DECISION SCIENCES

Theory and Practice

DECISION SCIENCES

Theory and Practice

Edited by

Raghu Nandan Sengupta

Aparna Gupta

Joydeep Dutta

CRC Press is an imprint of the
Taylor & Francis Group, an **informa** business

CRC Press
Taylor & Francis Group
6000 Broken Sound Parkway NW, Suite 300
Boca Raton, FL 33487-2742

First issued in paperback 2020

ISBN-13: 978-0-367-57437-6 (pbk)
ISBN-13: 978-1-4665-6430-5 (hbk)

Visit the Taylor & Francis Web site at
http://www.taylorandfrancis.com

and the CRC Press Web site at
http://www.crcpress.com

To my late mother, who would have been very happy to see whatever little I have achieved and to my wife, son, and father all of whom are really my greatest strengths

Raghu Nandan Sengupta

To Sree, for your friendship

Aparna Gupta

To my late father

Joydeep Dutta

Contents

About the Book

Preface and Introduction

This handbook is undoubtedly a bold and an honest attempt to cover many current, relevant, and essential topics related to *decision sciences* in a very scientific manner. Using this handbook, graduate students, researchers, as well as practitioners from engineering, statistics, sociology, economics, etc. will undoubtedly find a new and a much refreshing paradigm shift as to how these new and useful topics (along with relevant theorems/lemmas/corollaries, etc.) can be put to use beneficially. Starting from the basics to the advanced concepts (including intelligent use of different practical examples to supplement and complement every concept), we hope to make the readers well aware of the different theoretical as well as practical ideas, which are the focus of study in decision sciences nowadays. With an excellent bibliography/reference/journal list, good information about a variety of datasets, lucid as well as nicely illustrated pseudo-codes (which can be implemented in any existing computer code), and discussion of future trends in research, we hope this unique handbook will be able to create a niche for itself in the related academic community. Covering topics ranging from optimization, networks and games, multiobjective optimization, inventory theory, statistical methods, artificial neural networks, times series analysis, simulation modeling, decision support system, data envelopment analysis, queueing theory, etc., this reference book is unique and may be the first sincere attempt by leading experts in different fields of decision sciences to make this area more meaningful for different readers. Some other noteworthy features of this handbook are its vast as well as in-depth coverage of different topics, solved practical examples, good as well as unique datasets for a variety of examples in the areas of decision sciences, in-depth analysis of problems through charts, 3D diagrams, and discussion about software.

Aim and Scope

With a growing demand and urgent need for different mathematical, statistical, and quantitative tools for solving a variety of theoretical as well as practical decision analysis problems, there is an immediate and urgent need to come up with an interesting, in-depth, and updated handbook that will cater to the growing demand of academicians and practitioners alike. This firm conviction is the primary motivation for this handbook. This book is a reference manual that covers a whole gamut of interesting ideas, starting from optimization, multicriteria decision making, metaheuristic techniques, artificial neural network, simulation, etc. Furthermore, chapters related to networks and game theory, time series analysis, volatility modeling, queueing, artificial neural networks, decision support system, different statistical tools, etc. are some of the other interesting concepts/ideas covered in detail in this handbook.

The innovative idea of the handbook is not only limited to the coverage of the above interesting topics but also to the fact how these topics are dealt, such that postgraduate students as well as researchers from diverse fields such as engineering, economics, social sciences, statistics, mathematics, etc. can derive the maximum benefit by reading/referring this reference book. Apart from the emphasis on techniques, theorems, relevant proofs, etc., the authors make all the concepts clear by solving a variety of practical problems utilizing interesting datasets. The use of exhaustive lists of solved examples as well as exercises is an integral part of this handbook. By using novel presentation techniques, we offer a good practical flavor of all the different topics covered in this handbook. Innovative solution techniques, detailed pictorial analysis (tables, graphs, pie charts, histograms, etc.), pseudo-codes for real solvable programs, detailed analysis of results, etc. are the other interesting facets of this handbook.

The editors and authors of this book have also included the latest and future research trends in different areas covered in the handbook. A very good set of bibliography/ reference list as well as journal names, dataset library relevant to different areas of applications, information about technical software, URLs related to different research centers/universities/societies, and important facets/snippets on famous researchers in pertinent areas and fields of decision sciences are the other added attractions of this in-depth one-volume handbook.

Key Features

1. This reference book covers a variety of topics ranging from operations research to statistical tools, metaheuristic techniques, network and game theory, time series analysis, production planning, artificial neural networks, decision support system, simulation, etc., and is a genuine attempt to consolidate a whole gamut of tools, techniques, and methodologies developed and used in different areas of decision sciences, be it in biological sciences, physics, statistics, marketing, finance, civil engineering, computer science, mechanical engineering, sociology, etc.
2. Rather than being the sole authority in any particular topic, this reference book is an attempt to *complement as well as supplement* the existing diverse applications areas in decision sciences.
3. Apart from dealing with the theorems, proofs, and lemmas, the material also stresses on the practical aspects, rather than only theoretical concepts and proposition.
4. The handbook deals with a variety of solved practical problems, and analyzes the results through extensive discussion using charts, figures, and tables.
5. Apart from the solutions, the book adds flavor to the learning process by including the detailed pseudo-codes, programming concepts of each and every simulation-based study covered in this volume.
6. The emphasis and relevance of each topic is made more interesting with an extensive list of references, journal names, software, as well as URLs for relevant information.

The reference book also has detailed information on the latest trends in theoretical as well as practical research of the future, such that the readers of this handbook are well equipped with the general idea as to how to tackle interesting examples in areas of decision sciences in the years to come.

MATLAB® is a registered trademark of The MathWorks, Inc. For product information, please contact:

The MathWorks, Inc.
3 Apple Hill Drive
Natick, MA 01760-2098 USA
Tel: 508 647 7000
Fax: 508-647-7001
E-mail: info@mathworks.com
Web: www.mathworks.com

Acknowledgments

Writing an acknowledgment sometimes can be difficult when the list of people is large and all of them have been genuine and sincere in their personal efforts to see this project to its successful completion. I distinctively remember my meeting with Dr. Gagandeep Singh of CRC Press, Taylor & Francis in 2012, which resulted in the finalization of my unstructured idea of venturing into editing/writing a handbook with an emphasis on the different techniques of *decision sciences*. After that, much time has flown and I have grown four years *younger* at heart, but the relentless support from Dr. Singh in spite of my innumerable deadline shifts was amazing. A big thanks to him and his team at CRC Press, Taylor & Francis for their patience and willingness to bear with my innumerable requests on any account.

Dr. Aparna Gupta and Dr. Joydeep Dutta, my other two editors, were really very helpful as colleagues and friends who let me handle issues and take decisions on any account in the process of this work. I am confident of their individual faith in my judgment and I thank them for that. My other coauthors and contributors in this project, some of whom I never knew before the start of this work, have been really genuine academic friends, whose patience and willingness to listen to my demands were praiseworthy. Moreover, their individual subject knowledge, willingness to stick to the academic rigor as experts in their respective fields, and constructive suggestions to me on many accounts really improved the overall presentation of each and every chapter and also that of the handbook.

Finally, and most important of all, the people without whose coaxing, cajoling, and sometimes threatening support this work would not have been ever possible are my lovely wife, my adorable son, and my ever supportive father. I thank them from the bottom of my heart for their infinite patience in bearing with my sometimes nonchalant attitude. I now firmly believe that whatever I have been able to do would not have been ever possible without their unflinching support.

Lastly, on a personal note, we would definitely like to hear from our readers about this handbook and for the improvement of the same. Errors, if any, should be pointed out to us so that we may rectify it to make this handbook/reference book better.

Raghu Nandan Sengupta

I would first and foremost thank my coeditor Dr. Raghu Nandan Sengupta. Dr. Sengupta both masterminded this comprehensive and ambitious project, and then has patiently and diligently seen it through to its conclusion. I also thank my second coeditor Dr. Joydeep Dutta for his significant contributions to this project. Each chapter in this handbook has been a valuable contribution of experts in a range of topics relevant and useful in decision sciences. I thank each of the authors for offering their vast knowledge and expertise in helping this handbook achieve its goals and objectives. I wish to finally, and in the greatest magnitude, thank our editorial manager, Dr. Gagandeep Singh, and his editorial staff at Taylor & Francis group for their patience, encouragement, and support to successfully complete this massive project.

Aparna Gupta

I thank Dr. Raghu Nandan Sengupta for his infinite patience with my chapter and with the way he carried forward the whole project from the day of its inception to it successful completion. I also thank Dr. Aparna Gupta and the editor at CRC Press, Taylor & Francis for not writing me off in spite of the delays that happened due to various reasons. Foremost and most important, I thank my wife and daughter for always being my greatest support.

Joydeep Dutta

Editors

Raghu Nandan Sengupta completed his bachelor's degree in mechanical engineering from Birla Institute of Technology, Mesra, Ranchi, India and his FPM (PhD) from the Indian Institute of Management, Calcutta, India with specialization in operations management. His research interests are in sequential analysis, decision sciences, and optimization and its use in financial optimization. His research work has been published in journals such as *Metrika*, *European Journal of Operational Research*, *Sequential Analysis*, *Computational Statistics & Data Analysis*, *Communications in Statistics: Simulation & Computation*, *Quantitative Finance*, etc. He is a professor in the Department of Industrial and Management Engineering at the Indian Institute of Technology, Kanpur, India, and teaches courses such as probability and statistics, stochastic processes and their applications, financial risk management, etc. He is the recipient of the IUSSTF Fellowship 2008 and visited the Department of Operations Research and Financial Engineering at Princeton University, USA; the Erasmus Mundus Fellowship 2011 to University of Warsaw, Poland; and the second Erasmus Mundus Fellowship 2014 to Instituto Superior Técnico, University of Lisbon, Portugal. He is also an editorial board member of the *Foundations of Computing and Decision Sciences* and *Open Journal of Statistics*.

Aparna Gupta is a faculty member in the Finance and Accounting Department of the Lally School of Management at Rensselaer Polytechnic Institute (RPI). She is the director of the MS program on quantitative finance and risk analytics at Lally School and the director of Lally School's Center for Financial Studies. She holds a joint appointment at the Industrial and Systems Engineering Department in the School of Engineering at RPI. Dr. Gupta has developed and taught courses on corporate finance, financial computations, derivatives, risk management, and financial simulations at the Lally School of Management. She is the author of the book *Risk Management and Simulation*, published by CRC Press, Taylor & Francis in June 2013. Dr. Gupta's research interests are in risk management, financial engineering, and financial decision support. She has addressed a range of issues in risk management at the individual and institutional levels, including with financial network considerations. She conducts the U.S. National Science Foundation-funded research in financial innovation for risk management in network domains, such as electricity markets and communication network. Her research has been published in reputed journals such as *Insurance: Mathematics and Economics*, *Physica A*, *European Journal of Operational Research*, *Journal of Financial Engineering*, *Journal of Financial Stability*, *Annals of Operations Research*, *Computational Optimization and Applications*, *Journal of Computational Finance*, *Computer Networks*, *Electricity Journal*, etc. Dr. Gupta serves on the editorial board and as a reviewer for several business, finance, and management science journals, and is the past chair of the financial services section of the Institute for Operations Research and Management Science (INFORMS). She is also a member of AFA, WFA, FMA, IAQF, INFORMS, and IEEE. Dr. Gupta earned her doctorate from Stanford University and her BSc and MSc degrees in mathematics from the Indian Institute of Technology, Kanpur, India.

Joydeep Dutta earned his PhD in mathematics from the Indian Institute of Technology, Kharagpur, India with specialization in nonsmooth optimization. Consequently, he did his postdoctoral studies in the Indian Statistical Institute, India and the Autonomous University of Barcelona, Spain. His main research interests are in abstract convexity, nonsmooth optimization, and vector optimization. He has published over 45 papers in journals such as *Optimization*, *Mathematical Methods of Operations Research*, *Journal of Convex Analysis*, *Journal of Optimization Theory and Applications*, *Journal of Global Optimization*, *Operation Research Letters*, *SIAM Journal of Optimization*, *Mathematical Programming*, etc. He also has two books to his credit, which have been published by Alpha Science International, UK and by CRC Press of Taylor & Francis, respectively. He is currently serving as a professor in the economics group at the Department of Humanities and Social Sciences at the Indian Institute of Technology, Kanpur, India. Until June 2014, he was a professor in the Department of Mathematics of the same institute.

Contributors

Sunith Bandaru
Department of Production and Automation Engineering
University of Skövde
Skövde, Sweden

Emre Berk
Department of Management
Bilkent University
Ankara, Turkey

Amit Bhaya
Department of Electrical Engineering
Federal University of Rio de Janeiro
Rio de Janeiro, Brazil

Carlos A. Coello Coello
Departamento de Computación
CINVESTAV-IPN
Mexico City, Mexico

Kalyanmoy Deb
Department of Electrical and Computer Engineering
Michigan State University
East Lansing, Michigan

Joydeep Dutta
Department of Humanities and Social Sciences
Indian Institute of Technology Kanpur
Kanpur, India

Aparna Gupta
Lally School of Management
Rensselaer Polytechnic Institute
Troy, New York

Phalguni Gupta
National Institute of Technical Teachers' Training and Research
Kolkata, India
and
Department of Computer Science and Engineering
Indian Institute of Technology Kanpur
Kanpur, India

Ülkü Gürler
Department of Industrial Engineering
Bilkent University
Ankara, Turkey

Jussi Hakanen
Department of Mathematical Information Technology
University of Jyväskylä
Jyväskylä, Finland

Aditya K. Jagannatham
Department of Electrical Engineering
Indian Institute of Technology Kanpur
Kanpur, India

R. Krishnan
Indira Gandhi Institute of Development Research
Mumbai, India

Vimal Kumar
Department of Humanities and Social Sciences
Indian Institute of Technology Kanpur
Kanpur, India

Debasis Kundu
Department of Mathematics and Statistics
Indian Institute of Technology Kanpur
Kanpur, India

Adriana Lara
Departamento de Matemáticas de la Escuela Superior de Física y Matemáticas
Instituto Politécnico Nacional
San Pedro Zacatenco, Méjico

Dmitri G. Markovitch
Lally School of Management
Rensselaer Polytechnic Institute
Troy, New York

Lois S. Peters
Lally School of Management
Rensselaer Polytechnic Institute
Troy, New York

Deepu Philip
Department of Industrial and Management Engineering
Indian Institute of Technology Kanpur
Kanpur, India

Surya Prakash
Department of Computer Science and Engineering
School of Engineering
Indian Institute of Technology Indore
Indore, India

Subhash C. Ray
Department of Economics
University of Connecticut
Storrs, Connecticut

Raghu Nandan Sengupta
Department of Industrial and Management Engineering
Indian Institute of Technology Kanpur
Kanpur, India

Karthik Sindhya
Department of Mathematical Information Technology
University of Jyväskylä
Jyväskylä, Finland

Raghunath Tewari
Department of Computer Science and Engineering
Indian Institute of Technology Kanpur
Kanpur, India

Saúl Zapotecas-Martínez
Department of Electrical and Electronic Engineering
Shinshu University
Nagano, Japan

1

Convex Functions in Optimization

Joydeep Dutta

CONTENTS

ABSTRACT Convex functions are central to the study of optimization. The main property that makes it pivotal to the study of optimization is that every local minimizer of a convex function over a convex set is a global minimizer. The very large number of problems that arise in engineering, business, economics, and finance can be modeled as a convex optimization problem. In this chapter we first focus on the various examples and important properties of convex functions, which includes the study of subdifferentials and their computation. We also focus on the various important models of convex optimization problems and then on their optimality and duality properties. We end the chapter by a study of monotone operators, variational inequalities, and of the proximal point method for convex optimization problems.

1.1 Introduction

In the traditional academic setup, at least in my country, optimization is largely viewed as consisting of two major parts. They are linear programming and nonlinear programming. It is usually believed that linear programming problems are tractable while problems in nonlinear programming are hard. What is missing in the standard discourse in both linear and nonlinear programming is the fundamental role of convex sets and convex functions. In fact, all linear and affine functions are convex and thus linear programming is truly a part of convex programming in nonlinear programming the best results are obtained when the problem data are convex. Thus, convexity is an inescapable part of an optimizer's life. This fact is best echoed in the following quote of Rockafellar [1, p. 185] where he says:

> The great watershed in optimization isn't between linearity and nonlinearity but between convexity and nonconvexity.

Thus, any modern education and research in optimization has to be convex centric and very recently this has been clearer in the phenomenal applications of semidefinite programming, which by itself is a very important class of convex optimization problem. In this chapter, we are going to focus on the fundamental role of convexity or more precisely convex functions in optimization. There has been several texts and monographs on convexity and its links with optimization. I mention below some of my personal favorites:

1. J.M. Borwein and A.S. Lewis, *Convex Analysis and Nonlinear Optimization*, Springer (2nd edition), 2006.
2. J.V. Tiel, *Convex Analysis: An Introductory Text*, Wiley, 1984.
3. J.B. Hiriart-Urruty and C. Lemarachal, *Fundamentals of Convex Analysis*, Springer, 2003.
4. J.B. Hiriart-Urruty and C. Lemarachal, *Convex Analysis and Minimization Algorithms*, Vols. I and II, Springer, 1996.
5. R.T. Rockafellar, *Convex Analysis*, Princeton, 1970.
6. J.M. Borwein and J. Vanderwerff, *Convex Functions*, Cambridge University Press, 2010.
7. D.P. Bertsekas, *Convex Analysis and Optimization*, Athena Scientific Publishing, 2003.

I apologize if I have missed some good text or monograph on convexity. Most of these texts are mainly for graduate students or researchers who are interested in optimization. The above texts might not always be an easy read for the very beginner or people in other fields such as engineering. In this chapter, I would like to present a wide view of the applications of convexity to optimization. It won't be very detailed like a text book but the main details would be emphasized. I list down the various sections the reader will find in this chapter. An experienced reader can choose any section he/she wants. A reader new to convex analysis and optimization will find a good amount of material to develop a fair idea about the subject and then can proceed to the above-mentioned texts. I would also like to mention that this chapter is not for someone who does not have any basic idea of optimization. The sections in this chapter are as follows:

1. Convex Functions: Fundamental Properties.
2. What Is Convex Optimization?
3. Optimality and Duality in Convex Programming.

We would like to end this section by giving the most basic definitions of a convex set and a convex function. We will keep ourselves in finite dimensions. A set C subset of $\mathbb{R}^n$ is convex if for any x and y in C and $\lambda \in [0,1]$, we have $\lambda y + (1-\lambda)x \in C$. In words this means that for any pair of points in the set C, the line segment joining those points lies entirely in C. Thus, given a convex set $C \subseteq \mathbb{R}^n$, a function $f : C \to \mathbb{R}$ is said to be convex if for any $x, y \in C$ and $\lambda \in [0,1]$ we have

$$f(\lambda y + (1-\lambda)x) \leq \lambda f(y) + (1-\lambda) f(y).$$

Note that every convex function can be defined over $\mathbb{R}^n$ provided we allow functions to take values in the extended real line $\overline{\mathbb{R}} = \mathbb{R} \cup \{+\infty, -\infty\}$. For example, the above convex function f on C can be viewed as a convex function $h : \mathbb{R}^n \to \overline{\mathbb{R}}$ if we define $h(x) = f(x)$, when $x \in C$ and $h(x) = +\infty$, when $x \notin C$. The important point to note is why one has to consider $h(x) = +\infty$ when $x \notin C$.

Why can't we take $h(x) = -\infty$. This is due to the geometrical fact that for a finite-valued convex function the region above the graph known as the epigraph is convex. The epigraph is given as

$$epi f = \{(x, \alpha) \in \mathbb{R}^n \times \mathbb{R} : f(x) \leq \alpha\}.$$

A function f is convex if and only if the epigraph is convex. We urge the reader to prove this. Note that when we allow infinite values and if $f(x) = \infty$ then the vertical line passing through x consisting of the points (x, α) does not belong to the epigraph. However, if we allow $f(x) = -\infty$ then the whole line through x consisting of the points (x, α) is in the epigraph. This may violate the convexity of the epigraph. A simple sketch of a real-valued convex function on an interval will make this idea clear. If we extend the function outside the interval by setting a constant value $-\infty$, then the convexity of the epigraph is broken. Thus, for an extended-value convex functions certain terms have come into frequent use. The most common term is that of the domain of f or $dom f$, which consists of all points where the function value is strictly less than $+\infty$. Note that $f(x) = -\infty$ for all $x \in \mathbb{R}^n$ is a convex function and $dom f = \mathbb{R}^n$. In most cases we would be considering proper convex functions that do not take the value $-\infty$ at any point and have at least one point where it is finite. Proper convex functions are of great importance in optimization. We discuss this is more detail in the next section. Our symbols are standard and follow largely Borwein and Lewis [2].

1.2 Convex Functions: Fundamental Properties

1.2.1 Examples of Convex Functions

Convexity is a great property to use if one, somehow, knows beforehand the convexity of the function. Is it easy to detect whether a function is convex? The answer is, in general, it is not. One of the important rules for characterizing a twice continuously differentiable convex function is to check whether its Hessian matrix is positive semidefinite. This might not be an easy task always and further convex functions may not even be differentiable and thus the above tool may not be handy. It is now a growing theme in mathematics and other science when the analytical approach on paper is making no headway or giving any clue one has to turn to the computer for visualization using some computer algebra system (CAS) such as MAPLE or MATHEMATICA. We shall show the use of CAS using MAPLE. Now let us look at the real-valued convex functions on $\mathbb{R}$ or some subset of $\mathbb{R}$. Consider the function $f : \mathbb{R} \to \mathbb{R}$ given as $f(x) = x^6 + x^5 + x^4 + x^3 + x^2 + 1$. It is not so easy to believe that the function is convex even if it is an even-degree polynomial. If we attempt to prove that it is convex, we need to check whether the second-order derivative is nonnegative on $\mathbb{R}$. In this particular case, we have

$$f''(x) = 30x^4 + 20x^3 + 12x^2 + 6x + 12.$$

It is not easy to immediately check that the second derivative is nonnegative on $\mathbb{R}$. Thus, the first approach is to try and plot the graph of f, which is given in Figure 1.1. The graph definitely looks like the graph of a convex function; however, the region between $x = -2$ and $x = 2$ is not clear. Thus, let us zoom in on the graph in the interval $[-2, 2]$. This is what is shown in Figure 1.2. Now in the zoomed-in graph, we can see the picture more clearly and we can see that the graph is indeed that of a convex function and that provides us much confidence to declare that the function is convex. However, one might argue that such experimentation is really not a proof but still gives us much insight and further add to our confidence that we can actually minimize f'' using any computational software and find its minimum value as positive. We encourage the reader to actually check this out.

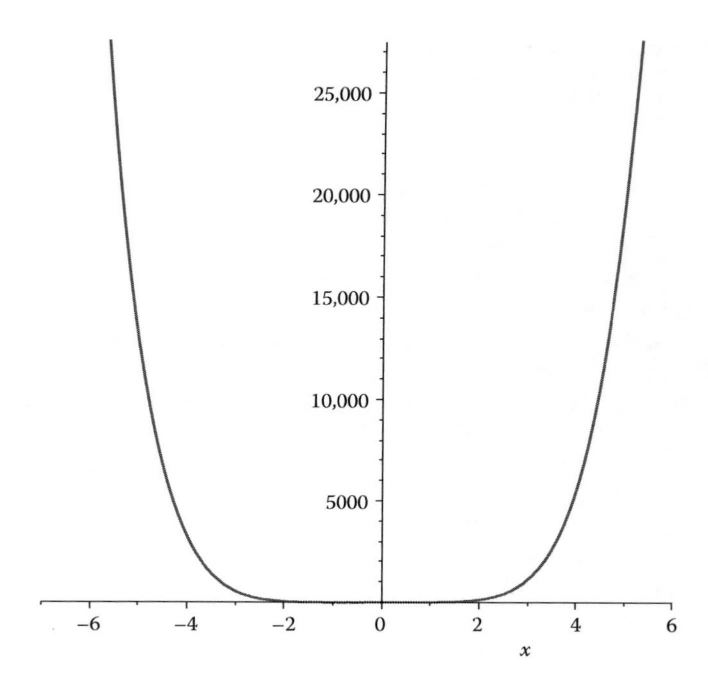

FIGURE 1.1
Graph of $x^6 + x^5 + x^4 + x^3 + x^2 + x + 1$.

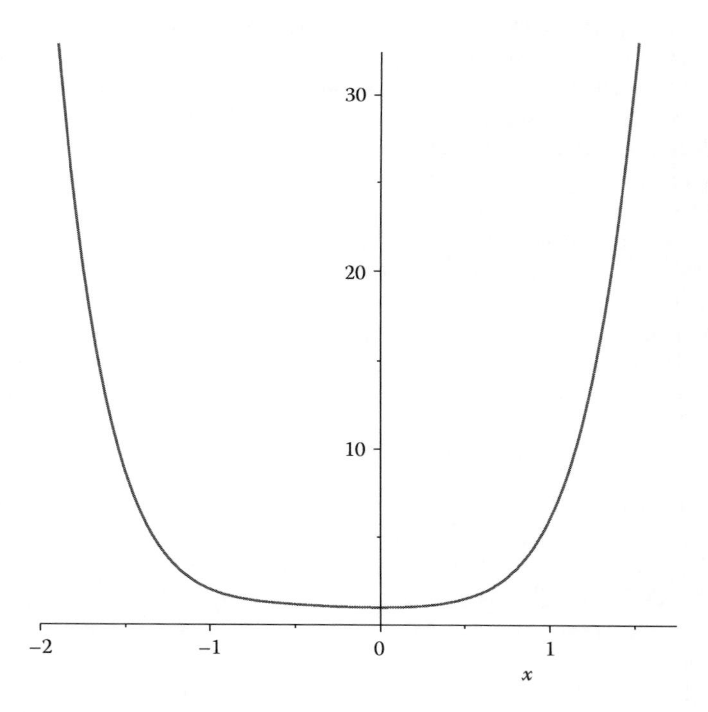

FIGURE 1.2
Graph of $x^6 + x^5 + x^4 + x^3 + x^2 + x + 1$ magnified.

Further it would be interesting to check whether the following function $\log(\sinh ax/\sinh x)$, $a \geq 1$ is convex. This example has been taken from Borwein and Lewis [2] and this also needs numerical experimentation using a CAS to guarantee that the function is convex. Using MAPLE for $a = 2$, we see that the second derivative is $1/(\cosh(x))^2$, which we see is positive and thus it is convex in this case. We would also like the reader to take note that the function $\log(\sinh ax/\sinh x)$, $a \geq 1$ has a removable discontinuity at $x = 0$. These two examples show that it might not always be easy to determine whether a real-valued function of a real variable is convex or not. Further we would like to mention that the convex functions of a real variable form a rich class of functions with many applications.

This richness of convex functions on the real line is demonstrated by the connection between convexity and the gamma function from classical analysis. This connection is given by the *Bohr–Mollerup Theorem*, which we will state and also provide a proof. Let us first begin by giving the definition of the gamma function. For $x \in \mathbb{R}$ and $x > 0$, the gamma function that is denoted as $\Gamma(x)$ is defined as

$$\Gamma(x) = \int_0^\infty t^{x-1} e^{-t}\, dt.$$

Integration by parts we conclude that $\Gamma(x+1) = x\Gamma(x)$ and from the very definition we have $\Gamma(1) = 1$. $\cosh x = (e^x + e^{-x})/2 > 0$. Any function whose logarithm is convex is called log convex. We will now show that Γ is a log-convex function. Consider $x, y > 0$ and consider $\lambda, \mu > 0$. Then, we have from the definition

$$\Gamma(\lambda x + \mu y) = \int_0^\infty t^{\lambda x + \mu y - 1} e^{-t}\, dt = \int_0^\infty (t^{x-1}e^{-t})^\lambda (t^{y-1}e^{-t})^\mu\, dt.$$

Now applying Hölder's inequality we have

$$\Gamma(\lambda x + \mu y) \leq \left(\int_0^\infty t^{x-1} e^{-t}\, dt \right)^\lambda \left(\int_0^\infty t^{y-1} e^{-t}\, dt \right)^\mu = \Gamma^\lambda(x)\Gamma^\mu(y).$$

This immediately shows that Γ is log convex. Using the arithmetic mean–geometric mean inequality, it is simple to show that every log-convex function is convex though the converse is not true. Hence, Γ is a convex function. Figure 1.3 presents a graph on gamma function.

Even from the graph, it is clear that Γ is a convex function on $(0, +\infty)$. In fact, the minimum value of Γ is attained at $x = 1.461632145\ldots$. This was known to the great Gauss. See, for example, Niculescu and Persson [3] for more details. Now, we shall present the Bohr–Mollerup theorem and its proof from Reference 3.

Theorem 1.1: Bohr–Mollerup

Let $f : (0, +\infty) \to \mathbb{R}$ be a given function, which satisfies the following conditions:

1. $f(x+1) = xf(x)$
2. $f(1) = 1$
3. f is log convex

Then $f = \Gamma$.

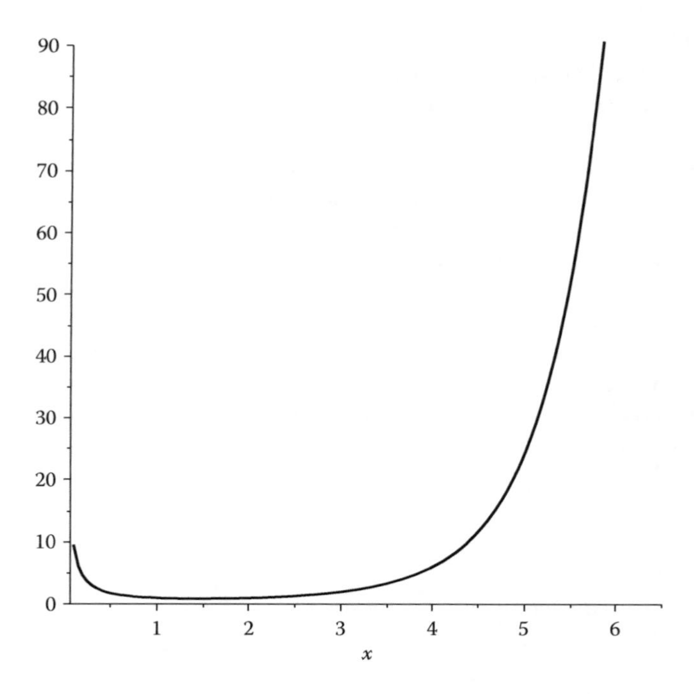

FIGURE 1.3
Graph of the gamma function.

Proof. It is clear that Γ satisfies all these properties as we have noted above. Now using (1) and (2) and using the principle of induction we see that $f(n+1) = n!$ for $n \in \mathbb{N}$, where $\mathbb{N}$ denotes as usual the set of natural numbers. Let us now consider $x \in (0, 1]$. Now using (1) and (3), we can deduce the following:

$$\begin{aligned} f(n+1+x) &= f((1-x)(n+1) + x(n+2)) \\ &\leq (f(n+1))^{1-x}(f(n+2))^x \\ &= (f(n+1))^{1-x}(n+1)^x(f(n+1))^x \\ &= (n+1)^x f(n+1) \\ &= (n+1)^x n! \end{aligned}$$

Further note that

$$\begin{aligned} n! &= f(n+1) \\ &= f(x(n+x) + (1-x)(n+1+x)) \\ &\leq (f(n+x))^x (f(n+1+x))^{1-x} \\ &= (n+x)^{-x}(f(n+1+x))^x(f(n+1+x))^{1-x} \\ &= (n+x)^{-x} f(n+1+x). \end{aligned}$$

Now observe that $f(n+1+x) = (n+x)(n-1+x)\cdots xf(x)$. Using this fact and doing some simple manipulations using the above inequalities, we have

$$\left(1+\frac{x}{n}\right)^x \leq \frac{(n+x)(n-1+x)\cdots xf(x)}{n!n^x} \leq \left(1+\frac{1}{n}\right)^n.$$

On passing to the limits as $n \to \infty$, we have

$$f(x) = \lim_{n\to\infty} \frac{n!n^x}{(n+x)(n-1+x)\cdots x}.$$

Now we shall show that even when $x > 1$ the three conditions given in the hypothesis lead to the same expression for $f(x)$. Consider $x > 1$ and consider an integer m such that $0 < x - m \leq 1$. Thus, repeatedly applying (1) we have

$$f(x) = (x-1)\cdots(x-m)f(x-m).$$

Since $x - m \leq 1$, we can now write

$$f(x) = (x-1)\cdots(x-m) \lim_{n\to\infty} \frac{n!n^x}{(n+x-m)(n-1+x-m)\cdots(x-m)}.$$

A little algebraic manipulation shows that

$$f(x) = \lim_{n\to\infty} \frac{n!n^x}{(n+x)(n-1+x)\cdots x} \lim_{n\to\infty} \left(\left(1+\frac{x}{n}\right)\left(1+\frac{x-1}{n}\right)\cdots\left(1+\frac{x-m+1}{n}\right)\right).$$

This shows that

$$f(x) = \lim_{n\to\infty} \frac{n!n^x}{(n+x)(n-1+x)\cdots x}.$$

This shows that for all $x > 0$ the function f is uniquely determined by the above expression. Hence, $f = \Gamma$ since Γ satisfies all the three properties. This proves the result. ■

Now let us turn our attention to convex functions from $\mathbb{R}^n$ to $\mathbb{R}$ or $f : C \to \mathbb{R}$, where $C \subset \mathbb{R}^n$ is a convex set. We just list a few examples:

1. $f(x) = \|x\|$, $f(x) = \|x\|_1$ and $f(x) = \|x\|_\infty$.
2. $f(x) = \frac{1}{2}\langle x, Qx\rangle$, Q is positive semi-definite.
3. $f(x) = -\sum_{i=1}^n \log x_i$, $x_i > 0$, $i = 1, \ldots, m$.
4. $f(x) = \max\{\langle a_i, x\rangle - b_i : i = 1, \ldots, m\}$, $a_i \in \mathbb{R}^n$ and $b_i \in \mathbb{R}$ for all $i = 1, \ldots, m$.
5. $f(x) = \max\{f_1(x), \ldots f_m(x)\}$, where for each $i = 1, \ldots, m$ the function f_i is convex.
6. $f(x) = -(x_1x_2, \ldots, x_n)^{1/n}$ where $x_i > 0$ for $i = 1, \ldots, m$.
7. *Indicator function.* Let C be a convex set then consider the function $\delta_C(x) = 0$ if $x \in C$ and $\delta_C(x) = +\infty$ if $x \notin C$.
8. *Support function.* Let C be a convex set then $\sigma_C(x) = \sup_{w\in C}\langle w, x\rangle$.

9. *Maximum eigenvalue*. Denote by S^n the space of all $n \times n$ real symmetric matrices and let $\lambda_1(A)$ denote the largest eigenvalue of the symmetric matrix A. This is given by the following:

$$\lambda_1(A) = \sup_{\|x\|=1} \langle x, Ax \rangle.$$

This is a convex function.

10. *Distance function.* Let C be a closed convex set. Then the distance function

$$d_C(x) = \inf_{y \in C} \|y - x\|$$

is convex.

11. *Sum of m-largest eigenvalues of a matrix.* Consider the space S^n of all $n \times n$ real symmetric matrices equipped with the inner product $\langle X, Y \rangle = \text{trace}(XY)$. Since the eigenvalues are real, we can order them as $\lambda_1(A) \geq \lambda_2(A) \geq \cdots \lambda_n(A)$. Now consider the function $f_m(A)$ that is the sum of the m largest eigenvalues of $A (m \leq n)$, that is,

$$f_m(A) = \sum_{i=1}^{m} \lambda_i(A).$$

It is interesting to note that f_m is a convex function.

12. Consider again S^n the space of all real symmetric matrices. Further let S^n_{++} denote the cone of all positive-definite matrices. Consider the following functions for $X \in S^n_{++}$:

$$f(X) = -\log(\det(X)) \quad g(x) = \log(\det(X^{-1})).$$

These functions are also convex.

One of the most important ways to construct a convex function is to take the maximum of a finite number of convex functions. This was mentioned in the above list of examples. However, this idea can be generalized to arbitrary index sets. Let $\{f_i : i \in I\}$ be an arbitrary family of convex functions where I is an arbitrary index set. Then, the function $f(x) = \sup_{i \in I} f_i(x)$ is also a convex function. This is a key to prove, for example, that the largest eigenvalue function is a convex function. Let us take a brief look at how it is done. Note that in S^n the inner product is defined as follows. Let X and Y be in S^n then $\langle X, Y \rangle = trace(XY)$, where $trace(A)$ is true for matrix A, which is nothing but the sum of the diagonal element of A. Once we know the definition of the inner product in S^n, one can write

$$\lambda_1(A) = \max_{\|x\|=1} \langle x, Ax \rangle = \max_{x^T x=1} \langle xx^T, A \rangle = \max_{U \in S^n_+, trace(U)=1} \langle U, A \rangle,$$

where S^n_+ is the cone of positive semidefinite matrices. We leave it to the reader to figure out from the above equalities that $\lambda_1(A)$ is indeed a convex function.

However, it is more trickier to show that the function defined by the sum of the m-largest eigenvalues is convex. This function $f_m(A)$ is given as follows in Horn and Johnson [4], Corollary 4.3.18,

$$f_m(A) = \max_{Z \in M^{n \times m}, Z^T Z=1} \langle Z, AZ \rangle,$$

where $M^{n \times m}$ is the set of all real $n \times m$ matrices. It is not so easy to see that this is convex. Note that if X and Y are $m \times n$ matrices then the inner product $\langle X, Y \rangle$ is given as $\langle X, Y \rangle = trace(X^T Y)$. In fact, there has been a complete paper on the convexity of this function and the associated issues. It was shown in Overton and Womersley [5] that

$$f_m(A) = \max_{U \in K_{m,n}} \langle U, A \rangle,$$

where $K_{m,n}$ is given as

$$K_{m,n} = \{U \in S^n : I - U \in S^n, trace(U) = m\}.$$

Once we have expressed $f_m(A)$ in the above format, then it is again simple to see why $f_m(A)$ is convex. However, the proof of the above format is not direct and we would recommend the reader for the very elegant proof in Overton and Womersley [5].

Let us now turn our attention to prove the convexity of $-\log(\det(X))$ over S^n_{++}. We shall prove this by proving the concavity of $\log(\det(X))$. We shall provide the proof from Boyd and Vandenberghe [6]. Hiriat-Urruty and Lemarechal have mentioned [7] that convexity is essentially a one-dimensional thing since a function f is convex if and only if $\varphi(t) = f(x + tv)$ is convex in $t \in \mathbb{R}$ for every fixed x and v. The same happens with concavity. With this in mind we shall show that for any $Z \in S^n_{++}$ and any $V \in S^n$, the function

$$\varphi(t) = \log(\det(Z + tV))$$

is concave for all $t \in \mathbb{R}$. Let us note the fact that if A is a positive-definite matrix then there exists another positive-definite matrix B such that $B^2 = A$. This is also symbolized as $B = A^{1/2}$. Thus, we can write $\varphi(t)$ as

$$\varphi(t) = \log(\det(Z^{1/2}(I + tZ^{-1/2}VZ^{-1/2})Z^{1/2})).$$

Note that the determinant of the product of matrices is the product of the determinants we have

$$\varphi(t) = \log(\det(Z)) + \log(\det(I + tZ^{-1/2}VZ^{-1/2})).$$

This can be further simplified as

$$\varphi(t) = \log(\det(Z)) + \sum_{i=1}^{n} \log(1 + t\lambda_i),$$

where λ_i is the ith eigenvalue of $Z^{-1/2}VZ^{-1/2}$. It is not difficult to calculate to obtain

$$\varphi''(t) = -\sum_{i=1}^{n} \frac{\lambda_i^2}{(1 + t\lambda_i)^2}.$$

Thus, $\varphi''(t) \leq 0$ for all $t \in \mathbb{R}$ establishing that $\varphi(t)$ is concave.

Before we continue our discussions on some more of the convex functions listed above, let us take a break and have a look at the graphs of some simple convex functions defined on $\mathbb{R}^2$. Let us begin by visualizing one of the most simple of all two dimensional convex functions. Let us look at Figure 1.4, $f(x, y) = x^2 + y^2$.

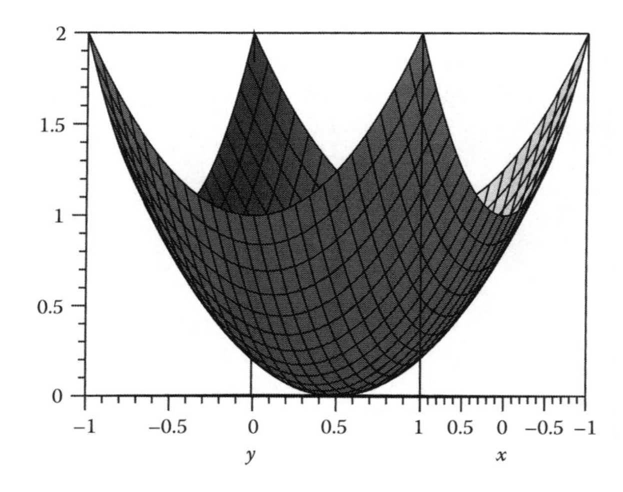

FIGURE 1.4
Graph of $x^2 + y^2$.

Though the graph shows that the function is a very nice one, it is smooth and has a unique minimizer at $(x^*, y^*) = (0, 0)$. However, this nice function is also an example of a convex function, which is not log convex. We shall produce the graph of $\log(x^2 + y^2)$ as shown in Figure 1.5 and that will make it clear.

We shall take a look at the graph of an interesting function that is important in the study of proximal point methods in convex optimization and is known usually as the negative entropy function. This is the function $f(x) = \sum_{i=1}^{n}(x_i \log x_i - x_i)$. In fact, to show that this function convex on $\mathbb{R}^n_{++}$ we have to just show that $t \log t - t$ is convex for any real $t > 0$. We will show this when we discuss conjugates of a convex function. We shall draw the graph in the two dimensional setting as shown in Figure 1.6.

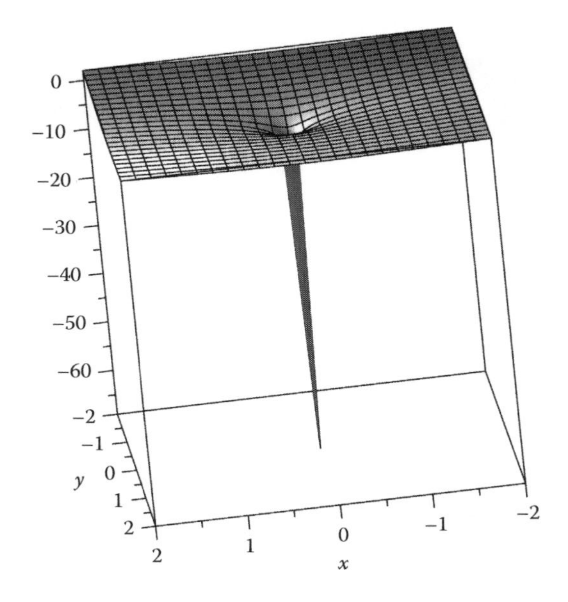

FIGURE 1.5
Graph of $\log(x^2 + y^2)$.

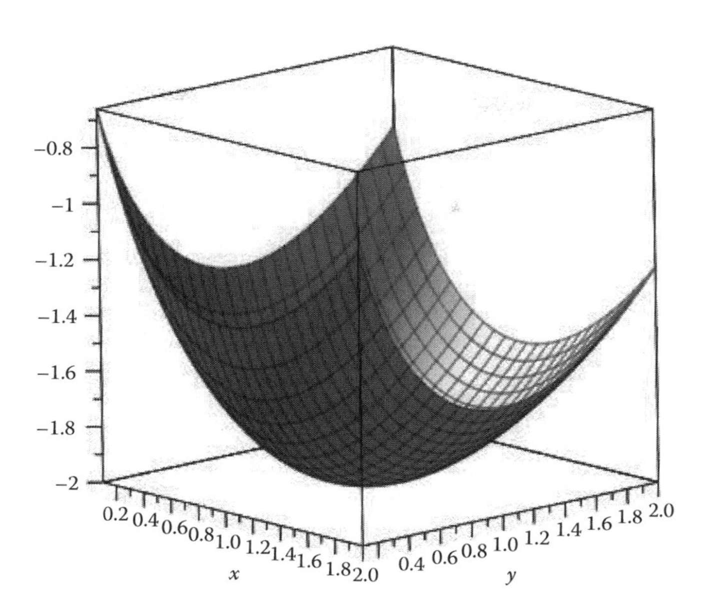

FIGURE 1.6
Graph of $(x\log x - x) + (y\log y - y)$.

Now let us take a look at a simple nondifferentiable convex function $f(x, y) = |x| + |y|, x, y \in \mathbb{R}$. We will see that it is composed of affine parts in the sense that its epigraph is a polyhedral set. A polyhedral set is a convex set formed by the intersection of a finite number of half-spaces. This function has a minimizer at $(0, 0)$ and is not differentiable there. Further it is not the only point where it is not differentiable. Observe the boundaries where two planes meet (Figure 1.7). Along those points too, the function does not have a derivative.

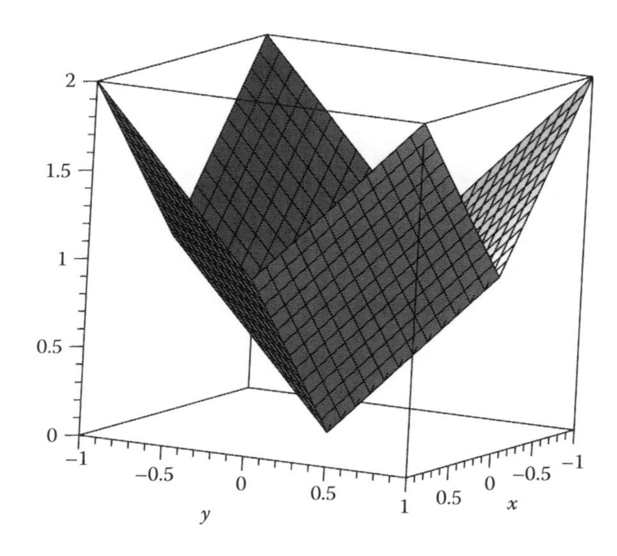

FIGURE 1.7
Graph of $|x| + |y|$.

Let us now again return to our discussions of the examples of convex function that we have stated above. Note that the support function and indicator functions are infinite-valued convex functions. Another important function is the distance function. How to show that the distance function is convex? There are two paths to do this. Let us first choose the path of *inf-convolution*. The inf-convolution of two proper convex functions f and g is given as

$$f \diamond g(x) = \inf_{y \in \mathbb{R}^n} (f(y) + g(x - y)).$$

We leave it to the reader to prove that $f \diamond g$ is convex. Now consider $f = \delta_C$, the indicator function and $g = \|.\|$. Then in that case we have

$$f \diamond g = \inf_{y \in C} \|x - y\|.$$

This shows that the distance function d_C is convex. The other route comes from the following information. If $h : \mathbb{R}^n \times \mathbb{R}^m \to \mathbb{R}$ is the convex, then the function $g(x) = \inf_{y \in C} h(x, y)$, where C is convex set of $\mathbb{R}^m$, is a convex function. This fact will immediately prove that the distance function is convex since $\|y - x\|$ is jointly convex in (x, y).

1.2.2 Continuity and Differentiability of Convex Functions

One of the most intriguing aspects of a convex function is related to continuity and differentiability. In fact, every convex function is continuous on the interior of its domain. However, problems can come at the boundary. Lower semicontinuity of extended-valued convex functions play, a major role in convex analysis. Instead of getting into a discussion on the lower semicontinuity of convex functions, let us refer the reader to the classic text *Convex Analysis* by R. T. Rockafellar [8]. Though in mathematics it is important to provide rigorous proofs of the claims one make, the ideas about lower-semicontinuous convex functions can be easily understood by visualizing the results using the epigraph of a convex function on the real line. For the purpose of optimization another important issue associated with convex functions should be our focus now. This issue of differentiability is one of the key necessities in optimization. A differentiable function $f : \mathbb{R}^n \to \mathbb{R}$ is convex if and only if for all $x, y \in \mathbb{R}^n$

$$f(y) - f(x) \geq \langle \nabla f(x), y - x \rangle.$$

We leave it to the reader to prove this fact. When we have a differentiable convex function, it is better to know at the very outset that it is continuously differentiable. Further there is another way of characterizing differentiable convex functions, which we believe is very fundamental. This is through the vehicle of monotonicity. A differentiable function is convex if and only if

$$\langle \nabla f(y) - \nabla f(x), y - x \rangle \geq 0; \quad \forall x,\ y \in \mathbb{R}^n. \tag{1.1}$$

The property expressed through Equation 1.1 is called monotonicity. The reader can easily prove this fact by noting a hint. It is important to keep in mind that when moving from monotonicity to convexity one should not forget to use the mean-value theorem.

However, in convex analysis differentiability is not such an easy thing to come by. A convex function need not always be differentiable throughout the domain. However, it is differentiable almost everywhere in the sense that the set of points at which a convex function is not differentiable forms a set of measure zero. This is not a very bad news but the drawback is that the point of minimizers of convex functions can well be at these nondifferentiable points. Just consider the simple function $f(x) = |x|$, where $x \in \mathbb{R}$. The only point of nondifferentiability is the point $x = 0$ and this is the unique minimizer of this function. How does one get over this issue? There can be two ways to view it. One from the theoretical point of view, and the other from the numerical point of view. The numerical point of view may gain importance by pointing to the fact that the real use of the derivative in optimization is the actual computation of the solution. Thus, one may be tempted to use finite-difference methods to provide a rough estimate of the derivative where the function is not differentiable. From the theoretical point of view, it is more prudent to search for an object that would replace the derivative and yet retain most of its important features. Of the several features of the derivative that we may want to replicate, the fundamental feature is the preservation of the Fermat's rule ($0 = \nabla f(x)$) at the minimizer in some form or the other. Unfortunately, finite-difference techniques when applied to nondifferentiable convex functions more often provide estimates that do not capture this fundamental feature. This can be very well gauged from the following famous quote of Claude Lemarechal [9, p. 539]:

> Actually, a mere finite differencing in nonsmooth optimization is a sin against mathematics.

The breakthrough came through the introduction of the notion of a subgradient of a convex function. Given a convex function $f : \mathbb{R}^n \to \mathbb{R}$, the subgradient of f at $x \in \mathbb{R}^n$ is a vector ξ, such that

$$f(y) - f(x) \geq \langle \xi, y - x \rangle, \quad \forall y \in \mathbb{R}^n.$$

The collection of all subgradients of a convex function f at x is called the subdifferential of f at x and is denoted by $\partial f(x)$. Further note that for a convex function $f : \mathbb{R}^n \to \mathbb{R}$ the set $\partial f(x)$ is a nonempty, convex, and compact set for each $x \in \mathbb{R}^n$. This is not so simple if we consider a convex function $f : \mathbb{R}^n \to [-\infty, +\infty]$. In this case, the subgradient is defined as follows. Let x be a point where $f(x)$ is finite. Then, a subgradient is defined in the same way as above. Note that when $f(x)$ is not finite we set $\partial f(x) = \emptyset$. The set $\partial f(x)$ is always convex. Note that $\partial f(x)$ in this setting need not be nonempty and compact unless we have $x \in \operatorname{int} dom f$. Using the above definition of the subdifferential, we can make computation of the subdifferential of some simple functions. For example, consider the function $f(x) = |x|$, $x \in \mathbb{R}$. Then, $\partial f(0) = [-1, +1]$. Now consider the function $f = \delta_C$, where C is a closed convex set. Then, the subdifferential of the indicator function is given as follows:

$$\partial \delta_C(x) = \{v \in \mathbb{R}^n : \langle v, y - x \rangle \leq 0, \forall y \in C\}, \quad \text{if } x \in C$$

and $\partial \delta_C(x) = \emptyset$, if $x \notin C$. Note that when $x \in C$ the set $\partial \delta_C(x)$ is a cone and this cone is called the normal cone to the closed convex set C at x and is denoted by $N_C(x)$ and plays a pivotal role as vehicle for expressing optimality conditions and further it is deeply connected with the notion of Lagrange multipliers. We shall discuss this in quite a bit of detail when we discuss the Karush–Kuhn–Tucker conditions or KKT conditions.

Parallel to the notion of the subdifferential is the notion of the one-sided directional derivative. Consider a proper convex function and let x be a point where $f(x)$ is finite. Then the one-sided directional derivative (or just the directional derivative) of the convex function f at x in the direction h is given as

$$f'(x, h) = \lim_{\lambda \downarrow 0} \frac{f(x + \lambda h) - f(x)}{\lambda},$$

where $\lambda \downarrow 0$ means that $\lambda > 0$ and $\lambda \to 0$. If $x \in \text{int}dom f$ then $f'(x, h)$ is finite for each h and is a convex and positively homogenous function (sublinear function, see below). Further one can also show that

$$f'(x,h) = \inf_{\lambda > 0} \frac{f(x + \lambda h) - f(x)}{\lambda},$$

Further one can also easily demonstrate that for a proper convex function for any $x \in dom f$ we have

$$\partial f(x) = \{\xi \in \mathbb{R}^n : f(y) - f(x) \geq \langle \xi, y - x \rangle, \quad \forall y \in \mathbb{R}^n\}.$$

It is a simple exercise to show that if f is differentiable at x then $\partial f(x) = \{\nabla f(x)\}$. This shows that when f is differentiable at x we have $f'(x, h) = \langle \nabla f(x), h \rangle$. There is an very beautiful connection between the subdifferential and the directional derivative. For a given $x \in dom f$ the support function of the subdifferential at x is the directional derivative. This is thus written as

$$f'(x,h) = \max_{\xi \in \partial f(x)} \langle \xi, h \rangle.$$

The notion of the directional derivative will play an important role as we compute subdifferential of some important convex functions.

The subdifferential took the center stage in convex analysis since it allowed for the development of calculus rules along the lines of differential calculus. However, there is one particular rule for the calculation of the subdifferential of a max function, which differentiates the convex calculus from the usual differential calculus. Note that if f and g are finite-valued convex functions on $\mathbb{R}^n$, then we have

1. $\partial(f + g)(x) = \partial f(x) + \partial g(x)$.
2. $\partial(\lambda f)(x) = \lambda \partial f(x)$, when $\lambda \geq 0$.

In the set of two rules above, the first one is called the addition rule or sum rule for subdifferentials, and the second one is called the scaling rule. It is interesting to note that the sum rule holds even if the functions f and g are extended valued. However, in the extended-valued case, the sum rule holds under additional assumptions. For example, it has been shown in Rockafellar [8] that if f and g are proper convex functions with $ri(dom f) \cap ri(dom g) \neq \emptyset$ then the sum rule holds at each point $x \in dom f \cap dom g$. One could also have a less stringent condition wherein one asks that if there exists at least one point $x_0 \in dom f \cap dom g$ where at least one of the functions is continuous. This will give us the sum rule if we additionally assume that f and g are proper and lower semicontinuous. In fact, the last stated condition is weak and if that is violated the sum rule may break down. The following example due to Phelps [10] illustrates this.

Example 1.1

Let $C = epi\,\varphi$ where $\varphi(x) = x^2$, $x \in \mathbb{R}$. Let L be the x-axis in $\mathbb{R}^2$. Let $f = \delta_C$ and $g = \delta_L$. In this case $dom f \cap dom g = \{0\}$. Further $\partial f(0) = \mu(0, 1)$ where $\mu \leq 0$ and $\partial g(0) = \alpha(0, 1)$ where $\alpha \in \mathbb{R}$. It is not difficult to see that $\partial(f + g)(0) = \mathbb{R}^2 \neq \partial f(0) + \partial g(0)$. Note that neither f nor g is continuous at x_0.

We shall now concentrate on the calculus rule for the composition of two convex functions. However, it is important to note that just arbitrary compositions of convex functions need not give us a convex function. In fact, in order to preserve convexity we need the outer composing function to be increasing in some sense. Consider a function $F : \mathbb{R}^n \to \mathbb{R}^m$ be a vector function given as

$$F(x) = (f_1(x), \ldots, f_m(x)),$$

where each f_i is a finite-valued convex function. Let us consider a convex function $g : \mathbb{R}^m \to \mathbb{R}$ which is increasing in the following sense. Let $y, z \in \mathbb{R}^m$. We shall say that $y \geq z$ if $y_i \geq z_i$ for all $i = 1, \ldots, m$. Now we say that g is increasing if $g(y) \geq g(z)$ whenever $y \geq z$. We can now easily show that $g \circ F$ is a convex function and through the following result we show how to compute the subdifferential of the composite convex function. We have

$$\partial(g \circ F)(x) = \left\{ \sum_{i=1}^{m} v_i \xi^i : (v_1, v_2, \ldots, v_m) \in \partial g(F(x)), \xi^i \in \partial f_i(x), i = 1, \ldots, m \right\}.$$

An immediate application of the above result is the computation of the subdifferential of the max function given by a finite number of convex functions. Let us consider the convex function $f(x) = \max\{f_1(x), \ldots, f_m(x)\}$ where each f_i is a convex function. In fact, f can be represented as a composite convex function with $g = \max\{y_1, \ldots, y_m\}$ and $F(x) = \{f_1(x), \ldots, f_m(x)\}$. Thus, we have

$$\partial f(x) = \text{conv}\{\partial g_i(x) : i \in J(x)\},$$

where $J(x) = \{i : f(x) = f_i(x)\}$. When each f_i is differentiable then the rule becomes

$$\partial f(x) = \text{conv}\{\nabla f_i(x) : i \in J(x)\}.$$

Further if each f_i is an affine function, that is, $f_i(x) = \langle a_i, x\rangle + b$ then the subdifferential at each x is a polyhedral function given as

$$\partial f(x) = \text{conv}\{a_i : i \in J(x)\}.$$

Let us now focus on computing subdifferentials of some interesting class of convex functions. We shall consider the maximum eigenvalue function and also the distance function. We shall first compute the subdifferential for the maximum eigenvalue function. Note that in order to do that we shall need to focus on an important class of convex functions called sublinear functions. A sublinear function is nothing but a positively homogeneous convex function. It is this simple to show that a finite-valued sublinear function satisfies the following two properties:

1. $p(x + y) \leq p(x) + p(y)$ (subadditivity)
2. $p(\lambda x) = \lambda p(x), \lambda \geq 0$ (positively homogenous)

Further it is simple to show that

$$\partial p(0) = \{v \in \mathbb{R}^n : p(x) \geq \langle \xi, x\rangle, \forall x \in \mathbb{R}^n\}.$$

This can be easily seen by observing that

$$p'(0, x) = \lim_{\lambda \downarrow 0} \frac{p(\lambda x) - p(0)}{\lambda} = p(x).$$

This shows that

$$p(x) = \max_{\xi \in \partial p(0)} \langle \xi, x \rangle.$$

This then allow us to compute $\partial p(0)$.

We shall begin by showing that $\lambda_1(A) : S^n \to \mathbb{R}$ is a positively homogeneous function. This can be seen by noting that for any $\mu \geq 0$

$$\lambda_1(\mu A) = \max_{\|x\|=1} \langle x, \mu A x \rangle = \mu \max_{\|x\|=1} \langle x, Ax \rangle = \mu \lambda_1(A).$$

Our goal now is to show that

$$\partial \lambda_1(0) = \left\{ Y \in S^n_+ : \text{trace} Y = 1 \right\},$$

where S^n_+ is the cone of all symmetric positive semidefinite. Noting the fact that $\lambda_1(A)$ is positively homogenous, we have

$$\partial \lambda_1(A) = \{ Y \in S^n : \lambda_1(A) \geq \langle Y, A \rangle, \forall A \in S^n \}.$$

Further we have

$$\lambda_1(A) = \lambda_1'(0, A) = \max_{Y \in \partial \lambda_1(0)} \langle Y, A \rangle.$$

Let us now compute $\lambda_1(A)$. Note that

$$\lambda_1(A) = \max_{x \in \mathbb{R}^n, \|x\|=1} \langle x, Ax \rangle.$$

This can be written in matrix norm format as

$$\lambda_1(A) = \max_{xx^T, \text{trace}(xx^T)=1} \langle xx^T, A \rangle.$$

Noting that $\text{conv}\{xx^T : \text{trace}(xx^T) = 1\} = \{Y \in S^n_+ : \text{trace}(Y) = 1\}$, we conclude that

$$\lambda_1(A) = \max_{Y \in S^n_+ : \text{trace} Y = 1} \langle Y, A \rangle.$$

Hence, we have

$$\max_{Y \in S^n_+ : \text{trace} Y = 1} \langle Y, A \rangle = \max_{Y \in \partial \lambda_1(0)} \langle Y, A \rangle.$$

We will show that $\{Y \in S^n_+ : \text{trace}(Y) = 1\}$ is a compact set. This is proved by noting that since $Y \in S^n_+$ we have $\text{trace}(YY) \leq \text{trace}(Y)\text{trace}(Y)$ and hence we have $\|Y\|^2 \leq 1$. It is a well-known fact of convex analysis that if the support functions of two compact convex sets are equal then the sets must be equal (see, e.g., Reference 11). This shows that

$$\partial \lambda_1(0) = \left\{ Y \in S^n_+ : \text{trace} Y = 1 \right\}.$$

Our aim now would be to compute $\partial\lambda_1(A)$ where A is not a zero matrix. To do so we need to compute the subdifferential of the convex function $\varphi(x) = \max_{y\in C} f(x, y)$ where for each $y \in C$ the function $x \mapsto f(x, y)$ is convex and C is a compact set. This is an important result in convex analysis and we give the detailed proof here following closely the proof given in Ruzyncski [12]. So let us compute $\partial\varphi(x_0)$ for any given x_0 in $\mathbb{R}^n$. We shall begin by stating the following assumptions:

1. $f(., y)$ is finite valued and convex in x for each fixed $y \in C$.
2. $f(x, .)$ is upper semicontinuous in y for each fixed $x \in \mathbb{R}^n$.
3. C is a compact set.

For the given point x_0 let us denote by $\hat{Y}(x_0)$ the following set:

$$\hat{Y}(x_0) = \text{argmax}_{y\in C} f(x_0, y) = \{y \in C : f(x_0, y) = \varphi(x_0)\}.$$

If conditions (1), (2), and (3) hold, we have

$$\partial\varphi(x_0) \supseteq \text{conv}\left(\bigcup_{y\in\hat{Y}(x_0)} \partial_x f(x_0, y)\right),$$

where $\partial_x f$ means that we are just considering the subdifferential with respect to the variable x. Further if the function $f(., y)$ is continuous at x_0 for all $y \in C$, then

$$\partial\varphi(x_0) = \text{conv}\left(\bigcup_{y\in\hat{Y}(x_0)} \partial_x f(x_0, y)\right).$$

Thus, let us go ahead and prove this fact. Let us first prove the inclusion under conditions (1), (2), and (3). Let $g \in \partial_x f(x_0, y_0)$ for some $y_0 \in \hat{Y}(x_0)$. Then, for any x we have

$$\varphi(x) \geq f(x, y_0) \geq f(x_0, y_0) + \langle g, x - x_0\rangle.$$

This shows that

$$\varphi(x) \geq \varphi(x_0) + \langle g, x - x_0\rangle.$$

Hence, we conclude that $g \in \partial\varphi(x_0)$. This shows that $\partial_x f(x_0, y_0) \subseteq \partial\varphi(x_0)$. Since $y_0 \in \hat{Y}(x_0)$ can be chosen arbitrarily, we can conclude that $\partial_x f(x_0, y) \subseteq \partial\varphi(x_0)$ for all $y \in \hat{Y}(x_0)$. Hence, we conclude that

$$\partial\varphi(x_0) \supseteq \text{conv}\left(\bigcup_{y\in\hat{Y}(x_0)} \partial_x f(x_0, y)\right).$$

We have proved the first assertion. We shall now establish the equality. So we will begin by proving that the set

$$\text{conv}\left(\bigcup_{y\in\hat{Y}(x_0)} \partial_x f(x_0, y)\right)$$

is a closed and bounded set and hence compact. Note that using (2) we can show that $\hat{Y}(x_0)$ is closed and since it is a subset of a compact set C the set $\hat{Y}(x_0)$ is a compact set. First of all note that from our first derivation of the set inclusion it is clear that

$$\left(\bigcup_{y\in\hat{Y}(x_0)} \partial_x f(x_0, y)\right) \subseteq \partial\varphi(x_0).$$

This shows that the set

$$\left(\bigcup_{y\in\hat{Y}(x_0)} \partial_x f(x_0, y)\right)$$

is bounded since $\partial\varphi(x_0)$ is a compact and convex set. We shall now have to show that the set on the left-hand side of the above inclusion is closed. Let $\xi^k \in \partial_x f(x_0, y^k)$, with $y^k \in \hat{Y}(x_0)$, be a convergent sequence and let us assume that $\xi^k \to \xi^*$. Since $\hat{Y}(x_0)$ is a compact set, we conclude without loss of generality that $y^k \to y^* \in \hat{Y}(x_0)$. Thus, we have

$$f(x, y^k) \geq f(x_0, y^k) + \langle \xi^k, x - x_0\rangle.$$

Noting that $f(x_0, y^k) = f(x_0, y^*)$ we have

$$f(x, y^k) \geq f(x_0, y^*) + \langle \xi^k, x - x_0\rangle.$$

Using hypothesis (2), which speaks about the upper semicontinuity of f in the variable y we conclude that as $k \to \infty$

$$f(x, y^*) \geq f(x_0, y^*) + \langle \xi^*, x - x_0\rangle.$$

Now this shows that $\xi^* \in \partial_x f(x_0, y^*)$. Hence, $\xi^* \in (\bigcup_{y\in\hat{Y}(x_0)} \partial_x f(x_0, y))$. This shows that $(\bigcup_{y\in\hat{Y}(x_0)} \partial_x f(x_0, y))$ is closed and bounded and hence compact and thus showing that $\text{conv}(\bigcup_{y\in\hat{Y}(x_0)} \partial_x f(x_0, y))$ is a compact convex set.

Consider on the contrary that there exists $\xi^0 \in \partial\varphi(x_0)$ such that

$$\xi^0 \notin \text{conv}\left(\bigcup_{y\in\hat{Y}(x_0)} \partial_x f(x_0, y)\right).$$

Thus, by standard separation theorem (see any good book on convex analysis) there exists $d \in \mathbb{R}^n$ with $d \neq 0$ such that

$$\langle \xi^0, d\rangle > \langle \xi, d\rangle, \quad \text{for all } \xi \in \partial_x f(x_0, y) \quad \text{and} \quad \text{for all } y \in \hat{Y}(x_0).$$

Let us consider a sequence $\{\lambda_k\} \subset \mathbb{R}_+$ with $\lambda_k \downarrow 0(\lambda_k > 0$ and $\lambda_k \to 0)$. Since φ is convex, we have

$$\frac{\varphi(x_0 + \lambda_k d) - \varphi(x_0)}{\lambda_k} \geq \langle \xi^0, d\rangle.$$

For each $k \in \mathbb{N}$ define the set

$$Y_k = \left\{ y \in C : \frac{\varphi(x_0 + \lambda_k d, y) - \varphi(x_0)}{\lambda_k} \geq \langle \xi^0, d \rangle \right\}.$$

We claim that for each k the set is nonempty and compact. Since $f(x, .)$ is upper semicontinuous for each x we can conclude that $\hat{Y}(x_0 + \lambda_k d)$ is nonempty. This means that $\hat{Y}(x_0 + \lambda_k d) \subseteq Y_k$, showing that $Y_k \neq \emptyset$. It is obvious that $Y_k \subset C$ and thus Y_k is bounded. To show that Y_k is compact, we have to now show that Y_k is closed. Let $y^r \in Y_k$ and $y^r \to \hat{y}$. Now $y_r \to \hat{y}$ means

$$\frac{f(x_0 + \lambda_k d, y^r) - \varphi(x_0)}{\lambda_k} \geq \langle \xi^0, d \rangle.$$

Now using the fact that $f(x, .)$ is upper semicontinuous we have as $k \to \infty$,

$$\frac{f(x_0 + \lambda_k d, \hat{y}) - \varphi(x_0)}{\lambda_k} \geq \langle \xi^0, d \rangle.$$

This shows that $\hat{y} \in Y_k$, showing that Y_k is closed.

Using the property of the directional derivative, we know that for $0 < \lambda_2 \leq \lambda_1$ we have

$$\frac{f(x_0 + \lambda_2 y, d) - f(x_0, y)}{\lambda_2} \leq \frac{f(x_0 + \lambda_1 d, y) - f(x_0, y)}{\lambda_1}.$$

Since $\varphi(x_0) \geq f(x_0, y)$ for every y, we have for $0 < \lambda_2 \leq \lambda_1$,

$$\frac{f(x_0, y) - \varphi(x_0)}{\lambda_2} \leq \frac{f(x_0, y) - \varphi(x_0)}{\lambda_1}.$$

Combining the above two inequalities we have for $0 < \lambda_2 \leq \lambda_1$,

$$\frac{f(x_0 + \lambda_2 d, y) - \varphi(x_0)}{\lambda_2} \leq \frac{f(x_0 + \lambda_1 d, y) - \varphi(x_0)}{\lambda_1}.$$

Thus, $Y_2 \subset Y_1$ for $0 < \lambda_1 < \lambda_2$. Hence, if $\{\lambda_k\}$ be a sequence of positive numbers decreasing to zero, then we have $Y_1 \supset Y_2 \supset Y_3 \ldots$. Further as $Y_1, Y_2, Y_3, \ldots$ are compact and as their diameters go to zero, we have $\bigcap_{k=1}^{\infty}$ is nonempty. Let $\tilde{y} \in Y_k$ for all $k \in \mathbb{N}$. Thus, for each k we have

$$\frac{f(x_0 + \lambda_k d, \tilde{y}) - \varphi(x_0)}{\lambda_k} \geq \langle \xi^0, d \rangle.$$

This implies that

$$f(x_0 + \lambda_k d, \tilde{y}) - \varphi(x_0) \geq \lambda_k \langle \xi^0, d \rangle.$$

Using the continuity of $f(., y)$ at point x_0, we have

$$f(x_0, \tilde{y}) - \varphi(x_0) \geq 0.$$

This shows that $\varphi(x_0) = f(x_0, \tilde{y})$. Hence, $\tilde{y} \in \hat{Y}(x_0)$. This simply means that

$$f'((x_0, \tilde{y}), d) \geq \langle \xi^0, d \rangle.$$

Then, there exists $\xi \in \partial_x f(x_0, \tilde{y})$ such that

$$\langle \xi, d \rangle \geq \langle \xi^0, d \rangle.$$

This contradicts the separation inequality. Hence, the result.

We can now apply this fact to compute $\partial\lambda_1(A)$ for A not equal to the zero matrix. It is easy to note that $\nabla\langle xx^T, A\rangle = xx^T$. The reader can check out that

$$\hat{Y}(A) = \{xx^T : x^T x = 1, Ax = \lambda_1(A)x\}.$$

What we have just said is that the maximizers of the functions $\langle x, Ax\rangle$ over norm $\|x\| = 1$ are the eigenvectors of $\lambda_1(A)$. Thus, we have

$$\partial\lambda_1(A) = \text{conv}\{xx^T : x^T x = 1, Ax = \lambda_1(A)x\}.$$

Now we shall see how to compute the subdifferential of a distance function to a closed convex set. Our approach is based on the approach in Hiriart-Urruty and Lemarechal [11]. Let C be a closed convex set. Then,

$$\partial d_C(x) = \begin{cases} N_C(x) \cap \mathbb{B}, & \text{if } x \in C \\ \left\{\frac{x - p_C(x)}{\|x - p_C(x)\|}\right\}, & \text{if } x \notin C, \text{ where } p_C(x) = \text{Proj}_C(x), \end{cases}$$

where $\text{Proj}_C(x)$ denotes the projection onto the closed convex set C from $x \in \mathbb{R}^n$. For more detail, see Hiriart-Urruty and Lemarechal [7] (Vol. I). When $x \notin C$, $d_C(x) > 0$. In fact, for each $x \notin C$, we can write

$$d_C(x) = \|x - p_C(x)\|.$$

Now since $x \neq p_C(x)$, $d_C(x)$ is differentiable at each $x \notin C$ and we have

$$d_C(x) = \sqrt{d_C^2(x)}.$$

Thus,

$$\nabla d_C(x) = \frac{1}{2\sqrt{d_C^2(x)}} \nabla d_C^2(x).$$

This simplifies to

$$\nabla d_C(x) = \frac{\nabla d_C^2(x)}{2 d_C(x)},$$

where $d_C^2(x) = \|x - p_C(x)\|^2$. Suppose we are able to prove that $\nabla d_C^2(x) = 2(x - p_C(x))$, then we have

$$\nabla d_C(x) = \frac{x - p_C(x)}{\|x - p_C(x)\|}.$$

To prove that, denote

$$\triangle = d_C^2(x+h) - d_C^2(x).$$

Now by definition

$$d_C^2(x) \le \|x - p_C(x+h)\|^2;$$

therefore,

$$\begin{aligned}
\triangle &\ge \|x+h - p_C(x+h)\|^2 - \|x - p_C(x+h)\|^2 \\
\triangle &\ge \|x - p_C(x+h)\|^2 - 2\langle x - p_C(x+h), h\rangle + \|h\|^2 - \|x - p_C(x+h)\|^2 \\
\triangle &\ge \|h\|^2 - 2\langle h, x - p_C(x+h)\rangle.
\end{aligned}$$

Further,

$$\triangle \le \|x+h - p_C(x)\|^2 - \|x - p_C(x)\|^2 = \|h\|^2 + 2\langle h, x - p_C(x)\rangle.$$

Thus,

$$\|h\|^2 + 2\langle h, x - p_C(x+h)\rangle \le \triangle \le \|h\|^2 + 2\langle h, x - p_C(x)\rangle.$$

Let

$$\begin{aligned}
\triangle &\ge \|h\|^2 + 2\langle h, x - p_C(x+h)\rangle \\
-\triangle &\le -\|h\|^2 + 2\langle h, p_C(x+h) - x\rangle \\
-\triangle &\ge -\|h\|^2 + 2\langle h, p_C(x+h) - p_C(x) + p_C(x) - x\rangle \\
&= -\|h\|^2 + 2\langle h, p_C(x+h) - p_C(x)\rangle + 2\langle h, p_C(x) - x\rangle.
\end{aligned}$$

Thus,

$$-\triangle \le -\|h\|^2 + 2\|h\|\|p_C(x+h) - p_C(x)\| + 2\langle h, p_C(x) - x\rangle.$$

Using the fact that p_C is nonexpansive, we have

$$\|p_C(x+h) - p_C(x)\| \le \|x+h-x\|,$$

that is,

$$\|p_C(x+h) - p_C(x)\| \le \|h\|.$$

Thus,

$$\begin{aligned}
-\triangle &\le -\|h\|^2 + 2\|h\|^2 + 2\langle h, p_C(x) - x\rangle \\
\triangle &\ge -\|h\|^2 + 2\langle h, x - p_C(x)\rangle
\end{aligned}$$

Thus,

$$\begin{aligned}
&-\|h\|^2 + 2\langle h, x - p_C(x)\rangle \le \triangle \le \|h\|^2 + 2\langle h, x - p_C(x)\rangle \\
&-\|h\|^2 \le d_C^2(x+h) - d_C^2(x) - \langle h, 2(x - p_C(x))\rangle \le \|h\|^2 \\
&-\|h\| \le \frac{d_C^2(x+h) - d_C^2(x) - \langle h, 2(x - p_C(x))\rangle}{\|h\|} \le \|h\|.
\end{aligned}$$

As $\|h\| \to 0$

$$\lim_{\|h\|\to 0} \frac{d_C^2(x+h) - d_C^2(x) - \langle h, 2(x - p_C(x))\rangle}{\|h\|} = 0.$$

Thus, we have

$$\nabla^2 d_C(x) = 2(x - p_C(x)).$$

Now as $x \in C$, let $s \in \partial d_C(x)$, we have

$$d_C(x') - d_C(x) \geq \langle s, x' - x\rangle, \quad \forall\, x \in \mathbb{R}^n.$$

As $x \in C \;\Rightarrow\; d_C(x) = 0$, then

$$d_C(x') \geq \langle s, x' - x\rangle, \quad \forall\, x \in \mathbb{R}^n. \tag{1.2}$$

For all $x' \notin C$, we have taken $x' = x + s$. Thus, we have from Equation 1.2

$$\begin{aligned} \|s\|^2 &\leq d_C(x+s) \leq \|x + s - x\| = \|s\| \\ \|s\|^2 &\leq d_C(x+s) \leq \|s\| \\ \|s\| &\leq 1 \Rightarrow s \in \mathbb{B} \quad \text{(the unit ball in } \mathbb{R}^n\text{)}. \end{aligned}$$

Thus, we have

$$\partial d_C(x) \subseteq N_C(x) \cap \mathbb{B}.$$

Let $s \in N_C(x) \cap \mathbb{B}$. Then, for all $x' \in \mathbb{R}^n$, we have

$$\langle s, x' - x\rangle = \langle s, x' - p_C(x')\rangle + \langle s, p_C(x') - x\rangle.$$

Now as $s \in N_C(x)$, we have

$$\begin{aligned} \langle s, p_C(x') - x\rangle &\leq 0 \\ \langle s, x' - x\rangle &\leq \langle s, x' - p_C(x')\rangle \\ &\leq \|s\|\|x' - p_C(x')\| \\ &\leq \|x' - p_C(x')\| = d_C(x') \; (\|s\| \leq 1). \end{aligned}$$

Thus,

$$d_C(x') \geq \langle s, x' - x\rangle, \quad \forall x \in \mathbb{R}^n.$$

Hence,

$$N_C(x) \cap \mathbb{B} \subseteq \partial d_C(x).$$

This establishes the result.

1.3 What Is Convex Optimization?

Convex optimization simply means minimizing a convex function over a convex set. More precisely, we can write the problem as

$$\min f(x) \qquad \text{(CP)}$$
$$\text{subject to} \quad x \in C,$$

where $f : \mathbb{R}^n \to \mathbb{R}$ is a convex function and C is a convex set. The set C need not be just an abstract convex set but can be viewed as represented by equality and inequality constraints. In general, we consider

$$C = \{x \in \mathbb{R}^n : g_i(x) \leq 0,\ i = 1, \ldots, m,\ \langle a_j, x\rangle - b_j = 0,\ j = 1, \ldots, k\},$$

where each g_i is a convex function, $a_j \in \mathbb{R}^n$ and $b_j \in \mathbb{R}$. Note that the equality constraints have to be affine or the set need not be convex. For example, consider

$$C_1 = \left\{(x_1, x_2) \in \mathbb{R}^2 : x_1^2 + x_2^2 - 1 = 0\right\},$$

which is the unit circle in two dimensions and naturally not a convex set though $g(x_1, x_2) = x_1^2 + x_2^2 - 1$ is a convex function. We would also like the reader to take note of the fact that a set C could be convex even if we do not have all the functions describing the inequality constraints to be convex. Consider the set C_2 given as

$$C_2 = \{(x_1, x_2) \in \mathbb{R}^2 : 1 - x_1x_2 \leq 0,\ -x_1 \leq 0,\ -x_2 \leq 0\}.$$

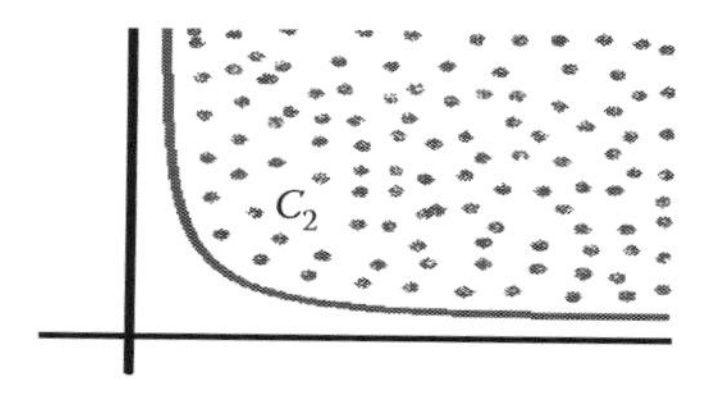

Though $1 - x_1x_2$ is not a convex function, the set C_2 is indeed convex. More about this issue can be found in the next section but let us now focus on the various important classes of convex optimization problems.

Before we begin looking into the important particular cases of CP. Let us just note that for simplicity of the exposition we consider f to be finite valued but there is a compelling reason theoretically as to why it should be so. In fact, one can consider f to be extended valued. However, most of the important examples of CP have their objective functions finite valued. Further, we should constantly remember the fact that for the problem (CP) every local minimizer is global.

1.3.1 Linear Programming Problem

Possibly, the most important class of convex optimization problems are the linear programming problems. These problems have a simple geometrical structure and have efficient algorithms to solve them.

A linear programming problem in its canonical form is given as

$$\min \langle c, x \rangle \qquad \text{(LP)}$$
$$\text{subject to} \quad Ax \geq b, \quad x \geq 0,$$

where $c \in \mathbb{R}^n$, usually called the cost vector; A is an $m \times n$ matrix called the constraints matrix and $b \in \mathbb{R}^m$. The symbol "$\geq$" in the above problem represents componentwise inequality. By subtracting a surplus variable, one can convert the inequality $Ax \geq b$ into an equality. Thus, in general, the linear programming problem in the standard form is given as

$$\min \langle c, x \rangle \qquad \text{(LPS)}$$
$$\text{subject to} \quad Ax = b, \quad x \geq 0.$$

In our discussion below let us just focus on the problem (LP). A remarkable feature of a linear programming problem is the following:

Theorem 1.2

If the problem (LP) is bounded below on the feasible set

$$\hat{C} = \{x \in \mathbb{R}^n : \ Ax \geq b, \ x \geq 0\},$$

then there exists a minimizer of (LP).

The question now is how to get this lower bound on the objective. Let us begin by observing that for any $x \in \hat{C}$, we have

$$Ax \geq b$$

and hence for any $y \in \mathbb{R}^m_+$, we have

$$\langle y, Ax \rangle \geq \langle b, y \rangle.$$

This shows that

$$\langle A^T y, x \rangle \geq \langle b, y \rangle.$$

Let $y \in \mathbb{R}^m_+$ be so chosen that $A^T y \leq c$. Then, for any $x \geq 0$, we have

$$\langle c, x \rangle \geq \langle A^T y, x \rangle = \langle b, y \rangle.$$

This shows that for any $x \geq 0$, with $Ax \geq b$, we have

$$\langle c, x \rangle \geq \max_{A^T y \leq c, \ y \geq 0} \langle b, y \rangle.$$

Thus,

$$\min_{Ax \geq b, x \geq 0} \langle c, x \rangle \geq \max_{A^T y \leq c, \ y \geq 0} \langle b, y \rangle.$$

The inequality is called the "weak duality." In fact, the problem

$$\begin{aligned} &\max \ \langle b, y \rangle && \text{(DP)} \\ &\text{subject to} \quad A^T y \le c, \quad y \ge 0. \end{aligned}$$

is called the dual problem of (LP). One of the far-reaching consequences of linear programming is the following fact.

Theorem 1.3: Strong Duality

If (LP) and (DP) are feasible, then their optimal values coincide, that is,

$$\min_{Ax \ge b,\, x \ge 0} \langle c, x \rangle = \max_{A^T y \le c,\, y \ge 0} \langle b, y \rangle.$$

Strong duality actually separates the linear programming problem from other classes of convex optimization problems. Strong duality does not come for free for nonlinear convex programming problems. The question now arises is whether there lies behind a minimization problem a maximization problem. We will discuss this later. Let us now provide an example of an important linear programming problem, the transportation problem.

Consider the following transportation network associated with a retailer.

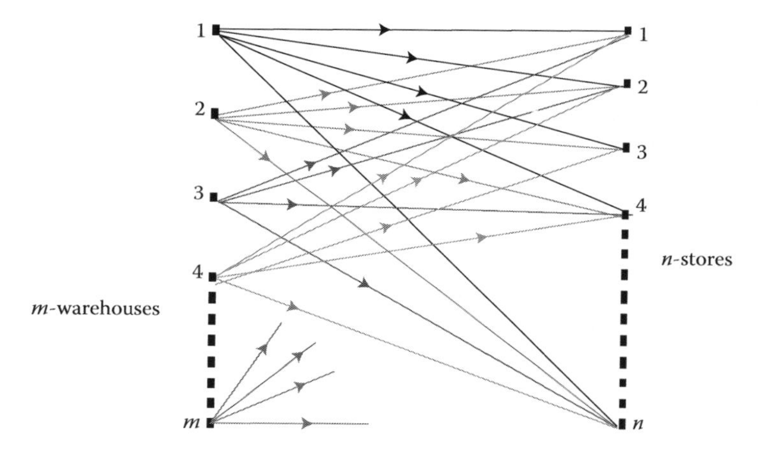

There are m warehouses, which cater to n stores. The problem is to transport the materials from the warehouses to the stores at minimum cost

$$a_i = \text{total supply at warehouse } i$$

$$b_j = \text{total demand at store } j.$$

We assume that $\sum_{i=1}^{m} a_i = \sum_{j=1}^{n} b_j$, that is, total supply equals total demands. So we have to decide on the optimal amount of material x_{ij} to be transported from warehouse i to store j where c_{ij} is the unit cost of transportation.

Thus, our problem is as follows:

$$\min \sum_{j=1}^{n} \sum_{i=1}^{m} c_{ij} x_{ij}$$

subject to

$$\sum_{j=1}^{n} x_{ij} = a_i, \quad i = 1, \ldots, m, \tag{1.3}$$

$$\sum_{i=1}^{m} x_{ij} = b_j, \quad j = 1, \ldots, n, \tag{1.4}$$

$$x_{ij} \geq 0, \quad i = 1, \ldots, m, \ i = 1, \ldots, n.$$

We will not explain how the constraints are constructed and we feel that it is self-explanatory. We leave it to the reader to show that an optimal solution to the transportation problem exists if and only if $a_i, b_j \geq 0$, $i = 1, \ldots, m$, $j = 1, \ldots, n$, and $\sum_{i=1}^{m} a_i = \sum_{j=1}^{n} b_j$ (demand = supply). We have in fact chosen the transportation problem with a purpose. The name "linear programming" was suggested to George B. Dantzig by the economist T.C. Koopmans. Every optimizer worth his/her salt would know the name of Dantzig as the inventor of the simplex method, which is the most popular way of solving a linear programming problem. But when we teach linear programming we hardly talk about the contribution of the great Russian mathematician L.V. Kantorovich (winner of the Nobel Prize in economics in 1975) who had been working on linear programming before Dantzig. Kantorovich actually worked on what is known as the optimal transport problem, which is an infinite-dimensional version of the transportation problem. Optimal transport was in fact first stated by Gaspard Monge in 1783 in France. Recently, optimal transport is back to the limelight after Cedric Villani, one of its great practitioners, who received the Fields Medal in 2010. Villani also wrote two lovely monographs on optimal transport [13,14]. Unfortunately, we are succumbing to the temptation of describing the Monge–Kantorovich transportation problem.

We shall now provide a motivation due to Villani [13] on how to develop the Monge form of the optimal transport problem.

The problem is the following. We are given a pile of sand and also a hole, which we have to fill up with sand (Figure 1.8). The first step is to normalize the mass to 1. We shall model the sandpile and hole by measure μ and ν, respectively, defined on some measure spaces X and Y. Thus, if $A \subset X$ and $B \subset Y$, then $\mu(A)$ measures the amount of sand in A and $\nu(B)$ measures the amount of sand that can be piled in B. Thus, μ and ν can be indeed thought of as probability measures. One goal is

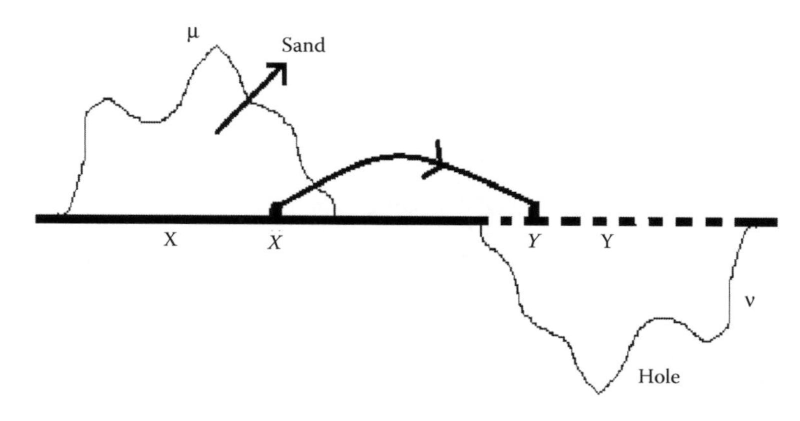

FIGURE 1.8
Transporting sand.

to transport the sand from the pile to the hole in minimum cost. Let $c : X \times Y \to \mathbb{R} \cup \{\infty\}$ be the cost function associated with moving sand from $x \in X$ to $y \in Y$. Monge posed the problem as

$$\min \mathrm{I(T)} = \int_X c(x, T(x)) d\mu(x)$$

over all measurable maps T such that $T\#\mu = \nu$, where $T\#\mu$ is the push-forward of μ by T, given as

$$T\#\mu(B) = \nu(B) = \mu[T^{-1}(B)],$$

where $B \subset Y$ is measurable. In fact, T is a measurable map from X to Y. Roughly, it would mean transport or push to μ by T to obtain ν.

The optimal transport problem in the Monge form has its own drawbacks. One of the major drawbacks is that it need not be feasible. Thus, we now pose the optimal transport problem in the Kantorovich form that has a striking resemblance with the transport problem in the finite-dimensional form. First let us discuss the "transference plan," which to an optimizer means the decision variable. The transference plan or the decision variable will be the probability measure π on the product space $X \times Y$. On a crude level $d\pi(x, y)$ can be viewed as the infinitesimal mass transferred from X to Y, thus the cost is given as

$$\int_{X \times Y} c(x, y)\, d\pi(x, y).$$

If we look at Equations 1.3 and 1.4, then it is natural to assume the constraints

$$\int_Y d\pi(x, y) = d\mu(x) \quad \text{and} \quad \int_X d\pi(x, y) = d\nu(y).$$

Note that

$$\int_X d\mu(x) = 1 = \int_Y d\nu(y) \quad \text{(total demand = total supply)}.$$

More precisely, we need the constraints

$$\Pi[A \times Y] = \mu[A],$$
$$\Pi[X \times B] = \nu[B],$$

for all measurable sets $A \subset X$ and $B \subset Y$. (Measurable set here truly means events and X and Y are probability spaces). Villani [13] notes that the above constraints can be equivalently stated as follows. There are functions $(\phi, \psi) \subset L^1(\nu) \times L^1(\mu)$ such that

$$\int_{X \times Y} [\phi(x) + \psi(y)]\, d\pi(x, y) = \int_X \phi(x)\, d\mu(x) + \int_Y \psi(y)\, d\nu(y).$$

Villani [13] defines the feasible set as follows:

$$\Pi(\mu, \nu) = \{\pi \in P(X \times Y) : \ \Pi(A \times Y) = \mu(A), \ \Pi(X \times B) = \nu(B)$$
$$\forall A \subset X, B \subset Y, \ \text{measurable}\}.$$

Thus, the Kantorovich optimal transport problem is given as

$$\min I(\pi) = \int_{X \times Y} c(x, y)\, d\pi(x, y)$$

$$\text{subject to} \quad \pi \in \Pi(\mu, \nu).$$

We would again like to note that the optimal transport problem is a linear programming problem in infinite dimensions. Can we construct a dual to it? The answer to this is yes and the dual problem of an optimal transport problem is usually called the Kantorovich dual problem or the Monge–Kantorovich dual problem. We will just call it Kantorovich duality following Villani [13] and this is given as follows:

$$\max_{(\phi,\psi)\in\Phi_C} J(\phi, \psi),$$

where

$$J(\phi, \psi) = \int_X \phi\, d\mu + \int_Y \psi\, d\nu, \quad \text{where } (\phi, \psi) \in L^1(\mu) \times L^1(\nu) \text{ and}$$

$$\Phi_C = \{(\phi, \psi) \in L^1(\mu) \times L^1(\nu) : \ \phi(x) + \psi(y) \leq c(x, y)\}.$$

The inequality $\phi(x) + \psi(y) \leq c(x, y)$ must be understood in an almost everywhere sense. We state below the famous Kantorovich duality theorem from Villani [13].

Theorem 1.4

Let X and Y be Polish spaces (Just think of a separable metric space or a Banach space). Let μ be a probability measure on X and ν, a probability measure on Y. Then,

$$\inf_{\pi\in\Pi(\mu,\nu)} I[\pi] = \sup_{(\phi,\psi)\in\Phi_C} J(\phi, \psi). \tag{1.5}$$

We shall not give a proof here since its detailed proof is quite involved. However, from Villani [13] we shall just provide a sketch of the proof, which is exciting in its own right. A very important conclusion that one can draw from the proof is that the infimum on the left-hand side of Equation 1.5 is attained. Further the value of $\sup_{(\phi,\psi)\in\Phi_C} J(\phi, \psi)$ is unchanged if we consider ϕ and ψ to be bounded and continuous functions on X and Y, respectively, that is, $(\phi, \psi) \in \Phi_C \cap (C_b(X) \times C_b(Y))$, that is,

$$\sup_{(\phi,\psi)\in\Phi_C} J(\phi, \psi) = \sup_{(\phi,\psi)\in\Phi_C\cap(C_b(X)\times C_b(Y))} J(\phi, \psi).$$

Note that $(C_b(X) \times C_b(Y)) \subset L^1(\mu) \times L^1(\nu)$. Hence, we have

$$\sup_{(\phi,\psi)\in\Phi_C\cap C_b} J(\phi, \psi) \leq \sup_{(\phi,\psi)\in\Phi_C} J(\phi, \psi) \leq \inf I(\pi),$$

where we write $\Phi_C \cap C_b := \Pi_C \cap (C_b(X) \times C_b(Y))$. Now let $(\phi, \psi) \in \Phi_C$ and let $\pi \in \Pi(\mu, \nu)$. Then, by definition of Π, we have

$$J(\phi, \psi) = \int_X \phi\, d\mu + \int_Y \psi\, d\nu = \int_{X\times Y} [\phi(x) + \psi(y)]\, d\pi(x, y),$$

where the above inequality is in the π-almost everywhere sense. To show this let us assume $N_x \subset X$ and $N_y \subset Y$ be such that $\mu[N_x] = 0 = \nu[N_y]$, and let

$$\phi(x) + \psi(y) \leq c(x, y)\ \forall (x, y) \in N_x^C \times N_y^C.$$

Here N_x^C and N_y^C denotes the compliments of the N_x and N_y respectively.

Now π has marginal ν and μ and we have

$$\pi[N_x \times Y] = \mu[N_x] = 0,$$
$$\pi[X \times N_x] = \nu[N_y] = 0$$

and $\pi[(N_x^C \times N_y^C)^C] = 0$. Thus,

$$\int_{X \times Y} [\phi(x) + \psi(y)]\, d\pi(x, y) \leq \int_{X \times Y} c(x, y)\, d\pi = I(\pi).$$

Hence,

$$\sup_{(\phi,\psi)\in\Phi_C} J(\phi, \psi) \leq \inf_{\pi\in\Pi(\mu,\nu)} I(\pi). \tag{1.6}$$

This result is often called weak duality in a manner analogous with the finite-dimensional linear programming problem. Then, we have equality in Equation 1.5 we call it the strong duality. Actually, the duality theorem of Kantorovich is the strong duality theorem. Note that we have

$$\inf_{\pi\in\Pi(\mu,\nu)} I(\pi) = \inf_{\pi\in M_{+(x,y)}} I(\pi) + \delta_{\Pi(\mu,\nu)}(\pi),$$

where $\delta_{\Pi(\mu,\nu)}$ is the indicator function of $\Pi(\mu, \nu)$ and $M_{+(x,y)}$ is the set of all nonnegative Borel measures on $X \times Y$. We can show that

$$\sup_{(\phi,\psi)\in C_b(x)\times C_b(y)} \left[\int_X \phi\, d\mu + \int_Y \psi\, d\nu - \int_{X\times Y} [\phi(x) + \psi(y)]\, d\pi(x, y)\right]$$
$$= \begin{cases} 0, & \text{if } \pi \in \Pi(\mu, \nu) \\ +\infty, & \text{otherwise.} \end{cases}$$

Hence, we have

$$\inf_{\pi\in\Pi(\mu,\nu)} I(\pi) = \inf_{\pi\in M_{+(x,y)}} \left[I(\pi) + \sup_{(\phi,\psi)} \left[\int_X \phi\, d\mu + \int_Y \psi\, d\nu - \int_{X\times Y} [\phi(x) + \psi(y)]\, d\pi(x, y)\right]\right]$$
$$= \inf_{\pi\in M_{+(x,y)}} \sup_{(\phi,\psi)} \left[I(\pi) + \int_X \phi\, d\mu + \int_Y \psi\, d\nu - \int_{X\times Y} [\phi(x) + \psi(y)]\, d\pi(x, y)\right].$$

Now let us take for granted a minimax principle and swap the infimum and the supremum to have

$$\inf_{\pi\in\Pi(\mu,\nu)} I(\pi) = \sup_{(\phi,\psi)} \inf_{\pi\in M_{+(x,y)}} \left[\int_{X\times Y} c(x,y)\,d\pi(x,y) + \int_X \phi\,d\mu + \int_Y \psi\,d\nu - \int_{X\times Y} [\phi(x)+\psi(y)]\,d\pi(x,y)\right]$$

$$= \sup_{(\phi,\psi)} \left[\int_X \phi\,d\mu + \int_Y \psi\,d\nu - \sup_{\pi\in M_{+(x,y)}} \int_{X\times Y} [\phi(x)+\psi(y)-c(x,y)]\,d\pi(x,y)\right].$$

Now let us try to calculate the supremum above. Consider $h(x,y) = \phi(x)+\psi(y)-c(x,y)$. Then we have two choices. First, we have $\phi(x)+\psi(y) \le c(x,y)$ for all (x,y) except a set of measure zero in the product space. Or else there is a set of nonzero measure N such that $\phi(x)+\psi(y) > c(x,y)\ \forall (x,y) \in N$. Let us take the second case, first let $(x_0,y_0) \in N$, then consider $\pi = \lambda\delta_{(x_0,y_0)}$, where $\delta_{(x_0,y_0)}$ is the Dirac mass at (x_0,y_0) and $\lambda > 0$. As $\lambda \to \infty$ we see that the supremum above becomes infinite. Now once we consider the first case, we have

$$\int_{X\times Y} [\phi(x)+\psi(y)-c(x,y)]\,d\pi(x,y) \le 0\ \forall \pi \in \Pi(\mu,\nu).$$

Now if we choose $\pi = 0$, we have

$$\int_{X\times Y} [\phi(x)+\psi(y)-c(x,y)]\,d\pi(x,y) = 0.$$

Hence,

$$\sup_{\pi\in M_{+(x,y)}} \int_{X\times Y} [\phi(x)+\psi(y)-c(x,y)]\,d\pi(x,y) = 0$$

if $(\phi,\psi) \in \Phi_C$. Thus, we have

$$\sup_{(\phi,\psi)} \left[\int_X \phi\,d\mu + \int_Y \psi\,d\nu - \sup_{\pi\in M_{+(x,y)}} \int_{X\times Y} [\phi(x)+\psi(y)-c(x,y)]\,d\pi(x,y)\right]$$

$$= \sup_{(\phi,\psi)\in\Phi_C} \left[\int_X \phi\,d\mu + \int_Y \psi\,d\nu\right]$$

$$= \sup_{(\phi,\psi)\in\Phi_C} J(\phi,\psi).$$

This implies that

$$\inf_{\pi\in\Pi(\mu,\nu)} I(\pi) = \sup_{(\phi,\psi)\in\Phi_C} J(\phi,\psi).$$

1.3.2 Quadratic and Conic Programming Problem

Every convex optimization problem can be viewed as a problem of minimizing a linear function over a convex set. Consider the convex optimization problem,

$$\min f(x)$$
$$\text{subject to} \quad x \in C.$$

Then, we can equivalently write it as

$$\min_{x,t} \ t$$
$$\text{subject to} \quad f(x) \leq t, \quad x \in C.$$

The feasible set of this problem is

$$C_1 = \{x \in \mathbb{R}^n : f(x) \leq t, \ x \in C\}$$

It is easy to see that C_1 is a convex set. Many important optimization problems that arise in application can be viewed as a linear programming problem.

Example 1.2: Basis Pursuit Problem

$$\min \|x\|_2$$
$$\text{subject to} \quad Ax = b.$$

The above problem is equivalent to

$$\min_{x,t} \ t$$
$$\text{subject to} \quad \|x\|_2 \leq t, \quad Ax = b.$$

Thus, we can also equivalently write the basis pursuit problem as

$$\min_{x,t} \ t$$
$$\text{subject to} \quad x_1^2 + x_2^2 + \cdots + x_n^2 \leq t^2, \quad Ax = b.$$

Thus, the basis pursuit problem is a differentiable optimization problem where we minimize a linear function subject to linear and quadratic constraints.

Example 1.3: l_∞-Residual Norm Minimization Problem

$$\min_x \|Ax - b\|_\infty,$$
$$A \in \mathbb{R}^{m \times n}, \quad b \in \mathbb{R}^m$$

equivalent problem

$$\min_{x,t} \ t$$
$$\|Ax - b\|_\infty \leq t.$$

Further we have $\|Ax - b\| \leq t$ implies that

$$|\langle a_i, x\rangle - b_i| \leq t, \quad \text{where } a_i \text{ is the } i\text{th row of } A \text{ and } b_i \text{ is the } i\text{th component of } b.$$

$$-t \leq \langle a_i, x\rangle - b_i \leq t$$

$$\langle a_i, x\rangle - t \leq b_i, \ \forall i,$$

$$\langle a_i, x\rangle + t \geq b_i, \ \forall i.$$

Thus, we have

$$\min_{x,t} \ t$$

$$\langle a_i, x\rangle - t \leq b_i, \quad i = 1, 2, \ldots, m$$

$$\langle a_i, x\rangle + t \geq b_i, \quad i = 1, 2, \ldots, m.$$

We now, in fact, arrive at a linear programming problem.

Example 1.4: Compressed Sensing Problem

Compressed sensing essentially means "making to do with less." This is, in fact, the title of a paper on compressed sensing by Kurt Bryan and Tanya Leise [15]. In compressed sensing, one has to detect very few faulty elements in a given set of elements in a given number of moves. The number of such moves is much less than the number of elements and this problem can be posed as finding a sparse solution of

$$Ax = b, \tag{1.7}$$

where A is an $m \times n$ matrix with $m \ll n$. In fact, one seeks to find a solution of a given level, that is, say k-space solution. This means at most that the k-components of the solution vector is nonzero. Emmanuel, Candes, and Terence Tao had shown in their seminal paper "Decoding by linear programming" (2006) that minimizing the 1-norm (or l_1-norm) over Equation 1.7 can lead to the sparse solution. The 1-norm problem is given as

$$\min \|x\|_1$$

$$\text{subject to} \quad Ax = b.$$

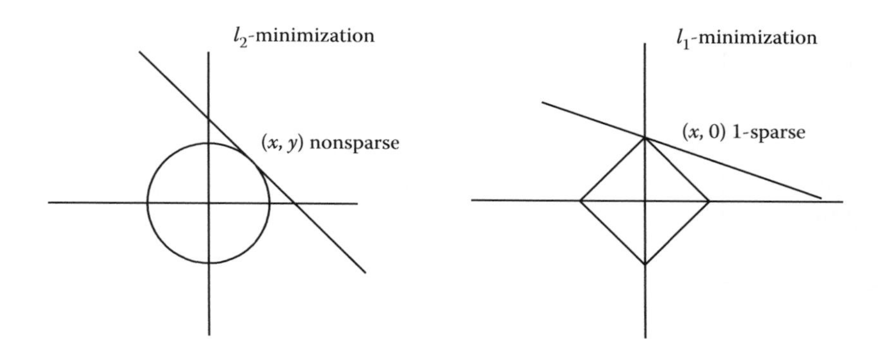

Note the sparse solutions are important in imaging sciences. The l_1-minimization problem can be viewed equivalently as

$$\min_{x,t} \sum_{i=1}^{n} t_i$$
$$\text{subject to} \quad |x_i| \leq t_i, \quad i = 1,2,\ldots,n, \quad Ax = b.$$

This is same as writing

$$\min_{x,t} \sum_{i=1}^{n} t_i$$
$$\text{subject to} \quad -t_i \leq x_i \leq t_i, \quad i = 1,2,\ldots,n, \quad Ax = b.$$

Any optimization problem in finite dimension is a linear programming problem in infinite dimensions.

Let K be a subset of $\mathbb{R}^n$ and let $\mu(K)$ be the space of all finite signal Borel measure on K. The positive cone (or nonnegative cone) $\mu(K)_+$ is the space of finite Borel measure (say) μ on K.

Consider the problems

$$f^* = \min f(x) \qquad \text{(P1)}$$
$$\text{subject to} \quad x \in K,$$

$$\rho = \min_{\mu \in M(K)_+} \int_K f d\mu \qquad \text{(P2)}$$
$$\text{subject to} \quad \int_K d\mu = 1.$$

Theorem 1.5: (Lasserre [16])

Problems (P1) and (P2) are equivalent, that is, $\rho = f^*$.

Proof. Let $f^* = -\infty$. Then, for $\mu < 0$ there exists $x \in K$ such that $f(x) \leq M$. Consider $\mu = \delta_x$ (the Dirac measure at x). Then, we have $\int_K f d\mu = f(x) \leq M$. Thus, $\rho = -\infty$. Consider the case $f^* > -\infty$ (i.e., f^* is finite). Then, for any $x \in K$, $f(x) \geq f^*$. Thus, for any $\mu \in \mu(K)_+$, we have

$$\int_K f\, d\mu \geq f^* \quad \text{thus } \rho \geq f^*.$$

To each $x \in K$, let us now associate the Dirac measure δ_x. Then, $\int f d\delta_x = f(x)$ is the feasible objective value of (P2) and thus $\rho \leq f(x)\ \forall\, x \in K \Rightarrow \rho \leq f^*$. Hence, $\rho = f^*$ as required. ■

Example 1.5: Quadratic Programming Models [QP-Models]

There are two types of quadratic programming models

- QP with affine constraints:

$$\min_x f(x) = \frac{1}{2}\langle x, Qx\rangle + \langle c, x\rangle + d$$
$$\text{subject to} \quad Ax = b, \quad x \geq 0,$$

where A is an $m \times n$ matrix and $x \geq 0$ means $x_1 \geq 0, \ldots, x_n \geq 0$, $c \in \mathbb{R}^n$, $d \in \mathbb{R}$, $Q \in S^n_+$. The set S^n_+ is the cone of all symmetric $n \times n$ positive semi-denite matrices.

- QP with quadratic constraints:

$$\min_x f(x) = \frac{1}{2}\langle x, Q_0 x\rangle + \langle c_0, x\rangle + d_0$$
$$\text{Subject to} \quad \frac{1}{2}\langle x, Q_i x\rangle + \langle c_i, x\rangle + d_i \leq 0, \quad i = 1, \ldots, m,$$
$$Q_0, Q_1, \ldots, Q_m \in S^n_+, \; c_0, c_1, \ldots, c_m \in \mathbb{R}^n \quad \text{and} \quad d_0, d_1, \ldots, d_m \in \mathbb{R}.$$

Let us end our discussion on quadratic programming by giving a proof of the famous Frank–Wolfe theorem. We shall state the proof due to Blum and Oettli [17].

Theorem 1.6: Frank–Wolfe Theorem

Let us consider a quadratic programming problem with affine constraints. If the objective function is bounded below on the feasible set, then the bound is attained.

Theorem 1.7: Frank–Wolfe Theorem [Blum and Oettli Format]

Let $x \in \mathbb{R}^n$ and let us consider

$$Q(x) = \langle x, Cx\rangle + \langle p, x\rangle, \; (C \in S^n).$$

Let $l_j(x) = \langle a_j, x\rangle + b_j$, $j \in J$ (J is a finite index set). Let $\hat{Q} = \inf_{x\in S} Q(x)$ where $S = \{x \in \mathbb{R}^n : l_j(x) \leq 0, \; \forall\, j \in J\}$ be finite. Then, there exists $\hat{x} \in S$ such that

$$Q(\hat{x}) = \hat{Q}.$$

(The proof due to Blum and Oettli is a direct proof and does not use the fact that a polyhedral set can be decomposed as the sum of a polytope and a polyhedral cone.)

Proof. Without loss of generality assume that $x = 0$ is feasible. This can be true after suitable transformation of the axis. Now $x = 0$ being feasible will only show that $b_j \leq 0 \; \forall \; j \in J$. For all $\rho \geq 0$ the sets $S_\rho = S \cap \{x : \|x\| \leq \rho\}$ is nonempty. In fact, S_ρ is a compact set since S is a closed set and $\{x : \|x\| \leq \rho\}$ is a compact set. The function $q(\rho) = \inf_{x\in S_\rho} Q(x)$ is monotone nonincreasing and $\lim_{\rho\to\infty} q(\rho) = \hat{Q}$. Since S_ρ is compact there exists $\xi \in S_\rho$ such that $Q(\xi) = q_\rho$. Thus, for each ρ the set

$$\{\xi : \xi \in S_\rho, \; Q(\xi) = q(\rho)\}$$

is a compact set and thus the function $\|\xi\|$ attains a minimum on this set. Thus, for any $\rho \geq 0$, there exists at least one point x_ρ such that $x_\rho \in S_\rho$, $Q(x_\rho) = q(\rho)$ and $\|x_\rho\| \leq \|\xi\| \; \forall \; \xi \in S_\rho$ with $Q(\xi) = q(\rho)$. Thus, x_ρ is the minimum norm solution of $\min_{x \in S_\rho} Q(x)$.

Our first claim is that $\|x\| < \rho$.

For $\rho > 0$ and sufficiently large, we will show that $\|x\| = \rho$ cannot hold for arbitrarily large values of ρ. Suppose that there is a sequence $\{\rho_k\}$ with $\rho_k \to +\infty$ *as* $k \to \infty$ and $\|x_{\rho_k}\| = \rho_k$. Since $x_{\rho_k} \in S$, we have $l_j(x_{\rho_k}) \leq 0 \; \forall \; j \in J$. Let

$$J_0 = \left\{ j \in J : \limsup_{k \to \infty} l_j(x_{\rho_k}) = 0 \right\}.$$

Since J is finite we can find $\varepsilon > 0$ such that

$$\limsup_{k \to \infty} l_j(x_{\rho_k}) \leq -2\varepsilon < 0, \quad \forall \; j \in J \backslash J_0.$$

It is possible to extract a subsequence $\{\rho_\nu\} \subseteq \{\rho_k\}$ such that $l_j(x_{\rho_\nu}) \to 0 \;\; \forall \; j \in J_0$ as $\nu \to \infty$ and $l_j(x_\nu) \leq -\varepsilon \; \forall \; j \in J \backslash J_0$.

Let us set $t_\nu = (x_{\rho_\nu}/\rho_\nu)$. Let us assume without loss of generality that $t_\nu \to t$ and $\|t\| = 1$ ($t \neq 0$) (note that $\|t_\nu\| = 1$ *as* $\|x_{\rho_\nu}\| = \rho_\nu$). Now

$$\begin{aligned} l_j(x_{\nu_\rho}) &= l_j(\rho_\nu, t_\nu) \\ &= \rho_\nu \langle a_j, t_\nu \rangle + b_j \end{aligned}$$

as $\nu \to \infty$, $\rho_\nu \to \infty$ and thus for $\forall \; j \in J_0 \;\; l_j(x_{\nu_\rho}) \to 0$, it implies that $\langle a_j, t \rangle = 0 \; \forall \; j \in J_0$. Further for $\forall j \in J \backslash J_0$, we must have $\langle a_j, t \rangle \leq 0$. Now

$$q(\rho_\nu) = Q(x_{\rho_\nu}) = Q(\rho_\nu t_\nu) = \rho_\nu \langle p, t_\nu \rangle + \rho_\nu^2 \langle t_\nu, c t_\nu \rangle.$$

Further $Q(x_{\rho_\nu}) \to \hat{Q}$ as $\rho_\nu \to \infty$. This will show that

$$\langle t, ct \rangle = 0.$$

Note that if $\langle t, ct \rangle = 2\alpha > 0$, then $\forall \; \nu \geq N$ (say) would have $\langle t_\nu, c t_\nu \rangle \geq \alpha_i$. Consequently, $Q(x_{\rho_\nu}) \geq \rho_\nu \beta + \rho_\nu^2 \alpha$, thus $Q(x_{\rho_\nu}) \to +\infty$ *as* $\nu \to +\infty$ (i.e., $\rho_\nu \to +\infty$), which is a contradiction. One can have a similar sort of contradiction if $\langle t, ct \rangle = 2\beta < 0$. It will show that $Q(x_{\rho_\nu}) \to \infty$ as $\nu \to \infty$.

So let us sum up till now. All we have is

$$\langle a_j, t \rangle = 0, \quad \forall \; j \in J_0 \langle a_j, t \rangle \leq 0, \quad \forall \; j \in J \backslash J_0 \tag{1.8}$$

and

$$\langle t, ct \rangle = 0. \tag{1.9}$$

Using Equation 1.8, we have

$$x_{\rho_\nu} + \lambda t \in S, \quad \forall \; \lambda \geq 0, \quad \forall \; \nu. \tag{1.10}$$

Further Equation 1.8 shows that

$$Q(x_{\rho_\nu} + \lambda t) = Q(x_{\rho_\nu}) + \lambda\,[\langle p,\, t\rangle + 2\langle x_{\rho_\nu}, ct\rangle], \quad \forall\,\lambda, \quad \forall\,\nu.$$

Now we cannot have $[\langle p, t\rangle + 2\langle x_{\rho_\nu}, ct\rangle] < 0$ for some ν. Otherwise as $\lambda \to \infty$ we have $Q(x_{\rho_\nu} + \lambda t) \to -\infty$. Note that we cannot have $\langle t_\nu, t\rangle \leq 0,\ \forall \nu$ (sufficiently large). Since $t_\nu \to t$, we shall have $\langle t, t\rangle \leq 0 \ \Rightarrow\ \|t\|^2 \leq 0 \ \Rightarrow\ \|t\| = 0$, which contradicts $\|t\| = 1$. Thus, there is a sufficiently large ν, that is ($\rho_\nu > 0$ and large) we have

$$\langle t_\nu, t\rangle > 0.$$

This means that $\langle x_{\rho_\nu}, t\rangle > 0$ and hence

$$\|x_{\rho_\nu} - \lambda t\|^2 = \|x_{\rho_\nu}\|^2 + \lambda^2\|t\|^2 - 2\lambda\langle x_{\rho_\nu}, t\rangle.$$

This show that $\exists \delta_1 > 0$, such that $\forall 0 < \lambda < \delta_1$ we have

$$\begin{aligned} &\lambda\,\|t\|^2 - 2\lambda\langle x_{\rho_\nu}, t\rangle < 0 \quad (\text{since } \langle x_{\rho_\nu}, .\rangle > 0) \\ &\|x_{\rho_\nu} - \lambda t\|^2 < \|x_{\rho_\nu}\|^2 \\ &\|x_{\rho_\nu} - \lambda t\| < \|x_{\rho_\nu}\|, \quad \forall\, 0 < \lambda < \delta_1. \end{aligned}$$

Hence,

$$l_j(x_{\rho_\nu} - \lambda t) = l_j(x_{\rho_\nu}) - \lambda\langle a_j, t\rangle = l_j(x_{\rho_\nu}) \leq 0, \quad \forall\, j \in J_0.$$

Further,

$$\begin{aligned} &l_j(x_{\rho_\nu}) = l_j(x_{\rho_\nu}) - \lambda\langle a_j, t\rangle = l_j(x_{\rho_\nu}) + \lambda\langle a_j, -t\rangle \leq -\varepsilon + \lambda\|a_j\|\|t\| \leq 0, \\ &\text{for } 0 \leq \lambda \leq \frac{\varepsilon}{\|a_j\|\|t\|}. \end{aligned}$$

Thus,

$$x_\nu - \lambda t \in S, \quad 0 < \lambda < \delta_2, \quad \delta_2 = \min\left\{\delta_1, \frac{\varepsilon}{\|a_j\|\|t\|}, j \in J\backslash J_0\right\}. \tag{1.11}$$

Thus, we have

$$Q(x_{\rho_\nu}) = Q(x_{\rho_\nu}) - \lambda[\langle p, t\rangle + 2\langle x_\nu, ct\rangle] \ \leq Q(x_{\rho_\nu}).$$

Put

$$\hat{\xi} = x_{\rho_\nu}, \quad 0 < \lambda < \min\{\delta_1, \delta_2\}.$$

Then, what we have obtained is as follows:

$$\hat{\xi} \in \mathbb{R}_{\rho_\nu}, \quad \|\hat{\xi}\| < \|x_{\rho_\nu}\|, \quad Q(\hat{\xi}) \leq Q(x_{\rho_\nu}).$$

This contradicts that x_{ρ_ν} is the minimum norm solution of

$$\min_{x \in S\rho_\nu} Q(x).$$

So, we have $\exists\ \bar{\rho} > 0$

$$\|x_\rho\| < \rho, \quad \forall\ \rho > \bar{\rho}. \tag{1.12}$$

We shall show that if Equation 1.12 is valid, then there exists ρ such that $q(\rho) = \hat{Q}$. Suppose, on the contrary

$$q(\rho) > \hat{Q}, \ \forall \rho. \tag{1.13}$$

We know that $q(\rho)$ is monotone and nonincreasing and in the limit hits $\hat{Q}$. Hence, we can find $\bar{\rho} < \rho_1 < \rho_2$ such that $q(\rho_1) > q(\rho_2)$. Since $\bar{\rho} < \rho_2$, we obtain from Equation 1.12 that $\|x_{\rho_2}\| < \rho_2$. Now $q(\rho_1) > q(\rho_2)$ shows that $\rho_1 < \|x_{\rho_2}\|$ (or else $q(\rho_1) = q(\rho_2)$ ......Think why!!).

Choose $\rho_3 = \|x_{\rho_2}\|$. Then, $\bar{\rho} < \rho_1 < \rho_3 < \rho_2$. Hence, $\bar{\rho} < \rho_3$ shows $\|x_{\rho_3} < \rho_3\|$. Further $\rho_2 > \rho_3$ shows that $Q(x_{\rho_2}) \leq Q(x_{\rho_3})$. This will make us to analyze two cases:

If $Q(x_{\rho_2}) = Q(x_{\rho_3})$, then by noting that $|x_{\rho_3}| < |x_{\rho_2}|$ we contradict that x_{ρ_3} is the minimum norm solution of

$$\min_{x \in S_{\rho_2}} Q(x).$$

If $Q(x_{\rho_2}) < Q(x_{\rho_3})$, and the fact that $|x_{\rho_2}| < \rho_3$ shows that x_{ρ_3} is not a solution of

$$\min_{x \in S_{\rho_3}} Q(x),$$

which is a contradiction.

These two contradictions show that Equation 1.13 is not true and hence we establish our claim. ■

Example 1.6: Second-Order Conic Programming Problem

What is conic programming? Consider the linear programming problem

$$\begin{aligned}&\min \langle c, x\rangle\\ &\text{subject to} \quad Ax = b,\\ &x \leq 0.\end{aligned}$$

This can be written as

$$\begin{aligned}&\min \langle c, x\rangle\\ &\text{subject to} \quad Ax = b,\\ &x \in \mathbb{R}^n_+.\end{aligned}$$

Note that $\mathbb{R}^n_+$ is a convex cone. We can generalize this by considering a convex cone $K \subset \mathbb{R}^n$ and consider the problem

$$\begin{aligned}&\min \langle c, x\rangle\\ &\text{subject to} \quad Ax = b,\\ &x \in K.\end{aligned}$$

Such a problem is called a conic programming problem. If $K = \mathcal{K}^{n-1}$, the second-order cone in $\mathbb{R}^n$ (defined below) then

$$\begin{aligned}&\min \langle c, x\rangle\\ &\text{subject to} \quad Ax = b,\\ &x \in \mathcal{K}^{n-1}\end{aligned}$$

is called the second-order cone programming problem (SOCP). Sometimes one also can have a problem as

$$\begin{aligned}&\min\ \langle c, x\rangle\\&\text{subject to}\quad Ax - b \in \mathcal{K}^{n-1}.\end{aligned}$$

This is also an SOCP.

1.3.2.1 More Details on the Second-Order Cone

The second-order cone $\mathcal{K}^n$

$$\mathcal{K}^n = \{(x, t) \in \mathbb{R}^n \times \mathbb{R} : \ \|x\|_2 \leq t\},$$

for example,

$$\mathcal{K}^2 = \left\{(x, t) \in \mathbb{R}^3 : \ \sqrt{x_1^2 + x_2^2} \leq t\right\}.$$

$\mathcal{K}^n$ is a closed convex cone but not polyhedral. In fact, we can write

$$\mathcal{K}^n = \bigcap_{\{u:\ \|u\|_2=1\}} \{(x, t) \in \mathbb{R}^n \times \mathbb{R} : \ \langle x, u\rangle \leq t\}.$$

Let us now look at the following variant of the second-order cone. This is called the rotated second-order cone.

$$\mathcal{K}_r^n = \{(x, y, z) : x \in \mathbb{R}^n, \ y, z \in \mathbb{R}; \ \langle x, x\rangle \leq 2yz, \ y \geq 0, \ z \geq 0\}.$$

$\mathcal{K}_r^n$ lives in $\mathbb{R}^{n+2}$ while $\mathcal{K}^n$ lives in $\mathbb{R}^{n+1}$. In fact, $\mathcal{K}_r^n$ can be viewed as a rotation of $\mathcal{K}^{n+1}$. Note that

$$\|x\|_2^2 \leq 2yz, \ y \geq 0, \ z \geq 0 \ \Leftrightarrow \left\|\begin{pmatrix} x \\ \frac{1}{\sqrt{2}}(y - z)\end{pmatrix}\right\| \leq \frac{1}{\sqrt{2}}(y + z).$$

Thus, $(x, y, z) \in \mathcal{K}_r^n$ if and only if $(w, t) \in \mathcal{K}^{n+1}$ where

$$w = \left\|\begin{pmatrix} x \\ \frac{1}{\sqrt{2}}(y - z)\end{pmatrix}\right\| \quad \text{and} \quad t = \frac{1}{\sqrt{2}}(y + z).$$

In fact,

$$\hat{R} = \begin{pmatrix} x \\ y \\ z\end{pmatrix} = \begin{pmatrix} I_n & 0 & 0 \\ 0 & \frac{1}{\sqrt{2}} & -\frac{1}{\sqrt{2}} \\ 0 & \frac{1}{\sqrt{2}} & \frac{1}{\sqrt{2}}\end{pmatrix} \quad \text{[check that } \hat{R} \text{ is invertible]}.$$

Thus, a linear transformation of $\mathcal{K}^{n+2}$ leads to $\mathcal{K}_r^n$. Thus, $\mathcal{K}_r^n$ is convex.

Let us now give some examples showing that some standard constraint systems can be represented as a rotated second-order cone.

1.3.2.1.1 Quadratic Constraints

Consider a quadratic constraint of the form

$$\langle x, Qx\rangle + \langle c, x\rangle \leq t, \quad Q \in S^n_+.$$

This is equivalent to the existence of w, y, z such that

$$\langle w, w\rangle \leq 2yz, \quad z = \frac{1}{2}, \quad w = Q^{1/2}x, \quad y = t - \langle c, x\rangle.$$

Hence, the feasible set generated by quadratic constraints given above can be viewed as the intersection of $\mathcal{K}^n_r$ with the affine sets.

$$z = \frac{1}{2}, \quad \{(x, w) : w = Q^{1/2}x\}, \quad \{(x, y) : y = t - \langle c, x\rangle\}$$

1.3.2.1.2 Second-Order Cone Constraints

These are of the form

$$(y, t) \in \mathcal{K}^m$$
$$\text{where } y \in \mathbb{R}^m \quad \text{and} \quad t \in \mathbb{R},$$

where $y = Ax + b$ (where A is an $m \times n$ matrix and $b \in \mathbb{R}^m$) and $t = \langle c, x\rangle + d$. Thus, second-order conic constrains is of the form

$$\|Ax + b\|_2 \leq \langle c, x\rangle + d.$$

In fact, the quadratic constraints that we have just studied can be written as

$$\left\| \begin{pmatrix} \sqrt{2}Q^{1/2}x \\ t - c^T x - \frac{1}{2} \end{pmatrix} \right\|_2 \leq t - c^T x + \frac{1}{2}.$$

SOCP in the standard form

$$\min_{x \in \mathbb{R}^n} \langle c, x\rangle$$
$$\|A_i x + b_i\| \leq c_i^T x + d, \quad i = 1, \ldots, m,$$

where $A_i \in \mathbb{R}^{m_i \times n}$ and $b_i \in \mathbb{R}^{m_i}$, $c_i \in \mathbb{R}^{n_i}$, $d_i \in \mathbb{R}$.

Thus, we can write

$$\min \langle c, x\rangle$$
$$(A_i x + b_i, c_i^T x + d_i) \in \mathcal{K}^{m_i}, \quad i = 1, \ldots, m.$$

In fact, Boyd and Vanderberghe [6] add an additional linear (or affine) constraints of the form $Fx = g$ to the above standard form of the SOCP.

Example 1.7: LASSO Problem

Consider the problem

$$\min_x \|Ax - b\|_2 + \lambda\|x\|_1,$$

where $A \in \mathbb{R}^{m\times n}$ and $b \in \mathbb{R}^m$, $\lambda > 0$. This is also called the LASSO problem, which plays an important role in image processing.

This can be written as the following SOCP:

$$\min_{x,t,u} t + \lambda \sum_{i=1}^{n} u_i$$
$$\text{subject to} \quad \|Ax - b\|_2 \le t \quad \text{and} \quad |x_i| \le u_i, \ \ i = 1, \ldots, n.$$

This is the standard form.

Example 1.8: SOCP Formulation in Robust Linear Programming

We will now consider a linear program whose data is uncertain in the sense that we know it takes value within some interval. What we will show is that such a problem can be equivalently posed as an SOCP where the problem data has no uncertainty.

Consider the linear programming problem

$$\min\langle c, x\rangle$$
$$\text{subject to} \quad \langle a_i, x\rangle \le b_i, \quad i = 1, \ldots, m.$$

Let us assume that the parameters c and b are fixed while that of a_i are known to lie in the given ellipsoid, that is,

$$a_i \in \xi_i = \{\bar{a}_i + P_i u \mid \|u\|_2 \le 1\} \quad \forall\, i = 1, \ldots, m.$$

For a nongenerate ellipsoid, we will have P_i nonsingular. If P_i is singular then, we say we have degenerate or flat ellipsoid.

To obtain a robust-linear programming problem, we need to find $x \in \mathbb{R}^n$ such that for each i $\langle a_i, x\rangle \le b_i, \forall\, a_i \in \xi_i$. Thus, we can reformulate the linear programming problem in a robust form and state it as

$$\min\langle c, x\rangle$$
$$\text{subject to} \quad \langle a_i, x\rangle \le b_i, \quad \forall\, a_i \in \xi_i\ i = 1, \ldots, m.$$

This is equivalent to

$$\min\ \langle c, x\rangle$$
$$\text{subject to} \quad \sup\{\langle a_i, x\rangle : a_i \in \xi_i\} \le b_i, \quad i = 1, \ldots, m.$$
$$\begin{aligned}\text{Further} \quad & \sup\{\langle a_i, x\rangle : a_i \in \xi_i\} \\ &= \sup\{\langle \bar{a}_i + P_i u, x\rangle; \|u\|_2 \le 1\} \\ &= \sup\{\langle \bar{a}_i, x\rangle + \langle u, P_i^T x\rangle; \|u\|_2 \le 1\} \\ &= \langle \bar{a}_i, x\rangle + \sup\{\langle u, P_i^T x\rangle; \|u\|_2 \le 1\} \\ &= \langle \bar{a}_i, x\rangle + \|P_i^T x\|.\end{aligned}$$

The robust linear program can be written as

$$\min\langle c, x\rangle$$
$$\langle a_i, x\rangle + \|P_i^T x\|_2 \leq b_i, \quad i = 1, \ldots, m,$$

that is,

$$\min\langle c, x\rangle$$
$$\|P_i^T x\|_2 \leq b_i - \langle a_i, x\rangle,$$

which is an SOCP in the standard form.

Example 1.9: Linear Programming with Probabilistic Constraints [SOCP Formulation]

We again consider a situation where the input vectors a_i are uncertain but this time they are probabilistic in nature, that is, each a_i is a normal random vector with mean $\bar{a}_i$ and variance and covariance matrix C_i. Thus, our linear programming problem is stated as

$$\min\langle c, x\rangle$$
$$\text{subject to} \quad P(\langle a_i, x\rangle \leq b_i) \geq \eta, \quad \forall\, i = 1, \ldots, m \quad \text{and} \quad \eta \geq \frac{1}{2}.$$

Let us consider fixed $i \in \{1, 2, \ldots, m\}$. Set $u_i = \langle a_i, x\rangle$ for each i and let σ^2 be its variance. Note that $u_i \sim N(\bar{u}_i, \sigma^2)$ where $\bar{u}_i = \langle \bar{a}_i, x\rangle$. Thus, we have

$$P(u \leq b_i) \geq \eta$$
$$P\left(\frac{u_i - \bar{u}}{\sigma} \leq \frac{b_i - \bar{u}}{\sigma}\right) \geq \eta.$$

Thus, $(u_i - \bar{u})/\sigma \sim N(0, 1)$. Hence,

$$\Phi\left(\frac{b_i - \bar{u}}{\sigma}\right) \geq \eta; \qquad \left(\text{Note that } \phi(z) = \frac{1}{\sqrt{2\pi}} \int\limits_{-\infty}^{z} e^{(-t^2/2)}\, dt\right).$$

Since Φ is nondecreasing and invertible, Φ^{-1} is nondecreasing, and hence

$$\frac{b_i - \bar{u}_i}{\sigma} \geq \Phi^{-1}(\eta).$$

Since $\eta \geq (1/2)$, $\Phi^{-1}(\eta) \geq 0$. Further

$$\bar{u}_i + \Phi^{-1}(\eta)\sigma \leq b_i.$$

Now,

$$\sigma^2 = \langle x, C_i x\rangle \quad \text{(check it)}.$$

Therefore,

$$\sigma = \sqrt{\langle {C_i}^{1/2} x, {C_i}^{1/2} x\rangle} \Rightarrow \sigma = \|{C_i}^{1/2} x\|_2.$$

Thus, we have

$$\bar{u}_i + \Phi^{-1}(\eta)\ \|C_i{}^{1/2}x\|_2 \le b_i.$$

$$\begin{aligned}&\min\langle c, x\rangle\\ &\text{subject to}\quad a_i^T x + \Phi^{-1}(\eta)\ \|C_i{}^{1/2}x\|_2 \le b_i,\quad \forall i = 1,\ldots,m,\end{aligned}$$

that is,

$$\begin{aligned}&\min\langle c, x\rangle\\ &\text{subject to}\quad \|C_i^{\frac{1}{2}}x\|_2 \le \frac{b_i - a_i^T x}{\Phi^{-1}(\eta)},\quad \forall i = 1,\ldots,m\quad \text{if } \phi_{-1}(\eta) \ge 0.\end{aligned}$$

Thus, this is an SOCP in the standard form.

One of the most detailed and fundamental paper on *SOCP* is by F. Alizadeh and D. Goldfarb [18]. This paper titled "Second-order cone programming," which is published in *Math. Prog. Series B* (Vol. 95, pp. 3–51, 2003). This paper deals in detail with the problem, the geometry, and the associated algebra of the second-order cone and algorithms. It is worthwhile to know some of the results in this paper.

In this paper, Alizadeh and Goldfarb consider the SOCP in the following form:

$$\begin{aligned}&\min\ \langle c, x\rangle,\\ &\text{subject to}\quad Ax = b,\quad x \in \mathcal{K}^n.\end{aligned}$$

Further they partition the variable x and the matrix A in the following way:

Let $x = (x_i, \ldots, x_r)$ where $x_i \in \mathbb{R}^{n_i}$ and $n_1 + n_2 + \cdots + n_r = n$, $C = (c_1, \ldots, c_r)$ where $c_r \in \mathbb{R}^{n_i}$ $A = [A_1|A_2|\cdots|A_r,]$, where $A_i \in \mathbb{R}^{m\times n_i}$. So each A_i is an $m \times n_i$ matrix.

$$\begin{aligned}Ax &= [A_1|A_2|\cdots|A_r,]\begin{pmatrix} x_1\\ x_2\\ \cdots\\ \cdots\\ \vdots\\ x_r\end{pmatrix}\\ &= A_1x_1 + A_2x_2 + \cdots + A_rx_r\\ \langle c, x\rangle &= \langle c_1, x_1\rangle + \langle c_2, x_2\rangle + \cdots + \langle c_r, x_r\rangle.\end{aligned}$$

Thus, the SOCP problem is given as

$$\begin{aligned}&\min\ \langle c_1, x_1\rangle + \langle c_2, x_2\rangle + \cdots + \langle c_r, x_r\rangle,\\ &\text{subject to}\quad A_1x_1 + A_2x_2 + \cdots + A_rx_r = b,\\ &x_i \in \mathcal{K}^{n_i}\quad \text{for } i = 1,\ldots,r.\end{aligned}$$

For a given $x \in \mathbb{R}^n$, partitioned as $(x_0, \bar{x}) = (x_0, x_1, \ldots, x_n)$ define the matrix

$$\text{Arrow matrix} \leftarrow \text{Arw}(x) = \begin{pmatrix} x_0, & \bar{x}^T\\ \bar{x}, & x_0 I\end{pmatrix}.$$

Further $x \in \mathcal{K}^n$ if and only if $\text{Arw}(x)$ is positive semidefinite, that is, $\text{Arw}(x) \in S^n_+$. Thus, we can write the SOCP as

$$\min \langle c_1, x_1\rangle + \langle c_2, x_2\rangle + \cdots + \langle c_r, x_r\rangle,$$
$$\text{subject to} \quad A_1x_1 + A_2x_2 + \cdots + A_rx_r = b,$$
$$\text{Arw}(x_i) \in S^{n_i}_+ \; i = 1, \ldots, r.$$

Thus, SOCP can be now viewed as a semidefinite programming problem (SDP), which we shall study after this. In fact, Alizadeh and Goldfarb [18] have shown that several interesting class of optimization problems can be modeled as an SOCP problem.

Example 1.10

Let us set $v_i = A_i x_i$, where $x_i \in \mathbb{R}^{n_i}$.
Consider the problem

$$\min \left(\max_{1 \leq i \leq r} \|v_i\| \right).$$

This can be written as

$$\min_{x,t,v} t,$$
$$A_i x_i + b_i = v_i, \quad i = 1, \ldots, r,$$
$$(v_i, t) \in \mathcal{K}^{n_i}, \quad i = 1, \ldots, r.$$

In fact, it can also be written as

$$\min_{x,t} t$$
$$\|A_i x_i + b_i\| \leq t, \quad \forall\, i = 1, 2, \ldots, r,$$

which is same as the standard SOCP.

Example 1.11: Semidefinite Programming

First state (or restate) several facts
S^n : Space of all $n \times n$ symmetric matrices.
S^n_+ : Cone of all $n \times n$ symmetric and positive semidefinite matrices.
$S^n_{++} := \text{int} S^n_+$ = Cone of all $n \times n$ symmetric and positive-definite matrices.
SDP can be viewed as a linear programming problem in symmetric matrices

$$\min \langle C, X\rangle,$$
$$\text{subject to} \quad \langle A_i, X\rangle = b_i, \quad i = 1, \ldots, n, \quad X \in S^n_+.$$

$\langle X, Y\rangle = \text{trace}(X, Y)$ is an inner product on S^n.

$$\text{Frobenius norm } \|X\|^2_F = \langle X, X\rangle = \text{trace}(X^2) > 0 \quad \text{if } X \neq 0, \; X \in S^n.$$

As S^n_+ is not a polyhedral cone, this is truly not a linear programming problem in S^n, but only a convex optimization problem.

Let K be a convex cone then the dual cone is given as

$$K^* = \{w \in \mathbb{R}^n : \langle w, x \rangle \geq 0, \ \forall \, x \in K\}.$$

If K is a closed convex cone, then $K^{**} = K$. K is called self-dual if $K^* = K$. We shall prove next that S_+^n is self-dual.

Proposition 1.1

S_+^n is self-dual.

Proof. Let $X \in S_+^n$. Then, for any $Y \in S_+^n$, we have

$$\begin{aligned} \text{trace}\langle X, Y \rangle &= \text{trace}(XY) \\ &= \text{trace}(X^{1/2}X^{1/2}Y) \\ &= \text{trace}(X^{1/2}YX^{1/2}) \geq 0. \end{aligned}$$

Note that $X^{1/2}YX^{1/2}$ is a p.s.d. matrix and thus $\text{tr}(X^{1/2}YX^{1/2}) \geq 0$. Thus, $X \in (S_+^n)^*$. Thus, $S_+^n \subseteq (S_+^n)^*$. Let $\hat{X} \in (S_+^n)^*$. Let $\hat{v}^{(1)}, \ldots, \hat{v}^{(n)}$ be the eigenvectors of $\hat{X}$ and let $\hat{v}^{(1)}, \ldots, \hat{v}^{(n)}$ be normalized.

Hence, $\hat{v}^{(i)} \, \hat{v}^{(i) \, T} \in S_+^n$. Thus,

$$\begin{aligned} 0 &\leq \left\langle \hat{X}, \hat{v}^{(i)} \, \hat{v}^{(i) \, T} \right\rangle \\ &= \text{tr}\left(\hat{X}, \ \hat{v}^{(i)} \, \hat{v}^{(i) \, T}\right) \\ &= \left\langle \hat{v}^{(i)}, \ \hat{X} \, \hat{v}^{(i)} \right\rangle \text{(This is the usual inner product in } \mathbb{R}^n\text{)}. \end{aligned}$$

Thus,

$$\left\langle \hat{v}^{(i)}, \ \hat{X} \, \hat{v}^{(i)} \right\rangle = \lambda_i(\hat{X}).$$

Thus, $\lambda_i(\hat{X}) \geq 0, \ \forall \, i$ and hence, $\hat{X} \in S_+^n$. Thus, $(S_+^n)^* \subseteq S_+^n$. This proves the proposition.

S_+^n is also a homogeneous cone. By that we mean for $X, V \in S_{++}^n \ \exists \, T \in \text{Aut}(S_+^n)$ such that

$$T(V) = X,$$

where $\text{Aut}(S_+^n)$ is the set of all automorphism of S_+^n into itself. ■

A cone which is self-dual and homogeneous is called a symmetric cone.

Sometimes the SDP is given as

$$\begin{aligned} &\min \ \langle c, x \rangle, \\ &\text{subject to} \quad A_0 + x_1 A_1 + x_2 A_2 + \cdots + x_n A_n \in S_+^n, \\ &\text{where} \quad A_0, A_1, \ldots, A_n \in S^n. \end{aligned}$$

This problem actually turns out to be the “dual” of the other.

An inequality of the above form, that is,

$$A_0 + x_1 A_1 + x_2 A_2 + \cdots + x_n A_n \in S_+^n$$

is called a linear matrix inequality (LMI). Let

$$C = \left\{ x \in \mathbb{R}^n : A_0 + \sum_{i=1}^{m} x_i A_i \succcurlyeq 0. \right\} \quad (X \in S_+^n \Leftrightarrow X \succcurlyeq 0)$$

The set C is convex. Further, if a convex set can be represented as above, then it is called LMI representable. Let us give some examples of LMI representable sets.

Example 1.12

Let $f(x) = \|x\|_2$. Consider epi$f$

$$\text{epi} f = \{(x,t) \in \mathbb{R}^n \times \mathbb{R} : \|x\|_2 \leq t\}.$$

Thus, epi$f = \mathcal{K}^n$. Then, we have

$$(x,t) \in \text{epi} f \Leftrightarrow M = \begin{pmatrix} t & x^T \\ x & tI \end{pmatrix} \succcurlyeq 0.$$

Note that if $t = 0$, then $M \succcurlyeq 0$ if and only if $x = 0$.

Note that if $t < 0$, then $M \notin S_+^n$ ($\langle e_1, Me_1 \rangle = t < 0$).

Now if $t > 0$. Then, by Schur complement, we have

$$\begin{aligned} M \succcurlyeq 0 &\Leftrightarrow tI - \frac{1}{t} xx^T \succcurlyeq 0 \\ &\Rightarrow t^2 I - xx^T \succcurlyeq 0 \\ &\Rightarrow \lambda_1(xx^T) \leq t^2 \qquad \text{(Think why!!)} \\ &\Rightarrow \|x\|_2^2 \leq t^2. \end{aligned}$$

Thus, we have epi$f$ to be LMI representable. There is an elaborate paper by Helton and Vinnikov [19], which describes the LMI representability of convex sets. See also Helton and Nie [21].

Another example of LMI representability

$$f : S^n \to \mathbb{R} \quad \text{and} \quad f(X) = \lambda_1(X) = \lambda_{\max}(X).$$

Now $(x,t) \in \text{epi} f = \text{epi}\lambda_1$ if and only if

$$\lambda_1(X) \leq t$$

This is same as $\langle w, X w \rangle \leq t\|w\|^2$.

Thus, showing that $\langle w, (tI - X)w \rangle \geq 0 \; \forall w \in \mathbb{R}^n$.

Thus, $(X,t) \in \text{epi}\lambda_1$ if and only if

$$tI - X \in S_+^n$$

and hence,

$$\text{epi}\lambda_1 = \{(X, t) \in S^n \times \mathbb{R} : tI - X \succcurlyeq 0\}.$$

Thus, $\text{epi}\lambda_1$ is an SDP representable set.

Linear, quadratic, and affine inequalities can be expanded in the LMI form. Thus, a linear programming problem under the above-mentioned feasible set system is an SDP problem. An important rule in studying SDP is the Schur complement rule, which we formulate below. Let

$$M = \begin{pmatrix} A & C^T \\ C & B \end{pmatrix}, \quad A, B \text{ symmetric.}$$

The following implications hold:

$$\begin{aligned} &\text{if } M \succcurlyeq 0 \Rightarrow A \succcurlyeq 0, \quad B \succcurlyeq 0, \\ \text{if } B = 0, \quad &\text{then } M \succcurlyeq 0 \Leftrightarrow A \succcurlyeq 0, \quad C = 0, \\ \text{if } A = 0, \quad &\text{then } M \succcurlyeq 0 \Leftrightarrow B \succcurlyeq 0, \quad C = 0. \end{aligned}$$

If B^{-1} exists then

$$M \succcurlyeq 0 \Leftrightarrow A - C^T B^{-1} C \succcurlyeq 0 \; (A - C^T B^{-1} C \text{ is Schur complement})$$

if equivalently

$$\begin{aligned} &M \succcurlyeq 0 \Leftrightarrow B - CA^{-1}C^T \succcurlyeq 0 \; (B - CA^{-1}C^T \text{ is Schur complement}), \\ &M \succ 0 \Leftrightarrow B \succ 0; \quad A - C^T B^{-1} C \succ 0, \\ &M \succ 0 \Leftrightarrow A \succ 0; \quad B - CA^{-1}C^T \succ 0. \end{aligned}$$

Let us end this section by showing how the SDP model can be used to study a different and difficult nonconvex problem.

Consider the QPQC problem: quadratic programming with quadratic constraints. Let

$$\begin{aligned} &p^* = \min \langle x, H_0 x\rangle + 2c_0^T x + d_0, \\ &\text{subject to} \quad \langle x, H_i x\rangle + 2c_i^T x + d_i \leq 0, \quad i = 1, \ldots, m, \\ &\langle x, H_j x\rangle + 2c_j^T x + d_j = 0, \quad j = 1, \ldots, k. \end{aligned}$$

Note that

$$\langle x, H_0 x\rangle = \langle H_0, xx^T\rangle = \text{tr}(H_0 xx^T).$$

This shows that

$$\begin{aligned} &p^* = \min_x \langle H_0, X\rangle + 2c_0^T x + d_0, \\ &\text{subject to} \quad \langle H_i, X\rangle + 2c_i^T x + d_i \leq 0, \quad i = 1, \ldots, m, \\ &\langle H_j, X\rangle + 2c_j^T x + d_j = 0, \quad j = 1, \ldots, k. \\ &X = xx^T. \end{aligned}$$

However, we can relax $X = xx^T$ to have $X - xx^T \in S_+^m$

$$X - xx^T \in S_+^m \Leftrightarrow \begin{pmatrix} X & x \\ x^T & I \end{pmatrix} \succcurlyeq 0.$$

Thus, we can relax the problem as

$$q^* = \min_{x,X} \langle H_0, X \rangle + 2c_0^T x + d_0,$$

$$\text{subject to} \quad \langle H_i, X \rangle + 2c_i^T x + d_i \leq 0, \quad i = 1, \ldots, m,$$

$$\langle H_j, X \rangle + 2c_j^T x + d_j = 0, \quad j = 1, \ldots, k$$

$$\begin{pmatrix} X & x \\ x^T & I \end{pmatrix} \succcurlyeq 0.$$

Thus, we have SDP which gives $q^* \leq p^*$ and q^* will be much easier to find compared with p^*. Note that by the term relax we mean that instead of considering the constraint $X - xx^T = 0$ we consider the constraint $X - xx^T \in S_+^n$. Note that the set of x satisfying $X - xx^T \in S_+^n$ is larger than the set of x satisfying $X - xx^T = 0$.

1.4 Optimality and Duality in Convex Programming

Optimality conditions and duality lie at the heart of any analysis of a convex optimization problem. We start by looking at the minimization of the convex function $f : \mathbb{R}^n \to \mathbb{R}$ over a convex set C. For the moment, let us consider f to be differentiable. We leave it to the readers to show that $\bar{x} \in C$ is a minimizer of f over C if and only if

$$\langle \nabla f(\bar{x}), x - \bar{x} \rangle \geq 0, \quad \forall\, x \in C.$$

Though the conditions may appear algebraic, in essence it is geometrical. If we recall the definition of the normal cone, the above optimality condition simply means

$$-\nabla f(\bar{x}) \in N_C(\bar{x}). \tag{1.14}$$

Now what happens if

$$C = \{x \in \mathbb{R}^n : g_i(x) \leq 0, \forall i = 1, \ldots, m\}, \tag{1.15}$$

where each $g_i : \mathbb{R}^n \to \mathbb{R}$ is a convex function and differentiable. If we could compute the normal cone $N_C(\bar{x})$ for each x, then we can have optimality condition (1.14) in its full generality. So, how do we compute $N_C(\bar{x})$ for each x? We have the following result:

Theorem 1.8

Let the convex set C be given by Equation 1.15 and let $\hat{x} \in \mathbb{R}^n$ such that $g_i(\hat{x}) < 0$, $\forall i$ that is, the Slater condition holds true. Then,

$$N_C(\bar{x}) = \left\{ \sum_{i=1}^{m} \lambda_i \nabla g_i(\bar{x}) : \lambda_i \geq 0,\ \forall\, i = 1, \ldots, m,\ \lambda_i g_i(\bar{x}) = 0\ \forall\, i = 1, \ldots, m \right\}.$$

How can we prove this? First, we shall show that the set

$$\left\{\sum_{i=1}^{m} \lambda_i \nabla g_i(\bar{x}) : \lambda_i \geq 0,\ \forall\, i = 1,\ldots,m,\ \lambda_i g_i(\bar{x}) = 0\, \forall\, i = 1,\ldots,m\right\} \subset N_C(\bar{x}). \tag{1.16}$$

Note that $\lambda_i g_i(\bar{x}) = 0,\ \forall\, i = 1,\ldots,m$ immediately tells us that $\lambda_i = 0$ when $g_i(\bar{x}) < 0$. Thus, we set

$$I(\bar{x}) = \{i \in \{1,\ldots,m\} : g_i(\bar{x}) = 0\},$$

which we shall call the set of active indices at $\bar{x}$. Now for $i \in I(\bar{x})$, the convexity of g_i shows that for all y

$$g_i(y) - g_i(\bar{x}) \geq \langle \nabla g_i(\bar{x}), y - \bar{x}\rangle.$$

Thus,

$$g_i(y) \geq \langle \nabla g_i(\bar{x}), y - \bar{x}\rangle, \quad \forall\, i \in I(\bar{x}).$$

If y is feasible, that is, $y \in C$ then for each $i \in I(\bar{x})$

$$\left\langle \sum_{i \in I(\bar{x})} \lambda_i \nabla g_i(\bar{x}), y - \bar{x}\right\rangle \leq 0, \quad \forall y \in C.$$

Since $\lambda_i = 0$ for $i \notin I(\bar{x})$, we have

$$\left\langle \sum_{i=1}^{m} \lambda_i \nabla g_i(\bar{x}), y - \bar{x}\right\rangle \leq 0, \quad \forall y \in C.$$

This shows that

$$\sum_{i=1}^{m} \lambda_i \nabla g_i(\bar{x}) \in N_C(\bar{x}).$$

Now let us look for the reverse; that is, we want to show that

$$N_C(\bar{x}) \subset \left\{\sum_{i=1}^{m} \lambda_i \nabla g_i(\bar{x}) : \lambda_i \geq 0,\ \lambda_i g_i(\bar{x}) = 0\, \forall\, i = 1,\ldots,m\right\}.$$

Let us set

$$S(\bar{x}) = \left\{\sum_{i=1}^{m} \lambda_i \nabla g_i(\bar{x}) : \lambda_i \geq 0,\ \lambda_i g_i(\bar{x}) = 0\, \forall\, i = 1,\ldots,m\right\}.$$

Note that we can write the above set as

$$S(\bar{x}) = \left\{\sum_{i \in I(\bar{x})} \lambda_i \nabla g_i(\bar{x}) : \lambda_i \geq 0\right\}.$$

This immediately shows that $S(\bar{x})$ is a finitely generated cone and hence closed. We leave it to the reader to show that $S(\bar{x})$ is convex!! Let us assume, on the contrary, that there exists $v \in N_C(\bar{x})$, but $v \notin S(\bar{x})$. Since $S(\bar{x})$ is a closed convex cone using the standard separation theorem [11], we have that there exists $w \neq 0,\ w \in \mathbb{R}^n$ such that

$$\langle w, \xi \rangle \leq 0 < \langle w, v \rangle, \quad \forall \xi \in S(\bar{x}).$$

(Warning!! we have skipped several steps here and this part of the proof can be skipped in first reading.)

Now the fact $\langle w, \xi \rangle \leq 0\, \forall \xi \in S(\bar{x})$ shows that

$$\sum_{i \in I(x)} \lambda_i \langle \nabla g_i(\bar{x}), w \rangle \leq 0; \quad \lambda_i \geq 0, \quad i \in I(\bar{x}).$$

This shows that

$$\langle \nabla g_i(\bar{x}), w \rangle \leq 0, \quad \forall\, i \in I(\bar{x}).$$

Consider the set $P = \{u \in \mathbb{R}^n : \langle \nabla g_i(\bar{x}), u \rangle < 0\ \forall\ i \in I(\bar{x})\}$. We urge the author to use Slater's condition to show that $P \neq \emptyset$. For any $u \in K$, we can show that for $\lambda > 0$ and sufficiently small we have $g_i(x + \lambda u) < 0,\ \forall\ i = 1, 2, \ldots, m$. We also leave this simple argument in analysis to the reader. This shows that $x + \lambda u \in C$. Hence, $u \in (1/\lambda)(C - x)$. Thus, $u \in$ cone $(C - x)$. Thus, $P \subseteq$ cone$(C - x)$. Note that

$$\bar{P} = \{u \in \mathbb{R}^n : \langle \nabla g_i(\bar{x}), u \rangle \leq 0\, \forall\, i \in I(\bar{x})\}.$$

Thus, $\bar{P} \subseteq \overline{\text{cone}(C - \bar{x})}$. Now $w \in \bar{P}$. This shows that $w \in \overline{\text{cone}(C - \bar{x})}$. Thus, $\langle v, w \rangle \leq 0$ as $v \in N_C(\bar{x})$. This brings us to a contradiction, and we conclude that

$$N_C(\bar{x}) = S(\bar{x}).$$

Voila! We have thus shown the celebrated KKT condition for a differentiable convex optimization problem. Let us formally state this below:

Theorem 1.9: KKT Conditions

Let f be a finite convex function on $\mathbb{R}^n$ and C be a convex set in $\mathbb{R}^n$ given as

$$C = \{x \in \mathbb{R}^n : g_i(x) \leq 0,\ \forall i = 1, \ldots, m\},$$

where each g_i is a finite convex function on $\mathbb{R}^n$. Consider that Slater's condition holds true, that is, there exists $\hat{x} \in \mathbb{R}^n$, such that $g_i(\hat{x}) < 0, i = 1, \ldots, m$. Then, $\bar{x} \in C$ is a global minimizer of the problem

$$\min f(x) \quad \text{subject to } x \in C,$$

if and only if there exists $\lambda_i \geq 0, i = 1, \ldots, m$ such that

1. $0 = \nabla f(\bar{x}) + \sum_{i=1}^{m} \lambda_i \nabla g_i(\bar{x})$
2. $0 = \lambda_i g_i(\bar{x}),\ i = 1, \ldots, m$

Condition (2) is called the complementary slackness condition, which means both inequalities $\lambda_i \geq 0$ and $g_i(x) \leq 0$ cannot hold with strict inequality at the same time.

Let us consider the case where f is not differentiable. It is tempting to say that $\bar{x}$ is a global minimizer of f over C then,

$$0 \in \partial f(\bar{x}) + N_C(\bar{x}). \tag{1.17}$$

Our intuition has prompted us to replace $\nabla f(\bar{x})$ with $\partial f(\bar{x})$ in Equation 1.14. In fact, Equation 1.14 shows that

$$0 \in \nabla f(\bar{x}) + N_C(\bar{x}).$$

Thus possibly Equation 1.17 makes sense. In my mind we would argue as follows. If $\bar{x}$ is a local minimizer, then there must be $\bar{\xi} \in \partial f(\bar{x})$ such that

$$0 \in \bar{\xi} + N_C(\bar{x}).$$

This simply means

$$0 \in \partial f(\bar{x}) + N_C(\bar{x}).$$

So we have replaced $\nabla f(\bar{x})$ by some element of $\bar{\xi} \in \partial f(\bar{x})$. Though such intuition is compelling by looking at the structures it must be backed up by proper mathematical reasoning.

Note that minimizing the convex function f over the set C is same as minimizing the convex function $f + \delta_C$ over $\mathbb{R}^n$. Now since $\text{dom} f = \mathbb{R}^n$ and $\text{dom}\delta_C = C$, we have $\text{ri}(\mathbb{R}^n) \cap \text{ri}\,(C) = \mathbb{R}^n \cap \text{ri}(C) = \text{ri}(C) \neq \emptyset$. So, if $\bar{x}$ minimizes f over C it minimizes $f + \delta_C$ over $\mathbb{R}^n$ and hence

$$0 \in \partial(f + \delta_C)(\bar{x}).$$

Since all the conditions for applying the sum rule hold, we have

$$0 \in \partial f(\bar{x}) + \partial \delta_C(\bar{x}).$$

Since $\delta_C(\bar{x}) = N_C(\bar{x})$, the above condition reduces to

$$0 \in \partial f(\bar{x}) + N_C(\bar{x})$$

just as we had predicted.

Let us now make the problem a bit more difficult. Let $f : \mathbb{R}^n \to \bar{\mathbb{R}}$ be a proper lower-semicontinuous convex function and C be a closed convex set. Then, we can write the optimality condition like we had done before. This is done using the following theorem:

Theorem 1.10: Rockefellar–Pschenichni Condition

Let $f : \mathbb{R}^n \to \mathbb{R} \cup \{+\infty\}$ be a proper lower-semicontinuous function and C be a closed convex set with a nonempty interior. Assume that $\text{dom} f \cap (\text{int} C) \neq \emptyset$. Then, $\bar{x}$ is a global minimizer of f over C if and only if

$$0 \in \partial f(\bar{x}) + N_C(\bar{x}). \tag{1.18}$$

Proof. As before we know that if $\bar{x}$ is a minimizer of f over C then,

$$0 \in \partial(f + \delta_C)(\bar{x}).$$

Now let $x_0 \in \text{dom} f \cap (\text{int}C)$. As C is closed, δ_C is a proper lower-semicontinuous convex function and δ_C is continuous on int$C$ and thus is continuous at $x_0 \in \text{dom} f \cap (\text{int}C) \subseteq \text{dom} f \cap \text{dom}\delta_C$. Hence, we can apply the sum rule to obtain

$$0 \in \partial f(\bar{x}) + N_C(\bar{x}).$$

This establishes the necessary part. The sufficiency of condition (1.18) is left to the reader. ∎

The above optimality condition is named after the famous American mathematician R.T. Rockafellar and a famous Russian mathematician, B.N. Pshenichni. Suppose, we make the problem more descriptive by assuming

$$C = \{x \in \mathbb{R}^n : g_i(x) \leq 0, \ \forall i = 1, \ldots, m\}, \tag{1.19}$$

where each g_i is convex but not differentiable throughout. How shall we compute $N_C(\bar{x})$ in this case. Again if the Slater CQ holds, then

$$\begin{aligned} N_C(\bar{x}) &= \left\{ v \in \mathbb{R}^n : v = \sum_{i=1}^{m} \lambda_i \xi_i; \ \xi_i \in \partial g_i(\bar{x}), \ \lambda_i g_i(\bar{x}) = 0, \ i = 1, \ldots, m \right\} \\ &= \text{cone}\{\partial g_i(\bar{x}) : i \in I(\bar{x})\} = \text{cone}\left(\bigcup_{i \in I(\bar{x})} \partial g_i(\bar{x}) \right). \end{aligned}$$

Thus, naturally in the nonsmooth setting the KKT conditions become

1. $0 \in \partial f(\bar{x}) + \sum_{i=1}^{m} \lambda_i \partial g_i(\bar{x})$
2. $\lambda_i g_i(\bar{x}) = 0, \ i = 1, \ldots, m$

Of course using the explicit computation of the normal cone above allows us to postulate the existence of $\lambda_i \geq 0, \ i = 1, \ldots, m$, which leads to the above conditions. Note that the normal cone is not polyhedral when g_i's are not differentiable. For the computation of the normal cone when g_i's need not be differentiable see, for example, Dhara and Dutta [21].

Let us go back and have a more historical look at the development of these optimality conditions. In fact, in 1948, Fritz John [22] first proposed a necessary condition for an inequality-constrained optimization problem. These conditions are called the Fritz John condition or John conditions. Fritz John by the way was a legendary name in partial differential equations. Let us see how shall we drive the Fritz John condition for the convex case. We shall as before minimize the convex function f over C, where C is given by Equation 1.19.

Consider the function $L : \mathbb{R}^n \times \mathbb{R}_+ \times \mathbb{R}^m_+ \to \mathbb{R}$ given as

$$L(x, \lambda_0, \lambda) = \lambda_0 f(x) + \sum_{i=1}^{m} \lambda_i g_i(x).$$

We call this the *fundamental Lagrangian function*. To derive the Fritz John conditions for a convex optimization problem with inequality constraints, we shall mention the well-known Gordan's theorem of the alternative.

Lemma 1.1: Gordan's Theorem of the Alternative

Let $X \subset \mathbb{R}^n$ be a convex set and let $f_i : X \to \mathbb{R},\ i = 1, \ldots, m$ be convex functions. Let us consider the following two systems:

$$\text{System 1} : f_1(x) < 0, f_2(x) < 0, \ldots, f_m(x) < 0, \quad x \in X$$

$$\text{System 2} : \exists \lambda \neq 0,\ \lambda \in \mathbb{R}^m_+ \quad \text{such that} \sum_{i=1}^{m} \lambda_i f_i(x) \geq 0, \quad \forall x \in X.$$

Any one of the system has a solution but not for both.

For a proof, see Rockafellar [8] or Craven [23].

Let x^* be a solution of the given convex optimization problem. Then,

$$f(x) - f(x^*) < 0$$
$$g_i(x) < 0, \quad \forall = 1, \ldots, m$$

has no solution in $\mathbb{R}^n$. Gordan's theorem of the alternative allows us to postulate the existence of $(\lambda_0^*, \lambda^*) \in \mathbb{R}_+ \times \mathbb{R}^m_+$ with $(\lambda_0^*, \lambda^*) \neq 0$ such that

$$\lambda_0^*(f(x) - f(x^*)) + \sum_{i=1}^{m} \lambda_i^* g_i(x) \geq 0, \quad \forall x \in \mathbb{R}^n. \tag{1.20}$$

Set $x = x^*$ to obtain $\sum_{i=1}^{m} \lambda_i^* g_i(x^*) \geq 0$. This shows that $\sum_{i=1}^{m} \lambda_i^* g_i(x^*) = 0$. Further from Equation 1.20, we obtain that for all x,

$$L(x, \lambda_0^*, \lambda^*) \geq L(x^*, \lambda_0^*, \lambda^*). \tag{1.21}$$

Now consider any $(\lambda_0, \lambda) \in \mathbb{R}_+ \times \mathbb{R}^m_+$ and we can conclude that

$$L(x^*, \lambda_0^*, \lambda^*) \geq L(x^*, \lambda_0, \lambda). \tag{1.22}$$

Combining Equation 1.21 and 1.22 and the fact that $\sum_{i=1}^{m} \lambda_i g_i(x) = 0$, we have

1. $L(x^*, \lambda_0, \lambda) \leq L(x^*, \lambda_0^*, \lambda^*) \leq L(x, \lambda_0^*, \lambda^*), \forall x \in \mathbb{R}^n,\ (\lambda_0, \lambda) \in \mathbb{R}_+ \times \mathbb{R}^m_+$
2. $\lambda_i g_i(x^*) = 0,\ i = 1, \ldots, m$

So let us put all the above discussion in the following theorem.

Theorem 1.11

Let x^* be a minimizer of a convex optimization problem with inequality constraints given by Equation 1.18. Then, there exists $(\lambda_0^*, \lambda^*) \in \mathbb{R}_+ \times \mathbb{R}^m_+$ such that $(\lambda_0^*, \lambda^*) \neq 0$ such that the conditions (1) and (2) above hold. Conversely, if $(x^*, \lambda_0^*, \lambda^*) \in \mathbb{R}^n \times \mathbb{R}_+ \times \mathbb{R}^m_+$ satisfies (1) and (2) with $\lambda_0^* > 0$; then x^* solves the convex optimization problem.

We leave the proof of the converse part to the readers. However, there is a crux in the converse. We must have $\lambda_0^* > 0$. Thus, if $\lambda_0^* = 0$ we cannot say anything about the converse. The KKT approach guarantees that $\lambda_0^* > 0$. In fact, we would encourage the readers to show that when Slater's conditions holds then $\lambda_0^* > 0$. Dividing by $\lambda_0^* > 0$ both sides of Equation 1.20, we can consider $\lambda_0^* = 1$. In fact, we can write

$$L(x, \lambda_0^*, \lambda^*) := L(x, \lambda^*) \ \text{if}\ \lambda_0^* = 1.$$

The set of multipliers $(\lambda_0^*, \lambda_1^*, \ldots, \lambda_m^*)$ is called the set of Fritz John multipliers of the problem. The main problem with Fritz John multipliers is λ_0^* could be zero. In that case the objective function ceases to have any role in the optimality conditions. In fact, using Equation 1.21 we can see that

$$x^* \text{ minimizes } L(x, \lambda_0^*, \lambda^*) \text{ over } \mathbb{R}^n.$$

This is equivalent to saying that

$$0 \in \partial L(x, \lambda_0^*, \lambda^*) \quad \text{[The subdifferential is with respect to } x\text{]}.$$

Hence,

$$0 \in \lambda_0^* \partial f_i(x^*) + \lambda_1^* \partial g_1(x^*) + \cdots + \lambda_m^* \partial g_m(x^*).$$

Slater's CQ guarantees that $\lambda_0^* > 0$, showing that

$$0 \in \partial f_i(x^*) + \frac{\lambda_1^*}{\lambda_0^*} \partial g_1(x^*) + \cdots + \frac{\lambda_m^*}{\lambda_0^*} \partial g_m(x^*).$$

The vectors $(\lambda_1^*/\lambda_0^*, \ldots, \lambda_m^*/\lambda_0^*)$ is called the set of KKT multipliers. In fact, in the literature one says that the set of Fritz John multipliers $(\lambda_0^*, \lambda_1^*, \ldots, \lambda_m^*)$ is normal if $\lambda_0^* > 0$ and abnormal if $\lambda_0^* = 0$. In fact, Slater's condition guarantees that there can never be any abnormal Fritz John multipliers. It will be interesting to have a little discussion over the debate on which optimality condition is better, Fritz John or KKT. In 1980, B. Pourciau [24] wrote a paper titled "Modern multipliers rules" which appear in the *American Mathematical Monthly*. In that paper, he advocated the importance of Fritz John conditions stating that it is more fundamental since it is necessary for a minimizer without any additional assumption. Pourciau says that KKT is a simple extension of Fritz John by adding an additional assumption on the constraints. In fact, researchers in optimal control view the Fritz John conditions as more fundamental. They try to use the assumption of the specific problem to prove that $\lambda_0^* > 0$. See, for example, the book *Optimization Insight and Applications* by Brinkhuis and Tikhomirov [25] and Clarke [26]. However, from the point of view of convex optimization, the KKT conditions are truly fundamental. KKT conditions are sufficient without any additional condition and thus can be used as a computational tool. In fact, the Fritz John conditions are only sufficient in the convex case under the naive assumption that $\lambda_0^* > 0$. The KKT conditions are necessary under very simple assumption. The Slater's conditions simply say that $\text{int} C \neq \emptyset$, that is, the set C is full dimensional. In most problems of interest, this can get satisfied. Further for the linear programming problem and the convex quadratic programming problems under affine/linear constraints, the KKT conditions are automatically necessary without any additional conditions. Thus, for two very important classes of convex optimization problem, the KKT conditions become a powerful computational tool. Modern interior point algorithms for linear programming problems are based on the solution of KKT conditions using Newton's method. Thus, the KKT condition is the truly celebrated optimality condition. The only advantage of the Fritz John conditions is that they can provide a negative certificate for optimality. It can certify that a given point is not a local minimizer. However, practically trying to do that is not really successful. Further for the class of nonlinear programming problems called mathematical programming problems with complementarity constraints (MPCC for short), every feasible point is a Fritz John point which essentially renders the Fritz John conditions ineffective. Thus, KKT conditions are truly important even in this nonconvex setting. We, however, won't give a comprehensive bibliography of the MPCC literature. One can look up the Mathscinet (of the AMS) for the recent work on this subject by C. Kanzow, Jane J. Ye, S. Scholtes, Danny Ralph, and their collaborators.

Let us now give two examples where Slater's conditions fail. Consider the following convex optimization problem with a bilevel structure.

$$\begin{aligned} &\min f(x), \\ &\text{subject to} \quad x \in S, \quad \text{where } S = \arg\min\{h(x) : x \in K\}, \end{aligned}$$

where $f_i : \mathbb{R}^n \to \mathbb{R}$, $h : \mathbb{R}^n \to \mathbb{R}$ are convex functions and K be a closed convex set. This problem was first studied by Solodov [27], and later analyzed by Dempe et al. [28]. In Dempe et al. [28], this problem was referred to as a simple bi-level problem. If $\alpha = \inf\{h(x) : x \in K\}$ is finite, then we can equivalently write the above problem as

$$\begin{aligned} &\min f(x), \\ &\text{subject to} \quad h(x) \leq \alpha, \quad x \in K. \end{aligned}$$

This reformulated problem never satisfies the Slater condition. Note that if $\hat{x} \in \text{int}K$ be such that $h(\hat{x}) < \alpha$, then it contradicts the fact $\alpha = \inf_{x \in K} h$.

For a moment let us change over to a first person narrative. I was fortunate to meet Harold W. Kuhn in 2010 at the European OR conference in Lisbon. He was 80 years old at the time. When I told him my own personal fascination about the KKT conditions, he told me the story about how it was developed. To begin with, Harold Kuhn along with Albert W. Tucker and David Gale developed the necessary and sufficient conditions for linear programming problem. Tucker then wanted to consider convex quadratic functions. Gale (a famous game theorist) left the team possibly losing interest. It was Kuhn who suggested Tucker that one should go straight for differentiable functions. It was Tucker who suggested the use of a theorem of the alternative to deduce the necessary condition. The paper Kuhn and Tucker [29] published in 1951 in a conference proceedings showed that in the convex case, optimality conditions are sufficient when the problem has convex data. To derive the necessary part, they introduced a constraint qualification based on the geometry of the feasible set. This was called the Kuhn–Tucker constraint qualification. Kuhn and Tucker actually went a step ahead and derived a necessary condition for the existence of Pareto minimizer of a multiobjective optimization problem. Their optimality conditions became famous as Kuhn–Tucker condition since it was immediately helpful in the study of optimization algorithms. It was in 1975, while reading a book on mathematical economics by Takayama, Kuhn came to know that W. Karush had developed a similar condition in his master's thesis. To correct this historical mistake, Kuhn wrote to Karush that his name should also be a part of the nomenclature. From the early eighties, the name Karush–Kuhn–Tucker conditions (KKT conditions) became vogue in the optimization community.

We shall now develop the KKT condition for a linear programming problem. To do so, we shall first present the inhomogenous form of Motzkin's theorem of alternative as given by Guler [30].

Theorem 1.12: Motzkin Theorem of the Alternative

Let $p_i,\ i = 1,\ldots,l,\ q_j,\ j = i,\ldots,m$ and $\gamma_k,\ k = 1,\ldots,n$ be vectors in $\mathbb{R}^n$. Further let $\alpha_i,\ i = 1,\ldots,l,\ \beta_j,\ j = i,\ldots,m,$ and $\gamma_k,\ k = 1,\ldots,n$ are scalars (real numbers). Then the linear system

$$\langle p_i, x\rangle < \alpha_i, \quad i = 1,\ldots,l$$
$$\langle q_j, x\rangle \le \beta_j, \quad i = 1,\ldots,m$$
$$\langle \gamma_k, x\rangle = \gamma_k, \quad k = 1,\ldots,n$$

is inconsistent if and only if there exists vectors $\lambda_0 \in \mathbb{R}$, $\lambda \in \mathbb{R}^l$, $\mu \in \mathbb{R}^m$, and $\delta \in \mathbb{R}^n$ such that

$$\sum_{i=1}^{l} \lambda_i p_i + \sum_{j=1}^{m} \mu_j q_j + \sum_{k=1}^{n} \delta_k \gamma_k = 0$$
$$\sum_{i=1}^{l} \lambda_i \alpha_i + \sum_{j=1}^{m} \mu_j \beta_j + \sum_{k=1}^{n} \delta_k \gamma_k + \lambda_0 = 0$$
$$(\lambda_0, \lambda, \mu) \in \mathbb{R}_+ \times \mathbb{R}_+^l \times \mathbb{R}_+^m, \quad 0 \ne (\lambda_0, \lambda).$$

The reader might have observed a slight difference between the presentation of Gordan's theorem of the alternative versus the one of Motzkin. If we carefully look at Motzkin's theorem of the alternative, there are in fact two systems and only one of them can have a solution and not both.

We have considered linear programming problem in the following form:

$$\min\ \langle c, x\rangle,$$
$$\text{subject to} \quad Ax \ge b, \quad x \ge 0.$$

However, in order to find the feasible set, it might be difficult to compute the feasible point by solving inequalities. Thus, one can introduce the surplus variable and convert the main inequality constraints ($Ax \ge b$) into equality ones. For example, the above problem can be written as

$$\min\ \langle c, x\rangle,$$
$$\text{subject to} \quad Ax - Iz = b, \quad x \ge 0, \quad z \ge 0.$$

Here z is the surplus variable which of course is nonnegative and I is an $m \times m$ identity matrix. Thus, one usually considers the linear programming problem in the standard form

$$\min\langle c, x\rangle \qquad \text{(LPS)}$$
$$\text{subject to} \quad Ax = b, \quad x \ge 0.$$

where A is an $m \times n$ matrix usually assumed to the full rank.

Theorem 1.13

Let us consider the linear programming problem in the standard form, that is, we consider the problem LPS. Then, $\bar{x} \in \mathbb{R}^n$ is a solution of LPS if and only if there exists a vector $\bar{\delta} \in \mathbb{R}^m$ and $\bar{s} \in \mathbb{R}_+^n$

such that

$$\begin{aligned} c + A^T \bar{\delta} &= \bar{s} \\ A\bar{x} &= b \\ \langle \bar{x}, \bar{s} \rangle &= 0 \\ \bar{x} \geq 0, &\quad \bar{s} \geq 0. \end{aligned}$$

Proof. Let $\bar{x}$ be a solution of LPS. Thus, $A\bar{x} = b$. Further let us write LPS as

$$\begin{aligned} &\min \langle c, x \rangle, \\ \text{subject to} \quad &\langle a_j, x \rangle = b_j, \quad j = 1, \ldots, m, \\ &x_i \geq 0, \quad i = i, \ldots, n. \end{aligned}$$

It is simple to see that if $\bar{x}$ is the solution of LPS, the following system

$$\begin{aligned} &\langle c, x \rangle < \langle c, \bar{x} \rangle, \\ &\langle -e_i, x \rangle \leq 0, \quad i = 1, \ldots, n, \\ &\langle a_j, x \rangle = b_j, \quad j = 1, \ldots, m \end{aligned}$$

has no solution in $\mathbb{R}^n$. Then, by using Motzkin's theorem of alternative, there exists scalars $\lambda_0 \geq 0$, $\lambda_1 \geq 0$, $\mu_i \geq 0$, $i = 1, \ldots, n$ and $\delta_j,\ j = 1, \ldots, m$ such that

$$\lambda_1 c + \sum_{i=1}^{n} \mu_i(-e_i) + \sum_{j=1}^{m} \delta_j a_j = 0 \tag{1.23}$$

$$\lambda_1 \langle c, \bar{x} \rangle + \sum_{i=1}^{n} \mu_i(0) + \sum_{j=1}^{m} \delta_j b_j + \lambda_0 = 0 \tag{1.24}$$

$$\text{with } (\lambda_0, \lambda_1, \mu) \in \mathbb{R}_+ \times \mathbb{R}_+ \times \mathbb{R}^n_+ \quad \text{and,} \quad (\lambda_0, \lambda_1) \neq 0.$$

We claim that $\lambda_0 = 0$. On the contrary, assume that $\lambda_0 \neq 0$. Hence, $\lambda_0 > 0$. Thus, from Equation 1.24 we have

$$\lambda_1 \langle c, \bar{x} \rangle + \sum_{j=1}^{m} \delta_j b_j = -\lambda_0 < 0$$

while from Equation 1.23, we have

$$\lambda_1 \langle c, \bar{x} \rangle + \sum_{j=1}^{m} \delta_j b_j = \sum_{i=1}^{m} \mu_i \langle e_i, x \rangle \geq 0.$$

This leads to a contradiction and hence $\lambda_0 = 0$, showing that $\lambda_1 > 0$. Thus, dividing Equations 1.23 and 1.24 throughout by λ_1, we have

$$c + \sum_{j=1}^{m} \bar{\delta}_j a_j = \sum_{i=1}^{n} \bar{\mu}_i e_i \tag{1.25}$$

$$\langle c, \bar{x} \rangle + \sum_{j=1}^{m} \delta_j b_j = 0, \tag{1.26}$$

where $\bar{\delta}_j = \delta_j/\lambda_1$ and $\bar{\mu}_i = \mu_i/\lambda_1$. Now from Equation 1.25, we have

$$\langle c, \bar{x} \rangle + \sum_{j=1}^{m} \bar{\delta}_j b_j = \langle \bar{s}, \bar{x} \rangle,$$

where $\bar{s} = \sum_{i=1}^{n} \bar{\mu}_i e_i$. Now using Equation 1.26, we have $\langle \bar{x}, \bar{s} \rangle = 0$. More compactly, Equation 1.25 can be written as

$$c + A^T \bar{\delta} = \bar{s}.$$

Thus, this completes the proof. ■

We can represent the above condition by setting $\bar{\delta} = -\bar{y} \in \mathbb{R}^m$. Hence, the conditions now become

$$\begin{aligned} &A^T \bar{y} + \bar{s} = c \\ &A\bar{x} = b \\ &\langle \bar{x}, \bar{s} \rangle = 0 \\ &\bar{x} \geq 0, \quad \bar{s} \geq 0. \end{aligned}$$

The solution of this system leads to the solution of both the problem LPS and its dual, which is given as

$$\begin{aligned} &\max \ \langle b, y \rangle, \\ &\text{subject to} \quad A^T y + s = c, \quad s \geq 0. \end{aligned}$$

We shall now turn our attention to the multiplier or KKT multiplier set of a convex optimization problem, which has its own interesting properties. Let $\bar{x}$ be a solution of the convex optimization problem (CP) with only inequality constraints. Then, we shall denote the KKT multiplier set at $\bar{x}$ as

$$\text{KKT}(\bar{x}) = \left\{ \lambda \in \mathbb{R}^m_+ : \ 0 \in \partial f(\bar{x}) + \sum_{i==1}^{m} \lambda_i \partial g_i(\bar{x}) \text{ and } \lambda_i g_i(\bar{x}) = 0, \ i = 1, \ldots, m \right\}.$$

If Slater's conditions hold, then $\text{KKT}(\bar{x})$ is compact. Further it may appear that if we consider another minimizer of the same problem, then the multipliers will change. The answer seems to be surprising. It doesn't. The key to understanding this fact needs an understanding of perturbation of the convex optimization problem. We shall thus consider the following perturbed optimization problem

$$\min f(x),$$
$$\text{subject to} \quad g_i(x) \leq b_i, \quad i = 1, \ldots, n,$$

where $b = (b_1, \ldots, b_m)^T$ is called the parameter vector and the above perturbed problem is useful in many practical situations. Consider the following value function

$$v(b) = \inf\{f(x) : g_i(x) \leq b_i,\ i = 1, \ldots, m\}.$$

A key fact is that v is a convex function in b and this time we need to allow the values $+\infty$ and $-\infty$. From Lemma 3.2.6 of Borwein and Lewis [2] if there exists $\hat{b} \in (\text{dom } v)$ such that $v(\hat{b}) > -\infty$, then v never takes the value $-\infty$. If $v(0)$ is finite, that is, the original problem has a finite infimal value, then $\text{KKT}(\bar{x}) = -\partial v(0)$ for any $\bar{x}$. This shows that $\text{KKT}(\bar{x})$ is independent of $\bar{x}$ and the result is a central fact in convex optimization.

Another key idea in convex analysis is the notion of the Fenchel conjugate. Let $f : \mathbb{R}^n \to \mathbb{R} \cup \{+\infty, -\infty\}$ be a given convex function. Then, the conjugate function $f^* : \mathbb{R}^n \to \mathbb{R} \cup \{+\infty\}$ is given as

$$f^*(x^*) = \sup_{x \in \mathbb{R}^n} \{\langle x^*, x\rangle - f(x)\}.$$

Note that f^* is a convex function and in fact it is a proper convex function if f is proper. Note that

$$-f^*(0) = \inf_{\mathbb{R}^n} f.$$

Further from the definition, we have $\forall (x, x^*) \in \mathbb{R}^n \times \mathbb{R}^n$

$$\langle x^*, x\rangle \leq f^*(x^*) + f(x). \quad \text{(Young–Fenchel inequality)}$$

In fact equality holds, that is,

$$\langle x^*, x\rangle = f^*(x^*) + f(x),$$

if and only if $x^* \in \partial f(x)$, we can also consider the bi-conjugate of f given as

$$f^{**}(x) = \sup_{x^* \in \mathbb{R}^n} \{\langle x^*, x\rangle - f^*(x^*)\}.$$

An immediate consequence of the above definition is that $f^{**}(x) \leq f(x)$ for all x. The following is a far-reaching result in the convex analysis. For more detail see, for example, Rockafellar [8].

Theorem 1.14

A function $f : \mathbb{R}^n \to \mathbb{R} \cup \{+\infty, -\infty\}$ is a proper and lower-semicontinuous convex function if and only if

$$f = f^{**}.$$

We will now show a simple application of this fact.

Corollary 1.1

Let $f : \mathbb{R}^n \to \mathbb{R} \cup \{+\infty, -\infty\}$ be a proper and lower-semicontinuous convex function then $f = f^*$ if and only if $f = (1/2)\|.\|^2$.

Proof. Let $f = (1/2)\|.\|^2$. Hence,

$$f^*(x^*) = \sup_{x \in \mathbb{R}^n} \left\{ \langle x^*, x \rangle - \frac{1}{2}\|x\|^2 \right\}.$$

Let $\phi(x) = \langle x^*, x \rangle - (1/2)\|x\|^2$. Hence, $\nabla\phi(x) = x^* - x$. Thus, $\nabla\phi(x) = 0$ iff $x^* = x$. Further $\nabla^2\phi(x) = -I$, showing that $\nabla^2\phi(x^*)$ is negative definite and hence $x = x^*$ is a strict local maximizer of ϕ. Note that $\lim_{\|x\| \to \infty} \phi(x) = -\infty$; hence there exists a global maximizer of ϕ. Since $x = x^*$ is the only critical point, we see that $x^* = x$ is the global minimizer of ϕ. Hence,

$$f^*(x^*) = \langle x^*, x^* \rangle - \frac{1}{2}\|x^*\|^2.$$

Therefore,

$$f^*(x^*) = \frac{1}{2}\|x\|^2.$$

For the converse, observe that for each fixed $x^* \in \mathbb{R}^n$ we have

$$f^*(x^*) + f(x) \geq \langle x^*, x \rangle, \quad \forall x \in \mathbb{R}^n.$$

In particular, setting $x = x^*$, we have

$$f^*(x^*) + f(x^*) \geq \langle x^*, x^* \rangle. \tag{1.27}$$

But as $f^* = f$, we have from Equation 1.27

$$f(x^*) \geq \frac{1}{2}\|x^*\|^2. \tag{1.28}$$

Now as Equation 1.28 holds for any $x^* \in \mathbb{R}^n$, we conclude that $f(x) \geq (1/2)\|x\|^2$, $\forall x \in \mathbb{R}^n$. Now from Equation 1.28, we have

$$-f^*(x^*) \leq -\frac{1}{2}\|x^*\|^2.$$

Therefore,

$$\langle x^*, x \rangle - f^*(x^*) \leq \langle x^*, x \rangle - \frac{1}{2}\|x^*\|^2.$$

Thus,

$$f^{**}(x) \leq \frac{1}{2}\|x\|^2.$$

Now using Theorem 3.7, we have $f(x) \leq (1/2)\|x\|^2$. Thus, we prove $f(x) = (1/2)\|x\|^2$, $\forall x \in \mathbb{R}^n$. ■

Let us now provide some interesting examples of computing the conjugate. Borwein and Lewis [2] have a good store of examples.

Example 1.13

1. Let $f(x) = \sqrt{1+x^2}$, $x \in \mathbb{R}$. Then, $f^*(x^*) - \sqrt{1-x^2}$, $x^* \in [-1,+1]$.
2. Let $f(x) = \ln x$, $x > 0$. Then, $f^*(x^*) = -1 - \log(-x^*)$, $x^* < 0$.
3. Let $f(x) = e^x$, $x \in \mathbb{R}$. Then,

$$f^*(x^*) = \begin{cases} x^* \log x^* - x^*, & x^* > 0 \\ 0, & x^* = 0. \end{cases}$$

   Using Theorem 3.7, we conclude that $f(x) = xlogx - x$, $x > 0$ is a convex function since it is the conjugate of the convex function e^x. The readers should try out the details.
4. The following interesting example is by Borwein and Lewis [2].

Let $f(x) = e^x$, $x \in \mathbb{R}$ and let f^* be its Fenchel conjugate. Define

$$g(z) = \inf_{x^* \in \mathbb{R}^{m+1}} \left\{ \sum_{i=1}^{m+1} f^*(x_i^*) : \sum_{i=1}^{m+1} x_i^* = 1, \sum_{i=1}^{m+1} x_i^* a^i = z \right\},$$

where $a^0, a^1, \ldots, a^m$ are the given points in $\mathbb{R}^n$, then it can be shown that

$$g^*(x^*) = 1 + \ln\left(\sum_{i=1}^{m+1} e^{\langle a^i, x^* \rangle} \right).$$

The conjugate function has been used to deduce Gordan's theorem of the alternative in Borwein and Lewis [2]. We would like to request the reader to check why $g(x)$ is convex.

A very key result in convex optimization is the Fenchel duality theorem, which brings out the fact that behind every minimization problem is a maximization problem and their optimal values coincide. This is a fundamental feature of optimization about which we have already mentioned during our discussion of linear programming problem. Let us now state and prove the Fenchel duality theorem as given in Villani [13]. Note that in Villani [13] it is given for a normed space. However, we will speak about it only in the finite-dimensional setting.

Theorem 1.15: Fenchel Duality Theorem

Let $f : \mathbb{R}^n \to \mathbb{R} \cup \{+\infty\}$ and $g : \mathbb{R}^n \to \mathbb{R} \cup \{+\infty\}$ be two convex functions. Assume that there exists $z_0 \in \mathbb{R}^n$ such that $f(z_0) < +\infty$ and $g(z_0) < +\infty$ and f is continuous at z_0. Then,

$$\inf_{x \in \mathbb{R}^n} \{f(x) + g(x)\} = \sup_{x^* \in \mathbb{R}^n} \{-f^*(-x^*) - g^*(x^*)\}. \tag{1.29}$$

Proof. If we observe carefully, one can rewrite Equation 1.29 as

$$\sup_{x^* \in \mathbb{R}^n} \inf_{x,y \in \mathbb{R}^n} \{f(x) + g(y) + \langle x^*, x - y \rangle\} = \inf_{x \in \mathbb{R}^n} \{f(x) + g(x)\}\ x \in \mathbb{R}^n. \tag{1.30}$$

Thus, it will be enough if we demonstrate that $\exists x^* \in \mathbb{R}^n$ such that Equation 1.30 holds. To see how Equation 1.29 is related to Equation 1.30, let us observe that

$$\sup_{x^*\in\mathbb{R}^n} \{-f^*(-x^*) - g^*(x^*)\} = \sup_{x^*\in\mathbb{R}^n} \left\{ - \sup_{x\in\mathbb{R}^n} \{\langle -x^*, x\rangle - f(x)\} - \sup_{y\in\mathbb{R}^n} \{\langle x^*, y\rangle - g(y)\} \right\}$$

$$= \sup_{x^*\in\mathbb{R}^n} \left\{ \inf_{x\in\mathbb{R}^n} \{\langle x^*, x\rangle + f(x)\} - \sup_{y\in\mathbb{R}^n} \{\langle x^*, y\rangle - g(y)\} \right\}.$$

$$\sup_{x^*\in\mathbb{R}^n} \{-f^*(-x^*) - g^*(x^*)\} = \sup_{x^*\in\mathbb{R}^n} \{\inf_x\{\langle x^*, x\rangle + f(x)\} + \inf_y\{g(y) - \langle x^*, y\rangle\}\}$$

$$= \sup_{x^*\in\mathbb{R}^n} \inf_{x,y\in\mathbb{R}^n} \{f(x) + g(y) + \langle x^*, x - y\rangle\}$$

when $x = y$ then both sides are equal. So what we indeed have to show is that $\forall x,\ y \in \mathbb{R}^n$, $\exists x^* \in \mathbb{R}^n$ such that

$$f(x) + g(y) - \langle x^*, x - y\rangle \geq m = \inf_{x\in\mathbb{R}^n} \{f(x) + g(x)\}.$$

Since $f(x_0) + g(x_0) < +\infty$, it is clear that m is finite. Let

$$C_1 = \{(x, \lambda) \in \mathbb{R}^n \times \mathbb{R} : f(x) < \lambda\},$$
$$C_2 = \{(y, \mu) \in \mathbb{R}^n \times \mathbb{R} : g(y) + \mu \leq m\}.$$

C_1 and C_2 are convex sets and C_1 is an open set since $f(z_0) < f(z_0) + 1$ (The reader has to put in same thought here). Further $C_1 \cap C_2 = \emptyset$ (the reader can check this out). Thus, by separation theorem, there exists $0 \neq l \in (\mathbb{R}^n \times \mathbb{R})$ such that

$$\inf_{c_1\in C_1} \langle l, c_1\rangle = \inf_{c_1\in \text{int}C_1} \langle l, c_1\rangle \geq \sup_{c_2\in C_2} \langle l, c_2\rangle.$$

Thus, there exists $w^* \in \mathbb{R}^n$ and $\alpha \in \mathbb{R}$ with $(w^*, \alpha) \neq (0, 0)$ such that

$$\langle w^*, x\rangle + \alpha\lambda \geq \langle w^*, y\rangle + \alpha\mu, \tag{1.31}$$

$\forall (x, \lambda) \in C_1$ and $(y, \mu) \in C_2$. We claim that $\alpha > 0$. Note that $\alpha \neq 0$. Since $\alpha = 0$, it can be shown that $\langle w^*, x - y\rangle \geq 0$. This will immediately show that $w^* = 0$. This will be a contradiction to the fact that $(w^*, \alpha) \neq (0, 0)$. We will now show that $\alpha < 0$ is also not a feasible case here. Let $\bar{x}, \bar{y}$, and $\bar{\mu}$ be given. Consider any $\lambda > 0$ such that $f(\bar{x}) < \lambda$. Then, from Equation 1.31, we have

$$\langle w^*, x_0 - y_0\rangle \geq \alpha(\bar{\mu} - \lambda). \tag{1.32}$$

If $\bar{\mu} \geq 0$ then by taking $\lambda > 0$ as large as we like, we will reach a value of $\hat{\lambda}$ of λ such that

$$\langle w^*, x_0 - y_0\rangle < \alpha(\bar{\mu} - \hat{\lambda}). \tag{1.33}$$

If $\bar{\mu} < 0$, we can argue in the same way and come to the same conclusion in Equation 1.33, which contradicts Equation 1.32. Hence, $\alpha > 0$. Dividing both sides of Equation 1.31 by α, we have

$$\left\langle \frac{w^*}{\alpha}, x \right\rangle + \lambda \geq \left\langle \frac{w^*}{\alpha}, y \right\rangle + \mu$$

Setting $w^*/\alpha = x^*$, we have

$$\langle x^*, x\rangle + \lambda \geq \langle x^*, y\rangle + \mu.$$

Hence, we have

$$\langle x^*, x\rangle + f(x) \geq \langle x^*, y\rangle + m - g(y).$$

This implies

$$f(x) + g(y) + \langle x^*, x - y\rangle \geq m.$$

Hence, we have established the result. ■

A key feature in the study of a convex optimization problem is the deep and fascinating link between optimality and duality. Optimality conditions analyze the nature of the conditions that an optimal point will satisfy and duality brings forth the fact that for every minimizing problem there is a maximization problem going at its back. The key to the study of these features is the Lagrangian function $L(x, \lambda)$, which for CP is given as

$$L(x, \lambda) = f(x) + \sum_{i=1}^{m} \lambda_i g_i(x),$$

where $x \in \mathbb{R}^n$ and $\lambda \in \mathbb{R}^n_+$.

We will now see how the idea of a dual problem arises naturally and how the extended-valued function framework is so natural to optimization. For the sake of brevity, we will just consider inequality constraints and the abstract constraint $x \in X$ and ignore the affine equality constraints. So our Lagrange function or Lagrangian is just

$$L(x, \lambda) = f(x) + \lambda_1 g_1(x) + \cdots + \lambda_m g_m(x).$$

Note the following:

$$\sup_{\lambda \geq 0 (\lambda \in \mathbb{R}^m_+)} L(x, \lambda) = \begin{cases} f(x), & \text{if } g_i(x) \leq 0, \forall\, i = 1, \ldots, m. \\ +\infty, & \text{otherwise} \end{cases}$$

Thus, our original problem can be written as

$$\inf_{x \in X} \sup_{\lambda \geq 0} L(x, \lambda).$$

Suppose for fun and most probably we are tempted to switch the min and the max. Then, we obtain a problem like

$$\sup_{\lambda \geq 0} \inf_{x \in X} L(x, \lambda).$$

The function $\theta(\lambda) = \inf_{x \in X} L(x, \lambda)$ is a concave function and thus the new problem tells us that

$$\sup_{\lambda \geq 0} \theta(\lambda) \qquad \text{(DP)}$$

can be viewed as a problem associated with CP as

$$\sup_{\lambda\geq 0}\inf_{x\in X} L(x,\lambda) \leq \inf_{x\in X}\sup_{\lambda\geq 0} L(x,\lambda).$$

Thus, the new problem DP provides a lower bound to CP and thus DP is often referred to as the dual problem. One might dismiss the approach as not so natural since we have switched min and max. (or sup and inf). In reality, can we do that?

In fact, we shall now see that these two types of problems will naturally arise if we consider our study from the game theory point of view and more precisely two-person zero-sum games. Let us consider a 2-person zero-sum game where X is the strategy set of player 1 and Y is the strategy set of player 2. So when player 1 chooses $x \in X$ and player 2 chooses $y \in Y$ and they are simultaneously revealed, then player 1 has to pay an amount $F(x, y)$ to player 2. So player 1 actually has now $-F(x, y)$ amount and thus the total payoff money with the players is zero.

Action of Player 1 Given a choice $x \in X$, what is the worst-case scenario, that is, what is the maximum amount one needs to pay to player 2.

So, for a given x the maximum amount he has to pay is

$$\varphi(x) = \sup_{y\in Y} F(x, y).$$

The idea is to choose an $x \in X$, which minimizes $\varphi(x)$, that is, player 1 problem is

$$\min_{x\in X} \varphi(x) = \min_{x\in X}\sup_{y\in Y} F(x, y).$$

Action of Player 2 For a given choice $y \in Y$, he thinks what is the least he can obtain from Player 1; that is

$$\psi(y) = \inf_{x\in X} F(x, y).$$

He would want to have $y \in Y$ such that it maximizes $\psi(y)$

$$\min_{y\in Y} \psi(y) = \max_{y\in Y}\inf_{x\in X} F(x, y).$$

So Player 2 plays the dual problem while Player 1 plays the so-called primal problem. A major question in game theory is when the following holds true:

$$\min_{x\in X}\max_{y\in Y} F(x, y) \stackrel{?}{=} \max_{y\in Y}\min_{x\in X} F(x, y).$$

Von-Neumann and Morgenstern answer this question in their famous book *Game Theory and Economic Behavior* (Ans: X, Y compact, $F(x, y)$ is convex in x and concave in y).

Thus, in our setting we can ask:

When is

$$\min_{x\in X}\sup_{\lambda\geq 0} L(x,\lambda) \stackrel{?}{=} \sup_{\lambda\geq 0}\min_{x\in X} L(x,\lambda).$$

Here $\lambda \in \mathbb{R}_+^m$, which is not compact and X need not be compact though L is convex in x and linear in λ. Note that the result of Von-Neumann and Morgenstern cannot be applied here because even we take X compact we cannot escape $Y = \mathbb{R}_+^m$. The key to this is Sion's famous minimax theorem, which we now state and give an outline of the proof (as given in a paper by Komiya [31]).

Suppose $\exists\ (x^*, y^*) \in X \times \mathbb{R}_+^m$ such that

$$L(x^*, \lambda^*) = \min_{x \in X} \sup_{\lambda \geq 0} L(x, \lambda) = \sup_{\lambda \geq 0} \min_{x \in X} L(x^*, \lambda^*).$$

Then, we have

$$L(x^*, \lambda) \leq L(x^*, \lambda^*) \leq L(x, \lambda^*), \quad \forall\, x \in X, \quad \lambda \in \mathbb{R}_+^m.$$

(x^*, λ^*) is a saddle point, a key object in convex optimization.

Theorem 1.16: Sion's Minimax Theorem

Let X be a subset of $\mathbb{R}^n$ and Y be a subset of $\mathbb{R}^m$. Let X be a compact convex set and Y is a convex set. Let $f : X \times Y \to \mathbb{R}$ be given and satisfies the following:

1. $f(x, .)$ is upper semicontinuous and concave on Y for each $x \in X$.
2. $f(., y)$ is lower semicontinuous and convex on X for each $y \in Y$.

Then,

$$\min_{x \in X} \sup_{y \in Y} f(x, y) = \sup_{y \in Y} \min_{x \in X} f(x, y).$$

Hitedoshi Komiya [31] provides an elementary proof of the result by first proving two lemmas.

Lemma 1.2

Consider the same assumptions as in Sion's minimax theorem. Then, for any y_1 and $y_2 \in Y$ and any real number α with $\alpha < \min_{x \in X}[\max f(x, y_1),\ f(x, y_2)]$, there is $y_0 \in Y$ such that

$$\alpha < \min_{x \in X} f(x, y_0).$$

Proof. On the contrary, assume that $\alpha \geq \min_{x \in X} f(x, y_0)$ for all $y \in Y$. Now choose β such that

$$\alpha < \beta < \min_{x \in X} \max\{f(x, y_1),\ f(x, y_2)\}.$$

For each $z \in [y_1, y_2]$ = line segment joining y_1 and y_2, define

$$Cz = \{x \in X : f(x, z) \leq \alpha\} \quad \text{and} \quad C'z = \{x \in X : f(x, z) \leq \beta\}.$$

Let $A = C'y_1$, $B = C'y_2$. Since $f(., z)$ is lsc, then Cz, $C'z$, A, and B are closed; $A \cap B = \phi$. Since $f(x, .)$ is concave, we have

$$f(x, z) \geq \min\{f(x, y_1),\ f(x, y_2)\}\ \forall\, x \in X\ \&\ z \in [y_1, y_2].$$

Since $C'z$ is convex (why!!), we have $C'z$ connected and we have $C'z \subset A \cup B$ (with $A \cap B = \phi$). Hence $Cz \subset C'z \subset A$ or $Cz \subset C'z \subset B$. Define $I = \{z \in [y_1, y_2] : Cz \subset A\}$ and $J = \{z \in [y_1, y_2] : Cz \subset B\}$, $I \cup J = [y_1, y_2]$; $I \cap J = \phi$. Let $\{z_n\}$ be a sequence in I with $\lim z_n \in [y_1, y_2]$. Let x be any point of Cz. Then, $f(x, z) \leq \alpha < \beta$.

By upper semicontinuity of $f(x, .)$, we have

$$\lim_{n\to\infty} \sup f(x, z_n) < \beta.$$

Thus, $\exists\ m \in \mathbb{N}$ such that $f(x, z_m) < \beta$, that is, $x \in C'z_m$. We have $C'z_m \subset A$. Since $Cz_m \subset A$. Then, $x \in A$. Thus, $Cz \subset A \Rightarrow z \in I$ and hence I is closed in $[y_1, y_2]$. Similarly, J is closed. But $I \cap J = \phi$ when $I \cup J = [y_1, y_2]$, which contradicts the connectness of $[y_1, y_2]$. ■

We simply state the following lemma, which is a consequence of Lemma 1.2.

Lemma 1.3

Under the same assumptions as Sion's minimax theorem for any finite $y_1, \ldots, y_n \in Y$ and any real number $\alpha < \min_{x\in X} \max_{1\leq i\leq n} f(x, y_i)$, there is $y_0 \in Y$ such that

$$\alpha < \min_{x\in X} f(x, y_0).$$

Proof of Sion's minimax theorem: (Based on the two lemma's and finite intersection property) we always have

$$\sup_{y\in Y} \min_{x\in X} f(x, y) \leq \min_{x\in X} \sup_{y\in Y} f(x, y).$$

Let $\alpha < \min_{x\in X} \sup_{y\in Y} f(x, y)$. Let $X_y = \{x \in X,\ y \in Y : f(x, y) \leq \alpha\} \subseteq X$. Then, $\cap_{y\in Y} X_y = \phi$ by FIP $\Rightarrow\ \exists\ y_1, \ldots, y_n \in Y,\ s.t.\ \cap_{i=1}^{n} X_{y_i} = \phi$. That is, $\alpha < \min_{x\in X} \max_{1\leq i\leq n} f(x, y)$. By Lemma 3.3 $\exists\ y_0 \in Y$ such that $\alpha < \min_{x\in X} f(x, y_0)$. Hence, $\alpha < \sup_{y\in Y} \min_{x\in X} f(x, y) \Rightarrow \min_{x\in X} \sup_{y\in Y} f(x, y) \leq \sup_{y\in Y} \min_{x\in X} f(x, y)$. Hence, the result. ■

Thus, if X is compact, we can apply Sion's minimax theorem to our case and hence

$$\min_{x\in X} \sup_{\lambda\geq 0} L(x, \lambda) = \sup_{\lambda\geq 0} \inf_{x\in X} L(x, \lambda).$$

$$\min_{x\in X,\ \forall\ i\ g_i(x)\leq 0} f(x) = \sup_{\lambda\geq 0} \theta(\lambda) \quad \text{and} \quad \theta(\lambda) = \inf_{x\in X} L(x, \lambda).$$

Thus, we have

$$\min -val(CP) = \max -val(DP).$$

This equality is what is called the strong duality. How to deal with the case when X is not compact? Moreover, in the above expression "min" can be replaced by "inf." To answer the case when X is not compact, we shall provide the following modified form of Theorem 3.4, which we call the saddle point theorem.

Theorem 1.17: Saddle Point Theorem

Let us consider the problem CP with only inequality constraints. Let the Slater constraints qualification hold. That is, $\exists\ \hat{x}$, such that $g_i(\hat{x}) < 0$ for $i = 1, \ldots, m$. Then, if x^* is a solution of CP, $\exists\ \lambda^* \in \mathbb{R}^m_+$ such that

1. $L(x^*, \lambda) \leq L(x^*, \lambda^*) \leq L(x, \lambda^*),\ \forall \lambda \in \mathbb{R}^m_+,\ x \in X$
2. $\lambda^* g_i(x^*) = 0,\ \forall\, i = 1, \ldots, m$

Conversely, if there exists $(x^*, \lambda^*) \in X \times \mathbb{R}^m_+$ such that (1) and (2) hold, then x^* is feasible for CP and is also the global minimizer of CP.

This is a famous result so we do not put in the proof here. The key of proving the saddle point conditions is as follows:

If x^* solves CP, then the following system

$$f(x) - f(x^*) < 0$$

$$g_i(x) < 0$$

$$x \in X$$

has no solution, $x \in \mathbb{R}^n$. Apply some separation theorem (or Gordan's theorem of a alternative) to conclude the result. Note that for the converse we do not need Slater's condition.
The saddle point conditions tell us that if x^* solves CP, $\lambda^* \in \mathbb{R}^m_+$ such that

i. $L(x^*, \lambda^*) = \min_{x \in X} L(x, \lambda^*)$
ii. $\lambda_i^* g_i(x^*) = 0,\ \forall\, i = 1, \ldots, m$

Suppose that we have the information that f and g_i are differentiable. Then, we can conclude that

$$(KKT) \begin{cases} \text{(i)}\ \ 0 \in \nabla_x L(x^*, \lambda^*) + N_X(x^*) \\ \\ \text{(ii)}\ \ \lambda_i^* g_i(x^*) = 0, \quad \forall\, i = 1, \ldots, m. \end{cases}$$

Recall that $N_X(x^*)$ is the normal cone to X at x^*. The above conditions are termed as KKT conditions.

Condition (ii) in the KKT system is called the complementary slackness condition. This tells that both g_i and λ_i cannot hold with strict inequality at the same time.
If $X = \mathbb{R}^n$, we have

$$\text{(i)}\ \ 0 = \nabla_x L(x^*, \lambda^*)$$

$$\text{(ii)}\ \ \lambda_i^* g_i(x^*) = 0, \quad \forall\, i = 1, \ldots, m$$

If the functions f and g_i are not differentiable, then we have

$$\text{(i)}\ \ 0 \in \partial_x L(x^*, \lambda^*) + N_X(x^*)$$

$$\text{(ii)}\ \ \lambda_i^* g_i(x^*) = 0, \quad \forall\, i = 1, \ldots, m$$

Let us now see the same application of the KKT conditions. Note that for CP obeying Slater's conditions KKT conditions are necessary. Further, they are always sufficient without any additional conditions.

Example 1.14: Lagrangian Dual of a Convex Quadratic Problem [32]

$$\min f(x) = \frac{1}{2}\langle x, Qx\rangle + \langle c, x\rangle$$
$$\text{subject to} \quad Ax = b.$$

(we have not written the optimality condition with $Ax = b$) where A is an $m \times n$ matrix with rank m $(m \leq n)$.

Let $C = \{x \in \mathbb{R}^n : Ax = b\}$.

If Q is p.d. $(Q \in S^n_{++})$, then there is a unique optimal solution of minimizing f over $C \neq \emptyset$. We shall see how to calculate it explicitly. Let $\bar{x}$ be the unique minimizer that exists, then

$$0 \in \nabla f(\bar{x}) + N_C(\bar{x}).$$

We claim that $N_C(\bar{x}) = Im(A^T)$.
Let $v \in Im(A^T)$, then $\exists\, \lambda \in \mathbb{R}^m$ such that $v = A^T\lambda$,

$$\begin{aligned}\text{for all } x \in C \Rightarrow \langle A^T\lambda, x - \bar{x}\rangle &= \langle \lambda, Ax - A\bar{x}\rangle \\ &= \langle \lambda, b - b\rangle = 0\end{aligned}$$

$\Rightarrow A^T\lambda \in N_C(\bar{x}). \Rightarrow \text{Im}A^T \subseteq N_C(\bar{x})$.

Let $v \in N_C(\bar{x})$. Thus,

$$\langle v, x - \bar{x}\rangle \leq 0, \quad \forall\, x \in C.$$

Let $x = \bar{x} + p$ where $p \in \text{Ker}(A)$, then

$$\begin{aligned}&\langle v, \bar{x} + p - \bar{x}\rangle \leq 0, \\ &\Rightarrow \langle v, p\rangle \leq 0, \quad \forall\, p \in \text{Ker}(A) \\ &\Rightarrow \langle v, p\rangle = 0 \Rightarrow\; v \in \text{Ker}(A)^{\perp} = \text{Im}A^T.\end{aligned}$$

Thus, we will have $\exists\, \lambda \in \mathbb{R}^m$ such that

$$\begin{aligned}&Q\bar{x} + c + A^T\lambda = 0, \\ &\Rightarrow Q^{-1}Q\bar{x} + Q^{-1}c + Q^{-1}A^T\lambda = 0 \\ &\Rightarrow \bar{x} + Q^{-1}c + Q^{-1}A^T\lambda = 0 \\ &\Rightarrow A\bar{x} + AQ^{-1}c + AQ^{-1}A^T\lambda = 0\end{aligned}$$

$AQ^{-1}A^T$ is in S^n_{++}

$$\langle AQ^{-1}A^Ty, y\rangle = \langle A^Ty, Q^{-1}A^Ty\rangle > 0, \quad \forall\, y \text{ with } A^Ty \neq 0.$$

If $A^T y = 0 \Rightarrow y = 0$ (since rank$(A) = m$).
Thus, if $y \neq 0$ $\langle A^T y, Q^{-1}A^T y\rangle > 0$.
Then, $\bar{x} = -Q^{-1}(A^T\lambda - c) = -Q^{-1}A^T\lambda + Q^{-1}c$ while λ is given as

$$\begin{aligned}AQ^{-1}A^T\lambda &= -AQ^{-1}c - b\\ &\Rightarrow \lambda = -(AQ^{-1}A^T)^{-1}(AQ^{-1}c + b)\\ &\Rightarrow \bar{x} = Q^{-1}A^T(AQ^{-1}A^T)^{-1}(AQ^{-1}c + b).\end{aligned}$$

Example 1.15: Bergstrom Inequality via Convex Minimization [33]

Let $A \in S^n_{++}$. Consider the problem

$$\begin{aligned}&\min\langle Ax, x\rangle\\ &\text{subject to} \quad x_1 \geq 1.\end{aligned}$$

Use this problem to show Bergstrom's inequality

$$\frac{\det(A+B)}{\det(A+B)_i} \geq \frac{\det A}{\det A_i} + \frac{\det B}{\det B_i} \quad (\text{where } A, B \in S^n_{++}),$$

where A_i is the submatrix of A after removing row i and column i.

$$\text{Here, } g(x) = -x_1 + 1.$$

Then, $C = \{x : g(x) \leq 0\}$ forms a closed convex set. This problem has a solution and the solution is unique. The KKT conditions give us (let $\bar{x}$ be the unique solution)

$$2A\bar{x} + \lambda\begin{pmatrix}-1\\0\\\vdots\\0\end{pmatrix} = 0.$$

Hence,

$$\lambda(g(\bar{x})) = 0.$$

We claim that $x_1 > 1$ is not true. Then in that case

$$\lambda = 0 \Rightarrow A(\bar{x}) = 0 \Rightarrow \bar{x} = 0,$$

which is a contradiction. Thus, $x_1 = 1$, we have

$$\bar{x} = \frac{\lambda}{2}A^{-1}\begin{pmatrix}1\\0\\\vdots\\0\end{pmatrix} = 0.$$

Let $A = [A^{-1}_{ij}]$, then $1 = \bar{x}_1 = (\lambda/2)A^{-1}_{1,1}$. Thus,

$$\lambda = \frac{2}{A^{-1}_{1,1}} \Rightarrow \bar{x} = \frac{1}{A^{-1}_{1,1}}A^{-1}\begin{pmatrix}1\\0\\\vdots\\0\end{pmatrix}.$$

Thus, $\langle A\bar{x}, \bar{x}\rangle = \lambda/2 = 1/A^{-1}_{1,1}$.

Further $A^{-1}_{1,1} = \det A_1/\det A$

$$\min_{x_1 \geq 1} \langle Ax, x\rangle = \min_{x_1 = 1} \langle Ax, x\rangle = \frac{\det A}{\det A_1}.$$

Further

$$\min_{x_1=1} \langle (A+B)x, x\rangle \geq \min_{x_1=1} \langle x, Ax\rangle + \min_{x_1=1} \langle x, Bx\rangle$$

$$\frac{\det(A+B)}{\det(A+B)_1} \geq \frac{\det A}{\det A_1} + \frac{\det B}{\det B_1}.$$

Further, for the general case consider the problem

$$\min \langle x, Ax\rangle$$
$$\text{subject to} \quad x_i \geq 1.$$

Let us turn to duality and how to construct the Lagrangian dual of some important class of problems.

Example 1.16: Lagrangian Dual of a Linear Programming Problem

$$\min \langle c, x\rangle$$
$$\text{subject to} \quad Ax \geq b, \quad x \geq 0.$$

Construct the Lagrangian as follows:

$$\begin{aligned}
L(x,\lambda) &= \langle c, x\rangle + \langle \lambda, b - Ax\rangle; \ \lambda \geq 0 \ (\lambda \in \mathbb{R}^m_+) \\
\theta(\lambda) &= \min_{x \in \mathbb{R}^n_+} \langle c, x\rangle + \langle \lambda, b - Ax\rangle \\
&= \min_{x \in \mathbb{R}^n} \langle b, \lambda\rangle + \langle c - A^T\lambda, x\rangle \\
\theta(\lambda) &= \langle b, \lambda\rangle \text{ if } \mathrm{A}^{\mathrm{T}}\lambda \leq \mathrm{c}, \ \lambda \geq 0 \\
&= -\infty \quad \text{otherwise.}
\end{aligned}$$

This leads to the dual problem

$$\max \langle b, \lambda\rangle$$
$$\text{subject to} \quad A^T\lambda \leq c, \quad \lambda \geq 0.$$

Example 1.17: Lagrangian Dual of an SDP

$$\min \langle C, X\rangle$$
$$\langle A_i, X\rangle = b_i, \quad i = 1, \ldots, m, \quad X \in S^n_+.$$

From the Lagrangian, it follows that.

$$L(X,\lambda) = \langle C, X\rangle + \lambda_i(b_i - \langle A_i, X\rangle)$$
$$\theta(\lambda) = \min_{x\in S_+^n} L(X,\lambda)$$
$$= \min_{x\in S_+^n} \left[\sum_{i=1}^{m} \lambda_i b_i + \left\langle C - \sum_{i=1}^{m} \lambda_i A_i, X\right\rangle\right].$$

If $C - \sum_{i=1}^{m} \lambda_i A_i \geq 0$, that is, $C - \sum_{i=1}^{m} \lambda_i A_i \in S_+^n$.
We have

$$\left[\sum_{i=1}^{m} \lambda_i b_i + \left\langle C - \sum_{i=1}^{m} \lambda_i A_i, X\right\rangle\right] \geq \sum_{i=1}^{m} \lambda_i b_i.$$

For $X = 0$, we have

$$\left[\sum_{i=1}^{m} \lambda_i b_i + \left\langle C - \sum_{i=1}^{m} \lambda_i A_i, X\right\rangle\right] \geq \sum_{i=1}^{m} \lambda_i b_i.$$

Thus, $\theta(\lambda) = \langle b, \lambda\rangle$, if $C - \sum_{i=1}^{m} \lambda_i A_i \in S_+^n$.
If

$$C - \sum_{i=1}^{m} \lambda_i A_i \notin S_+^n, \quad \text{then there exists } X \in S_+^n \text{ such that}$$
$$\left\langle C - \sum_{i=1}^{m} \lambda_i A_i, X\right\rangle < 0 \quad \text{(this is because } ((S_+^n)^* = S_+^n).$$

Consider $\mu > 0$. Thus,

$$\left\langle C - \sum_{i=1}^{m} \lambda_i A_i, \mu X\right\rangle < 0, \quad \forall\, \mu > 0.$$

Hence,

$$\sum_{i=1}^{m} \lambda_i b_i + \left\langle C - \sum_{i=1}^{m} \lambda_i A_i, \mu X\right\rangle \to -\infty.$$

If $C - \sum_{i=1}^{m} \lambda_i A_i \notin S_+^n$.
Then,

$$\theta(\lambda) = \begin{cases} \langle b, \lambda\rangle : C - (\lambda_1 A_1 + \cdots + \lambda_m A_m) \succeq 0 \\ -\infty : \text{otherwise.} \end{cases}$$

Thus, the dual problem is

$$\max \langle b, \lambda\rangle$$
$$\text{subject to} \quad C - (\lambda_1 A_1 + \cdots + \lambda_m A_m) \succeq 0, \quad \lambda \geq 0.$$

1.5 Miscellaneous Notes

Though we have converted quite a large ground, we have still left with some important issues associated with convex optimization problems. We have not spoken about how to devise algorithms for convex optimization problems, and which are the current applications of convex optimization? One might argue that if the convex function is differentiable and one just bothers about unconstrained optimization. Then standard algorithms are available. But these algorithms do not exploit the convexity of the functions and even for differentiable convex functions it is worthwhile to build separate algorithms. See, for example, Nesterov [34]. For the time being let us turn ourselves to the issue of solving a nonsmooth convex minimization problem. In this case, we will be interested in the unconstrained optimization problem

$$\min_{x \in \mathbb{R}^n} f(x), \qquad \text{(CP2)}$$

where $f : \mathbb{R}^n \to \mathbb{R}$ is a convex function, which need not be differentiable. In effect, we need to know how to solve the differentiable inclusion

$$0 \in \partial f(x). \tag{1.34}$$

There are several popular methods about how to solve the problem (CP2) or the inclusion (1.34). We list a few of them:

1. Descent methods
2. Subgradient methods (projection methods)
3. Bundle methods
4. Proximal point methods

We shall briefly discuss the descent methods, subgradient methods, and proximal point methods. Discussing the bundle methods is not possible due to lack of space. For the study of bundle methods, the best source still remains to be the wonderful two volume monograph by Hiriart-Urruty and Lemarechal.

To begin our discussion let us ask ourselves, what do we mean by descent in the nonsmooth setting? The notion of descent is very intuitive and Hiriart-Urruty and Lemarechal describe it as follows:

A descent direction $d \in \mathbb{R}^n$ of a convex function $f : \mathbb{R}^n \to \mathbb{R}$ at x satisfies the following:

$$\exists\, t > 0, \quad \text{such that } f(x + td) < f(x).$$

The question is of course how to find d. By analogy with the differentiable case, we can intuitively conclude that d will be a descent direction, if

$$f'(x, d) < 0;$$

or equivalently,

$$\langle \xi, d \rangle < 0, \quad \text{for all } \xi \in \partial f(x).$$

It is, however, not so easy to figure out a descent direction in this case. See Hiriart-Urruty and Lemarechal [7] (Vol. I) for details. Once a descent direction is obtained, it is quite simple to set up a descent algorithm. However, just as in differentiable case we can define the steepest descent direction d of f at x if d solves the problem

$$\min f'(x,d)$$
$$\text{subject to} \quad \|d\| = 1,$$

where $\|.\|$ is any norm on $\mathbb{R}^n$. This, of course, is an idealized scenario and is not so easy to solve. However, we can pose the problem equivalently as

$$\min f'(x,d)$$
$$\text{subject to} \quad \|d\| \leq 1. \tag{1.35}$$

Of course one must follow some approximate techniques to solve the above problem, which is a nondifferentiable constraint optimization problem. Hiriart-Urruty and Lemarechal [7] (Vol. I) again remains in our opinion the best source.

So let us now write down the steepest descent algorithm as given in Hiriart-Urruty and Lemarechal [7] (Vol. I).

Steepest Descent Algorithm

Step 1: If $0 \in \partial f(x^k)$, stop (stopping criteria).

Step 2: For some norm on $\mathbb{R}^n$ solve Equation 1.35 to obtain d^k (the direction-finding step).

Step 3: Find $t_k > 0$ such that for $x^{k+1} = x^k + t_k d^k$; we have $f(x^{k+1}) < f(x^k)$.

Step 4: Set $k := k + 1$ and loop to step 1.

Let us now briefly discuss the subgradient methods. The key to the subgradient methods is the assumption on the step lengths $\lambda_k > 0$, which also satisfies the facts.

$$\lambda_k \to 0, \quad \sum_{k=0}^{\infty} \lambda_k = \infty.$$

Of course $\lambda_k = 1/k$ is the prototype example. The iteration step for $k \geq 0$ is given as

$$x^{k+1} = x^k - \lambda_k g^k, \tag{1.36}$$

where $g^k \in \partial f(x^k)$. In fact, any subgradient of $\partial f(x^k)$ will do. Sometimes the scheme can be written as

$$x^{k+1} = x^k - \lambda_k \frac{g^k}{\|g^k\|}, \quad k > 0.$$

We shall briefly discuss some issues related to its convergence. We shall present the analysis from Nesterov [35]. First one needs to understand why one needs the criteria, $\lambda \downarrow 0$. Note that if we are considering say a strongly convex function, the point in the neighborhood of a minimizer will not have zero in their subgradients. This compels us to assume that $\lambda \downarrow 0$. Let us now analyze the situation using the iteration scheme (1.36). Let us additionally assume that

$$\|g^k\|_2 \leq L$$

for any $x \in \mathbb{R}$. Thus, we have

$$\begin{aligned}\|x - x^{k+1}\|_2^2 &= \|x - x^k + \lambda_k g^k\|_2^2 \\ &= \|x - x^k\|_2^2 + 2\lambda_k \langle g^k, x - x^k\rangle + (\lambda_k)^2 \|g^k\|_2^2 \\ &\leq \|x - x^k\|_2^2 + 2\lambda_k \langle g^k, x - x^k\rangle + (\lambda_k)^2 L^2. \qquad (1.37)\end{aligned}$$

Let x^0 be the starting point of the iteration. Consider any $x \in \mathbb{R}^n$, with $1/2\|x - x^0\|^2 \leq D$. Now for each $i = 0, 1, 2, \ldots$, we have by the definition of a subgradient

$$f(x) - f(x^i) \geq \langle g^i, x - x^i\rangle.$$

Hence,

$$f(x) \geq f(x^i) + \langle g^i, x - x^i\rangle.$$

Hence,

$$f(x) \geq \frac{\sum_{i=1}^k \lambda_i f(x^i) + \sum_{i=1}^k \lambda_i \langle g^i, x - x^i\rangle}{\sum_{i=1}^k \lambda_i}.$$

Thus,

$$f(x) \geq \frac{\sum_{i=1}^k \lambda_i [f(x^i) + \langle g^i, x - x^i\rangle]}{\sum_{i=1}^k \lambda_i}. \qquad (1.38)$$

Using Equation 1.37, we can write for $i = 0, 1, 2, \ldots, k$

$$-\frac{1}{2}\lambda_i^2 L^2 + \|x - x^{i+1}\|^2 - \|x - x^i\|^2 \leq \lambda_i \langle g^i, x - x^i\rangle.$$

Summing from $i = 0$ to $i = k$, we have

$$-\frac{1}{2}\sum_{i=0}^k \lambda_i^2 L^2 + \|x - x^{k+1}\|^2 - \|x - x^0\|^2 \leq \sum_{i=0}^k \lambda_i \langle g^i, x - x^i\rangle.$$

Hence, we have

$$-\frac{1}{2}\sum_{i=0}^k \lambda_i^2 L^2 - \|x - x^0\|^2 \leq \sum_{i=0}^k \lambda_i \langle g^i, x - x^i\rangle.$$

Thus,

$$-D - \frac{1}{2}\sum_{i=0}^k \lambda_i^2 L^2 \leq \sum_{i=0}^k \lambda_i \langle g^i, x - x^i\rangle.$$

From Equation 1.38, we have

$$f(x) \geq \frac{\left[\sum_{i=0}^{k} \lambda_i f(x^i) - D - \frac{1}{2} L^2 \sum_{i=0}^{k} \lambda_i^2\right]}{\sum_{i=0}^{k} \lambda_i}. \tag{1.39}$$

Suppose we have chosen D large enough such that the minimizer of f, if exists, and satisfies $\|x^* - x^0\| \leq 2D$. Then, if $f(x^*) = f^*$, we have

$$f^* = f_D^* = \min \left\{ f(x) : \frac{1}{2} \|x - x^0\|^2 \leq D \right\}.$$

From Equation 1.39, we have

$$f(x) \geq \bar{f}_k - w_k,$$

for all x such that $\frac{1}{2}\|x - x^0\|^2 \leq D$, where

$$\bar{f}_k = \frac{\sum_{i=0}^{k} \lambda_i f(x^i)}{\sum_{i=0}^{k} \lambda_i} \quad \text{and} \quad w_k = \frac{2D + L^2 \sum_{i=0}^{k} \lambda_i^2}{2 \sum_{i=0}^{k} \lambda_i}.$$

Thus,

$$\bar{f}_k - f_D^* \leq w_k, \quad \text{that is,} \quad \bar{f}_k - f^* \leq w_k.$$

For the convergence of the subgradient process, one thus needs $w_k \to 0$ and this is generated by the assumption, $\sum_{i=0}^{k} \lambda_i = \infty$. This process is interesting since it immediately gives us a stopping criterion. We can stop the process if $w_k \leq \epsilon$, where $\epsilon > 0$ is some precision parameter. For more details on the subgradient method, see the lecture videos on advanced convex optimization by Stephen Boyd available on YouTube.

It is important to have a careful look at the expression $\bar{f}_k$. Thus, at each x^k, we don't just consider the function value at x^k but some weighted sum or a convex combination of all the previous function values and that at k. The expression $\bar{f}_k$ is often called esergodic sum.

Many researchers use the term subgradient method for an approach based on projection which also encompasses the case when f is minimized over a closed convex set C. Let us very briefly mention what we mean by the term "point of projection." Given a closed convex set C and $x \in \mathbb{R}^n$, we already know that the distance of C from x is measured by the convex function $d_C : \mathbb{R}^n \to \mathbb{R}$, where

$$d_C(x) = \inf_{y \in C} \|y - x\|.$$

It can be shown that there exists $\hat{y} \in C$ such that

$$d_C(x) = \|\hat{y} - x\|,$$

then $\hat{y}$ is called the projection of x on C, denoted as $\hat{y} = \text{Proj}_C(x)$. The vector $x - \hat{y}$ is often called the normal vector to C at $\hat{y}$. Now if $x^* \in C$ is a minimizer of f over C, then one has

$$0 \in \partial f(x^*) + N_C(x^*).$$

Now this means that there exists $\xi \in \partial f(x^*)$ such that

$$-\xi \in N_C(x^*).$$

Since $N_C(x^*)$ is nothing but the cone of normals at x^*, we see that there exists $\hat{x}$ such that

$$-\lambda\xi = \hat{x} - x^*, \quad \text{for some } \lambda > 0 \quad \text{proj}_C(\hat{x}) = x^*.$$

Thus, $\hat{x} = x^* - \lambda\xi$. Hence,

$$\begin{aligned}\text{proj}_C(\hat{x}) &= \text{proj}_C(x^* - \lambda\xi)\\ x^* &= \text{proj}_C(x^* - \lambda\xi).\end{aligned}$$

This actually holds for any $\lambda > 0$ and the reader can easily see this as an if and only if condition for x^* to be a minimizer of f over C. This leads to the iterative scheme

$$x^{k+1} = \text{proj}_C(x^k - \lambda_k \xi^k), \quad \xi^k \in \partial f(x^k), \quad \lambda_k > 0. \tag{1.40}$$

When $C = \mathbb{R}^n$, we have

$$x^{k+1} = x^k - \lambda_k \xi^k, \quad \xi^k \in \partial f(x^k), \quad \lambda_k > 0.$$

Thus, when $C = \mathbb{R}^n$, we get back the subgradient scheme discussed earlier. Thus, the terms subgradient scheme and projected subgradient scheme are used interchangeably. The iteration scheme (1.40) is called the projected subgradient method. If f is differentiable, Equation 1.40 reduces to

$$x^{k+1} = \text{proj}_C(x^k - \lambda_k \nabla f(x^k)), \quad \lambda_k > 0. \tag{1.41}$$

We shall now provide a very basic convergent result for scheme (1.41) when $\lambda_k = \lambda$ for all $k \in \mathbb{N}$. We shall provide the result from Ruszczynksi [12].

Theorem 1.18

Let $f : \mathbb{R}^n \to \mathbb{R}$ be a differentiable convex function and $C \subset \mathbb{R}^n$ a closed convex set. Let ∇f be Lipschitz over $\mathbb{R}^n$ with Lipschitz rank $L \geq 0$, that is,

$$\|\nabla f(x) - \nabla f(y)\| \leq L\|x - y\|, \quad \forall\, x, \quad y \in \mathbb{R}^n.$$

Furthermore, assume that the set

$$\{x \in C : \ f(x) \leq f(x^0)\}$$

is bounded, where x^0 is the starting solution. Further let us also assume that

$$0 < \lambda < \frac{1}{L}.$$

Then, the iterate $\{x^k\}$ generated by Equation 1.41 forms a bounded sequence and every limit point is a minimizer of f over C.

Let us now come to the last part of our discussion. This is centered on the proximal point method. The basic idea is simple. View the convex optimization problem as a sequence of a strongly convex optimization problem. In effect, we speak of an iterative scheme given as

$$x^{k+1} = \arg\min_x \left\{ f(x) + \frac{1}{2}\|x - x^*\|^2 \right\}.$$

Hence,

$$0 \in \partial f(x^{k+1}) + (x^{k+1} - x^k).$$

This implies that

$$-(x^{k+1} - x^k) \in \partial f(x^{k+1})$$

or

$$x^k \in (\partial f + I)(x^{k+1}).$$

Now $(\partial f + I)$ is a set-valued map. Its inverse is given as

$$(\partial f + I)^{-1}(x) = \{z \in \mathbb{R}^n : x \in (\partial f + I)(z)\}.$$

Thus,

$$x^{k+1} \in (\partial f + I)^{-1}(x^k).$$

Surprisingly, $(\partial f + I)^{-1}$ turns out to be single valued and thus we can write

$$x^{k+1} = (\partial f + I)^{-1}(x^k).$$

In fact, one can also use an iterative scheme

$$x^{k+1} \in (\lambda_k \partial f + I)^{-1}(x^k), \quad \lambda_k > 0.$$

Further this marvellous property of $(\lambda_k \partial f + I)^{-1}$ being single valued comes from the fact that ∂f is a maximal monotone operator. This means that ∂f is a monotone operator, that is, for all $v \in \partial f(y)$ and $w \in \partial f(x)$

$$\langle v - w, y - x \rangle \geq 0, \quad \forall y,\ x \in \mathbb{R}^n.$$

By maximal we mean that there is no other set-valued monotone operator $T : \mathbb{R}^n \rightrightarrows \mathbb{R}^n$, whose graph property contains the graph of ∂f. By the graph of set-valued map T, we mean the set

$$\text{gph}T = \{(x, y) \in \mathbb{R}^n \to \mathbb{R}^n : y \in T(x)\}$$

In fact, to use the proximal point method f need not be a finite convex function, but it is proper and lower semicontinuous. In fact, one can consider the more general problem

$$0 \in T(x),$$

where T is a maximal monotone operator. Then, the proximal point method is given as

$$x^{k+1} \in (I + \lambda_k T)^{-1}(x^k).$$

A very good source for proximal point methods is the monograph by Burachick and Iusem [36]. However, it is worthwhile to study the first paper in this direction by Rockafellar [37]. The expression $(I + \lambda T)^{-1}$ is called resolvent of T and for the fact that it is often a singleton it is referred to as Minty's lemma. See Rockafellar and Wets [38] for more details. Let us wind up our discussion here but we would like to refer a more application-oriented reader to a monograph, *Optimization in Machine Learning*, edited by S. Sra, S. Nowozin and S.J. Wright (MIT Press 2012). Machine learning at present is one of the principal areas where convex function plays an important role.

Acknowledgment

We are grateful to Dr. Sanjeev Gupta, who typed the major part of the manuscript and also pointed out several anomalies.

References

1. R.T. Rockafellar, Lagrange multipliers and optimality, *SIAM Review*, Vol. 35, pp. 182–238, 1993.
2. J.M. Borwein and A.S. Lewis, *Convex Analysis and Nonlinear Optimization*, Springer, (2nd edition), New York, 2006.
3. C. Niculescu and L.-E. Persson, *Convex Functions and Their Applications*, Springer, New York, 2006.
4. R.A. Horn and C.R. Johnson, *Matrix Analysis*, Cambridge University Press, Cambridge, 1985.
5. M.L. Overton and R.S. Womersley, Optimality conditions and duality theory for minimizing sums of largest eigenvalues of symmetric matrices, *Mathematical Programming*, Vol. 62, pp. 321–357, 1993.
6. S. Boyd and L. Vandenberghe, *Convex Optimization*, Cambridge University Press, New York, 2004.
7. J.B. Hiriart-Urruty and C. Lemarechal, *Convex Analysis and Minimization Algorithms*, Vols. I and II, Springer, Berlin, 1993.
8. R.T. Rockafellar, *Convex Analysis*, Princeton University Press, New Jersey, 1970.
9. C. Lemarechal, Nondifferentiable optimization, *Handbook in Operations Research and Management Sciences*, pp. 529–572, North Holland, Amsterdam, 1989.
10. R.R. Phelps, *Monotone Operators and Differentiability*, Lecture Notes in Mathematics, 1364, Springer, Berlin, Heidelberg, 1993.
11. J.B. Hiriart-Urruty and C. Lemarechal, *Fundamentals of Convex Analysis*, Springer, Berlin, 2001.
12. A. Ruszczynski, *Nonlinear Optimization*, Princeton University Press, New Jersey, 2006.
13. C. Villani, *Topics in Optimal Transport, Old and New, Graduate Studies in Mathematics,* AMS, Providence, Rhode Island, 2003.
14. C. Villani, *Optimal Transport, Old and New*, Springer, Berlin, Heidelberg, 2008.
15. K. Brayan and T. Leise, Making to do with less: An introduction to compressed sensing, *Siam Review*, Vol. 55, pp. 547–566, 2013.
16. J.B. Lasserre, *Moments, Positive Polynomials and Their Applications*, Vol. 1, Imperial College Press, London, 2009.
17. E. Blum and W. Oettli, Direct proof of the existence theorem for quadratic programming, *Operational Research*, Vol. 20, pp. 165–167, 1972.
18. F. Alizadeh and D. Goldfarb, Second-order cone programming, *Mathematical Programming*, Vol. 95, pp. 3–51, 2003.

19. W.J. Helton and V. Vinnikov, Linear matrix inequality representation of sets. *Communications on Pure and Applied Mathematics*, Vol. 60, no. 5, pp. 654–674, 2007.
20. J.W. Helton and J. Nie, Semidefinite representation of convex sets, *Mathematical Programming*, Vol. 122, pp. 21–64, 2010.
21. A. Dhara and J. Dutta, *Optimality Conditions in Convex Optimization: A Finite Dimensional View,* CRC/Taylor and Francis, Boca Raton, Florida, 2011.
22. F. John, Extremum problems with inequalities as side conditions. In *Studies and Essays, Courant Anniversary Volume* (K.O. Friedrichs, O.E. Neugebauer and J.J. Stoker, eds.), Wiley (Interscience), New York, pp. 187–204, 1948.
23. B.D. Craven, *Mathematical Programming and Control Theory,* Chapman and Hall, London, 1978.
24. B. Pourciau, Modern mulipliers rules, *American Mathematical Monthly*, Vol. 87, pp. 433–452, 1980.
25. J. Brinkhuis and V. Tikhomirov, *Optimization Insight and Applications*, Princeton University Press, New Jersey, 2005.
26. F.H. Clarke, *Functional Analysis, Calculus of Variations and Optimal Contro*, Springer, New York, 2013.
27. M. Solodov, An explicit descent method for bilevel convex optimization, *Journal of Convex Analysis* , Vol. 14, pp. 227–237, 2007.
28. S. Dempe, N. Dinh, and J. Dutta, Optimality conditions for a simple convex bilevel programming problem. Variational analysis and generalized differentiation in optimization and control, *Springer Optimization and Its Applications*, Vol. 47, pp. 149–161, 2010.
29. H.W. Kuhn and A.W. Tucker, Nonlinear programming. *Proceedings of the Second Berkeley Symposium on Mathematical Statistics and Probability*, University of California Press, Berkeley and Los Angeles, pp. 481–492, 1951.
30. O. Guler, *Foundation of Optimization,* Springer, New York, 2010.
31. H. Komiya, Elementary proof of sions minimax theorem, *Kodai Mathematical Journal*, Vol. 11, pp. 5–7, 1980.
32. G.C. Calafiore and L. El. Ghaoui, *Optimization Model*, Cambridge University Press, London, 2014.
33. D. Aze and J.B. Hiriart-Ururty, *Analyse Variationnelle et Optimisation (in French)*, Cepadues, Toulouse, France, 2009.
34. Y. Nesterov, *Introductory Lectures on Convex Optimization: A Basic Course*, Kluwer Academic Publisher, Boston, Massachusetts, 2004.
35. Y. Nesterov, Primal dual subgradient methods for convex problems, *Mathematical Programming Series B*, Vol. 120, pp. 221–259, 2009.
36. R.S. Burachik and A.N. Iusem, *Set-Valued Mappings and Enlargement of Maximal Monotone Operators*, Springer, New York, 2008.
37. R.T. Rockafellar, Monotone operator and proximal point algorithms, *SIAM Journal on Control and Optimization* , Vol. 14, pp. 877–898, 1976.
38. R.T. Rockafellar and R.J.B. Wets, *Variational Analysis*, Springer, New York, 1997.

2

Introduction to Game Theory

Aditya K. Jagannatham and Vimal Kumar

CONTENTS

ABSTRACT The focus of this chapter is game theory and its various applications. This chapter is divided into two parts dealing with static games and dynamic games. The first part of the chapter begins with a mathematical formulation of games followed by the description of a strategic-form game. This chapter subsequently describes the various solutions and related concepts, including domination, rationalizability, Nash equilibrium, and their refinements. An elaborate description of the concept and procedure to evaluate a mixed-strategy Nash equilibrium follows, in which the players use randomized strategies or randomly mix various possible actions. This also includes a discussion of a special form of games termed as *zero-sum* games. The next topic deals with games in which the players are not certain of the characterization of the game, which leads to the formulation of Bayesian games. This gives rise to the novel paradigm of Bayesian–Nash equilibria to analyze the behavior in such games. An offshoot of the Bayesian game framework is auction theory, which is subsequently developed separately. An analysis of various auction forms such as first-price, second-price, and all-pay auctions is described next, along with the central principle of *revenue equivalence* that governs such auctions. The next section deals with the Vickrey–Clarke–Groves (VCG) procedure for a competitive resource allocation and its application in the design of pricing mechanisms. This chapter then proceeds to cover a different form of games, termed as *potential* games, which arise in the context of networks. An application of this is illustrated in the context of road planning and traffic congestion, followed by elaboration of the key properties of such games. The second part of the chapter deals with dynamic games. This part begins with the description and mathematical formulation of an extensive-form game. The description emphasizes the graph-theoretic representation of extensive-form games followed by the technique of backward induction and the concept of a sub-game-perfect equilibrium.

2.1 Definition of a Game

A *game* can be defined as "a competitive activity in which players or agents contend for limited resources according to a set of rules." This definition of a game can be applied to several challenging problems in real-life situations. For instance, in the context of business, the different firms can be modeled as players and the consumers in the market, or basically the *market share*, as the limited resources that these firms are contending for. In the case of politics, the players are the different political parties or individuals associated with political parties and the limited resources are the votes/voters. Further, the different players can be inanimate objects as well. For example, in a wireless communication scenario, the players are the different devices and the frequency spectrum or bandwidth is the limited resource. A *reward* or *payoff* to each player can be associated with every outcome of the game, which is proportional to his share of allocation of the limited resource in the specific outcome.

Another important concept in game theory is the notion of a strategy. A strategy is simply a plan or a sequence of maneuvers employed by a player of the game toward obtaining a goal. Such strategies can be employed in the context of business, making them business strategies, or they can be political strategies and so on. This naturally implies that when a wise player plans a strategy, he would consider the probable strategies other players would employ and try to come up with an optimum strategy that maximizes his reward or his payoff.

A key assumption of every player in game theory is the notion of *rationality*. Rationality assumes that each player tries to maximize his or her payoff irrespective of the payoffs to others. This is significantly different from conventional resource allocation scenarios where a central controller assigns or enforces jointly optimal or Pareto optimal allocations, which maximize the reward of a group, sometimes at the cost of a reduced reward for the individual users. In the course of our discussion, we will always assume rational players.

2.2 Example of a Basic Game: Prisoner's Dilemma

In this section, we introduce the notation and basic ideas of game theory through a simple and ubiquitous example termed the *Prisoner's Dilemma*. Consider two suspects who are held up for the purpose of interrogation for a major crime that has been committed. There is enough evidence to convict them for a minor offense. However, there is not enough evidence to press charges for the major crime as there are no witnesses of the crime. The aim of the interrogator therefore is to use one of them as a witness against the other. The game table in Figure 2.1 can be employed to model the game in this scenario. In the game table, P_1, P_2 denote prisoner 1 and 2, respectively. Each prisoner can choose one of two actions, that is, to either to confess (C) or deny (D) the crime. For the table, P_1 is also termed the *row player* and P_2 is also termed the *column player*. The rewards or payoffs to each player for every possible outcome of the game are listed in the table. For instance, if P_1, P_2 both deny the crime, that is, both choose D, then there is no evidence to convict either of them for the major crime. Hence, as charges can only be pressed for the minor crime, each receives a mild prison sentence of 1 year. This can be interpreted as a negative reward of 1 or a payoff of -1 to each player. Hence, the net payoff of the players corresponding to outcome (D, D) is given as $(-1, -1)$, where the first entry gives the payoff of the row player P_1 and the second entry gives the payoff of the column player P_2.

P_1 \ P_2	C	D
C	(−3, −3)	(0, −4)
D	(−4, 0)	(−1, −1)

FIGURE 2.1
Game table for Prisoner's Dilemma.

However, if one of the players, for instance, P_1, chooses to confess to the major crime, then, as an incentive for confessing, he is exempted from serving time in the prison. As a result, his reward is 0 years in prison. However, now since P_2 can be convicted of the major crime, he receives a harsher prison sentence of 4 years, that is, a reward of -4. Hence, the outcome (C, D) with P_1 confessing and P_2 denying leads to the reward of $(0, -4)$. The reader can now observe a key property of games: the payoff for P_2 is affected by the action choice of P_1. It is equal to either -1 or -4 depending on whether P_1 chooses to confess C or deny D, respectively. Similarly, outcome (D, C), with P_1 choosing to deny and P_2 choosing to confess, yields payoffs $(-4, 0)$. Finally, when both P_1 and P_2 choose to confess, that is, corresponding to outcome (C, C), the payoff is a 3-year prison sentence to each of them, that is, $(-3, -3)$. Thus, the game table succinctly captures the game framework.

Let n denote the total number of players of the game. As illustrated by this example, a formal description of a game involves the following quantities. A set of players $\mathcal{P}$. In the above game, $\mathcal{P} = \{P_1, P_2\}$. A set of rules $\mathcal{R}$, which are often implicit. For instance, an implicit rule in the game of Prisoner's Dilemma is that neither of the prisoners can escape. The set of actions $\mathcal{A}_i, 1 \leq i \leq n$, denotes the action choices available to the ith user. For instance, in this example, $\mathcal{A}_1 = \mathcal{A}_2 = \{C, D\}$. However, in general, the action sets of all users need not be identical as in this example and can be different from each other. The set of outcomes $\mathcal{A}$ is the Cartesian product of the action sets $\mathcal{A}_i$ of the users:

$$\mathcal{A} = \mathcal{A}_1 \times \mathcal{A}_2 \times \cdots \times \mathcal{A}_n. \tag{2.1}$$

Thus, in this example, $\mathcal{A} = \{C, D\} \times \{C, D\} = \{(C, C), (C, D), (D, C), (D, D)\}$. Finally, the payoff or utility function $u_i(\cdot), 1 \leq i \leq n$, for each user i. The quantity $u_i(\mathbf{a})$ yields his reward as a function of the outcome $\mathbf{a} \in \mathcal{A}$. For instance, in this example, $u_1((D, C)) = -4$. Another convenient way of representing the utility functions in game theory, which will be very useful in future discussions, is to denote the utility of user i as $u_i(a_i, a_{-i})$, where a_i is the action of the ith user and a_{-i} is defined as

$$a_{-i} = (a_1, a_2, \ldots, a_{i-1}, a_{i+1}, \ldots, a_n),$$

denotes the actions of all the users other than the ith user. For instance, using this notation, the payoffs in this example can be described as $u_1(C, D) = 0 = u_2(C, D)$, since $u_1(C, D)$ denotes player 1 choosing C and player 2 choosing D, whereas $u_2(C, D)$ denotes player 2 choosing C and player 1 choosing D. Also, $u_1(D, C) = -4 = u_2(D, C)$. Thus, $u_2(C, D) \neq u_2((C, D))$. The use of the ordered pair (C, D) versus simply C, D in the argument of $u_i(\cdot)$ distinguishes both these notations.

2.3 Best Response Dynamic

Consider again the Prisoner's Dilemma game table in Figure 2.1. Observe that if P_2 confesses, in other words chooses action C, then P_1 is better off confessing (C) rather than denying (D). This is because denying results in a harsher punishment of 4 years, leading to a payoff of -4, while confessing results in a punishment of 3 years, that is, reward of -3. Thus, given the fixed action C of player P_2, confessing is the preferred action of P_1 compared to denying. Hence, confessing is termed as the *best response* of P_1 to the fixed action C of player P_2. This can be represented as

$$u_1(C, C) \geq u_1(D, C).$$

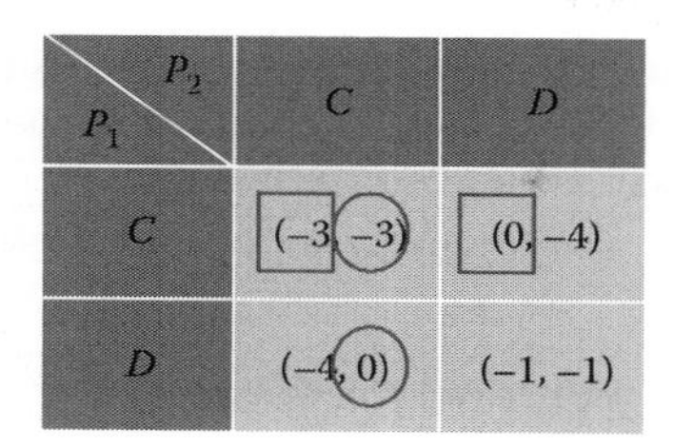

FIGURE 2.2
Best responses in Prisoner's Dilemma.

The *best response* $B_i(a_{-i})$ of the ith player, with respect to the fixed action a_{-i} of all the other users, can be defined as

$$B_i(a_{-i}) = \left\{a_i \in \mathcal{A}_i | u_i(a_i, a_{-i}) \geq u_i\left(a_i', a_{-i}\right), \quad \forall a_i' \in \mathcal{A}_i\right\}.$$

Similarly, for $a_{-1} = D$, the best response of player P_1 is given as C, since $u_1(C, D) = 0 \geq -1 = u_1(D, D)$. The best responses of P_1 are highlighted by squares in the game table in Figure 2.2 for each of the actions of P_2. Similarly, the best responses of P_2 are marked by circles in the same figure.

2.4 Nash Equilibrium

The concept of *Nash equilibrium* is one of the most important and central concepts of game theory. The idea of Nash equilibrium can be illustrated in the context of the Prisoner's Dilemma as follows. Consider superposition of the best responses of P_1, P_2, illustrated in Figure 2.2. One can readily see that there is one entry in this table that stands apart from the others. This is the entry corresponding to the outcome (C, C), where the best response functions of both the players intersect. In other words, each player is playing his best response to the other player's action. The Nash equilibrium outcome can be formally defined as follows. An action profile $\mathbf{a}^\star = [a_1, a_2, \ldots, a_n]$ of an n player game is a Nash equilibrium if for each player P_i, we have,

$$u_i\left(a_i^\star, a_{-i}^\star\right) \geq u_i\left(a_i, a_{-i}^\star\right), \quad \forall a_i \in \mathcal{A}_i, \quad 1 \leq i \leq N.$$

Thus, the action $a_i^\star$ of each user i is his best response to the actions $a_{-i}^\star$ of all other users. Another way of stating the same property is that given the action profile $a_{-i}^\star$ of all users other than the ith user, there is no other action a_i, which yields a strictly higher payoff for user i compared to $a_i^\star$. Thus, no user can increase his payoff by *unilaterally* deviating from the Nash equilibrium.

Another interesting interpretation of the Nash equilibrium is that it is a *self-enforcing* contract. That is to say that if all the users agree on the action profile $\mathbf{a}^\star$, there is no incentive for anyone to deviate from this during game play. For instance, in the example of the Prisoner's Dilemma, if both P_1, P_2 agree on the action $a_i = C$, then each will act as previously agreed during game play, since a deviation by any player does not increase his payoff. Finally, the Nash equilibrium also represents a *no-regret* outcome for a game, in the sense that having played $a_i^\star$ to the profile $a_{-i}^\star$ of all the other users, the ith user has no regret, since any other action would not have increased his payoff. Thus, at the outcome (C, C) in the Prisoner's Dilemma, each prisoner has no regret, as choosing D would only have decreased his payoff. Finally, observe the important point that the Nash equilibrium is not necessarily the *optimal* outcome of the game. For instance, the outcome (D, D) yields a higher payoff to each user compared to (C, C). However, the outcome (D, D) is not an equilibrium since each player can unilaterally deviate to increase his payoff. A formal proof of the existence of the Nash equilibrium is beyond the scope of this overview chapter. However, the result below guarantees the existence of a Nash equilibrium for a certain class of games.

Theorem 2.1

A Nash equilibrium exists for the n player game with utility functions $u_i(\cdot), 1 \leq i \leq n$ and action sets $\mathcal{A}_i$, if each utility function $u_i(\cdot)$ is continuous quasi-concave over $\mathcal{A}_i$ and each action set $\mathcal{A}_i$ is a nonempty compact convex set.

Proof. Can be found in Reference 1. ■

2.5 Other Games Similar to Prisoner's Dilemma

There are several other games that arise in different contexts and are similar to the Prisoner's Dilemma game introduced previously. For instance, consider a scenario of students collaborating on a project. The game table for this situation is given in Figure 2.3, where S_1, S_2 denote the students who are the players. Each student can either choose to work W or be lazy L. The payoffs for each outcome are given in the corresponding table entries, from which it can be readily seen that this game is similar to the Prisoner's Dilemma. Each player prefers to be lazy while the partner works on the project. It can be seen that if S_1 chooses to work, being lazy yields a higher payoff to S_2. Further, the same is also true if S_1 chooses to be lazy himself. Hence, for both action choices of S_1, lazy is the best response of S_2. Since the game is symmetric, the same holds true for the best response characteristic of S_1. Thus, the outcome (L, L) is where the best responses of both the players intersect. Therefore, it represents the Nash equilibrium of the game. It can again be seen that similar to the (C, C) option in the Prisoner's Dilemma game, it yields a lower payoff for both the players compared to the outcome (W, W).

Yet another example similar to Prisoner's Dilemma is the Arms Race scenario, which can be modeled with the game table on the right in Figure 2.3. The players are the countries represented by C_1, C_2. Each country would like to have the strategic defense advantage over the other. If C_1 spends on defense (D), C_2 is better off spending on defense (D) too. On the other hand, in case C_1 chooses to spend on health (H), C_2 is yet again better off spending on defense since it gains a strategic advantage. Thus, it can be seen again the outcome (D, D) is the Nash equilibrium, which leads

S_1 \ S_2	W	L
W	(2, 2)	(0, 3)
L	(3, 0)	(1, 1)

C_1 \ C_2	D	H
D	(1, 1)	(15, −10)
H	(−10, 15)	(10, 10)

FIGURE 2.3
Student project collaboration (left) and arms race (right) game tables.

to a significant arms buildup and is significantly lower in payoff compared to the ideally desirable (H, H) outcome where each country focuses on the health of its citizens.

2.6 Examples of Other Games

2.6.1 Coordination Games

Consider the example of a Deer Hunt illustrated in the game table in Figure 2.4. The two hunters are denoted by the players H_1, H_2. Each of the hunters can choose to hunt for the deer, indicated by action D, or to hunt for a rabbit, represented by R. However, the deer being larger in size, it requires the coordinated efforts of both the players, yielding a net payoff of 2 to each player, represented by the outcome (D, D). However, if one of the hunters chooses to go after the rabbit, it yields him a payoff of 1, while the other player who chooses to hunt for the deer receives 0, since he is now without a partner to hunt for the deer. Finally, if both the players H_1 and H_2 choose to hunt for the rabbit, it yields both of them an equal reward of 1 each corresponding to the outcome (R, R). However, this payoff is lower than $(2, 2)$ when both choose to make a concerted effort toward hunting the deer.

For obvious reasons, this game is termed as the *coordination game*, since the outcome depends on the level of coordination by the players. The best response of player H_1 corresponding to D by player H_2 is D, while the best response for R of H_2 is R. This is intuitively reasonable since choosing to hunt for deer is desirable for H_1 if H_2 also chooses to hunt for deer, while rabbit is better for H_1 if H_2 chooses to hunt for the rabbit. The best responses of H_1 are shown by boxes in Figure 2.4. The best responses of H_2 are identical to those of H_1 and are indicated by circles in the same figure. Thus, it can now be seen that the best responses intersect for two outcomes, viz., (D, D) and (R, R). Both these represent the Nash equilibrium outcomes for the Deer Hunt game. In this game, observe that there is a "good" equilibrium and a "bad" equilibrium. The good one corresponds to (D, D) where both receive a payoff of 2 and the bad one is the (R, R) outcome where each receives a payoff of 1. At this point, we cannot distinguish between both the equilibrium outcomes. However, one can see that there is a tendency to drift toward the (R, R) outcome, since if one player chooses R and the second chooses to go for the deer D, the first one still receives a payoff of 1. Thus, the action R appears to be a *safe* option, yielding a guaranteed payoff of 1 to the user. However, the option D needs the player to place faith in the other player that he will indeed choose D, failing which, the first player is left with a payoff of 0. Thus, this is indeed a case of coordination, where each player needs to rally the other player or all the players in a group to hunt for the deer.

H_1 \ H_2	D	R
D	(2, 2)	(0, 1)
R	(1, 0)	(1, 1)

FIGURE 2.4
Deer Hunt—A coordination game.

Again, several other coordination games can be modeled similar to the Deer Hunt game. For instance, consider working for a technology start-up company. The players are two employees E_1, E_2 with action possibilities work W or quit Q. While if both the employees work together toward the success of the company, it yields them each a reward of 2 on the company's success. However, an employee who develops cold feet and decides to quit, thus ensuring a comfortable position in a larger established firm, receives a smaller payoff of 1, while the other player who continues in the start-up receives 0, since his effort alone is not enough for the success of the company. Finally, both employees receive a payoff of 1 each if they choose to quit simultaneously for a larger firm. Thus, (W, W) and (Q, Q) are the equilibrium outcomes. In such scenarios, one can see the necessity of a strong leader, such as a project manager, to keep the employees motivated toward working in the company toward the goal of succeeding. Yet another example similar to this scenario is one of paying taxes. All citizens choosing to pay taxes P yields a higher payoff compared to not paying taxes N. On the other hand, it is worse if only a few choose to pay taxes, since they lose valuable pay, while the level of services is subpar since not many can be supported when only a few choose to pay taxes. The task of formally developing a game table similar to this scenario of Deer Hunt is left as an exercise to the reader.

2.6.2 Battle of the Sexes

Another popular example in game theory is that of the Battle of the Sexes (BoS), the game table for which is shown in Figure 2.5. In this context, the players are a dating couple comprising of a boy and a girl, who are deciding on a plan for the evening. The boy prefers watching the game (G), while the girl prefers going out for a movie (M). Hence, with the row player and column players representing the boy and girl, respectively, an outcome of (G, G) yields a payoff of $(10, 5)$, while (M, M) yields a payoff of $(5, 10)$. However, the outcomes (G, M) and (M, G) are not desirable since it would result in them not spending the day together, leading to payoffs of $(0, 0)$. One can also argue that the outcome (M, G) is worse compared to (G, M) since it leads to each player choosing the least desired action, with the boy choosing a movie and girl choosing to watch the game. However, for the purpose of analysis, this will not make a difference.

The best responses for each of the players are shown in the game table. They can be explained as follows. If the girl chooses to watch the game, it is best for the boy to go with the game, while if she chooses to go for a movie, the best response of the boy is to go along. Similarly for the girl, the best response is to choose whatever action is chosen by the boy, as indicated by the circles in the game table. One can observe that the best response characteristics intersect for the (G, G) and (M, M) outcomes. Thus, this game has two Nash equilibria, similar to the coordination game in Section 2.6.1. However, a fundamental difference with respect to the coordination game is that each of the players *prefers a different equilibrium*, with the boy preferring (G, G), and the girl preferring (M, M). What then will happen when such a game arises in real-life scenarios? Hidden in this game

Boy \ Girl	Game	Movie
Game	(10, 5)	(0, 0)
Movie	(0, 0)	(5, 10)

FIGURE 2.5
Battle of the Sexes.

is another kind of equilibrium termed as a *mixed-strategy* equilibrium, where each of the players is compromising for a fraction of the time, or in other words, the boy and girl choose one of the actions randomly with a certain frequency for each. This concept of a mixed-strategy equilibrium will be introduced and explored in detail in the future sections.

2.7 A Game of Matching Pennies

Finally, we consider the game of *Matching Pennies*. Each of the two players P_1, P_2 possesses a coin and they simultaneously choose to show either heads or tails. Player P_1 wins if he chooses the same face as P_2, leading to a payoff of 1 for him and -1 for his opponent P_2. However, if they choose different faces, that is, P_1 choosing heads and P_2 choosing tails or vice versa, P_1 loses, leading to a payoff of $(-1, 1)$. The game table for this is shown in Figure 2.6. The best responses for each player are once again shown in the game table. It can now be naturally seen that the best response of player P_1 for each action of P_2 is to choose the same action as P_2, while that of P_2 is to choose a different action from P_1. Thus, it can be seen that the best responses do not intersect, resulting in no Nash equilibrium, very different from the other games demonstrated previously, which possessed one or more Nash equilibrium outcomes. However, once again, we will demonstrate in future sections that a Nash equilibrium does indeed exist, but only when each of the players randomly chooses a face H or T with equal probability, which basically constitutes a mixed strategy.

2.8 Strategy Domination

Consider the game table shown in Figure 2.7 for an $N = 2$ player game with the row and column players denoted by P_1, P_2, respectively. While P_1 can choose from actions $\alpha_i, 1 \leq i \leq 4$, the actions available to P_2 are $\beta_i, 1 \leq i \leq 4$. On closer examination of the game table, it can be readily seen that the action β_4 always yields a higher reward to P_2 compared to action β_1, irrespective of the action α_i of P_1. In other words, action β_1 cannot be part of an equilibrium profile for P_2 since he can always increase his payoff by deviating to β_4. Such an action is termed as a dominated action. An action $\tilde{a}_i$ is *strictly dominated* by another action a_i if

$$u_i(a_i, a_{-i}) > u_i(\tilde{a}_i, a_{-i}), \quad \forall a_{-i} \in A_{-i}.$$

On the other hand, action $\tilde{a}_i$ is *weakly dominated* by another action a_i if $u_i(a_i, a_{-i}) \geq u_i(\tilde{a}_i, a_{-i})$ for all $a_{-i} \in A_{-i}$ and $u_i(a_i, a_{-i}) > u_i(\tilde{a}_i, a_{-i})$ for some $a_{-i} \in A_{-i}$. In such a situation, one can

P_1 \ P_2	Heads	Tails
Heads	(1, −1)	(−1, 1)
Tails	(−1, 1)	(1, −1)

FIGURE 2.6
Matching Pennies.

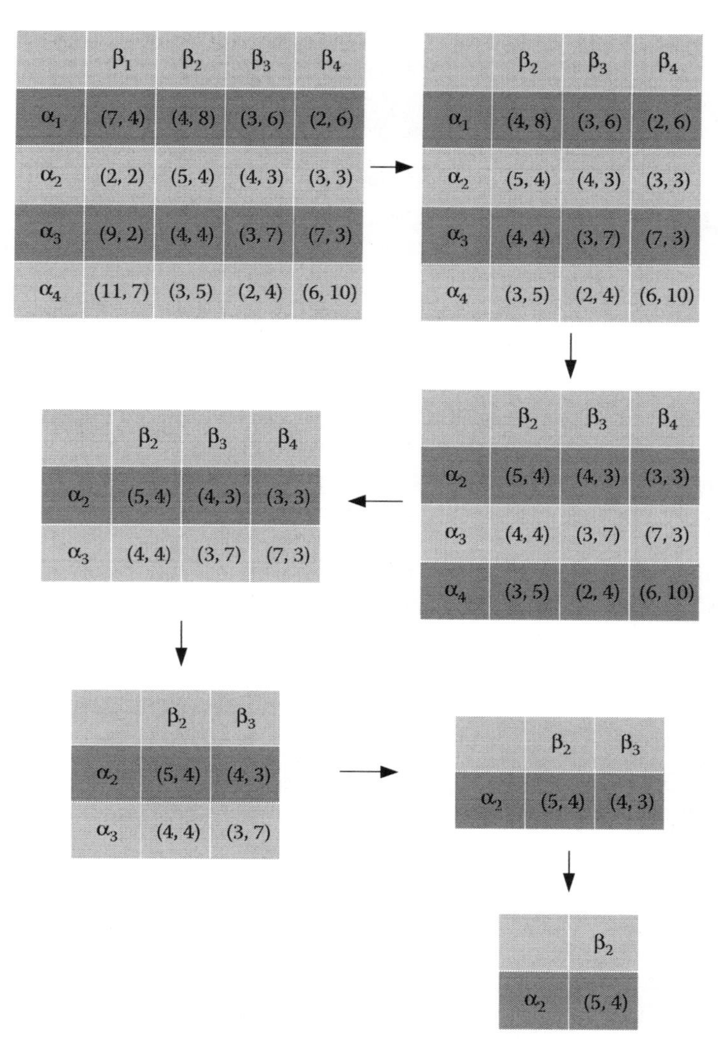

FIGURE 2.7
Successive dominated strategy removal.

progressively reduce the game table by removing the dominated strategies. If the successive removal of dominated strategies (SRDS) eventually leads to a set of indifferent outcomes for players, the game is said to be dominance solvable. For instance, a sequence of steps for the reduction of the initial game table is shown in Figure 2.7. Since action β_1 is dominated by β_4, the game table can be reduced by the removal of the column corresponding to β_1. In the reduced table, α_1 can be removed as it is dominated by α_2. Subsequently, α_4 is removed as it is dominated by α_3, β_4 can be removed as it is dominated by β_3, α_3 can be removed as it is dominated by α_2, and finally β_3 can be removed as it is dominated by β_2. Thus, in the end, only the outcome (α_2, β_2) remains, which can be readily seen to be the Nash equilibrium of the game.

Consider another example of a competition between two retailers to choose among high, low, and matched pricing. The payoffs for different pricing choices are given in Figure 2.8. It can be seen for both the players that action Match weakly dominates High. Once the dominated action High is

	High	Low	Match
High	(17, 17)	(9, 22)	(17, 17)
Low	(22, 9)	(12, 12)	(12, 12)
Match	(17, 17)	(12, 12)	(17, 17)

FIGURE 2.8
Removal of dominated strategies in retail game.

removed, in the reduced game table, Match dominates Low. Removal of Low leads to the Nash equilibrium outcome of (Match, Match) for this game. However, it is also important to note that the order of elimination plays a significant role in the final outcome in the elimination of weakly dominated strategies, meaning to say that different orders can lead to entirely different outcomes. For instance, consider the game in Figure 2.9 where the row player can choose from actions U, M, D, while the column player can choose from L, R. It can be seen from the table that D weakly dominates U and removal of U leads to the reduced game table on the left. In this table, subsequently, R dominates L and D dominates M. This thus finally leads to the Nash equilibrium outcome (D, R). However, if one begins by removal of the weakly dominated strategy M initially, leading to the reduced game table on the right, followed by removal of the dominated strategies M and R, the final remaining outcome is the Nash equilibrium (D, L). Thus, the two different orders of elimination lead to different equilibrium outcomes. Consider yet another example regarding the order of elimination of

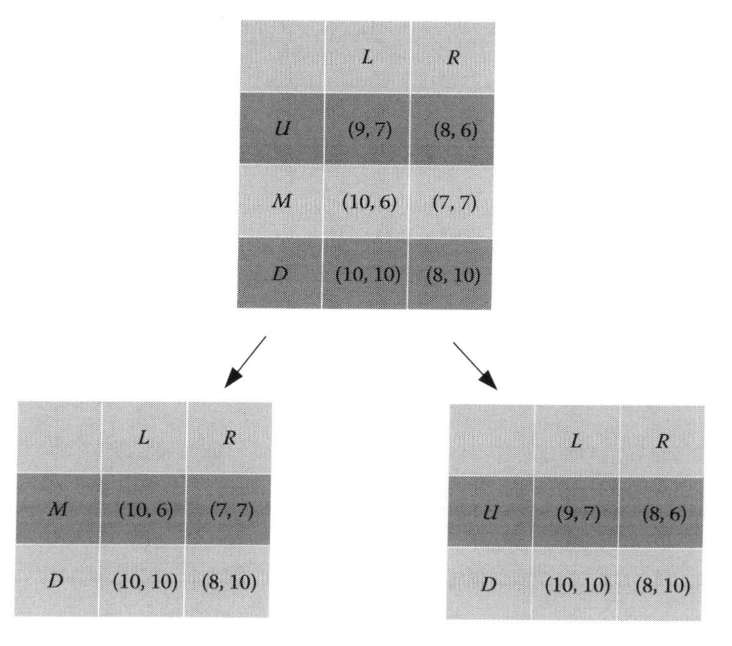

	L	R
U	(9, 7)	(8, 6)
M	(10, 6)	(7, 7)
D	(10, 10)	(8, 10)

	L	R
M	(10, 6)	(7, 7)
D	(10, 10)	(8, 10)

	L	R
U	(9, 7)	(8, 6)
D	(10, 10)	(8, 10)

FIGURE 2.9
Order of removal of weakly dominated strategies (I).

	L	C	R
U	(4, 4)	(2, 4)	(2, 3)
M	(4, 2)	(3, 3)	(2, 4)
D	(3, 2)	(4, 2)	(2, 2)

FIGURE 2.10
Order of removal of weakly dominated strategies (II).

weakly dominated strategies shown in Figure 2.10, where the actions U, M, D are available to the row player and L, C, R are available to the column player. It can be observed that C dominates L, followed by M dominating U and R dominating C in the reduced action tables. Thus, elimination of the dominated strategies L, U, C leads to the equilibrium outcomes (M, R), (D, R) with payoffs $(2, 2)$, $(2, 4)$, respectively. On the other hand, following another possible elimination order L, U, M leads to the set of outcomes (D, C) and (D, R) with payoffs (4, 2) and (2, 2), respectively. Thus, they lead to different outcomes. Also, both the orders miss the equilibrium outcome (U, L) with payoff $(2, 2)$. Thus, the elimination of weakly dominated strategies can also potentially lead to missing Nash equilibrium outcomes.

2.8.1 Example: Competition between Firms

Consider a competition between two firms that are selling a similar product. The number of units sold at price p is given as $4 - p$ and the firm with the lower price captures the market. Further, if both firms set equal price, each meets half of the demand. The possible values of the price p are $p \in \{1, 2, 3, 4\}$ and the production cost is 0. The game table for this game is given in Figure 2.11. This can be understood as follows. Consider the outcome $(1, 1)$, where each firm sets a price of

	1	2	3	4
1	(1.5, 1.5)	(3, 0)	(3, 0)	(3, 0)
2	(0, 3)	(2, 2)	(4, 0)	(4, 0)
3	(0, 3)	(0, 4)	(1.5, 1.5)	(3, 0)
4	(0, 3)	(0, 4)	(0, 3)	(0, 0)

FIGURE 2.11
Game of competition between firms.

$p = 1$. At this price, both the firms sell half the demand of $4 - p = 3$, that is, 1.5 units, to receive an payoff of $1.5p = 1.5$. On the other hand, at a price point of $(2, 1)$, firm 2 captures the entire market since it has the lower price, receiving a payoff of $(4 - p)\,p = 3$, while firm 1 receives 0. The payoff for this outcome is given by $(0, 3)$. The rest of the entries in the table can be explained similarly. The best response dynamic of both the firms is also shown in the table. From the intersection of best responses, it can be seen that the outcome $(1, 1)$ is the Nash equilibrium of the game. Also, it can be seen that the action $p = 2$ weakly dominates $p = 3, 4$ for both the players. Removing this, in the reduced game table, $p = 1$ in turn dominates $p = 2$ and subsequent removal of this action once again leaves the Nash equilibrium outcome $(1, 1)$, where each firm sets price $p = 1$.

2.9 Cournot Competition: Strategic Substitutes

A popular example in game theory is one involving a competition between two firms F_1, F_2 termed as a *duopoly*. Consider two firms producing quantities s_1, s_2 of two identical products, which are also termed as *strategic substitutes*. The price per unit of the product in such a scenario can be modeled as

$$p\,(s_1, s_2) = \alpha - \beta\,(s_1 + s_2),$$

from which it can be seen that the price per unit decreases with increasing total quantity $s_1 + s_2$ of the good. Let the cost toward producing quantity s_i, $1 \leq i \leq 2$, be modeled as

$$c\,(s_i) = \gamma s_i + \delta,$$

where δ is the fixed cost associated with the manufacture of the product. Hence, the profit or utility of F_1 for quantities s_1, s_2 produced by F_1, F_2, respectively, is given as

$$\begin{aligned} u_1\,(s_1, s_2) &= s_1 p\,(s_1, s_2) - (\gamma s_1 + \delta) \\ &= s_1\,(\alpha - \beta\,(s_1 + s_2)) - (\gamma s_1 + \delta)\,. \end{aligned}$$

Observe from this expression that the net utility of F_1 depends not only on the quantity s_1 it produces, but also on the quantity s_2 produced by F_2. Also, it can be seen that unlike the earlier games where a finite set of actions were available to each player, in this game, the action of each firm is the quantity s_i produced by it, which belongs to a continuous set. Hence, the best response of F_1 for a given quantity s_2 of F_2 can be obtained by maximizing the net utility of F_1, which is derived by setting the first-order derivative $(\partial/\partial s_1)u_1\,(s_1, s_2)$ to 0 as

$$\begin{aligned} 0 &= \alpha - \beta s_2 - 2\beta s_1 - \gamma, \\ s_1^\star = B_1\,(s_2) &= \frac{\alpha - \gamma - \beta s_2}{2\beta}, \end{aligned}$$

where $B_1\,(s_2)$ denotes the best response of F_1 to the quantity s_2 produced by F_2. Similarly, the best response of F_2 to quantity s_1 produced by F_1 is given as

$$s_2^\star = B_2\,(s_1) = \frac{\alpha - \gamma - \beta s_1}{2\beta}.$$

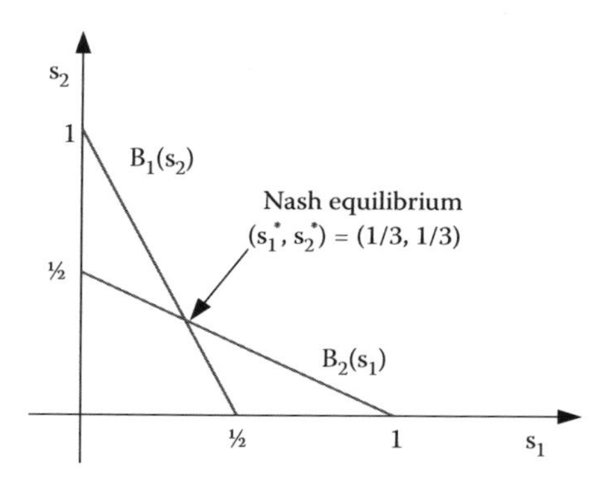

FIGURE 2.12
Best responses in a Cournot duopoly with $\alpha = 2$, $\beta = 1$, and $\gamma = 1$.

At Nash equilibrium, the quantities $s_1^\star, s_2^\star$ are best responses to each other and can be computed as the solution to the system of equations

$$s_1^\star = B_1\left(s_2^\star\right) = \frac{\alpha - \gamma - \beta s_2^\star}{2\beta},$$

$$s_2^\star = B_2\left(s_1^\star\right) = \frac{\alpha - \gamma - \beta s_1^\star}{2\beta}.$$

Solving these equations, the Nash equilibrium quantities $s_1^\star, s_2^\star$ are given as

$$s_1^\star = s_2^\star = \frac{\alpha - \gamma}{3\beta}.$$

This Nash equilibrium quantity is also termed as the *Cournot quantity*. For instance, with $\alpha = 2$, $\beta = 1$, and $\gamma = 1$, the equilibrium quantity is $s_1^\star = s_2^\star = \frac{1}{3}$. The best responses and the Nash equilibrium dynamic for this game are shown in Figure 2.12.

2.10 Regulation versus Competition: Tragedy of Commons

In this section, we look at a practical example of a game dealing with the utilization of shared societal resources such as fisheries and forests. Consider a scenario with N fisherman, where the ith fisherman chooses action a_i, which indicates the extent of his effort in fishing. The total effort of all the fishermen therefore equals $\sum_{i=1}^{N} a_i$. Let the utility or payoff of the ith fisherman be given by

$$\begin{aligned} u_i\left(a_i, a_{-i}\right) &= a_i \left(2000 - \sum_{j=1}^{N} a_j\right), \\ &= a_i \left(2000 - a_i - \sum_{\substack{j=1 \\ j \neq i}}^{N} a_j\right). \end{aligned} \tag{2.2}$$

The first term in this expression for utility indicates that the payoff to player i increases with his effort a_i. However, as the collective effort of all the fishermen increases, it leads to a net depletion of resources. Hence, the second term indicates this decreasing tendency of the payoff with the increased aggregate effort of the fishermen. Hence, the best response effort of fisherman i can be found as

$$\frac{\partial}{\partial a_i} u_i\left(a_i, a_{-i}\right) = 0,$$

$$2000 - a_i - \sum_{\substack{j=1 \\ j \neq i}}^{N} a_j - a_i = 0,$$

$$a_i^\star = \frac{1}{2}\left(2000 - \sum_{\substack{j=1 \\ j \neq i}}^{N} a_j\right).$$

At the Nash equilibrium of this game, each player i is playing his best response $a_i^\star$. Hence, for Nash equilibrium, we have

$$a_i^\star = \frac{1}{2}\left(2000 - \sum_{\substack{j=1 \\ j \neq i}}^{N} a_j^\star\right). \tag{2.3}$$

Also, from the symmetry of the payoff functions of the fisherman, we have $a_j^\star = a_i^\star$. Hence, it follows that, at Nash equilibrium,

$$2a_i^\star = 2000 - (N-1)\, a_i^\star,$$

$$a_i^\star = \frac{2000}{N+1}.$$

Thus, the Nash equilibrium outcome of efforts is given as

$$\mathrm{NE} = \left(\frac{2000}{N+1}, \frac{2000}{N+1}, \ldots, \frac{2000}{N+1}\right).$$

Finally, the utility of each player at Nash equilibrium is given as

$$u_i\left(a_i^\star, a_{-i}^\star\right) = \frac{2000}{N+1}\left(2000 - N \times \frac{2000}{N+1}\right)$$

$$= \frac{2000^2}{(N+1)^2}. \tag{2.4}$$

Consider now a different approach that regulates the total fishing effort toward payoff maximization. The net payoff of all the fishermen together is given as

$$\begin{aligned}
u(a_1, a_2, \ldots, a_N) &= \sum_{i=1}^{N} u_i(a_i, a_{-1}) \\
&= \sum_{i=1}^{N} a_i \left(2000 - \sum_{j=1}^{N} a_j\right) \\
&= \underbrace{\left(\sum_{i=1}^{N} a_i\right)}_{U} \left(2000 - \sum_{j=1}^{N} a_j\right) \\
&= U(2000 - U),
\end{aligned}$$

where $U = \sum_{i=1}^{N} a_i$ denotes the total fishing effort. The net payoff function can be readily seen to be maximized for $U = 2000 - U$, leading to the total optimal effort of $U = 1000$. Dividing this effort equally among all the fishermen, the effort of each fisherman is given as $1000/N$. Hence, the payoff to each fisherman at this *regulated* optimal outcome is given as

$$\begin{aligned}
u_i\left(a_i^R, a_{-i}^R\right) &= \frac{1000}{N}(2000 - 1000) \\
&= \frac{1000^2}{N}.
\end{aligned} \tag{2.5}$$

The outcome of the fishermen in the game versus regulated scenario can be better visualized by considering a simple example scenario. Let $N = 1000$ denote the total number of fishermen. From Equation 2.4, the payoff at Nash equilibrium to each of the fishermen is

$$\text{NE payoff} = \frac{2000^2}{1001^2} \approx 4.$$

As compared to this, the payoff for the regulated scenario is given from Equation 2.5 as

$$\text{Regulated payoff} = \frac{1000^2}{1000} = 1000,$$

thus clearly showing that the imposing regulations lead to an overall increase in the payoff of the players, due to tolerable consumption levels. This is also known as the *Tragedy of Commons*, since it indicates the poor utilization and reckless consumption of the common societal resources in the absence of appropriate regulatory framework.

2.11 Mixed Strategies

Consider again the Matching Pennies game shown in Figure 2.6. As described already in Section 2.7, there is no Nash equilibrium for this game in *pure* strategies, that is, when each user is choosing a single action. However, consider now a scenario where each user is randomly choosing one of the two possible actions. For instance, the row player P_1 can choose heads (H) with probability p and tails (T) with probability $(1-p)$. This is termed as a *mixed* strategy, since the player is mixing the two possible actions. Similarly, the column player can randomly choose between H,T with probabilities q and $1-q$, respectively. The Nash equilibrium in such a game can be found as follows. Consider the payoffs of the column player to the mixed strategy $(p,1-p)$ of the row player. If the column player chooses H always, his payoff on an average can be computed as follows. Since the row player is mixing H, T with probability p and $1-p$, with probability p his payoff is -1 and with probability $1-p$ his payoff is 1. Hence, his payoff to his pure strategy H is

$$u_2(H,(p,1-p)) = p \times (-1) + (1-p) \times 1 = 1-2p,$$

where the notation $u_2(H,(p,1-p))$ represents the utility of player 2 for action H, while player 1 is employing the mixed strategy $(p,1-p)$. Similarly, if player 2 chooses T, with probability p his payoff is 1 and with probability $1-p$ his payoff is -1. Hence, his average payoff is given as

$$u_2(T,(p,1-p)) = p \times (1) + (1-p) \times (-1) = 2p-1.$$

Now comes the critical part, which forms the key to the equilibrium analysis of this game. If the payoff of player P_2 is higher for one of the fixed strategies H or T, then he will choose a fixed strategy and not mix his actions. For instance, consider $p = 0.4$. Then, the payoffs of player P_2 are

$$u_2(H,(0.4,0.6)) = 0.2, u_2(T,(0.4,0.6)) = -0.2.$$

Since the payoff to choosing H is strictly greater than that of choosing T, player P_2 will always choose T as a best response. Similarly, if $p = 0.6$, then the payoffs to player P_2 are -0.2 and 0.2 for actions H and T, respectively, and the best response is to always choose T in this case. Therefore, there is no need for player P_2 to *mix* actions, since mixing, that is, averaging will only lower his payoff. Thus, for player P_2 to be mixing his actions H,T with probabilities q and $1-q$, it must be the case that both of them yield the same payoff. Therefore, it follows that

$$\begin{aligned} u_2(H,(p,1-p)) &= u_2(T,(p,1-p)) \\ 1-2p &= 2p-1 \\ p &= \frac{1}{2}. \end{aligned}$$

Thus, fundamentally, for a mixed-strategy equilibrium in this Matching Pennies game, it must be the case that player P_1 is mixing H,T with probabilities $p = 0.5$ and $1-p = 0.5$, respectively. Similarly, it can now be readily seen that repeating this analysis from the perspective of player P_1, his payoffs for choosing H versus T to the mixing strategy $(q,1-q)$ of player P_2 are given as

$$\begin{aligned} u_1(H,(q,1-q)) &= q \times 1 + (1-q) \times (-1) = 2q-1, \\ u_1(T,(q,1-q)) &= q \times (-1) + (1-q) \times 1 = 1-2q. \end{aligned} \tag{2.6}$$

Thus, if player P_1 is mixing, it must be the case his payoffs from both H and T are equal. Therefore, we have

$$\begin{aligned} u_1(H,(q,1-q)) &= u_1(T,(q,1-q)) \\ 2q-1 &= 1-2q \\ q &= \frac{1}{2}. \end{aligned}$$

Thus, the equilibrium strategy is mixing with probability $p^\star = \frac{1}{2}$ for the row player and mixing with probability $q^\star = \frac{1}{2}$ for the column player. Thus, the Nash equilibrium outcome of this game can now be represented as

$$\left(\underbrace{\left(\frac{1}{2},\frac{1}{2}\right)}_{\sigma_1^\star},\underbrace{\left(\frac{1}{2},\frac{1}{2}\right)}_{\sigma_2^\star}\right),$$

where $\sigma_1^\star, \sigma_2^\star$ are the equilibrium mixed strategies employed by players P_1, P_2, respectively. Naturally, unlike the pure strategies dealt with in the initial part of this chapter, these strategies $\sigma_1^\star, \sigma_2^\star$ are not actions, but probability distributions on their action sets $\mathcal{A}_1, \mathcal{A}_2$. We will formally define this framework for mixed-strategy equilibria in the succeeding sections.

2.11.1 Intuition for Mixed-Strategy Equilibrium

Consider the mixed-strategy equilibrium of $(\sigma_1^\star, \sigma_2^\star)$ derived for the Matched Pennies game above. The average payoff to player P_1 from employing the mixed strategy $\sigma_1^\star$ can be computed as follows. His payoffs for choosing the actions H, T can be computed as illustrated by the equations in Equation 2.6 as

$$\begin{aligned} u_1(H,\sigma_2^\star) &= u_1\left(H,\left(\frac{1}{2},\frac{1}{2}\right)\right) = 2\times\frac{1}{2}-1 = 0, \\ u_1(T,\sigma_2^\star) &= u_1\left(T,\left(\frac{1}{2},\frac{1}{2}\right)\right) = 1-2\times\frac{1}{2} = 0. \end{aligned}$$

Thus, his net payoff for the mixed strategy $\sigma_1^\star = \left(\frac{1}{2},\frac{1}{2}\right)$ is given as

$$\begin{aligned} u_1(\sigma_1^\star,\sigma_2^\star) &= \sigma_1^\star(H)\,u_1(H,\sigma_2^\star) + \sigma_1^\star(T)\,u_1(T,\sigma_2^\star) \\ &= \frac{1}{2}\times 0 + \frac{1}{2}\times 0 \\ &= 0. \end{aligned}$$

Thus, the average payoff to player P_1 from his strategy $\sigma_1^\star$ against the strategy $\sigma_2^\star$ of player P_2 is 0. Now, similar to idea of *unilateral deviation* introduced in the context of pure strategies, let us consider a deviation by player P_1. Let player P_1 deviate to any strategy

$$\sigma_1^d(H) = p, \sigma_1^d(T) = 1-p,$$

where the superscript d denotes a deviation from the Nash strategy $\sigma_1^\star$. The average payoff to player P_1 for this strategy deviation is given as

$$\begin{aligned} u_1\left(\underbrace{(p, 1-p)}_{\sigma_1^d}, \sigma_2^\star\right) &= \sigma_1^d(H)\, u_1\left(H, \sigma_2^\star\right) + \sigma_1^d(T)\, u_1\left(T, \sigma_2^\star\right) \\ &= p \times 0 + (1-p) \times 0 \\ &= 0. \end{aligned}$$

Therefore, at Nash equilibrium, the payoff $u_1\left(\sigma_1^d, \sigma_2^\star\right)$ to player P_1 for any deviation σ_1^d is exactly equal to the Nash payoff $u_1\left(\sigma_1^\star, \sigma_2^\star\right)$. A similar line of reasoning holds for player P_2. It follows therefore, akin to the scenario of Nash equilibrium in pure strategies, that there is no incentive to any player for unilateral deviation from his Nash strategy.

2.12 Battle of the Sexes: Another Example of Mixed-Strategy Equilibrium

Let us go back to the BoS game introduced in Section 2.6.2. We show the game table for this game again in Figure 2.13 for convenience. It was mentioned therein that in addition to the (G, G) and (M, M) pure strategy equilibrium outcomes in the game, there is yet another mixed-strategy Nash equilibrium outcome. In the wake of the above discussion on mixed-strategy Nash equilibria, let us derive this mixed-strategy Nash equilibrium for the BoS game. Let the boy employ the mixed strategy $\sigma_b = (p, 1-p)$, that is, he randomly mixes his actions G and M with probabilities p and $1-p$, respectively. Similarly, let the girl employ the mixture $\sigma_g = (q, 1-q)$. The payoffs to the girl for the pure strategies G and M, respectively, are given as

$$\begin{aligned} u_g(G, \sigma_b) &= P_b(G)\, u_g(M, G) + P_b(M)\, u_g(G, M) \\ &= p \times 5 + (1-p) \times 0 \\ &= 5p, \\ u_g(M, \sigma_b) &= P_b(G)\, u_g(M, G) + P_b(M)\, u_g(M, M) \\ &= p \times 0 + (1-p) \times 10 \\ &= 10\,(1-p)\,. \end{aligned} \tag{2.7}$$

Girl / Boy	Game	Movie
Game	(10, 5)	(0, 0)
Movie	(0, 0)	(5, 10)

FIGURE 2.13
Battle of the Sexes.

The above quantities can be reasoned as follows. If the girl chooses the pure strategy G against the mixed strategy σ_b of the boy, then with probability p, they will both end up at the movies, yielding a payoff on an average of $5p$ to the girl, and with probability $1-p$, they will end up with the girl watching the game and the boy at the movies, resulting in an average payoff of $0 \times (1-p) = 0$ to the girl. Hence, the net average payoff to the girl for the pure strategy of G is given as $u_g(G, \sigma_b) = 5p + 0 = 0$. The other quantity $u_g(M, \sigma_b)$ can be derived similarly. Now, if the girl is mixing G and M, it must be the case that her payoffs from both are equal. Hence, we have

$$u_g(G, \sigma_b) = u_g(M, \sigma_b),$$
$$5p^\star = 10\left(1 - p^\star\right),$$
$$p^\star = \frac{2}{3}.$$

Similarly, consider now the game from the perspective of the boy. His average payoffs for the pure strategies G, M to the mixed strategy $q, (1-q)$ by the girl can be derived as

$$u_b\left(G, \sigma_g\right) = 10q,$$
$$u_b(M, \sigma_b) = 5(1-q).$$

Hence, since he is mixing G and M, we have $10q^\star = 5(1-q^\star)$, yielding the Nash equilibrium mixture $q^\star = \frac{1}{3}, 1 - q^\star = \frac{2}{3}$. Thus, the Nash equilibrium mixture outcome is given as

$$\left(\sigma_1^\star, \sigma_2^\star\right) = \left(\left(\frac{2}{3}, \frac{1}{3}\right), \left(\frac{1}{3}, \frac{2}{3}\right)\right).$$

2.12.1 Mixed-Strategy Equilibrium through Best Responses

Let us now analyze and derive the Nash equilibrium outcome for the BoS game employing the approach of best response characteristics of each of the players. Consider again a mixture $\sigma_b = (p, 1-p)$ and $\sigma_g = (q, 1-q)$ for the boy and girl, respectively. Now, consider the game from the girl's perspective. As illustrated in Equation 2.7 above, her payoffs to the pure strategies G, M, respectively, are

$$u_g(G, \sigma_b) = 5p, u_g(M, \sigma_b) = 10(1-p).$$

Now, if $u_g(G, \sigma_b) > u_g(M, \sigma_b)$, that is, her payoff to choosing the game G is strictly greater than that of choosing the movie M, she will choose G, since any mixture is only bound to reduce her payoff. This occurs for

$$5p > 10(1-p) \Rightarrow p > \frac{2}{3}.$$

In this scenario, the girl therefore chooses G with certainty or probability $q = 1$. Thus, her best response to any mixture $\sigma_b = (p, 1-p)$, with $p > \frac{2}{3}$ is $\sigma_g = (1, 0)$. On the other hand, if $u_g(G, \sigma_b) < u_g(M, \sigma_b)$, her best response would be to choose M with certainty, resulting in a best response of $\sigma_g = (0, 1)$, that is, $q = 0$. However, if $u_g(G, \sigma_b) = u_g(M, \sigma_b)$, something interesting occurs. Her payoffs to both G and M are exactly equal. Hence, any mixture yields the exact same payoff as any other. This occurs when $p = \frac{2}{3}$. This can be seen as follows. For $p = \frac{2}{3}$, we have

$$u_g\left(G, \left(\frac{2}{3}, \frac{1}{3}\right)\right) = 5p = \frac{10}{3},$$
$$u_g\left(M, \left(\frac{2}{3}, \frac{1}{3}\right)\right) = 10\,(1 - p) = \frac{10}{3}.$$

Thus, the payoff of the girl to any mixture $\sigma_g = (q, 1 - q)$ is

$$\begin{aligned} u_g\left(\sigma_g, \left(\frac{2}{3}, \frac{1}{3}\right)\right) &= q \times \frac{10}{3} + (1 - q) \times \frac{10}{3} \\ &= \frac{10}{3}. \end{aligned}$$

Thus, in this case, the best response of the girl is to choose any $\sigma_g = (q, 1 - q)$ with $0 \leq q \leq 1$, resulting in an identical payoff of $\frac{10}{3}$ for every such strategy. Thus, the best response $\sigma_g^b = \left(q^b, 1 - q^b\right)$ can be summarized as

$$\sigma_g^b = \left(q^b, 1 - q^b\right) = \begin{cases} (0, 1), & \text{if } 0 \leq p < \frac{2}{3} \\ (q, 1 - q), 0 \leq q \leq 1, & \text{if } p = \frac{2}{3} \\ (1, 0), & \text{if } \frac{2}{3} < p \leq 1 \end{cases}$$

The best response mixture probability q^b for the girl is shown in Figure 2.14 as a function of the mixture probability p of the boy. Employing a similar approach as above, one can derive the best response mixture $\sigma_b^b = \left(p^b, 1 - p^b\right)$ as a function of the mixture $\sigma_g = (q, 1 - q)$ of the girl. The average payoffs of the boy for the fixed strategies G, M, respectively, to the mixture $(q, 1 - q)$ are given as

$$u_b\,(G, (q, 1 - q)) = 10q,$$
$$u_b\,(M, (q, 1 - q)) = 5\,(1 - q).$$

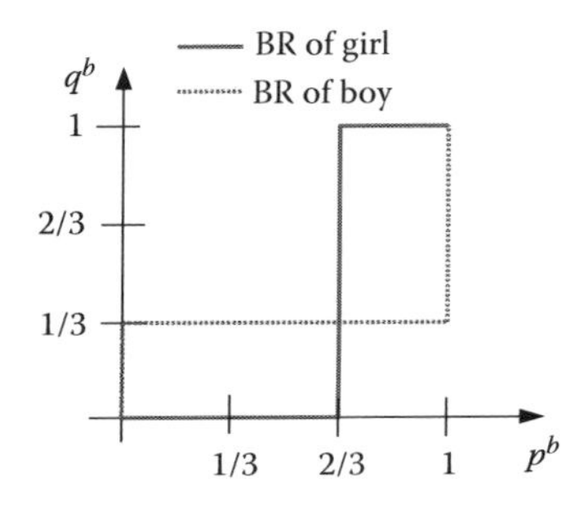

FIGURE 2.14
Battle of the Sexes: Best responses.

Therefore, if $10q > 5(1-q)$, that is, $q > 1/3$, the best response of the boy is to choose G with certainty, leading to $\sigma_b^b = (1,0)$. Otherwise, if $10q < 5(1-q)$, then he chooses M with certainty, implying $p^b = 0$ and $\sigma_b^b = (0,1)$. Finally, in the scenario $10q = 5(1-q)$ or $q = \frac{1}{3}$, any $0 \leq p^b \leq 1$ is the best response as the payoff for each p^b is equal to $\frac{10}{3}$. Thus, the best response characteristic of the girl can be derived as

$$\sigma_b^b = \left(p^b, 1-p^b\right) = \begin{cases} (0,1), & \text{if } 0 \leq q < \frac{1}{3} \\ (p, 1-p), 0 \leq p \leq 1, & \text{if } q = \frac{1}{3} \\ (1,0), & \text{if } \frac{1}{3} < q \leq 1 \end{cases}$$

The best responses q^b as a function of p and p^b as a function of q are shown in Figure 2.14. The dotted line represents the best response function of the boy while the solid line represents the best response of the girl. It can be seen that the best responses intersect at three points. The two points corresponding to $p^b = q^b = 1$ and $p^b = q^b = 0$ correspond to the pure strategy Nash equilibria (G,G) and (M,M), respectively. And now it can be clearly seen that the remaining point $p^b = \frac{2}{3}$ and $q^b = \frac{1}{3}$ corresponds to the mixed-strategy Nash equilibrium

$$\left(\sigma_1^\star, \sigma_2^\star\right) = \left(\underbrace{\left(\frac{2}{3}, \frac{1}{3}\right)}_{\text{Mixture of boy}}, \underbrace{\left(\frac{1}{3}, \frac{2}{3}\right)}_{\text{Mixture of girl}} \right).$$

2.13 Formal Definition of a Mixed-Strategy Equilibrium

Let the mixed strategy employed by player i be denoted by σ_i, which randomly chooses action $a_i \in \mathcal{A}_i$ with probability $\sigma_i(a_i)$. For instance, in the BoS game, $\sigma_b(G) = p$ and $\sigma_b(M) = 1-p$. Let each $\sigma_i \in \Sigma_i$, where Σ_i is the space of probability distributions over the action set $\mathcal{A}_i$ of the ith user. Let $\boldsymbol{\sigma} = (\sigma_1, \sigma_2, \ldots, \sigma_n)$ denote the mixed-strategy profile employed by the n players, where $\boldsymbol{\sigma} \in \boldsymbol{\Sigma}$ defined as $\boldsymbol{\Sigma} = \Sigma_1 \times \Sigma_2 \times \cdots \times \Sigma_n$, the Cartesian product of the mixed strategies of all the users. This is similar to the outcome space $\mathcal{A}$ in the context of pure strategies. The average payoff to the ith user for the mixed-strategy profile $\boldsymbol{\sigma}$ is given as

$$u_i(\boldsymbol{\sigma}) = \sum_{\mathbf{a} \in \mathcal{A}} \left(\prod_{j=1}^{n} \sigma_j(a_j) \right) u_i(\mathbf{a}). \tag{2.8}$$

where $\mathbf{a} = (a_1, a_2, \ldots, a_n)$ denotes an action profile such that $a_i \in \mathcal{A}_i$. Thus, the above computation essentially weights the $u_i(\mathbf{a})$ to each user i for the action profile $\mathbf{a}$, with the probability that $\mathbf{a}$ is chosen randomly corresponding to the mixed-strategy profile $\boldsymbol{\sigma}$. Further, similar to the case of pure strategies, the quantity $u_i(\boldsymbol{\sigma})$ can be more naturally represented in the context of game theory as

$$\begin{aligned} u_i(\boldsymbol{\sigma}) &= u_i((\sigma_1, \sigma_2, \ldots, \sigma_n)) \\ &= u_i\left(\sigma_i, \underbrace{(\sigma_1, \sigma_2, \ldots, \sigma_{i-1}, \sigma_{i+1}, \ldots, \sigma_n)}_{\boldsymbol{\sigma}_{-i}}\right) \\ &= u_i(\sigma_i, \boldsymbol{\sigma}_{-i}), \end{aligned}$$

where $\boldsymbol{\sigma}_{-i}$ represents the strategy profile of all the users other than the ith user. A strategy profile $\boldsymbol{\sigma}^\star = \left(\sigma_1^\star, \sigma_2^\star, \ldots, \sigma_n^\star\right) \in \boldsymbol{\Sigma}$ is a mixed-strategy Nash equilibrium if

$$u_i\left(\sigma_i^\star, \boldsymbol{\sigma}_{-i}^\star\right) \geq u_i\left(\sigma_i, \boldsymbol{\sigma}_{-i}^\star\right),$$

for all mixed strategies $\sigma_i \in \Sigma_i$ of the ith user and for all users $1 \leq i \leq n$. This basically means that no user can benefit by unilaterally deviating from his Nash strategy $\sigma_i^\star$ to any strategy $\sigma_i \in \Sigma_i$. Observe that this also naturally implies that $u_i\left(\sigma_i^\star, \boldsymbol{\sigma}_{-i}^\star\right) \geq u_i\left(a_i, \boldsymbol{\sigma}_{-i}^\star\right)$, for every action $a_i \in \mathcal{A}_i$, since choosing the pure action a_i corresponds to the trivial $\sigma_i = (0, 0, \ldots, 0, 1, 0, \ldots, 0)$ with the lone 1 in the ith position.

2.14 Mixed-Strategy Nash Equilibrium for Two-Player Zero-Sum Games

Deriving the mixed-strategy Nash equilibrium for a general two-player game with more than two possible actions for each player, that is, $|\mathcal{A}_i| > 2$, where $|\cdot|$ denotes the cardinality of the set, is not straightforward. Hence, in this section, we develop the framework to derive the mixture Nash equilibrium for a special class of games termed as *zero-sum* games. Consider the game table for the Matching Pennies game shown in Figure 2.6. It can be seen from the utility functions therein that for any outcome, the sum of the payoffs to each player is identically equal to zero, implying that if one of the player wins, the other has to lose. Such a game is termed as a *zero-sum* game. Consider the mixed strategies $\mathbf{p} = [p_1, p_2]^T$ and $\mathbf{q} = [q_1, q_2]^T$ employed by the row and column players, respectively. Now, since $\mathbf{p}, \mathbf{q}$ are probability distributions, $p_1 + p_2 = 1 = q_1 + q_2$ and $p_1, p_2, q_1, q_2 \geq 0$. Then, from the expression in Equation 2.8 for the payoff to each player from a chosen mixed-strategy profile, it can be seen that the average payoff to player P_1 under the strategy profile $(\mathbf{p}, \mathbf{q})$ is given as

$$\begin{aligned} u_1(\boldsymbol{p}, \boldsymbol{q}) &= p_1 u_1(H, \boldsymbol{q}) + p_2 u_1(T, \boldsymbol{q}) \\ &= p_1 \{q_1 u_1(H, H) + q_2 u_1(H, T)\} + p_2 \{q_1 u_1(T, H) + q_2 u_1(T, T)\} \\ &= p_1 q_1 u_1(H, H) + p_1 q_2 u_1(H, T) + p_2 q_1 u_1(T, H) + p_2 q_2 u_1(T, T) \\ &= p_1 q_1 - p_1 q_2 - p_2 q_1 + p_2 q_2 \\ &= \begin{bmatrix} p_1 & p_2 \end{bmatrix} \underbrace{\begin{bmatrix} 1 & -1 \\ -1 & 1 \end{bmatrix}}_{\mathbf{U}} \begin{bmatrix} q_1 \\ q_2 \end{bmatrix} \\ &= \mathbf{p}^T \mathbf{U} \mathbf{q}. \end{aligned}$$

It can also be seen that the entry u_{ij} of the matrix $\mathbf{U}$ is the payoff $u_1\left(a_1 = \alpha_i^{(1)}, a_2 = \alpha_j^{(2)}\right)$ to player P_1 corresponding to the outcome $\left(\alpha_i^{(1)}, \alpha_j^{(2)}\right)$, where $1 \leq i, j \leq 2$, with $\alpha_1^{(k)}, \alpha_2^{(k)}$ denoting H, T, respectively. It can be readily verified that in the zero-sum game above, the average payoff to player P_2 is given as

$$u_2(\mathbf{q}, \mathbf{p}) = -u_1(\mathbf{p}, \mathbf{q}) = -\mathbf{p}^T\mathbf{U}\mathbf{q}.$$

Now, consider a general game table of size $K \times L$, that is, one in which player P_1 can choose one out of K possible actions in the set $\mathcal{A}_1 = \left\{\alpha_1^{(1)}, \alpha_2^{(1)}, \ldots, \alpha_K^{(1)}\right\}$, while P_2 can choose from L possible actions belonging to the set $\left\{\alpha_1^{(2)}, \alpha_2^{(2)}, \ldots, \alpha_L^{(2)}\right\}$. In other words, $|\mathcal{A}_1| = K$, while $|\mathcal{A}_2| = L$. The mixtures $\mathbf{p}, \mathbf{q}$ of P_1, P_2 are now probability vectors of dimension K, L, respectively. Let $\mathbf{p} = [p_1, p_2, \ldots, p_K]$ and $\mathbf{q} = [q_1, q_2, \ldots, q_L]$, where each $p_i, q_j \geq 0$ and $\sum_{i=1}^{K} p_i = \sum_{j=1}^{L} q_j = 1$, with p_i, q_j corresponding to the probabilities with which P_1, P_2 choose actions $\alpha_i^{(1)}, \alpha_j^{(2)}$, respectively. The average payoff to P_1 can now be generalized for the $K \times L$ dimensional action space $\mathcal{A}_1 \times \mathcal{A}_2$ as $u_1(\mathbf{p}, \mathbf{q}) = \mathbf{p}^T\mathbf{U}\mathbf{q} = -u_2(\mathbf{q}, \mathbf{p})$, where the matrix $\mathbf{U}$ is

$$\mathbf{U} = \begin{bmatrix} u_1\left(\alpha_1^{(1)}, \alpha_1^{(2)}\right) & u_1\left(\alpha_1^{(1)}, \alpha_2^{(2)}\right) & \ldots & u_1\left(\alpha_1^{(1)}, \alpha_{L-1}^{(2)}\right) & u_1\left(\alpha_1^{(1)}, \alpha_L^{(2)}\right) \\ u_1\left(\alpha_2^{(1)}, \alpha_1^{(2)}\right) & u_1\left(\alpha_2^{(1)}, \alpha_2^{(2)}\right) & \ldots & u_1\left(\alpha_2^{(1)}, \alpha_{L-1}^{(2)}\right) & u_1\left(\alpha_2^{(1)}, \alpha_L^{(2)}\right) \\ \vdots & \vdots & \ddots & \vdots & \vdots \\ u_1\left(\alpha_{K-1}^{(1)}, \alpha_1^{(2)}\right) & u_1\left(\alpha_{K-1}^{(1)}, \alpha_2^{(2)}\right) & \ldots & u_1\left(\alpha_{K-1}^{(1)}, \alpha_{L-1}^{(2)}\right) & u_1\left(\alpha_{K-1}^{(1)}, \alpha_L^{(2)}\right) \\ u_1\left(\alpha_K^{(1)}, \alpha_1^{(2)}\right) & u_1\left(\alpha_K^{(1)}, \alpha_2^{(2)}\right) & \ldots & u_1\left(\alpha_K^{(1)}, \alpha_{L-1}^{(2)}\right) & u_1\left(\alpha_K^{(1)}, \alpha_L^{(2)}\right) \end{bmatrix}. \tag{2.9}$$

Thus, player P_1 wishes to maximize his payoff $\mathbf{p}^T\mathbf{U}\mathbf{q}$ while player P_2 wishes to minimize $\mathbf{p}^T\mathbf{U}\mathbf{q}$, since his payoff is $-\mathbf{p}^T\mathbf{U}\mathbf{q}$. Therefore, the Nash equilibrium of the two-player zero-sum game can be defined as follows. The mixed strategies $\check{\mathbf{p}}, \check{\mathbf{q}}$ form the mixed-strategy Nash equilibrium if

$$\begin{aligned} \check{\mathbf{p}}^T\mathbf{U}\check{\mathbf{q}} &\geq \mathbf{p}^T\mathbf{U}\check{\mathbf{q}}, \forall \mathbf{p}, \\ \check{\mathbf{p}}^T\mathbf{U}\check{\mathbf{q}} &\leq \check{\mathbf{p}}^T\mathbf{U}\mathbf{q}, \forall \mathbf{q}. \end{aligned} \tag{2.10}$$

In other words, given the strategy $\check{\mathbf{q}}$ of P_2, the strategy $\check{\mathbf{p}}$ maximizes the payoff of P_1 and similarly given $\check{\mathbf{p}}$ of P_1, $\check{\mathbf{q}}$ maximizes the payoff, that is, minimizes $\check{\mathbf{p}}^T\mathbf{U}\mathbf{q}$, of P_2. Next, we describe the minimax framework to deduce the Nash equilibrium of the above two-person zero-sum game.

2.15 Minimax Framework

One can define the fundamental quantities V_r and V_c associated with the above zero-sum game as follows:

$$\begin{aligned} V_r &= \max_{\mathbf{p}} \min_{\mathbf{q}} \mathbf{p}^T\mathbf{U}\mathbf{q}, \\ V_c &= \min_{\mathbf{q}} \max_{\mathbf{p}} \mathbf{p}^T\mathbf{U}\mathbf{q}. \end{aligned}$$

The above quantities V_r, V_c can then be readily described as the assured payoff of the row and column players, that is, P_1, P_2, respectively. Below, we prove results that underlie the basic framework of zero-sum games.

Theorem 2.2

The assured payoffs V_r, V_c of the row and column players, respectively, satisfy the fundamental property $V_c \geq V_r$.

Proof. From the definitions above, it follows that

$$\max_{\mathbf{p}} \mathbf{p}^T \mathbf{U}\mathbf{q} \geq \max_{\mathbf{p}} \min_{\mathbf{q}} \mathbf{p}^T \mathbf{U}\mathbf{q},$$

$$\max_{\mathbf{p}} \mathbf{p}^T \mathbf{U}\mathbf{q} \geq V_r,$$

where the last inequality above follows from the definition of V_r. Similarly, we also have

$$\max_{p} \mathbf{p}^T \mathbf{U}\mathbf{q} \geq V_r,$$

$$\min_{q} \max_{p} \mathbf{p}^T \mathbf{U}\mathbf{q} \geq V_r,$$

$$\Rightarrow V_c \geq V_r.$$

This completes the proof of the theorem. ■

The next two results further shed light on the structure of the Nash equilibrium strategies.

Theorem 2.3

If $\mathbf{p}^\star, \mathbf{q}^\star$ is the Nash equilibrium profile of the zero-sum game, then

$$V_c = V_r = \left(\mathbf{p}^\star\right)^T \mathbf{U}\mathbf{q}^\star.$$

Proof. If $\mathbf{p}^\star, \mathbf{q}^\star$ is the Nash equilibrium of the zero-sum game, it follows from Equation 2.10 that

$$\left(\mathbf{p}^\star\right)^T \mathbf{U}\mathbf{q}^\star = \min_{\mathbf{q}} \left(\mathbf{p}^\star\right)^T \mathbf{U}\mathbf{q}.$$

Further, from the definition of V_r, we have $V_r = \max_{\mathbf{p}} \min_{\mathbf{q}} \mathbf{p}^T \mathbf{U}\mathbf{q}$. It naturally follows that $V_r \geq \left(\mathbf{p}^\star\right)^T \mathbf{U}\mathbf{q}^\star$. Thus, we have

$$\left(\mathbf{p}^\star\right)^T \mathbf{U}\mathbf{q}^\star = \max_{\mathbf{p}} \left(\mathbf{p}\right)^T \mathbf{U}\mathbf{q}^\star.$$

Hence, $V_c \leq \left(\mathbf{p}^\star\right)^T \mathbf{U}\mathbf{q}$ and $V_r \geq V_c$. Also, from the result proved above in Theorem 2.2, we have $V_c \geq V_r$. Hence, it follows that

$$V_r = V_c = \left(\mathbf{p}^\star\right) \mathbf{U}\mathbf{q}^\star,$$

thus proving the theorem. ■

Finally, we prove that if $V_c = V_r$, the mixed strategies $\mathbf{p}^\star, \mathbf{q}^\star$ indeed are the Nash equilibrium strategies.

Theorem 2.4

If $V_c = V_r$ and $\mathbf{p}^\star = \arg\max_{\mathbf{p}} \min_{\mathbf{q}} \mathbf{p}^T \mathbf{U} \mathbf{q}$ and $\mathbf{q}^\star = \arg\min_{\mathbf{q}} \max_{\mathbf{p}} \mathbf{p}^T \mathbf{U} \mathbf{q}$, then $\mathbf{p}^\star$ and $\mathbf{q}^\star$ are the Nash equilibrium strategies of P_1, P_2 in the zero-sum game.

Proof. From the definition of V_r, we have $V_r = \max_{\mathbf{p}} \min_{\mathbf{q}} \mathbf{p}^T \mathbf{U} \mathbf{q}$. Let $\tilde{\mathbf{p}}$ and $\tilde{\mathbf{q}}$ be mixture strategies such that

$$\begin{aligned} \left(\mathbf{p}^\star\right)^T \mathbf{U}\tilde{\mathbf{q}} &= V_r = V_c = \min_{\mathbf{q}} \left(\mathbf{p}^\star\right)^T \mathbf{U}\tilde{\mathbf{q}}, \\ \tilde{\mathbf{p}}^T \mathbf{U}\mathbf{q}^\star &= V_c = V_r = \max_{\mathbf{p}} \mathbf{p}^T \mathbf{U}\mathbf{q}^\star. \end{aligned} \tag{2.11}$$

Observe also that

$$V_c = \left(\mathbf{p}^\star\right)^T \mathbf{U}\tilde{\mathbf{q}} = \max_{\mathbf{p}} \mathbf{p}^T \mathbf{U}\mathbf{q}^\star \geq \left(\mathbf{p}^\star\right)^T \mathbf{U}\mathbf{q}^\star. \tag{2.12}$$

Also, $(\mathbf{p}^\star)^T \mathbf{U}\tilde{\mathbf{q}} = \min_{\mathbf{q}} (\mathbf{p}^\star)^T \mathbf{U}\mathbf{q} \leq (\mathbf{p}^\star)^T \mathbf{U}\mathbf{q}^\star$. Hence, we have $(\mathbf{p}^\star)^T \mathbf{U}\tilde{\mathbf{q}} = (\mathbf{p}^\star)^T \mathbf{U}\mathbf{q}^\star = \min_{\mathbf{q}} (\mathbf{p}^\star)^T \mathbf{U}$ from which it can be readily deduced that $(\mathbf{p}^\star)^T \mathbf{U}\mathbf{q}^\star \leq (\mathbf{p}^\star)^T \mathbf{U}\mathbf{q}, \forall \mathbf{q}$. Therefore, it follows that

$$\begin{aligned} \left(\mathbf{p}^\star\right)^T \mathbf{U}\mathbf{q}^\star &= \max_{\mathbf{p}} \mathbf{p}^T \mathbf{U}\mathbf{q}^\star, \\ \left(\mathbf{p}^\star\right)^T \mathbf{U}\mathbf{q}^\star &\geq \mathbf{p}^T \mathbf{U}\mathbf{q}^\star, \forall \mathbf{p}. \end{aligned} \tag{2.13}$$

Hence, $\mathbf{p}^\star, \mathbf{q}^\star$ is the Nash equilibrium. ■

Thus, the above minimax framework can be conveniently employed to find the Nash equilibrium strategies for zero-sum games. Next, we illustrate a systematic procedure to find the solution to such zero-sum games through linear programming (LP)-based optimization problems.

2.16 Solving Zero-Sum Games through Linear Programming

Consider the mixed strategy $\mathbf{p} = [p_1, p_2, \ldots, p_k]$ of player P_1. From the structure of the payoff matrix $\mathbf{U}$ given in Equation 2.9, his payoff to the action a_j of player P_2 can be seen to be given as

$$u_1\left(\mathbf{p}, a_j\right) = \sum_{i=1}^{k} p_i u_1\left(a_i, a_j\right).$$

Thus, the payoffs of player P_1 for different choices $a_1, a_2, \ldots, a_l$ of player P_2 are given by

$$\sum_{i=1}^{k} p_i u_1\left(a_i, a_1\right), \quad \sum_{i=1}^{k} p_i u_1\left(a_i, a_2\right), \ldots, \sum_{i=1}^{k} p_i u_1\left(a_i, a_l\right).$$

Thus, since player P_2 desires to minimize the payoff of player P_1, in turn maximizing his own payoff, he would choose action $a_{\tilde{j}}$ such that

$$\tilde{j} = \arg\min_{j} \sum_{i=1}^{k} p_i u_1 \left(a_i, a_j\right).$$

Player P_1 thus desires to maximize the minimum among the above payoffs. Let this minimum payoff be denoted by z. Hence, the optimal strategy $\mathbf{p}^\star$ is the solution to the optimization problem

$$\begin{aligned}
&\max z \\
&\text{s.t.} \sum_{i=1}^{k} p_i u_1 \left(a_i, a_j\right) \geq z, \\
&\sum_{i=1}^{k} p_i u_1 \left(a_i, a_2\right) \geq z, \\
&\vdots \\
&\sum_{i=1}^{k} p_i u_1 \left(a_i, a_l\right) \geq z, \\
&\sum_{i=1}^{k} p_i = 1, \\
&p_i \geq 0, 1 \leq i \leq k,
\end{aligned} \tag{2.14}$$

where the last two constraints follow from the fact that $\mathbf{p}$ is a probability mixture and hence contains nonnegative elements with total probability equal to unity. The above optimization problem can be readily seen to be an LP problem. Thus, the solution to the two player zero-sum game can be obtained by solving an appropriate linear program. Similarly, the corresponding linear program to derive the equilibrium mixture $\mathbf{q}$ corresponding the column player P_2 can be formulated similar to above.

2.16.1 Zero-Sum Game Example

Consider a zero-sum game between the row and column players with payoff table given in Figure 2.15, with actions T, B available to the row player and L, M, R available to the column player. Consider the row player mixing T, B with probabilities $p, 1-p$. The payoffs to the row player for different action choices of the column player are given as

$$\begin{aligned}
u_1\left((p, 1-p), L\right) &= 0 + 5\left(1-p\right) = 5\left(1-p\right), \\
u_1\left((p, 1-p), M\right) &= -2p + 4\left(1-p\right) = 4 - 6p, \\
u_1\left((p, 1-p), R\right) &= 2p - 3\left(1-p\right) = 5p - 3.
\end{aligned}$$

The payoffs for each of the pure actions L, M, R of the column player as a function of the mixing probability p are shown in plot in Figure 2.16. The minimum payoff for each value of p is marked

	L	M	R
T	(0, 0)	(−2, 2)	(2, −2)
B	(5, −5)	(4, −4)	(−3, 3)

FIGURE 2.15
Zero sum—Minimax example.

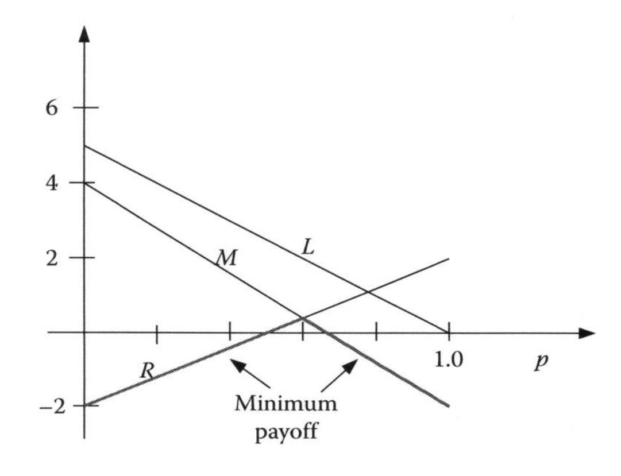

FIGURE 2.16
Zero-sum—Minimax payoffs.

using the thick curve. It can be seen that the maxmin point corresponds to the intersection of the payoffs corresponding to the pure strategies M, R given by $4 - 6p = 5p - 3$, leading to a value of $p = \frac{7}{11}$. Therefore, at Nash equilibrium, the row player is mixing T, B with probabilities $\frac{7}{11}, \frac{4}{11}$, respectively. The probability mixture of the column player can now be found employing the procedure similar to the one illustrated in the section on mixed-strategy games. Consider the column player mixing M, R with probabilities $q, 1 - q$, respectively. Equating the payoffs of the row player to T, B, we have

$$\begin{aligned} -2q + 2(1-q) &= 4q - 3(1-q) \\ q &= \frac{5}{11}. \end{aligned} \tag{2.15}$$

Hence, the column player is mixing M, R with probabilities $\frac{5}{11}, \frac{6}{11}$, respectively. Therefore, the mixed-strategy Nash equilibrium for the zero-sum game above is given as

$$\text{NE} = \Big(\underbrace{\left(\frac{7}{11}, \frac{4}{11}\right)}_{\text{Row player}}, \underbrace{\left(0, \frac{5}{11}, \frac{6}{11}\right)}_{\text{Column player}}\Big).$$

2.17 Bayesian Games

In the games we have considered so far, a critical assumption is that each player knows the other player's preferences. However, the above assumption might not hold true in several scenarios. For instance, in a market game involving several firms, the firms may not know the cost to other users. In an auction, the bidders might not know the valuation of others. Such scenarios can be effectively modeled using the framework of a Bayesian game, which we will further elaborate in this section. Let us consider again the BoS game introduced in Section 2.6.2 with a slight modification. The boy and girl are deciding between watching a game (G) or movie (M). However, in this scenario, the girl might be in one of two different moods, either interested (I) or uninterested (U), while the boy is always interested (I). The payoff tables corresponding to each mood or *type* of the girl are shown in Figure 2.17. The exact mood of the girl is unknown to the boy. The probability that the girl is interested or uninterested is $P_G(I) = \frac{1}{2}$, $P_G(U) = \frac{1}{2}$. Since there is only one type for the boy, we indicate this default type by I.

In such a situation, one has to consider an equilibrium action for a girl of each type. For instance, if the girl of each type chooses G, and the boy also chooses G, the payoff to the boy is

$$\begin{aligned} u_{b(I)}(G,(G,G)) &= P_G(I)\, u_{b(I)}(G,G,I) + P_G(U)\, u_{b(I)}(G,G,U) \\ &= \frac{1}{2} \times 10 + \frac{1}{2} \times 10 \\ &= 10, \end{aligned}$$

where the notation $u_{b(I)}(G,(G,G))$ denotes the payoff of the boy of type I for his action G and the strategy (G,G) for the girl of interested and uninterested types. Further, $u_{b(I)}(G,G,I)$, which employs the notation $u_{i(\theta_i)}(a_i, a_{-i}, \theta_{-i})$, denotes the payoff to the boy in the interested state of the girl, with both the boy and her choosing G. However, if the boy chooses M, while the girl of both types chooses (G,G), the payoff to the boy is

$$\begin{aligned} u_{b(I)}(M,(G,G)) &= P_G(I)\, u_{b(I)}(M,G,I) + P_G(U)\, u_{b(I)}(M,G,U) \\ &= \frac{1}{2} \times 0 + \frac{1}{2} \times 0 \\ &= 0. \end{aligned}$$

Similarly, the average payoffs for different actions of the boy for the strategy (G,M) employed by the girl, that is, one where the girl of interested type chooses G and uninterested type chooses M is

Girl = I, $P_G(I) = 1/2$

Boy \ Girl	Game	Movie
Game	(10, 5)	(0, 0)
Movie	(0, 0)	(5, 10)

Girl = U, $P_G(U) = 1/2$

Boy \ Girl	Game	Movie
Game	(10, 0)	(0, 10)
Movie	(0, 5)	(5, 0)

FIGURE 2.17
Game tables for Bayesian BoS with girl of type I or U.

given as

$$\begin{aligned}
u_{b(I)}(G,(G,M)) &= P_G(I)\,u_{b(I)}(G,G,I) + P_G(U)\,u_{b(I)}(G,M,U) \\
&= \frac{1}{2}\times 10 + \frac{1}{2}\times 0 \\
&= 5, \\
u_{b(I)}(M,(G,M)) &= P_G(I)\,u_{b(I)}(M,G,I) + P_G(U)\,u_{b(I)}(M,M,U) \\
&= \frac{1}{2}\times 0 + \frac{1}{2}\times 5 \\
&= \frac{5}{2}.
\end{aligned}$$

Following the procedure above, one can also compute the payoffs of the boy for the other strategy profiles, that is, (M,G), (M,M) that can be employed by the girl. These are summarized in Table 2.1. The Nash equilibrium for the above Bayesian game can now be deduced as follows. Consider the outcome $(G,(G,M))$, where the boy chooses action G and the girl chooses strategy profile (G,M). From Table 2.1, it can be seen that G is the best response of the boy to the strategy profile (G,M) employed by the girl, given his belief of $P(I) = P(U) = \frac{1}{2}$, since deviating to M decreases his payoff from 5 to $\frac{5}{2}$. Further, it can also be seen that the best response of girl of type I to G of boy is G, while the best response of girl to type U is M. Thus, the strategy profile (G,M) represents the best response of girl of each type to action G of the boy. Thus, G and (G,M) are the best responses of the boy and girl, respectively, to the strategies of each other and hence constitutes the Nash equilibrium of this Bayesian game. Therefore, the Nash equilibrium of the above game is given by

$$\text{NE} = \left(\underbrace{G}_{\text{Boy}}, \underbrace{\left(\underbrace{G}_{\text{Type } I}, \underbrace{M}_{\text{Type } U} \right)}_{\text{Girl}} \right).$$

Thus, for the boy, the best response has to be over the strategy pair of the girl, with respect to his belief of the different types. For the girl, her strategy for each of her types should be the best response to the strategy of the boy. The unique aspect of the above Bayesian game can be summarized as follows. Given strategy (G,M) and belief $P(I) = P(U) = \frac{1}{2}$, action G is the best response of the boy. Further, given G of the boy, G is the best response of girl of type I while M is the best response of girl of type U. This idea can be further reinforced through the following observation. For instance,

TABLE 2.1

Payoff to Boy for Various Strategy Profiles of Girl

	(G,G)	(G,M)	(M,G)	(M,M)
G	10	5	5	0
M	0	$\frac{5}{2}$	$\frac{5}{2}$	5

Note: For instance, in the strategy (G,M), the girl of interested and uninterested type chooses G and M, respectively.

$(G,(M,G))$ is not a Nash equilibrium of the above game, since even though G is the best response of the boy to strategy (M,G) of the girl, the strategy M is not the best response of the girl of type I to action G of the boy.

2.18 Another Example: Yield versus Fight

In this section, we present another example of a Bayesian in a two-person confrontation. Each person P_1, P_2 has two choices, either to fight (F) or to yield (Y). However, person two can be either of type strong (S) with probability $P_2(S) = \frac{1}{4}$ or type weak (W), also with probability $P_2(W) = \frac{3}{4}$. Player P_1 is of default type normal (N). The game tables for this Bayesian game are shown in Figure 2.18. The payoffs of P_1 for the different strategy profiles employed by P_2 are given in Table 2.2. For instance, the payoffs to P_1 for his action Y against strategy (F,Y) employed by P_2 can be computed as follows:

$$\begin{aligned} u_{1(N)}(Y,(F,Y)) &= P_2(S)\,u_{1(N)}(Y,F,S) + P_2(W)\,u_{1(N)}(Y,Y,W) \\ &= \frac{1}{4}\times 0 + \frac{3}{4}\times 0 \\ &= 0. \end{aligned}$$

Similarly, the payoff to P_1 for action F versus strategy (F,Y) of P_2 is given as

$$\begin{aligned} u_{1(N)}(F,(F,Y)) &= P_2(S)\,u_{1(N)}(F,F,S) + P_2(W)\,u_{1(N)}(F,Y,W) \\ &= \frac{1}{4}\times(-4) + \frac{3}{4}\times(4) \\ &= 2. \end{aligned}$$

P_2 = Strong

P_1 \ P_2	Yield	Fight
Yield	(0, 0)	(0, 4)
Fight	(4, 0)	(–4, 4)

P_2 = Weak

Boy \ Girl	Yield	Fight
Yield	(0, 0)	(0, 4)
Fight	(4, 0)	(4, –4)

FIGURE 2.18
Game tables for Bayesian yield versus fight.

TABLE 2.2
Payoff to P_1 for Various Strategy Profiles of P_2 in the Two-Person Confrontation Bayesian Game

	(F,Y)	(F,F)	(Y,F)	(Y,Y)
Y	0	0	–	–
F	2	2	–	–

Similarly, one can compute the payoffs to P_1 for other strategy profiles and these are listed in Table 2.2. At this point, one can make an interesting observation. Observe that the action Y of P_2 of type strong is not a best response to either of the actions Y, F of P_1. Hence, in any Nash equilibrium, P_2 of type S will never choose Y since it is dominated by F. Thus, the strategy profiles $(Y, F), (Y, Y)$ are not employed in any Nash equilibrium and can be safely omitted. Hence, these payoffs are not computed and thereby simply omitted in the table. It can be readily seen that the outcome $(F, (F, Y))$ is a Nash equilibrium of the above game. This can be clearly explained as follows. Action F is the best response of P_1 to the strategy (F, Y) employed by P_2, given his belief of $P(S) = \frac{1}{4}, P(W) = \frac{3}{4}$, since F yields him a payoff of 2 while Y yields him a payoff of 0. Further, F is the best response of P_2 of type S while Y is the best response of P_2 of type W to the action F of P_1. Thus, F, (F, Y) are best responses to each other and constitute the Nash equilibrium.

2.19 Cournot Duopoly with Cost Uncertainty

In this section, we consider the Bayesian version of the Cournot game introduced in Section 2.9, with cost uncertainty. Let the price per unit be given as

$$p(s_1, s_2) = \alpha - (s_1 + s_2),$$

where β from Section 2.9 has been set equal to 1. Further, let $\delta = 0$, and the cost per unit be equal to γ for firm F_1. Firm F_2 can be of one of two types: one with a high (H) cost of γ and another with a low (L) cost equal to $\frac{1}{2}\gamma$. Let the probability of each type be given as $P(H) = P(L) = \frac{1}{2}$. Let firm F_1 produce quantity s_1, while firm F_2 of types H and L produces quantities s_2^H, s_2^L, respectively. Payoff to firm F_2 of type L is known as

$$u_2^L\left(s_2^L, s_1\right) = \left(\alpha - \left(s_2^L + s_1\right)\right)s_2^L - \frac{1}{2}\gamma s_2^L,$$

which can be differentiated and set equal to 0 to yield a best response quantity

$$\left(s_2^L\right)^\star = \frac{1}{2}\left(\alpha - \frac{1}{2}\gamma - s_1\right). \tag{2.16}$$

Similarly, the payoff and best response of the firm F_2 of type H are given as

$$\begin{aligned} u_2^H\left(s_2^H, s_1\right) &= \left(\alpha - \left(s_2^H + s_1\right)\right)s_2^H - \gamma s_2^H, \\ \left(s_2^H\right)^\star &= \frac{1}{2}(\alpha - \gamma - s_1). \end{aligned} \tag{2.17}$$

Finally, the payoff to firm F_1 averaged with respect to the probabilities of firm F_2 of types L and H is given as

$$\begin{aligned} u_1\left(s_1, \left(s_2^L, s_2^H\right)\right) &= \frac{1}{2}\left(\left(\alpha - \left(s_1 + s_2^L\right)\right)s_1 - \gamma s_1\right) + \frac{1}{2}\left(\left(\alpha - \left(s_1 + s_2^H\right)\right)s_1 - \gamma s_1\right) \\ &= ((\alpha - s_1)s_1 - \gamma s_1) - \frac{1}{2}\left(s_2^L + s_2^H\right)s_1. \end{aligned}$$

The best response can once again be found by differentiating the above cost function and setting equal to zero, and is given as

$$\alpha - \gamma - 2s_1 - \frac{1}{2}\left(s_2^L + s_2^H\right) = 0,$$

$$s_1^\star = \frac{1}{2}\left(\alpha - \gamma - \frac{1}{2}\left(s_2^L + s_2^H\right)\right).$$

Finally, the Nash equilibrium is given as the intersection of best responses from Equations 2.16 and 2.17 and the above equation as

$$s_1^\star = \frac{1}{2}(\alpha - \gamma) - \frac{1}{4}\left(\frac{1}{2}\left(\alpha - \frac{1}{2}\gamma - s_1^\star\right) + \frac{1}{2}\left(\alpha - \gamma - s_1^\star\right)\right)$$

$$s_1^\star = \frac{1}{2}(\alpha - \gamma) - \frac{1}{8}\left(\alpha - \frac{1}{2}\gamma\right) - \frac{1}{8}(\alpha - \gamma) + \frac{1}{4}s_1^\star$$

$$\frac{3}{4}s_1^\star = \frac{1}{4}\alpha - \frac{5}{16}\gamma$$

$$s_1^\star = \frac{1}{3}\left(\alpha - \frac{5}{4}\gamma\right).$$

Finally, from Equations 2.16 and 2.17, the Nash equilibrium quantities $\left(s_2^L\right)^\star, \left(s_2^H\right)^\star$ for firm F_2 of types L and H, respectively, are given as

$$\left(s_2^L\right)^\star = \frac{1}{2}\left(\left(\alpha - \frac{1}{2}\gamma\right) - \frac{1}{3}\left(\alpha - \frac{5}{4}\gamma\right)\right) = \frac{1}{3}\alpha - \frac{1}{24}\gamma,$$

$$\left(s_2^H\right)^\star = \frac{1}{3}\alpha - \frac{7}{24}\gamma.$$

2.20 Bayesian Game with Multiplayer Uncertainty

Consider now a generalization of the BoS game shown in Section 2.17 with multiplayer uncertainty. Figure 2.19 shows the payoff tables for the BoS game, where both the boy and girl can be of the interested and uninterested types. The probabilities that the boy is interested or uninterested are given as $P_b(I) = \frac{2}{3}$, $P_b(U) = \frac{1}{3}$, respectively, while the corresponding probabilities for the girl are $P_g(I) = P_g(U) = \frac{1}{2}$, similar to the earlier example. It can be seen now that there are four game tables corresponding to each of the type combinations $(I, I), (I, U), (U, I), (U, U)$ for the boy and the girl.

Tables 2.3 through 2.6 give the payoff for each type of each player for various strategy profiles adopted by the other player. In the earlier example, the computation of the payoffs for type I of the boy for the different strategy profiles of the girl of I, U types has been elaborated in great detail. Below, we simply choose one case to illustrate the corresponding computation for type I of the girl for various strategy profiles of the boy. For instance, consider the payoff $u_{g(I)}(G, (G, M))$, that is, the payoff to girl of type I for her action G, versus the profile (G, M) for the boy, where G, M are

Boy = I, Girl = I

Boy \ Girl	Game	Movie
Game	(10, 5)	(0, 0)
Movie	(0, 0)	(5, 10)

Boy = I, Girl = U

Boy \ Girl	Game	Movie
Game	(10, 0)	(0, 10)
Movie	(0, 5)	(5, 0)

Boy = U, Girl = I

Boy \ Girl	Game	Movie
Game	(0, 5)	(10, 0)
Movie	(5, 0)	(0, 10)

Boy = U, Girl = U

Boy \ Girl	Game	Movie
Game	(0, 0)	(10, 10)
Movie	(5, 5)	(0, 0)

FIGURE 2.19
Game tables for Bayesian BoS with both boy and girl of types I or U.

TABLE 2.3
Payoff to Boy of Type I for Various Strategy Profiles of Girl

	(G, G)	(G, M)	(M, G)	(M, M)
G	10	5	5	0
M	0	$\frac{5}{2}$	$\frac{5}{2}$	5

TABLE 2.4
Payoff to Boy of Type U for Various Strategy Profiles of Girl

	(G, G)	(G, M)	(M, G)	(M, M)
G	0	5	5	10
M	5	$\frac{5}{2}$	$\frac{5}{2}$	0

TABLE 2.5
Payoff to Girl of Type I for Various Strategy Profiles of Boy

	(G, G)	(G, M)	(M, G)	(M, M)
G	5	$\frac{10}{3}$	$\frac{5}{3}$	0
M	0	$\frac{10}{3}$	$\frac{20}{3}$	10

TABLE 2.6

Payoff to Girl of Type U for Various Strategy Profiles of Boy

	(G,G)	(G,M)	(M,G)	(M,M)
G	0	$\frac{5}{3}$	$\frac{10}{3}$	5
M	10	$\frac{20}{3}$	$\frac{10}{3}$	0

the actions of interested and uninterested types, respectively. This is given as

$$\begin{aligned} u_{g(I)}(G,(G,M)) &= P_b(I)\,u_{g(I)}(G,G,I) + P_b(U)\,u_{g(I)}(G,M,U) \\ &= \frac{2}{3}\times 5 + \frac{1}{3}\times 0 \\ &= \frac{10}{3}. \end{aligned}$$

The rest of the entries in the tables can be computed in a fashion similar to the one illustrated above. It can be seen that the strategy profile $((G,G),(G,M))$, with (G,G), (G,M) denoting the actions of the (I,U) types of the boy and girl, respectively, constitutes a Nash equilibrium for this game. This can be seen as follows. Given profile (G,M) of the girl, action G is the best response for a boy of types I,U. Thus, (G,G) is a best response for the boy to strategy (G,M) of the girl. Also, given strategy (G,G) of the boy, action G is the best response of a girl of type I, while M is the best response for a girl of type U, thus leading to a best response strategy of (G,M) for the girl. Thus, the strategies (G,G) and (G,M) are mutual best responses for the boy and the girl, respectively, indicating that the profile $((G,G),(G,M))$ is indeed the Nash equilibrium. Similarly, it can be easily verified that another Nash equilibrium of the game is given by the profile $((M,G),(M,M))$.

2.21 Bayesian Nash Equilibrium: Formal Definition

Based on the above examples of Bayesian games, the concept of a Bayesian Nash equilibrium can be formally defined as follows. Let $\theta_i \in \Theta_i$ denote the type θ_i of player i, belonging to the set Θ_i, consisting of all possible types for player i. Naturally, in this scenario, a Nash equilibrium profile has to specify one action for each player of each type. The action profile $a_i(\theta_i)$ is a Nash equilibrium if

$$a_i(\theta_i) = \arg\max_{a_i \in \mathcal{A}_i} \sum_{\theta_{-i}\in\Theta_{-i}} P_{-i}(\theta_{-i})\,u_{i(\theta_i)}(a_i, a_{-i}, \theta_{-i}).$$

This basically implies the action $a_i(\theta_i)$ for each player i of each type θ_i is indeed the best response of player i of type θ_i, averaged with respect to his belief $p_{-i}(\theta_{-i})$ for the rest of the players of each of their types, with their equilibrium strategies fixed. The above definition is significantly complex and abstract in nature. Hence, to clarify it further for the first-time reader, we illustrate it in the context of the BoS game with multiplayer uncertainty described above. In this case, the action sets $\mathcal{A}_1 = \mathcal{A}_2 = (G,M)$, while the type sets $\Theta_1 = \Theta_2 = (I,U)$. The probabilities $p_1(I) = \frac{2}{3}$, $p_1(U) = \frac{1}{3}$ and $p_2(I) = p_2(U) = \frac{1}{2}$. To verify that the profile $((G,G),(G,M))$ is a Nash equilibrium, let us begin by considering the equilibrium action G for the boy of type I, that is, $a_1(I) = G$. For this to be the best response for type I of the boy, it must be true that

$$\begin{aligned} u_{1(I)}(G,(G,M)) &= P_2(I)\,u_{1(I)}(G,G,I) + P_2(U)\,u_{1(I)}(G,M,U) \\ &\geq u_{1(I)}(M,(G,M)) \\ &= P_2(I)\,u_{1(I)}(M,G,I) + P_2(U)\,u_{1(I)}(M,M,U), \end{aligned}$$

which can indeed be seen to be true since $u_{1(I)}(G,(G,M)) = 5 \geq \frac{5}{2} = u_{1(I)}(M,(G,M))$. At equilibrium, this must hold for the actions of all players of all types, that is, $a_1(I) = G$, $a_1(U) = G$, $a_2(I) = G$, and $a_2(U) = M$.

2.22 Mixed Bayesian Nash Equilibrium

Consider again the Bayesian game in Figure 2.17, where the girl can be of either the interested (I) or uninterested (U) type, while the boy is of the default type I. In this section, we illustrate a mixed-strategy Nash equilibrium for this Bayesian game, similar to the mixed-strategy equilibrium for strategic games developed earlier. Consider the boy, that is, player 1, mixing actions G and M with probabilities p, $1-p$, respectively. In this scenario, the girl of type I mixes only if her payoffs from G and M are equal, that is,

$$u_{g(I)}(G,(p,1-p),I) = u_{g(I)}(M,(p,1-p),I). \tag{2.18}$$

These payoffs can be computed for the girl of type I as

$$\begin{aligned} u_{g(I)}(G,(p,1-p),I) &= p \times u_{g(I)}(G,G,I) + (1-p) \times u_{g(I)}(G,M,I) \\ &= p \times 5 + (1-p) \times 0 \\ &= 5p, \\ u_{g(I)}(M,(p,1-p),I) &= p \times u_{g(I)}(M,G,I) + (1-p) \times u_{g(I)}(M,M,I) \\ &= p \times 0 + (1-p) \times 10 \\ &= 10(1-p). \end{aligned}$$

Hence, from Equation 2.18, the girl of type I employs a mixed strategy only if

$$\begin{aligned} 5p &= 10(1-p) \\ \Rightarrow p &= \frac{2}{3}. \end{aligned}$$

Further, observe that for $p = \frac{2}{3}, (1-p) = \frac{1}{3}$, the payoffs of the girl of type U are given as

$$\begin{aligned} u_{g(U)}\left(G, \left(\frac{2}{3}, \frac{1}{3}\right), I\right) &= \frac{2}{3} \times u_{g(U)}(G, G, I) + \frac{1}{3} \times u_{g(U)}(G, M, I) \\ &= \frac{2}{3} \times 0 + \frac{1}{3} \times 5 \\ &= \frac{5}{3}, \\ u_{g(U)}\left(M, \left(\frac{2}{3}, \frac{1}{3}\right), I\right) &= \frac{2}{3} \times u_{g(U)}(M, G, I) + \frac{1}{3} \times u_{g(U)}(M, M, I) \\ &= \frac{2}{3} \times 10 + \frac{1}{3} \times 0 \\ &= \frac{20}{3}. \end{aligned}$$

Hence, it is clear that M is the best response of type U of the girl to the mixed strategy $\left(\frac{2}{3}, \frac{1}{3}\right)$ of the boy. One can now compute the mixing probabilities $(q, 1-q)$ of the girl of type I by inspecting the payoffs of the boy for actions G, M. His payoff to the pure strategy G is

$$\begin{aligned} & u_{b(I)}(G, ((q, 1-q), (0, 1))) \\ &= P_g(I)\, u_{b(I)}(G, (q, 1-q), I) + P_g(U)\, u_{b(I)}(G, (0, 1), U) \\ &= \frac{1}{2}(q \times 10 + (1-q) \times 0) + \frac{1}{2} \times 0 \\ &= 5q. \end{aligned}$$

Similarly, his payoff to the pure strategy M can be computed as

$$\begin{aligned} &= u_{b(I)}(M, ((q, 1-q), (0, 1))) \\ &= P_g(I)\, u_{b(I)}(M, (q, 1-q), I) + P_g(U)\, u_{b(I)}(M, (0, 1), U) \\ &= \frac{1}{2}(q \times 0 + (1-q) \times 5) + \frac{1}{2} \times 5 \\ &= 5 - \frac{5}{2}q. \end{aligned}$$

Finally, since the boy is mixing, it must be the case that his payoff from the pure strategy choices G, M are equal, that is,

$$\begin{aligned} 5q &= 5 - \frac{5}{2}q \\ q &= \frac{2}{3}. \end{aligned}$$

Hence, the mixed-strategy Nash equilibrium for the Bayesian game above is given as

$$\text{NE} = \left(\underbrace{\left(\frac{2}{3}, \frac{1}{3}\right)}_{\text{Boy}}, \underbrace{\left(\underbrace{\left(\frac{2}{3}, \frac{1}{3}\right)}_{\text{Type } I}, \underbrace{(0, 1)}_{\text{Type } U} \right)}_{\text{Girl}} \right).$$

Thus, it can be readily seen from the above equilibrium outcome that the Nash strategy profile for mixed-strategy Bayesian game specifies a *mixed* strategy for each player of each type. To further elaborate on this, at the Nash equilibrium above, the boy is mixing G, M with $p = \frac{2}{3}, 1 - p = \frac{1}{3}$, respectively, while the girl of type I is mixing G, M with probabilities $\frac{2}{3}, \frac{1}{3}$ and the girl of type U is always choosing the pure strategy M. Interestingly, another mixed-strategy Nash equilibrium for the above game can be derived as follows. Consider instead the girl of type U is mixing G, M with probabilities $q, 1 - q$, respectively, while the boy is mixing with $p, 1 - p$. For this to hold, her payoffs from the pure strategies G, M must be equal, that is,

$$\begin{aligned} u_{g(U)}\left(G, (p, 1-p), I\right) &= u_{g(U)}\left(M, (p, 1-p), I\right) \\ p \times 0 + (1-p) \times 5 &= p \times 10 + (1-p) \times 0 \\ 5(1-p) &= 10p \\ p &= \frac{1}{3}. \end{aligned}$$

For the above value of $p = \frac{1}{3}$, the payoffs to the girl of type I for the different strategies G, M are $u_{g(I)}\left(G, (p, 1-p)\right) = \frac{5}{3}, u_{g(I)}\left(M, (p, 1-p)\right) = \frac{20}{3}$. Hence, the best response of the girl of type I is to choose M. Finally, to find the mixing probabilities $(q, 1-q)$ for the girl of type U, we compute the payoffs of the boy for each for the pure strategies G, M, and then impose the condition that they must be equal at equilibrium. Therefore, we have

$$\begin{aligned} u_{b(I)}\left(G, ((0,1), (q, 1-q))\right) &= u_{b(I)}\left(M, ((0,1), (q, 1-q))\right) \\ \Rightarrow \frac{1}{2} \times 0 + \frac{1}{2} \times 10 \times q &= \frac{1}{2} \times 5 + \frac{1}{2} \times 5 \times (1-q) \\ \Rightarrow 5q &= \frac{5}{2} + \frac{5}{2}(1-q) \\ \Rightarrow q &= \frac{2}{3}. \end{aligned}$$

Hence, yet another mixed-strategy Nash equilibrium for the Bayesian game above is given as

$$\text{NE} = \left(\underbrace{\left(\frac{1}{3}, \frac{2}{3}\right)}_{\text{Boy}}, \underbrace{\left(\underbrace{(0, 1)}_{\text{Type } I}, \underbrace{\left(\frac{2}{3}, \frac{1}{3}\right)}_{\text{Type } U} \right)}_{\text{Girl}} \right).$$

Below, we conclude this section on Bayesian games with the formal definition for mixed-strategy equilibrium in such scenarios.

2.23 Bayesian Mixed-Strategy Nash Equilibrium: Formal Definition

Consider a Bayesian where $\theta_i \in \Theta_i$ denotes the type of player i and $a_i \in \mathcal{A}_i$ denotes his action. Let $\mathcal{N}$ denote the set of players. Let $\sigma_i \in \Sigma_i$ denote the set of probability distributions over the action set $\mathcal{A}_i$. The mixed-strategy profile $\sigma_i(\theta_i) \in \Sigma_i$ is a Nash equilibrium for the Bayesian game above if $\forall \theta_i \in \Theta_i, \forall i$,

$$\sigma_i(\theta_i) = \arg\max_{\tilde{\sigma} \in \Sigma_i} \sum_{\theta_{-i} \in \Theta_{-i}} p_{-i}(\theta_{-i}) \sum_{\mathbf{a} \in \mathcal{A}} \prod_{j \in \mathcal{N} \setminus \{i\}} \sigma_j\left(a_j, \theta_j\right) \tilde{\sigma}(a_i)\, u_{i(\theta_i)}(a_i, a_{-i}, \theta_{-i}).$$

2.24 Introduction to Fundamentals of Probability Distributions

In this section, we introduce the reader to basic concepts in probability distributions, which form the basis for concepts in Bayesian auctions described in future sections. Let X denote a random variable that takes values randomly over a continuum of values, also termed as an *interval*. In this context, one can define the *probability density function* $f_X(x)$ for the random variable X, from which one can compute the probability that X lies in an infinitesimal interval $[x, x + \Delta x]$ of width Δx as

$$\mathrm{P}(x < X \leq x + \Delta x) = f_X(x)\Delta x.$$

The probability that the random variable X assumes a value in the interval $[a, b]$ is given as

$$\mathrm{P}(a < X \leq b) = \int_a^b f_X(x)dx.$$

Naturally, since $-\infty < X < \infty$, we must have

$$\mathrm{P}(-\infty < X \leq \infty) = \int_{-\infty}^{\infty} f_X(x)dx = 1.$$

Another related quantity, the *cumulative distribution function* (CDF) of the random variable X, denoted by $F_X(x)$, can be defined as

$$F_X(x) = P(X \leq x) = \int_{-\infty}^{x} f_X(x)\, dx.$$

Consider, for example, the probability distribution shown in Figure 2.20 such that $f_X(x) = 1$ if $0 \leq x \leq 1$ and 0 otherwise. It is easy to verify that this is a valid probability density function, since

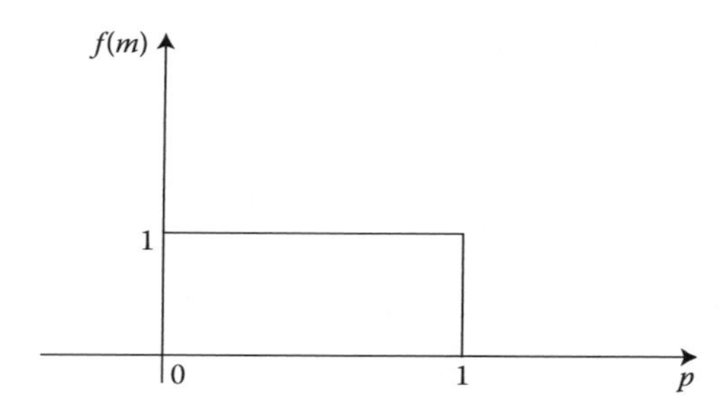

FIGURE 2.20
Uniform distribution.

$f_X(x) \geq 0, \forall x$, and

$$\int_{-\infty}^{\infty} f_X(x)\, dx = \int_{0}^{1} 1 dx = 1.$$

Also, to elucidate this concept further, the probability that the above random variable takes a value in the interval $\left[\frac{1}{3}, \frac{1}{2}\right]$ is given as

$$\mathrm{P}\left(\frac{1}{3} \leq X \leq \frac{1}{2}\right) = \int_{\frac{1}{3}}^{\frac{1}{2}} 1 dx = \frac{1}{2} - \frac{1}{3} = \frac{1}{6}.$$

The CDF for the above density function can be readily computed as follows. For $0 \leq x \leq 1$,

$$F_X(x) = \mathrm{P}(X \leq x) = \int_{-\infty}^{x} f_X(x)\, dx = \int_{0}^{x} 1 dx = x.$$

Therefore, the CDF is given as

$$F_X(x) = \begin{cases} 0 & x \leq 0 \\ x & 0 \leq x \leq 1 \\ 1 & x \geq 1 \end{cases}$$

2.25 Bayesian Auction

Consider an N-player first-price auction with valuations $V_1, V_2, \ldots, V_N$ for the N players. Without loss of generality, it can be assumed that

$$V_1 > V_2 > \cdots > V_N.$$

If the valuations are known, it can be shown that the Nash equilibrium for this is $(V_2, V_2, V_3, \ldots, V_N)$. How can one model the scenario when the valuations are private and unknown? To characterize the Nash equilibrium in such scenarios is precisely the intention of this discussion. In the absence of exact knowledge of the valuations, they can assumed to be random in nature. For instance, consider each valuation to be *uniformly* distributed in the interval [0, 1], which can be denoted as $V_i \sim \mathcal{U}[0, 1]$. Profit or payoff to player i is

$$\text{Payoff of } i\text{th player} = \begin{cases} v_i - b_i & \text{if } b_i \text{ is highest} \\ 0 & \text{otherwise} \end{cases}$$

Let $\mathrm{P_{win}}(b_i)$ denote the probability of winning for a bid b_i of the ith user. The expected revenue as a function of the bid b_i is given as

$$\begin{aligned} \text{Expected revenue}(b_i) &= \mathrm{P_{win}}(b_i) \times \text{revenue}(b_i) + \mathrm{P_{lose}}(b_i) \times 0 \\ &= \mathrm{P_{win}}(b_i)(v_i - b_i). \end{aligned} \tag{2.19}$$

Theorem 2.5

The Nash equilibrium bid $b_i^\star$ of the ith player for this Bayesian first-price auction is given as

$$b_i^\star = \frac{N-1}{N} V_i.$$

Proof. This can be demonstrated as follows. Let each player $j \neq i$ bid $b_j^\star = ((N-1)/N)V_j$. We then demonstrate that the bid $b_i^\star = ((N-1)/N)V_i$ is the best response for the ith player. To prove this, consider any bid b_i for the ith player. As can be seen from Equation 2.19, to compute the payoff to the ith player for bid b_i, we need to deduce the corresponding probability of winning $\mathrm{P_{win}}(b_i)$. This can be derived as follows. The ith player wins the auction if $b_j^\star < b_i$ for all $j \neq i$. This in turn implies that

$$\begin{aligned} &\frac{N-1}{N} V_j < b_i \\ &\Rightarrow V_j < \frac{N}{N-1} b_i. \end{aligned}$$

Given that V_j is uniformly distributed in [0, 1], as computed in the example earlier in this section, the probability that $V_j < (N/(N-1))b_i$ is given as $\mathrm{P}\left(V_j < (N/(N-1))b_i\right) = (N/(N-1))b_i$. Further, for b_i to be the winning bid, this has to hold for each $j \neq i$. Hence, the net probability of winning $\mathrm{P_{win}}(b_i)$ can be computed as

$$\begin{aligned} \mathrm{P_{win}}(b_i) &= \mathrm{P}\left(V_1, V_2, \ldots, V_{i-1}, V_{i+1}, \ldots, V_N < \frac{N}{N-1} b_i\right) \\ &\overset{(a)}{=} \prod_{j=1, j\neq i}^{N} \mathrm{P}\left(V_j < \frac{N}{N-1} b_i\right) \\ &= \left(\frac{N}{N-1} b_i\right)^{N-1}, \end{aligned}$$

where the equality (a) follows above from the fact that the valuations V_j are *independent* random variables, and therefore the joint probability is given by the product of the individual probabilities. Hence, the payoff to player i for bid b_i is given as

$$\begin{aligned} u_i\left(b_i, b_{-i}^{\star}\right) &= \mathrm{P_{win}}\left(b_i\right)\left(V_i - b_i\right) \\ &= \left(\frac{N-1}{N}\right)^{N-1} \underbrace{b_i^{N-1}\left(V_i - b_i\right)}_{g(b_i)}. \end{aligned}$$

Since the factor $((N-1)/N)^{N-1}$ is a constant, the above payoff is maximum at the value $b_i^{\star}$ for which $g\left(b_i\right)$ is maximized. This can be obtained by differentiating $g\left(b_i\right)$ and setting finding the zeros as

$$\begin{aligned} 0 &= \left.\frac{d}{db_i} g\left(b_i\right)\right|_{b_i^{\star}} \\ &= (N-1)\left(b_i^{\star}\right)^{N-2}\left(V_i - b_i^{\star}\right) - \left(b_i^{\star}\right)^{N-1} \\ \Rightarrow b_i^{\star} &= (N-1)\left(V_i - b_i^{\star}\right) \\ \Rightarrow b_i^{\star} &= \frac{N-1}{N} V_i. \end{aligned}$$

Thus, $b_i^{\star} = ((N-1)/N)V_i$ is indeed the best response of player i, when each of the players $j \neq i$ is bidding $b_j^{\star} = (N/(N-1))V_j$. Hence, it constitutes the Nash equilibrium for the above Bayesian first-price auction in which the valuations of the players are independent random variables and uniformly distributed in [0, 1]. ■

2.26 Second-Price Bayesian Auction

Consider now a second-price auction with valuations $V_1, V_2, \ldots, V_N$ for the N users, with each valuation V_i distributed uniformly randomly. The result below gives the Nash equilibrium for this second-price auction.

Theorem 2.6

The Nash equilibrium of the second-price auction Bayesian game above is given by

$$\left(V_1, V_2, \ldots, V_N\right),$$

that is, each player i bids $b_i^{\star} = V_i$.

Proof. This can be seen as follows. Let each of the other players $j \neq i$ bid $b_j = V_j$. If the valuation V_i of the ith player is not the highest, he can increase his bid to the highest valuation to win the auction. However, this gives him a negative payoff and hence there is no incentive to deviate in this scenario. If his valuation V_i is indeed the highest, bidding $b_i = V_i$ will win the auction for him and he pays the second highest valuation. Increasing his bid has no effect, while decreasing his bid might lose the auction for him. Hence, there is no incentive to deviate in this case either. Therefore, the best response of the ith player is to bid $b_i^\star = V_i$. The above is indeed the Nash equilibrium of the game. Further, the attentive reader will realize that nowhere in this deduction process have we used the fact that the valuations are distributed uniformly randomly. Thus, this result holds for any arbitrary distribution of the individual valuations. ■

2.27 Revenue in First-Price Auction

In this section, we will compute the expected revenue for the first-price Bayesian auction described above at equilibrium.

Theorem 2.7

The expected revenue in a Bayesian first-price auction with N players and valuations distributed uniformly randomly in $[0, 1]$ is $((N-1)/(N+1))$.

Proof. It can be readily seen that the revenue of the first-price auction is given as

$$\text{Revenue} = \text{E}\left\{b^{\max}\right\},$$

where $b^{\max} = \max\{b_1, b_2, \ldots, b_N\}$ is the maximum among the bids of the different players. As has been demonstrated above for the first-price Bayesian auction with uniform distribution, the equilibrium bid $b_i^\star$ of the ith player is

$$b_i^\star = \frac{N-1}{N} V_i.$$

Therefore, the winning bid $b^{\max} = ((N-1)/N)V^{\max}$, where $V^{\max} = \max\{V_1, V_2, \ldots, V_N\}$ is the maximum among the valuations. The cumulative distribution $F_{V^{\max}}(v)$ is given as

$$\begin{aligned} F_{V^{\max}}(v) &= \text{P}\left(V^{\max} < v\right) \\ &\overset{(a)}{=} \text{P}\left(V_1, V_2, \ldots, V_N < v\right) \\ &\overset{(b)}{=} \prod_{i=1}^{N} \text{P}\left(V_i < v\right) \\ &= v^N. \end{aligned} \tag{2.20}$$

The equality (a) above follows from the fact that the maximum valuation $V^{\max} < v$ implies that each of the valuations $V_i < v$ and equality (b) follows from the independence of the different valuations, as already described in sections above. Therefore, the probability density function $f_{V^{\max}}(v)$ can be derived as

$$f_{V^{\max}}(v) = \frac{d}{dv} F_{V^{\max}}(v) = N v^{N-1}.$$

The expected revenue of the first-price Bayesian auction can therefore be computed as

$$\begin{aligned}
\text{Expected revenue} &= \mathrm{E}\left\{\frac{N-1}{N}V^{\max}\right\} \\
&= \int_0^1 \frac{N-1}{N} v f_{V^{\max}}(v)\, dv \\
&= \frac{N-1}{N}\int_0^1 vNv^{N-1} dv \\
&= \frac{N-1}{N+1} v^N\Big|_0^1 \\
&= \frac{N-1}{N+1}.
\end{aligned}$$

■

2.28 Expected Revenue in Second-Price Bayesian Auction

In the second-price Bayesian auction, the bid $b_i^\star$ of each player at equilibrium is $b_i^\star = V_i$. The winner is the one with the highest bid and pays an amount equal to that of the second highest bid, which in turn is equal to the second highest valuation at equilibrium. Hence, denoting the second highest bid and valuation by $b_{[2]}$, $V_{[2]}$, respectively, we have

$$\text{Expected revenue} = \mathrm{E}\left\{b_{[2]}\right\} = \mathrm{E}\left\{V_{[2]}\right\}.$$

We now derive the expected revenue for the second-price Bayesian auction.

Theorem 2.8

The expected revenue of the second-price auction is also $(N-1)/(N+1)$.

Proof. Similar to the procedure in the case of the first-price Bayesian auction, we begin by determining the CDF of the second highest valuation. This cumulative distribution $F_{V_{[2]}}(v)$ can be computed as follows. The second highest valuation $V_{[2]} < v$ if either all the valuations are less than v or all the valuations except one are less than v. The probability of the first event is v^N as illustrated previously in Equation 2.20. The probability of the second event is

$$\begin{aligned}
\mathrm{P}(\text{exactly one } V_i \geq v) &= \sum_{i=1}^{N} \mathrm{P}(V_i \geq v; V_1, V_2, \ldots, V_{i-1}, V_{i+1}, \ldots, V_N < v) \\
&= \sum_{i=1}^{N} v^{N-1}(1-v) \\
&= Nv^{N-1}(1-v).
\end{aligned}$$

Therefore, the CDF $F_{V_{[2]}}(v)$ is finally given as

$$\begin{aligned} F_{V_{[2]}}(v) &= v^N + N v^{N-1}(1-v) \\ &= v^{N-1}(v + N(1-v)). \end{aligned}$$

The probability density function of the second highest valuation $f_{V_{[2]}}(v)$ can now be computed as

$$\begin{aligned} f_{V_{[2]}}(v) &= \frac{d}{dv} F_{V_{[2]}}(v) \\ &= (N-1) v^{N-2}(v + N(1-v)) + v^{N-1}(1-N) \\ &= (N-1) v^{N-2}(v + N(1-v) - v) \\ &= N(N-1) v^{N-2}(1-v). \end{aligned}$$

The expected revenue of the second-price Bayesian auction can now be computed as the expected value of the random variable $V_{[2]}$, given as

$$\begin{aligned} \text{Expected revenue} &= \mathrm{E}\left\{V_{[2]}\right\} \\ &= \int_0^1 f_{V_{[2]}}(v)\, v dv \\ &= \int_0^1 N(N-1) v^{N-2}(1-v) \\ &= N(N-1)\left(\frac{1}{N} v^N - \frac{1}{N+1} v^{N+1}\right)\Bigg|_0^1 \\ &= N(N-1)\left(\frac{1}{N} - \frac{1}{N+1}\right) \\ &= N(N-1)\frac{1}{N(N+1)} \\ &= \frac{N-1}{N+1}. \end{aligned}$$

■

Thus, surprisingly, the expected revenue of the second-price auction is also $(N-1)/(N+1)$, equal to that of the first-price auction. However, this is not a coincidence and follows from the *revenue equivalence* principle that states that the revenues from first- and second-price auctions are equal under a certain set of conditions. This is stated below.

Theorem 2.9

Consider an auction in which the bidders' valuations are independent random variables and the item being auctioned is sold to the bidder with the highest valuation, while the expected payoff of the bidder with the lowest valuation is zero. At the Nash equilibrium of such an auction, the expected revenue of each bidder and the auctioneer are fixed for any auction mechanism.

A proof of the revenue equivalence principle is beyond the scope of this discussion and the interested reader is referred to advanced references such as References 2 and 3. Next, we illustrate the example of yet another auction mechanism, the *all-pay* auction.

2.29 Example: All-Pay Auction

Another example of an auction is the "all-pay" auction in which the bidder pays his bid irrespective of winning or losing. This can be thought of as a "lobbying" scenario in which the bid is paid irrespective of the outcome. Consider an N-player all-pay auction scenario in which the valuations $V_i, 1 \leq i \leq N$, are drawn uniformly from the interval [0, 1]. The lemma below gives the Nash equilibrium bidding strategy for this auction.

Lemma 2.1

The Nash equilibrium bid $b_i^\star$ for the all-pay auction is given as

$$b_i^\star = \frac{N-1}{N} V_i^N, \quad 1 \leq i \leq N. \tag{2.21}$$

Proof. To demonstrate this, let us start with the assumption that the ith bidder is bidding $b_i = b$, while all bidders other than the ith bidder are bidding

$$b_j = \frac{N-1}{N} V_j^N, \quad \forall j \neq i. \tag{2.22}$$

The payoff $\pi(b)$ to player i as a function of the bid b is given as

$$\pi(b) = \Pr(\text{win}|\, b)\,(V_i - b) + \Pr(\text{lose}|\, b)\,(-b). \tag{2.23}$$

Observe above that unlike the first- and second-price auctions, the payoff on losing is $-b$ since the bid cannot be recovered in an all-pay auction. Thus, it is akin to a *sunk* cost in a business. The probability of winning of the ith player as a function of his bid b can be computed as follows. The player i wins if his bid is the highest among all the bids. Hence, we have

$$\begin{aligned}
\Pr(\text{win}|\, b) &= \Pr(b_1 \leq b, b_2 \leq b, \ldots, b_{i-1} \leq b, b_{i+1} \leq b, \ldots, b_N \leq b) \\
&= \prod_{\substack{j=1 \\ j\neq i}}^{N} \Pr(b_j \leq b) \\
&= \prod_{\substack{j=1 \\ j\neq i}}^{N} \Pr\left(\frac{N-1}{N} V_j^N \leq b\right) \\
&= \prod_{\substack{j=1 \\ j\neq i}}^{N} \Pr\left(V_j \leq \left(\frac{N}{N-1}\right)^{\frac{1}{N}} b^{\frac{1}{N}}\right).
\end{aligned} \tag{2.24}$$

Substituting the above probability for win for bid b in the payoff equation of player i in Equation 2.23 yields

$$\pi(b) = \left(\frac{N}{N-1}\right)^{\frac{1}{N}} b^{\frac{1}{N}} (V_i - b) + \left(1 - \left(\frac{N}{N-1}\right)^{\frac{1}{N}} b^{\frac{1}{N}}\right)(-b)$$

$$= \left(\frac{N}{N-1}\right)^{\frac{1}{N}} b^{\frac{1}{N}} V_i - b.$$

To find the best response, that is, to maximize the payoff above, differentiate with respect to b and set equal to zero as

$$\frac{\partial}{\partial b}\pi(b) = 0,$$

$$\left(\frac{N}{N-1}\right)^{\frac{1}{N}} \left(\frac{1}{N} - 1\right) b^{\left(\frac{1}{N}-1\right)} V_i - 1 = 0, \tag{2.25}$$

$$b_i^{\star} = V_i^N \frac{N-1}{N}.$$

Thus, the above results clearly shows that if each player other than player i is bidding $b_j = ((N-1)/N)V_j^N$, the best response for player i is to bid $b_i^{\star} = ((N-1)/N)V_i^N$. Hence, it is the Nash equilibrium bidding strategy. ■

Lemma 2.2

The expected revenue of the all-pay auction equals $(N-1)/(N+1)$.

Proof. The expected revenue to the auctioneer can be calculated as follows. Observe that each player pays his bid $((N-1)/N)V_i^N$ irrespective of the outcome in an all-pay auction at Nash equilibrium. Hence, the expected payment of each player is given as

$$\mathrm{E}\left\{\frac{N-1}{N}V_i^N\right\} = \frac{N-1}{N}\int_0^1 V_i^N dV_i$$

$$= \frac{N-1}{N}\frac{1}{N+1}$$

$$= \frac{1}{N}\frac{N-1}{N+1}.$$

Finally, the expected revenue of the auctioneer is given as

$$\text{Expected revenue} = N \times \frac{1}{N}\frac{N-1}{N+1} = \frac{N-1}{N+1},$$

which can be seen equal to that of the first- and second-price auctions and also follows the *revenue equivalence principle* described earlier. ■

2.30 Vickrey–Clarke–Groves Procedure

In this section, we introduce the VCG auction mechanism for competitive resource allocation. This procedure is named after its inventors William Vickrey, Edward H. Clarke, and Theodore Groves. Consider N agents competing for a set of M resources. Let $\theta_1, \theta_2, \ldots, \theta_M$ denote these resources, while $Q_1, Q_2, \ldots, Q_M$ denote the quantities available for the corresponding resources. Let $\mathbf{q}^l$, the allocation vector of the lth player, be defined as

$$\mathbf{q}^l = \begin{bmatrix} q_1^l \\ q_2^l \\ \vdots \\ q_M^l \end{bmatrix},$$

where q_j^l is the quantity of the jth commodity allocated to him. Let $r_i\left(\mathbf{q}^i\right)$ denote the reported payoff function to the ith agent corresponding to the allocation vector $\mathbf{q}^i$. The optimal resource allocation that maximizes the net payoff to all the users can be computed as the solution to the optimization problem

$$\begin{aligned} \max \sum_{i=1}^{N} r_i\left(\mathbf{q}^i\right), \\ \text{s.t.} \quad \sum_{i=1}^{N} \mathbf{q}^i \preceq \mathbf{q}_0 \\ \mathbf{q}^i \succeq 0, 1 \leq i \leq N, \end{aligned} \tag{2.26}$$

where $\mathbf{q}_0 = [Q_1, Q_2, \ldots, Q_M]^T$ denotes the total quantity vector for the M commodities and the symbols $\preceq$, $\succeq$ denote componentwise inequalities, that is, $\mathbf{x} \preceq \mathbf{y}$ implies that every element in $\mathbf{x}$ is less than or equal to the corresponding element in $\mathbf{y}$. The problem with the above allocation process is that the users might distort or misrepresent their utility functions to subvert the resource allocation process and gain unfair advantage. Let the true utility function of each user be $u_i\left(\mathbf{q}^i\right)$, which is not known to the central allocation agent. Each agent reports a utility function $r_i\left(\mathbf{q}^i\right)$. In this context, how should the centralized controller or distribution agent allocate the various commodities, while not be subject to willful manipulation? This is the aim of the VCG mechanism, which is described next.

Initially, the controller solicits the bids or utility functions $r_1(\mathbf{q}), r_2(\mathbf{q}), \ldots, r_N(\mathbf{q})$ of all the N agents. The controller then computes the optimal allocation $\mathbf{q}_1^\star, \mathbf{q}_2^\star, \ldots, \mathbf{q}_N^\star$ as the solution to the optimization problem in Equation 2.26. Let $\mathbf{q} = \left(\mathbf{q}^1, \mathbf{q}^2, \ldots, \mathbf{q}^N\right)$ and the function $P_k(\mathbf{q})$ be defined as

$$P_k(\mathbf{q}) = \sum_{j=1, j \neq k}^{N} r_j\left(\mathbf{q}^i\right),$$

and the factor G_{-k} be defined such that $G_{-k} : r_{-k} \rightarrow R$, that is, it maps the reported utility function of all other agents other than the kth agent to a real number. For instance, one possible definition of

G_{-k} could be as the solution to the optimization problem

$$\max \sum_{i=1, i \neq k}^{N} r_i\left(\mathbf{q}^i\right),$$

$$\text{s.t.} \quad \sum_{i=1, i \neq k}^{N} \mathbf{q}^i \preceq \mathbf{q}_0$$

$$\mathbf{q}^i \succeq 0, 1 \leq i \leq N, i \neq k.$$

The final step in the VCG step, which is the key step in the VCG auction, is to charge the agent k a price given as

$$\text{Price paid by agent } k = G_{-k}() - P_k\left(\mathbf{q}^\star\right).$$

The above price charged to player k can be interpreted as follows. G_{-k} is the net utility to all the players in the *absence* of any allocation to player k. The quantity $P_k(\mathbf{q}^\star)$ is the net allocation to all the players, taking into consideration the reported utility function of agent k. Thus, the quantity $G_{-k}() - P_k(\mathbf{q}^\star)$ can be intuitively seen to be the system cost of allocation of resources to agent k. This is the price he is expected to pay.

2.31 Illustration of the Optimality of the VCG Procedure

In this section, we demonstrate that the VCG procedure indeed encourages the agents to report their true utility functions, by penalizing an agent who misrepresents his payoffs.

Lemma 2.3

The net payoff of the kth player is maximum when the reported utility function $r_k(\cdot)$ equals the true utility function $u_k(\cdot)$.

Proof. This can be readily seen as follows. The true utility of the kth agent corresponding to an allocation $\mathbf{q}^\star$ is given as $u_k\left(\mathbf{q}_k^\star\right)$. Hence, his net payoff after deducting the price paid can be computed as

$$\begin{aligned}
\text{Net payoff of } k\text{th player} &= u_k\left(\mathbf{q}_k^\star\right) - \text{price} \\
&= u_k\left(\mathbf{q}_k^\star\right) - \left(G_{-k} - P_k\left(\mathbf{q}^\star\right)\right) \\
&= \underbrace{u_k\left(\mathbf{q}_k^\star\right) + \sum_{j=1, j \neq k}^{N} r_j\left(\mathbf{q}_j^\star\right)}_{T_1} - \underbrace{G_{-k}}_{T_2}.
\end{aligned}$$

Note that the quantity T_2 in the above expression is independent of the utility function reported by the kth user. Hence, in order to maximize his net payoff, he wishes to maximize the term T_1 in the above expression at $\mathbf{q}^\star$. However, $\mathbf{q}^\star$ maximizes the cost function

$$\mathbf{q}^\star = \arg\max \sum_{j=1}^{N} r_j\left(\mathbf{q}^j\right) = r_k\left(\mathbf{q}^k\right) + \sum_{j=1, j\neq k}^{N} r_j\left(\mathbf{q}^j\right),$$

from which it can be seen that T_1 is maximized at $\mathbf{q}^\star$ only if the reported utility function $r_i\,()$ coincides with the true utility function $u_i\,()$. ■

Thus, this scheme punishes malicious users through its ingenious pricing mechanism.

2.32 VCG Example: Second-Price Auction

Consider a single-item auction scenario in which N players are bidding for a single item. Let $\mathbf{q} = [q_1, q_2, \ldots, q_N]$ denote the allocation of the auction, such that

$$q_i = \begin{cases} 1 & \text{if item is allocated to the } i\text{th user} \\ 0 & \text{otherwise} \end{cases}$$

Hence, $\mathbf{q}$ satisfies the property $\sum_{i=1}^{N} q_i = 1$ with $q_i \in \{0, 1\}$. Let the valuation of each player be V_i. Hence, the true utility $u_i\,(q_i) = V_i q_i$. If his bid is b_i, then the reported utility $r_i\,(q_i) = b_i q_i$. The VCG procedure then involves the following sequence of steps. The first step of the VCG procedure is to solve the optimization problem:

$$\max_{\mathbf{q}} \sum_{i=1}^{N} r_i\,(q_i) = \sum_{i=1}^{N} b_i q_i,$$

$$\text{s.t.} \quad \sum_{i=1}^{N} q_i = 1, q_i \in \{0, 1\}.$$

It can be readily seen that the solution $\mathbf{q}^\star$ to the above problem is given as $\mathbf{q}^\star = \left[q_1^\star, q_2^\star, \ldots, q_N^\star\right]^T$, where

$$q_i^\star = \begin{cases} 1 & \text{if } b_i = b_{\max} \\ 0 & \text{else} \end{cases}$$

Let the highest bid be b_k, corresponding to that of the kth player. To compute the price paid by the players, we need to compute the factor $G_{-i}\,()$, given as

$$G_{-i}\,() = \max_{\mathbf{q}} \sum_{j=1, j\neq i}^{N} r_j\,(q_j) = \sum_{j=1, j\neq i}^{N} b_j q_j.$$

Consider the user $l \neq k$, that is, not the highest bid. The factor G_{-l} is given as

$$G_{-l} = \max_{\mathbf{q}} \sum_{j=1, j\neq l}^{N} b_j q_j = b_{\max}.$$

The price paid by the user $l \neq k$ is derived as

$$\begin{aligned}\text{Price to the } l\text{th user} &= G_{-l} - \sum_{j=1, j\neq l}^{N} b_j q_j^\star \\ &= b_{\max} - \sum_{j=1, j\neq l}^{N} b_j q_j^\star \\ &= b_{\max} - b_{\max} \\ &= 0.\end{aligned}$$

Consider now the kth user with the highest bid. The factor G_{-k} is given as

$$G_{-k} = \max_{\mathbf{q}} \sum_{j=1, j\neq k}^{N} b_j q_j = b_{[2]},$$

where $b_{[2]}$ is the second highest bid. This naturally follows since the summation above involves all the terms except $b_k q_k$ corresponding to the highest bid b_k. Therefore, the maximum occurs at the second highest bid $b_{[2]}$. The price to be paid by this player is then given as

$$\begin{aligned}\text{Price paid by the } k\text{th player} &= b_{[2]} - \sum_{j=1, j\neq k}^{N} b_j q_j^\star \\ &= b_{[2]} - \sum_{j=1, j\neq k}^{N} b_j \times 0 \\ &= b_{[2]}.\end{aligned}$$

Thus, the price paid by the winner is $b_{[2]}$. Therefore, the VCG-based procedure applied to single-item auctions basically leads to a mechanism in which the object is allocated to the highest bidder at the second highest bid. This is basically the second-price auction. Thus, it can be seen that the standard second-price auction is an optimal VCG mechanism.

2.33 Games on Networks: Potential Games

In this section, we consider a special class of games on networks, termed as *potential* games. Consider the packet data network shown in Figure 2.21a, consisting of a source node $\mathcal{S}$, destination node $\mathcal{D}$, with a total of $N = 12$ flows. Let $l_i \in \mathcal{L}$ denote the ith, with $\mathcal{L}$ denoting the set of all links. For instance, in Figure 2.21a, $\mathcal{L} = \{l_1, l_2, l_3, l_4\}$. Also, $p_i \subset \mathcal{L}$ denotes the path taken by a flow from source to destination. For instance, $p_1 = \{l_1, l_2\}$ is a path for player 1 from source to destination. The quantity $P = (p_1, p_2, \ldots, p_N)$ denotes the collection of paths taken by N flows in the network. Let $n_i(P)$ denote the number of players j such that $l_i \in p_j$, that is, number of flows that are using the ith link. Note here that because of the unique nature of this game, we are only concerned with the total number of flows using a particular link, rather than the specific identities of the flows that

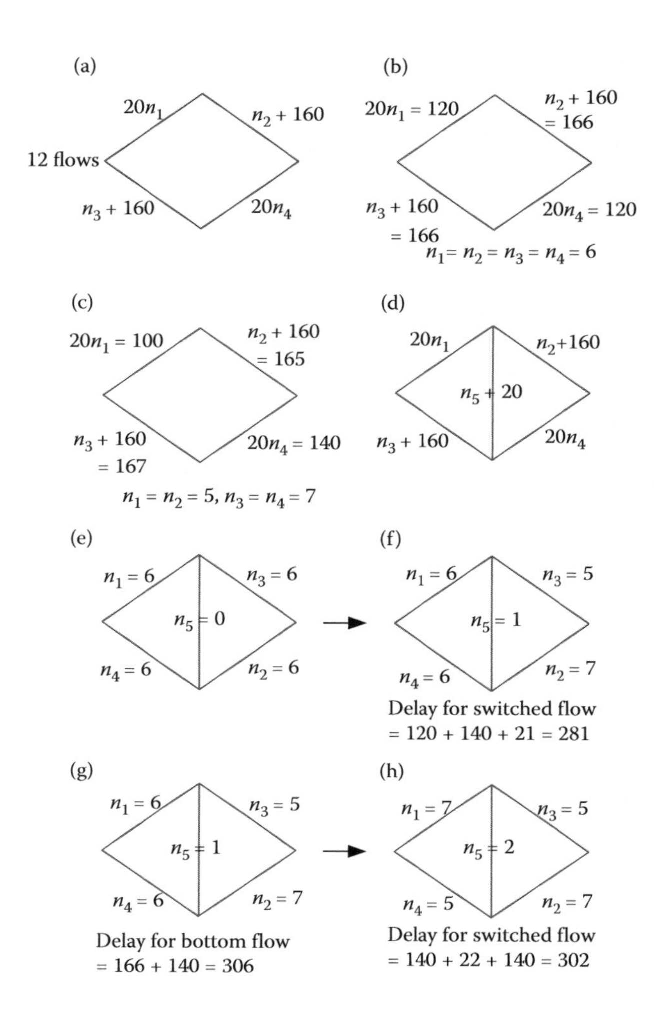

FIGURE 2.21
Potential game.

are using the link. Obviously, any such allocation of flows to link must satisfy the network flow constraints, that is, the total number of flows into a node must equal the total number of flows leaving the node.

We model this scenario of route selection as a *congestion* game, in which the payoff to each player or flow is the negative of the delay associated with the path. Thus, the lower the delay, the higher the payoff. For the example shown in Figure 2.21a, let the delays associated with the links $l_i, 1 \leq i \leq 4$, be given by $20n_1, n_2 + 160, n_3 + 160, 20n_4$, where n_i is the number of flows on link i. Consider the routing of flows show in Figure 2.21b, in which 6 flows employ the route $\{l_1, l_2\}$ and the others employ route $\{l_3, l_4\}$. It can be seen that the net delay d_j for each flow is $d_j = 120 + 166 = 286$ for $1 \leq j \leq 12$. Is this route allocation a Nash equilibrium? Let us check by considering a deviation as shown in Figure 2.21c, where a flow from the top switches to the bottom. The delay for this flow increases to $140 + 167 = 307$. Thus, there is no incentive for any flow to unilaterally switch routes, and therefore this is indeed a Nash equilibrium routing for the congestion game above.

Consider now a modification to the network with the addition of a new link l_5 as shown in Figure 2.21d, with delay $d_5 = n_5 + 20$ for the link. How does the addition of this new link affect the network? Consider one car switching from the top switching to the middle as shown in Figure 2.21e and f. Thus, the new delay for the flow is given as $120 + 21 + 140 = 281$. Thus, the delay decreases from 286 to 281. Hence, one of the flows from the top indeed deviates unilaterally to increase its payoff, that is, to decrease its delay. Is another switch possible? Consider one flow from the bottom switching to a path that flows from the top through the middle as shown in Figure 2.21g and h. Before the switch, the delay for this flow on the bottom is $166 + 140 = 306$. After the switch, its delay is $140 + 22 + 140 = 302$. Thus, since it can reduce its delay by switching, it will indeed switch toward maximizing its payoff. The delays of the different flows in the network are now as follows. The delays for the flows exclusively on the top or bottom links is $140 + 165 = 305$, while that for the ones through the middle is $140 + 22 + 140 = 302$. It can be seen that this is indeed the Nash equilibrium for this network flow game. For instance, consider another flow switching from top to bottom. The delay associated with this is either $166 + 160 = 326$ or $166 + 23 + 166 = 355$. Both the cases result in an increasing delay and hence there is no incentive to deviate. How about a flow switching from bottom to top. The delays in this case are either $160 + 166 = 326$ or $160 + 23 + 160 = 343$. Once again, both lead to an increasing delay. Thus, this is the Nash equilibrium flow routing for this game. Further, observe a very interesting, counterintuitive aspect of the above game. Addition of a new link increases the delay of all flows from 286 without the link to 302 or 305 at equilibrium with the new link! This is indeed a surprising outcome and is termed as the *Braess's paradox*. Further, another key aspect of the above game is that each agent iteratively updating his strategy leads to a convergence to the Nash equilibrium. Below, we demonstrate that this is a unique feature of such congestion games, which belong to the class of potential games and develop a framework to formally characterize such games.

2.34 Network Routing Potential Game

As per the framework developed above for the network routing game, let $d_i\left(n_i\left(P\right)\right)$ denote the delay of link i, which in turn depends on the quantity $n_i\left(P\right)$, that is, the number of paths in P that use link l_i. Naturally, the payoff to flow j can be defined as the negative of the total delay associated with the corresponding path, which can be formally expressed as

$$\begin{aligned}
\text{Payoff to flow } j &= u_j\left(P\right) \\
&= u_j\left(p_j, p_{-j}\right) \\
&= -\sum_{i \in p_j} d_i\left(n_i\left(P\right)\right) \\
&= -\sum_{i \in p_j} d_i\left(n_i\left(p_{-j}\right) + 1\right),
\end{aligned} \tag{2.27}$$

where the last equality above arises from the fact the $n_i\left(P\right)$, which is the total number of flows using the link l_i in the path of flow j, is equal to the total number of flows other than j using this link plus one, that is, $n_i\left(p_{-j}\right) + 1$. Let the current path assignment be P. Consider flow j switching to path

$\tilde{p}_j$. The new payoff to agent j is given as

$$u_j\left(\tilde{p}_j, p_{-j}\right) = -\sum_{i\in\tilde{p}_j} d_i\left(n_i\left(p_{-j}\right)+1\right). \tag{2.28}$$

From the payoffs to agent j associated with $p_j, \tilde{p}_j$, given in Equations 2.27 and 2.28, agent j will consider switching only if

$$\begin{aligned}
&u_j\left(\tilde{p}_j, p_{-j}\right) - u_j\left(\tilde{p}_j, p_{-j}\right) > 0 \\
\Rightarrow &\sum_{i\in p_j} d_i\left(n_i\left(p_{-j}\right)+1\right) - \sum_{i\in\tilde{p}_j} d_i\left(n_i\left(p_{-j}\right)+1\right) > 0 \\
\Rightarrow &\sum_{i\in p_j\setminus\tilde{p}_j} d_i\left(n_i\left(p_{-j}\right)+1\right) - \sum_{i\in\tilde{p}_j\setminus p_j} d_i\left(n_i\left(p_{-j}\right)+1\right) > 0.
\end{aligned}$$

Further, the flow allocation P represents a Nash equilibrium for this routing game if it satisfies

$$u_j\left(p_j, p_{-j}\right) - u_j\left(\tilde{p}_j, p_{-j}\right) \geq 0, \forall\tilde{p}_j,$$

for each flow j, that is, no flow benefits from a unilateral path deviation. For this game, one can define a *potential* function $\mathcal{V}(P)$ associated with the flow allocation P as

$$V(P) = -\sum_{i\in\mathcal{L}}\sum_{l=0}^{n_j(P)} d_j(l).$$

Observe that the potential function defined above is a *global* function, and is not specific to a particular flow j. The above expression $V(P)$ can now be further simplified for a particular flow j as

$$V\left(p_j, p_{-j}\right) = -\sum_{i\in\mathcal{L}}\sum_{l=0}^{n_i(p_{-j})} d_j(l) - \sum_{i\in p_j} d_i\left(n_i\left(p_{-j}\right)+1\right),$$

$$V\left(\tilde{p}_j, p_{-j}\right) = -\sum_{i\in\mathcal{L}}\sum_{l=0}^{n_i(p_{-j})} d_j(l) - \sum_{i\in\tilde{p}_j} d_i\left(n_i\left(p_{-j}\right)+1\right).$$

It can now be readily seen that $V\left(\tilde{p}_j, p_{-j}\right) - V\left(p_j, p_{-j}\right)$ can be simplified as

$$\begin{aligned}
V\left(\tilde{p}_j, p_{-j}\right) - V\left(p_j, p_{-j}\right) &= \sum_{i\in p_j} d_i\left(n_i\left(p_{-j}\right)+1\right) - \sum_{i\in\tilde{p}_j} d_i\left(n_i\left(p_{-j}\right)+1\right) \\
&= u_j\left(\tilde{p}_j, p_{-j}\right) - u_j\left(\tilde{p}_j, p_{-j}\right).
\end{aligned}$$

This is the unique property associated with the potential function. A game is termed as an *exact* potential game if there exists a function $V : \mathcal{P} \rightarrow \mathbb{R}$, such that

$$V\left(\tilde{p}_j, p_{-j}\right) - V\left(p_j, p_{-j}\right) = u_j\left(\tilde{p}_j, p_{-j}\right) - u_j\left(p_j, p_{-j}\right).$$

A game is term an *ordinal* potential game if

$$V(\tilde{p}_j, p_{-j}) - V(p_j, p_{-j}) > 0 \Rightarrow u_j(\tilde{p}_j, p_{-j}) - u_j(p_j, p_{-j}) > 0.$$

It can be readily seen from the definition above that every exact potential game is also an ordinal potential game. An important property of ordinal potential games can be described as follows.

Lemma 2.4

Let V denote the potential function for an ordinal potential game Γ and $P^\star$ be a maximizer of V, that is, $P^\star = \arg\max_{P \in \mathcal{P}} V(P)$. The $P^\star$ is a Nash equilibrium of the game.

Proof. This can be deduced as follows:

$$\begin{aligned} V\left(p_j^\star, p_{-j}^\star\right) - V\left(p_j, p_{-j}^\star\right) > 0, \quad \forall p_j, j \\ \Rightarrow u_j\left(p_j^\star, p_{-j}^\star\right) - u_j\left(p_j, p_{-j}^\star\right) > 0, \quad \forall p_j, j \\ \Rightarrow u_j\left(p_j^\star, p_{-j}^\star\right) > u_i\left(p_j, p_{-j}^\star\right), \quad \forall p_j, j. \end{aligned}$$

■

The above property of potential games allows one to readily compute the Nash equilibrium of the game through maximization of the potential function. In general, it can be proven that ordinal potential games have a Nash equilibrium. The detailed proof of this principle is beyond the scope of this discussion and can be found in Reference 4. A sequence of outcomes $\left\{P^{(0)}, P^{(1)}, P^{(2)}, \ldots\right\}$ is termed as a *path* (not to be confused with the earlier usage of the term *path*, which is specific to the context of a routing game) in such a game if for each $k > 0$, there exists a player $j(k)$ such that $p_{-j(k)}^{(k-1)} = p_{-j(k)}^{(k)}$, that is, at each stage, at most one player changes his strategy. Such a path is termed an *improvement* path if for each $k > 0$,

$$u_{j(k)}\left(p_{j(k)}^{(k)}, p_{-j(k)}^{(k-1)}\right) > u_{j(k)}\left(p_{j(k)}^{(k-1)}, p_{-j(k)}^{(k-1)}\right).$$

In other words, an improvement path is a player who changes strategies that always improves his utility. The *best* and *better* reply dynamics can be employed in such games to construct improvement paths. In a best reply dynamic in a potential game, user $j(k)$ at stage k chooses a strategy $p_{j(k)}$ such that

$$p_{j(k)} = \arg\max_{p_j \in \mathcal{P}_j} u_{j(k)}\left(p_j, p_{-j(k)}^{(k-1)}\right),$$

that is, agent $j(k)$ chooses his best strategy, while the strategies of all the other players are fixed. Similarly, a better reply dynamic can be formulated as

$$p_{j(k)} \in \left\{p_j \in \mathcal{P}_j \middle| u_{j(k)}\left(p_j, p_{-j(k)}^{(k-1)}\right) > u_{j(k)}\left(p_{j(k)}^{(k-1)}, p_{-j(k)}^{(k-1)}\right)\right\}.$$

The unique aspect of such potential games is that the above low complexity best and better response dynamic updates can be employed by the agents to converge to the Nash equilibrium outcome.

2.35 Dynamic Games

One defining trait of all the strategic-form games discussed so far is that players decide their moves only once and at the beginning. And, when they move, they do so without knowing the actual moves of other players. This particular way of abstraction for modeling strategic interactions is quite useful for many engineering, managerial, business, political, and real-life problems. However, such abstractions grossly ignore the sequential/temporal nature and informational features of interactions, which may play a very important role in some of the games. The following examples demonstrate the need of having a more detailed framework that takes into account the informational and temporal structure of strategic interactions:

1. A variation of the BoS: The boy first decides whether to watch a movie (M) or play a game (G). The girl, first, observes the action taken by the boy and, then, decides to watch a movie (M) or play a game (G). The ability to observe the action of the boy certainly influences the move made by the girl.
2. A variation of Matching Pennies: Player P_1 shows either head or tail. Player P_2 observes the action taken by P_1 and then decides to show either head or tail. In this game also, the way the game is played makes a big difference in the nature of a strategic interaction.
3. Stackelberg competition: Unlike a Cournot competition, in which the firms decide to produce the respective quantities simultaneously, in a Stackelberg competition, one of the firms, called the leader, moves first. The other firm, called the follower, observes the move of the leader and, then, decides its move. This particular sequential nature of the interaction strongly influences the actual move of the follower.
4. Repeated games: In many real-life situations, players strategically interact many times and not just once. For example, two firms may compete against each other on a day-to-day basis. The repeated nature of the interaction may alter the underlying game fundamentally. A repeated interaction helps in building reputation and can be instrumental in achieving cooperative behavior that is frequently observed in real life but not so often predicted by the models using one-shot games/static games.
5. Signaling games: In a two-player signaling game, one player has more information than the other. The player having more information moves first, and after observing his action, the uninformed agent takes his strategic action. In these situations, the uninformed agent cares about the informed agents' action not only because of payoffs but also because the action conveys something about the type of the player.

It is clear from the above examples that there are indispensable differences between dynamic games and static games that are represented in strategic-form or normal-form games. A game is considered as dynamic in two different cases: first, if the interaction among players is sequential, a player moves after observing the action taken by another player, and second, if a static game is repeated again and again and players observe the outcome of the static game and then interact again. Games such as Stackelberg competition and signaling games fall into the first category while repeated games are dynamic games of the second category.

2.36 Extensive-Form Games

A static game is represented in normal form, which is also called strategic form. This form requires three ingredients: a list of players, a strategy set for each player containing all her strategies, and the preference of each player over all the strategy profiles or payoff for each player associated with all the strategy profiles. In case of two players, a strategic form can be nicely represented in a bimatrix table. In dynamic games, some of the players get to observe what other players have done in the past. Players who know the past moves of other players are able to condition their actions on the moves of those other players. This fact considerably changes the nature of a strategic interaction among players. In a sense, dynamic games are more complex as they have a rich temporal and information structure. The strategic form is unable to capture the richness of a dynamic game. To capture the full richness of a dynamic game, the following details would be required:

1. A list of players participating in the strategic interaction
2. Timings/instances when players get to move
3. Actions available to the player when she has a potential move in the game
4. Knowledge/information that a player has when she has a potential move in the game
5. Payoffs for all the players when a game ends in all of the several possible ways

All these details can be conveniently depicted with the help of a game tree.

2.36.1 Game Tree

Consider an interaction between Charlie Brown and Lucy from the comic strip *Peanuts*. A typical interaction between Charlie Brown and Lucy can be described as: First, Lucy convinces Charlie Brown to kick a football. Then, Lucy holds the football and Charlie Brown attempts to kick it. Charlie runs to give a powerful kick to the football. And, at the last moment, Lucy withdraws the ball, and as a consequence, Charlie Brown ends up flying up into the air and eventually lands on his back.* If this interaction is modeled as a game, one can give the following details: the game has two players: Charlie Brown (CB) and Lucy (L). The interaction begins when Lucy asks Charlie Brown to kick a football. So, in the beginning of the interaction, Lucy has two actions available to her: (1) ask [A] or (2) don't ask [DA] Charlie Brown to kick the ball. If Lucy does not ask Charlie Brown, the game ends. Otherwise, the game proceeds further. After Lucy asks CB, he can either accept her offer to kick or reject her offer. If he rejects, the interaction (game) ends right there. If he agrees to kick the ball, Lucy can either leave the ball on the ground [LG] or takes away [TA] the ball at the last moment. In both the conditions, the game ends. Any end of the game should be associated with the payoffs for both the players, or both the players should have preference relations defined over the set of all outcomes (possible ends). Let us assume that Lucy's ordering of the outcomes from best to worst is Charlie Brown falls on the ground, Lucy does not ask Charlie Brown to kick the ball, Lucy asks but Charlie Brown rejects her offer, Lucy keeps the ball on the ground, and Charlie kicks it. Let us further assume that Charlie Brown's preference ordering is just the reverse of Lucy's. For a simple representation of Lucy's and Charlie Brown's preferences, payoffs should be associated with different outcomes. A higher number indicates a preferred outcome. In this example, let us say, the best outcome is associated with 4, the second-best outcome with 3, the third-best outcome with 2, and the worst outcome with 1.

* The situation is humorous as Lucy always comes up with a new logic to convince Charlie Brown to make another attempt to kick a football.

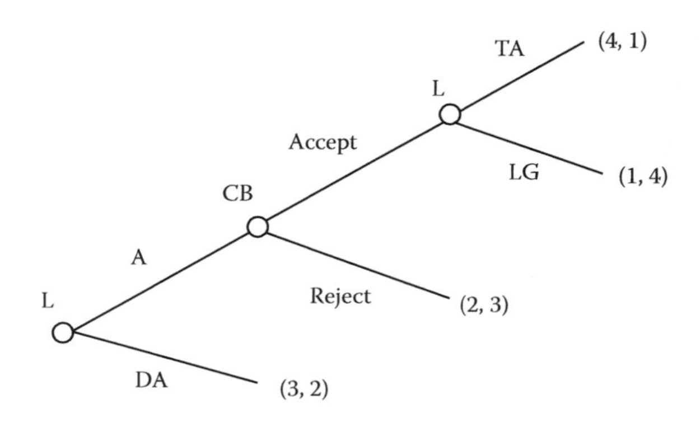

FIGURE 2.22
Example of a game tree.

The diagram given in Figure 2.22 that represents the dynamic interaction between Lucy and Charlie Brown is called a game tree. The diagram represents the following details: list of players: Lucy and Charlie Brown; timings/instances when players get to move: Lucy moves first and Charlie Brown takes an action if Lucy asks him to kick a ball and so on; actions available to players when she or he has a potential move in the game, for example: Lucy, in the beginning of the game can either ask [A] or don't ask [DA] Charlie Brown to kick a ball and the payoffs associated with the possible end of the game that indicate preference ordering of all players defined over all outcomes such as (4, 1) are written after Lucy takes away [TA] the ball. In this game tree, numbers associated with a possible outcome are an ordered pair in which the first number shall represent the payoff to Lucy and the second number shall represent the payoff to Charlie Brown. The discussion on how to represent the information in such figures shall be taken up in the next section.

Formally, a game tree is a directed graph G that is defined as a pair $G = (V, E)$, where V is a finite set of nodes and E is a finite set of branches. An element e of the set E connects a node $v \in V$ to another node $v' \in V$; in other words, $E \subseteq V \times V$. The branch e is said to be leading to the node v' and leading away from the node v. Furthermore, the following three conditions are satisfied on G to be characterized as a game tree:

1. There exists an initial node v_0 in V such that there is no edge leading to it.
2. For every other node $v \in V - \{v_0\}$, there exists a unique branch leading to it.
3. For a unique pair of nodes in a game tree, there exists a unique path* that starts from one of these two nodes and ends at the other node.

The graph in Figure 2.23 is a game tree while the graph in Figure 2.24 is not a game tree as for node E, and two branches S and T are leading to it. In other words, if a graph contains a "loop," then it cannot be a game tree.

In simple words, a game tree consists of nodes and branches. There are three different types of nodes

1. An initial node that represents the unique beginning of the game
2. Decision nodes that indicate that it is the player's turn to take an action

* A path in a graph $G = (V, E)$ is an alternating sequence of nodes and branches starting and ending with nodes, such that two neighboring nodes in the sequence are connected by the branch given in the middle of those two nodes and no branch appears more than once in the sequence.

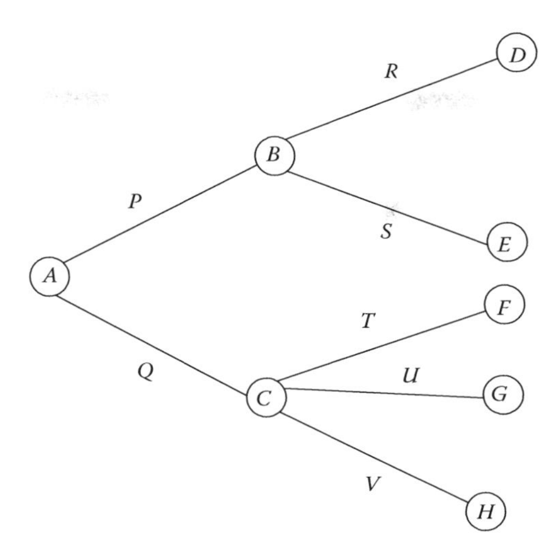

FIGURE 2.23
Another example of a game tree.

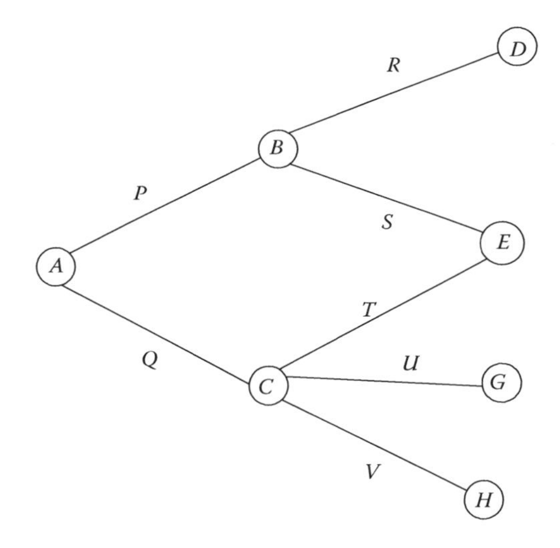

FIGURE 2.24
Not an example of a game tree.

3. Terminal nodes that do not precede any other nodes and indicate the possible ends of the game

An initial node is also a decision node. A branch represents a possible move at the decision node. In Figure 2.23, A is the initial node, A, B, and C are decision nodes, and D, E, F, G, and H are terminal nodes. Sometimes, circles representing nodes are omitted from diagrammatic representations. For example, in Figure 2.22, terminal nodes are not represented using circles.

2.36.2 Information

Consider the variation of the Matching Pennies game discussed at the beginning of this section. The game is represented in Figures 2.25 and 2.26.

Compare the variation with the original Matching Pennies game. The only difference is that in the variation, player 2 knows the actual move of player 1 when it is his turn to move. In the original game, both the players move simultaneously. The question is how to capture and represent the difference in the level of information that player 2 has in these two different strategic settings.

The original game in which both the players move simultaneously can be thought as a game in which player 1 moves first but his move is not observed by player 2.* So, in the game tree representation of the original Matching Pennies game, a dotted line can be drawn between two

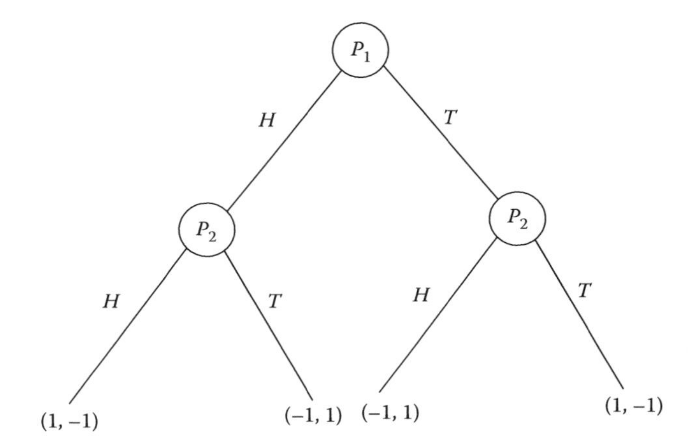

FIGURE 2.25
Variation of Matching Pennies game.

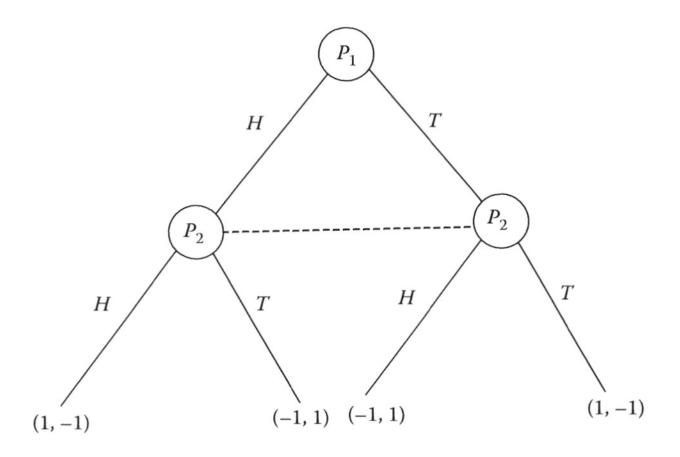

FIGURE 2.26
Matching Pennies game.

* The strategic interaction will remain the same even if we consider that player 2 moves first but his move is not observed by player 1.

decision nodes corresponding to player 2 indicating that he is not able to distinguish which node among these two nodes has reached when it is his turn to move.

A key idea to understand the notion of information in any extensive-form game is of *history*. In a game tree, there exists a unique path that connects two nodes. Therefore, there exists a unique path from the initial node to any node under consideration. This unique path is called a history, which gives the sequence of events describing the progression of the game from the initial node to that node. Now, think of a player, who is supposed to move but does not know the precise history of the game, meaning the player knows that it is his turn to move and also that he has reached one of the nodes from a subset of nodes, but he does not know the precise identity of that specific node. An information set is defined as a collection of all such nodes. For example, in the game of Matching Pennies given in Figure 2.9, player 2, when it is his turn to move, does not know the exact history of the game but knows that it is his turn to move, so both the nodes at which he has the potential move are in the same information set. In the variation of Matching Pennies, given in Figure 2.25, when it is player 2's turn to move, he knows the exact history of the game. Therefore, the two nodes at which he has potential moves are in different information sets.

Consider all decision nodes of a game. A player has to make a move at each decision node. One can define a function that assigns a decision node to a particular player who is supposed to move at that node. This function is called the player function. If the set of all decision nodes is indicated by X, and the set of all players is indicated by I, then the player function P is $P : X \to I$. The set of all decision nodes assigned to player i is $X_i = \{x \in X : P(x) = i\}$. The player function partitions the set X of all decision nodes as a set $\{X_1, X_2, X_3, \ldots, X_I\}$ such that $\bigcup X_i = X$ and $X_i \bigcap X_J = \emptyset, \forall i,\ j \in I$. Player i owns all the nodes of the set X_I in the sense that it is player i's turn to move if any of the nodes in the set X_i is reached.

The set X_i whose nodes belong to player i can further be partitioned as a collection of information sets of h_i's denoted as H_i. An information set h_i has the following properties:

1. If h_i contains only one node x, then player i who gets to move when the information set h_i is reached in the game knows that he is at node x.
2. If there are more than one node in h_i, for example, the information set h_i contains exactly two different nodes x and x', then player i who gets to move when this information set is reached does not know whether he is at x or x'.
3. If two different nodes x and x' are in h_i, then branches leading away from x and x' are the same. In other words, at nodes belonging to the same information set h_i, player i has the same actions available to him.

2.36.2.1 Example

In a game described by the game tree given in Figure 2.27, the set of all nodes is $\{A, B, C, D, E, T, U, V, W, X, Y, Z\}$. A is the initial node. The set of all decision nodes X is $\{A, B, C, D, E\}$. Partition of the set of all decision nodes induced by the player function of the game is $\{X_1, X_2, X_3\}$, where $X_1 = \{A\}$, $X_2 = \{B, C\}$ and $X_3 = \{D, E\}$. Both X_1 and X_2 contain only one information set; X_1 as it is singleton and X_2 as player 2 does not know whether player 1 takes action on P or Q so nodes B and C in X_2 belong to the same information set. X_3 contains two different information sets and both are singletons. A collection of all information sets of player i is denoted as H_i. Therefore, $H_1 = \{\{A\}\}$, $H_2 = \{\{B, C\}\}$, and $H_3 = \{\{D\}, \{E\}\}$. Notice that H_1 contains a single information set that itself is a singleton set, H_2 contains a single information set that has two nodes, and H_3 contains two information sets, and each of them is singleton.

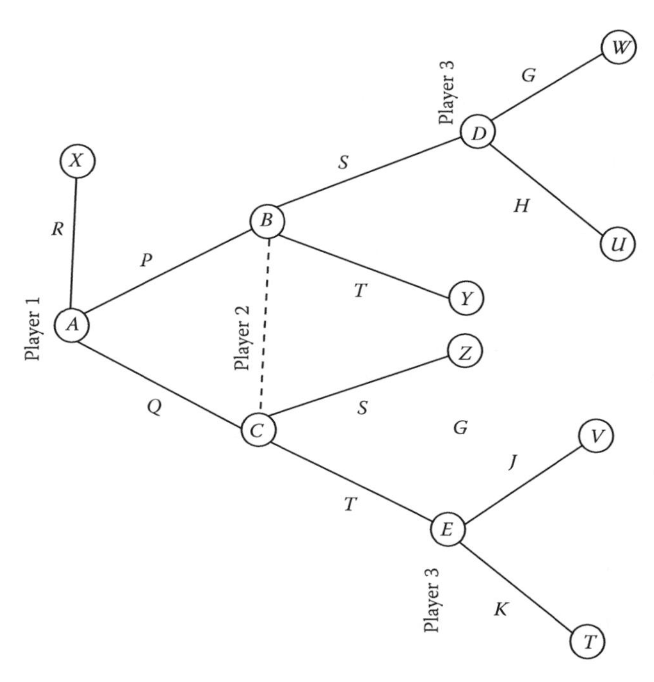

FIGURE 2.27
Example to understand the information set.

Definition 2.1

A game in which all the information sets are singletons is called a game of perfect information as whenever it is a player's turn to move in this game, he knows the exact history of the progression of the game. A game in which at least one information set is not singleton is called a game of imperfect information.

2.36.3 Actions and Strategies

In normal-form games, the terms *actions* and *strategies* are used interchangeably. In an extensive-form game, the term *strategy* has a slightly different meaning than the term *action*. Consider the Matching Pennies game given in Figures 2.28 and 2.29. In this game, both the players have one potential play each. A strategy of player 1 should describe what he would do whenever he is called upon to play in this game. Player 1 moves only at the beginning of the game and he can choose either H or T. Therefore, player 1's strategy set is $S_1 = \{H, T\}$. Player 2 is called upon to make a move only once in the game and he can choose either H or T; therefore, player 2's strategy set is $S_2 = \{H, T\}$. In contrast, consider the variation of the Matching Pennies game given in Figure 2.25. In this variation, player 2, when it is his turn to move, knows the action taken by player 1. Therefore, he can condition his action on the action taken by player 1. Consider one strategy for player 2: take action T if player 1 has taken action H and take action H if player 1's action is T. This strategy of player 2 is different from the strategy: take action H if player 1 has taken action H and take action H if player 1's action is T. In this case, the strategy is much more than a simple action. In fact, a strategy can be called "a complete plan of action." The strategy of a player specifies what action he would take whenever he is called upon to move. More formally, the strategy of a player is a list of actions specifying which action he would take at each of his information set.

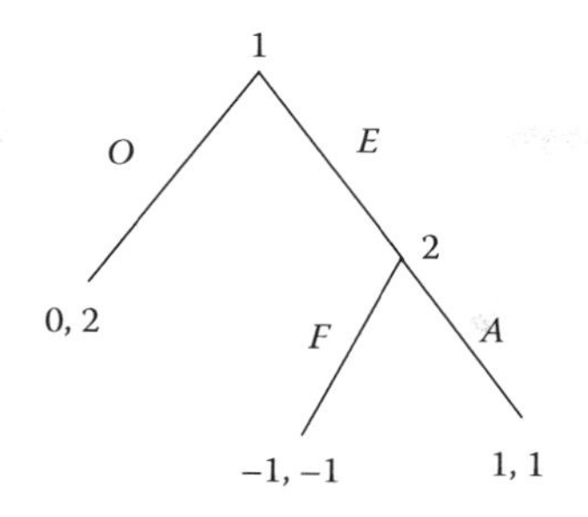

FIGURE 2.28
Extensive-form representation of the entry game.

P_1 \ P_2	A	F
E	1, 1	−1, −1
O	0, 2	0, 2

FIGURE 2.29
Normal-form representation of the entry game.

Consider the game described in Figure 2.27. In this game, players 1 and 2 have one information set each, while player 3 has two information sets:

$$S_1 = \{P, Q, R\},$$
$$S_2 = \{S, T\},$$
$$S_3 = \{GJ, HJ, GK, HK\},$$

where

$GJ =$ Player 3 plays G if player 2 plays S and player 3 plays J if player 2 plays T.

$HJ =$ Player 3 plays H if player 2 plays S and player 3 plays J if player 2 plays T.

$GK =$ Player 3 plays G if player 2 plays S and player 3 plays K if player 2 plays T.

$HK =$ Player 3 plays H if player 2 plays S and player 3 plays K if player 2 plays T.

A strategy is sometimes called "a complete plan of action" describing all the actions that a player needs to make in a strategic interaction. This description is nice and simple but not apt in some games. Consider the game described in Figure 2.22. In this game, there are three decision nodes: the initial node at which Lucy moves, the decision node at which Charlie Brown moves, and the other decision node at which Lucy moves. By definition, Lucy's strategy should describe her moves at the initial node as well as at the other node at which she has a potential move. Charlie Brown's strategy should contain only one action. Lucy's and Charlie Brown's strategy sets are as follows:

S_L = {Ask at the beginning and Take Away if CB tries to kick,

Ask at the beginning and leave the ball on the ground if CB tries to kick,

Don't Ask at the beginning and Take Away if CB tries to kick,

Don't Ask at the beginning and leave the ball on the ground if CB tries to kick}

S_{CB} = {Accept, Reject}

Notice that if Lucy decides not to ask [DA] Charlie Brown to kick the ball, then the game ends right there and Lucy does not get to move again. So, one of her complete plan of action is don't ask Charlie Brown to kick the ball, which is not a strategy as this plan does not assign an action to the information set containing the other decision node at which Lucy has a move. So, why is strategy defined as a list of actions specifying which action the player would take at each of his information set? Lucy will decide not to ask Charlie Brown to kick the ball only if not asking [DA] gives her a better payoff than asking [A]. Lucy's payoff from asking further depends on the action that Charlie Brown will take after being asked to kick the ball. Charlie Brown will accept or reject the offer after comparing the payoffs from both these actions. But, Charlie Brown's payoff from accepting depends on Lucy's action, namely, whether she keeps the ball on the ground or takes away the ball. The strategy that Charlie Brown will use depends on what Lucy does or may do on all her information sets. If "don't ask" is Lucy's strategy, it may not be possible for Charlie Brown to figure out his best response. If Charlie Brown is unable to decide his best response, then Lucy also cannot figure out her response. Therefore, a strategy needs to be defined as a list of actions specifying which action the player would take at each of his information set.

Definition 2.2

If A_i is the set of all actions available to player i in the game and $A_i(h_i)$ is the set of all actions available to player i in an information set h_i, then a (pure) strategy of player i in an extensive-form game can be formally defined as a function $s_i : H_i \rightarrow A_i$ such that $s_i(h_i) \in A_i(h_i)$ for each $h_i \in H_i$.

2.36.3.1 Behavior Strategies and Mixed Strategies

The concept of mixed strategies in normal-form games can naturally be extended in the context of extensive-form games.

Definition 2.3

A mixed strategy for player i, σ_i, in an extensive-form game is defined as a probability distribution over his pure strategies $s_i \in S_i$.

The interpretation of a mixed strategy in an extensive-form game is exactly the same as in the case of a mixed strategy in a normal-form game. A mixed strategy describes the different probabilities that the player assigns to different pure strategies. Therefore, using a mixed strategy in an extensive-form game implies that the player selects a particular strategy randomly at the beginning of the game and follows that particular strategy throughout the game. A mixed strategy does not give the player an opportunity to randomize at each information set where he has a potential play in the game. A player in an extensive-form game may like to randomize at a particular information set irrespective of the actions taken at his earlier information sets. The idea of randomizing at each information set independently is not captured by the notion of mixed strategy. The concept of behavioral strategy in the extensive-form game captures the idea in which players do pick an action randomly at each information set, and this random selection is independent of what they do at other information sets.

Definition 2.4

A behavioral strategy for player i in an extensive-form game is defined as a function $\sigma_i : H_i \rightarrow \Delta(A_i)$ such that $\sigma_i(h_i) \in \Delta(A_i(h_i)), \forall h_i \in H_i$.

Definition 2.5

A game is of perfect recall if players

- Never forget a decision that they take in the game till the conclusion of the strategic interaction
- Never forget any information that they once know in the game

In a game of perfect recall, mixed and behavioral strategies are equivalent in the sense that for one mixed strategy of player i, there exists a behavioral strategy of the same player i, such that both of them lead to the same probability distribution over outcomes for any mixed or behavioral strategies of other players. In this chapter, all the games considered are of perfect recall.

2.36.4 Nash Equilibrium of Extensive-Form Games

The extensive-form description of a strategic interaction in comparison to a normal-form/strategic-form description of the same game is relatively rich in details. Apart from describing the list of players, strategy sets, and payoffs of all the players, the extensive form also contains the timing of moves and knowledge that a player has at the time of his move. Therefore, a strategic interaction that is represented in the extensive form can very easily be represented in its normal form by suppressing the timing of each player's move.

Consider the following strategic interaction between two players. Player 1 is a potential entrant to a market that has a monopolist incumbent, player 2. At the beginning of the strategic interaction, player 1 has two possible actions: enter into market $[E]$ and remain out $[O]$. If the potential entrant decides to remain out, the strategic interaction ends and player 1 gets 0 and player 2 gets 2 as payoff. If player 1 enters the market, player 2 gets to move. Player 2 can accommodate $[A]$ or fight $[F]$ the entry of player 1. If player 2 accommodates, both the players get a payoff of 1, and if player 2 decides to fight, both the players get -1.

Nash equilibria of this entry game can be obtained using the normal-form representation of the game. The game has two Nash equilibria in pure strategies: (O, F) and (E, A). In equilibrium (O, F), it is player 2's threat of fight that keeps player 1 away from entering. If this equilibrium gets a player out, then player 1 will remain out of the market and player 1 will not get an opportunity to fight. This Nash equilibrium is not credible and intuitive, as a careful observation of the extensive-form representation clearly indicates that if player 1 enters the market, player 2 will strictly prefer to accommodate player 1. Therefore, player 1 should definitely enter the market as entering will ensure a payoff of 1 as opposed to a payoff of 0, which he gets if he remains out. This Nash equilibrium is obtained because the normal form of the game ignores the sequential nature of entry game. Some Nash equilibria obtained using a normal form of the game, which treats strategies as choices made once and for all at the beginning of the game, may look illogical in the later stages of the game. Nash equilibria that are reasonable not only at the beginning of the game but also at each information set of the game are called "sequentially rational."

References

1. M. Osborne and A. Rubinstein, *A Course in Game Theory*. The MIT Press, Cambridge, 1994.
2. M. Osborne, *An Introduction to Game Theory*. Oxford University Press, USA, 2003.
3. V. Krishna, *Auction Theory*. Academic Press, 2009.
4. A. MacKenzie, *Game Theory for Wireless Engineers (Synthesis Lectures on Communications)*. Morgan & Claypool Publishers, 2006.

3

Multi-Objective Optimization

Kalyanmoy Deb, Karthik Sindhya, and Jussi Hakanen

CONTENTS

ABSTRACT Basic concepts and principles of multi-objective optimization (MO) methods and multiple criteria decision making (MCDM) approaches will be the focus area of this chapter. Algorithms are classified based on the role of decision makers, and four different classifications have been made. Algorithms such as the weighting method, epsilon-constraint method, achievement scalarizing function method, and normal boundary intersection method are discussed in detail. Moreover, evolutionary multi-objective optimization (EMO) methods are discussed highlighting certain recent advanced topics. Methods for handling constraints and many objectives (as many as 15 objectives) are also discussed. Under *a priori* methods, lexicographic and goal programming methods are presented as well. Finally, a number of existing and recent interactive methods such as the reference point method, GUESS method, STOM method, and the NIMBUS method are discussed.

3.1 Introduction

A multi-objective optimization problem (MOOP) deals with more than one objective function. In most practical decision-making problems, multiple objectives or multiple criteria are evident. Owing to lack of suitable solution methodologies, traditionally, an MOOP has been mostly cast and solved as a single-objective optimization problem. However, there exist a number of fundamental differences between the working principles of single- and multi-objective optimization (MO) algorithms. In a single-objective optimization problem, the task is to find one solution (except in some specific multimodal optimization problems, where multiple optimal solutions are sought), which optimizes the sole objective function. Extending the idea to MO, it may be wrongly assumed that the task in MO is to find an optimal solution corresponding to each objective function. In this chapter, we will discuss the principles of MO and present optimality conditions for any solution to be optimal in the presence of multiple objectives.

MOOPs, by nature, give rise to a set of Pareto-optimal solutions, which need a further processing to arrive at a single preferred solution. To achieve the first task, it becomes quite a natural proposition to use an evolutionary optimizer (EO), because the use of a population in an iteration helps an EO to simultaneously find multiple nondominated solutions, which portrays a tradeoff among objectives, in a single run of the algorithm.

In this chapter, we begin with a brief introduction to MO, its principle, and a few important terminologies that will be useful to read the rest of the chapter. We describe the optimality conditions briefly. Thereafter, we discuss a number of traditional and evolutionary multi-objective optimization (EMO) methods, including interactive methods. Conclusions are then drawn at the end.

3.2 MO Basics

An MOOP has a number of objective functions, which are to be minimized or maximized simultaneously. As in the single-objective optimization problem, an MOOP usually has a number of constraints, which any feasible solution (including the optimal solution) must satisfy. In the following, we state the MOOP in its general form:

$$\left.\begin{array}{lll} \text{min/max } f_i(\mathbf{x}), & & i = 1, 2, \ldots, k; \\ \text{s.t. } g_j(\mathbf{x}) \geq 0, & & j = 1, 2, \ldots, J; \\ h_p(\mathbf{x}) = 0, & & p = 1, 2, \ldots, H; \\ x_i^{(L)} \leq x_i \leq x_i^{(U)}, & & i = 1, 2, \ldots, n. \end{array}\right\} \tag{3.1}$$

A solution $\mathbf{x}$ is a vector of n decision variables: $\mathbf{x} = (x_1, x_2, \ldots, x_n)^T$. The last set of constraints are called variable bounds, restricting each decision variable x_i to take a value within a lower $x_i^{(L)}$ and an upper $x_i^{(U)}$ bound. These bounds constitute a *decision variable space* $\mathcal{D}$ or simply the decision space. Throughout this book, we use the terms *point* and *solution* interchangeably to mean a solution vector $\mathbf{x}$. Usually, for an MOOP several optimal solutions exist with different tradeoffs and are termed as Pareto-optimal solutions.

Associated with the problem are J inequality and K equality constraints. The terms $g_j(\mathbf{x})$ and $h_k(\mathbf{x})$ are called constraint functions. The inequality constraints are treated as "greater-than-equal-to" types, although a "less-than-equal-to" type inequality constraint is also taken care of in the above formulation. In the latter case, the constraint must be converted into a "greater-than-equal-to" type constraint by multiplying the constraint function by -1 [16]. A solution $\mathbf{x}$ that does not satisfy *all* of the $(J + K)$ constraints and *all* of the $2N$ variable bounds stated above is called an *infeasible solution*. On the other hand, if any solution $\mathbf{x}$ satisfies all constraints and variable bounds, it is known as a *feasible solution*. Therefore, we realize that in the presence of constraints, the entire decision variable space $\mathcal{D}$ need not be feasible. The set of all feasible solutions is called the *feasible region* or $\mathcal{S}$. In this book, sometimes we will refer to the feasible region as simply the search space.

There are k objective functions $\mathbf{f}(\mathbf{x}) = (f_1(\mathbf{x}), f_2(\mathbf{x}), \ldots, f_k(\mathbf{x}))^T$ considered in the above formulation. Each objective function can be either minimized or maximized. The duality principle [16,57,58], in the context of optimization, suggests that we can convert a maximization problem into a minimization one by multiplying the objective function by -1. The duality principle has made the task of handling mixed type of objectives much easier. Many optimization algorithms are developed to solve only one type of optimization problem such as the minimization problems. When an objective is required to be maximized using such an algorithm, the duality principle can be used to transform the original objective for maximization into an objective for minimization.

One of the striking differences between single-objective and MO is that in MO the objective functions constitute a multi-dimensional space, in addition to the usual decision variable space. This additional space is called the *objective space*, $\mathcal{Z}$. For each solution $\mathbf{x}$ in the decision variable space, there exists a point in the objective space, denoted by $\mathbf{f}(\mathbf{x}) = \mathbf{z} = (z_1, z_2, \ldots, z_k)^T$. The mapping takes place between an n-dimensional solution vector and a k-dimensional objective vector. Figure 3.1 illustrates these two spaces and a mapping between them.

MO is sometimes referred to as *vector optimization* because a vector of objectives, instead of a single objective, is optimized.

3.2.1 Principles of MO

It is clear from the above discussion that, in principle, the search space in the context of multiple objectives can be divided into two nonoverlapping regions, namely, one which is optimal and the other which is nonoptimal. In the presence of multiple Pareto-optimal solutions, it is difficult to prefer one solution over the other without any further information about the problem. If higher-level information is satisfactorily available, this can be used to make a *biased* search. However, in the absence of any such information, all Pareto-optimal solutions are equally important. Hence, in

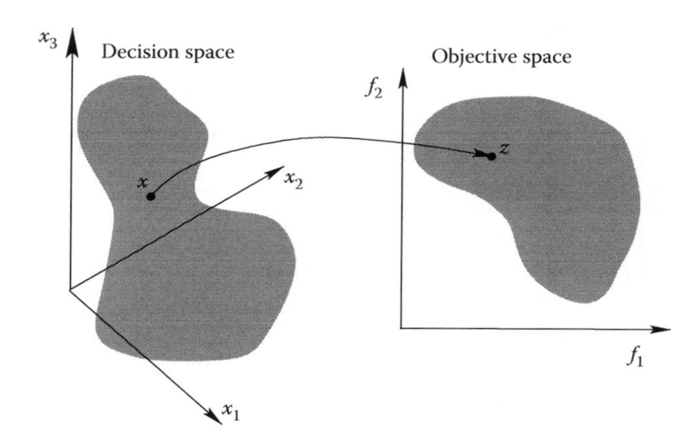

FIGURE 3.1
Representation of the decision variable space and the corresponding objective space.

light of the ideal approach, it is important to find as many Pareto-optimal solutions as possible in a problem. Thus, it can be conjectured that there are two goals in an MO.

1. To find a set of solutions as close as possible to the Pareto-optimal front
2. To find a set of solutions as diverse as possible

The first goal is mandatory in any optimization task. Converging to a set of solutions, which are not close to the true optimal set of solutions is not desirable. It is only when solutions converge close to the true optimal solutions that one can be assured of their near-optimality properties. This goal of MO is common to the similar optimality goal in a single-objective optimization.

On the other hand, the second goal is entirely specific to MO. In addition to being converged close to the Pareto-optimal front, they must also be sparsely spaced in the Pareto-optimal region. We can be assured of having a good set of tradeoff solutions among objectives only with a diverse set of solutions. Since multi-objective evolutionary algorithms (MOEAs) deal with two spaces—decision variable space and objective space—"diversity" among solutions can be defined in both of these spaces. For example, two solutions can be said to be diverse in the decision variable space if their Euclidean distance in the decision variable space is large. Similarly, two solutions are diverse in the objective space, if their Euclidean distance in the objective space is large. Although in most problems diversity in one space usually means diversity in the other space, this may not be so in all problems. In such complex and nonlinear problems, it is then the task to find a set of solutions having a good diversity in the desired space [17].

3.2.2 Special Solutions

We first define some special solutions, which are often used in MO algorithms.

3.2.2.1 Ideal Objective Vector

For each of the k conflicting objectives, there exists one different optimal solution. An objective vector constructed with these individual optimal objective values constitutes the ideal objective vector.

Definition 3.1

The ith component of the ideal objective vector $\mathbf{z}^*$ is the constrained minimum solution of the following problem:

$$\left.\begin{array}{ll} \min & f_i(\mathbf{x}) \\ \text{s.t.} & \mathbf{x} \in \mathcal{S}. \end{array}\right\} \tag{3.2}$$

Thus, if the minimum solution for the ith objective function is the decision vector $\mathbf{x}^{*(i)}$ with function value f_i^*, then the ideal vector is as follows:

$$\mathbf{z}^* = \mathbf{f}^* = (f_1^*, f_2^*, \ldots, f_k^*)^T.$$

In general, the ideal objective vector corresponds to a nonexistent solution. This is because the minimum solution of Equation 3.2 for each objective function need not be the same solution. The only way an ideal objective vector corresponds to a feasible solution is when the minimal solutions to all objective functions are identical. In this case, the objectives are not conflicting with each other and the minimum solution to any objective function would be the only optimal solution to the MOOP. Figure 3.2 shows the ideal objective vector ($\mathbf{z}^*$) in the objective space of a hypothetical two-objective minimization problem.

It is interesting to ponder the question, "If the ideal objective vector is nonexistent, what is its use?" In most algorithms, which are seeking to find Pareto-optimal solutions, the ideal objective vector is used as a reference solution (we are using the word "solution" corresponding to the ideal objective vector loosely here, realizing that an ideal vector represents a nonexistent solution). It is also clear from Figure 3.2 that solutions closer to the ideal objective vector are better. Moreover, many algorithms require the knowledge of the lower bound on each objective function to normalize objective values in a common range.

3.2.2.2 Utopian Objective Vector

The ideal objective vector denotes an array of the lower bound of all objective functions. This means that for every objective function there exists at least one solution in the feasible search space sharing an identical value with the corresponding element in the ideal solution. Some algorithms may require

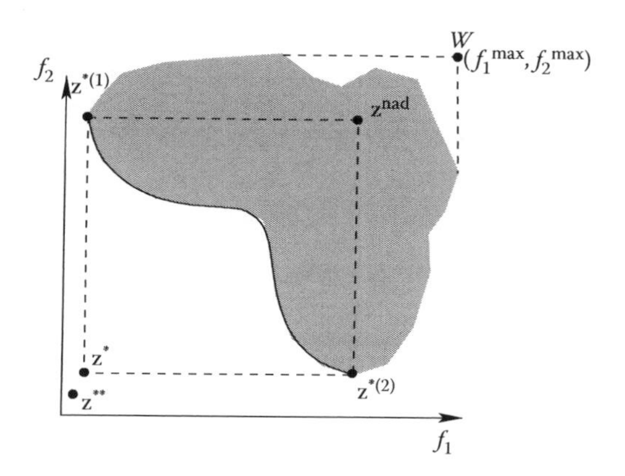

FIGURE 3.2
Ideal, utopian, and nadir objective vectors.

a solution, which has an objective value strictly better than (and not equal to) that of any solution in the search space. For this purpose, the utopian objective vector is defined as follows (for objectives to be minimized).

Definition 3.2

A utopian objective vector $\mathbf{z}^{**}$ has each of its components marginally smaller than that of the ideal objective vector, or $\mathbf{z}_i^{**} = \mathbf{z}_i^* - \epsilon_i$ with $\epsilon_i > 0$ for all $i = 1, 2, \ldots, k$.

Figure 3.2 shows a utopian objective vector. Like the ideal objective vector, the utopian objective vector also represents a nonexistent solution.

3.2.2.3 Nadir Objective Vector

Unlike the ideal objective vector that represents the lower bound of each objective in the entire feasible search space, the nadir objective vector, $\mathbf{z}^{\text{nad}}$, represents the upper bound of each objective in the entire Pareto-optimal set, and not in the entire search space. A nadir objective vector must not be confused with a vector of objectives (marked as "W" in Figure 3.2) found by using the worst feasible function values, $f_i^{\max}$, in the entire search space.

Although the ideal objective vector is easy to compute (except in complex multimodal objective problems), the nadir objective vector is difficult to compute in practice. However, for well-behaved problems (including linear MOOPs), the nadir objective vector can be derived from the ideal objective vector using the *payoff table* method described in Miettinen [52]. For the two objectives (Figure 3.2), if $\mathbf{z}^{*(1)} = \left(f_1(\mathbf{x}^{*(1)}), f_2(\mathbf{x}^{*(1)})\right)^T$ and $\mathbf{z}^{*(2)} = \left(f_1(\mathbf{x}^{*(2)}), f_2(\mathbf{x}^{*(2)})\right)^T$ are coordinates of the minimum solutions of f_1 and f_2, respectively, in the objective space, then the nadir objective vector can be estimated as $\mathbf{z}^{\text{nad}} = \left(f_1(\mathbf{x}^{*(2)}), f_2(\mathbf{x}^{*(1)})\right)^T$.

The nadir objective vector may represent an existent or a nonexistent solution, depending on the convexity and continuity of the Pareto-optimal set. To normalize each objective in the entire range of the Pareto-optimal region, the knowledge of nadir and ideal objective vectors can be used as follows:

$$f_i^{\text{norm}} = \frac{f_i - z_i^*}{z_i^{\text{nad}} - z_i^*}. \tag{3.3}$$

3.2.3 Concept of Domination in MO

Most MO algorithms use the concept of domination. In these algorithms, two solutions are compared on the basis of whether one dominates the other solution or not. We will describe the concept of domination in the following paragraph.

We assume that there are k objective functions. To cover both minimization and maximization of objective functions, we use the operator $\triangleleft$ between two solutions i and j as $i \triangleleft j$ to denote that the solution i is better than solution j on a particular objective. Similarly, $i \triangleright j$ for a particular objective implies that solution i is worse than solution j on this objective. For example, if an objective function is to be minimized, the operator $\triangleleft$ would mean the "$<$" operator, whereas if the objective function is to be maximized, the operator $\triangleleft$ would mean the "$>$" operator. The following definition covers mixed problems with minimization of some objective functions and maximization of the rest of them.

Definition 3.3

A solution $\mathbf{x}^{(1)}$ is said to dominate the other solution $\mathbf{x}^{(2)}$, if both conditions 1 and 2 are true:

1. Solution $\mathbf{x}^{(1)}$ is no worse than $\mathbf{x}^{(2)}$ in all objectives, or $f_j(\mathbf{x}^{(1)}) \not\triangleright f_j(\mathbf{x}^{(2)})$ for all $j = 1, 2, \ldots, k$.
2. Solution $\mathbf{x}^{(1)}$ is strictly better than $\mathbf{x}^{(2)}$ in at least one objective, or $f_{\bar{j}}(\mathbf{x}^{(1)}) \triangleleft f_{\bar{j}}(\mathbf{x}^{(2)})$ for at least one $\bar{j} \in \{1, 2, \ldots, k\}$.

If any of the above condition is violated, solution $\mathbf{x}^{(1)}$ does not dominate solution $\mathbf{x}^{(2)}$. If $\mathbf{x}^{(1)}$ dominates solution $\mathbf{x}^{(2)}$ (or mathematically $\mathbf{x}^{(1)} \preceq \mathbf{x}^{(2)}$), it is also customary to write any of the following:

- $\mathbf{x}^{(2)}$ is dominated by $\mathbf{x}^{(1)}$.
- $\mathbf{x}^{(1)}$ is nondominated by $\mathbf{x}^{(2)}$.
- $\mathbf{x}^{(1)}$ is noninferior to $\mathbf{x}^{(2)}$.

Let us consider a two-objective optimization problem with five different solutions shown in the objective space, as illustrated in Figure 3.3. Let us also assume that objective function 1 needs to be maximized while objective function 2 needs to be minimized. Five solutions with different objective function values are shown in this figure. Since both objective functions are of importance to us, it is usually difficult to find one solution which is best with respect to both the objectives. However, we can use the above definition of domination to decide which solution is better among any two given solutions in terms of both the objectives. For example, if solutions 1 and 2 are to be compared, we observe that solution 1 is better than solution 2 in objective function 1 and solution 1 is also better than solution 2 in objective function 2. Thus, both of the above conditions for domination are satisfied and we may write that solution 1 dominates solution 2. We take another instance of comparing solutions 1 and 5. Here, solution 5 is better than solution 1 in the first objective and solution 5 is no worse (in fact, they are equal) than solution 1 in the second objective. Thus, both the above conditions for domination are also satisfied and we may write that solution 5 dominates solution 1.

It is intuitive that if solution $\mathbf{x}^{(1)}$ dominates another solution $\mathbf{x}^{(2)}$, then solution $\mathbf{x}^{(1)}$ is better than $\mathbf{x}^{(2)}$ in the parlance of MO. Since the concept of domination allows a way to compare solutions with multiple objectives, most MO methods use this domination concept to search for nondominated solutions.

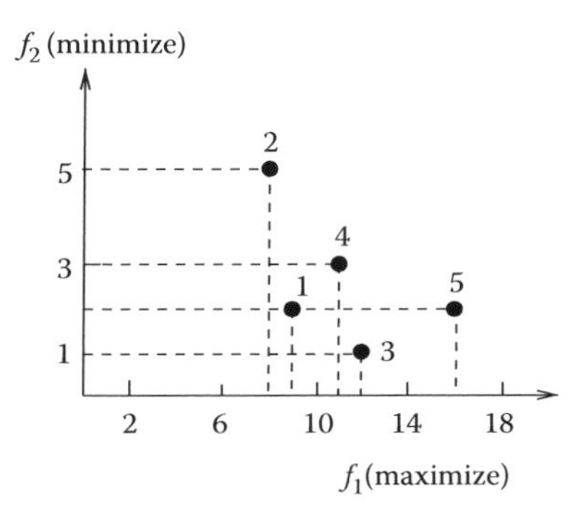

FIGURE 3.3
A population of five solutions.

3.2.4 Properties of Dominance Relation

Definition 3.3 defines the dominance relation between any two solutions. There are three possibilities that can be the outcome of the dominance check between solutions 1 and 2, that is, (i) solution 1 dominates solution 2, (ii) solution 1 gets dominated by solution 2, or (iii) solutions 1 and 2 do not dominate each other. Let us now discuss the different binary relation properties [11] of the dominance operator.

Reflexive: The dominance relation is *not reflexive*, since any solution p does not dominate itself according to Definition 3.3. The second condition of dominance relation in Definition 3.3 does not allow this property to be satisfied.

Symmetric: The dominance relation is also *not symmetric*, because $p \preceq q$ does not imply $q \preceq p$. In fact, the opposite is true. That is, if p dominates q, then q does not dominate p. Thus, the dominance relation is *asymmetric*.

Antisymmetric: Since the dominance relation is not symmetric, it cannot be antisymmetric as well.

Transitive: The dominance relation is *transitive*. This is because if $p \preceq q$ and $q \preceq r$, then $p \preceq r$.

There is another interesting property that the dominance relation possesses. If solution p does not dominate solution q, this does not imply that q dominates p.

In order for a binary relation to qualify as an ordering relation, it must be at least transitive [5]. Thus, the dominance relation qualifies as an ordering relation. Since the dominance relation is not reflexive, it is a *strict partial order*. In general, if a relation is reflexive, antisymmetric, and transitive, it is loosely called a *partial order* and a set on which a partial order is defined is called a *partially ordered set*. However, it is important to note that the dominance relation is not reflexive and is not antisymmetric. Thus, the dominance relation is not a partial order relation in its general sense. The dominance relation is only a strict partial order relation.

3.2.5 Pareto-Optimality

Continuing with the comparisons in the previous subsection, let us compare solutions 3 and 5 in Figure 3.3, because this comparison shows an interesting aspect. We observe that solution 5 is better than solution 3 in the first objective, while solution 5 is worse than solution 3 in the second objective. Thus, the first condition is not satisfied for both these solutions. This simply suggests that we can neither conclude that solution 5 dominates solution 3 nor can we say that solution 3 dominates solution 5. When this happens, it is customary to say that solutions 3 and 5 are nondominated with respect to each other. When both objectives are important, it cannot be said which of the two solutions, in solutions 3 and 5, is better.

For a given finite set of solutions, we can perform all possible pairwise comparisons and find which solution dominates which and which solutions are nondominated with respect to each other. At the end, we expect to have a set of solutions, any two of which do not dominate each other. This set also has another property. For any solution outside of this set, we can always find a solution in this set, which will dominate the former. Thus, this particular set has a property of dominating all other solutions, which do not belong to this set. In simple terms, this means that the solutions of this set are better compared with the rest of solutions. This set is given a special name. It is called the *nondominated set* for the given set of solutions. In the example problem, solutions 3 and 5 constitute the nondominated set of the given set of five solutions. Thus, we define a set of nondominated solutions as follows.

Definition 3.4: Nondominated Set

Among a set of solutions P, the nondominated set of solutions P' are those that are not dominated by any member of the set P.

When set P is the entire search space, or $P = \mathcal{S}$, the resulting nondominated set P' is called the *Pareto-optimal set*. Figure 3.4 marks the Pareto-optimal set with continuous curves for four different scenarios with two objectives. Each objective can be minimized or maximized. In the top-left figure, the task is to minimize both objectives f_1 and f_2. The solid curve marks the Pareto-optimal solution set. If f_1 is to be minimized and f_2 is to be maximized for a problem having the same search space, the resulting Pareto-optimal set is different and is shown in the top-right figure. Here, the Pareto-optimal set is a union of two disconnected Pareto-optimal regions. Similarly, the Pareto-optimal sets for two other cases ([maximizing f_1, minimizing f_2] and [maximizing f_1, maximizing f_2]) are also shown in the bottom-left and bottom-right figures, respectively. In any case, the Pareto-optimal set always consists of solutions from a particular edge of the feasible search region.

It is important to note that an multi-objective evolutionary algorithm (MOEA) can be easily used to handle all of the above cases by simply using the domination definition. However, to avoid any confusion, most applications use the (discussed earlier) to convert a maximization problem into a minimization problem and treat every problem as a combination of minimizing all objectives.

Like global and local optimal solutions in the case of single-objective optimization, there could be global and local Pareto-optimal sets in MO.

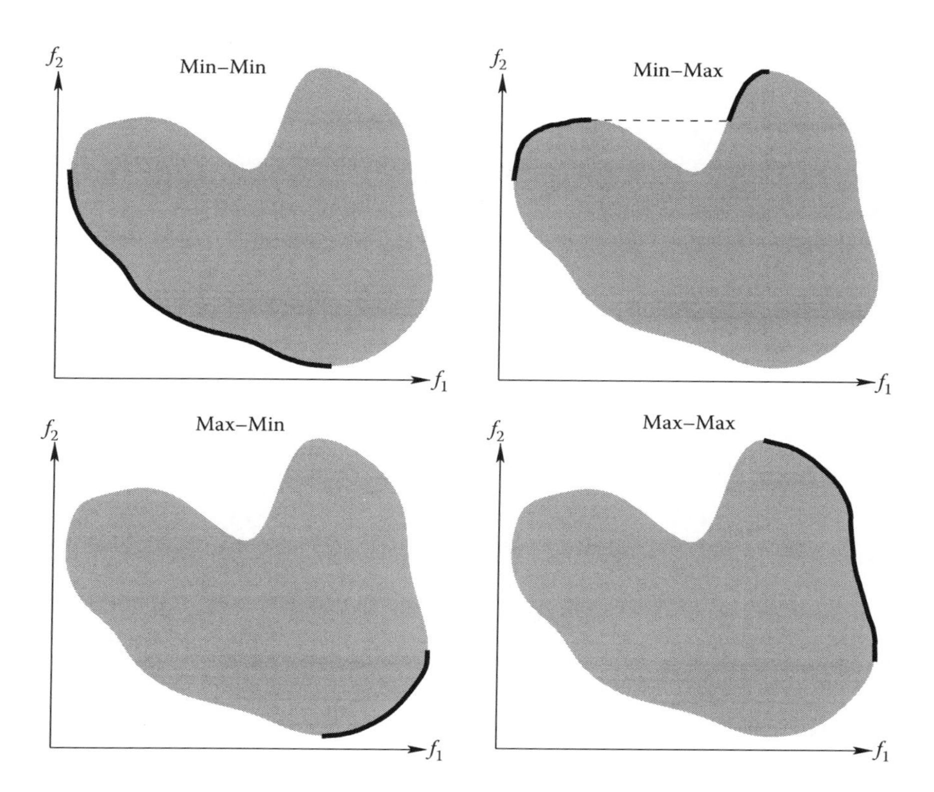

FIGURE 3.4
Pareto-optimal solutions are marked with continuous curves for four combinations of two types of objectives.

Definition 3.5: Global Pareto-Optimal Set

The nondominated set of the entire feasible search space $\mathcal{S}$ is the global Pareto-optimal set.

On many occasions, the global Pareto-optimal set is simply referred to as the Pareto-optimal set. Since solutions of this set are not dominated by any feasible member of the search space, they are optimal solutions of the MOOP. We define a local Pareto-optimal set as follows [17,52].

Definition 3.6

If for every member **x** in a set $\underline{P}$ there exists no solution **y** (in the neighborhood of **x** such that $\|\mathbf{y}-\mathbf{x}\|_\infty \leq \epsilon$, where ϵ is a small positive number) dominating any member of the set $\underline{P}$, then solutions belonging to the set $\underline{P}$ constitute a local Pareto-optimal set.

Figure 3.5 shows two local Pareto-optimal sets (marked with continuous curves). When any solution (say "B") in this set is perturbed locally in the decision variable space, no solution can be found dominating any member of the set. It is interesting to note that for continuous search space problems, the local Pareto-optimal solutions need not be continuous in the decision variable space and the above definition will still hold good. Zitzler [73] added a neighborhood constraint on the objective space in the above definition to make it more generic.

By the above definition, it is also true that a global Pareto-optimal set is also a local Pareto-optimal set.

3.2.6 Strong Dominance and Weak Pareto-Optimality

The dominance relationship between the two solutions defined in Definition 3.3 is sometimes referred to as a *weak* dominance relation. This definition can be modified and a strong dominance relation can be defined as follows.

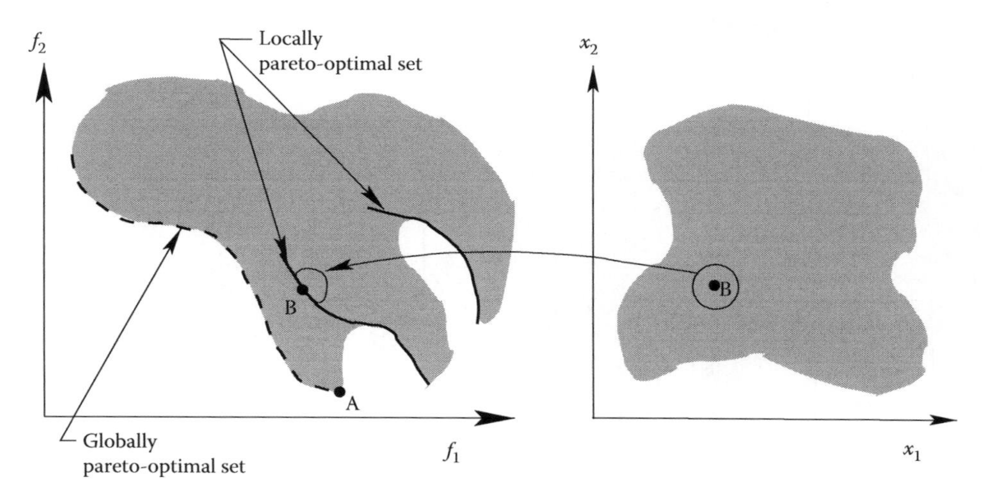

FIGURE 3.5
Local and global Pareto-optimal solutions.

Definition 3.7

Solution $\mathbf{x}^{(1)}$ *strongly* dominates a solution $\mathbf{x}^{(2)}$ (or $\mathbf{x}^{(1)} \prec \mathbf{x}^{(2)}$), if solution $\mathbf{x}^{(1)}$ is strictly better than solution $\mathbf{x}^{(2)}$ in all k objectives.

Referring to Figure 3.3, we now observe that solution 5 does not strongly dominate solution 1, although we have seen earlier that solution 5 weakly dominates solution 1. However, solution 3 strongly dominates solution 1, since solution 3 is better than solution 1 in both objectives. Thus, if solution $\mathbf{x}^{(1)}$ strongly dominates solution $\mathbf{x}^{(2)}$, solution $\mathbf{x}^{(1)}$ also weakly dominates solution $\mathbf{x}^{(2)}$, but not vice versa. The strong dominance operator has the same properties as that described in Section 3.2.4 for the weak dominance operator.

The above definition of strong dominance can be used to define a *weakly nondominated set.*

Definition 3.8: Weakly Nondominated Set

Among a set of solutions P, the weakly nondominated set of solutions P' are those that are not strongly dominated by any other member of the set P.

The above definition suggests that a weakly nondominated set found from a set of P solutions contains all members of the nondominated set obtained using Definition 3.4 from the same set P. In other words, for a given population of solutions, the cardinality of the weakly nondominated set is greater than or equal to the cardinality of the nondominated set obtained using the usual Definition 3.4. The definition of a globally or locally weakly Pareto-optimal set can also be defined similarly using the definition of the weak nondominated set.

3.3 Optimality Conditions

In this section, we outline the theoretical Pareto-optimality conditions for a constrained MOOP. Here, we state the optimality conditions for the MOOP given in Equation 3.1. We assume that all objectives and constraint functions are continuously differentiable. As in single-objective optimization, there exist first- and second-order optimality conditions for MO. Here, we state the first-order necessary condition and a specific sufficient condition. Interested readers may refer to more advanced classical books on MO [5,28,52] for a more comprehensive treatment of different optimality conditions.

The following condition is known as the necessary condition for Pareto-optimality.

Theorem 3.1

A necessary condition for $\mathbf{x}^*$ to be Pareto optimal is that there exist vectors $\boldsymbol{\lambda} \geq \mathbf{0}$ and $\mathbf{u} \geq \mathbf{0}$ (where $\boldsymbol{\lambda} \in \mathbb{R}^k$, $\mathbf{u} \in \mathbb{R}^J$ and $\boldsymbol{\lambda}, \mathbf{u} \neq \mathbf{0}$) such that the following conditions are true:

1. $\sum_{m=1}^{k} \lambda_m \nabla f_m(\mathbf{x}^*) - \sum_{j=1}^{J} u_j \nabla g_j(\mathbf{x}^*) = \mathbf{0}$ and
2. $u_j g_j(\mathbf{x}^*) = 0$ for all $j = 1, 2, \ldots, J$

For proof, readers may refer to Cunha and Polak [13]. Miettinen [52] argues that the above theorem is also valid as the necessary condition for a solution to be weakly Pareto optimal. Those readers familiar with the Kuhn–Tucker necessary conditions for single-objective optimization will immediately recognize the similarity between the above conditions and that of the single-objective optimization. The difference is in the inclusion of a λ-vector with the gradient vector of the objectives.

For an unconstrained MOOP, the above theorem requires the following condition:

$$\sum_{m=1}^{k} \lambda_m \nabla f_m(\mathbf{x}^*) = \mathbf{0}$$

to be necessary for a solution to be Pareto optimal. Writing the above vector equation in matrix form, we have the following necessary condition for a k-objective and n-variable unconstrained MOOP:

$$\begin{bmatrix} \frac{\partial f_1}{\partial x_1} & \frac{\partial f_2}{\partial x_1} & \cdots & \frac{\partial f_k}{\partial x_1} \\ \frac{\partial f_1}{\partial x_2} & \frac{\partial f_2}{\partial x_2} & \cdots & \frac{\partial f_k}{\partial x_2} \\ \cdots & \cdots & \cdots & \cdots \\ \frac{\partial f_1}{\partial x_n} & \frac{\partial f_2}{\partial x_n} & \cdots & \frac{\partial f_k}{\partial x_n} \end{bmatrix} \begin{bmatrix} \lambda_1 \\ \lambda_2 \\ \vdots \\ \lambda_k \end{bmatrix} = \begin{bmatrix} 0 \\ 0 \\ \vdots \\ 0 \end{bmatrix}. \tag{3.4}$$

For nonlinear objective functions, the partial derivatives are expected to be nonlinear. For a given λ-vector, the nonexistence of a corresponding Pareto-optimal solution can be checked using the above equation. If the above set of equations are not satisfied, a Pareto-optimal solution corresponding to the given λ-vector does not exist. However, the existence of a Pareto-optimal solution is not guaranteed by the above necessary condition. That is, any solution that satisfies Equation 3.4 is not necessarily a Pareto-optimal solution.

For problems with $n = k$ (identical number of decision variables and objectives), the Pareto-optimal solutions must satisfy the following:

$$\begin{vmatrix} \frac{\partial f_1}{\partial x_1} & \frac{\partial f_2}{\partial x_1} & \cdots & \frac{\partial f_k}{\partial x_1} \\ \frac{\partial f_1}{\partial x_2} & \frac{\partial f_2}{\partial x_2} & \cdots & \frac{\partial f_k}{\partial x_2} \\ \cdots & \cdots & \cdots & \cdots \\ \frac{\partial f_1}{\partial x_n} & \frac{\partial f_2}{\partial x_n} & \cdots & \frac{\partial f_k}{\partial x_n} \end{vmatrix} = 0. \tag{3.5}$$

The determinant of the partial derivative matrix must be zero for Pareto-optimal solutions. For a two-variable, two-objective MOOP, the above condition reduces to the following:

$$\frac{\partial f_1}{\partial x_1}\frac{\partial f_2}{\partial x_2} = \frac{\partial f_1}{\partial x_2}\frac{\partial f_2}{\partial x_1}. \tag{3.6}$$

The following theorem offers sufficient conditions for a solution to be Pareto optimal for convex functions.

Theorem 3.2

Let the objective functions be convex and the constraint functions of the problem shown in Equation 3.1 be nonconvex. Let the objective and constraint functions be continuously differentiable at a feasible solution $\mathbf{x}^*$. A sufficient condition for $\mathbf{x}^*$ to be Pareto optimal is that there exist vectors $\boldsymbol{\lambda} > \mathbf{0}$ and $\mathbf{u} \geq \mathbf{0}$ (where $\boldsymbol{\lambda} \in \mathbb{R}^k$ and $\mathbf{u} \in \mathbb{R}^J$) such that the following equations are true:

1. $\sum_{m=1}^{k} \lambda_m \nabla f_m(\mathbf{x}^*) - \sum_{j=1}^{J} u_j \nabla g_j(\mathbf{x}^*) = \mathbf{0}$
2. $u_j g_j(\mathbf{x}^*) = 0$ for all $j = 1, 2, \ldots, J$.

For a proof, see Miettinen [52]. If the objective functions and constraints are not convex, the above theorem does not hold. However, for pseudo-convex and nondifferentiable problems, different necessary and sufficient conditions do exist [2].

3.4 Classification Based on the Role of the Decision Maker

Several methods have been proposed in the literature for solving multi-objective optimization problems and are classified in different ways. Often solving an MOOP means supporting the DM in finding his/her most preferred Pareto-optimal solution(s). Hence, the classification based on the role of the DM is popular [52] and consists of the following methods:

- *No-preference methods.* In no-preference methods, preference information from the DM is not considered and an MOOP is solved to obtain a Pareto-optimal solution, which can be considered as some sort of compromise solution. The DM can subsequently consider or reject the proposed solution. Since no-preference methods do not consider the preferences of the DM, they are usually considered when either a DM is unavailable or he/she has no preference information to provide. The method of global criterion [68,70] is one of the most common examples for no-preference methods.
- *A priori methods.* In *a priori* methods, the DM is asked to provide his/her preference information beforehand, which is subsequently used to formulate a scalarizing function. The resulting scalarizing problem defined using the formulated scalarizing function is subsequently solved to obtain a Pareto-optimal solution that satisfies the DM's preferences. Usually, it is difficult for the DM to express *a priori* his/her preference information, which may be too optimistic or too pessimistic. Lexicographic ordering [31] and goal programming [6–8] are examples of *a priori* methods.
- *A posteriori methods.* In *a posteriori* methods, a representative set of Pareto-optimal solutions is presented to the DM and later on the DM selects one among them based on his/her preferences. Although finding a representative set of Pareto-optimal solutions beforehand looks attractive, it is difficult/impractical to try to represent the entire Pareto front in multi-objective optimization problems with more than two objectives, as many real-world multi-objective optimization problems are computationally expensive. The weighting method [34] and the ϵ-constraint methods [38] are commonly used as *a posteriori* methods by solving several single-objective optimization problems sequentially to obtain a representation of the entire Pareto-optimal front.
- *Interactive methods.* In interactive methods, the DM provides preference information iteratively and thus directs the solution process toward his/her preferred Pareto-optimal solution(s). Here, a constant interaction is involved between the interactive method and the DM. Initially, a Pareto-optimal solution is provided to the DM based on which the DM can specify his/her preference information. Using this preference information, Pareto-optimal

solution(s) are generated. The procedure is iterated until the DM is satisfied with the Pareto-optimal solution and does not wish to continue further. There are several interactive methods in the literature such as STOM [55], etc.

3.5 No-Preference Methods

3.5.1 Method of Global Criterion

The most well-known no-preference method is *the method of global criterion* [68,70], which is that is also known as *compromise programming*. The general idea is to find a Pareto-optimal solution that is, in some sense, closest to the ideal objective vector. To measure the closeness, different metrics can be used, typically, L^p-metrics for $1 \leq p \leq \infty$. By using different metrics, different solutions can be obtained.

The scalarized problem to be solved in the method of global criterion when $p < \infty$ is

$$\begin{aligned} \min \ & \left(\sum_{i=1}^{k} |f_i(x) - z_i^*|^p \right)^{1/p} \\ \text{s.t.} \ & x \in S. \end{aligned} \tag{3.7}$$

It can be shown that the solution of problem (3.7) is Pareto optimal [52]. When $p = \infty$, the problem is of the form

$$\begin{aligned} \min \ & \max_{i=1,\ldots,k} \left[|f_i(x) - z_i^*| \right] \\ \text{s.t.} \ & x \in S. \end{aligned} \tag{3.8}$$

Note that problem (3.8) is nondifferentiable even if the objective and constraint functions are differentiable. It can be shown that the solution of problem (3.8) is weakly Pareto optimal [52]. If the scales of the objective functions are totally different, a normalization should be used. Instead of using $f_i(x) - z_i^*$ in Equations 3.7 and 3.8, it can be substituted, for example, with $(f_i(x) - z_i^*)/(z_i^{nad} - z_i^*)$. Note that in this case, a nadir objective vector is also needed.

If all the objective and constraint functions are differentiable and the ideal objective vector is known globally, problem (3.8) can be written in a corresponding differentiable form

$$\begin{aligned} \min_{x,\delta} \ & \delta \\ \text{s.t.} \ & f_i(x) - z_i^* \leq \delta, \quad \text{for all } i = 1, \ldots, k, \\ & x \in S, \quad \delta \in \mathbb{R}. \end{aligned} \tag{3.9}$$

This means that one additional variable and k inequality constraints are added. Note that the minimization in problem (3.9) is with respect to both x and δ.

An example of the performance of the method of global criterion can be seen from Figure 3.6 for different values of p. Note that for this problem, the values of $p = 1$ and $p = \infty$ will produce the same Pareto-optimal solution $z = (0, 1)^T$.

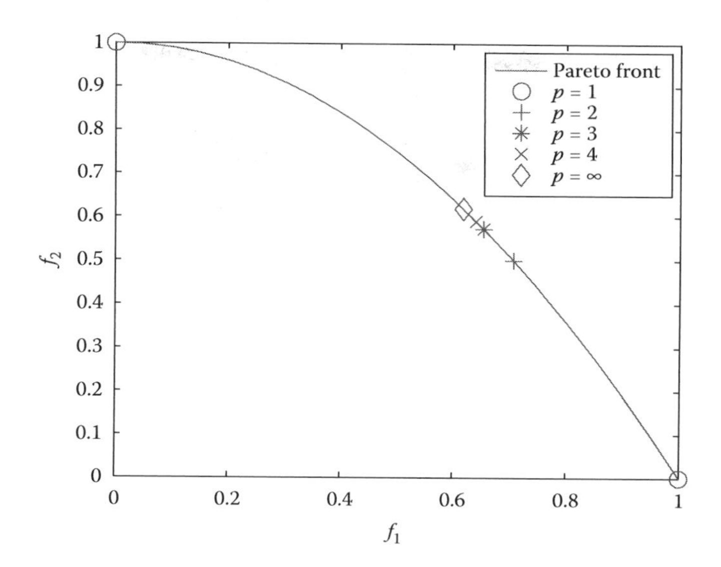

FIGURE 3.6
Pareto-optimal solutions produced by the method of global criterion for different values of p in the case of the ZDT2 problem.

3.6 *A Posteriori* Methods

3.6.1 Weighting Method

Probably the most widely used and well-known MO method is the *weighting method*, where a weighted sum of the objective functions is optimized. Reasons for its popularity include that it is very easy to implement by using traditional single-objective optimization methods and it retains the linearity of the optimization problem if all the objective and constraint functions are linear. The weighting method is presented, for example, in References 34 and 69.

The scalarized single-objective optimization problem to be solved in the weighting method is

$$\begin{aligned} \min \ & \sum_{i=1}^{k} w_i f_i(x) \\ \text{s.t.} \ & x \in S, \end{aligned} \tag{3.10}$$

where $w = (w_1, \ldots, w_k)^T$ is the vector of weights such that $w_i \geq 0$ for all $i = 1, \ldots, k$ and $\sum_{i=1}^{k} w_i = 1$. It can be shown that a solution to problem (3.10) is weakly Pareto optimal and it is Pareto optimal if all the weights are positive, that is $w_i > 0$ for all $i = 1, \ldots, k$ (see, e.g., Reference 52). Different (weakly) Pareto-optimal solutions can be obtained using different weight vectors w.

The weighting method can be used either as an *a posteriori* or as an *a priori* method. In the former case, different weight vectors are used to produce a set of Pareto-optimal solutions, which is then shown to the DM for selecting a most preferred solution. In the latter case, the preferences of the DM are represented by a weight vector w and the corresponding solution is obtained by solving problem (3.10).

However, the weighting method has well-known weaknesses (see, e.g., References 14, 52, and 55). First, it is not able to produce any Pareto-optimal solutions that lie in the nonconvex part of the Pareto-optimal front because the level sets of the objective function in Equation 3.10 are linear

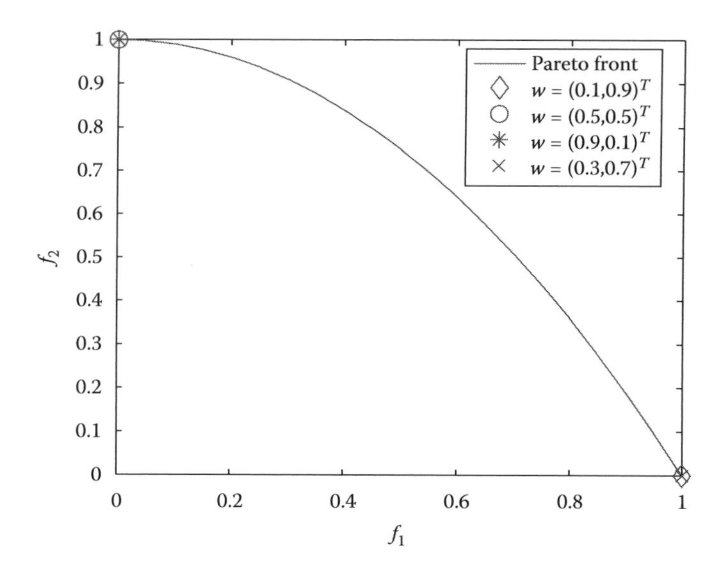

FIGURE 3.7
Pareto-optimal solution is produced by the weighting method for different weight vectors w. Note that since the ZDT2 problem has a nonconvex Pareto front, only the extreme solutions can be found.

subspaces in the objective space. On the other hand, when representing preferences of the DM with weights, the solution obtained with the weighting method does not necessarily reflect the preferences expressed as demonstrated, for example, in Reference 64 (an example originally formulated by Professor P. Korhonen). Furthermore, a set of evenly distributed weights does not guarantee an evenly distributed set of Pareto-optimal solutions even for convex problems [14].

Figure 3.7 illustrates the performance of the weighting method when applied to the ZDT2 problem. Four different weight vectors are used, namely, $(0.1, 0.9)^T$, $(0.5, 0.5)^T$, $(0.9, 0.1)^T$, and $(0.3, 0.7)^T$. Since the Pareto-optimal front of the ZDT2 problem is nonconvex, only the two extreme Pareto-optimal solutions can be found with the weighting method. Therefore, the weighting method is not a suitable method to be used for this problem. The extreme Pareto-optimal solution corresponding to the objective that has a bigger weight is found. Furthermore, this example emphasizes that the weighting method should only be used for problems having convex Pareto front.

3.6.2 ϵ-Constraint Method

Another example of a widely used MO method that retains the linearity of the optimization problem to be solved is the ϵ-*constraint method* [38], where one of the objective functions is selected to be optimized and the other objective functions are converted into constraint functions by specifying bounds for them. When all the objective functions are to be minimized, the scalarized problem of the ϵ-constraint method is

$$\begin{aligned} &\min\ f_j(x) \\ &\text{s.t.}\ \ f_i(x) \leq \epsilon_i \quad \text{for all } i \neq j, \quad x \in S, \end{aligned} \tag{3.11}$$

where ϵ_i is an upper bound for the value of the objective function f_i. It can be shown that a solution $x^* \in S$ of problem (3.11) is weakly Pareto optimal. Furthermore, a solution $x^* \in S$ is Pareto optimal if and only if it is a solution of problem (3.11) for every $j = 1, \ldots, k$, where $\epsilon_i = f_i(x^*)$

for $i = 1, \ldots, k$, $i \neq j$. In addition, a solution $x^* \in S$ is Pareto optimal if it is a unique solution of Equation 3.11 for some j with $\epsilon_i = f_i(x^*)$ for $i = 1, \ldots, k$, $i \neq j$. Finally, any unique solution of Equation 3.11 is Pareto optimal for any given upper bounds ϵ_i. The proofs can be found, for example, in Reference 52. Different (weakly) Pareto-optimal solutions are obtained by specifying different upper bounds ϵ_i and/or selecting different objective function f_j to be optimized.

When compared with the weighting method, the ϵ-constraint method is able to find Pareto-optimal solutions also in the nonconvex parts of the Pareto-optimal set. On the other hand, the ϵ-constraint method is computationally more demanding than the weighting method since $k - 1$ inequality constraints are added to the scalarized problem to be optimized. It can also be noted from the above results that any Pareto-optimal solution can be found with the ϵ-constraint method using the appropriate objective function f_j to be optimized and appropriate values for the upper bounds ϵ_i. However, those appropriate values are not known beforehand.

The drawbacks of the ϵ-constraint method include that in order to guarantee that a Pareto-optimal solution is obtained, the upper bounds need to be set properly into the values that are typically not known before optimization. In addition, selection of the objective function f_j to be optimized has also an unpredictable effect to the solutions obtained.

The ϵ-constraint method can also be used either as an *a posteriori* or as an *a priori* method. In the former case, different Pareto-optimal solutions are computed for the DM by using different upper bounds (or changing the function to be optimized). In the latter case, the DM selects the function to be optimized as well as specifies the upper bounds to be used.

An illustration of the ϵ-constraint method for the ZDT2 problem is shown in Figure 3.8. The objective function f_1 is minimized while different upper bounds are used for f_2. As can be seen from the figure and from the theory above, the ϵ-constraint method is able to find Pareto-optimal solutions also in the nonconvex part of the Pareto-optimal front. When compared with the weighting method, this is a clear advantage of the ϵ-constraint method. Due to the symmetry of the Pareto-optimal front of the ZDT2 problem, a similar behavior can be assumed if the objective f_2 is selected to be minimized and the upper bounds are given to f_1.

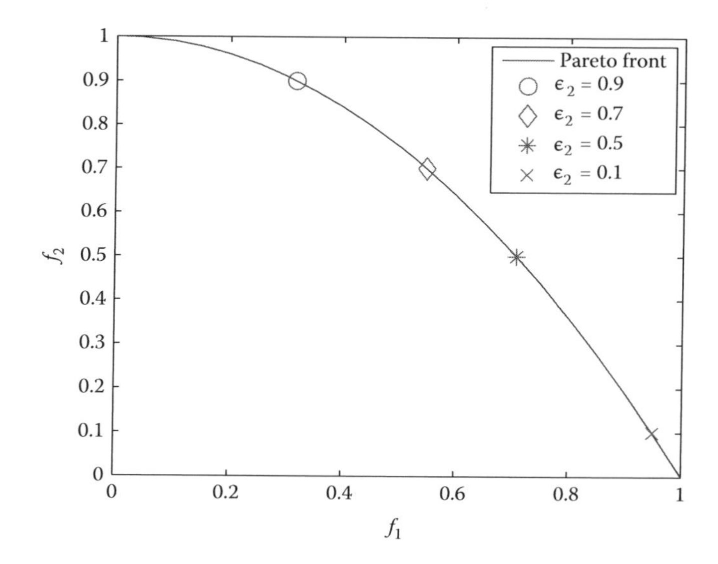

FIGURE 3.8
The Pareto-optimal solutions produced by the ϵ-constraint method for the ZDT2 problem. The objective f_1 is minimized while different upper bounds are used for f_2.

3.6.3 Method of Weighted Metrics

In the *method of weighted metrics*, the idea is to find a Pareto-optimal solution closest to the ideal objective vector with respect to different metrics like in the method of global criterion. The difference here is that the metric used can be augmented with weights to get different Pareto-optimal solutions. The method of weighted metrics is also sometimes called compromise programming [70].

The scalarized problem to be solved is of the form

$$\begin{aligned} &\min \left(\sum_{i=1}^{k} w_i |f_i(x) - z_i^*|^p \right)^{1/p} \\ &\text{s.t.} \quad x \in S \end{aligned} \tag{3.12}$$

for the L^p metric when $p < \infty$. We assume that $w_i \geq 0$ for all $i = 1, \ldots, k$ and $\sum_{i=1}^{k} w_i = 1$. When $p = \infty$, the problem is of the form

$$\begin{aligned} &\min \max_{i=1,\ldots,k} \left[w_i |f_i(x) - z_i^*| \right] \\ &\text{s.t.} \quad x \in S \end{aligned} \tag{3.13}$$

and it was first introduced in Reference 3. It can be shown that the solution of problem (3.12) is Pareto optimal if it is unique or all the weighting coefficients are positive [52]. Further, the solution of problem (3.13) is weakly Pareto optimal if all the weighting coefficients are positive [52].

Note that problem (3.13) is nondifferentiable like problem (3.8) in the method of global criterion as mentioned before. Similarly, if all the objective and constraint functions are differentiable and the ideal objective vector is known globally, the corresponding differentiable form of problem (3.13) can be formulated.

3.6.4 Achievement Scalarizing Function Approach

Wierzbicki introduced the idea of achievement scalarizing functions, for example, in References 66 and 67. The achievement (scalarizing) function measures the distance between the reference point and Pareto-optimal solutions and produces a new Pareto-optimal solution closest to the reference point. A *reference point* $\bar{z} = (\bar{z}_1, \ldots, \bar{z}_k)^T \in \mathbb{R}^k$ consists of *aspiration levels* $\bar{z}_i$ that are desirable values for the objective functions f_i typically specified by the DM.

The most widely used approach utilizing achievement (scalarizing) function is the *reference point method* where the scalarized subproblem to be solved is of the form

$$\begin{aligned} &\min \max_{i=1,\ldots,k} [w_i(f_i(x) - \bar{z}_i)] + \rho \sum_{i=1}^{k} w_i f_i(x) \\ &\text{s.t.} \quad x \in S, \end{aligned} \tag{3.14}$$

where $w = (w_1, \ldots, w_k)^T$ is the vector of weights used for scaling. Typical choice for the weights is $w_i = 1/(z_i^{nad} - z_i^{uto})$. The second term of the objective function in Equation 3.14 is called an *augmentation term* (with the small *augmentation coefficient* $\rho > 0$) and it is used to guarantee Pareto-optimality of the solution. Note that problem (3.14) is nondifferentiable. The corresponding differentiable formulation (if all the objective and constraint functions are differentiable) is

$$\begin{aligned} &\min_{x,\delta} \delta + \rho \sum_{i=1}^{k} w_i f_i(x) \\ &\text{s.t.} \quad w_i(f_i(x) - \bar{z}_i) \leq \delta, \quad 1 \leq i \leq k, \\ &\qquad x \in S, \quad \delta \in \mathbb{R}. \end{aligned} \tag{3.15}$$

It can be shown (see, e.g., Reference 52) that the solution of problem (3.14) is Pareto optimal and, without the augmentation term, the solution is weakly Pareto optimal. Different Pareto-optimal solutions can be obtained using different reference points and any Pareto-optimal solution can be found using a suitable reference point. The advantages of the reference point method include that the solution is always Pareto optimal independently of whether the reference point used is feasible or not. In addition, the use of reference points in expressing the preferences of the DM has been found intuitive and cognitively valid [48].

In the reference point method, the DM provides a reference point and a corresponding solution of problem (3.14) is obtained and shown to the DM. In addition to k, other Pareto-optimal solutions around are computed by shifting the reference point slightly along each coordinate axis in the objective space. Therefore, the DM can choose between $k + 1$ Pareto-optimal solutions.

3.6.5 Normal Boundary Intersection

The main idea of the *normal boundary intersection (NBI)* method [15] is to produce an evenly distributed set of Pareto-optimal solutions. First, an evenly distributed set of points is generated in the convex hull of individual minima (*CHIM*) in the objective space. In mathematical notation, CHIM is defined by $CHIM = \{\Phi\beta : \beta \in \mathbb{R}^k, \sum_{i=1}^{k} \beta_i = 1, \beta_i \geq 0\}$, where Φ is the $k \times k$ matrix where the ith column is $f(x^{*,i}) - z^*$ and $x^{*,i}$ is the point where f_i achieves its minimum. To begin with, the points $x^{*,i}$, $i = 1, \ldots, k$ need to be computed by optimizing each objective function independently. Vector $\beta = (\beta_1, \ldots, \beta_k)^T$ denotes the coordinates of the points in CHIM. After a set of evenly distributed points in CHIM has been generated (how to do it is discussed later), each point is projected toward the Pareto-optimal set by solving parameterized scalar subproblems of the form

$$\begin{aligned} &\max_{x,t} \; t \\ &\text{s.t.} \quad \Phi\beta + t\hat{n} = f(x) - z^*, \; x \in S, \end{aligned} \tag{3.16}$$

where $\beta_i \geq 0$, $i = 1, \ldots, k$, $\sum_{i=1}^{k} \beta_i = 1$ and $\hat{n}$ is the unit normal to CHIM pointing toward the origin. The equality constraint in problem (3.16) guarantees that a point in CHIM defined by the coordinates β is projected along the normal $\hat{n}$. Note that maximization in problem (3.16) is with respect to both x and t. The solution of problem (3.16) is not always Pareto optimal because it can also lie in a strongly nonconvex part of the Pareto-optimal set as illustrated in Figure 3.9.

The normal vector in CHIM pointing toward the origin can be computed in practice, for example, by $\hat{n} = -\Phi e$, which gives a quasi-normal that is an approximation of the normal. Vector $e \in \mathbb{R}^k$ is the column vector of all ones and a minus sign is used to guarantee that the obtained vector points toward the origin since $\Phi_{i,j} \geq 0$ for each $i \neq j$ by its definition.

In Reference 15, an approach to generate an evenly distributed set of points in CHIM is introduced. Let us denote by $\delta > 0$ the distance between two consecutive values for β_i, that is, a step size. Here for the sake of simplicity, we consider only the case where the step size is equal for all β_i (see Reference 15 for details in the case of different step sizes). Let us also assume that $p = 1/\delta$ is an integer. In that case, the number of points the approach produces is

$$\binom{k+p-1}{p},$$

which means that there are as many scalarized problems to be solved. Now, the possible values for β_1 are $[0, \delta, 2\delta, \ldots, 1]$. If a fixed value for $\beta_1 = m_1\delta$ is given, the possible values for β_2 are $[0, \delta, 2\delta, \ldots, l_2\delta]$, where $l_2 = (1 - \beta_1)/\delta = (1 - m_1\delta)/\delta = p - m_1$. By induction, we get that

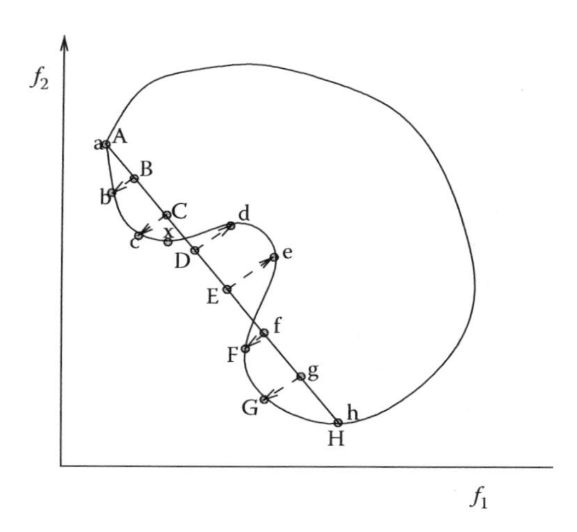

FIGURE 3.9
Illustration of the NBI method.

possible values for β_j corresponding to $\beta_i = m_i\delta$, $i = 1, \ldots, j-1$ for $j = 2, \ldots, k-1$ are

$$\left[0, \delta, 2\delta, \ldots, \left(p - \sum_{i=1}^{j-1} m_i\right)\delta\right]$$

and $\beta_k = 1 - \sum_{i=1}^{k-1} \beta_i$.

For every point β generated on CHIM, a scalarized problem (3.16) needs to be solved except for k extreme points where one of the components is equal to one and others are zero (these are the individual optima computed in the beginning). For solving the large number of scalarized subproblems efficiently, a parallelization strategy is presented in Reference 15.

An example of applying the NBI method to the ZDT2 problem is shown in Figure 3.10. In addition to the extreme solutions $(0, 1)^T$ and $(1, 0)^T$, four solutions were computed by solving problem (3.16) with $p = 5$ ($\delta = 1/5$). Since the ZDT2 problem is smooth and symmetric, the obtained solutions are all Pareto optimal and evenly distributed along the Pareto-optimal set.

3.6.6 Normalized Normal Constraint Method

An improvement of the NBI method called the *normalized normal constraint method (nnc)* is presented in Reference 51. The basic idea is to replace the equality constraint in the scalar subproblem (3.16) of the NBI method with the corresponding inequality constraint that reduces the possibility of producing dominated solutions. In addition, a *Pareto filter* is presented in Reference 51 to exclude the dominated solutions from the solutions produced by NNC.

The scalarized subproblem to be solved in the NNC method is of the form

$$\begin{aligned} &\min_x \; \bar{f}_k(x) \\ &\text{s.t.} \;\; N_i(\bar{f}(x) - X_j)^T \leq 0, \quad 1 \leq i \leq k-1 \quad x \in S, \end{aligned} \tag{3.17}$$

where $N_i = \bar{f}(x^{*,k}) - \bar{f}(x^{*,i})$ is the direction from the minimum of the ith objective function to the minimum of the kth objective function in objective space. In other words, the vectors N_i,

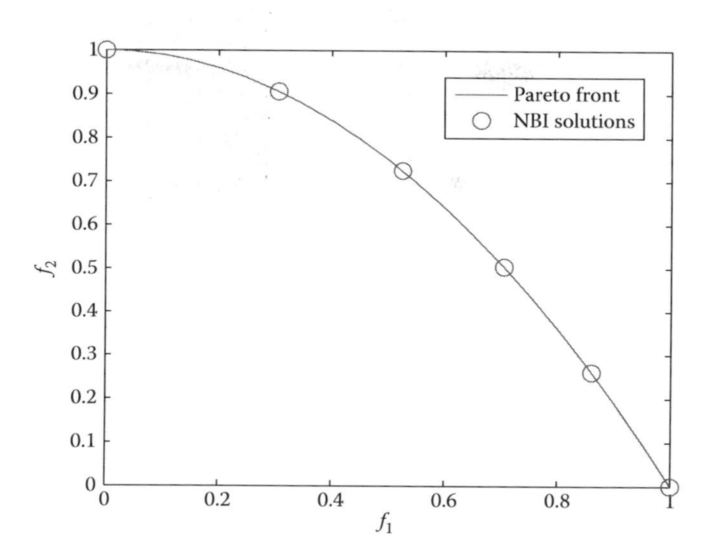

FIGURE 3.10
Pareto-optimal solutions produced by the NBI method for the ZDT2 problem. The values used for the coordinates β were $(0.2, 0.8)^T$, $(0.4, 0.6)^T$, $(0.6, 0.4)^T$, and $(0.8, 0.2)^T$.

$i = 1, \ldots, k$, span CHIM. In addition, $X_j = \sum_{i=1}^{k} \beta_{ij} \bar{f}(x^{*,i})$ denotes the jth point in CHIM generated similarly than in the NBI method. Function $\bar{f} = (\bar{f}_1, \ldots, \bar{f}_k)^T$ is a normalized objective function where $\bar{f}_i(x) = (f_i(x) - z_i^*)/(z_i^{nad} - z_i^*)$.

3.6.7 Evolutionary Multi-Objective Optimization

EMO method is *a posteriori* method and is developed based on the principles of evolutionary algorithms (EAs) [35,39]. Details about EAs for solving single-objective optimization problems can be found in the above-mentioned literature. Here, we describe their extension for solving multi-objective optimization problems.

3.6.7.1 A Brief History of EMO Methodologies

During the early years, EA researchers have realized the need of solving multi-objective optimization problems in practice and mainly resorted to using weighted-sum approaches to convert multiple objectives into a single goal [32,59].

However, the first implementation of a real MOEA (vector-evaluated GA or VEGA) was suggested by David Schaffer in the year 1984 [62]. Schaffer modified the simple three-operator genetic algorithm [39] (with selection, crossover, and mutation) by performing independent selection cycles according to each objective. The selection method is repeated for each individual objective to fill up a portion of the mating pool. Then, the entire population is thoroughly shuffled to apply crossover and mutation operators. This is performed to achieve the mating of individuals of different subpopulation groups. The algorithm worked efficiently for some generations but in some cases suffered from its bias toward some individuals or regions (mostly individual objective champions). This does not fulfill the second goal of EMO, discussed earlier.

Ironically, no significant study was performed for almost a decade after the pioneering work of Schaffer, until a revolutionary 10-line sketch of a new nondominated sorting procedure suggested by David E. Goldberg in his seminal book on GAs [35]. Since an EA needs a fitness function for reproduction, the trick was to find a single metric from a number of objective functions. Goldberg's suggestion was to use the concept of *domination* to assign more copies to nondominated individuals in a population. Since diversity is the other concern, he also suggested the use of a *niching* strategy [36] among solutions of a nondominated class. Niching is an operator that is often used to control the diversity of a population by not allowing a single solution to take over the entire population. Getting this clue, at least three independent groups of researchers developed different versions of MOEAs during 1993–1994 [33,40,63]. These algorithms differ in the way a fitness assignment scheme is introduced to each individual.

These EMO methodologies gave a good head start to the research and application of EMO, but suffered from the fact that they did not use an elite-preservation mechanism in their procedures. Inclusion of elitism in an EO provides a monotonically nondegrading performance [60], as previously found best solutions are preserved. The second-generation EMO algorithms implemented an elite-preserving operator in different ways and gave birth to elitist EMO procedures, such as NSGA-II [20], strength Pareto EA (SPEA) [75], Pareto-archived ES (PAES) [44], and others. More recently, EMO algorithms have been broadly classified as domination-, decomposition-, and indicator-based algorithms. The domination-based algorithms follow Goldberg's suggestion of using dominance concept explained above. In decomposition-based algorithms, an MOOP is converted into a number of single-objective optimization problems and solved simultaneously, for example, MOEA/D [71] and indicator-based algorithms use the value of a performance metric such as hypervolume as fitness values of the individuals in an EMO algorithm, for example, SMS-MOEA [29]. Since the domination-based EMO algorithms are the commonly used procedures, we describe one of these algorithms in detail.

3.6.7.2 Elitist EMO: NSGA-II

The NSGA-II procedure [20] is one of the popularly used EMO procedures, which attempt to find multiple Pareto-optimal solutions in an MOOP and has the following three features:

1. It uses an elitist principle.
2. It uses an explicit diversity-preserving mechanism.
3. It emphasizes nondominated solutions.

At any generation t, the offspring population (say, Q_t) is first created using the parent population (say, P_t) and the usual genetic operators. Thereafter, the two populations are combined together to form a new population (say, R_t) of size $2N$. Then, the population R_t is classified into different nondominated classes. Thereafter, the new population is filled by points of different nondominated fronts, one at a time. The filling starts with the first nondominated front (of class one) and continues with points of the second nondominated front, and so on. Since the overall population size of R_t is $2N$, not all fronts can be accommodated in N slots available for the new population. All fronts that could not be accommodated are deleted. When the last allowed front is being considered, there may exist more points in the front than the remaining slots in the new population. This scenario is illustrated in Figure 3.11. Instead of arbitrarily discarding some members from the last front, the points which will make the diversity of the selected points the highest are chosen.

The crowded-sorting of the points of the last front, which could not be accommodated fully, is achieved in the descending order of their *crowding distance values* and points from the top of the ordered list are chosen. The crowding distance d_i of point i is a measure of the objective space around i, which is not occupied by any other solution in the population. Here, we simply calculate

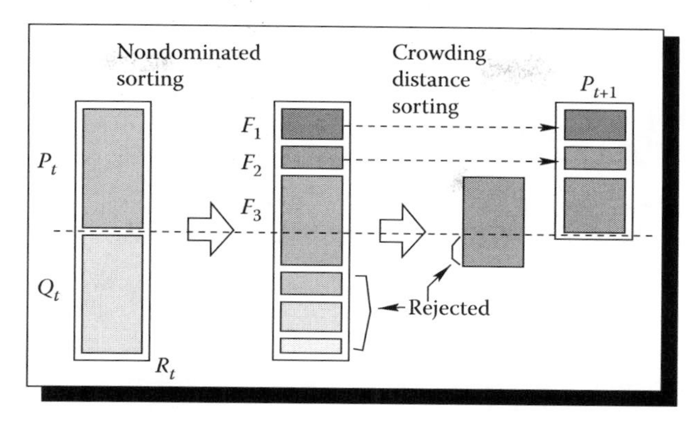

FIGURE 3.11
Schematic of the NSGA-II procedure.

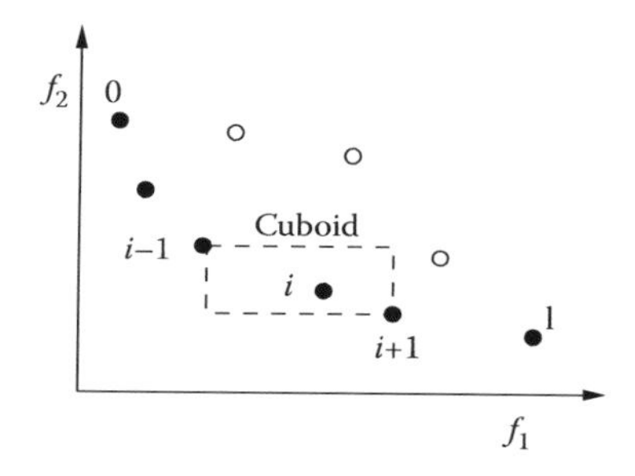

FIGURE 3.12
Crowding distance calculation.

this quantity d_i by estimating the perimeter of the cuboid (Figure 3.12) formed using the nearest neighbors in the objective space as the vertices (we call this the *crowding distance*).

Sample Results: Here, we show the results from several runs of the NSGA-II algorithm on two-test problems. The first problem (ZDT2) is a two-objective, 30-variable problem with a concave Pareto-optimal front:

$$\text{ZDT2}: \begin{cases} \text{Min } f_1(\mathbf{x}) = x_1, \\ \text{Min } f_2(\mathbf{x}) = s(\mathbf{x})\left[1 - (f_1(\mathbf{x})/s(\mathbf{x}))^2\right], \\ \text{where} \quad s(\mathbf{x}) = 1 + \frac{9}{29}\sum_{i=2}^{30} x_i \\ \qquad\quad 0 \le x_1 \le 1, \\ \qquad\quad -1 \le x_i \le 1, \quad i = 2, 3, \ldots, 30. \end{cases} \tag{3.18}$$

The second problem (KUR), with three variables, has a disconnected Pareto-optimal front:

$$\text{KUR}: \begin{cases} \text{Min } f_1(\mathbf{x}) = \sum_{i=1}^{2}\left[-10\exp(-0.2\sqrt{x_i^2 + x_{i+1}^2})\right], \\ \text{Min } f_2(\mathbf{x}) = \sum_{i=1}^{3}\left[|x_i|^{0.8} + 5\sin(x_i^3)\right], \\ \qquad -5 \le x_i \le 5, \quad i = 1, 2, 3. \end{cases} \tag{3.19}$$

NSGA-II is run with a population size of 100 and for 250 generations. The variables are used as real numbers and an SBX recombination operator [19] with $p_c = 0.9$ and distribution index of $\eta_c = 10$ and a polynomial mutation operator [18] with $p_m = 1/n$ (n is the number of variables) and

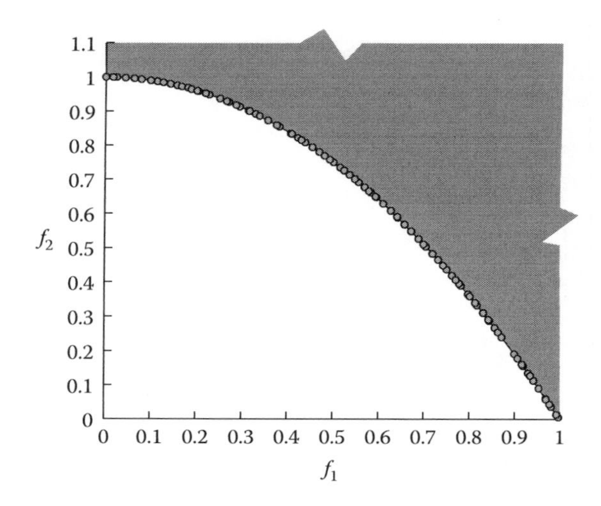

FIGURE 3.13
NSGA-II on ZDT2.

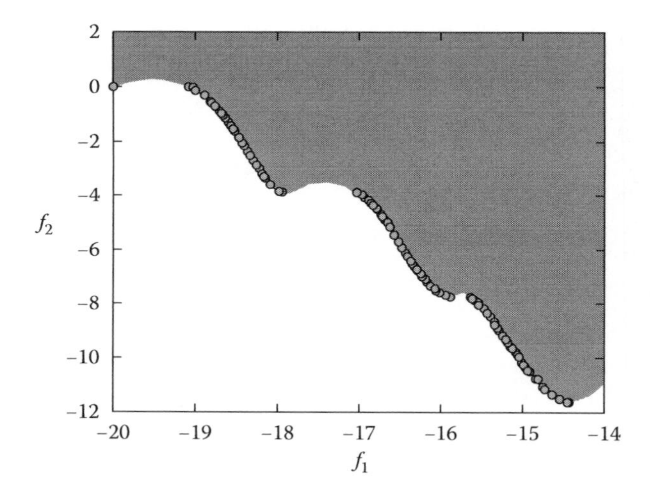

FIGURE 3.14
NSGA-II on KUR.

distribution index of $\eta_m = 20$ are used. Figures 3.13 and 3.14 show that NSGA-II converges to the Pareto-optimal front and maintains a good spread of solutions on both test problems.

There also exist other competent EMOs, such as SPEA and its improved version SPEA2 [74], PAES and its improved versions PESA and PESA2 [12], multi-objective messy GA (MOMGA) [65], multi-objective micro-GA [10], neighborhood constraint GA [49], ARMOGA [61], and others. Besides, there exists other EA-based methodologies, such as particle swarm EMO [9,54], ant-based EMO [37,50], and differential evolution-based EMO [1].

3.6.7.3 Constraint Handling in EMO

The constraint handling method modifies the binary tournament selection, where two solutions are picked from the population and the better solution is chosen. In the presence of constraints, each solution can be either feasible or infeasible. Thus, there may be at most three situations: (i) both solutions are feasible, (ii) one is feasible and other is not, and (iii) both are infeasible. We consider each case by simply redefining the domination principle as follows (we call it the *constrained-domination* condition for any two solutions $\mathbf{x}^{(i)}$ and $\mathbf{x}^{(j)}$):

Definition 3.9

A solution $\mathbf{x}^{(i)}$ is said to "constrained-dominate" a solution $\mathbf{x}^{(j)}$ (or $\mathbf{x}^{(i)} \preceq_c \mathbf{x}^{(j)}$), if any of the following conditions are true:

1. Solution $\mathbf{x}^{(i)}$ is feasible and solution $\mathbf{x}^{(j)}$ is not.
2. Solutions $\mathbf{x}^{(i)}$ and $\mathbf{x}^{(j)}$ are both infeasible, but solution $\mathbf{x}^{(i)}$ has a smaller constraint violation, which can be computed by adding the normalized violation of all constraints:

$$\mathrm{CV}(\mathbf{x}) = \sum_{j=1}^{J} \max\left(0, -\bar{g}_j(\mathbf{x})\right) + \sum_{k=1}^{K} \mathrm{abs}(\bar{h}_k(\mathbf{x})).$$

   The normalization of a constraint $g_j(\mathbf{x}) \geq g_{j,r}$ can be achieved as $\bar{g}_j(\mathbf{x}) \geq 0$, where $\bar{g}_j(\mathbf{x}) = g_j(\mathbf{x})/g_{j,r} - 1$.
3. Solutions $\mathbf{x}^{(i)}$ and $\mathbf{x}^{(j)}$ are feasible and solution $\mathbf{x}^{(i)}$ dominates solution $\mathbf{x}^{(j)}$ in the usual sense (Definition 3.3).

The above change in the definition requires a minimal change in the NSGA-II procedure described earlier. Figure 3.15 shows the nondominated fronts on a six-member population due to the introduction of two constraints (the minimization problem is described as CONSTR elsewhere [18]). In the absence of the constraints, the nondominated fronts (shown by dashed lines) would have been `((1,3,5), (2,6), (4))`, but in their presence, the new fronts are `((4,5), (6), (2), (1),`

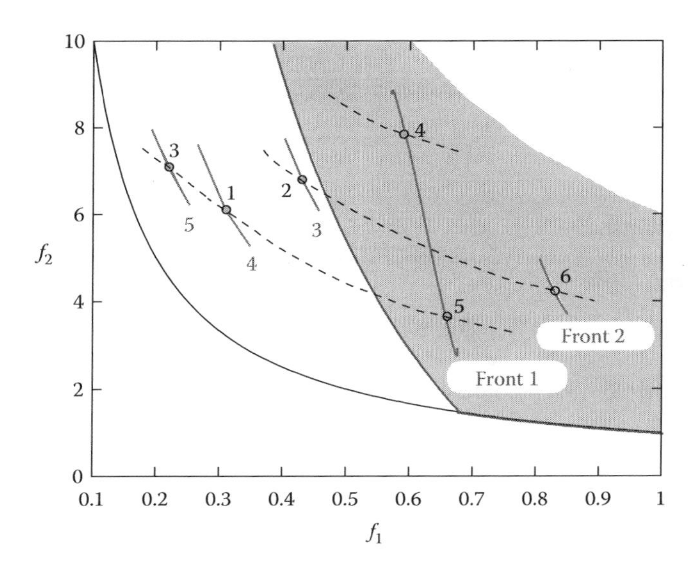

FIGURE 3.15
Nonconstrained-domination fronts.

(3)). The first nondominated front consists of the "best" (i.e., nondominated and feasible) points from the population and any feasible point lies on a better nondominated front than on an infeasible point.

3.6.7.4 Many-Objective EMO

Owing to the increased popularity of EMO methods in practice and demand for addressing more than two or three objectives, EMO researchers have suggested new and improved EMO algorithms for many-objective ($k > 3$) problems recently. The dominance-based algorithms suffer a (convergence) failure for many-objective problems that is mainly due to the high levels of incomparability between solutions, in a dominance sense. Many years of research resulted in one common belief:

> For solving many-objective optimization problems, it is too much to demand from a population-based optimization algorithm to simultaneously converge to the Pareto-optimal front as well as spread its population members across the entire Pareto-optimal front.

To beat the higher dimension in a problem, prompted EMO researchers to provide a set of guided directions or points *a priori* and the task from an EMO is reduced to find a single Pareto-optimal solution *near* each guided direction or point. In 2009, Zhang et al. [71] suggested the MOEA/D algorithm that uses a prespecified set of reference directions originating from the ideal point for this purpose. In an MOEA/D population, members are first assigned a particular reference direction based on the proximity and distance from the ideal point along the reference direction. Thereafter, restricting the search within neighboring reference directions, MOEA/D is able to find a widely distributed and well-converged solutions. Later, in 2014, Deb and Gupta [23] suggested NSGA-III in which a set of reference points are supplied *a priori*. On the basis of a simpler and parameterless strategy for member association with a reference point and nondomination-cum-niching methodology, they were able to solve up to 15-objective optimization problems. These methods are flexible in that they can be used to find a partial front, constrained problems [42], and real-world problems. Results using NSGA-III on an unconstrained three-objective problem and a 15-objective problem are shown in Figures 3.16 and 3.17, respectively. It is interesting to note the uniformity in NSGA-III's final population on the three-objective efficient front and on a 10-objective parallel coordinate plot (PCP).

3.7 *A Priori* Methods

3.7.1 Lexicographic Ordering

In *lexicographic ordering* [31], the objective functions are first ordered based on their order of importance by the DM. Let us denote the order by $\{i_1, \ldots, i_k\}$. Then, the most important objective function is optimized with respect to the feasible region S, that is, a problem

$$\begin{aligned} &\min\ f_{i_1}(x) \\ &\text{s.t.}\ \ x \in S \end{aligned} \tag{3.20}$$

is solved. If the solution of problem (3.20) is not unique, then another optimization is performed by solving problem

$$\begin{aligned} &\min\ f_{i_2}(x) \\ &\text{s.t.}\ \ f_{i_1} \leq c^*_{i_1}, \quad x \in S, \end{aligned} \tag{3.21}$$

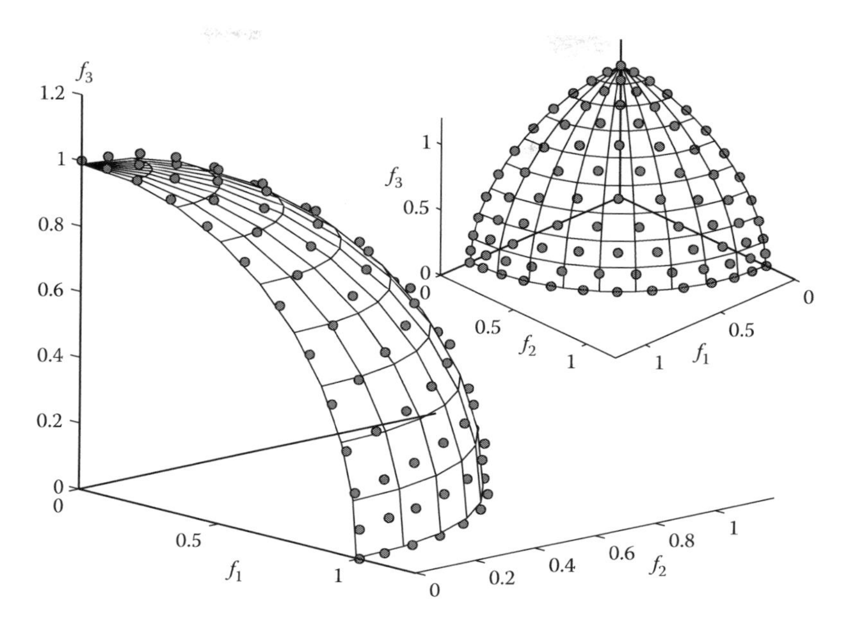

FIGURE 3.16
NSGA-III solutions on a three-objective problem.

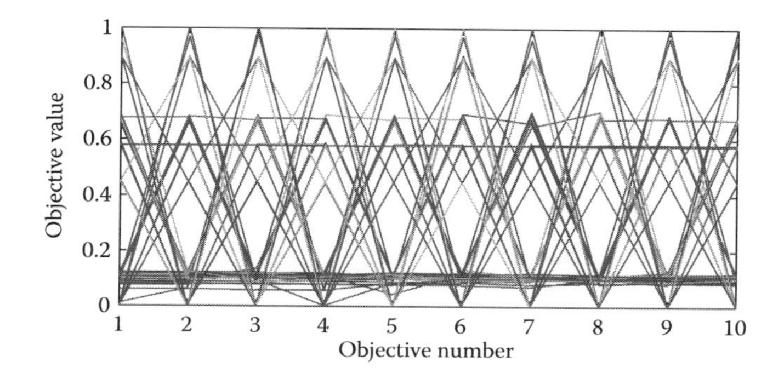

FIGURE 3.17
NSGA-III solutions on a 10-objective PCP. The horizontal axis represents the objective functions while the objective values are presented in the vertical axis. Each piecewise linear path represents a single solution in the final population. If two paths intersect each other, then they do not dominate each other.

where $c_{i_1}^*$ is the optimal objective function value of problem (3.20). In other words, problem (3.21) optimizes the second most important objective function while making sure that the value of the most important objective function is not worse than its optimal value. If the solution of problem (3.21) is not unique, then the third most important objective is optimized with constraints for the most important and the second most important objectives. The solution process continues until the obtained solution is unique.

It can be shown that the solution obtained by lexicographic ordering is Pareto optimal (see, e.g., Reference 52). Different Pareto-optimal solutions can be obtained using different importance orders for the objective functions.

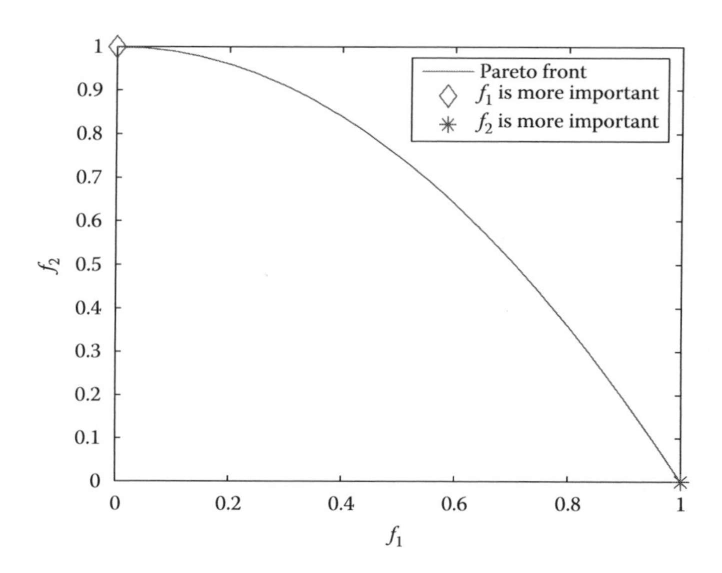

FIGURE 3.18
Pareto-optimal solutions produced by lexicographic ordering for the ZDT2 problem.

The difficulty in lexicographic ordering is that, in many times, the solution process ends after optimizing the first or the second most important objectives due to the additional constraints introduced after each step. In practice, this can be improved by allowing a small deviation from the optimal values, that is, using an upper bound $c^*_{i_j} + \delta$ instead of $c^*_{i_j}$ where $\delta > 0$ is a small constant.

Pareto-optimal solutions of applying lexicographic ordering to the ZDT2 problem are shown in Figure 3.18. Both the possible orders of importance are used.

3.7.2 Goal Programming

In *goal programming* [6–8], the DM specifies the desired aspiration levels $\bar{z}_i$, $i = 1, \ldots, k$, for the objective functions. An objective function together with an aspiration level forms a *goal* $f_i(x) \leq \bar{z}_i$. The idea of goal programming is to minimize the deviation between the objective function value and the aspiration level. It is sufficient to study the *deviational variables* $\delta_i = \bar{z}_i - f_i(x)$, which can be presented as the difference between two positive variables, $\delta_i = \delta_i^- - \delta_i^+$. Now, we can minimize the deviational variables and, in the case of minimization problems, it is sufficient to minimize *positive deviations* δ_i^+ for all $i = 1, \ldots, k$. There are several possibilities to formulate the minimization problem in goal programming, and as an example, we present one formulation based on weighting coefficients w_i. For information about the different formulations, see Reference 52 and references therein. The minimization problem is of the form

$$\begin{aligned} &\min_{x,\delta^+} \sum_{i=1}^{k} w_i \delta_i^+ \\ &\text{s.t.} \quad f_i(x) - \delta_i^+ \leq \bar{z}_i, \quad \text{for all } i = 1, \ldots, k, \\ &\qquad \delta_i^+ \geq 0 \quad \text{for all } i = 1, \ldots, k, \quad x \in S, \end{aligned} \tag{3.22}$$

where $\delta_i^+, i = 1, \ldots, k$, and x are the variables.

It can be shown that the solution of the goal programming problem (3.22) is Pareto optimal if either the aspiration levels form a Pareto-optimal reference point or all the deviational variables δ_i^+ have positive values at the optimum [52]. Therefore, the main drawback is that the goal has to be selected correctly.

3.8 Interactive Methods

Among the four different types of methods that classify MO methods described earlier, interactive methods are most interesting as they actively involve the DM in the solution process. A pseudo-code of an interactive method involves the following steps:

1. *Initialization*. Calculate and show to the DM the ideal and nadir points, which provide the DM information about the bounds of the Pareto-optimal front. Set $it = 1$, where *it* is the iteration number.
2. *Starting solution*. An unbiased compromise solution is calculated and shown to the DM as a starting feasible solution.
3. *Preference information*. The DM investigates the presented feasible solution and expresses his/her preference information, if possible.
4. *Generation of new solutions*. The preference information provided by the DM is subsequently used to generate new Pareto-optimal solution(s) and shown to the DM.
5. *Iterative decision*. If the DM can choose his/her preferred solution(s) among the solutions presented to him/her and wishes to stop, the interactive method is terminated, else set $it = it + 1$ and the DM presents his/her new preference information in step 3.

As can be seen above, in interactive methods the DM can progressively build or refine his/her preferences and finally obtain solution(s) that is most satisfactory. In fact, interactive methods are most useful when computationally expensive problems (wherein evaluation of objective functions involves considerable computational cost) are involved as only solution(s) that are desirable to the DM that are generated.

Different types of interactive methods are proposed in the literature and they mainly differ in the following aspects:

1. Preference information provided by the DM
2. Type of information provided to the DM, for example, ideal and nadir points, tradeoff information, etc. to enable him/her to express preferences
3. Type of scalarizing function used
4. Mathematical assumptions behind the method, for example, some interactive methods are only applicable to linear multi-objective optimization problems, etc.

In practice, most interactive methods are equally good as different methods appeal to different DMs. Their efficacy depends on the DM's ability to provide the desirable preference information and find his/her preferred Pareto-optimal solution. In addition, a graphical user interface through which a DM interacts with an interactive method may also enhance the efficacy of most interactive methods. An exhaustive description of different interactive methods can be found in Reference 52. Here, we present briefly a few interactive methods enumerating the steps in the pseudo-code pertaining to individual methods to provide an overall understanding of the existing methods in the literature.

3.8.1 Reference Point Method

The reference point method utilizing an achievement scalarizing function described as an *a priori* method in Section 3.6.4 was extended as an interactive method by Wierzbicki [66].

Let us consider an instance of the pseudo-code as a reference point algorithm. Step 2 of the pseudo-code is not considered in the reference point algorithm. Here, the DM is directly asked to provide his/her preference information as a reference point $\bar{\mathbf{z}}^{it} \in \mathbb{R}^k$ in step 3. This reference point is used to generate k other reference points as follows: $\bar{\mathbf{z}}(i) = \bar{\mathbf{z}}^{it} + d^{it}\hat{\mathbf{e}}^i$. Here, $d^{it} = \|\bar{\mathbf{z}}^{it} - \mathbf{z}^{\mathbf{it}}\|$ and $\hat{\mathbf{e}}^i$ is the ith unit vector for $i = 1, \ldots, k$. These $k + 1$ reference points are used to formulate $k + 1$ scalarized problems (see problem (3.14)) and solved using any appropriate mathematical programming technique. The $k + 1$ Pareto-optimal solutions thus obtained are shown to the DM in step 4. In step 5, if the DM is not satisfied with any of the provided Pareto-optimal solutions he/she provides a new reference point $\bar{\mathbf{z}}^{it+1}$ by revisiting step 3, else the reference point algorithm is terminated.

The idea of generating $k + 1$ reference points is to provide the DM with a set of Pareto-optimal solutions that might reflect her/his preferences instead of a single solution. If the reference point is farther from the Pareto-optimal front, a wide spread set of solutions is generated and if the reference point is in close proximity to the Pareto-optimal front, the Pareto-optimal solutions generated are close to each other. The reference point algorithm is widely considered to be simple as the DM has to provide only the reference point and he/she can direct the solution process in any direction by specifying new reference points. However, sometimes the convergence to final solution desirable to the DM can be time consuming as the DM is not provided any extra support to set new reference points and obtain improved solutions.

3.8.2 GUESS Method

The GUESS method proposed by Buchanan [4] is similar to the reference point method. An instance of pseudo-code is considered as a GUESS method, with some changes. As in reference point algorithm, step 2 is also not considered in the GUESS method. However, in step 3, the DM can provide the upper and lower bounds to each of the objective functions, if he/she desires in addition to a reference point $\bar{\mathbf{z}}^{it} \in \mathbb{R}^k$ satisfying the bounds. The reference point $\bar{\mathbf{z}}^{it}$ is subsequently used to formulate a scalarized problem as follows:

$$\begin{aligned} \max \quad & \min_{i=1,\ldots,k} \left[\frac{z_i^{nad} - f_i(\mathbf{x})}{z_i^{nad} - \bar{\mathbf{z}}^{it}} \right] \\ \text{s.t.} \quad & \mathbf{x} \in S \end{aligned} \tag{3.23}$$

and solved to obtain a weakly Pareto-optimal solution and shown to the DM in step 4. If the DM is satisfied with the solution shown in step 4, the GUESS method is terminated, else the DM is asked to provide a new reference point $\bar{\mathbf{z}}^{it+1}$ by revisiting step 3.

The GUESS method is extremely simple to use, but provide very little support to the DM to set subsequent reference points as her/his preference information. In fact, the operation of the GUESS method can be considered similar to the what-if analysis, wherein the DM investigates the changes in the obtained solutions by setting different reference points. As indicated by Buchanan, the DM usually stops the GUESS method if the obtained solution is similar to her/his reference point, which in turn can cause premature convergence.

3.8.3 STOM Method

The STOM method presented in this section was proposed by Nakayama [55]. This method attempts to find a satisficing solution to the DM's preferences. The basic idea behind the STOM method is similar to the reference point and the GUESS methods. As with other methods, an instance of the pseudo-code is considered as the STOM method.

In step 1, an utopian objective vector $\mathbf{z}^{**}$ is calculated, which may or may not be shown to the DM. Step 2 is also not considered in the STOM method. The DM is asked to provide a reference point $\bar{\mathbf{z}}^{it} \in \mathbb{R}^k$ and $\bar{z}_i^{it} < z_i^{**}$ $\forall$ $i = 1, \ldots, k$. Next in step 4, a scalarized problem is formulated as follows:

$$\begin{aligned} \min \max_{i=1,\ldots,k} & \left[w_i^{it}(f_i(\mathbf{x}) - z_i^{**})\right] + \rho \sum_{i=1}^{k} w_i^{it} f_i(\mathbf{x}) \\ \text{s.t.} \quad & \mathbf{x} \in S, \end{aligned} \tag{3.24}$$

where $w_i^{it} = 1/(\bar{z}_i^{it} - z_i^{**})$ $\forall$ $i = 1, \ldots, k$ and solved to obtain a Pareto-optimal solution, $\mathbf{z^{it}}$. The solution $\mathbf{z^{it}}$ is subsequently shown to the DM. In step 5, the DM is asked to classify the objectives as the objective function whose values

1. Have to be improved ($I^{<}$)
2. Could be impaired ($I^{>}$)
3. Are acceptable as they are ($I^{=}$)

If the DM chooses not to improve any of the objective function values (i.e., $I^{<} = \emptyset$), the STOM method is terminated and $\mathbf{z^{it}}$ is declared to be the most preferred solution to the DM. Next, if the DM wishes to improve the present solution by providing new preference information the algorithm moves back to step 3.

In step 3, for $it > 1$ the following procedure is followed to obtain a new reference point $\bar{\mathbf{z}}^{it+1}$. If the DM wishes to improve some of the objective function values, the aspiration levels for those objectives $\bar{z}_i^{it+1}$ (for $i \in I^{<}$) are obtained from the DM. For those objective functions $i \in I^{=}$, set $\bar{z}_i^{it+1} = z_i^{it}$. An important highlight of the STOM method is the use of tradeoff information to calculate the values for $\bar{z}_i^{it+1}$ (for $i \in I^{>}$) as

$$\bar{z}_i^{it+1} = z_i^{it} + \frac{1}{N(\lambda_i^{it} + \rho)w_i^{it}} \sum_{j \in I^{<}} (\lambda_j^{it} + \rho) w_j^{it} (z_i^{it} - \bar{z}_i^{it+1}),$$

where λ_i^{it} is the Karush–Kuhn–Tucker multipliers corresponding to the current Pareto-optimal solution and N the cardinality of the class $I^{>}$. For further information regarding the calculation of $\bar{z}_i^{it+1}$ (for $i \in I^{>}$), see References 52 and 55.

The STOM method is similar to the GUESS method with the exception of using the tradeoff information. The tradeoff information is in fact very useful when a large number of objectives are involved, as they avoid extra cognitive load on the DM to specify new aspiration levels for objective functions belonging to the class $I^{>}$.

3.8.4 Reference Direction Approach

Korhonen and Laakso [45–47] proposed a new reference direction approach and termed it visual interactive approach. Unlike the reference point method, in reference direction approach a Pareto-optimal point is generated using a reference direction (a vector formed from the current point to the reference point provided by the DM). Let us consider an instance of the pseudo-code as a visual interactive approach.

The background information such as the bounds of the Pareto-optimal front, which are part of step 1 is not considered. In step 2, a starting solution $\mathbf{z^1} \in \mathbb{R}^\mathbf{k}$ that can be arbitrary is considered to start. Unlike most interactive methods, the starting point in step 2 may not be Pareto optimal and even can be infeasible. In step 3, the DM is asked to provide a reference point $\bar{\mathbf{z}}^{it} \in \mathbb{R}^k$ and then a preference direction $\mathbf{d^{it+1}}$ is set as $\bar{\mathbf{z}}^{it} - \mathbf{z^{it}}$. Next, a set of weakly Pareto-optimal solutions Z^{it+1} that are solutions to the following problem (3.25) are obtained:

$$\begin{aligned} \min\ & s_{\bar{\mathbf{z}},\mathbf{w}}(\mathbf{z}) = \max_{i \in I} \frac{z_i - \bar{z}_i}{w_i} \\ \text{s.t.}\ & \bar{\mathbf{z}} = \mathbf{z}^{it} + t\mathbf{d}^{it+1} \\ & \mathbf{z} \in Z \text{ is Pareto optimal,} \end{aligned} \tag{3.25}$$

where $I = \{i | w_i > 0\} \subset 1, \ldots, k$, $\mathbf{w}$ is a weighing vector, $\mathbf{z} \in Z$ is the sought after desirable objective vector, $\bar{\mathbf{z}} \in \mathbb{R}^k$ and $t > 0$ takes on different discrete values. In step 5, the DM picks one solution $\mathbf{z}^{it+1}$ as his/her preferred solution among the solutions in Z^{it+1}. In addition, in step 5, the newly chosen solution $\mathbf{z}^{it+1}$ is compared against the solution chosen in the previous iteration $\mathbf{z}^{it}$, if the solutions are different the DM is asked to specify a new reference point in step 3, else optimality conditions are checked to decide about the termination of the algorithm.

The reference direction approach uses a value function $U : \mathbb{R}^k \to \mathbb{R}$ to capture the DM's preferences. To evaluate the optimality conditions the approach considers some assumptions such as

1. The existence of a pseudo-concave value function U
2. Compact and convex feasible region S
3. Differentiable constraint functions

Then, the objective vector $\mathbf{z}^{it+1}$ can be considered to be optimal with respect to U if the following condition is satisfied:

$$U(\mathbf{z}^{it+1}) \geq U(\mathbf{z}^{it+1} + \beta_j \mathbf{d}(j)) \quad \forall \beta_j \geq 0 \quad \text{and} \quad j = 1, \ldots, p,$$

where $\mathbf{d}(j)$ represents all possible feasible directions at $\mathbf{z}^{it+1}$. For further discussions on the optimality conditions, see Reference 52. If, in step 5, the optimality conditions are satisfied, $\mathbf{z}^{it+1}$ is declared to the final solution chosen by the DM, else a new search direction ($\mathbf{d}^{it+1}$) obtained during the optimality checking is considered and step 4 is revisited to generate new weakly Pareto-optimal solutions. It must be noted that this approach is particularly suitable for multi-objective linear programming problems, as in this case it is possible to verify the optimality conditions and find the direction of improvement.

3.8.5 Synchronous NIMBUS Algorithm

A novel approach in interactive MO was proposed by Miettinen and Mäkelä [53], which extended the previously proposed NIMBUS method [52] involving classification of objectives as preference information of the DM. The novelty of this approach is the use of several scalarizing functions utilizing identical preference information provided by the DM to yield different Pareto-optimal solutions in a synchronous way. The authors termed the algorithm following this approach as the synchronous

NIMBUS algorithm. The working of this algorithm is presented next utilizing an instance of the pseudo-code presented earlier.

The synchronous NIMBUS algorithm [53] follows the pseudo-code very closely. In step 3, for the starting solution, the DM is asked to classify the objective functions and specify her/his aspiration levels and upper bounds. It is interesting to note that even though the preference information is in the form of classification of objectives they can be easily converted to a reference point. In synchronous NIMBUS algorithm, the DM can classify the objective functions into up to five different classes, that is,

1. $\mathbf{I}^{<}$: For every $i \in I^{<}$, the value of the objective functions must be improved from the current value. The reference point z_i can be set as z_i^*.
2. $\mathbf{I}^{\leq}$: For every $i \in I^{\leq}$, the value of the objective functions must be improved to a desired value (z_i^{asp}) specified by the DM.
3. $\mathbf{I}^{=}$: For every $i \in I^{=}$, the value of the objective functions must stay the same as that of the current solution ($f_i(x^c)$).
4. $\mathbf{I}^{\geq}$: For every $i \in I^{\geq}$, the value of the objective functions must be impaired up to a desired value (ϵ_i) specified by the DM.
5. $\mathbf{I}^{\diamond}$: For every $i \in I^{\diamond}$, the value of the objective functions can be temporarily allowed to change freely. During implementation, the reference point z_i can be set as z_i^{nad}.

In addition, the DM can also request up to four solutions to be shown in step 4. The four solutions are generated using four different scalarized problems, that is,

1. NIMBUS variant:

$$\begin{aligned}
&\min \max_{i \in I^{<}, j \in I^{\leq}} \left[\frac{f_i(x) - z_i^*}{z_i^{nad} - z_i^{**}}, \frac{f_j(x) - z_j^{asp}}{z_j^{nad} - z_j^{**}} \right] + \rho \textstyle\sum_{i=1}^{k} \frac{f_i(x)}{z_i^{nad} - z_i^{**}} \\
&\text{s.t.} \quad f_i(x) \leq f_i(x^c) \quad \text{for all } i \in I^{<} \cup I^{\leq} \cup I^{=} \\
&\qquad\; f_i(x) \leq \epsilon_i \quad \text{for all } i \in I^{\geq} \\
&\qquad\; x \in S
\end{aligned}$$

2. STOM-based subproblem (3.24)
3. GUESS scalarized subproblem

$$\begin{aligned}
&\min \max_{i \notin I^{\diamond}} \frac{f_i(x) - z_i^{nad}}{z_i^{nad} - \bar{z}_i} + \rho \textstyle\sum_{i=1}^{k} \frac{f_i(x)}{z_i^{nad} - \bar{z}_i} \\
&\text{s.t.} \quad x \in S
\end{aligned}$$

4. Achievement scalarizing function-based subproblem (3.14).

This algorithm provides the DM flexibility to save all or some of the solutions shown to him/her in a repository for future use in subsequent iterations, if needed. In step 5, the DM is asked to choose one solution among the solutions shown to him/her or from the repository as his/her preferred current solution. In addition, the DM also has the flexibility to choose two solutions shown to him/her or from the repository in step 4 and request for more Pareto-optimal solutions between them. Here, the algorithm generates solutions between the two solutions and projects them on to the Pareto-optimal front using the scalarized subproblem (3.14).

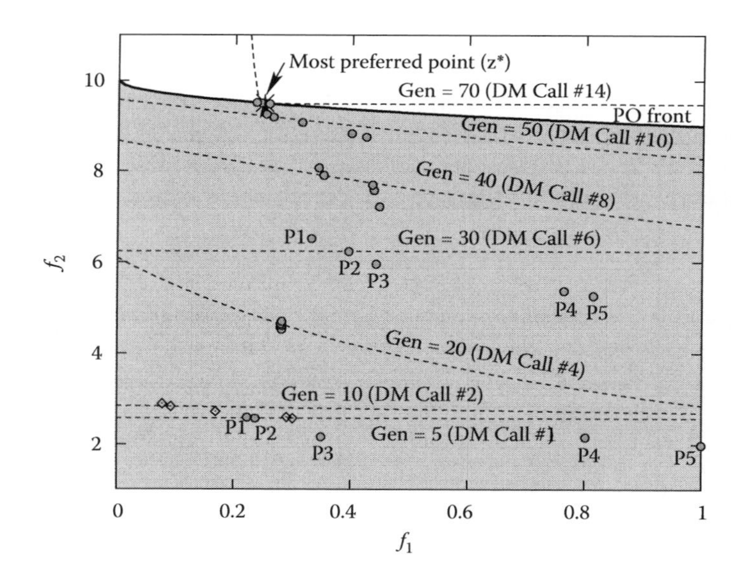

FIGURE 3.19
Progressively interactive EMO method is illustrated on a test problem.

Among different interactive methods, the synchronous NIMBUS algorithm is commonly used. By not limiting to a single scalarized subproblem, the algorithm provides the DM a better view of the solutions possible for his/her preference information.

3.8.6 Interactive EMO

Interactive methods for EMO have also been suggested [56]. A recent study [27] used the following strategy. After a few generations, four or five nondominated yet well-distributed solutions are presented to the decision maker(s) for a feedback on preference. On the basis of partial or complete pairwise comparison, the decision-maker's preferences are modeled using a predefined mathematical structure. For the next few generations, the EMO procedure is modified based on the modeled preference function. The domination and termination conditions of the EMO procedure are changed accordingly. This process is continued until two consecutive modeling efforts do not produce an identical solution. Figure 3.19 shows the progress of such a progressively interactive EMO method (PI-EMO) on a two-objective test problem involving an artificial decision maker. It can be seen that the PI-EMO is able to concentrate on a single Pareto-optimal solution at the end of the simulation, rather than the usual method of finding the whole front first and then choosing a single preferred solution. The PI-EMO method was also found to be *robust* in terms of limited inconsistencies in the preference information simulating the uncertainties involved with human decision makers.

3.9 Conclusions

Most practical search and optimization problems involve more than one objective, simply because no product, process, or system can be evaluated with a single criterion. Since, often, there are conflicting criteria, such as minimizing cost and maximizing quality of a product, multi-objective optimization problems give rise to a set of tradeoff optimal (known as Pareto optimal) solutions as the final outcome. In this chapter, we have discussed the principles of MO and its difference with its more commonly practiced single-objective counterpart. After defining the theoretical optimality conditions that make solutions optimal for different objectives simultaneously, we have discussed a number of numerical optimization methods. The methods have been classified into four broad classes:

1. No-preference methods in which no preference information from the decision maker is considered and a neutral compromise Pareto-optimal solution is generated.
2. *A priori* methods in which preference information about multiple objectives is used before optimization to form a composite objective function, so that a single-objective optimization method can be applied directly to find the corresponding optimal solution.
3. *A posteriori* methods, which first find a set of tradeoff Pareto-optimal solutions using generative numerical methods or a population-based evolutionary optimization approach and then preference information is used to choose a single preferred solution.
4. Interactive methods in which decision makers provide preference information after every few iterations of the optimization algorithm to eventually converge close to the most desired (yet unknown beforehand) Pareto-optimal solution.

Recent many-objective optimization methods have been discussed to address a large number (10+) of objectives. The multi-objective literature stores many other practical aspects, which are not discussed in this chapter. Some of these are handling uncertainties in decision variables [21,22], dynamic optimization problems [25], bilevel problems [26], computationally expensive problems using meta-modeling methods [30], multi-scenario problems involving multiple operating or loading conditions [72], combinatorial optimization problems [43], scheduling problems [24,41], etc.

Owing to the practical significance of MO, research and application in MO is progressing at a fast pace. Many commercial softwares are now available enabling MO. Different but contemporary research groups on the broad areas of MO, multiple criteria decision-making (MCDM), and mathematical optimization community are collaborative well through various workshops and conference meetings. Hopefully, this chapter provides a brief introduction to the growing activities of MO so that novice as well as expert researchers are motivated to contribute and choose this area as their career direction of research and application.

Acknowledgment

Some part of this chapter is excerpted from the first author's 2001 Wiley book [18].

References

1. B. V. Babu and M. L. Jehan. Differential evolution for multi-objective optimization. In *Proceedings of the 2003 Congress on Evolutionary Computation (CEC'2003)*, volume 4, pages 2696–2703, Canberra, Australia, December 2003. IEEE Press.
2. D. Bhatia and S. Aggarwal. Optimality and duality for multiobjective nonsmooth programming. *European Journal of Operational Research*, 57(3):360–367, 1992.

3. V. J. Jr. Bowman. On the relationship of the tchebycheff norm and the efficient frontier of multiple-criteria objectives. In H. Thiriez and S. Zionts, editors, *Multiple Criteria Decision Making*, Lecture Notes in Economics and Mathematical Systems, pages 76–85. Springer–Verlag, Berlin, Heidelberg, 1976.
4. J. T. Buchanan. A naive approach for solving MCDM problems: The GUESS method. *Journal of the Operational Research Society*, 48(2):202–206, 1997.
5. V. Chankong and Y. Y. Haimes. *Multiobjective Decision Making Theory and Methodology*. New York: North-Holland, 1983.
6. A. Charnes and W. W. Cooper. *Management Models and Industrial Applications of Linear Programming*, volume 1. New York: John Wiley & Sons, 1961.
7. A. Charnes and W. W. Cooper. Goal programming and multiple objective optimizations. *European Journal of Operational Research*, 1:39–54, 1977.
8. A. Charnes, W. W. Cooper, and R. O. Ferguson. Optimal estimation of executive compensation by linear programming. *Management Science*, 1:138–151, 1955.
9. C. A. C. Coello and M. S. Lechuga. MOPSO: A proposal for multiple objective particle swarm optimization. In *Congress on Evolutionary Computation (CEC'2002)*, volume 2, pages 1051–1056, Piscataway, New Jersey, May 2002. IEEE Service Center.
10. C. A. C. Coello and G. Toscano. A micro-genetic algorithm for multi-objective optimization. Technical Report Lania-RI-2000-06, Laboratoria Nacional de Informatica AvRyerkerk, M., Averill, R., Deb, K., and Goodman, E. 2012. Optimization for variable-size problems using genetic algorithms. *Proceedings of the 14th AIAA/ISSMO Multidisciplinary Analysis and Optimization Conference* (Indianapolis, USA). AIAA 2012-5569, Reston, VA: AIAA. Avanzada, Xalapa, Veracruz, Mexico, 2000.
11. T. H. Cormen, C. E. Leiserson, and R. L. Rivest. *Introduction to Algorithms*. New Delhi: Prentice-Hall, 1990.
12. D. W. Corne, J. D. Knowles, and M. Oates. The Pareto envelope-based selection algorithm for multiobjective optimization. In *Proceedings of the Sixth International Conference on Parallel Problem Solving from Nature VI (PPSN-VI)*, pages 839–848, Paris, France, 2000.
13. N. O. Da Cunha and E. Polak. Constrained minimization under vector-evaluated criteria in finite dimensional spaces. *Journal of Mathematical Analysis and Applications*, 19(1):103–124, 1967.
14. I. Das and J. E. Dennis. A closer look at drawbacks of minimizing weighted sums of objectives for Pareto set generation in multicriteria optimization problems. *Structural Optimization*, 14(1):63–69, 1997.
15. I. Das and J. E. Dennis. Normal-boundary intersection: A new method for generating the pareto surface in nonlinear multicriteria optimization problems. *SIAM Journal on Optimization*, 8(3):631–657, 1998.
16. K. Deb. *Optimization for Engineering Design: Algorithms and Examples*. New Delhi: Prentice-Hall, 1995.
17. K. Deb. Multi-objective genetic algorithms: Problem difficulties and construction of test problems. *Evolutionary Computation Journal*, 7(3):205–230, 1999.
18. K. Deb. *Multi-Objective Optimization Using Evolutionary Algorithms*. Chichester, United Kingdom: Wiley, 2001.
19. K. Deb and R. B. Agrawal. Simulated binary crossover for continuous search space. *Complex Systems*, 9(2):115–148, 1995.
20. K. Deb, S. Agrawal, A. Pratap, and T. Meyarivan. A fast and elitist multi-objective genetic algorithm: NSGA-II. *IEEE Transactions on Evolutionary Computation*, 6(2):182–197, 2002.
21. K. Deb and H. Gupta. Introducing robustness in multi-objective optimization. *Evolutionary Computation Journal*, 14(4):463–494, 2006.
22. K. Deb, S. Gupta, D. Daum, J. Branke, A. Mall, and D. Padmanabhan. Reliability-based optimization using evolutionary algorithms. *IEEE Transactions on Evolutionary Computation*, 13(5):1054–1074, 2009.
23. K. Deb and H. Jain. An evolutionary many-objective optimization algorithm using reference-point based non-dominated sorting approach, Part I: Solving problems with box constraints. *IEEE Transactions on Evolutionary Computation*, 18(4):577–601, 2014.
24. K. Deb, P. Jain, N. Gupta, and H. Maji. Multi-objective placement of electronic components using evolutionary algorithms. *IEEE Transactions on Components and Packaging Technologies*, 27(3):480–492, 2004.

25. K. Deb, U. B. Rao, and S. Karthik. Dynamic multi-objective optimization and decision-making using modified NSGA-II: A case study on hydro-thermal power scheduling bi-objective optimization problems. In *Proceedings of the Fourth International Conference on Evolutionary Multi-Criterion Optimization (EMO-2007)*, Matsushima, Japan, 2007.
26. K. Deb and A. Sinha. An efficient and accurate solution methodology for bilevel multi-objective programming problems using a hybrid evolutionary-local-search algorithm. *Evolutionary Computation Journal*, 18(3):403–449, 2010.
27. K. Deb, A. Sinha, P. Korhonen, and J. Wallenius. An interactive evolutionary multi-objective optimization method based on progressively approximated value functions. *IEEE Transactions on Evolutionary Computation*, 14(5):723–739, 2010.
28. M. Ehrgott. *Multicriteria Optimization*. Berlin: Springer, 2000.
29. M. Emmerich, N. Beume, and B. Naujoks. An EMO algorithm using the hypervolume measure as selection criterion. In C. A. C. Coello, H. A. Aguirre, and E. Zitzler, editors, *Evolutionary Multi-Criterion Optimization*, Lecture Notes in Computer Science, pages 62–76. Berlin, Heidelberg: Springer–Verlag, 2005.
30. M. Emmerich, K. C. Giannakoglou, and B. Naujoks. Single and multiobjective evolutionary optimization assisted by Gaussian random field metamodels. *IEEE Transactions on Evolutionary Computation*, 10(4):421–439, 2006.
31. P. C. Fishburn. Lexicographic orders, utilities and decision rules: A survey. *Management Science*, 20:1442–1471, 1974.
32. L. J. Fogel, A. J. Owens, and M. J. Walsh. *Artificial Intelligence Through Simulated Evolution*. New York: Wiley, 1966.
33. C. M. Fonseca and P. J. Fleming. Genetic algorithms for multiobjective optimization: Formulation, discussion, and generalization. In *Proceedings of the Fifth International Conference on Genetic Algorithms*, pages 416–423. San Mateo, California: Morgan Kaufmann, 1993.
34. S. Gass and T. Saaty. The computational algorithm for the parametric objective function. *Naval Research Logistics Quarterly*, 2:39–45, 1955.
35. D. E. Goldberg. *Genetic Algorithms for Search, Optimization, and Machine Learning*. Reading, Massachusetts: Addison-Wesley, 1989.
36. D. E. Goldberg and J. Richardson. Genetic algorithms with sharing for multimodal function optimization. In *Proceedings of the First International Conference on Genetic Algorithms and Their Applications*, pages 41–49, Hillsdale, New Jersey, 1987.
37. M. Gravel, W. L. Price, and C. Gagné. Scheduling continuous casting of aluminum using a multiple objective ant colony optimization metaheuristic. *European Journal of Operational Research*, 143(1):218–229, 2002.
38. Y. Y. Haimes, L. S. Lasdon, and D. A. Wismer. On a bicriterion formulation of the problems of integrated system identification and system optimization. *IEEE Transactions on Systems, Man and Cybernetics*, 1:296–297, 1971.
39. J. H. Holland. *Adaptation in Natural and Artificial Systems*. Ann Arbor, Michigan: MIT Press, 1975.
40. J. Horn, N. Nafploitis, and D. E. Goldberg. A niched Pareto genetic algorithm for multi-objective optimization. In *Proceedings of the First IEEE Conference on Evolutionary Computation*, pages 82–87, Piscataway, New Jersey, 1994.
41. H. Ishibuchi and T. Murata. A multi-objective genetic local search algorithm and its application to flowshop scheduling. *IEEE Transactions on Systems, Man and Cybernetics—Part C: Applications and Reviews*, 28(3):392–403, 1998.
42. H. Jain and K. Deb. An evolutionary many-objective optimization algorithm using reference-point based non-dominated sorting approach, Part II: Handling constraints and extending to an adaptive approach. *IEEE Transactions on Evolutionary Computation*, 18(4):602–622, 2014.
43. A. Jaszkiewicz. Genetic local search for multiple objective combinatorial optimization. *European Journal of Operational Research*, 137(1):50–71, 2002.
44. J. D. Knowles and D. W. Corne. Approximating the non-dominated front using the Pareto archived evolution strategy. *Evolutionary Computation Journal*, 8(2):149–172, 2000.

45. P. Korhonen and J. Laakso. A visual interactive method for solving the multiple-criteria problem. In M. Grauer and A. P. Wierzbicki, editors, *Interactive Decision Analysis*, Lecture Notes in Economics and Mathematical Systems, pages 146–153. Springer–Verlag, Berlin, Heidelberg, 1984.
46. P. Korhonen and J. Laakso. On developing a visual interactive multiple criteria method—An outline. In Y. Y. Haimes and V. Chankong, editors, *Decision Making with Multiple Objectives*, Lecture Notes in Economics and Mathematical Systems, pages 272–281. Springer–Verlag, Berlin, Heidelberg, 1985.
47. P. Korhonen and J. Laakso. A visual interactive method for solving the multiple-criteria problem. *European Journal of Operational Research*, 24(2):277–287, 1986.
48. O. Larichev. Cognitive validity in design of decision aiding techniques. *Journal of Multi-Criteria Decision Analysis*, 1(3):127–138, 1992.
49. D. H. Loughlin and S. Ranjithan. The neighborhood constraint method: A multiobjective optimization technique. In *Proceedings of the Seventh International Conference on Genetic Algorithms*, pages 666–673, East Lansing, Michigan, 1997.
50. P. R. McMullen. An ant colony optimization approach to addressing a JIT sequencing problem with multiple objectives. *Artificial Intelligence in Engineering*, 15:309–317, 2001.
51. A. Messac, A. Ismail-Yahaya, and C. A. Mattson. The normalized normal constraint method for generating the pareto frontier. *Structural and Multidisciplinary Optimization*, 25(2):86–98, 2003.
52. K. Miettinen. *Nonlinear Multiobjective Optimization*. Boston: Kluwer Academic Publishers, 1999.
53. K. Miettinen and M. M. Mäkelä. Synchronous approach in interactive multiobjective optimization. *European Journal of Operational Research*, 170(3):909–922, 2006.
54. S. Mostaghim and J. Teich. Strategies for finding good local guides in multi-objective particle swarm optimization (MOPSO). In *2003 IEEE Swarm Intelligence Symposium Proceedings*, pages 26–33, Indianapolis, Indiana, April 2003. IEEE Service Center.
55. H. Nakayama. Aspiration level approach to interactive multi-objective programming and its applications. In P. M. Pardalos, Y. Siskos, and C. Zopounidis, editors, *Advances in Multicriteria Analysis*, pages 147–174. Dordrecht: Kluwer Academic Publishers, 1995.
56. R. C. Purshouse, K. Deb, M. M. Mansor, S. Mostaghim, and R. Wang. A review of hybrid evolutionary multiple criteria decision making methods. In *Proceedings of the IEEE Congress on Evolutionary Computation (CEC)*, pages 1147–1154, Piscataway, New Jersey, 2014.
57. S. S. Rao. *Optimization: Theory and Applications*. New York: Wiley, 1984.
58. G. V. Reklaitis, A. Ravindran, and K. M. Ragsdell. *Engineering Optimization Methods and Applications*. New York: Wiley, 1983.
59. R. S. Rosenberg. *Simulation of Genetic Populations with Biochemical Properties*. PhD thesis, Ann Arbor, Michigan: University of Michigan, 1967.
60. G. Rudolph. Convergence analysis of canonical genetic algorithms. *IEEE Transactions on Neural Network*, 5(1):96–101, 1994.
61. D. Sasaki, M. Morikawa, S. Obayashi, and K. Nakahashi. Aerodynamic shape optimization of supersonic wings by adaptive range multiobjective genetic algorithms. In *Proceedings of the First International Conference on Evolutionary Multi-Criterion Optimization (EMO 2001)*, pages 639–652, Zurich, Switzerland, 2001.
62. J. D. Schaffer. *Some Experiments in Machine Learning Using Vector Evaluated Genetic Algorithms*. PhD thesis, Nashville, Tennessee: Vanderbilt University, 1984.
63. N. Srinivas and K. Deb. Multi-objective function optimization using non-dominated sorting genetic algorithms. *Evolutionary Computation Journal*, 2(3):221–248, 1994.
64. L. Tanner. Selecting a text-processing system as a qualitative multiple criteria problem. *European Journal of Operational Research*, 50:179–187, 1991.
65. D. Van Veldhuizen and G. B. Lamont. Multiobjective evolutionary algorithms: Analyzing the state-of-the-art. *Evolutionary Computation Journal*, 8(2):125–148, 2000.
66. A. P. Wierzbicki. A mathematical basis for satisficing decision making. *Mathematical Modelling*, 3(25):391–405, 1982.
67. A. P. Wierzbicki. On the completeness and constructiveness of parametric characterizations to vector optimization problems. *OR Spectrum*, 8(2):73–87, June 1986.
68. P. L. Yu. A class of solutions for group decision problems. *Management Science*, 19(8):936–946, 1973.

69. L. Zadeh. Optimality and non-scalar-valued performance criteria. *IEEE Transactions on Automatic Control*, 8:59–60, 1963.
70. M. Zeleny. Compromise programming. In J. L. Cochrane and M. Zeleny, editors, *Multiple Criteria Decision Making*, pages 262–301. Columbia, South Carolina: University of South Carolina Press, 1973.
71. Q. Zhang and H. Li. MOEA/D: A multiobjective evolutionary algorithm based on decomposition. *IEEE Transactions on Evolutionary Computation,* 11(6):712–731, 2007.
72. L. Zhu, K. Deb, and S. Kulkarni. Multi-scenario optimization using multi-criterion methods: A case study on byzantine agreement problem. In *Proceedings of World Congress on Computational Intelligence (WCCI-2014)*, pages 2601–2608, Piscataway, New Jersey, 2014. IEEE Press.
73. E. Zitzler. *Evolutionary Algorithms for Multiobjective Optimization: Methods and Applications*. PhD thesis, Zürich, Switzerland: Swiss Federal Institute of Technology ETH, 1999.
74. E. Zitzler, M. Laumanns, and L. Thiele. SPEA2: Improving the strength Pareto evolutionary algorithm for multiobjective optimization. In K. C. Giannakoglou et al., editors, *Evolutionary Methods for Design Optimization and Control with Applications to Industrial Problems*, pages 95–100. Barcelona, Spain: International Center for Numerical Methods in Engineering (CIMNE), 2001.
75. E. Zitzler and L. Thiele. Multiobjective evolutionary algorithms: A comparative case study and the strength Pareto approach. *IEEE Transactions on Evolutionary Computation*, 3(4):257–271, 1999.

4

Hybridizing MOEAs with Mathematical-Programming Techniques

Saúl Zapotecas-Martínez, Adriana Lara, and Carlos A. Coello Coello

CONTENTS

ABSTRACT In this chapter, we present hybridization techniques that allow us to combine evolutionary algorithms with mathematical-programming techniques for solving continuous multiobjective optimization problems. The main motivation for this hybridization is to improve the performance by coupling a global search engine (a multiobjective evolutionary algorithm [MOEA]) with a local search engine (a mathematical-programming technique). The chapter includes a short introduction to multiobjective optimization concepts, as well as some general background about mathematical-programming techniques used for multiobjective optimization and state-of-the-art MOEAs. Also, a general discussion of memetic algorithms (which combine global search engines with local search engines) is provided. Then, the chapter discusses a variety of hybrid approaches in detail, including combinations of MOEAs with both gradient and non-gradient methods.

4.1 Introduction

This chapter presents methods, and the general background, to the coupling of multiobjective evolutionary algorithms (MOEAs) with mathematical-programming methods. The goal of this mixture of techniques is to improve the efficiency of MOEAs when solving continuous multiobjective optimization problems (MOPs). Introductory descriptions for both techniques (MOEAs and classical approaches) are also included. The chapter makes a distinction between gradient-based and nongradient-based methods. Different hybridization options are presented, to provide the reader with a richer companion of algorithms as well as discussions about them.

4.1.1 Multiobjective Background Concepts

The traditional optimization literature [1,2] focuses on solving a single-objective optimization problem (SOP), in which there is only one objective function to deal with; that is, it focuses on optimizing

$$f : \mathbb{R}^n \to \mathbb{R}$$

subject to certain domain constraints. In this case, the solution is a single point—or a set of points with the same image value—such that its image is a unique maximum or minimum value in $\mathbb{R}$. In contrast, when dealing with an MOP, we are interested in finding the best values for a function

$$\mathbf{F} : \mathbb{R}^n \to \mathbb{R}^m$$

subject to certain domain constraints. In this case, there is not a single-point solution. This can be noticed, for example, in Figure 4.1, in which the functions are, in general, in conflict with each

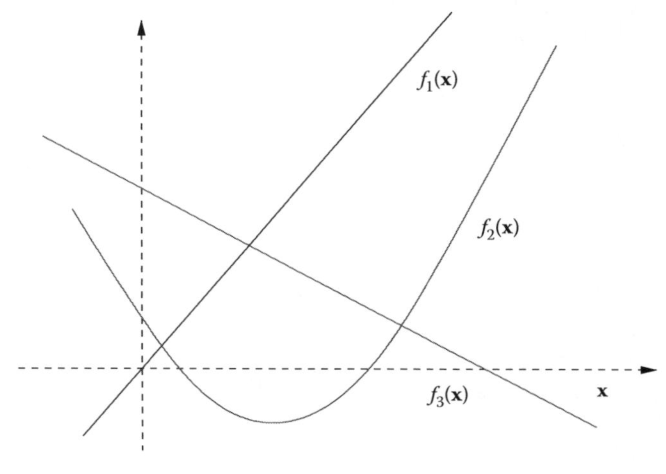

FIGURE 4.1
It shows function $f_3(\mathbf{x})$ in conflict with functions $f_1(\mathbf{x})$ and $f_2(\mathbf{x})$.

other—that is, while one increases, the other decreases. In the following, we are going to deepen into this idea, after explaining some basic concepts.

If an objective function is to be maximized (e.g., if it corresponds to a certain profit), then, it is possible to restate it in terms of minimization, using the *duality principle* when multiplying the function by -1. Therefore, an MOP can be stated in general as

$$\text{Minimizing } \mathbf{F}(\mathbf{x}) := [f_1(\mathbf{x}), f_2(\mathbf{x}), \ldots, f_m(\mathbf{x})]^T \tag{4.1}$$

subject to

$$g_i(\mathbf{x}) \leq 0 \quad i = 1, 2, \ldots, k \tag{4.2}$$

$$h_j(\mathbf{x}) = 0 \quad j = 1, 2, \ldots, p \tag{4.3}$$

where $\mathbf{x} = [x_1, x_2, \ldots, x_n]^T \in \mathbb{R}^n$ is the vector of *decision variables* (also known as *decision parameters* vector or *solution* vector), $f_i : \mathbb{R}^n \rightarrow \mathbb{R}$, where $i = 1, \ldots, m$ are the *objective functions* (or *objectives,* in short), and $g_i, h_j : \mathbb{R}^n \rightarrow \mathbb{R}$, where $i = 1, \ldots, k,\ j = 1, \ldots, p$ are the constraint functions of the problem. If functions g_i and h_i are not present, then, we are dealing with an *unconstrained* MOP. Solving the above problem is known as solving an MOP. For most of the following sections, we will assume unconstrained MOPs.

Even though MOPs can be defined over other domains—such as discrete sets, for example—in the methods presented in this chapter, we are only interested in continuous domains, which are contained in $\mathbb{R}^n$. When all the objective functions and the constraint functions are linear, problem (4.1) is called a *linear* MOP, and there are several techniques to solve it. If at least one of the functions is nonlinear, the problem is then called a *nonlinear* MOP. If all the objective functions are convex, and also the feasible region is convex, then, the problem is known as a *convex* MOP. In this chapter, we are dealing with nonlinear problems, which are either convex or not. For the remainder of this chapter, if conditions such as differentiability or continuous differentiability are assumed for the f_i functions, then, this will be pointed out for each particular method.

In general, a solution of an MOP consists of an entire set of points that fulfills certain properties. We state the definitions given below to establish a way to decide when a point can be considered as a part of this particular solution set.

Definition 4.1

Given two vectors $\mathbf{x}, \mathbf{y} \in \mathbb{R}^n$, we say that $\mathbf{x}$ **dominates*** $\mathbf{y}$ (denoted by $\mathbf{x} \prec \mathbf{y}$) if $f_i(\mathbf{x}) \leq f_i(\mathbf{y})$ for $i = 1, \ldots, m$, and $\mathbf{F}(\mathbf{x}) \neq \mathbf{F}(\mathbf{y})$.

Definition 4.2

We say that a vector of decision variables $\mathbf{x} \in \mathcal{X} \subseteq \mathbb{R}^n$ is **nondominated** with respect to $\mathcal{X}$ (where $\mathcal{X}$ is the feasible region), if there does not exist another $\mathbf{x}' \in \mathcal{X}$ such that $\mathbf{x}' \prec \mathbf{x}$ [3].

Definition 4.3

We say that a vector of decision variables $\mathbf{x}^* \in \mathcal{X} \subset \mathbf{R}^n$ is **Pareto optimal** if it is nondominated with respect to $\mathcal{X}$.

* Additional concepts, such as a weak optimality, are also very interesting, but they are not included here since they are not necessary for the rest of this chapter.

Definition 4.4

The Pareto optimal set or **Pareto set** $\mathcal{P}^*$ is defined by

$$\mathcal{P}^* = \{\mathbf{x} \in \mathcal{X} \mid \mathbf{x} \text{ is Pareto optimal}\}$$

Definition 4.5

The **Pareto front** $\mathcal{PF}^*$ is defined by

$$\mathcal{PF}^* = \{F(\mathbf{x}) \in \mathbb{R}^k \mid \mathbf{x} \in \mathcal{P}^*\}$$

Remarks

It is important to note that, in the specialized literature, these terms and definitions are not necessarily unique. For example, some authors use the term *efficient solution* to refer to Pareto optimal points. Also, other equivalent definitions for efficiency are available; for example, the general approach based on *ordering cones* (see [3,4]). In particular, a point x' is a nondominated point if there is no $x \in \mathcal{X}$ such that $F(x') - F(x) \in \mathbb{R}^k_+ \{0\}$.

We thus wish to determine the Pareto optimal set from the set $\mathcal{X}$ of all the decision variable vectors that satisfy the constraints of the problem. However, note that in practice, just a finite representation of the Pareto optimal set is normally achievable. Assuming $\mathbf{x}^*$ as a Pareto optimal point of Equation 4.1, there exists [5] a vector $\boldsymbol{\alpha} \in \mathbb{R}^k$, with $0 \leq \alpha_i$, where $i = 1, \ldots, k$ and $\sum_{i=1}^k \alpha_i = 1$ such that

$$\sum_i^k \alpha_i \nabla f_i(\mathbf{x}^*) = 0 \tag{4.4}$$

A point $\mathbf{x}^*$ that satisfies Equation 4.4 is called a Karush–Kuhn–Tucker (KKT) point.

Example 4.1

Consider the following unconstrained MOP. Minimize

$$f_1 = x^2 + y^2$$
$$f_2 = (x + 4)^2 + y^2 \tag{4.5}$$

with $x, y \in \mathbb{R}$. The Pareto set of this problem is the line segment $[(0, 0), (-4, 0)] \in \mathbb{R}^2$. Figure 4.2 shows the image of uniformly generated points in the domain of the problem. The axes correspond to the values regarding each function, what is known as the *objective function space*.

4.2 Solving an MOP

The most common procedures to solve Equation 4.1 can be classified [6,7] into four classes: *no-preference methods, a posteriori methods, a priori methods, and interactive methods,* according to the stage, and level, at which the decision maker intervenes. In this chapter, we focus on the *a posteriori* approach, in which the goal is to obtain the best approximation of the entire set of optima. This solution will then be presented *a posteriori* to the decision maker—who will then select the most suitable solution out of it. In the following section, we present a brief review of some classical techniques to mention the important aspects that will then be compared with respect to MOEAs.

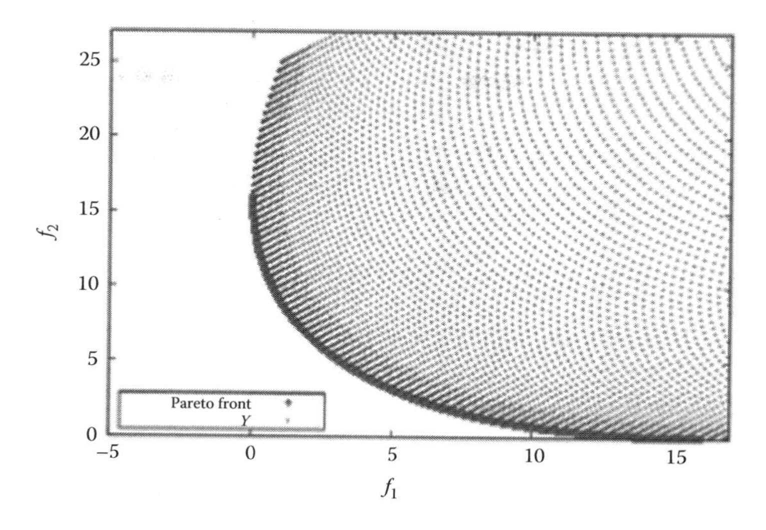

FIGURE 4.2
It emphasizes the Pareto front corresponding to Example 4.1. The points of Y are the images of uniformly generated vectors from the domain.

4.2.1 Scalarizing Methods

The traditional approaches developed for solving MOPs are part of the mathematical-programming literature, and are known as *classical methods*. Up to the early 1980s, most of the computational methods to solve MOPs consisted of minimizing only one function, either using the other objective functions as constraints of the problem, or simply by taking a combination of all the objectives [8]. The most common way to tackle an MOP is by *scalarization* that means reducing the problem to an SOP. One example of this approach is the following method:

Weighted sum method: This method consists of transforming the vector of function values into a scalar value using an aggregating function over the vector function, getting the following problem:

$$\text{Minimize } g(\mathbf{x}|\mathbf{w}) = \sum_{i=0}^{m} w_i f_i(\mathbf{x}) \tag{4.6}$$

where $\mathbf{x} \in \mathcal{X}$ and $\mathbf{w}$ is a weighting vector, that is, $w_i \geq 0$ for all $i \in \{1, \ldots, m\}$ and $\sum_{i=1}^{m} w_i = 1$.

After this reformulation, the solution set—or point—found corresponds only to one single point in objective space; this point is related to each weight combination. The main drawback of this approach is that the weights distribution does not necessarily correspond to the distribution of the final points in objective space. Besides, there are points that cannot be generated as a combination of weights in nonconvex cases—see [9].

Tchebycheff approach: This approach also transforms the vector of function values into a scalar optimization problem that is of the form

$$\text{Minimize } g(\mathbf{x}|\mathbf{w}, \mathbf{z}^\star) = \max_{1 \leq i \leq k} \{w_i |f_i(\mathbf{x}) - z_i|\} \tag{4.7}$$

where $\mathbf{x} \in \mathcal{X}$, $\mathbf{z}^\star = [z_1, \ldots, z_k]^T$, such that $z_i = \min\{f_i(\mathbf{x})|\mathbf{x} \in \mathcal{X}\}$ and $\mathbf{w}$ is a weighting vector.

For each Pareto optimal point $\mathbf{x}^\star$ there exists a weighting vector $\mathbf{w}$ such that $\mathbf{x}^\star$ is the optimal solution of Equation 4.7 and each optimal solution of Equation 4.7 is a Pareto optimal solution of Equation 4.1. Therefore, one is able to obtain different Pareto optimal solutions by altering the weight vector $\mathbf{w}$. One weakness of this approach is that its aggregation function is not smooth for a continuous MOP. There exist, in general, many scalarization methods that transform the MOP into a "classical" SOP. It is worth noting that a certain selection of suitable SOPs can lead to a reasonably good approximation of the entire Pareto set—see, for example, [10].

ϵ-Constraint method: In the *ϵ-constraint method* [11,12], one of the objectives is chosen for minimization while the rest of the objectives conform to a set of constraints limited by user-specified bounds ϵ_i, that is

$$\begin{aligned}&\text{Minimize} f_j\\&\text{subject to}\quad f_i \leq \epsilon_i \quad \text{for all } i \in \{1,\ldots,m\}, \quad i \neq j\end{aligned}$$

The ϵ-constraint problem should be solved using multiple different values for ϵ_i, if several Pareto optimal solutions are desired. This method can deal with convex and nonconvex functions; but, choosing the ϵ_i values is still an issue since there is no warranty that a feasible optimum exists for a specific ϵ_i. An in-depth analysis of this method can be found in [6].

When a method iteratively explores solutions from a neighborhood, it is classified as a local search (LS) procedure. When solving an MOP according to the *a posteriori* approach, the use of population-based heuristics, such as the evolutionary algorithms (EAs) presented in the following section, can be seen as an advantage since a global exploration of the space is possible. Other reasons for which this sort of approach is interesting is due to the fact that no previous knowledge regarding the MOP is necessary for the algorithm. Besides, at the end of the run, an EA, being a population-based algorithm, generates an entire set of (ideally) "well-distributed" solutions. The idea of having a good distribution of solutions is to produce the best-possible discrete approximation of the real Pareto front. In particular for a large number of objective functions, producing such an approximation is a very challenging problem.

4.3 Multiobjective Evolutionary Algorithms

The limitations of traditional mathematical-programming methods to solve MOPs have motivated the development of new strategies to deal with these types of problems. EAs are stochastic search and optimization methods that simulate the natural evolution process. At the end of the 1960s, Rosenberg [13] proposed the use of genetic algorithms (GAs) to solve MOPs. However, it was until 1984, when David Schaffer [14] introduced the first implementation of what is now called an MOEA. Since then, many researchers [15–18] have developed a wide variety of MOEAs. MOEAs are particularly well suited to solve MOPs because they operate over a set of potential solutions (they are based on a population). This feature allows them to generate several elements of the Pareto optimal set (or a good approximation of them) in a single run. Furthermore, MOEAs are less susceptible to the shape or continuity of the Pareto front than the traditional mathematical-programming techniques, require little domain information, and are relatively easy to implement and use. As pointed out by Coello et al. [19] and Zitzler and Thiele [20], finding an approximation to the Pareto front is by itself a biobjective problem whose objectives are

1. Minimize the distance of the generated vectors to the true Pareto front
2. Maximize the diversity of the achieved Pareto front approximation

Therefore, the fitness assignment scheme must consider these two objectives.

4.3.1 Nondominated Sorting Genetic Algorithm

The nondominated sorting genetic algorithm (NSGA) was proposed by Srinivas and Deb [21] and is a variation of Goldberg's approach [22]. The NSGA is based on several layers of classifications of the individuals. Before the selection is performed, the population is ranked on the basis of nondomination: all nondominated individuals are classified into one category (with a dummy fitness value, which is proportional to the population size, to provide an equal reproductive potential for these individuals). To maintain the diversity of the population, these classified individuals are shared with their dummy fitness values. Then, this group of classified individuals is ignored and another layer of nondominated individuals is considered. The process continues until all individuals in the population are classified. A stochastic remainder-proportionate selection is adopted for this technique. Since individuals in the first front have the maximum fitness value, they always get more copies than the rest of the population. This allows for a better search of the different nondominated regions and results in convergence of the population toward such regions. Fitness sharing, by its part, helps to distribute the population over this region. As a result, one might think that this MOEA converges rather quickly; however, a computational bottleneck occurs with the fitness-sharing mechanism. An improved version of the NSGA algorithm, called the nondominated sorting genetic algorithm II (NSGA-II), was proposed by Deb et al. [15]. The NSGA-II builds a population of competing individuals, ranks and sorts each individual according to its nondomination level, it applies evolutionary operators to create a new offspring pool, and then combines the parents and offspring before partitioning the new combined pool into fronts. For each ranking level, a crowding distance is estimated by calculating the sum of the Euclidean distances between the two neighboring solutions from either side of the solution along each of the objectives. Once the nodomination rank and the crowding distance is calculated, the next population is stated by using the crowded-comparison operator ($\prec_n$). The crowded-comparison operator guides the selection process at the various stages of the algorithm toward a uniformly spread-out Pareto optimal front. Assuming that every individual in the population has two attributes: (1) nondomination rank (i_{rank}) and (2) crowding distance ($i_{distance}$), the partial order $\prec_n$ is defined as

$$\begin{aligned} i \prec_n j : \text{if } (i_{rank} < j_{rank}) \quad & \text{or} \\ ((i_{rank} = j_{rank}) \quad & \text{and} \quad (i_{distance} > j_{distance})) \end{aligned} \tag{4.8}$$

That is, between two solutions with different nondomination ranks, we prefer the solution with the lower (better) rank. Otherwise, if both solutions belong to the same front, then, the solution that is located in a lesser-crowded region is preferred. Algorithm 4.1 presents the outline of the NSGA-II, which (in the last decade) has been the most popular MOEA, and it is frequently adopted to compare the performance of newly introduced MOEAs.

Algorithm 4.1: General Framework of NSGA-II

Input:
N: the population size;
T_{max}: the maximum number of generations;
Output:
A: the final approximation to the Pareto optimal front;

```
1  begin
2      t = 0;
3      Generate a random population P_t of size N;
4      Evaluate the population P_t;
5      while t < T_max do
6          Generate the offspring population Q_t by using binary tournament and genetic
           operators (crossover and mutation);
7          Evaluate the offspring population Q_t;
8          R_t = P_t ∪ Q_t;
9          Rank R_t by using nondominated sorting to define F; // F = (F_1, F_2, ...), all
           nondominated fronts of R_t
10         P_{t+1} = ∅ and i = 1;
11         while (|P_{t+1}| + |F_i| ≤ N) do
12             Assign crowding distance to each front F_i;
13             P_{t+1} = P_{t+1} ∪ F_i;
14             i = i + 1;
15         end
16         Sort F_i by using the crowded-comparison operator;
17         P_{t+1} = P_{t+1} ∪ F_i[1 : (N − |P_{t+1}|)];
18         t = t + 1;
19     end
20     A = P_t;
21 end
```

4.3.2 Strength Pareto Evolutionary Algorithm

The strength Pareto evolutionary algorithm (SPEA) was introduced by Zitzler and Thiele [20]. This evolutionary approach integrates some successful mechanisms from other MOEAs, namely, a secondary population (external archive) and the use of Pareto ranking. SPEA uses an external archive containing nondominated solutions that were previously found. At each generation, nondominated individuals are copied to the external nondominated set. In SPEA, the fitness of each individual in the primary population is computed using the individuals of the external archive. First, for each individual in this external set, a strength value is computed. The strength, $S(i)$, of individual i is determined by

$$S(i) = \frac{n}{\overline{N} + 1}$$

where n is the number of solutions, in the archive, dominated by i, and $\overline{N}$ is the size of the archive.

Finally, the fitness of each individual in the primary population is equal to the sum of the strengths of all the external members that dominate it. This fitness assignment considers both closeness to the true Pareto front and uniform distribution of solutions at the same time. Thus, instead of using niches based on the distance, Pareto dominance is used to ensure that the solutions are properly distributed along the Pareto front. Since the size of the archive may be too large, the authors employed a technique that prunes the contents of the external nondominated set so that its size remains below a certain threshold. There is also a revised version of SPEA, called strength Pareto evolutionary algorithm 2 (SPEA2) [18]. SPEA2 has three main differences with respect to its predecessor: (1) it incorporates a fine-grained fitness assignment strategy that takes into account, for each individual, the number of individuals that dominate it and the number of individuals to which it dominates; (2) it uses a nearest-neighbor density estimation technique that guides the search more efficiently; and (3) it has an enhanced archive truncation method that guarantees the preservation of boundary solutions. The outline of SPEA2 is shown in Algorithm 4.2.

In detail, let $\overline{P}_t$ and P_t be the external archive and the population at the t generation. Each individual i in the external archive $\overline{P}_t$ and the population P_t is assigned a strength value $S(i)$, representing the number of solutions it dominates: $S(i) = |j|j \in P_t + \overline{P}_t \wedge i \prec j|$, where $|\cdot|$ denotes the cardinality of a set, $+$ stands for a multiset union, and the symbol $\prec$ corresponds to the Pareto dominance relation. On the basis of the S values, the raw fitness $R(i)$ of an individual i is calculated:

$$R(i) = \sum_{j \in P \cup \overline{P}, j \prec i} S(j)$$

Algorithm 4.2: General Framework of SPEA2

Input:
N: the population size;
$\overline{N}$: the archive size;
T_{max}: the maximum number of generations;
Output:
A: the final approximation to the Pareto optimal front.

1 **begin**
2 $t = 0$;
3 Generate a random population P_t of size N;
4 $\overline{P}_t = \emptyset$; `// the external archive`
5 **while** $(t < T_{max})$ **do**
6 Calculate the fitness values of individuals in P_t and $\overline{P}_t$;
7 Copy all nondominated individuals in P_t and $\overline{P}_t$ to P_{t+1}. If the size of P_{t+1} exceeds $\overline{N}$, then reduce P_{t+1} by means of the truncation operator; otherwise if the size of P_{t+1} is less than $\overline{N}$, then fill P_{t+1} with dominated individuals in P_t and $\overline{P}_t$;
8 **if** $(t + 1 < T_{max})$ **then**
9 Perform binary tournament selection with replacement on $\overline{P}_{t+1}$ in order to fill the mating pool;
10 Apply recombination and mutation operators to the mating pool and set P_{t+1} to the resulting population;
11 **end**
12 $t = t + 1$;
13 **end**
14 Set A as the set of decision vectors represented by the nondominated individuals in $\overline{P}_t$;
15 **end**

Although the raw fitness assignment provides a sort of niching mechanism based on the concept of Pareto dominance, it may fail when most individuals do not dominate each other. Therefore, the additional density information is incorporated to discriminate between individuals having identical raw fitness values. The density estimation technique used in SPEA2 is an adaptation of the kth nearest-neighbor method, and it is calculated by

$$D(i) = \frac{1}{\sigma_i^k + 2}$$

where $k = \sqrt{|N| + |\overline{N}|}$, and σ_i^k denotes the distance of i to its kth nearest neighbor in $P_t + \overline{P}_t$. N and $\overline{N}$ represent the population size and the archive size, respectively. Finally, the fitness value $F(i)$ of an individual i is calculated by

$$F(i) = R(i) + D(i) \tag{4.9}$$

During environmental selection, the first step is to copy all nondominated individuals. If the nondominated front fits exactly into the archive ($|\overline{P}_{t+1}| = N$), then, the environmental selection step is completed. Otherwise, there can be two situations: Either the archive is too small ($|\overline{P}_{t+1}| < N$) or too large ($|\overline{P}_{t+1}| > N$). In the first case, the best $\overline{N} - |\overline{P}_{t+1}|$ dominated individuals in the previous archive and population are copied to the new archive. This can be implemented by sorting the multiset $P_t + \overline{P}_t$ according to the fitness values and copying the first $\overline{N} - |\overline{P}_{t+1}|$ individuals i with $F(i) \geq 1$ from the resulting ordered list to $\overline{P}_{t+1}$. In the second case, when the size of the current nondominated (multi)set exceeds $\overline{N}$, an archive truncation procedure is invoked that iteratively removes individuals from $\overline{P}_{t+1}$ until $|\overline{P}_{t+1}| = \overline{N}$. Here, at each iteration, that individual i is chosen for removal for which $i \leq_d j$ for all $j \in \overline{P}_{t+1}$ with

$$\begin{aligned} i \leq_d j \iff & \forall 0 < k < |\overline{P}_{t+1}| : \sigma_i^k = \sigma_j^k \quad \vee \\ & \exists 0 < k < |\overline{P}_{t+1}| : [(\forall 0 < l < k : \sigma_i^l = \sigma_j^l) \wedge \sigma_i^k < \sigma_j^k] \end{aligned}$$

where σ_i^k denotes the distance from i to its kth nearest neighbor in $\overline{P}_{t+1}$. In other words, the individual that has the minimum distance to another individual is chosen at each stage; if there are several individuals with a minimum distance, then, the tie is broken by considering the second-smallest distances and so forth.

4.3.3 Multiobjective Evolutionary Algorithm Based on Decomposition

The multiobjective evolutionary algorithm based on decomposition (MOEA/D) was introduced by Zhang and Li [17]. MOEA/D explicitly decomposes the MOP into several scalar optimization subproblems. It is well known that a Pareto optimal solution to an MOP, under certain conditions, could be an optimal solution of a scalar optimization problem in which the objective is an aggregation of all the objective functions. Therefore, an approximation of the Pareto optimal front can be decomposed into a number of scalar objective optimization subproblems. This is a basic idea behind many traditional mathematical-programming methods for approximating the Pareto optimal front. Several methods for constructing aggregation functions can be found in [4,6]. This basic idea of decomposition is used by MOEA/D, and it solves these subproblems simultaneously by evolving a population of solutions. At each generation, the population is composed of the best solution found so far (i.e., since the start of the run of the algorithm) for each subproblem. The neighborhood relations among these subproblems are defined based on the distances between their aggregation coefficient vectors.

The optimal solutions to two neighboring subproblems should be very similar. Each subproblem (i.e., each scalar aggregation function) is optimized in MOEA/D by using information only from its neighboring subproblems. To obtain a good representation of the Pareto optimal front, a set of evenly spread weighting vectors needs to be previously generated.

Considering N as the number of scalar optimization subproblems and

$$W = \{\mathbf{w}_1, \ldots, \mathbf{w}_N\}$$

as the set of weighting vectors that defines such subproblems, MOEA/D finds the best solution to each subproblem along the evolutionary process. Assuming the Tchebycheff approach (4.7), the fitness function of the ith subproblem is stated by $g(\mathbf{x}|\mathbf{w}_i, \mathbf{z})$. MOEA/D defines a neighborhood of each weighting vector $\mathbf{w}_i$ as a set of its closest weighting vectors in W. Therefore, the neighborhood of the ith subproblem consists of all the subproblems with the weighting vectors from the neighborhood of $\mathbf{w}_i$ and it is denoted by $B(\mathbf{w}_i)$. At each generation, MOEA/D maintains (1) a population of N points $P = \{\mathbf{x}_1, \ldots, \mathbf{x}_N\}$, where $\mathbf{x}_i \in \mathcal{X}$ is the current solution to the ith subproblem; (2) $FV^1, \ldots, FV^N$, where FV^i is the F-value of $\mathbf{x}_i$, that is, $FV^i = \mathbf{F}(\mathbf{x}_i)$ for each $i = 1, \ldots, N$; and (3) an external archive EP, which is used to store the nondominated solutions found during the search. In contrast to NSGA-II and SPEA2 that use density estimators (crowding distance and neighboring solutions, respectively), MOEA/D uses a well-distributed set of weighting vectors for guiding the search, and therefore, multiple solutions along the Pareto optimal set are maintained. With that, the diversity in the population of MOEA/D is implicitly maintained. For an easier interpretation of MOEA/D, it is outlined in Algorithm 4.3.

4.3.4 Memetic Algorithms

The term memetic algorithm (MA) was first introduced in 1989 by Pablo Moscato [23]. The term "memetic" has its roots in the word "meme" introduced by Richard Dawkins in 1976 [24] to denote the unit of imitation in cultural transmission. The essential idea behind MAs is the combination of LS refinement techniques with a strategy based on a population, such as EAs. The main difference between genetic and MAs is the approach and view of the information's transmission techniques. In GAs, the genetic information carried out by genes is usually transmitted intact to the offspring; meanwhile, in the MA, the base units are the so-called "memes" and they are typically adapted by the individual transmitting information. While GAs are good at exploring the solution space from a set of candidate solutions, MAs explore from a single point, allowing to exploit solutions that are close to the optimal solutions. The main design goal of such an approach will be the efficiency of the final algorithm (i.e., the MA approach should perform a reduced number of objective function evaluations as compared to state-of-the-art EAs) on standard test functions. During the last few years, MAs have been successfully applied to find solutions of SOPs. In the multiobjective case, the initial efforts were done over discrete domain problems, followed by an increasing interest for the continuous case. Some important decisions should be taken when mixing these techniques, they are

1. In which moment, of the evolutionary iteration should the LS be performed?
2. How often should the LS be applied along the entire evolutionary process?
3. From which solutions should the LS be started?

While answering the above questions, it is necessary to keep in mind the evident trade-off between the computational cost and the benefits of the LS procedure. This makes it important to focus on the

Algorithm 4.3: General Framework of MOEA-D

Input:
a stopping criterion;
N: the number of the subproblems considered in MOEA/D;
W: a set of uniformly distributed weighting vectors $\{\mathbf{w}_1, \ldots, \mathbf{w}_N\}$;
T: the number of weight vectors in the neighborhood of each weighting vector.
Output:
EP: the nondominated solutions found during the search;
P: the final population found by MOEA/D.

1 **begin**
2 **Step 1.** INITIALIZATION:
3 $EP = \emptyset$;
4 Generate an initial population $P = \{\mathbf{x}_1, \ldots, \mathbf{x}_N\}$ randomly;
5 $FV^i = \mathbf{F}(\mathbf{x}_i)$;
6 $B(\mathbf{w}_i) = \{\mathbf{w}_{i_1}, \ldots, \mathbf{w}_{i_T}\}$ where $\mathbf{w}_{i_1}, \ldots, \mathbf{w}_{i_T}$ are the T closest weighting vectors to $\mathbf{w}_i$, for each $i = 1, \ldots, N$;
7 $\mathbf{z} = (+\infty, \ldots, +\infty)^T$;
8 **while** *stopping criterion is not satisfied* **do**
9 **Step 2.** UPDATE: (the next population)
10 **for** $\mathbf{x}_i \in P$ **do**
11 REPRODUCTION: Randomly select two indexes k, l from $B(\mathbf{w}_i)$, and then generate a new solution $\mathbf{y}$ from $\mathbf{x}_k$ and $\mathbf{x}_l$ by using genetic operators.
12 MUTATION: Apply a mutation operator on $\mathbf{y}$ to produce $\mathbf{y}'$.
13 UPDATE OF $\mathbf{z}$: For each $j = 1, \ldots, k$, if $z_j < f_j(\mathbf{x})$, then set $z_j = f_j(\mathbf{y}')$.
14 UPDATE OF NEIGHBORING SOLUTIONS: For each index $j \in B(\mathbf{w}_i)$, $g(\mathbf{y}'|\mathbf{w}_j, \mathbf{z}) > g(\mathbf{x}_i|\mathbf{w}_j, \mathbf{z})$, then set $\mathbf{x}_j = \mathbf{y}'$ and $FV^j = \mathbf{F}(\mathbf{y}')$.
15 UPDATE OF *EP*: Remove from *EP* all the vectors dominated by $\mathbf{F}(\mathbf{y}')$. Add $\mathbf{F}(\mathbf{y}')$ to *EP* if no vectors in *EP* dominate $\mathbf{F}(\mathbf{y}')$.
16 **end**
17 **end**
18 **end**

development of suitable methods to perform LS. Because of this, mathematical methods based on particular geometrical properties of the MOP would produce good results. In particular, methods that use gradient information will produce excellent approximations; but in practice, this is not always the ideal strategy, since they imply a high computational cost. From this point of view, other choices have shown advantages in practice, producing results of a comparative quality. There is no specific method to design an MA. However, Algorithm 4.4 shows a general framework of what an MA should contain.

Hybridization of MOEAs with LS algorithms has been investigated for more than one decade [25]. Some of the first memetic MOEAs for models on discrete domains were presented in [26,27]. These include multiobjective genetic local search (MOGLS), and Pareto memetic algorithm (PMA) [28]; these two approaches use scalarization functions. In [29], a proposal employing a Pareto ranking selection (called memetic Pareto archived evolution strategy [M-PAES]) was introduced. Also, Murata et al. [30] proposed an LS process with a generalized replacement rule based on the dominance relation. In [31], the cross-dominant multiobjective memetic algorithm (CDMOMA) was proposed as an adaptation of the NSGA-II, and two LS engines: a multiobjective implementation

Algorithm 4.4: General Scheme of an MA

Input:
N: the population size;
T_{max}: the maximum number of generations;
Output: A: the final population

```
1  begin
2      t = 0;
3      Generate a random population P_t of size N;
4      Evaluate the population P_t;
5      while t < T_max do
6          Generate the offspring population Q_t by using stochastic operators;
7          Evaluate the offspring population Q_t;
8          Select a set of solutions R_t from Q_t;
9          forall the r^t ∈ R_t do
10             IMPROVE: r^t = i(r^t) // using an improvement mechanism;
11         end
12         Set P_{t+1} as the set of N solutions selected from P_t ∪ Q_t ∪ R_t according to any rule
           of selection (commonly using the fitness of each individual);
13         t = t + 1;
14     end
15     A = P_t;
16 end
```

of Rosenbrock's algorithm [32], which performs very small movements, and the Pareto domination multiobjective simulated annealing (PDMOSA) [33], which performs a more global exploration. A memetic version of the coevolutionary multiobjective differential evolution (CMODE) was proposed in [34] and was called CMODE-MEM. Most of this work has been proposed for combinatorial problems. For the continuous case—that is, continuous objectives defined on a continuous domain—the first attempts started, to the authors' best knowledge, with Goel and Deb [35], where a neighborhood search was applied to NSGA-II. This is a very simple scheme and the authors found that the added computational work had a severe impact on the efficiency of the algorithm. In the following sections, we provide an in-depth analysis of the continuous-domain case.

4.4 Methods Based on Descent Directions

Many mathematical-programming techniques are traditionally based on differentiable properties of the objective functions. These tools are popular since they have strong foundations, and have been intensively taught to engineers through many decades. Exploiting the particular knowledge about the problem is a very common feature of almost all traditional optimization techniques. For example, some of the methods developed in [36] use linear properties of the functions and have a well-founded theoretical basis. Other methods, such as those explained in [6], succeed on solving nonlinear MOPs, when using just the differentiability properties of the functions. One drawback of all these (traditional) methods is precisely that they cannot be applied to a wide variety of MOPs, because of their special assumptions. On the other hand, it is worth noting that when certain conditions are fulfilled, they are indeed very efficient. Because of the nature of the differentiability theory, these techniques can be seen as LS procedures; that is, they can only guarantee—in most cases—the approximation of local optima points. Talking specifically about differentiable problems, in this section, we mainly focus on those methods that use gradient-based descent directions to perform line search. The remarkable features of these classic methods are the following:

- They generate new candidate solutions in an iterative way. For each iteration, at least an approximation of the gradient function is required; second-order information (Hessian matrix) is also necessary, in some cases.
- They take an advantage of a starting point for triggering the search, and they normally produce a single solution per run (in fact, in some cases, different starting points may lead to the same final objective value).
- They can only guarantee that the final point is optimum in a local neighborhood.

For MOPs where the objective functions are continuously differentiable, the use of gradient information seems to be a natural choice. However, due to the local nature of gradient information, its combination with a global search engine such as a population-based heuristic is an obvious choice. Several proposals, such as those developed in [37–41], are based on this hypothesis, that is, they attempt to show that the combination of gradient-based methods and MOEAs can boost performance.

Directing the search, from a particular solution, toward a special improvement direction is a widely used idea in optimization. This procedure is known as *line search.* It starts with a particular (non-optimal) vector solution $\boldsymbol{x_0}$, and is composed of two goals:

1. First, it is necessary to find a promising search direction $\boldsymbol{\nu} \in \mathbb{R}^n$ in which the function value improves. This is commonly based on information related to the gradient of the objective function.
2. Once the direction $\boldsymbol{\nu}$ is obtained, it is necessary to state a suitable step size t^* to move from $\boldsymbol{x_0}$ in this direction, such that this movement best improves the objective value of the new iterative solution $\boldsymbol{x_1}$.

For example, when minimizing

$$f : \mathbb{R}^n \to \mathbb{R}$$

to solve an SOP, the well-known *steepest descent method* is based on an iteration given by

$$\boldsymbol{x^{(1)}} = \boldsymbol{x^{(0)}} - t^{*(0)} \nabla \boldsymbol{f(x)}$$

When trying to use a similar procedure for the multiobjective optimization case, the first goal is to find a descent direction to be equivalent to the role played by the gradient of a function in the SOP case. Let us remember that, for MOPs, m-conflicting objective functions are involved. So, to do this, the gradient of each of the objective functions should be combined. The idea is then to find a direction

$$\boldsymbol{\nu} \in \mathbb{R}^n$$

such that, moving along $\boldsymbol{\nu}$ decrements the value of each objective function $f_i(\boldsymbol{x})$ simultaneously. This is stated in the next definition.

Definition 4.6

Let $f_1, \ldots, f_m$ be the objective functions of an MOP. A multiobjective descent direction (MDD) is a direction $\mathbf{v} \in \mathbb{R}^n$ such that the directional derivatives $D_{\mathbf{v}} f_i(\boldsymbol{x})$ with respect to $\mathbf{v}$, at the point $\boldsymbol{x} \in \mathbb{R}^n$, are nonpositive, that is, $D_{\mathbf{v}} f_i(\boldsymbol{x}) = \langle \nabla f_i(\boldsymbol{x}), \mathbf{v} \rangle \leq 0$ for all $i \in 1, \ldots, m$ without allowing all of them to be equal to zero.

Then, if there exists an MDD at a certain point $\boldsymbol{x} \in \mathbb{R}^n$, then, we can guarantee by the above definition that x is not a Pareto optimal point. In this way, assuming conflicting functions, when performing a small movement over an MDD $\mathbf{v}$, we will get a local improvement, at least in one objective function, and hopefully in several (or all) of them simultaneously—without compromising the values of the rest of the functions, in that point. The computation of an MDD is a procedure that regularly implies the solution of another optimization problem. Different proposals are available to set this problem; we explain some of these methods in the subsequent sections.

We will start by first noticing that, given $i \in \{1, \ldots, m\}$, any direction $\mathbf{v}$ such that $\langle \mathbf{v}, \nabla f_i(\boldsymbol{x}) \rangle < 0$ is a descent direction for f_i at $\boldsymbol{x}$. Figure 4.3 shows how, for any point $\boldsymbol{x}$, the search space is split into two semispaces, one of them corresponding to descent directions. From Definition 4.6, it follows that, considering the intersection of these semispaces provides a number of MDDs (see Figure 4.4).

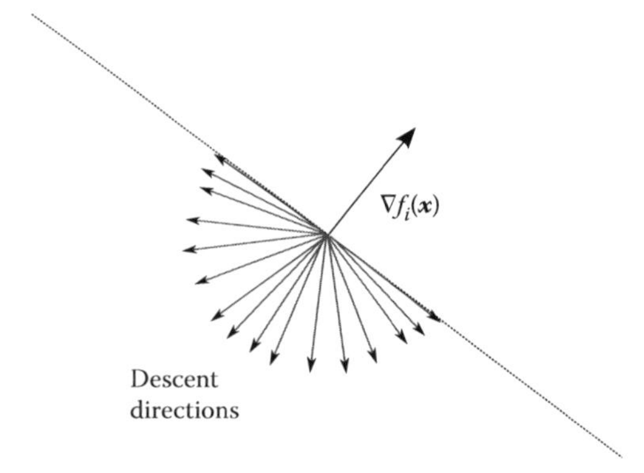

FIGURE 4.3
It illustrates the descent directions for a particular function f_i.

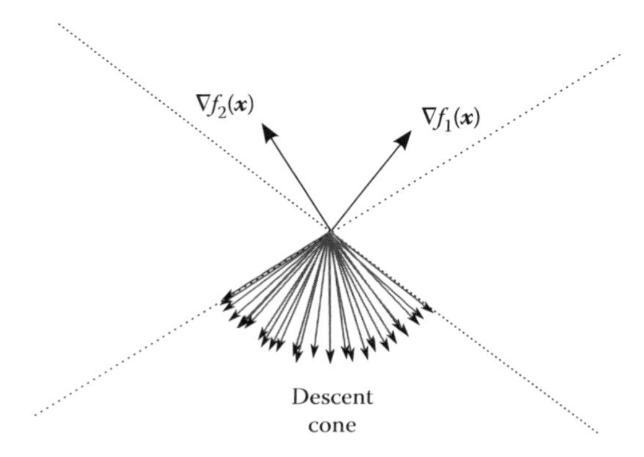

FIGURE 4.4
It illustrates the descent cone that encloses MDDs.

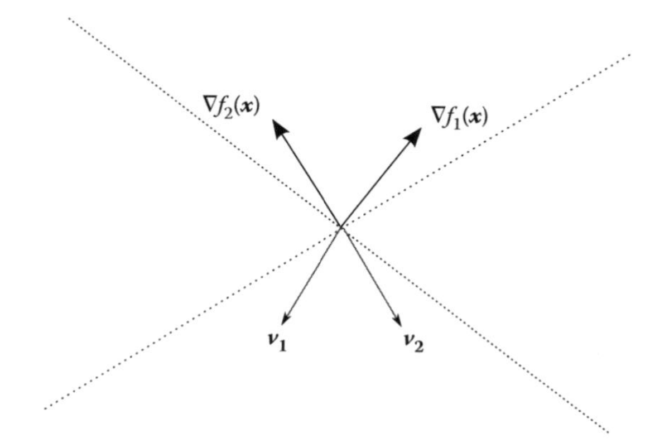

FIGURE 4.5
It illustrates two MDDs $\mathbf{v_1}$ and $\mathbf{v_2}$ for which the simultaneous objective function improvement is incomparable.

It is worth noting that the problem of finding an appropriate descent direction is again an MOP. To explain this, let us assume that two different vectors $\mathbf{v_1}$ and $\mathbf{v_2}$ are descent directions. And also that

$$D_{\mathbf{v_1}} f_1(\boldsymbol{x}) < D_{\mathbf{v_2}} f_1(\boldsymbol{x})$$

and

$$D_{\mathbf{v_2}} f_2(\boldsymbol{x}) < D_{\mathbf{v_1}} f_2(\boldsymbol{x})$$

hold. Then, even when both directions get simultaneous improvements, it is not possible to decide which one is the best—at least not without setting a preference between the objective functions. This is illustrated in Figure 4.5. Since finding an MDD led us to face an MOP again, it should be clear that the "best" direction is not defined in the MOP context.

Given a point $\boldsymbol{x} \in \mathbb{R}^n$ that is not a KKT point, solving a quadratic-programming problem leads to a descent direction, as the next theorem says.

Theorem 4.1

Given an unconstrained MOP, as in Equation 4.1, and $q : \mathbb{R}^n \to \mathbb{R}^n$, it can be defined by [41]

$$q(\boldsymbol{x}) = \sum_{i=1}^{m} \hat{\alpha}_i \boldsymbol{\nabla f_i}(\boldsymbol{x}) \tag{4.10}$$

where $\hat{\boldsymbol{\alpha}}$ is a solution of

$$\min_{\boldsymbol{\alpha} \in \mathbb{R}^m} \left\{ \left\| \sum_{i=1}^{m} \alpha_i \boldsymbol{\nabla f_i}(\boldsymbol{x}) \right\|_2^2 ; \alpha_i \geq 0, i = 1, \ldots, m, \sum_{i=1}^{m} \alpha_i = 1 \right\} \tag{4.11}$$

then, either $q(\boldsymbol{x}) = 0$, or it is the case that $-q(\boldsymbol{x})$ is a descent direction for F at x.

Proof. See [42]. ■

An alternative way to get an MDD can be found in [43]; the direction computed in that way is called by its authors as the *steepest descent direction*. The method works for convex Pareto fronts as well as for concave Pareto fronts. In this case, an MDD $\mathbf{v}$ can be computed using the information provided by the Jacobian $J_{F(\mathbf{x})}$ of the problem F evaluated on $\mathbf{x}$. It is, then, necessary to solve the following quadratic-programming problem:

$$\text{Minimize } \alpha + \frac{1}{2}\|\mathbf{v}\|^2 \tag{4.12}$$

$$\text{subject to} \quad (J_{F(\mathbf{x})}\mathbf{v})_i \leq \alpha, \quad \text{for all } i \in \{1, \ldots, m\}$$

The above problem produces a solution

$$(\mathbf{v}^*, \alpha^*)$$

where the descent direction that we are looking for is $\mathbf{v}^*$. This method triggers automatically, similar to Theorem 4.1, by using a condition that verifies if $\mathbf{x}$ fulfills condition (4.4), which is the case when

$$\alpha^* = 0$$

In practice, it is necessary to set a tolerance parameter τ, such that $\tau < 0$ to stop the descent when

$$\tau \leq \alpha^*$$

holds; note that, by construction, $\alpha^* \leq 0$.

For two-objective problems, there is a simple way to compute an MDD with no cost, other than the gradient approximation for each function. In this case, the descent direction $\overline{\nabla}_x$ can be obtained by

$$\overline{\nabla}_{\boldsymbol{x}} = -\left(\frac{\nabla f_1(\boldsymbol{x})}{\|\nabla f_1(\boldsymbol{x})\|} + \frac{\nabla f_2(\boldsymbol{x})}{\|\nabla f_2(\boldsymbol{x})\|}\right) \tag{4.13}$$

where $\|\cdot\| = \|\cdot\|_2$ is the Euclidean norm, and $\nabla f_1(\boldsymbol{x}), \nabla f_2(\boldsymbol{x})$ are the gradients of the objective functions, at solution x. The reader can refer to [44] for details and a proof that Equation 4.13 leads to an MDD. Also, for a possible hybridization of this procedure with NSGA-II, it is important to note that this simple formula cannot be generalized for problems with more than two objectives.

Once the MDD is set, a regular line search can be performed. The computation of a step length in the multiobjective context is an open problem, since each objective function will have its own optimal step size (see [45]). In practice, an Armijo-like rule, or a sequential quadratic approximations approach have been commonly used. Then, after computing a reasonable step length t, and getting $\mathbf{x}^{(1)} = \mathbf{x} + t\mathbf{v}$, we are in a condition to repeat the movement by calculating a new descent direction from $\mathbf{x}^{(1)}$ or, if this is not possible, we can assume that a critical point has been achieved. Other ways to calculate MDD have been proposed [37–39,46]. A study of the efficiency of each method is the subject of ongoing research. But, descent directions are not the only interesting directions during the search; sometimes, it is also necessary to perform movements along the Pareto front, or specifically directed toward a particular region (see [47]).

To show the coupling of this LS procedure with an MOEA, we chose the widely adopted NSGA-II as our global search engine. Nevertheless, the coupling with other MOEAs is also possible. In Algorithm 4.5, we present a simple version of an MOEA hybridized with a gradient-based LS procedure. The parameters N and G in Algorithm 4.5 represent the population size and the maximum number

Algorithm 4.5: Memetic NSGA-II Using Gradient-Based Line Search

Input:
G: maximum number of generations;
Output:
P: final approximation to the Pareto front;

```
1  begin
2      Generate a Random Population P of size N;
3      Evaluate Objective Function Values;
4      Fast Non-Dominated Sort;
5      Crowding Distance Assignment;
6      for i ← 1, ..., G do
7          Generate Offspring Population P_offs;
8          Set P ← P ∪ P_offs;
9          Fast Non-Dominated Sort;
10         Crowding Distance Assignment;
11         Apply Local Search using (i, P);
12     end
13 end
```

Algorithm 4.6: Gradient-Based Line Search Procedure

Input:
i : generation number;
ξ : rule to establish the frequency of application of Local Search (LS);
η : rule to select the individuals to which the LS is applied;
P : population before applying the LS;
Output:
P : population after applying the LS;

```
1  begin
2      if i fullfils ξ then
3          Conform the set E ⊂ P according to η, with randomly selected individuals from the
           non-dominated set R_1 ⊂ P;
4          for a ∈ E do
5              if local improvement is possible then
6                  Apply line search to obtain a';
7                  Replace a ← a';
8                  Set a' ∈ R_1 ⊂ P;
9                  Set the crowding distance of a' as ∞;
10             end
11         end
12     end
13 end
```

of generations. The procedures "Fast Non-Dominated Sort," "Crowding Distance Assignment," and "Generate Offspring Population" correspond to the well-known components of the NSGA-II. Algorithm 4.5 places the LS inside the NSGA-II just after the reproduction and the ranking–crowding process. LS is applied only to nondominated individuals, but not to all of them. In Algorithm 4.6, the LS procedure is described.

4.5 Gradient-Based Numerical Continuation

A method to deal with both, unconstrained and constrained MOPs, is shown by Hillermeier [48]. He introduces a homotopy approach using differential geometry and parametric-optimization concepts to extend the MOP with m objective functions on an n-dimensional space to an auxiliary problem. That is, when having a Pareto point $\boldsymbol{x}$, a necessary condition for the convex case (see [6]) is the existence of

$$\boldsymbol{\alpha} \in \mathbb{R}^m \quad \text{with} \sum_{i=1}^{m} \alpha_i = 1$$

such that

$$\sum_{i=1}^{m} \alpha_i \nabla f_i(\boldsymbol{x}) = \mathbf{0}$$

On the basis of this, it is possible to construct a suitable function

$$\widetilde{F} : \mathbb{R}^{n+m} \longrightarrow \mathbb{R}^{m+1}$$

defined by

$$\widetilde{F}(\boldsymbol{x}, \boldsymbol{\alpha}) = \begin{pmatrix} \sum_{i=1}^{m} \alpha_i \nabla f_i(\boldsymbol{x}) \\ \sum_{i=1}^{m} \alpha_i - 1 \end{pmatrix} \tag{4.14}$$

Then, for every Pareto point

$$\boldsymbol{x}$$

there exists a vector

$$\boldsymbol{\alpha^*} \in \mathbb{R}^m$$

such that

$$\widetilde{F}(\boldsymbol{x}, \boldsymbol{\alpha^*}) = \mathbf{0}$$

With this auxiliary function, it is possible to show [48] that the Pareto set of the original problem forms, under certain assumptions, an $m-1$ dimensional manifold.* In this method, all the functions are assumed to be twice continuously differentiable, too. Hillermeier states that this method is scalable to problems of a high dimensionality. Even when this procedure is capable of computing a set of Pareto optimal points, it only finds connected components. It is worth noting that this drawback can be avoided if the method is complemented with a stochastic technique—such as an MOEA.

An interesting way to use LS with stochastic methods is to refine the solutions obtained by a population-based algorithm. As an example, multiobjective continuation methods have been used [50] as a recovering technique to complement the application of a multiobjective particle swarm

---

* For a general treatment of this subject, the reader can refer to [49].

algorithm [19]. It is worth noting that the particle swarm heuristic could be replaced, in this algorithm, by any other MOEA.

When using the continuation part of [50], it is assumed that a Pareto optimal point $\boldsymbol{x_0}$ is given. Then, a linearization* of the Pareto front, at the solution $\boldsymbol{x_0}$, is computed using second-order information. In this way, a predictor–corrector method is used to follow the frontier of the Pareto front, and a new candidate point $\boldsymbol{x_0'}$ is obtained. The possible error in the prediction is corrected by the solution of Equation 4.11.

It is worth noting that, as with Hillermeier's method, this predictor–corrector method can only detect connected components of the Pareto set; then, the entire procedure can be seen as an LS technique. This observation is an important motivation to combine these techniques with a stochastic heuristic, when looking for global solutions of an MOP. A key element of this mix is the archiving method. This is used to manage the solutions generated by both techniques, in different moments. It is because of this feature that the convergence of the entire method can be ensured, in a theoretical sense. Also, the data structure to store the information is based on data trees—that reduce the computational complexity of the information access. This information is related to specially designed boxes that contain, and manage, the approximation points.

As another example, in [52], an LS engine called HCS2—Hill climber with side step 2—it consists of a hill climber part and a one-step continuation. The hill climber part is performed by an approach based on Fliege's method [43]. Alternatively, Shaffler's method (Equation 4.11) can also be used. In both cases, a stopping criterion is automatically given, when the MDD cannot be computed since a Pareto optimal point has been reached. This is done by a certain tolerance value, for computational purposes, to assess when a quantity is small enough—and can then be considered as zero. When this condition is fulfilled, a side step is performed—assuming that the Pareto front has been reached, and a movement over the frontier is suitable. The side-step part of the operator is performed by a technique based on the continuation proposed in [48]. Again, it uses second-order information and intends to generate two more points at the Pareto front with an affordable computational cost. Figure 4.6 shows the points obtained by the repeated execution of the HCS2, as a stand-alone procedure, over a convex problem (in convex problems, the local optima are also global [53]). In [52], a second type of LS engine is proposed—named as HCS1—but since it does not use gradient information, it is not described in this section. Each of these LS engines were combined with the Nondominated Sorting Genetic Algorithm II (NSGA-II) and SPEA2. The four MAs were tested on two and three objective-standard test functions; and those experiments showed [52] that the memetic approach outperformed the plain MOEA, in certain problems with a moderate number of local Pareto points.

Finally, it is worth mentioning that, in [50,54–56], other hybrids can be found, in which heuristic methods are coupled with particular multiobjective continuation methods.

4.6 Reference Point Methods

Rigoni and Poles [57] proposed a hybrid technique that combines the robustness of MOGA-II [58] with the accuracy and speed of NBI-NLPQLP. In [59], the proposed LS process employs quadratic approximations for all the objective functions.

An important work that explicitly emphasizes the benefits of using LS to obtain accurate solutions was presented in [60]. The authors combined the Nondominated Sorting Genetic Algorithm

* A linearization of the Pareto front can also be found in Eichfelder's work (see, e.g., [51]).

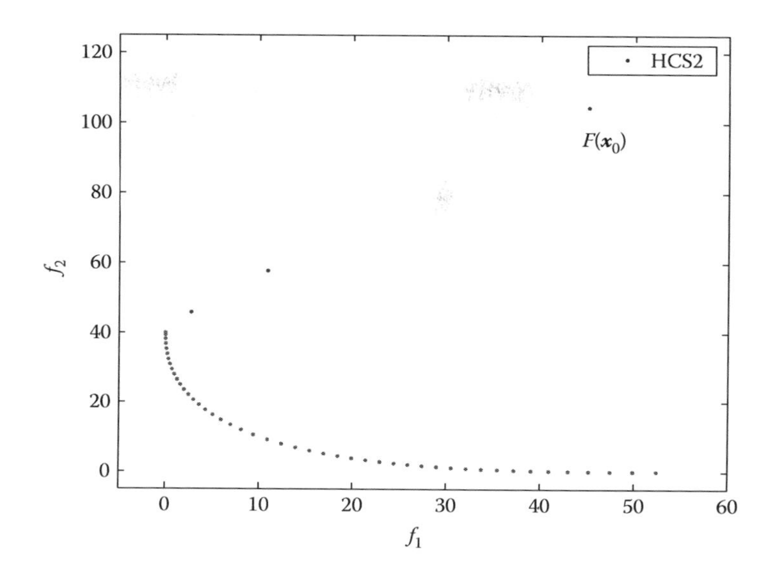

FIGURE 4.6
It shows a plot of the set of points obtained by the local operator HCS2 as a stand-alone procedure.

II (NSGA-II) with a reference point method. Applying this method, the authors were able to accelerate the convergence of NSGA-II. In this work, the authors proposed a concurrent approach, instead of a serial approach—in which the LS is applied after the population-based algorithm finishes. In the concurrent approach, both techniques interact over the same generation of individuals, at different times of the evolutionary process. This avoids the problem of having to set, *a priori*, a specific number of resources to be spent by each technique.

Then, applying LS to a specific individual $\boldsymbol{y}$ from the population consists of solving the next problem* for $\boldsymbol{x}$

$$\text{Minimize} \max_{i=1}^{m} \frac{f_i(\boldsymbol{x}) - z_i}{f_i^{max} - f_i^{min}} + \rho \sum_{j=1}^{m} \frac{f_j(\boldsymbol{x}) - z_j}{f_j^{max} - f_j^{min}} \tag{4.15}$$

where $\boldsymbol{x}$ is subject to $\boldsymbol{x} \in S$, with S being the feasible region. In this case, to take advantage of the population information into the procedure, f_i^{max} and f_i^{min} are the maximum and minimum values taken from the entire population, at a certain generation, respectively. Also, $\boldsymbol{z}$ is the so-called reference point defined by

$$\boldsymbol{z} := F(\boldsymbol{y})$$

with $\boldsymbol{y}$ being the individual from the population that has been chosen to be used as a departing point for the LS procedure. A suggested value is

$$\rho = 10^{-2}$$

Like other gradient-based local searchers, this method allows to use the KKT conditions to terminate the LS. Anyway, the authors proposed to check a secondary-stopping condition, when the LS had

* For an explanation and survey on reference point methods, the interested reader can refer to [51].

elapsed to more than 25 function calls. This avoids consuming too many resources for the execution of the LS.

Besides, to balance the cost of both techniques, the LS is applied with a certain probability p_{ls} at each generation. The function used [60] is defined as follows: starting from zero at the beginning of the evolutionary process (*generation* = 0), the probability of applying LS increases up to 0.01 in $0.5N - 1$ generations. Here, N is the population size. Then, the probability drops until reaching zero after spending $0.5N$ generations. This way of managing the probability has several motivations. One is to linearly increase the amount of resources spent by the LS. Another one is to avoid the application of LS at the beginning of the process, which is particularly convenient for the specific LS method chosen in this case. A final reason is to have a restart mechanism to prevent the loss of diversity during the evolutionary process.

This method was tested with the Zitzler–Deb–Thiele (ZDT) benchmark problems with good results. It showed that the application of this kind of LS brings benefits in terms of the speed of convergence and the final accuracy of the solutions obtained. In these experiments, the chosen MOEA to be hybridized was NSGA-II.

4.7 Other Gradient-Based Approaches

In [37], the problem of finding an improvement direction for the multiobjective case is stated as a multiobjective problem as well. Therefore, the solution (i.e., the "multiobjective gradient") must be a set of suitable movement directions. When performing a search over any of these directions, some of the objective functions from the original MOP decrease simultaneously, while the others can either decrease or just maintain the same value. In other words, the aim of this method is to describe a set of descent directions rather than a single one. An analytical way to calculate this set of directions is also shown. This method only requires a few matrix operations and the solution of a linear optimization problem. Once the set of descent directions is settled, one of such directions is randomly selected. The method mentioned above is called combined-objectives repeated line search (CORL) in [37]. A later improvement for CORL (by combining it with other criteria) can be found in [46]. In both cases, the hybridization of CORL is made by a combination with an estimation of distribution algorithm (EDA) called $\mathbb{MIDEA}$ [61]. The hybridization is made in the following manner: at the end of the generational cycle, that is, after the variation operators have been applied to the population, the algorithms choose a specific set of individuals to perform the LS over them. Thus, an improvement on the fitness for each one of them is obtained.

A revision of this method was presented by Harada et al. [39]. They used the ideas introduced in [43] to build what they called the Pareto descent method (PDM). These researchers proposed PDM as an option to deal with particular constrained MOPs, when the solution lies on the boundary between the feasible and the infeasible regions. In these cases, it is necessary to find different descent directions. The time complexity [62] of the PDM algorithm is also polynomial—as its authors refer to—because the basic operations of this method consist of solving systems of linear equations (one system for each new direction). Same as with other gradient-based LS engines, PDM performs successfully on MOPs with no local Pareto fronts. Harada et al. [39] compared PDM against CORL and also against a simple weighted linear-aggregating function. They also evaluated a randomized generator of solutions, similar to the mutation mechanism used in evolution strategies—they concluded that this last method was the worst performer. PDM does not show dramatic improvements over the other methods, except in the specific case when CORL has trouble (on a three-objective problem). In this particular situation, PDM does not obtain a set of descent directions—as CORL does—but it

TABLE 4.1

Test Functions Used in [41] to Test Gradient-Based MAs

Problem	Functions	Domain
ZDT 1	$f_1(x) = x_1$ $f_2(x) = g(x)(2 - \sqrt{f_1(x)/g(x)})$ $g(x) = 1 + \frac{9}{n-1}\sum_{i=2}^{n} x_i^2$	$[0,1] \times [-1,1]^n$
ZDT 2	$f_1(x) = x_1$ $f_2(x) = g(x)(2 - (f_1(x)/g(x))^2)$ $g(x) = 1 + \frac{9}{n-1}\sum_{i=2}^{n} x_i^2$	$[0,1] \times [-1,1]^n$
ZDT 3	$f_1(x) = x_1$ $f_2(x) = g(x)(2 - \sqrt{f_1(x)/g(x)}$ $-(f_1(x)/g(x))\sin(10\pi f_1))$ $g(x) = 1 + \frac{9}{n-1}\sum_{i=2}^{n} x_i^2$	$[0,1] \times [-1,1]^n$
ZDT 4	$f_1(x) = x_1$ $f_2(x) = g(x)(2 - \sqrt{f_1(x)/g(x)})$ $g(x) = 1 + 10(n-1)$ $+\sum_{i=2}^{n}(x_i^2 - 10\cos 4\pi f_1)$	$[0,1] \times [-5,5]^n$
ZDT 6	$f_1(x) = 1 - e^{-4x_1}$ $f_2(x) = g(x)(2 - (f_1(x)/g(x))^2)$ $g(x) = 1 + \frac{9}{n-1}\sum_{i=2}^{n} x_i^2$	$[0,1] \times [-1,1]^n$

offers a good alternative. There is no MA based on PDM; this is left by the authors as future work. In the same work, the authors stated that their method could have scalability issues when more than three objectives are used.

Shukla [41] introduced the use of two stochastic gradient-based techniques to improve the mutation mechanism of NSGA-II. These two techniques are Schäffler's stochastic method [42] and Timmel's [63] method. Both hybrid algorithms were competitive in some modified versions, described in Table 4.1, of the well-known ZDT test problems, outperforming the plain NSGA-II. However, the ZDT4 problem could not be properly solved by any of these hybrids.

In this case, only the NSGA-II was able to converge to the true Pareto front of ZDT4, since all the hybrids got trapped in local Pareto fronts. It is clear that the hybrids proposed by Shukla are relatively straightforward approaches that could be improved, but they also illustrate the local nature of the gradient-based information and its possible limitations. As an example, Figure 4.7 shows the set of points generated by the HCS2, a more sophisticated LS engine that is obviously trapped in false fronts at the same test problem indicated before (ZDT4). The reason for this lies on the use of gradient-based procedures, and the high multimodality of these particular problems. These problems show the importance of an efficient balance between such deterministic methods and their stochastic counterpart (the MOEA).

Sequential quadratic programming: In [40], a gradient-based local algorithm (sequential quadratic programming [SQP]) was used in combination with NSGA-II and SPEA [20] to solve the ZDT benchmark suite [64]. The authors stated that if there are no local Pareto fronts, then, the hybrid MOEA has a faster convergence toward the true Pareto front than the original approach. Furthermore, they found that the hybridization technique does not decrease the solution diversity.

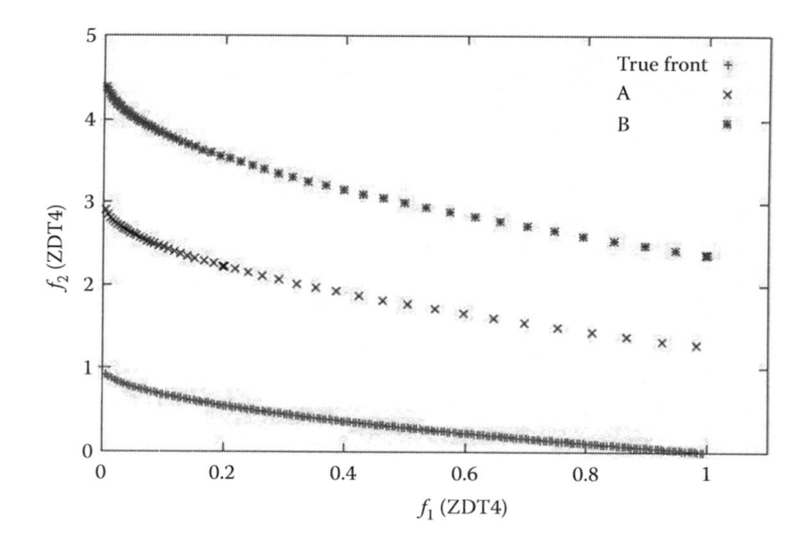

FIGURE 4.7
It shows the set of points generated by the repeated application of the operator HCS2, starting from two different points A and B. In this test problem (ZDT4), local Pareto fronts are formed from different points, which are far away from the global one.

4.8 Approaches Based on Nelder and Mead's Algorithm

The multiobjective memetic algorithms (MOMAs) presented in the earlier sections require the gradient information of the functions. Therefore, their use is limited to certain types of problems. This has motivated the development of new approaches that couple direct search methods (i.e., methods that do not require gradient information) with an MOEA. Among the mathematical-programming techniques available for this coupling, Nelder and Mead's algorithm [65] is perhaps the most obvious choice, since it is the most popular direct search method adopted for solving unconstrained optimization problems. Nelder and Mead's method, also known as the nonlinear simplex search (NSS), has been used extensively to solve parameter estimation problems as well as other optimization problems since its inception, in 1965. The search done by the NSS is based on geometric operations (reflection, expansion, contraction, and shrinkage) on a set of points, which define an n-dimensional polygon called a "simplex." Mathematically, an n-simplex can be defined as follows.

Definition 4.7

A simplex or n-simplex Δ is the convex hull of a set of $n+1$ affinely independent points Δ_i ($i = 1, \ldots, n+1$), in some Euclidean space of dimension n.

If all the vertices of the simplex are mutually equidistant, then, the simplex is said to be regular. Thus, in two dimensions, a regular simplex is an equilateral triangle (see Figure 4.8), while in three dimensions, a regular simplex is a regular tetrahedron.

The NSS expands or focuses the search adaptively on the basis of the topography of the fitness landscape. The full algorithm is defined stating three scalar parameters to control the movements performed in the simplex: reflection (ρ), expansion (χ), contraction (γ), and shrinkage (σ). According

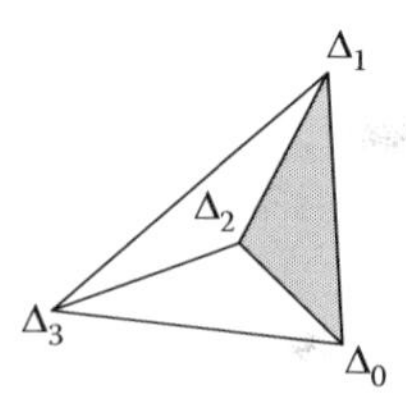

FIGURE 4.8
n-Simplex with $n = 2$.

to Nelder and Mead [65], these parameters should satisfy

$$\rho > 0, \quad \chi > 1, \quad \chi > \rho, \quad 0 < \gamma < 1, \quad \text{and} \quad 0 < \sigma < 1 \tag{4.16}$$

The nearly universal choices used in the Nelder and Mead algorithm are

$$\rho = 1, \quad \chi = 2, \quad \gamma = \frac{1}{2}, \quad \text{and} \quad \sigma = \frac{1}{2} \tag{4.17}$$

At each iteration of the NSS, the $n + 1$ vertices Δ_i's of the simplex represent solutions that are evaluated and sorted according to $f(\Delta_1) \leq f(\Delta_2) \leq \cdots \leq f(\Delta_{n+1})$.

Considering $\Delta = \{\Delta_1, \Delta_2, \ldots, \Delta_{n+1}\}$ as the simplex with the vertices sorted according to the function value, the transformations performed by the NSS into the simplex are defined as

1. Reflection: $\mathbf{x}_r = (1 + \rho)\mathbf{x}_c - \rho\Delta_{n+1}$ (see Figure 4.9)
2. Expansion: $\mathbf{x}_e = (1 + \rho\gamma)\mathbf{x}_c - \rho\chi\Delta_{n+1}$ (see Figure 4.10)
3. Contraction:
   a. *Outside*: $\mathbf{x}_{oc} = (1 + \rho\gamma)\mathbf{x}_c - \rho\gamma\Delta_{n+1}$
   b. *Inside*: $\mathbf{x}_{ic} = (1 - \gamma)\mathbf{x}_c + \gamma\Delta_{n+1}$ (see Figure 4.11)

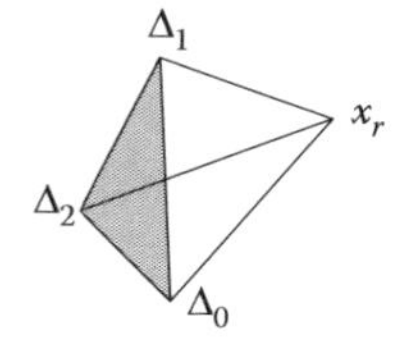

FIGURE 4.9
Reflection.

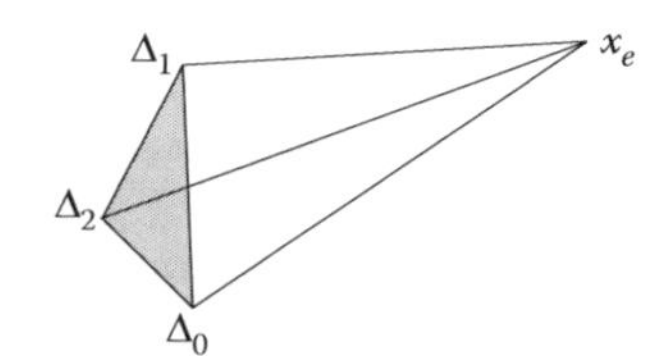

FIGURE 4.10
Expansion.

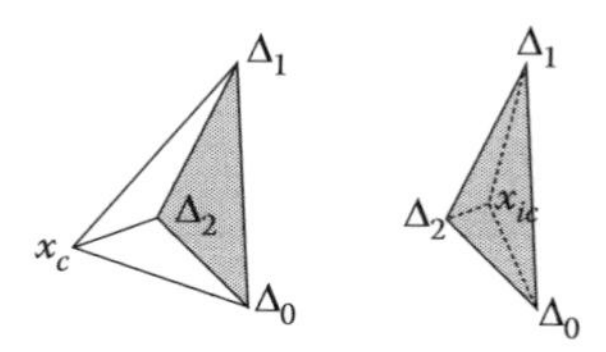

FIGURE 4.11
Inside and outside contraction.

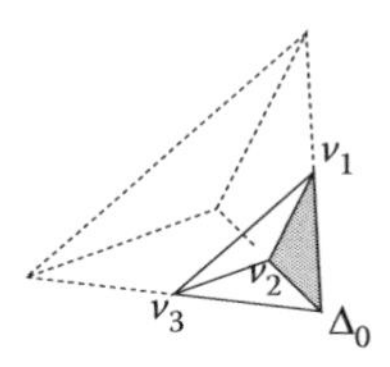

FIGURE 4.12
Shrinkage.

4. Shrinkage: Each vertex of the simplex is transformed by the geometric shrinkage defined by $\Delta_i = \Delta_1 + \sigma(\Delta_i - \Delta_1)$, where $i = 2, \ldots, n+1$, and the new vertices are evaluated—see Figure 4.12

where $\mathbf{x}_c = (1/n)\sum_{i=1}^{n} \Delta_i$ is the centroid of the n-best points (all vertices except for Δ_{n+1}), Δ_1 and Δ_{n+1} are the best and the worst solutions identified within the simplex, respectively. At each iteration, the simplex is modified by one of the above movements, according to the following rules:

1. If $f(\Delta_1) \leq f(\mathbf{x}_r) \leq f(\Delta_n)$, then $\Delta_{n+1} = \mathbf{x}_r$
2. If $f(\mathbf{x}_e) < f(\mathbf{x}_r) < f(\Delta_1)$, then $\Delta_{n+1} = \mathbf{x}_e$, otherwise $\Delta_{n+1} = \mathbf{x}_r$
3. If $f(\Delta_n) \leq f(\mathbf{x}_r) < f(\Delta_{n+1})$ and $f(\mathbf{x}_{oc}) \leq f(\mathbf{x}_r)$, then $\Delta_{n+1} = \mathbf{x}_{oc}$, otherwise perform a shrinkage
4. If $f(\mathbf{x}_r) \geq f(\Delta_{n+1})$ and $f(\mathbf{x}_{ic}) < f(\Delta_{n+1})$, then $\Delta_{n+1} = \mathbf{x}_{ic}$, otherwise perform a shrinkage

In the last few years, some MOMAs that combine the NSS with different state-of-the-art MOEAs have been reported in the specialized literature. In the following sections, we present several hybrid approaches that have reported significant improvements with respect to the original MOEA adopted.

4.8.1 Multiobjective GA-Simplex Hybrid Algorithm

Koduru et al. [66] proposed a hybrid GA using fuzzy dominance and Nelder and Mead's algorithm. The simplex search algorithm is adopted to improve solutions in the population of a GA. This memetic approach is used to estimate the parameters of a gene regulatory network for flowering time control in rice. The MA minimizes the difference between the model behavior and real-world data. Because of the nature of the data, a multiobjective approach is stated.

To understand the fuzzy dominance relation, the following definitions are introduced. Assuming a minimization problem with n decision variables and considering $\mathcal{X} \subset \mathbb{R}^n$ as the solution space, that is, the set of all possible solution vectors, the fuzzy i-dominance by a solution is defined as follows.

Definition 4.8

Given a monotonically nondecreasing function $\mu_i^{dom} : \mathcal{X} \rightarrow [0,1], i = \{1,\ldots,n\}$ such that $\mu_i^{dom}(0) = 0$, a solution $\mathbf{u} \in \mathcal{X}$ is said to ***i*-dominate** solution $\mathbf{v} \in \mathcal{X}$, if and only if $f_i(\mathbf{u}) < f_i(\mathbf{v})$. This relationship will be denoted as $\mathbf{u} \prec_i^F \mathbf{v}$. If $\mathbf{u} \prec_i^F \mathbf{v}$, then, the degree of fuzzy i-dominance is equal to $\mu_i^{dom}(f_i(\mathbf{v}) - f_i(\mathbf{u})) \equiv \mu_i^{dom}(\mathbf{u} \prec_i^F \mathbf{v})$. Fuzzy dominance can be regarded as a fuzzy relationship $\mathbf{u} \prec_i^F \mathbf{v}$ between $\mathbf{u}$ and $\mathbf{v}$ [67].

Definition 4.9

The solution $\mathbf{u} \in \mathcal{X}$ is said to **fuzzy dominate** solution $\mathbf{v} \in \mathcal{X}$ if and only if $\forall i \in \{1,\ldots,k\}, \mathbf{u} \prec_i^F \mathbf{v}$. This relationship will be denoted as $\mathbf{u} \prec^F \mathbf{v}$. The degree of fuzzy dominance can be defined by invoking the concept of a fuzzy intersection [67]. If $\mathbf{u} \prec^F \mathbf{v}$, the degree of fuzzy dominance $\mu^{dom}(\mathbf{u} \prec^F \mathbf{v})$ is obtained by computing the intersection of the fuzzy relationships $\mathbf{u} \prec_i^F \mathbf{v}$ for each i. The fuzzy intersection operation is carried out using a family of functions called t-norms, denoted by $\bigcap$. Hence

$$\mu^{dom}(\mathbf{u} \prec^F \mathbf{v}) = \bigcap_{i=1}^{k} \mu_i^{dom}(\mathbf{u} \prec \mathbf{v}) \tag{4.18}$$

where k is the number of objective functions.

Definition 4.10

Given a population of solutions $P \subset \mathcal{X}$, a solution $\mathbf{v} \in P$ is said to be fuzzy dominated in P iff it is fuzzy dominated by any other solution $\mathbf{u} \in P$. In this case, the degree of fuzzy dominance can be computed by performing a union operation over every possible $\mu^{dom}(\mathbf{u} \prec^F \mathbf{v})$, carried out using t-conorms, which are denoted by $\bigcup$. Hence, the degree of fuzzy dominance of a solution $\mathbf{v} \in P$ in the set P is given by

$$\mu^{dom}(P \prec^F \mathbf{v}) = \bigcup_{\mathbf{u} \in P} \mu^{dom}(\mathbf{u} \prec^F \mathbf{v}) \tag{4.19}$$

To calculate the fuzzy dominance relationship between two solution vectors, trapezoidal membership functions are used. Therefore,

$$\mu_i^{dom}(\mathbf{u} \prec_i^F \mathbf{v}) = \begin{cases} 0 & \text{if } f_i(\mathbf{v}) - f_i(\mathbf{u}) < 0 \\ \dfrac{f_i(\mathbf{v}) - f_i(\mathbf{u})}{p_i} & \text{if } 0 \leq f_i(\mathbf{v}) - f_i(\mathbf{u}) < p_i \\ 1 & \text{otherwise} \end{cases} \tag{4.20}$$

where p_i determines the length of the linear region of the trapezoid for the objective function f_i. The t-norm and t-conorms are defined as $x \cap y = xy$ and $x \cup y = x + y - xy$. Both are standard forms of operators [67].

At each generation t, a fraction of the next population P_{t+1} is obtained by applying genetic operators and the rest of the population is stated by using the NSS as it is illustrated in Figure 4.13. At the beginning of each iteration, the fuzzy dominances of all solutions in the current population P_t are calculated according to Equation 4.20. Then, the fuzzy dominance of the population is stored as a two-dimensional array, each entry of which is a fuzzy dominance relationship between two solution vectors. The first part of the population is obtained by evolving a set B of $N - (n+1)$ solutions

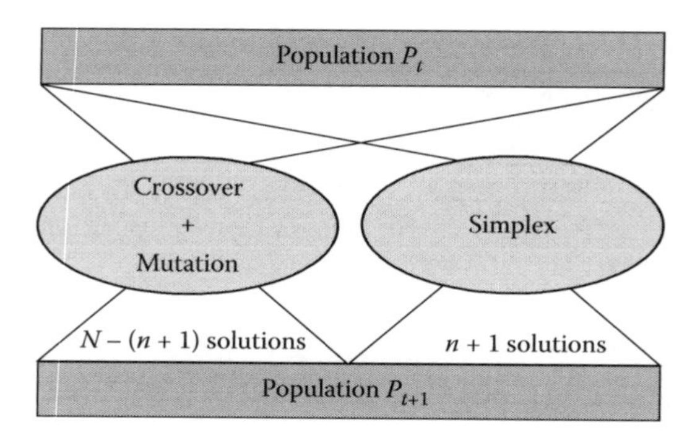

FIGURE 4.13
Offspring population generated by the multiobjective GA-simplex hybrid algorithm.

chosen randomly from the population P_t, where N denotes the population size and n denotes the number of decision variables of the MOP. The subpopulation B is evolved by applying genetic operators (crossover and mutation) and the fuzzy dominance relation is used as a measure of fitness during the selection into the GA. The resulting offspring population Q_1 is inserted as a part of the next population P_{t+1}.

The second part of the population is generated by the NSS. The simplex is built by selecting a sample set S of $n+1$ solutions from the current population P_t and then, the centroid $\mathbf{c}$ of the sample S is calculated. Any solution $\mathbf{u} \in S$ at a distance $\|\mathbf{c}-\mathbf{u}\| > \rho_{simplex}$ is rejected and replaced with another one taken in a random way from the population P_t, where $\rho_{simplex}$ represents the radius parameter of the simplex and $\|\cdot\|$ denotes the Euclidean norm. This process is repeated until either all the sample solutions fit within the radius $\rho_{simplex}$, or the total replacements exceed r_{max}. After selecting the initial vertices of the simplex, the NSS is performed for a total of α times. For each solution in the simplex, fuzzy dominance is calculated among solutions of the simplex and the vertices are sorted according to the fuzzy dominance relation. From the solutions obtained by the NSS, a set Q_2 of the best $n+1$ solutions (according to the fuzzy dominance relation) are selected and they are inserted into the next population P_{t+1}. The evolutionary process of this MOMA is carried out during T_{max} generations. Algorithm 4.7 shows the general framework of the multiobjective GA-simplex hybrid algorithm.

The authors suggested the use of $\alpha = 10$ as the maximum number of iterations for the NSS. The coefficients for the reflection, expansion, and contraction movements of the NSS were defined as $\rho = 1, \chi = 1.5$, and $\gamma = 0.5$, respectively. The shrinkage step was omitted in this memetic approach. The hybrid GA was tested using a population size of $N = 100$ and it was compared against a well-known state-of-the art MOEA:SPEA. A more-detailed description of this algorithm can be found in [66].

4.8.2 Multiobjective Hybrid Particle Swarm Optimization Algorithm

Koduru et al. [68] hybridized a multiobjective particle swarm optimization (PSO) algorithm with Nelder and Mead's method. The NSS was used as an LS engine for finding nondominated solutions in the neighborhood defined by the solution to be improved. This bioinspired technique evolves a set of solutions called swarm P for approximating solutions toward the Pareto optimal front. Each particle $\mathbf{x}_i$ in the swarm possesses a flight velocity that is initially set to zero. The swarm is evolved by updating both the velocity $\mathbf{v}_i^{t+1}$ and the position of each particle $\mathbf{x}_i^{t+1}$ according to the following

Algorithm 4.7: Multiobjective GA-Simplex Hybrid Algorithm

Input:
N: the population size;
T_{max}: the maximum number of generations;
Output:
P: the final approximation to the Pareto optimal front;

```
1  begin
2      t = 0;
3      Generate a random population P_t of size N;
4      Evaluate the population P_t;
5      while t < T_max do
6          // Select the solutions for performing the evolutionary process;
7          B = x_i ∈ P_t such that: x_i is randomly chosen from P_t and |P_t| = N − (n + 1);
8          Q_1 = MUTATION(CROSSOVER(B)); // Apply genetic operators
9          S = x_i ∈ P_t such that: |S| = n + 1; // Defining the simplex
10         for j = 0 to j < α do
11             Execute Nonlinear Simplex Search (NSS) using the initial simplex S;
12         end
13         Define Q_2 as the best n + 1 solutions (according to fuzzy dominance) found by the
           simplex search;
14         P_{t+1} = Q_1 ∪ Q_2;
15         t = t + 1;
16     end
17     return P_t
18 end
```

equations:

$$\mathbf{v}_i^{t+1} = w(\mathbf{v}_i^t + c_1 r_1(\mathbf{x}_{pb,i} - \mathbf{x}_t^t) + c_2 r_2(\mathbf{x}_{gb,i} - \mathbf{x}_i^t)) \tag{4.21}$$

and the new particle position is updated according to Equation 4.21 [69]

$$\mathbf{x}_i^{t+1} = \mathbf{x}_i^t + \mathbf{v}_i^{t+1} \tag{4.22}$$

where $w \geq 0$ represents the constriction coefficient, $c_1, c_2 \geq 0$ are the constraints on the velocity, and r_1, r_2 are two random variables having a uniform distribution in the range $(0, 1)$. $\mathbf{v}_i, \mathbf{x}_{pb,i}$, and $\mathbf{x}_{gb,i}$ represent the velocity, the personal best, and the global best position for the ith particle, respectively. Since, at the beginning of the search, a particle does not have the previous position, the best personal position is initialized with the same position as the particle, that is, $\mathbf{x}_{pb,i} = \mathbf{x}_i$. To avoid getting stuck into a local minimum, a turbulence factor is implemented into the velocity update (see Equation 4.21), which is similar to a mutation operator in GAs. The modified update equation is given by

$$\mathbf{v}_i^{t+1} = w(\mathbf{v}_i^t + c_1 r_1(\mathbf{x}_{pb,i} - \mathbf{x}_t^t) + c_2 r_2(\mathbf{x}_{gb,i} - \mathbf{x}_i^t)) + \exp(-\delta t) \cdot u \tag{4.23}$$

where δ is the turbulence coefficient and u is a uniformly distributed random number in $[-1, 1]$. The negative exponential term ensures that the turbulence in the velocities is higher at the initial generations, which promotes a more-explorative behavior. Later on, the behavior of the algorithm will become more exploitative.

The nondominated solutions found during the evolutionary process are stored in an external archive denoted as A. This set of nondominated solutions is updated during the evolutionary process by selecting the best N solutions from the union of the current population P and the external archive A, according to the fuzzy dominance relation. In the previous implementation of fuzzy dominance [66], the membership functions $\mu_i^{dom}(\cdot)$ used to compute the fuzzy i-dominances were defined to be zero for negative arguments. Therefore, whenever $f_i(\mathbf{u}) > f_i(\mathbf{v})$, the degree of fuzzy dominance $\mathbf{u} \prec_i^F \mathbf{v}$ is necessarily zero. In this memetic approach, nonzero values are allowed. The membership functions used are trapezoidal, yielding nonzero values whenever their arguments are to the right of a threshold ε. Mathematically, the memberships $\mu_i^{dom}(\mathbf{u} \prec^F \mathbf{v})$ are defined as

$$\mu_i^{dom}(\delta f_i) = \begin{cases} 0, & \delta f_i \leq -\varepsilon \\ \dfrac{\delta f_i}{\delta_i}, & -\varepsilon < \delta f_i < \delta_i - \varepsilon \\ 1, & \delta f_i \geq \delta_i - \varepsilon \end{cases} \tag{4.24}$$

where $\delta f_i = f_i(\mathbf{v}) - f_i(\mathbf{u})$. Given a population of solutions $P \subset \mathcal{X}$, a solution $\mathbf{v} \in P$ is said to be fuzzy dominated in P iff it is fuzzy dominated by any other solution $\mathbf{u} \in S$. In this way, each solution can be assigned a single measure to reflect the magnitude by which it dominates the others in a population. Better solutions within a set will have a lower fuzzy dominance value, although, unlike in [66], a nondominated solution may not necessarily be assigned zero values. To compare multiple solutions having similar fuzzy dominance values, the crowding distance of NSGA-II is used [15].

At each generation of the multiobjective PSO, NSS is executed. Considering an MOP with n decision variables, the set of solutions P is divided into separate clusters, where each cluster consists of proximally located solutions and it is generated by using a variant of the k-means algorithm [70]. The clusters are disjoint, with $n + 1$ points on each one. Each cluster represents the simplex from which the local simplex search is performed. At each iteration of the LS procedure, NSS performs l movements (reflection, expansion, or contraction) into the simplex before finishing its execution. The solutions found by the NSS are used to update both the population P and the archive A by using the fuzzy dominance relation. Algorithm 4.8 shows the general framework of the multiobjective PSO.

The authors suggested the use of $k = 9$ as the number of centers for the k-means algorithm, and $l = 2$ for the number of movements (reflection, expansion, or contraction) into the simplex. The simplex search was tested using $\rho = 1$, $\chi = 1.5$, and $\gamma = 0.5$ for the reflection, expansion, and contraction, respectively. In this hybrid multiobjective Particle Swarm Optimization (PSO), the use of the shrinkage transformation is omitted. The population size was set to $N = 100$ and the external archive was limited to a maximum of 100 particles. The proposed hybrid was used to solve artificial test functions and a molecular genetic model plant problem having between 3 and 10 decision variables and two objective functions. For a more-detailed description of this algorithm, see [68].

4.8.3 Nonlinear Simplex Search Genetic Algorithm

Zapotecas and Coello [71] presented a hybridization between the well-known NSGA-II and Nelder and Mead's method. The proposed nonlinear simplex search genetic algorithm (NSS-GA) combines the explorative power of NSGA-II with the exploitative power of Nelder and Mead's method, which acts as an LS engine. The general framework of the proposed MOMA is shown in Algorithm 4.9. This evolutionary approach evolves a population P_t by using the genetic operators of the NSGA-II (simulated binary crossover [SBX] and polynomial-based mutation [PBM]) and then, the LS mechanism is applied. The general idea of the LS procedure is to intensify the search toward better

Algorithm 4.8: Multiobjective Hybrid PSO

Input:
T_{max}: maximum number of generations;
Output:
A: final approximation to the Pareto optimal front;

1 **begin**
2 $t = 0$;
3 Generate a set of particles P_t of size N `// using a uniform distribution`;
4 Initialize all velocities $\mathbf{v}_i^t$, to zero;
5 **while** $t < t_{max}$ **do**
6 Evaluate the set of particles P_t;
7 Evaluate the fuzzy dominance in the population P_t according to Equation 4.24;
8 Update the archive A;
9 Update each particle $\mathbf{x}_i \in P_t$ including its personal best and global best;
10 Randomly initialize k cluster centers;
11 Assign each particle $\mathbf{x}_i$ to a cluster using the k-means algorithm;
12 For each cluster, apply the simplex search algorithm.;
13 Update the velocities $\mathbf{v}_i^{t+1}$ according to Equation 4.23;
14 Update the positions $\mathbf{x}_i \in P_t$ according to Equation 4.22;
15 $t = t + 1$;
16 **end**
17 **return** P_t
18 **end**

solutions for each objective function and the maximum bungle (sometimes called the knee) of the Pareto optimal front. The main goal of the NSS is to obtain Λ, which is defined as

$$\Lambda = \lambda_1 \cup \lambda_2 \cup \cdots \lambda_k \cup \Upsilon$$

where λ_i is a set of the best solutions found for the ith objective function of the MOP. Υ is a set of the best solutions found by minimizing an aggregating function that approximates solutions to the knee of the Pareto optimal front.

The LS mechanism is applied at each $n/2$ generations, where n denotes the number of decision variables of the MOP. Initially, the LS is focused on minimizing, separately, the objective functions f_i's of the MOP. Once the separate functions are minimized, an aggregating function is used for approximating solutions to the maximum bungle of the Pareto front. The initial search point from which the LS starts is defined according to the following rules:

- Minimizing separate functions. In the population P, an individual $\mathbf{x}_\Delta \in P^\star$ is chosen such that

$$\mathbf{x}_\Delta = \mathbf{x}_l | \mathbf{x}_l = \min_{\forall \mathbf{x}_l \in P^\star} \{f_i(\mathbf{x}_l)\}$$

where $P^\star$ is a set of nondominated solutions within the population P. In other words, the selected individual is the best nondominated solution.

Algorithm 4.9: The NSS-GA

```
   Input:
   t_max: maximum number of generations;
   Output:
   P: final approximation to the Pareto front;
1  begin
2      t = 0;
3      Randomly initialize a population P_t of size N;
4      Evaluate the fitness of each individual in P_t;
5      while t < t_max do
6          Q_t = MUTATION(CROSSOVER(B)); // Apply the genetic operators of NSGA-II
7          R_t = P_t ∪ Q_t;
8          Assign to P* the N best individuals from R_t // According to the crowding
           comparison operator;
9          if (t mod n/2 = 0) then
10             Get Λ set by minimizing each function of the Multiobjective Optimization
               Problem (MOP) and using the aggregating function (Equation 4.26).
11             R*_t = P*_t ∪ Λ;
12             Assign to P_{t+1} the N best individuals from R*_t // According to the crowding
               comparison operator;
13         else
14             P_{t+1} = P*;
15         end
16         t = t + 1;
17     end
18     return P_t
19 end
```

- Minimizing an aggregating function. An individual $\mathbf{x}_\Delta \in P^\star$ is chosen such that it minimizes

$$G(\mathbf{x}_\Delta) = \sum_{i=1}^{k} \frac{|z_i - f_i(\mathbf{x}_\Delta)|}{|z_i|} \tag{4.25}$$

  where $\mathbf{z}^\star = (z_i^\star, \ldots, z_k^\star)$ is the utopian vector defined by the minimum values f_i^* of the k objective functions until the current generation. In this way, the LS minimizes the aggregating function defined by

$$g(\mathbf{x}) = ED(\mathbf{F}(\mathbf{x}), \mathbf{z}^\star) \tag{4.26}$$

  where ED is the Euclidean distance between the vector of objective functions $\mathbf{F}(\mathbf{x})$ and the utopian vector $\mathbf{z}^\star$.

The selected solution $\mathbf{x}_\Delta$ is called the "simplex-head," which is the first vertex of the n-simplex. The remaining n vertices are created in two phases:

1. Reducing the search domain: A sample of s solutions that minimize the objective function to be optimized is identified, and then, the average ($\mathbf{m}$) and standard deviation (σ) of these

decision variables is computed. On the basis of that information, the new search space is defined as

$$\mathbf{L}_{bound} = \mathbf{m} - \sigma$$
$$\mathbf{U}_{bound} = \mathbf{m} + \sigma$$

where $\mathbf{L}_{bound}$ and $\mathbf{U}_{bound}$ are the vectors that define the lower and upper bounds of the new search space, respectively. In this work, the authors proposed to use $s = 0.20 \times N$, where N is the population size of the EA—that is, 20% of the population size.

2. Building the vertices: Once the new search domain has been defined, the remaining vertices are determined by using either the Halton [72] or the Hammersley [73] sequence (each has a 50% probability of being selected) in the new bounds $\mathbf{L}_{bound}$ and $\mathbf{U}_{bound}$, which are previously defined.

Once the simplex is defined, the NSS method is executed during a determined number of iterations defined by the following stopping criteria. The LS is stopped if (1) it does not generate a better solution after $n + 1$ iterations, or (2) if after performing $2(n + 1)$ iterations, the convergence is less than ϵ. The knowledge of the LS is introduced to the population of the NSGA-II by using the crowding comparison operator [15] using the union of the current population P and the set of solutions found by the LS Λ, that is, $P \cup \Lambda$.

The authors suggested a population size of $N = 100$, and the simplex was controlled using $\rho = 1$, $\chi = 2$, and $\gamma = 0.5$ for the reflection, expansion, and contraction coefficients, respectively. The shrinkage step is not employed in this approach. The threshold for the convergence in the simplex search was set to $\epsilon = 1 \times 10^{-3}$. The hybrid algorithm was tested over artificial test functions having between 10 and 30 decision variables, and two and three objective functions. A more-detailed description of this hybrid algorithm can be found in [71].

4.8.4 Hybrid Nondominated Sorting Differential Evolution Algorithm

Zhong et al. [74] hybridized Nelder and Mead's method with differential evolution (DE) [75] by using nondominated sorting. The proposal of Zhong et al. adopts NSS as its LS engine to obtain nondominated solutions along the evolutionary process according to the Pareto dominance relation. The sorting strategy adopted in this work involves the evaluation of the fitness function of each solution, and the dominance relation among the individuals of the population is defined according to their fitness cost. Throughout the search, the nondominated solutions are stored in a separate set A which, at the end of the search, will constitute an approximation of the Pareto optimal set.

At each iteration t, DE generates an offspring population Q_t by evolving each solution $\mathbf{x}_i$ of the current population P_t. The DE/best/2 strategy is employed to generate the trial vector $\mathbf{v}_i$

$$\mathbf{v}_i = \mathbf{x}_i^{best} + F \cdot (\mathbf{x}_{r_0} - \mathbf{x}_{r_1}) + F \cdot (\mathbf{x}_{r_2} - \mathbf{x}_{r_3}) \tag{4.27}$$

where $\mathbf{x}_{r_0}, \mathbf{x}_{r_1}, \mathbf{x}_{r_2}$, and $\mathbf{x}_{r_3}$ are different solutions taken from P_t and $\mathbf{x}_i^{best}$ is a solution randomly chosen from the set of nondominated solutions A. The trial vector $\mathbf{v}_i$ is then used for generating the new solution $\mathbf{x}_i'$ by using a binary crossover

$$\mathbf{x}_i'(j) = \begin{cases} \mathbf{v}_i(j) & \text{if } r < CR \\ \mathbf{x}_i(j) & \text{otherwise} \end{cases} \tag{4.28}$$

where r is a random number having a uniform distribution, $j = 1, \ldots, n$ is the jth parameter of each vector, and CR represents the crossover ratio. After the whole offspring population Q_t is generated, the nondominated sorting of $P_t \cup Q_t$ is used for obtaining the set of N solutions (N is the number of solutions in P_t) for the next population P_{t+1}.

In the LS mechanism, a simplex S is built by randomly selecting a nondominated solution from A, and the other n points (where n is the number of decision variables) are randomly chosen from the current population P_t. If the population P_t cannot provide enough points to create the simplex, other points are selected from A. After the simplex is built, the vertices of the simplex are stored according to the nondominated sorting differential evolution (NSDE). The sorting strategy involves evaluating the fitness value of each solution, and the dominance relation among the solutions in the simplex is built according to their fitness cost. NSS performs any movement in the sorted simplex. The movements in the simplex are performed according to the Nelder and Mead algorithm. However, for the comparisons among the solutions, the dominance relation is used instead of a function cost. The shrinkage step is performed if either inside or outside, a contraction fails. In this case, all the vertices into the transformed simplex S are sorted to obtain the solutions that are nondominated. Considering m as the number of the current nondominated solutions in the simplex, the shrinkage step is performed according to the following description.

If $m > 1$, then, there exist different converging directions, which could help to maintain the diversity of the solutions. Then, new simplexes $S_1, S_2, \ldots, S_m$ with m nondominated solutions, each as the respective guiding points, are generated. The new simplexes are stored within a bounded array. If the total number exceeds the storing space of the array, then, no more new simplexes are accepted. Then, these simplexes iterate to shrink the Pareto fronts in the array order as it is described in the following: If $m \leq 1$ or $S \in S_1, \ldots, S_m$, set the Pareto point m correspondingly in the simplex S_m as the guiding point x_1. The vertices in the simplex are relocated according to

$$\mathbf{v}_i = \mathbf{x}_1 + \sigma(\mathbf{x}_i - \mathbf{x}_1), \quad i = 2, \ldots, n+1$$

where σ is the shrinkage coefficient. The new form of the simplex uses $\mathbf{x}_1, \mathbf{v}_2, \ldots, \mathbf{v}_{n+1}$ as vertices to form the new simplex.

The Euclidean distance among the centroid and the vertices of the simplex is used for assessing the convergence at each simplex. After the convergence has taken place in all simplexes of the array, the nondominated solutions found by the simplex searches are introduced into the population of the EA according to the nondominated sorting strategy used in NSDE.

The authors suggested a population size of $N = 20 \times k \times n$ where n and k represent the number of decision variables and the number of objective functions of the MOP, respectively. The DE/best/2/bin strategy was used with a crossover ratio $CR = 0.8$ and a weighting factor $F = 0.5$. The NSS was performed using $\rho = 1$, $\chi = 2$, $\gamma = 0.5$, and $\sigma = 0.5$ for the reflection, expansion, contraction, and shrinkage movements, respectively. The Euclidean distance criterion to assess the convergence was set as 1×10^{-12}. For a more-detailed description of this multiobjective evolutionary technique, the interested reader can refer to [74].

4.8.5 Multiobjective Memetic Evolutionary Algorithm Based on Decomposition

Zapotecas and Coello [76] presented an MA using Nelder and Mead's algorithm that was coupled to Multiobjective Evolutionary Algorithm based on Decomposition (MOEA/D) (Algorithm 4.10) [17]. The LS engine is based on the multiobjective nonlinear simplex search (MONSS) framework presented in [77]. The memetic EA exploits the promising neighborhoods of the nondominated solutions found by MOEA/D. Considering P as the set of solutions found by MOEA/D in any generation, this approach assumes that if a solution $\mathbf{p} \in P$ is nondominated, then, there exists another

Algorithm 4.10: Multiobjective Memetic Evolutionary Algorithm Based on Decomposition

Input:
a stopping criterion;
N: The number of the subproblems considered in the memetic algorithm;
W: A well-distributed set of weighting vectors $\{\mathbf{w}_1, \ldots, \mathbf{w}_N\}$;
T: The number of weight vectors in the neighborhood of each weighting vector;
R_{ls}: The maximum number of evaluations for the local search;
A_{ls}: The action range for the local search.
Output:
P: the final approximation to the Pareto optimal front found by the memetic algorithm.

```
1  begin
2      Compute the Euclidean distances between any two weighting vectors and then work out
       the T closest weighting vectors to each weighting vector. For each i = 1, ..., N, set
       B(i) = {i_1, ..., i_T} where w_{i_1}, ..., w_{i_T} are the T closest weighting vectors to w_i;
3      Generate an initial population P = {x_1, ..., x_N} randomly. Set FV^i = F(x_i);
       z = (+∞, ..., +∞)^T;
4      while the stopping criteria is not satisfied do
5          Perform an iteration of Multiobjective Evolutionary Algorithm Based on
           Decomposition (MOEA/D) (see Algorithm 4.3);
6          if the percentage of nondominated solutions in P is less than 50% then
7              while there are enough resources and the simplex has not collapsed do
8                  Select a solution from P as the initial search solution (p_ini);
9                  Build the simplex;
10                 Select the search direction for the NSS;
11                 Perform an iteration of NSS for obtaining p_new;
12                 Update the population P using the new solution p_new;
13             end
14         end
15     end
16 end
```

nondominated solution $\mathbf{q} \in \Omega$ such that $\|\mathbf{p} - \mathbf{q}\| < \delta$ for any small $\delta \in \mathbb{R}_+$. In other words, the probability that $\mathbf{q}$ is nondominated with respect to $\mathbf{p}$ in the neighborhood defined by δ is equal to one, which implies that $\mathbf{q}$ is also nondominated. This property is considered to obtain new nondominated solutions departing from nondominated solutions located in the current population P. Since MOEA/D decomposes an MOP into several scalarization subproblems and such subproblems are solved during the evolutionary process, if all solutions in P are nondominated, then, it assumes that the minimum value to each subproblem has been achieved. Considering that at the end of the evolutionary process, if the population converges to a particular region of the search space (the place where the nondominated solutions are contained), then, the performance of the LS engine should be better when the diversity in the population is higher, that is, when having a low number of nondominated solutions. The LS procedure is then applied, when the percentage of nondominated solutions in P is less than a certain threshold.

Considering $P^\star \subseteq P$ as the set of nondominated solutions found by MOEA/D in any generation, and assuming that all the nondominated solutions in $P^\star$ are equally efficient, the solution $\mathbf{p}_{ini}$ that starts the LS is randomly taken from $P^\star$. Solution $\mathbf{p}_{ini}$ represents not only the initial search point, but also the simplex head from which the simplex is built. Let $\mathbf{w}_{ini}$ be the weighting vector that defines the subproblem for which the initial search solution $\mathbf{p}_{ini}$ is minimum. Let $S(\mathbf{w}_{ini})$ be the neighborhood of the n-closest weighting vectors to $\mathbf{w}_{ini}$ (where n is the number of decision variables of the MOP). Since the dimensionality of the simplex depends on the number of decision variables of the MOP, the population size of the MOEA needs to be larger than the number of decision variables. Then, the simplex defined as

$$\Delta = \{\mathbf{p}_{ini}, \mathbf{p}_1, \ldots, \mathbf{p}_n\}$$

is built in two different ways by using a probability $P_s = 0.3$, according to the two following strategies:

1. *Neighboring solutions:* The remaining n solutions $\mathbf{p}_i \in P$ $(i = 1, \ldots, n)$ are chosen, such that $\mathbf{p}_i$ minimizes each subproblem defined by each weighting vector in $S(\mathbf{w}_{ini})$. This is the same strategy employed for constructing the simplex used in MONSS [77].
2. *Sample solutions:* The remaining n solutions $\mathbf{p}_i \in \Omega$ $(i = 1, \ldots, n)$ are generated by using a low-discrepancy sequence. The Hammersley sequence [73] is adopted to generate a well-distributed sampling of solutions in a determined search space. In an analogous way as in [71], this approach uses a strategy based on the genetic analysis of a sample from the current population for reducing the search space. However, here, the average ($\mathbf{m}$) and standard deviation (σ) of the solutions that minimize each subproblem defined by the weighting vectors in $S(\mathbf{w}_{ini})$ are computed. In this way, the new bounds are defined by

   $$\mathbf{L}_{bound} = \mathbf{m} - \sigma$$
   $$\mathbf{U}_{bound} = \mathbf{m} + \sigma$$

   where $\mathbf{L}_{bound}$ and $\mathbf{U}_{bound}$ are the vectors that define the lower and upper bounds of the new search space, respectively.

   Once the search space has been reduced, the n-remaining solutions are generated by means of the Hammersley sequence using the bounds $\mathbf{L}_{bound}$ and $\mathbf{U}_{bound}$.

Let $B(\mathbf{w}_{ini})$ be the neighborhood of the T-closest weighting vectors to $\mathbf{w}_{ini}$, such that $\mathbf{w}_{ini}$ defines the subproblem for which the initial search solution $\mathbf{p}_{ini}$ is minimum. Let $D(\mathbf{w}_{ini})$ be the $A_r = 5$ closest weighting vectors to $\mathbf{w}_{ini}$. NSS focuses on minimizing a subproblem defined by the weighting vector $\mathbf{w}_{obj}$, which is defined according to the following rules:

1. The farthest weighting vector in $B(\mathbf{w}_{ini})$ to $\mathbf{w}_{ini}$, if it is the first iteration of the LS
2. Otherwise, a random weighting vector taken from $D(\mathbf{w}_{ini})$ is employed

At each iteration of the LS, the vertices of the simplex Δ are sorted according to their value for the subproblem that it tries to minimize. In this way, a movement into the simplex is performed for generating the new solution $\mathbf{p}_{new}$ by using Nelder and Mead's method. The new solution generated by the NSS replaces one or more solutions of the population according to: Let $B(\mathbf{w}_{obj})$ and $W = \{\mathbf{w}_1, \ldots, \mathbf{w}_N\}$ be the neighborhood of the T-closest weighting vectors to $\mathbf{w}_{obj}$, and the well-distributed set of all weighting vectors, respectively. The set of weighting vectors is defined by

$$Q = \begin{cases} B(\mathbf{w}_{obj}), & \text{if } r < 0.9 \\ W & \text{otherwise} \end{cases}$$

where r is a random number having a uniform distribution.

The population P is updated by replacing at most R_{ls} solutions from P such that $g(\mathbf{p}_{new}|\mathbf{w}_i, z) < g(\mathbf{x}_i|\mathbf{w}_i, z)$, where $\mathbf{w}_i \in Q$ and $\mathbf{x}_i \in P$ is the solution $\mathbf{x}_i$ that minimizes the subproblem defined by $\mathbf{w}_i$. The LS is performed by a maximum number of fitness function evaluations E_{ls}. If the NSS overcomes this maximum number of evaluations, then, the simplex search is stopped and the evolutionary process of MOEA/D continues. The search could be inefficient if the simplex has been deformed so that it has collapsed into a region where there are no local minima. Therefore, if the NSS does not find a minimum value in $n+1$ iterations, then, a reinitialization of the simplex is performed.

The authors used $\rho = 1$, $\chi = 2$, and $\gamma = 0.5$ for the reflection, expansion, and contraction movements of the NSS. The population size was set as $N = 100$ for two-objective problems and 300 for three-objective problems. The action range for the NSS was set as $A_r = 5$. The number of solutions to be replaced was set as $R_{ls} = 15$. The total number of iterations of the NSS was set as $E_{ls} = 300$. The proposed MA was tested using artificial functions between 10 and 30 decision variables, with two and three objectives.

4.9 Other Direct Search Approaches

In the last few years, several algorithms that hybridize Nelder and Mead's algorithm with an MOEA have been presented. Such implementations have used solutions from the population of the MOEA as the vertices to define the simplex. Since the solutions in the population are evaluated, the vertices of the simplex do not need to be evaluated. This reduces the number of function evaluations that an MOMA based on the NSS could require. Nevertheless, in the specialized literature, there are other direct search methods that do not require a simplex to obtain an optimal point of a function, but their use has been somewhat scarce in MOMAs during the last several years. Most of these methods start the search from a single point (which implies less information of the fitness landscape), and generate a solution for each iteration, whereupon, the convergence to an optimal value could require several function evaluations and the hybridization with an MOEA could be inefficient. However, in recent years, the coupling of this type of direct search methods with an MOMA has attracted a lot of interest from some researchers. In the following sections, we present some strategies that have been used for coupling other direct search methods (different from Nelder and Mead's method) with an MOEA.

4.9.1 Multiobjective Meta-Model-Assisted Memetic Algorithm

Zapotecas and Coello [78] presented a strategy that combines an MOEA with a direct search method assisted by a surrogate model. The LS mechanism adopts the Hooke and Jeeves algorithm [79] as its LS engine that is assisted by a surrogate model based on support vector regression (SVR).

Multiobjective meta-model-assisted memetic algorithm (MOMAMA) generates a sample S of size $2N$ (where N is the population size) that is randomly distributed in the search space using the Latin hypercube sampling method [80]. The surrogate model is trained using the set of solutions $D = S$ that is evaluated with the real fitness function. The initial population P_0 is defined by N solutions randomly chosen from S. The nondominated solutions found throughout the search are stored in an external archive A. The external archive is always maintained with at most N nondominated solutions, and it is pruned by performing Pareto ranking [15]. A technique based on clustering is used for selecting the best nondominated solutions when the external archive contains more than N solutions.

At each iteration t, recombination takes place between each individual $\mathbf{x}_i \in P_t$ and an individual that can be chosen from either R_t or A, according to the next rule

$$parent^1 = \mathbf{x}_i \in P_t \quad \forall i = 1, \ldots, N$$
$$parent^2 = \begin{cases} \mathbf{y} \in R_t, & \textbf{if } \left(g < 1 - \frac{|A|}{2N}\right) \\ \mathbf{y} \in A, & \textbf{otherwise} \end{cases} \tag{4.29}$$

where g is a uniformly distributed random number within (0, 1) and $\mathbf{y}$ is a solution randomly chosen from A or R_t. After the parents are recombined, the mutation operator is applied to each child for obtaining the offspring population Q_t. The next population P_{t+1} is then stated by selecting N individuals from $P_t \cup Q_t$ according to Pareto ranking [15].

The LS mechanism uses the surrogate model to find new solutions nearby the solutions provided by the MOEA. The LS is guided by a set W of n_w weighting vectors that define a scalar optimization problem using the following augmented-weighting Tchebycheff problem:

$$\min_{\mathbf{x} \in \mathbb{R}^n} \max_{i=1,\ldots,k} \{w_i |f_i(\mathbf{x}) - z_i^\star|\} + \alpha \sum_{i=1}^{k} |f_i(\mathbf{x}) - z_i^\star| \tag{4.30}$$

where α is a sufficiently small positive scalar and $\mathbf{z}^\star$ represents the utopian vector.

The Hooke and Jeeves algorithm approximates solutions to the Pareto optimal set by solving the n_w Tchebycheff problems. For each weighting vector $\mathbf{w}_j \in W$, a set of solutions λ_j that consists of all solutions evaluated into the metamodel is found. The initial search point $\mathbf{x}_s^1$ for solving the problem corresponding to the weighting vector $\mathbf{w}_1$ is stated as the $\mathbf{x}^\star \in P_t \cup A$ that minimizes Equation 4.30. The remaining initial search points $\mathbf{x}_s^j$ $(j = 2, \ldots, n_w)$ are defined by the local optimal solution found by the Hooke and Jeeves algorithm for the weighting vector $\mathbf{w}_{j-1} \in W$.

Let $\Lambda = \bigcup_{j=1}^{n_w} \lambda_j$ be the set of solutions found so far by the LS mechanisms. Considering that the probability that, given a nondominated solution $\mathbf{q}^*$, there exists another nondominated solution $\mathbf{p}$ in its neighborhood is equal to one, that is

$$P(\exists \mathbf{p} \in \mathbb{R}^n : \|\mathbf{q}^* - \mathbf{p}\| < \delta \text{ and } \mathbf{q}^* \nprec \mathbf{p}) = 1 \tag{4.31}$$

for any small $\delta \in \mathbb{R}_+$, the proposed approach generates more approximate solutions by using DE [75].

The initial population is given by $\mathcal{G}_0 = \Lambda$. Each new individual is stored in an external archive $\mathcal{L}$ according to the dominance rule. The archiving strategy can make that the set of solutions $\mathcal{L}$ increases or decreases its size. The next population for the DE algorithm is then defined by $\mathcal{G}_{g+1} = \mathcal{L}$.

Since all the solutions in the archive $\mathcal{L}$ are nondominated, we can say that DE has converged (at least locally) when it has obtained N different nondominated solutions from the evolutionary process. That is

$$\textbf{if } |\mathcal{L}| = N \textbf{ then} \quad \text{stop the DE algorithm} \tag{4.32}$$

The solutions set R_t obtained by the LS mechanism is given by $R_t = \mathcal{L}$. However, this stopping criterion is not always satisfied. Thus, R_t set can be defined by selecting N individuals from $\Lambda \cup \mathcal{L}$

Algorithm 4.11: The MOMAMA

Input:
T_{max}: Maximum number of generations;
Output:
A: Final approximation to the Pareto optimal front;

```
1  begin
2      A = ∅;
3      Generate S of size 2N // using the Latin Hyper-cubes method;
4      EVALUATE(S) // using the real fitness function;
5      P = {x_i ∈ S} such that: x_i is randomly chosen from S and |P_t| = N;
6      R_t = S \ P_t;
7      A =UPDATEARCHIVE(R_t, A);
8      D = S;
9      while t < T_max do
10         A =UPDATEARCHIVE(P_t, A);
11         Q_t =CREATEOFFSPRING(P_t, R_t, A_t); // apply genetic operators
12         EVALUATE(Q_t); // using the real fitness function
13         D = D ∪ Q_t;
14         Retrain the meta model using the training set D;
15         P_{t+1} =SELECTNEXTPOPULATION(P_t, Q_t); // using Pareto ranking
16         R_{t+1} =SURROGATELOCALSEARCH(P_t, A); // using the Hooke and Jeeves
           algorithm, and Differential Evolution (DE) into the surrogate model
17         t = t + 1;
18     end
19 end
```

using Pareto ranking [15] after a certain number of iterations. The final set R_t is then evaluated using the real fitness function and it is used for generating the offspring population Q_t of the MA. After applying the LS, the surrogate model is retrained by using the data set D with the set of local optimal solutions Q_t, that is, the training set is defined by $D = D \cup Q_t$. Algorithm 4.11 shows the general scheme of the proposed MOMAMA. A more-detailed description of this evolutionary approach can be found in [78].

4.9.2 Hybrid MOEA Based on the S-Metric

Koch et al. [81] presented a hybrid algorithm that combines the exploratory properties of the S-metric selection evolutionary multiobjective optimization algorithm (SMS-EMOA) [82] with the exploitative power of the Hooke and Jeeves algorithm [79] that is used in the LS engine. SMS-EMOA optimizes a set of solutions according to the S-metric or hypervolume indicator [83], which measures the size of the space dominated by the population. This performance measure is integrated into the selection operator of SMS-EMOA that aims for maximization of the S-metric and thereby guides the population toward the Pareto optimal front. A $(\mu + 1)$ (or steady-state) selection scheme is applied: At each generation, SMS-EMOA discards the individual that contributes the least to the S-metric value of the population. The invoked variation operators are not specific for the SMS-EMOA but are taken from the literature, namely PBM and SBX, with the same parameterization as in the NSGA-II [15]. At each iteration, the Hooke and Jeeves algorithm performs an exploratory move along the coordinate axes. Afterward, the vectors of the last exploratory moves are combined to a projected direction that can accelerate the descent of the search vector. When the exploratory moves lead to no improvement in any coordinate direction, step sizes are reduced by a factor η. The search terminates after a number of predefined function evaluations or, alternatively, when the step size falls below a constant value $\varepsilon > 0$.

The Hooke and Jeeves algorithm was conceived for minimizing SOPs; therefore, its use to deal with MOPs is not possible without modifications. Koch et al. adopted a scalar function by using the weighting sum approach developed in [84]. Besides, the proposed MOMA introduces a probability function $p_{ls}(t)$ for extending the idea presented in Sindhya et al. [60] who linearly oscillated the probability for starting the LS procedure. The probability function adopted in this work is given by

$$p_{ls}(t) = \frac{p_{max} \cdot \Phi(t \mod (\alpha\mu))}{\Phi(\alpha\mu - 1)} \tag{4.33}$$

where the parameter μ refers to the population size of the MOEA and $\alpha \in (0, 1]$ is a small constant value—in the experiments, the authors suggested to use $\alpha = 0.05$. The probability function oscillates with period $\alpha \cdot \mu$ and is linearly decreasing in each period. The auxiliary function Φ determines the type of reduction, that is, linear, quadratic, or logarithmic, and has to be defined by the user. Algorithm 4.12 shows the general framework of the proposed hybrid SMS-EMOA.

Using the same outlined Algorithm 4.12, Koch et al. hybridized S-Metric Selection Evolutionary Multiobjective Optimization Algorithm (SMS-EMOA) with other mathematical-programming techniques. The multiobjective Newton method [85] and the steepest descent method [43] are also hybridized with the SMS-EMOA. Koch et al. emphasize the importance of the probability function p_{ls} that controls the frequency of the LS during the optimization process. Three different functions

Algorithm 4.12: Hybrid SMS-EMOA

Input:
T_{max}: Maximum number of generations;
Output:
A: Final approximation to the Pareto optimal front;

```
1  begin
2      t = 0;
3      Generate a population P_t of size N; // using uniform distribution
4      Evaluate the population P_t;
5      while t < T_max do
6          Select μ parents of P_t;
7          Create population Q_t with λ offspring;
8          for i=1 to λ do
9              Choose random variable r ∈ [0, 1];
10             if r ≤ p_ls(t) then
11                 Local search for Q_t[i];
12             end
13         end
14         Evaluate λ offspring;
15         Create population P_{t+1} out of P_t and Q_t;
16         t = t + 1;
17     end
18 end
```

using Equation 4.33 and a constant probability p_{ls} were adopted. The hybrid variants using Equation 4.33 obtained a value of $\alpha = 0.5$ as proposed by Sindhya et al. [60] and the subsequent functions were used.

1. $p_{ls}(t)$ with $\Phi(x) = x$ (in Equation 4.33)
2. $p_{ls}(t)$ with $\Phi(x) = x^2$ (in Equation 4.33)
3. $p_{ls}(t)$ with $\Phi(x) = \log(x)$ (in Equation 4.33)
4. $p_{ls}(t)$ with $\Phi(x) = 0.01$

Each hybridization with the above probability functions was tested on the ZDT test suite. The hybrid SMS-EMOA started with a population size of $N = 100$, the SBX recombination operator proposed by Sindhya et al. [60] was adopted with a probability of 1.0, and the polynomial mutation operator was adopted with a probability of 1.0. According to the results reported, the hybrid using the multiobjective Newton method achieved better results than those obtained by both the hybrid SMS-EMOA using the Hooke and Jeeves algorithm, and using the steepest descent method. More details of this hybridization can be found in [81].

4.10 Conclusions

This chapter has provided a detailed description of the hybridization of MOEAs with mathematical-programming methods. The main motivation for this sort of coupling is, evidently, to improve the performance that any of these two types of techniques can provide, when considered separately. Our discussion has included the use of both gradient-based and nongradient-based methods for the LS engine. To make the chapter self-contained, the most important required background has also been included.

The development of hybrid MOEAs raises several challenges that include the design of efficient and effective LS engines (particularly in continuous search spaces), the proper balance between the global and the LS engines, and the appropriate choice of the LS engine to be adopted, among others [19]. It is not possible to say, so far, that a certain hybrid configuration is the best choice for a particular problem. However, the empirical and theoretical work present in the current literature provides a good start to guide practitioners.

The review of approaches provided in this chapter clearly illustrates the interest that hybrid MOEAs have attracted in the last few years. Such an interest is expected to raise in the next few years, since this is, indeed, a very promising research line within evolutionary multiobjective optimization due to the potential benefits that these hybrid MOEAs could bring about (particularly, when dealing with complex real-world problems).

Acknowledgments

The first author acknowledges CONACyT for providing him a scholarship to pursue his PhD studies at the Computer Science Department of CINVESTAV-IPN. The second author acknowledges the support from IPN SIP20152082 project. The third author gratefully acknowledges the support from CONACyT project no. 221551.

Acronyms

CDMOMA	Cross Dominant Multiobjective Memetic Algorithm
CMODE	Coevolutionary Multiobjective Differential Evolution
CORL	Combined-Objectives Repeated Line Search
DE	Differential Evolution
DTLZ	Deb-Thiele-Laumanns-Zitzler
EA	Evolutionary Algorithm
EC	Evolutionary Computation
EDA	Estimation of Distribution Algorithm
EMOA	Evolutionary Multi-objective Optimization Algorithm
GA	Genetic Algorithm
KKT	Karush-Kuhn-Tucker
LS	Local Search
MA	Memetic Algorithm
MDD	Multiobjective Descent Direction
MOEA	Multiobjective Evolutionary Algorithm
MOEA/D	Multiobjective Evolutionary Algorithm Based on Decomposition
MOGA	Multiobjective Genetic Algorithm
MOGLS	Multiobjective Genetic Local Search
MOMA	Multiobjective Memetic Algorithm
MOMAMA	Multiobjective Meta-model Assisted Memetic Algorithm
MONSS	Multiobjective Nonlinear Simplex Search Algorithm
MOP	Multiobjective Optimization Problem
M-PAES	Memetic Pareto Archived Evolution Strategy
NSDE	Nondominated Sorting Differential Evolution
NSGA	Nondominated Sorting Genetic Algorithm
NSGA-II	Nondominated Sorting Genetic Algorithm II
NSS	Nonlinear Simplex Search
NSS-GA	Nonlinear Simplex Search Genetic Algorithm
PBM	Polynomial-Based Mutation
PDM	Pareto Descent Method
PDMOSA	Pareto Domination Multiobjective Simulated Annealing
PMA	Pareto Memetic Algorithm
PSO	Particle Swarm Optimization
SBX	Simulated Binary Crossover
SMS-EMOA	S-Metric Selection Evolutionary Multiobjective Optimization Algorithm
SOP	Single-objective Optimization Problem
SPEA	Strength Pareto Evolutionary Algorithm
SPEA2	Strength Pareto Evolutionary Algorithm 2
SQP	Sequential Quadratic Programming
SVR	Support Vector Regression
ZDT	Zitzler-Deb-Thiele

References

1. S. S. Rao. *Engineering Optimization*. John Wiley & Sons Inc., USA, 3rd edition, 1996.
2. A. Ravindran, K. M. Ragsdell, and G. V. Reklaitis. *Engineering Optimization. Methods and Applications*. John Wiley & Sons, Inc., Hoboken, New Jersey, USA, 2006.
3. J. Jahn. *Vector Optimization: Theory, Applications, and Extensions*. Springer-Verlag, Berlin, Heidelberg, 2010.
4. M. Ehrgott. *Multicriteria Optimization*. Springer, Berlin, 2nd edition, June 2005.
5. H. W. Kuhn and A. W. Tucker. Nonlinear programming. In *Proceedings of the Second Berkeley Symposium on Mathematics, Statistics, and Probability*. Statistical Laboratory of the University of California, Berkeley, pp. 481–492, 1951.
6. K. Miettinen. *Nonlinear Multiobjective Optimization*. Kluwer Academic Publishers, Boston, Massachusetts, 1999.
7. K. M. Miettinen. Some methods for nonlinear multi-objective optimization. In E. Zitzler et al., editor, *Evolutionary Multi-Criterion Optimization*, volume 1993/2001 of *Lecture Notes in Computer Science*, pp. 1–20. Springer, Berlin, Heidelberg, 2001.
8. H. Mukai. Algorithms for multicriterion optimization. *IEEE Transactions on Automatic Control*, 25(2):177–186, 1980.
9. I. Das and J. E. Dennis. A closer look at drawbacks of minimizing weighted sums of objectives for Pareto set generation in multicriteria optimization problems. *Structural and Multidisciplinary Optimization*, 14(1):63–69, 1997.
10. I. Das and J. E. Dennis. Normal-boundary intersection: A new method for generating Pareto optimal points in multicriteria optimization problems. *SIAM Journal on Optimization*, 8(3):631–657, 1998.
11. Y. Y. Haimes, L. S. Lasdon, and D. A. Wismer. On a bicriterion formulation of the problems of integrated system identification and system optimization. *IEEE Transactions on Systems, Man, and Cybernetics*, 1(3):296–297, 1971.
12. C. Vira and Y. Y. Haimes. *Multiobjective Decision Making: Theory and Methodology*. Number 8. North-Holland, Amsterdam, The Netherlands, 1983.
13. R. S. Rosenberg. *Simulation of Genetic Populations with Biochemical Properties*. PhD thesis, University of Michigan, Ann Arbor, Michigan, USA, 1967.
14. J. David Schaffer. *Multiple Objective Optimization with Vector Evaluated Genetic Algorithms*. PhD thesis, Vanderbilt University, 1984.
15. K. Deb, S. Agrawal, A. Pratap, and T. Meyarivan. A fast and elitist multiobjective genetic algorithm: NSGA-II. *IEEE Transactions Evolutionary Computation*, 6(2):182–197, 2002.
16. J. D. Knowles and D. W. Corne. The Pareto archived evolution strategy: A new baseline algorithm for multiobjective optimisation. In *1999 Congress on Evolutionary Computation*, pp. 98–105, Washington, DC, July 1999. IEEE Service Center.
17. Q. Zhang and H. Li. MOEA/D: A multiobjective evolutionary algorithm based on decomposition. *IEEE Transactions on Evolutionary Computation*, 11(6):712–731, 2007.
18. E. Zitzler, M. Laumanns, and L. Thiele. SPEA2: Improving the strength Pareto evolutionary algorithm. In K. Giannakoglou, D. Tsahalis, J. Periaux, P. Papailou, and T. Fogarty, editors, *EUROGEN 2001. Evolutionary Methods for Design, Optimization and Control with Applications to Industrial Problems*, pp. 95–100, Athens, Greece, 2002.
19. C. A. Coello Coello, G. B. Lamont, and D. A. Van Veldhuizen. *Evolutionary Algorithms for Solving Multi-Objective Problems*. Springer, New York, 2nd edition, September 2007. ISBN 978-0-387-33254-3.
20. E. Zitzler and L. Thiele. Multiobjective evolutionary algorithms: A comparative case study and the strength Pareto approach. *IEEE Transactions on Evolutionary Computation*, 3(4):257–271, 1999.
21. N. Srinivas and K. Deb. Multiobjective optimization using nondominated sorting in genetic algorithms. *Evolutionary Computation*, 2(3):221–248, 1994.
22. D. E. Goldberg. *Genetic Algorithms in Search, Optimization and Machine Learning*. Addison-Wesley Publishing Company, Reading, Massachusetts, 1989.

23. P. Moscato. On *Evolution, Search, Optimization, Genetic Algorithms and Martial Arts: Towards Memetic Algorithms*. Technical Report Caltech Concurrent Computation Program, Report 826, California Institute of Technology, Pasadena, California, USA, 1989.
24. R. Dawkins. *The Selfish Gene*. Oxford University Press, UK, 1990.
25. J. D. Knowles. *Local-Search and Hybrid Evolutionary Algorithms for Pareto Optimization*. PhD thesis, The University of Reading, Department of Computer Science, Reading, UK, 2002.
26. H. Ishibuchi and T. Murata. Multi-objective genetic local search algorithm. In T. Fukuda and T. Furuhashi, editors, *Proceedings of the 1996 International Conference on Evolutionary Computation*, pp. 119–124, Nagoya, Japan, 1996. IEEE.
27. H. Ishibuchi and T. Murata. Multi-objective genetic local search algorithm and its application to flowshop scheduling. *IEEE Transactions on Systems, Man and Cybernetics—Part C: Applications and Reviews*, 28(3):392–403, 1998.
28. A. Jaszkiewicz. Do multiple-objective metaheuristics deliver on their promises? A computational experiment on the set-covering problem. *IEEE Transactions on Evolutionary Computation*, 7(2):133–143, 2003.
29. J. Knowles and D. Corne. M-PAES: A memetic algorithm for multiobjective optimization. In *2000 Congress on Evolutionary Computation*, volume 1, pp. 325–332, Piscataway, New Jersey, July 2000. IEEE Service Center.
30. T. Murata, S. Kaige, and H. Ishibuchi. Generalization of dominance relation-based replacement rules for memetic EMO algorithms. In E. Cantú-Paz et al., editor, *Genetic and Evolutionary Computation—GECCO 2003. Proceedings, Part I*, Chicago, Illinois, pp. 1234–1245. Springer. Lecture Notes in Computer Science, Vol. 2723, July 2003.
31. A. Caponio and F. Neri. Integrating cross-dominance adaption in multi-objective memetic algorithms. In C.-K. Goh, Y.-S. Ong, and K. C. Tan, editors, *Multi-Objective Memetic Algorithms*, pp. 325–351. Springer, Studies in Computational Intelligence, Vol. 171, 2009.
32. H. H. Rosenbrock. An automatic method for finding the greatest or least value of a function. *The Computer Journal*, 3(3):175–184, 1960.
33. B. Suman. Study of simulated annealing based algorithms for multiobjective optimization of a constrained problem. *Computers and Chemical Engineering*, 28:1849–1871, 2004.
34. O. Soliman, L. T. Bui, and H. Abbass. A memetic coevolutionary multi-objective differential evolution algorithm. In C.-K. Goh, Y.-S. Ong, and K. C. Tan, editors, *Multi-Objective Memetic Algorithms*, pp. 325–351. Springer, Studies in Computational Intelligence, Vol. 171, 2009.
35. T. Goel and K. Deb. Hybrid methods for multi-objective evolutionary algorithms. In L. Wang, K. C. Tan, T. Furuhashi, J.-H. Kim, and X. Yao, editors, *Proceedings of the 4th Asia-Pacific Conference on Simulated Evolution and Learning (SEAL'02)*, volume 1, pp. 188–192, Orchid Country Club, Singapore, November 2002. Nanyang Technical University.
36. J. Johannes. *Mathematical Vector Optimization in Partially Ordered Linear Spaces*. Frankfurt am Main; New York: Lang, 1986.
37. P. A. N. Bosman and E. D. de Jong. Exploiting gradient information in numerical multi-objective evolutionary optimization. In H.-G. Beyer et al., editor, *2005 Genetic and Evolutionary Computation Conference (GECCO'2005)*, volume 1, pp. 755–762, New York, USA, June 2005. ACM Press.
38. M. Brown and R. E. Smith. Effective use of directional information in multi-objective evolutionary computation. In *Genetic and Evolutionary Computation (GECCO) 2003*, volume 2723/2003 of *Lecture Notes in Computer Science*, pp. 778–789. Springer, Berlin/Heidelberg, 2003.
39. K. Harada, J. Sakuma, and S. Kobayashi. Local search for multiobjective function optimization: Pareto descent method. In *GECCO '06: Proceedings of the 8th Annual Conference on Genetic and Evolutionary Computation*, pp. 659–666, New York, NY, USA, 2006. ACM Press.
40. X. Hu, Z. Huang, and Z. Wang. Hybridization of the multi-objective evolutionary algorithms and the gradient-based algorithms. In *Proceedings of the 2003 Congress on Evolutionary Computation (CEC'2003)*, volume 2, pp. 870–877, Canberra, Australia, December 2003. IEEE Press.
41. P. K. Shukla. On gradient based local search methods in unconstrained evolutionary multi-objective optimization. In S. Obayashi, K. Deb, C. Poloni, T. Hiroyasu, and T. Murata, editors, *Evolutionary*

Multi-Criterion Optimization, 4th International Conference, EMO 2007, pp. 96–110, Matshushima, Japan, March 2007. Springer. Lecture Notes in Computer Science, Vol. 4403.
42. S. Schäffler, R. Schultz, and K. Weinzierl. Stochastic method for the solution of unconstrained vector optimization problems. *Journal of Optimization Theory and Applications*, 114(1):209–222, 2002.
43. J. Fliege and B. Fux Svaiter. Steepest descent methods for multicriteria optimization. *Mathematical Methods of Operations Research*, 51(3):479–494, 2000.
44. A. Lara, C. A. Coello Coello, and O. Schütze. A painless gradient-assisted multi-objective memetic mechanism for solving continuous bi-objective optimization problems. In *IEEE Congress on Evolutionary Computation (CEC 2010)*, Barcelona, Spain, pp. 1–8. IEEE Press, 2010.
45. A. Lara, O. Schütze, and C. A. Coello Coello. New challenges for memetic algorithms on continuous multi-objective problems. In *GECCO 2010 Workshop on Theoretical Aspects of Evolutionary Multiobjective Optimization*, pp. 1967–1970, Portland, Oregon, USA, July 2010. ACM.
46. P. A. N. Bosman and E. D. de Jong. Combining gradient techniques for numerical multi-objective evolutionary optimization. In M. Keijzer et al., editor, *2006 Genetic and Evolutionary Computation Conference (GECCO'2006)*, volume 1, pp. 627–634, Seattle, Washington, USA, July 2006. ACM Press. ISBN 1-59593-186-4.
47. A. Lara, O. Schütze, and C. A. Coello Coello. Gradient-based local search to hybridize multi-objective evolutionary algorithms. In E. Tantar, A.-A. Tantar, P. Bouvry, P. D. Moral, P. Legrand, C. A. C. Coello, and O. Schütze, editors, *EVOLVE: A Bridge between Probability, Set Oriented Numerics and Evolutionary Computation*, volume 447 of *Studies in Computational Intelligence*, Chapter 9, pp. 305–332, Springer-Verlag, Heidelberg, Germany, 2013. ISBN 978-3-642-32725-4.
48. C. Hillermeier. *Nonlinear Multiobjective Optimization: A Generalized Homotopy Approach*. Birkhäuser, Basel, 2000.
49. S. Schecter. Structure of the first-order solution set for a class of nonlinear programs with parameters. *Mathematical Programming*, 34(1):84–110, 1986.
50. O. Schütze, S. Mostaghim, M. Dellnitz, and J. Teich. Covering Pareto sets by multilevel evolutionary subdivision techniques. In C. M. Fonseca, P. J. Fleming, E. Zitzler, K. Deb, and L. Thiele, editors, *Evolutionary Multi-Criterion Optimization. Second International Conference, EMO 2003*, pp. 118–132, Faro, Portugal, April 2003. Springer. Lecture Notes in Computer Science, Vol. 2632.
51. G. Eichfelder. *Adaptive Scalarization Methods in Multiobjective Optimization*. Springer, Berlin, Germany, 2008.
52. A. Lara, G. Sanchez, C. A. Coello Coello, and O. Schütze. HCS: A new local search strategy for memetic multi-objective evolutionary algorithms. *IEEE Transactions on Evolutionary Computation*, 14(1):112–132, 2010.
53. R. Tyrell Rockafellar. *Convex Analysis*, volume 28. Princeton University Press, USA, 1997.
54. K. Harada, J. Sakuma, and S. Kobayashi. Uniform sampling of local Pareto-optimal solution curves by Pareto path following and its applications in multi-objective GA. In D. Thierens, editor, *2007 Genetic and Evolutionary Computation Conference (GECCO'2007)*, volume 1, pp. 813–820, London, UK, July 2007. ACM Press.
55. O. Schütze. *Set Oriented Methods for Global Optimization*. PhD thesis, University of Paderborn, 2004. <http://ubdata.uni-paderborn.de/ediss/17/2004/schuetze/>.
56. O. Schütze, C. A. Coello Coello, S. Mostaghim, E.-G. Talbi, and M. Dellnitz. Hybridizing evolutionary strategies with continuation methods for solving multi-objective problems. *Engineering Optimization*, 40(5):383–402, 2008.
57. E. Rigoni and S. Poles. NBI and MOGA-II, two complementary algorithms for multi-objective optimization. In *Dagstuhl Seminar Proceedings 04461. Practical Approaches to Multi-Objective Optimization*, Dagstuhl, Germany, pp. 1–22, 2005.
58. S. Poles, E. Rigoni, and T. Robič. MOGA-II performance on noisy optimization problems. In B. Filipič and J. Šilc, editors, *Bioinspired Optimization Methods and Their Applications. Proceedings of the International Conference on Bioinspired Optimization Methods and Their Applications, BIOMA 2004*, pp. 51–62. Jožef Stefan Institute, Ljubljana, Slovenia, October 2004.
59. E. F. Wanner, F. G. Guimaraes, R. H. C. Takahashi, and P. J. Fleming. A quadratic approximation-based local search procedure for multiobjective genetic algorithms. In *2006 IEEE Congress*

on Evolutionary Computation (CEC'2006), pp. 3361–3368, Vancouver, BC, Canada, July 2006. IEEE.
60. K. Sindhya, K. Deb, and K. Miettinen. A local search based evolutionary multi-objective optimization approach for fast and accurate convergence. In G. Rudolph, T. Jansen, S. Lucas, C. Poloni, and N. Beume, editors, *Parallel Problem Solving from Nature–PPSN X*, pp. 815–824. Springer. Lecture Notes in Computer Science, Vol. 5199, Dortmund, Germany, September 2008.
61. P. A. N. Bosman and D. Thierens. The naive MIDEA: A baseline multi-objective EA. In C. A. Coello Coello, A. Hernández Aguirre, and E. Zitzler, editors, *Evolutionary Multi-Criterion Optimization. Third International Conference, EMO 2005*, pp. 428–442, Guanajuato, México, March 2005. Springer. Lecture Notes in Computer Science, Vol. 3410.
62. T. H. Cormen, C. E. Leiserson, R. L. Rivest, and C. Stein. *Introduction to Algorithms*. MIT Press, USA, 2001.
63. G. Timmel. Ein stochastisches suchverfahren zur bestimmung der optimalen kompromisslösung bei statistischen polykriteriellen optimierungsaufgaben. Technical Report, TH Illmenau, 1980.
64. E. Zitzler, K. Deb, and L. Thiele. Comparison of multiobjective evolutionary algorithms: Empirical results. *Evolutionary Computation*, 8(2):173–195, 2000.
65. J. A. Nelder and R. Mead. A simplex method for function minimization. *The Computer Journal*, 7:308–313, 1965.
66. P. Koduru, S. Das, S. Welch, and J. L. Roe. Fuzzy dominance based multi-objective GA-simplex hybrid algorithms applied to gene network models. In K. Deb et al., editor, *Genetic and Evolutionary Computation–GECCO 2004. Proceedings of the Genetic and Evolutionary Computation Conference. Part I*, pp. 356–367, Seattle, Washington, USA, June 2004. Springer-Verlag, Lecture Notes in Computer Science, Vol. 3102.
67. J. M. Mendel. Fuzzy logic systems for engineering: A tutorial. *Proceedings of the IEEE*, 83(3):345–377, 1995.
68. P. Koduru, S. Das, and S. M. Welch. Multi-objective hybrid PSO using ϵ-fuzzy dominance. In D. Thierens, editor, *2007 Genetic and Evolutionary Computation Conference (GECCO'2007)*, volume 1, pp. 853–860, London, UK, July 2007. ACM Press.
69. J. Kennedy and R. C. Eberhart. Particle swarm optimization. In *Proceedings of the IEEE International Conference on Neural Networks*, Perth, Australia, pp. 1942–1948, 1995.
70. J. B. MacQueen. Some methods for classification and analysis of multivariate observations. In *Proceedings of the Fifth Berkeley Symposium on Mathematical Statistics and Probability*, volume 1, pp. 281–297. Statistical Laboratory of the University of California, Berkeley, 1967.
71. S. Zapotecas-Martínez and C. A. Coello Coello. A proposal to hybridize multi-objective evolutionary algorithms with non-gradient mathematical programming techniques. In G. Rudolph, T. Jansen, S. Lucas, C. Poloni, and N. Beume, *Parallel Problem Solving from Nature—PPSN X*, volume 5199, pp. 837–846. Springer, Berlin, Germany. Lecture Notes in Computer Science, September 2008.
72. J. H. Halton. On the efficiency of certain quasi-random sequences of points in evaluating multi-dimensional integrals. *Numerische Mathematik*, 2:84–90, 1960.
73. J. M. Hammersley. Monte-Carlo methods for solving multivariable problems. *Annals of the New York Academy of Science*, 86:844–874, 1960.
74. X. Zhong, W. Fan, J. Lin, and Z. Zhao. Hybrid non-dominated sorting differential evolutionary algorithm with Nelder–Mead. In *Intelligent Systems (GCIS), 2010 Second WRI Global Congress on*, Wuhan, China, volume 1, pp. 306–311, December 2010.
75. R. M. Storn and K. V. Price. *Differential Evolution—A Simple and Efficient Adaptive Scheme for Global Optimization over Continuous Spaces.* Technical Report TR-95-012, ICSI, Berkeley, CA, March 1995.
76. S. Zapotecas-Martínez and C. A. Coello Coello. A direct local search mechanism for decomposition-based multi-objective evolutionary algorithms. In *2012 IEEE Congress on Evolutionary Computation (CEC'2012)*, pp. 3431–3438, Brisbane, Australia, June 2012. IEEE Press.
77. S. Zapotecas-Martínez and C. A. Coello Coello. *MONSS: A Multi-Objective Nonlinear Simplex Search Algorithm.* Technical Report TR:2011-09-09, Centro de Investigación y de Estudios Avanzados del IPN (CINVESTAV-IPN), México, D.F., MÉXICO, September 2011.

78. S. Zapotecas-Martínez and C. A. Coello Coello. A memetic algorithm with non gradient-based local search assisted by a meta-model. In R. Schaefer, C. Cotta, J. Kołodziej, and G. Rudolph, editors, *Parallel Problem Solving from Nature—PPSN XI*, volume 6238, pp. 576–585, Kraków, Poland, September 2010. Springer, Lecture Notes in Computer Science.
79. R. Hooke and T. A. Jeeves. "Direct search" solution of numerical and statistical problems. *Journal of the ACM*, 8(2):212–229, 1961.
80. M. D. McKay, R. J. Beckman, and W. J. Conover. A comparison of three methods for selecting values of input variables in the analysis of output from a computer code. *Technometrics*, 21(2):239–245, 1979.
81. P. Koch, O. Kramer, G. Rudolph, and N. Beume. On the hybridization of SMS-EMOA and local search for continuous multiobjective optimization. In *Proceedings of the 11th Annual Conference on Genetic and Evolutionary Computation*, GECCO '09, pp. 603–610, New York, NY, USA, 2009. ACM.
82. N. Beume, B. Naujoks, and M. Emmerich. SMS-EMOA: Multiobjective selection based on dominated hypervolume. *European Journal of Operational Research*, 181(3):1653–1669, 2007.
83. E. Zitzler and L. Thiele. Multiobjective optimization using evolutionary algorithms—A comparative study. In A. E. Eiben, editor, *Parallel Problem Solving from Nature V*, pp. 292–301, Amsterdam, September 1998. Springer-Verlag.
84. K. Deb and T. Goel. A hybrid multi-objective evolutionary approach to engineering shape design. In *Proceedings of the First International Conference on Evolutionary Multi-Criterion Optimization*, EMO '01, pp. 385–399, London, UK, 2001. Springer-Verlag.
85. J. Fliege, L. M. Graña Drummond, and B. F. Svaiter. Newton's method for multiobjective optimization. *SIAM Journal on Optimization*, 20(2):602–626, 2009.

5

Other Decision-Making Models

Raghu Nandan Sengupta

CONTENTS

ABSTRACT This chapter will deal with different methodologies of decision making when the parametric as well as the nonparametric form of the objective functions are not known as this is quite possible in practical situations. The chapter starts with the definition of the utility theory with relevant examples. It then goes on to discuss the concept of Pareto principle and dwells into the idea of multiattribute utility theory (MAUT) and the relevance of goal programming with respect to MAUT. After building on this background, the basic tenets along with relevant examples for topics ranging from analytical network process, analytical hierarchy process, elimination and choice-translating reality method, preference-ranking organization method for enrichment evaluation, technique for order preference by similarity to ideal solution, and VIseKriterijumska Optimizacija I Kompromisno Resenje potentially all pairwise ranking of all possible alternatives, measuring attractiveness by a categorical based evaluation technique, decision trees, are discussed. Finally, the chapter ends with a list of softwares and urls, relevant to the topics covered in the chapter.

5.1 Introduction

Multi-criteria decision-making (MCDM) problems are those classes of problems where the decision maker has a range of decisions/alternatives in front of him/her, one among which should be chosen depending on the available attributes/decision criteria/goals for each of these decisions/alternatives. MCDM is generally categorized as continuous or discrete depending on the domain of the available decisions/alternatives. Hwang and Yoon (1981) classify MCDM methods under two general categories, namely, *multiobjective decision making* (MODM) and *multiattribute decision making* (MADM). MODM has decision variables that are determined in a continuous or integer domain with either an infinite or a large number of decisions/alternatives. On the other hand, MADM is used to solve problems with a discrete decision space and a predetermined or a limited number of decisions/alternatives. The commonality between different MCDM techniques, using the notions/concepts of decisions/alternatives and attributes/decision criteria/goals, can be found in Chen and Hwang (1992). Few good references that give an excellent theoretical foundation of the different MCDM methods, along with their comparative strengths and weaknesses, are Belton and Stewart (2002), Ehrgott et al. (2010), Figueira et al. (2005), Gal et al. (1999), Greco et al. (2010), Kahraman (2008), Larichev and Olson (2001), Parlos (2000), Pomerol and Barba-Romero (2000), Tzeng and Huang (2011), and Zeleny (1982). In solving MCDM problems, conflicts occur; hence, obtaining solutions by satisfying all the available attributes/decision criteria/goals at the same time may not be possible; hence, the concept of *Pareto optimality* (Kuhn and Tucker 1951; Pareto 1896; Zadeh 1963) is used. Furthermore, in making decisions, rather than the quantitative criteria, we use the concept of a utility function, Arrow (1971), Berger (1985), Ingersoll (1987), Marshall (1920), Morgenstern (1976), and Neumann and Morgenstern (2007). To present the different MCDM methods in a logical manner, we discuss the core concepts of the utility theory and their significance from the point of view of MCDM, and then go into each of the different MCDM techniques in some details.

For the interested readers, we state a simple general formulation of an MCDM problem that may be stated as follows: *optimize*: $\{f_1(\boldsymbol{x}), \ldots, f_k(\boldsymbol{x})\}$, s.t.: $g_j(\boldsymbol{x}) \geq a_j$ where $j \in J$ and $\boldsymbol{x} \in \boldsymbol{X}$. As we see, one needs to optimize (maximize/minimize) k number of objective functions, $f_i(\cdot), i = 1, \ldots, k$ in $\boldsymbol{x}$, given constraint functions, $g_i(\cdot)$ and constants a_j, where $j = 1, \ldots, J$. It is important to note what $\boldsymbol{X}$ is. If $\boldsymbol{x}$ is defined explicitly (by a set of alternatives), then, the resulting problem is called a multiple-criteria evaluation problem, while if $\boldsymbol{x}$ is defined implicitly (by a set of constraints), then, the resulting problem is called a multiple-criteria design problem.

While solving the above problem, it is assumed that under certain conditions, a preference for any decision/alternative can be represented by a utility function that assigns a numerical value to each of them. But in general, these conditions may not be exactly satisfied. To ensure a preference/ranking in such cases, one uses the multi-criteria decision-analysis (MCDA) method where a certain outranking criterion is used to rank the decisions/alternatives. The outranking relation is a binary relation, S, defined on the set of decisions/alternatives, $\boldsymbol{A}$, such that $A_i S A_j$, where $A_i, A_j \in \boldsymbol{A}$. Moreover, MCDA is nonparametric in nature as the attributes/decision criteria/goals, based on which the selection among decisions/alternatives is made, may not be quantified in any explicit functional form, but can definitely be stated/compared by the decision maker when making choices between attributes/decision criteria/goals.

5.1.1 Utility Theory

Imagine that a decision maker faces a choice among a set of risky alternatives or lotteries. We assume that the decision maker has a rational preference over these risky decisions/alternatives or lotteries, based on which he/she ranks the preferences. To analyze this, let us assume that the risky decisions/alternatives result in one out of a number of consequences/outcomes, where the consequences/outcomes are c_i, where $i = 1, \ldots, n$, that is, $\boldsymbol{c} = (c_1, \ldots, c_n)$. Also, assume that $p_i(a) = Pr\{s|F(a,s) = c_i\} \geq 0$, where $p_i(a)$ denotes the probability for a consequence/outcome, c_i, given the action is $a \in A$ in state $s \in S$, while $F(.,.)$ is a functional mapping from $A \times S$ to C. Utilizing this, one can formally define a lottery $L = (p_1, \ldots, p_n)$ with consequences, $(c_1, \ldots, c_n)$, such that $\sum_{i=1}^{n} p_i(a) = 1$. Furthermore, we can also have a compound lottery $\boldsymbol{L} = (L_1, \ldots, L_p;\ p_1, \ldots, p_p)$, where each L_j, $j = 1, \ldots, p$ is itself made up of simple lotteries as described above, that is, $L_j = (q_j^1, \ldots, q_j^n)$, here $\sum_{i=1}^{n} q_j^i = 1$. In general, one might think that the preferences of the individual may be represented by a set of indifference curves, or an indifference map. This is where we first encounter the concept of a utility function denoted by $u(c_i)$, where the utility function $u(\cdot)$ is defined over consequences/outcomes, such that one can evaluate the decisions/alternatives/lottery by the mathematical expectation or expected value of this utility function. This underlying $u(\cdot)$ function is termed as Bernoulli utility function or von Neumann–Morgenstern utility function. To understand the general notion of an average, we use the term "expected utility," that is, $E(U) = \sum_{i=1}^{n} p_i(a)u(c_i)$. Similarly, one can define the variance as $Var(U) = \sum_{i=1}^{n}\{u(c_i) - E(U)\}^2 p_i(a)$.

A simple diagram (Figure 5.1) helps us to illustrate this concept. One should remember that it can be recursively extended as required where each compound lottery is again broken up into simple lotteries.

5.1.1.1 Axioms of Utility Theory

The four axioms of von Neumann–Morgenstern utility theorem that are important for us to understand are as follows:

Axiom 1 (completeness): Under Axiom 1, we assume that an individual has well-defined preferences, that is, for any lottery, L, M, exactly one of the following holds, that is, (i) $L \prec M$ (M is preferred over L), (ii) $L \succ M$ (L is preferred over M), and (iii) $L \sim M$ (one is indifferent between L and M).

Axiom 2 (transitivity): Under Axiom 2, we assume that for an individual, if $L \preceq M$ and $M \preceq N$, then, $L \preceq N$, that is, the preference is consistent across any three options, L, M, and N.

Axiom 3 (continuity): Under Axiom 3, we assume that for an individual, if $L \preceq M \preceq N$, then, there exists $p \in [0, 1]$, such that $pL + (1 - p)N \sim M$.

Axiom 4 (independence): Under Axiom 4, we assume that for an individual, if $L \prec M$, then, for any N and $p \in [0, 1]$, $pL + (1 - p)N \prec pM + (1 - p)N$.

5.1.1.2 Example of Utility Functions

Consider some wealth W, where this wealth, W, implies some numerical concept used to define the input variable based on which the utility may be calculated. Thus, from the perspective of economics

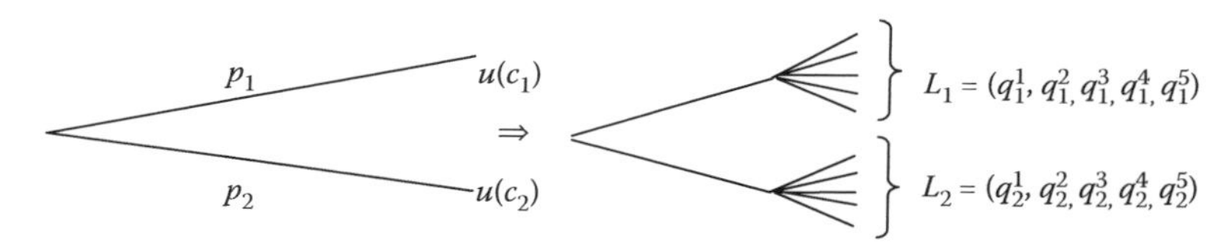

FIGURE 5.1
Simple example of a lottery.

and in general, one may define a utility as $U(W)$, and in general, the following utility functions are used:

- Quadratic: $U(W) = W - bW^2$, where b is a positive constant
- Logarithmic: $U(W) = ln(W)$
- Exponential: $U(W) = -exp(-aW)$, where a is a positive constant
- Power: $U(W) = cW^c$, where $c \neq 0$ and $c \leq 1$

Furthermore, a utility function is said to exhibit a hyperbolic absolute risk aversion (HARA) *iff* the level of risk tolerance, that is, the reciprocal of absolute risk aversion $A(W)$, is a linear function of wealth W. A utility function $U(W)$ that has this property is the HARA utility function that is of the form, $U(W) = ((1-\gamma)/\gamma)((aW/(1-\gamma)) + b)^\gamma$, where $a > 0$, $((aW/(1-\gamma)) + b) > 0$, while a, b, and γ are parameters for the HARA utility function.

One can derive different utility functions from the HARA case under certain conditions that are stated below.

Case 1: If $\gamma \to 1$, then, the utility is a linear utility function.

Case 2: A variant of HARA utility function is the quadratic utility function, which is obtained if we consider $\gamma > 2$ as the parameter value in the HARA function. Note that this quadratic utility is an increasing function when $W < (1/2\gamma)$.

Case 3: In the case of HARA function, if we consider $a > 0$, $b = 1$, and $\gamma \to -\infty$, then, we obtain the exponential utility function. It is true that the value of an exponential utility function is negative. But as utility functions use the concept of cardinal ranking, what is important is the *relative* value of the utility function; hence, the negative value of the exponential utility function does not matter when making a relative decision.

Case 4: The power utility function is obtained if we consider $\gamma \leq 1$, $\gamma \neq 0$, and $a = 1 - \gamma$ as the parameter values in the general form of the HARA utility function. Power utility function takes the form of the linear utility if the value of parameter γ approaches 1.

Case 5: If we consider $a = 1$ and $\gamma \to 0$ in the HARA utility function, then, we obtain the logarithmic utility function of the form $ln(W)$. If there is any positive probability of obtaining an outcome of 0 for W, then, the expected utility of the logarithmic utility function will be $-\infty$.

Note: The actual value of the expected utility is of no use, except when comparing with other alternatives. Hence, we use an important concept of a *certainty equivalent*, which is the amount of certain wealth (under risk-free conditions) that has the utility level exactly equal to this expected utility value. We define $U(C) = E\{U(W)\}$, where C is the certainty value.

Example 5.1

Suppose you face two options. Under option # 1, you toss a coin and if the head appears, you win Rs. 10, while if the tail appears, you win Rs. 0. Under option # 2, you get an amount of Rs. M. Also, assume that your utility function is of the form $U(W) = W - 0.04W^2$. The utility function signifies that after you win, the functional form of the utility function that quantifies the benefit is W–$0.04W^2$. For the first option, the expected utility value would be Rs. 3, while the second option has an expected utility of Rs. M–$0.04M^2$. To find the certainty equivalent, we should have $E\{U(M)\} = M–0.04M^2 = 3$. Thus, $M = 3.49$, that is, $C = 3.49$, as $E\{U(3.49)\} = E\{U(W)\}$.

Example 5.2*

In finance, one considers the returns, $r = log_e(P_{t+1}/P_t)$, where P_t is the closing/adjusted closing price of a particular stock/script. If we consider the utility function of the returns as quadratic, then, the distribution of r is normal, that is, $r \sim N(\mu_r, \sigma_r^2)$, where $E(r) = \mu_r$ and $Var(r) = \sigma_r^2$. But in practical terms, returns are extreme value distributed (EVD) (a few good references for EVD are Castillo et al. 2004; Coles 2001; Embrechts et al. 1997; Kotz and Nadarajah 2000; Leadbetter et al. 1983; Resnick 1987). Continuing with our discussion, let us assume that the returns of the portfolio are EVD, that is, $r \sim GEV(\mu, \sigma, \xi)$, where $\mu \in \mathbb{R}$ is the location parameter, $\sigma > 0$ is the scale parameter, and finally $\xi \in \mathbb{R}$ is the shape parameter of the generalized EVD. If one considers the Gumbel distribution (a class of EVD), as the actual return distribution, then, the probability density function of returns is $f(r) = (1/\sigma)\exp[((r-\mu)/\sigma) - \exp((r-\mu)/\sigma)]$. One can then find the probability density function of its corresponding utility, $U(r)$, using a simple Jacobian transformation. Finally, based on the simple utility maximization concept, our portfolio optimization problem may be formulated as

$$\text{Maximize: } E\{U(r_{Portfolio})\},$$

$$\text{s.t: } Var\{U(r_{Portfolio})\} \leq \sigma^{2*}_{portfolio},$$

$$\sum_{i=1}^{N} w_i = 1,$$

$$0 \leq w_i \leq 1,$$

where w_i, $i = 1, \ldots, N$ are the weight of the ith stock/script in the portfolio, thus formed, while $\sigma^{2*}_{portfolio}$ is the stipulated value of the risk below which the portfolio variance should always lie. If one considers the data from National Stock Exchange (NSE) http://www.nse-india.com from Jan-01-2000 to Dec-31-2010, then, after finding the w_i's, where $i = 1, \ldots, N$, one can easily obtain the risk versus return graph that is shown in Figure 5.2.

Though not exactly the shape of an efficient frontier, yet, the graph in Figure 5.2 depicts the risk return profile as seen in a variety of such problems.

5.1.1.3 Stochastic Dominance Principle

5.1.1.3.1 First-Order Stochastic Dominance

When a lottery A_i dominates A_j in the sense of first-order stochastic dominance, then, the decision maker prefers A_i to A_j regardless of what the utility function, $u(x)$, is, as long as it is weakly increasing. Thus, in terms of cumulative distribution functions (cdfs), $F_{A_i}(x)$ and $F_{A_j}(x)$, A_i dominates A_j which implies $F_{A_i}(x) \geq F_{A_j}(x)$ for all x, with a strict inequality at some x. Remember that, here, x is an outcome and $x \in X$. Furthermore, A_i first-order stochastically dominates A_j, *iff* every expected utility maximizer with an increasing utility function prefers A_i over A_j.

* The results of Examples 5.2, 5.4, and 5.5 are parts of different un-published masters theses of students in the Industrial and Management Engineering, Indian Institute of Technology, Kanpur, India, who have worked under the guidance of the author, Raghu Nandan Sengupta.

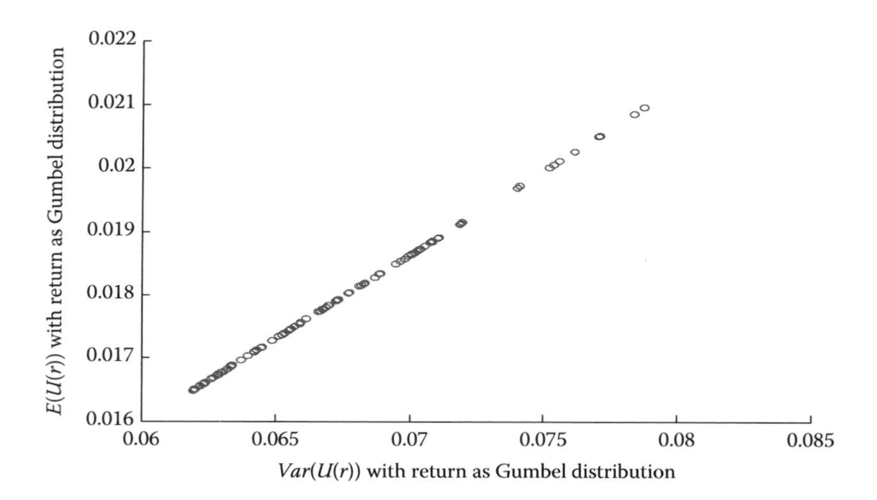

FIGURE 5.2
Expected utility versus the variance of utility, considering that the assets return distribution is Gumbel distributed.

Example 5.3

Consider the example below that is as follows:

Outcome from Die	**1**	**2**	**3**	**4**	**5**	**6**
Lottery A_1	1	1	2	2	2	2
Lottery A_2	1	1	1	2	2	2
Lottery A_3	3	3	3	1	1	1

Here, A_1 as well A_3 dominates A_2 in the first-order sense, while nothing can be said about the dominance characteristics between A_1 and A_3.

5.1.1.3.2 Second-Order Stochastic Dominance

When a lottery A_i dominates A_j in the sense of second-order stochastic dominance, then, the decision maker prefers A_i to A_j as long as he/she is risk averse and the utility function, $u(x)$, is weakly increasing. In terms of cdfs, $F_{A_i}(x)$ and $F_{A_j}(x)$, A_i are second-order stochastically dominant over A_j *iff* $\int_{-\infty}^{x}\{F_{A_j}(x) - F_{A_i}(x)\}dx \geq 0$ for all x, with a strict inequality at some x. Remember that, here, x is an outcome and $x \in X$. Expressed using the utility function, the second-order stochastic dominance can be expressed as $E[u(A_i)] \geq E[u(A_j)]$.

There are higher-order stochastic dominance characteristics also, but, we do not go into those as one can refer to any good book for a thorough understanding of this importance concept.

5.1.2 Concept of Pareto Optimality and Its Significance

The term "*Pareto optimality*" is named after Vilfredo Pareto (1848–1923), an Italian economist who used the concept in his studies of economic efficiency and income distribution. It is true that as per this concept, there can be noninferior solutions as introduced by Vilfredo Pareto in 1896, though the formalization of this idea that leads to its mathematical understanding may be credited to Francis Ysidro Edgeworth (1845–1926). The author is best known for the Edgeworth box, but it must be remembered that in the field of multiobjective optimization, the concept is known as Pareto optimality rather than Edgeworth optimality. The origin of the term "Pareto optimality" goes back to the following text from Pareto (1906, Chapter VI, Section 33). *We will begin by defining a term which is desirable to use in order to avoid prolixity. We will say that the members of a collectivity enjoy maximum ophelimity in a certain position when it is impossible to find a way of moving from that position very slightly in such a manner that the ophelimity enjoyed by each of the individuals of that collectivity increases or decreases. That is to say, any small displacement in departing from that position necessarily has the effect of increasing the ophelimity which certain individuals enjoy, and decreasing that which others enjoy, of being agreeable to some and disagreeable to others.*

There are other sources of references for Pareto optimality, for example, Koopmans (1951) who may also be credited to have formally studied production as a resource allocation problem, where the combination of different activities represents the output of commodities as a function of various factors. Another classic reference of Pareto optimality concept used in optimization is the seminal paper by Kuhn and Tucker (1951). The reference about Pareto optimality can also be found in the work of Kenneth J. Arrow who worked in the domain of a multiobjective optimization.

Let us consider a multiobjective optimization problem of the form $Optimize_{\forall x \in X}$: $\{f_1(\boldsymbol{x}), \ldots, f_k(\boldsymbol{x})\}$, where $\boldsymbol{x} \in \mathbb{R}^n$. The individual objective functions are $f_1(\boldsymbol{x}), \ldots, f_k(\boldsymbol{x})$, such that $f_i : X \rightarrow \mathbb{R}, i = 1, \ldots, k$. There may be instances when the objective functions are at least partially in conflict, that is, we do not have any decision vector, $\boldsymbol{x}$, which optimizes all the objective functions simultaneously. To make things more formal, we state a few results (properties of Pareto optimal solutions) that may be considered as relevant and important to this area.

1. *Property of dominance*: A vector $\boldsymbol{x} \in X$ is said to dominate a vector $\boldsymbol{y} \in X$ if $f_i(\boldsymbol{x}) \geq f_i(\boldsymbol{y})$, $i \in \{1, \ldots, k\}$, and $f_j(\boldsymbol{x}) > f_j(\boldsymbol{y})$ for at least one $j \in \{1, \ldots, k\}$.
2. *Strong Pareto optimality*: A vector $\boldsymbol{x}^* \in X$ is defined as a strong Pareto optimal if there exists a vector $\boldsymbol{x}^*$ such that it dominates the vector $\boldsymbol{x} \in X$. An objective vector $z^* = f(\boldsymbol{x}^*)$ is called a strong Pareto optimal if the corresponding vector $\boldsymbol{x}^*$ is a strong Pareto optimal. The set of strong Pareto optimal decision vectors $\boldsymbol{x}^* \in X$ is denoted by $\mathcal{P} \subseteq X$.
3. *Weak Pareto optimality*: A vector $\boldsymbol{x}^* \in X$ is defined as a weak Pareto optimal if there exists no other vector $\boldsymbol{x} \in X$ such that $f_i(\boldsymbol{x}) < f_i(\boldsymbol{x}^*)$, $i \in \{1, \ldots, k\}$. An objective vector $z^* = f(\boldsymbol{x}^*)$ is called weakly Pareto optimal if the corresponding vector $\boldsymbol{x}^*$ is weakly Pareto optimal. The set of weakly Pareto optimal vectors is denoted by $\mathcal{P}_W$.

Thus, we are interested to solve $Optimize_{\forall x \in X} : \{f_1(\boldsymbol{x}), \ldots, f_k(\boldsymbol{x})\}$ to find $\mathcal{P} \subseteq X$. In case we have multiobjectives to solve, then, finding the Pareto optimal points becomes difficult. Figure 5.3 is a hypothetical example in two-dimensional (2-D) space (two objectives) that illustrates the Pareto optimality plots for four different combinations of the objective functions (nomenclatured as Obj_1 and Obj_2). The bounded area depicts the hypothetical feasible set and the boundary is considered to be the optimal set of feasible points depending on whether it is maximization or minimization for Obj_1 or Obj_2. The Pareto optimal points (A–B) is for the case when we consider the (i) minimization of Obj_1 along with minimization of Obj_2. Similarly, C–D, E–F, and G–H are for the instances when one considers (ii) the maximization of Obj_1 along with minimization of Obj_2, (iii) maximization of Obj_1 along with maximization of Obj_2, and finally (iv) minimization of Obj_1 along with maximization of Obj_2. The four distinct Pareto optimal plots may not be exact in shape and size, but it should give a good feel to the reader about the set of Pareto optimal point curves one would obtain depending on the four different combinations, when one has two objective functions as shown.

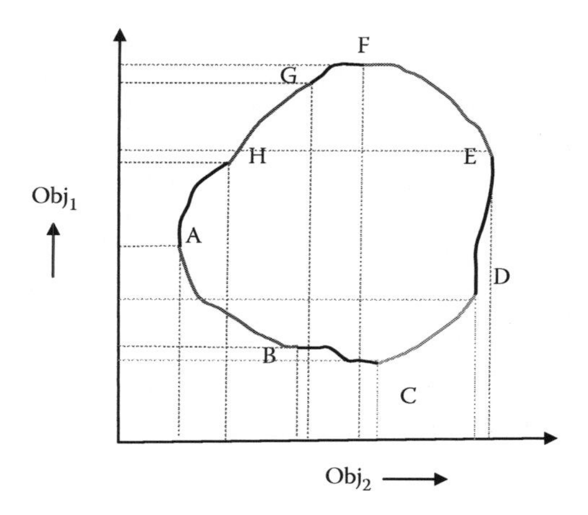

FIGURE 5.3
Hypothetical example to illustrate the set of Pareto optimal points for four different combinations of two different objective functions.

Example 5.4

Consider the mean–variance (MV) model in portfolio optimization, where we have N number of stocks/scripts where their individual returns are r_i, average returns are $\bar{r}_i$, and variance–covariance are σ_{i_1,i_2}, where $i_1, i_2 = 1, \ldots, N$. Then, the MV optimization problem is of the form

$$\text{Minimize: } \sum_{i_2=1}^{N} \sum_{i_1=1}^{N} w_{i_1} w_{i_2} \sigma_{i_1,i_2},$$

$$\text{s.t.: } \sum_{i_1=1}^{N} w_i \bar{r}_i \geq r^*$$

$$\sum_{i=1}^{N} w_i = 1,$$

$$0 \leq w_i \leq 1,$$

where r^* is some predetermined value of the return over which we would definitely like the return of the portfolio to be. Utilizing data from DAX (Deutscher Aktienindex) (i.e., German stock index) http://www.dax-indices.com for the time period from January 1, 2012 to January 31, 2015 and then solving the problem gives us the risk return graph (or a set of Pareto points, Figure 5.4) where one can easily understand that maximizing the return would also increase the risk (i.e., variance). The set of points obtained is the loci that illustrate the relationship between the return and variance that would be the Pareto optimal set of points for the MV optimization problem.

Example 5.5

Continuing with another example in finance, let us now illustrate the following multiobjective problem that is as follows:

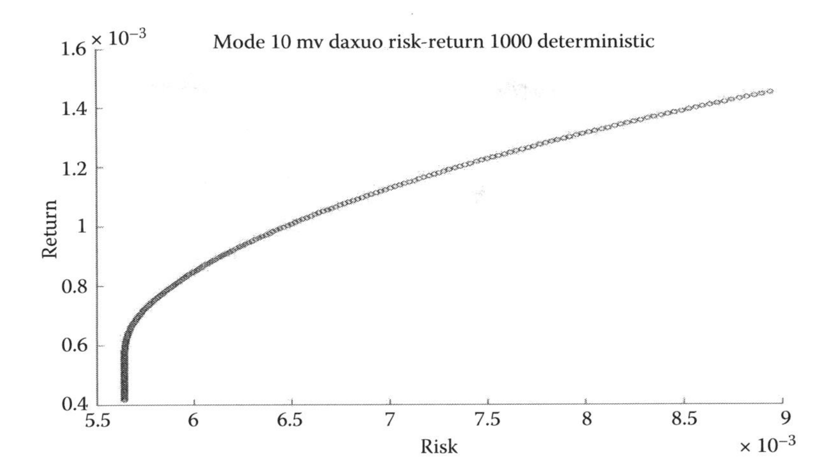

FIGURE 5.4
Risk return graph or the Pareto optimal curve, Example 5.4.

$$\textit{Maximize } r^* \quad \textit{Minimize } \sigma^{2^*},$$

$$\text{s.t}: \left\{\sum_{i=1}^{N}(1+r_i)w_i\right\} \geq 1,$$

$$\sum_{i=1}^{N} r_i w_i \geq r^*,$$

$$\sum_{i_2=1}^{N}\sum_{i_1=1}^{N} w_{i_1}w_{i_2}\sigma_{i_1,i_2} \leq \sigma^{2^*},$$

$$\sum_{i=1}^{N} w_i = 1,$$

$$0 \leq w_{i,\min} \leq w_i \leq w_{i,\max} \quad \forall i, i_1, i_2 = 1, \ldots, N,$$

where, as usual, r_i, w_i, σ_{i_1,i_2}, r^*, and σ^{2^*} have their usual meaning as already discussed.

Utilizing data from Dow Jones Industrial Average (DJIA) http://www.djaverages. com/ for the time frame from January 1, 2012 to January 31, 2015, one obtains the Pareto plots for the expected return and risk (variance) for one of the iteration runs (Figure 5.5). To end this section, we give a few relevant references such as Aleskerov et al. (2007), Berger (1985), Fishburn (1970, 1988, 1989), Kreps (1988), and Neumann and Morgenstern (2007) that one may read to get a good idea about the utility theory and the related topics.

5.1.3 Multiattribute Utility Theory

We know that single-attribute utility (SAU) functions, $u(x)$, where x is the variable, are obtained by using a set of lottery questions based on the certainty equivalence. SAUs are monotonic functions, where the best outcome is set at 1, and the worst outcome is set at 0. SAU functions are developed to describe the decision maker's compromise between the best and the worst alternatives based on the lottery questions. On the other hand, as a concept, multiattribute utility theory (MAUT) tends

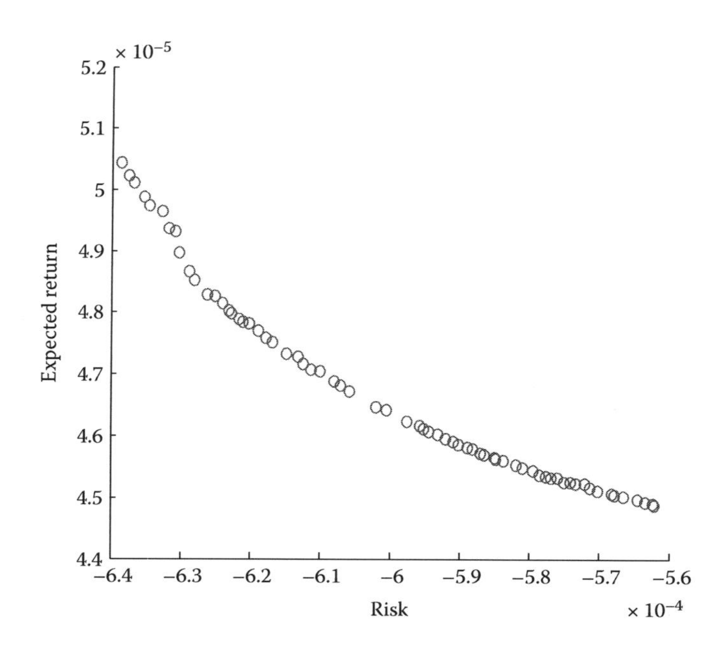

FIGURE 5.5
Risk return graph or the Pareto optimal point curve.

to reduce the complex problem of assessing a multiattribute utility function into one of assessing a series of unidimensional utility functions. MAUT stitches or joins the unidimensional utility functions in a way such that the multiattribute utility function can be portrayed objectively. The main problem of MAUT is to assess the form $u(x_1, \ldots, x_n)$. A simple method could be to derive a function such that $u(x_1, \ldots, x_n) = f\{u_1(x_1), \ldots, u_n(x_n)\}$. One of the simplest decomposition is the additive function where $u(x_1, \ldots, x_n) = \sum_{i=1}^{n} \lambda_i u_i(x_i)$, while λ_i's are the corresponding weights. Another parameter that is also used is the scaling factor, k. The scaling factor assures that the compound utility function, u, assumes values in the interval between 0 and 1. Thus, if $\sum_{i=1}^{n} \lambda_i = 1$, then $k = 0$. There are other functional forms of describing the relationship between MAUT and the unidimensional utility functions and for the convenience of the reader, Table 5.1 gives a few of the commonly used utility decompositions as utilized in the context of MAUT. In the context of Table 5.1, one should note the following:

1. When $\sum_{i=1}^{n} \lambda_i = 1$, then $k = 0$, such that the multiplicative decomposition reduces to the weighted additive decomposition concept.
2. The special case of the quasi additive form is the multiplicative form.
3. The special case of multilinear decomposition is the additive and the multiplicative functions.

The reader should note two important properties of MAUT that are (i) preferential independence and (ii) utility independence, which can be explained as stated below:

- Preferential independence: The pair of attributes X and Y is preferentially independent of attribute Z if the value trade-off between X and Y is not affected by a given level of Z.

TABLE 5.1

Common Utility Decompositions

Model	General Representation of $u(x_1, \cdots, x_n)$	Calculations
Additive	$\sum_{i=1}^{n} u_i(x_i)$	• n number of utility functions
Weighted additive	$\sum_{i=1}^{n} \lambda_i u_i(x_i)$	• n number of utility functions • n number of scaling constraints
Multiplicative/log additive	$\sum_{i=1}^{n} \lambda_i u_i(x_i) + \sum_{i=1}^{n} \sum_{j>i}^{n} k\lambda_i \lambda_j u_i(x_i) u_j(x_j) + \sum_{i=1}^{n} \sum_{j>i}^{n} \sum_{k>j}^{n} k^2 \lambda_i \lambda_j \lambda_k u_i(x_i) u_j(x_j) u_k(x_k)$ $+ \cdots + k^{n-1} \lambda_1 \lambda_2 \lambda_3 \cdots \lambda_n u_1(x_1) u_2(x_2) u_3(x_3) \cdots u_n(x_n)$	• n number of utility functions • n number of scaling constraints
Quasiadditive	$\sum_{i=1}^{n} \lambda_i u_i(x_i) + \sum_{i=1}^{n} \sum_{j>i}^{n} \lambda_{ij} u_i(x_i) u_j(x_j) + \sum_{i=1}^{n} \sum_{j>i}^{n} \sum_{k>j}^{n} \lambda_{ijk} u_i(x_i) u_j(x_j) u_k(x_k)$ $+ \cdots + \lambda_{123 \cdots n} u_1(x_1) u_2(x_2) u_3(x_3) \cdots u_n(x_n)$	• n number of utility functions • $(2^n - 1)$ number of scaling constraints
Bilateral	$\sum_{i=1}^{n} \lambda_i u_i(x_i) + \sum_{i=1}^{n} \sum_{j>i}^{n} \lambda_{ij} f_i(x_i) f_j(x_j) + \sum_{i=1}^{n} \sum_{j>i}^{n} \sum_{k>j}^{n} \lambda_{ijk} f_i(x_i) f_j(x_j) f_k(x_k)$ $+ \cdots + \lambda_{123 \cdots n} f_1(x_1) f_2(x_2) f_3(x_3) \cdots f_n(x_n)$	• $2n$ number of utility functions • $(2^n - 1)$ number of scaling constraints
Hybrid	$\sum_{i=1}^{n} \lambda_i u_i(x_i) + \sum_{i=1}^{n} \sum_{j>i}^{n} \lambda_{ij} u_i(x_i) f_j(x_j) + \sum_{i=1}^{n} \sum_{j>i}^{n} \sum_{k>j}^{n} \lambda_{ijk} u_i(x_i) f_j(x_j) f_k(x_k)$ $+ \cdots + \lambda_{123 \cdots n} u_1(x_1) f_2(x_2) f_3(x_3) \cdots f_n(x_n)$	• $f(n)$ number of scaling constraints, where $f(\cdot)$ is some function and depends on n

(Continued)

TABLE 5.1 ***(Continued)***

Common Utility Decompositions

Model	General Representation of $u(x_1, \cdots, x_n)$	Calculations
Quasipyramid	$\sum_{i=1}^{n} \lambda_i u_i(x_i) + \sum_{i=1}^{n} \sum_{j>i}^{n} \lambda_{ij} u_{ij}(x_i, x_j) + \sum_{i=1}^{n} \sum_{j>i}^{n} \sum_{k>j}^{n} \lambda_{ijk} u_{ijk}(x_i, x_j, x_k) + \cdots + \lambda_{123\cdots n} u_1(x_1) u_2(x_2) u_3(x_3) \cdots u_n(x_n)$	• $\{n(n-1)/2\}$ number of utility functions (with nonseparable interactions) • $f(n)$ number of scaling constraints, where $f(\cdot)$ is some function and depends on n
Semicube	$\sum_{i=1}^{n} \lambda_i u_i(x_i) + \sum_{i=1}^{n} \sum_{j>i}^{n} \lambda_{ij} u_{ij}(x_i, x_j) + \sum_{i=1}^{n} \sum_{j>i}^{n} \sum_{k>j}^{n} \lambda_{ijk} u_{ijk}(x_i, x_j, x_k) + \cdots + \lambda_{123\cdots n} f_1(x_1) f_2(x_2) f_3(x_3) \cdots f_n(x_n)$	• $f(n)$ Number of scaling constraints, where $f(\cdot)$ is some function and depends on n
Interdependent variable	$\sum_{i=1}^{n} \lambda_i u_i(x_i) + \sum_{i=1}^{n} \sum_{j>i}^{n} \lambda_{ij} u_{ij}(x_i, x_j) + \sum_{i=1}^{n} \sum_{j>i}^{n} \sum_{k>j}^{n} \lambda_{ijk} u_{ijk}(x_i, x_j, x_k) + \cdots + \sum_{i=1}^{n} \sum_{j>i}^{n} \cdots \sum_{n-1>n-2}^{n} \lambda_{ijk\cdots n-1} u_{ijk\cdots n-1}(x_i, x_j, x_k, \cdots, x_{n-1})$	• $f(n)$ number of scaling constraints, where $f(\cdot)$ is some function and depends on n
Multilinear	$\sum_{i=1}^{n} \lambda_i u_i(x_i) + \sum_{i=1}^{n} \sum_{j>i}^{n} \lambda_{ij} u_i(x_i) u_j(x_j) + \sum_{i=1}^{n} \sum_{j>i}^{n} \sum_{k>j}^{n} \lambda_{ijk} u_i(x_i) u_j(x_j) u_k(x_k) + \cdots + \lambda_{123\cdots n} \lambda_1 \lambda_2 \lambda_3 u_1(x_1) u_2(x_2) u_3(x_3) \cdots u_n(x_n)$	• n number of one-attribute utility functions • (2^n-2) number of scaling constants

- Utility independence: Attribute X is utility independent of attribute Y when conditional preferences of lotteries on X, given Y, do not depend on the particular level of Y.

Some references that can be used as a good starting point to understand MAUT are Dyer et al. (1992), Keeney and Raiffa (1993), Tzeng and Huang (2011), and Zeleny (1982).

5.1.3.1 Use of Goal Programming in MAUT

Goal programming (GP) is an analytical approach devised to address decision-making problems, where targets have been assigned to all the attributes and where the decision maker is interested in minimizing the nonachievement of the corresponding goals. In other words, the decision maker seeks a satisfactory and sufficient solution with this strategy. The purpose of GP is to minimize the deviations between the achievement of goals and their aspiration levels. Mathematically, it may be represented as $min \sum_{i=1}^{k} w_i |f_i(x) - a_i|$, s.t.: $g_j(\boldsymbol{x}) \geq b_j$ where $j \in J$, $\boldsymbol{x} \in \boldsymbol{X}$, a_i is the ith objective goal ($i = 1, \ldots, k$), and b_j, ($j = 1, \ldots, J$) is the constant corresponding to the jth constraint. Few good references that should give a good idea about GP are Charnes and Cooper (1961), Jones and Tamiz (2010), Lee (1972), Romero (1991), Schniederjans (1995), and Trzaskalik and Michnik (2002).

One can use GP to address the issues of MAUT. Without the formal solution, let us state a problem that can be modeled as a GP problem. Assume k number of criterion functions denoted by $f_1(\boldsymbol{x}), \ldots, f_{k_1}(\boldsymbol{x})$, $f_{k_1+1}(\boldsymbol{x}), \ldots, f_{k_2}(\boldsymbol{x})$, $f_{k_2+1}(\boldsymbol{x}), \ldots$, and $f_{k_3}(\boldsymbol{x})$ where $\boldsymbol{x} \in X$ and is $\mathbb{R}^n$. It is obvious that $k_1 + k_2 + k_3 = k$. Furthermore, for simplicity, consider that k_1, k_2, and k_3 are distinct, such that $k_1 \cap k_2 = k_1 \cap k_3 = k_2 \cap k_3 = \varnothing$, and $k_1 \cup k_2 \cup k_3 = k$. As per the problem, it is stated that (i) for $f_1(\boldsymbol{x}), \ldots, f_{k_1}(\boldsymbol{x})$, the respective values of these functions should be at least as large as $b_1, \ldots, b_{k_1}$, (ii) for $f_{k_1+1}(\boldsymbol{x}), \ldots, f_{k_2}(\boldsymbol{x})$, the respective functional values are somewhere in the sets $(b_{k_1+1,L}, b_{k_1+1,U}), \ldots, (b_{k_2,L}, b_{k_2,U})$, and finally (iii) for the $f_{k_2+1}(\boldsymbol{x}), \ldots, f_{k_3}(\boldsymbol{x})$, the functional values should be at most $b_{k_2+1}, \ldots, b_{k_3}$. Remember that for any deviation from the restrictions imposed on the functional values, it would entail some assignment of weights that are (i) $w_1, \ldots, w_{k_1}$, (ii) $w_{k_1+1,L}, w_{k_1+1,U}, \ldots, w_{k_2,L}$, and $w_{k_2,U}$, and (iii) $w_{k_2+1}, \ldots, w_{k_3}$, respectively. Using this information, we can write down the formulation as

$$
\begin{aligned}
minimize_{\boldsymbol{x} \in X} :& \{w_1 y_1^- + \cdots + w_{k_1} y_{k_1}^-\} + \{w_{k_1+1,L} y_{k_1+1,L}^- + \cdots + w_{k_2,L} y_{k_2,L}^-\} \\
& + \{w_{k_1+1,U} y_{k_1+1,U}^+ + \cdots + w_{k_2,U} y_{k_2,U}^+\} + \{w_{k_2+1} y_{k_2+1}^+ + \cdots + w_{k_3} y_{k_3}^+\}. \\
\text{s.t} :& f_i(\boldsymbol{x}) - y_i^+ + y_i^- = b_i. \\
& f_j(\boldsymbol{x}) - y_{jL}^+ + y_{jL}^- = b_{jL} \\
& f_s(\boldsymbol{x}) - y_{sU}^+ + y_{sU}^- = b_{sU} \\
& f_l(\boldsymbol{x}) - y_{lU}^+ + y_{lU}^- = b_{lU} \\
& y_i^+, y_i^-, y_{jL}^+, y_{jL}^-, y_{sU}^+, y_{sU}^-, y_{lU}^+, y_{lU}^- \geq 0.
\end{aligned}
$$

Here, $i = 1, \ldots, k_1$, $j = k_1 + 1, \ldots, k_2$, $s = k_1 + 1, \ldots, k_2$, and $l = k_2 + 1, \ldots, k_3$.

With some simplification, one can write this as a maximization problem where the objective function may be interpreted as the sum of piecewise linear functions. In other words, we replace the nonlinear part with piece wise linear functions and then solve the problem. Though it gives approximate results, yet, the results are intuitive. When utilizing this method, one should be cautious as to the properties of MAUT and what decomposition function one is using.

5.1.4 Multicriteria Decision Making

As a decision-making process, the MCDM technique involves five main steps involving the numerical analysis of different decisions/alternatives. These steps may be summarized as stated:

1. First, decide and finalize the set of n **attributes/decision criteria/goals** $(C_1, \ldots, C_n)$ and their evaluation process so that they relate to the goal that is required to be achieved by the decision maker.
2. Next, determine the relevant m **decisions/alternatives** $(A_1, \ldots, A_m)$ such that the goal may be achieved using these n **decisions/alternatives**. In simple words, these **decisions/alternatives** represent the different choices of actions available to the decision maker.
3. Next, attach numerical measures to the relative importance of the **attributes/decision criteria/goals** so that one is able to portray the effect of these **attributes/decision criteria/goals** on the **decisions/alternatives**.
4. The fourth step involves processing the numerical values to determine a ranking of each **decision/alternative**.
5. Finally, if the analysis has to be repeated, depending on the answers that are not in line with practical situations, then, redo steps 1–4.

In MCDM problems, other important concepts that are important and should be remembered so that these concepts are taken into consideration when doing such an analysis are (i) the presence of conflict among attributes/decision criteria/goals, (ii) presence of incommensurate units, (iii) decision weights, w_j, where $j = 1, \ldots, n$, and (iv) the decision matrix

$$\boldsymbol{A} = \begin{bmatrix} a_{11} & \cdots & a_{1n} \\ \vdots & \ddots & \vdots \\ a_{m1} & \cdots & a_{mn} \end{bmatrix}.$$

Under the MCDM technique, the framework for the MCDM problem is as follows. We have a set of m decisions/alternatives denoted by A_i, where $i = 1, \ldots, m$, along with a set of n attributes/decision criteria/goals denoted by C_j, where $j = 1, \ldots, n$, associated with each of the decision/alternative. We assume that the decision maker has determined (in an absolute or relative sense) the performance values a_{ij}, where $i = 1, \ldots, m$, $j = 1, \ldots, n$, for each decision/alternative in terms of each attribute/decision criterion/goal, that is, a decision matrix

$$\boldsymbol{A} = \begin{bmatrix} a_{11} & \cdots & a_{1n} \\ \vdots & \ddots & \vdots \\ a_{m1} & \cdots & a_{mn} \end{bmatrix},$$

along with criteria weights w_j. Given this, our aim is to devise rules of MCDM to compare decisions/alternatives $(A_1, \ldots, A_m) \in \boldsymbol{D}$, utilizing attributes/decision criteria/goals $(C_1, \ldots, C_n) \in \boldsymbol{C}$, where each of the ith decision/alternative is characterized by its own set of n attributes/criteria denoted by $(A_{i,1}, \ldots, A_{i,n})$. Thus, for two decisions/alternatives, A_i and A_j, where $(A_i, A_j \in \boldsymbol{D})$, a comparison of their respective kth attribute/decision criterion/goal is characterized by $g(A_{i,k}, A_{j,k})$, where $k \in n$, which signifies a functional mapping portraying the benefit that accrues from the kth attribute/decision criterion/goal, if one takes the ith decision/alternative rather than the

jth one. Thus, the preference for A_i over A_j is based on the fact that whether the collective functional mapping, $\sum_{\forall k \in n} g(d_{i,k}, d_{j,k})$, is positive or otherwise. In case one compares the jth decision/alternative with the ith one, then, the benefit is denoted by $g(d_{j,k}, d_{i,k})$. It is usually true that $g(d_{i,k}, d_{j,k}) \neq g(d_{j,k}, d_{i,k})$. After finding $\sum_{\forall k \in n} g(d_{i,k}, d_{j,k})$, a sort of outranking concept is utilized where we use *pair* wise comparison to make a decision to choose between decisions/alternatives. During the comparison mapping, one constructs a *concordance* coefficient, $c_{ij} = \sum_{\forall k \in n_1} g(d_{i,k}, d_{j,k})$, where $n_1 \in n$, which signifies a conglomeration of all *positive* functional mapping for those attributes/decision criteria/goals when A_i is considered over A_j, while the *discordance* coefficient $dc_{ij} = \sum_{\forall k \in \{n-n_1\}} g(d_{i,k}, d_{j,k})$ signifies the so-called *negative* worth emanating from those attributes/decision criteria/goals when one is forced to choose A_i and not A_j. An important point to remember for the weights is the fact that depending on the concept of utility, we may need to normalize them, such that the *normalized* weights, w_j^*, should add up to 1, that is, $\sum_{j=1}^{n} w_j^* = 1$.

According to the definition of Zimmermann (1996), let $\boldsymbol{D} = \{A_i : i = 1, \ldots, m\}$ be a finite set of decisions/alternatives and $\boldsymbol{C} = \{C_j : j = 1, \ldots, n\}$ be the finite set of attributes/decision criteria/goals according to which the desirability of an action is judged. Then, we need to determine an optimal decision/alternative, $A^* \in \boldsymbol{D}$ with the highest degree of desirability with respect to all the relevant attributes/decision criteria/goals.

5.1.5 Methods of MCDM

The classification of MCDM is done according to the type of data that can be (i) deterministic, (ii) stochastic, and (iii) fuzzy. If the classification is made according to the number of decision makers, then, we have the single decision maker MCDM and the group of decision makers in MCDM. With respect to the relevance of the chapter in the overall context of the title that is *Decision Sciences: Theory and Practice*, we discuss a few of the MCDM methods that have a practical relevance in real sense, and they are

1. Weighted sum model (WSM)
2. Weighted product model (WPM)
3. Analytical network process (ANP) and analytical hierarchy process (AHP)
4. Elimination and choice-translating reality (ELECTRE)
5. Preference-ranking organization method for enrichment evaluation (PROMETHEE)
6. Technique for order preference by similarity to ideal solution (TOPSIS)
7. VIseKriterijumska Optimizacija I Kompromisno Resenje (VIKOR)
8. Potentially all pairwise rankings of all possible alternatives (PAPRIKA)
9. Measuring attractiveness by a categorical-based evaluation technique (MACBETH)
10. Statistical decision trees.

5.1.5.1 Weighted Sum Model

Example 5.6

Consider Ms Y. Takada who is working with Mitsubishi after her bachelors, has calls for admissions for a master of business administration (MBA) degree from six different universities/institutes, namely University of Toronto (A_1), Harvard Business School (HBS) (A_2),

National University of Singapore (NUS) (A_3), Hong Kong University of Science and Technology (HKUST) (A_4), Indian Institute of Management (IIM) Calcutta (A_5), and Indian School of Business (ISB) (A_6). Her decision of enrolling into the MBA program is based on a few factors that are cost of education (C_1), quality of the faculty (C_2), international exposure (C_3), the prospect of future placement (C_4), and academic rigor (C_5). For this problem, the universities/institutes are the decision/alternatives (A_i, $i = 1, 2, 3, 4, 5, 6$, $m = 6$), while the factors are the attributes/decision criteria/goals (C_j, $j = 1, 2, 3, 4$, $n = 5$), based on which Ms Takada makes her choice. We solve this problem using the simplest of the method called the WSM. Let us suppose that the attribute/decision criterion/goal and the decision/alternative matrix, $\boldsymbol{A}$, be denoted as shown in Table 5.2.

If we refer to MAUT, and Table 5.1, then, the values inside the cells corresponding to A_i and C_j imply the utility that Ms Takada assigns to her combination of decisions/alternatives and attributes/decision criteria/goals. Thus, for each of the attributes/decision criteria/goals, she tries to analyze what is the benefit/utility she gets in case she prefers a particular decision/alternative. Though we consider hypothetical values, it must be remembered that the values inside the cell typically depict the preference value based on the decision maker's utility function. If we now check the weights, then, one may conclude that they throw some light on the level of importance that a particular attribute/decision criteria/goal has for the decision maker. Here, for Ms Takada, the cost of education (C_1) and international exposure (C_3) are of the highest importance. Using the formulae $A^*_{WSM} = \max_{\forall i} \sum_{j=1}^{n} a_{ij} w_j$, we get the values as $A_3(= 82) \succ A_5(= 80) \succ A_2(= 78) \equiv A_4(= 78) \succ A_6(= 76.5) \succ A_1(= 73.5)$. Thus, NUS is the best decision/alternative for Ms Takada. One should also note that the decision between HBS and HKUST is not possible as the scores are the same.

Note:

- A question that may be asked is how to normalize the values of a_{ij} that are used in WSM. A simple concept could be $A^*_{WSM} = \max_{\forall i} \sum_{j=1}^{n} \bar{a}_{ij}, w_j$, where $\bar{a}_{ij} = a_{ij} / \max_{\forall i} a_{ij}$ or $\bar{a}_{ij} = a_{ij} / \min_{\forall i} a_{ij}$.
- The ratio w_{j_1}/w_{j_2}, where $j_1, j_2 = \{1, \ldots, n\}$ is called the rate of substitution between attributes/decision criteria/goals j_1 and j_2 with other things being equal. If the utility function $u_{j_1}(.)$ decreases by δ_{j_1}, then, the corresponding utility $u_{j_2}(.)$ has to be increased by $\delta_{j_1}(w_{j_1}/w_{j_2})$. The idea is simply the marginal rate of substitution and can be easily

TABLE 5.2

Decision Matrix for Ms Y. Takada (Example 5.6)

	Criteria				
Alternatives ↓	**Cost of Education (C_1)[a]**	**Quality of Faculty (C_2)[a]**	**International Exposure (C_3)[a]**	**Prospect of Future Placement (C_4)[a]**	**Academic Rigor (C_5)[a]**
Weights ⟶	0.3	0.1	0.3	0.2	0.1
University of Toronto (A_1)	60	85	75	80	85
HBS (A_2)	50	100	90	85	90
NUS (A_3)	70	80	85	100	75
HKUST (A_4)	75	65	80	90	70
IIM Calcutta (A_5)	80	70	80	90	70
ISB (A_6)	65	85	75	90	80

[a] Implies that the points assigned are scaled on a score of 0–100.

expressed by

$$\frac{w_{j_1}}{w_{j_2}} = \frac{\left(\dfrac{\partial \sum_{j=1}^{n} \bar{a}_{ij} w_j}{\partial \bar{a}_{ij_1}}\right)}{\left(\dfrac{\partial \sum_{j=1}^{n} \bar{a}_{ij} w_j}{\partial \bar{a}_{ij_2}}\right)}.$$

- To further illustrate the principal methods of vector normalization, one can use Table 5.3. We should remember that the concept of normalization is used to make comparison on the same scale, while the type of normalization depends on the underlying utility function.

  If

$$\boldsymbol{U}(\boldsymbol{W}) = \begin{bmatrix} U(W_{11}) & \cdots & U(W_{1n}) \\ \vdots & \ddots & \vdots \\ U(W_{m1}) & \cdots & U(W_{mn}) \end{bmatrix},$$

then the normalized matrix can be represented as

$$\boldsymbol{U}(\boldsymbol{W})_{Normalized} = \begin{bmatrix} \dfrac{U(W_{11})}{f\{U(W_{11}) \otimes \cdots \otimes U(W_{m1})\}} & \cdots & \dfrac{U(W_{1n})}{f\{U(W_{1n}) \otimes \cdots \otimes U(W_{mn})\}} \\ \vdots & \ddots & \vdots \\ \dfrac{U(W_{m1})}{f\{U(W_{11}) \otimes \cdots \otimes U(W_{m1})\}} & \cdots & \dfrac{U(W_{mn})}{f\{U(W_{1n}) \otimes \cdots \otimes U(W_{mn})\}} \end{bmatrix},$$

where

$$f^{-1}\left[\frac{U(W_{11})}{f\{U(W_{11}) \otimes \cdots \otimes U(W_{m1})\}}\right] \otimes \cdots \otimes f^{-1}\left[\frac{U(W_{m1})}{f\{U(W_{11}) \otimes \cdots \otimes U(W_{m1})\}}\right] = 1.$$

This sum being 1 will be true for all other columns if we do column-wise normalization. In case it is row-wise normalization, then, the sum should be taken along the rows. Thus, for (i) $U(W) = W$, $\otimes$ is the additive sum $+$, $f(.)$ is additive, that is, Σ for any $j = 1, \ldots, n$ while for (ii) $U(W) = W^2$, $\otimes$ is the additive sum $+$, $f(.)$ is $\sqrt{\square}$, while $f^{-1}(.)$ is $(.)^2$.

TABLE 5.3

Principal Methods of Normalization Used in MCDM

Properties of $\bar{a}_{ij}$	**Procedure # 1**	**Procedure # 2**	**Procedure # 3**	**Procedure # 4**	**Procedure # 5**
Definition	$\frac{a_{ij}}{\max_{\forall i} a_{ij}}$	$\frac{a_{ij}}{\min_{\forall i} a_{ij}}$	$\frac{a_{ij} - \min a_{ij}}{\max a_{ij} - \min a_{ij}}$	$\frac{a_{ij}}{\sum_{\forall i} a_{ij}}$	$\frac{a_{ij}}{\sqrt{\sum_{\forall i} a_{ij}^2}}$
Normalized vector	$0 < \bar{a}_{ij} \leq 1$	$1 \leq \bar{a}_{ij}$	$0 \leq \bar{a}_{ij} < 1$	$0 < \bar{a}_{ij} < 1$	$0 < \bar{a}_{ij} < 1$
Modulus	Variable	Variable	No	Yes	Yes
Proportionality conserved	Yes	Yes	No	Yes	Yes
Interpretation	% of maximum	% of minimum	% of range	% of total	ith component of the unit vector

5.1.5.2 Weighted Product Model

The WPM is similar to WSM. The difference between WSM and WPM being, that in place of addition, we use the multiplicative concept. The decisions/alternatives are compared by multiplying a number of ratios, one for each attribute/decision criteria/goal. Each ratio is raised to the power equivalent to the relative weight of the corresponding criterion. Thus, to compare two decisions/alternatives, namely, A_{i_1} and A_{i_2}, i_1, $i_2 \in \{1, \ldots, m\}$, we need to calculate $R\left(A_{i_1}/A_{i_2}\right) = \prod_{j=1}^{n}\left(a_{i_1 j}/a_{i_2 j}\right)^{w_j}$. It is a dimensionless quantity and if $R\left(A_{i_1}/A_{i_2}\right) > 1$, then, $A_{i_1} \succ A_{i_2}$; furthermore, instead of the actual values, one can also use the relative values (using the concept of normalization as mentioned in Table 5.3). If we use the WPM model for Example 5.6, then, we obtain the ranking matrix values using the formulae $\prod_{j=1}^{n}\left(a_{i_1 j}/a_{i_2 j}\right)^{w_j}$, using which the ranks are now $A_3 \succ A_5 \succ A_4 \succ A_6 \succ A_2 \succ A_1$. For the convenience of the reader, the values based on which we arrive at the result are provided in a matrix notion, that is, Table 5.4.

An interesting thing to note is the fact that the ranking of the institutes where Ms Takada would like to study is different for the WSM and WPM methods. An alternative approach to WPM is to take only the products and not the ratios, that is, $\prod_{j=1}^{n}\left(a_{i_1 j}\right)^{w_j}$.

Note: In statistics, we know the importance of loss functions that are used to estimate the distribution parameters. Different references can be found regarding squared error loss (SEL) function, linear exponential (LINEX) loss function, lin-lin loss function, etc., and work has been done where combinations of different loss functions have also been considered. Keeping in mind that, one may attempt to use the weighted aggregated sum product assessment (WASPAS) method for ranking of alternatives where we obtain the ranking based on

$$\lambda \sum_{j=1}^{n} \bar{a}_{ij} w_j + (1 - \lambda) \prod_{j=1}^{n} (\bar{a}_{ij})^{w_j},$$

where $\lambda \in [0, 1]$ is the weight assigned to the WSM.

5.1.5.3 ANP and AHP

5.1.5.3.1 Analytical Network Process

Many decision problems cannot be structured hierarchically because of the presence of dependence and interaction between higher-level and lower-level elements. This results in decisions/alternatives (A_i) and attributes/decision criteria/goals (C_j) being interlinked. As one can guess by the name, hierarchy is a linear top-down structure, while a network spreads out in all directions and involves cycles between clusters and loops within the same cluster. The feedback structure in networks does

TABLE 5.4

WPM Values for Example 5.5

i ⟶	1	2	3	4	5	6
$R(A_i/A_1)$	1	1.034680474	1.116104035	1.065686539	1.094602827	1.042380414
$R(A_i/A_2)$	0.966481948	1	1.078694401	1.029966801	1.057913872	1.007441853
$R(A_i/A_3)$	0.895973824	0.927046621	1	0.954827243	0.980735481	0.933945566
$R(A_i/A4)$	0.938362233	0.97090508	1.047309874	1	1.027133953	0.978130413
$R(A_i/A_5)$	0.913573376	0.945256534	1.019642931	0.973582849	1	0.952290994
$R(A_i/A_6)$	0.959342661	0.99261312	1.070726214	1.02235856	1.050099189	1

not have the linear top-to-bottom form of a hierarchy but looks more like an actual network. In other words, ANP is a generalization of the AHP, where ANP is an influence network of clusters and nodes contained within the clusters, and priorities are established as in AHP using pairwise comparisons and judgment. In many practical situations, decision problems cannot be structured hierarchically because they involve the interaction and dependence of higher-level elements in a hierarchy on lower-level elements. Not only does the importance of the criteria determine the importance of the alternatives as in a hierarchy, but also the importance of the alternatives themselves determines the importance of the criteria. This sort of feedback enables us to factor the future into the present to determine the final result. As a decision-making process, so far, ANP has been neglected because of the linear structure used in traditional approaches and their inability to deal with feedbacks when choosing decisions/alternatives.

As already mentioned, ANP, Saaty (1996, 2004), is a more general form of the AHP and is used in MCDA. In ANP, one has a goal (which is to be achieved) and for that, we have some alternatives/decisions, a combination of which helps us to attain the goal. Each alternative/decision has some inherent properties that we classify as an attribute/decision criterion/goal. Thus, put simply, ANP models the problem as a network and uses pairwise comparison.

A few good references for pairwise comparison and calculating weights are the eigenvector method (Saaty 2003; Saaty and Hu 1998), modified eigenvector method (Cogger and Yu 1985), direct least-squares method (Chu et al. 1979), weighted least-squares method (Chu et al. 1979), logarithmic least-squares method (Crawford and Williams 1985), logarithmic least absolute-values method (Cook and Kress 1988), logarithmic GP method (Bryson 1995), and fuzzy preference programming method (Mikhailov 2004). For a detailed study of these techniques, one may check Golany and Kress (1993), Ishizaka and Lusti (2006), and Srdjevic (2005). Khademi et al. (2014) developed an algorithm that may be used to derive the appropriate criteria set based on the predetermined alternative and control criteria sets of the ANP problem. Furthermore, the authors show how the network of the ANP should be constructed systematically to give the correct overall weights.

According to Saaty and Vargas (2006), in ANP, a source node is an origin of the paths of influence (importance) and never a destination of such paths. A sink node is a destination of the paths of influence and never an origin of such paths. Thus, a full network can include source nodes; intermediate nodes that fall on paths from source nodes, lie on cycles, or fall on paths to sink nodes; and finally sink nodes, while some ANP networks can contain only source and sink nodes. Pictorially, an ANP structure may be represented as shown in Figure 5.6, where we have the source component, C_4, sink component, C_2, and transient components, C_1, C_2, C_3, and C_5. Moreover, we have a cycle between C_2 and C_5, while C_1 and C_3 are loops.

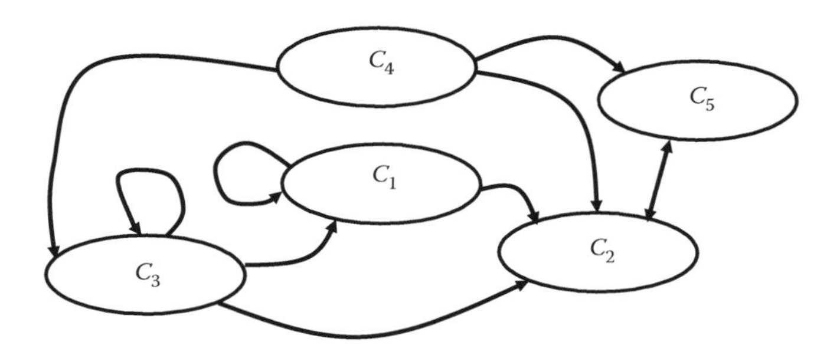

FIGURE 5.6
Illustrative ANP diagram. (With kind permission from Springer Science+Business Media: *Decision Making with the Analytic Network Process: Economic, Political, Social and Technological Applications with Benefits, Opportunities, Costs and Risks*, 2006, Saaty, T. L. and Vargas, L. G.)

For a better understanding, let us denote a component of a decision network by C_i, where $i = 1, \ldots, m$ and assume that it has n_i elements denoted by $e_{i1}, \ldots, e_{im_i}$. Thus, the elements for C_1 are $e_{11}, \ldots, e_{1m_1}$, for C_2, they are $e_{21}, \ldots, e_{2m_2}, \ldots$, and finally for C_m, they are $e_{m1}, \ldots, e_{mm_m}$. Given this, we would like to derive the priority vector from paired comparisons where these comparisons depict the influences of a given set of elements in a component on any element in the system. When making the comparison, we should keep in mind whether the influence is flowing from the parent element to the elements being compared, or the other way around. Once we decide on what is the flow of comparison, we should follow that throughout the comparison process. The resulting supermatrix (which denotes these relationships) is what is used for ANP and in general represents the influence of elements in the network on other elements in that network.

Consider that Mr S. Agarwal is planning to buy a car. He has choices that are Volkswagen (VW), Jaguar, Bayerische Motoren Werke (BMW), Audi, Porsche, and Toyota. To buy the car of his choice, he considers the following, which he thinks are the important factors to be considered by him during the decision process. The factors are price/cost, safety features, capacity/space, and other factors. Price/cost according to him is affected by the loan that he can get, the resale value of the car, and the average cost of maintenance. Under safety features, the important factors that are of concern to Mr Agarwal are crash proof worthiness, braking capacity, comfort, and other general safety features. Finally, under other factors, he may consider the color of the car, the status symbol of maintaining a particular car, and finally the choice of his family members. If we assume the values $e_{i1}, \ldots, e_{im_i}$, then, the supermatrix may be represented by Table 5.5.

The above supermatrix is of 18 × 18. In case we denote it with submatrices, then, they would be represented as shown above. In case the cost of maintenance under the price category consists of

TABLE 5.5

Illustration of the Supermatrix for ANP

		Car						Price			Safety				Space		Other Factors		
		C1	C2	C3	C4	C5	C6	P1	P2	P3	S1	S2	S3	S4	Sp1	Sp2	O1	O2	O3
Car	C1	$e_{1,1}$	$e_{1,2}$	$e_{1,3}$	$e_{1,4}$	$e_{1,5}$	$e_{1,6}$	$e_{1,7}$	$e_{1,8}$	$e_{1,9}$	$e_{1,10}$	$e_{1,11}$	$e_{1,12}$	$e_{1,13}$	$e_{1,14}$	$e_{1,15}$	$e_{1,16}$	$e_{1,17}$	$e_{1,18}$
	C2	$e_{2,1}$	$e_{2,2}$	$e_{2,3}$	$e_{2,4}$	$e_{2,5}$	$e_{2,6}$	$e_{2,7}$	$e_{2,8}$	$e_{2,9}$	$e_{2,10}$	$e_{2,11}$	$e_{2,12}$	$e_{2,13}$	$e_{2,14}$	$e_{2,15}$	$e_{2,16}$	$e_{2,17}$	$e_{2,18}$
	C3	$e_{3,1}$	$e_{3,2}$	$e_{3,3}$	$e_{3,4}$	$e_{3,5}$	$e_{3,6}$	$e_{3,7}$	$e_{3,8}$	$e_{3,9}$	$e_{3,10}$	$e_{3,11}$	$e_{3,12}$	$e_{3,13}$	$e_{3,14}$	$e_{3,15}$	$e_{3,16}$	$e_{3,17}$	$e_{3,18}$
	C4	$e_{4,1}$	$e_{4,2}$	$e_{4,3}$	$e_{4,4}$	$e_{4,5}$	$e_{4,6}$	$e_{4,7}$	$e_{4,8}$	$e_{4,9}$	$e_{4,10}$	$e_{4,11}$	$e_{4,12}$	$e_{4,13}$	$e_{4,14}$	$e_{4,15}$	$e_{4,16}$	$e_{4,17}$	$e_{4,18}$
	C5	$e_{5,1}$	$e_{5,2}$	$e_{5,3}$	$e_{5,4}$	$e_{5,5}$	$e_{5,6}$	$e_{5,7}$	$e_{5,8}$	$e_{5,9}$	$e_{5,10}$	$e_{5,11}$	$e_{5,12}$	$e_{5,13}$	$e_{5,14}$	$e_{5,15}$	$e_{5,16}$	$e_{5,17}$	$e_{5,18}$
	C6	$e_{6,1}$	$e_{6,2}$	$e_{6,3}$	$e_{6,4}$	$e_{6,5}$	$e_{6,6}$	$e_{6,7}$	$e_{6,8}$	$e_{6,9}$	$e_{6,10}$	$e_{6,11}$	$e_{6,12}$	$e_{6,13}$	$e_{6,14}$	$e_{6,15}$	$e_{6,16}$	$e_{6,17}$	$e_{6,18}$
Price	P1	$e_{7,1}$	$e_{7,2}$	$e_{7,3}$	$e_{7,4}$	$e_{7,5}$	$e_{7,6}$		$e_{3\times3}$			$e_{3\times4}$			$e_{3\times2}$			$e_{3\times3}$	
	P2	$e_{8,1}$	$e_{8,2}$	$e_{8,3}$	$e_{8,4}$	$e_{8,5}$	$e_{8,6}$												
	P3	$e_{9,1}$	$e_{9,2}$	$e_{9,3}$	$e_{9,4}$	$e_{9,5}$	$e_{9,6}$												
Safety	S1	$e_{10,1}$	$e_{10,2}$	$e_{10,3}$	$e_{10,4}$	$e_{10,5}$	$e_{10,6}$		$e_{4\times3}$			$e_{4\times4}$			$e_{4\times2}$			$e_{4\times3}$	
	S2	$e_{11,1}$	$e_{11,2}$	$e_{11,3}$	$e_{11,4}$	$e_{11,5}$	$e_{11,6}$												
	S3	$e_{12,1}$	$e_{12,2}$	$e_{12,3}$	$e_{12,4}$	$e_{12,5}$	$e_{12,6}$												
	S4	$e_{13,1}$	$e_{13,2}$	$e_{13,3}$	$e_{13,4}$	$e_{13,5}$	$e_{13,6}$												
Space	Sp1	$e_{14,1}$	$e_{14,2}$	$e_{14,3}$	$e_{14,4}$	$e_{14,5}$	$e_{14,6}$		$e_{2\times3}$			$e_{2\times4}$			$e_{2\times2}$			$e_{2\times3}$	
	Sp2	$e_{15,1}$	$e_{15,2}$	$e_{15,3}$	$e_{15,4}$	$e_{15,5}$	$e_{15,6}$												
Other factors	O1	$e_{16,1}$	$e_{16,2}$	$e_{16,3}$	$e_{16,4}$	$e_{16,5}$	$e_{16,6}$		$e_{3\times3}$			$e_{3\times4}$			$e_{3\times2}$			$e_{3\times3}$	
	O2	$e_{17,1}$	$e_{17,2}$	$e_{17,3}$	$e_{17,4}$	$e_{17,5}$	$e_{17,6}$												
	O3	$e_{18,1}$	$e_{18,2}$	$e_{18,3}$	$e_{18,4}$	$e_{18,5}$	$e_{18,6}$												

two more subgroups, then, that would also be expanded with the corresponding values of a 2×2 matrix. One should remember that this supermatrix is not symmetric and also some e_{ij} values may be zero depending on whether there is any linkage between the nodes.

Example 5.7*

To understand ANP, we discuss a very interesting problem in all its technical details. The model is available in the *Super Decision Software* http://www.superdecisions.com/ and the ANP package can be downloaded/purchased. The file used for this analysis is ***NationalMissileDefense.mod***. The model was presented by Thomas L. Saaty at the 6th International Symposium on AHP in Bern, Switzerland, 2001. This model is a complete multilayer structure with benefits, opportunities, costs, and risks (BOCR) merit nodes in the top-level network and control criteria in their attached subnets. The bottom-level decision subnets contain the alternatives that are attached to the control criteria. The top-level model also has an attached ratings component for evaluating the importance of the BOCR characteristics. An overview of the model structure is given in Figure 5.7.

The background of the problem is related to the United States government decision dilemma it faces in making the crucial choice of whether or not to commit plans regarding the deployment of a national missile defense (NMD) system. Different expert opinions in politics, military, and academia have different views regarding the deployment of the NMD decision; hence, the need of ANP (as one of the many methods which may be used to solve such problems) to consider all these opinions collectively and rationally.

The following alternatives and criteria for evaluating the decision may be identified and they are

1. Deploy NMD fully
2. Amend the antiballistic missile (ABM) treaty
3. Spend more resources on research and development in the area of missile defense technology
4. Terminate the NMD program

For the evaluation process in the software model, one needs to create the following, which are

Top-level network: It has the BOCR nodes and the strategic criteria used to evaluate their importance for this decision.

Control criteria networks: Each of the BOCR in this stage has a subnet attached to it containing its control criteria, which are generally hierarchical. Subnets are also created depending on the importance of the criterion.

Decision networks: A decision subnet is created for each high-value control criterion, where the alternatives are clustered in each decision subnet.

Next, one undertakes the assessment criteria step to determine the priorities of the BOCR merits that are then found and shown in Figure 5.8. After the initial set of runs, one obtains Table 5.6 that shows the 23 control criteria under the BOCRs and their corresponding priorities.

One should note that the merit values obtained in the first column (Table 5.6) are found using the calculations/priorities command in the ratings. The criteria and subcriteria values are also obtained by going into the control criteria subnet (e.g., benefits), using the computations, unweighted supermatrix command, and reading the values from it. The global priorities values are then obtained using

* The general analysis and discussion for Example 5.7 has been taken from http://www. superdecisions.com/

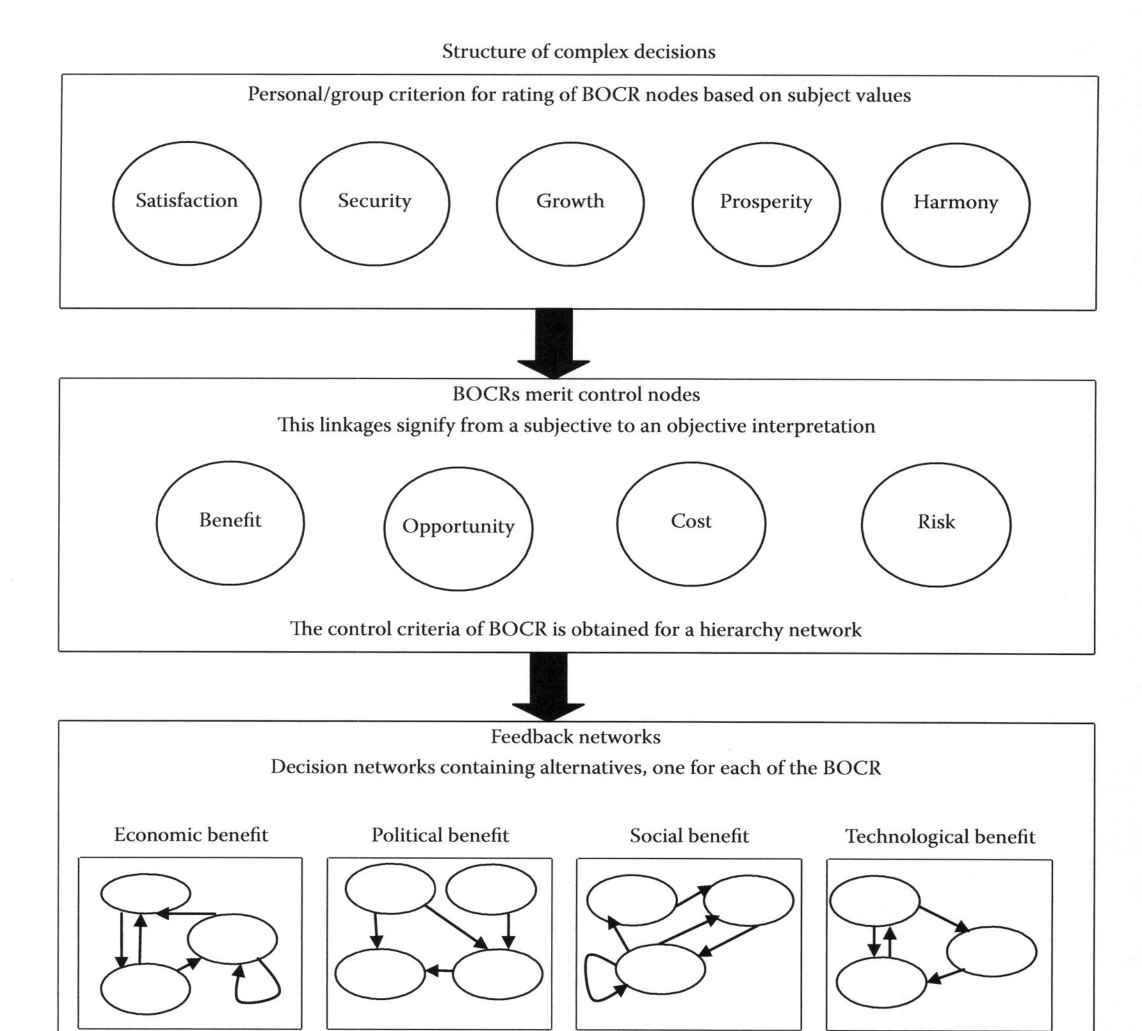

FIGURE 5.7
Schematic diagram of the structure for Example 5.7 using AHP.

Excel. For example, the first one, 0.006, is obtained by multiplying $0.264 \times 0.157 \times 0.141$. The 23 criteria are in the control criteria subnets attached to the BOCR, and they were prioritized through pairwise comparisons. Among these 23 criteria, the sum of the priorities of nine of them, the security threat, arms sales, technical failure, military capability, technological advancement, sunk cost, spin off, arms race, and further investment, account for more than 0.76 of the total. To economize the effort, we used these nine to do the analysis. We renormalize their priorities within their respective merits and continue likewise.

Table 5.7 shows the final synthesis of the alternatives for each of the BOCR merits and the overall result, using the reciprocals of the synthesized priorities of costs and risks.

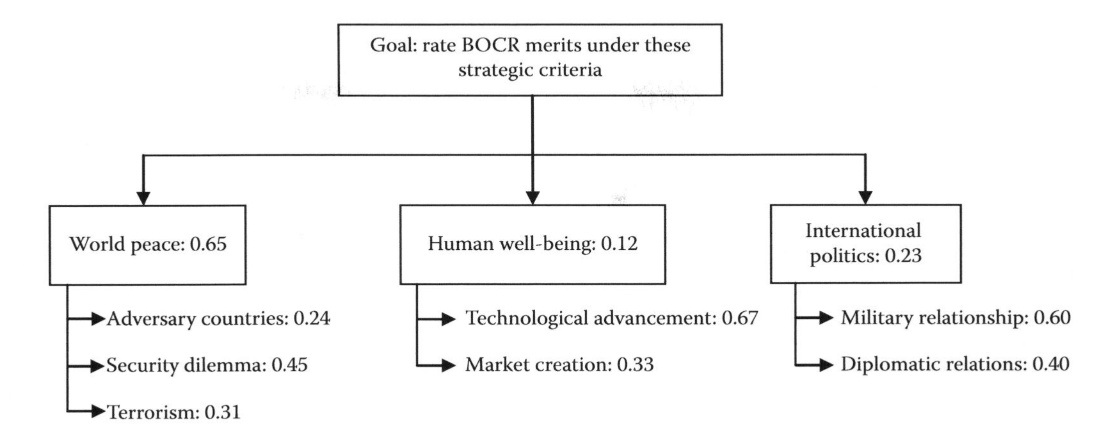

FIGURE 5.8
Strategic criteria for rating BOCRs.

TABLE 5.6
Criteria and Their Priorities

Merits	Criteria	Subcriteria	Local Priorities	Global Priorities
Benefits (0.264)	Economic (0.157)	Local economy	0.141	0.006
		Defense industry	0.859	0.036
	Political (0.074)	Bargaining power	0.859	0.017
		U.S. military leadership	0.141	0.003
	Security (0.481)	Deterrence	0.267	0.034
		Military capability	0.590	**0.076**
		Antiterrorism	0.143	0.018
	Technology (0.288)	Technological advancement	0.834	**0.064**
		Technological leadership	0.166	0.013
Opportunities (0.185)		Arms sales	0.520	**0.094**
		Spin-off	0.326	**0.059**
		Space development	0.051	0.009
		Protection of allies	0.103	0.019
Costs (0.363)	Security (0.687)	Security threat (vulnerability to the security threat)	1.000	**0.248**
	Economic (0.228)	Sunk cost	0.539	**0.044**
		Further investment	0.461	**0.038**
	Political (0.085)	ABM treaty	0.589	0.018
		Foreign relations	0.411	0.013
Risks (0.188)		Technical failure	0.430	**0.082**
		Arms race	0.268	**0.051**
		Increased terrorism	0.052	0.010
		Environmental damage	0.080	0.015
		U.S. reputation	0.170	0.032

TABLE 5.7
Final Outcome for the NMD Problem

	Benefits (0.264)	Opportunities (0.184)	invCosts (0.363)	invRisks (0.188)	Final Outcome Using Additive (Norm.)	Final Outcome Using Multiplicative (Norm.)
Deploy NMD	0.434	0.473	0.306	0.116	**0.331**	**0.493**
Global defense	0.357	0.290	0.305	0.178	0.291	0.379
R & D	0.161	0.151	0.236	0.289	0.212	0.110
Termination	0.049	0.085	0.153	0.417	0.165	0.018

The final outcome is calculated in two ways, using both an additive and a multiplicative formula. The effect of using the multiplicative formula is that the weights of the BOCR cancel out due to the form of the formula; so, in effect, they are equal to 0.25. The reason that the Deploy option is so much better under the multiplicative formula is that costs are very high for that alternative. Thus, when costs are weighted by 0.363, as they are in the additive formula, it drags down the value of the Deploy option.

To end this discussion about ANP, one can summarize that this method is generally composed of four major steps that are (i) model construction and problem structuring, (ii) pairwise comparison matrices and priority vectors, (iii) supermatrix formation, and finally (iv) selection of the best alternatives. In the literature, one comes across few algorithms for solving the ANP-modeled problem. Finally, to understand the whole range of a variety of ANP applications, one can refer to the book by Saaty and Ozdemir (2005).

5.1.5.3.2 Analytical Hierarchy Process

The AHP with its *independence* assumptions on the upper levels from the lower levels and the independence of the elements in a level is a special case of the ANP. AHP was also developed by Saaty (1994, 2005), Saaty and Vargas (2000), and is an MADM approach that simplifies complex and not that well-structured problems by arranging the decision attributes and alternatives. Figure 5.9 shows the general structure for AHP.

As stated by Saaty and Vargas (2006), the fundamental scale of values to represent the intensities of judgments is shown in Table 5.8. The scale has been derived through a stimulus response theory

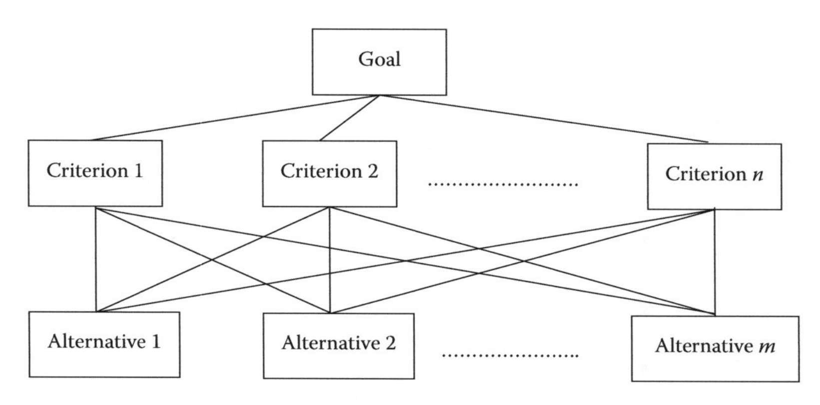

FIGURE 5.9
General structure for AHP.

TABLE 5.8

Scaling of Absolute Numbers as Used in ANP and AHP

Intensity of Importance	Definition	Explanation
1	Equal importance	Two activities equally contribute to the objective
2	Weak	
3	Moderate importance	Experience and judgment slightly favor one activity over another
4	Moderate plus	
5	Strong importance	Experience and judgment strongly favor one activity over another
6	Strong plus	
7	Very strong or demonstrated importance	An activity is favored very strongly over another; its dominance is demonstrated in practice
8	Very, very strong	
9	Extreme importance	The evidence favoring one activity over another is of the highest-possible order of affirmation

and validated for effectiveness, not only in many applications by a number of people, but also through a theoretical justification of what scale one must use in the comparison of homogeneous elements. For the convenience of the readers, Table 5.8 gives the fundamental scale of absolute numbers.

Note: If activity i has one of the above nonzero numbers assigned to it when compared with activity j, then, j has the reciprocal value when compared with i. This is a genuine and a reasonable assumption.

Example 5.8

Consider that we have the first matrix related to the decisions/alternatives ($A_i, i = 1, 2, 3$) and attributes/decision criteria/goals ($C_j, j = 1, 2, 3, 4$). Consider that we use the column-wise linear normalization concept, and the data are shown in Table 5.9 (refer to part (a)).

If we use the concept of $\sum_{j=1}^{3} \bar{a}_{ij} w_j$, for the data given in Table 5.9a, then, $A_2 \succ A_1 \succ A_3$. Now, a ranking inconsistency can occur when AHP is used. Consider another alternative, say A_4, is added in the problem. The root cause for inconsistencies is due to the fact that the relative values of each criterion sum up to one, due to which the ranking, using $\sum_{j=1}^{4} \bar{a}_{ij} w_j$, now results in $A_1 \succ A_2 \equiv A_4 \succ A_3$. Rather than using the linear normalization, if we utilize normalization using the maximum value, then, we get $A_2 \equiv A_4 \succ A_1 \succ A_3$. Similarly, different results are obtained depending on the normalization concept used. Hence, in AHP or any other ranking process, there should be consistency in the ranking process that implies coherency in the judgment process for the decision maker. Given a problem, any decision maker has to make a judgment in many practical situations and the property of consistency may not always hold true as illustrated in Example 5.8.

5.1.5.3.3 Concept of Consistency in Ranking

Continuing further, consistency in raking implies that a rational human being under whatever circumstances will arrive at the same set of conclusions regarding the priorities among the decisions/alternatives, in spite of the fact that he/she is using separate methodologies when analyzing

TABLE 5.9

Hypothetical Example to Illustrate (a) AHP and (b) Revised AHP

	Criteria		
	C_1	C_2	C_3
Alternatives ↓	$w_1 = (1/3)$	$w_2 = (1/3)$	$w_3 = (1/3)$
(a) AHP			
A_1	$\frac{1}{11}$	$\frac{9}{11}$	$\frac{8}{18}$
A_2	$\frac{9}{11}$	$\frac{1}{11}$	$\frac{9}{18}$
A_3	$\frac{1}{11}$	$\frac{1}{11}$	$\frac{1}{18}$
Relative weights	1	1	1
(b) Revised AHP			
A_1	$\frac{1}{20}$	$\frac{9}{12}$	$\frac{8}{27}$
A_2	$\frac{9}{20}$	$\frac{1}{12}$	$\frac{9}{27}$
A_3	$\frac{1}{20}$	$\frac{1}{12}$	$\frac{1}{27}$
A_4	$\frac{9}{20}$	$\frac{1}{12}$	$\frac{9}{27}$
Relative weights	1	1	1

the decisions/alternatives. By the word separate methodology, we mean that the fundamental principle of utility or the normalization concept being used is not the same. The idea of consistency is a fundamental concept based on which any decision-making process is undertaken, when pairwise comparison is used. Without going into the details, we state a few important properties with respect to the consistency concept and they are

1. For consistency, we should always have $a_{ij} \times a_{jk} = a_{ik}$, where $i, j, k = 1, \ldots, n$
2. A 2×2 matrix, $A_{2\times 2}$, is always consistent
3. Consistency implies that all rows/columns are linearly independent
4. $A_{m\times n} w_{n\times 1} = m w_{n\times 1}$ iff

$$A_{m\times n} = \begin{bmatrix} a_{11} & \cdots & a_{1n} \\ \vdots & \ddots & \vdots \\ a_{m1} & \cdots & a_{mn} \end{bmatrix}$$

is consistent. As an example, consider

$$A_{m\times n} = \begin{bmatrix} 1 & \cdots & \frac{w_1}{w_m} \\ \vdots & \ddots & \vdots \\ \frac{w_m}{w_1} & \cdots & 1 \end{bmatrix},$$

then, it is consistent with

$$\begin{bmatrix} 1 & \cdots & \frac{w_1}{w_m} \\ \vdots & \ddots & \vdots \\ \frac{w_m}{w_n} & \cdots & 1 \end{bmatrix} \begin{bmatrix} w_1 \\ \vdots \\ w_m \end{bmatrix} = \begin{bmatrix} m w_1 \\ \vdots \\ m w_m \end{bmatrix} = m \begin{bmatrix} w_1 \\ \vdots \\ w_m \end{bmatrix}.$$

To check the consistency, we use the following method that is:

1. From the matrix $\boldsymbol{A}$, the normalized principal eigenvector is obtained.
2. This normalized principal eigenvector is also called the priority vector, and it shows the relative weights among the attributes that are to be compared.
3. For the consistency check, we calculate the principal eigenvalue, where the principal eigenvalue is obtained from the summation of products between each element of the eigenvector and the sum of columns of the reciprocal matrix.

Note: Saaty gave a measure of consistency, called the consistency index (CI) that is the deviation or degree of consistency using the following formula: $CI = (\lambda_{max} - n)/(n - 1)$. Remember that using CI, random CI, and the consistency ratio, one can decide whether a decision may be termed as consistent or not.

Example 5.9

Mrs Rosanwala wants to buy an apartment (given her budgetary constraints as well as her liking/preference). The attributes/decision criteria/goals being considered by her are (i) cost, (ii) location, (iii) availability of amenities, and (iv) the number of rooms. The cost attribute/decision criterion/goal is further subdivided into (i) purchase price, (ii) ease of availability of the loan from banks, (iii) maintenance cost, and (iv) resale value, while the attribute/decision criterion/goal, that is the number of rooms is further broken down into (i) the size of the living room and (ii) size of the master bedroom. The numbers of decision/alternatives are in different localities of Mumbai that are Dadar, Bandra, Andheri, Juhu, Powai, and Goregaon. Thus, the hierarchy for Mrs Rosanwala's decision of buying an apartment can be visualized as shown in Figure 5.10.

Furthermore, based on her personal judgment, we assume the ranking criteria scaling scores as cost versus location: 3 vs $\frac{1}{3}$, cost versus availability of amenities: 7 vs $\frac{1}{7}$, cost versus number of rooms: 3 vs $\frac{1}{3}$, location versus availability of amenities: 9 vs $\frac{1}{9}$, location versus number of rooms; 1 vs 1, and availability of amenities versus number of rooms: $\frac{1}{7}$ vs 7. On the basis of the first stage of the scaling points assigned, one gets the scores for the primary criteria as cost: 0.504, location: 0.237, availability of amenities: 0.042, and number of rooms: 0.217. To double check

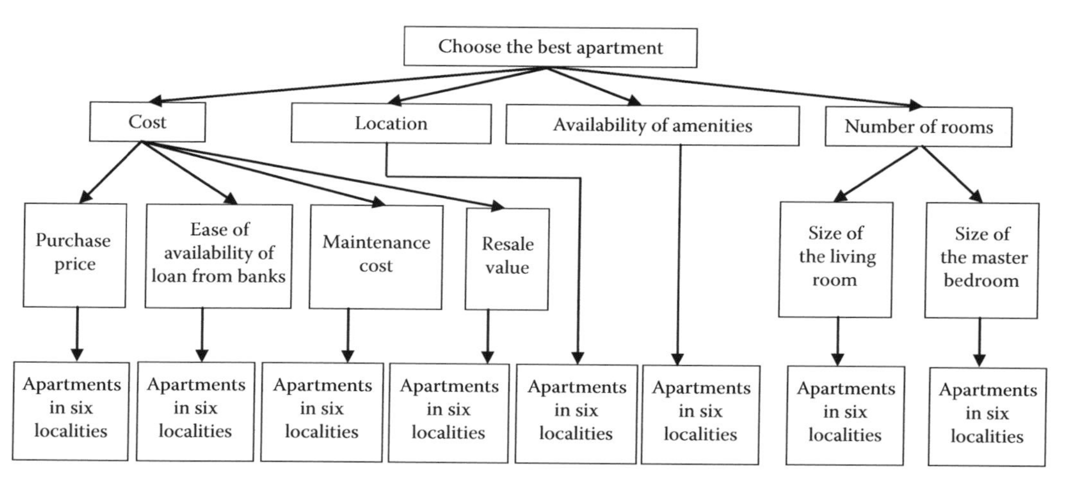

FIGURE 5.10
AHP hierarchy for Mrs Rosanwala (Example 5.9).

the validity of the values, one needs to check that they sum up to 1. In the next stage, consider that Mrs Rosanwala is able to compare the subcriteria related to the cost and number of rooms. On the basis of her information, we obtain the weights as purchase price = 0.246, ease of availability of the loan from banks = 0.127, maintenance cost = 0.050, resale value = 0.081, size of the living room = 0.036, and size of the master bedroom = 0.181, whereby the normalization function used is $\left(\prod_{j=1}^{n} a_{ij}\right)^{n}$.

Next, we need to compare the six decisions/alternatives with respect to the criteria and as the first step consider the cost criterion. Assume that the cost (in INR) of an average apartment at the above-mentioned locations is Dadar = $6,311,600$, Bandra = $9,637,900$, Andheri = $8,554,450$, Juhu = $6,417,000$, Powai = $5,883,800$, and Goregaon = $7,949,950$, while the budgetary constraint for Mrs Rosanwala is 7,750,000. Hence, any apartment's cost below her budgetary constraint is definitely manageable for her as she does not have to take any loan, while a cost above is financially a burden as she has to take a loan from the bank at an existing interest rate of 10.40%. On the basis of the above information, Table 5.10 shows the local and global rating for the localities based on price only. In deriving the scores, we assume that Mrs Rosanwala can exceed her budget by an amount of INR 3,10,000, which means that she is comfortable if the budget increases to a maximum of INR 3,10,000. Now, the cost for a decision/alternative, in this case an apartment in a location, if it is more than INR 8,060,000, is too high to be considered, but Mrs Rosanwala does not immediately discard the decision/alternative but puts a low score on those decision/alternatives that exceed the cost price of INR 8,060,000. On the other hand, a cost price of INR 7,440,000 or INR 4,650,000, for the apartment, would be a respective, yes and a strong yes from her side. One should remember that the local priorities show how much the cost price of an apartment contributes to the subcriterion of purchase price, while on the other hand, the global priorities illustrate how much the cost price of each apartment contributes to the overall goal of choosing the best apartment for Mrs Rosanwala.

Furthermore, the location is very important for Mrs Rosanwala, as she is a working lady; hence, she enquires from her friends residing in different localities, plus gets information from brokers to arrive at a set of scaling scores (uses the basic information provided in Table 5.8) when comparing localities. The information of pairwise comparison for the location is as follows: Dadar versus Bandra: 1 vs 1, Dadar versus Andheri: 5 vs (1/5), Dadar versus Juhu: 7 vs (1/7), Dadar versus Powai: 9 vs (1/9), Dadar versus Goregaon: (1/3) vs 3, Bandra versus Andheri: 5 vs (1/5), Bandra versus Juhu: 7 vs (1/7), Bandra versus Powai: 9 vs (1/9), Bandra versus Goregoan: (1/3) vs 3, Andheri versus Juhu: 2 vs (1/2), Andheri versus Powai: 9 vs (1/9), Andheri versus Goregaon: (1/8) vs 8, Juhu versus Powai: 2 vs (1/2), Juhu versus Goregaon: (1/8) vs 8, and Powai versus Goregaon: (1/9) vs 9. Using this scaling score, one obtains the priority score based on location and the same is given in Table 5.11.

A similar set of scaling scores are available for the number of rooms and they are Dadar versus Bandra: 1 vs 1, Dadar versus Andheri: $\frac{1}{2}$ vs 2, Dadar versus Juhu: 1 vs 1, Dadar versus Powai:

TABLE 5.10

Priority List Based on Cost Price for Example 5.9

Locality in Mumbai	Local Priority	Global Priority
Dadar	0.242	0.060
Bandra	0.027	0.007
Andheri	0.027	0.007
Juhu	0.242	0.060
Powai	0.362	0.089
Goregaon	0.100	0.025
Total	1.000	0.248

TABLE 5.11

Priority Score Based on Location for Example 5.9

Locality in Mumbai	Local Priority	Global Priority
Dadar	0.215	0.051
Bandra	0.215	0.051
Andheri	0.083	0.020
Juhu	0.038	0.009
Powai	0.025	0.006
Goregaon	0.424	0.100
Total	1.000	0.237

3 vs $\frac{1}{3}$, Dadar versus Goregaon: $\frac{1}{2}$ vs 2, Bandra versus Andheri: $\frac{1}{2}$ vs 2, Bandra versus Juhu: 1 vs 1, Bandra versus Powai: 3 vs $\frac{1}{3}$, Bandra versus Goregoan: $\frac{1}{2}$ vs 2, Andheri versus Juhu: 2 vs $\frac{1}{2}$, Andheri versus Powai: 6 vs $\frac{1}{6}$, Andheri versus Goregaon: 1 vs 1, Juhu versus Powai: 3 vs $\frac{1}{3}$, Juhu versus Goregaon: $\frac{1}{2}$ vs 2, and Powai versus Goregaon: $\frac{1}{6}$ vs 6. On the basis of this set of information, the list of local and global priorities for the attribute/decision criterion/goal (i.e., the number of rooms) is illustrated in Table 5.12.

Without going into the detailed calculations for each and every attribute/decision criterion/goal and decision/alternative, one gets the final matrix as shown in Table 5.13.

Giving all the information above, Mrs Rosanwala would buy an apartment in the area of Goregaon as the global score/priority is 0.220 for that decision/alternative. A similar ranking can also be done for all the other localities.

5.1.5.4 Elimination and Choice-Translating Reality

ELimination Et Choix Traduisant la REalité (ELECTRE), in French, that is, ELECTRE in English, Roy (1968) is one of the MCDM tools based on the concepts of the *outranking method.* In course of time, six different variants of ELECTRE (ELECTRE I, II, III, IV, Tri, and IS) methodology have been developed, whereby each variant has some unique properties and hence variations when compared with the other. Good texts related to ELECTRE as a ranking method can be found in Figueira et al. (2005), Roy (1996), and Rogers et al. (2000). For our understanding, we concentrate

TABLE 5.12

Priority Score Based on Number of Rooms for Example 5.9

Locality in Mumbai	Local Priority	Global Priority
Dadar	0.136	0.025
Bandra	0.136	0.025
Andheri	0.273	0.049
Juhu	0.136	0.025
Powai	0.046	0.008
Goregaon	0.273	0.049
Total	1.000	0.181

TABLE 5.13

Final Decision Matrix for Mrs Rosanwala for Example 5.9

	Cost						Number of Rooms		
Locality	Purchase Price	Ease of Availability of Loan	Maintenance Cost	Resale Value	Location	Availability of Amenities	Size of Living Room	Size of Master Bed room	Total
Dadar	0.060	0.024	0.018	0.018	0.051	0.015	0.003	0.025	0.213
Bandra	0.007	0.027	0.016	0.008	0.051	0.015	0.003	0.025	0.150
Andheri	0.007	0.017	0.004	0.004	0.020	0.002	0.006	0.049	0.109
Juhu	0.060	0.020	0.005	0.034	0.009	0.007	0.006	0.025	0.165
Powai	0.089	0.019	0.004	0.009	0.006	0.001	0.006	0.008	0.143
Goregaon	0.025	0.020	0.003	0.009	0.100	0.003	0.011	0.049	0.220
Total	0.246	0.127	0.050	0.081			0.036	0.181	1.000
	0.504				0.237	0.042	0.217		1.000
	1.000								

on the basic concepts of ELECTRE I, and afterward give a comparison of these variants for the appreciation of the readers.

The first question that invariably comes to one's mind is, what do we understand by the concept of *outranking*? Stated simply, the method of *outranking* is the idea in which one compares one decision/alternative against the others and ranks them based on some set principle that is already decided. In doing so, one assures that at the end of the pairwise-ranking process, we have a unique ranking system among all the decisions/alternatives. Intuitively, one starts with a set $\boldsymbol{A} = \{A_1, A_2, \ldots, A_m\}$, where $\#\boldsymbol{A} = m$ of decisions/alternatives, such that *each* a_i, where $\forall i = 1, \ldots, m$, has $j = 1, \ldots, n$, set of attributes/decision criteria/goals. Given this, one wants to accomplish the task of ranking based on the *collective/cumulative* effect of these n attributes/decision criteria/goals by comparing any two different decisions/alternatives, say A_i and A_j, where $\forall i \neq j = 1, \ldots, m$. At the end of this comparison process, we end up with the kernel (may be called the best choice), set as $\boldsymbol{A}_1$ (an ordered set), where $\boldsymbol{A}_1 \subset \boldsymbol{A}$ and $\#\boldsymbol{A}_1 \leq \#\boldsymbol{A}$. An advantage of the ELECTRE method is the fact that the final result, $\boldsymbol{A}_1$, when presented to the decision maker, makes it easier for him/her to take the final view, rather than what he/she would have taken only with $\boldsymbol{A}$, as in that case, it is difficult for the person to judge the n number of alternatives in a rational manner so as to choose the so called best set of elements among the n numbers originally there. ELECTRE II, unlike ELECTRE I, is able to rank the elements of the complete set $\boldsymbol{A} = \{A_1, A_2, \ldots, A_m\}$, such that a more broader view can be taken by the decision maker, with even the outranked decisions now being shown in the overall ranking process.

With this brief background, let us give a formal approach about how this outranking methodology is utilized in the ELECTRE I (from now to be referred to as ELECTRE) ranking system. In case the ith decision/alternative, A_i, *outranks* the jth alternative, A_j, then, we denote this as $A_i \rightarrow A_j$, which means that the risk (or loss whatever one wants to say) for A_i is not as much as A_j or A_i is as good as A_j, and not as worse as A_j. How one decides the so-called relative ranking between the decisions/alternatives, A_i, and A_j is a matter of prime importance to us. It is also worthwhile to mention that here is where the collective/cumulative effect of all the attributes/decision criteria/goals comes into play. Remember that the ELECTRE-ranking system is not *transitive*, that is, even when $A_i \rightarrow A_j$ and $A_j \rightarrow A_k$, it does not imply that $A_i \rightarrow A_k$. To overcome this, one uses the concept of *concordance* and *discordance*, which is a sort of the level of *liking* and *disliking*, respectively, which one states clearly when comparing two attributes/decision criteria/goals based on his/her concept of utility/net worth. So, if we say that we like A_i with respect to A_j when one chooses A_i in place of A_j, then, one assigns some score or points to quantify the level of liking, which, as mentioned before, is the concordance level. If due to some reason one is forced to choose A_i with respect to A_j, then, that level of disliking is objectively stated using the concept of discordance.

5.1.5.4.1 Basic Concepts/Steps in ELECTRE

1. To deal with *outranking relations* by using *pairwise* comparisons among decisions/alternatives for each of the attribute/decision criterion/goal separately.
2. The outranking between two decisions/alternatives A_i and A_j is denoted by $A_i \rightarrow A_j$, which generally implies that even if the ith alternative is not dominating the jth one, yet, the decision maker may choose the ith one.
3. Decisions/alternatives are dominated if there is another decision/alternative that excels them in one or more attributes/decision criteria/goals and equals in the remaining attributes/decision criteria/goals.
4. Pairwise comparison of decisions/alternatives for each attribute/decision criterion/goal is accomplished using some matching index, $g_j(A_i)$, where $i = 1, \ldots, m$ and $j = 1, \ldots, n$. This $g_j(A_i)$ may be compared with a monetary/numeric value.
5. One should also specify the threshold, $\varepsilon = g_j(A_{i_1}) - g_j(A_{i_2})$, which would dictate the level of liking/disliking. Remember that this threshold depends on the decision maker's choice (one should remember that the concept of the utility function is important here).
6. The decision maker usually assigns weights or an importance to the attributes/decision criteria/goals. This is done to express one's preference/relative importance of one attribute/decision criterion/goal over the other. This concept also comes from utility functions or the net value perspective of any decision-making process and is the driving force in the choices/ranking process being done by a decision maker.

Note: The ELECTRE method is convenient where there are few criteria and a large number of alternatives.

ELECTRE as an MCDM method works on the simple steps that are (i) normalizing the decision matrix (using any relevant criteria of normalization as mentioned in Table 5.3), (ii) weighting the normalized decision matrix, (iii) determining the concordance and discordance sets, (iv) constructing the concordance and discordance matrices, (v) determining the concordance and discordance dominance matrices, (vi) determining the aggregate dominance matrix, and finally (vii) eliminating the less-favorable decisions/alternatives and arriving at an answer. Using this, let us solve a hypothetical example (Example 5.10).

Example 5.10

Consider

$$\boldsymbol{A} = \begin{bmatrix} 2 & 1 & 2 \\ 3 & 4 & 1 \\ 1 & 3 & 2 \end{bmatrix},$$

$$\boldsymbol{W} = \begin{bmatrix} \frac{1}{4} & 0 & 0 \\ 0 & \frac{1}{2} & 0 \\ 0 & 0 & \frac{1}{4} \end{bmatrix}.$$

We solve this problem using the ELECTRE method considering (i) the linear normalization, (ii) concordance principle that is based on the concept that a concordance set C_{kl} of two decisions/alternatives A_k and A_l, where $m \geq k, l \geq 1$, is defined as the *set of all attributes/decision criteria/goals* for which A_k is preferred to A_l; hence, the concordance set, $c_{kl} = \{j : x_{kl} \geq x_{lj}\}$ for $j = 1, \ldots, n$, and finally (iii) discordance principle, based on the concept, that a discordance set D_{kl} of two decisions/alternatives A_k and A_l, where $m \geq k, l \geq 1$, is defined as the *set of all attributes/decision criteria/goals* for which A_k is not preferred to A_l; hence, the discordance set is $d_{kl} = \{j : x_{kl} < x_{lj}\}$ for $j = 1, \ldots, n$. Using the basic concepts/steps (refer to Section 5.1.5.4.1), we obtain the concordance and discordance matrices as

$$\boldsymbol{C} = \begin{bmatrix} - & 0.25 & 0.50 \\ 0.75 & - & 0.75 \\ 0.25 & 0.25 & - \end{bmatrix} \text{ and } \boldsymbol{D} = \begin{bmatrix} - & 1.00 & 1.00 \\ 0.31 & - & 0.64 \\ 0.37 & 1.00 & - \end{bmatrix}.$$

One can now use threshold values for concordance and discordance matrices given as $c^* = (1/m(m-1)) \sum_{k=1}^{m} \sum_{l(\neq k)=1}^{m} c_{kl}$ and $d^* = (1/m(m-1)) \sum_{k=1}^{m} \sum_{l(\neq k)=1}^{m} d_{kl}$ respectively, using which we obtain the concordance and discordance dominance matrices that are

$$\boldsymbol{C}_{dom} = \begin{bmatrix} - & 0 & 1 \\ 1 & - & 1 \\ 1 & 0 & - \end{bmatrix} \text{ and } \boldsymbol{D}_{dom} = \begin{bmatrix} - & 1 & 1 \\ 0 & - & 0 \\ 0 & 1 & - \end{bmatrix}.$$

Finally, $\boldsymbol{C}_{dom}$ and $\boldsymbol{D}_{dom}$ yield the ultimate matrix for the ELECTRE process that is

$$\begin{bmatrix} - & 0 & 1 \\ 0 & - & 0 \\ 0 & 0 & - \end{bmatrix}.$$

If we look at the result, we may conclude that $A_1 \succ A_3$, while for A_2, one is not sure so as to pass any judgment like this. We should remember that the normalization used here is linear in nature, and in case different normalizations are used, one can obtain a different set of rankings between A_1, A_2, and A_3. For a better appreciation of this ranking method, we state the general pseudocode for the ELECTRE (Figure 5.11).

5.1.5.4.2 ϵ ELECTRE

A variant of the ELECTRE method may be considered whereby we assume that the concordance and discordance sets are *not mutually exhaustive*. This implies that we incorporate the idea of the indifference principle (a common notion in utility theory), where for any two decisions/alternatives A_k and A_l, we now have $C_{kl} \cup D_{kl} \cup I_{kl}$ as the universal set of all possible rankings between A_k and A_l for all attributes/decision criteria/goals. It means that we divide the choices into three distinct sets called the *concordance* set, *discordance* set, and *indifference* set, that is, $\boldsymbol{I}$. Thus, the general indifference concept is given as (i) $|y_{kl} - y_{lj}| \leq \epsilon_j$ or (ii) $y_{kl} - y_{lj} \leq \epsilon_{j,1}$ and $y_{lj} - y_{kl} \leq \epsilon_{j,2}$ for $k, l = 1, \ldots, m$ and $j = 1, \ldots, n$, where ϵ_j implies equal dispersion on both the sides to signify the indifference, while $\epsilon_{j,1}$ along with $\epsilon_{j,2}$ signifies an unequal indifference depending on how less happy a person is with gains but on the other hand very sad with an equal quantum of losses. Using deviations between C_{kl} and I_{kl} and those between D_{kl} and I_{kl} as unequal, it is possible to bring the characteristics of a human being into the picture who is very sad for a minor loss but never happy even if his/her gains are huge.

```
1:  DEFINE: [a_ij], i = 1,...,m, j = 1,...,n i.e., normalized matrix;
    w_j, j = 1,...,n, i.e., the weights; c_kl = {j : x_kl ≥ x_lj}, j = 1,...,n,
    i.e., concordance values; d_kl = {j : x_kl < x_lj}, j = 1,...,n, i.e.,
    discordance values; c* = 1/(m(m−1)) Σ_{k=1}^{m} Σ_{l(≠k)=1}^{m} c_kl; d* = 1/(m(m−1)) Σ_{k=1}^{m}
    Σ_{l(≠k)=1}^{m} d_kl; C concordance matrix; D discordance matrix, F comparison
    matrix.
2:  INPUT: [a_ij], i = 1,...,m; j = 1,...,n i.e., normalized matrix, w_j,
                    j = 1,...,n i.e., the weights
3:  START If: j = 1 : n
4:     START If: k,l = 1 : m
5:     CALCULATE: c_kl and d_kl
6:     END if
7:  END if
8:  CALCULATE: c* = 1/(m(m−1)) Σ_{k=1}^{m} Σ_{l(≠k)=1}^{m} c_kl, d* = 1/(m(m−1)) Σ_{k=1}^{m}
                    Σ_{l(≠k)=1}^{m} d_kl, C, D, F
9:  REPORT: C, D, F
10: END
```

FIGURE 5.11
Pseudocode for the general ELECTRE.

Example 5.11

Consider that we have three decisions/alternatives and corresponding to each decision, there are three attributes/decision criteria/goals. The corresponding matrix

$$\mathbf{A}_{nonnormalized} = \begin{bmatrix} 100 & 200 & 200 \\ 200 & 50 & 250 \\ 150 & 150 & 200 \end{bmatrix},$$

while $w_1 = w_2 = w_3 = \frac{1}{3}$. Utilizing the logarithmic utility function and linear normalization, one would finally obtain

$$\begin{bmatrix} - & 1 & 0 \\ 1 & - & - \\ - & - & - \end{bmatrix}.$$

This result shows that $A_1 \approx A_2$, while for A_3, one is not sure. An interesting notion is the fact that the off-diagonal elements are nondeterminable, due to the fact that depending on the ϵ values, one may not be able to exactly differentiate between different decisions/alternatives even when one compares two decisions/alternatives from their respective positive as well as negative perspectives. To understand the differences between the different ELECTRE methods in all their technicalities, one can refer to the work of Roy (1996).

5.1.5.5 Preference-Ranking Organization Method for Enrichment Evaluations

PROMETHEEs are a family of outranking methods developed by Brans (1982) and refined later by Brans and Vincke (1985). The application areas of PROMETHEE are as diverse as environment management, hydrology and water management, chemistry, logistics and transportation, etc. Good references are Brans et al. (1984), Brans and Vincke (1985), Brans and Mareschal (2005), and Behzadian et al. (2010). For the interested readers, it would be appreciated if they have a look at the exhaustive bibliography available at http://www.promethee-gaia.net/files/BiblioPromethee.pdf, and as of September-23-2015 the number is 571. One can also look at http://www.promethee-gaia.net, that is, the website that gives all the necessary information regarding the resources, software, methods, conferences, etc., relevant to PROMETHEE. In literature, one observes a chronological development of the PROMETHEE method and as the names suggest, they are PROMETHEE I (partial outranking), PROMETHEE II (complete outranking), PROMETHEE III (ranking based on intervals), PROMETHEE IV (ranking based on a continuous case), PROMETHEE V (MCDA including segmentation constraints), and PROMETHEE VI (representation of the human brain).

To motivate our reader into this discussion of PROMETHEE, consider A_i, where $i = 1, \ldots, m$, as the m number of decisions/alternatives, while C_j, where $j = 1, \ldots, n$, are the n number of attributes/decision criteria/goals, each of which is associated with each individual decision/alternative. The basic data may be presented as shown (Table 5.14), where w_j, $j = 1, \ldots, n$, are nonnegative weights (normalized) associated with each attribute/decision criteria/goal such that $\sum_{j=1}^{n} w_j = 1$. One may consider that these weights express the importance of the corresponding criterion, C_j.

As mentioned, the preference structure of PROMETHEE is based on pairwise comparison, whereby the comparison is given by the concept of deviation. This idea of deviation is characterized by preference, whereby small deviations signify small or no preferences while large deviations imply large preferences. For an ease of understanding, let us discuss the preference function that is denoted by $P_j\left(A_{i_1}, A_{i_2}\right) = F_j\left[d_j\left(A_{i_1}, A_{i_2}\right)\right] = F_j\left[C_j\left(A_{i_1}\right) - C_j\left(A_{i_2}\right)\right]$, where $i_1, i_2 = 1, \ldots, m$ and is between 0 and 1, that is, $0 \leq P_j\left(A_{i_1}, A_{i_2}\right) \leq 1$. In case the criterion is to be maximized, then, the graph of the preference function versus deviation is as shown in Figure 5.12 and $P_j\left(A_{i_1}, A_{i_2}\right) > 0 \Rightarrow P_j\left(A_{i_2}, A_{i_1}\right) = 0$. Logically, for the minimization case, the preference function would be reversed such that $P_j\left(A_{i_1}, A_{i_2}\right) = F_j\left[-d_j\left(A_{i_1}, A_{i_2}\right)\right] = F_j\left[C_j\left(A_{i_2}\right) - C_j\left(A_{i_1}\right)\right]$. In literature, one comes across the pair $\{C_j(.), P_j(.)\}$ that is termed as the *generalized criterion* associated with criterion $C_j(.)$, and in general, there are six types of generalized criterion as illustrated in detail in Figure 5.13.

The parameters d^{**} and d^{*} signify the threshold of a strict indifference and strict preference, respectively, while s is any value in-between d^{**} and d^{*}. For the case when the deviation, $d_j\left(A_{i_1}, A_{i_2}\right)$, is in-between d^{**} and d^{*}, the decision maker is indifferent but for values greater

TABLE 5.14

Basic Data Illustrating Decisions/Alternatives and Attributes/Decision Criteria/Goals for PROMETHEE

$C \longrightarrow$ / $\downarrow A$	C_1		C_n
A_1	$C_1(A_1)$	.	$C_n(A_1)$
.	.	.	.
.	.	.	.
.	.	.	.
A_m	$C_1(A_m)$	...	$C_n(A_m)$
	w_1	.	w_n

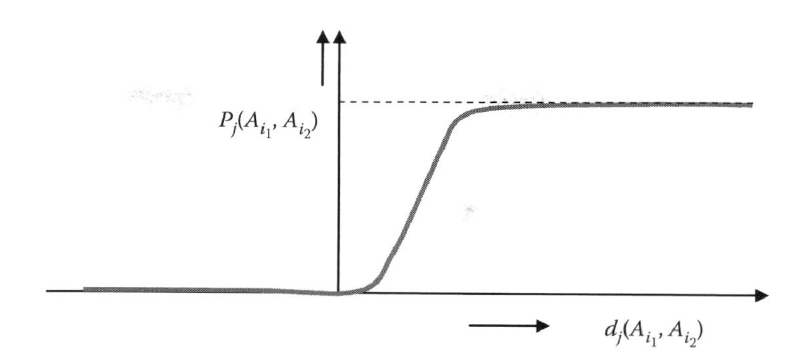

FIGURE 5.12
Preference function plotted against deviation.

than d^*, the decision maker has a strong preference for the decision/alternative A_{i_1}, while for less than d^{**} value, the decision maker has a strong preference for the decision/alternative A_{i_2}. From a mathematical point of view, s can also be termed as the point of inflexion, that is, the point at which the preference changes.

5.1.5.5.1 PROMETHEE I (Partial Outranking)

Let $A_{i_1}, A_{i_2} \in \boldsymbol{D}$. Furthermore, consider $\sum_{j=1}^{n} P_j\left(A_{i_1}, A_{i_2}\right) w_j$ and $\sum_{j=1}^{n} P_j\left(A_{i_2}, A_{i_1}\right) w_j$, which express the sum of the preference liking of A_{i_1} over A_{i_2} and A_{i_2} over A_{i_1}, respectively. Logically, $\sum_{j=1}^{n} P_j\left(A_{i_1}, A_{i_2}\right) w_j \neq \sum_{j=1}^{n} P_j\left(A_{i_2}, A_{i_1}\right) w_j$. Below, we state a few properties for PROMETHEE I that are

- $\sum_{j=1}^{n} P_j(A_i, A_i) w_j = 0$, for $i = 1, \ldots, m$
- $0 \leq \sum_{j=1}^{n} P_j\left(A_{i_1}, A_{i_2}\right) w_j \leq 1$ and $0 \leq \sum_{j=1}^{n} P_j\left(A_{i_2}, A_{i_1}\right) w_j \leq 1$, for $i_1, i_2 = 1, \ldots, m$
- $0 \leq \sum_{j=1}^{n} P_j\left(A_{i_1}, A_{i_2}\right) w_j + \sum_{j=1}^{n} P_j\left(A_{i_2}, A_{i_1}\right) w_j \leq 1$, for $i_1, i_2 = 1, \ldots, m$
- $\phi^+(A_i) = (1/(n-1)) \sum_{x \in \boldsymbol{D}} P_j(A_i, x) w_j$ is termed as the positive outranking flow, that is, sum of the *liking* deviation based on the fact that A_i is chosen.
- $\phi^-(A_i) = (1/(n-1)) \sum_{x \in \boldsymbol{D}} P_j(x, A_i) w_j$ is termed as the negative outranking flow, that is, sum of the *liking* deviation based on the fact that A_i is *not* chosen, that is, other alternatives are chosen.
- A_{i_1} *preferred* to A_{i_2} iff (i) $\phi^+\left(A_{i_1}\right) > \phi^+\left(A_{i_2}\right)$ and $\phi^-\left(A_{i_1}\right) < \phi^-\left(A_{i_2}\right)$ or (ii) $\phi^+\left(A_{i_1}\right) = \phi^+\left(A_{i_2}\right)$ and $\phi^-\left(A_{i_1}\right) < \phi^-\left(A_{i_2}\right)$ or (iii) $\phi^+\left(A_{i_1}\right) > \phi^+\left(A_{i_2}\right)$ and $\phi^-\left(A_{i_1}\right) = \phi^-\left(A_{i_2}\right)$.
- A_{i_1} is *indifferent* to A_{i_2} iff $\phi^+\left(A_{i_1}\right) = \phi^+\left(A_{i_2}\right)$ and $\phi^-\left(A_{i_1}\right) = \phi^-\left(A_{i_2}\right)$.
- A_{i_1} is *incompatible* to A_{i_2} iff (i) $\phi^+\left(A_{i_1}\right) > \phi^+\left(A_{i_2}\right)$ and $\varphi^-\left(A_{i_1}\right) > \phi^-\left(A_{i_2}\right)$ or (ii) $\phi^+\left(A_{i_1}\right) < \phi^+\left(A_{i_2}\right)$ and $\phi^-\left(A_{i_1}\right) < \phi^-\left(A_{i_2}\right)$.

To build an appropriate multicriteria concept for PROMETHEE methods, some requirements that are important and that should be considered are as follows: (i) the amplitude of the deviations between the evaluations of the alternatives within each criterion should be taken into account, (ii) the evaluations of each criterion expressed in their own units (i.e., scaling effects) should be completely eliminated, (iii) in pairwise comparisons, an appropriate multicriteria method should provide

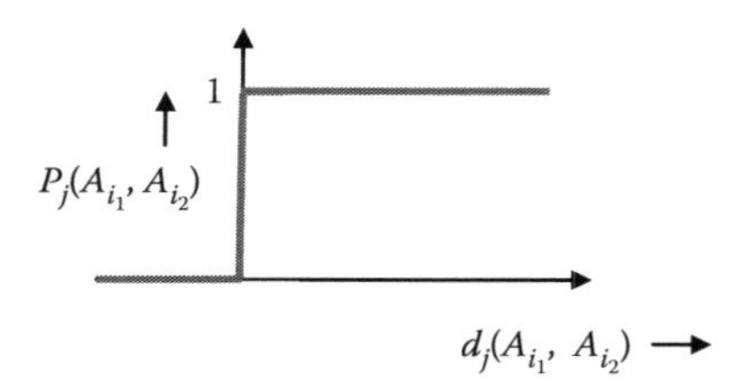

$$P_j(A_{i_1}, A_{i_2}) = \begin{cases} 0 & d_j(A_{i_1}, A_{i_2}) \le 0 \\ 1 & d_j(A_{i_1}, A_{i_2}) > 0 \end{cases}$$

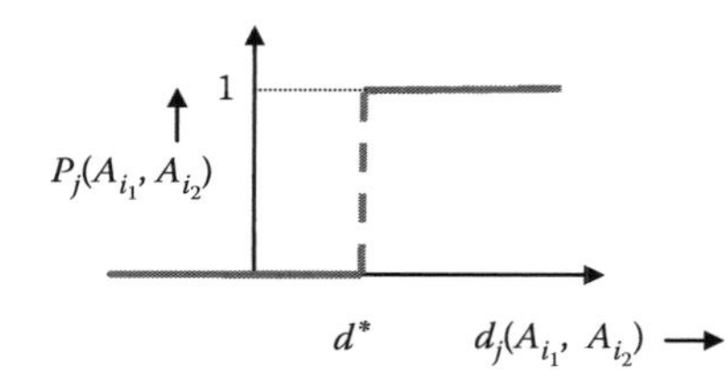

$$P_j(A_{i_1}, A_{i_2}) = \begin{cases} 0 & d_j(A_{i_1}, A_{i_2}) \le d^* \\ 1 & d_j(A_{i_1}, A_{i_2}) > d^* \end{cases}$$

The parameter is d^*

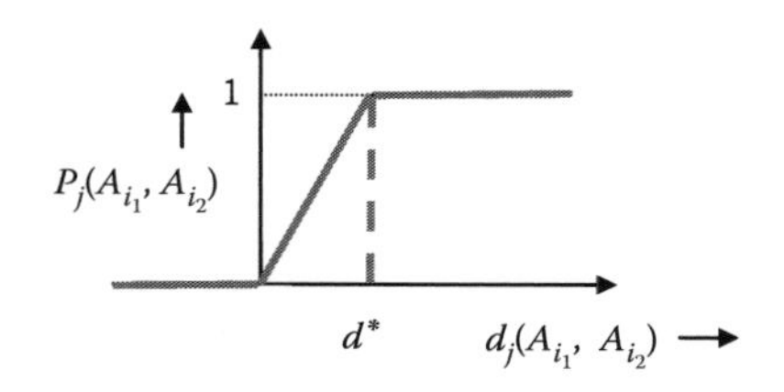

$$P_j(A_{i_1}, A_{i_2}) = \begin{cases} 0 & d_j(A_{i_1}, A_{i_2}) \le 0 \\ \frac{d_j(A_{i_1}, A_{i_2})}{d^*} & 0 \le d_j(A_{i_1}, A_{i_2}) < d^* \\ 1 & d_j(A_{i_1}, A_{i_2}) > d^* \end{cases}$$

The parameter is d^*

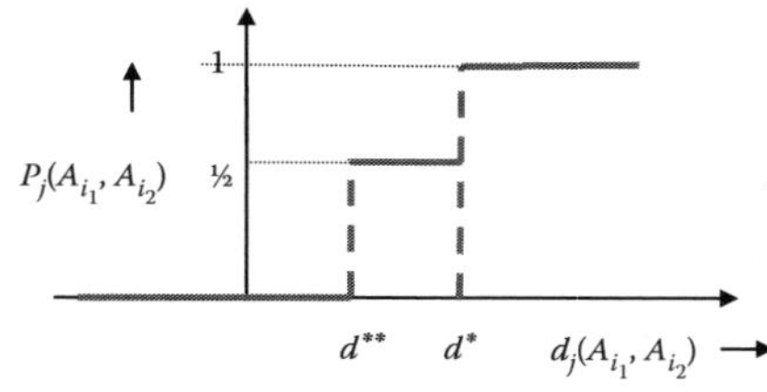

$$P_j(A_{i_1}, A_{i_2}) = \begin{cases} 0 & d_j(A_{i_1}, A_{i_2}) \le d^{**} \\ \frac{1}{2} & (A_{i_1}, A_{i_2}) \le d^* \\ 1 & d_j(A_{i_1}, A_{i_2}) > d^* \end{cases}$$

The parameters are d^{**} and d^*

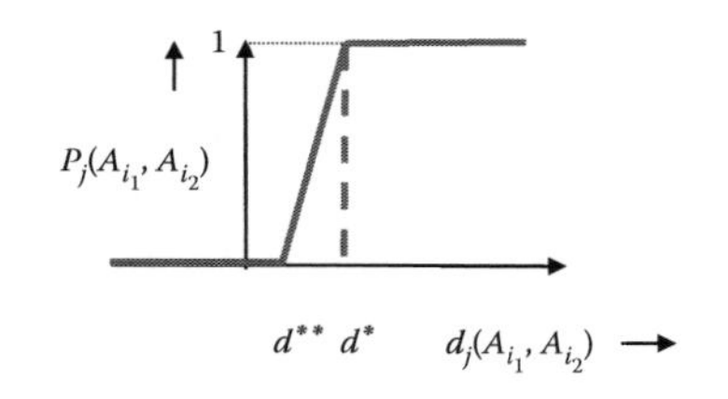

$$P_j(A_{i_1}, A_{i_2}) = \begin{cases} 0 & d_j(A_{i_1}, A_{i_2}) \le d^{**} \\ \frac{d_j(A_{i_1}, A_{i_2}) - d^{**}}{d^* - d^{**}} & d^{**} < d_j(A_{i_1}, A_{i_2}) \le d^* \\ 1 & d_j(A_{i_1}, A_{i_2}) > d^* \end{cases}$$

The parameters are d^{**} and d^*

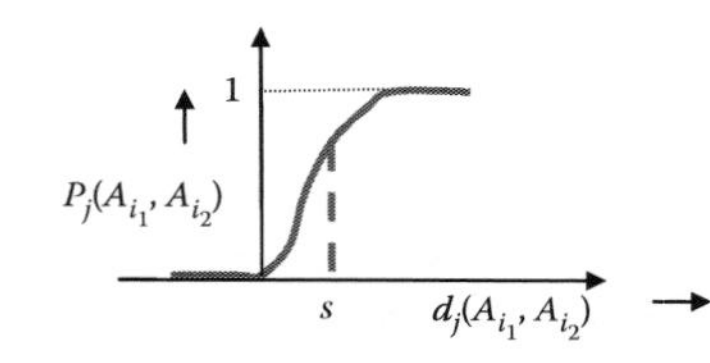

$$P_j(A_{i_1}, A_{i_2}) = \begin{cases} 0 & d_j(A_{i_1}, A_{i_2}) \le 0 \\ 1 - e^{-\frac{d^2}{2s^2}} & d_j(A_{i_1}, A_{i_2}) > 0 \end{cases}$$

The parameter is s

FIGURE 5.13
Types of a generalized criterion used in PROMETHEE.

the following information: that is, whether $A_{i_1} \succ A_{i_2}$, $A_{i_1} \approx A_{i_2}$; $A_{i_1} \not\asymp A_{i_2}$ (where $\not\asymp$ implies incompatibility), (iv) for different multicriteria methods with different additional information and different calculation procedures, the solutions they propose can be different, (v) any appropriate procedure should not include technical parameters that has no significance for the decision maker, (vi) an appropriate method should provide information on the conflicting nature of the criteria, and finally (vii) for the multicriteria method, allocating weights of a relative importance to the criteria is important, as this assignment of weights signifies a major part of the multicriteria decision process.

5.1.5.5.2 PROMETHEE II (Complete Outranking)

PROMETHEE II consists of preference and indifference ranking. It is often the case that the decision maker requests a complete ranking and the net outranking flow for any decision/alternative, A_{i_1}, where $i_1 = 1, \ldots, m$, can then be considered as

$$\Phi\left(\mathrm{A}_{i_1}\right) = \frac{1}{m-1} \sum_{A_{i_2} \in \boldsymbol{A}} \sum_{j=1}^{n} P_j\left(A_{i_1}, A_{i_2}\right) w_{j+} - \frac{1}{m-1} \sum_{A_{i_2} \in \boldsymbol{A}} \sum_{j=1}^{n} P_j\left(A_{i_1}, A_{i_2}\right) w_{j-},$$

where the suffix $+/-$ implies the positive/negative outranking flow. The positive outranking/negative outranking flow expresses how an alternative is outranking/underranking all the others, and its power (in terms of value) characterizes the outranking/underranking property. Thus, a higher/lower value implies whether a decision/alternative is better/worse with respect to other decisions/alternatives. The complete ranking is easy to use but the analysis of the incompatibilities often helps the decision maker to finalize a proper decision. This idea of net flow (somewhat akin to network flows in graphs) thus provides a complete ranking and may be compared with a utility function. One advantage of this net flow concept is the fact that it is built on clear and simple preference information (weights and preference functions). In real-world applications, one recommends decision makers to consider both PROMETHEE I and PROMETHEE II and one can refer to the existing literature in PROMETHEE to get a good idea about these methods.

5.1.5.5.3 Geometrical Analysis for Interactive Assistance

It is possible to denote the ranking process, also using the graphical method, whereby the geometrical analysis for interactive assistance (GAIA) plane concept is used. The concept of GAIA used in PROMETHEE is akin to the concept of principal component analysis (PCA) used in multivariate statistical analysis. If we look at the aggregate function, then, each decision/alternative, A_i, where $i = 1, \ldots, m$, is represented by a vector

$$\left[\Phi_1\left(A_i\right), \Phi_2\left(A_i\right), \ldots, \Phi_n\left(A_i\right)\right],$$

where

$$\Phi_j\left(\mathrm{A}_{i_1}\right) = \sum_{A_{i_2} \in \boldsymbol{A}} w_j \left\{P_j\left(A_{i_1}, A_{i_2}\right) - P_j\left(A_{i_2}, A_{i_1}\right)\right\}.$$

Thus, we have a space of n dimension where m points are denoted as vectors. Diagrammatically, it may be represented as shown in Figure 5.14 where we have $n = 3$, while the dots denote the decisions/alternatives. As per the concept of PCA, we denote **PC # 1** (principal component # 1) in a way such that it minimizes the distance function of each point on the line (the aggregate function of all decision/alternatives); hence, **PC # 1** is found in a way such that it goes through the maximum

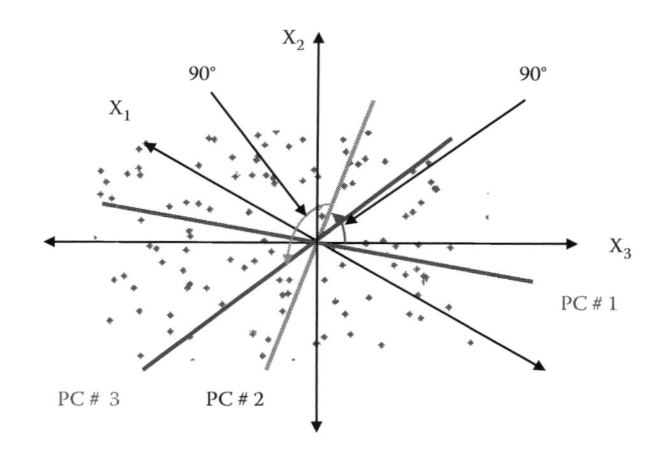

FIGURE 5.14
Hypothetical example illustrating the concept of PCA as used in PROMETHEE.

variation in the data. In a similar manner, **PC # 2** and **PC # 3**, and other principal component lines are drawn one after the other. Simple optimization techniques may also be used where one minimizes the total variation subject to the condition that the variation of $w_j \left\{P_j\left(A_{i_1}, A_{i_2}\right) - P_j\left(A_{i_2}, A_{i_1}\right)\right\}$ is minimized. One should remember that the variation one uses is that of the preference function, and hence, it depends on the utility function used. For the interest of the readers, we mentioned one good book (Jolliffe 2002) in PCA that the reader may refer to get some initial idea of PCA and hence use it in the context of the PROMETHEE MCDM technique.

To end this discussion, we state the general pseudocode for the PROMETHEE method (Figure 5.15).

1: **DEFINE**: $[a_{ij}]$, $i = 1,\ldots,m$, $j = 1,\ldots,n$, i.e., normalized matrix; w_j, $j = 1,\ldots,n$, i.e., weights; $P_j(A_{i_1}, A_{i_2})$ for $i_1, i_2 = 1,\ldots,m$ i.e., preference function; $\phi^+(A_i) = \frac{1}{n-1}\sum_{x\in \boldsymbol{D}} P_j(A_i, x)w_j$ i.e., positive outranking flow; $\phi^-(A_i) = \frac{1}{n-1}\sum_{x\in \boldsymbol{D}} P_j(x, A_i)w_j$, i.e., negative outflow ranking
2: **INPUT**: $[a_{ij}]$, $i = 1,\ldots,m$, $j = 1,\ldots,n$, i.e., normalized matrix;
   w_j, $j = 1,\ldots,n$, i.e., weights
3: **START If**: $i = 1 : m$
4:    **START If**: $j = 1 : m$
5:    **CALCULATE**: $P_j(A_{i_1}, A_{i_2})$, $\phi^+(A_i)$ and $\phi^-(A_i)$
6:    **END if**
7: **END if**
8: **CALCULATE**: $\phi^+(A_i) - \phi^-(A_i)$
9: **REPORT**: $\phi^+(A_i) - \phi^-(A_i)$
10: **END**

FIGURE 5.15
Pseudocode for the general PROMETHEE.

5.1.5.6 TOPSIS Method

TOPSIS method (Hwang and Yoon 1981; Chen and Hwang 1992) is widely used to solve various MCDM problems. This method was developed for the ***INTEGRated Human Exploration Mission Simulation FacilITY*** (INTEGRITY) project in the Johnson Space Center to assess the priority of a set of human spaceflight mission simulators. In TOPSIS, we first choose one positive ideal solution (PIS) and one negative ideal solution (NIS) of the original ranking problem. After that, one finds the distances from each decisions/alternatives, A_i, where $i = 1, \ldots, m$ to PIS and NIS, that is, $d(A_i, PIS)$ and $d(A_i, NIS)$ respectively, given which, we calculate $r_i = d(A_i, NIS)/(d(A_i, NIS) + d(A_i, PIS))$. After that, we rank the ratios, r_i, to get the best alternative. The concept of distance, d, as used in TOPSIS is Euclidean in nature and hence the ratio can be expressed as

$$r_i = \frac{\sqrt{(A_i - NIS)^2}}{\sqrt{(A_i - NIS)^2} + \sqrt{(A_i - PIS)^2}}.$$

Here, the basic premise being the Euclidean distance portrays the concept of the utility function, $U(W)$ that is quadratic. This ensures that our main motivation that is to minimize the dispersion, that is, $Var(X)$, is met.

5.1.5.6.1 Algorithms for TOPSIS and Its Variants

Assume decisions/alternatives as A_i, where $i = 1, \ldots, m$, while if the attributes/decision criteria/goals are C_j, where $j = 1, \ldots, n$, then, the pseudocodes for the working principle of TOPSIS (Figure 5.16), A-TOPSIS (Figure 5.17), and M-TOPSIS (Figure 5.18) are as follows.

5.1.5.7 VIKOR Method

The idea of VIKOR, that is, a multicriteria optimization and compromise solution, an MCDM technique, was developed by Serafim Opricovic during his PhD work. Though there are other papers, yet, we would like to mention the following: that is, Opricovic and Tzeng (2004, 2007), which may be considered as the initial academic work based on which one can understand the working principle of VIKOR. VIKOR as an MCDM technique is an outranking method and solves discrete decision problems where the criteria are conflicting and noncommensurable (of different units). In this method, the decision maker likes a solution that is closest to the ideal, and hence, the decisions/alternatives are evaluated/compared/ranked accordingly. While ranking, the decisions/alternatives, rather than the best solution, are the target as finding out the ideal solution is not always feasible, but rather, the closest to the ideal is what is practically possible.

Two of the MCDM methods, that is, VIKOR and TOPSIS, are based on an aggregating function that represents the concept of closeness of the solution to the ideal solution. In VIKOR, we follow the linear normalization, while in TOPSIS, it is vector normalization. Normalization is used to eliminate the units of criterion functions and thus ensure a level-playing field for a different criterion. In VIKOR, we determine a maximum group utility for the majority and a minimum of an individual regret for the opponent. While in TOPSIS, a solution with the shortest distance to the ideal solution and the greatest distance from the NIS is required to be found. While doing this, we do not consider the relative importance of these distances.

Assume that you have m decisions/alternatives, A_i, where $i = 1, \ldots, m$ and n attributes/decision criteria/goals C_j, where $j = 1, \ldots, n$ and consider $C_j(A_i)$ as the value of the jth attributes/decision

```
1:  DEFINE: X_{m×n} (matrix consisting of priority scores assigned to
    decisions/alternatives), A_i, based on attributes/decision criteria/
    goals, C_j; w_j (weight for the attributes/decision criteria/goals)
    such that ∑_{j=1}^{n} w_j = 1; B (benefit matrix); C (cost matrix); r_{i,j} =
    x_{i,j} / sqrt(∑_{i=1}^{m} x_{i,j}^2); AIS = (v_1^+, ..., v_m^+) = [{max_{∀i}(v_{i,j} | j ∈ B)}, {min_{∀i}(v_{i,j} | j ∈ C)}]
    (negative ideal solution); PIS = (v_1^-, ..., v_m^-) = [min_{∀i}{(v_{i,j} | j ∈ B)},
    {max_{∀i}(v_{i,j} | j ∈ C)}] (positive ideal
    solution); S_i^+ = sqrt(∑_{j=1}^{n} (v_{i,j} - v_j^+)^2); S_i^- = sqrt(∑_{j=1}^{n} (v_{i,j} - v_j^-)^2); T_i =
    (S_i^- / (S_i^+ + S_i^-)) (relative closeness); M = (S_i^+, S_i^-) (separation measure). Here
    i = 1, ..., m and j = 1, ..., n
2:  INPUT: X_{m×n} (matrix consisting of priority scores assigned to
    decisions/alternatives), A_i, based on attributes/decision criteria/
    goals, C_j; w_j (weight for the attributes/decision criteria/goals)
    such that ∑_{j=1}^{n} w_j = 1; B (benefit matrix); C (cost matrix). Here
    i = 1, ..., m and j = 1, ..., n
3:  START if: i = 1 : m
4:     START if: j = 1 : n
5:     CALCULATE: r_{i,j} = x_{i,j} / sqrt(∑_{i=1}^{m} x_{i,j}^2); v_{i,j} = w_j r_{i,j} where i = 1, ..., m and j = 1, ..., n
6:     END if
7:  END if
8:  CALCULATE: v_i^+; v_i^-; AIS; PIS; S_i^+; S_i^-; M; T_i
9:  REPORT: AIS; PIS; M; T_i
10: END
```

FIGURE 5.16
Pseudocode for TOPSIS.

criteria/goals for the ith alternative such that

$$L_{p,i} = \left[\sum_{j=1}^{n}\left\{\frac{w_j\left(C_j(A)^+ - C_j(A_i)\right)}{\left(C_j(A)^+ - C_j(A)^-\right)}\right\}^p\right]^{1/p},$$

where $C_j(A)^+ = \max_i C_j(A_i)$ and $C_j(A)^- = \min_i C_j(A_i)$. Remember that $L_{1,i}$ and $L_{\infty,i}$ are used to formulate the ranking measure, where $p = 1, 2, \ldots$, is an integer and denotes the distance measure used. Hence, $p = 1$ signifies the Manhattan norm while $p = \infty$ denotes the infinity norm. Finally, Figure 5.19 illustrates the pseudocode for the VIKOR method that would give us a clear idea of the working principle of this MCDM method.

5.1.5.8 Potentially All Pairwise Rankings of All Possible Alternatives

PAPRIKA is a method for MCDM or a conjoint analysis based on decision-makers' preferences as expressed using pairwise rankings of alternatives. PAPRIKA method specifically applies to additive

```
1:  DEFINE: X_{m×n} (matrix consisting of priority scores assigned to
    decisions/alternatives), A_i, based on attributes/decision criteria/
    goals, C_j; w_j (weight for the attributes/decision criteria/goals) such
    that Σ_{j=1}^{n} w_j = 1; B (benefit matrix); C (cost matrix); r_{i,j} = x_{i,j} / sqrt(Σ_{i=1}^{m} x_{i,j}^2);
    AIS = (v_1^+, ..., v_m^+) = [max_{∀i}{(r_{i,j} | j ∈ B)}, {min_{∀i}(r_{i,j} | j ∈ C)}] (negative ideal
    solution); PIS = (v_1^-, ..., v_m^-) = [min_{∀i}{(r_{i,j} | j ∈ B)}, {max_{∀i}(r_{i,j} | j ∈ C)}] positive
    ideal solution); S_i^+ = sqrt(Σ_{j=1}^{n} w_j(r_{i,j} - v_j^+)^2); S_i^- = sqrt(Σ_{j=1}^{n} w_j(r_{i,j} - v_j^-)^2);
    T_i = (S_i^- / (S_i^+ + S_i^-)) (relative closeness); M = (S_i^+, S_i^-) (separation measure).
    Here i = 1, ..., m and j = 1, ..., n
2:  INPUT: X_{m×n} (matrix consisting of priority scores assigned to
    decisions/alternatives), A_i, based on attributes/decision criteria/
    goals, C_j; w_j (weight for the attributes/decision criteria/goals) such
    that Σ_{j=1}^{n} w_j = 1; B (benefit matrix); C (cost matrix). Here i = 1, ..., m
    and j = 1, ..., n
3:  START if: i = 1 : m
4:     START if: j = 1 : n
5:     CALCULATE: r_{i,j} = x_{i,j} / sqrt(Σ_{i=1}^{m} x_{i,j}^2); v_{i,j} = w_j r_{i,j} where i = 1, ..., m and j = 1, ..., n
6:     END if
7:  END if
8:  CALCULATE: v_i^+; v_i^-; AIS; PIS; S_i^+; S_i^-; M; T_i
9:  REPORT: AIS; PIS; M; T_i
10: END
```

FIGURE 5.17
Pseudocode for A-TOPSIS.

multiattribute value models with performance categories—also known as points/scoring/point-count/linear systems, or models. One should remember that this method is based on the fundamental principle that an overall ranking of all possible alternatives representable by a given value model—that is, all possible combinations of the categories on the criteria—are defined when all pairwise rankings of the alternatives vis-à-vis each other are known (provided the rankings are consistent). So, if there are n possible alternatives, then, there are $n(n-1)/2$ pairwise rankings, which may be very large if n is a big number. PAPRIKA solves this problem by ensuring that the number of pairwise rankings that the decision makers need to perform is kept to a minimum, so that the method is practicable. It does this by (for each undominated pair explicitly ranked by decision makers) identifying (and eliminating) all undominated pairs that are implicitly ranked as corollaries of this and other explicitly ranked pairs (via the transitivity property of additive value models). PAPRIKA method's closest theoretical antecedent is pairwise trade-off analysis, a precursor to the adaptive conjoint analysis in marketing research. Like the PAPRIKA method, pairwise trade-off analysis is based on the idea that undominated pairs that are explicitly ranked by the decision maker can be used to implicitly rank other undominated pairs. Before discussing the general pseudocode of the PAPRIKA method (Figure 5.20), it would be relevant to state the reference for PAPRIKA that is Hansen and Ombler (2008).

```
1: DEFINE: X_{m×n}(matrix consisting of priority scores assigned to
   decisions/alternatives), A_i, based on attributes/decision criteria/
   goals, C_j; w_j (weight for the attributes/decision criteria/goals) such
   that Σ_{j=1}^{n} w_j = 1; B (benefit matrix); C (cost matrix); r_{i,j} = x_{i,j} / sqrt(Σ_{i=1}^{m} x_{i,j}^2);
   AIS = (v_1^+, ..., v_m^+) = [max_{∀i}{(v_{i,j} | j ∈ B)}, {min_{∀i}(v_{i,j} | j ∈ C)}] (negative ideal
   solution); PIS = (v_1^-, ..., v_m^-) = [min_{∀i}{(v_{i,j} | j ∈ B)}, {max_{∀i}(v_{i,j} | j ∈ C)}]
   (positive ideal solution); S_i^+ = sqrt(Σ_{j=1}^{n} (v_{i,j} - v_j^+)^2); S_i^- = sqrt(Σ_{j=1}^{n} (v_{i,j} - v_j^-)^2);
   S = (min(S_i^+), max(S_i^-)) (separation measure);
   T_i = sqrt({S_i^+ - min(S_i^+)}^2 + {S_i^- - max(S_i^-)}^2). Here i = 1, ..., m and j = 1, ..., n
2: INPUT: X_{m×n}(matrix consisting of priority scores assigned to
   decisions/alternatives), A_i, based on attributes/decision criteria/
   goals, C_j; w_j (weight for the attributes/decision criteria/goals)
   such that Σ_{j=1}^{n} w_j = 1; B (benefit matrix); C (cost matrix). Here
   i = 1, ..., m and j = 1, ..., n
3: START if: i = 1 : m
4:    START if: j = 1 : n
5:    CALCULATE: r_{i,j} = x_{i,j} / sqrt(Σ_{i=1}^{m} x_{i,j}^2); v_{i,j} = w_j r_{i,j} where i = 1, ..., m and j = 1, ..., n
6:    END if
7: END if
8: CALCULATE: v_i^+; v_i^-; AIS; PIS; S_i^+; S_i^-; S; T_i
9: REPORT: AIS; PIS; M; T_i
10: END
```

FIGURE 5.18
Pseudocode for M-TOPSIS.

5.1.5.9 Measuring Attractiveness by a Categorical-Based Evaluation Technique

Measuring the attractiveness through a category-based evaluation technique is the goal of the MACBETH approach that was designed by Carlos António Bana e Costa, from the University of Lisbon, in cooperation with Professor Jean-Claude Vansnick and Dr. Jean-Marie De Corte, from the Université de Mons. MACBETH permits the evaluation of options against multiple attributes/decision criteria/goals. The key distinction between MACBETH and other MCDA methods is that it needs only qualitative judgments about the difference of attractiveness between two elements at a time, to generate numerical scores for the options in each attribute/decision criterion/goal and to weight the attribute/decision criterion/goal. MACBETH is an interactive approach that uses a semantic judgment about the differences in attractiveness of several stimuli to help a decision maker quantify the relative attractiveness of each. The process begins with the elicitation of the key aspects that the decision maker considers to be the attribute/decision criterion/goal by which the attractiveness of any potential career option should be appraised. A tree is created in the MACBETH decision-support system listing the attribute/decision criterion/goal, after which the options are introduced into the model. Next, we create a value scale for each of the attribute/decision criterion/goal. The decision maker then ranks the options, as well as the previously established neutral and good reference levels, in order of their attractiveness in terms of a decided monetary reward. Qualitative judgments

```
1:  DEFINE: X_{m×n} (matrix consisting of priority scores assigned to
    decisions/alternatives), A_i, based on attributes/decision criteria/
    goals, C_j; w_j (weight for the attributes/decision criteria/goals)
    such that Σ_{j=1}^{n} w_j = 1; C_j(A_i) (function relationship between
    attributes/decision criteria/goals for each decisions/alternatives);
    C_j(A)^+ = max_i C_j(A_i); C_j(A)^- = min_i C_j(A_i);
    L_{p,i} = [Σ_{j=1}^{n} {w_j(C_j(A)^+ - C_j(A_i)) / (C_j(A)^+ - C_j(A)^-)}^p]^{1/p}.
    Here i = 1,...,m; j = 1,...,n and p = 1,2,...,∞ (distance norm)
2:  INPUT: X_{m×n} (matrix consisting of priority scores assigned to
    decisions/alternatives), A_i, based on attributes/decision criteria/
    goals, C_j; w_j (weight for the attributes/decision criteria/goals)
    such that Σ_{j=1}^{n} w_j = 1; C_j(A_i) (function relationship between
    attributes/decision criteria/goals for each decisions/alternatives).
    Here i = 1,...,m and j = 1,...,n.
3:  START if: i = 1 : m
4:      START if: j = 1 : n
5:      CALCULATE: C_j(A_i); C_j(A)^+ = max_i C_j(A_i); C_j(A)^- = min_i C_j(A_i);
        L_{p,i} = [Σ_{j=1}^{n} {w_j(C_j(A)^+ - C_j(A_i)) / (C_j(A)^+ - C_j(A)^-)}^p]^{1/p} where i = 1,...,m; j = 1,...,n and
        p = 1,2,...,∞ (distance norm)
6:      END if
7:  END if
8:  CALCULATE: C_j(A)^+; C_j(A)^-; L_{p,i}
9:  REPORT: C_j(A)^+; C_j(A)^-; L_{p,i}
10: END
```

FIGURE 5.19
Pseudocode for VIKOR.

regarding the difference of attractiveness between options are elicited from the decision maker, who then responds with *no*, *very weak*, *weak*, *moderate*, *strong*, *very strong*, or an *extreme* rating, after which the ranking is done. Remember that this qualitative response is notionally somewhat similar to the AHP scale that as we know is quantitative (Table 5.8) in nature. For a better understanding of the MACBETH MCDM method, we urge the readers to have a look at the following pseudocode (Figure 5.21).

Few good references for MACBETH are Bana e Costa and Chagas (2004), and Bana e Costa and Vansnick (1994, 1995, 1997).

5.1.5.10 Decision Trees

According to Breiman et al. (1984) and Quinlan (1993), the study of decision tree analysis and its algorithms is a nonparametric approach to data modeling. The advantages such as the use of optimization algorithms, likelihood theory, and Bayesian theory in formulating decision trees within a statistical framework are reasons for the popularity of use of the decision tree analysis. A good starting point for this area is Jordan and Jacobs (1994). A decision tree consists of nodes that form a *rooted tree*. It is a directed tree with a node called a *root* that has no incoming edges. All other nodes have exactly one incoming edge. Furthermore, a node with outgoing edges is called an internal/test

```
1:  DEFINE: X_{m× n}(matrix consisting of priority scores assigned to
    decisions/alternatives),A_i,based on attributes/decision criteria/
    goals, C_j;w_j (weight for the attributes/decision criteria/goals)
    such that ∑_{j=1}^{n} w_j = 1;C_j(A_i) (function relationship between
    attributes/decision criteria/goals for each decisions/alternatives);
    D(ranking matrix). Here i = 1,...,m and j = 1,...,n
2:  INPUT: X_{m× n}(matrix consisting of priority scores assigned to
    decisions/alternatives),A_i,based on attributes/decision criteria/
    goals, C_j;w_j (weight for the attributes/decision criteria/goals)
    such that ∑_{j=1}^{n} w_j = 1;C_j(A_i) (function relationship between
    attributes/decision criteria/goals for each decisions/alternatives).
    Here i = 1,...,m and j = 1,...,n
3:  START if: i = 1 : m
4:      START if: j = 1 : n
5:      CALCULATE: C_j(A_{i1}) − C_j(A_{i2}) for all dominated pairs only
6:      END if
7:  END if
8:  CALCULATE: ∑_{∀ j∈ X1}{C_j(A_{i1}) − C_j(A_{i2})},where X1∈ X is the set of
        dominated pairs;max[∑_{∀ j∈ X1} {C_j(A_{i1}) − C_j(A_{i2})}]
9:  REPORT: ∑_{∀ j∈ X1}{C_j(A_{i1}) − C_j(A_{i2})},where X1∈ X is the set of
        dominated pairs;max[∑_{∀ j∈ X1} {C_j(A_{i1}) − C_j(A_{i2})}]
10: END
```

FIGURE 5.20
Pseudocode for PAPRIKA.

node that splits the decision space into two or more subspaces. In case the node is the terminal decision, it is called a *leaf*. A decision tree may incorporate the nominal attribute (NoA) as well as numeric attributes (NuAs). Thus, a typical decision tree may look like that as shown in Figure 5.22.

Usually, the tree complexity is measured by one of the following metrics: the total number of nodes, total number of leaves, tree depth, and number of attributes used. The induction of an optimal decision tree from a given data is considered to be a hard task. It has been shown that finding a minimal decision tree consistent with the training set is NP hard (Hancock et al. 1996). Moreover, it has been shown that constructing a minimal binary tree with respect to the expected number of tests required for classifying an unseen instance is NP complete (Hyafil and Rivest 1976). Even finding the minimal equivalent decision tree for a given decision tree (Zantema and Bodlaender 2000) or building the optimal decision tree from decision tables is known to be NP hard (Naumov 1991). Roughly speaking, the method of induction of an optimal decision tree from a given data is of two types and can be divided into two groups: top down and bottom up with a clear preference in the literature to the first group.

Example 5.12

An oil company while evaluating an oil basin for drilling has three alternatives, which are (i) drill, (ii) conduct a seismic test before drilling for which the cost incurred to find the nature of the underlying oil basin is 20,000, and finally (iii) do nothing. If the company drills, then it is likely to find the oil basin as (i) dry, (ii) wet, or (iii) soaking. A dry well yields nothing,

```
1:  DEFINE: X_{m×n} (matrix consisting of priority assigned to decisions/
    alternatives), A_i, based on attributes/decision criteria/goals, C_j; C_j(A_i)
    (function relationship between attributes/decision criteria/goals for
    each decisions/alternatives); D comparison matrix; ⊗ the comparison
    monetary function. Here i = 1,...,m and j = 1,...,n
2:  INPUT: X_{m×n} (matrix consisting of priority assigned to decisions/
    alternatives), A_i, based on attributes/decision criteria/goals, C_j; C_j(A_i)
    (function relationship between attributes/decision criteria/goals for
    each decisions/alternatives); ⊗ the comparison monetary function. Here
    i = 1,...,m and j = 1,...,n
3:  START if: i = 1 : m
4:     START if: j = 1 : n
5:     CALCULATE: {C_j(A_i)}_(1) ⊗ {C_j(A_i)}_(n), {C_j(A_i)}_(2) ⊗ {C_j(A_i)}_(n), ...,
       {C_j(A_i)}_(n−1) ⊗ {C_j(A_i)}_(n) (the last column of comparison matrix is
       filled from top to bottom); {C_j(A_i)}_(1) ⊗ {C_j(A_i)}_(n), {C_j(A_i)}_(1) ⊗
       {C_j(A_i)}_(2), ..., {C_j(A_i)}_(1) ⊗ {C_j(A_i)}_(n) (the first row column of
       comparison matrix is filled from left to right) and {C_j(A_i)}_(1) ⊗
       {C_j(A_i)}_(2), {C_j(A_i)}_(2) ⊗ {C_j(A_i)}_(3), ..., {C_j(A_i)}_(n−1) ⊗ {C_j(A_i)}_(n) (the
       other elements in the top half of the principal diagonal of the
       comparison matrix is filled)
6:     END if
7:  END if
8:  CALCULATE: D
9:  REPORT: D
10: END
```

FIGURE 5.21
Pseudocode for MACBETH. *Note:* $\{C_j(A_i)\}_{(1)}, \ldots, \{C_j(A_i)\}_{(n)}$ is the ordered pair signifying most attractive to least attractive.

while a wet well provides a moderate quantity of oil and a soaking well generates a substantial quantity of oil. If the oil company conducts seismic tests, then, it can learn about the underlying structure of the oil basin before deciding whether to drill for oil or not. The underlying oil basin structure may be one of the following types, which are (i) no structure, (ii) an open structure, or (iii) a closed structure. If no structure is found, then, the prospect of finding oil is bleak. If an open structure is discovered, then, the prospect of finding oil is fair, while finally, if the structure is closed, then, the prospect of finding oil is bright. The following joint probability values (Table 5.15) between the underlying geological structure and the oil-bearing state are also given.

Finally, the oil company also has the following set of information regarding the net present value (NPV) of the three states, which are (i) NPV(dry state) $= -0.6$ million; (ii) NPV(wet state) $= 0.8$ million, and (iii) NPV(soaking state) $= 2.4$ million. With these sets of information, one can find the best course of action for the oil company for the situation as depicted in Figure 5.23.

Thus, *Expected Mean Value at* $C_1 = 0.5$. Similarly, $EMV(C_3) = -0.16$, $EMV(C_4) = 0.37$, and $EMV(C_5) = 1.51$. Hence, at the decision points D_2, D_3, and D_4, we have

- At D_2 : D_{21} (Drill) and $EMV(D_{21}) = -0.16$; D_{22} (Do not drill) and $EMV(D_{22}) = 0$

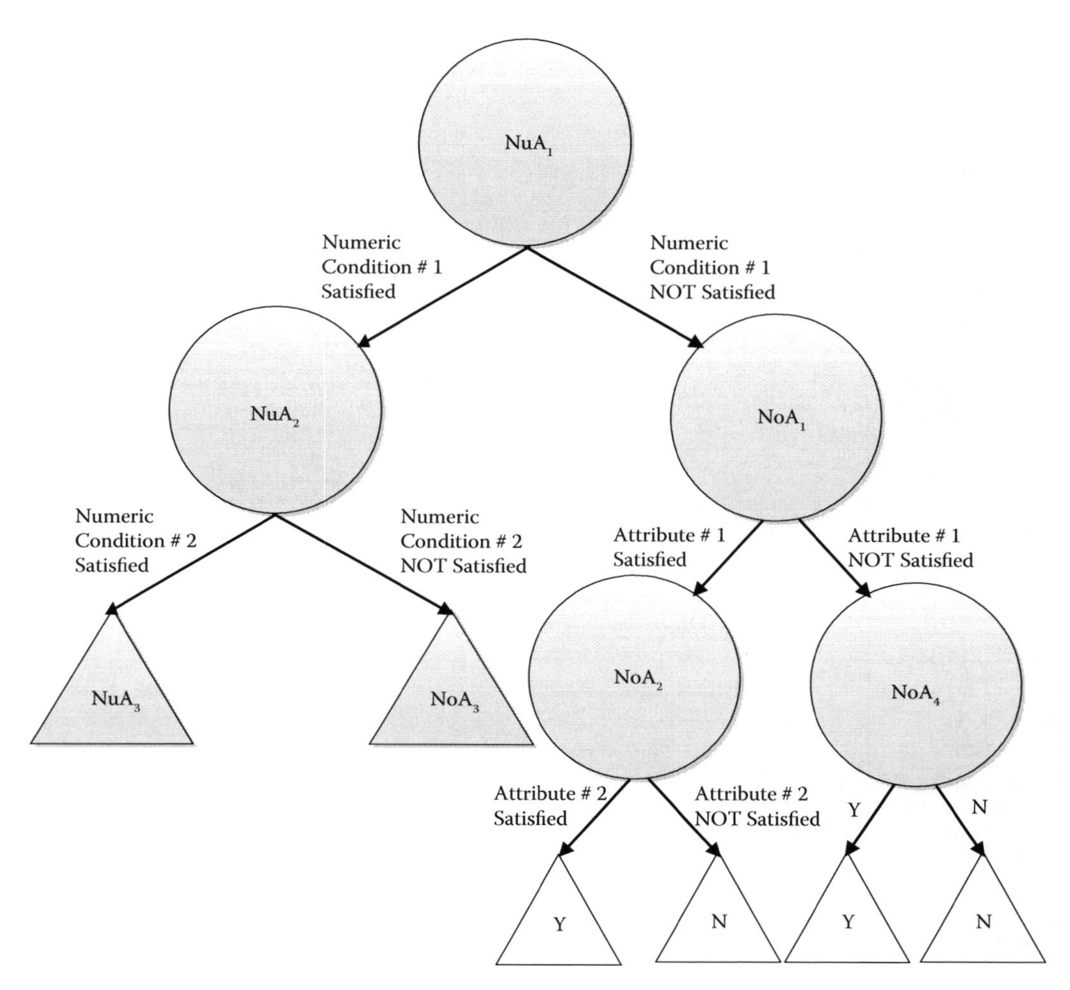

FIGURE 5.22
Typical decision tree.

TABLE 5.15
Joint Probability Distribution for the Oil Exploration (Example 5.12)

Oil-Bearing State	Underlying Geological Structure: No Structure	Open Structure	Closed Structure	Marginal Probability of the State
Dry	0.32	0.15	0.03	0.50
Wet	0.04	0.10	0.11	0.25
Soaking	0.04	0.05	0.16	0.25
Marginal probability of the geological structure	0.40	0.30	0.30	1.00

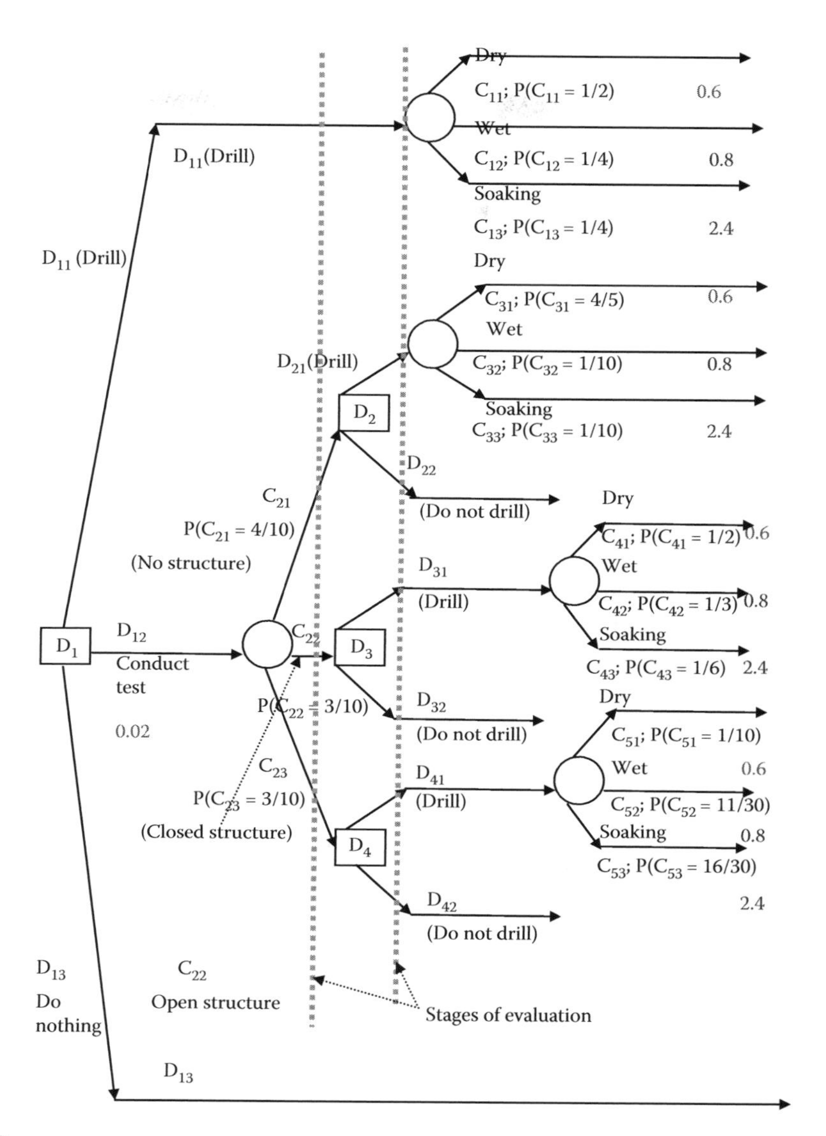

FIGURE 5.23
Decision tree for an oil company (Example 5.12).

- At D_3 : D_{31} (Drill) and $EMV(D_{31}) = 0.37$; D_{32} (Do not drill) and $EMV(D_{32}) = 0$
- At D_4 : D_{41} (Drill) and $EMV(D_{41}) = 1.51$; D_{42} (Do not drill) and $EMV(D_{42}) = 0$

Furthermore $EMV(C_2) = 0.56$. Hence, at the decision point D_1, we have

- D_{11} (Drill) and $EMV(D_{11}) = 0.50$
- D_{12} (Conduct seismic test) and $EMV(D_{12}) = 0.544$
- D_{13} (Do nothing) and $EMV(D_{13}) = 0$

On the basis of the evaluation of alternatives, the set of decision strategies is shown in Table 5.16.

TABLE 5.16

Set of Decision Strategies (Example 5.12)

Path	Probability	NPV
$D_{12} \rightarrow C_{12} \rightarrow D_{22}$	0.40	$-20,000$
$D_{12} \rightarrow C_{22} \rightarrow D_{31} \rightarrow C_{41}$	0.15	$-620,000$
$D_{12} \rightarrow C_{22} \rightarrow D_{31} \rightarrow C_{42}$	0.10	$+780,000$
$D_{12} \rightarrow C_{22} \rightarrow D_{31} \rightarrow C_{43}$	0.05	$+2,380,000$
$D_{12} \rightarrow C_{23} \rightarrow D_{41} \rightarrow C_{51}$	0.03	$-620,000$
$D_{12} \rightarrow C_{23} \rightarrow D_{41} \rightarrow C_{52}$	0.11	$+780,000$
$D_{12} \rightarrow C_{23} \rightarrow D_{41} \rightarrow C_{53}$	0.16	$+2,380,000$

Note: So, the expected value of the NPV is $E(NPV) = -20,000 \times 0.40 - 620,000 \times 0.15 + 780,000 \times 0.10 + 2,380,000 \times 0.05 = 544,000$.

5.1.6 Other Methods

Considering the paucity of space plus the fact that the chapter is a part of the set of topics related to decision sciences, we end this discussion with a mention of few of the existing methods in decision analysis such as (i) aggregated indices randomization method (AIRM) (developed by Aleksey Krylov), (ii) dominance-based rough set approach (DRSA) method, (iii) the evidential reasoning approach method, (iv) gray relational analysis (GRA) method, (v) inner product of vectors (IPV) method, (vi) new approach to appraisal (NATA) method, (vii) nonstructural fuzzy decision-support system (NSFDSS) method, (viii) superiority and inferiority ranking (SIR) method, (ix) value analysis (VA) method, (x) value engineering (VE) method, etc. The interested reader can check with the existing references to study these different MDCM methods in detail.

Finally, at the end of this chapter, we add a few relevant urls, data-set information, and software information for the areas covered in this chapter. We hope that this would help the reader to get the information that is essential for the successful implementation of different MCDM/MCDA techniques, which are widely used nowadays.

References

Aleskerov, F., Bouyssou, D., and Monjardet, B. 2007. *Utility Maximization, Choice and Preference (Studies in Economic Theory)*, Springer, Berlin, ISBN: 978-3540341826.

Arrow, K. J. 1971. *Essays in the Theory of Risk-Bearing*, Markham Publishing Company, Chicago, ISBN: 0841020019.

Bana e Costa, C. A. and Chagas, M. P. 2004. A career choice problem: An example of how to use MACBETH to build a quantitative value model based on qualitative value judgments, *European Journal of Operational Research*, **153**, 323–331.

Bana e Costa, C. A. and Vansnick, J.-C. 1994. MACBETH: An interactive path towards the construction of cardinal value functions, *International Transactions in Operational Research*, **1**, 489–500.

Bana e Costa, C. A. and Vansnick, J.-C. 1995. General overview of the MACBETH approach in advances in multicriteria analysis. In P. M. Pardalos, Y. Siskos, and C. Zopounidis (Editors), *Non-Convex Optimization and Its Applications*, Kluwer Academic Publishers, Dordrecht, 93–100, ISBN: 0-7923-3671-2.

Bana e Costa, C. A. and Vansnick, J.-C. 1997. A theoretical framework for measuring attractiveness by a categorical based evaluation technique (MACBETH) in multicriteria analysis. In J. Clímaco (Editor),

Proceedings of the XIth International Conference on MCDM, August 1–6, 1994, Coimbra, Portugal, Springer-Verlag, 15–24, ISBN: 978-3-642-64500-6.

Behzadian, M., Kazemzadeh, R. B., Albadvi, A., and Aghdasi, M. 2010. PROMETHEE: A comprehensive literature review on methodologies and applications, *European Journal of Operational Research*, **200**, 198–215.

Belton, V. and Stewart, T. J. 2002. *Multiple Criteria Decision Analysis: An Integrated Approach*, Kluwer Academic, Dordrecht, ISBN: 978-1-4615-1495-4.

Berger, J. O. 1985. *Utility and Loss in Statistical Decision Theory and Bayesian Analysis*, Springer-Verlag, New York, ISBN: 3-540-96098-8.

Brans, J. P. 1982. L'ingénierie de la décision: élaboration d'instruments d'aide à la décision. La méthode PROMETHEE. In R. Nadeau and M. Landry (Editors), L'aide à la décision: Nature, Instruments et Perspectives d'Avenir. Quebec, Presses de l'Université Laval.

Brans, J. P. and Vincke, Ph. 1985. A preference ranking organization method, *Management Science*, **31**, 647–656.

Brans, J. P. and Mareschal, B. 2005. PROMETHEE methods (Chapter 5, pp. 163–195). In J. Figueira, S. Greco, and M. Ehrgott (Editors), *Multiple Attributes Decision Analysis: State of the Art Survey*, Springer, New York, ISBN: 9780387230818.

Brans, J. P., Mareschal, B., and Vincke, Ph. 1984. *PROMETHEE: A New Family of Outranking Methods in Multicriteria Analysis*. In (OR'84: J. P. Brans, Editor), 408–421, North Holland.

Breiman, L., Friedman, J. H., Olshen, R. A., and Stone, C. J. 1984. *Classification and Regression Trees*, Wadsworth International Group, Belmont, CA, ISBN: 978-0412048418.

Bryson, N. 1995. A goal programming method for generating priority vectors, *The Journal of the Operational Research Society*, **46**, 641–648.

Castillo, E., Hadi, A. S., Balakrishnan, N., and Sarabia, J. M. 2004. *Extreme Value and Related Models with Applications in Engineering and Science*, John Wiley & Sons, New York, ISBN: 978-0471671725.

Charnes, A. and Cooper, W. W. 1961. *Management Models and Industrial Applications of Linear Programming*, John Wiley & Sons, New York, ISBN: 9780471148500.

Chen, S.-J. and Hwang, C.-L. 1992. *Fuzzy Multiple Attribute Decision Making: Methods and Applications in Lecture Notes in Economics and Mathematical Systems*, **375**, Springer-Verlag, Berlin, ISBN: 978-3-642-46768-4.

Chu, A., Kalaba, R., and Springam, K. 1979. A comparison of two methods for determining the weights of belonging to fuzzy sets, *Journal of Optimization Theory and Applications*, **27**, 531–541.

Cogger, K. O. and Yu, P. L. 1985. Eigenweight vectors and least distance approximation for revealed preference in pairwise weight ratios, *Journal of Optimization Theory and Applications*, **46**, 483–491.

Coles, S. 2001. *An Introduction to Statistical Modeling of Extreme Values*, Springer-Verlag, New York, ISBN: 978-1-4471-3675-0.

Cook, W. D. and Kress, M. 1988. Deriving weights from pairwise comparison ratio matrices: An axiomatic approach, *European Journal of Operational Research*, **37**, 355–362.

Crawford, G. and Williams, C. 1985. A note on the analysis of subjective judgment matrices, *Journal of Mathematical Psychology*, **29**, 387–405.

Dyer, J. S., Fishburn, P. C., Steuer, R. E., Wallenius, J., and Zionts, S. 1992. Multiple criteria decision making, Multiattribute utility theory: The next ten years, *Management Science*, **38**, 645–654.

Ehrgott, M., Figueira, J. R., and Greco, S. (Editors). 2010. *Trends in Multiple Criteria Decision Analysis*, Springer, Berlin, ISBN: 978-1-4419-5903-4.

Embrechts, P., Klüppelberg, C., and Mikosch, T. 1997. *Modelling Extremal Events for Insurance and Finance*, Springer-Verlag, Berlin, ISBN: 978-3-642-33483-2.

Figueira, J., Greco, S., and Ehrgott, M. 2005. *Multiple Attributes Decision Analysis: State of the Art Survey*, Springer, New York, ISBN: 9780387230818.

Fishburn, P. C. 1970. *Utility Theory for Decision Making*, John Wiley & Sons, New York, ISBN: 0-88275-736-9.

Fishburn, P. C. 1988. *Nonlinear Preference and Utility Theory*, Johns Hopkins University Press, Baltimore, ISBN: 978-0801835988.

Fishburn, P. C. 1989. Generalizations of expected utility theories: A survey of recent proposals, *Annals of Operations Research*, **19**, 3–28.

Gal, T., Stewart, T. J., and Hanne, T. 1999. *Multicriterion Decision Making*, Springer-Verlag, New York, ISBN: 978-1-4613-7283-7.

Golany, B. and Kress, M. 1993. A multicriteria evaluation of methods for obtaining weights from ratio-scale matrices, *European Journal of Operational Research*, **69**, 210–220.

Greco, S., Ehrgott, M., and Figueira, J. R. 2010. *Trends in Multiple Criteria Decision Analysis*, Springer-Verlag, Berlin, ISBN: 9781441959041.

Hancock, T. R., Jiang, T., Li, M., and Tromp, J. 1996. Lower bounds on learning decision lists and trees, *Information and Computation*, **126**, 114–122.

Hansen, P. and Ombler, F. 2008. A new method for scoring additive multi-attribute value models using pairwise rankings of alternatives, *Journal of Multi-Criteria Decision Analysis*, **15**, 87–105.

Hwang, C.-L. and Yoon, K. 1981. *Multiple Attribute Decision Making Methods and Applications: A State-of-the-Art Survey in Lecture Notes in Economics and Mathematical Systems*, **186**, Springer-Verlag, Berlin, ISBN: 978-3-540-10558-9.

Hyafil, L. and Rivest, R. L. 1976. Constructing optimal binary decision trees is NP complete, *Information Processing Letters*, **5**, 15–17.

Ingersoll, J. E. Jr. 1987. *Theory of Financial Decision Making*, Rowman and Littlefield, Totowa, NJ, ISBN: 0-8476-7359-6.

Ishizaka, A. and Lusti, M. 2006. How to derive priorities in AHP: A comparative study, *Central European Journal of Operations Research*, **14**, 387–400.

Jolliffe, I. T. 2002. *Principal Component Analysis*, Springer-Verlag, New York, ISBN: 978-0387954424.

Jones, D. and Tamiz, M. 2010. *Practical Goal Programming*, Springer-Verlag, New York, ISBN: 978-1-4419-5770-2.

Jordan, M. I. and Jacobs, R. A. 1994. Hierarchical mixtures of experts and the EM algorithm, *Neural Computation*, **6**, 181–214.

Kahraman, C. 2008. *Fuzzy Multi-Criteria Decision Making: Theory and Applications with Recent Developments*, Springer, New York, ISBN: 978-0-387-76812-0.

Keeney, R. L. and Raiffa, H. 1993. *Decisions with Multiple Objectives: Preferences and Value Tradeoffs*, Cambridge University Press, Cambridge and New York, ISBN: 0-521-44185-4.

Khademi, N., Behnia, K. and Saedia, R. 2014. Using analytic hierarchy/network process (AHP/ANP) in developing countries: Shortcomings and suggestions, *Energy Economy*, **59**, 2–29.

Koopmans, T. C. 1951. Analysis of production as an efficient combination of activities. In T. C. Koopmans (Editor), *Activity Analysis of Production and Allocation, Cowles Commission for Research in Economics Monograph No. 13*, John Wiley & Sons, New York, 33–97.

Kotz, S. and Nadarajah, S. 2000. *Extreme Value Distributions: Theory and Applications*, World Scientific Press, Singapore, ISBN: 978-1-86094-224-2.

Kreps, D. M. 1988. *Notes on the Theory of Choice*, West View Press, Boulder, CO, ISBN: 0-8133-7553-3.

Kuhn, H. W. and Tucker, A. W. 1951. In J. Neyman (Editor), *Nonlinear Programming in Proceedings of the Second Berkeley Symposium on Mathematical Statistics and Probability*, Statistical Laboratory of the University of California, Berkeley, USA, 481–492.

Larichev, O. I. and Olson, D. L. 2001. *Multiple Attributes Analysis in Strategic Siting Problems*, Kluwer Academic Publishers, Boston, ISBN: 0792373790.

Leadbetter, M. R., Lindgren, G., and Rootzén, H. 1983. *Extremes and Related Properties of Random Sequences and Processes*, Springer-Verlag, New York, ISBN: 0-387-90731-9.

Lee, S. M. 1972. *Goal Programming for Decision Analysis*, Auerback, Philadelphia, ISBN: 9780877691440.

Marshall, A. 1920. *Principles of Economics*, Macmillan, London, ISBN: 1-57392-140-8.

Mikhailov, L. 2004. Group prioritization in the AHP by fuzzy preference programming method, *Computers and Operations Research*, **31**, 293–301.

Morgenstern, O. 1976. Some reflections on utility. In A. Schotter (Editor), *Selected Economic Writings of Oskar Morgenstern*, New York University Press, New York, 65–70, ISBN: 0-8147-7771-6.

Naumov, G. E. 1991. NP-completeness of problems of construction of optimal decision trees, *Soviet Physics*, **36**, 270–271.

von Neumann, J. and Morgenstern, O. 2007. *Theory of Games and Economic Behavior*, Princeton University Press, Princeton, ISBN: 9780691130613.

Opricovic, S. and Tzeng, G.-H. 2004. Compromise solution by MCDM methods: A comparative analysis of VIKOR and TOPSIS, *European Journal of Operational Research*, **156**, 445–455.

Opricovic, S. and Tzeng, G.-H. 2007. Extended VIKOR method in comparison with outranking methods, *European Journal of Operational Research*, **178**, 514–529.

Pareto, V. 1896. Cours d'Économie Politique Professé a l'Université de Lausanne, Vol. I. http://www.institutcoppet.org/wp-content/uploads/2012/05/Cours-d%C3%A9conomie-politique-Tome-I-Vilfredo-Pareto.pdf.

Parlos, P. M. (Editor). 2000. *Multi-Criteria Decision Making Methods: A Comparative Study*, Kluwer Academic Publishers, Boston, ISBN: 0-7923-6607-7.

Pomerol, J.-C. and Barba-Romero, S. 2000. *Multicriterion Decision in Management: Principles and Practice*, Kluwer Academic Publishers, Norwell, ISBN: 0-7923-7756-7.

Quinlan, J. R. 1993. *Programs for Machine Learning*, Morgan Kaufmann, San Mateo, CA, ISBN: 1-55860-238-0.

Resnick, S. I. 1987. *Extreme Values, Regular Variation and Point Processes*, Springer-Verlag, New York, ISBN: 0-387-96481-9.

Rogers, M., Bruen, M., and Maystre, L.-Y. 2000. *ELECTRE and Decision Support Methods and Applications in Engineering and Infrastructure Investment*, Springer-Verlag, New York, ISBN: 978-1-4757-5057-7.

Romero, C. 1991. *Handbook of Critical Issues in Goal Programming*, Pergamon Press, Oxford, ISBN: 0-08-040661-0.

Roy, B. 1968. Classement et choix en présence de points de vue multiples (la méthode ELECTRE), *La Revue d'Informatique et de Recherche Opérationelle (RIRO)*, **8**, 57–75.

Roy, B. 1996. *Multicriteria Methodology for Decision Aiding: Nonconvex Optimization and Its Applications*, Kluwer Academic Publishers, Dordrecht, ISBN: 978-1-4757-2500-1.

Saaty, T. L. 1994. How to make a decision: The analytic hierarchy process, *Interfaces*, **24**, 19–43.

Saaty, T. L. 1996. *Decision Making with Dependence and Feedback: The Analytic Network Process*, RWS Publications, Pittsburgh, ISBN: 0-9620317-9-8.

Saaty, T. S. 2003. Decision-making with the AHP: Why is the principal eigenvector necessary, *European Journal of Operational Research*, **145**, 85–91.

Saaty, T. L. 2004. Fundamentals of the analytic network process—Dependence and feedback in decision-making with a single network, *Journal of Systems Science and Systems Engineering*, **13**, 129–157.

Saaty, T. L. 2005. *Theory and Applications of the Analytic Network Process: Decision Making with Benefits, Opportunities, Costs and Risks*, RWS Publications, Pittsburgh, ISBN: 1-888603-06-2.

Saaty, T. L. and Ozdemir, M. S. 2005. *The Encyclicon*, RWS Publications, Pittsburgh, ISBN: 978-1-888603-09-5.

Saaty, T. L. and Hu, G. 1998. Ranking by the eigenvector versus other methods in the analytic hierarchy process, *Applied Mathematical Letters*, **11**, 121–125.

Saaty, T. L. and Vargas, L. G. 2000. *Models, Methods, Concepts and Applications of the Analytic Hierarchy Process*, Kluwer Academic Publishers, Boston, ISBN: 0-7923-7267-0.

Saaty, T. L. and Vargas, L. G. 2006. *Decision Making with the Analytic Network Process: Economic, Political, Social and Technological Applications with Benefits, Opportunities, Costs and Risks*, Springer Publishers, New York, ISBN: 0-387-33859-4.

Schniederjans, M. 1995. *Goal Programming: Methodology and Applications*, Springer-Verlag, New York, ISBN: 978-1-4615-2229-4.

Srdjevic, B. 2005. Combining different prioritization methods in the analytic hierarchy process synthesis, *Computers and Operations Research*, **32**, 1897–1919.

Trzaskalik, T. and Michnik, J. (Editor). 2002. *Multiple Objective and Goal Programming: Recent Developments (Advances in Intelligent and Soft Computing)*, Physica-Verlag, Heidelberg, ISBN: 978-3-7908-1812-3.

Tzeng, G.-H. and Huang, J.-J. 2011. *Multiple Attribute Decision Making: Methods and Applications*, CRC Press, Taylor & Francis Group, Boca Raton, ISBN: 978-1-4398-6157-8.

Zadeh, L. A. 1963. Optimality and non-scalar-valued performance criteria, *IEEE Transactions on Automatic Control*, **8**, 59–60.

Zantema, H. and Bodlaender, H. L. 2000. Finding small equivalent decision trees is hard, *International Journal of Foundations of Computer Science*, **11**, 343–354.

Zeleny, M. 1982. *Multi Criterion Decision Making*, McGraw-Hill, New York, ISBN: 978-0070727953.
Zimmermann, H.-J. 1996. *Fuzzy Set Theory and Its Applications*, Kluwer Academic Publishers, Boston, ISBN: 0792374355.

Relevant Information Related to Data Sets, Softwares, etc.

S No.	Description	URL
1	Wikipedia definition of MCDA	http://en.wikipedia.org/wiki/Multiple-criteria_decision_analysis
2	MCDM decision-making software	https://en.wikipedia.org/wiki/Decision-making_software
3	International Society on Multiple Criteria Decision Making	http://www.mcdmsociety.org/
4	Multicriteria decision aid for business	http://www.decision-drive.com/
5	Society for Judgment and Decision Making	http://www.sjdm.org/
6	MCDM Books	http://www.mcdmsociety.org/books.html
7	*Journal of Multi-Criteria Decision Analysis*	http://onlinelibrary.wiley.com/journal/10.1002/%28ISSN%291099-1360;jsessionid=E9AD71395B9D174753DD1FE0DDABEBAA.f04t01
8	Softwares related to MCDM: International Society on Multiple Criteria Decision Making	http://www.mcdmsociety.org/soft.html
9	MCDA softwares	http://www.cs.put.poznan.pl/ewgmcda/index.php/software
10	Smart Picker Pro	http://www.smart-picker.com/
11	Multiobjective combinatorial optimization collection of problems	http://xgandibleux.free.fr/MOCOlib/
12	Multiobjective Optimization Library	http://home.ku.edu.tr/~moolibrary/
13	MCDMlib: test problems for multiobjective optimization	http://www.mcdmsociety.org/MCDMlib.html#MOCOlib
14	EURO Working Group on Multicriteria Decision Aiding	http://www.cs.put.poznan.pl/ewgmcda/
15	PROMETHEE-GAIA software and related information	http://www.promethee-gaia.net
16	Wikipedia	http://en.wikipedia.org/wiki/Preference_ranking_organization_method_for_enrichment_evaluation
17	Visual PROMETHEE	http://visual-promethee.software.informer.com/
18	Promethee 1.3 software	http://promethee.irsn.org/doku.php
19	Decision Lab 2000	http://decision-lab-2000.software.informer.com/
20	Logical Decisions	http://www.logicaldecisions.com/
21	Expert Choice	http://expertchoice.com/
22	Qualiflex	http://www.swmath.org/software/9592
23	Advanced Information Management	http://www.aimsoftware.com/
24	Criterium Decision Plus	http://www.infoharvest.com/ihroot/index.asp
25	Decision Pad	http://www.apian.com/downloads/

Continued

S No.	Description	URL
26	1000Minds	https://www.1000minds.com/
27	Web-Based Multi-Criteria Decision Making Software	http://www.visadecisions.com/
28	A Multicriteria Decision Support System	http://www.m-macbeth.com/en/m-home.html
29	Multiple Criteria Decision Support Software	http://www.isy.vcu.edu/~hweistro/mcdmchapter.htm
30	Software for MACBETH	http://www.umh.ac.be/~smq

Relevant Information Related to the Use of Software for Few of the MCDM/MCDA Techniques

Software	Supported MCDA Methods	Pairwise Comparison	Sensitivity Analysis	Group Evaluation	Web Based
WISED	MACBETH	Yes	Yes	Yes	Yes
Super Decisions	AHP, ANP	Yes	Yes	No	Yes
PriEsT	AHP	Yes	Yes	No	No
M-MACBETH	MACBETH	Yes	Yes	Yes	No
Logical Decisions	AHP	Yes	Yes	Yes	No
Intelligent Decision System	Evidential reasoning approach, Bayesian inference, Dempster–Shafer theory, and utility	Yes	Yes	Yes	Available on request
Hiview3		No	Yes	Yes	No
Expert Choice	AHP	Yes	Yes	Yes	Yes
Decision Lens	AHP, ANP	Yes	Yes	Yes	Yes
DecideIT	MAUT	Yes	Yes	Yes	Yes
D-Sight	PROMETHEE, utility	Yes	Yes	Yes	Yes
Criterium DecisionPlus	AHP, Smart	Yes	Yes	No	No
Analytica		No	Yes	No	Yes
Altova MetaTeam	WSM	No	No	Yes	Yes
Ahoona	WSM, utility	No	No	Yes	Yes
1000Minds	PAPRIKA	Yes	Yes	Yes	Yes

6

Queueing Theory

Ülkü Gürler and Emre Berk

CONTENTS

ABSTRACT Queueing theory is concerned with the quantitative modeling of dynamic systems that generate waiting lines, and the analysis of the behavior of such systems in the short and long time spans. In this chapter, we present a brief overview of the history of queueing theory and the basic concepts of it. The aim of this chapter is to expose the reader to one of the most instrumental and widely applicable topics of stochastic analysis, to provide the basic concepts of it, and to stimulate further interest in it. Some motivating examples of stochastic processes are discussed followed by the main topics in a standard course of stochastic processes. These include the renewal processes, Markov chains, and continuous-time Markov chains. Queueing models are then discussed in more detail, where the focus is mainly restricted to Markovian queues. A more complicated topic of queueing networks is also briefly discussed. Most of the topics discussed are also supported by illustrative examples. The methods are presented mainly under the assumption of a single queue, where a single type of service is provided by possibly several servers under the FIFO (first in first out) service protocol and under the traditional assumption that once a customer enters a queue, she/he stays there until the service has been completed. In an extensive section, we pointed out relaxation of such assumptions and extensions in several directions regarding the queue discipline, service protocol, customer behavior, several service types, estimation of major system parameters, as well as the current research interests in the field. In an appendix, we included a brief review of the background material in probability theory.

6.1 Introduction and Brief History

Waiting lines in front of ticket offices, bank counters, and ATM machines are a common sight. Queueing theory is concerned with the quantitative modeling of dynamic systems that generate such lines, their analysis, their short- or long-term behavior, and their improvement. Besides physical manifestations of actual customers waiting in lines in real life, a *waiting line* in queueing theory may refer to more abstract applications, as well. For example, the jobs that wait for processing in the buffer of a computer solver, the phone calls on hold at a call center, or the parts that wait for the next operation in a production environment are also considered as waiting lines. Queues are not desirable; the amount of time during waiting is in general a loss. People waiting in a line become unproductive during that time; semifinished products waiting in line for operations incur holding costs and/or may experience degradation in quality. Although highly undesirable, it is still almost impossible to completely avoid queues due to: (i) shortage of resources, such as the limited number of pay booths on a highway or the small number of counters in a bank. When a desired service cannot be provided immediately due to lack of resources, queues naturally arise and conversely if the resources were unlimited, there would be no queues. Although theoretically it may be possible to remove the shortages by employing more counters or building more booths, and in general increasing the resources, economically it may not be feasible to do so, or it may be impossible to do so as in the case of limited natural resources. When it is possible to increase the resources or the capacity by investment, the trade-off is between the acquisition cost for increased capacity and the opportunity costs incurred due to delays resulting from long lines, and (ii) due to the variability and uncertainty inherent in such systems, such as the variability in service times arising from the nature of the jobs or human factors, uncertainty of the arrival times for services, and machine failures. Various situations where queues are formed are described by *queueing models*. Analysis of these models allows us to understand the performance of the underlying systems and indicate how to improve their performances by balancing the above-mentioned costs in a rational way.

The earliest models studied in queueing theory were about telephone traffic congestion. The Danish mathematician A.K. Erlang was a pioneer researcher who published his results in his seminal work "The Theory of Probabilities and Telephone Conversations" in 1909. He observed that telephone arrivals were characterized by a Poisson input and exponential or constant holding (service) times. He provided analytical results for constant service times with up to three service channels and for exponential service times with an arbitrary number of channels. His papers that have appeared in the next two decades introduced the most important concepts of the topic. An early account of Erlang's work is published by Brockmeyer et al. (1948).

The works of Erlang (1909, 1917) have stimulated a stream of research. O'Dell (1920), Molina (1927), and Fry (1928) provided extensions to Erlang's results. Pollaczek (1930, 1934, 1965) made significant contributions. In particular, Khintchine (1932) and Pollaczek independently established the well-known Pollaczek–Khintchine formula for a single server queue with Poisson arrivals and general service time. Starting from 1937, Palm (1937, 1938, 1943, 1947) contributed to the theory with a series of papers and emphasized the impact of variable traffic intensity. Other researchers with important contributions include Crommelin (1932, 1934), Feller (1949a,b), and Kolmogorov (1931). Much of the basic queueing theory regarding the main models has been studied in the 1950s and 1960s, significant contributors being Kendall (1951, 1953), Lindley (1952), and Bailey (1954). Further details can be found in Syski (1960), Saaty (1961), and Bhat (1968). The advances in computer technology and traffics encountered in computer networks stimulated the research in queueing networks pioneered by Jackson (1957) and later contributions are made by Kingman (1969) and Whitt (1984a,b) among others.

6.2 Basic Concepts of Queueing Theory

In its generality, a queueing system is composed of one (or more) server(s), customers who are receiving service and those who may be waiting for their turn for service. All customers leave upon completion of their service. The term "customer" is a generic term and may refer to humans that arrive for service as well as parts/components in a production line that arrive for assembly. A server (also called a channel) is a resource, which is kept busy by a customer. The service time of a server is a nonnegative time interval over which both the customer and the server are busy. Typically, queueing systems experience congestion that will occur from time to time due to the scarcity of resources and the underlying uncertainties in the system mentioned above. Queueing theory investigates this congestion phenomenon—its sources, characteristics, and means to alleviate it.

Most of the queueing systems are characterized by five basic components: (1) arrival process of customers, (2) service process of servers, (3) number of servers (channels), (4) system capacity, and (5) queue discipline. Further characterizations are also possible by the number of service stages indicating whether a single stage or several stages (such as series of medical tests following one another) of services are undertaken, or by the number of input sources indicating whether the arrivals are generated by a finite or an infinite source. However, the first five characteristics are essential to most of the basic queueing systems.

6.2.1 Arrival (Input) Process

Arrival (input) process describes how the customer arrivals occur in the system. The intensity of arrivals is measured by the average number of arrivals per unit time, called the arrival rate, or by the average time between successive arrivals called the interarrival time. The dynamics of the arrival process is expressed commonly by the following patterns:

- Unit or batch arrivals (customers arrive one at a time or in groups)
- Renewal arrivals (times between arrivals are independent random variables)
- Stationary arrivals (parameters of arrival process do not change over time)
- Independent increments (number of arrivals in non-overlapping time intervals are independent)

When a customer arrives at the service facility, s/he may enter the system and wait until the service is provided, or may decide not to enter the system if the length of the queue is deemed long. The latter type of behavior is called *balking*. A customer who enters the system may leave it after waiting a certain amount of time; this is called *reneging*. Further, a customer who enters a queue in a multiple-channel system may switch from one line to the other; this is called *jockeying*. In this chapter, such customers are called *impatient*.

6.2.2 Service Process

Service process describes the service time distribution, which can be any nonnegative probability distribution. The characteristics regarding service time distributions are similar to those of inter-arrival time distributions. Service may be provided to a single customer such as a patient in a healthcare facility, or to a batch of customers such as a busload of tourists on a tour. Service rates may or may not depend on the congestion in the system. The former is referred to as state-dependent service rate. Service rate can be stationary or may change in time. For instance, if learning is in effect, service rate may increase over time, or if there is deterioration in the server's condition such as fatigue, service rate may decrease.

6.2.3 Number of Servers

Number of servers refers to the number of parallel service channels that can provide service to customers simultaneously. When there are several channels, how many queues are formed in front of the servers is a concern. Each server may have a queue of its own or, a single queue may be formed from which the customers are allocated to the servers as they become available. The second structure is more effective in several measures.

6.2.4 Queue Discipline

Queue discipline refers to the rule by which the next customer to be served is selected. The most common one is the first come first served (or first in first out) protocol denoted by $FCFS \equiv FIFO$. Another discipline is last come first served $LCFC$, which may be applicable in storage systems where the last arrived item to the storage may be served the next arriving customer. Service-in-random-order ($SIRO$) may apply to stocks of durable products where storage times are not relevant to the serving order. Another set of protocols regarding the queue discipline take into account the *priority* of customers. There are two general rules for prioritizing the customers. In *preemptive* queues, the customers with the highest priority are allowed to enter the service immediately (such as in hospital emergency rooms), whereas in *non-preemptive* queues, the priority customer waits until the service of the presently served customer is finished.

6.2.5 System Capacity

System capacity refers to the maximum number of customers that can be present in a system at a given time mostly due to physical constraints. Such systems can be considered as those where the arriving customers are forced to balk when the maximum capacity is reached.

Kendall (1953) introduced a notation to represent the main characteristics of a queueing system using a series of symbols separated by slashes as A/B/C/D/E/F, described as follows:

A. Specification of the arrival process
   M: Markovian—exponential
   D: Deterministic
   E_k: Erlang (k)
   GI : General, independent
   G: General arrival
B. Specification of the service process (the same notation for the arrival process is used)
C. Number of parallel service channels
D. Capacity (servers + queue length) of the system (maximum number of customers in the system at any point in time)
E. Queue discipline
   FCFS/LCFS/SIRO (RSS: random service selection)
F. Size of the input source

The above notation provides a detailed description of the queueing systems; however, for the most part, the first three letters are used. To illustrate, $M/E_k/3/7/FCFS/\infty$ in the above notation refers to a system with Poisson arrivals, Erlang(k) service times with three servers and a system capacity of seven customers who are served on a first come first served basis, and the source population of arrivals has an infinite size. As FIFO is the most common discipline and infinite source population is very common, sometimes the E and F segments are dropped in such cases for brevity, so that an $M/E_k/3/7/FCFS/\infty$ system is simply denoted as $M/E_k/3/7$. Furthermore, when arrivals are Poisson, the size of the source population is always infinite so that the last segment becomes redundant anyway.

6.2.6 Performance Measures

Whatever queueing model is used, we are interested in the performance of the system measured with respect to some criteria. Commonly used performance measures are the queue length (the number of customers waiting for service) at an arbitrary time t, the number of busy servers, waiting time in the queue, and the waiting time in the service. In certain systems, the delay (as a measure of customer satisfaction) may be more important whereas, in others, server utilization may be significant (as servers/machines may be considerably expensive). The analysis of the system and such performance measures for finite values of t is referred to as *transient* analysis. Except for some special cases such as an $M/M/1$ system, exact results are difficult to obtain in transient analysis. It is, therefore, common to provide the analysis of the system and the assessment of its performance in the *steady state*, when the system reaches a stable condition after a long time has passed since it started operation. Steady-state analysis is also of practical importance since the initial behavior of a system may have undesirable fluctuations, breaks, delays, etc. due to lack of experience and other factors and therefore may not reflect the operating characteristics in the long run. Therefore, in many cases, the decision makers would like to base their system assessments on the long-term behavior. There are, however, situations where the initial behavior is also important. For example, for setting the warranty periods for a product, the failure behavior in the early stages is more important. In this chapter, we will also consider the steady-state performances of the systems that will be studied. Whether a given system reaches steady state or not is another crucial aspect in the analysis of stochastic systems. These concepts will be further discussed in the context of specific queueing models. Before discussing specific queueing models, we introduce stochastic processes and provide some specific examples. The material in this chapter requires a good understanding of the basic concepts in probability. We, therefore, provide a brief review of probability in the appendix. The readers are recommended to read the material in the appendix to refresh their background unless they have a working knowledge of probability.

6.3 Stochastic Processes: Basic Concepts and Poisson Process

6.3.1 Introduction

The behavior of dynamic systems that evolve over time is characterized by a set of infinitely many variables. Consider, for example, the blood screening tests conducted in a medical lab and every time a rare disease is encountered it is recorded as an "event." At any time t, the health officials may be interested in the total number of events detected by time t. This example can be described by a sequence of random variables $N(t)$ defined as the number of events counted in the interval $(0, t]$, $t \geq 0$. Another simple example is coin tossing. Suppose an unfair coin is tossed indefinitely many of the previous ones. Then, the number of heads obtained can be described by a sequence of random variables $N(n), n = 1, 2, \ldots,$ which denotes the number of heads that appeared in n tossings. Obviously, $N(t)$ counts the number of rare disease that appeared by time t where t is an index considered over a compact interval, whereas $N(n)$ counts the number of heads obtained by the nth tossing where n is an index defined over a discrete set.

Formally stating, a *stochastic process* $\{X(t), t \in T\}$ is a collection of random variables taking on values over the set S. The set T is the index (time) space and S is the state space of the process. For a fixed value of $t \in T$, $X(t)$ is a random variable, whereas for a fixed outcome in the sample space, $\{X(t), t \in T\}$ is a realization of the stochastic process, which is also called a *sample path*. Stochastic processes represent dynamic systems where $X(t)$ corresponds to the state of the system at time t and they are classified according to the structure of the sets T and S. If both T and S are discrete sets, then $X(t)$ is said to be a discrete-time, discrete-state stochastic process; if T is discrete and S is a continuous set, then $X(t)$ is said to be a discrete-time, continuous-state process, and so on and so forth. Note here that the "time" index above is a generic term, and it does not necessarily correspond to chronological time; it may refer to the length of a manufactured metal rod, where $X(t)$ may be defined as the number of defects observed on a rod of length t. In the blood screening test example above, $\{N(t), t \geq 0\}$ is an example of a continuous-time, discrete-state stochastic process with $T = [0, \infty]$ and $S = (0, 1, 2, \ldots)$, whereas in the coin tossing example, $\{N(n), n \geq 1\}$ is an example of a discrete-time, discrete-state stochastic process with $T = [0, 1, 2, \ldots]$ and the same S set. Consider now the cash status of a financial agency at the beginning of review periods where a review period can be a day, week, etc. Suppose random amounts of deposits and withdrawals take place also at random instances in time. However, we consider the cash level only at the beginning of the review periods. Since cash is considered as an uncountable entity, the cash level $\{X(n), n = 0, 1, 2, \ldots\}$ at the beginning of the nth period constitutes a stochastic process with discrete-time and continuous-state space possibly described by $S = (-\infty, \infty)$. If we consider the same process at arbitrary time instances, rather than the beginning of review periods, then the corresponding process would be a continuous-time continuous-state stochastic process. As T is an arbitrary set, a stochastic process may be composed of an uncountable number of random variables. Hence, characterization of their joint behavior is beyond the concept of joint distribution of random variables. On the other hand, the joint distributions of $X(t_1), X(t_2), \ldots, X(t_n)$ for $t_i \in T, i \geq 1$ are called finite-dimensional distributions of $\{X(t), t \in T\}$. Stochastic processes are characterized by finite-dimensional distributions. As an example, if all finite-dimensional distributions of a stochastic process are multivariate normal, then the corresponding process is called a Gaussian process. The definition of a stochastic process given above is rather general, and it is almost impossible to derive further properties to provide practical and theoretical insights.To study such properties and obtain their performance measures, some structures are imposed on them. *Stationarity* and *independence* of increments are two commonly imposed properties. A stochastic process is said to have independent increments property if $X(t_1) - X(t_0), X(t_2) - X(t_1), \ldots, X(t_n) - X(t_{n-1})$ are independent random variables for all $t_i \in T, i \leq n, n \geq 2$. This property implies that the behavior of the process over non-overlapping intervals is independent of each other. A stochastic process is said to have stationary increments property if the random vectors $(X(t_1) - X(t_0), X(t_2) - X(t_1), \ldots, X(t_n) - X(t_{n-1}))$ and $(X(t_1 + h) - X(t_0 + h), X(t_2 + h) - X(t_1 + h), \ldots, X(t_n + h) - X(t_{n-1} + h))$ have identical probability distributions for all $t_i \in T, i \leq n, n \geq 2$, and all choices of $h > 0$. This property implies that the joint distribution of the increments of the process may depend on the length of the increments but not on the location. Note that the term *homogeneity* is also used to describe stationarity in stochastic processes. The stationary and independent increments assumptions are strong structural assumptions that yield important analytical results and insights. The homogeneity assumption is easier to relax in some processes; however, violence of independent increments assumption may result in complicated models. On the other hand, nonhomogeneous processes are commonly observed in real life when the underlying stochastic mechanisms change in time, for instance, if there is seasonality. Therefore, nonhomogeneous processes are widely studied in time-series analysis. For dependent increments, financial risk models and ruin probabilities in insurance are two frequently encountered areas. For some early and more recent discussions of these extensions, see, for example, Priestley (1965), Pfeifer (1982), Hughes et al. (1999), Asmussen (1999), and Muller and Pflug (2001). Further properties of special stochastic processes will be discussed in more detail in the following sections. In particular, we will introduce the Poisson process, renewal processes, and Markov chains as fundamental processes before we move on to the basic queueing models.

6.3.2 Poisson Process

We next discuss one of the most well-known processes that has both theoretical and practical importance, the Poisson process named after the French mathematician Simeon Denis Poisson (1781–1840) who derived the Poisson distribution and the limiting case of the binomial distribution with large number of trials with small *success* probabilities. A discrete-time stochastic process $X(t), t \geq 0$ is a Poisson process with rate $\lambda > 0$, if

1. $X(0) = 0$.
2. $\{X(t), t \geq 0\}$ has independent increments. That is, for $0 < t_1 < t_2 < \cdots < t_k$, the increments $X(t_1) - X(0), X(t_2) - X(t_1), \ldots, X(t_k) - X(t_{k-1})$ are independent random variables for all $k \geq 2$.
3. For all $s, t, \geq 0$ it holds that

$$P(X(t+s) - X(s) = n) = \frac{e^{-\lambda t}(\lambda t)^n}{n!}$$

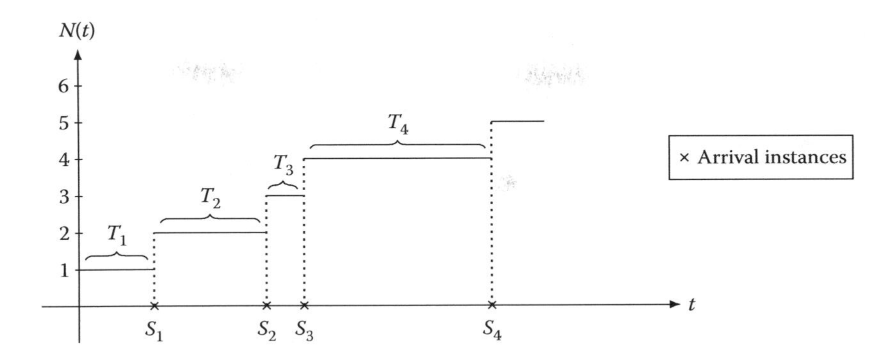

FIGURE 6.1
Number of events in $(0, t]$: A sample path from a Poisson process.

The above definition indicates that the distribution of number of events over an interval merely depends on the length of the interval, not on its location. Hence, the Poisson process has the *stationarity* or *homogeneity* property discussed earlier. Let T_n be the time between nth and $(n+1)$th events. Then $P(T_1 > t) = P(0 \text{ events in}(0,t]) = e^{-\lambda t}$, implying that the waiting time until the first event is exponential with rate λ. Using the independent increments property, we argue that $P(T_2 > t \mid T_1 = s) = P(0 \text{ events in}(s, s+t] \mid T_1 = s) = e^{-\lambda t}$. Hence, T_2 is also an exponential random variable with the same rate which is independent of T_1. Generalizing this argument, we conclude that the sequence $\{T_n, n \geq 1\}$ are i.i.d. exponential random variables with rate λ. Let $S_n = \sum_0^{n-1} T_k$ denote the time of the nth arrival (see Figure 6.1). The result that a Poisson process has independent and identically distributed interarrival times then implies that S_n has a gamma distribution with shape parameter n and scale parameter λ.

6.3.2.1 Superposition of Poisson Processes

Let $X_1(t), t \geq 0$ be a Poisson process with rate λ_1 and $X_2(t), t \geq 0$ be another Poisson process independent of $X_1(t)$ with rate λ_2. Define $Y(t) = X_1(t) + X_2(t)$, which denotes the total number of arrivals in $(0, t]$. Then,

$$P(Y(t) = k) = P(X_1(t) + X_2(t) = k) = \sum_{m=0}^{k} P(X_1(t) = m, X_1(t) + X_2(t) = k)$$

$$= \sum_{m=0}^{k} P(X_1(t) = m) P(X_2(t) = k - m) = \sum_{m=0}^{k} \frac{e^{-\lambda_1 t}(\lambda t)^m}{m!} \frac{e^{-\lambda_2 t}(\lambda_2 t)^{k-m}}{(k-m)!}$$

$$= \frac{e^{(\lambda_1+\lambda_2)t}[(\lambda_1+\lambda_2)t]^k}{k!} \sum_{m=0}^{k} \binom{k}{m} \left(\frac{\lambda_1}{\lambda_1+\lambda_2}\right)^m \left(\frac{\lambda_2}{\lambda_1+\lambda_2}\right)^{k-m}$$

where the sum adds up to one as the sum of binomial probabilities. This argument leads to the following result.

Theorem 6.1

$\{Y(t), t \geq 0\}$ is a Poisson process with rate $\lambda = \lambda_1 + \lambda_2$.

6.3.2.2 Decomposition of a Poisson Process

Let $\{X(t), t \geq 0\}$ be a Poisson process with rate λ. Also assume that each arrival (birth) is of Type 1 with probability p, independent of the other births, and of Type 2 with probability $1 - p$. Let $\{X_1(t), t \geq 0\}$ denote the number of Type 1 births $\in (0, t]$ and $\{X_2(t), t \geq 0\}$ be that of Type 2 births $\in (0, t]$. Then, we have

Theorem 6.2

$\{X_1(t), t \geq 0\}$ and $\{X_2(t), t \geq 0\}$ are independent Poisson processes with rates $\lambda_1 = \lambda p$ and $\lambda_2 = \lambda(1 - p)$, respectively.

6.3.2.3 Uniform Distribution and Poisson Process

Let $\{X(t), t \geq 0\}$ be a Poisson process with rate λ and let S_1 be the time of the first event. If we know that $X(t) = 1$, the distribution of S_1 is obtained as follows:

$$P(S_1 \leq u \mid X(t) = 1) = \begin{cases} 0 & \text{if } u \leq 0 \\ 1 & \text{if } u \geq t \end{cases}$$

For $0 < u < t$

$$\begin{aligned} P(S_1 \leq u \mid X(t) = 1) &= \frac{P(S_1 \leq u; X(t) = 1)}{P(X(t) = 1)} \\ &= \frac{P(X(u) = 1; X(t) - X(u) = 0)}{P(X(t) = 1)} \\ &= \frac{e^{-\lambda u}(\lambda u)^1}{1!} \cdot \frac{e^{\lambda(t-u)}[\lambda(t-u)]^0}{0!} \cdot \frac{1!}{e^{-\lambda t}(\lambda t)^1} = \frac{\lambda u}{\lambda t} = \frac{u}{t} \end{aligned}$$

from which we conclude that the conditional distribution of S_1 given $X(t) = 1$ is $U(0, t)$. This is an important result, which can be generalized to n events over the interval $(0, t]$. In particular, it can be shown that the joint conditional distribution of the times of the n events, given that $X(t) = n$ is similar to the joint distribution of the order statistics of size n from the uniform distribution over $(0, t]$. This result is formally stated as follows: Let $\{S_i, i \geq 1\}$ denote the time of the ith event and let $f_{S_1,S_2,\ldots,S_n}(s_1, s_2, \ldots, s_n | X(t) = n)$ denote the joint conditional density of $S_1, S_2, \ldots, S_n$ given that n events have occurred in $(0, t]$. Then it holds that

$$f_{S_1,S_2,\ldots,S_n}(s_1, s_2, \ldots, s_n | X(t) = n) = \begin{cases} n!/t^n & \text{if } 0 < s_1 < s_2 < \cdots < s_n < t \\ 0 & \text{otherwise.} \end{cases}$$

This result facilitates the calculation of the expectations related to the event times by using this connection with the uniform distribution.

6.4 Renewal Processes

In this section, we will briefly introduce some basic concepts from renewal theory, which may be used in other chapters as well since it provides some fundamental results used both in theory and applications. Renewal processes in general refer to dynamic systems which change states randomly in time and are *renewed* at a sequence of time instances in the infinite horizon. A typical example is a device whose lifetime is described by a random variable and upon failure, the device is replaced (renewed) by a new and identical device. Number of renewals over an interval of length t, total number of renewals over the interval $(0,t]$, or the renewal rate in a long-term horizon (expected number of renewals per unit time in the limit) are some concepts of interest. For a formal exposition, let $\{X_n, n \geq 1\}$ be a sequence of nonnegative i.i.d. random variables with d.f. F and mean $\mu = E(X_n) < \infty$ such that $F(0) \neq 1$, to avoid a degenerate case. The X_n's are considered as the renewal intervals corresponding to the times between the $(n-1)$th and nth event. Let $S_0 = 0$ and define $S_n = X_1 + \cdots + X_n$ as the time of the nth renewal (event) with d.f. F_n. Assuming that, in a finite length interval, only finite number of events can occur (assured by finiteness of μ), we define $N(t) = \max\{n : S_n \leq t\}$. The process $\{N(t), t \geq 0\}$ is called a *renewal process* and it counts the number of events that occur by time t. The function $M(t) = E[N(t)]$ is called the *renewal function.* It can be shown that

$$M(t) = \sum_{n=1}^{\infty} F_n(t) < \infty$$

In case $M(t)$ is continuous and differentiable, the derivative $m(t)$ of it is called the *renewal density.*

Example 6.1

Suppose $X_i, i \geq 1$ are independent and each has identical uniform distribution over the unit interval (0,1). Then, $F_n(t)$ is the distribution function of sum of n uniform random variables, which has a continuous piecewise linear form over the intervals $i \leq t \leq i+1$, $i = 1, \ldots, n-1$ and $F_n(t) = 1$ for $t > n$. Then, $M(t)$ does not have a simple closed form but it is quite suitable for computations.

Example 6.2

Suppose $X_i, i \geq 1$ are independent and each has identical exponential distribution with parameter λ. Then from the properties of i.i.d. exponential random variables, S_n has an Erlang distribution with parameters n and λ. In this case, $N(t)$ corresponds to a Poisson process and we have $M(t) = \lambda t$ and $m(t) = \lambda$ for all t.

The calculation of $M(t)$ is usually based on the Laplace transforms. In particular, if $f*(s)$ is the Laplace transform of the p.d.f. $f(t)$ of $F(t)$ (assuming F is absolutely continuous) and $m*(s)$ is that of $m(t)$, then it can be shown that there is a one-to-one relationship between $m*(s)$ and $f*(s)$ given by $m*(s) = f*(s)/(1 - f*(s))$. Usually, first $m*(s)$ is obtained from the above relationship and then it is converted by using the inverse transform, which is then integrated to obtain $M(t)$.

Example 6.3

Suppose the interarrival density is given as $f(t) = 2(e^{-t} - e^{-2t})$ when $t > 0$ and zero otherwise. Then, $f*(s) = 2/(s+1) - 2/(s+2)$ from which we get $m*(s) = 2/(s^2 + 3s)$. Inverting this, we get $m(t) = 2(1 - e^{-3t})/3$ and

$$M(t) = \int_0^t m(u)du = \frac{2}{3}t + \frac{2e^{-3t}}{9} - \frac{2}{9}$$

Letting $N(\infty) = \lim_{t\to\infty} N(t)$, it holds that $N(\infty) = \infty$ and $P(N(\infty) < \infty) = 0$. That is, in a finite interval, finite number of events can occur and as time span increases, the number renewals converge to infinity, both of which are intuitive results and they hold in the setting that we described above. The main results in the limiting cases regarding renewal processes require a careful and sometimes tedious analysis. We present below some important results, avoiding the technicalities in their proofs (see Ross (1996) for a formal treatment). Before providing these results, let us give two definitions. A nonnegative random variable with d.f. F is called *lattice* if there exists a $d \geq 0$ such that $\sum_n P(X = nd) = 1$. The largest d with this property is called the period of X. If X is lattice, F is also called lattice. We also need a stronger integrability condition as follows: Let h be a nonnegative function. For any $a > 0$, let $M_n(a)$ be the supremum and $m_n(a)$ be the infimum of $h(t)$ over the interval $(n-1)a \leq t \leq na$. Then, h is said to be *directly Riemann integrable* if $\lim_{a\to 0} am_n(a) = \lim_{a\to 0} aM_n(a) < \infty$ for all a.

Theorem 6.3

1. $N(t)/t \to 1/\mu$ as $t \to \infty$ with probability one.
2. $m(t)/t \to 1/\mu$ as $t \to \infty$ (the elementary renewal theorem).
3. (Blackwells's theorem) If F is not *lattice*, then $m(t + a) - m(t) \to a/\mu$ as $t \to \infty$ for all $a \geq 0$. If F is lattice with period d then E(number of renewals at nd) $\to d/\mu$ as $n \to \infty$.
4. (The key renewal theorem) If F is not lattice and $h(t)$ is directly Riemann integrable, then

$$\lim_{t\to\infty} \int_0^t h(t - x)dM(x) = \frac{1}{\mu} \int_0^\infty h(t)dt$$

where

$$M(x) = \sum_{n=1}^{\infty} F_n(x) \quad \text{and} \quad \mu = \int_0^\infty [1 - F(t)]dt$$

The results in the above theorem provide results for the long-term behavior of renewal processes. The first part states that the rate of renewals converges to the reciprocal of the mean interarrival time in long term. The second part indicates that the same rate holds for the expected rate of renewals as well. The third part implies that in the limit the $M(t)$ function becomes linear. The fourth part is very important to calculate the limiting values of functions, where h may refer to some kind of cost or reward.

6.4.1 Renewal Reward Processes

Renewal processes provide a basis for the analysis of another very general process called the *renewal reward processes*. Let $\{N(t), t \geq 0\}$ be a renewal process with interarrival time sequence $\{X_n, n \geq 1\}$ with d.f. F. Suppose at each renewal, a random reward (or cost) is earned. Let $\{R_n, n \geq 1\}$ denote the sequence of rewards earned at the nth renewal, which are assumed to be i.i.d. random variables. The reward R_n at the nth renewal may, however, depend on the nth renewal interval X_n. Let $E(R) = E(R_n)$ and $E(X) = E(X_n)$ denote expected reward and expected renewal interval, respectively. The total reward earned by time t is then expressed as

$$R(t) = \sum_{n=1}^{N(t)} R_n$$

We state below a fundamental theorem known as the *renewal reward theorem (RRT)*.

Theorem 6.4

If $E(R) < \infty$ and $E(X) < \infty$, then

1. $R(t)/t \to E(R)/E(X)$ as $t \to \infty$ with probability one
2. $E[R(t)]/t \to E(R)/E(X)$ as $t \to \infty$

The RRT is very important for assessing the long-term expected costs or benefits per unit time, of operating a stochastic system, which can be characterized as a renewal process. For instance, consider an inventory system with i.i.d. random weekly demands, where the stocks are raised to S at the beginning of each week. Suppose also that both holding costs for stocked items and lost sales costs for unsatisfied demands are incurred. If we would like to find the expected cost per unit time of this business over a long time horizon, the above theorem applies. Every time the inventory level is raised to S will be a renewal instance with renewal intervals corresponding to a week. Hence, the expected cost rate in the long run will be the expected cost per week.

Example 6.4

Suppose the customers arrive at a bus-stop according to a renewal process with mean interarrival time μ. The bus capacity is N, so that when there are a total of N customers, the bus departs. The cost of keeping n customers waiting is nc per unit time for $c > 0$ and on every time a bus departs, a cost of K is incurred. We observe here that this process is a renewal reward process where a renewal occurs every time a bus departs from the bus-stop. The long run expected cost rate is then calculated as the ratio of the expected cost of a renewal interval to the expected length of that. Let T be a renewal interval which is the sum of N independent interarrival times with mean μ. Then, it is easy to see that $E(T) = N\mu$. Expected cost, $E(CT)$, of a renewal interval can be obtained as

$$E(CT) = E[c(X_1 + 2X_2 + \cdots + (N-1)X_{N-1}) + K] = \frac{c\mu N(N-1)}{2} + K$$

From the above expressions, we get the long run expected lost rate as

$$E(TC)/E(T) = \frac{c(N-1)}{2} + \frac{K}{N\mu}$$

We note from the above expression that to the first term increases and the second term decreases in N. Hence, a unique value of N that minimizes the cost rate above can be obtained to select the optimal bus capacity.

6.5 Markov Chains and Continuous-Time Markov Chains

6.5.1 Markov Chains

An important class of stochastic processes are Markov chains (MCs), named after the Russian mathematician Andrei Andreevich Markov (1856–1922). An MC is a discrete-time discrete-state process, for which the future behavior, given the present, only depends on the present, not on the past. Consider a discrete-time stochastic process $\{X_n, n = 0, 1, \ldots\}$ that takes on a finite or a countably infinite number of values on the set S. If $X_n = i$, then, the process is in state i at time n.

A discrete-time stochastic process is an MC if $P\{X_n = j_n \mid X_{n-1} = j_{n-1}, X_{n-2} = j_{n-2}, \ldots, X_0 = j_0\} = P\{X_n = j_n \mid X_{n-1} = j_{n-1}\}$ for $n \geq 0$ and all state sequences $j_n, j_{n-1}, \ldots, j_0$.

$P\{X_{n+1} = j \mid X_n = i\} \equiv p_{ij}(n)$ are called one-step transition probabilities of the MC, where $p_{ij}(n) \geq 0$ and $\sum_{n=1}^{\infty} p_{ij}(n) = 1 \forall ij$. An MC is *stationary* (time-homogeneous) if $P\{X_{n+1} = j \mid X_n = i\} = p_{ij}$, independent of n. For a stationary MC, $P(X_{n+1} = j \mid X_n = i) = p_{ij} = P(X_1 = j \mid X_0 = i)$. Hence, without loss of generality, we condition on the initial state. The $n - step$ transition probability is accordingly defined as $p_{ij}^{(n)} = P(X_n = j \mid X_0 = i)$. The *transition matrix*, $P = \{p_{ij}\}$, of an MC is the matrix whose elements are the one-step transition probabilities given as below:

$$P = \begin{bmatrix} p_{00} & p_{01} & p_{02} & \cdots \\ p_{10} & p_{11} & p_{12} & \cdots \\ p_{20} & p_{21} & p_{22} & \cdots \\ \vdots & & & \vdots \end{bmatrix}$$

For a transition matrix P, it holds that $\sum_j p_{ij} = 1, \forall i$.

Example 6.5

Consider a sequence of independent Bernoulli trials and let X_n denote the number of successes in n trials. The sequence of random variables $\{X_n, n \geq 1\}$ is an MC (called a Bernoulli process) with

$$P = \begin{bmatrix} q & p & 0 & 0 & & \cdots \\ 0 & q & p & 0 & & \cdots \\ 0 & 0 & q & p & 0 & \cdots \\ \vdots & & & & & \end{bmatrix}$$

Example 6.6

Now, let $\{X_n, n \geq 1\}$ be the time of the n'th success in a Bernoulli process. Then, $\{X_n, n \geq 1\}$ is also an MC with

$$p_{ij} = p(X_{n+1} = j \mid X_n = i) = \begin{cases} pq^{j-i-1} & j \geq i+1 \\ 0 & \text{otherwise} \end{cases}$$

6.5.1.1 Unconditional State Probabilities

Sometimes we are interested in the unconditional probability of being in some state j at time n, regardless of the initial states. For calculation of unconditional probabilities, we need the probability distribution of the initial state. Let $\pi_i^{(0)} = p(X_0 = i)$, $\pi^{(0)} = (\pi_0^{(0)}, \pi_1^{(0)}, \pi_2^{(0)}, \ldots)$ denote the *initial probability distribution* and $\pi_i^{(n)} = P(X_n = i)$, $\pi^{(n)} = (\pi_0^{(n)}, \pi_1^{(n)}, \pi_2^{(n)}, \ldots)$ denote the unconditional probabilities at time n.

6.5.1.2 Chapman–Kolmogorov Equations

From total probability law, we know that in order to be in state j after $n + m$ transitions, starting in state i, the process enters some state k after n transitions. This implies

$$p_{ij}^{(n+m)} = \sum_{k=0}^{\infty} p_{ik}^{(n)} p_{kj}^{(m)}$$

which follows from observing

$$p_{ij}^{(n+m)} = P(X_{n+m} = j \mid X_0 = i) = \sum_{k=0}^{\infty} p(X_n = k, X_{n+m} = j \mid X_0 = i)$$
$$= \sum_{k=0} P(X_n = k \mid X_0 = i) P(X_{n+m} = j \mid X_n = k, X_0 = i) = \sum_{k=0} p_{ik}^{(n)} p_{kj}^{(m)}$$

where the last equality follows from the Markov property.

6.5.1.3 Matrix Notation

Let $P^{(n)} = \left\{p_{ij}^{(n)}\right\}$ denote the matrix of n-step transition probabilities. Then, from matrix algebra, we have $P^{(n+m)} = \left\{p_{ij}^{(n+m)}\right\} = \left\{\sum_{k=0}^{\infty} p_{ik}^{(n)} p_{kj}^{(m)}\right\} = \{(P^{(n)} \times P^{(m)})_{ij}\} = P^{(n)} P^{(m)}$. That is, the $(n+m)$-step transition matrix is obtained as the product of the n- and m-step transition matrices. Using this argument, we deduce that the n-step transition matrix is the nth power of the 1-step transition matrix P, that is, $P^{(n)} = P^n$.

Example 6.7

Let $\{X_n, n \geq 1\}$ denote the evolution of weather condition, where the states are represented by the weather status of two consecutive days denoted as either rainy (R) or not rainy ($\bar{R}$). Then, $S = \{RR, \bar{R}R, R\bar{R}, \bar{R}\bar{R}\}$ and we have

$$P = \begin{matrix} RR \\ \bar{R}R \\ R\bar{R} \\ \bar{R}\bar{R} \end{matrix} \begin{bmatrix} 0.7 & 0 & 0.3 & 0 \\ 0.5 & 0 & 0.5 & 0 \\ 0 & 0.4 & 0 & 0.6 \\ 0 & 0.2 & 0 & 0.8 \end{bmatrix}, \quad P^2 = \begin{matrix} RR \\ \bar{R}R \\ R\bar{R} \\ \bar{R}\bar{R} \end{matrix} \begin{bmatrix} 0.49 & 0.12 & 0.21 & 0.18 \\ 0.35 & 0.20 & 0.15 & 0.30 \\ 0.20 & 0.12 & 0.20 & 0.48 \\ 0.10 & 0.16 & 0.10 & 0.64 \end{bmatrix}$$

Given that it rained on Monday and Tuesday, the probability that it will rain on Thursday is obtained as $R_W R_{TH} + \bar{R}_W R_{TH} = 0.49 + 0.12 = 0.61$.

6.5.1.4 Unconditional State Probabilities

Recall that $\pi_i^{(n)} = P(X_n = i)$ is the unconditional probability of being in state i at time n. Then, $\pi_i^{(n)} = \sum_j P(X_0 = j) P(X_n = i \mid X_n = j)$ and $\pi^{(n)} = \pi^{(0)} P^{(n)}$.

6.5.1.5 Expected Number of Visits to a State

In many applications, we are interested in the expected number of visits made to a state in n-transitions. Let $e_{ij}^{(n)} = E$[number of visits made to state j in the first n-transitions $\mid X_0 = i$]. Assuming $X_0 = i$, define the indicator variable $I_{ij}^{(k)}$:

$$I_{ij}^{(k)} = \begin{cases} 1 & \text{if } X_k = j \\ 0 & \text{otherwise} \end{cases}$$

Then, $E[I_{ij}^{(k)}] = P(X_k = j \mid X_0 = i)$ and $e_{ij}^{(n)} = \sum_{k=0}^{n} E[I_{ij}^{(k)}] = \sum_{k=0}^{n} p_{ij}^{(k)}$, from which we have

$$e_{ij}^{(n)} = \begin{cases} 1 + \sum_{k=0}^{n} p_{jj}^{k} & \text{if } i = j \\ \sum_{k=1}^{n} p_{ij}^{(k)} & \text{if } i \neq j \end{cases}$$

The matrix version of the same equation is $E^{(n)} = I + P + P^{(2)} + \cdots + P^{(n)} = \sum_{k=0}^{n} P^{(k)}$.

6.5.1.6 Sojourn Times

The sojourn time at state i is a random variable N_i that counts the number of consecutive epochs that the process is in state i, given that it starts in state i. The initial entrance is also counted; hence $N_i \geq 1$. Then, $P(N_i = k) = P$ (number of visits to state $i = k \mid X_0 = i) = (p_{ii})^{k-1}(1 - p_{ii}), k = 1, 2, \ldots$. Hence, $N_i \sim Geom(p = 1 - p_{ii})$ and $E[N_i] = 1/(1 - p_{ii})$. We conclude that the sojourn times of an MC have geometric distributions.

6.5.1.7 First Passage and Return Probabilities

The first passage time from state i to state j, denoted as N_{ij}, is the random variable that corresponds to the number of transitions necessary to reach state j for the first time starting from state i. Let $f_{ij}^{(n)} = P(N_{ij} = n)$ denote the probability that starting at state i, the first visit to state j will be at the nth transition. We can write

$$\begin{aligned} P(N_{ij} = n) &= P(X_n = j \mid X_0 = i) - P(\text{all visits to } j \text{ in earlier transitions} \mid x_0 = i) \\ &= p_{ij}^{(n)} - \sum_{k=1}^{n-1} P(\text{first visit to } j \text{ at } k\text{th transition} \mid X_0 = i) P(X_n = j \mid X_k = j) \\ &= p_{ij}^{(n)} - \sum_{k=1}^{n-1} f_{ij}^{(k)} p_{jj}^{(n-k)} = f_{ij}^{(n)} \end{aligned}$$

which provides an iterative way to calculate $f_{ij}^{(n)}$ values, with $f_{ij}^{(1)} = p_{ij}$.

Example 6.8

$$f_{12}^{(4)} = p_{12}^{(4)} - f_{12}^{(1)} p_{22}^{(3)} - f_{12}^{(2)} p_{22}^{(2)} - f_{12}^{(3)} p_{22}^{(1)}.$$

6.5.1.8 Limiting Probabilities

Recall that the n-step transition probability from i to j is $p_{ij}^{(n)} = P(X_n = j \mid X_0 = i)$. Usually, the behavior of the process in the long run is of interest. In mathematical terms, we are interested in the behavior of the above probabilities as $n \to \infty$ and whether the process reaches a "stable" state or not. Stability may correspond to different notions. We will distinguish between *limiting distribution* and *stationary distribution*.

Suppose first that $\lim_{n\to\infty} p_{ij}^{(n)} = \lim_{n\to\infty} P(X_n = j \mid X_0 = i)$ exists for all i and is independent of the initial state i.

The *limiting or steady-state probabilities*, or *limiting (steady-state) distribution* for a Markov chain is the set of limit values

$$\pi_j = \lim_{n\to\infty} p_{ij}^{(n)}$$

The steady-state vector is the row vector, $\pi = [\pi_0, \pi_1, \pi_2, \ldots]$.

As illustrated by the following example, such a limit may not always exist.

Example 6.9

(Alternating renewal process) Consider an MC with the following transition matrix:

$$P = \begin{bmatrix} 0 & 1 \\ 1 & 0 \end{bmatrix} \quad P^{(2)} = \begin{bmatrix} 1 & 0 \\ 0 & 1 \end{bmatrix} \quad P^{(3)} = \begin{bmatrix} 0 & 1 \\ 1 & 0 \end{bmatrix} \cdots$$

Obviously, a limiting distribution does not exist since the unconditional state probabilities will always depend on the initial state.

6.5.1.9 Calculation of Limiting Probabilities

Suppose the limiting steady-state distribution of an MC exists. Next, we discuss how to find this distribution. Recall that from Chapman–Kolmogorov equations, we have $P^{(n)} = P^{(n-1)}P$ and $P^{(n)} = \left\{p_{ij}^{(n)}\right\} = \{P(X_n = j \mid X_0 = i)\}$. Suppose $\lim_{n\to\infty} p_{ij}^{(n)} = \pi_j$ exists independent of the initial state. Let $\pi = [\pi_1, \pi_2, \ldots]$ be the row vector of these limiting probabilities. Since we should have $\lim_{n\to\infty} P^{(n)} = \lim_{n\to\infty} P^{(n-1)}P$, letting $\lim_{n\to\infty} P^{(n)} = P^0$ denote the matrix of limiting probabilities, we have $P^0 = P^0 P$. However, note that the rows of the P^0 matrix are identical and are all π, since the limiting probabilities are independent of the initial states. Therefore, the limiting probabilities satisfy $\pi = \pi P$ and $\pi \underline{1} = \underline{1}$, where $\underline{1}$ is a column of 1's. These two equations are called the *stationary equations*. The vector of π of limiting probabilities is obtained by solving the stationary equations and the solution is called the *stationary distribution* (or the steady-state distribution) of the MC.

We have the following interpretations of steady-state probabilities: π_j is (i) the probability of finding the process in state j after a long time has passed since the process has started, (ii) the fraction of time that the process spends in state j in the long run, (iii) the fraction of the processes, which are in state j after a long time has passed, if identical processes operate simultaneously, π_j, for example, if a number of identical machines operate simultaneously and that randomly change states, π_j will be the fraction of the machines which are in state j at any time, and (iv) the reciprocal of the expected time between two consecutive visits to state j, that is, $1/\pi_j$ = expected time between two visits to state j.

It is possible that the stationary equations have a solution but the limiting distribution does not exist. The above example of alternating renewal process is an example. Solution of the above equations for this process is $(1/2, 1/2)$. However, as illustrated in the example, this process does not have a limiting distribution.

Next, we address the question of when an MC has a limiting distribution. To this end, we first introduce some definitions. State j of an MC is *accessible* from state i if for some $n \geq 0$, $p_{ij}^{n} > 0$. Two states that is all states communicate with each other. Let f_{ij} = be the probability that the process ever visits state j, starting in state i. State j is said to be *recurrent* if $f_{jj} = 1$, *transient*

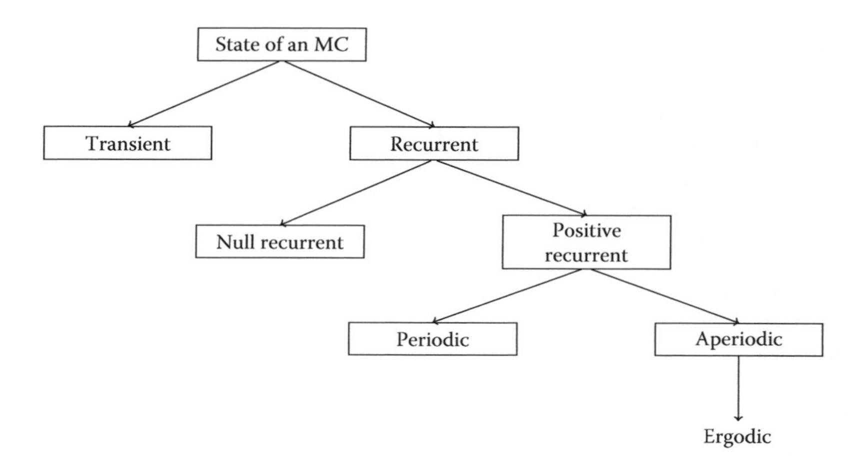

FIGURE 6.2
Classification of the states of a Markov chain.

otherwise. Let μ_{jj} be the expected number of transitions needed to return to state j. State j is said to be *positive recurrent* if $\mu_{jj} < \infty$; otherwise, it is called *null-recurrent*. State j is said to have a *period d* if $p^n_{jj} > 0$ only when n is divisible by d; otherwise, it is *aperiodic* (see Figure 6.2). Positive (null) recurrency and periodicity are class properties; hence, if one state in a class possesses these properties, all the others do as well.

Having introduced these basic definitions, we have the following result (see Ross (1996), p. 175):

Theorem 6.5

An irreducible aperiodic Markov chain belongs to one of the following two classes:

1. The states are all transient or all null recurrent. In this case, $\lim_{n\to\infty} p^{(n)}_{ij} = 0$ for all i, j and there is no stationary distribution.
2. The states are all positive recurrent, $\lim_{n\to\infty} \pi_j = p^{(n)}_{ij} > 0$. In this case, $\{\pi_j, j \geq 0\}$ is the unique stationary distribution.

An irreducible, aperiodic, positive recurrent MC is *ergodic*. From the above theorem, we conclude that if an MC is ergodic, both limiting and stationary distributions exist and they are identical. If the chain is irreducible, positive recurrent, and periodic, the stationary equations have a nondegenerate solution and the solution π_j satisfies $\pi_j = 1/\mu_{jj}$. In this case, limiting distribution does not exist and π_j values correspond to the proportion of time that the process spends in state j.

6.5.2 Continuous-Time Markov Chains

An MC is stochastic process, which makes transitions at every discrete-time unit. Modeling time in discrete units is a matter of choice usually for mathematical convenience. We next consider a process with Markov property and continuous-time index.

A continuous-time stochastic process $\{X(t), t \geq 0\}$ with state space $S = \{0, 1, 2, \}$ is a *continuous-time Markov chain (CTMC)* if for all $s, t \geq 0$ it holds that

$$P\{X(t+s) = j \mid X(s) = i, \quad X(u) = x(u), \quad 0 \leq u < s\} = P\{X(t+s) = j \mid X(s) = i\}$$

A CTMC is *stationary* if for all states i and j it holds that $P(X(t+s)=j \mid X(s)=i) = P(X(t)=j \mid X(0)=i)$. The transition probability function of a CTMC is defined by $P_{ij}(t) = P(X(t)=j \mid X(0)=i)$, which is the probability of finding the process in state j after t time units have passed, given that initially the process was in state i. Transition functions satisfy (i) $P_{ij}(0)=0, i \neq j$, and (ii) $\sum_j P_{ij}(t)=1$ for $t>0$.

There are alternative ways to characterize a CTMC. We will start by considering the time the process spends in state i before it makes a transition to state j. Let T_{ij} be the time the process spends in state i before making a transition to state j. Then, using the Markov property, it can be shown that $T_{ij} \sim Exp(\lambda_{ij})$ and $E[T_{ij}] = 1/\lambda_{ij}$ for some $\lambda_{ij}>0$, which is the rate of transition from state i to state j. It can be shown that T_{ij}'s are independent random variables. Let $T_i = \min_j(T_{ij})$ be the time the process spends in state i before making a transition to another state, that is the sojourn time in state i. The distribution of T_i is obtained as follows, using the independence of T_{ij}'s

$$P(T_i > t) = P\left(\min_j(T_{ij}) > t\right) = \prod_j P(T_{ij} > t) = exp\left\{-\sum_j \lambda_{ij}\right\}$$

Hence, $T_i \sim Exp(\lambda_i = \sum_j \lambda_{ij})$. If $\lambda_i = \infty$, the state i is called *instantaneous* implying that the process immediately leaves this state when enters it. Throughout this chapter, we will assume that the departure rates of the states are finite. Given that the process is in state i, the probability that the next state is j is expressed as $P_{ij} = P(\min_k\{T_{ik}\} = T_{ij}) = \lambda_{ij}/\sum_k \lambda_{ik}$. This discussion leads to an alternative characterization of CTMC. A CTMC is a continuous-time stochastic process, which stays in state i an exponential amount of time with rate $\lambda_i > 0$ and makes a transition to state j, with probability $P_{ij} = \lambda_{ij}/\lambda_i$. Let $\lambda_{ii} = -\sum_{j \neq i} \lambda_{ij}$. The *infinitesimal generator matrix* Λ is defined as $\Lambda = \{\lambda_{ij}\}$, which plays a significant role in expressing the key characteristics of a CTMC.

Example 6.10

Let $S = \{0,1,2\}$. Then,

$$\Lambda = \begin{array}{c} \\ 0 \\ 1 \\ 2 \end{array} \begin{array}{c} \begin{array}{ccc} 0 & 1 & 2 \end{array} \\ \begin{pmatrix} -(\lambda_{01}+\lambda_{02}) & \lambda_{01} & \lambda_{02} \\ \lambda_{10} & -(\lambda_{10}+\lambda_{12}) & \lambda_{12} \\ \lambda_{20} & \lambda_{21} & -(\lambda_{20}+\lambda_{21}) \end{pmatrix} \end{array}$$

Every CTMC contains an MC, which is composed of the sequence of transitions that CTMC makes. Let X_n denote the state that the CTMC visits at the nth transition. Then, $\{X_n, n \geq 0\}$ is an MC with 1-step transition probability matrix $P = \{P_{ij}\}$. In general, finding the values of $P_{ij}(t)$ for an arbitrary process can be difficult. Usually, a system of differential equations called the *Kolomogorov differential equations (KDE)* are used to find these functions. Before providing KDE, we introduce two results.

Lemma 6.1: Chapman–Kolmogorov Equations

For all $s,t>0$, it holds that $P_{ij}(t+s) = \sum_{k\neq j} P_{ik}(t)P_{kj}(s) - P_{ij}(t)\lambda_j$.

The above results are obtained by conditioning on an intermediate state k and using the Markov property.

Lemma 6.2

For a CTMC, it holds that

1. $\sum_{h\to 0} \frac{1 - P_{ii}(h)}{h} = \lambda_i$
2. $\sum_{h\to 0} \frac{P_{ij}(h)}{h} = \lambda_{ij} \quad i \neq j$

6.5.2.1 Kolmogorov Differential Equations

For any CTMC, the $P_{ij}(t)$ function satisfies

$$P'_{ij}(t) = \frac{d}{dt}P_{ij}(t) = \sum_k P_{ik}(t)\lambda_{kj}$$

The basic steps for obtaining KDE are as follows. Using Chapman–Kolmogorov equations, we write $P_{ij}(t+s) = \sum P_{ik}(t)P_{kj}(s)$. Letting $s = \Delta t$, we write $P_{ij}(t + \Delta t) = \sum P_{ik}(t)P_{kj}(\Delta t)$ and subtracting $P_{ij}(t)$ from both sides, we get

$$P_{ij}(t+\Delta t) - P_{ij}(t) = \sum_{k\neq j} P_{ik}(t)P_{kj}(\Delta t) - P_{ij}(t)(1 - P_{jj}(\Delta t))$$

Letting $\Delta t \to 0$ and invoking the above lemma, the result is obtained. A more compact expression for KDE can be obtained as follows. Let $P(t) = \{p_{ij}(t)\}$. Then, the KDE are written in the matrix form as

$$P'(t) = P(t)\Lambda$$

Example 6.11

Let $S = \{1, 2\}$ and suppose

$$\Lambda = \begin{bmatrix} -3 & 3 \\ 2 & 2 \end{bmatrix}$$

Then, the KDE yield
$P'_{11}(t) = P_{11}(t)\lambda_{11} + P_{12}(t)\lambda_{21} = -3P_{11}(t) + 2P_{12}(t)$
$P'_{12}(t) = P_{11}(t)\lambda_{12} + P_{12}(t)\lambda_{22} = 3P_{11}(t) - 2P_{12}(t)$
$P'_{21}(t) = P_{11}(t)\lambda_{21} + P_{22}(t)\lambda_{21} = -3P_{21}(t) + 2P_{22}(t)$
$P'_{22}(t) = 3P_{21}(t) - 2P_{22}(t)$. The initial conditions $P_{ij}(0) = 0, i \neq j$, and $P_{ii}(0) = 1$ imply $P_{11}(t) = \frac{2}{5} + \frac{3}{5}e^{-5t}$, $P_{12}(t) = \frac{3}{5} - \frac{3}{5}e^{-5t}$, $P_{21}(t) = \frac{2}{5} - \frac{2}{5}e^{-5t}$, $P_{22}(t) = \frac{3}{5} + \frac{2}{5}e^{-5t}$

6.5.2.2 Unconditional State Probabilities

Let $P_j(t)$ be the unconditional probability of being in state j at time t. Then, $P_j(t) = P(X(t) = j) = \sum_i P(X(0) = i)P(X(t) = j \mid X(0) = i) = \sum_i P_i(0)P_{ij}(t)$, which is written in matrix form as $P(t) = P(0)P(t)$.

6.5.2.3 Steady-State Probabilities

KDE provide a tool to solve for the $P_{ij}(t)$ functions, which may be quite difficult especially if the number of states is large. In many cases, it is also of interest to know the behavior of the process as t gets large. Suppose the limiting probabilities $\pi_j = \lim_{t\to\infty} P_{ij}(t)$ exist for the system. Taking limits with KDE and assuming the limit and differentiation operations can be interchanged, we have

$$\lim_{t\to\infty} P'_{ij}(t) = \lim_{t\to\infty} \sum_k P_{ik}(t)\lambda_{kj}$$

$$\lim_{t\to\infty} P'_{ij}(t) = \sum_k \pi_k \lambda_{kj}$$

Since the limits as $t \to \infty$ will be independent of t, the left-hand side of the above expression will be zero and we observe that π_j values satisfy $0 = \sum_k \pi_k \lambda_{kj}$. These equations are referred to as the *balance equations* and the limiting probabilities of CTMC are used solving the balance equations with the boundary condition $\sum \pi_k = 1$.

6.5.3 Birth and Death Processes

A special case of CTMC $\{X(t), t > 0\}$ is a birth and death (B&D) process where the process makes a transition either to one step up or one step down as given in Figure 6.3. If a transition is made to one step up, it is referred to as a birth, otherwise it is a death. For a B&D process, a double subscript is not necessary; therefore, λ_j is used to denote the rate of moving from j to $(j+1)$ and μ_i to denote the rate of moving from i to $(i-1)$.

The infinitesimal generator matrix is given as

$$\Lambda = \begin{array}{c} \\ 0 \\ 1 \\ 2 \\ \vdots \end{array} \begin{array}{c} \begin{array}{cccccc} 0 & 1 & 2 & 3 & \cdots \end{array} \\ \begin{pmatrix} -\lambda_0 & \lambda_0 & 0 & & \cdots \\ \mu_1 & -(\lambda_1+\mu_1) & \lambda_1 & 0 & \cdots \\ 0 & \mu_2 & -(\lambda_2+\mu_2) & \lambda_2 & \cdots \\ \vdots & & & & \end{pmatrix} \end{array}$$

Solution to the equations $0 = \pi\Lambda$ and $\sum \pi_k = 1$ should satisfy

$$\pi_j = \frac{\lambda_0 \cdot \lambda_1 \cdots \lambda_{j-1}}{\mu_1 \cdots \mu_j} \pi_0$$

and

$$\sum_{i=0}^{\infty} \pi_j = 1$$

FIGURE 6.3
Birth–death process.

Hence, a stationary distribution exists if and only if the following convergence holds:

$$\sum_{i=0}^{\infty}\prod_{j=0}^{i-1}\frac{\lambda_j}{\mu_{j-1}} < \infty$$

Example 6.12

Suppose there are three machines; each has exponential lifetime with rate $\lambda = 0.01$, there is a single repairman, where repair takes exponential time with rate $\mu = 1$. Let $X(t)$ be the number of the failed machines at time t. $X(t)$ can be considered as a B&D process, $X(t) \in \{0, 1, 2, 3\}$. We have $\lambda_0 = 3\lambda$, since all three machines are working. Whenever one of them fails, the system changes state, that is, the minimum of the lifetimes makes the system change its state. Hence, $\lambda_1 = 2\lambda$, $\lambda_2 = \lambda$. Since there is a single repairman, the μ's are all same $\pi_1 = (3\lambda/\mu)\pi_0$, $\pi_2 = 6(\lambda/\mu)^2\pi_0$, $\pi_3 = 6(\lambda/\mu)^3\pi_0$, $\pi_0 + \pi_1 + \pi_2 + \pi_3 = 1$, resulting in $\pi_0[1 + 3(\lambda/\mu) + 6(\lambda/\mu)^2 + 6(\lambda/\mu)^3] = 1$, which gives $\pi_0 = 1/[1 + 0.03 + 6(0.01)^2 + 6(0.01)^3]$.

A very important special case of a B&D process is a pure birth process with identical arrival rates, which is the Poisson process introduced in Section 6.2. In this special case, $\lambda_i = \lambda$, $\mu_i = 0, i \geq 1$. Letting $X(t), t \geq 0$ denote the number of individuals in the system at time t with $X(0) = 0$, we immediately observe that a Poisson process will count the number of births (events) that have arrived by time t and therefore a Poisson process is also a counting process. In a B&D process, the amount of time the process spends in a state is exponentially distributed. Therefore, a B&D is a special case of a renewal process with the renewal interval sequence $X_n, n \geq 1$ having exponential distribution with rate λ. Assume that initially the system is empty $X(0) = 0$. It is sufficient to consider the transition probabilities $P_{0j}(t) \equiv P_j(t)$. The differential equations satisfy $dP_0(t)/dt = -\lambda P_0(t)$ implying $P_0(t) = e^{-\lambda t}$, $dP_1(t)/dt = \lambda P_0(t) - \lambda P_1(t)$, implying $P_1(t) = \lambda t e^{-\lambda t}$. In general, we have $dP_j(t)/dt = \lambda P_{j-1}(t) - \lambda P_j(t)$. Solving the general differential equations results in

$$P_{0j}(t) = P(j \text{ events} \in (0, t]) = e^{-\lambda t}(\lambda t)^j/j!$$

which is the Poisson distribution discussed in the appendix, corresponding to the distribution of the number of events over an interval of length t.

6.6 Queueing Models

6.6.1 Introduction

Consider a queueing system where $X(t)$ denotes the number of customers in the system at time t, λ denotes the arrival rate (the expected number of arrivals per unit time), and μ denotes the service rate (the expected number of completed servers by a single server per unit time). As mentioned earlier, transient analysis for most systems is complicated, and intractable for many. Furthermore, the steady-state analysis captures the intricacies and the fundamental peculiarities of the underlying queueing models. This allows for gaining insights regarding the operating characteristics of the systems, which is also useful for designing service or production systems that are characterized as queuing systems. Therefore, our analysis throughout this chapter is limited to the behavior of the systems in steady state. Approaches for transient analysis are mentioned in Chapter 8.

We adopt the following steady-state notation: π_j denotes the probability that there are j customers in the system, b_j the probability that j servers are busy, q_j the probability that j customers are in the queue. Furthermore, L denotes the expected number of customers in the system, B denotes the expected number of busy servers, L_q denotes mean number of customers in the queue, W denotes the mean waiting time in the system, and W_q denotes the mean waiting time in the queue. Lastly, $U = B/s$ is the mean utilization with s servers in the system and R is the mean throughput rate (mean number of completed services per unit time). Then, we have $L = \sum_{j=1}^{\infty} j\pi_j$, which is also written as

$$L = L_q + B$$

where $B = \sum_{j=1}^{\infty} jb_j$. If there are s servers, L_q is the expected number of customers not served, written as $L_q = \sum_{j=s}^{\infty} (j - s)\pi_j$ and $W = W_q + 1/\mu$. We have $q_0 = \pi_0 + \pi_1 + \cdots + \pi_s = P(0$ customers in queue), $q_j = P(j$ customers in queue) $= \pi_{s+j}$ and

$$b_j = p(j \text{ servers are busy}) = \begin{cases} \pi_j & \text{if } j < s \\ \pi_s + \pi_s + 1 + \cdots & j = s \end{cases}$$

If $s = 1$, then $U = B = 1 - \pi_0$. $R = \lambda$ for infinite capacity systems and adjusted as $R = \lambda(1 - \pi_N)$ for systems with capacity $N(< \infty)$ where arrivals see time averages, referred to as the ASTA property.

6.6.1.1 Little's Law

A fundamental result on the queueing systems is a very simple relation between the expected number of customers in the system, the mean waiting time, and the arrival rate, known after the person who first described it as the Little's law. (A number of proofs offering various insights have been provided over the years starting with Little (1961) followed by Jewell (1967), Eilon (1969), and Stidham (1974)).

According to Little's law, in any queueing system for which steady-state behavior exists and arrivals are state independent, it holds that

$$L = RW$$

and that, for infinite capacity systems,

$$L = \lambda W$$

The latter states that the expected number of customers in the system in the steady state is the expected number of arrivals during the expected time spent in the system by a customer. The intuition behind the above result may be explained as follows: the expected number of customers that departing customers see in the system must be equal to the expected number of arrivals during the expected stay of customers in the system. The same reasoning also applies to the number of customers in the queue $L_q = \lambda W_q$ and $L_q = RW_q$. If the arrivals are state dependent with λ_j when there are j customers in the system, then the above formulas can still be used with λ replaced by

$$\bar{\lambda} = \sum_{j=0}^{\infty} \lambda_j \pi_j$$

provided that ASTA property holds. Similarly, $B = \lambda/\mu = \lambda E$ [service time] for finite-capacity systems, and $B = R/\mu$ for infinite-capacity systems.

After the above preliminaries, we introduce a Markovian queueing model. A Markovian queueing model is a system where both interarrival times and service times are exponential. Exponential distribution has both advantages and disadvantages. However, it is the most commonly used distribution to represent the "time" until certain tasks/events completed/occur in a stochastic system. Both advantages and disadvantages arise mainly from the "memoryless property," which provides significant mathematical convenience, but at the same time restricts possible applications of such models. Even if the Markovian assumption may not hold in real cases, the insights and results obtained from Markovian systems are, nonetheless, surprisingly good.

6.6.2 The M/M/1 Queue

Consider a service system with a single server, Poisson arrivals with rate λ, and exponential service times with service rate μ. This system is a B&D process as depicted in Figure 6.4.

Let $X(t)$ be the number of customers in the system at time t. The balance equations result in

$$\begin{aligned}
\lambda\pi_0 &= \mu\pi_1 \\
(\lambda+\mu)\pi_1 &= \lambda\pi_0 + \mu\pi_2 \\
(\lambda+\mu)\pi_2 &= \lambda\pi_1 + \mu\pi_3 \\
&\vdots \\
(\lambda+\mu)\pi_n &= \lambda\pi_{n-1} + \mu\pi_{n+1}
\end{aligned}$$

Hence, $\pi_j = (\lambda/\mu)^j\pi_0 = \rho^j\pi_0$, where $\rho = \lambda/\mu$ is called the *traffic intensity.*

$$\sum_{j=0}^{\infty}\pi_i = \pi_0 + \pi_1 + \pi_2 + \cdots = \pi_0(1+\rho+\rho^2+\rho^3+\cdots) = 1$$

These expressions imply that the $M/M/1$ queue will have a stationary distribution only if $\rho = \lambda/\mu < 1$, in which case $\pi_0 = 1-\rho$ and $\pi_j = \rho^j(1-\rho), j \geq 1$.

Then, the steady-state distribution of the number of customers in the system, N, for an $M/M/1$ queue has a geometric distribution with $p = 1-\rho$. Then,

$$\begin{aligned}
L = E[N] &= \sum_{j=0}^{\infty} j\pi_j = \sum_{j=0}^{\infty} j\rho^j(1-\rho) \\
&= \rho(1-\rho)\sum_{j=0}^{\infty} j\rho^{j-1} = \frac{\rho}{(1-\rho)}
\end{aligned}$$

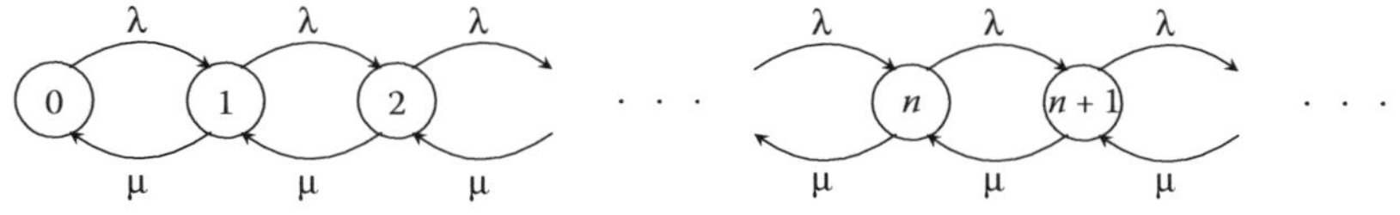

FIGURE 6.4
$M/M/1$ queueing system.

For the queue length, we have $q_0 = \pi_0 + \pi_1 = (1-\rho) + \rho(1-\rho) = 1-\rho^2$, and for $j \geq 1$, $q_j = \pi_{j+1} = \rho^{j+1}(1-\rho)$. The expected queue length is obtained as

$$\begin{aligned} L_q &= \sum_{j=0}^{\infty} j q_j = \sum_{j=0}^{\infty} j \rho^{j+1}(1-\rho) \\ &= (1-\rho)\rho^2 \sum_{j=0}^{\infty} j \rho^{j-1} = (1-\rho)\rho^2 \frac{1}{(1-\rho)^2} = \frac{\rho^2}{(1-\rho)} \end{aligned}$$

For the expected number of busy servers, we have $B = 0\pi_0 + 1(1-\pi_0) = \rho$. From the above expressions, we can verify Little's formula $L = \lambda W$, which implies $W = L/\mu = \rho/\lambda(1-\rho) = 1/\mu(1-\rho)$. Similarly, $W_q = L_q/\lambda = \rho^2/\lambda(1-\rho) = \rho/(1-\rho)\mu$. It is also of interest to know the expected queue length. Let $L_{nq} = \mathrm{E}$ [number of waiting customers (NWC)$|NWC > 0$]. Then, we can show that $L_{nq} = 1/(1-\rho) = \mu/(\mu-\lambda)$.

6.6.2.1 Probability Distribution of Waiting Times

Above we have obtained W and W_q, which are the expected waiting times in the system and in the queue, respectively. For the $M/M/1$ queue, we can also find the probability distribution of the waiting times explicitly. However, for most other models, it is not so easy. Let Z be the total waiting time of a customer (including the service time). Then,

$$\begin{aligned} P(Z > t) &= \sum_{k=0}^{\infty} P(Z > t \mid k \text{ customers in system upon arrival}) \\ &\quad P(k \text{ customers in system upon arrival}) \\ &= \sum_{k=0}^{\infty} \pi_k P(Z > t \mid k \text{ customers in system upon arrival}) \\ &= \sum_{k=0}^{\infty} \pi_k \sum_{j=0}^{k} \frac{e^{-\mu t}(\mu t)^j}{j!} \\ &= e^{-\mu t}(1-\rho) \sum_{j=0}^{\infty} \sum_{k=j}^{\infty} \rho^k \frac{(\mu t)^j}{j!} \\ &= e^{-\mu t}(1-\rho) \sum_{j=0}^{\infty} \frac{(\mu t)^j}{j!} \sum_{k=j}^{\infty} \rho^k \\ &= e^{-\mu t} e^{\mu \rho t} = e^{-\mu t(1-\rho)} \end{aligned}$$

where the third equality above follows from the memoryless property of the exponential distribution. The above result verifies that Z has an exponential distribution with parameter $\mu(1-\rho)$. Let Z_q denote the waiting time in the queue. Since there is a positive probability that the system can be idle at any time, $Z_q = 0$ with probability $P(Z_q = 0) = \pi_0$. For $t > 0$

$$\begin{aligned} P(Z_q > t) &= \sum_{n=1}^{\infty} \pi_n P(T_1 + T_2 + \cdots + T_n > t) \\ &= \rho \sum_{n=1}^{\infty} (\rho)^{n-1}(1-\rho) P(T_1 + T_2 + \cdots + T_n > t) \\ &= \rho \sum_{k=0}^{\infty} \pi_k P(T_1 + T_2 + \cdots + T_{k+1} > t) \\ &= \rho P(Z > t) = \rho e^{-\mu(1-\rho)t} \end{aligned}$$

Hence, Z_q is a mixture random variable with positive probability at $t = 0$.

Example 6.13

Consider an $M/M/1$ queue with $\lambda = 2$ and $\mu = 3$. Then, the probability that an arriving customer waits more than 3.2 units of time is

$$P(Z > 3.2) = e^{-3\times(3.2)(1-2/3)} = e^{-3.2}$$

whereas the probability that an arriving customer spends more than 3.2 in the system if there are already three customers in the system at the arrival instance $P(T_q > 3.2|$ already three customers) $= \sum_{k=0}^{3} \frac{e^{-3.2\mu}(3.2\mu)^k}{k!}$.

Example 6.14

Suppose now $\lambda = 15$ and $\mu = 20$ per hour. Then $\rho = 15/20 = 0.75$ so that utilization is 75%. For the steady-state distribution, we find $\pi_0 = 0.25$ which implies that 25% of time the server is idle. For j=1,2,..., $\pi_j = (0.75)^j(0.25)$. The expected number in the system is $\rho(1-\rho) = L = 3$ and the expected queue length is $L_q = 2.25$. The expected total time in the system is $W = 12$ min and that in the queue is $W_q = 9$ min. The probability that the time T a customer spends in the system exceeds a quarter of an hour is $P(T > 0.25) = e^{(11-0.75)20\times0.25} = e^{-1.25} = 0.286$. To compare the system performance measures, now suppose the arrival rate has increased to $\lambda = 18$/h. Then, $\rho = 18/20 = 0.9$ which implies that the utilization has increased significantly and the proportion of idle time decreased to $\pi_0 = 0.1$ and steady-state system size probabilities are given as $\pi_j = (0.9)^j(0.1)$ for $j \geq 1$. Then, the expected system size is $L = 0.9/0.1 = 9$ which increases three times and the total expected time spent in the system is $W = 30$ min, where the expected waiting time in the queue increases to L_q. This example illustrates that performance measures deteriorate in a nonlinear manner as the arrival rate increases while the service rate stays constant.

6.6.3 M/M/s Queue

Now suppose that there are s identical and independently operating servers, each with service rate μ. Then, the system can again be modeled as a B&D process with rates

$$\mu_n = \begin{cases} n\mu & \text{if } n \leq s \\ s\mu & \text{if } n \geq s \end{cases}$$

A representation of the system is given in Figure 6.5.

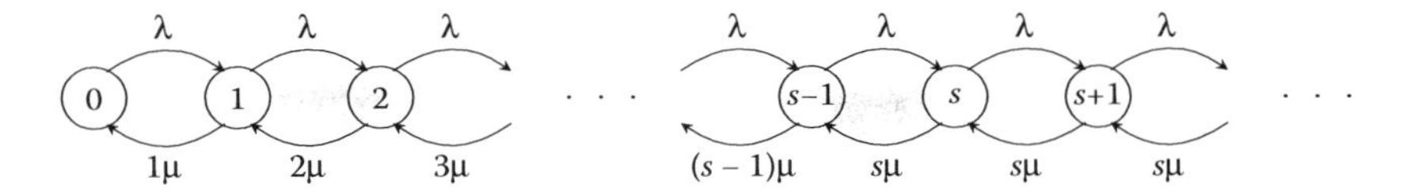

FIGURE 6.5
$M/M/s$ queueing system.

Let $c_n = (\lambda_0 \cdots \lambda_{n-1})/(\mu_1 \cdots \mu_n)$. Then, from the balance equations of B&D processes, we have $\pi_n = c_n \pi_0$ and $\pi_0 = [\sum_{n=0}^{\infty} c_n]^{-1}$. Hence, the condition for the existence of the steady-state distribution is $\sum_{n=0}^{\infty} c_n < \infty$. For the M/M/s queue, we have

$$c_n = \begin{cases} \left(\dfrac{\lambda}{\mu}\right)^n \dfrac{1}{n!} & n = 1, \ldots, s \\ \left(\dfrac{\lambda}{\mu}\right)^s \dfrac{1}{s!} \left(\dfrac{\lambda}{s\mu}\right)^{n-s} & n = s, s+1, \ldots \end{cases}$$

and $\pi_0 = [\sum_{n=0}^{\infty} c_n]^{-1}$ and

$$\pi_n = \begin{cases} \pi_0 \left(\dfrac{\lambda}{\mu}\right)^n \dfrac{1}{n!} & 0 \leq n \leq s \\ \pi_0 \left(\dfrac{\lambda}{\mu}\right)^n \dfrac{1}{s! s^{n-s}} & n \geq s \end{cases}$$

To derive the performance measures, let $\rho = \lambda/\mu$.

$$\begin{aligned} L_q &= \sum_{n=s}^{\infty} (n-s)\pi_n = \sum j \pi_{j+s} \\ &= \sum j \left(\frac{\lambda}{\mu}\right)^s \frac{1}{s!} \rho^j \pi_0 = \pi_0 \left(\frac{\lambda}{\mu}\right)^s \frac{1}{s!} \sum_{j=0}^{\infty} j \rho^j \\ &= \pi_0 \frac{(\lambda/\mu)^s \rho}{s!(1-\rho)^2} \end{aligned}$$

For $s \leq 4$, π_0 and the corresponding L_q values are given in Table 6.1.

For larger values of s, the approximate formula $L_q \equiv \rho^{\sqrt{2(s+1)}}/(1-\rho)$ can be used. For instance, for $s = 6$, $\rho = 0.9$, the approximation yields $L_q = 6.74$, whereas the exact value is 6.66.

Having found the expected queue length, we can find the expected waiting time in the queue from Little's law as

$$W_q = \frac{L_q}{\lambda} = \pi_0 \frac{(\lambda/\mu)^s}{\mu s!(1-\rho)^2}$$

Next, using the relation $W = W_q + 1/\mu$, we obtain the expected waiting time in the system as

$$W = \frac{1}{\mu} + \pi_0 \frac{(\lambda/\mu)^s}{\mu s!(1-\rho)^2}$$

TABLE 6.1

π_0 and L_q Values for the $M/M/s$ System

s	π_0	L_q
1	$1-\rho$	$\frac{\rho^2}{1-\rho}$
2	$\frac{1-\rho}{1+\rho}$	$\frac{2\rho^3}{1-\rho^2}$
3	$\frac{2(1-\rho)}{2+4\rho+3\rho^2}$	$\frac{9\rho^4}{2+2\rho-\rho^2-3\rho^3}$
4	$\frac{3(1-\rho)}{3+9\rho+12\rho^2+8\rho^3}$	$\frac{32\rho^5}{3+6\rho+3\rho^2-4\rho^3-8\rho^5}$

Finally, again invoking Little's law, the expected number in the system is given by

$$L = \lambda W = \frac{\lambda}{\mu} + \pi_0 \frac{(\lambda/\mu)^{s+1}}{s!(1-\rho)^2}$$

6.6.3.1 Waiting Time Distribution

It can be shown that for the total time in the system, Z, it holds that

$$P(Z > t) = e^{-\mu t}\left[\frac{1+\pi_0(\lambda/\mu)^s}{s!(1-\rho)}\right]\left[\frac{1-e^{-\mu t(s-1-(\lambda/\mu))}}{s-1-(\lambda/\mu)}\right]$$

The probability that waiting time in the queue will be zero is given by

$$P(Z_q = 0) = \sum_{n=0}^{s-1} \pi_n = \sum_{n=0}^{s-1}\left(\frac{\lambda}{\mu}\right)^n \frac{1}{s!s^{n-s}}$$

$$= \frac{s^{s-1}}{(s-1)!}\frac{1-\rho^s}{1-\rho}$$

In general, it can be shown that

$$P(Z_q > t) = [1 - P(Z_q = 0)]e^{-s\mu(1-\rho)t}$$

Example 6.15

Suppose there are three servers and traffic ratio is 0.7, that is, $s = 3$ and $\rho = 0.7$. Then from Table 6.1, the fraction of time the system is idle and the expected queue length are obtained as

$$\pi_0 = \frac{2(1-0.7)}{2+4\times 0.7+3(0.7)^2} = \frac{0.6}{6.27} = 0.09$$

and

$$L_q = \frac{9\times(0.7)^4}{2+1.4-0.49-3\times 0.343} = \frac{2.16}{1.88} = 1.14$$

6.6.4 M/M/s/N Queue

We next consider a single server Markovian queue with limited queue capacity of N and s serves as depicted in Figure 6.6, which is also a special case of a B&D process. The parameters of this process are given as $\lambda_j = \lambda$ for $0 \le j \le N$, $\lambda_j = 0$ otherwise, and $\mu_j = j\mu$ for $0 \le j < s$, $\lambda_j = s\mu, s \le j \le N$. Using the balance equations, we have

$$\pi_j = \begin{cases} \left(\dfrac{\lambda}{\mu}\right)^j \dfrac{\pi_0}{n!} & \text{if } 0 \le j < s \\ \left(\dfrac{\lambda}{\mu}\right)^j \dfrac{\pi_0}{s^{j-s} s!} & \text{if } s \le j \le N \end{cases}$$

The boundary condition $\sum_{j=0}^{N} \pi_j = 1$, which is equivalent to

$$\sum_{j=0}^{s-1} \left(\frac{\lambda}{\mu}\right)^j \frac{\pi_0}{n!} + \sum_{j=s}^{N} \left(\frac{\lambda}{\mu}\right)^j \frac{\pi_0}{s^{j-s} s!}$$

results in the following:

$$\pi_0 = \begin{cases} \left[\sum_{j=0}^{s-1} \left(\dfrac{\lambda}{\mu}\right)^j \dfrac{1}{n!} + \dfrac{(\lambda/\mu)^s}{s!} \dfrac{1-\rho^{N-s+1}}{1-\rho} \right]^{-1} & \text{if } \rho \ne 1 \\ \left[\sum_{j=0}^{s-1} \left(\dfrac{\lambda}{\mu}\right)^j \dfrac{1}{n!} + \dfrac{(\lambda/\mu)^s}{s!} (N-s+1) \right]^{-1} & \text{if } \rho = 1 \end{cases}$$

where $\rho = \lambda/s\mu$. The values of $\pi_j, j \ge 1$ are obtained from the previous expressions. The expected queue length is obtained as

$$L_q = \begin{cases} \pi_0 \dfrac{(s\rho)^s \rho}{s!(1-\rho)^2} [1 - \rho^{N-s+1} - (1-\rho)(N-s+1)\rho^{N-s}] & \text{if } \rho \ne 1 \\ \pi_0 \dfrac{(s\rho)^s \rho}{s!} \dfrac{(N-s)(N-s+1)}{s} & \text{if } \rho = 1 \end{cases}$$

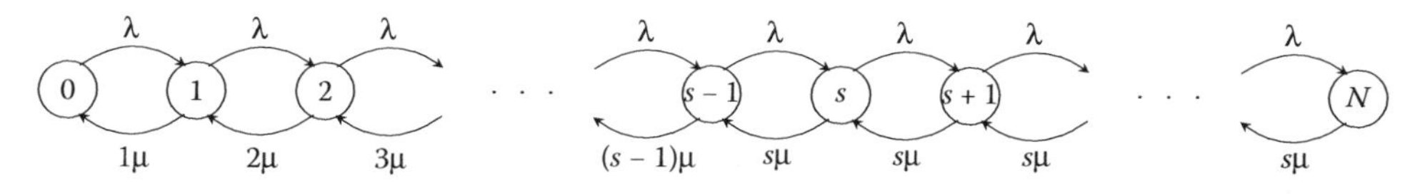

FIGURE 6.6
M/M/s/N queueing system.

The other performance measures are obtained as follows:

$$L = L_q + s - \pi_0 \sum_{j=0}^{s-1} \frac{(s-j)(\rho s)^j}{j!}$$

$$W = \frac{L}{\lambda(1-\pi_N)}$$

$$W_q = W - \frac{1}{\mu} = \frac{L_q}{\lambda(1-\pi_N)}$$

6.6.5 M/M/1/N Queue

This is the special case of the previous model with $s = 1$. From the previous results, we obtain $\pi_j = (\lambda/\mu)^j \pi_0$ for $j = 1, \ldots, N$ and $\pi_0(1 + \rho + \rho^2 + \cdots + \rho^N) = 1$. Then,

$$\pi_0 = \begin{cases} \dfrac{1-\rho}{1-\rho^{N+1}} & \text{if } \rho \neq 1 \\ \dfrac{1}{N+1} & \text{if } \rho = 1 \end{cases}$$

and

$$\pi_j = \left(\frac{\lambda}{\mu}\right)^j \pi_0 = \begin{cases} \rho^j \dfrac{1-\rho}{1-\rho^{N+1}} & \text{if } \rho \neq 1 \\ \dfrac{1}{N+1} & \text{if } \rho = 1 \end{cases}$$

For the expected number of customers in the system, we have

$$L = \begin{cases} \dfrac{\rho}{1-\rho}\left[\dfrac{1-\rho^N}{1-\rho^{N+1}}\right] - \dfrac{N\rho^{N+1}}{1-\rho^{N+1}} & \text{if } \rho \neq 1 \\ \dfrac{N}{2} & \text{if } \rho = 1 \end{cases}$$

and for the expected queue length, we have

$$L_q = \begin{cases} \dfrac{\rho}{1-\rho} - \dfrac{\rho(1+N\rho^N)}{1-\rho^{N+1}} & \text{if } \rho \neq 1 \\ \dfrac{N(N-1)}{2(N+1)} & \text{if } \rho = 1 \end{cases}$$

6.6.6 M/M/s/s Queue

We now consider a lost system with multiple servers where the arrivals balk when all servers are busy, as depicted in Figure 6.7. This is a special case of the previously studied system.

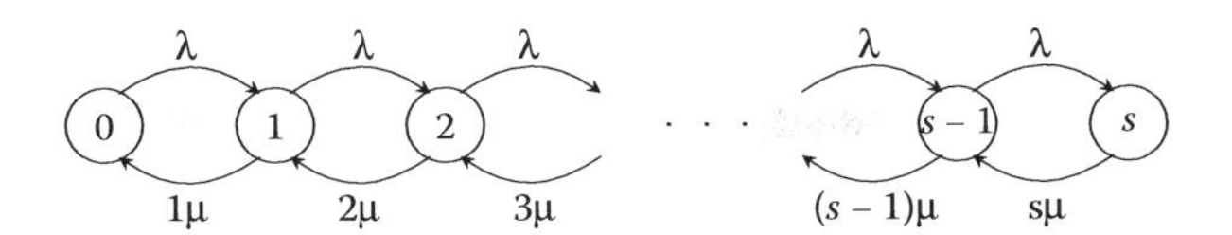

FIGURE 6.7
M/M/s/s queueing system.

The balance equations are written as

$$\lambda\pi_0 = \mu\pi_1$$
$$(\lambda + \mu)\pi_1 = \lambda\pi_0 + 2\mu\pi_2$$
$$(\lambda + 2\mu)\pi_2 = \lambda\pi_1 + 3\mu\pi_3$$
$$\vdots$$
$$(\lambda + n\mu)\pi_n = \lambda\pi_{n-1} + (n+1)\mu\pi_{n+1}$$

Solving the system of equations above, we get

$$\pi_j = \pi_0 \left(\frac{\lambda}{\mu}\right)^j \cdot \frac{1}{j!} \quad j = 1, 2, \ldots, s$$

and $\sum_{j=0}^{s} \pi_j = 1$. Letting $\rho = \lambda/\mu$, we have

$$\pi_j = \frac{\rho^j / j!}{\sum_{k=0}^{s} \rho^k / k!}$$

for $j = 1, 2, \ldots, s$. Consequently, the probability that all servers are busy is given by

$$\pi_s = \frac{\rho^s / s!}{\sum_{k=0}^{s} \rho^k / k!}$$

The above formula is known as the Erlang's loss formula. Erlang first derived it in 1917 to assess the percentage of lost customers when all the channels are busy in a telephone system. This formula later turned out to be of great importance when it was discovered that it is valid for *any* M/G/s/s queue with *general* service time distributions. This result indicates that the steady-state probabilities for such systems are characterized only by the mean service time and is independent of G.

6.6.7 M/M/∞ Queue

Next, we consider a queueing system with infinite number of servers as depicted in Figure 6.8. Population of living organisms is a good example, where each arrival to the system (a birth) starts service immediately which takes a random amount of time (the service duration corresponding to a lifetime).

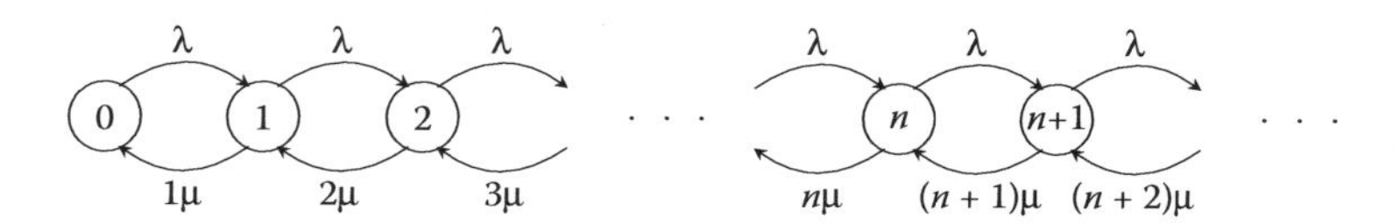

FIGURE 6.8
$M/M/\infty$ queueing system.

From the balance equations, we find a relationship similar to that for the $M/M/s/s$ model, except that in this case, capacity is not finite. The limiting probabilities satisfy

$$\pi_j = \left(\frac{\lambda}{\mu}\right)^j \frac{1}{j!} \quad j = 1, 2, \ldots$$

and

$$1 = \pi_0 \left[1 + \frac{\rho^2}{2} + \frac{\rho^3}{3!} + \frac{\rho^4}{4!} + \cdots\right]$$

We easily obtain

$$\pi_j = \frac{e^{-\rho}\rho^j}{j!} \quad j = 0, 1, \ldots$$

Hence, the steady-state distribution of the number of customers in the system has a Poisson distribution with rate ρ ($= \lambda/\mu$). Erlang obtained these results in 1917 and conjectured that these would also hold for arbitrary service times. The conjecture was later verified by Takacs (1969) implying that $M/M/C/C \equiv M/G/C/C$ and $M/M/\infty \equiv M/G/\infty$.

In designing queueing systems with multiple servers, it is interesting to know whether it is better to merge or separate the lines feeding the servers. Consider s identical servers. In the first system, A, there is a single queue and the customers in the queue are served by the first server that becomes available. In the second system, B, customers randomly choose, upon arrival, any one of these servers and form a queue with no reneging. Everything else being equal, system A is always better in all performance measures. Another question of interest is whether it is better to have more servers or faster servers. To answer that, suppose the arrival rate is μ and we consider two queueing systems. In system A, there are $s = 2$ servers each with service rate μ, and in system B, there is one server ($s = 1$) with service rate 2μ. For the first system, the traffic intensity is $\rho_A = \lambda/\mu$ and for the second $\rho_B = \lambda/2\mu$, and $\pi_{0A} = (1-\rho)/(1+\rho) = (\mu-\lambda)/(\mu+\lambda)$, whereas $\pi_{0B} = 1 - \lambda/2\mu = (2\mu-\lambda)/2\mu$, resulting in $\pi_{0B} > \pi_{0A}$. It follows that $L_{qA} < L_{q,B}$, $W_{qA} < W_{qB}$ and $W_A > W_B$. We observe that neither of the systems dominates the other in all performance measures. For example, although the overall waiting time in the system is shorter in system B, waiting time in the queue is shorter in system A on the average.

6.6.8 M/M/1/N/N Queue

In the next system, we consider a model where customer arrivals come from a finite population. Consider, for example, a production facility with N machines and a single server for repairs. Different than the previous models, in this system, the arrival rate is state-dependent. The system is depicted in Figure 6.9.

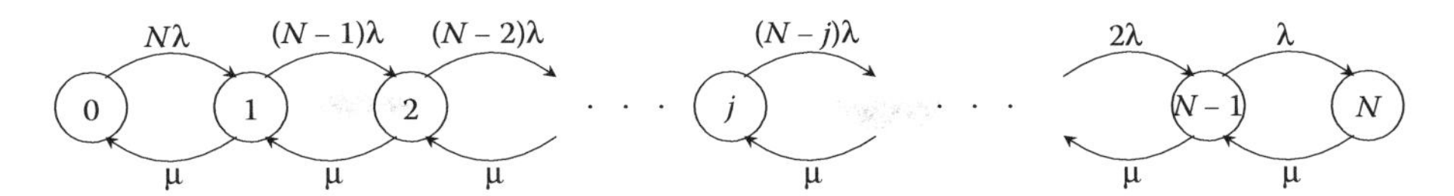

FIGURE 6.9
$M/M/1/N/N$ queueing system.

The balance equations yield

$$
\begin{aligned}
N\lambda\pi_0 &= \mu\pi_1 \\
(N-1)\lambda + \mu\pi_1 &= N\lambda\pi_0 + \mu\pi_2 \\
\pi_j &= \pi_0 \frac{N!}{(N-j)!}\left(\frac{\lambda}{\mu}\right)^j \\
(N-j)\lambda + \mu\pi_j &= (N-j+1)\lambda\pi_{j-1} + \mu\pi_{j+1} \\
\lambda\pi_{N-1} &= \mu\pi_N, j = N \\
\pi_j &= \pi_0 \frac{N!}{(N-j)!}\left(\frac{\lambda}{\mu}\right)^j
\end{aligned}
$$

Letting $\rho = \lambda/\mu$, we have $\pi_0 = [1 + N\rho + N(N-1)\rho^2 + \cdots + \rho^N N!]^{-1}$ and $\pi_j = \pi_0(N!/(N-j)!)\rho^j$. The expressions for L, L_q, and W are not simple; however, generalization to s servers $(s \leq N)$ is straightforward.

In the preceding sections, we have considered only the Markovian systems where both the interarrival and the service times were exponential. We will next relax this assumption and state some results of non-Markovian queueing models, where the interarrival or service times may not be exponential.

6.6.9 M/G/1 Queue

Consider a single server queue with Poisson arrivals and i.i.d. service times denoted by Y with a general distribution function G, mean, $E[Y] = 1/\mu$, and variance $\text{Var}(Y) = \sigma^2$. Also, suppose the Laplace transform of G is given by

$$
G^*(z) = \int_0^\infty e^{-zt} dG(t)
$$

Let $\{X(t), t \geq 0\}$ denote the number of customers in the system at time t. In this system, because the service times are not Markovian and the memoryless property does not hold, the future system behavior when the server is busy depends on what portion of the service has been completed up to that instance. This introduces significant complications to the analysis. However, if the system is observed at the instances of service completions, then the difficulty of this dependency is overcome. In this case, the stochastic process $\{X(t_i), i \geq 1\}$, where t_i corresponds to the ith service completion (departure) time has the Markovian property. Let $\{X_n, n \geq 1\}$ be the number of customers left in the system at the departure instance of the nth customer, then $\{X_n, n \geq 1\}$ is a Markov chain with transition probabilities given by

$$p_{ij} = P(X_{n+1} = j | X_n = i) = \begin{cases} \int_0^\infty \frac{e^{-\lambda t} \lambda t^{j-i+1}}{(j-i+1)!} dG(t) & \text{if } j \geq i-1, i \geq 1 \\ 0 & \text{otherwise} \end{cases}$$

Note that if the system becomes idle at a departure instance, then it stays idle until a new arrival occurs. Hence, $p_{0j} = p_{1j}$. The MC defined above is the *embedded Markov chain* of the process $\{X(t), t \geq 0\}$. Let $\rho = \lambda/\mu$ denote the expected number of arrivals during mean service time. The following are known to hold: (i) The steady-state distribution $\{\pi_n, n \geq 0\}$ of the embedded Markov chain exists if $\rho < 1$ and it coincides with the steady-state solution of the M/M/1 queue. (ii) The steady-state probability that there are n customers left behind at departure instances is the same as the steady-state probability that there are n customers in the system at any instance of time. (See, e.g., Gross and Harris (1985) for more formal statements of these results.) We provide the following results regarding the performance measures, omitting the proofs. Let A denote the number of arrivals during a service time and $w_n = P(A = n)$ denote the probability that n customers have arrived during a service time. Then, the one-step transition probability of the embedded chain is written as $p_{ij} = w_{j-i+1}$. We write w_n as

$$w_n = \int_0^\infty \frac{e^{-\lambda t} \lambda t^n}{(n)!} dG(t)$$

Then, $\{w_n, n \geq 0\}$ defines a discrete probability distribution. Let $\phi_w(z)$ denote the corresponding probability generating function, given by

$$\phi_w(z) = \sum_{i=0}^\infty w_i z^i = \sum_{i=0}^\infty z^i \int_0^\infty \frac{e^{-\lambda t} \lambda t^i}{i!} dG(t) \tag{6.1}$$

$$= \int_0^\infty e^{-(\lambda - \lambda z)t} dG(t) = G * (\lambda - \lambda z) \tag{6.2}$$

from which $E(A) = \phi'_w(1) = -\lambda G *' (0) = \lambda/\mu = \rho$ is obtained. Similarly, let $\phi_\pi(z) = \sum_{i=0}^\infty \pi_i z^i$ be the probability generating function of $\{\pi_n, n \geq 0\}$. Then, it can be shown that

$$\phi_\pi(z) = \frac{(1-\rho)(1-z)\phi_w(z)}{\phi_w(z) - z} = \frac{(1-\rho)(1-z)G * (\lambda - \lambda z)}{G * (\lambda - \lambda z) - z}$$

The above formula is known as the Pollaczek–Knitchine formula. The probability generating function above can be used to obtain the steady-state probabilities of the embedded MC by taking the derivatives and evaluating at $z = 0$ as described in the appendix. In particular, we can obtain

$$\pi_0 = 1 - \rho$$

$$L_q = \frac{\lambda^2 \sigma^2 + \rho^2}{2(1-\rho)}$$

$$L = \rho + L_q$$

The formula given for L_q above is also known as the Pollaczek–Knitchine formula for the mean queue length. This expression was developed in the early 1930s independently of each other by Pollaczek and Khintchine and implies that L_q increases with the variance of the service time distribution.

We next consider the waiting times in the system. We know that $T = T_q + Y$, where T is the total time in the system with $E(T) = W$ and T_q is the time spent in the queue with $E(T_q) = W_q$. Let F_T and F_{T_q} denote the distribution functions of T and T_q, respectively, with corresponding Laplace transforms given by F_T^* and $F_{T_q}^*$. Since the service times are assumed independent of the waiting times in the queue, the Laplace transform of F_T is written as $F_T^*(z) = F_{T_q}^*(z)G*(z)$. We first find $F_T^*(z)$ by exploiting the relationship between the number of customers in the system and the total waiting time. In particular, we have

$$
\begin{aligned}
w_i &= P(\text{a departing customer leaves } i \text{ in the system}) \\
&= \int_0^\infty P(\text{a departing customer leaves } i \text{ in the system} | T = t)\, dF_T(t) = \int_0^\infty \frac{e^{-\lambda t}\lambda t^i}{i!} dF_T(t)
\end{aligned}
$$

Hence

$$
\begin{aligned}
\phi_w(z) &= \sum_{i=0}^{\infty} w_i z^i = \sum_{i=0}^{\infty} z^i = \int_0^\infty \frac{e^{-\lambda t}\lambda t^i}{i!} dF_T(t) \\
&= \int_0^\infty e^{-(\lambda - \lambda z)t} dF_T(t) = F *_T (\lambda - \lambda z) \\
&= \frac{(1-\rho)(1-z)G*(\lambda - \lambda z)}{G*(\lambda - \lambda z) - z}
\end{aligned}
$$

according to the previous expression given for $\phi_w(z)$. Rewriting the above by transforming the argument, we have

$$
F *_T (z) = \frac{(1-\rho)zG*(z)}{z - \lambda(1 - G*(z))}
$$

Now we get the Laplace transform of the waiting time distribution in the queue as

$$
F_{T_q}^*(z) = \frac{z(1-\rho)}{z - \lambda(1 - G*(z))}
$$

This expression is also known as the Pollaczek–Knitchine formula for Laplace transforms. These results can be exploited to obtain the mean waiting time in the queue and in the system as follows:

$$
\begin{aligned}
W_q &= \frac{L_q}{\lambda} \\
W &= W_q + \frac{1}{\mu}
\end{aligned}
$$

Example 6.16

Suppose the service time has an Erlang distribution with shape parameter 2 and scale parameter $\alpha = 2$. Then $E[X] = 2/\alpha$ and $Var(X) = \sigma^2 = 2/\alpha^2 = 1/2$. We have

$$L_q = \frac{(2\lambda^2/\alpha^2) + \rho^2}{2(1-\rho)}$$
$$= \frac{(\lambda^2/2) + \lambda^2}{2(1-\lambda)} = \frac{3\lambda^2}{4(1-\lambda)} = 0.75\left(\frac{\lambda^2}{1-\lambda}\right)$$

Now, suppose we have an M/M/1 system with same mean service time. Then,

$$L_q = \frac{\rho^2}{1-\rho} = \frac{\lambda^2}{1-\rho}$$

Note that the variance has increased the queue length by 60% on average.

6.6.9.1 M/D/1 Queue

In this model, the arrivals are Poisson but the service times are assumed to be a fixed positive quantity with value $1/\mu$. The necessary and sufficient condition for the existence of the steady-state probability distribution $\pi_j j \geq 0$ is that $\rho = \lambda/\mu < 1$. Under this condition, the probability generating function is given by

$$\phi_\pi(z) = \frac{(1-\rho)(1-z)}{1 - ze^{\rho(1-z)}}$$

The steady-state probabilities are obtained from the above function as follows (Saaty (1961), p. 155):

$$\pi_0 = 1 - \rho$$
$$\pi_1 = (1-\rho)(e^\rho - 1)$$
$$\vdots$$
$$\pi_j = (1-\rho)\sum_{k=1} j(-1)^{j-k} e^{k\rho}\left[\frac{(k\rho)^{j-k}}{(j-k)!} + \frac{(k\rho)^{j-k-1}}{(j-k-1)!}\right]$$

where the last equation is valid for $j \geq 2$, and for $k = j$, the term in the square bracket is ignored.

6.6.9.2 M/E_k/S Queue

There is an important class of queues where the service times are characterized by an Erlang distribution, for which some closed form expressions are available. Note that for an $M/D/1$ queue, service time has no variation, whereas for an $M/M/1$ queue, the variance of the service times is $1/\mu^2$. Between these two extremes lies a big region where many distributions used in practice fall. One important class is Erlang $(k, k\mu)$ distribution, for which we chose the scale parameter as $k\mu$ in order to get clean expressions for the mean and coefficient of variation. Recall that an Erlang distribution with integer shape parameter k and scale parameter $k\mu$ can be viewed as the distribution of the sum of k i.i.d. exponential random variables with rate μ, which is denoted as E_k. Erlang has exploited this fact in an ingenious way to analyze the system behavior of the queues with Poisson arrivals and service times with E_k distributions. The approach used for the analysis is called the *method of stages.* The idea behind this approach is the following: Since the service time is E_k, we consider a service time as a duration composed of k exponential stages. Since exponential distribution has the memoryless property, the amount of time that passed in the current stage is unimportant. This observation allows to describe the state (number of customers on the system) of the $M/E_k/1$ system in terms of the total stages in the system at an arbitrary instance. If there are n customers in the system, $(n-1)$ of them waiting and one of them being served at the ith stage of the k Erlang stages, then the total number of stages (which can be considered as the workload remaining in the system) at that instance, m, is given by $m = kn - i + 1$. Letting P_m denote the probability that there are m stages in the system in the steady state and π_j denote the probability that there are j customers in the system, we have the following result:

$$\pi_j = \sum_{m=(j-1)k+1}^{jk} P_m$$

Let $(z_1, z_2, \ldots, z_k)$ be the unique set of roots that solve $kz - \lambda(z + z^2 + z^3 + \cdots + z^k) = 0$, which is an expression obtained from the p.g.f. of the distribution $\{P_m, m \geq 1\}$ of number of stages in the system. Then, the solution for the steady-state probabilities of number of stages in the system is given by

$$a_i = \prod_{n=2, n\neq i}^{k} \frac{1}{(1 - z_i/z_n)}$$

$$P_m = (1-\rho)\sum_{i=1}^{k} a_i/z_i^m$$

The above are then employed to obtain $\pi_j, j \geq 0$.

6.6.10 G/M/1 Queue

We next consider a system where the interarrival times are independent and identically distributed with a general distribution function G with mean $1/\lambda$, arrival rate λ, and Laplace transform $G*$. Service times of the single server is exponential with rate μ. Similar to the M/G/1 queue, the approach of embedded MCs is used. We consider the system immediately before an arrival instance, and define $\{X_n, n \geq 0\}$ as the number of customers that the nth arriving customer sees in the system. Then $\{X_n, n \geq 0\}$ is an MC with transition probabilities given by

$$p_{ij} = P(X_{n+1} = j | X_n = i) = \begin{cases} \int_0^\infty \frac{e^{-\mu t}\mu t^{i+1-j}}{(i+1-j)!} dG(t) & \text{if } i+1 \geq j \geq 1 \\ 0 & \text{otherwise} \end{cases}$$

This embedded MC has a steady-state distribution $\{\pi_n, n \geq 0\}$ provided that $\rho = \lambda/\mu < 1$. Let A denote the number of completed services during an interarrival time and $w_n = P(A = n)$ denote the probability that n services have been completed between two successive arrivals. The one-step transition probability of the embedded chain is written as $p_{ij} = w_{j-i+1}$. We write w_n as

$$w_n = \int_0^\infty \frac{e^{-\mu t} \mu t^n}{n!} dG(t)$$

Similar to the M/G/1 model, $\{w_n, n \geq 0\}$ defines a discrete probability distribution. Let $\phi_w(z)$ denote the corresponding probability generating function, given by $\phi_w(z) = \sum_{i=0}^{\infty} w_i z^i$. It can be shown that this function satisfies $\phi_w(z) = z$ which has a unique root in (0,1). Let us denote this root by r. Then the steady-state distribution of the system size immediately before an arrival occurrence is given by $\{\pi_n = (1-r)r^n, \ n \geq 0\}$, which is a geometric distribution which is similar to the steady-state distribution of the M/M/1 queue where ρ is replaced by r. This similarity allows to use the results for the performance measures of the M/M/1 queue for those of the G/M/1 queue at the arrival instances, by replacing ρ with r. It is important to emphasize again that the above results are valid only at the instances immediately prior to an arrival to the G/M/1 queue. Unlike the results of the M/G/1 queue, these results do not hold for an arbitrary time instance at the steady state. As the distribution of the number of completed services between two consecutive arrivals is geometric and the service time is exponential, the waiting time in the queue, T_q, is a random sum of exponential variables, particularly a sum of a geometric number of exponential service times. We have $P(T_q = 0)b = 1 - r$ and $P(T_q \leq t) = 1 - re^{-\mu(1-r)t}$ for $t > 0$, from which we obtain the expected waiting time in the queue as $E(T_q) = W_q = r/[\mu(1-r)]$.

6.6.10.1 E_k/M/1 Queue

In this model, the arrival process is no longer Poisson. The interarrival times are independent E_k variables with mean λ and service times are independent exponential variables with rate μ. The analysis of this system is carried out in a fashion similar to that for the $M/E_k/1$ queue by means of the method of stages. In this model, interarrival distribution has k stages. Hence, the state description of the system will be the total number of arrival stages in the system. If there are n customers in the system, and the customer that will arrive next is in the ith stage of the k Erlang stages of the interarrival distribution ($i-1$ stages of the interarrival distribution are completed), then the total number of stages, m, is given by $m = kn + i - 1$. Letting P_m denote the probability that there are m stages in the system in the steady state and π_j denote the probability that there are j customers in the system, we have the following:

$$\pi_j = \sum_{m=jk}^{k(j+1)-1} P_m$$

Let z_0 be the unique root of that $kz^{k+1}\rho - (1 + k\rho)z^k + 1 = 0$, which satisfies $|z_0| > 1$. Then the solution for the steady-state probabilities of number of stages and the number of customers in the system are given by

$$P_m = \begin{cases} \dfrac{1}{k}(1 - z_0^{-(m+1)}) & \text{if } 0 \leq m < k \\ \rho(z_0 - 1)z_0^{k-(m+1)} & \text{if } m \geq k \end{cases}$$

The solution for π_j are then obtained as

$$p_j = \begin{cases} 1-\rho & \text{if } j=0 \\ \rho(z_0^r - 1)z_0^{-kj} & \text{if } j>0 \end{cases}$$

6.6.11 GI/G/s Queue

Now suppose that the interarrival times are independent with a general distribution function and the service times are also general and independent. Such a system is highly complicated and exact analysis is not available. However, certain useful approximations exist which will be considered herein. Specifically, we will introduce approximations for the expected queue length L_q and upper and lower bounds for the expected waiting time in the queue in the steady state. Let X refer to the random interarrival time and Y to the random service time with respective means $\mu_X = E[X]$ and $\mu_Y = E[Y]$, variances $\sigma_X^2 = Var(X)$ and $\sigma_Y^2 = Var(Y)$, and coefficient of variations $c_X^2 = \sigma_X^2/\mu_X^2$ and $c_Y^2 = \sigma_Y^2/\mu_Y^2$. For this system, letting $\rho = \lambda/(s\mu)$, the following approximation known as Whitt's approximation (Whitt 1989) is given for the expected queue length:

$$L_q \equiv \left(\frac{c_X^2 + c_Y^2}{2}\right)\left(\frac{\rho^{\sqrt{2(s+1)}}}{1-\rho}\right)$$

For the expected waiting time in the queue, the following bounds are provided by Kingman (1962):

$$\frac{\lambda^2\sigma_Y^2 + \rho(\rho-2)}{2\lambda(1-\rho)} \le W_q \le \frac{\lambda}{2(1-\rho)}\left(\sigma_X^2 + \sigma_Y^2\right)$$

Note that although the upper bound is always nonnegative, the lower bound may get negative values so that it becomes uninformative. In such cases, the lower bound is naturally zero.

6.6.12 Queues with Phase-Type Distributions

6.6.12.1 Phase-Type Distributions

An important avenue to the analysis of queueing systems is provided by the works of Neuts who provided matrix analytic methods for studying stochastic systems, basics of which can be found in Neuts (1981, 1984, 1989). This approach allows to represent systems with complicated input and output structures in a compact way and enables to derive and extend results in a structured manner. See also Breuer and Baum (2005) for matrix methods in queueing. The family of phase-type distributions (PTD) is a nice example to illustrate the matrix-analytic approach which also provides a flexible family for modeling interarrival or service times in queueing models. PTDs were introduced in the 1970s (see, e.g., Neuts (1975, 1978, 1979)) as a versatile class of stochastic processes which are then extensively studied. The following exposition is taken from a study of the author (Balcioglu and Gürler 2011). We start with a general definition. A PTD of order k is considered to be the distribution of the time until absorption in a continuous-time Markov chain (CTMC) with states $\{1,\ldots,k+1\}$, $(k+1)$st (or state 0) being the single absorbing state. The general structure of PTD is depicted in Figure 6.10, where $c_{0,i}$ denotes the initial branching probabilities, μ_i is the rate of the sojourn time in state i, and $c_{i,j}$ denotes the transition probability from state i to state j satisfying $c_{i,1} + \cdots + c_{i,k} + c_{i,k+1} = 1$, for $i = 1,\ldots,k$.

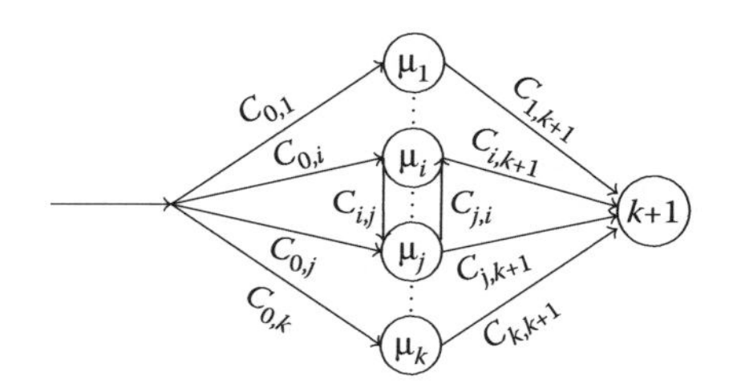

FIGURE 6.10
A k-stage phase-type distribution.

A PTD r.v. can be expressed in matrix notation as follows. Let $\hat{\mathbf{C}} = (\mathbf{C}, 0)$ with $\mathbf{C} = (c_{0,1}, c_{0,2}, \ldots, c_{0,k})$ be the initial probability row vector satisfying $\mathbf{CE} = 1$ with $\mathbf{E}$ being a $k \times 1$ column vector of 1's. The $(k+1) \times (k+1)$ infinitesimal generator $\hat{\mathbf{G}}$ of this CTMC chain is then

$$\hat{\mathbf{G}} = \begin{pmatrix} \mathbf{G} & \mathbf{G_0} \\ \mathbf{0} & 0 \end{pmatrix} \tag{6.3}$$

where $\mathbf{G}$ is a $k \times k$ nonsingular (with the inverse given by G^{-1}) matrix with entries as $G_{ii} = -\mu_i$ for $i = 1, \ldots, k$ and $G_{ij} = c_{i,j}\mu_i$ for $i \neq j$ and the $k \times 1$ $\mathbf{G_0}$ vector satisfies $\mathbf{GE} + \mathbf{G_0} = \mathbf{0}$. The pair $(\mathbf{C}, \mathbf{G})$ is called a *representation* of the PTD.

Two special PTDs, namely, the Coxian and the hyper-exponential distributions are commonly used for modeling purposes.

In a Coxian distribution, once the sojourn time in phase i with exponential rate μ_i is over, the only possible transition types are either to the next state $i+1$ with probability $c_{i,i+1} = a_i$ or to the absorbing state $k+1$ with probability $c_{i,k+1} = 1 - a_i$ for $i = 1, \ldots, k-1$ and $c_{k,k+1} = 1$. The graphical representation of a Coxian distribution is provided in Figure 6.11. In many cases, the probability a_0 is usually assumed to be 1.

Example 6.17

Two-stage Coxian distributions with $a_1 \leq 1$ are practical for modeling or approximating distributions with squared coefficient of variation (variance to the squared mean ratio) $c^2 \geq 0.5$. For the two-stage Coxian distribution, we have $\mathbf{C} = (1, 0)$ and from Equation 6.3,

$$\hat{\mathbf{G}} = \begin{pmatrix} -\mu_1 & a_1\mu_1 & (1-a_1)\mu_1 \\ 0 & -\mu_2 & \mu_2 \\ 0 & 0 & 0 \end{pmatrix}$$

A k-stage hyper-exponential random variable, on the other hand, is a proper mixture of exponential random variables (see Figure 6.12). With an initial branching probability of $c_{0,i} = p_i$, it enters state i to stay for an exponential duration with rate μ_i, where $p_1 + p_2 + \cdots + p_k = 1$. Once the sojourn

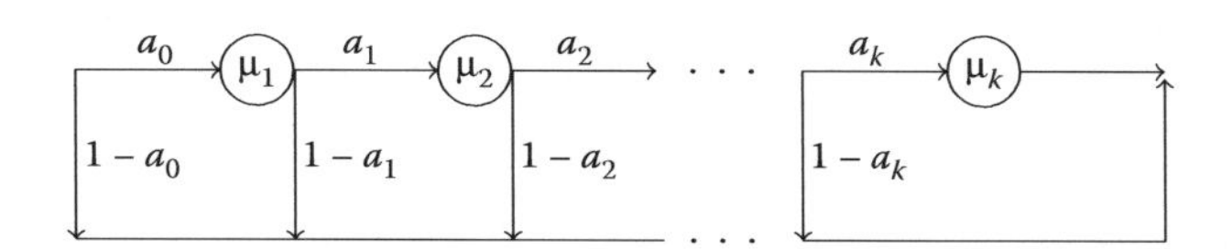

FIGURE 6.11
A k-stage Coxian distribution.

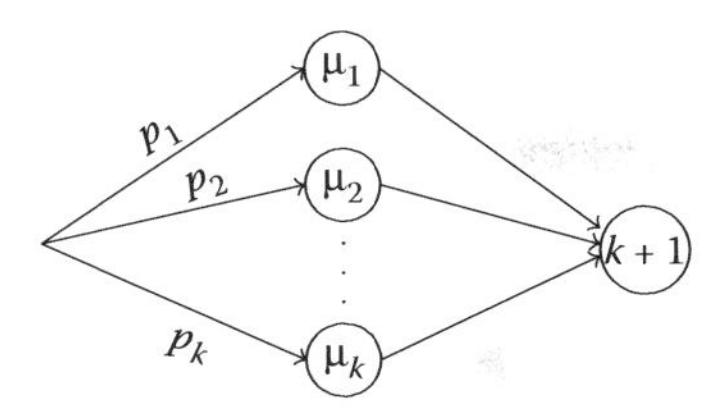

FIGURE 6.12
A k-stage hyper-exponential distribution.

time in that state is over, it enters the absorbing state. The hyper-exponential r.v. is popularly used to model or approximate distributions with squared coefficient of variation $c^2 \geq 1$. In many practical settings, the special case of two-stage hyper-exponential is deemed sufficient for accurate modeling.

Neuts (1982) provides the steady-state distribution of the queues where the service time or interarrival times have PTD and the following discussion summarizes his results.

6.6.12.2 M/PT/1 Queue

A single server queue with Poisson arrivals with rate λ and PTD service time represented by (C, G) as described above can be considered as a continuous-time Markov chain with state space $S = \{0, (i, j), i \geq 1, 1 \leq j \leq k\}$, where (i, j) is a state where there are $i \geq$ customers in the system and the stage of the service time is $j \geq 1$ and state 0 corresponds to an empty queue. The infinitesimal generator of the Markov chain is given by the following partitioned form:

$$\Lambda = \begin{array}{c} \\ 0 \\ \bar{1} \\ \bar{2} \\ \bar{3} \end{array} \begin{array}{c} \begin{array}{ccccc} 0 & \bar{1} & \bar{2} & \bar{3} & \cdots \end{array} \\ \begin{pmatrix} -\lambda_0 & \lambda C & 0 & & \cdots \\ G_0 & G - \lambda I & \lambda I & 0 & \cdots \\ 0 & G_0 C & G - \lambda I & \lambda I & \cdots \\ \vdots & & & & \end{pmatrix} \end{array}$$

The following matrix

$$R = \lambda(\lambda I - \lambda EC - G)^{-1}$$

plays a basic role in the main results obtained for queues with PTD. Let F be the distribution function of the service time given as $F(x) = 1 - Cexp(Gx)E$ for $x \geq 0$ with mean $\tau = -CG^{-1}E$. We denote the steady-state probability vector of the Markov chain with infinitesimal generator matrix Λ by $\bar{\pi}$ which is also partitioned to $\pi_0, \bar{\pi}_1, \bar{\pi}_2$, where the row vectors $\pi_i, \bar{i} \geq 1$ are of dimension k. Then, letting $\rho = \lambda\tau$, the following results are obtained:

1. *Limiting distribution:* Provided that $\rho < 1$, it holds that
   a. $\pi_0 = 1 - \rho$
   b. $\bar{\pi}_i = (1 - \rho)CR^i, \ i \geq 1$
2. *Waiting time distribution:* The steady-state distribution of the waiting time in an $M/PT/1$ queue is a PTD with representation $(\bar{\beta}, L)$, where $\bar{\beta} = (1 - \rho)CR(I - R)^{-1}$ and $L = G + \rho G^0 \beta$.

Example 6.18

As a special case, consider the queue $M/E_3/1$. Then the PTD is represented as (C, T), where $C = (1, 0, 0)$ and

$$G = \begin{pmatrix} -\mu & \mu & 0 \\ 0 & -\mu & \mu \\ 0 & 0 & -\mu \end{pmatrix}$$

The mean is $\tau = 3/\mu$ and R is the inverse of the matrix

$$\lambda^{-1}(\lambda I - \lambda EC - T) = \lambda^{-1} \begin{pmatrix} -\mu & \mu & 0 \\ -\lambda & \lambda+\mu & -\mu \\ -\lambda & 0 & \lambda+\mu \end{pmatrix}$$

For the waiting time distribution, we have $\bar{\beta} = \lambda\mu^{-1}(1, 1, 1)$ and

$$L = \begin{pmatrix} -\mu & \mu & 0 \\ 0 & -\mu & \mu \\ \lambda & \lambda & \lambda-\mu \end{pmatrix}$$

The mean of the waiting time is given by $\bar{\beta}L^{-1}E$.

Results for the bounded queue $PT/M/s/k+s$ and the unbounded queue $PT/M/s$ are also available in Neuts (1982), and more general queues in Asmussen (2001) for the interested readers, which we do not further discuss in the present context.

6.7 Queueing Networks

In real applications, we often encounter queues as parts of larger systems, such as the service networks in different departments which are connected to each other in a hospital. The system in such applications can be described as a network of subsystems, where each subsystem consists of a single queue, possibly with several servers. Such networks are difficult to analyze unless some assumptions are made to reduce the models to simpler forms. We will consider below some commonly known networks, namely, the series networks, Jackson networks, and cyclic networks. Before considering the special cases, we first introduce the general setting. Consider a service network of k service facilities, referred to as *nodes*, each facility having a number of identical servers with independent exponential service times. In particular, let node i have c_i identical servers with exponential service times and service rates μ_i, $i = 1, \ldots, k$. A customer served by a server at node i goes to node j with probability r_{ij}, where r_{i0} refers to the exit probability from the system after the service is completed at node i. Furthermore, all the nodes may receive arrivals from outside of the system according to Poisson processes. Let γ_i denote the external rate of arrivals to state i. There is no capacity restriction for the customers waiting to make a transition from one state to another.

6.7.1 Series Networks

In series networks, customers enter the first node, proceed through each node and leave the system from the last node. The underlying parameters are given as

$$\gamma_i = \begin{cases} \lambda & \text{if } i = 1 \\ 0 & \text{otherwise} \end{cases}$$

and

$$r_{ij} = \begin{cases} 1 & \text{if } j = i+1, 1 \le i \le k-1 \\ 1 & \text{if } j = 0, i = k \\ 0 & \text{otherwise} \end{cases}$$

Note that the first node behaves like an $M/M/c_1$ queue. In order to study the network behavior, we also need the departure process from node 1. Quite interestingly, the departure process is found to be the same as the arrival process. This means that all the nodes have the same departure processes, which is the arrival process to node 1 (interdeparture times are exponential with rate λ). With this result, the system steady-state performance measures can be obtained from k independently and operating identical $M/M/c_1$ queues.

6.7.2 Open Jackson Networks

Open Jackson networks are general networks where arrivals from outside to any node are possible and the customer may leave the system from any node, as well. The system state for the open Jackson networks is defined as the number of customers at each node. Let N_i denote the number of customers in state i at the steady state. Then the joint distribution of the N_is is given as

$$P(N_1 = n_1, N_2 = n_2, \ldots, N_k = n_k) \equiv p_{n_1,n_2,\ldots,n_k}$$

Total arrival rate to node i is the sum of the external arrivals and the arrivals from within the network nodes, which is given as

$$\lambda_i = \gamma_i + \sum_{j=1}^{k} r_{ji}\lambda_j.\| \tag{6.4}$$

Now, as before, we define the traffic rate at node i as $\rho_i = \lambda_i/\mu_i, i = 1, \ldots, k$. Solving the balance equations, the following steady-state result is obtained:

$$p_{n_1,n_2,\ldots,n_k} = \prod_{i=1}^{k} \frac{\rho_i^{n_i}}{w_i(n_i)} p_{0i}$$

where

$$w_i(n_i) = \begin{cases} n_i! & \text{if } n_i \le c_i \\ 1 & \text{if } j = 0, i = k \\ c_i^{n_i - c_i} c_i! & \text{otherwise} \end{cases}$$

and for node $i, i = 1, \ldots, k$, p_{0i} is obtained from the normalization $\sum_{n_i=1}^{\infty} p_{0i}\rho_i^{n_i}/w_i(n_i) = 1$. The above result shows that the network behaves as if it is composed of independent nodes in the steady state, where node i behaves like an $M/M/c_i$ queue. This result should be taken with caution however. In particular, although the number of customers in the nodes behaves independently at the steady state, this does not hold for the waiting times at each node. Because, in general Jackson networks, the arrivals are not truly Poisson, the distribution of the joint waiting times at the nodes does not turn out to be the product of the marginal distribution of the waiting times.

Example 6.19

Consider a five-node network (see Figure 6.13) with $c_1 = 3, c_2 = 2, c_3 = 2, c_4 = 2, c_5 = 1$ identical servers and corresponding service rates $\mu_1 = 1.2, \mu_2 = 1.8, \mu_3 = 1.8, \mu_4 = 1.2, \mu_5 = 6.4$, respectively, at nodes $1, 2, \ldots, 5$. Only nodes 1 and 3 receive external arrivals with $\gamma_1 = 3, \gamma_2 = 2, \gamma_i = 0, i = 3, 4, 5$. The positive transition probabilities are as follows: $r_{12} = 1, r_{24} = 2/3, r_{24} = 1/3, r_{35} = 1, r_{40} = 1, r_{50} = 4/5, r_{53} = 1/5$, and the others are zero. Then the arrival rates to each node are calculated using the expressions given in (6.4) above as $\lambda_1 = 3, \lambda_2 = 3, \lambda_3 = 3.1, \lambda_4 = 2, \lambda_5 = 5.5$.

As mentioned above, node i behaves like an $M/M/c_i$ queue in the steady state independently of each other. Hence, we can calculate the performance measures at each node separately referring to the results of an $M/M/s$ queue. To be brief, we will calculate the probabilities that each node is empty at any time in the steady state and the expected queue lengths (Table 6.1).

From Table 6.2, we see that the system is highly loaded. In particular, at a given instance in the steady state, the average total queue length in the system L_T is about 22, which is obtained as the sum of the waiting customers, i.e., $L_T = \sum_{i=1}^{5} L_{qi} = 21.09$. The probability π_0 that the entire system is empty is given by the product of the idleness probabilities of the nodes as $\pi_0 = \prod_{i=1}^{5} \pi_{0i} = 6 \times 10^{-6}$, whereas the probability that an arriving customer finds the entering node

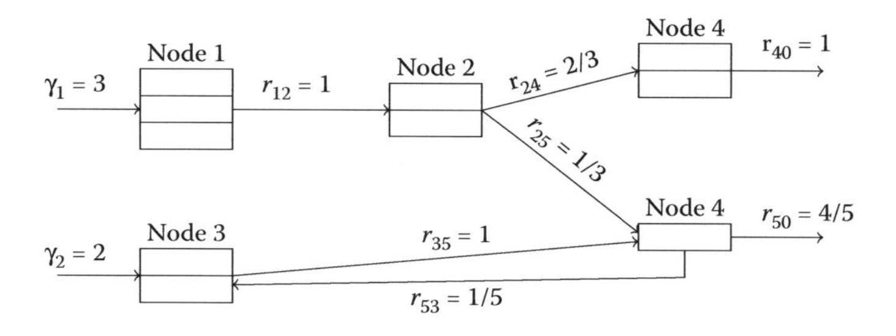

FIGURE 6.13
An open Jackson network with five nodes.

TABLE 6.2
Traffic Ratio, Percentage of Idle Times, and Mean Queue Length for the Five Nodes of an Open Jackson Network

Node i	c_i	λ_i	μ_i	ρ_i	π_{0i}	L_{qi}
1	3	3	1.2	0.833	0.045	3.5
2	2	3	1.8	0.833	0.091	4.53
3	2	3.1	1.8	0.861	0.075	5.65
4	2	2	1.2	0.833	0.091	4.53
5	1	5.5	6.4	0.786	0.215	2.88

empty is $3/5 \times 0.045 + 2/5 \times 0.075$ =0.057; only about 6% of the arrivals find the entry of the system from a free node. Obviously, the node performances or the overall system performance can be improved by either increasing the service rates or the number of servers, which requires a realistic cost assessment.

The product-form solution for the stationary probability distribution for a single class Jackson network also holds for the Kelly networks (Kelly 1975, Chen and Yao 2001a,b), where customers of different classes may have different priority levels at different service nodes and each node is quasireversible.

6.7.3 Closed Jackson Networks

A special case of Jackson networks where no customers enter ($\gamma_i = r_{i0} = 0$ for all i) or leave the system (resulting in a fixed population size, N, circulating within the system) is called the Gordon–Newell (Gordon and Newell 1967) network, or the closed Jackson network. A typical example is a repair facility in a system with N machines which are subject to random failures. A failed item can go to repair which may consist of several nodes and then return to the service system. The steady-state solution of this system is obtained from the balance equations similar to the previous case, which is given as

$$p_{n_1,n_2,\ldots,n_k} = \prod_{i=1}^{k} C(N) \frac{\rho_i^{n_i}}{w_i(n_i)}$$

where $C(N)$ is the normalizing constant obtained from the k-fold summation

$$\sum_{n_1+n_1+\cdots+n_k=N} C(N) \prod_{i=1}^{k} \frac{\rho_i^{n_i}}{w_i(n_i)} = 1$$

which satisfies

$$C(N) = \left[\sum_{n_1+n_1+\cdots+n_k=N} \prod_{i=1}^{k} \frac{\rho_i^{n_i}}{w_i(n_i)} \right]^{-1}$$

where $w_i(n_i)$ is defined as in open Jackson networks.

Example 6.20

Consider a repair facility with two servers with mean service times 2 and 3, respectively. There are N machines in the system which fail randomly in time. In this case, we can write

$$p_{n_1,n_2} = \frac{1}{3^{N+1} - 2^{N+1}} 2^{n_1} \times 2^{n_2}$$

6.7.4 BCMP Network

An important extension of the results on the Jacksonian networks has been provided by Baskett et al. (1975) for a network with general service time, regimes, and customer routing subject to particular service disciplines. The so-called BCMP network has the following properties: Customer arrival process may be single or multiple Poisson streams which may also be dependent on the total number of customers in the network. A service center is subject to any one of the following four service conditions:

1. FCFS discipline where all customers have the same negative exponential service time distribution for which the service rate can be dependent on the number of customers at the center.
2. A single server with a service discipline of processor sharing (that is, the service rate is equally shared by all n customers in service).
3. The number of servers is at least as large as the maximum number of customers that can be queued at the center.
4. Service discipline last come first served (LCSF) with preemptive resume.

The last three conditions also require that the service time distributions have rational Laplace transforms, which subsume the cases for the exponential, hyper- and hypo-exponential distributions, and the phase-type distributions as approximations of any other distribution.

For a BCMP network of m queues, which is open, closed, or mixed in which each queue is of the four types above, the equilibrium state probabilities has a Jacksonian-type product form, $\pi(x_1, x_2, \ldots, x_m) = CP(S)\pi_1(x_1)\pi_2(x_2)\cdots\pi_m(x_m)$, where S denotes the state of the network, x_i is the configuration of customers (typically, the number and current stage of service for each class) at center i, and C is the normalizing constant.

6.7.5 G Networks

Another type of network is the so-called G-networks first proposed by Gelenbe (1991), where the service times are no longer exponential and "negative" customers or triggers may arrive to remove the customers in a nonempty queue, which mimic disposal of congestion through batch removals. They are used for modeling system with specific control functions such as traffic rerouting. The G-networks also have a stationary solution form superficially resembling a Jacksonian network solution. It has been observed that G-networks can be used to approximate quite general input–output behaviors. They have been used in packet networks in communications and neural networks. See, for example, Gelenbe (1993a,b) and Gelenbe et al. (1999).

6.7.6 Finite Buffer Networks

Queueing networks with finite buffers (i.e., limited number of customers in line) have received much attention in the past years. Owing to the finiteness of the buffers, *blocking* may occur in such systems. Blocking refers to the instance when an arrival from node i (or from outside) is not permitted at node j because the buffer at j is full; node j is said to be blocked after service. A before-service-blocking would be the case when the service of a customer is not allowed to start at node i until there is room available in its destination buffer j. A node that is empty is usually considered starved. Such networks have been used to model primarily manufacturing systems. In the industrial jargon, buffers are sometimes called *kanban*. Except in special cases, queueing networks with blocking do not have product form solutions. Therefore, exact solutions can only be obtained by means of numerical techniques. A number of approximate solution approaches have been proposed, mostly on decomposition techniques. (For open networks, see Perros (1989) and Dallery and Gershwin (1991); for closed networks, see Onvural (1990) and Dallery and Frein (1993)). In a related vein, the difficulties arising in optimal control of queueing networks can be found in Papadimitriou and Tsitsiklis (1999).

6.8 Remarks and Extensions

In the previous sections, we have introduced basic queueing models commonly used in theory and practice. In this section, we state some important extensions to the queueing models described in the previous sections and provide some related references.

6.8.1 Single Stage Queues

So far, we emphasized the Markovian models with unit arrivals (customers arriving one by one). If, instead of a single customer, a possibly random number of customers arrive at each arrival instance, it is called a system with *batch* or *bulk* arrivals. Some useful references for such systems may be cited as the following. The book by Chaudhry and Templeton (1983) inspects bulk arrival queues, Lucantoni (1993) provides a tutorial for systems with batch arrivals and general service times, Armero and Conesa (1998) consider inference and prediction in bulk arrival queues, Chaudhry and Gupta (1992) provide exact analysis for waiting-time distribution of a single-server bulk arrival queue, Chen et al. (2010) consider Markovian bulk arrival and bulk service queues, and El-Sherbiny (2008) analyzes a bulk arrival queue with reneging, balking, and state-dependent servers.

6.8.1.1 Transient Solution

In most of the models, it is assumed that the interarrival and service processes are either exponential, or can be expressed as an *Erlang(k)* random variable, for which some closed form analytical expressions can be obtained and most of the results were given for the steady state of the system. Except for the Poisson process and simple CTMC models, transient behavior is generally difficult to obtain. When transient solutions are needed, Kolmogorov differential equations are solved, usually first solving the Laplace transforms and then inverting them to obtain the values of the transition probability functions $p_{ij}(t)$ for finite values of t. For transient solutions and applications in advanced systems, see, for example, Bertsimas and Nakazato (1992), Abdallah and Marie (1993), Csenki (2002), Gürler et al. (1997), Chassioti and Worthington (2004), and Balcioglu and Gürler (2011).

6.8.1.2 Queue Discipline

Another commonly made assumption in the analysis of queueing models is the FIFO queue discipline. However LCFS or SIRO models are also encountered in real applications. In an early study, Doshi (1983) considers a system with a hybrid discipline, Ritter and Wacker (1991) consider a general queue with general arrival and services with LCFS and priority discipline, Takagi and Kudoh (1997) and Flatto (1997) study the moments of the waiting time in a SIRO queues. Alfa and Fitzpatrick (1999) consider the waiting time distribution for FIFO/LIFO queues.

6.8.1.3 Priority Queues

The queue discipline may change if a *priority* customer arrives at the system. Such queues are first discussed by Cobham (1954) and Takács (1964); Jaiswal and Thiruvengadam (1967), Jaiswal (1968), and Balachandran (1971) provided some early results, while Williams (1980), Bagchi and Sullivan (1985), Ritter and Wacker (1991), Takagi and Takahashi (1991), Kao and Wilson (1999), and more recently Iravani and Balcioglu (2008) present further extensions in multiserver, general priority functions, general input and service distributions, batch Poisson arrivals, two priority classes, and impatient customers, respectively.

6.8.1.4 Impatient Customers

In most of the queueing models, it is assumed that the patients wait in line until their service is completed, once they enter the queueing system. In real life, however, customers may be *impatient* and may leave the queue after waiting for a while before their service starts. Although the idea of impatient customers has been introduced quite early, models with impatient queues have been extensively studied more recently partly due to the applications arising in call centers and other telecommunication networks. Barrer (1957) provides an initial work in this area. De Kok and Tijms (1985) study queueing systems with impatient customers while Boots and Tijms (1999) and Mandelbaum and Momcilovic (2012) consider multiserver queues. Bae and Kim (2010) consider the stationary workload of such queues. A review of the models with impatient customers can be found in Wang et al. (2010). For applications in call centers, readers may refer to Garnett et al. (2002), Gans et al. (2003), and Zeltyn and Mandelbaum (2005) among others.

6.8.1.5 Retrial Queues

Impatient customers who leave the system after waiting for a while may come back to the system with a certain probability after some time passes. Queueing systems with such customers are referred to as *retrial* queues. The interested readers may refer to the book by Falin and Templeton (1997) and Kulkarni and Liang (1997) as useful resources for the retrial queues as well as the review papers by Yang and Templeton (1987), Falin (1990), Artalejo (1999a,b), and Gómez-Corral (2006).

6.8.1.6 Polling Models

There is another category of queueing models called the *Polling models* where one or more servers provide service to several queues in a cyclical manner. For research on polling models, see a special issue of *Queueing Systems* edited by Boxma and Takagi (1992), as well as Gupta and Günalay (1997), Takagi (1997), Hirayama et al. (2004), and Boon et al. (2011).

6.8.1.7 Vacation Models

These models refer to queueing systems where service breaks occur due to reasons such as machine breakdowns, service disruptions, and cyclic server (polling) queues. For bibliographies on this topic, see Doshi (1986) and Alfa (2003).

6.8.1.8 Estimation

In all the queueing models described above, it is assumed that the system parameters such as arrival and service rates or distributional parameters are known. In practice, however, it is more realistic that these are not exactly known and have to be estimated. The first theoretical treatment of the estimation problem was given by Clarke (1957) who derived maximum likelihood estimates of arrival and service rates in an $M/M/1$ queueing system. Billingsley (1961) provided inference for Markov processes in general and Wolff (1965) derived likelihood ratio tests and maximum likelihood estimates for B&D processes. Also see Cox (1955) for a comprehensive survey of statistical problems related to queues. Cox also provides a broad guideline for inference investigations in non-Markovian queues. An early paper on inferencing for a non-Markovian system is by Goyal and Harris (1972). Significant progress has occurred in developing statistical procedures for various queueing systems, including Basawa and Prabhu (1981, 1988) who considered estimation of parameters in $GI/G/1$, and Armero (1994) who used Bayesian techniques for inference in Markovian queues. Bhat et al.

(1997) provide a survey of statistical inference problems. Bhat and Basawa (2002) provide MLE methods for queueing systems.

6.8.2 Network of Queues

Closed networks with batch arrivals are analyzed in Henderson et al. (1990); Onvural (1990) provides a survey of closed blocking networks, Bitran and Dasu (1992) provide a survey of open networks in manufacturing processes, Boucherie and Van Dijk (1991) consider networks with state-dependent multiple job transitions, Bolch et al. (2006) discuss queueing networks with computer science applications, and Creemers and Lambrecht (2010) provide modeling approaches for a hospital queueing network. We may also mention Meyn (2001, 2003, 2008) for multiclass networks with feedback regulation and Stidham (2002) for analysis and design of queueing systems. There is a growing literature on the queueing networks where both arrival and service processes are described by Markovian arrival processes (MAP), which introduce quite complicated systems that are however computationally tractible. See, for example, Takine (1999), Asmussen and Moller (2001), Wu et al. (2011), and Baek et al. (2013). An extremely popular application of closed queueing networks has been the CONWIP model in production systems, first proposed by Spearman et al. (1990). Other related work are by Gelenbe et al. (1998), Gong et al. (2008) for applications in supply chains, Graham (1978) for computer networks, and Koizimu et al. (2005), and Lam and Wong (1982) for other applications. (See also Framinan et al. (2003) for a review.)

6.8.3 Further Resources

For a more in-depth analysis of the systems and the methodologies as well as further extensions, we refer the readers to the following references among many others: Saaty (1961), Bhat (1969, 1984), Kleinrock (1975, Vol. 1, Vol. 2), Gross and Harris (1985), Cohen (1992), Prabhu (1997), Parlar (2000) (where computational aspects are emphasized), Buzacott and Shantikumar (1993) for queueing networks in manufacturing, Bhat (2008), Kleinrock and Gail (1996), Stidham (2009) (for optimal design of queueing systems), Haviv (2013), and Gross et al. (2013). Shantikumar et al. (2007) provide a survey and the open problems in queueing models. On queueing networks, we may recommend Gelenbe and Pujolle (1998), Walrand (1993), Chen and Yao (2001a,b), Bolch et al. (2006), and Kelly and Yudovina (2014).

Notation

FCFS	First come first served
FIFO	First in first out
SIRO	Service in random order
LCFC	Last come first served
E_k	Erlang k
$\{X(t), t \in T\}$	A stochastic process with time index T
$M(t)$	Expected number renewals in $(0, t]$
$\{R_n, n \geq 1\}$	Sequence of rewards earned at the nth renewal
p_{ij}	One-step transition probability of a homogeneous Markov chain (MC) from state i to j
$P = \{p_{ij}\}$	One-step transition matrix of an MC
P^n	n-Step transition matrix of an MC

$\pi_i^{(0)}$	Initial probability distribution of an MC
$\pi_i^{(n)}$	Unconditional probability distribution of an MC at the nth transition
N_i	Sojourn time of a Markov chain in state i
N_{ij}	First passage time from state i to state j of an MC
$f_{ij}^{(n)}$	Probability distribution of N_{ij}
π	Vector of limiting probabilities of an MC
T_{ij}	Sojourn time of an MC in state i before making a transition to state j
T_i	Sojourn time of an MC in state i
b_j	Probability that j servers are busy
q_j	Probability that j customers are in the queue
L	Expected number of customers in the system
L_q	Expected number of customers in the queue
W	Expected waiting time in the system
W_q	Expected waiting time in the queue
B	Expected number of busy servers
U	Expected utilization
R	Expected throughput
PTD	Phase-type distribution
$f*$	Laplace transform of a function f
$M/M/s$	A queue with exponential interarrival and service times with s servers
$M/D/1$	A queue with exponential interarrival and deterministic service times with a single server.
$M/M/s/N$	A queue with exponential interarrival and service times with s servers and queue capacity N
$M/M/1/N/N$	A queue with exponential interarrival and service times with s servers, queue capacity N, and source population of size N
$G_1I/G_2/s$	A queue with independent interarrival times with d.f. G_1 and service times with d.f. G_2 and s servers
$U(a,b)$	Uniform distribution over the interval (a,b)
$E(X)$	Expectation of a random variable X
$E(X\|Y)$	Conditional expectation of a random variable X given Y
$M_X(t)$	Moment generating function of random variable X
$\xi_X(z)$	Probability generating function of random variable X

Appendix A: Review of Probability Concepts

We start with describing the premises of probability. A *random experiment* is an operation or process whose outcome is not known in advance. *Sample space* is the collection of all possible outcomes of a random experiment, denoted by Ω. An *event* is any subset of a sample space, denoted by uppercase letters $A, B, \ldots$. *Probability* is a set function from the subsets of the sample space to the interval

$[0, 1]$, which satisfies the probability axioms: (i) $0 \leq P(A)$ for $A \subset \Omega$, (ii) $P(\Omega) = 1$, and (iii) if $E_1, E_2, \ldots$ is a sequence of mutually exclusive events ($E_m \cap E_n = \emptyset$, $\forall\, m \neq n$), it holds that

$$P\left(\bigcup_{n=1}^{\infty} E_n\right) = \sum_{n=1}^{\infty} P(E_n)$$

The sample space of a random experiment identifies all possible outcomes but which ones will occur as a result of the experiment is totally unknown. If some information about the outcome is revealed, this information may alter the probabilities of the events, which is defined as follows. The *conditional probability* of event E given the event A is denoted by $P(E|A)$ and defined as $P(E|A) = P(E \cap A)/P(A)$. Two events A and B are said to be *independent* if $P(A \cap B) = P(A)P(B)$. Events $A_1, A_2, \ldots$ are independent if $P(A_{i_1} \cap A_{i_2} \cap \ldots A_{i_n}) = P(A_{i_1})P(A_{i_2}) \ldots P(A_{i_n})$ for every finite set of indices $i_1, i_2, \ldots, i_n$.

A.1 Partitioning and Bayes' Rule

A set of mutually exclusive events $\mathbf{B} = \{B_1, \ldots, B_n\}$ is called a *partition of* Ω if $\bigcup_{n=1}^{n} B_i = \Omega$. Suppose $\mathbf{B}$ is a partition of Ω and $A \subset \Omega$. Then, according to total probability law,

$$P(A) = \sum_{i=1}^{\infty} P(A \cap B_i) = \sum_{n=1}^{\infty} P(B_i)P(A \mid B_i)$$

and the Bayes' formula for conditional probability of event B_j is given as

$$P(B_j|A) = \frac{P(A \cap B_j)}{P(A)} = \frac{P(B_j)P(A \mid B_j)}{\sum_{i=1}^{n} P(B_i)p(A \mid B_i)}$$

A *random variable* (r.v.) X is a real-valued function from the sample space to the real line. That is, for $w \in \Omega$, $X(w)$ is a rule that assigns a real value to the outcome w. Usually, the dependence on w is suppressed and $X = X(w)$ is used for brevity. In many cases, the identity function $X(w) = w$ is encountered if the sample space directly represents the quantity of interest. The randomness of X is inherited from the random experiment, the sample space of which is the domain of the function X. We assign probabilities to X by referring to the probabilities of the outcomes in the sample space that give rise to X. To this end, the *distribution function* (d.f.), also called the cumulative d.f. of an r.v. X is defined as $F(x) = P(X \leq x) = P(w \in \Omega : X(w) \leq x)$, for $x \in R$. Some basic properties of $F(x)$ are the following: (i) $F(x)$ is a nondecreasing function of x, (ii) $\lim_{x \to \infty} F(x) = 1$, (iii) $\lim_{x \to -\infty} F(x) = 0$, and (iv) $F(x)$ is a right continuous function. That is, for $\epsilon > 0$, $F(x^+) = \lim_{\epsilon \to 0} F(x + \epsilon) = F(x)$.

Random variables are classified as *discrete, continuous,* or *mixture* random variables. If the set of possible values is a countable set D, then X is called a discrete r.v. In that case, the d.f. $F(x)$ is a step function, and $P(x) = P(X = x) = F(x) - F(x^-)$ is called the *probability mass function (p.m.f.)* of X that satisfies $0 \leq P(z) \leq 1$ and $\sum_{x \in D} P(x) = 1$. In the above expression, $F(x^-) = \lim_{\epsilon \to 0} F(x - \epsilon)$. In general, $F(x) = \sum_{x_i \leq x, x_i \in D} P(x_i)$, where x_i's are the mass (or jump) points of F. If $F(x)$ is an absolutely continuous function, then X is a continuous random variable. In that case, there exists a nonnegative function f called the *probability density function (p.d.f.)* of X that satisfies $f(x) = dF(x)/dx$ at the points where $F(x)$ is differentiable. In all other cases, X is a mixture random variable which has both discrete and continuous parts.

A.2 Expected Values

The expected value or the mean of a random variable X with d.f. F is defined as the following Riemann–Stieltjes integral:

$$E[X] = \int_{-\infty}^{\infty} u dF_X(u)$$

provided that $E[X] = \int_{\infty}^{\infty} |u| dF(u) < \infty$. When X is continuous, we can write $E[X] = \int_{-\infty}^{\infty} uf(u)du$, and when X is discrete, $E[X] = \sum_{x \in D} xP(x)$. For the mixture random variables, we have $E[X] = \int_{x \in C} xf(x)dx + \sum_{x \in D} xP(x)$, where C and D are the set of continuity and jump points of F, respectively, and $f(x)$ is the derivative of F at the continuity points of F where it is also differentiable. The expected value of a function g of a random variable is given as

$$E[X] = \int_{-\infty}^{\infty} g(x)dF(x) = \int_{-\infty}^{\infty} xdH(x)$$

where H is the d.f. of the r.v. $g(X)$ and the above integrals are equal provided that they exist in the absolute sense.

A.3 Moments of Random Variables

For a positive integer k, $E[X^k]$ is called the moment of order k of X. Moments have both theoretical and practical importance. The first four moments are related to the mean, variance, skewness, and kurtosis of a random variable, respectively. The variance of a random variable X with mean μ is $Var(X) = E[(X - \mu)^2] = E(X^2) - E^2(X)$ and is a measure of the dispersion of the values of X around the mean μ. For positive random variables, a useful measure is the *coefficient of variation*, defined as $\eta_X = \sqrt{Var[X]}/E[X]$.

A.4 Moment Generating Function

The moments of a random variable can be conveniently obtained using the moment generating function (m.g.f.), $M_X(t)$, provided that it exists. The m.g.f. of an r.v. X is defined as a special expectation as $M_X(t) = E(e^{tX})$, for a real $t > 0$. If the m.g.f. exists, then the moments of all orders can be obtained through the relationship

$$E(X^k) = \frac{d^k M_X(t)}{dt^k}|_{t=0}$$

Furthermore, the m.g.f. uniquely defines the distribution of a random variable, which makes it useful for identifying the distributions of functions of random variables, especially in the limiting cases that involve infinitely many random variables. When the m.g.f. does not exist, similar results are obtained using the characteristic function $\xi_X(t)$ defined on the complex domain as $\xi_X(t) = E(e^{itX})$, where $i = \sqrt{-1}$, which is always finite despite the disadvantage of being a complex function.

A.5 Probability Generating Function

Another useful generating function for nonnegative discrete random variables is the *probability generating function*, also known as the *factorial moment generating function*, ξ, defined as $\xi(z) =$

$E(z^X)$ provided that it is finite for some $|z| \leq 1$. Letting $\xi^{(k)}(z)$ denote the kth derivative of ξ with respect to z, it can be easily verified that

$$\frac{\xi^{(k)}(z)}{k!}|_{z=0} = P(X = k)$$

and

$$\xi^{(k)}(z)|_{z=1} = E[X(X-1)\dots(X-k+1)]$$

where the last expression is known as the factorial moment of order k.

A.6 Laplace Transform

The Laplace transform is an integral transform of a function particularly useful in solving differential equations. For a continuous nonnegative function f, the Laplace transform is defined as

$$l(s) = \int_0^\infty f(t)e^{-st}dt$$

When f is a probability density function of a random variable X with distribution function F, an alternative expression is given as

$$F*(s) = \int_0^\infty e^{-st}f(x)dx = \int_0^\infty e^{-sx}dF(x)$$

Note in this case that $F*(s) = M_X - s$ and that $F*(s)$ is finite for all $s > 0$ unlike the m.g.f. which may not exist.

A.7 Some Important Distributions

An experiment that has only two possible outcomes, usually referred to as *success* and *failure* with probabilities p and $q = 1 - p$, respectively, is also called a Bernoulli trial. A *Bernoulli* or an indicator random variable X takes only the values 1 or 0 with respective probabilities p and $q = 1 - p$, which is a random variable that maps the outcomes of a Bernoulli trial to the set $(1, 0)$ on R. A sequence of Bernoulli random variables obtained from independent Bernoulli trials with the same probability p of success is called a Bernoulli sequence. A *binomial* r.v. X corresponds to the number of successes in n Bernoulli trials. The notation $X \sim Bin(n, p)$ is used and the p.m.f. is given by

$$P(x) = P(X = x) = \binom{n}{x} p^x(1-p)^{n-x} \quad x = 0, 1, \dots, n$$

The *geometric* r.v. corresponds to the number of Bernoulli trials needed until the first success is obtained. The corresponding p.m.f. is

$$P(x) = P(X = x) = \begin{cases} q^{x-1}p & \text{if } x \geq 1 \\ 0 & \text{otherwise} \end{cases}$$

Geometric r.v. has a conveniently expressed c.d.f. given by $F(x) = p(X \leq x) = \sum_{i=0}^{x} pq^{i-1} = 1 - q^x$. Geometric random variable has the *memoryless property* defined as follows: A random variable is said to have the memoryless property if $P\{T > t + u \mid T > u\} = P(T > t)$ for all $u, v > 0$. Equivalently, this property is expressed as satisfying $P(T > t + u) = P(T > t)P(T > u)$ for all $u, v > 0$. It can be shown that the geometric r.v. is the only discrete r.v. with this property. The notation $X \sim Geom(p)$ is used for a geometric random variable X with parameter p.

Another commonly used discrete random variable is the *Poisson* random variable, which under some technical assumptions corresponds to the number of events that occur randomly over a fixed time interval. For fixed $\alpha > 0$, the p.m.f. of a Poisson r.v. is given by

$$P(x) = \frac{e^{-\alpha}\alpha^x}{x!} \qquad x = 0, 1, \ldots$$

Poisson random variable is a special case of the Poisson process discussed in Section 6.3.2 and the underlying technical assumptions will be discussed there. The notation for a Poisson r.v. with parameter λ is $X \sim Poisson(\lambda)$.

Next, we introduce some commonly used continuous random variables. A *uniform* r.v. represents an even distribution of probability over an interval (a, b) with p.d.f.

$$f(x) = \begin{cases} \dfrac{1}{b-a} & \text{if } a < x < b \\ 0 & \text{otherwise} \end{cases}$$

and the c.d.f.

$$F(x) = \int_a^x f(u)du = \begin{cases} 0 & \text{if } x < a \\ \dfrac{x-a}{b-a} & a < x < b \\ 1 & x \geq b \end{cases}$$

The notation for a uniform r.v. with parameters a and b is $X \sim U(a, b)$. The special case $U(0, 1)$ is of particular importance.

An *exponential* random variable is a nonnegative r.v. commonly used for modeling lifetimes of components, duration of events, or the time until a specific event occurs. For some fixed $\lambda > 0$, the p.d.f. of an exponential r.v. is given as

$$f(x) = \begin{cases} \lambda e^{-\lambda x} & x > 0 \\ 0 & \text{otherwise} \end{cases}$$

and the corresponding c.d.f. is

$$F(x) = \begin{cases} 0 & x < 0 \\ 1 - e^{-\lambda x} & x \geq 0 \end{cases}$$

The notation for an exponential r.v. with parameter λ is $X \sim Exp(\lambda)$. Exponential distribution is a key distribution in stochastic modeling and particularly in the context of queueing theory. Hence, its properties will be further elaborated.

A.8 Some Properties of Exponential Distributions

1. Exponential distribution possesses memoryless property since $P(X > t+u \mid X > u) = P(X > t+u; X > u)/P(X > u) = P(X > t+u)P(X > u) = e^{-\alpha(t+u)}/e^{-\alpha(u)} = e^{-\alpha t} = P(T > t)$.
2. The minimum of several independent (will be defined for jointly distributed r.v.'s) exponential r.v.'s is exponential. Suppose X_i's are independent random variables having exponential distribution with parameters $\alpha_i, i = 1, \ldots, n$. Then, $T = \min(X_1, \ldots, X_n)$ has exponential distribution with parameter $\alpha = \sum_i \alpha_i$. This follows from noting

$$\begin{aligned} P(T > t) &= P(T = min(X_1, \ldots, X_n) > t) = P(X_1 \geq t, X_2 > t, \ldots, X_n > t) \\ &= P(X_1 > t)P(X_2 \geq t) \cdots P(X_n > t) \\ &= e^{-\alpha_1 t}e^{-\alpha_2 t} \cdots e^{-\alpha_n t} = e^{-(\alpha_1+\alpha_2+\cdots+\alpha_n)t} \end{aligned}$$

3. Consider an arrival process where independent unit arrivals occur randomly. If interarrival times are identical and exponential with common rate λ, then the number of arrivals in a unit interval has Poisson distribution with parameter λ.
4. Exponential distribution is the unique continuous distribution with constant hazard rate. The hazard rate, h, of an r.v. X with d.f. F is defined as

$$h(t) = \lim_{\Delta t \to 0} \frac{P(X < t + \Delta t \mid X > t)}{\Delta t} = \lim_{\Delta t \to 0} \frac{[F(t+\Delta t) - F(t)]}{1 - F(t)}$$

If X is continuous with p.d.f. f, then the hazard rate is given as $h(t) = f(t)/1 - F(t)$. In the special case of exponential distribution with rate α, $h(t) = \alpha e^{-\alpha t}/e^{-\alpha t} = \alpha$. The uniqueness property can be shown using mathematical arguments.

A *gamma* r.v. is a nonnegative continuous random variable with parameters $\gamma > 0$ and $\lambda > 0$. The p.d.f. is given as

$$f(x) = \begin{cases} \frac{1}{\Gamma(\gamma)} \lambda e^{-\lambda x} \lambda^{\gamma} x^{\gamma-1} & x > 0 \\ 0 & \text{otherwise} \end{cases}$$

The corresponding c.d.f. does not have a simple form, but if γ is an integer, it can be obtained from the sum of Poisson probabilities as follows:

$$F(x) = P(X \leq x) = 1 - \sum_{i=0}^{\gamma-1} e^{-\lambda x}(\lambda x)i/i!$$

The notation for a gamma r.v. is $X \sim gamma(\gamma, \lambda)$, where γ is called the shape parameter and λ is called the scale parameter. The special case with $\gamma = 1$ is the exponential distribution. If $\gamma = k$, where k is a positive integer, the distribution is known as Erlang(k,) after the discovery of Erlang that it appears in the context of telephone call arrivals.

A *normal* r.v. X is a continuous random variable with parameters $\mu \in R$ and $\sigma^2 > 0$. The p.d.f. for $x \in R$ is given as

$$f(x) = \frac{1}{\sqrt{2\pi\sigma}} e^{-(x-\mu)^2/\sigma^2}$$

Normal distribution is symmetric about μ which is the mean of the distribution and the variance is σ^2. The common notation to denote a normal r.v. is $N(\mu, \sigma^2)$. The c.d.f. of normal density does not have a closed form. The probabilities for normal distribution is obtained using the fact that linear combinations of normal random variables also have normal distributions. As a special case, if X is $N(\mu, \sigma^2)$, then the linear transformation, known as the Z transformation defined as $Z = (X - \mu)/\sigma$, has an $N(0, 1)$ distribution. Since $P(X \leq x) = P(Z \leq x - \mu/\sigma) = F_Z(x - \mu/\sigma) \equiv \Phi(x - \mu/\sigma)$, the probabilities relating to X can be calculated using the c.d.f. of Z which is usually denoted as Φ, for which probability integral tables are provided in standard text books in probability. The importance of normal distribution lies in the fact that the sum of n random variables tend to have a normal distribution as n tends to infinity, known as the central limit theorem (CLT), which is a fundamental result that provides a means for approximating the distributions of sum of random variables when their exact distributions are not known or are too complicated.

A.9 Jointly Distributed Random Variables

In many practical situations, we are interested in the joint behavior of several random variables, rather than a single variable.

Most of the concepts that we have developed to characterize random variables are generalized to vector random variables as well; however, the issues usually become more complicated and in higher dimensions.

The joint distribution function of the random vector $\mathbf{X} = (X_1, X_2, \ldots, X_n)$ is as follows for any $(x_1, x_2, \ldots, x_n) \in R_n$:

$$F_{\mathbf{X}}(x_1, x_2, \ldots, x_n) = P(X_1 \leq x_1, X_2 \leq x_2, \ldots, X_n \leq x_n)$$

The above function is bounded between zero and one, nondecreasing and right continuous in each component. The random variables $X_1, X_2, \ldots, X_n$ are said to be jointly continuous if there is a nonnegative function $f(x_1, x_2, \ldots, x_n) \in R_n$ such that for any $B \in R_n$ it holds that

$$P(X_1, X_2, \ldots, X_n) \in B = \int \ldots \int_B f(x_1, x_2, \ldots, x_n) dx_1 dx_2 \cdots dx_n$$

and $\int \ldots \int_{R_n} f(x_1, x_2, \ldots, x_n) dx_1 dx_2 \cdots dx_n = 1$. This function is called the joint p.d.f. of X. The *marginal* joint d.f. of any subset S of X can be obtained by letting $x_i \to \infty, i : x_i \notin S$ in $F_{\mathbf{X}}(x_1, x_2, \ldots, x_n)$. That is, if $n = 6$, the marginal joint d.f. of $S = (X_1, X_2, X_6)$ is $F_{X_1, X_2, X_6}(x_1, x_2, x_6) = F_{\mathbf{X}}(x_1, x_2, \infty, \infty, \infty, x_6)$. If S consists of a single variable, the marginal is a *univariate marginal*, if it consists of two variables, it is a *bivariate marginal*, and so on.

We say that the random variables $X_1, X_2, \ldots, X_n$ are *independent* if

$$F_{\mathbf{X}}(x_1, x_2, \ldots, x_n) = F_{X_1}(x_1) F_{X_2}(x_2) \cdots F_{X_n}(x_n) \quad \forall (x_1, x_2, \ldots, x_n) \in R_n$$

A sequence of independent random variables with a common distribution is called i.i.d. (independent, identically distributed) random variables. Independent random variables and their sums play a central role in probability theory and several properties of them are obtained using the moment generating functions using the following result. Suppose $X_1, X_2, \ldots, X_n$ are independent random variables with respective m.g.f.'s $M_{X_1}(t), M_{X_2}(t), \ldots, M_{X_n}(t)$. Then, the m.g.f. of $S = X_1 + X_2 + \cdots + X_n$ is given by $M_S(t) = M_{X_1}(t) \times M_{X_2}(t) \times \cdots \times M_{X_n}(t)$.

A.10 Conditional Probability Distributions and Conditional Expectation

The concept of conditional probabilities can be extended to the conditional distributions when random variables are involved. To avoid messy notation, we illustrate the idea for two random variables. Let X, Y be two discrete random variables with joint p.m.f. $P_{X,Y}(x, y) = P(X = x, Y = y)$ and the marginal p.m.f.'s $P_X(x)$ and $P_Y(y)$. Then the *conditional p.m.f.* of X given $Y = y$ is $P_{X|Y}(x|y) = P_{X,Y}(x, y)/P_Y(y)$ provided that $P_Y(y) > 0$. If X, Y are continuous r.v.'s with joint p.d.f. $f_{X,Y}(x, y)$, and marginal p.d.f.'s $f_X(x)$ and $f_Y(y)$, then the conditional p.d.f. of X given Y is $f_{X|Y}(x|y) = f_{X,Y}(x, y)/f_Y(y)$ provided that $f_Y(y) > 0$. A conditional p.m.f. or p.d.f. satisfies all the properties of a p.m.f. or a p.d.f. Hence, quantities such as means, variances, moments, and moment generating functions can all be obtained from these conditional distributions. Of particular importance is the *conditional expectation*. Suppose X, Y are r.v.'s and g is a real-valued function. Then the conditional expectation of $G(X)$ given Y is $E[g(X) \mid Y = y] = \sum_x g(x)P_{X|Y}(x \mid y)$ for discrete X, Y and $E[g(X) \mid Y = y] = \int_x g(x)f_{X|Y}(x \mid y)dx$ for continuous X, Y. An important result regarding conditional expectations is the iterated expectations stated as follows: For any two random variables X and Y, it holds that

$$E[X] = E_Y[E[X \mid Y]]$$

This result indicates that $E(X)$ can be calculated by first conditioning X on another convenient r.v. Y to obtain the conditional expectation $E[X|Y]$ and the taking another expectation with respect to Y. This method of finding $E[X]$ proves to be a useful tool for calculating expectations which are difficult to evaluate directly.

References

Abdallah, H., and Marie, R. 1993. The uniformized power method for transient solutions of Markov processes. *Computers & Operations Research*, 20(5):515–526.

Alfa, A., and Fitzpatrick, G. 1999. Waiting time distribution of a FIFO/LIFO Geo/D/1 queue. *INFOR*, 37(2):149–159.

Alfa, A. S. 2003. Vacation models in discrete time. *Queueing Systems*, 44(1):5–30.

Armero, C. 1994. Bayesian inference in Markovian queues. *Queueing Systems*, 15(1–4):419–426.

Armero, C., and Conesa, D. 1998. Inference and prediction in bulk arrival queues and queues with service in stages. *Applied Stochastic Models and Data Analysis*, 14(1):35–46.

Artalejo, J. 1999a. Accessible bibliography on retrial queues. *Mathematical and Computer Modelling*, 30(3):1–6.

Artalejo, J. 1999b. A classified bibliography of research on retrial queues: Progress in 1990–1999. *Top*, 7(2):187–211.

Asmussen, S. 1999. *On the Ruin Problem for Some Adapted Premium Rules*. Centre for Mathematical Physics and Stochastics, University of Aarhus.

Asmussen, S., and Moller J. R. 2001. Calculation of the steady state waiting time distribution in GI/PH/c and MAP/PH/c queues. *Queueing Systems*, 37:9–29.

Bae, J., and Kim, S. 2010. The stationary workload of the G/M/1 queue with impatient customers. *Queueing Systems*, 64(3):253–265.

Baek, J. W., Lee, H. W., Lee, S. W., and Ahn, S. 2013. A MAP-modulated fluid flow model with multiple vacations. *Annals of Operations Research,* 202(1):19–34.

Bagchi, U., and Sullivan, R. S. 1985. Dynamic, non-preemptive priority queues with general, linearly increasing priority function. *Operations Research*, 33(6):1278–1298.

Bailey, N. T. 1954. A continuous time treatment of a simple queue using generating functions. *Journal of the Royal Statistical Society. Series B (Methodological)*, 15:288–291.

Balachandran, K. 1971. Queue length dependent priority queues. *Management Science*, 17(7):463–471.

Balcioglu, B., and Gürler, Ü. 2011. On the use of phase-type distributions for inventory management with supply disruptions. *Applied Stochastic Models in Business and Industry*, 27(6):660–675.

Barrer, D. 1957. Queuing with impatient customers and indifferent clerks. *Operations Research*, 5(5):644–649.

Basawa, I., and Prabhu, N. 1981. Estimation in single server queues. *Naval Research Logistics Quarterly*, 28(3):475–487.

Basawa, I. V., and Prabhu, N. U. 1988. Large sample inference from single server queues. *Queueing Systems*, 3(4):289–304.

Baskett, F., Chandy, K. Mani, Muntz, R. R., and Palacios, F. G. 1975. Open, closed and mixed networks of queues with different classes of customers. *Journal of the ACM*, 22(2):248–260, doi: 10.1145/321879.321887.

Bertsimas, D. J., and Nakazato, D. 1992. Transient and busy period analysis of the GI/G/1 queue: The method of stages. *Queueing Systems*, 10(3):153–184.

Bhat, U. N. 1968. *Elements of Applied Stochastic Processes*. Wiley, New York.

Bhat, U. N. 1984. *A Study of the Queueing Systems M/G/1 and GI/M/1*. Springer-Verlag, New York.

Bhat, U. N. 2008. *An Introduction to Queueing Theory: Modeling and Analysis in Applications*. Birkhauser, Boston.

Bhat, U., and Basawa, I. 2002. Maximum likelihood estimation in queueing systems. In: *Advances on Methodological and Applied Aspects of Probability and Statistics*, (Ed. N. Balakrishnan), Taylor & Francis, New York, 13–29.

Bhat, U. N. 1969. Sixty years of queueing theory. *Management Science*, 15(6):B–280.

Bhat, U. N., Miller, G. K., and Rao, S. S. 1997. Statistical analysis of queueing systems. In: Dshalalow, J.H. (ed.), *Frontiers in Queueing Theory*, CRC Press, Boca Raton, FL, 351–394.

Billingsley, P. 1961. *Statistical Inference for Markov Processes*, volume 2. University of Chicago Press, Chicago.

Bitran, G. R., and Dasu, S. 1992. A review of open queueing network models of manufacturing systems. *Queueing Systems*, 12(1–2):95–133.

Bolch, G., Greiner, S., de Meer, H., and Trivedi, K. S. 2006. *Queueing Networks and Markov Chains: Modeling and Performance Evaluation with Computer Science Applications*. John Wiley & Sons, New York.

Boon, M. A. A., Van der Mei, R. D., and Winands, E. M. M. 2011. Applications of polling systems. *Surveys in Operations Research and Management Science*, 16(2):67–82.

Boots, N. K., and Tijms, H. 1999. A multiserver queueing system with impatient customers. *Management Science*, 45(3):444–448.

Boucherie, R. J., and Van Dijk, N. M. 1991. Product forms for queueing networks with state-dependent multiple job transitions. *Advances in Applied Probability*, 23:152–187.

Boucherie, R. J., and van Dijk, N. M. 2011. *Queueing Networks*. Springer, New York.

Boxma, O., and Takagi, H. 1992. Special issue on polling systems. *Queueing Systems*, 11(1):2.

Breuer, L., and Baum, D. 2005. *An Introduction to Queueing Theory and Matrix-Analytic Methods*. Springer, Dordrecht, Netherlands.

Brockmeyer, E., Halstrm, H., Jensen, A., and Erlang, A. K. 1948. The life and works of A. K. Erlang.

Buzacott, J. A., and J. G. Shantikumar. 1993. *Stochastic Models of Manufacturing Systems*. Prentice-Hall, Englewood Cliffs, NJ.

Chassioti, E., and Worthington, D. 2004. A new model for call centre queue management&star. *Journal of the Operational Research Society*, 55(12):1352–1357.

Chaudhry, M., and Gupta, U. 1992. Exact computational analysis of waiting-time distributions of single-server bulk-arrival queues: MX/G/1. *European Journal of Operational Research*, 63(3): 445–462.

Chaudhry, M., and Templeton, J. G. 1983. *A First Course in Bulk Queues*. Wiley, New York.

Chen, A., Pollett, P., Li, J., and Zhang, H. 2010. Markovian bulk-arrival and bulk-service queues with state-dependent control. *Queueing Systems*, 64(3):267–304.

Chen, H., and Yao, D. D. 2001a. *Fundamentals of Queueing Networks: Performance, Asymptotics, and Optimization*, volume 46, Springer, New York.

Chen, H., and Yao, D. 2001b. Kelly networks in fundamentals of queueing networks. *Stochastic Modelling and Applied Probability* 46:69–96. ISBN 978-1-4419-2896-2.
Clarke, A. B. 1957. Maximum likelihood estimates in a simple queue. *The Annals of Mathematical Statistics*, 28(4):1036–1040.
Cobham, A. 1954. Priority assignment in waiting line problems. *Journal of the Operations Research Society of America* 2:70–76.
Cohen, J. W. 1992. *The Single Server Queue*. North-Holland Series in Applied Mathematics and Mechanics, North-Holland.
Cox, D. R. 1955. The statistical analysis of congestion. *Journal of the Royal Statistical Society: Series A*, 118:324–335.
Creemers, S., and Lambrecht, M. 2010. Queueing models for appointment-driven systems. *Annals of Operations Research*, 178(1):155–172.
Crommelin, C. D. 1932. Delay probability formulae when the holding times are constant. *P.O. Elec. Engrs. J.*, 25:41–50.
Crommelin, C.D. 1934. Delay probability formulae. *P.O. Elec. Engrs. J.*, 26:266–274.
Csenki, A. 2002. Transient analysis of semi-Markov reliability models: A tutorial review with emphasis on discrete parameter approaches. In: *Stochastic Models in Reliability and Maintenance*, (Ed. S. Osaki), Springer, Berlin, 219–251.
Dallery, Y., and Frein, Y. 1993. On decomposition methods for tandem queueing networks with blocking. *Operations Research*, 41(2):386–399.
Dallery, Y., and Gershwin, S. B. 1991. Manufacturing flow line systems: A review of models and analytical results. *Technical Report MASI No. 91-18*, Universite Pierre et Marie Curie, Paris, France.
De Kok, A. G., and Tijms, H. 1985. A queueing system with impatient customers. *Journal of Applied Probability*, 22:688–696.
Doshi, B. 1983. An M/G/1 queue with a hybrid discipline. *Bell System Technical Journal*, 62(5):1251–1271.
Doshi, B. 1986. Queueing systems with vacations: A survey. *Queueing Systems*, 1(1):29–66.
Eilon, S. 1969. A simpler proof of L=AW. *Operations Research*, 17(5):915–917.
El-Sherbiny, A. A. 2008. The non-truncated bulk arrival queue MX/M/1 with reneging, balking, state-dependent and an additional server for longer queues. *Applied Mathematical Sciences*, 2:747–752.
Erlang, A. K. 1909. The theory of probabilities and telephone conversations. *Nyt Tidsskrift for Mathematik Series B*, 20:33–39.
Erlang, A. K. 1917. Solutions of some problems in the theory of probabilities of significance in automatic telephone exchanges. *Electroteknikeren*, 13:5–13. Reproduced in Brockmeyer et al. 1948. The life and works of AK Erlang, 138–155.
Falin, G. 1990. A survey of retrial queues. *Queueing Systems*, 7(2):127–167.
Falin, G., and Templeton, J. G. 1997. *Retrial Queues*, volume 75, Chapman & Hall, London.
Feller, W. 1949a. On the theory of Markov processes with particular reference to applications. *First Berkeley Symposium on Mathematical Statistics and Probability*, 403–432. University of California Press, Berkeley, CA.
Feller, W. 1949b. Fluctuation theory of recurrent events. *Transactions of the American Mathematical Society*, 48:98–119.
Flatto, L. 1997. The waiting time distribution for the random order service $m/m/1$ queue. *The Annals of Applied Probability*, 7(2):382–409.
Framinan, J. M., González, P. L., and Ruiz-Usano, R. 2003. The CONWIP production control system: Review and research issues. *Production Planning & Control: The Management of Operations*, 14(3):255–265.
Fry, T. C. 1928. *Probability and Its Engineering Uses*, D. Van Nostrand Co. Inc., New York, 48.
Gans, N., Koole, G., and Mandelbaum, A. 2003. Telephone call centers: Tutorial, review, and research prospects. *Manufacturing & Service Operations Management*, 5(2):79–141.
Garnett, O., Mandelbaum, A., and Reiman, M. 2002. Designing a call center with impatient customers. *Manufacturing & Service Operations Management*, 4(3):208–227.
Gelenbe, E. 1991. Product-form queueing networks with negative and positive customers. *Journal of Applied Probability*, 28(3):656–663.
Gelenbe, E. 1993a. G-Networks with triggered customer movement. *Journal of Applied Probability*, 30(3): 742–748.

Gelenbe, E. 1993b. G-Networks with signals and batch removal. *Probability in the Engineering and Informational Sciences*, 7:335–342.

Gelenbe, E., Pujolle, G., and Nelson, J. 1998. *Introduction to Queueing Networks*. Wiley, Chichester.

Gelenbe, E., Zhi-Hong, M., and Yan, D. L. 1999. Function approximation with spiked random networks. *IEEE Transactions on Neural Networks*, 10(1):3–9.

Gómez-Corral, A. 2006. A bibliographical guide to the analysis of retrial queues through matrix analytic techniques. *Annals of Operations Research*, 141(1):163–191.

Gong, Q., Lai, K. K., and Wang, S. 2008. Supply chain networks: Closed Jackson network models and properties. *International Journal of Production Economics*, 113(2):567.

Gordon, W. J., and Newell, G. F. 1967. Closed queuing systems with exponential servers. *Operations Research*, 15(2):254.

Goyal, T. L., and Harris, C. M. 1972. Maximum-likelihood estimates for queues with state-dependent service. *Sankhyā: The Indian Journal of Statistics, Series A*, 34:65–80.

Graham, G. S. 1978. Queueing network models of computer system performance. *ACM Computing Surveys (CSUR)*, 10(3):219–224.

Gross, D., and Harris, C. M. 1985. *Fundamentals of Queueing Theory*. John Wiley & Sons, New York.

Gross, D., Shortle, J. F., Thompson, J. M., and Harris, C. M. 2013. *Fundamentals of Queueing Theory*. John Wiley & Sons, New York.

Gupta, D., and Günalay, Y. 1997. Recent advances in the analysis of polling systems. In: *Advances in Combinatorial Methods and Applications to Probability and Statistics*, (Ed. N. Balakrishnan), Birkhäuser, Boston, 339–360.

Gürler, Ü., and Parlar, M. 1997. An inventory problem with two randomly available suppliers. *Operations Research*, 45(6):904–918.

Haviv, M. 2013. Queues. *A Course in Queueing Theory*, Springer, New York.

Henderson, W., Pearce, C. E. M., Taylor, P. G., and van Dijk, N. M. 1990. Closed queueing networks with batch services. *Queueing Systems*, 6(1):59–70.

Hirayama, T., Hong, S. J., and Krunz, M. M. 2004. A new approach to analysis of polling systems. *Queueing Systems*, 48(1–2):135–158.

Hughes, J. P., P. Guttorp, and S. P. Charles. 1999. A non-homogeneous hidden Markov model for precipitation occurrence. *Journal of the Royal Statistical Society: Series C*, 48(1):15–30.

Iravani, F., and Balcioglu, B. 2008. On priority queues with impatient customers. *Queueing Systems*, 58(4): 239–260.

Jackson, J. R. 1957. Networks of waiting lines. *Operations Research*, 5(4):518–521.

Jaiswal, N., and Thiruvengadam, K. 1967. Finite-source priority queues. *SIAM Journal on Applied Mathematics*, 15(5):1278–1293.

Jaiswal, N. K. 1968. *Priority Queues*, volume 50, Academic Press, New York.

Jewell, W. S. 1967. A simple proof of L=AW. *Operations Research*, 15(6):1109–1116.

Kao, E. P., and Wilson, S. D. 1999. Analysis of nonpreemptive priority queues with multiple servers and two priority classes. *European Journal of Operational Research*, 118(1):181–193.

Kelly, F., and Yudovina, E. 2014. *Stochastic Networks*. Cambridge University Press, Cambridge, UK.

Kelly, F. P. 1975. Networks of queues with customers of different types. *Journal of Applied Probability*, 12(3):542–554.

Kendall, D. G. 1951. Some problems in the theory of queues. *Journal of the Royal Statistical Society. Series B*, 13(2):151–185.

Kendall, D. G. 1953. Stochastic processes occurring in the theory of queues and their analysis by the method of imbedded Markov chain. *Annals of Mathematical Statistics*, 24:338–354.

Khintchine, A. 1932. Mathematisches uber die Erwartung wor ienemoffentlichen schalter. *Mat. Sbornik*, 39: 73–84.

Kingman, J. F. C. 1962. Some inequalities for the queue GI/G/1. *Biometrika* 49:315–324.

Kingman, J. F. C. 1969. Markov population processes. *Journal of Applied Probability*, 6:1–18.

Kleinrock, L. 1975. *Queueing Systems, Volume 1: Theory*. Wiley-Interscience, New York.

Kleinrock, L. 1976. *Queueing Systems, Volume II: Computer Applications* Wiley, New York.

Kleinrock, L., and Gail, R. 1996. *Queueing Systems: Problems and Solutions*. Wiley, New York.

Koizimu, N., Kuno, E., and T. E. Smith. 2005. Modeling patient flows using a queuing network with blocking. *Health Care Management Science*, 8:49–60.
Kolmogorov, A.N. 1931. Sur le probleme d'attante. *Matematicheskii Sbornik*, 38:101–106.
Kulkarni, V., and Liang, H. 1997. Retrial queues revisited. *Frontiers in Queueing: Models and Applications in Science and Engineering*, 7:19.
Lam, S. S., and Wong, J. W. 1982. Queueing network models of packet switching networks. *Performance Evaluation*, 2:161–180.
Lindley, D. V. 1952. The theory of queues with a single server. *Proceedings of the Cambridge Philosophical Society*, 48:277–289.
Little, J. D. C. 1961. A proof of the queuing formula: L=AW. *Operations Research*, 9(3):383–387.
Lucantoni, D. M. 1993. The BMAP/G/1 queue: A tutorial. In: *Performance Evaluation of Computer and Communication Systems*, (Eds. M. C. Calzarossa and S. Tucci), Springer, Berlin, 330–358.
Mandelbaum, A., and Momcilovic, P. 2012. Queues with many servers and impatient customers. *Mathematics of Operations Research*, 37(1):41–65.
Meyn, S. P. 2001. Sequencing and routing in multiclass queueing networks part I: Feedback regulation. *SIAM Journal on Control and Optimization*, 40(3):741–776.
Meyn, S. P. 2003. Sequencing and routing in multiclass queueing networks part II: Workload relaxations. *SIAM Journal on Control and Optimization*, 42(1):178–217.
Meyn, S. P. 2008. *Control Techniques for Complex Networks*. Cambridge University Press, Cambridge.
Molina, E. C. 1927. Application of the theory of probability to telephone trunking problems. *Bell System Technical Journal*, 6(3):461–494.
Muller A., and Pflug, G. 2001. Asymptotic ruin probabilities for risk processes with dependent increments. *Insurance: Mathematics and Economics*, 28:381–392.
Neuts, M. F. 1975. Computational uses of the method of phases in the theory of queues. *Computers & Mathematics with Applications* 1:151–166.
Neuts, M. F. 1978. Renewal processes of phase type. *Naval Research Logistics Quarterly*, 25:445–454.
Neuts, M. F. 1979. A versatile Markovian point process. *Journal of Applied Probability*, 16:764–779
Neuts, M. F. 1981. *Matrix-Geometric Solutions in Stochastic Models: An Algorithmic Approach*. Courier Dover Publications, New York.
Neuts, M. F. 1982. Explicit steady-state solutions to some elementary queueing models. *Operations Research*, 30.3: 480–489.
Neuts, M. F. 1984. Matrix-analytic methods in queuing theory. *European Journal of Operational Research*, 15(1):2–12.
Neuts, M. F. 1989. *Structured Stochastic Matrices of M/G/1 Type and Their Applications*, volume 701, Marcel Dekker, New York.
O'Dell, G. F. 1920. The influence of traffic in automatic exchange design. *The Post Office Electrical Engineers' Journal*, 13:209–223.
Onvural, R. O. 1990. Survey of closed queueing networks with blocking. *ACM Computing Surveys (CSUR)*, 22(2):83–121.
Palm, C. 1937. Inhomogeneous telephone traffic in full availability groups. *Ericsson Tech*, 5:39.
Palm, C. 1938. Analysis of the Erlang traffic formulae for busy-signal arrangements. *Ericsson Tech*, 6:39–58.
Palm, C. 1943. Intensity fluctuations in telephone traffic. *Ericsson Tech*, 44:1–188.
Palm, C. 1947. Waiting times when traffic has variable mean intensity. *Ericsson Reviews*, 24:102–107.
Papadimitriou, C.H., and Tsitsiklis, J.N. 1999. The complexity of optimal queuing network control *Mathematics of Operations Research*, 24(2):293–305.
Parlar, M. 2000. *Interactive Operations Research with Maple*, Methods and Models. Birkhäuser, Basel.
Perros, H.G. 1989. Open queueing networks with blocking. In: *Stochastic Analysis of Computer and Communication Systems* (Ed. H. Takagi), North-Holland, Amsterdam.
Pfeifer, D. 1982. Characterizations of exponential distributions by independent non-stationary record increments. *Journal of Applied Probability*, 19:127–135.
Priestley, M. B. 1965. Evolutionary spectra and non-stationary processes. *Journal of the Royal Statistical Society. Series B*, 27:204–237
Pollaczek, F. 1930. Über eine Aufgabe der Wahrscheinlichkeitstheorie. I. *Mathematische Zeitschrift*, 32(1): 64–100.

Pollaczek, F. 1934. Über das warteproblem. *Mathematische Zeitschrift*, 38(1):492–537.
Pollaczek, F. 1965. Concerning an analytic method for the treatment of queueing problems. *Proceedings of Symposium on Congestion Theory*, (Eds. W. L. Smith and W. E. Wilkinson), University of North Carolina Press, Chapel Hill, NC, 1–42.
Prabhu, N. 1987. A bibliography of books and survey papers on queueing systems: Theory and applications. *Queueing Systems*, 2(4):393–398.
Prabhu, N. U. 1997. *Foundations of Queueing Theory*. Kluwer Academic Publishers, New York.
Ritter, G., and Wacker, U. 1991. The mean waiting time in a G/G/m/∞ queue with the lcfs-p service discipline. *Advances in Applied Probability*, 23:406–428.
Ross, S. M. 1996. *Stochastic Processes*. John Wiley & Sons, New York.
Saaty, T. L. 1961. *Elements of Queueing Theory with Applications*. Dove Publications Inc., New York.
Shanthikumar, J. G., Shengwei, D., and Zhang, M. T. 2007. Queueing theory for semiconductor manufacturing systems: A survey and open problems. *IEEE Transactions on Automation Science and Engineering*, 4(4):513–522.
Spearman, M. L., Woodruff, D. L., and Hoop, W. J. 1990. Conwip: A pull alternative to Kanban. *International Journal of Production Research*, 28(5):879–894.
Stidham, Jr, S. 1974. A last word on L=AW. *Operations Research*, 22(2):417–421.
Stidham, Jr, S. 2002. Analysis, design, and control of queueing systems. *Operations Research*, 50(1):197–216.
Stidham, Jr, S. 2009. *Optimal Design of Queueing Systems*. CRC Press, Chapman & Hall, Boca Raton.
Syski, R. Introduction to congestion theory in telephone systems, Oliver and Boyd, London. 1960. *Lecture Notes in Economics and Mathematical Systems*.
Takács, L. 1964. Priority queues. *Operations Research*, 12(1):63–74.
Takagi, H. 1997. Queueing analysis of polling models: Progress in 1990–1994. *Frontiers in Queueing: Models and Applications in Science and Engineering*, 7:119.
Takagi, H., and Kudoh, S. 1997. Symbolic higher-order moments of the waiting time in an M/G/1 queue with random order of service. *Stochastic Models*, 13(1):167–179.
Takagi, H., and Takahashi, Y. 1991. Priority queues with batch Poisson arrivals. *Operations Research Letters*, 10(4):225–232.
Takine, T. 1999. The nonpreemptive priority MAP/G/1 queue. *Operations Research*, 47(6):917–927.
Walrand, J. 1993. Queueing networks, in *Handbooks in Operations Research and Management Science*, Vol. 2: Stochastic Models, (Eds. D. P. Heyman and M. J. Sobel), North Holland, New York, 519–603.
Wang, K., Li, N., and Jiang, Z. 2010. Queueing system with impatient customers: A review. In: *Proceedings of 2010 IEEE International Conference on Service Operations and Logistics, and Informatics*, QingDao, China, 82–87.
Whitt, W. 1989. An interpolation approximation for the mean workload in a GI/G/1 queue. *Operations Research*, 37:936–952.
Whitt, W. 1984a. The amount of overtaking in a network of queues. *Networks*, 14:411–426.
Whitt, W. 1984b. Open and closed models for networks of queues. *AT&T Bell Laboratories Technical Journal*, 63:1911–1979.
Williams, T. 1980. Nonpreemptive multi-server priority queues. *Journal of the Operational Research Society*, 31:1105–1107.
Wolff, R. W. 1965. Problems of statistical inference for birth and death queuing models. *Operations Research*, 13(3):343–357.
Wu, J., Liu, Z., and Yang, G. 2011. Analysis of the finite source MAP/PH/N retrial G-queue operating in a random environment. *Applied Mathematical Modelling*, 35(3):1184–1193.
Yang, T., and Templeton, J. G. C. 1987. A survey on retrial queues. *Queueing Systems*, 2(3):201–233.
Zeltyn, S., and Mandelbaum, A. 2005. Call centers with impatient customers: Many-server asymptotics of the M/M/n+ G queue. *Queueing Systems*, 51(3–4):361–402.

7

Inventory Theory

Emre Berk and Ülkü Gürler

CONTENTS

ABSTRACT Inventory theory is concerned with management of the quantity and timing of the replenishment of assets typically stored to satisfy future demands. In this chapter, we introduce the fundamentals of inventory theory and the basic models that constitute the technical core of supply chain management. The purpose of the chapter is to expose the reader to the basic concepts, models, and theoretical results that would serve as a foundation to build upon and to introduce the existing literature. The exposition follows the common classification of inventory models on the basis of echelon structures (single vs. multiple locations/levels), demand processes (deterministic vs. random demands), problem horizon lengths (finite vs. infinite), and perishability (nonperishables vs. perishables). Specifically, the chapter introduces the basic terminology and inventory-related costs and proceeds to construct the continuous and periodic review models with deterministic demands. The models considered herein establish the fundamental trade-offs between the cost components. In each category, the stylistic models are presented first to develop intuition, followed by relaxations of the basic assumptions that result in more realistic models and richer insights. Stochastic demands are treated extensively as they constitute not only the more general contexts but also a vaster portion of the existing literature. For the continuous-review, stochastic demand setting, three different modeling approaches are illustrated in detail to provide a foundation for different solution methodologies encountered in the literature. Special cases are also treated in detail to the same end. The multiple-item and multiple-echelon inventory systems are discussed concisely but all the basic models and results are presented. For such systems, extensive references are given. The modeling approaches assume a basic familiarity with probability and stochastic processes; in cases of specialized techniques, sufficient intuition is provided for the uninitiated. The emphasis is on the model development and codification of the existing knowledge. Where available, optimal control policies are established and presented. In the absence of such policies, commonly used approximations and/or heuristics are given. To illustrate some of the heuristic methods and models, simple numerical examples are also provided. The chapter concludes with some discussion of issues and practices encountered in the implementation of the discussed models.

7.1 Introduction

An inventory typically refers to a stockpile of goods (physical units) stored in the anticipation of future demands. However, any asset that exists to satisfy some need in the future may be also viewed as inventories. For example, seats on a plane for a particular flight, movie tickets, or workers. Inventory control is management of the quantity and timing of the replenishment of such assets (stored goods or otherwise) while satisfying demands in accordance with the decision-maker's objective(s). The objectives are, typically, to minimize some cost measure and/or achieve a service level expressed as an ability (or probability thereof) to satisfy demands. Inventory control policies are decision rules that focus on the trade-off between the costs and benefits of alternative solutions to questions of when and how much to order. Inventories can be found in both retail and production settings; depending on their location on a supply chain, they may be raw materials, work-in-progress (WIP), or finished goods. In practice, each item carried out in a stock (currently or in the past) is assigned a specific code and referred to as a stock-keeping unit (SKU); typical retail/production inventories may involve tens of thousands of SKUs.

The main reasons for carrying out inventories are (i) uncertainties in future demands and the available supply quantities, (ii) existence of and the possible variability in delivery lead times, (iii) opportunities for economies of scale, (iv) speculative objectives when future price movements are anticipated, and (v) other uncertainties in production environments due to machine failures and/or quality defects. Of these, economies of scale and the presence and/or uncertainty of delivery lead times constitute the most important reasons. Scale economies can be achieved through ordering/producing in larger quantities by either spreading fixed ordering/setup costs over larger volumes, or exploiting quantity discounts offered by suppliers.

An inventory system incurs costs in five categories: (i) ordering or setup costs; (ii) replenishment costs; (iii) inventory-holding costs; (iv) shortage costs that are associated with not being able to satisfy the demands immediately; and, finally, (v) costs due to decay, perishing, and obsolescence of items. Of these, ordering/setup costs are fixed, and the remainder are variable costs. (There is also the cost of monitoring the inventories and implementing an inventory control policy. However, such costs are typically assumed as sunk and, hence, ignored in the literature.)

Ordering costs are the costs incurred in a retail setting every time a replenishment action is taken regardless of the size of the order. They include the bureaucratic and legal paperwork costs, the flat fee component of transportation costs, and the fixed portion of quality control activities for incoming materials. Their counterpart in production environments—setup costs—refers to the costs incurred for preparing machinery for production, such as cleaning, changeover/retooling, and calibration. In retail settings, it is denoted by K, and in production, it is denoted by S.

Replenishment costs refer to all costs incurred on a unit basis to acquire one unit of an item from the supplier or to the point of storage: In retail, they entail unit-purchasing costs, and the variable components of transportation costs and the incoming quality inspections. It will be denoted by c.

A unit-holding cost, h, is the cost incurred to hold one unit of an item in a stock for a prespecified time period, typically expressed in years. It consists of two main parts: (i) the cost of capital is the opportunity cost of (i.e., the interest rate on) the capital tied up in an inventory and is computed as the cost of capital of the firm i multiplied by the unit replenishment cost, c. (ii) Unit-warehousing cost, w, is the cost of holding one unit physically in a stock; and, refers to the cost of operating the warehouse or other storage spaces, materials-handling equipment and labor costs, and pilferage. Hence, $h = ic + w$.

Shortage costs can be expressed either explicitly on a unit basis, or implicitly through the desired service levels. Whenever the on-hand stock becomes zero, the inventory system is said to experience a *stock-out*. A stock-out period may be of a random duration in the presence of demand and/or delivery uncertainties. A demand arriving during a stock-out period is either backordered (to be fulfilled in the future when outstanding orders arrive) or lost. The backordering may be partial or full; in case of partial backordering, a portion of demand is backordered while the rest is lost. When shortages are explicitly costed, each demand that is not immediately met, typically, incurs a unit shortage cost, $\hat{\pi}$. Additionally, there may be time-weighted backordering cost per unit, π, which is incurred for the time period during which the demand remains backordered.

Service levels may be defined either through the lead-time service level or fill rate. The lead-time service level is the probability that demand does not exceed the stock during a lead-time period (i.e., the probability of not incurring a stock-out). This measure, although popular in textbooks, is not a particularly useful measure because there may be more than one outstanding orders at any time. The fill rate is the fraction of demand that is immediately satisfied by an on-hand inventory.

Perishing refers to items becoming unfit for use when they reach a certain age, which may be fixed or random. Decay differs in that, only a prescribed portion of an on-hand stock deteriorates over time; it is most appropriate for items that are volatile. Obsolescence refers to the phenomenon where the product goes out of fashion at some time; it implies a terminus for the planning horizon. In the literature, all three forms assume variable costs. Some practitioners regard the costs associated with perishing or decaying of items and obsolescence as holding costs but they should be treated separately. In most of the inventory models in the literature, it is assumed that the items do not perish.

In the above discussion, it is assumed, as is customary, that holding and shortage are costed in a linear fashion. However, there is also a growing body of knowledge on non-linear-holding cost structures where larger inventories and/or shortages are penalized more. (See Snyder.) Also, the acquisition costs may be nonlinear in the order size, as in the case of quantity discounts.

Inventory control is performed via policies that dictate the quantity and timing of replenishments. Inventory control aims at balancing the above costs with each other. In many simple inventory models, there are two fundamental trade-offs: One between ordering and holding costs; and another between holding and shortage costs. The trade-offs emerge in the control policy parameters.

Inventory control policies can be classified into two classes depending on how inventory levels are monitored and orders are placed. *Periodic-review* policies give ordering decisions at prespecified points in time, separated typically by equal intervals—for example, days, weeks; inventory levels may be monitored at those instances prior to ordering decisions. *Continuous-review* policies monitor inventory levels in real time and may give orders at any point in time. Although common in practice, periodic-review systems are losing their prominence due to advances in information technologies such as bar codes and RFID and novel management practices such as a collaborative replenishment. In general, in periodic-review control, order sizes would be variable while order timing is fixed, whereas, in a continuous review, order sizes would be fixed but order timing is variable.

Firms prioritize their inventory control efforts by employing a Pareto analysis on their stocks. A small percentage of the SKUs may account for a very large share of the total sales value, which is computed as the annual demand multiplied by the unit price; these are called class A items. Aside from the class A items, and in the opposite direction, there exists a large percentage of SKUs that tend to constitute a much smaller portion of the sales value (class C items). The remaining 20%–30% of the items in the middle are called class B items. This categorization is referred to as the *ABC analysis*. The criticality of items in production may also be taken into account despite their low unit values in this classification. The ABC analysis may help the management to direct more of its attention to managing inventories of class A items and may lead to a greater cost-effectiveness. This cost-based approach has been extended to multiple criteria (Flores and Whybark 1986, 1987).

Firms are able to reduce the overall inventory levels by employing advanced forecasting techniques (which reduce effective demand uncertainties) and lean design practices such as standardization, modularity, and interchangeability of parts. At a strategic level, supplier selection, and reconfiguration of supply networks may also lead to substantial savings in inventory costs.

In this chapter, we consider a typical breakdown of inventory models on the basis of (i) the number of locations and levels (single location, multilocation serial, and multilocation arborescent), (ii) demand variability (deterministic and stochastic), (iii) lead times (negligible, deterministic, and stochastic), (iv) replenishment mode (retail and production), and (v) inventory monitoring (continuous review and periodic review).

Herein, we concentrate on the models with a single decision maker. As such, we consider the *supply chain management*, which is the interaction of multiple agents, out of the scope of the work herein. The emphasis is on the basic models and fundamental insights derived from such models, with occasional discussions of the most relevant and recently emerging extensions.

7.2 Deterministic Demand Models

Inventory systems with a deterministic demand are controlled through periodic or continuous-review policies. For expositional convenience and building a solid understanding of the fundamental trade-offs, we begin our discussion with continuous-review models.

7.2.1 Continuous Review: Economic Order Quantity Model

A basic inventory system under a continuous review is the so-called EOQ (economic order quantity) model that dates back to the work by F. Harris in 1913 (see also Erlenkotter 1989, 1990; Harris 1990).

The assumptions of the basic EOQ setting are as follows:

A1. Demand is known and constant at D units per unit time

A2. Shortages are not permitted

A3. Lead time for the delivery is instantaneous

A4. Orders arrive in one batch

A5. Units in the stock do not undergo any change in their usefulness nor in their value (i.e., do not perish/decay or become obsolete)

A6. Each order incurs a fixed cost K, each unit is acquired at a cost of c, and each unit held in the stock for one unit time incurs a cost of h

A7. Costs do not vary over time nor depend on the replenished quantity per order

A8. The planning horizon is infinitely long

A9. The objective is to minimize the average cost per unit time

Assumption (A2) implies that shortage costs per unit and/or per unit per unit time are prohibitively high, and, coupled with (A3) and a positive unit-holding cost rate in (A6), dictate that it is optimal to place orders only when the inventory level hits zero. Assumptions (A3) and (A4) imply that the replenishment rate of the system is infinite, and, hence, there are no staggered deliveries. Assumption (A5) refers to the item being nonperishable, and Assumption (A8) implies that the inventory system perpetually repeats itself. Due to stationary costs and infinite horizon, a stationary control policy exists. Assumption (A9) ignores the time value of money, that is, discounting of cashflows. It is implicitly assumed that inventory levels can be continuously monitored (at no cost).

This stylized inventory model is of a historical and practical importance. It is the oldest inventory model analyzed (Harris 1913) and provides an excellent example of the first fundamental trade-off (between fixed-ordering costs and holding costs) of inventory management.

The only decision to make here is to determine how much to order, since the timing of replenishment (at instances of zero stock) is already dictated by not allowing any shortages. Let Q denote the size of the order. Then, the control policy is formally stated as follows.

EOQ Policy: Monitor inventory levels continuously, and order Q units whenever the inventory level hits zero.

A depiction of the inventory levels over an item is provided in Figure 7.1, which illustrates the sawtooth-shape characteristic for the EOQ model.

Despite its deterministic nature, a renewal-theoretic approach is convenient herein for the analysis of the model. The inventory levels take on values between Q and 0; and, since deliveries are instantaneous, there are no orders outstanding at any time. Hence, the inventory level fully describes the state of the inventory system and, thereby, the instances at which it takes on a particular value (chosen arbitrarily over $[0, Q]$) constitute regeneration points. Define the *cycle time* as the time between two consecutive instances when the inventory level is raised to Q immediately after an order placement, denoted by T. Note that $T = Q/D$. Since all cycles are identical by definition, the average cost per unit time is the total cost incurred in a single cycle divided by the cycle length.

In each cycle, there is a single replenishment order incurring K, and the acquisition cost of cQ. The inventory-level process at some time t since the start of a cycle, $IL(t)$ is simply $IL(t) = Q - Dt$ for $t \in [0, Q/D)$, which results in the total holding cost incurred in a cycle given by

$$h \int_0^T IL(t)dt = h\left(QT - \frac{1}{2}DT^2\right) = \frac{1}{2}hTQ.$$

Then, the total average cost per unit time is

$$C(Q) = \left(K + cQ + \frac{1}{2}hTQ\right)/T = KD/Q + \frac{1}{2}hQ + cD.$$

Note that the acquisition cost component is typically ignored in the above expression since it may be assumed as sunk (due to all demands being met). $C(Q)$ is convex in Q with a global finite minimizer, Q^*, which is found by solving the first-order condition (FOC). It is commonly known as the EOQ given by

$$Q^* = \sqrt{2KD/h}, \tag{7.1}$$

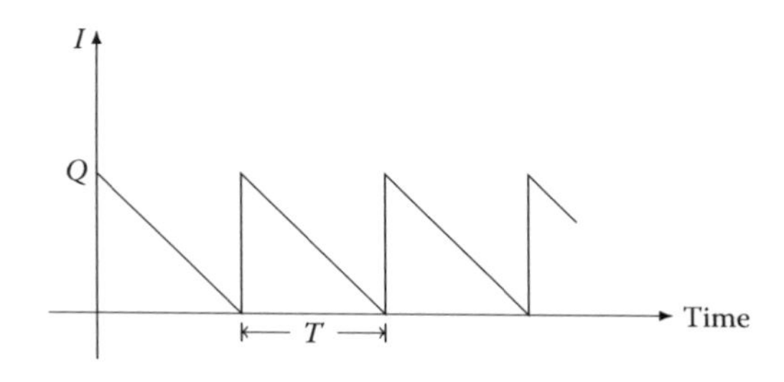

FIGURE 7.1
EOQ model.

resulting in the optimal cost rate

$$C^*(Q^*) = \sqrt{2KDh} + cD. \tag{7.2}$$

An interesting property of the optimal solution is that the average cost rate associated with ordering CO and that with an inventory-holding COH are equal to each other at the EOQ ($CO = COH = 0.5\sqrt{2KDh}$), and only at the EOQ. That is, the first fundamental trade-off between ordering and inventory-holding costs emerges.

Another interesting property of the EOQ solution is its robustness (conversely, the flatness/insensitivity of the corresponding optimal cost rate function around the optimal) to values of the cost parameters. Considering only the inventory-related components, the cost rate ensuing from using a suboptimal order size $\hat{Q}$ is given by $C(\hat{Q}) = (1/2)[(Q^*/\hat{Q}) + (\hat{Q}/Q^*)]C(Q^*)$. For example, an error of 100% in Q (i.e., $\hat{Q} = 2Q^*$) results in a change in the average cost rate, $[C(\hat{Q}) - C(Q^*)]/C(Q^*)\%$, of only 25%. The practical implications of this have been the reason for the popularity of the EOQ model and its usage in settings where system parameter values need to be estimated highly.

7.2.2 Relaxations

Some of the stylized assumptions for the EOQ model can be relaxed easily while retaining the structural properties of the model and of the optimal solution. We consider these relaxations next; others resulting in special extensions of the basic model will be discussed separately.

7.2.2.1 Demand Structure

1. Suppose demand is not continuous but discrete with interdemand times being constant at $1/D$ time units. This corresponds to the realistic case of demands being for an integral number of units and the order size being an integer. The classical treatment with continuous demands is an approximation of this setting.

   In this case, the replenishment policy is modified slightly so that an order of size Q units is placed at the occurrence of the first demand after the on-hand inventory is depleted. Thus, the inventory level within a cycle is $[0, Q - 1]$ with a cycle length of Q/D time units. Then, the total cost per unit time is given by

   $$C(Q) = KD/Q + \left[\frac{Q(Q-1)h}{2D}\right]/[Q/D] + \left[\frac{cQ}{Q/D}\right] = KD/Q + h(Q-1)/2 + cD, \tag{7.3}$$

   with $Q \geq 1$. The optimal order size that minimizes the above cost rate, Q^*, is the largest positive integer Q for which

   $$Q(Q-1) < \frac{2KD}{h}. \tag{7.4}$$

   Note that the optimal order size occurs again when the holding and ordering cost rates are balanced, and that, when unity is negligible in comparison with Q and the order size can vary continuously, we get the EOQ formula.

2. Suppose demand is not deterministic but random per unit time with a known stationary distribution with a mean of $\overline{d}$. *Ceteris paribus*, the optimal solution is given by the EOQ formula with $\overline{d}$ substituted for the constant demand rate, D.

3. Suppose demand is deterministic but nonstationary. Nonstationarity can be of two types: (i) Demand rate depends on the elapsed chronological time since a time origin, and (ii) demand rate depends on the time elapsed since the last replenishment. For the first type, if the changes in the demand rate are small and/or rapid around a time-average $\overline{d}$, then, the EOQ model where the demand rate is taken as the time average provides a good approximation. If the rate evolves over time relatively slowly, then, it is a practical approximation to use the current value of demand at the instance of zero stock for computing the order size via the EOQ formula. The second type of demand nonstationarity can occur either as a *decay* through which a fraction of stock perishes/is lost per unit time or as a variant of *stock-dependent demands* that have been empirically observed in retail settings. (This model is discussed below under perishing.)

7.2.2.2 Cost Parameters Vary

Cost parameters may vary in two ways: (i) over time and (ii) as a function of the order size.

Suppose some of the cost parameters change over time—either at an announced time in the future or continuously through inflation. Consider that the announced cost changes first. Clearly, after the new costs are in effect, the optimal order size is the new EOQ computed with the new parameter values. The question is, what is the optimal ordering policy between the announcement of the cost change and the time, say τ when the new cost takes effect? It has been shown that either (1) a special order (the size of which is only a function of old and new cost parameters) is placed immediately before τ, or (2) no special order is placed and an order computed via the EOQ formula with the new costs is placed when the inventory level hits zero for the first time after τ, or (3) a sequence of orders are placed to be modified so that the inventory level hits zero immediately before τ (Taylor and Bradley 1985). In the case of inflation, a practical solution is to modify the EOQ orders at certain time intervals to capture the changes in the cost parameters similar to the treatment of time-varying demand rates above.

Suppose some of the cost parameters change with the order size. A typical example is where the supplier charges less per unit for larger orders to create an incentive to increase the lot size. It is called quantity discount pricing. The ensuing benefits to the supplier are faster depletion of accumulated stocks and stimulated demand. The models in the literature focus on the first case while ignoring the effects on the possibly increased demand. The most commonly used quantity discount price schedules are (1) an *all-units discount* schedule with discrete prices in which all units in the order are purchased at the lower unit cost corresponding to the order size (Hadley and Whitin 1963) or with price being a continuous function of the order size (Ladany and Sternlieb 1974; Rosenblatt and Lee 1985), and (2) an *incremental unit discount* that is applied only to the additional units beyond each (discrete) price break point (Hadley and Whitin 1963; Hax and Candea 1984). The quantity discount schedules impact the average unit acquisition cost directly, and the unit-holding cost rate indirectly.

For continuous-discount schedules, where $c(Q)$ denotes the total acquisition cost of Q units, the optimal order size needs to be obtained numerically by solving the following transcendental equation:

$$hQ^2/2 - c'(Q)DQ - [K + c(Q)]D. \tag{7.5}$$

For the all-units discount schedule with a finite number of price breakpoints, $b_0, b_1, \ldots, b_m$ with corresponding unit prices c_j for $b_j < Q \leq b_{j+1}\ \forall j$, it is either optimal to order the EOQ quantity (computed with $h = w + ic_m$) at the lowest-feasible price level c_m, or to order as much as one of the breakpoints of higher prices, b_j with $j < m$. The optimal order size is obtained algorithmically by starting with the lowest price level and proceeding upward as long as the total cost rate is non-decreasing (see Hadley and Whitin 1963).

For the incremental quantity discount schedule, the total acquisition cost of Q units, $AC(Q)$, for $b_j < Q \leq b_{j+1}$ is given by $AC(Q) = AC(b_j) + c_j(Q - b_j)$ for $j = 0, 1, \ldots, m$ where $AC(0) = 0$ and $b_0 = 0$. In this case, the optimal order size can never occur at a price break point and the optimal order size is given by a feasible Q_j, that is, $b_j < Q_j \leq b_{j+1}$, which results in the lowest cost rate, where

$$Q_j = \sqrt{\frac{2D(K + AC(b_j) - c_j b_j)}{h_j}}, \tag{7.6}$$

and h_j is the unit-holding cost rate computed at the corresponding price level.

7.2.2.3 Problem Horizon

Suppose the planning horizon is of a finite length. Then, it is optimal to place a finite number of orders in equal sizes while the inventory levels at the beginning and end of the horizon are set at zero (Schwarz 1972). The optimal order size is $D\tau/n^*$ where τ denotes the length of the problem horizon and n^* is the smallest integer satisfying

$$\tau \leq \sqrt{\frac{2n(n+1)K}{Dh}}.$$

Although the order size deviates from the EOQ formula, the inventory levels still undergo a saw-tooth pattern. If τ is selected as a duration of 5-EOQ's worth of supply, then, the cost penalty of operating with the EOQ formula is only 1% (Schwarz 1977). This result is related to the concepts of protection- or forecast horizons in lot sizing (Lundin and Morton 1975).

7.2.2.4 Optimization Objective

Suppose that the time value of money is taken into account so that the present value of total costs is minimized. The optimal policy structure for the basic model is retained for an infinitely long problem horizon. As the cost capital tied up in an inventory is explicitly taken into account, the unit-holding cost rate now consists only of the out-of-pocket physical storage costs, w. In each cycle, a cash outflow of $K + cQ$ occurs at the instance corresponding to the cycle start and a continuous cash outflow of $wIL(t)$ occurs within the cycle for carrying the inventory. Let α denote the continuous-discount factor. Then, the discounted total cost is given by

$$\begin{aligned} C(Q) &= \sum_{n=0}^{\infty} e^{-\alpha n Q/D}\left[K + cQ + \int_{u=0}^{Q/D} w(Q - Du)e^{-\alpha u}du\right] \\ &= \frac{K + (w/\alpha^2)[(\alpha Q - D) + De^{-\alpha Q/D}] + cQ}{1 - e^{-\alpha Q/D}}. \end{aligned}$$

This complex cost expression requires numerical solution techniques to obtain the exact optimal order size. However, using a McLauren series approximation, it is possible to get an EOQ-type approximate solution as $Q^* \approx \sqrt{2Kd/(w + \alpha c)}$ (Jesse et al. 1983, Porteus 1985b; Zipkin 2000). The approximation has been found to work remarkably well in comparison with the exact numerical solution.

7.2.2.5 Perishing and Obsolescence

Perishing refers to items becoming unfit for use when they reach a certain age, which may be fixed or random. Perishing may be modeled as all-units perishing or individual perishing within an order (Raafat 1991). Consider all-units perishing. Suppose the units in each order have identical constant lifetimes, say τ; that is, they perish after they have stayed in the stock for τ time units. In this case, the optimal order size is given by the minimum of the EOQ formula ignoring perishability (implying that the lifetime constraint is not binding) and $D\tau$ (implying that the lifetime constraint is binding).

For deterministic shelflives, all-units perishing and individual unit perishing are identical. In the case of all-units perishing with a random lifetime, τ having a cumulative density function $F(\tau)$, the cycle time becomes a random variable, $T = min\{Q/d, \tau\}$ with its expected value $E[T]$ computed as $\int_{\tau} min\{Q/d, \tau\} dF(\tau)$. Then, the total expected holding cost $E[OHC]$ in a cycle is given by

$$E[OHC] = hint_{\tau} \int_{0}^{min\{Q/D,\tau\}} IL(t)dt \; dF(\tau),$$

and the expected total cost rate $EC(Q)$ is

$$EC(Q) = (K + cQ + E[OHC])/E[T].$$

If there is an (additional out-of-pocket) cost per unit that perishes π, the expected perishing cost in a cycle is given by $\pi(Q - DE[T])$ and needs to be included in the cycle costs.

A special case of individual perishing occurs when a constant fraction θ of stock perishes/is lost per unit time. This phenomenon has been observed in volatile chemicals, pharmaceuticals, and foodstuffs. Owing to the ensuing structure of the inventory level (as shown below), this type of perishing is called *exponential decay*. In the presence of decay, the inventory is depleted by external demand at rate D and by the decay process itself so that $dI(t)/dt = -D - \theta I(t)$. Hence, the inventory level when t time units have elapsed since the beginning of an order cycle is given by $IL(t) = Qe^{-\theta t} - (D/\theta)(1 - e^{-\theta t})$. It is more convenient to treat the cycle length T as the decision variable and compute the corresponding order size as $Q = (D/\theta)(e^{\theta T} - 1)$. The cost rate is given by

$$\begin{aligned} C(T) = {} & [K + (c+\pi)((D/\theta)(e^{\theta T} - 1))]/T - \pi D \\ & + w(D\theta)(e^{\theta T} - \theta T - 1)/(\theta T) + ic((D/\theta)(e^{\theta T} - 1))/2, \end{aligned} \tag{7.7}$$

where π is the perishing cost per unit, w is the physical storage cost rate, and i is the cost of capital (see Zipkin 2000 for an excellent discussion on how to assess unit-holding cost rates in the cost rate construction). This complex cost expression requires numerical solution techniques to obtain the exact optimal order size. However, using a McLauren series approximation for a small θ, it is possible to get an EOQ-type approximate solution as $Q^* \approx \sqrt{2KD/(w + ic + \theta(c+\pi))}$.

The exponential decay model is interesting for another reason. The structure of this perishable model mimics that of a nonperishable inventory system with stock-level-dependent demands. Stock-level-dependent demands have been observed and well documented in the economics literature. For this, the demand rate is set to be $D + \theta I(t)$ when t time units have elapsed since the start of the current cycle and the inventory level is $I(t)$, and $\pi(< 0)$ corresponds to the unit profit margin on a unit sold.

The oldest work on exponential decay (outlined above) is by Ghare and Schrader (1963). Covert and Phillip (1973) extend the analysis to Weibull distributions, which mimics the reliability of electronic components well. Comprehensive reviews for decaying inventories can be found in Raafat (1991) and Goyal and Giri (2001).

Obsolescence refers to the phenomenon where the product goes out of fashion at some time. It differs from perishability in that the problem horizon ends with the instance of obsolescence whereas the latter assumes an infinitely repeating replenishment process. If the exact timing τ of obsolescence (in the future) is known, then, the problem reduces to a finite horizon problem as discussed above. For random obsolescence times, the problem may be modeled via dynamic programming (DP) with stochastic horizon lengths (Jain and Silver 1994).

7.2.2.6 Delivery Times and Modes

Suppose there is a constant lead time of $L(>0)$ time units. In this case, the order quantity Q given by the EOQ formula in Equation 7.1 remains optimal but the replenishment policy is modified in the timing decision so that an order of size Q is placed at L time units before the inventory level hits zero. In inventory management parlance, it is more appropriate to express the timing of ordering in terms of inventory levels, as well. Therefore, we have the control policy formally stated as follows.

Policy: Monitor inventory levels continuously, and order Q units whenever the inventory level hits DL.

Note that DL is the so-called *reorder point* that corresponds to the demand that the system would face during the lead time. (The above assumes that demand during the lead time does not exceed the order quantity. Otherwise, the reorder point is $DL \mod Q$.) Random lead times are discussed further below under the general *lotsize-reorder point* model. Among the random lead times, we should also mention the models with supply disruptions, where the system undergoes a renewal process that results in either total supply unavailability or ample supply with a prescribed lead time (see Parlar and Berkin 1991; Berk and Arreola-Risa 1994; Gürler and Parlar 1997; Balcioglu and Gürler 2011, and, for more general settings, Ozekici and Parlar 1999; Schmitt et al. 2010). Another way to relax the infinite replenishment rate assumption is to directly assume a *finite* replenishment rate. This results in a special inventory model that has been studied in its own right—the economic production quantity, EPQ (also known as the economic manufacturing quantity, EMQ) model. While the EOQ model captures the fundamental trade-off in retail settings (where replenishment is from upstream-stocking installations), the EPQ model is most appropriate for production settings with a limited capacity to replenish the stock of the desired item. This is discussed in a separate subsection below.

7.2.2.7 Nonlinear Holding Costs

In certain inventory systems, there may be implicitly or explicitly nonlinear holding costs. The implicit setting refers to the case of storage or ordering/shipment limitations, W. Owing to the convex nature of the cost rate function, the optimal order size is given by the minimum of the EOQ and the storage limit W.

The explicit optimal solution is available for the case where the unit-holding cost is a convex function $h(\cdot)$ of the amount of stock on hand. In this case, the optimal order size solves the following transcendental equation:

$$Q^* = \sqrt{Kd/h'(Q^*)}.$$

7.2.2.8 Allowed Shortages: Backordering and Lost Sales

A very important class of inventory models are those with possible shortages. When shortage costs are finite, it may be possible and desirable to allow for backordering and/or lost sales; such shortages are also called *planned shortages* to distinguish them from the shortages that occur (undesirably) in stochastic demand settings. Backordering implies that, in cases of stock-outs, a customer is willing to wait until the demand is fulfilled (when the next available order is received), whereas lost sales occur when the customer is not willing to wait at all.

For this general model, the fundamental changes in the assumptions are as follows:

A1′. Demand is known and constant but depends on inventory levels. Whenever there is a positive stock, demand is d units per unit time; in cases of stock-out, the α fraction of demand is backordered while the remainder is lost

A2′. Shortages are permitted

A6′. In addition to the ordering, holding, and acquisition cots, the backordering cost is charged at b per unit of demand satisfied immediately and at $\hat{b}$ per unit of demand kept backordered for one unit of time. Each unit of lost demand incurs a cost of π

The inventory model is called a *full backordering* model for $\alpha = 1$, a *lost sales* model for $\alpha = 0$, and a *partial backordering* model for $0 < \alpha < 1$.

In case of allowable backorders, the inventory level may drop below zero. For that reason, it is convenient to refer to the inventory levels for such systems as *net* inventory levels to differentiate them from merely the on-hand stock. Net inventory levels enable us to fully describe the state of the inventory system. In this case, the replenishment policy is general in its form specifying both the replenishment quantity and its timing in terms of the net inventory level. For the sake of convenience, we adopt a slightly different approach herein and base-reordering decisions on the total accumulated shortages instead of the total demand backordered or lost as it is typically the case. The replenishment policy is formally stated as follows.

Policy: Order Q units of stock whenever r units of shortage accumulate.

For the (full or partial) backordering models, the reorder point as defined herein would correspond to ordering when the net inventory level drops to $-\alpha r$ (i.e., αr units of demand are backordered in a cycle). For the full lost sales model, reordering may be said to occur after a stock-out period of length r/D time units (i.e., a total of r units of demand have been lost in a cycle).

When backordering is present, the inventory level fully describes the state of the inventory system, and the instances at which it takes on a particular value (chosen arbitrarily over $[\alpha r, Q]$) constitute regeneration points. However, for a full lost sales system, the inventory level, by definition, stays at zero during the entire stock-out period. Therefore, the inventory control policy cannot be solely based on the inventory level. The system state needs to be described through, say, on-hand stock and accumulated shortages at any time. The regeneration points can be chosen accordingly. In practice, the system-state representation does not present much difficulty in the deterministic setting herein but it points at the difficulties encountered in stochastic demand settings with lost sales.

The analysis with allowable shortages rests on a renewal cycle defined by two consecutive instances of order decisions. Define the cycle time, denoted by T, as the time between two consecutive instances when an order of size Q is placed. Since all cycles are identical by definition, the average cost per unit time is the total cost incurred in a single cycle divided by the cycle length (see Figure 7.2).

In each cycle, there is a single replenishment order incurring K, and the acquisition cost of cQ. Each cycle consists of two parts—one of length T_1 during which there is a positive stock and the other with a duration of T_2 in which shortages are incurred, where $T_1 + T_2 = T$. Noting that only

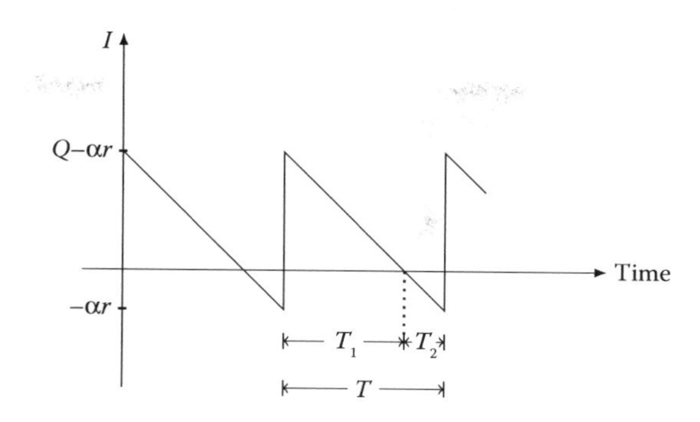

FIGURE 7.2
EOQ model with allowable shortages.

the α fraction of shortages are backordered, $T_1 = (Q - \alpha r)/D$ and $T_2 = r/D$. The inventory-level process at some time t since the start of the cycle is $IL(t) = Q - \alpha r - Dt$ for $t \in [0, (Q - \alpha r)/D]$, and $IL(t) = -\alpha D(t - (Q - \alpha r)/D)$ for $t \in [(Q - \alpha r)/D, Q/D + (1 - \alpha)r/D]$. The total holding cost incurred in a cycle, $HC(Q, r)$, is given by

$$HC(Q,r) = h \int_0^T \mathbf{1}_{IL(t)>0} IL(t)dt = h \int_0^{(Q-\alpha r)/D} (Q - \alpha r - Dt)dt = \frac{h}{2D}(Q - \alpha r)^2. \quad (7.8)$$

The total shortage cost in a cycle, $SC(Q, r)$, is obtained by considering the inventory level in the second part. It is given by

$$\begin{aligned} SC(Q,r) &= \pi(1 - \alpha)r + b\alpha r + \hat{b} \int_0^T \mathbf{1}_{IL(t)<0} IL(t)dt \\ &= \pi(1 - \alpha)r + b\alpha r + \hat{b} \int_0^{r/D} \alpha D t dt \\ &= (\pi(1 - \alpha) + b\alpha)r + \frac{\hat{b}\alpha}{2D} r^2. \end{aligned} \quad (7.9)$$

Note that the shortage cost $SC(Q, r)$ within a cycle is nonlinear in the planned shortages. Even though the construction above assumes time-weighted backordering costs, the model is identical to an inventory system in which shortages are costs in a polynomial form $\beta_0 + \beta_1 r + \beta_2 r^2$ where $\beta = 0$, $\beta_1 = \pi(1 - \alpha) + b\alpha$ and $\beta_2 = \hat{b}\alpha/2D$. Thus, the conclusions to be drawn about the operational characteristics of this particular deterministic inventory system with allowable shortages hold for similar systems with such nonlinear shortage costs, as well.

The total average cost per unit time is the total cycle cost divided by the cycle length as before

$$C(Q,r) = \frac{K + cQ + HC(Q,r) + SC(Q,r)}{Q/D + r(1 - \alpha)/D}. \quad (7.10)$$

Using a standard calculus, the FOCs on the order size and reorder point give, respectively,

$$(Q - \alpha r)^2 h/2D + (Q - \alpha r)hr/D - (\pi - c(1 - \alpha))r - \hat{b}\alpha r^2/2D - K = 0, \quad (7.11)$$

and

$$[\alpha(h\alpha^2 - \alpha(h - \hat{b}) - \hat{b})/2]r^2 - [\alpha Q(\alpha h + \hat{b})]r$$
$$+ [Q^2(1 + \alpha)h/2 + (K + cQ)(1 - \alpha)D - D\pi Q] = 0. \quad (7.12)$$

Depending on the system parameters, the optimal solution for the general problem setting is found as one of the following: (i) It is not worthwhile to carry any inventories and, thereby, not satisfying any of the demand, ($r^* \to \infty$, $Q^* = 0$). (ii) It is optimal not to incur any shortages at all, ($r^* = 0$, $Q^* = \sqrt{2KD/h}$). Finally, (iii) it is optimal to accumulate a finite number of shortages before ordering, ($r^* = \tilde{r}$, $Q^* = \tilde{Q}$) where $\tilde{Q}$ and $\tilde{r}$ are the unique positive pair that solves

$$\tilde{Q} = D\pi/h + (1 + \hat{b}/h]\tilde{r}, \quad (7.13)$$

and

$$[\alpha\hat{b}(\alpha\hat{b} + h)]\tilde{r}^2 + [2D(\pi\hat{b}\alpha + hc(1 - \alpha))]\tilde{r} + [\pi^2 D^2 - 2KDh] = 0. \quad (7.14)$$

A fundamental result about a deterministic inventory system is that, if shortage costs are only incurred per unit with $\hat{b} = 0$ (i.e., shortage costs are linear in the planned shortages), then, the inventory system either operates as an EOQ model or does not operate at all. Only in the presence of time-weighted shortage costs (i.e., nonlinear shortage costs), it may be optimal to operate while allowing for shortages. To elucidate this important result, below, we consider the special cases in detail.

Case 1. Suppose $0 < \alpha \leq 1$ and $\hat{b} = 0$ while $b > 0$ and $\pi > 0$. This corresponds to the case where part of the demand that is not satisfied immediately is backordered while the rest (if any) is lost and shortage costs are charged per unit only. For $\alpha b + (1 - \alpha)\pi = \sqrt{2Kh/D}$, there is no unique r^*. Hence, without the loss of generality, $r^* = 0$. This results in the basic EOQ model and the optimal order size is given by $Q^* = \sqrt{2KD/h}$. For $\alpha b + (1 - \alpha)\pi > \sqrt{2Kh/D}$, $r^* = 0$, which again reduces to the EOQ model. Finally, for $\alpha b + (1 - \alpha)\pi < \sqrt{2Kh/D}$, it is never optimal to order; hence, $r^* = \infty$ and $Q^* = 0$. In this last instance, all backordered demands are backordered forever! In practice, this corresponds to a voluntary lost sales model on the part of the decision maker resulting in a total cost rate of $(\alpha b + (1 - \alpha)\pi)D$.

Case 2. Suppose $\alpha = 0$ and $\pi > 0$ while, by definition, $b = \hat{b} = 0$. This corresponds to the case where all demand that is not satisfied immediately is lost and, thereby, the shortage cost is charged only on a unit basis. The analysis and conclusion for this case is similar to that above. If $\pi - c < \sqrt{2Kh/D}$, it is optimal not to operate the inventory system and to lose all demands; that is, $r^* \to \infty$ and $Q^* = 0$. Otherwise (i.e., if the inventory system is to be operated at all), then, it is never optimal to incur any stock-outs; that is, $r^* = 0$ that implies the EOQ model.

Case 3. Finally, suppose that $\hat{b} > 0$ and $0 < \alpha \leq 1$ while $b \geq 0$ and $\pi > 0$, if applicable. This corresponds to the general case where there may be a mixture of backorders and lost sales, and backorders become more costly as they accumulate. For $\alpha b + (1 - \alpha)\pi \geq \sqrt{2Kh/D}$, $r^* = 0$ and the model reduces to the EOQ model. Otherwise, the optimal order size and reorder point are given by the unique positive pair $Q^* = \tilde{Q}$ and $r^* = \tilde{r}$ as defined above.

A thorough analytical treatment and a detailed discussion of the special cases of a full backordering and lost sales is in Hadley and Whitin (1963), and also in Zipkin (2000).

Note that the shortage cost expression within a cycle, $SC(Q,r)$, is nonlinear in the planned shortages. Even though the construction above assumes time-weighted backordering costs, the model is identical to an inventory system in which shortages are costed via a polynomial of the form $\beta_0 + \beta_1 r + \beta_2 r^2$ where $\beta = 0$, $\beta_1 = \pi(1-\alpha) + b\alpha$, and $\beta_2 = \hat{b}\alpha/2D$. Thus, the conclusions drawn about the operational characteristics of an inventory system with allowable shortages hold for this particular cost structure, as well.

7.2.2.9 Waiting Time Limits

In the above analysis, it is assumed that there are no restrictions on the number of backorders at any point in time. However, such limitations may occur due to customers' unwillingness to wait for more than some threshold. Incidentally, lost sales occur because (some) customers have zero waiting time tolerance. For the case of a homogeneous population of waiting customers with a maximum waiting tolerance of τ, the above results are optimal if the obtained reorder points do not exceed $\alpha\tau D$; otherwise, $r^* = \alpha\tau D$ and the optimal order size is found from Equation 7.11.

7.2.3 Continuous Review: EPQ Model—Finite Replenishment Rates

Another way to relax the infinite replenishment rate assumption is to directly assume a finite replenishment rate. This results in a special inventory model that has been studied in its own right—the EPQ (also known as the EMQ) model. While the EOQ model captures the fundamental trade-off in retail settings (where replenishment is done from upstream-stocking installations), the EPQ model is most appropriate for production settings with a limited capacity to replenish the stock of the desired item.

In the EPQ model, an order of the stock does not arrive instantaneously from an external source but, instead, the stock is produced at a finite rate of ρ per unit time, with $\rho > D$ to ensure no shortages. The production facility is assumed to undergo a setup that incurs a fixed cost K each time the production is restarted. Let Q denote the amount to be produced whenever a production decision is made. The control policy is formally stated as follows.

EPQ Policy: Monitor the inventory levels continuously, and produce Q units whenever the inventory level hits zero.

A depiction of the inventory levels over an item is provided in Figure 7.3.

The main difference in system evolution lies in that in the EOQ model, inventory levels are immediately raised to their maximum levels whereas in the EPQ model, the stock gradually accumulates at a net rate of $\rho - D$. Each cycle consists of two parts: In the first part, both production and

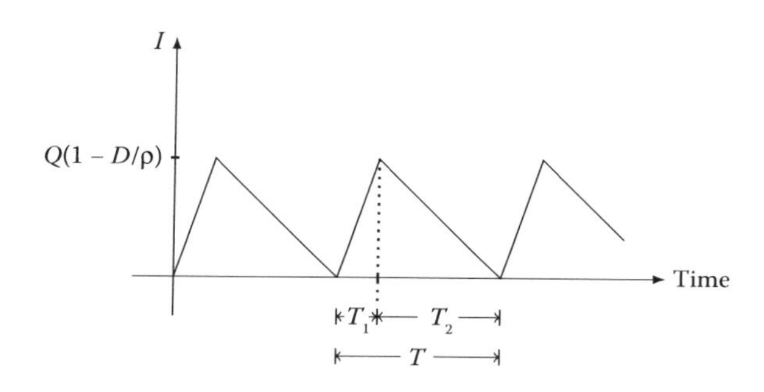

FIGURE 7.3
EPQ model.

depletion occur simultaneously, and in the second part, production has stopped and the accumulated inventory is depleted. The total time of production in a cycle is Q/ρ, so that the maximum stock in a cycle is $(\rho - D)Q/\rho$; the total cycle length is, as before, $T = Q/D$. The inventory-level process at some time t since the start of a cycle is $IL(t) = (\rho - D)t$ for $0 \leq tQ/\rho$ and $IL(t) = (\rho - D)Q/\rho - D(t - Q/\rho)$ for $Q/\rho \leq t \leq T$. The total inventory held in a cycle is $(\rho - D)QT/2\rho$. The total cost rate is convex in Q resulting in the optimal lot size Q^*, called the EPQ, given by

$$Q^* = \sqrt{2KD/h}\sqrt{\rho/(\rho - D)}, \tag{7.15}$$

resulting in the optimal cost rate

$$C^*(Q^*) = \sqrt{2KDh}\sqrt{(\rho - D)/\rho} + cd. \tag{7.16}$$

Note that the EPQ is a multiple of the EOQ where the correction factor is related to the ratio of the rates of net inventory accumulation and replenishment. Thus, as ρ goes to infinity, the EPQ reduces to the EOQ, as expected. Also, the EPQ is at least as large as the EOQ for the same cost parameters; $\sqrt{\rho/(\rho - D)} \geq 1$. This is due to the impact of gradual accumulation of the stock versus instantaneous replenishment: The effective unit-holding cost rate in the presence of the finite replenishment rate, $\hat{h} = h(\rho - D)/\rho$, is lower than that with the instantaneous replenishment, h. Except for the use of an effective holding-cost rate $\hat{h}$ instead of the actual unit cost rate h, the EPQ model is identical to the EOQ model in construction, analysis, and interpretation. Hence, the relaxations about the other assumptions on the operational setting hold for the EPQ model, as well.

7.2.3.1 Setup Times

The classical EPQ model assumes instantaneous setups. However, in many manufacturing settings, setups involve activities such as cleaning, retooling, calibration, and pilot runs to ensure the desired quality levels with the associated nonnegligible setup times. Let $L(> 0)$ denote the setup time. Then, the optimal production quantity is the maximum of the EPQ formula ignoring the setup time (which implies that the setup time is less than the time it takes to deplete the stock after production has stopped) and $LD/(1 - D/\rho)$ (which implies that the setup time constraint is binding). Thus, the EPQ model serves as a building block for the economic lot-sizing problem (ELSP), where a number of different items are produced on the same machine under a, typically, cyclical sequence. The shadow prices of idle times resulting from the cyclical schedule constitute the basis of the imputed setup costs in this case; and, the replenishment quantity of each item may be computed in line with the EPQ formula (see Elmaghraby 1978 for a general exposition of the ELSP and a comprehensive review).

7.2.4 Periodic Review: The Wagner–Whitin Model

7.2.4.1 Exact Analysis

The periodic-review inventory models are important due to (1) their applicability to production-planning settings and (2) their amenability to provide a framework to obtain the optimal policy structures.

The problem horizon is finite with a length of T periods. Let $\{d_1, d_2, \ldots, d_T\}$ be the known requirements (net demands) for a single product over the next T periods. Periods may be of equal or unequal lengths. The beginning of each period constitutes a replenishment opportunity. Let $\{y_1, y_2, \ldots, y_T\}$ be the order sizes placed at these instances; $y_t \leq 0\ \forall t$.

The original formulation of the model is due to Wagner and Whitin (1958). The model assumptions are as follows:

A1. Shortages are not permitted. (A forward-solution algorithm does not exist when backordering is allowed. However, Aksen et al. (2003) have shown for allowable lost sales that (i) either all or none of the demands in a period are lost, and (ii) the results for no shortage hold for the resulting net demands.)

A2. The starting and ending inventories are zero.

A3. The order lead time is assumed to be negligible. (This assumption is easily relaxed for constant lead times; the order obtained under the no-lead-time assumption is translated backward in time by the number of periods corresponding to the lead time.)

A4. A fixed cost K_t is incurred per order placed in period t, and a unit-holding cost h_t is incurred per unit at the end of each period t.

A5. Unit acquisition cost is stationary, $c_t = c \; \forall t$.

A6. Nonspeculative cost structures hold: (i) $c_{t+1} \leq c_t + h_t$ and $K_{t+1} \geq K_t$.
This assumption implies that (i) it is more beneficial to order a unit in a period as close to that in which it will be used to satisfy the requirement, and (ii) it is beneficial to batch (increase the order sizes). Out of these two assumptions, emerges the fundamental trade-off between ordering and holding costs.

A7. The objective is to minimize the total costs over the finite horizon. As the individual ordering and holding-cost components may be nonstationary, the formulation subsumes the discounted cost objective, as well.

With a fixed-ordering cost and a linear acquisition cost, the total ordering cost is concave. Likewise, a linear holding cost also results in a concave inventory-holding cost. Although the basic model assumes concave total costs, the results hold under the more general assumption of holding costs being concave in the ending inventory and the total ordering costs being concave in the order quantity in each period. The decision is to determine the nonnegative order vector, $\mathbf{y} = \{y_1, y_2, \ldots, y_T\}$. The model is formally stated as follows:

$$min_{\mathbf{y} \geq \mathbf{0}} \sum_{t=1}^{T} [(\delta_t K_t + c y_t) + h_t x_t],$$

subject to

$$x_t = \sum_{j=1}^{t} (y_t - d_t),$$

$$x_t \geq 0; \quad t = 1, \ldots, T,$$

$$x_0 = 0,$$

where δ_t is an indicator variable of whether an order is placed in period t ($\delta_t = 1$) or not ($\delta_t = 0$), and x_t is the ending inventory in period t.

The optimal order policy rests on the following property:

In an optimal policy,

$$y_t x_{t-1} = 0, \quad t = 1, 2, \ldots, T. \tag{7.17}$$

This means that ordering occurs only in periods when the starting inventory (immediately before an ordering decision) is zero. The implication of this result is that the optimal policy is the one in which demand integrality is preserved. That is, order sizes correspond to the exact requirements such that $y_t \in \{0, d_t, d_t + d_{t+1}, \ldots, d_t + d_{t+1} + \cdots + d_T\}$ $\forall t$. Then, the optimal policy can be found by finding the shortest path through an acyclic network with nodes labeled as $1, 2, \ldots, T+1$ and arcs (i, j) connecting all pairs of nodes with $i < j$. The length of the arc-connecting nodes i and j, c_{ij} is the total cost of satisfying the requirements through period $j-1$ by an order placed in period i, where $i < j \leq T+1$ with $T+1$ denoting a dummy ending period and

$$c_{ij} = K\delta_i + c\sum_{k=i}^{j-1} d_k + \sum k = i^{j-2} h_k \left(\sum_{m=k+1}^{j-1} d_m \right).$$

With the preprocessing of c_{ij}, the optimal order policy is found by solving either a forward or a backward dynamic program in $O(T)$ time. The backward formulation is

$$f_i^* = min_{j>i}(c_{ij} + f_j^*); \quad i = 1, \ldots, T,$$

$$f_{T+1}^* = 0,$$

where f_i^* corresponds to the cost of following an optimal policy when i periods remain in the problem subhorizon, and f_1^* gives the optimal total cost solution. For example, suppose $T = 2$, $d_1 = 60, d_2 = 100, K_1 = 150, K_2 = 140, c = 7$, and $h_1 = h + 2 = 1$. Then, $c_{11} = 150 + 7 \times 60 = 570$, $c_{12} = 150 + 7 \times 100 + 7 \times 60 + 1 \times 100 = 1370$, and $c_{22} = 140 + 7 \times 100 = 840$ resulting in $f_3^* = 0$, $f_2^* = 8400 + 0 = 840$, and $f_1^* = min\{c_{12} + f_3^*, c_{11} + f_2^*\} = min\{1370 + 0.570 + 840\} = 1310$. The optimal policy is to replenish all demands in period 1.

A similar result was established simultaneously in Manne (1958) in an intermediate step of a proof in its appendix for a linear-programming formulation of the production-planning problem. This basic algorithm can be improved to yield a solution in $O(TlogT)$ (Federgruen and Tzur 1991; Wagelmans et al. 1992). (For reviews, see Wolsey 1995; Drexl and Kimms 1997; Jans and Degraeve 2008.) Recently, Kian et al. (2014) have considered polynomial-type convex production costs. For a review, see Brahimi et al. (2006). For network flow representations of lot-sizing problems in detail, see Johnson and Montgomery (1974).

7.2.4.2 Lot-Sizing Heuristics

Despite the benign nature of the optimal policy and its computation algorithm, there has been considerable work in developing approximate solutions to the basic Wagner–Whitin model. This interest is partly due to the reticence of practitioners to incorporate the seemingly sophisticated DP methodology in their planning systems, and partly due to the desire to provide more stable solutions in rolling-horizon environments in which future requirements undergo a periodic revision arising from cancelations of tentative orders and placement of new firm orders by customers. Interestingly, order schedules obtained heuristically may dominate the optimal DP solutions in a rolling-horizon setting; the improvements may be as substantial. Although some heuristics may allow for nonstationary cost parameters, typically, they have been developed and tested for stationary costs.

The (forward) heuristics developed for the single-item setting are Silver and Meal (1973), part period balancing (DeMatteis 1968), least unit cost, economic order interval, McLauren's moment (Vollmann et al. 1997), least unit cost (Narasimhan and McLeavy 1995), and Groff's algorithm (Groff 1979).

A comprehensive comparison of the currently available heuristics reveals that the Silver–Meal heuristic is the best contender (Simpson 2001). The Silver–Meal heuristic is based on computing the average cost per time period as a function of the number of periods whose requirements are covered by the current order. Consider the subproblem horizon over the current period labeled conveniently as period 1 (with zero-starting inventory) through period n for $1 \leq n \leq T$. Let $y_1 = \sum_{i=1}^{n} d_i$, and $C(n) = K + h\sum_{i=1}^{n} d_i$ denote the total costs over the next n periods. The heuristic employs the stopping rule that $C(n) > C(n-1)$ for the first time for $n = \hat{n}$, and sets $y_1 = \sum_{i=1}^{\hat{n}} d_i$. The forward procedure is started again at period $\hat{n}$ and continues until the end of the problem horizon.

For DP-based heuristics (especially useful with concave costs), see Johnson and Montgomery (1974).

7.2.4.3 Capacitated Lot Sizing

In the presence of finite capacities, the feasibility of the problem needs to be ensured: $\sum_{i=1}^{t} \leq \sum_{i=1}^{t} R_i$ where R_t denotes the capacity in period t. If fixed-ordering costs are negligible, then, capacities being finite do not pose any difficulty as the optimal solution is still a lot-for-lot schedule implemented on the preprocessed demand $\hat{d}_t$ so that all $\hat{d}_t \leq R_t$ and $\sum_{i=1}^{t} \hat{d}_i \geq \sum_{i=1}^{t} d_i$ for all t over the horizon. With finite fixed costs, however, the problem becomes extremely difficult: Even establishing that a feasible solution is NP hard. Nonetheless, some structural results are possible that allow one to represent the problem as a network flow problem (Florian and Klein 1971). Heuristic solution methods based on Lagrangian relaxation procedures are popular due to the fact that subproblems are easily solvable. There is also a vast body of work on the problems with setup times within a limited period length, which impose an implicit setup cost. Another implicit capacity limitation occurs when the process can be kept ready/warm so that a minor instead of a major setup would be needed for the next run. Such systems have been analyzed in Robinson and Sahin (2001), Toy and Berk (2006, 2013), and Berk et al. (2008). For comprehensive reviews on capacitated lot-sizing problems, see Karimi et al. (2003) and Allahverdi et al. (2008).

7.3 Stochastic Demands

7.3.1 Single Period: The Newsvendor Model

The newsvendor model is the building block for stochastic inventory models in discrete time. It considers the stocking decision for a product that can be replenished once at the start of a selling season and needs to be disposed off at its end. It is also called the newsboy problem, the Christmas tree problem, and the single-period inventory model. It is directly applicable when a product becomes obsolete quickly such as a fresh produce, fashion goods, and newspapers (hence, the name). The earliest analysis of the newsboy problem is by Arrow et al. (1951). It illustrates the second fundamental trade-off in inventory theory—the one between holding (or, excess, overage) costs and shortage (or, underage) costs.

The model assumptions are as follows:

A1. The planning horizon consists of a single period

A2. Demand, X is random but is fully characterized by its known probability density function (pdf) $f(x)$ and cumulative distribution function (cdf) $F(x)$

A3. A single replenishment is possible at the beginning of the period

A4. There is no initial stock

A5. Each unit of stock left over at the end of the period incurs a cost c_e and each unit of unsatisfied demand incurs a cost c_u

A6. The objective is to minimize the expected total costs

The decision is to determine the optimal stocking level so that the total expected costs are minimized. Let y denote the stocking level at the beginning of the period immediately after ordering. The decision variable y is also called the order-up-to level. It is implicitly assumed that there is no delivery lead time, or that the period starts after a positive lead time during which all demands are lost as sunk cost.

The number of units of stock left over at the end of the period is $y - min(y, X)$, and, likewise, the number of unmet demands by the end of the period is $X - min(y, X)$. Then, the total cost incurred within the period for a given initial stock, y, is

$$c_e[y - min(y, X)] + c_u[X - min(y, X)], \tag{7.18}$$

which is a random variable. The expected value of the above total cost is

$$G(y) = \int_{x=0}^{\infty} \{c_e[y - min(y, X)] + c_u[X - min(y, X)]\} f(x)dx \tag{7.19}$$

$$= \int_{x=0}^{y} c_e(y - x) f(x)dx + \sum_{x=y}^{\infty} c_u(x - y) f(x)dx. \tag{7.20}$$

The expected total cost $G(y)$ is convex in y so that its unique finite global optimizer y^* is given by

$$F(y^*) = \frac{c_u}{c_e + c_u}. \tag{7.21}$$

The critical ratio on the right-hand side is called the *desired service level*. It corresponds to the probability that all demands will be satisfied within the period with the optimal initial stocking level; and equivalently, to the probability that demand within the period will not exceed y^*. The second fundamental trade-off in inventory theory emerges in this marginal cost balance: The optimal order-up-to-level y^* is such that the marginal expected cost of holding one less unit of stock is equal to the marginal expected cost of not being able to satisfy one more unit of demand.

Remarks

The above analysis implicitly assumes that demand is purely continuous but similar results hold when X is discrete or of a mixed nature. In case of discrete demand values, if the exact equality is not possible, then, the convention is to set y^* to the smallest demand value for which the demand satisfaction probability exceeds the critical ratio. The most commonly used distributions are normal for the continuous demand, and Poisson and a negative binomial for discrete demand; the last providing a better fit empirically for retail data is Agrawal and Smith (1996).

The cost parameters in the critical ratio can be computed in practice as follows. Let c denote the unit acquisition cost of the product, p denote the revenue per unit sold, h denote the unit-holding cost, s denote the salvage value of a left-over unit, and $\hat{p}$ denote the cost of goodwill loss and similar penalties associated with each unmet demand. Then, $c_e = c - s - h$ and $c_u = p - c + \hat{p}$. For other details of the interpretation of costs in practice, we refer to Hillier and Lieberman (1995) and Nahmias (2014). Lowe et al. (1988) consider the uncertainty about the underage and overage costs expressed as an interval for each and independent of the uncertainty in demand. The optimal solution structure still holds with the costs being replaced by their expected values.

7.3.2 Relaxations

7.3.2.1 Positive Initial Stock and Ordering Cost

The basic newsvendor model indicates the structure of the optimal control policy for discrete-time problems with horizons longer than one period, as well. To see this, first, suppose that there is an initial on-hand inventory at the start of the period (immediately before a replenishment decision is made), $z \geq 0$. Then, the optimal solution becomes *Order* $y^* - z$ *if* $z < y$, *and do not order, otherwise.* This policy is known as the critical number policy. Its logic lies in that the optimal ordering decision is based on the net demand process, $max(0, X - z)$. Hence, in inventory systems in which the excess inventory may be carried out into the following time period, with the appropriate choice of excess and shortage costs incurred at the end of a single period, the newsvendor solution provides either a good myopic approximation or captures the true optimal policy. In the additional presence of a fixed-ordering cost, K, the optimal solution is again a critical number policy, stated as follows. *If* $z < \hat{y}$ *order* $y^* - z$, *otherwise do not order, where* $\hat{y}$ *is the smallest solution of* $G(\hat{y}) = G(y^*) + K$. The logic of the policy is that $G(y^*) + K$ is the possible minimum expected total cost with ordering, but the on-hand inventory levels below $\hat{y}$ result in the same expected cost with no ordering. Thus, the newsvendor model brings forth the decision of when to order—the concept of a reorder point for systems with demand uncertainty.

7.3.2.2 Optimization Objective

The above analysis has been carried out with a cost-minimization objective. An equivalent analysis under profit maximization is possible along similar lines (see Hadley and Whitin 1963; Berk and Gürler 2015 and see below). In the literature, there have been a number of formulations of the classical newsvendor problem with different objectives. Below, we review some of the important ones. Scarf (1958) provides a version of the problem under the *minimax* criterion, which is also called the *distribution-free newsvendor* problem. In this variant, only the first two moments of the demand distribution are assumed as known (μ and σ^2). Let p, c, s, and π denote the unit-selling price, unit acquisition cost, unit salvage value, and unit shortage cost, respectively. Furthermore, define the markup and discount factors $m = p/c - 1$ and $d = 1 - s/c$. The optimal stocking level that maximizes the expected profit against the worst distribution of demand is given by

$$Q^* = \mu + \frac{\sigma}{2}\left[\sqrt{(\pi/c + m)/d} - \sqrt{d/(\pi/c + m)}\right].$$

For the distribution-free newsvendor problem, see also Gallego (1992), Gallego and Moon (1993), and Moon and Gallego (1994) and Alfares and Elmorra (2005) and Bouakiz and Sobel (1992) consider an exponential utility criterion. Although their model is formulated as an infinite horizon problem, their findings subsume the single-period setting as well.

The classical newsvendor model assumes a risk-neutral decision maker; hence, the rationale of working with the expected costs or profits. However, risk aversion has recently received attention in operations management contexts as well. Ahmed et al. (2007) and Chen et al. (2007) discuss a number of alternative objectives incorporating preferences toward the risk. Of these, value-at-risk (VaR) and conditional value-at-risk (CVaR) formulations deserve particular attention (Gotoh and Takano 2007; Jammernegg and Kischka 2012). The latter concept has been introduced by Rockafellar and Uryasev (2000). The VaR analysis for profit maximization states that the expected profit is maximized while ensuring that the probability of profit being below a threshold z_0 does not exceed some prespecified probability, β. Letting p, c, and s denote the unit revenue, unit acquisition cost, and unit salvage value, the optimal solution is as follows. If the unconstrained solution of the classical newsvendor problem satisfies this condition on the cdf of profit, then, this solution is also the solution of the VaR formulation. Otherwise, if the condition is feasible,

$$Q^* = \frac{F^{-1}(\beta)(p-s) - z_0}{c-s}.$$

In CVaR optimization, given the risk-aversion level of the decision maker, η, $(0 < \eta \leq 1)$, there are two decision variables—the stocking level, Q and an objective threshold value, ω. We consider the CVaR optimization problem on the profit with the following notation: The unit acquisition cost is c, unit revenue is p, each left-over unit has a salvage value of s, and each unmet demand incurs a cost of π. The optimization problem is formally stated as follows:

$$Max_{Q,\omega} = CVaR(Q,\omega|\eta),$$

where

$$CVaR(Q,\omega|\eta) = \omega + \frac{1}{\eta}\left\{\int_0^\infty [(p-c)Q - (p-s)[Q-x]^+ - \pi[x-Q]^+ - \omega]^- dF(x)\right\}.$$

The objective function is jointly concave in Q and ω, and the optimal solution (Q^*, ω^*) is given by

$$Q^* = \left(\frac{\pi}{p-s+\pi}\right) F^{-1}\left(\frac{(p-s+\pi)-(c-s)}{p-s+\pi}\right) + \left(\frac{p-s}{p-s+\pi}\right) F\left(\frac{p-c+\pi}{p-s+\pi}\right),$$

and

$$\omega^* = \left(\frac{(p-s)(p-c+\pi)}{p-s+\pi}\right) F^{-1}\left(\frac{p-c+\pi}{p-s+\pi}\right) - \left(\frac{\pi(c-s)}{p-s+\pi}\right) F^{-1}\left(\frac{(p-s+\pi)-(c-s)}{p-s+\pi}\right).$$

Note that the problem reduces to the risk-neutral newsvendor problem when $\eta = 1$, and that, for $\eta < 1$, the optimal stocking level under the CVaR criterion is less than that for the classical newsvendor problem.

7.3.2.3 Parameter Updating

If the item is a standard product but has very short lifetimes (a single period), then, the inventory problem may be viewed as a sequence of independent newsvendor problems solved myopically. In markets for relatively new products, demand distributions or their parameters may be unknown or, at best, reasonable estimates. As new sales data come in, decision makers update these estimates. If lost demands are fully observable, then, stocking decisions do not impact parameter updating, and each period's problem can be solved independently with the given estimates. If lost demands are not observable, however, then, the censoring of data interlinks the stocking decisions in earlier periods with the accuracy of the parameter estimates to be used currently. For the frequentist approach, we refer the reader to Nahmias (1994) and Agrawal and Smith (1996). Bayesian methodology provides an opportunity to incorporate prior information about demand parameters (from prior market research or pilot markets) into current observations about the demand. Scarf (1959) is the earliest work on Bayesian updating in the inventory context. Hill (1997) and Eppen and Iyer (1997) consider fully observed lost sales and build on the multi-period-setting studies by Azoury (1985), who established that the total sales are a sufficient statistic to update the demand parameters and that the posterior distributions retain the structure of their conjugate priors. With censored demands, this is not possible in general. Lariviere and Porteus (1999) showed that the total sales are still a sufficient statistic for the special case of the so-called newsvendor distribution. Ding et al. (2002) considered the impact of censoring on updating and stocking decisions for other general demand distributions (see also Lu et al. 2005). Berk et al. (2007) provided an effective two-parameter approximation that retains the conjugacy of the posteriors. In a different vein, Graves (1999) combines the demands generated from an autoregressive-integrated moving-average (ARIMA) process and updates the ordering policy.

7.3.3 Periodic Review

The periodic-review inventory models have been typically formulated as dynamic programs. The earliest attempt is Bellman et al. (1955). The formulation rests on a succession of newsvendor-type models for each of the periods considered over the problem horizon $\{1, 2, \ldots, T\}$. The model assumptions are as follows:

A1. The planning horizon consists of T periods
A2. The demand in period n, X_n is random but is fully characterized by its known pdf $f_n(x)$ and cdf $F_n(x)$
A3. Replenishment is possible at the beginning of each period at a unit acquisition cost c_n
A4. Order lead time in each period is zero
A5. There is no initial stock
A6. Each unit of stock left over at the end of a period incurs a cost h_n and each unit of unsatisfied demand incurs a cost p_n
A7. The objective is to minimize the expected total costs (appropriately discounted) over the problem horizon

The decision is to determine the optimal stocking levels in each period so that the total discounted expected costs are minimized. To avoid ill-defined costs, it is assumed that the (time-dependent) discount factor $0 \leq \alpha_n \leq 1 \; \forall n$ where $\alpha_n = 1$ only if $T < \infty$.

Let y denote the stocking level at the beginning of a period immediately after ordering and define $C_n(z_n)$ as the minimum expected discounted cost when n periods remain until the end of the problem horizon and z_n is the current inventory level. Then, the DP formulation of the problem is as follows:

$$C_n = \min_{y \geq z_n} \{G_n(y) - c_n z_n + \alpha_n \int_0^\infty C_{n-1}[s(y,x)] f_n x dx\} \quad \text{for } n \geq 1,$$

where

$$G_n(y) = c_n y + \int_{x=0}^{y} h_n(y-x) f_n(x)dx + \int_{x=y}^{\infty} p_n(x-y) f_n(x)dx.$$

The function $s(y,x)$ is called the transfer function, which describes the evolution of inventory levels z_n from period to period. If excess demand is backordered, then, $s(y,x) = y - x$ whereas if it is lost, then, $s(y,x) = max(y-x,0)$. $C_T(0)$ corresponds to the optimal cost over the horizon and the initial condition is generally taken as $C_0(z) = 0\ \forall z$. For $n = 1$, the problem reduces to the classical newsvendor model.

In its general form, the above inventory control problem subsumes speculative acquisitions, as well: If acquisition prices are nonstationary and/or the discount factor is nonstationary, then, it may be possible to postpone or accelerate replenishment decisions to replenish the stock at effectively lower costs. To eliminate this additional trade-off, the basic model considers only stationary unit acquisition costs and discount factors: $c_n = c$ and $\alpha_n = \alpha\ \forall n$. Then, with $C_0(z) = 0\ \forall z$, the optimal policy has the following structure: *In any period* $1 \leq n \leq T$, *order up to* y_n^* *if* $z_n < y_n^*$; *do not order, otherwise.* For holding and shortage costs that are nonincreasing in time, $y_1^* \geq y_2^* \geq \cdots y_T^*$.

If demand is backordered, then, y_1^* solves the newsvendor model and $y_\infty^* = (p-(1-\alpha)c)/(p+h)$ for $T = \infty$. As there are no restrictions on the replenishment quantities and order lead times are zero, similar results hold for fully lost sales.

The seminal results on the optimality of the periodic-review (s,S) policies are found in Iglehart (1963) and Porteus (1971). Porteus (1985a) provides a numerical comparison of simpler periodic-review policies with respect to this class. Federgruen and Zipkin (1984) provide a discussion of the computational issues and Zheng and Federgruen (1991) give a simple algorithm to compute the optimal policy parameters. For nonlinear costs, see Cetinkaya and Parlar (1998, 2004).

7.3.3.1 Positive-Order Lead Time

When a positive-order lead time is included, it does not suffice to consider only inventory levels. As Karlin and Scarf (1958) proved, as long as demand is backordered, the optimal policy depends only on the sum of on-hand and on-order stock. Furthermore, the optimal policy has the same structure as the zero lead time case though the critical numbers are computed differently.

When excess is (partially or fully) lost rather than backordered, the optimal policy is extremely complex. It is a nonlinear function of the vector of outstanding orders. Suppose order lead time is a constant L for all periods. Define the system state in period n as the L-dimensional vector of replenishment quantities z_i to be received in the i periods for $0 \leq i \leq T-1$ where $i = 0$ corresponds to the order already received, hence the on-hand stock z_0. Then, the functional equation of the DP formulation is written as follows. For $n \leq 1$

$$C_n(z_0, z_1, \ldots, z_{L-1}) = \min_{y \geq 0} \left\{ c_n y + G_n(z_0) + \alpha_n \int_{0}^{\infty} C_{n-1}[s(y, z_0, z_1, \ldots, z_{L-1}, x)] f_n x dx \right\},$$

where y denotes the size of the current order, $G_n(z_0)$, denoting the expected holding and shortage cost in period n if the inventory level at the start of the period z_0 is given as

$$G_n(z_0) = \int_{x=0}^{y} h_n(y-x) f_n(x)dx + \int_{x=y}^{\infty} p_n(x-y) f_n(x)dx,$$

and the transfer function $s(y, z_0, z_1, \ldots, z_{L-1}, x) = (z_0 + z_1 - x, x_2, \ldots, x_{L-1}, y)$ for $0 \le x \le z_0$ and $s(y, z_0, z_1, \ldots, z_{L-1}, x) = (z_1, z_2, \ldots, z_{L-1}, y)$.

Although the optimal policy can be obtained numerically for small-scale problems, approximations are important. For an infinite-horizon problem with stationary parameters and demand processes, Morton (1969) proposes the following approximate solution for the optimal order quantity in a period:

$$y^* = min\{y_1^*, y_2^*, \ldots, y_{L+1}^*\},$$

where y_i^* solves

$$F^{(i)}(y_i^*) = \frac{p - \alpha^{-L}c}{p + h - \alpha^{-L+1}c},$$

and $F^{(i)}$ denotes the cdf of the ith convolution of the per-period demand cdf $F(\cdot)$. The logic of the solution rests on the basic newsvendor trade-off considering the entirety of demand over the order lead time with an effective (discounted) acquisition cost. Nahmias (1979) proposes other approximations. Morton (1971) demonstrated that myopic policies (which ignore the future periods) rendering the finite problem to a newsvendor problem are quite efficient. Sobel and Zhang (2001) consider a system facing a mixture of (backorderable) random demands and deterministic (fully lost) demands.

Lead-time uncertainty presents additional difficulties due to the possibility that orders may cross. For example, shipments that are transported independently may cross each other on the road; a shipment that left the upstream supplier may be received earlier than the one that left afterward. As the next order's delivery time is the minimum of those of all outstanding orders, order crossing gives rise to the peculiar case of lead times getting stochastically smaller as the number of outstanding orders increases. But, if orders are placed with one supplier and follow identical shipment procedures, then, it is unlikely that they would cross in time. In this case, however, the order lead times become dependent random variables. To see the complications arising from order crossing, consider the case of constant lead times with duration L. In this case, to compute the net inventory level at time $i + L$, it suffices to take into account only those orders that are to be received within the next L time periods. However, if orders can cross, then, all the outstanding orders at time i and those that may be placed in periods $i + 1$ to $i + L - 1$ must be taken into account. That is, replenishment decisions to be made in the future affect the decision in the current epoch, as well.

Kaplan (1970) overcame such difficulties in a novel way and incorporated random lead times into the DP formulation. Let β_i denote the probability that all orders placed at i or more time periods ago arrive in the current period. This event guarantees that orders do not cross by definition as the arrival of an order implies that of an older one. The evolution of inventory levels over two consecutive periods can then be computed by making use of β_i and the vector of outstanding orders. Hence, the critical number policy structure is still optimal. Bashyam and Fu (1998) propose a constrained optimization problem allowing for orders to cross resulting in close-to-optimal solutions.

Veinott (1965) showed that, under reasonable assumptions about the salvage value and the transfer function, a T period multiperiod dynamic inventory problem can be decomposed into T single-period problems. To this end, we construct a forward formulation of the problem.

Let z_n denote the starting period in period n and y_n denote the order-up-to level in period n. Then, the expected discounted cost for a T-period problem may be written as

$$\tilde{C}_T = \sum_{n=1}^{T} (\Pi_{i=1}^{n-1}\alpha_i)[E\{c(y_n - z_n)\} + G_n(y_n)] - (\Pi_{i=1}^{T}\alpha_i)cz_{T+1}.$$

This formulation assumes that the stock remaining at the end of the horizon z_{T+1} can be returned at the original purchase price of c. If excess demand is backordered, then, $z_{n+1} = y_n - X_n$ for $1 \leq n \leq T$. Then

$$\tilde{C}_T = \sum_{n=1}^{T} (\Pi_{i=1}^{n-1} \alpha_i)[c y_n (1 - \alpha_n) + G_n(y_n)] - c z_1 - \sum_{n=1}^{T} (\Pi_{i=1}^{n} \alpha_i) c \overline{d_n},$$

where $\overline{d_n}$ is the expected demand in period n.

If there are no constraints on y_n, then, the optimal policy is the value of y_n^* that minimizes $c y_n (1 - \alpha_n) + G_n(y_n)$. Hence, the solution of the optimal policy for the multiperiod problem is as easy as that for the basic newsvendor problem. Similar results apply when there is a positive-order lead time and the excess demand is backordered.

The key assumption in the above result is that the unit acquisition cost is stationary over the problem horizon. If all parameters of the problem are stationary, then, there is a stationary-ordering policy regardless of the length of the horizon!

In this case, consider the setting where orders can only be integer multiples of a certain base size, M. Suppose $G_n(y) = G(y)$ is strictly quasi-convex, then, there is a unique value k satisfying $G(k) = G(k + M)$. Then, the optimal ordering policy is as follows. *If the starting inventory level* $z < k$, *order that number of base sizes to raise the inventory level to* $\hat{y}$ *where* $k \leq \hat{y} \leq k + M$; *do not order otherwise.* Note that the order-up-to level, in this case, depends on the starting inventory level; and, hence, the optimal policy is characterized by an interval.

7.3.3.2 Positive-Setup Costs

A setup cost for ordering results in that the cost of acquiring u units is $K\mathbf{1}_{u>0} + cu$ where $\mathbf{1}_{u>0}$ is the indicator function. That is, the acquisition cost is discontinuous at zero. For a single-period problem, it poses no difficulty. Above, we showed that the objective function is convex and the optimal policy is a critical number policy of the form: *If* $z < s$ *order up to* S, *otherwise do not order.* This policy structure is called the (s, S) policy. For a multiperiod dynamic inventory problem with positive-ordering costs, the optimal policy would still be an (s, S) policy if the objective function were convex throughout the problem horizon. But this is not the case. Nevertheless, the (s, S) policy is optimal if the total cost function has a special property—the so-called K-convexity.

Definition

A differentiable function $g(x)$ is K-convex if $K + g(x + y) - g(x) - y g'(x) \geq 0$.

There are a number of interpretations of this property. (1) A K-convex function with a global minimum at $x = S$ may be multimodal (i.e., it may contain many humps) but the steepness of its contour is bounded over its domain. (2) A K-convex function with a global minimum at $x = S$ may be multimodal but attains the value of $g(S) + K$ at most at two points over its domain. That is, a tall traveler (with a height of K units) walking on this hilly terrain will always be able to see over the hills! This has the important implication that an order is given when the inventory position falls within only a single region. Otherwise, the control policy would consist of multiple ordering regions. Clearly, a k-convex function is also K-convex for $0 \leq k \leq K$. Therefore, in a multiperiod problem, as long as positive-order costs are nondecreasing, $K_n \leq K_{n+1} \leq K$ $1 \leq n \leq T$, the $C_n(\cdot)$ is K-convex.

In his seminal work, Scarf (1960) showed inductively that $C_n(\cdot)$ is K-convex for the full backordering case and, hence, established that the optimal policy is an (s, S)-type policy where the values of s and S may change from one period to the next. Iglehart (1963) proved that a stationary policy exists for stationary-problem parameters for the infinite-horizon problem (see also Scarf 2002). The optimal values of the policy parameters can be computed numerically by successive approximations.

7.3.3.3 Variants

In some settings, especially where demands are generated by additive processes over time (e.g., Poisson or normally distributed), the review-period lengths may be selected by the decision maker. In such cases, the order-up-to policy becomes the (R, T) policy and the periodic-review (s, S) policy becomes the (r, R, T) policy. The above expressions for the operating characteristics hold. Additionally, the period length T is a decision variable. Under the (R, T) policy, in essence, the review interval length functions as a reordering point. The (r, R, T) policy provides further flexibility with an additional decision variable. If order quantities are restricted to integer multiples of some pre-determined quantity, then, the (nQ, r, T) policy is employed, under which inventories are monitored every T time units, and an order of size that is an integer multiple of Q is placed whenever the inventory position crosses r. The minimum order quantity Q may also be a decision variable. For detailed derivations of these models, see Hadley and Whitin (1963). As $T \to 0$, these periodic-review models reduce to the lotsize-reorder point models discussed later under a continuous review.

The above analysis assumes an end-of-period costing for holding inventories or shortages. An exact costing method also exists where demand occurrence instances are taken into account. An on-hand item incurs carrying costs from the beginning of a period until such an instance at which it is withdrawn from the stock to satisfy the demand; similarly, shortage costs within a period are computed over the exact interval from demand occurrence until the end of the period. (Outstanding backorders incur time-based costs throughout the period lengths, when lead times are integer multiples of period lengths.) Although it appears to be complex, the method results in relatively simple expressions in the case of Poisson demands since the demand arrival times are uniformly distributed and conditioned on the number of demand occurrences within a period.

7.3.3.4 Perishable Inventories

The optimal policy structure for perishable inventories under a periodic review has been shown independently by Fries (1975) and Nahmias (1975) for items with constant lifetimes ($m \geq 2$ periods). (Note that for $m = 1$, the problem reduces to the newsvendor problem.) The special case of two-period shelflives is tractable and yields the fundamental insights. The main difference between the works is that the former bases the analysis on the outdating function that contains the information vector. Nahmias (1977a) shows that the two approaches give the same policy in an appropriately selected infinite-horizon setting. Random lifetimes have been considered by Nahmias (1978, 1977b). For multiple-product settings (as blood banks), see the seminal work by Nahmias and Pierskalla (1976). A comprehensive review of perishables is Nahmias (1982).

7.3.4 Continuous Review

7.3.4.1 Basic Model

Under a continuous review, random demands, by themselves, do not present much difficulty if replenishment lead times are negligible. The analysis and formulas of the deterministic models with zero lead times still hold with a slight modification in the demand rate. Instead of a constant demand rate, one needs to work with the mean demand per unit time, say $\overline{d}$. It suffices to simply substitute

$\overline{d}$ in lieu of D in the previous discussions of the deterministic inventory systems, and the results are superb approximations if not exact representations of the stochastic systems. But, this simple approach does not work when lead times are nonnegligible.

Positive lead times occur, primarily, due to shipment times and the delays encountered because of the stock-outs at upper echelons. Shipment times are practically constant although there may be instances of uncertainty arising from weather and traffic conditions especially over long distances. The delays experienced at upper echelons are the main reason for lead-time uncertainty. Such delays can be explicitly modeled in multiechelon systems but, for single-location models, they are usually taken as exogeneous random processes. Naturally, in real life, lead times do not cross—that is, an order placed later cannot be received earlier than a preceding order. However, this necessitates the use of the exact process that generates the lead times (as one would in the transient solution of a queuing system). This is either impossible or intractable and, hence, undesirable in the analysis. Therefore, all single-location models with positive lead times assume that lead times are generated exogeneously (independent of the order-process history) but may be dependent on the size of the current order.

The assumptions of the basic continuous-review stochastic demand model are as follows:

A1. The demand times constitute a renewal process with $H_Z(z)$ and $h_Z(z)$, $z \geq 0$, denoting the (absolutely continuous) cdf and pdf of the interdemand time Z with mean $1/\mu$, and demand sizes are i.i.d. integer-valued random variables with $G_Y(j)$ and $g_Y(j)$, $j \geq 1$, denoting the cdf and probability mass function (pmf) of the batch size Y with mean $\overline{b}$. (For hybrid and/or discrete-time distributions, the integration operations in the analysis below are to be replaced by summations at the possible jumps in the distributions.)

A2. Replenishment lead times are independent of the order history and are of length LT that may be random with $H_{LT}(l)$ and $h_{LT}(l)$, $l \geq 0$, denoting the cdf and pdf of the lead time LT with mean L.

A3. Orders arrive in one batch.

A4. Units do not undergo any change in their usefulness nor in their value (i.e., do not perish/decay or become obsolete).

A5. Shortages are permitted. With compound demands, demand integrality may or may not be preserved in case of a shortage: In one scenario, the entire quantity demanded would be fully backordered (or, lost) and, in another scenario, only the quantity in excess of an on-hand stock would be backordered (or, lost). The first scenario, which is not too common, leads to a more complex model and its system behavior is reported as not being fundamentally different from that of the latter when the underlying demand process is a Poisson process (Zipkin 2000). So, it is assumed that all of the on-hand stock is used to satisfy the arriving demand and the unmet portion, if any, is fully backordered to be satisfied later piecemeally as orders are received.

A6. Each order incurs a fixed cost K; each unit held in the stock for one unit time incurs a cost of h; each unit of demand that is not immediately satisfied incurs a cost of $\hat{\pi}$ per unit and a cost of π per unit per unit time while it remains backordered; and, a cost of π_s per unit time while the system remains out of stock. (The unit acquisition costs are ignored due to full backordering.)

All cost parameters are nonnegative. The stock-out cost rate π_s represents the effective costs that are incurred during a stock-out period: In cases when a desired service level is prescribed in terms of the fraction of time that the system is out of stock, it represents the shadow price of this constraint. In cases when each shortage incurs a unit cost, it represents the shortage cost incurred per unit time during stock-outs, which is the unit cost of shortage times the average demand rate for Markovian demand processes.

A7. Costs do not vary over time nor depend on the replenished quantity per order.

A8. The planning horizon is infinitely long.

A9. The objective is to minimize the expected cost per unit time.

A10. The inventory system is continuously monitored under a *reasonable* (stationary) control policy.

Letting $NI(t)$ denote the net inventory level at time t, the corresponding on-hand inventory denoted by $OH(t)$ is $[NI(t)]^+$, the corresponding number of backorders denoted by $NBO(t)$ is $[-NI(t)]^+$, and the probability that the system is out of stock at time t, $P_{out}(t)$ is $Pr[NI(t) \leq 0]$. Define the inventory position at some time t, denoted by $IP(t)$, as the on-hand inventory plus the total amount in outstanding orders denoted by $OO(t)$ minus the backorders; $IP(t) = OH(t) + OO(t) - NBO(t)$. Let $BOD(t)$ denote the amount of demand that is backordered during the (infinitesimally small) interval $(t, t + dt]$. It has been shown by Stidham (1974) that the inventory system operated under a *reasonable* stationary control policy is ergodic; that is, there exists a stationary distribution of the above entities as $t \to \infty$. The same notation is retained for all entities at a steady state while suppressing the time argument; for example, *NI*, *OH*, *etc.*

The expected cost rate at a steady state, $E[TC]$, is given by

$$E[TC] = K\mu_D/\hat{Q} + hE[OH] + \pi E[NBO] + \hat{pi}E[BOD] + \pi_s P_{out}, \tag{7.22}$$

where μ_D is the mean demand rate ($\mu_D = \mu\bar{b}$) and $\hat{Q}$ denotes the average order size.

7.3.4.2 Exact Analysis: A General Framework

For the exact analysis, replenishment lead times are assumed to be of a constant length L; $h_{LT}(L) = 1$ and zero, otherwise. Suppose that, under the employed inventory control policy, $IP(t)$ is a random variable that takes on integer values in the finite interval $[IP_{min}(t), IP_{max}(t)]$ denoted by $\Omega_{IP(t)}$ for all t. The analysis rests on (i) obtaining the steady-state distribution of the inventory position (at some arbitrary time t) and (ii) relating it to the distribution of the inventory levels at a steady state by means of the demand process that occurs over the time segment $(t, t + L]$.

Consider the inventory system at some arbitrary time instant $t - L$ (> 0). Suppose that the inventory position at this instant is $IP(t - L) = k$ with $k \in \Omega_{IP(t-L)}$ and that the time elapsed since the last arrival of a demand is $Z(t - L) = z$ where $0 \leq z \leq t - L$. Next, consider the time instant t. If no demands occur during $(t - L, t]$, then, all the orders that are outstanding at time $t - L$ (if any) will have joined the stock clearing the backorders (if any) so that the inventory position at time $t - L$ becomes the on-hand inventory at time t. If demands occur, however, inventories will be accordingly depleted and no order that might have been placed in the meantime will be able to arrive (as the times elapsed since order placements will be less than L by a construct). Hence, the net inventory $NI(t) = IP(t - L) - D(t - L, t)$.

For all j,

$$Pr[NI(t) = j] = \sum_{k \in \Omega_{IP(t-L)}} \int_{z=0}^{t-L} Pr[D(t - L, t) = k - j | z \leq Z(t - L) \leq z + dz]$$
$$\times\, Prob[IP(t - L) = k, z \leq Z(t - L) \leq z + dz]. \tag{7.23}$$

Then, for all j (≥ 0), $Pr[OH(t) = j] = Pr[NI(t) = j]$ and $Pr[NBO(t) = j] = Pr[NI(t) = -j]$.

In the above construction so far, the inventory control policy has not been specified. The optimal policy structure depends on the objective function. For the basic setting, under a continuous review, the objective is taken as minimizing the expected cost rate, although recently, a desire has emerged to consider other performance measures. For general renewal demands, the optimal continuous-review control policy remains an open question. However, for some special cases—primarily the Poisson demand process—the optimal control policy classes are known. Despite their suboptimality, these policies are deemed reasonable to be employed in practice also for other settings because of their ease and simplicity. All of them restrict replenishment epochs to demand arrival instances and are stationary IP-based policies with the following general structure:

General IP-based policy: Monitor inventory positions continuously, and raise the inventory position to IP_k whenever the inventory position hits k for the first time since the last replenishment epoch, where $-\infty < IP_{min} \leq k \leq s < IP_k \leq IP_{max} < \infty$.

Under this general IP-based policy class, by construct, the $\{IP(t)\}$ and $\{Z(t)\}$ processes are independent. At a steady state, $Pr[Z \leq z] = \int_{\eta=0}^{z}[1 - H(\eta)]d\eta$. Letting $\Delta = IP_{max} - IP_{min}$, we have the following fundamental result.

Proposition 7.1

At a steady state, for $0 \leq i < \Delta$,

$$Pr[IP = i] = p(IP_{max} - i), \tag{7.24}$$

where $p(i) = (m_{i+1} - m_i)/m_\Delta$ with $m_0 = 0$ and $m_j = 1 + \sum_{i=1}^{j} g_Y(i)m_{j-1}$ for $j \geq 1$.

There are four policies of this structure, which are commonly used and are of interest to both theoreticians and practitioners: The (Q, r), (s, S), $(S - 1, S)$, and (r, Nq) policies, which are discussed separately below.

Another performance measure of an inventory system is the length of time an order remains backordered and its distribution. This entity is particularly important when shortage costs are nonlinear in the waits. Such costs are incurred when customers are sensitive to the length of wait for their demands to be fulfilled, as would be the case for spare-parts supply contracts.

Kruse (1981) considered the waiting-time distribution of demands in an (s, S) continuous-review system, where demand times constitute a renewal process, demand sizes are i.i.d. integer values random variables, and the lead times are constant. A similar analysis for an $(S - 1, S)$ inventory system with unit Poisson demands and general i.i.d. lead times is provided, in the exact form, by Kruse (1980) using the results of the $M/G/\infty$ queue and, as an approximation, by Higa et al. (1975). Sherbrooke (1975) provides an exact waiting-time distribution for constant lead times. Below, we provide a generalization of these results under a stationary IP-based policy.

Consider the distribution of wait for each unit in the demand. Suppose that a demand of size y arrives at time t, where $Z(t) = 0$. Each unit in the demand is indexed by j from 1 to y. The jth unit will wait for at most τ time units iff the demands preceding the jth unit are less than the inventory position at time $t + \tau - L$, $IP(t + \tau - L)$. So,

$$Pr[j^{\text{th}} \text{ unit waits} \leq \tau] = \sum_{k=IP_{min}}^{IP_{max}} Pr[IP(t + \tau - L) = k, D(t + \tau - L, t) \leq k - j | Z(t) = 0].$$

The number of demand occurrences over the given time segment at a steady state forms an ordinary renewal process (Richards 1975; Sahin 1979; Stidham 1974). Hence, letting $F_{W_j}(\cdot)$ denote the steady-state distribution of a function of the waiting time for the jth unit of a demand

$$F_{W_j}(\tau) = \sum_{k=IP_{min}}^{IP_{max}} Prob(IP = k) \sum_{i=0}^{k-j} p(i); \quad 0 \leq \tau < L, \tag{7.25}$$

$$F_{W_j}(L) = 1, \tag{7.26}$$

where

$$p(i) = \sum_{n=1}^{i} g^{(n)}(i)[H^{(n)}(L-\tau) - H^{(n+1)}(L-\tau)]; \quad i \leq 1,$$

and $p(0) = 1 - H(L - \tau)$.

A demand (of any size) is immediately satisfied with probability $\sum_{j=1}^{\infty} g_Y(j) F_{W_j}(0)$. The expected (maximum) wait of the entire demand (of any size), $E[W_e]$, is given by

$$E[W_e] = \sum_{j=1}^{\infty} g_Y(j) \sum_{\tau=0}^{\infty} [1 - F_{W_j}(\tau)] d\tau.$$

The expected total wait that a demand (of any size) experiences $E[W_{tot}]$ is given by

$$E[W_{tot}] = \sum_{j=1}^{\infty} j g_Y(j) \sum_{\tau=0}^{\infty} [1 - F_{W_j}(\tau)] d\tau,$$

when it is satisfied only when sufficient inventory has accumulated, and by

$$E[W_{tot}] = \sum_{y=1}^{\infty} g_Y(y) \sum_{j=1}^{y} \sum \tau = 0^{\infty} [1 - F_{W_j}(\tau)] d\tau,$$

when backordered demands are fulfilled as orders arrive. The expected amount of demand backordered per unit time, $E[BOD]$, is given by

$$E[BOD] = \mu \sum_{y}^{\infty} y g_Y(y) [1 - F_{W_j}(0)],$$

when it is satisfied only when sufficient inventory has accumulated, and by

$$E[BOD] = \mu \sum_{y}^{\infty} g_Y(y) \sum_{j=1}^{y} [1 - F_{W_j}(0)],$$

when backordered demands are fulfilled as orders arrive.

7.3.4.3 The (Q,r) Model with Poisson Demands

The (Q,r) or lotsize-reorder policies are widely used in the industry and have been extensively studied in the literature. The employed policy is formally stated as follows.

The (Q,r) Policy: The inventory position, *IP* is monitored continuously, and an order of size Q is placed whenever the *IP* hits r.

The policy parameter Q is the order size and r is the reorder point. This policy is also known as the *lotsize-reorder point* policy. It is appropriate for unit-sized demands, and it is optimal for Poisson demands.

Corollary 7.1

Under the (Q,r) policy and with unit demands, at a steady state, the inventory position *IP* is uniformly distributed over $[r+1, r+Q]$.

Suppose that (unit-sized) demands are generated by a Poisson process with mean rate μ. Let $f_{DL}(\cdot)$ and $F_{DL}(\cdot)$ denote the pmf and cdf of demand that occurs at a steady state over a time segment of length L (i.e., the lead-time demand). Owing to the memoryless property of the exponential distribution, $f_{DL}(n)$ is the Poisson probability that exactly n demands occur with rate $\mu_{DL} = L\mu$. Thus,

$$\begin{aligned} Pr[NI = j] &= \frac{1}{Q}\sum_{k=r+1}^{r+Q} f_{DL}(k-j) \\ &= \frac{1}{Q}[F_{DL}(r+Q-j) - F_{DL}(r+1-j)] \quad \text{for } j \leq r+Q. \end{aligned} \tag{7.27}$$

The expected number of backorders at a steady state is given by

$$E[NBO] = \frac{1}{Q}[\beta(r) - \beta(r+Q)], \tag{7.28}$$

where $\beta(v) = \sum_{i=1}^{\infty} iF_{DL}(i+v+1)$.

The expected on-hand inventory at a steady state is given by

$$E[OH] = (Q+1)/2 + r - \mu_{DL} + E[NBO], \tag{7.29}$$

where μL corresponds to the expected amount on the order at a steady state, $E[OO]$.

The probability that the system is out of stock at a steady state is

$$P_{out} = \frac{1}{Q}[\omega(r) - \omega(r+Q)], \tag{7.30}$$

where $\omega(v) = \sum_{i=v+1}^{\infty} F_{DL}(i)$.

Owing to the PASTAs (Poisson arrivals see time averages) property, the expected number of demands backordered per unit time is μP_{out}. Assuming that each shortage incurs a unit cost of $\hat{\pi}$, we have $\pi_o = \hat{\pi}\mu$.

Noting that each order is of size Q, the expected cost per unit time can be written as follows:

$$\begin{aligned} E[TC] = {} & K\mu/Q + h[(Q+1)/2 + r - \mu_{DL}] + \pi\mu\frac{1}{Q}[\omega(r) - \omega(r+Q)] \\ & + (\bar{\pi} + h)\frac{1}{Q}[\beta(r) - \beta(r+Q)]. \end{aligned} \tag{7.31}$$

In the derivations above, we assumed that the reorder point r is positive. This is the most common setting; however, if need be, the above expressions can be used with the additional convention that $F_{DL}(i) = 0$ if $i < 0$. The above expressions are not amenable to obtain analytical expressions for the optimal policy parameter values. However, when it is negligibly unlikely that the lead-time demand will be larger than $r + Q$ (thereby, rendering $\omega(r + Q)$ and $\beta(Q + r)$ to be zero), we have, at an optimality,

$$Q^*(Q^* - 1) < \frac{2\mu}{h}[K + \pi\omega(r^*) + (\bar{\pi} + h)\beta(r^*)/\mu]. \tag{7.32}$$

$$\left[1 - \frac{(\bar{\pi} + h)}{\mu\pi}(r^* - \mu_{DL})\right] F_{DL}(r^*) + frac(\bar{\pi} + h)r^*\mu\pi f_{DL}(r^*) > \frac{Qh}{\pi\mu}. \tag{7.33}$$

The above set of equations are solved iteratively to arrive at the optimal values by starting with $\tilde{Q}$ that solves Equation 7.32 with r set at zero. The convergence has been observed to be quite fast.

For large-order sizes, it is possible to use a continuous approximation so that Equation 7.32 yields

$$Q^* = \sqrt{\frac{2\mu}{h}[K + \pi\omega(r^*) + (\bar{\pi} + h)\beta(r^*)/\mu]}, \tag{7.34}$$

to be used conjunctively with Equation 7.33.

For large lead-time demands, it may be convenient to approximate the Poisson probabilities by a normal distribution as well as to treat the policy parameters and the inventory-related entities as continuous variables (Hadley and Whitin 1963; Zipkin 2000). Federgruen and Zheng (1992) have provided a simple and very efficient algorithm for determining the optimal policy parameters; the algorithm is linear in the optimal order quantity. Song et al. (2010) consider the effects of lead-time variability with compound Poisson demands. They show that the base-stock model properties can be partially extended to this general setting.

7.3.4.4 The (s, S) Model with Compound Poisson Demands

The continuous-review (s, S) policy is a natural extension of the optimal policy class obtained with a periodic review. The policy parameter s denotes the reorder point and S is the *target stock level*. The order size under this policy is random. This policy is optimal for compound Poisson demands. (This policy is also called the (r, s) policy in the literature.) When demands are unit sized, this policy reduces to the (Q, r) policy with $S = Q + r$ and $s = r$, as expected. The policy is formally stated as follows.

The (s, S) Policy: The inventory position, *IP*, is monitored continuously, and an order is placed to raise *IP* to S whenever the *IP* hits/crosses s with $(S \geq s + 1)$.

Corollary 7.2

At a steady state, for $0 \leq i < S - s$,

$$Pr[IP = i] = p(S - i), \tag{7.35}$$

where $p(i) = (m_{i+1} - m_i)/m_{S-s}$ with $m_0 = 0$ and $m_j = 1 + \sum_{i=1}^{j} g_Y(i)m_{j-1}$ for $j \geq 1$.

The ergodicity of the system has been shown by Stidham (1974, 1977) and, hence, we know that stationary distributions exist. The stationary distribution of the inventory position was derived by Sivazlian (1974) for unit demand size and extended to a random demand size by Richards (1975). Archibald and Silver (1978) also studied the system for the compound demand process. The most comprehensive analysis of the model is in Sahin (1979) where transient and stationary distributions are derived for the inventory position and levels using a renewal-theoretic approach. The objective-function behavior is investigated in Sahin (1982) and in Zipkin (1986) with service-level measures. The (s, S) system with nonlinear costs has been studied in Stidham (1986). Browne and Zipkin (1991) considered continuous demands. Nahmias (1976) provides service-levels comparisons with other control policies under service constraints and proposes optimization efficiencies. Urbach (1977) analyzed the system with no more than one order that was outstanding and the lead times are random.

7.3.4.5 The (r, Nq) Model with Compound Poisson Demands

This policy class is meaningful especially when there is some "natural" order size, such as a truckload. When demands are unit sized, this policy also reduces to the (Q, r) policy. Furthermore, the best (r, Nq) policy is often close to the optimal in the presence of compound Poisson demands (Zipkin 2000). The policy is formally stated as follows.

The (r, Nq) Policy: The inventory position, IP is monitored continuously, an order is placed whenever the IP hits or crosses r, and each order is the smallest positive-integer multiple of q so as to bring the inventory position above r.

The policy parameter r denotes the reorder point and q denotes the *basic batch size*. Under this policy, the order size is also random, where N indicates the random number of basic batches in an order.

Corollary 7.3

Under the (r, Nq) policy and with unit demands, at a steady state, the inventory position IP is uniformly distributed over $[r+1, r+Q]$.

Note that the inventory position is uniformly distributed under this policy as well. Hence, the operating characteristics are those under the (Q, r) policy with unit Poisson demands. To get the expected fixed-ordering costs, consider the general cost structure where each order incurs a major cost K and each basic batch size in the order further contributes to a minor cost k. Then, the expected cost per unit time can be written as follows:

$$E[TC] = \left\{ K\mu_D \left[E[Y] - \sum_{j=q}^{\infty} \bar{G}(j) \right] /[qE[Y]] + k\mu_D/q \right\}$$
$$+ h[(q+1)/2 + r - \mu_{DL}] + \pi\mu_D \frac{1}{q}[\omega(r) - \omega(r+q)]$$
$$+ (\bar{\pi} + h)\frac{1}{q}[\beta(r) - \beta(r+q)] + c\mu_D. \quad (7.36)$$

7.3.4.6 The $(S-1, S)$ Model with Poisson Demands

This is a special case of the (s, S) policy with $s = S - 1$, and is considered appropriate when fixed-ordering costs are negligible so that the inventory position is always maintained at its maximum value, S. That is, at a steady state, $Pr[IP = S] = 1$ and zero, otherwise. This policy is also called the *one-for-one* policy, indicating the possibility of demand sizes larger than one. The policy is formally stated as follows.

The $(S-1, S)$ *Policy:* The inventory position, IP, is monitored continuously, and an order is placed to raise IP to S whenever a demand occurs.

The $(S-1, S)$ policy deserves special attention for a number of reasons. It may be optimal to order units one at a time when the demand rate for an item is low and/or the item is very expensive (rendering the ordering costs negligible in comparison to holding costs). It clearly illustrates the modeling machinery employed in the general framework, and, hence is a useful instructional tool. It occurs commonly in practice, and, due to its simplicity, acts as a building block for modeling more complex inventory systems, such as random lead times, multiechelon structures, customer preferences for delivery timing, etc.

Below, we develop the operating characteristics of the inventory system in two ways to illustrate different modeling approaches.

Approach I

With Poisson demands, the $(S-1, S)$ is a special case of the (Q, r) policy. Hence, the expressions can be readily obtained from the general framework by setting $Q = 1$ and $r = S - 1$ as follows:

$$E[NBO] = \sum_{j=S+1}^{\infty} (j-S) f_{DL}(j) = \mu_{DL}[1 - F_{DL}(S-1)] - S[1 - F_{DL}(S)], \tag{7.37}$$

$$E[OH] = \sum_{j=0}^{S} (S-j) f_{DL}(j) = S - \mu_{DL} + E[NBO], \tag{7.38}$$

$$P_{out} = 1 - F_{DL}(S-1), \tag{7.39}$$

$$E[DOB] = \mu P_{out}. \tag{7.40}$$

Thus,

$$\begin{aligned} E[TC] = h[S - \mu_{DL}] + [\hat{\pi}\mu + \pi_o][1 - F_{DL}(S-1)] \\ + [\pi + h][\mu_{DL}[1 - F_{DL}(S-1)] - S[1 - F_{DL}(S)]]. \end{aligned} \tag{7.41}$$

Simplifying

$$\begin{aligned} E[TC] = h[S - \mu_{DL}] + [\{[h+\pi]\mu_{DL} + \pi_o + \hat{\pi}\mu\}[1 - F_{DL}(S-1)]] \\ - [[h+\pi]S[1 - F_{DL}(S)]]. \end{aligned} \tag{7.42}$$

The optimal base-stock level, S^*, is the largest S for which

$$[1 - F_{DL}(S-1)] + \frac{\hat{pi}\mu + \pi_o}{h+\pi} f_{DL}(S-1) > \frac{h}{\pi + h}.$$

Since lead-time demand has a unimodal probability function, a linear search on S can be terminated at the first reversal of the inequality. When $\pi_o = \hat{\pi} = 0$, the optimization problem for the base-stock level becomes a newsvendor-type problem as a trade-off purely between holding and backordering costs (over an infinitesimal time interval at a steady state). The optimal S^* is the smallest S such that

$$F_{DL}(S) \geq \frac{\pi}{\pi + h}.$$

When $\pi = 0$, the optimal is the largest S such that

$$\frac{h}{\pi_o + \hat{\pi}\mu}[1 - F_{DL}(S-1)] + f_{DL}(S-1) > \frac{h}{\pi_o + \hat{\pi}\mu}.$$

Remarks

Note that the expected cost rate expression and the optimality results bear an equivalence with the classical newsvendor problem. In this case, the "single period" is the lead-time period and the trade-off is made on the basis of the demand during a time segment whose length is equal to a lead time. This simple analogy enables one to consider random lead times. If they are not allowed to cross, then, one can use the steady-state distribution of an exogeneous lead time to arrive at the lead-time demand distribution, $f_{DL}(\cdot)$. If orders are allowed to cross, then, we use a queuing analogy. The $(S-1, S)$ inventory system with Poisson demands and constant lead times can be, equivalently, represented by an $M/D/\infty$ queuing system with the number of customers in the latter corresponding to the outstanding orders. Similarly, when orders are allowed to cross, the operating characteristics of the inventory system with *random lead times* with a mean of L can be derived from an equivalent $M/G/\infty$ queuing system. Hence, with random lead times with a probability distribution $f_{LT}(\cdot)$, lead-time demand distribution $f_{rDL}(j)$ can be obtained as $f_{rDL}(j) = \int_{u=0}^{\infty} Po(j;\mu u) f_{LT}(u)du$ with a mean of μL as before.

Approach II

The second approach rests on the *method of supplementary variables* introduced and developed independently by Cox (1955) and Takács (1955). (For a general exposition of the technique, see Gnedenko and Kovalenko 1989; Keilson and Kooharian 1960.) To obtain the operating characteristics (of our interest) for the inventory system under the $(S-1, S)$ policy with constant lead times, it suffices to know the age of the Sth most recent order at any time. This immediately follows from the first-in first-out (FIFO) issuance discipline. Suppose the age of this order at some time is x_1. If $x_1 \geq L$, then, it implies that this unit is on the shelf and has been so for $x_1 - L$ time units. Otherwise, it is outstanding and the system resides in a stock-out period since all the more recent orders must be also outstanding.

To illustrate the approach, consider the S-dimensional stochastic process $\xi = \{\xi_1, \xi_2, \ldots, \xi_S\}$ where ξ_S denotes, at a steady state, the time elapsed since the most recent order was placed (i.e., the most recent order's age), ξ_{S-1} denotes that of the second-most recent order, etc. with $\xi_1 \geq \xi_2 \geq \cdots \geq \xi_S \geq 0$. Let $p_\xi(x_1, x_2, \ldots, x_S)$ denote the pdf of ξ at a steady state. (Owing to the ergodicity of the system, this stationary distribution exists; and, for convenience, $p(\cdot)$ is assumed to be absolutely continuous and differentiable.) For $x_S > 0$,

$$\sum_{i=1}^{S} \frac{\partial}{\partial x_i} p_\xi(x_1, x_2, \ldots, x_S) = -\mu p_\xi(x_1, x_2, \ldots, x_S), \tag{7.43}$$

and

$$\sum_{i=1}^{S} \frac{\partial}{\partial x_i} p_{\xi}(x_2, \ldots, x_S, 0) = \mu \int_{x_1=x_2}^{\infty} p_{\xi}(x_1, x_2, \ldots, x_S). \tag{7.44}$$

The unique solution of this set of partial differential equations (pde's) is given by

$$p_{\xi}(x_1, x_2, \ldots, x_S) = C_0 e^{-\mu x_1}; \quad x_1 \geq x_2 \geq \cdots \geq x_S \geq 0, \tag{7.45}$$

where C_0 is the normalizing constant so that $p(\cdot)$ integrates to unity over its domain. The marginal distribution $p_{X_1}(x_1)$ denotes the pdf of the age of the Sth oldest order or, conversely, the age of the unit that is to be used to satisfy the next demand. Carrying out the integration, we get

$$p_{X_1}(x_1) = C_0 e^{-\mu x_1} \frac{(x_1)^{S-1}}{(S-1)!}, \tag{7.46}$$

with $C_0 = \mu^S$. That is, the age of the oldest unit in the system follows an $Erlang(\mu, S)$ distribution as noted before (see Kruse 1980). Using the PASTA property,

$$E[NBO] = \mu \int_{x_1=0}^{\infty} [L - x_1]^+ p_{X_1}(x_1) dx_1,$$

$$E[OH] = \mu \int_{x_1=0}^{\infty} [x_1 - L]^+ p_{X_1}(x_1) dx_1,$$

and

$$P_{out} = \int_{x_1=0}^{L} p_{X_1}(x_1) dx_1.$$

As before, $E[BO] = \mu P_{out}$. The expected cost rate can be obtained with these components as before.

This technique has been first employed in inventory theory by Schmidt and Nahmias (1985) for perishables, but it is applicable to other Markovian demand settings, as well. (See the discussion on lost sales below.)

Song (1994) studies the effect of lead time on a base-stock model. She shows that a stochastically larger lead time requires a higher optimal base-stock level while a stochastically larger lead time may not necessarily result in a higher optimal average cost. However, a more variable lead time always leads to a higher optimal average cost. The effect of lead-time variability on optimal policies depends on the inventory cost structure: A more variable lead time requires a higher optimal base-stock level if and only if the unit penalty (holding) cost rate is high (low).

7.3.4.7 Approximate Analysis for the (Q, r) Model

The approximate analysis and the ensuing models are important in that they use the same renewal-theoretic approach as the deterministic models and provide simple extensions of the basic deterministic models to the random demand cases. There are three basic approximations of the (Q, r) model.

Approximation I

This approximation retains the construction of the exact formulation but replaces the Poisson demand expressions with normally distributed demands. It is appropriate when lead-time demands are large.

Approximation II

The second approximate analysis rests on the fundamental assumption that there is *at most* one outstanding order at any time. That is, it is assumed that, once an order is placed, demand during the lead-time period will not exceed the order quantity. The assumption does not, in fact, restrict the demand process in the analysis; rather, the analysis ignores such an event assuming that it may occur only with a very small probability. Naturally, so long as this probability is indeed negligibly small, the approximate treatment works very well. Moreover, this assumption further simplifies the modeling because one can work with the net inventory equivalently as with the inventory position (see Figure 7.4 for a depiction).

For this approximate analysis, it is typical to consider continuous-demand processes, as well. Let the demand per unit time (say a day), X, have a pdf $f_D(x)$ with μ_D and σ_D denoting its mean and standard deviation, and let the demand during a lead-time period, Y, have a pdf $f_{DL}(x)$ with μ_{DL} and σ_{DL} denoting its mean and standard deviation. The lead time is allowed to be random. All unmet demand is fully backordered. The shortage cost consists of backordering costs incurred only on a unit basis; $\pi_o = \pi = 0, \hat{\pi} > 0$.

The approximate analysis uses the renewal-theoretic approach for constructing the model and deriving the relevant operating characteristics. Without a loss of generality, define a *cycle* as the time between two consecutive instances when the inventory level hits the reorder point r (see Figure 7.4). By definition, during each cycle, exactly Q units of demand occur so that the expected cycle length is $E[CL] = Q/\mu_D$. In each cycle, there are two time segments to consider: the lead-time period and the time after the order is received until the next order is placed. The lead-time period starts with r units on hand and ends (immediately before the outstanding order is received) with $r - \mu_{DL}$ units on hand, on an average. The expected left-over inventory amount at the end of a lead-time period corresponds to the *safety stock*, which we will denote by SS. The remainder of the cycle, on an average, starts (immediately after the outstanding order is received) with $r - \mu_{DL} + Q$ units and ends, by definition, with r units of the inventory. Then, the average expected on-hand inventory level within a cycle is $E[OH] = [r + r - \mu_{DL}]/2 + [r - \mu_{DL} + Q - r]/2 = Q/2 + r - \mu_{DL}$. Under the one-outstanding-order assumption, in each cycle, there is only one replenishment incurring the fixed cost K and the variable purchasing cost cQ. Also, the shortage-related costs can be incurred *only* within

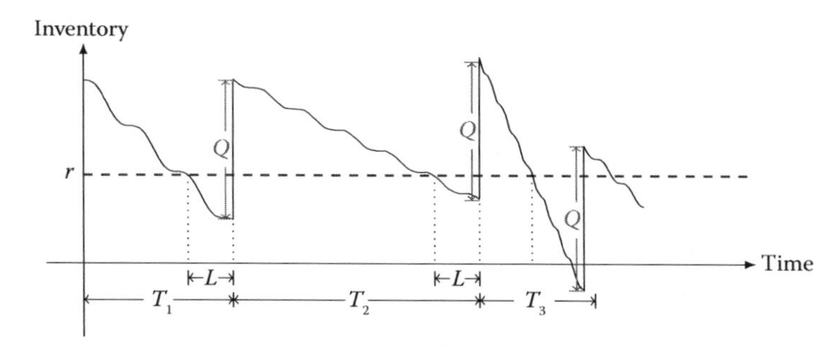

FIGURE 7.4
(Q, r) inventory policy with a constant lead time.

a lead-time period. The amount of backordered demands at the end of a lead-time period, *NBO*, is given by $[Y - r]^+$ with the corresponding expected value $E[NBO]$ as follows:

$$E[NBO] = \int_{y=0}^{\infty} \{[y - r]^+ f_{DL}(y)dy\} = \int_{y=r}^{\infty} y f_{DL}(y)dy - r\bar{F}_{DL}(r). \tag{7.47}$$

The expected cost rate per unit time, $E[TC]$, is given by

$$E[TC] = K\mu_D/Q + h(Q/2 + r - \mu_{DL}) + \pi\mu_D/QE[NBO]. \tag{7.48}$$

Carrying out the joint optimization over the policy parameters, we obtain the optimal values solving the following set of equations arising from the FOCs:

$$Q^* = \sqrt{\frac{2\mu_D(K + \pi E[NBO])}{h}}, \tag{7.49}$$

and

$$\bar{F}_{DL}(r^*) = \frac{Q^* h}{\pi\mu_D}. \tag{7.50}$$

The optimal values jointly satisfying the above expressions are obtained computationally in an iterative fashion, typically solving for an initial reorder point via Equation 7.50 and proceeding with the corresponding order quantity from Equation 7.49. The convergence has been found to be quite rapid.

As in the case of deterministic demands with allowable shortages, it is not necessarily true that a nonnegative optimal reorder point is feasible—implying that all demands should be backordered forever! However, for realistic shortage costs, the expected cost rate is jointly convex in the policy parameters.

For the special case of normally distributed demands during the lead time, the expected cost rate can be conveniently rewritten as

$$E[TC] = K\mu_D/Q + h(Q/2 + r - \mu_{DL}) + (h + \pi\mu_D/Q)\left[(\mu_{DL} - r)\bar{F}_{DL}\left(\frac{r - \mu_{DL}}{\sigma_{DL}}\right) + \sigma f_{DL}\left(\frac{r - \mu_{DL}}{\sigma_{DL}}\right)\right]. \tag{7.51}$$

Approximation III

The last approximation is a simpler variant of Approximation II with normally distributed lead-time demands and, is frequently encountered in introductory-learning resources (e.g., Stevenson 2014). It decouples the quantity and timing decisions for reorders, and computes the policy parameter for each decision based on the two fundamental trade-offs: Q is obtained via the EOQ formula using the mean demands rate, and r is obtained via the newsvendor model on the basis of a desired *lead-time service level*, α, which is defined as the probability that lead demand and time demand does not exceed the reorder point

$$Q = \sqrt{2K\mu_D/h},$$

and

$$r = \mu_{DL} + SS,$$

where SS denotes the safety stock and is given by $z_\alpha \sigma_{DL}$. The concept of a safety stock refers to the amount of inventory that is in excess of the average lead-time demand when an order is placed. It is viewed as a hedge against the possible surges of demand while the order is outstanding. Notice the close resemblance to the construct of the classical newsvendor problem: When the period's demand is normally distributed and the desired service level is computed as $SL = \alpha$, the optimal newsvendor-stocking level is the same as r above.

Let $E(z_\alpha)$ be the loss function value for a unit-normal variate evaluated at z_α; $E(z_\alpha) = \int_{z_\alpha}^{\infty} u\phi_1(u)du$. The expected number of units short in a cycle is given by $E(z_\alpha)\sigma_{DL}$ where $E(z_\alpha)$ is the loss function value for a unit-normal variate evaluated at z_α. Noting that there are μ_D/Q cycles on average per unit time (say, a year), we get the *annual service level*, SL_{ann} that is defined as the fraction of demand that is immediately satisfied, as follows:

$$SL_{ann} = [1 - (1/Q)E(z_\alpha)\sigma_{DL}].$$

The corresponding expected cost rate is given by

$$E[TC] = \sqrt{2K\mu h} + hSS + \hat{\pi}(\mu_D/Q)E(z_\alpha)\sigma_{DL} + c\mu.$$

7.3.4.8 The (Q, r) Model with Fully or Partially Lost Sales

In the presence of fully or partially lost sales and lead times, the inventory problem becomes extremely difficult. Except for the $(S-1, S)$, there is no exact treatment of the setting. For the (Q, r) model, some approximations have been suggested. The approximate analysis in the presence of lost sales differs little from the fully backordered case. When unmet demand is (partially) lost, the main difficulty in the treatment arises from the time segment within a cycle during which the system is out of stock. Given that the analysis is of a heuristic nature anyhow, this time segment is ignored assuming that it constitutes a negligibly small portion of a lead-time period. Let α denote the fraction of demand that is backordered in case of a stock-out at unit cost π; the remainder of excess demand is lost at unit cost $\hat{\pi}$.

Retaining our definition of a cycle and proceeding in a fashion similar to the full backordering case, the average expected on-hand inventory within a cycle is given by

$$E[OH] = \left[r + \int_{y=0}^{r} [(r-y)f_{DL}(y)dy]\right]/2 + \left[\int_{y=0}^{r} [(r-y)f_{DL}(y)dy] + Q + r\right]/2$$

$$= Q/2 + r - \mu_{DL} + \int_{y=r}^{\infty} yf_{DL}(y)dy - r\bar{F}_{DL}(r), \tag{7.52}$$

the expected amount of backordered demand is given by

$$E[NBO] = \alpha \int_{y=r}^{\infty} [(y-r)f_{DL}(y)dy]$$

$$= \alpha \int_{y=r}^{\infty} [yf_{DL}(y)dy] - r\bar{F}_{DL}(r), \tag{7.53}$$

and the expected amount of lost demand is given by $E[LS] = [(1-\alpha)/\alpha]E[NBO]$. Then, the expected total cost rate is

$$E[TC] = K\mu_D/Q + h(Q/2 + r - \mu_{DL}) + (h + \alpha\pi + (1-\alpha)\bar{\pi}) \times \left(\int_{y=r}^{\infty} yf_{DL}(y)dy - r\bar{F}_{DL}(r) \right). \tag{7.54}$$

Carrying out the joint optimization over the policy parameters, we obtain the optimal values as

$$Q^* = \sqrt{\frac{2\mu_D(K + \pi E[NBO] + \bar{\pi}E[LS])}{h}}, \tag{7.55}$$

and

$$\bar{F_{DL}}(r^*) = \frac{Q^*h}{(\alpha\pi + (1-\alpha)\bar{\pi})\mu_D + Q^*h}. \tag{7.56}$$

The optimal values jointly satisfying the above expressions are obtained computationally in an iterative fashion, typically solving for an initial reorder point via Equation 7.56 and proceeding with the corresponding order quantity from Equation 7.55.

For the special case of normally distributed demands during the lead time, the expected cost rate can be conveniently rewritten as

$$E[TC] = K\mu_D/Q + h(Q/2 + r - \mu_{DL}) + (h + \pi\mu_D/Q)\left[(\mu_{DL} - r)\bar{F}_{DL}\left(\frac{r-\mu_{DL}}{\sigma_{DL}}\right) + \sigma f_{DL}\left(\frac{r-\mu_{DL}}{\sigma_{DL}}\right)\right]. \tag{7.57}$$

For the $(S-1, S)$ model, Smith (1977) proposes an approximate solution. All the three exact works employ the age distribution of the oldest unit in the system. When unmet demands are fully lost, from Schmidt and Nahmias (1985) (as the shelf life of the perishable item, $\tau \to \infty$), we have

$$p_{X_1}(x_1) = C_0 e^{-\mu x_1}; \quad x_1 \geq L,$$

$$= C_0; \quad \text{otherwise}, \tag{7.58}$$

where C_0 is the normalizing constant. Moinzadeh (1989) considers the partial backordering case where unmet demand is backordered with probability β and lost otherwise. Modifying his choice of the stochastic process a bit, we get

$$p_{X_1}(x_1) = C_0 e^{-\mu x_1}; \quad x_1 \geq L,$$
$$= C_0 e^{-\beta \mu x_1}; \quad \text{otherwise}, \tag{7.59}$$

where C_0 is the normalizing constant. Finally, Perry and Posner (1998) consider the case when customers have different impatience levels for backorders so that, if the remaining lead time of the oldest item in the system exceeds their tolerance, the demand is lost, otherwise, it is backordered. They use the order-crossing methodology but their setting can be modeled equally well with the method of supplementary variables.

For an insightful analysis and discussions on lost-sales inventory systems, we refer to Zipkin (2008a,b). For a recent comprehensive review, see Bijvank and Vis (2011).

7.3.4.9 Perishable Inventories with Random Demands

Kalpakam and Sapna (1996) consider a perishable inventory model with Markovian lead times, exponential decay, and demands being generated by a renewal process. The exponential lead-time assumption greatly facilitates the analysis as orders are allowed to cross. The only study that considers decay in the presence of random demands and constant lead times is Nahmias and Wang (1979) who modify the approximate (Q, r) model above to account for the decay. This heuristic treatment results in a worst-case performance of less than 2.8% cost error.

With constant shelflives, the analysis becomes quickly intractable. The only exact treatment is by Schmidt and Nahmias (1985) for Poisson demands and the constant shelf life denoted by τ. They derive the steady-state age distribution of the oldest unit in the system given by

$$p_{X_1}(x_1) = C_0 e^{\mu x_1}; \quad x_1 \geq \tau$$
$$= C_0; \quad x_1 < \tau, \tag{7.60}$$

where C_0 is the normalizing constant. The expected holding cost is again found by the PASTA property as for the nonpersihable case above. The steady-state perishing rate is given by $x_1 \rightarrow \tau p_{X_1}(x_1)$. Perry and Posner (1998) provide another variant with lead-time-dependent arrivals.

Under a continuous review, the earliest work is by Weiss (1980) with zero lead times. The results were generalized by Liu and Lian (1999) for renewal demands (see also Gürler and Ozkaya 2003). Gürler and Ozkaya (2008) extend their model to the case where the lifetime of a batch is random. With constant lead times, Berk and Gürler (2008) analyze a (Q, r) model in the absence of the optimal policy. The analysis assumes that $r < Q$, but their structural results can be extended to the case of $r > Q$, as well. An interesting model is in Tekin et al. (2001) where each batch in a stock is kept unperishable until it starts being used. The control policy is a modified (Q, r) policy that incorporates two reorder points—one on the basis of inventory levels and the other on the remaining age of the current batch: the (Q, r, T) policy. A thorough review and analysis on perishable inventories can be found in Nahmias (1982, 2011), Karaesmen et al. (2011), and Bakker et al. (2012).

7.3.4.10 Multiple Modes of Supply

Replenishment may be possible not only from a single supplier with a single-shipment mode but from different suppliers offering different lead times and/or utilizing different modes of shipment such as rail or airfreight. Therefore, inventory systems with multiple modes of supply have also received attention in the literature. The optimal replenishment policy with multiple delivery modes is unknown. Hence, models under simple control policies have been investigated. Moinzadeh and Nahmias (1988) are among the early works that consider dual modes of supply with constant delivery times and consider an approximate analysis of the system. Moinzadeh and Schmidt (1991) propose a modified one-for-one policy based on on-hand inventory levels for the problem and provide an exact analysis of the system. Mohebbi and Posner (1999) consider a lost-sales system. Expediting lead times can also be viewed as multiple modes of replenishment (e.g., Mamani and Moinzadeh 2014). For a review, see Minner (2003). (See also Chiang and Gutierrez 1996; Tagaras and Vlachos 2001 for periodic-review treatments of multiple supply modes.)

7.4 Joint-Replenishment Models for Multiple Items

When more than one item is carried in the stock, it may be optimal to replenish the items not independently but jointly. The benefits ensue from the economies of scale in fixed-ordering costs. Typically, fixed costs of ordering consist of two parts: A major ordering cost (K) that is incurred whenever an order is placed, and a minor ordering cost (a_i) that is incurred on an item basis for those included in the current order where $1 \leq i \leq m$ with m denoting the total number of items considered for joint replenishment. The joint-replenishment problem arises in both deterministic and stochastic demand settings. The joint-replenishment models with deterministic demands are an encountered distribution/transportation scheduling.

A commonly used policy is the cyclic joint-replenishment policy so that an order is placed at every T time units and to include the ith item into every kth replenishment; it is the $(T; k_1, \ldots, k_m)$ policy. If $k_{min} = min_i\{k_1, \ldots, k_m\} = 1$, then, the policy is called a strict cyclic policy. It is established that the optimal solution within the class of cyclic strategies does not necessarily belong to the class of strict cyclic strategies. Goyal (1974) proposed an algorithm to find the optimal solution within the cyclic class of strategies (see also Van Eijs 1993 on optimality). The infinite-horizon version with constant demands, with the objective of minimizing the average cost per time period, has been studied by Jackson et al. (1985). They develop a heuristic algorithm that is always within 6% of the optimal solution. The infinite-horizon case with constant demands, but in a more general production and distribution model, has been considered by Roundy (1985). In the finite-horizon case with nonconstant demands, Veinott (1969) and Zangwill (1966) propose DP algorithms. However, these proposed algorithms are applicable to small-scale problems. Hence, attention has been focused on developing heuristics. Among these, Silver (1976) is a very simple and efficient heuristic and Kaspi and Rosenblatt (1983) provide an improvement on this heuristic.

The earliest study on inventory control in the presence of random demands and joint-replenishment economies is due to Balintfy (1964) in the single-location multiple-product setting where the continuous-review can-order $(\mathbf{s}, \mathbf{c}, \mathbf{S})$ policy is introduced. Given the complexity of the can-order policy structure, several authors have developed approximations to this policy class (Silver 1965, 1974; Thompstone and Silver 1975; Silver 1981; Federgruen et al. 1984; Van Eijs 1994; Schultz and Johansen 1999; Melchiors 2002; Johansen and Melchiors 2003). Other continuous-review policies proposed for the problem are the following: The $(Q, \mathbf{S})$ policy, where all items are raised to their order-up-to levels when a total of Q demands arrive since the previous-order instance (Renberg and Planche 1967; Pantumsinchai 1992); the $Q(\mathbf{s}, \mathbf{S})$ policy, where the inventories are reviewed only when a total of Q demands have accumulated since the last-review instance and the items whose inventory positions at or below their reorder levels are raised to their order-up-to levels

(Nielsen and Larsen 2005); and, the $(Q, \mathbf{S}, T)$ policy, where the inventory positions of the items are raised to their order-up-to levels whenever a total of at least Q demands have accumulated or T time units have elapsed since the last-ordering instance (Özkaya et al. 2006; Larsen 2011).

Under a periodic review, Atkins and Iyogun (1988) introduce two base-stock periodic-review policies for unit Poisson demands. The first policy $(\mathbf{S}, T)$ imposes the same review-period length T for all items, and the inventory levels of all items are raised to their order-up-to levels defined by $\mathbf{S}$. The second policy, MP, is a modified periodic policy that utilizes item-specific review-period lengths based on a lower bound. Viswanathan (1997) recommends a new periodic-policy class, $P(\mathbf{s}, \mathbf{S})$, under which independent, periodic-review (s, S) policies are used for each item with a common review interval, T. Lately, Cachon (2001) proposed and studied another periodic-review policy, $(Q, \mathbf{S}|T)$, under which the system is reviewed every T time units and if a total of at least Q demands have accumulated since the last-order instance, then, the inventory positions of all items are raised to their order-up-to levels. It is reported that it is a commonly used policy by retailers.

On the deterministic-demand settings, Aksoy and Erenguc (1988) and Goyal and Satir (1989) provide comprehensive reviews. The next 15 years are discussed in Adelman and Klabjan (2005) and Khouja and Goyal (2008).

7.5 Multiechelon Inventory Systems

Multiechelon inventory systems represent the environments where units flow successively from an ample supplier through a number of locations to, finally, reside at a final location to satisfy demands that occur at that location. There are two basic structures in multiechelon inventory systems that are the building blocks for complex systems: *convergent* and *divergent* (or arborescent, tree-like) structures (Diks et al. 1996). A convergent system characterizes the assembly/production process of a finished good, and is typically seen in most upstream portions of a supply chain. For the analysis of convergent multiechelon systems, we refer to Federgruen and Axsäter (1993) and Van Houtum et al. (1996).

An arborescent (tree-like) system consists of N echelons where each echelon may consist of more than one location. A typical arborescent structure is depicted in Figure 7.5a. Echelon 0, which is assumed to be an ample supplier, supplies the locations $(1, i)$ $(i = 1, \ldots, n_1)$ on echelon 1, and each location $(i-1, j)$ on echelon $(i-1)$ supplies location j on echelon i $(1 < i < N)$, and $j = 1, \ldots, n_i$. Echelon N is the echelon closest to the customer. This type of a multiechelon structure is typically seen for sales/distribution networks finished products. Most work in the literature is restricted to two-echelon $(N = 2)$ systems. Herein, we concentrate on divergent multiechelon inventory systems.

A special case of a divergent structure of importance on its own is the serial system. A *serial* system consists of a chain of N inventory echelons that have a one-to-one supply relationship; that is, echelon $j-1$ supplies echelon j only. Echelon 1 is supplied by echelon 0, which is assumed to be an ample supplier, and echelon N is the echelon closest to the customer. A typical serial structure is depicted in Figure 7.5b.

Similar to single-location systems, multiechelon systems have been studied under periodic- and continuous-review policy classes. Under a continuous review, it is mostly assumed that the demand process is a (compound) Poisson process. Under a periodic review, demands can have arbitrary distributions. Herein, we focus on continuous-review control policies while referring to the seminal periodic-review models (see also Diks et al. 1996 for periodic-review models).

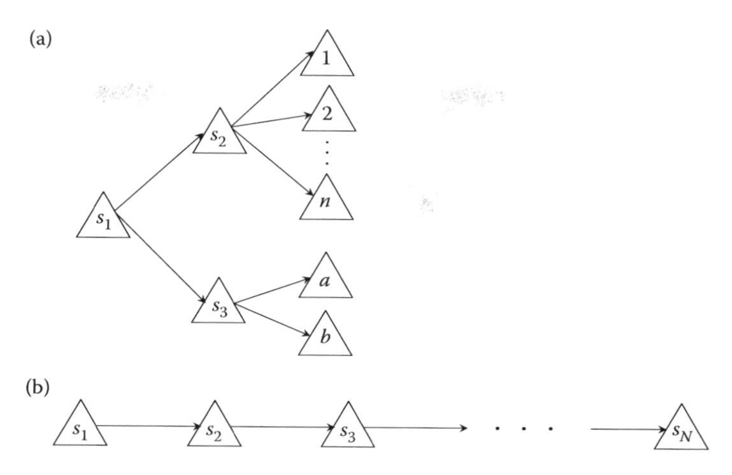

FIGURE 7.5
Illustration of a multiechelon inventory system.

The replenishment decisions at locations in multiechelon inventory systems typically use only the information on the stock at that location, which is the so-called *installation* stock. However, installation stock policies may not be optimal for multiechelon systems. The *echelon* stock policies based on the echelon inventory position have been shown to be optimal. The echelon inventory position (or stock) is defined as the sum of inventory positions (or stock) of the installation plus of all the downstream locations. A discussion of the differences and performance impact of the installation and echelon stock-based policies can be found in Axsäter and Rosling (1993) and Chen and Zheng (1994a,b).

Herein, we briefly discuss periodic-review echelon stock policies and focus on continuous-review installation stock policies that are, in essence, driven by and based on the modeling and optimization principles of stochastic single-location models.

7.5.1 Periodic-Review Echelon Stock Policies

The concept of echelon stock was introduced by Clark (1958) and employed by Clark and Scarf (1960) to establish the optimality of periodic-review echelon order-up-to policies in a serial system with no fixed costs, i.i.d. demands, and deterministic lead times in finite-horizon setting. The proof (developed for a two-echelon system) involves a decomposition of the multiechelon system into a series of single-location systems. It rests on defining a penalty function for the upper echelon if it is unable to immediately ship the order placed by the retailers. Federgruen and Zipkin (1984) extend the results to the stationary infinite-horizon setting. Chen and Zheng (1994a) provide an alternative proof that is also valid in continuous time. Rosling (1989) shows that a general-assembly system can be converted into an equivalent serial system. Chen and Zheng (1994b) consider positive fixed costs and propose an echelon stock (s, nQ) policy. Echelon stock policies require some kind of rationing rule to allocate the available stock to downstream installations. Eppen and Schrage (1981) introduced a fair-share rationing policy that maintains all lowest-echelon installations at a balanced position so that all such installations have the same stock-out probability. De Kok et al. (1994) provide the more general appropriate-share rationing policy ensuring that a prespecified target level is attained. Lagodimos (1992) considers a priority-rationing policy (see Diks et al. 1996 for a review).

7.5.2 Periodic-Review Installation Stock Policies

We briefly discuss periodic-review installation policies in the context of serial systems. Owing to its applicability in serial-assembly systems with *Kanban* practice, serial systems have also received some attention on their own. Gallego et al. (2007) develop simple approximate methods for the analysis and heuristics for control of the two-stage distribution system consisting of one warehouse and multiple retailers with stochastic demand. The analyses rest on decomposing the system into newsvendor-type subsystems. Bounds on the optimal policy and various reasonable and efficient heuristics are also provided. Shang and Song (2003) provide a simple decomposition heuristic based on the newsvendor model for a serial system operating under the desired service levels. Gallego and Özer (2005) propose a new computational algorithm and a different heuristic for the system. Shang and Song (2006) develop a closed-form approximation for the optimal base-stock levels. Shang (2008) incorporates fixed-ordering costs and proposed a simple heuristic for the serial system.

7.5.3 Continuous-Review Installation Stock Policies: One-for-One Replenishment

7.5.3.1 The Exact Analysis

The system consists of a single warehouse and N (≥ 1) retailers. The warehouse is supplied by an ample supplier. Installation i experiences a constant shipment lead time of L_i, $i = 0, 1, 2, \ldots, N$ where 0 denotes the warehouse. Demands at retailer i are generated according to a Poisson process with mean rate λ_i and the intensity at the warehouse is $\lambda_0 = \sum_{i=1}^{n} \lambda_i$. All locations employ $(S-1, S)$ policies with S_i denoting the order-up-to level at location i. We first provide an exact analysis, and, then discuss various approximate approaches.

For an inventory location i replenished under the $(S-1, S)$ policy, the outstanding orders at time t are given by

$$OO_i(t) = NBO_0^i(t - L_i) + D^i(t - L_i, t],$$

where $NBO_0^i(t - L_i)$ denotes the orders placed by retailer i already backordered at the warehouse at time $t - L_i$. If the upper-echelon location is an ample supplier, then, $OO_i(t)$ merely corresponds to the demands that occur within the immediately preceding time interval of lead-time length. If the upper supplier is subject to stock-outs, this is no longer the case. Hence, in a multiechelon setting, $OO_i(t)$ becomes dependent on the inventory situation at the upper location. By definition, $NI_i(t) = S_i - OO_i(t)$ for all i. Hence, the relevant operating characteristics can be obtained via $NI(t)$ (as discussed for single-location models) if the steady-state distribution of OO_i is available. The exact and approximate analyses differ in how one expresses this particular distribution.

Let $f_{OO}^i(\cdot)$ and $F_{OO}^i(\cdot)$ denote the probability functions of OO_i at a steady state. As the warehouse is supplied by an ample supplier with a deterministic transportation lead time, $f_{OO}^0(\cdot)$ is Poisson probability with mean $\lambda_0 L_0$. The expected number of backorders at the warehouse is given by

$$E[NBO_0] = \sum_{j=S_0+1}^{\infty} (j - S_0) f_{OO}^0(j).$$

Similarly, the expected number of the on-hand inventory is given by

$$E[OH_0] = \sum_{j=0}^{S_0} (S_0 - j) f_{OO}^0(j),$$

and the stock-out probability at the warehouse P^0_{out} is given by

$$P^0_{out} = \sum_{j=S_0+1}^{\infty} f^0_{OO}(j),$$

and the expected number of demands backordered is given by

$$E[BO]_0 = \lambda_0 P^0_{out}.$$

Next, consider the retailers. Suppose L_i's are identical. Following Graves (1985), let $OO(t)$ denote the sum of the outstanding orders for all retailers at time t. Then

$$OO(t) = \sum_{i=1}^{N} OO_i(t) = NBO_0(t - L_i) + D(t - L_i, t],$$

where $D(t - L_i, t]$ is the total (Poisson) demand that occurs at the retailers between $t - L_i$ and t and $NBO_0(t) = [-NI_0(t)]^+ = [OO_0(t) - S_0]^+$. The distribution of $OO(t)$ is obtained by convolving the backorders at the warehouse and the demand process at the particular retailer. The conditional distribution of $OO_i(t)$ is binomial due to the *FIFO* issuance policy and superimposed Poisson demand streams for a given $OO(t)$.

$$Pr[OO_i(t) = j] = \sum_{k=j}^{\infty} Pr[OO(t) = k]\binom{k}{j}\left(\frac{\lambda_i}{\lambda_0}\right)^j\left(\frac{\lambda_i - \lambda_0}{\lambda_0}\right)^{k-j}.$$

Letting $t \to \infty$, we get the stationary results. Then, using the steady-state distribution $f^i_{OO}(\cdot)$, we obtain the relevant measures for the retailers similar to the warehouse, and write the total expected cost rate for the system as follows:

$$E[TC] = h_0 E[OH_0] + \sum_{i=0}^{N} \{h_i E[OH_i] + \pi_i E[NBO_i] + \hat{\pi} E[BO_i]\}.$$

The first derivation of exact probabilities was by Simon (1971), which was extended to more than two echelons by Kruse (1979) and to compound Poisson demands by Shanker (1981). Svoronos and Zipkin (1991) considered the exogeneous random lead times that preserve the replenishment order.

Although the derivation of the exact probability distributions discussed above is straightforward, the convolutions render the computational effort to be extensive. The following observation has led to different approaches to compute the system characteristics: For a single-location system, $NI = S - DL$ where DL denotes the lead-time demands, whereas, for all i in a multiechelon system, $NI_i = S_i - OO_i$. Then, the multiechelon system may be viewed *as if* it is a collection of single-location models such that each location has an *effective* stochastic-replenishment lead time during which the demands follow the distribution of OO_i. The demands that a retailer faces are still generated by a Poisson process with rate λ_i but the effective lead time $\tilde{L}_i$ is the sum of transportation times (L_i) and (possible) waits (or, delays, retards) at the upper echelon, W_{i-1} which are random.

Axsäter (1990a) bases his exact derivation approach on this fact. For illustration, consider a two-echelon system. A unit ordered by retailer i at time t is used to satisfy the S_i^{th} consecutive demand at this location, which arrives at time $t + Z_i$, and is received by the retailer at time $t + L_i + W_0$.

If $t + Z_i < t + L_i + W_0$, then, the arriving demand finds the retailer out of stock and is backordered for $W_0 - Z_i$ time units; otherwise, the demand is immediately satisfied and the unit has remained in the stock at the retailer for $Z - W_0$ time units. The *conditional* expected cost incurred by a unit for a given $W_0 = w_0$ is denoted by $E[C_i](S_i|w_0)$ and is given by

$$E[C_i](S_i|w_0) = h_i \int_{z=0}^{L_i+w_0} [L_i + w_0 - z] f_{Z_i}(z)dz + \pi_i \int_{z=L_i+w_0}^{\infty} [z - L_i - w_0] f_{Z_i}(z)dz, \quad (7.61)$$

for $S_i > 0$ and,

$$E[C_i](0|w_0) = \pi_i [L_i + w_0], \quad (7.62)$$

where Z_i is $Erlang(\lambda_i, S_i)$ distributed.

The item that is issued from the warehouse to satisfy the order placed at time t corresponds to the unit that was ordered by the warehouse before the last S_0 orders were placed at the warehouse at time $t - Z_0$ where Z_0 is $Erlang(\lambda_0, S_0)$ distributed. The retard at the warehouse is $W_0 = [L_0 - Z_0]^+$. Therefore, the expected cost incurred at retailer i, $E[TC_i(S_i, S_0)]$ is given by

$$E[TC_i(S_i, S_0)] = \int_{z=0}^{L_0} E[C_i](S_i|L_0 - z) f_{Z_0}(z)dz + E[C_i](S_i|0)(1 - F_{Z_0}(L_0)),$$

for $S_i, S_0 > 0$, and

$$E[TC_i(S_i, 0)] = E[C_i](S_i|L_0),$$

for $S_i > 0$.

Given the Poisson demand processes, Axsäter (1990a) provides the recursive relationship

$$\begin{aligned} E[TC_i(S_i, S_0 - 1)] = {} & \frac{\lambda_i}{\lambda_0} E[TC_i(S_i - 1, S_0)] + \frac{\lambda_0 - \lambda_i}{\lambda_i} E[TC_i(S_i, S_0)] \\ & + \frac{\lambda_i}{\lambda_0}(1 - F_{Z_0}(L_0))(E[C_i](S_i|0) - E[C_i](S_i - 1|0)), \end{aligned} \quad (7.63)$$

for $S_i, S_0 > 0$ and, for $S_0 > 0$,

$$E[TC_i(0, S_0)] = F_{Z_0}(L_0)\pi_i L_0 - F'_{Z_0}(L_0)\frac{\pi_i S_0}{\lambda_0} + \pi_i L_i,$$

where $F'_{Z_0}(\cdot)$ stands for the cdf of $Erlang(\lambda_0, S_0 + 1)$. Also, for large S_0 values, $E[TC_i(S_i, S_0)] \approx E[C_i](S_i|0)$. Hence, starting with a large warehouse target stock level, the expected cost at retailer i can be computed without difficulty. Finally, similar to the retailer costs, the expected cost at the warehouse (consisting solely of holding costs) is given by

$$E[C_0](S_0) = h_0 \int_{z=L_0}^{\infty} [z - L_0] f_{Z_0}(z)dz.$$

Thus, all the relevant measures of the inventory system can be easily computed. This approach is amenable to consider other cost structures and extensible to the case with $N > 2$. For an exposition of multiechelon inventory-modeling techniques, see Sherbrooke (2004).

There are two simple approximations that have received attention in the literature due to their historic import and modeling technique: The *METRIC* and *two-parameter* approximations, both of which provide approximations to the distribution of outstanding orders at retailer i, which is OO_i.

7.5.3.2 The METRIC Approach

For a single-location inventory system with Poisson demands with rate μ, constant lead time L, and an ample supplier, the number of outstanding orders at any time corresponds to the number of customers being served (i.e., the occupancy level) in an $M/D/\infty$ queuing system, where the service time is L and the arrival rate is μ. Palm's theorem (Palm 1938) dictates that the steady-state occupancy level in the $M/G/\infty$ system is independent of the shape of the service-time distribution and that it is Poisson distributed with mean μL. Therefore, if replenishment lead times are random but orders are allowed to cross, the single-location results for Markovian arrivals and constant-replenishment lead times can be directly extended to stochastic lead times working with their mean values. In a multiechelon inventory system, however, the lead times that lower echelons experience are *not* independent because they depend on the inventory levels at the upper echelons and the demand-process history.

The *METRIC* approximation rests on *assuming* that successive lead times experienced by retailers are not correlated and applying Palm's theorem to replace the true stochastic lead time with a constant one equal to its mean $\overline{L_i}$. That is, $f^i_{OO}(\cdot)$ is approximated by a Poisson distribution with rate $\lambda_i \overline{L_i}$.

As the warehouse is supplied by an ample supplier with a deterministic-transportation lead time L_0, $f^0_{OO}(\cdot)$ is truly Poisson probability with mean $\lambda_0 L_0$. Therefore, the expressions obtained above for the warehouse already hold in the *METRIC* approach.

From Little's Law, the average waiting time due to stock-outs at the warehouse is

$$E[W_0] = E[NBO_0]/\lambda_0.$$

The *average replenishment lead time* that retailer i experiences is

$$\overline{L_i} = L_i + E[W_0].$$

Each retailer takes on a single-location inventory system as we did for the warehouse. Therefore, for $i = 1, 2, \ldots, N$,

$$E[NBO_i] = \sum_{j=S_0+1}^{\infty} (j - S_i) Po(j; \lambda_i \overline{L_i}).$$

From Little's Law, the average waiting time due to stock-outs at the warehouse is

$$E[W_i] = E[NBO_i]/\lambda_i.$$

Similarly, the expected number of the on-hand inventory is given by

$$E[OH_i] = \sum_{j=0}^{S_i} (S_i - j) Po(j; \lambda_0 \overline{L_i}),$$

and the stock-out probability at the warehouse P^i_{out} is given by

$$P^i_{out} = \sum_{j=S_0+1}^{\infty} (j - S_i) Po(j; \lambda_i \overline{L_i}),$$

$$E[BO]_i = \lambda_i P^i_{out}.$$

The expected total cost for the system is

$$E[TC] = h_0 E[OH_0] + \sum_{i=0}^{N} h_i E[OH_i] + \sum_{i=0}^{N} \pi_i E[NBO_i] + \sum_{i=0}^{N} [\hat{\pi}\lambda_i] P^i_{out}. \quad (7.64)$$

An extension to more than two levels is straightforward: Obtain the exact operating characteristics for the uppermost echelon (warehouse), and, using the mean values of the waits at the lower echelons, approximate the outstanding-order distributions with Poisson distributions. As Palm's theorem holds for a compound Poisson demand and batch service, the approximation can be extended to such inventory systems, as well. Muckstadt (1973) extended the approximation to a hierarchical or indentured-parts structure (called the *MODMETRIC* model) to treat groups of items. A similar approach has been used by Andersson and Melchiors (2001) to approximate the two-echelon system with Poisson demands and lost sales.

Note that the holding and shortage costs at retailer i are independent of the inventory positions of the other retailers. The costs at the warehouse depend only on S_0, and the costs at retailer i depend on its target stock level S_i as well as S_0. Thus, for a given S_0, the cost optimization for each retailer can be performed separately with respect to S_i's.

The expected cost rate at each retailer has the same properties as the newsvendor problem; and, hence, it is convex in S_i. Let $S^*_i(S_0)$ denote the optimal target stock level for retailer i for a given S_0. $E[TC]$ is not necessarily convex in S_0. However, this does not pose a great difficulty as tight upper and lower bounds can be found and the optimization is carried out within this band.

It is reported that the *METRIC* approximation will, in general, work well as long as the demand at each retailer is low relative to the total demand (i.e., for a small λ_i/λ_0), which would be the case with many small retailers served by a common warehouse. The errors arising from the approximation may result in an incorrect allocation of the stock if it is used for optimization. Graves (1985) reported the wrong stockage decisions in around 11.5% of the considered cases.

7.5.3.3 The Two-Moment Approximation

The *METRIC* approximation represents the number of outstanding orders at retailer i as a Poisson process with mean $\lambda_i \overline{L_i}$, which is a single-parameter characterization. It is expected that a characterization that captures higher moments would provide a better approximation. Graves (1985) proposed that the distribution of the outstanding orders would be approximated by a negative-binomial distribution (which is characterized by the parameters p and r) that has the same mean and variance of the true distribution: $E[OO_i] = (\lambda_i \overline{L_i})$ and $Var(OO_i)$ where

$$Var(OO_i) = (\lambda_i/\lambda_0)^2 Var(NBO_0) + \left(\frac{\lambda_i}{\lambda_0}\right)\left(\frac{\lambda_0 - \lambda_i}{\lambda_0}\right) E[NBO_0] + \lambda_i L_i.$$

The average backorder level at the warehouse $E[NBO_0]$ is obtained exactly as before according to the equation below and $Var(NBO_0)$ is (through a recurve relation in terms of $Var(NBO_0)$ for S_0 and $Var(NBO_0)$ for $S_0 - 1$) given by

$$Var(NBO_0) = \sum_{S_0+1} \infty (j - S_0)^2 Po(j; \lambda_0 L_0) - (E[NBO_i])^2.$$

Then, $f^i_{OO}(\cdot) \approx \hat{f}^i_{OO}(\cdot)$ where

$$\hat{f}^i_{OO}(j) = \binom{r + j - 1}{j} p^r (1 - p)^j, \quad j = 0, 1, 2, \ldots,$$

so that $E[OO_i] = r(1 - p)/p$ and $Var(OO_i) = r(1 - p)/P^2$. The relevant entities for the retailers are found as before with the approximate distribution.

The numerical comparison reported by Graves (1985) indicates that the two-parameter approximation overestimates backorders and thereby the waiting times while the *METRIC* approximation underestimates these measures. In terms of finding the optimal inventory levels, the two-parameter approximation makes considerably fewer errors. The negative-binomial approximation has been independently studied by Slay (1984), as well. Sherbrooke (1986) extended the approach to the multi-indenture setting (i.e., the *VARI-METRIC* model). Using a similar approach, Lee (1987) considered *lateral transshipments* whereby groups of lower-echelon locations are allowed to pool and share their stocks in cases of stock-outs (see also Axsäter 1990b; Aggarwal and Moinzadeh 1994; Moinzadeh and Aggarwal 1997).

7.5.4 Continuous-Review Installation Stock Policies: Batch-Ordering Policies

Multiechelon inventory systems under batch-ordering policies pose a great analytical difficulty. To see this, suppose that N retailers face Poisson demands with rate λ_i and employ the (Q_i, r_i) policies independently. The orders that a retailer places at the warehouse constitute an $Erlang(\lambda_i, Q_i)$ renewal process. As retailers act independently, the overall process that the warehouse faces is a superposition of N-such Erlang processes instead of the benign Poisson process obtained above under one-for-one policies. Two special cases are relatively tractable and are of a theoretical interest: (i) The retailers employ $(S_i - 1, S_i)$ policies and only the warehouse batches its orders via the (Q, r) policy. For this system, the operating characteristics and the corresponding retard at the warehouse can be obtained by using the results for the single-location model with Poisson demand with rate λ_0 and the methodology employed above. (ii) The one-warehouse-one-retailer setting (i.e., a serial system) has received considerable attention especially in supply chain management for the contract design and game-theoretic studies. We discuss this model below.

Consider the one-warehouse-one-retailer setting. Both echelons employ the lotsize-reorder point policy (Q_0, r_0) by the warehouse and (Q_1, r_1) by the retailer. It is assumed that r_0 and Q_0 are integer multiples of the retailer's lotsize and that all deliveries are in full batches. Consider the batch ordered by the warehouse. It consists of Q_0 subbatches of size Q_1. Similar to the argumentation used for the waiting times in the single location (s, S) by Kruse (1981), the jth $(j = 1, 2, \ldots, Q_0)$ subbatch in this order will be released from the warehouse when $(r_0 + 1)Q_0$ demands have occurred. These dynamics are similar to an item ordered by the warehouse and are used to satisfy the S_0^{th} consecutive demand at the warehouse and, similarly, an item ordered by the retailer is used to satisfy the S_1^{th} consecutive demand at the retailer. Thus, each subbatch may be viewed as operating under a unique one-for-one policy. That is, batch-ordering policies in a one-warehouse-one-retailer system can be analyzed by one-for-one policies. Following Axsäter (1993a,b), the holding and backordering costs per unit time can be computed by taking the average over the entire batch as

$$C = (1/(Q_0 Q_1)) \sum_{j=r_0+1}^{r_0+Q_0} \sum_{r_1+1}^{r_1+Q_1} C(jQ_1, k), \tag{7.65}$$

where $C(S_0, S_1)$ is the expected cost rate for a two-echelon system under one-for-one policies with the warehouse target level S_0 and S_1. The average number of orders per unit time at the warehouse is given by $\lambda_0/(Q_0 Q_1)$ and at the single retailer, it is given by λ_1/Q_1 (see also Gallego and Zipkin 1999). For multiechelon repairable-item inventories, see Moinzadeh and Lee (1986). For a novel approach on unit decomposition, see Muharremoglu and Tsitsiklis (2008).

For arborescent systems, the case of identical retailers with $Q_i = Q = 1$ is treated in an exact analogy with Equation 7.65. For $Q_i = Q > 1$, let p_{ij} denote the probability that the jth demand will trigger the ith retailer order. Then, in a similar fashion,

$$C = (1/(Q_0 Q_1)) \sum_{j=r_0+1}^{r_0+Q_0} \sum_{r_1+1}^{r_1+Q_1} \sum_{i=j}^{\infty} p_{ij} C(i, k). \tag{7.66}$$

The computational effort needed to compute p_{ij} is, however, prohibitively high requiring approximations. Noting that exactly every Qth demand generates a retailer order in a serial system, one simple approximation is $p_{ij} = 1$ if $i = jQ$ and zero otherwise. Another is to assume that each demand triggers a retailer order with probability $1/Q$ resulting in a negative-binomial distribution, which is reasonable when each retailer's demand process contributes little to the overall demand process. Axsäter (1993a) suggests a mixture of the two with weights $1/N$ and $(N - 1/N)$, respectively. Other approximations include Deuermeyer and Schwarz (1981), Schwarz et al. (1985), Lee and Moinzadeh (1987a,b), Axsäter (1997), and Svoronos and Zipkin (1988). Of these, the last two have been reported to be very accurate (Axsäter 1993b). See also Axsäter (1998).

7.5.5 Multiechelon Inventory-Practice Issues

MRP and ERP Systems: In all the models considered so far, demands are independent. For items facing dependent demands such as semifinished goods used in manufacturing, material requirements planning (MRP) systems (which are production-inventory scheduling softwares that make use of computerized files and data-processing equipment) have received widespread application since the 1970s (Orlicky 1975). They recognize the implications of dependent demands in multiechelon manufacturing (which includes lumpy production requirements). Integrating the bills of materials, the given production requirements of end products, and the inventory records file, MRP systems generate a complete list of a production-inventory schedule for parts, subassemblies, and end products, taking into account the lead-time requirements. MRP has proved to be a useful tool for manufacturers, especially in assembly operations. Over time, MRP has evolved into and been absorbed in enterprise resources planning (ERP) systems that consider the entirety of the firm's activities.

Just-in-time: While MRP systems were being developed in the United States, some Japanese manufacturers achieved a widely acclaimed success with a different system. By producing components "just in time" (JIT) to be used in the next step of the production process, and by extending this concept throughout the production line so that even the finished goods are delivered JIT to be sold, they obtained substantial reductions in inventories. One of the key factors for establishing JIT is altering the manufacturing process to drastically reduce the setup times and simplifying the ordering and procurement process so that ordering costs are cut down. The idea is to enable the producer to operate with small lot sizes, which get produced when the need arises (and not before). The practice in its origin is known as the *Toyota production system* (Monden 2011). For a review of the current philosophy and practice, see Hutchins (1999). Once JIT is established, an information system is used to determine the timing and quantities of production. Card signals—that is, visible records (in Japanese, *Kanban*)—are used to specify withdrawals from the preceding production stages, and to order for production, the number, and type of items required. Because small batches of production have become economical, the production orders can be filled JIT. Advocates of Kanban characterize it as a pull process and criticize MRP as a push system. Though Kanban is a simple idea and yields an adaptive–flexible production system, its appropriateness hinges on whether setup and ordering costs have been drastically reduced so as to allow for small production batches. Lean-manufacturing systems and supply chains may and have suffered severely from disruptions due to climate, labor disputes, and terrorism.

Bullwhip effect: In multiechelon systems, the variability of demands (orders) increases as one moves away from the retailer echelons. This is called the *bullwhip effect* implying that a small perturbation progresses very fast and grows in size as the physical wave cracks at the tip of the whip. The main reason for this well-established phenomenon are the batching policies employed, speculative purchasing due to price promotions, and observed unavailability of stocks and forecasting techniques. The information distortion can be eliminated through more transparency and collaborative-forecasting replenishment activities along the supply chain (Lee et al. 2004; Cachon et al. 2007). The information on where the outstanding orders are in a supply chain may also be used for constructing better control policies; see Aggarwal and Moinzadeh (1994), Moinzadeh and Aggarwal (1997), Moinzadeh (2002), and Jain and Moinzadeh (2005).

Vendor-managed inventory: Another approach to reduce the bullwhip effect is the vendor-managed inventory (VMI) practice (Waller et al. 1999). The name refers to the control of lower-echelon stocks by the upper echelons. In a two-echelon system, the warehouse (or producer) will act as the central decision maker to replenish the stocks of the retailers, as well. Following the seminal work by Cetinkaya and Lee (2000), a large literature has emerged on the practice and theory of such systems. Disney and Towill (2003) provide a recent analysis of VMI practice on serial systems.

Game-theoretic concerns: The models herein deal with centralized systems with a single decision maker. In the last two decades, multiple actors in multiechelon settings have received a lot of attention resulting in a vast body of knowledge in supply chain management. A key characteristic of these models is the game-theory applications within the inventory management context. The single-location models presented herein constitute the so-called response functions of individual actors. That is, the decision variables of the other actors are taken as the given parameters and optimization is carried out accordingly. Then, all actors are assumed to behave optimally simultaneously, resulting in an equilibrium (Nash-equilibrium) solution. Although the existence of equilibrium may not be guaranteed in general, under mild conditions that preserve cost convexity, the models herein would result in such equilibrium solutions. The strength of this framework rests in that, if a Nash equilibrium exits, then, a decentralized system where actors behave independently would achieve the optimal solution for a centralized system with a single decision maker. Game-theoretic models have significantly contributed to our understanding of competition in multiechelon systems and the designing of supply contracts. The seminal works in this area include Cachon (1999), Cachon and Zipkin (1999), and Cachon and Netessine (2004).

References

Adelman, D., and D. Klabjan. Duality and existence of optimal policies in generalized joint replenishment. *Mathematics of Operations Research* 3, 2005: 28–50.

Aggarwal, P. K., and K. Moinzadeh. Order expedition in multi-echelon production/distribution systems. *IIE Transactions* 26.2, 1994: 86–96.

Agrawal, N., and S. A. Smith. Estimating negative binomial demand for retail inventory management with lost sales. *Naval Research Logistics* 43, 1996: 839–861.

Ahmed, S., U. Cakmak, and A. Shapiro. Coherent risk measures in inventory problems. *European Journal of Operational Research* 182.1, 2007: 226–238.

Aksen, D., K. Altinkemer, and S. Chand. The single-item lot-sizing problem with immediate lost sales. *European Journal of Operational Research* 147.3, 2003: 558–566.

Aksoy, Y., and S. Erenguc. Multi-item models with coordinated replenishments: A survey. *International Journal Operations and Production Management* 8(1), 1988: 63–73.

Alfares, H. K., and H. H. Elmorra. The distribution-free newsboy problem: Extensions to the shortage penalty case. *International Journal of Production Economics* 93, 2005: 465–477.

Allahverdi, A. et al. A survey of scheduling problems with setup times or costs. *European Journal of Operational Research* 187.3, 2008: 985–1032.

Andersson, J., and P. Melchiors. A two-echelon inventory model with lost sales. *International Journal of Production Economics* 69.3, 2001: 307–315.

Archibald, B. C., and E. A. Silver. (s, S) policies under continuous review and discrete compound Poisson demand. *Management Science* 24.9, 1978: 899–909.

Arrow, K. J., T. Harris, and J. Marschak. Optimal inventory policy. *Econometrica*, 19, 1951: 250–272.

Atkins, D. R., and P. O. Iyogun. Periodic versus can-order policies for coordinated multi-item inventory systems. *Management Science* 34, 1988: 791–796.

Axsäter, S. Simple solution procedures for a class of two-echelon inventory problems. *Operations Research* 38.1, 1990a: 64–69.

Axsäter, S. Modelling emergency lateral transshipments in inventory systems. *Management Science* 36.11, 1990b: 1329–1338.

Axsäter, S. Exact and approximate evaluation of batch-ordering policies for two-level inventory systems. *Operations Research* 41.4, 1993a: 777–785.

Axsäter, S. Continuous review policies for multi-level inventory systems with stochastic demand. *Handbooks in Operations Research and Management Science* 4, 1993b: 175–197.

Axsäter, S. Simple evaluation of echelon stock (R, Q) policies for two-level inventory systems. *IIE Transactions* 29.8, 1997: 661–669.

Axsäter, S. Evaluation of installation stock based (R, Q)-policies for two-level inventory systems with Poisson demand. *Operations Research* 46.3(supplement-3), 1998: S135–S145.

Axsäter, S., and K. Rosling. Notes: Installation vs. echelon stock policies for multilevel inventory control. *Management Science* 39.10, 1993: 1274–1280.

Azoury, K. S. Bayes solution to dynamic inventory models under unknown demand distribution. *Management Science* 31, 1985: 1150–1160.

Bakker, M., J. Riezebos, and R. H. Teunter. Review of inventory systems with deterioration since 2001. *European Journal of Operational Research* 221.2, 2012: 275–284.

Balcioglu, B., and Ü. Gürler. On the use of phase-type distributions for inventory management with supply disruptions. *Applied Stochastic Models in Business and Industry* 27.6, 2011: 660–675.

Balintfy, J. L. On a basic class on inventory problems. *Management Science* 10, 1964: 287–297.

Bashyam, S., and M. C. Fu. Optimization of (s, S) inventory systems with random lead times and a service level constraint. *Management Science* 44.12—Part-2, 1998: S243–S256.

Bellman, R., I. Glicksberg, and O. Gross. On the optimal inventory equation. *Management Science* 2.1, 1955: 83–104.

Berk, E., and A. Arreola-Risa. Note on future supply uncertainty in EOQ models. *Naval Research Logistics* 41.1, 1994: 129–132.

Berk, E., and Ü. Gürler. Newsboy inventory problem. *Wiley StatsRef: Statistics Reference Online*. 2015: 1–10.

Berk, E., and Ü. Gürler. Analysis of the (Q, r) inventory model for perishables with positive lead times and lost sales. *Operations Research* 56.5, 2008: 1238–1246.

Berk, E., Ü. Gürler, and R. A. Levine. Bayesian demand updating in the lost sales newsvendor problem: A two-moment approximation. *European Journal of Operational Research* 182.1, 2007: 256–281.

Berk, E., A. O. Toy, and O. Hazir. Single item lot-sizing problem for a warm/cold process with immediate lost sales. *European Journal of Operational Research* 187.3, 2008: 1251–1267.

Bijvank, M., and I. F. A. Vis. Lost-sales inventory theory: A review. *European Journal Operational Research* 215, 2011: 1–13.

Bouakiz, M., and M. J. Sobel. Inventory control with an exponential utility criterion. *Operations Research* 40.3, 1992: 603–608.

Brahimi, N., S. Dauzere-Peres, N. M. Najid, and A. Nordli. Single item lot sizing problems. *European Journal of Operational Research* 168(1), 2006: 1–16.

Browne, S., and P. Zipkin. Inventory models with continuous, stochastic demands. *The Annals of Applied Probability* 1.3, 1991: 419–435.

Cachon, G. P. Competitive supply chain inventory management. In S. Tayur, R. Ganeshan, and M. Magazine (eds.), *Quantitative Models for Supply Chain Management*. Springer, USA, 1999: 111–146.

Cachon, G. P. Managing a retailer's shelf space, inventory and transportation. *Manufacturing Service Operations Management* 3, 2001: 211–229.

Cachon, G. P., and S. Netessine. Game theory in supply chain analysis. In D. Simchi-Levi, S. D. Wu, and Z. J. Shen (Max) (eds.), *Handbook of Quantitative Supply Chain Analysis: Modeling in the eBusiness Era*. Kluwer, Springer, USA, 2004: 13–65.

Cachon, G. P., T. Randall, and G. M. Schmidt. In search of the bullwhip effect. *Manufacturing and Service Operations Management* 9.4, 2007: 457–479.

Cachon, G. P., and P. H. Zipkin. Competitive and cooperative inventory policies in a two-stage supply chain. *Management Science* 45.7, 1999: 936–953.

Cetinkaya, S., and C.-Y. Lee. Stock replenishment and shipment scheduling for vendor-managed inventory systems. *Management Science* 46.2, 2000: 217–232.

Cetinkaya, S., and M. Parlar. Optimal myopic policy for a stochastic inventory problem with fixed and proportional backorder costs. *European Journal of Operational Research* 110.1, 1998: 20–41.

Cetinkaya, S., and M. Parlar. Computing a stationary base-stock policy for a finite horizon stochastic inventory problem with non-linear shortage costs. *Stochastic Analysis and Applications* 22.3, 2004: 589–625.

Chen, R., and Y. Zheng. Lower bounds for multi-echelon stochastic inventory systems. *Management Science* 40, 1994a: 1426–1443.

Chen, F., and Y.-S. Zheng. Evaluating echelon stock (R, nQ) policies in serial production/inventory systems with stochastic demand. *Management Science* 40.10, 1994b: 1262–1275.

Chen, X. et al. Risk aversion in inventory management. *Operations Research* 55.5, 2007: 828–842.

Chiang, C., and G. J. Gutierrez. A periodic review inventory system with two supply modes. *European Journal of Operational Research* 94.3, 1996: 527–547.

Clark, A. A Dynamic, Single-Item, Multi-Echelon Inventory Model, RM-2297, Santa Monica, California, The RAND Corporation. December, 1958.

Clark, A. J., and H. Scarf. Optimal policies for a multi-echelon inventory problem. *Management Science* 6.4, 1960: 475–490.

Covert, R. P., and G.C. Phillip. An EOQ model for items with Weibull distribution deterioration. *AIIE Transactions* 5.4, 1973: 323–326.

Cox, D. R. The analysis of non-Markovian stochastic processes by the inclusion of supplementary variables. *Mathematical Proceedings of the Cambridge Philosophical Society*. 51(03). Cambridge University Press, 1955: 433–441.

De Kok, A. G., A. G. Lagodimos, and H. P. Seidel. Stock Allocation in a Two-Echelon Distribution Network under Service Constraints. TUE/BDK/LBS 94-03, Eindhoven University of Technology, The Netherlands, 1994.

DeMatteis, J. J. An economic lot-sizing technique, I: The part-period algorithm. *IBM Systems Journal* 7.1, 1968: 30–38.

Deuermeyer, B. L., and L. B. Schwarz. A model for the analysis of system service level in warehouse–retailer distribution systems: The identical retailer case. In L. B. Schwarz (ed.) *Multi-Level Production Inventory Control System, Studies in Management Science*. Vol. 16. North-Holland, Amsterdam, 1981: 163–193.

Diks, E. B., A. G. De Kok, and A. G. Lagodimos. Multi-echelon systems: A service measure perspective. *European Journal of Operational Research* 95.2, 1996: 241–263.

Ding, X., M. L. Puterman, and A. Bisi. The censored newsvendor and the optimal acquisition of information. *Operations Research* 50.3, 2002: 517–527.

Disney, S. M., and D. R. Towill. The effect of vendor managed inventory (VMI) dynamics on the bullwhip effect in supply chains. *International Journal of Production Economics* 85.2, 2003: 199–215.

Drexl, A., and A. Kimms. Lot sizing and scheduling survey and extensions. *European Journal of Operational Research* 99.2, 1997: 221–235.

Elmaghraby, S. E. The economic lot scheduling problem (ELSP): Review and extensions. *Management Science* 24.6, 1978: 587–598.

Eppen, G., and L. Schrage. Centralized ordering policies in a multi-warehouse system with lead times and random demand. *Multi-Level Production/Inventory Control Systems: Theory and Practice* 16, 1981: 51–67.

Eppen, G. D., and A. V. Iyer. Improved fashion buying with Bayesian updates. *Operations Research* 45, 1997: 805–819.

Erlenkotter, D. Note an early classic misplaced: Ford W. Harris's economic order quantity model of 1915. *Management Science* 35.7, 1989: 898–900.

Erlenkotter, D. Ford Whitman Harris and the economic order quantity model. *Operations Research* 38.6, 1990: 937–946.

Federgruen, A., and S. Axsäter. Logistics of production and inventory. In S. C. Graves, A. G. H. Rinnooy Kan, and P. H. Zipkin (eds.), *Handbooks in Operations Research and Management Science 4*. Elsevier (North-Holland), Amsterdam, 1993: 757.

Federgruen, A., H. Greoenevelt, and H. C. Tijms. Coordinated replenishments in a multi-item inventory system with compound Poisson demands and constant lead times. *Management Science* 30, 1984: 344–357.

Federgruen, A., and M. Tzur. A simple forward algorithm to solve general dynamic lot sizing models with n periods in 0 (n log n) or 0 (n) time. *Management Science* 37.8, 1991: 909–925.

Federgruen, A., and Y.-S. Zheng. An efficient algorithm for computing an optimal (r, Q) policy in continuous review stochastic inventory systems. *Operations Research* 40.4, 1992: 808–813.

Federgruen, A., and P. Zipkin. Computational issues in an infinite horizon, multi-echelon inventory model. *Operations Research* 32, 1984: 818–836.

Flores, B. E., and D. C. Whybark. Multiple criteria ABC analysis. *International Journal of Operations and Production Management* 6.3, 1986: 38–46.

Flores, B. E., and D. C. Whybark. Implementing multiple criteria ABC analysis. *Journal of Operations Management* 7.1, 1987: 79–85.

Florian, M., and M. Klein. Deterministic production planning with concave costs and capacity constraints. *Management Science* 18.1, 1971: 12–20.

Fries, B. E. Optimal ordering policy for a perishable commodity with fixed lifetime. *Operations Research* 23.1, 1975: 46–61.

Gallego, G. A minmax distribution free procedure for the (Q, r) inventory model. *Operations Research Letters* 11, 1992: 55–60.

Gallego, G., and I. Moon. The distribution free newsboy problem: Review and extensions. *Journal of the Operational Research Society* 44.8, 1993: 825–834.

Gallego, G., and O. Özer. A new algorithm and a new heuristic for serial supply systems. *Operations Research Letters* 33.4, 2005: 349–362.

Gallego, G., O. Özer, and P. Zipkin. Bounds, heuristics, and approximations for distribution systems. *Operations Research* 55.3, 2007: 503–517.

Gallego, G., and P. Zipkin. Stock positioning and performance estimation in serial production–transportation systems. *Manufacturing and Service Operations Management* 1, 1999: 77–87.

Ghare, P. M., and G. F. Schrader. A model for exponentially decaying inventory. *Journal of Industrial Engineering* 14.5, 1963: 238–243.

Gnedenko, B. V., and I. N. Kovalenko. *Introduction to Queueing Theory*. Birkhauser Boston Inc., 1989.

Gotoh, J., and Y. Takano. Newsvendor solutions via conditional value-at-risk minimization. *European Journal of Operational Research* 179.1, 2007: 80–96.

Goyal, S. K. Determination of optimum packaging frequency of items jointly replenished. *Management Science* 21, 1974: 436–443.

Goyal, S. K., and B. C. Giri. Recent trends in modeling of deteriorating inventory. *European Journal of Operational Research* 134.1, 2001: 1–16.

Goyal, S. H., and A. T. Satir. Joint replenishment inventory control: Deterministic and stochastic models. *European Journal of Operational Research* 38, 1989: 2–13.

Graves, S. C. A multi-echelon inventory model for a repairable item with one-for-one replenishment. *Management Science* 31.10, 1985: 1247–1256.

Graves, S. C. A single-item inventory model for a nonstationary demand process. *Manufacturing and Service Operations Management* 1.1, 1999: 50–61.

Groff, G. K. A lot sizing rule for time-phased component demand. *Production and Inventory Management* 20.1, 1979: 47–53.

Gürler, Ü., and B. Y. Ozkaya. A note on continuous review perishable inventory systems: Models and heuristics. *IIE Transactions* 35, 2003: 321–323.

Gürler, Ü., and B. Y. Ozkaya. Analysis of the (s, S) policy for perishables with a random shelf life. *IIE Transactions* 40.8, 2008: 759–781.

Gürler, Ü., and M. Parlar. An inventory problem with two randomly available suppliers. *Operations Research* 45.6, 1997: 904–918.

Hadley, G., and T. M. Whitin. *Analysis of Inventory Systems*. Prentice-Hall, Englewood Cliffs, NJ, 1963.

Harris, F. W. How many parts to make at once. *Factory, the Magazine of Management* 10(2), 1913: 135–136, 152.

Harris, F. W. How many parts to make at once. *Operations Research* 38.6, 1990: 947–950.

Hax, A. C., and D. Candea. *Production and Operations Management*. Prentice-Hall, Englewood Cliffs, NJ, 1984.

Higa, I., A. M. Feyerherm, and A. L. Machado. Waiting time in an (S–1, S) inventory system. *Operations Research* 23.4, 1975: 674–680.

Hill, R. M. Applying Bayesian methodology with a uniform prior to the single period inventory model. *European Journal of Operations Research* 98, 1997: 555–562.

Hillier, F. S., and G. J. Lieberman. *Introduction to Operations Research*. Holdenday Inc., San Francisco, CA, 1995.

Hutchins, D. *Just in Time*. Gower Publishing Ltd., Brookfield Vermont, USA, 1999.

Iglehart, D. L. Optimality of (s, S) policies in the infinite horizon dynamic inventory problem. *Management Science* 9.2, 1963: 259–267.

Jackson, P., W. Maxwell, and J. Muckstadt. The joint replenishment problem with a powers-of-two restriction. *IIE Transactions* 17.1, 1985: 25–32.

Jain, A., and K. Moinzadeh. A supply chain model with reverse information exchange. *Manufacturing and Service Operations Management* 7.4, 2005: 360–378.

Jain, K., and E. A. Silver. Lot sizing for a product subject to obsolescence or perishability. *European Journal of Operational Research* 75.2, 1994: 287–295.

Jammernegg, W., and P. Kischka. Newsvendor problems with VaR and CVaR consideration. *Handbook of Newsvendor Problems*. Springer, New York, 2012: 197–216.

Jans, R., and Z. Degraeve. Modeling industrial lot sizing problems: A review. *International Journal of Production Research* 46.6, 2008: 1619–1643.

Jesse, R. R., A. Mitra, and J. F. Cox. EOQ formula: Is it valid under inflationary conditions? *Decision Sciences* 14.3, 1983: 370–374.

Johansen, S.G., and P. Melchiors. Can-order policy for the periodic review joint replenishment problem. *Journal of the Operational Research Society* 54, 2003: 283–290.

Johnson, L.A., and D.C. Montgomery. *Operations Research in Production Planning, Scheduling and Inventory Control*. John Wiley and Sons, New York, 1974.

Kalpakam, S., and K. P. Sapna. A lost sales (S1, S) perishable inventory system with renewal demand. *Naval Research Logistics* 43.1, 1996: 129–142.

Kaplan, R. S. A dynamic inventory model with stochastic lead times. *Management Science* 16.7, 1970: 491–507.

Karaesmen, I. Z., A. Scheller-Wolf, and B. Deniz. Managing perishable and aging inventories: Review and future research directions. In K. G. Kempf, P. Keskinocak, and R. Uzsoy (eds.), *Planning Production and Inventories in the Extended Enterprise*. Springer, USA, 2011: 393–436.

Karlin, S., and H. Scarf. Inventory models of the Arrow–Harris–Marschak type with time lag. *Studies in the Mathematical Theory of Inventory and Production* 1, 1958: 155.

Karimi, B., S. M. T. Fatemi Ghomi, and J. M. Wilson. The capacitated lot sizing problem: A review of models and algorithms. *Omega* 31.5, 2003: 365–378.

Kaspi, M., and M. J. Rosenblatt. An improvement of Silver's algorithm for the joint replenishment problem. *AIIE Transactions* 15.3, 1983: 264–267.

Keilson, J., and A. Kooharian. On time dependent queuing processes. *The Annals of Mathematical Statistics* 31.3, 1960: 104–112.

Khouja, M., and S. Goyal. A review of the joint replenishment problem literature: 1989–2005. *European Journal of Operational Research* 186.1, 2008: 1–16.

Kian, R., U. Gürler, and E. Berk. The dynamic lot-sizing problem with convex economic production costs and setups. *International Journal of Production Economics* 155, 2014: 361–379.

Kruse, W. K. An exact n echelon inventory model: The simple Simon method. No. IRO-TR-79-2. Army Inventory Research Office, Philadelphia, PA, 1979.

Kruse, W. K. Waiting time in an S–1, S inventory system with arbitrarily distributed lead times. *Operations Research* 28.2, 1980: 348–352.

Kruse, W. K. Technical note waiting time in a continuous review (s, S) inventory system with constant lead times. *Operations Research* 29.1, 1981: 202–207.

Ladany, S., and A. Sternlieb. The interaction of economic ordering quantities and marketing policies. *AIIE Transactions* 6.1, 1974: 35–40.

Lagodimos, A. G. Multi-echelon service models for inventory systems under different rationing policies. *International Journal of Production Research* 30.4, 1992: 939–956.

Lariviere, M. A., and L. E. Porteus. Stalking information: Bayesian inventory management with unobserved lost sales. *Management Science* 45, 1999: 346–363.

Larsen, C. Note to editor. *Naval Research Logistics* 58, 2011: 410–410.

Lee, H. L. A multi-echelon inventory model for repairable items with emergency lateral transshipments. *Management Science* 33.10, 1987: 1302–1316.

Lee, H. L., and K. Moinzadeh. Operating characteristics of a two-echelon inventory system for repairable and consumable items under batch ordering and shipment policy. *Naval Research Logistics* 34.3, 1987a: 365–380.

Lee, H. L., and K. Moinzadeh. Two-parameter approximations for multi-echelon repairable inventory models with batch ordering policy. *IIE Transactions* 19.2, 1987b: 140–149.

Lee, H. L., P. Venkata, and S. Whang. Information distortion in a supply chain: The bullwhip effect. *Management Science* 50.12 Supplement, 2004: 1875–1886.

Liu, L., and Z. Lian. (s, S) continuous review models for products with fixed lifetimes. *Operations Research* 47.1, 1999: 150–158.

Lowe, T. J., L. B. Schwarz, and E. J. McGavin. The determination of optimal base-stock inventory policy when the costs of under- and oversupply are uncertain. *Naval Research Logistics* 35.4, 1988: 539–554.

Lu, X., J.-S. Song, and K. Zhu. On the censored newsvendor and the optimal acquisition of information. *Operations Research* 53.6, 2005: 1024–1026.

Lundin, R. A., and T. E. Morton. Planning horizons for the dynamic lot size model: Zabel vs. protective procedures and computational results. *Operations Research* 23.4, 1975: 711–734.

Mamani, H., and K. Moinzadeh. Lead time management through expediting in a continuous review inventory system. *Production and Operations Management* 23.1, 2014: 95–109.

Manne, A. S. Programming of economic lot sizes. *Management Science* 4.2, 1958: 115–135.

Minner, S. Multiple-supplier inventory models in supply chain management: A review. *International Journal of Production Economics* 81, 2003: 265–279.

Moinzadeh, K. Operating characteristics of the (S–1, S) inventory system with partial backorders and constant resupply times. *Management Science* 35.4, 1989: 472–477.

Moinzadeh, K. A multi-echelon inventory system with information exchange. *Management Science* 48.3, 2002: 414–426.

Moinzadeh, K., and P. K. Aggarwal. An information based multiechelon inventory system with emergency orders. *Operations Research* 45.5, 1997: 694–701.

Moinzadeh, K., and H. L. Lee. Batch size and stocking levels in multi-echelon repairable systems. *Management Science* 32.12, 1986: 1567–1581.

Moinzadeh, K., and S. Nahmias. A continuous review model for an inventory system with two supply modes. *Management Science* 34.6, 1988: 761–773.

Moinzadeh, K., and C. P. Schmidt. An (S–1, S) inventory system with emergency orders. *Operations Research*, 39.2, 1991: 308–321.

Mohebbi, E., and M. J. M. Posner. A lost-sales continuous-review inventory system with emergency ordering. *International Journal of Production Economics* 58.1, 1999: 93–112.

Monden, Y. *Toyota Production System: An Integrated Approach to Just-in-Time*. CRC Press, Boca Raton, FL, 2011.

Melchiors, P. Calculating can-order policies for the joint replenishment problem by the compensation approach. *European Journal of Operational Research* 141, 2002: 587–595.

Moon, I. and G. Gallego. Distribution free procedures for some inventory models. *Journal of Operational Research Society* 45, 1994: 651–658.

Morton, T. E. Bounds on the solution of the lagged optimal inventory equation with no demand backlogging and proportional costs. *SIAM Review* 11.4, 1969: 572–596.

Morton, T. E. The near-myopic nature of the lagged-proportional-cost inventory problem with lost sales. *Operations Research* 19.7, 1971: 1708–1716.

Muckstadt, J. A. A model for a multi-item, multi-echelon, multi-indenture inventory system. *Management Science* 20.4—Part-I, 1973: 472–481.

Muharremoglu, A., and J. N. Tsitsiklis. A single-unit decomposition approach to multiechelon inventory systems. *Operations Research* 56.5, 2008: 1089–1103.

Nahmias, S. Optimal ordering policies for perishable inventory II. *Operations Research* 23.4, 1975: 735–749.

Nahmias, S. On the equivalence of three approximate continuous review inventory models. *Naval Research Logistics Quarterly* 23.1, 1976: 31–36.

Nahmias, S. Technical note comparison between two dynamic perishable inventory models. *Operations Research* 25.1, 1977a: 168–172.

Nahmias, S. On ordering perishable inventory when both demand and lifetime are random. *Management Science* 24.1, 1977b: 82–90.

Nahmias, S. The fixed-charge perishable inventory problem. *Operations Research* 26.3, 1978: 464–481.

Nahmias, S. Simple approximations for a variety of dynamic leadtime lost-sales inventory models. *Operations Research* 27.5, 1979: 904–924.

Nahmias, S. Perishable inventory theory: A review. *Operations Research* 30.4, 1982: 680–708.

Nahmias, S. Demand estimation in lost sales inventory systems. *Naval Research Logistics* 41, 1994: 739–757.

Nahmias, S. *Production and Operations Analysis*. McGraw-Hill/Irwin, New York, 2014.

Nahmias, S. *Perishable Inventory Systems*. International Series in Operations Research and Management Science, Vol. 160. Springer, New York, 2011.

Nahmias, S., and W. P. Pierskalla. A two-product perishable/nonperishable inventory problem. *SIAM Journal on Applied Mathematics* 30.3, 1976: 483–500.

Nahmias, S., and S. S. Wang. A heuristic lot size reorder point model for decaying inventories. *Management Science* 25.1, 1979: 90–97.

Narasimhan, S., and D. W. McLeavy. *Production Planning and Inventory Control*, 2nd ed. Prentice-Hall, Englewood Cliffs, 1995.

Nielsen, C., and C. Larsen. An analytical study of the $Q(s, S)$ policy applied to the joint replenishment problem. *European Journal of Operational Research* 163, 2005: 721–732.

Orlicky, J. *Material Requirements Planning: The New Way of Life in Production and Inventory Management*. McGraw-Hill, New York, 1975.

Ozekici, S., and M. Parlar. Inventory models with unreliable suppliers in a random environment. *Annals of Operations Research* 91, 1999: 123–136.

Özkaya, B. Y., Ü. Gürler, and E. Berk. The stochastic joint replenishment problem: A new policy, analysis and insights. *Naval Research Logistics* 53, 2006: 525–546.

Palm, C. Analysis of the Erlang traffic formula for busy-signal arrangements. *Ericsson Technics* 5.9, 1938: 39–58.

Pantumsinchai, P. A comparison of three joint ordering policies. *Decision Science* 23, 1992: 111–127.

Parlar, M., and D. Berkin. Future supply uncertainty in EOQ models. *Naval Research Logistics* 38.1, 1991: 107–121.

Perry, D., and M. J. M. Posner. An (S–1, S) inventory system with fixed shelf life and constant lead times. *Operations Research* 46.3—Supplement-3, 1998: S65–S71.

Porteus, E. L. On the optimality of generalized (s, S) policies. *Management Science* 17.7, 1971: 411–426.

Porteus, E. L. Numerical comparisons of inventory policies for periodic review systems. *Operations Research* 33.1, 1985a: 134–152.

Porteus, E. L. Undiscounted approximations of discounted regenerative models. *Operations Research Letters* 3.6, 1985b: 293–300.

Raafat, F. Survey of literature on continuously deteriorating inventory models. *Journal of the Operational Research Society* 24.1, 1991: 27–37.

Renberg, B., and R. Planche. Un Modéle pour la gestion simultanée des n articles d'un stock. *Revue Francaise d'Informatique et de Recherche Opérationelle* 6, 1967: 47–59.

Richards, F. R. Technical note comments on the distribution of inventory position in a continuous-review (s, S) inventory system. *Operations Research* 23.2, 1975: 366–371.

Robinson, E. P., and F. Sahin. Economic production lot sizing with periodic costs and overtime. *Decision Sciences* 32(3), 2001: 423–452.

Rockafellar, R. T., and S. Uryasev. Optimization of conditional value-at-risk. *Journal of Risk* 2(3), 2000: 21–41.

Rosenblatt, M. J., and H. L. Lee. Improving profitability with quantity discounts under fixed demand. *IIE Transactions* 17.4, 1985: 388–395.

Rosling, K. Optimal inventory policies for assembly systems under random demands. *Operations Research* 37.4, 1989: 565–579.

Roundy, R. 98%-Effective integer-ratio lot-sizing for one-warehouse multi-retailer systems. *Management Science* 31.11, 1985: 1416–1430.

Sahin, I. On the stationary analysis of continuous review (s, S) inventory systems with constant lead times. *Operations Research* 27.4, 1979: 717–729.

Sahin, I. On the objective function behavior in (s, S) inventory models. *Operations Research* 30.4, 1982: 709–724.

Scarf, H. A min-max solution of an inventory problem. In K. Arrow, S. Karlin, and H. Scarf (eds.) *Studies in the Mathematical Theory of Inventory and Production*. Stanford University Press, California, 1958: 201–209.

Scarf, H. Bayes solutions of the statistical inventory problem. *The Annals of Mathematical Statistics* 30.2, 1959: 490–508.

Scarf, H. E. The optimality of (s, S) policies in the dynamic inventory problem. In K. J. Arrow, S. Karlin, and P. Suppes (eds.) *Mathematical Methods in the Social Science*, Stanford University Press, Stanford, 1960, Chapter 13.

Scarf, H. E. Inventory theory. *Operations Research* 50.1, 2002: 186–191.

Schmidt, C. P., and S. Nahmias. (S–1, S) policies for perishable inventory. *Management Science* 31.6, 1985: 719–728.

Schmitt, A. J., L. V. Snyder, and Z.-J. M. Shen. Inventory systems with stochastic demand and supply: Properties and approximations. *European Journal of Operational Research* 206.2, 2010: 313–328.

Schwarz, L. B. Economic order quantities for products with finite demand horizons. *AIIE Transactions* 4.3, 1972: 234–237.

Schwarz, L. B. Technical note: A note on the near optimality of 5-EOQ's worth forecast horizons. *Operations Research* 25.3, 1977: 533–536.

Schwarz, L. B., B. L. Deuermeyer, and R. D. Badinelli. Fill-rate optimization in a one-warehouse N-identical retailer distribution system. *Management Science* 31.4, 1985: 488–498.

Shang, K. H. Note: A simple heuristic for serial inventory systems with fixed order costs. *Operations Research* 56.4, 2008: 1039–1043.

Shang, K. H., and J.-S. Song. Newsvendor bounds and heuristic for optimal policies in serial supply chains. *Management Science* 49.5, 2003: 618–638.

Shang, K. H., and J.-S. Song. A closed-form approximation for serial inventory systems and its application to system design. *Manufacturing and Service Operations Management* 8.4, 2006: 394–406.

Shanker, K. Exact analysis of a two-echelon inventory system for recoverable items under batch inspection policy. *Naval Research Logistics Quarterly* 28.4, 1981: 579–601.

Sherbrooke, C. C. Technical note waiting time in an (S–1, S) inventory system constant service time case. *Operations Research* 23.4, 1975: 819–820.

Sherbrooke, C. C. VARI-METRIC: Improved approximations for multi-indenture, multi-echelon availability models. *Operations Research* 34.2, 1986: 311–319.

Sherbrooke, C. C. *Optimal Inventory Modeling of Systems: Multi-Echelon Techniques*. Vol. 72. Springer, New York, 2004.

Schultz, H., and S.G. Johansen. Can-order policies for coordinated inventory replenishment with Erlang distributed times between ordering. *European Journal of Operational Research* 113, 1999: 30–41.

Silver, E. A. Some characteristics of a special joint-order inventory model. *Operations Research* 13, 1965: 319–322.

Silver, E. A. A control system for coordinated inventory replenishment. *International Journal of Production Research* 12, 1974: 647–671.

Silver, E. A. A simple method of determining order quantities in joint replenishments under deterministic demand. *Management Science* 22, 1976: 351–1361.

Silver, E. A. Establishing reorder points in the (S, c, s) coordinated control system under compound Poisson demand. *International Journal of Production Research* 19, 1981: 743–750.

Silver, E. A., and H. C. Meal. A heuristic for selecting lot size quantities for the case of a deterministic time-varying demand rate and discrete opportunities for replenishment. *Production and Inventory Management* 14.2, 1973: 64–74.

Simon, R. M. Stationary properties of a two-echelon inventory model for low demand items. *Operations Research* 19.3, 1971: 761–773.

Simpson, N. C. Questioning the relative virtues of dynamic lot sizing rules. *Computers and Operations Research* 28.9, 2001: 899–914.

Sivazlian, B. D. A continuous-review (s, S) inventory system with arbitrary interarrival distribution between unit demand. *Operations Research* 22.1, 1974: 65–71.

Slay, F. M. *VARI-METRIC: An Approach to Modelling Multi-Echelon Resupply When the Demand Process Is Poisson with a Gamma Prior*. Logistics Management Institute, Washington, DC, Report AF301-3, 1984.

Smith, S. A. Optimal inventories for an (S–1, S) system with no backorders. *Management Science* 23.5, 1977: 522–528.

Sobel, M. J., and R. Q. Zhang. Inventory policies for systems with stochastic and deterministic demand. *Operations Research* 49.1, 2001: 157–162.

Song, J.-S. The effect of leadtime uncertainty in a simple stochastic inventory model. *Management Science* 40.5, 1994: 603–613.

Song, J.-S., H. Zhang, Y. Hou, and M. Wang. The effect of lead time and demand uncertainties in (r, q) inventory systems. *Operations Research* 58.1, 2010: 68–80.

Stevenson, W. *Operations Management*, 12th ed. McGraw-Hill/Irwin, New York, 2014.

Stidham Jr., S. Stochastic clearing systems. *Stochastic Processes and Their Applications* 2.1, 1974: 85–113.

Stidham Jr., S. Cost models for stochastic clearing systems. *Operations Research* 25.1, 1977: 100–127.

Stidham Jr., S. Clearing systems and (s, S) inventory systems with nonlinear costs and positive lead times. *Operations Research* 34.2, 1986: 276–280.

Svoronos, A., and P. Zipkin. Estimating the performance of multi-level inventory systems. *Operations Research* 36.1, 1988: 57–72.

Svoronos, A., and P. Zipkin. Evaluation of one-for-one replenishment policies for multiechelon inventory systems. *Management Science* 37.1, 1991: 68–83.

Tagaras, G., and D. Vlachos. A periodic review inventory system with emergency replenishments. *Management Science* 47.3, 2001: 415–429.

Takács, L. Investigation of waiting time problems by reduction to Markov processes. *Acta Mathematica Hungarica* 6.1, 1955: 101–129.

Taylor, S. G., and C. E. Bradley. Optimal ordering strategies for announced price increases. *Operations Research* 33.2, 1985: 312–325.

Tekin, E., Ü. Gürler, and E. Berk. Age-based vs. stock level control policies for a perishable inventory system. *European Journal of Operational Research* 134.2, 2001: 309–329.

Thompstone, R. M., and E. A. Silver. A coordinated inventory control system under compound Poisson demand and zero lead time. *International Journal of Production Research* 13, 1975: 581–602.

Toy, A. O., and E. Berk. Dynamic lot sizing problem for a warm/cold process. *IIE Transactions* 38.11, 2006: 1027–1044.

Toy, A. O., and E. Berk. Dynamic lot sizing for a warm/cold process: Heuristics and insights. *International Journal of Production Economics* 145.1, 2013: 53–66.

Urbach, R. Inventory average costs: Non-unit order sizes and random lead times. No. IRO-TR-77-5. DRC Inventory Research Office, Philadelphia, PA, 1977.

Van Eijs, M. J. G. A note on the joint replenishment problem under constant demand. *Journal of the Operational Research Society* 44.2, 1993: 185–191.

Van Eijs, M. J. G. On the determination of the control parameters of the optimal can-order policy. *Zeitschrift für Operations Research* 39, 1994: 289–304.

Van Houtum, G. J., K. Inderfurth, and W. H. M. Zijm. Materials coordination in stochastic multi-echelon systems. *European Journal of Operational Research* 95.1, 1996: 1–23.

Veinott Jr., A. F. The optimal inventory policy for batch ordering. *Operations Research* 13.3, 1965: 424–432.

Veinott Jr., A. F. Minimum concave-cost solution of Leontief substitution models of multi-facility inventory systems. *Operations Research* 17.2, 1969: 262–291.

Viswanathan, S. Note. Periodic review (s, S) policies for joint replenishment inventory systems. *Management Science* 43, 1997: 1447–1454.

Vollmann, T. E., W. L. Berry, and D. C. Whybark. *Manufacturing Planning and Control Systems*, 4th ed. McGraw-Hill/Irwin, Boston, 1997.

Wagelmans, A., S. Van Hoesel, and A. Kolen. Economic lot sizing: An O (n log n) algorithm that runs in linear time in the Wagner–Whitin case. *Operations Research* 40.1—Supplement-1, 1992: S145–S156.

Wagner, H. M., and T. M. Whitin. Dynamic version of the economic lot size model. *Management Science* 5.1, 1958: 89–96.

Waller, M., M. E. Johnson, and T. Davis. Vendor-managed inventory in the retail supply chain. *Journal of Business Logistics* 20, 1999: 183–204.

Weiss, H. J. Optimal ordering policies for continuous review perishable inventory models. *Operations Research* 28.2, 1980: 365–374.

Wolsey, L. A. Progress with single-item lot-sizing. *European Journal of Operational Research* 86.3, 1995: 395–401.

Zangwill, W. I. A deterministic multiproduct, multi-facility production and inventory model. *Operations Research* 14.3, 1966: 486–507.

Zheng, Y.-S., and A. Federgruen. Finding optimal (s, S) policies is about as simple as evaluating a single policy. *Operations Research* 39.4, 1991: 654–665.

Zipkin, P. Inventory service-level measures: Convexity and approximation. *Management Science* 32.8, 1986: 975–981.

Zipkin, P. *Foundations of Inventory Management*. Vol. 2. McGraw-Hill, New York, 2000.

Zipkin, P. On the structure of lost-sales inventory models. *Operations Research* 56.4, 2008a: 937–944.

Zipkin, P. Old and new methods for lost-sales inventory systems. *Operations Research* 56.5, 2008b: 1256–1263.

8

Statistical Methods

Raghu Nandan Sengupta and Debasis Kundu

CONTENTS

ABSTRACT The chapter of Statistical Methods starts with the basic concepts of data analysis and then leads into the concepts of probability, important properties of probability, limit theorems, and inequalities. The chapter also covers the basic tenets of estimation, desirable properties of estimates, before going on to the topic of maximum likelihood estimation, general methods of moments, Baye's estimation principle. Under linear and nonlinear regression different concepts of regressions are discussed. After which we discuss few important multivariate distributions and devote some time on copula theory also. In the later part of the chapter, emphasis is laid on both the theoretical content as well as the practical applications of a variety of multivariate techniques like Principle Component Analysis (PCA), Factor Analysis, Analysis of Variance (ANOVA), Multivariate Analysis of Variance (MANOVA), Conjoint Analysis, Canonical Correlation, Cluster Analysis, Multiple Discriminant Analysis, Multidimensional Scaling, Structural Equation Modeling, etc. Finally, the chapter ends with a good repertoire of information related to softwares, data sets, journals, etc., related to the topics covered in this chapter.

8.1 Introduction

Many people are familiar with the term *statistics*. It denotes recording of numerical facts and figures, for example, the daily prices of selected stocks on a stock exchange, the annual employment and unemployment of a country, the daily rainfall in the monsoon season, etc. However, statistics deals with situations in which the occurrence of some events cannot be predicted with certainty. It also provides methods for organizing and summarizing facts and for using information to draw various conclusions.

Historically, the word *statistics* is derived from the Latin word *status* meaning *state*. For several decades, statistics was associated solely with the display of facts and figures pertaining to economic, demographic, and political situations prevailing in a country. As a subject, statistics now encompasses concepts and methods that are of far-reaching importance in all enquires/questions that involve planning or designing of the experiment, gathering of data by a process of experimentation or observation, and finally making inference or conclusions by analyzing such data, which eventually helps in making the future decision.

Fact finding through the collection of data is not confined to professional researchers. It is a part of the everyday life of all people who strive, consciously or unconsciously, to know matters of interest concerning society, living conditions, the environment, and the world at large. Sources of factual information range from individual experience to reports in the news media, government records, and articles published in professional journals. Weather forecasts, market reports, costs of living indexes, and the results of public opinion are some other examples. Statistical methods are employed extensively in the production of such reports. Reports that are based on sound statistical reasoning and careful interpretation of conclusions are truly informative. However, the deliberate or inadvertent misuse of statistics leads to erroneous conclusions and distortions of truths.

8.2 Basic Concepts of Data Analysis

In order to clarify the preceding generalities, a few examples are provided:

Socioeconomic surveys: In the interdisciplinary areas of sociology, economics, and political science, such aspects are taken as the economic well-being of different ethnic groups, consumer expenditure patterns of different income levels, and attitudes toward pending legislation. Such studies are typically based on data oriented by interviewing or contacting a representative sample of person selected by statistical process from a large population that forms the domain of study. The data are then analyzed and interpretations of the issue in questions are made. See, for example, a recent monograph by Bandyopadhyay et al. (2011) on this topic.

Clinical diagnosis: Early detection is of paramount importance for the successful surgical treatment of many types of fatal diseases, say, for example, cancer or AIDS. Because frequent in-hospital checkups are expensive or inconvenient, doctors are searching for effective diagnosis process that patients can administer themselves. To determine the merits of a new process in terms of its rates of success in detecting true cases avoiding false detection, the process must be field tested on a large number of persons, who must then undergo in-hospital diagnostic test for comparison. Therefore, proper planning (designing the experiments) and data collection are required, which then need to be analyzed for final conclusions. An extensive survey of the different statistical methods used in clinical trial design can be found in Chen et al. (2015).

Plant breeding: Experiments involving the cross fertilization of different genetic types of plant species to produce high-yielding hybrids are of considerable interest to agricultural scientists. As a simple example, suppose that the yield of two hybrid varieties are to be compared under specific climatic conditions. The only way to learn about the relative performance of these two varieties is to grow them at a number of sites, collect data on their yield, and then analyze the data. Interested readers may refer to the edited volume by Kempton and Fox (2012) for further reading on this particular topic.

In recent years, attempts have been made to treat all these problems within the framework of a unified theory called decision theory. Whether or not statistical inference is viewed within the broader framework of decision theory depends heavily on the theory of probability. This is a mathematical theory, but the question of subjectivity versus objectivity arises in its applications and in its interpretations. We shall approach the subject of statistics as a science, developing each statistical idea as far as possible from its probabilistic foundation and applying each idea to different real-life problems as soon as it has been developed.

Statistical data obtained from surveys, experiments, or any series of measurements are often so numerous that they are virtually useless, unless they are condensed or reduced into a more suitable form. Sometimes, it may be satisfactory to present data just as they are, and let them speak for themselves; on other occasions, it may be necessary only to group the data and present results in the form of tables or in a graphical form. The summarization and exposition of the different important aspects of the data is commonly called descriptive statistics. This idea includes the condensation of the data in the form of tables, their graphical presentation, and computation of numerical indicators of the central tendency and variability.

There are mainly two main aspects of describing a data set:

1. Summarization and description of the overall pattern of the data by
   a. Presentation of tables and graphs
   b. Examination of the overall shape of the graphical data for important features, including symmetry or departure from it
   c. Scanning graphical data for any unusual observations, which seems to stick out from the major mass of the data
2. Computation of the numerical measures for
   a. A typical or representative value that indicates the center of the data
   b. The amount of spread or variation present in the data

Summarization and description of the data can be done in different ways. For a univariate data, the most popular methods are histogram, bar chart, frequency tables, box plot, or the stem and leaf plots. For bivariate or multivariate data, the useful methods are scatter plots or Chernoff faces. A wonderful exposition of the different exploratory data analysis techniques can be found in Tukey (1977), and for some recent development, see Theus and Urbanek (2008).

A typical or representative value that indicates the center of the data is the average value or the mean of the data. But since the mean is not a very robust estimate and is very much susceptible to the outliers, often, median can be used to represent the center of the data. In case of a symmetric distribution, both mean and median are the same, but in general, they are different. Other than mean or median, trimmed mean or the Windsorized mean can also be used to represent the central value of a data set. The amount of spread or the variation present in a data set can be measured using the standard deviation or the interquartile range.

8.3 Probability

The main aim of this section is to introduce the basic concepts of probability theory that are used quite extensively in developing different statistical inference procedures. We will try to provide the basic assumptions needed for the axiomatic development of the probability theory and will present some of the important results that are essential tools for statistical inference. For further study, the readers may refer to some of the classical books in probability theory such as Doob (1953) or Billingsley (1995), and for some recent development and treatment, readers are referred to Athreya and Lahiri (2006).

8.3.1 Sample Space and Events

The concept of probability is relevant to experiments that have somewhat uncertain outcomes. These are the situations in which, despite every effort to maintain fixed conditions, some variation of the result in repeated trials of the experiment is unavoidable. In probability, the term "experiment" is not restricted to laboratory experiments but includes any activity that results in the collection of data pertaining to the phenomena that exhibit variation. The domain of probability encompasses all phenomena for which outcomes cannot be exactly predicted in advance. Therefore, an experiment is the process of collecting data relevant to phenomena that exhibits variation in its outcomes. Let us consider the following examples:

Experiment (a). Let each of 10 persons taste a cup of instant coffee and a cup of percolated coffee. Report how many people prefer the instant coffee.

Experiment (b). Give 10 children a specific dose of multivitamin in addition to their normal diet. Observe the children's height and weight after 12 weeks.

Experiment (c). Note the sex of the first 2 new born babies in a particular hospital on a given day.

In all these examples, the experiment is described in terms of what is to be done and what aspect of the result is to be recorded. Although each experimental outcome is unpredictable, we can describe the collection of all possible outcomes.

Definition

The collection of all possible distinct outcomes of an experiment is called the sample space of the experiment, and each distinct outcome is called a simple event or an element of the sample space. The sample space is denoted by Ω.

In a given situation, the sample space is presented either by listing all possible results of the experiments, using convenient symbols to identify the results or by making a descriptive statement characterizing the set of possible results. The sample space of the above three experiments can be described as follows:

Experiment (a). $\Omega = \{0, 1, \ldots, 10\}$.

Experiment (b). Here, the experimental result consists of the measurements of two characteristics, height and weight. Both of these are measured on a continuous scale. Denoting the measurements of gain in height and weight by x and y, respectively, the sample space can be described as $\Omega = \{(x, y);\ x$ nonnegative, y positive, negative or zero.$\}$

Experiment (c). $\Omega = \{BB, BG, GB, GG\}$, where, for example, BG denotes the birth of a boy first and then followed by a girl. Similarly, the other symbols are also defined.

In our study of probability, we are interested not only in the individual outcomes of Ω but also in any collection of outcomes of Ω.

Definition

An event is any collection of outcomes contained in the sample space Ω. An event is said to be simple, if it consists of exactly one outcome, and compound, if it consists of more than one outcome.

Definition

A sample space consisting of either a finite or a countably infinite number of elements is called a discrete sample space. When the sample space includes all the numbers in some interval (finite or infinite) of the real line, it is called continuous sample space.

8.3.2 Axioms, Interpretations, and Properties of Probability

Given an experiment and a sample space Ω, the objective of probability is to assign to each event A, a number $P(A)$, called probability of the event A, which will give a precise measure of the chance that A will occur. To ensure that the probability assignment will be consistent with our intuitive notion of probability, all assignments should satisfy the following axioms (basic properties) of probability:

- Axiom 1: For any event $A, 0 \leq P(A) \leq 1$.
- Axiom 2: $P(\Omega) = 1$.
- Axiom 3: If $\{A_1, A_2, \ldots\}$ is an infinite collection of mutually exclusive events, then

$$P(A_1 \cup A_2 \cup A_3 \ldots) = \sum_{i=1}^{\infty} P(A_i).$$

Axiom 1 reflects the intuitive notion that the chance of A occurring should be at least zero, so that negative probabilities are not allowed. The sample space is by definition an event that must occur when the experiment performed (Ω) contains all possible outcomes. So, Axiom 2 says that the maximum probability of occurrence is assigned to Ω. The third axiom formalizes the idea that if we wish the probability that at least one of a number of events will occur, and no two of the events can occur simultaneously, then the chance of at least one occurring is the sum of the chances of individual events.

Consider an experiment in which a single coin is tossed once. The sample space is $\Omega = \{H, T\}$. The axioms specify $P(\Omega) = 1$, so to complete the probability assignment, it remains only to determine $P(H)$ and $P(T)$. Since H and T are disjoint events, and $H \cup T = \Omega$, Axiom 3 implies that $1 = P(\Omega) = P(H) + P(T)$. So, $P(T) = 1 - P(H)$. Thus, the only freedom allowed by the axioms in this experiment is the probability assigned to H. One possible assignment of probabilities is $P(H) = 0.5$, $P(T) = 0.5$, while another possible assignment is $P(H) = 0.75$, $P(T) = 0.25$. In fact, letting p represent any fixed number between 0 and 1, $P(H) = p$, $P(T) = 1 - p$ is an assignment consistent with the axioms.

8.3.3 Borel σ-Field, Random Variables, and Convergence

The basic idea of probability is to define a set function whose domain is a class of subsets of the sample space Ω, whose range is $[0, 1]$, and it satisfies the three axioms mentioned in the previous subsection. If Ω is the collection of finite number or countable number of points, then it is quite easy to define the probability function always, for the class of all subsets of Ω, so that it satisfies Axioms 1–3. If Ω is not countable, it is not always possible to define for the class of all subsets of Ω. For example, if $\Omega = \mathbb{R}$, the whole real line, then the probability function (from now onward, we call it as a probability measure) is not possible to define for the class of all subsets of Ω. Therefore, we define a particular class of subsets of $\mathbb{R}$, called Borel σ-field (it will be denoted by $\mathcal{B}$); see Billingsley (1995) for details, on which probability measure can be defined. The triplet $(\Omega, \mathcal{B}, P)$ is called the probability space, while Ω or $(\Omega, \mathcal{B})$ is called the sample space.

Random variable: A real-valued point function $X(\cdot)$ defined on the space $(\Omega, \mathcal{B}, P)$ is called a random variable of the set $\{\omega : X(\omega) \leq x\} \in \mathcal{B}$, for all $x \in \mathbb{R}$.

Distribution function: The point function

$$F(x) = P\{\omega : X(\omega) \leq x\} = P(X^{-1}(-\infty, x]),$$

defined on $\mathbb{R}$, is called the distribution function of X.

Now, we will define three important concepts of convergence of a sequence of random variables. Suppose $\{X_n\}$ is a sequence of random variables, and X is also a random variable, and all are defined of the same probability space $(\Omega, \mathcal{B}, P)$.

Convergence in probability or weakly: The sequence of random variables $\{X_n\}$ is said to converge to X in probability (denoted by $X_n \xrightarrow{p} X$) if for all $\epsilon > 0$,

$$\lim_{n\to\infty} P(|X_n - X| \geq \epsilon) = 0.$$

Almost sure convergence or strongly: The sequence of random variables $\{X_n\}$ is said to converge to X strongly (denoted by $X_n \xrightarrow{a.e.} X$), if

$$P\left(\lim_{n\to\infty} X_n = X\right) = 1.$$

Convergence in distribution: The sequence of random variables $\{X_n\}$ is said to converge to X in distribution (denoted by $X_n \xrightarrow{d} X$), if

$$\lim_{n\to\infty} F_n(x) = F(x),$$

for all x, such that F is continuous at x. Here, F_n and F denote the distribution functions of X_n and X, respectively.

8.3.4 Some Important Results

In this subsection, we present some of the most important results of probability theory that have direct relevance in statistical sciences. The books by Chung (1974) or Serfling (1980) are referred for details.

The characteristic function of a random variable X with the distribution function $F(x)$ is defined as follows:

$$\phi_X(t) = E\left(e^{itX}\right) = \int_{-\infty}^{\infty} e^{itx} dF(x), \quad \text{for} \quad t \in \mathbb{R},$$

where $i = \sqrt{-1}$. The characteristic function uniquely defines a distribution function. For example, if $\phi_1(t)$ and $\phi_2(t)$ are the characteristic functions associated with the distribution functions $F_1(x)$ and $F_2(x)$, respectively, and $\phi_1(t) = \phi_2(t)$, for all $t \in \mathbb{R}$, then $F_1(x) = F_2(x)$, for all $x \in \mathbb{R}$.

Chebyshev's theorem: If $\{X_n\}$ is a sequence of random variables, such that $E(X_i) = \mu_i$, $V(X_i) = \sigma_i^2$, and they are uncorrelated, then

$$\lim_{n\to\infty} \frac{1}{n^2}\sigma_i^2 = 0 \Rightarrow \left[\frac{1}{n}\sum_{i=1}^{n} X_i - \frac{1}{n}\sum_{i=1}^{n} \mu_i\right] \xrightarrow{p} 0.$$

Khinchine's theorem: If $\{X_n\}$ is a sequence of independent and identically distributed random variables, such that $E(X_1) = \mu < \infty$, then

$$\lim_{n\to\infty} \frac{1}{n}\sum_{i=1}^{n} X_i \xrightarrow{p} \mu.$$

Kolmogorov theorem 1: If $\{X_n\}$ is a sequence of independent random variables, such that $E(X_i) = \mu_i$, $V(X_i) = \sigma_i^2$, then

$$\sum_{i=1}^{\infty} \frac{\sigma_i^2}{i^2} < \infty \Rightarrow \left[\frac{1}{n}\sum_{i=1}^{n} X_i - \frac{1}{n}\sum_{i=1}^{n} \mu_i\right] \stackrel{a.s.}{\to} 0.$$

Kolmogorov theorem 2: If $\{X_n\}$ is a sequence of independent and identically distributed random variables, then a necessary and sufficient condition that

$$\frac{1}{n}\sum_{i=1}^{n} X_i \stackrel{a.s.}{\to} \mu$$

is that $E(X_1) < \infty$, and it is equal to μ.

Central limit theorem: If $\{X_n\}$ is a sequence of independent and identically distributed random variables, such that $E(X_1) = \mu$, and $V(X_1) = \sigma^2 < \infty$, then

$$\frac{1}{\sigma\sqrt{n}}\sum_{i=1}^{n}(X_i - \mu) \stackrel{d}{\to} Z.$$

Here, Z is a standard normal random variable with mean zero and variance 1.

Example 8.1

Suppose $X_1, X_2, \ldots$ is a sequence of i.i.d. exponential random variable with the following probability density function for $x > 0$:

$$f(x) = \begin{cases} e^{-x} & \text{if } x \geq 0, \\ 0 & \text{if } x < 0. \end{cases}$$

In this case, $E(X_1) = V(X_1) = 1$. Therefore, by the weak law of large numbers (WLLN) of Khinchine, it immediately follows that

$$\frac{1}{n}\sum_{i=1}^{n} X_i \stackrel{p}{\to} 1,$$

and by Kolmogorov's strong law of large numbers (SLLN),

$$\frac{1}{n}\sum_{i=1}^{n} X_i \stackrel{a.e.}{\to} 1.$$

Further, by the central limit theorem (CLT), we have

$$\frac{1}{\sqrt{n}}\sum_{i=1}^{n}(X_i - 1) \stackrel{d}{\to} Z \sim N(0, 1).$$

8.4 Estimation

8.4.1 Introduction

The topic of parameter estimation deals with the estimation of some parameters from the data that characterizes the underlying process or phenomenon. For example, one is posed with the data taken repeatedly of the same temperature. These data are not equal, although the underlying *true* temperature was the same. In such a situation, one would like to obtain an estimate of the true temperature from the given data.

We may also be interested in finding the coefficient of resolution of a steel ball from the data on successive heights to which the ball rose. One may be interested in obtaining the face flow speed of vehicles from data on speed and density. All these estimation problems come under the purview of parameter estimation. The question of estimation arises because one always tries to obtain knowledge on the parameters of the population from the information available through the sample. The estimate obtained depends on the sample collected. Further, one could generally obtain more than one sample from a given population, and therefore the estimates of the same parameter could be different from one another. Most of the desirable properties of an estimate are defined keeping in mind the variability of the estimates.

In this discussion on the said topic, we will look into desirable properties of an estimator, and some methods for obtaining estimates. We will also see some examples that will help to clarify some of the salient features of parameter estimation. Finally, we will introduce the ideas of interval estimation, and illustrate its relevance to real-world problems.

8.4.2 Desirable Properties

The desirable properties of an estimator are defined keeping in mind that the estimates obtained are random. In the following discussion, T will represent an estimate while θ will represent the true parameter value of a parameter. The properties that will be discussed are the following:

Unbiasedness: The unbiasedness property states that $E(T) = \theta$. The desirability of this property is self-evident. It basically implies that on an average the estimator should be equal to the parameter value.

Minimum variance: It is also desirable that any realization of T (i.e., any estimate) may not be far off from the true value. Alternatively stated, it means that the probability of θ being near to θ should be high, or as high as possible. This is equivalent to saying that the variance of T should be minimal. An estimator that has the minimum variance in the class of all unbiased estimators is called an efficient estimator.

Sufficiency: An estimator is sufficient if it uses all the information about the population parameter, θ, that is available from the sample. For example, the sample median is not a sufficient estimator of the population mean, because median only utilizes the ranking of the sample values and not their relative distance. Sufficiency is important because it is a necessary condition for the minimum variance property (i.e., efficiency).

Consistency: The property of consistency demands that an estimate be very close to the true value of the parameter when the estimate is obtained from a large sample. More specifically, if $\lim_{n\to\infty} P(|T - \theta| < \epsilon) = 1$, for any $\epsilon > 0$, however small it might be, the estimator T is said to be a consistent estimator of the parameter θ. It may be noted that if T has a zero bias, and the variance of T tends to zero, then T is a consistent estimator of θ.

Asymptotic properties: The asymptotic properties of estimators relate to the behavior of the estimators based on a large sample. Consistency is thus an asymptotic property of an estimator. Other asymptotic properties include asymptotic unbiasedness and asymptotic efficiency.

As the nomenclature suggests, asymptotic unbiasedness refers to the unbiasedness of an estimator based on a large sample. Alternatively, it can be stated as follows:

$$\lim_{n\to\infty} E(T) = \theta.$$

For example, an estimator whose $E(T) = \theta - (1/n)$ is an asymptotically unbiased estimator of θ. For small samples, however, this estimator has a finite negative bias.

Similarly, asymptotic efficiency suggests that an asymptotically efficient estimator is the minimum variance unbiased estimator of θ for large samples. Asymptotic efficiency may be thought of as the large sample equivalent of best unbiasedness, while asymptotic unbiasedness may be thought of as the large sample equivalent of unbiasedness property.

Minimum mean square error: The minimum mean square error (MSE) property states that the estimator T should be such that the quantity MSE defined below is minimum:

$$MSE = E(T - \theta)^2.$$

Alternatively written,

$$MSE = Var(T) + (E(T) - \theta)^2.$$

Intuitively, it is appealing because it looks for an estimator that has small bias (may be zero) and small variance. This property is appealing because it does not constrain an estimator to be unbiased before looking at the variance of the estimator. Thus, the minimum MSE property does not give higher importance to unbiasedness than to variance. Both the factors are considered simultaneously.

Robustness: Another desirable property of an estimator is that the estimator should not be very sensitive to the presence of outliers or obviously erroneous points in the data set. Such an estimator is called a robust estimator. The robust property is important because, loosely speaking, it captures the reliability of an estimator. There are different ways in which robustness is quantified. Influence function and breakdown point are two such methods. Influence functions describe the effect of one outlier on the estimator. Breakdown point of an estimator is the proportion of incorrect observations (for example, arbitrarily large observations) an estimator can handle before giving an incorrect (that is arbitrarily large) result.

8.4.3 Methods of Estimation

One of the important questions in parameter estimation is, how does one estimate (the method of estimation) the unknown parameters so that the properties of the resulting estimators are in reasonable agreement with the desirable properties? There are many methods that are available in the literature, and needless to say that none of these methods provide estimators that satisfy all the desirable properties. As we will see later, some methods provide good estimators under certain assumptions and others provide good estimators with minor modifications. Although salient, one important aspect of developing a method for estimation that one should bear in mind the amount and complexity of the computation requirement associated with the methodology.

We will elaborate on four different methods, namely, (a) the method of maximum likelihood, (b) the method of least squares, (c) the method of moments, and (d) the method of minimum absolute deviation.

The method of maximum likelihood: Suppose $\boldsymbol{x} = \{x_1, \ldots, x_n\}$ is a random sample from a population that is characterized by m parameters $\boldsymbol{\theta} = (\theta_1, \ldots, \theta_m)$. It is assumed that the population has the probability density function (PDF) or probability mass function (PMF) as $f(x; \boldsymbol{\theta})$. The principle of maximum likelihood estimation consists of choosing as an estimate of $\boldsymbol{\theta}$ a $\widehat{\boldsymbol{\theta}}(\boldsymbol{x})$ that maximizes the likelihood function, which is defined as follows:

$$L(\boldsymbol{\theta}; x_1, \ldots, x_n) = \prod_{i=1}^{n} f(x_i; \boldsymbol{\theta}).$$

Therefore,

$$L(\widehat{\boldsymbol{\theta}}; x_1, \ldots, x_n) = \sup_{\boldsymbol{\theta}} L(\boldsymbol{\theta}; x_1, \ldots, x_n),$$

or in other words

$$\widehat{\boldsymbol{\theta}} = \text{argmax } L(\boldsymbol{\theta}, x_1, \ldots, x_n).$$

The notation "argmax" means that $L(\boldsymbol{\theta}, x_1, \ldots, x_n)$ achieves the maximum value at $\widehat{\boldsymbol{\theta}}$, and $\widehat{\boldsymbol{\theta}}$ is called the maximum likelihood estimator (MLE) of $\boldsymbol{\theta}$.

To motivate the use of the likelihood function, we begin with a simple example, and then provide with a theoretical justification.

Let $X_1, \ldots, X_n$ be a random sample from a Bernoulli distribution with parameter θ, which has the following probability mass function:

$$p(x) = \begin{cases} \theta^x (1-\theta)^x & x = 0, 1 \\ 0 & \text{otherwise,} \end{cases}$$

where $0 < \theta < 1$. Then

$$P(X_1 = x_1, \ldots, X_n = x_n) = \theta^{\sum_{i=1}^n x_i} (1-\theta)^{n - \sum_{i=1}^n x_i},$$

where x_i can be either 0 or 1. This probability, which is the joint probability mass function of $X_1, \ldots, X_n$, as a function of θ, is the likelihood function $L(\theta)$ defined above, that is,

$$L(\theta) = \theta^{\sum_{i=1}^n x_i} (1-\theta)^{n - \sum_{i=1}^n x_i}, \quad 0 < \theta < 1.$$

Now we may ask what should be the value of θ that would maximize the above probability $L(\theta)$ to obtain the specific observed sample $x_1, \ldots, x_n$. The value of θ that maximizes $L(\theta)$ seems to be a good estimate of θ as it provides the largest probability for this particular sample. Since

$$\widehat{\theta} = \text{argmax } L(\theta) = \text{argmax } \ln L(\theta),$$

it is often easier to maximize $l(\theta) = \ln L(\theta)$, rather than $L(\theta)$. In this case

$$l(\theta) = \ln L(\theta) = \left(\sum_{i=1}^n x_i\right) \ln \theta + \left(n - \sum_{i=1}^n x_i\right) \ln(1-\theta),$$

provided θ is not equal to 0 or 1. So we have

$$\frac{dl(\theta)}{d\theta} = \frac{\sum_{i=1}^{n} x_i}{\theta} - \frac{n - \sum_{i=1}^{n} x_i}{1-\theta} = 0.$$

Therefore,

$$\widehat{\theta} = \frac{\sum_{i=1}^{n} x_i}{n}.$$

Now we provide the theoretical justification to use the maximum likelihood estimator as a reasonable estimator of θ. Suppose θ_0 denotes the true value of θ, then Theorem 8.1 provides a theoretical reason for maximizing the likelihood function. It says that the maximum of $L(\theta)$ asymptotically separates the true model at θ_0 from models at $\theta \neq \theta_0$. We will state the main result without proof. For details, interested readers are referred to Lehmann and Casella (1998).

Regularity conditions A1. The PDFs are distinct, that is, for $\theta \neq \theta' \Rightarrow f(x;\theta) \neq f(x;\theta')$.

A2. The PDFs have common support for all θ.

A3. The point θ_0 is an interior point of the parameter space Ω.

Note that the first assumption states that the parameters identify the PDFs. The second assumption implies that the support of the random variables does not depend on the parameter. Now, based on the above assumptions, we have the following important result regarding the existence and uniqueness of the maximum likelihood estimator of θ.

Theorem 8.1

Let θ_0 be the true parameter value, then under Assumptions A1–A3

$$\lim_{n\to\infty} P_{\theta_0}\left[L(\theta_0, X_1, \ldots, X_n) > L(\theta, X_1, \ldots, X_n)\right] = 1, \quad \text{for all } \theta \neq \theta_0.$$

Theorem 8.1 states that the likelihood function is asymptotically maximized at the true value θ_0. The following theorem provides the consistency property of the maximum likelihood estimator under some suitable regularity conditions.

Theorem 8.2

Let $X_1, \ldots, X_n$ be a random sample from the probability density function $f(x;\theta)$, which satisfies Assumptions A1–A3. Further, it is assumed that $f(x;\theta)$ is differentiable with respect to $\theta \in \Omega$. If θ_0 is the true parameter value, then the likelihood equation

$$\frac{\partial}{\partial\theta} L(\theta) = 0 \Leftrightarrow \frac{\partial}{\partial\theta} l(\theta) = 0$$

has a solution $\widehat{\theta}_n$, such that $\widehat{\theta}_n$ converges to θ_0 in probability.

Finally, we state the asymptotic normality results based on certain regularity conditions. The details of the regularity conditions and the proof can be obtained in Lehmann and Casella (1998).

Theorem 8.3

Let $X_1, \ldots, X_n$ be a random sample from the probability density function $f(x; \theta)$, which satisfies the regularity conditions as stated in Lehmann and Casella (1998). Then

$$\sqrt{n}\left(\widehat{\theta}_n - \theta_0\right) \xrightarrow{D} N_p(\mathbf{0}, \boldsymbol{I}^{-1}).$$

Here, $\xrightarrow{D}$ means converges in distribution, and $\boldsymbol{I}$ is the Fisher information matrix.

Theorem 8.3 can be used for the construction of confidence intervals and also for testing purposes. One point should be mentioned that although the MLE is the most popular estimator, it may not be in the explicit form always. Let us consider the following example:

Example 8.2

Suppose $X_1, \ldots, X_n$ are i.i.d. random variables with the following PDF:

$$f(x; \theta) = \frac{e^{-(x-\theta)}}{(1+e^{-(x-\theta)})^2}; \quad -\infty < x < \infty, \quad -\infty < \theta < \infty. \tag{8.1}$$

It may be mentioned that Equation 8.1 is the PDF of a logistic distribution. The logarithm of the likelihood function can be written as

$$l(\theta) = \sum_{i=1}^{n} \ln f(x_i; \theta) = n\theta - \sum_{i=1}^{n} x_i - 2\sum_{i=1}^{n} \ln\left(1 + e^{-(x_i-\theta)}\right). \tag{8.2}$$

The MLE of the unknown parameter θ can be obtained by maximizing Equation 8.2 with respect to the unknown parameter θ. Setting the first partial derivative of Equation 8.2 equals to zero, we obtain

$$\frac{d}{d\theta} l(\theta) = n - 2\sum_{i=1}^{n} \frac{e^{-(x_i-\theta)}}{1+e^{-(x_i-\theta)}} = 0. \tag{8.3}$$

Rearranging Equation 8.3, it becomes

$$\sum_{i=1}^{n} \frac{e^{-(x_i-\theta)}}{1+e^{-(x_i-\theta)}} = \frac{n}{2}. \tag{8.4}$$

The MLE of θ, $\widehat{\theta}$, can be obtained by solving Equation 8.4. Unfortunately, it cannot be obtained in explicit form. It can be shown that the solution exists and it is unique. It can be obtained as follows. The first derivative of the left-hand side of Equation 8.4 is

$$\frac{d}{d\theta} \sum_{i=1}^{n} \frac{e^{-(x_i-\theta)}}{1+e^{-(x_i-\theta)}} = \sum_{i=1}^{n} \frac{e^{-(x_i-\theta)}}{\left(1+e^{-(x_i-\theta)}\right)^2} > 0.$$

Therefore, the left-hand side of Equation 8.4 is a strictly increasing function of θ, and it goes to zero, as θ goes to ∞ or $-\infty$. Hence, Equation 8.4 has a unique solution, and it has to be obtained numerically. The standard numerical analysis method like bisection or Newton's method may be used to compute $\widehat{\theta}$. It can be easily shown that the PDF of the logistic distribution satisfies all the regulatory conditions. Hence, we can conclude that as $n \to \infty$, $\widehat{\theta} \to \theta_0$; here, θ_0 is the true value of the parameter. Moreover,

$$\sqrt{n}(\widehat{\theta} - \theta_0) \xrightarrow{D} N(0, 1.0/I(\theta_0)), \tag{8.5}$$

where

$$I(\theta_0) = -E\left[\frac{d^2}{d\theta^2} \ln f(x;\theta_0)\right] = \int_{-\infty}^{\infty} \frac{e^{-2(x-\theta_0)}}{(1+e^{-(x-\theta_0)})^4} dx = \frac{1}{6}.$$

Therefore, using Equation 8.5, for large sample size, we can obtain an approximate 95% confidence interval of θ_0 as $\left(\widehat{\theta} - 1.96 \times \sqrt{6/n}, \widehat{\theta} + 1.96 \times \sqrt{6/n}\right)$.

Now we will provide some of the numerical methods that can be used to compute the MLEs. In most of the cases, the MLEs have to be obtained by solving a set of nonlinear equations, or solving a multidimensional optimization problem. Some of the standard general-purpose algorithms can be used to compute the MLEs. For example, genetic algorithm of Goldberg (1989), simulated annealing of Kirkpatrick et al. (1983), downhill simplex method of Nelder and Mead (1965), etc. can be used to compute the MLEs of the unknown parameters by maximizing the likelihood function.

Another very important method that can be used very successfully to compute the MLEs of the unknown parameters, particularly if some of the data are missing or censored, is known as the expectation maximization (EM) algorithm introduced by Dempster et al. (1977); see the excellent book by McLachlan and Krishnan (1997) in this respect. The EM algorithm has two steps: (i) E-Step and (ii) M-Step. In E-Step, pseudo-likelihood function has been obtained by replacing the missing values with their corresponding expected values, and M-Step involves maximizing the pseudo-likelihood function. Although, EM algorithm has been used mainly when the complete data are not available, but it has been used in many cases in case of complete sample also by treating it as a missing value problem.

Before describing another very popular estimator, we will mention one very useful property of the MLE, and that is known as invariance property, which may not be true for most of the other estimators. It can be stated as follows. If $\widehat{\theta}$ is the MLE of θ, and $h(\theta)$ is a "nice" function, then $h(\widehat{\theta})$ is the MLE of $h(\theta)$. Moreover, similar to Equation 8.5, in this case, we have

$$\sqrt{n}(g(\widehat{\theta}) - g(\theta_0)) \xrightarrow{D} N(0, (g'(\theta_0))^2/I(\theta_0)). \tag{8.6}$$

Hence, similar to θ_0, using Equation 8.6, an asymptotic confidence interval of $g(\theta_0)$ can also be obtained along the same line.

8.4.4 Method of Moment Estimators

The method of moment estimators is the oldest method of finding point estimators. Dating goes back to Karl Pearson in the late 1800. It is very simple to use and most of the time it provides some sort of estimate. In many cases, it may happen that it can be improved upon; however, it is a good place to start with when other methods may be very difficult to implement.

Let $X_1, \ldots, X_n$ be a random sample from a PDF or PMF $f(x|\theta_1, \ldots, \theta_k)$. The method of moment estimators are found by equating the first k sample moments to the corresponding k population moments. The method of moment estimators are obtained by solving the resulting systems of equations simultaneously. To be more precise, for $j = 1, \ldots, k$, define

$$\begin{aligned} m_1 &= \frac{1}{n}\sum_{i=1}^{n} X_i^1, \quad \mu_1 = E(X^1); \\ m_2 &= \frac{1}{n}\sum_{i=1}^{n} X_i^2, \quad \mu_2 = E(X^2); \\ &\vdots \\ m_k &= \frac{1}{n}\sum_{i=1}^{n} X_i^k, \quad \mu_1 = E(X^k). \end{aligned} \tag{8.7}$$

Usually, the population moment μ_j, will be a function of $\theta_1, \ldots, \theta_k$, say $\mu_j(\theta_1, \ldots, \theta_k)$. Hence, the method of moment estimators $\widetilde{\theta}_1, \ldots, \widetilde{\theta}_k$ of $\theta_1, \ldots, \theta_k$, respectively, can be obtained by solving the following k-equations simultaneously:

$$\begin{aligned} m_1 &= \mu_1(\theta_1, \ldots, \theta_k); \\ m_2 &= \mu_2(\theta_1, \ldots, \theta_k); \\ &\vdots \\ m_k &= \mu_k(\theta_1, \ldots, \theta_k). \end{aligned} \tag{8.8}$$

The justification of the method of moment estimators mainly comes from the SLLN, and also from the CLT. Owing to SLLN, under some very mild conditions, it can be shown that the method of moment estimators are always consistent estimators of the corresponding parameters. Further, because of the CLT, asymptotically, the method of moment estimators follow multivariate normal distribution, whose covariance matrix can be easily obtained. For illustrative purposes, we provide a simple example where the method of moment estimators can be obtained explicitly. But it may not be the case always. Most of the times, we need to solve a system of nonlinear equations to compute the method of moment estimators.

Example 8.3

Suppose $X_1, \ldots, X_n$ are i.i.d. from a two-parameter exponential distribution with the following PDF for $\theta > 0$, and $\mu \in \mathbb{R}$:

$$f(x; \mu, \theta) = \begin{cases} \frac{1}{\theta} e^{-(1/\theta)(x-\mu)} & \text{if } x > \mu \\ 0 & \text{if } x \leq \mu. \end{cases}$$

In this case, using the same notation as above, we obtain

$$m_1 = \frac{1}{n} X_i, \quad m_2 = \frac{1}{n}\sum_{i=1}^{n} X_i^2, \quad \mu_1 = \theta + \mu, \quad \mu_2 = (\mu + \theta)^2 + \theta^2.$$

Hence, in this case, the method of moment estimators of θ and μ can be easily obtained as

$$\widetilde{\theta} = \sqrt{m_2 - m_1^2} = \sqrt{\frac{1}{n}\sum_{i=1}^{n} X_i^2 - \left(\frac{1}{n}X_i\right)^2} \quad \text{and}$$

$$\widetilde{\mu} = m_1 - \widetilde{\theta} = \frac{1}{n}X_i - \sqrt{\frac{1}{n}\sum_{i=1}^{n} X_i^2 - \left(\frac{1}{n}X_i\right)^2}.$$

Note that the method of moment estimators are different from the MLEs. Finally, we will wind up this section with another estimator, which is becoming extremely popular in recent days.

8.4.5 Bayes Estimators

The Bayesian approach to statistics is philosophically different from the classical approach that we have just mentioned. First, let us describe the Bayesian approach to statistics. The main difference between the classical approach and the Bayesian approach is the following. In the classical approach, the parameter θ is assumed to be unknown but it is assumed to be a fixed quantity. A random sample is drawn from a population that is characterized by the parameter θ, and based on the random sample, a knowledge about the parameter θ is obtained. On the other hand, in the Bayesian approach, the parameter θ is not assumed to be a fixed quantity, and it is considered to be a quantity whose variation can be described by a probability distribution, known as the prior distribution. This prior distribution is purely a subjective distribution, and it depends on the choice of the experimenter. The most important aspect of the prior distribution is that it is formulated before any sample is observed. A sample is then obtained from the population indexed by the parameter θ, and then the prior distribution is updated based on the information of the present sample. The updated information about the parameter θ is known as the posterior distribution.

If we denote the prior distribution by $\pi(\theta)$ and the sampling distribution by $f(\boldsymbol{x}|\theta)$, then the posterior distribution of θ given the sample $\boldsymbol{x}$ becomes

$$\pi(\theta|\boldsymbol{x}) = f(\boldsymbol{x}|\theta)\pi(\theta)/m(\boldsymbol{x}).$$

Here, $m(\boldsymbol{x})$ is the marginal distribution of $\boldsymbol{x}$, and it can be obtained as

$$m(\boldsymbol{x}) = \int f(\boldsymbol{x}|\theta)\pi(\theta)d\theta.$$

Note that the posterior distribution of θ provides the complete information regarding the unknown parameter θ. The posterior distribution of θ can be used to obtain a point estimate of θ or to construct confidence interval (known as credible interval in the Bayesian terminology) of θ. The most popular point estimate of θ is the posterior mean. It has some other nice interpretation also. Other point estimates like median or mode can also be used in this case. Let us consider the following example to illustrate the Bayesian methodology.

Example 8.4

Let $X_1, \ldots, X_n$ be a random sample from an exponential distribution with parameter θ, and it has the following PDF:

$$f(x|\theta) = \begin{cases} \theta e^{-\theta x} & \text{if } x > 0 \\ 0 & \text{if } x \le 0. \end{cases}$$

Suppose the prior $\pi(\theta)$ on θ has a gamma distribution with the known shape and scale parameters as $a > 0$ and $b > 0$, respectively. Therefore, $\pi(\theta)$ for $\theta > 0$ has the following form:

$$\pi(\theta|a,b) = \frac{b^a}{\Gamma(a)}\theta^{a-1}\, e^{-b\theta},$$

and zero otherwise. Therefore, for $\boldsymbol{x} = (x_1, \ldots, x_n)$, the posterior distribution of θ can be obtained as

$$\pi(\theta|\boldsymbol{x}) = \frac{b^a}{m(\boldsymbol{x})\Gamma(a)}\theta^{n+a-1}\, e^{-\theta(b+\sum_{i=1}^{n} x_i)} \tag{8.9}$$

for $\theta > 0$, and zero otherwise. Here

$$m(\boldsymbol{x}) = \int_0^\infty \frac{b^a}{\Gamma(a)}\theta^{n+a-1}\, e^{-\theta(b+\sum_{i=1}^{n} x_i)}d\theta = \frac{b^a}{\Gamma(a)} \times \frac{\Gamma(n+a)}{(b+\sum_{i=1}^{n} x_i)^{n+a}}.$$

Therefore,

$$\pi(\theta|\boldsymbol{x}) = \frac{(b+\sum_{i=1}^{n} x_i)^{n+a}}{\Gamma(n+a)}\theta^{n+a-1}e^{-\theta(b+\sum_{i=1}^{n} x_i)}.$$

Hence, the posterior distribution of θ becomes a gamma distribution with the shape and scale parameters as $n + a$ and $(b + \sum_{i=1}^{n} x_i)$, respectively. As we have mentioned before, the posterior distribution of θ provides complete information about the unknown parameter θ. If we want a point estimate of θ, the posterior mean can be considered as one such estimate and in this case it will be

$$\widehat{\theta}_{Bayes} = \frac{\Gamma(n+a)}{b+\sum_{i=1}^{n} x_i}.$$

Similarly, an associated confidence $100(1-\alpha)\%$ credible interval of θ, say (L, U), can also be constructed using posterior distribution as follows. Choose L and U such that

$$\int_L^U \pi(\theta|\boldsymbol{x})d\theta = 1 - \alpha.$$

It is clear from the above discussions that integration techniques play a significant role in Bayesian inference. Here, we provide some illustration of some of the Monte Carlo techniques used for integration in Bayesian inference. We provide it with an example. Suppose $(X_1, \ldots, X_n)$, a random sample, is drawn from an $N(\theta, \sigma^2)$, where σ^2 is known. Then, $Y = \bar{X}$ is a sufficient statistic. Consider the Bayes model:

$$Y|\theta \sim N(\theta, \sigma^2/n)$$

$$\Theta \sim h(\theta) \propto \exp\{-(\theta-a)/b\}/(1+\exp\{-[(\theta-a)/b]^2\}); \quad -\infty < \theta < \infty. \tag{8.10}$$

Here, a and $b > 0$ are known hyperparameters. The distribution of Θ is known as the logistic distribution with parameter a and b. The posterior PDF is

$$h(\theta|y) = \frac{\frac{1}{\sqrt{2\pi\sigma/n}}\exp\left\{-\frac{(y-\theta)^2}{2\sigma^2/n}\right\}e^{-(\theta-a)/b}/(1+e^{[-(\theta-a)/b]^2})}{\int_{-\infty}^{\infty}\frac{1}{\sqrt{2\pi\sigma/n}}\exp\left\{-\frac{(y-\theta)^2}{2\sigma^2/n}\right\}e^{-(\theta-a)/b}/(1+e^{[-(\theta-a)/b]^2})d\theta}.$$

Based on the squared error loss function, the Bayes estimate of θ becomes the mean of the posterior distribution. It involves computing two integrations, which cannot be obtained explicitly. In this case, Monte Carlo simulation technique can be used quite effectively to compute the Bayes estimate and the associated credible interval of θ. Consider the following likelihood function as a function of θ:

$$w(\theta) = \frac{1}{\sqrt{2\pi\sigma/n}} \exp\left\{-\frac{(y-\theta)^2}{2\sigma^2/n}\right\}.$$

Therefore, the Bayes estimate can be written as

$$\delta(y) = \frac{\int_{-\infty}^{\infty} \theta w(\theta) b^{-1} e^{-(\theta-a)/b}/(1+e^{[-(\theta-a)/b]^2}) d\theta}{\int_{-\infty}^{\infty} w(\theta) b^{-1} e^{-(\theta-a)/b}/(1+e^{[-(\theta-a)/b]^2}) d\theta} = \frac{E(\Theta w(\Theta))}{E(w(\Theta))},$$

where the expectation is taken with Θ having a logistic prior distribution. The computation of $\delta(y)$ can be carried out by the simple Monte Carlo technique as follows: Generate independently $\Theta_1, \Theta_2, \ldots, \Theta_M$ from the logistic distribution with PDF (8.10). This generation is straightforward, as the inverse function of the logistic distribution can be expressed in explicit form. Then compute

$$T_M = \frac{M^{-1}\sum_{i=1}^{M} \Theta_i w(\Theta_i)}{M^{-1}\sum_{i=1}^{M} w(\Theta_i)}.$$

By WLLN (Khinchine's theorem), it immediately follows that $T_M \xrightarrow{p} \delta(y)$, as $M \to \infty$. By bootstrapping this sample, the confidence interval of $\delta(y)$ can also be obtained, see, for example, Davison and Hinkley (1997). There are several very good books available on Bayesian theory and methodology. The readers are referred to Gilks et al. (1996) or Ghosh et al. (2006) for an easy reading on different Bayesian methodologies.

8.5 Linear and Nonlinear Regression Analysis

One of the most important problems in statistical analysis is to find the relationships, if any, that exist in a set of variables when at least one is random. In a regression problem, typically, one of the variables, usually called the dependent variable, is of particular interest, and it is denoted by y. The other variables $x_1, x_2, \ldots, x_k$, usually called explanatory variables or independent variables, are mainly used to predict or explain the behavior of y. If the prior experience or the plots of the data suggest some relationship between y and x_i, then we would like to express this relationship via some function, f, namely

$$y \approx f(x_1, x_2, \ldots, x_k). \tag{8.11}$$

Now, using the functional Equation 8.11, from the given $x_1, x_2, \ldots, x_k$, we might be able to predict y. For example, y could be the price of a used car of a certain make, x_1, the number of previous owners, x_2, the age of the car, and x_3 the mileage. As expected, the relationship (8.11) can never be exact, as data will always contain unexplained fluctuations or noise, and some degree of measurements error is usually present.

Explanatory variables can be random or fixed (i.e., controlled). Consider an experiment conducted to measure the yield (y) of wheat at different specified levels of density planting (x_1), and fertilizer application (x_2). In this case, both x_1 and x_2 are fixed. If at the time of planting, soil pH (x_3) was also measured on each plot, then x_3 would be random.

In both linear and nonlinear regression analysis, it is assumed that the mathematical form of the relationship (8.11) is known, except for some unknown constants or coefficients, called parameters, the relationship being determined by a known underlying physical process or governed by some accepted scientific laws. Therefore, mathematically, Equation 8.11 can be written as

$$y \approx f(x_1, x_2, \ldots, x_k, \theta), \tag{8.12}$$

where the function f is entirely known, except for the parameter vector $\theta = (\theta_1, \theta_2, \ldots, \theta_p)$, which is unknown, and based on the observation it needs to be estimated. In both linear and nonlinear regression analysis, it is often assumed that the noise present is additive in nature. Hence, mathematically, the model can be written in the following form:

$$y = f(x_1, x_2, \ldots, x_k, \theta) + \epsilon, \tag{8.13}$$

where ϵ is the noise random variable, and $E(\epsilon) = 0$.

In case of linear regression model, the function f is assumed to be linear, and the model has the following form:

$$y = \beta_0 + \beta_1 x_1 + \cdots + \beta_{p-1} x_{p-1} + \epsilon. \tag{8.14}$$

Here, x_i's can include squares, cross products, higher powers, and even transformations of the original measurements. For example

$$y = \beta_0 + \beta_1 x_1 + \beta_2 x_2 + \beta_3 x_1 x_3 + \beta_4 x_2^2 + \epsilon$$

or

$$y = \beta_0 + \beta_1 e^{x_1} + \beta_2 \ln x_2 + \beta_3 \sin x_3 + \epsilon$$

are both linear models. The important requirement is that the expression should be linear in the parameters. On the other hand, if the function f is not linear in the parameters, it is called a nonlinear regression model. For example, the following models:

$$y = \beta_0 + \beta_1 e^{\beta_2 x_1} + \beta_3 e^{\beta_4 x_2} + \epsilon$$

and

$$y = \beta_0 + \beta_1 \sin(\beta_2 x_1) + \beta_3 \cos(\beta_4 x_1) + \epsilon$$

are nonlinear regression models.

As can be observed from the above examples, linear models are very flexible, and so it is often used in the absence of a theoretical model f. Nonlinear models tend to be used either when they are suggested by theoretical consideration to build a known nonlinear behavior into a model. Even when a linear approximation works well, a nonlinear model may still be used to retain a clear interpretation of the parameters.

The main aim of this section is to consider both the linear and nonlinear regression models and discuss different inferential issues associated in both of them. It may be mentioned that the linear regression models are very well studied in the statistical literature and they are quite well understood also. On the other hand, the nonlinear regression analysis is much less understood, although it has a huge scope of applications. There are several unanswered questions, and lots of scope for future research.

8.5.1 Linear Regression Analysis

It is quite convenient to represent the linear regression model in the following matrix notation:

$$\boldsymbol{y} = \boldsymbol{X}\boldsymbol{\beta} + \boldsymbol{\epsilon}, \tag{8.15}$$

here, $\boldsymbol{y} = (y_1, y_2, \ldots, y_n)^T$ is the vector of n observations, $\boldsymbol{X}$ is the known $n \times p$ design matrix, where $p < n$ and the rank of the matrix $\boldsymbol{X}$ is p. It will be denoted by

$$\boldsymbol{X} = \begin{bmatrix} x_{11} & \ldots & x_{1p} \\ \vdots & \ddots & \vdots \\ x_{p1} & \ldots & x_{pp} \end{bmatrix},$$

and $\boldsymbol{\epsilon} = (\epsilon_1, \epsilon_2, \ldots, \epsilon_p)^T$ is the noise random vector. For simplicity, we will assume that ϵ_i's are independent identically distributed normal random variables with mean zero and variance σ^2. The problem is to provide the statistical inferences of the unknown parameters $\boldsymbol{\beta}$ and σ^2, based on the observation vector $\boldsymbol{y}$ and the design matrix $\boldsymbol{X}$. It is immediate from the above assumptions that $\boldsymbol{y}$ has n-variate normal distribution with the mean vector $\boldsymbol{X}\boldsymbol{\beta}$ and dispersion matrix $\sigma^2\boldsymbol{I}_n$. Here, $\boldsymbol{I}_n$ denotes the $n \times n$ identity matrix.

The joint probability density function of $\boldsymbol{y}$, given $\boldsymbol{\beta}$ and σ^2, is

$$\begin{aligned} p(\boldsymbol{y}|\boldsymbol{\beta}, \sigma^2) &= (2\pi\sigma^2)^{-n/2} \exp\left\{-\frac{(\boldsymbol{y} - \boldsymbol{X}\boldsymbol{\beta})^T(\boldsymbol{y} - \boldsymbol{X}\boldsymbol{\beta})}{2\sigma^2}\right\} \\ &= (2\pi\sigma^2)^{-n/2} \exp\left\{-\frac{||\boldsymbol{y} - \boldsymbol{X}\boldsymbol{\beta}||^2}{2\sigma^2}\right\}, \end{aligned} \tag{8.16}$$

here, for the vector $\boldsymbol{a} = (a_1, a_2, \ldots, a_n)^T$, $||\boldsymbol{a}|| = \sqrt{a_1^2 + a_2^2 + \cdots + a_n^2}$. The likelihood function, or more simply, the likelihood $l(\boldsymbol{\beta}, \sigma|\boldsymbol{y})$, for $\boldsymbol{\beta}$ and σ is identical in form to the joint probability density (8.16), except that $l(\boldsymbol{\beta}, \sigma|\boldsymbol{y})$ is regarded as a function of the parameters conditional of the observed data, rather than as a function of the responses conditional of the values of the parameters. Suppressing the constant $(2\pi)^{-n/2}$, the likelihood can be written as

$$l(\boldsymbol{\beta}, \sigma|\boldsymbol{y}) \propto \sigma^{-n} \exp\left\{-\frac{||\boldsymbol{y} - \boldsymbol{X}\boldsymbol{\beta}||^2}{2\sigma^2}\right\}. \tag{8.17}$$

Therefore, the MLEs of the unknown parameters can be obtained by maximizing Equation 8.17 with respect to $\boldsymbol{\beta}$ and σ. It immediately follows that the likelihood (8.17) is maximized with respect to $\boldsymbol{\beta}$, when the residual sum of squares $S(\boldsymbol{\beta}) = ||\boldsymbol{y} - \boldsymbol{X}\boldsymbol{\beta}||^2$ is minimum. Thus, the MLE $\widehat{\boldsymbol{\beta}}$ is the value of $\boldsymbol{\beta}$ that minimizes $S(\boldsymbol{\beta})$ can be obtained as

$$\widehat{\boldsymbol{\beta}} = \left(\boldsymbol{X}^T\boldsymbol{X}\right)^{-1}\boldsymbol{X}^T\boldsymbol{y}. \tag{8.18}$$

$\widehat{\boldsymbol{\beta}}$ is the least squares estimate of $\boldsymbol{\beta}$ also. In deriving Equation 8.18, it is assumed that the matrix $\boldsymbol{X}$ is of full rank. If the rank of the matrix is less than p, then clearly $\left(\boldsymbol{X}^T\boldsymbol{X}\right)^{-1}$ does not exist. In this case, although $\widehat{\boldsymbol{\beta}}$ exists, it is not unique. It is not pursued any further here. Moreover, the MLE of σ^2

can be obtained as

$$\widehat{\sigma^2} = \frac{||y - X\widehat{\beta}||^2}{n}. \tag{8.19}$$

For detailed treatments on linear regression models and for their applications, the readers are referred to Arnold (1981) and Rao (2008).

Least squares estimates can also be derived using sampling theory, since the least squares estimator is the minimum variance unbiased estimator of β, or by using a Bayesian approach with a noninformative prior density on β and σ. Interestingly, all three methods of inference, the likelihood approach, the sampling theory approach, and the Bayesian approach, produce the same point estimate of β.

However, it is important to realize that the MLE of least squares estimates are appropriate when the model and the error assumptions are correct. Expressed in another way, using the least squares estimates, we assume

1. The expectation function is correct.
2. The response is expectation function plus noise.
3. The noise is independent of the expectation function.
4. Each error component has a normal distribution.
5. Each error component has mean zero.
6. The errors have equal variances.
7. The errors are independently distributed.

When these assumptions appear reasonable, we can go to make further inferences about the least squares estimates.

Least squares estimator or the MLE has a number of desirable properties. For example:

1. The least squares estimator $\widehat{\beta}$ is normally distributed. This mainly follows as the least squares estimator is a linear function of y, which in turn is a linear function of ϵ. Since ϵ is assumed to be normally distributed, $\widehat{\beta}$ is also normally distributed.
2. $\widehat{\beta}$ is an unbiased estimator of β, that is, $E(\widehat{\beta}) = \beta$.
3. $\text{Var}(\widehat{\beta}) = \sigma^2(X^T X)^{-1}$; that is, the covariance matrix of the least squares estimator depends on the error variances and design matrix X.
4. A $100(1 - \alpha)\%$ joint confidence set for β is the ellipsoid

$$\left(\beta - \widehat{\beta}\right)^T X^T X \left(\beta - \widehat{\beta}\right) \leq ps^2 F_{p,n-p,\alpha}, \tag{8.20}$$

where

$$s^2 = \frac{S(\widehat{\beta})}{n - p}$$

is the residual mean square or an unbiased estimator of σ^2, and $F_{p,n-p,\alpha}$ is the upper α quantile for Fisher's F distribution with p and $n - p$ degrees of freedom.

5. A $100(1-\alpha)\%$ marginal confidence interval for the parameter β_j, for $j = 1, 2, \ldots, p$, is

$$\widehat{\beta}_j \pm \text{se}(\widehat{\beta})_j t_{n-p,\alpha/2}, \tag{8.21}$$

where $t_{n-p,\alpha/2}$ is the upper $\alpha/2$ quantile for Student's t distribution with $n-p$ degrees of freedom, and the standard error of the parameter estimator is

$$\text{se}(\widehat{\beta})_j = s\sqrt{\left\{\left(\boldsymbol{X}^T\boldsymbol{X}\right)^{-1}\right\}_{jj}},$$

with $\left\{\left(\boldsymbol{X}^T\boldsymbol{X}\right)^{-1}\right\}_{jj}$ equal to the jth diagonal term of the matrix $\left(\boldsymbol{X}^T\boldsymbol{X}\right)^{-1}$.

6. A $100(1-\alpha)\%$ confidence interval for the expected response at $\boldsymbol{x}_0$ is

$$\boldsymbol{x}_0^T\widehat{\beta} \pm t_{n-p,\alpha/2}\sqrt{\boldsymbol{x}_0^T\left(\boldsymbol{X}^T\boldsymbol{X}\right)^{-1}\boldsymbol{x}_0}. \tag{8.22}$$

The least squares estimators are the most popular method to estimate the unknown parameters of a linear regression model, and it has several desirable properties. For example, it can be obtained in explicit form; in case of i.i.d. normal error distributions, the MLEs and the LSEs coincide, but it has certain disadvantages. For example, LSEs are not robust, that is, even in the presence of a very small number of outliers, the estimators can change drastically, which may not be desirable. Moreover, in the presence of heavy tail errors, the LSEs do not behave well. Owing to this reason, least absolute deviation (LAD) estimator can be used, which is more robust, and behaves very well in the presence of heavy tail error. The LAD estimator of $\boldsymbol{\beta}$ can be obtained by minimizing

$$|\boldsymbol{y} - \boldsymbol{X}\boldsymbol{\beta}|, \tag{8.23}$$

with respect to $\boldsymbol{\beta}$, where for $\boldsymbol{a} = (a_1, \ldots, a_n)^T$ and $\boldsymbol{b} = (b_1, \ldots, b_n)^T$, $|\boldsymbol{a} - \boldsymbol{b}| = |a_1 - b_1| + \cdots + |a_n - b_n|$. The estimator $\widetilde{\boldsymbol{\beta}}$, which minimizes Equation 8.23, cannot be obtained in explicit form. It has to be obtained numerically. Several numerical methods are available to solve this problem. One important technique is to convert this problem to a linear programming problem, and then solve the linear programming problem, by some efficient linear programming problem solvers; see, for example, an excellent book by Kennedy and Gentle (1980) in this respect. Regarding the theoretical development of the LAD estimators, see Huber (1981).

Another important aspect in a linear regression problem is to estimate the number of predictors. It is a fairly difficult problem, and it can be treated as a model selection problem. Classically, the problem was solved by using stepwise regression method, but recently, different information theoretic criteria such as Akaike information criterion (AIC) or Bayesian information criterion (BIC) have been used to solve this problem. Recently, the least absolute shrinkage and selection operator (LASSO) method proposed by Tibshirani (1996) has received considerable amount of attention in the last one decade.

8.5.1.1 Bayesian Inference

Before closing this section, we will briefly discuss the Bayesian inference of the linear regression model. The Bayesian marginal posterior density for $\boldsymbol{\beta}$, assuming a noninformative prior density for $\boldsymbol{\beta}$ and σ of the form,

$$p(\boldsymbol{\beta}, \sigma) \propto \sigma^{-1} \tag{8.24}$$

is

$$p(\beta|\sigma) \propto \left\{1 + \frac{(\beta - \widehat{\beta})^T X^T X (\beta - \widehat{\beta})}{\nu s^2}\right\}^{-n/2}. \quad (8.25)$$

It is in the form of a p-variate Student's t density with the location parameter $\widehat{\beta}$, scaling matrix $s^2(X^T X)^{-1}$, and $\nu = n - p$ degrees of freedom.

Furthermore, the marginal posterior density for a single parameter β_j, say, is a univariate Student's t density with location parameter $\widehat{\beta}_j$, scale parameter $s^2 \left\{(X^T X)^{-1}\right\}_{jj}$, and the degrees of freedom $n - p$. The marginal posterior density for the mean of y at x_0 is a univariate Student's t density with the location parameter $x_0^T \widehat{\beta}$, scale parameter $s^2 x_0^T (X^T X)^{-1} x_0$, and the degrees of freedom $n - p$.

A $100(1 - \alpha)\%$ highest posterior density (HPD) region of content is defined as a region R in the parameter space such that $P(\beta \in R) = 1 - \alpha$, and for $\beta_1 \in R$, and $\beta_2 \notin R$, $P(\beta_1 \in R) \geq P(\beta_2 \in R)$. For linear models with a noninformative prior, an HPD region is therefore given by the ellipsoid defined in Equation 8.20. Similarly, the marginal HPD regions for β_j and $x_0^T \beta$ are numerically identical to the sampling theory regions (8.21) and (8.22), respectively.

Example 8.5

We consider the data of the maximum January temperature (in degrees Fahrenheit) for 62 cities in the United States (from 1931 to 1960) along with their latitude (degrees), longitude (degrees), and altitude (feet). The data have been taken from Mosteller and Tukey (1977). We want to relate the maximum January temperature with the other three variables. We write the model in the following form:

$$\textit{Max Temp} = \beta_0 + \beta_1 \times \textit{Latitude} + \beta_2 \times \textit{Longitude} + \beta_3 \times \textit{Altitude} + \epsilon.$$

The following summary measures are obtained for the design matrix X:

$$\mathbf{X}^T\mathbf{X} = \begin{bmatrix} 62.0 & 2365.0 & 5674.0 & 56{,}012.0 \\ 2365.0 & 92{,}955.0 & 217{,}285.0 & 2{,}244{,}586.0 \\ 5674.0 & 217{,}285.0 & 538{,}752.0 & 5{,}685{,}654.0 \\ 56{,}012.0 & 2{,}244{,}586.0 & 5{,}685{,}654.0 & 1.772 \times 10^8 \end{bmatrix},$$

$$\left(\mathbf{X}^T\mathbf{X}\right)^{-1} = \begin{bmatrix} 94{,}883.1914 & -1342.5011 & -485.0209 & 2.5756 \\ -1342.5011 & 37.8582 & -0.8276 & -0.0286 \\ -485.0209 & -0.8276 & 5.8951 & -0.0254 \\ 2.5756 & -0.0286 & -0.0254 & 0.0009 \end{bmatrix},$$

$$X^T y = (2739.0, 99{,}168.0, 252{,}007.0, 2{,}158{,}463.0)^T.$$

It gives

$$\widehat{\beta} = (100.8260, -1.9315, 0.2033, -0.0017)^T \quad \text{and} \quad s = 6.05185.$$

Therefore, based on the normality assumption on the error random variables, and the priors discussed above, $\widehat{\beta}$ can be taken as the least squares estimators and Bayes estimators of β. Further, 100

$(1-\alpha)\%$ credible interval for β can be obtained as

$$\left\{\beta : (\beta - \widehat{\beta})^T \boldsymbol{X}^T \boldsymbol{X}(\beta - \widehat{\beta}) \leq ps^2 F_{p,n-p}(\alpha)\right\}.$$

8.5.2 Nonlinear Regression Analysis

The basic problem in the nonlinear regression analysis can be expressed as follows. Suppose that we have n observations $\{(\boldsymbol{x}_i, y_i); i = 1, 2, \ldots, n\}$ from a nonlinear model with a known functional relationship f. Thus

$$y_i = f(\boldsymbol{x}_i, \theta^*) + \epsilon_i; \quad i = 1, 2 \ldots, n, \tag{8.26}$$

where ϵ_i's are assumed to be independent and identically distributed normal random variables with mean zero, and variance σ^2, $\boldsymbol{x}_i$ is a $k \times 1$ vector, and the true value θ^* of θ is known to belong to Θ, a subset of $\mathcal{R}^p$. The problem is to estimate the unknown parameter θ^*, based on the above observations $\{(\boldsymbol{x}_i, y_i); i = 1, 2, \ldots, n\}$.

Among the different methods, the most popular one is the least squares estimator, denoted by $\widehat{\theta}$, and it can be obtained by minimizing the error sum of squares $S(\theta)$, where

$$S(\theta) = \sum_{i=1}^{n} (y_i - f(\boldsymbol{x}_i, \theta))^2, \tag{8.27}$$

over $\theta \in \Theta$. Unlike least squares estimator, a simple analytical solution of $\widehat{\theta}$ does not exist, and it has to be obtained numerically. Several numerical methods are available, which can be used to compute $\widehat{\theta}$, but all the methods are iterative in nature. Hence, each iterative method requires some kind of initial guesses to start the process. This is one of the major problems in computing the least squares estimator $\widehat{\theta}$. The least squares surface $S(\theta)$ may have several local minima; hence, a very careful choice of the initial guess is required to compute $\widehat{\theta}$.

For completeness purposes, we provide one numerical method that can be used to compute $\widehat{\theta}$, but it should be mentioned that it may not be the best method in all possible cases. We use the following notation: $\theta = (\theta_1, \theta_2, \ldots, \theta_p)^T$, $f_i(\theta) = f(\boldsymbol{x}_i, \theta)$, for $i = 1, \ldots, n$, $\boldsymbol{f}(\theta) = (f_1(\theta), f_2(\theta), \ldots, f_n(\theta))^T$. The main idea of the proposed method is to approximate the nonlinear surface $\boldsymbol{f}(\theta)$ near $\widehat{\theta}$ by a linear surface as follows. Suppose $\theta^{(a)}$ is an approximation to the least squares estimate $\widehat{\theta}$. For θ close to $\widehat{\theta}$, by using Taylor series approximation, $f(\theta)$ can be written as follows:

$$\boldsymbol{f}(\theta) \approx \boldsymbol{f}(\theta^{(a)}) + \boldsymbol{F}_{\bullet}{}^{(a)}(\theta - \theta^{(a)}), \tag{8.28}$$

here, $\boldsymbol{F}_{\bullet}{}^{(a)} = \boldsymbol{F}_{\bullet}(\theta^{(a)})$, and

$$\boldsymbol{F}_{\bullet}(\theta) = \begin{bmatrix} \dfrac{\partial f_1(\theta)}{\partial \theta_1} & \cdots & \dfrac{\partial f_1(\theta)}{\partial \theta_p} \\ \cdots & \ddots & \cdots \\ \dfrac{\partial f_n(\theta)}{\partial \theta_1} & \cdots & \dfrac{\partial f_n(\theta)}{\partial \theta_p} \end{bmatrix}.$$

Applying this to the residual vector

$$\boldsymbol{r}(\theta) = \boldsymbol{y} - \boldsymbol{f}(\theta) \approx \boldsymbol{r}(\theta^{(a)}) - \boldsymbol{F}_{\bullet}{}^{(a)}(\theta - \theta^{(a)}),$$

in $S(\theta) = \boldsymbol{r}^T(\theta)\boldsymbol{r}(\theta)$, leads to

$$S(\theta) \approx \boldsymbol{r}^T(\theta^{(a)})\boldsymbol{r}(\theta^{(a)}) - 2\boldsymbol{r}^T(\theta^{(a)})\boldsymbol{F}_{\bullet}{}^{(a)}(\theta - \theta^{(a)}) + (\theta - \theta^{(a)})^T\boldsymbol{F}_{\bullet}{}^{(a)T}\boldsymbol{F}_{\bullet}{}^{(a)}(\theta - \theta^{(a)}). \tag{8.29}$$

The right-hand side of Equation 8.29 is minimized with respect to θ, when

$$\theta - \theta^{(a)} = \left(\boldsymbol{F}_{\bullet}{}^{(a)T}\boldsymbol{F}_{\bullet}{}^{(a)}\right)^{-1}\boldsymbol{F}_{\bullet}{}^{(a)T}\boldsymbol{r}(\theta^{(a)}) = \delta^{(a)}.$$

This suggests that given a current approximation $\theta^{(a)}$, the next approximation can be obtained as

$$\theta^{(a+1)} = \theta^{(a)} + \delta^{(a)}.$$

This provides an iterative scheme for computing $\widehat{\theta}$. This particular iterative scheme is known as Gauss–Newton method. It forms the basis of a number of least squares algorithm used in the literature. The Gauss–Newton algorithm is convergent, that is, $\theta^{(a)} \to \widehat{\theta}$, as $a \to \infty$, provided that $\theta^{(1)}$ is close to θ^*, and n is large enough.

Finally, we will conclude this section to provide some basic properties of the least squares estimators without any formal proof. Under certain regularity conditions on the function f, if the error random variables are i.i.d. normal random variables with mean zero and variance σ^2, for large n, we have the following results:

1. $(\widehat{\theta} - \theta^*) \sim N_p(\boldsymbol{0}, \sigma^2\boldsymbol{C})$, where $\boldsymbol{C} = \boldsymbol{F}_{\bullet}{}^T(\theta^*)\boldsymbol{F}_{\bullet}(\theta^*)$.
2. $S(\widehat{\theta})/\sigma^2 \sim \chi^2_{n-p}$.
3. $\dfrac{(S(\theta^*) - S(\widehat{\theta}))/p}{S(\widehat{\theta})/(n-p)} \sim F_{p,n-p}$.

Here, χ^2_{n-p} and $F_{p,n-p}$ denote the chi square distribution with $n - p$ degrees of freedom, and F distribution with p and $n - p$ degrees of freedom, respectively. The result (1) provides the consistency and asymptotic normality properties of the least squares estimators; moreover, (2) and (3) can be used for the construction of the confidence interval or confidence set for σ^2 and θ.

Example 8.6

The data set has been obtained from Osborne (1972), and it represents the concentration of a chemical during the different time point of an experiment. We fit the following nonlinear model:

$$y_t = \alpha_0 + \alpha_1 e^{\beta_1 t} + \alpha_2 e^{\beta_2 t} + \epsilon_t.$$

We start with the initial guesses as $\alpha_0 = 0.5$, $\alpha_1 = 1.5$, $\alpha_2 = -1.0$, $\beta_1 = -0.01$, and $\beta_2 = -0.02$. Using Newton–Raphson method, we finally obtained the least squares estimators as

$$\widehat{\alpha}_0 = 0.3754, \quad \widehat{\alpha}_1 = 1.9358, \quad \widehat{\alpha}_2 = -1.4647, \quad \widehat{\beta}_1 = -0.01287, \quad \widehat{\beta}_2 = -0.02212.$$

8.6 Introduction to Multivariate Analysis

Multivariate analysis (MVA) is the study based on the statistical principle of multivariate statistics, and involves the observation and analysis of more than one statistical outcome variable at a time. To motivate our readers, we present three different examples of MVA below.

Example 8.7

Consider as a dietician in the hospital you are interested to study the physical features and find out the relevant parameters, such as body density (BD), body mass index (BMI), etc., of patients who undergo treatment in the hospital. Your job is to decide on the right diet plan based on the data/information such as percent body fat, age (years), weight (kg), height (cm), etc. of the patients. For the study, you use the past data (http://wiki.stat.ucla.edu/socr/index.php/SOCR_Data_ BMI_Regression), which consists of a sample set of 252 patients, as this enables you to do a detailed study/analysis of the different characteristics, such as body fat index, height, and weight, using a three-dimensional (3-D) scatter plot, as illustrated in Figure 8.1.

Example 8.8

As the next example, assume you are the real estate agent in the state of California, USA, and your job is to forecast the median house value (MHV). To facilitate better forecasting, you have with you 20,640 data points consisting of information such as MHV, median income (MI), housing median age (HMA), total rooms (TR), total bedrooms (TB), population (P), households (H), etc. Information about the data set can be obtained in the paper by Kelley and Barry (1997). If one fits the multiple linear regression (MLR) model (as used by the authors) to this data, one obtains the ordinary least square (OLS) regression coefficient vector, $\hat{\beta} = (\hat{\beta}_0 = 11.4939,\ \hat{\beta}_1 = 0.4790,\ \hat{\beta}_2 = -0.0166,\ \hat{\beta}_3 = -0.0002,\ \hat{\beta}_4 = 0.1570,\ \hat{\beta}_5 = -0.8582,\ \hat{\beta}_6 = 0.8043,\ \hat{\beta}_7 = -0.4077,\ \hat{\beta}_8 = 0.0477)$, using which we can forecast the 20,641th MHV as $\widehat{log_e(MHV)}_{20,641} = \hat{\beta}_0 + \hat{\beta}_1 \times MI_{20,641} + \hat{\beta}_2 \times MI^2_{20,641} + \hat{\beta}_3 \times MI^3_{20,641} + \hat{\beta}_4 \times log_e(MA_{20,641}) + \hat{\beta}_5 \times log_e(TR_{20,641}/P_{20,641}) + \hat{\beta}_6 \times log_e(TB_{20,641}/P_{20,641}) + \hat{\beta}_7 \times log_e(P_{20,641}/H_{20,641}) + \hat{\beta}_8 \times log_e(H_{20,641})$. For example, if one wants to forecast the 20,621th reading, which is 11.5129255, then the forecasted value is 12.3302108, which results in an error of 0.8172853.

Example 8.9

As a third example, consider that Professor Manisha Kumari, a faculty member in the Finance Group at the Indian Institute of Management, Indore, India, is interested to study the change

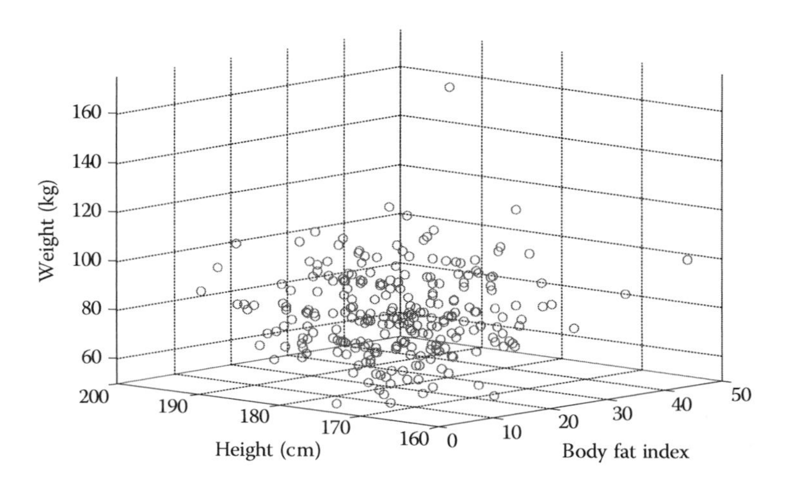

FIGURE 8.1
A three-dimensional scatter plot of body fat index, height, and weight of 252 patients, Example 8.7.

in the prices of seven stocks, namely, Bajaj Auto, Maruti Suzuki Indian Limited, Tata Motors, Steel Authority of India, Tata Steel, Infosys Limited, and Tata Consultancy Services Limited, for the time period January 1, 2014 to December 31, 2014. She utilizes the prices of these seven stocks from National Stock Exchange (NSE), which is available at http://in.finance.yahoo.com or http://www.nse-india.com. A closer look convinces her that the price for the first three scripts (Bajaj Auto [#1], Maruti Suzuki Indian Limited [#2], and Tata Motors [#3]), the next two (Steel Authority of India [#4] and Tata Steel [#5]), and the last two (Infosys Limited [#6] and Tata Consultancy Services Limited [#7]) moves in tandem as separate groups as they are from the automobile, steel, and information technology sectors, respectively. Her surmise is valid as the companies that are in the same sector tend to vary together as economic conditions change and this fact is also substantiated by the factor analysis (FA) performed by her (Figure 8.2).

In all these three examples, what is important to note is the fact that given a multidimensional data set, $X_{n \times p}$, of size $(n \times p)$, the users, be it the dietician, the real estate agent, or the faculty member, are all interested to draw some meaningful conclusions from this data set, $X_{n \times p}$. The study of multivariate statistics leads us to analyze multivariate distributions. As is apparent, such studies are of prime importance in many areas of our practical life. It is interesting to note that Francis Galton (1822–1911) may be credited as the first person who worked in the area of multivariate statistical analysis. In his work *Natural Inheritance* (1889), the author summarized the ideas of regression considering bivariate normal distribution. Other notable early researchers whose contributions in the area of multivariate statistics are worth mentioning are Theodore Wilbur Anderson (1918–), Steven F. Arnold (1944–2014), Debabrata Basu (1924–2001), Morris L. Eaton (1939–), Ronald Aylmer Fisher (1890–1962), Narayan Chandra Giri (1928–2006), Ramanathan Gnanadesikan (1932), Harold Hotelling (1895–1973), Norman Lloyd Johnson (1917–2004), Maurice George Kendall (1907–1983), C. G. Khatri (1931–1989), Samuel Kotz (1930–2010), Paruchuri R. Krishnaiah (1932–1987), Anant M. Kshirsagar (1931), Prasanta Chandra Mahalanobis (1893–1972), Calyampudi Radhakrishna Rao (1920–), Samarendra Nath Roy (1906–1964), George Arthur Frederick Seber (1938–), Samuel Stanley Wilks (1906–1964), and many others. Thus, the study of the body of methodologies to investigate simultaneous measurements on many variables is termed as MVA.

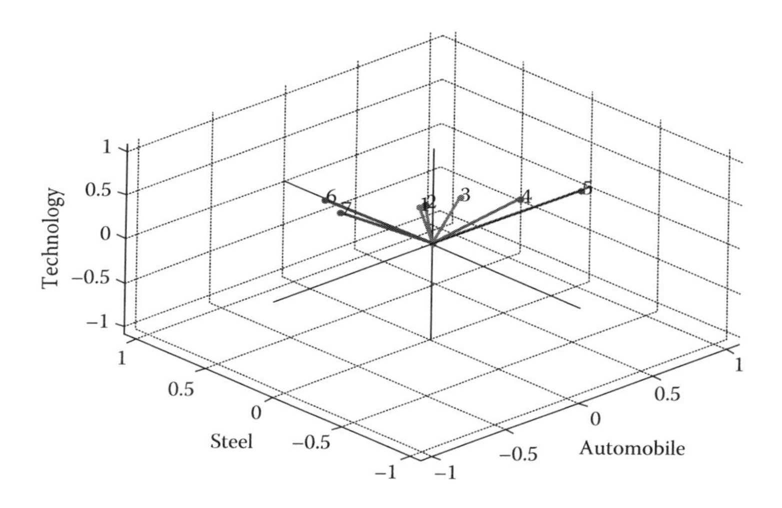

FIGURE 8.2
Illustration of factor analysis (FA) method considering seven stocks from NSE, India for the time period January 1, 2014 to December 31, 2014, Example 8.9.

While studying multivariate methods, the objectives that are of prime importance are: data reduction or structural simplification, sorting and grouping, investigation of the dependence among variables, prediction and hypothesis construction, and subsequent testing of the same.

To start with, let us define $\boldsymbol{X}_{n\times p} = (\boldsymbol{X}_1, \ldots, \boldsymbol{X}_p)$ or $(X_{i,j})$, $i = 1, \ldots, n$ and $j = 1, \ldots, p$ as an $(n \times p)$-dimensional matrix of random variables, where n signifies the number of readings and p signifies the dimension, corresponding to different factors in a random variable that are of interest to us. A few important definitions that are useful to understand MVA are:

1. Mean value vector: $\boldsymbol{\mu}_{p\times 1} = (\mu_1, \ldots, \mu_p)'$, while the sample counterpart is $\bar{\boldsymbol{X}}_{p\times 1} = (\bar{X}_1, \ldots, \bar{X}_p)'$.
2. Variance–covariance matrix:
$$\boldsymbol{\Sigma}_{p\times p} = \begin{pmatrix} \sigma_{1,1} & \cdots & \sigma_{1,p} \\ \vdots & \ddots & \vdots \\ \sigma_{p,1} & \cdots & \sigma_{p,p} \end{pmatrix},$$
while the sample counterpart is
$$\boldsymbol{S}_{p\times p} = \begin{pmatrix} s_{1,1} & \cdots & s_{1,p} \\ \vdots & \ddots & \vdots \\ s_{p,1} & \cdots & s_{p,p} \end{pmatrix}.$$
3. Correlation coefficient matrix:
$$\boldsymbol{\rho}_{p\times p} = \begin{pmatrix} 1 & \cdots & \rho_{1,p} \\ \vdots & \ddots & \vdots \\ \rho_{p,1} & \cdots & 1 \end{pmatrix},$$
while the sample counterpart is
$$\boldsymbol{R}_{p\times p} = \begin{pmatrix} 1 & \cdots & r_{1,p} \\ \vdots & \ddots & \vdots \\ r_{p,1} & \cdots & 1 \end{pmatrix}.$$
4. Mean: $E(X_j) = \mu_j = \sum_{\forall x_j} x_j Pr(X_j = x_j)$, or $E(X_j) = \mu_j = \int_{x_{j,\min}}^{x_{j,\max}} x_j f(x_j) dx_j = \int_{x_{j,\min}}^{x_{j,\max}} x_j dF_{X_j}(x_j)$, while the sample counterpart is $\bar{X}_j = (1/n)\sum_{i=1}^{n} X_{i,j}$, for $j = 1, \ldots, p$.
5. Covariance: $Covar\left(X_{j_1}, X_{j_2}\right) = E\left[\left\{X_{j_1} - E\left(X_{j_1}\right)\right\}\left\{X_{j_2} - E\left(X_{j_2}\right)\right\}\right] = \sigma_{j_1,j_2} = \sum_{\forall x_{j_1}, x_{j_2}} \left\{X_{j_1} - E\left(X_{j_1}\right)\right\}\left\{X_{j_2} - E\left(X_{j_2}\right)\right\} Pr\left(X_{j_1} = x_{j_1}, X_{j_2} = x_{j_2}\right)$, or $Covar\left(X_{j_1}, X_{j_2}\right) = E\left[\left\{X_{j_1} - E\left(X_{j_1}\right)\right\}\left\{X_{j_2} - E\left(X_{j_2}\right)\right\}\right] = \sigma_{j_1,j_2} = \int_{x_{j_2,\min}}^{x_{j_2,\max}} \int_{x_{j_1,\min}}^{x_{j_1,\max}} \left\{X_{j_1} - E\left(X_{j_1}\right)\right\} \left\{X_{j_2} - E\left(X_{j_2}\right)\right\} f\left(x_{j_1}, x_{j_2}\right) dx_{j_1}\, dx_{j_2} = \int_{x_{j_2,\min}}^{x_{j_2,\max}} \int_{x_{j_1,\min}}^{x_{j_1,\max}} \left\{X_{j_1} - E\left(X_{j_1}\right)\right\}\left\{X_{j_2} - E\left(X_{j_2}\right)\right\} dF_{X_{j_1}, X_{j_2}}\left(x_{j_1}, x_{j_2}\right)$, while the sample counterpart is $s_{j_1,j_2} = (1/(n-1))\sum_{i=1}^{n}\left(X_{i,j_1} - \bar{X}_{j_1}\right)\left(X_{i,j_2} - \bar{X}_{j_2}\right)$, for j_1, $j_2 = 1, \ldots, p$.

6. Correlation coefficient: $corr\left(X_{j1}, X_{j2}\right) = \rho_{j1,j2} = Covar\left(X_{j1}, X_{j2}\right) / \sqrt{Var\left(X_{j1}\right)} \sqrt{Var\left(X_{j2}\right)}$, while the sample counterpart is $r_{j1,j2} = \left(s_{j1,j2} / \sqrt{s_{j1,j1}} \sqrt{s_{j2,j2}}\right)$, for j_1, $j_2 = 1, \ldots, p$.
7. Co-skewness:

$$E\left[\left\{X_{j1} - E\left(X_{j1}\right)\right\}\left\{X_{j2} - E\left(X_{j2}\right)\right\}\left\{X_{j3} - E\left(X_{j3}\right)\right\}\right]$$
$$= \sum_{\forall x_{j1}, x_{j2}, x_{j3}} \left\{X_{j1} - E\left(X_{j1}\right)\right\}\left\{X_{j2} - E\left(X_{j2}\right)\right\}\left\{X_{j3} - E\left(X_{j3}\right)\right\}$$
$$\times Pr\left(X_{j1} = x_{j1}, X_{j2} = x_{j2}, X_{j3} = x_{j3}\right)$$

or

$$\int_{x_{j3,\min}}^{x_{j3,\max}} \int_{x_{j2,\min}}^{x_{j2,\max}} \int_{x_{j1,\min}}^{x_{j1,\max}} \left\{X_{j1} - E\left(X_{j1}\right)\right\}\left\{X_{j2} - E\left(X_{j2}\right)\right\}\left\{X_{j3} - E\left(X_{j3}\right)\right\}$$
$$\times f\left(x_{j1}, x_{j2}, x_{j3}\right) dx_{j1}\, dx_{j2}\, dx_{j3},$$

for $j_1, j_2, j_3 = 1, \ldots, p$.

Note: Co-skewness is related to skewness as covariance is related to variance.

8. Skew relation:

$$\frac{E\left[\left\{X_{j1} - E\left(X_{j1}\right)\right\}\left\{X_{j2} - E\left(X_{j2}\right)\right\}\left\{X_{j3} - E\left(X_{j3}\right)\right\}\right]}{\sqrt{E\left\{X_{j1} - E\left(X_{j1}\right)\right\}^2} \sqrt{E\left\{X_{j2} - E\left(X_{j2}\right)\right\}^2} \sqrt{E\left\{X_{j3} - E\left(X_{j3}\right)\right\}^2}},$$

for $j_1, j_2, j_3 = 1, \ldots, p$.
9. Co-kurtosis:

$$E\left[\left\{X_{j1} - E\left(X_{j1}\right)\right\}\left\{X_{j2} - E\left(X_{j2}\right)\right\}\left\{X_{j3} - E\left(X_{j3}\right)\right\}\left\{X_{j4} - E\left(X_{j4}\right)\right\}\right]$$
$$= \sum_{\forall x_{j1}, x_{j2}, x_{j3}, x_{j4}} \left\{X_{j1} - E\left(X_{j1}\right)\right\}\left\{X_{j2} - E\left(X_{j2}\right)\right\}\left\{X_{j3} - E\left(X_{j3}\right)\right\}\left\{X_{j4} - E\left(X_{j4}\right)\right\}$$
$$\times \Pr\left(X_{j1} = x_{j1}, X_{j2} = x_{j2}, X_{j3} = x_{j3}, X_{j4} = x_{j4}\right)$$

or

$$\int_{x_{j4,\min}}^{x_{j4,\max}} \int_{x_{j3,\min}}^{x_{j3,\max}} \int_{x_{j2,\min}}^{x_{j2,\max}} \int_{x_{j1,\min}}^{x_{j1,\max}} \left\{X_{j1} - E\left(X_{j1}\right)\right\}\left\{X_{j2} - E\left(X_{j2}\right)\right\}\left\{X_{j3} - E\left(X_{j3}\right)\right\}$$
$$\times \left\{X_{j4} - E\left(X_{j4}\right)\right\} f\left(x_{j1}, x_{j2}, x_{j3}, x_{j4}\right) dx_{j1}\, dx_{j2}\, dx_{j3}\, dx_{j4},$$

for $j_1, j_2, j_3, j_4 = 1, \ldots, p$.

Note: Co-kurtosis is related to kurtosis as covariance is related to variance.

10. Kurtic relation:

$$\frac{E\left[\left\{X_{j_1}-E\left(X_{j_1}\right)\right\}\left\{X_{j_2}-E\left(X_{j_2}\right)\right\}\left\{X_{j_3}-E\left(X_{j_3}\right)\right\}\left\{X_{j_4}-E\left(X_{j_4}\right)\right\}\right]}{\sqrt{E\left\{X_{j_1}-E\left(X_{j_1}\right)\right\}^2}\sqrt{E\left\{X_{j_2}-E\left(X_{j_2}\right)\right\}^2}\sqrt{E\left\{X_{j_3}-E\left(X_{j_3}\right)\right\}^2}\sqrt{E\left\{X_{j_4}-E\left(X_{j_4}\right)\right\}^2}},$$

for $j_1, j_2, j_3 j_4 = 1, \ldots, p$.

To represent the multivariate data, different graphical techniques can also be used, some of which are: scatter diagram/scatter plot/marginal dot diagram, multiple scatter plot, box plot, 3-D scatter plot, linked scatter plot, rotated plot, growth plot, Chernoff faces, stars, etc. Another important concept that is used in the study of multivariate statistics is the idea of distance measure. A few examples are: Euclidean distance, Bhattacharyya distance, Mahalanobis distance, Pitman closeness criterion, Bregman divergence, Kullback–Leibler distance, Hellinger distance, Chernoff bound, Rényi entropy, and Cook's distance. An interested reader can refer many good references for a better understanding of MVA, a few examples of which are: Anderson (2003), Arnold (1981), Bock (1975), Cooley and Lohnes (1971), Dillon and Goldstein (1984), Eaton (1983), Everitt and Dunn (2001), Giri (2004), Gnanadesikan (2011), Hair et al. (2005), Härdle and Simar (2007), Jobson (1991), Johnson and Wichern (2002), Kendall (1980), Kotz et al. (2000), Kshirsagar (1972), Mardia et al. (1979), Morrison (1990), Muirhead (2005), Press (1982), Rao (2008), Roy (1957), Roy et al. (1971), Seber (2004), Srivastava and Khatri (1979), Takeuchi et al. (1982), and Tatsuoka (1988). Sen (1986) gives a good review of textbooks, papers, monographs, and other related materials in the area of multivariate statistics.

For the interest of the readers, we consider a few multivariate distributions such as multinomial distribution, multivariate normal distribution (MND), multivariate Student t-distribution, Wishart distribution, and multivariate extreme value distribution (MEVD) before discussing copula theory. After that, we cover different multivariate techniques that are widely used. One may note that other multivariate distributions such as Dirichlet distribution, Hotelling distribution, multivariate gamma distribution, multivariate beta distribution, multivariate exponential distribution, etc. are not considered due to paucity of space. Moreover, the distributions discussed here are based on their general relevance and practicality.

8.7 Joint and Marginal Distribution

The joint distribution of $(X_1, \ldots, X_p)$ may be expressed as $F_{X_1,\ldots,X_p}(x_1, \ldots, x_p) = Pr(X_1 \leq x_1, \ldots, X_p \leq x_p)$. If one thinks from the marginal distribution point of view, then $F_{X_1,\ldots,X_p}(x_1, \ldots, x_p)$ consists of $(2^p - 2)$ number of marginal distributions of which $\binom{p}{1}$ are univariate, $\binom{p}{2}$ are bivariate,..., and finally $\binom{p}{p-1}$ are $(p-1)$ variate. If $X_1, \ldots, X_p$ are pairwise independent, then the joint distribution function $F_{X_1,\ldots,X_p}(x_1, \ldots, x_p) = F_{X_1}(x_1) \times \cdots \times F_{X_p}(x_p)$, where $F_{X_j}(x_j)$ is the corresponding marginal distribution of X_j, $j = 1, \ldots, p$. In case $F_{X_1,\ldots,X_p}(x_1, \ldots, x_p) = F_{X_1,\ldots,X_{j_1}}\left(x_1, \ldots, x_{j_1}\right) \times F_{X_{j_1+1},\ldots,X_{j_2}}\left(x_{j_1+1}, \ldots, x_{j_2}\right) \times F_{X_{j_2+1},\ldots,X_p}\left(x_{j_2+1}, \ldots, x_p\right)$, then one can similarly add that $\left(X_1, \ldots, X_{j_1}\right), \left(X_{j_1+1}, \ldots, X_{j_2}\right)$, and $\left(X_{j_2+1}, \ldots, X_p\right)$ are independent and $F_{X_1,\ldots,X_{j_1}}\left(x_1, \ldots, x_{j_1}\right)$, $F_{X_{j_1+1},\ldots,X_{j_2}}\left(x_{j_1+1}, \ldots, x_{j_2}\right)$, and $F_{X_{j_2+1},\ldots,X_p}\left(x_{j_2+1}, \ldots, x_p\right)$ are the joint distributions of $\left(X_1, \ldots, X_{j_1}\right)$, $\left(X_{j_1+1}, \ldots, X_{j_2}\right)$, and $\left(X_{j_2+1}, \ldots, X_p\right)$, respectively.

If the distribution is of discrete type, then the total mass of the distribution of $(X_1, \ldots, X_p)$ is concentrated at the points in a way such that $\sum_{\forall j_1} \cdots \sum_{\forall j_p} Pr\left\{X_1 = x_{1,j_1}, \ldots, X_p = x_{p,j_p}\right\} = 1$, while for the continuous case, we have $\int_{x_p=-\infty}^{x_p=+\infty} \cdots \int_{x_1=-\infty}^{x_1=+\infty} f(x_1, \ldots, x_p)\, dx_1 \cdots dx_p = 1$. The corresponding joint distribution functions would be given as $F_{X_1,\ldots,X_p}(x_1, \ldots, x_p) = \sum_{X_1 \le x_1} \cdots \sum_{X_p \le x_p} Pr\{X_1 \le x_1, \ldots, X_p \le x_p\}$ or $F_{X_1,\ldots,X_p}(x_1, \ldots, x_p) = \int_{x_p=-\infty}^{x_p} \cdots \int_{x_1=-\infty}^{x_1} f(x_1, \ldots, x_p) dx_1 \cdots dx_p$ as the case may be.

Considering $\boldsymbol{X}_{n\times p} = (\boldsymbol{X}_1, \ldots, \boldsymbol{X}_p)$, we may be interested to measure the degree to which one of the variable, say X_j, is dependent on the remaining $(p-1)$ number of variables, that is, $(X_1, \ldots, X_{j-1}, X_{j+1}, \ldots, X_p)$ taken jointly. This is called *multiple correlation* and the measure is given by *multiple correlation coefficient*

$$\rho_{j,(1,\ldots,j-1,j+1,\ldots,p)} = \frac{Covar\left(X_j, X_{1,\ldots,j-1,j+1,\ldots,p}\right)}{\sqrt{Var(X_j)} \times \sqrt{Var\left(X_{j,(1,\ldots,j-1,j+1,\ldots,p)}\right)}} = \left(1 - \frac{R}{R_{j_1,j_2}}\right)^{1/2},$$

where R_{j_1,j_2} is the cofactor of ρ_{j_1,j_2} in the determinant R of the correlation matrix

$$\boldsymbol{R} = \begin{pmatrix} \rho_{1,1} & \cdots & \rho_{1,p} \\ \vdots & \ddots & \vdots \\ \rho_{p,1} & \cdots & \rho_{p,p} \end{pmatrix}.$$

Another way of representing the multiple correlation coefficient is

$$\rho^2_{j,(1,\ldots,j-1,j+1,\ldots,p)} = 1 - \left\{\frac{Var\left(\varepsilon_{j,(1,\ldots,j-1,j+1,\ldots,p)}\right)}{Var(X_j)}\right\},$$

where $\varepsilon_{j,(1,\ldots,j-1,j+1,\ldots,p)}$ is the residual of X_j corresponding to its multiple regression on $X_1, \ldots, X_{j-1}, X_{j+1}, \ldots, X_p$. This multiple correlation coefficient may be interpreted as the *maximum* correlation between X_j and a linear function of $X_1, \ldots, X_{j-1}, X_{j+1}, \ldots, X_p$, say $X_j = \alpha + \beta_1 X_1 + \cdots + \beta_{j-1} X_{j-1} + \beta_{j+1} X_{j+1} + \cdots + \beta_p X_p + \varepsilon$. On the other hand, *partial correlation coefficient* between X_{j_1} and X_{j_2} is denoted by $\rho_{(j_1,j_2|1,\ldots,j_1-1,j_1+1,\ldots,j_2-1,j_2+1,\ldots,p)} = (-1)^{j_1+j_2}\left(R_{j_1,j_2} / \left(R_{j_1,j_1} R_{j_2,j_2}\right)^{1/2}\right)$, where R_{j_1,j_1}, R_{j_2,j_2}, and R_{j_1,j_2} have the usual definition as already mentioned while discussing multiple correlation coefficient, $\rho_{j,(1,\ldots,j-1,j+1,\ldots,p)}$. Remember that partial correlation coefficients are related to the partial regression coefficients by the formula

$$\beta_{(j_1,j_2|1,\ldots,j_1-1,j_1+1,\ldots,j_2-1,j_2+1,\ldots,p)} = \rho_{(j_1,j_2|1,\ldots,j_1-1,j_1+1,\ldots,j_2-1,j_2+1,\ldots,p)} \times \frac{\sigma_{(j_1|1,\ldots,j_1-1,j_1+1,\ldots,j_2-1,j_2+1,\ldots,p)}}{\sigma_{(j_2|1,\ldots,j_1-1,j_1+1,\ldots,j_2-1,j_2+1,\ldots,p)}}.$$

Example 8.10

Consider the data related to cigarettes given in Mendenhall and Sincich (2006). The data set contains measurements related to brand, tar content (mg), nicotine content (mg), weight (g), and carbon monoxide content (mg) for $n = 25$ brands of cigarettes. The data set can be accessed at http://www.amstat.org/publications/jse/datasets/cigarettes. dat.txt. Considering the variables,

```
1:  DEFINE: n, p, A, R, determinant of R (i.e., det(R)), cofactors of (j1,j2) (i.e., R_{j1,j2}) from R
2:  INPUT: n, p, A
3:  START If: j1 = 1:p
4:     START If: j2 = 1:p
5:     CALCULATE: R, determinant of R (i.e., det(R)), cofactors of (j1,j2) (i.e., R_{j1,j2}) from R
6:     END if
7:  END if
8:  CALCULATE: (1 - det(R)/R_{j1j2})^(1/2)
9:  REPORT: (1 - det(R)/R_{j1j2})^(1/2)
10: END
```

FIGURE 8.3
Pseudo-code used for calculating multiple correlation coefficient vector.

$p = 4$, that is, X_1, X_2, X_3, and X_4 as the tar content (mg), nicotine content (mg), weight (g), and carbon monoxide content (mg), respectively, one obtains

$$\boldsymbol{R} = \begin{bmatrix} 1 & 0.9766 & 0.4908 & 0.9575 \\ 0.9766 & 1 & 0.5002 & 0.9259 \\ 0.4908 & 0.5002 & 1 & 0.4640 \\ 0.9575 & 0.9259 & 0.4640 & 1 \end{bmatrix}.$$

Furthermore, utilizing $\boldsymbol{R}$, we get the multiple correlation coefficient vector as

$$\begin{pmatrix} 0.9867 \\ 0.9774 \\ 0.5001 \\ 0.9584 \end{pmatrix},$$

while the corresponding R^2 values are 0.9720, 0.9554, 0.5366, and 0.9174. Simple calculations would also yield the partial correlation coefficient matrix as

$$\begin{pmatrix} 1 & -0.8199 & -0.0141 & -0.6556 \\ -0.8199 & 1 & -0.1092 & 0.1465 \\ -0.0141 & -0.1092 & 1 & 0.0072 \\ -0.6556 & 0.1465 & 0.0072 & 1 \end{pmatrix}.$$

To double verify the calculation, we may also calculate the partial regression coefficients. The pseudo-codes for calculating the multiple correlation coefficient vector and the partial correlation coefficient matrix are given in Figures 8.3 and 8.4, respectively.

```
1:  DEFINE: n, p, A, R, cofactors of (j1,j2) (i.e., R_{j1,j2}) from R
2:  INPUT: n, p, A
3:  START If: j1 = 1:p
4:     START If: j2 = 1:p
5:     CALCULATE: R, cofactors of (j1,j2) (i.e., R_{j1,j2}) from R
6:     END if
7:  END if
8:  CALCULATE: (-1)^(j1+j2) R_{j1j2} / (R_{j1j1} R_{j2j2})^(1/2)
9:  REPORT: (-1)^(j1+j2) R_{j1j2} / (R_{j1j1} R_{j2j2})^(1/2)
10: END
```

FIGURE 8.4
Pseudo-code used for calculating partial correlation coefficient matrix.

8.8 Multinomial Distribution

Suppose $X_1, \ldots, X_p$ be p jointly distributed random variables each of which is discrete, nonnegative, and integer valued. Then, the joint probability mass function of $X_1, \ldots, X_p$ is called the *multinomial* distribution and is of the form

$$\binom{n}{x_1, \ldots, x_p} \times p_1^{x_1} \times \cdots \times p_p^{x_p},$$

where $x_i \in (0, 1, \ldots n-1, n)$, $\sum_{j=1}^{p} x_j = n$, and $\sum_{j=1}^{p} p_j = 1$.

For the multinomial distribution, it can be easily proved that

1. $E(X_j) = np_j, j = 1, \ldots, p$.
2. $Var(X_j) = np_j(1 - p_j), j = 1, \ldots, p$.
3. $Covar\left(X_{j_1}, X_{j_2}\right) = -np_{j_1}p_{j_2}, j_1 \neq j_2 = 1, \ldots, p$.
4. $$corr\left(X_{j_1}, X_{j_2}\right) = -\left\{\frac{p_{j_1}p_{j_2}}{\left(1 - p_{j_1}\right)\left(1 - p_{j_2}\right)}\right\}^{1/2}, \quad j_1 \neq j_2 = 1, \ldots, p.$$
5. Moment generating function (MGF) $= \left\{\sum_{j=1}^{p} p_j e^{t_j}\right\}^n$.
6. Characteristic function (CF) $= \left\{\sum_{j=1}^{p} p_j e^{it_j}\right\}^n$, where $i^2 = -1$.
7. Probability generating function (PGF) $= \left\{\sum_{j=1}^{p} p_j z_j\right\}^n$ for $(z_1, \ldots, z_p) \in C^p$.

For a better appreciation of the multinomial distribution, consider the polynomial coefficients of the expansion of the multinomial expansion, $\{p_1x_1 + \cdots + p_px_p\}^n$. A closer look at this expansion will make it immediately evident how the polynomial coefficients of the multinomial expansion correspond to the multinomial distribution discussed above. Another interesting analogy for multinomial distribution can be made from the *Pascal pyramid* (Figure 8.5).

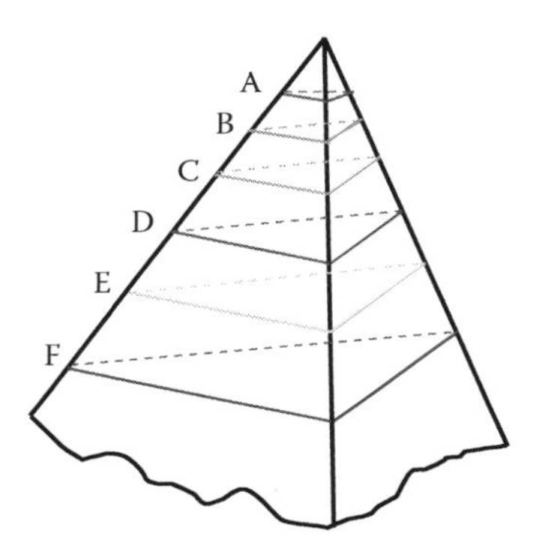

FIGURE 8.5
Pascal pyramid considered to depict the coefficient of the multinomial distribution.

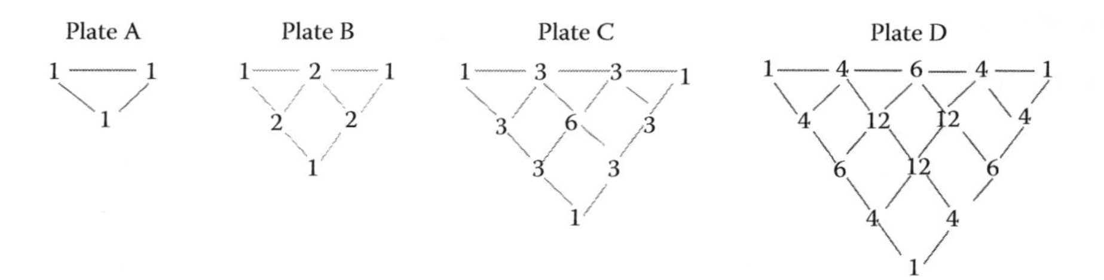

FIGURE 8.6
Depiction of the flat plates of Pascal pyramid to signify the concept of multinomial distribution.

If we view the slices (as represented by A, B, C, D, E, F, and so on) of the *Pascal pyramid* as flat triangular plates, then the numbers depicted on them are as shown in Figure 8.6.

When $p = 2$, we have the binomial distribution, while for $p = 3$, one obtains the trinomial distribution, and so on. It can be shown that the *marginal distribution* of X_j, $j = 1, \ldots, p$ is binomially distributed with parameters n and p_j, and is given by

$$Pr(X_j = x_j) = \binom{n}{x_j} p_j^{x_j}(1 - p_j)^{n-x_j},$$

where $(p_1 + \cdots + p_{j-1}) + p_j + (p_{j+1} + \cdots + p_p) = 1$ and $\{(x_1 + \cdots + x_{j-1}) + x_j + (x_{j+1} + \cdots + x_p)\} = n$. If one considers the *conditional distribution* of X_j, then given $X_1 = x_1, \ldots, X_{j-1} = x_{j-1}, X_{j+1} = x_{j+1}, \ldots, X_p = x_p$, the *conditional distribution* is

$$Pr(X_j = x_j | X_1 = x_1, \ldots, X_{j-1} = x_{j-1}, X_{j+1} = x_{j+1}, \ldots, X_p = x_p)$$

$$= \left(n / \left(j! \{n - \sum_{i \in (J-j)} x_i\}! \right) \right) p_j^{x_j}(1 - p_j)^{n-x_j},$$

Example 8.11

Consider the use of contraceptive among married women in El Salvador in 1985, and the data for the same for a sample of 3165 respondents is shown in Table 8.1.

TABLE 8.1

Data Related to Use of Contraceptive among Married Women in El Salvador, 1985

	Contraceptive Method			
Age (Years)	**Sterilization**	**Other Method**	**None**	**All**
15–19	3	61	232	296
20–24	80	137	400	617
25–29	216	131	301	648
30–34	268	76	203	547
35–39	197	50	188	435
40–44	150	24	164	338
45–49	91	19	183	284
All	1005	489	1671	3165

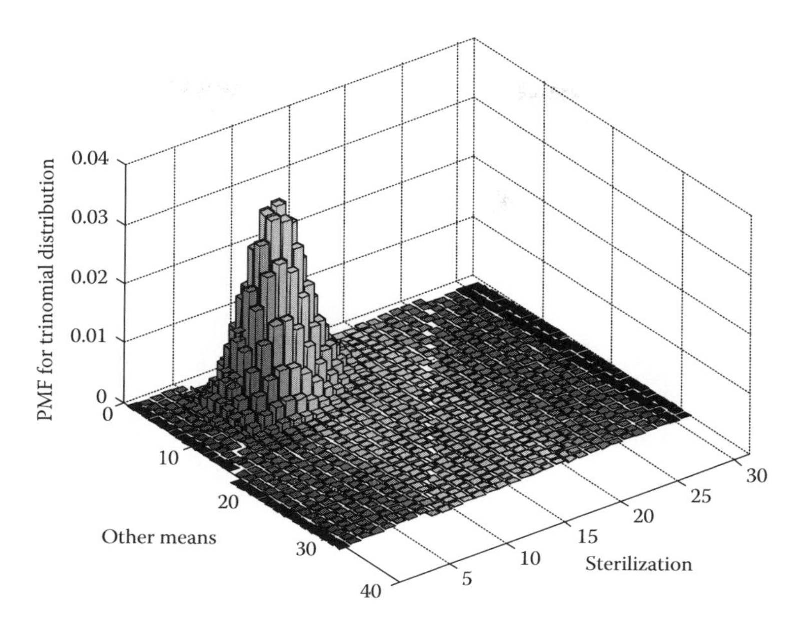

FIGURE 8.7
PMF for the trinomial distribution utilizing the data for the contraceptive use by married women in El Salvadore, 1985, Example 8.11.

Consider X_1, X_2, and X_3 as the random variable signifying the case of sterilization, other method, and none. Then, the joint multinomial distribution may be written as

$$f_{X_1,X_2,X_3}(x_1, x_2, x_3) = Pr(X_1 = x_1, X_2 = x_2, X_3 = x_3)$$
$$= \binom{n}{x_1, x_2, x_3} \times \left(\frac{1005}{3165}\right)^{x_1} \times \left(\frac{489}{3165}\right)^{x_2} \times \left(\frac{1671}{3165}\right)^{x_3},$$

where $(x_1 + x_2 + x_3) = n$. If we assume $n = 30$, then the PMF of the trinomial distribution and the general pseudo-code used for generating the same are given in Figures 8.7 and 8.8, respectively.

In case one is interested to generate the multivariate multinomial random variable, then the MATLAB® code is `mnrnd(n,pp,m)`, where n is the dimension of the multinomial distribution, $\boldsymbol{p}_p = (p_1, \ldots, p_p)$, and m is the number of observations, that is, the sample size we wish to generate. Hence, if $p = (0.1, 0.2, 0.3, 0.4), n = 100$ then $E(X) = (\mu_1, \mu_2, \mu_3, \mu_4) = (10, 20, 30, 40)$,

```
1:  DEFINE: p, p_j, x_j, j = 1,...,p, f(x) = (n choose x_1,...,x_p) × p_1^{x_1} × .....× p_p^{x_p}, x-axis, y-axis, z-axis
2:  INPUT: p, p_j, j = 1,...,p
3:  START If: i = 1:n
4:     CALCULATE: f(x)
5:  END if
6:  PLOT: (x-axis, y-axis, z-axis)
7:  REPORT: f(x)
8:  END
```

FIGURE 8.8
Pseudo-code used for Example 8.11.

$Var(\boldsymbol{X}) = (9, 16, 21, 24)$. Other values such as $Covar\left(X_{i_1}, X_{i_2}\right)$, $corr\left(X_{i_1}, X_{i_2}\right)$, PGF, MGF, and CF can be calculated accordingly.

8.9 Multivariate Normal Distribution

We say $\boldsymbol{X}_p \sim N_p(\boldsymbol{\mu}, \boldsymbol{\Sigma})$ is a nonsingular MND when its density function is given by

$$f_{X_1,\ldots,X_p}(x_1, \ldots, x_p) = \frac{1}{(2\pi)^{p/2}} |\boldsymbol{\Sigma}|^{-1/2} e^{\{-(1/2)(\boldsymbol{X}-\boldsymbol{\mu})'\Sigma^{-1}(\boldsymbol{X}-\boldsymbol{\mu})\}},$$

where $-\infty < X_i < +\infty$, $E(\boldsymbol{X}) = \boldsymbol{\mu}$, $Covar(\boldsymbol{X}) = \boldsymbol{\Sigma}$ along with the fact that

$$\boldsymbol{\Sigma} = \begin{bmatrix} \sigma_{11} & \cdots & \sigma_{1p} \\ \vdots & \ddots & \vdots \\ \sigma_{p1} & \cdots & \sigma_{pp} \end{bmatrix} > \boldsymbol{0}.$$

The study of MND is useful as many other multivariate statistics are approximately normal regardless of the parent distribution because of the CLT effect.

For the MND, it can be easily proved that

1. $E(X_j) = \mu_j$, $j = 1, \ldots, p$.
2. $Var(X_j) = \sigma^2$, $j = 1, \ldots, p$.
3. $Covar\left(X_{j_1}, X_{j_2}\right) = \sigma_{j_1, j_2}$, $j_1, j_2 = 1, \ldots, p$.
4. $corr\left(X_{j_1}, X_{j_2}\right) = \rho_{j_1, j_2}$, $j_1, j_2 = 1, \ldots, p$.
5. $\text{MGF} = e^{(\boldsymbol{\mu}'\boldsymbol{t}+(1/2)\boldsymbol{t}'\boldsymbol{\Sigma}\boldsymbol{t})}$, $t \in \mathbb{R}$.
6. $\text{CF} = e^{(i_m\boldsymbol{\mu}'\boldsymbol{t}+(1/2)\boldsymbol{t}'\boldsymbol{\Sigma}\boldsymbol{t})}$, where $i_m^2 = -1$.

Furthermore, we state a few results without proofs. Given $\boldsymbol{X}_p \sim N_p(\boldsymbol{\mu}, \boldsymbol{\Sigma})$, we have the following:

1. Contours of consistent density for the p-dimensional normal distribution are ellipsoids defined by $\boldsymbol{X}$, such that $(\boldsymbol{X} - \boldsymbol{\mu})', \boldsymbol{\Sigma}^{-1}(\boldsymbol{X} - \boldsymbol{\mu}) = c^2$, where c is a constant. These ellipsoids are centered at $\boldsymbol{\mu}$ and have axes $\pm c\sqrt{\lambda_j} e_j$, where $\boldsymbol{\Sigma} e_j = \lambda_j e_j$, $j = 1, \ldots, p$. Here, $(\lambda, \boldsymbol{e})$ is the eigen value–eigen vector for $\boldsymbol{\Sigma}$ corresponding to the pair $(1/\lambda, \boldsymbol{e})$ for $\boldsymbol{\Sigma}^{-1}$. Remember $\boldsymbol{\Sigma}^{-1}$ is positive definite.
2. If $\boldsymbol{\Sigma}$ is positive definite, so that $\boldsymbol{\Sigma}^{-1}$ exists, then $\boldsymbol{\Sigma}\boldsymbol{e} = \lambda\boldsymbol{e}$ implies $\boldsymbol{\Sigma}^{-1}\boldsymbol{e} = (1/\lambda)\boldsymbol{e}$.
3. All subsets of the components of $\boldsymbol{X}$ have MNDs.
4. Zero covariance implies that the corresponding components are independently distributed.
5. The conditional distributions of the components are multivariate normal.
6. The q linear combinations of the components of $\boldsymbol{X}$ are also normally distributed. Thus, if $\boldsymbol{X}$ is distributed as $N_p(\boldsymbol{\mu}, \boldsymbol{\Sigma})$, then q linear combinations $\boldsymbol{A}_{q\times p}\boldsymbol{X}_{p\times 1} \sim N_q(\boldsymbol{A}\boldsymbol{\mu}, \boldsymbol{A}'\boldsymbol{\Sigma}\boldsymbol{A})$. Also, $\boldsymbol{X}_{p\times 1} + \boldsymbol{d}_{p\times 1}$, where $\boldsymbol{d}_{p\times 1}$ is a vector of constants, is distributed as $N_p(\boldsymbol{\mu} + \boldsymbol{d}\boldsymbol{\Sigma})$.
7. $(\boldsymbol{X}_p - \boldsymbol{\mu}) \sim N_p(\boldsymbol{0}, \boldsymbol{\Sigma})$.

8. $X \sim N_p(\mu, \sigma^2\mathbf{I})$, provided $X_j \sim N(\mu_j, \sigma^2)$, $j = 1, \ldots, p$, are mutually independent univariate normal distributions.
9. The solid ellipsoid of x values satisfying $(x - \mu)'\Sigma^{-1}(x - \mu) \leq \chi_p^2(\alpha)$ has probability $(1 - \alpha)$, that is, $Pr\left\{(x - \mu)'\Sigma^{-1}(x - \mu) \leq \chi_p^2(\alpha)\right\} = (1 - \alpha)$.
10. If X is distributed as $N_p(\mu, \Sigma)$, then any linear combination of variables $a'X = \sum_{j=1}^{p} a_j X_j$ is distributed as $N_p(a'\mu, a'\Sigma a)$. Conversely, if $a'X$ is distributed as $N_p(a'\mu, a'\Sigma a)$ for every a, then X must be $N_p(\mu, \Sigma)$.
11. Suppose $X \sim N_p(\mu, \Sigma)$, and

$$X_p = \begin{bmatrix} X_{k\times 1} \\ \cdots \\ X_{(p-k)\times 1} \end{bmatrix}, \quad \mu_p = \begin{bmatrix} \mu_{k\times 1} \\ \cdots \\ \mu_{(p-k)\times 1} \end{bmatrix}, \quad \Sigma_{p\times p} = \begin{bmatrix} \Sigma_{11} & \vdots & \Sigma_{12} \\ \cdots & \cdots & \cdots \\ \Sigma_{21} & \vdots & \Sigma_{22} \end{bmatrix},$$

then $X_k \sim N_k(\mu_k, \Sigma_{11})$ and $X_{p-k} \sim N_{p-k}(\mu_{p-k}, \Sigma_{22})$. The converse of this is also true.

An interesting and important concept in MND is something to do with circles and ellipses. Consider $p = 2$, then

$$f_{X_1,X_2}(x_1, x_2) = \frac{1}{2\pi\sqrt{\sigma_{11}\sigma_{22}\left(1 - \rho_{12}^2\right)}} \times \exp\left[-\frac{1}{2\left(1 - \rho_{12}^2\right)}\left\{\left(\frac{x_1 - \mu_1}{\sqrt{\sigma_{11}}}\right)^2 + \left(\frac{x_2 - \mu_2}{\sqrt{\sigma_{22}}}\right)^2 \right.\right.$$
$$\left.\left. -2\rho_{12}\left(\frac{x_1 - \mu_1}{\sqrt{\sigma_{11}}}\right)\left(\frac{x_2 - \mu_2}{\sqrt{\sigma_{22}}}\right)\right\}\right].$$

If we consider three different values of ρ_{12}, that is, negative, zero, and positive, and consider, $\sigma_{11} = \sigma_{22}$, then we obtain the contours as shown in Figure 8.9.

Example 8.12

Let us consider the data presented by Galton (1886), which shows a cross-tabulation of 963 adult children (486 sons and 476 daughters) born to 205 families, by their height and their midparent's

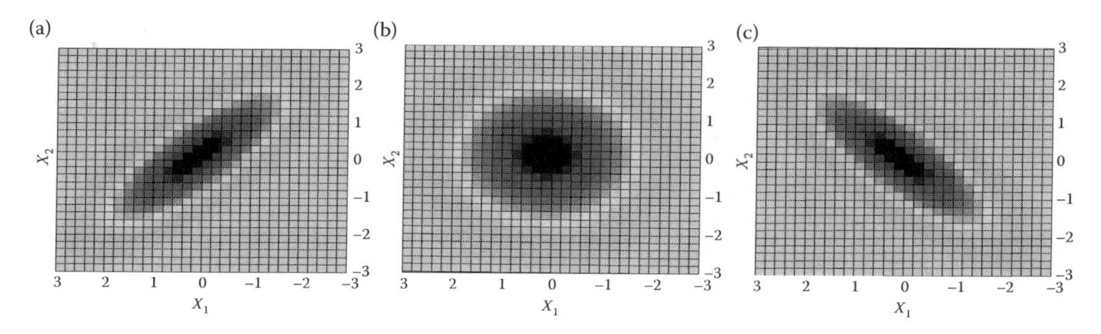

FIGURE 8.9
Contour plots for bivariate normal distribution considering ρ_{12} as (a) negative, (b) zero, and (c) positive.

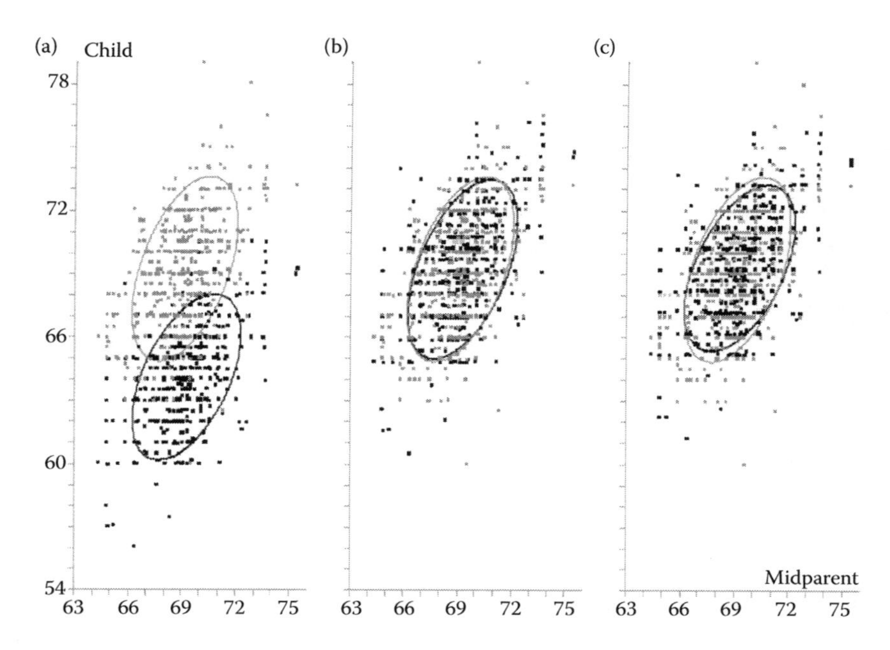

FIGURE 8.10
Ellipsoidal plots for Example 8.12 (refer Hanley, 2004).

height. The author visually smoothed the bivariate frequency distribution and showed that the contours formed concentric and similar ellipses, thus setting the stage for correlation, regression, and the bivariate normal distribution. The data is recorded in class intervals of width 1.0″. Furthermore, he used noninteger values for the center of each class interval because of the strong bias toward integral inches. All of the heights of female children were multiplied by 1.08 before tabulation to compensate for sex differences. One can also refer to Hanley (2004), along with the source materials at http://www.medicine.mcgill.ca/epidemiology/hanley/galton/ to have a better understanding about this study and the corresponding data analysis. The related ellipsoidal plots for this problem are shown in Figure 8.10.

The use of the basic concept of ellipsoids may be found in the area of reliability-based design optimization (RBDO), where this concept is used to depict the most reliable area (search space) within which the optimization solution is feasible, depending on the level of confidence. Though not exhaustive, a few good references in the area of RBDO are: Ben-Tal et al. (2009), Ben-Tal and Nemirovski (1998, 1999, 2002), and Bertsimas and Sim (2003, 2004, 2006).

8.10 Multivariate Student t-Distribution

The joint probability distribution function for the multivariate Student t-distribution (standard form) is

$$f_{Y_1,\ldots,Y_p}(y_1,\ldots,y_p) = \frac{\Gamma\{(v+p)/2\}}{(\pi v)^{p/2}\Gamma(v/2)|\mathbf{R}|^{1/2}}\left\{1+\frac{\mathbf{y}'\mathbf{R}^{-1}\mathbf{y}}{v}\right\}^{-(v+p)/2},$$

where $\boldsymbol{Y} = (Y_1,\ldots,Y_p)$, v is the degree of freedom for univariate t-distribution, and $Y_j = X_j/(S_j/\sqrt{v})$. Remember $X_1,\ldots,X_p$ have a joint standard multinormal distribution with

$E(\boldsymbol{X}) = \boldsymbol{0}$, $Covar(\boldsymbol{X}) = \mathbf{R}$ (the correlation matrix), and $S_j = (1/(n-1))\sum_{i=1}^{n}(X_{ij} - \bar{X}_j)^2$. Given this, the following results for this distribution can be easily derived:

1. $E(\boldsymbol{Y}) = \boldsymbol{0}$.
2. $Median(\boldsymbol{Y}) = \boldsymbol{0}$.
3. $Mode(\boldsymbol{Y}) = \boldsymbol{0}$.
4. $Covar(\boldsymbol{Y}) = (v/(v-2))\boldsymbol{R}$.

In its nonstandard form, it can be expressed as

$$f_{X_1,\ldots,X_p}(x_1,\ldots,x_p) = \frac{\Gamma\{(v+p)/2\}}{(\pi v)^{p/2}\Gamma(v/2)|\boldsymbol{\Sigma}|^{1/2}}\left\{1 + \frac{(\boldsymbol{x}-\boldsymbol{\mu})'\boldsymbol{\Sigma}^{-1}(\boldsymbol{x}-\boldsymbol{\mu})}{v}\right\}^{-(v+p)/2},$$

where $E(\boldsymbol{X}) = \boldsymbol{\mu}(v > 1)Covar(\boldsymbol{X}) = (v/(v-2))\boldsymbol{\Sigma}(v > 2)$, mode is $\boldsymbol{\mu}$, and finally median is also $\boldsymbol{\mu}$. When the p random variables are independent then $Y_1^2 + \cdots + Y_k^2$ has a joint multivariate F-distribution with parameters $(k-1)$ and $(v+k)$, and for the case when $v = 1$ one obtains the multivariate Cauchy distribution. One should remember that as v tends to infinity the joint distribution of $Y_1, \ldots, Y_p$ tends to the multinormal distribution with $E(\boldsymbol{Y}) = \boldsymbol{0}$ and $Covar(\boldsymbol{Y}) = \boldsymbol{R}$. Furthermore, the conditional probability distribution for the independent case is given by

$$f_{Y_{k+1},\ldots,Y_p|Y_1,\ldots,Y_k}(Y_{k+1},\ldots,Y_p|Y_1,\ldots,Y_k) = \frac{\Gamma\left(\frac{v+p}{2}\right)\left\{\frac{\left(1+v^{-1}\sum_{j=1}^{k}y_j^2\right)(v+k)}{v}\right\}^{(p-k)/2}}{\{\pi(v+k)\}^{(p-k)/2}\Gamma\left(\frac{v+k}{2}\right)} \times \left[1 + \frac{1}{v\left(1+v^{-1}\sum_{j=1}^{k}y_j^2\right)}\sum_{j=k+1}^{p}y_j^2\right]^{-(v+p)/2}.$$

Provided all the marginals have the same degrees of freedom, v, the marginal probability distribution is of the form

$$\frac{\Gamma((v+p)/2)|\boldsymbol{\Sigma}^{-1}|^{1/2}}{\sqrt{v^p\pi^p}\Gamma(v/2)}\left\{1 + \frac{1}{v}\sum_{j_1=1}^{p}\sum_{j_2=1}^{p}\sum_{j_1,j_2}^{-1} y_{j_1,j_2}\right\},$$

where

$$\boldsymbol{\Sigma}^{-1} = \begin{bmatrix} \Sigma_{1,1}^{-1} & \cdots & \Sigma_{1,p}^{-1} \\ \vdots & \ddots & \vdots \\ \Sigma_{p,1}^{-1} & \cdots & \Sigma_{p,p}^{-1} \end{bmatrix}.$$

In case one is interested to obtain the noncentral multivariate t-distribution of

$$Y_j = \left(\frac{U_j + \delta_j}{S_j/\sqrt{v}}\right),$$

where δ_j is the noncentrality parameters for t_j distribution (univariate t-distribution with parameter v), $\boldsymbol{U} \sim MVN(\boldsymbol{0}, \boldsymbol{R})$ and $\boldsymbol{U} = (U_1, \ldots, U_p)$, then it is given by

$$f_{Y_1,\ldots,Y_p}(y_1, \ldots, y_p) = \frac{e^{-(1/2)\boldsymbol{\delta}'\boldsymbol{R}^{-1}\boldsymbol{\delta}}}{(\pi v)^{p/2}\Gamma(v/2)|\boldsymbol{R}|^{1/2}}(1 + v^{-1}\mathbf{y}'\boldsymbol{R}^{-1}\mathbf{y})^{-(v+p)/2}$$

$$\times \sum_{j=0}^{\infty} \frac{\Gamma((v+p+j)/2)}{j!} \left\{ \frac{2(\boldsymbol{\delta}'\boldsymbol{R}^{-1}\mathbf{y})^2}{v(1 + v^{-1}\mathbf{y}'\boldsymbol{R}\mathbf{y})} \right\}^{(1/2)j}$$

Here, $\boldsymbol{R}$ is the correlation matrix for the standardized multinormal variables $\boldsymbol{U} = (U_1, \ldots, U_p)$. The use of multivariate Student t-distribution can be found in areas ranging from constructing simultaneous confidence intervals for the expected values of a number of normal populations to the study of stepwise linear multiple regression analysis. The multivariate Behrens–Fisher distribution may also be created using multivariate Student t-distribution. The use of multivariate Student t-distribution is nowadays utilized in finance whereby one uses the Student t-copula to find the dependence structure of p number of financial scripts (Cherubini et al. 2004). Other areas where multivariate Student t-distribution is used are multiple decision problems, discriminant and cluster analysis, speech recognition, etc. One may refer to the following texts, viz., Johnson and Kotz (1972) and Kotz and Nadarajah (2004) to get a good idea about multivariate Student t-distribution.

Example 8.13*

A copula, $C(u_1, \ldots, u_p)$, is a multivariate probability distribution for which the marginal probability distribution of each variable, u_j, $j = 1, \ldots, p$ is uniform, that is, [0,1]. They are used to describe the dependence between random variables, $X_1, \ldots, X_p$. As per the fundamental theorem of Sklar, every distribution $F_{X_1,\ldots,X_p}(x_1, \ldots, x_p)$ with marginals $F_{X_1}(x_1), \ldots, F_{X_p}(x_p)$ may be written using the copula function as $F_{X_1,\ldots,X_p}(x_1, \ldots, x_p) = C\left\{F_{X_1}(x_1), \ldots, F_{X_p}(x_p)\right\}$. Alternatively, $C(u_1, \ldots, u_p) = F_{U_1,\ldots,U_p}\left\{F_{u_1}^{-1}(u_1), \ldots, F_{u_p}^{-1}(u_p)\right\}$.

To illustrate the application of multivariate t-distribution, let us consider the following two scripts, namely, TATA STEEL and SBI from NSE, India (http://www.nse-india.com/) for the time period January 1, 2015 to May 29, 2015. If we draw the two-dimensional (2-D) copula (Figure 8.11) considering bivariate t-distributions between the returns, r, of the pair of stocks, then one obtains the PDF graphs. The Pearson correlation coefficient between the two scripts is found out to be

$$\begin{pmatrix} 1.0000 & 0.3513 \\ 0.3513 & 1.0000 \end{pmatrix}.$$

One should use the closing price, P_{it}, of each day, say t, to find

$$r = log_e\left(\frac{P_{i,t+1}}{P_{i,t}}\right),$$

and we use this for our calculations.

* The results of Examples 8.13 and 8.15 are part of different unpublished master's theses of students in Industrial and Management Engineering, Indian Institute of Technology, Kanpur, India, who have worked under the guidance of the first author, Raghu Nandan Sengupta.

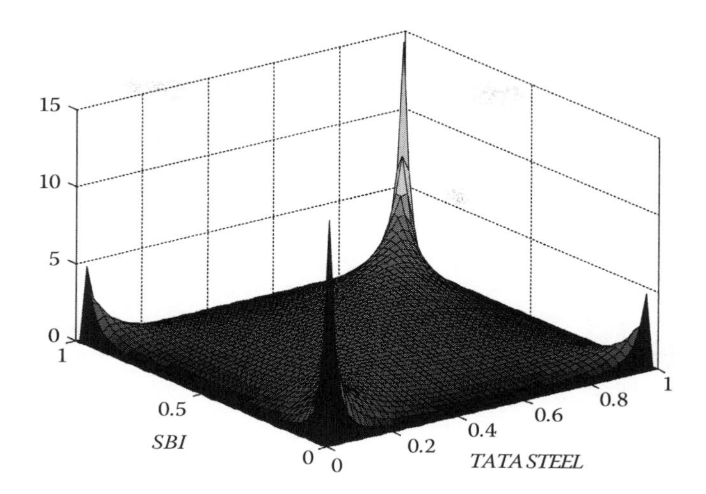

FIGURE 8.11
PDF of *C(TATA STEEL,SBI)* utilizing the concept of bivariate t-distribution, Example 8.13.

8.11 Wishart Distribution

In statistics, the Wishart distribution is a generalization of the chi-squared distribution in multiple dimensions. It was first formulated by John Wishart (1898–1956) (Wishart, 1928). Suppose $\boldsymbol{X}_k \sim N_p(\mu_k \boldsymbol{\Sigma})$, $k = 1, \ldots, \upsilon$ be independently distributed, then $\boldsymbol{W} = \sum_{k=1}^{\upsilon} X_k X_k' \sim W_p(\upsilon, \boldsymbol{\Sigma})$, where the parameters $\Sigma > 0$ is of size $(p \times p)$ and is positive definite, while $\upsilon > (p-1)$ is the degree of freedom. The Wishart distribution arises as the distribution of the sample covariance matrix for a sample from an MND. If $\boldsymbol{X}_{n\times p}$ is an $(n \times p)$ matrix of random variables, then the PDF is given by

$$f_{X_1,\ldots,X_p}(x_1,\ldots,x_p) = \frac{1}{2^{\upsilon p/2}|\Sigma|^{\upsilon/2}\Gamma_p(\upsilon/2)}|\boldsymbol{X}|^{(\upsilon-p-1)/2}e^{-(1/2)tr(\Sigma^{-1}\boldsymbol{X})}.$$

We now state a few relevant properties of the Wishart distribution:

1. $E(\boldsymbol{W}) = \upsilon\boldsymbol{\Sigma} + \boldsymbol{M}'\boldsymbol{M}$, where $\boldsymbol{M}' = (\mu_1, \ldots, \mu_\upsilon)$.
2. Rank of $\boldsymbol{W} = \min(\upsilon, p)$.
3. If $\boldsymbol{W}_k \sim W_p(k, \boldsymbol{\Sigma}), k = 1, \ldots, \upsilon$, then $\sum_{k=1}^{\upsilon} \boldsymbol{W}_k \sim W_p\left(\sum_{k=1}^{\upsilon} k, \boldsymbol{\Sigma}, \mathbf{M}\right)$, where $\boldsymbol{M}' = \left[\boldsymbol{M}_1 \vdots \cdots \vdots \boldsymbol{M}_\upsilon\right]$.
4. If $\boldsymbol{W} \sim W_p(\upsilon, \boldsymbol{\Sigma})$ and $\boldsymbol{C}$ is any $(p \times q)$ matrix of constants, then $\boldsymbol{C}'\boldsymbol{W}\boldsymbol{C} \sim W_q(\upsilon, \boldsymbol{C}'\boldsymbol{\Sigma}\mathbf{C}, \mathbf{MC})$, where $\boldsymbol{M}' = (\mu_1, \ldots, \mu_\upsilon)$.
5. $E(\boldsymbol{W}) = \upsilon\boldsymbol{\Sigma}$.
6. $\text{Mode}(\boldsymbol{W}) = (\upsilon - p - 1)$.
7. $\text{Var}(\boldsymbol{W}_{i,j}) = \upsilon\left(\sigma_{i,j}^2 + \sigma_{i,i}\sigma_{j,j}\right)$, where $\sigma_{i,j}$ is the ith row and jth column element of $\boldsymbol{\Sigma}$.

Let us now consider the inverse Wishart distribution, denoted by $IW_p(.)$. It is the multivariate extension of the inverse gamma distribution. If one considers that the Wishart distribution generates

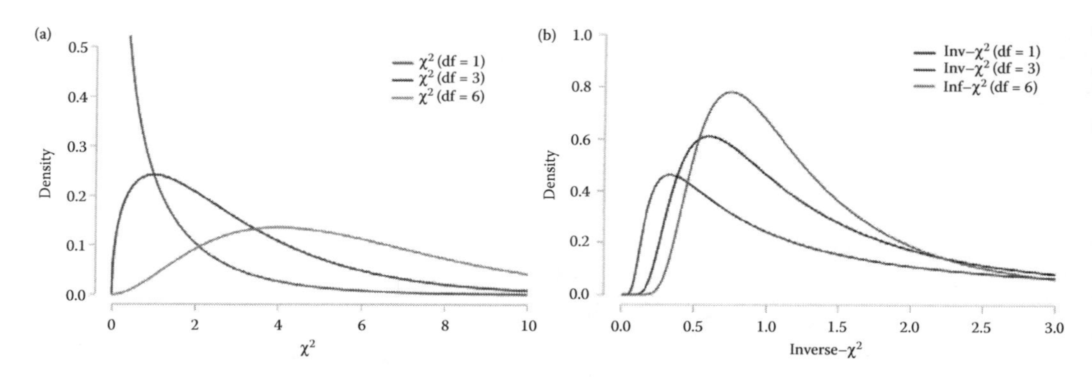

FIGURE 8.12
PDF for (a) χ^2_υ and (b) inverse-χ^2_υ considering $\upsilon = 1, 3, 6$, Example 8.14.

the sum of squares matrices, then the inverse Wishart distribution can be imagined as that which generates random covariance matrices. Hence, if $\boldsymbol{W} \sim W_p(\upsilon, \boldsymbol{\Sigma})$, then $\boldsymbol{W}^{-1} \sim IW_p(\upsilon, \boldsymbol{\Sigma}^{-1})$. The use of inverse Wishart distribution can be found in Bayesian statistics where it is used as a prior on the variance/covariance matrix, $\boldsymbol{\Sigma}$, of an MND. If we consider the inverse gamma distribution as the conjugate prior of the variance parameter, σ^2, for the univariate normal distribution, then the inverse Wishart distribution can be said to extend this conjugacy to the MND case. Another important point worth mentioning is the fact that using Helmert transformation, the Wishart distribution can be expressed as two distributions, one for the sample means and another for the sample variances–covariances. These transformations are orthogonal in nature, which makes it intuitive to understand that the sample means and the sample variances–covariances are independent of each other. Thus, as in univariate theory, the sample mean vector and the sample variance–covariance are also independently distributed for the multidimensional case.

A few good references for Wishart distribution are: Anderson (2003), Chatfield and Collins (1980), Cuadras and Rao (1993), Dempster (1969), and Eaton (1983).

Example 8.14

Let us illustrate the Wishart and inverse Wishart distributions in the simple case where we consider their univariate counterpart, which are χ^2_υ and inverse-χ^2_υ (Figure 8.12). In Figure 8.12, the degrees of freedom considered are 1, 3, and 6. Remember that for the multivariate case, one can make deductions about the Wishart and inverse Wishart distributions in a similar manner as we can do for χ^2_υ and inverse-χ^2_υ cases in the univariate setup.

8.12 Multivariate Extreme Value Distribution

Multivariate extreme gives us the picture of the asymptotic behavior of componentwise maxima of *i.i.d.* observations. The main problem one faces is how to define MEVD. This problem arises due to the fact that there does not exist any strict ordering principle for multivariate observations. Though we use concepts of ordering such as marginal ordering (M-ordering), reduced (aggregate) ordering (R-ordering), partial ordering (P-ordering), conditional (sequential) ordering (C-ordering), etc. to accomplish this task, yet the ordering problem does occur in many cases.

Let $\boldsymbol{Y}_i = (Y_{i1}, \ldots, Y_{ip})$ be *i.i.d.* such that $M_{\max,j} = \max\{Y_{1,j}, \ldots, Y_{n,j}\}$, where $i = 1, \ldots, n$ and $j = 1, \ldots, p$. As per definition, n is the number of observations, while p is the dimension. Given this, we are interested to find the normalizing scaling constants $a_{n,j}$ and $b_{n,j}$ such that $Pr\{((\boldsymbol{M}_{\max} - \boldsymbol{b}_n)/\boldsymbol{a}_n) \le \boldsymbol{x}\} \to G(\boldsymbol{x})$ as $n \to \infty$. Here, three important points about $G(\boldsymbol{x})$ should be mentioned:

1. If $G(\boldsymbol{x})$ is MEVD, then each of its marginal must be one of the univariate extreme value distributions (EVDs) and hence can be represented in general extreme value (GEV) form.
2. The form of the limiting distribution is invariant under monotonic transformation for each of the component.
3. Each marginal distribution can be transformed into specified forms, and one of these transformed specified forms is the Fréchet form given by $Pr\{X_j \le x\} = e^{-x^{-\alpha}}, x > 0, j = 1, \ldots, p$ and $\alpha > 0$. When $\alpha = 1$, we have the unit Fréchet.

In general, we are interested to find the scalar transformations of each, X_j, $j = 1, \ldots, p$. Using the same concept as used in the univariate case, we intend to find $M_{\max,1}, \ldots, M_{\max,p} = \{\max_{1 \le i \le n}, X_{i,1}, \ldots, \max_{1 \le i \le n} X_{i,p}\}$ as $n \to \infty$. Utilizing simple scalar transformation, one needs to find

$$Pr\left\{\frac{(M_{n,1} - b_{n,1})}{a_{n,1}} \le x_1, \ldots, \frac{(M_{n,p} - b_{n,p})}{a_{n,p}} \le x_p\right\}$$
$$= F^n\{(a_{n,1}x_1 + b_1), \ldots, (a_{n,p}x_p + b_p)\} \to G(x_1, \ldots, x_p),$$

when $n \to \infty$. In case if

$$Pr\left\{\frac{(M_{n,1} - b_{n,1})}{a_{n,1}} \le x_1, \ldots, \frac{(M_{n,p} - b_{n,p})}{a_{n,p}} \le x_p\right\}$$
$$= F^n\{(a_{n,1}x_1 + b_1), \ldots, (a_{n,p}x_p + b_p)\} \to G(x_1, \ldots, x_p)$$

holds, for some suitable choices of $a_{n,j}$ and $b_{n,j}$, $j = 1, \ldots, p$, then we say that $G(x_1, \ldots, x_p)$ is a MEVD and F is in the domain of attraction of G. A question that automatically arises is what are the normalizing scaling constants, $a_{n,j}$ and $b_{n,j}$, $j = 1, \ldots, p$, in their general form. For the convenience of the readers, we state below the scaling constants, $a_{n,j}$ and $b_{n,j}$ for the case when $n \to \infty$ and $j = 1, \ldots, p$.

1. For Type I distribution: $a_{n,j} = F_{X_j}^{-1}(1 - 1/n)$ and $b_{n,j} = F_{X_j}^{-1}(1 - 1/ne) - F_{X_j}^{-1}$ $(1 - 1/n)$.
2. For Type II distribution: $a_{n,j} = 0$ and $b_{n,j} = F_{X_j}^{-1}(1 - 1/n)$.
3. For Type III distribution: $a_{n,j} = F_{X_j}^{-1}(1)$ and $b_{n,j} = F_{X_j}^{-1}(1) - F_{X_j}^{-1}(1 - 1/n)$.

The two extreme forms of the limiting multivariate distribution correspond to (i) the case of the asymptotic total independence between componentwise maxima for which $G(x_1, \ldots, x_p) = G_1(x_1) \cdots G_p(x_p)$ and (ii) the case of asymptotic total dependence between componentwise maxima for which $G(x_1, \ldots, x_p) = \min\{G_1(x_1), \ldots, G_p(x_p)\}$.

Remember that $G(x_1, \ldots, x_p) = G_1(x_1) \cdots G_p(x_p)$ holds true if and only if

1. $G(0, \ldots, 0) = G_1(0) \cdots G_p(0) = e^{-p}$, provided $G_j's$ are Gumble type with $G_j(x_j) = exp\{-exp(-x_j)\}$ for $j = 1, \ldots, p$.
2. $G(1, \ldots, 1) = G_1(1) \cdots G_p(1) = e^{-p}$, provided $G_j's$ are Fréchet type with $G_j(x_j) = exp\left\{-x_j^{-\alpha_j}\right\}$ and $\alpha_j > 0$ for $j = 1, \ldots, p$.

3. $G(-1,\ldots,-1) = G_1(-1)\cdots G_p(-1) = e^{-p}$, provided $G_j's$ are Weibull type with $G_j(x_j) = exp\{-(-x_j)^{\alpha_j}\}$ and $\alpha_j > 0$ for $j = 1,\ldots,p$.

Before we end the discussion regarding MEVD, we give a few examples of MEVD considering $p = 2$.

1. Logistic MEVD:

$$exp\left[-\left(\frac{1-\psi_1}{x_1}\right)-\left(\frac{1-\psi_2}{x_2}\right)-\left\{\left(\frac{\psi_1}{x_1}\right)^q+\left(\frac{\psi_2}{x_2}\right)^q\right\}^{1/q}\right],$$

where $0 \le \psi_1, \psi_2 \le 1$, and $q > 1$ have three usual meanings. In case $\psi_1 = 1$ and $\psi_2 = \alpha$, then we obtain the biextremal distribution of the form

$$exp\left[-\left(\frac{1-\alpha}{x_2}\right)-\left\{\left(\frac{1}{x_1}\right)^q+\left(\frac{\alpha}{x_2}\right)^q\right\}^{1/q}\right],$$

while for $\psi_1 = \psi_2 = \alpha$, one obtains the Gumble distribution.

2. Negative logistic MEVD:

$$exp\left[-\left(\frac{1}{x_1}\right)-\left(\frac{1}{x_2}\right)-\alpha\left\{\left(\frac{1}{x_1}\right)^q+\left(\frac{1}{x_2}\right)^q\right\}^{1/q}\right],$$

where $0 \le \psi_1, \psi_2 \le 1, q < 0$, and α have three usual meanings.

3. Bilogistic MEVD:

$$exp\left[-\int_0^1 \max\left\{\frac{(q_1-1)s^{-1/q_1}}{q_1x_1},\frac{(q_2-1)s^{-1/q_2}}{q_2x_2}\right\}ds\right],$$

where $q_1, q_2 > 1$.

Note: A general multivariate case can be thought as

$$exp\left[-\int_0^1 \max\left\{\frac{(q_1-1)s^{-1/q_1}}{q_1x_1},\ldots,\frac{(q_p-1)s^{-1/q_p}}{q_px_p}\right\}ds\right],$$

where $q_j > 1, j = 1,\ldots,p$.

4. Negative bilogistic MEVD:

$$exp\left[-\int_0^1 \max\left\{\frac{(q_1-1)s^{-1/q_1}}{q_1x_1},\frac{(q_2-1)s^{-1/q_2}}{q_2x_2}\right\}ds\right],$$

where $q_1, q_2 < 1$.

5. Gaussian MEVD: If one considers the bivariate extremes for the normal distribution, then one obtains

$$exp\left[-\left(\frac{1}{x_1}\right)\Phi\left\{a-s\left(\frac{x_1}{x_1+x_2}\right)\right\}-\left(\frac{1}{x_2}\right)\Phi\left\{s\left(\frac{x_1}{x_1+x_{12}}\right)\right\}\right],$$

where $s(w) = (a^2 + 2log_e w - 2log_e(1-w))/2a$ and $a = ((x_1 - x_2)/\sigma)^2$. Here, Φ and σ imply the standard normal cumulative deviate and the standard deviation, respectively, for the normal distribution, based on which the Gaussian MEVD is formulated.

A few good references for MEVD are: Coles (2001), de Haan and Resnick (1977), Kotz and Nadarajah (2000), Marshall and Olkin (1967, 1983), Pickards (1981), Sibuya (1960), and Tiago de Oliveira (1958, 1975).

Example 8.15

Figure 8.13 illustrates the EVD (for ease of illustration, we show the univariate EVD case only) using the *positive* values of returns, $r = log_e(P_{i,t+1}/P_{i,t})$, of the indices of four countries, namely, Nikkei (Japan), Nifty (India), FTSE (the United Kingdom), and KOSPI (Korea), for a time period of 10 years from 2003 to 2012. The values of *shape* (μ), *scale* (σ), and *location* (ξ) parameters for the EVD for the four indices are (i) 0.092, 0.0075, 0.0158; (ii) 0.1745, 0.0084, 0.0169; (iii) 0.1475, 0.0068, 0.0118; and (iv) 0.1571, 0.0074, 0.0163, respectively. Considering returns as *negative*,, one can also calculate the values of μ, σ, and ξ and draw similar EVD graphs for these four indices. We leave that to the readers to work on them so that they get a better understanding of the concepts about which we have discussed. Finally, considering the overall

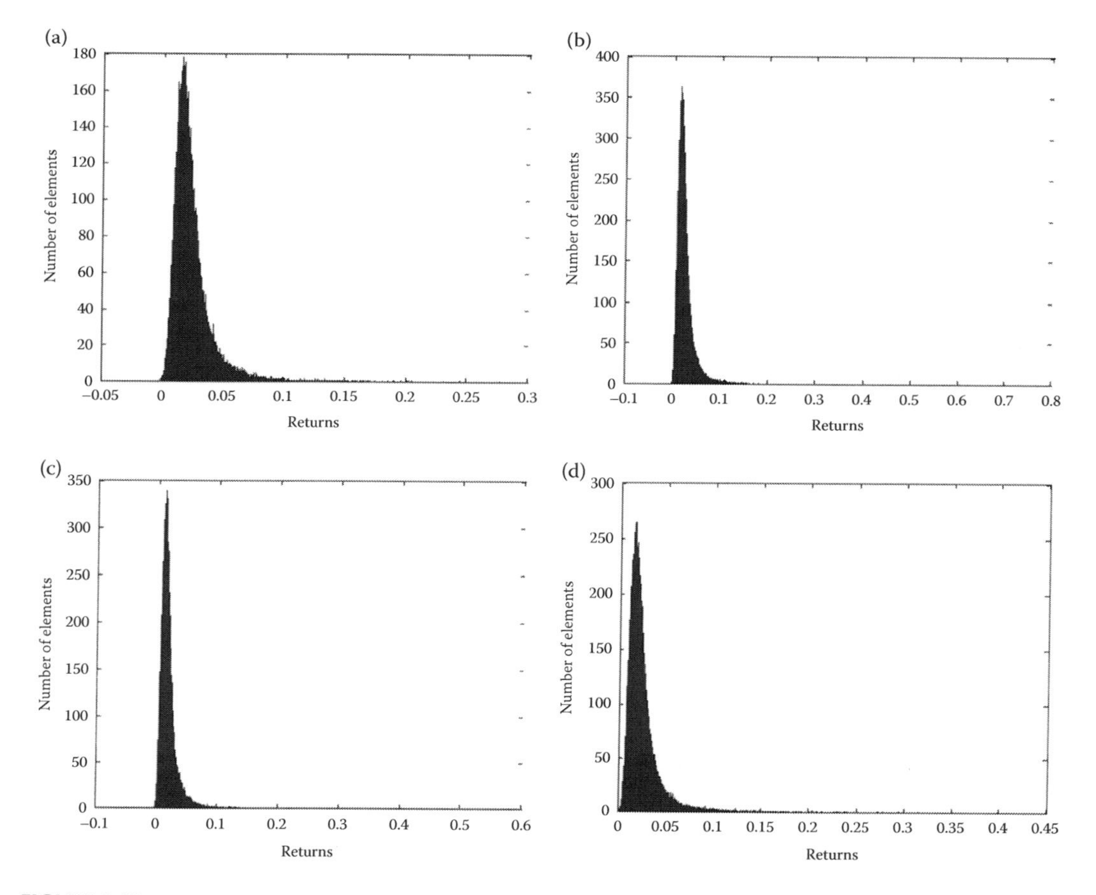

FIGURE 8.13
PDF for (a) Nikkei, (b) Nifty, (c) FTSE, and (d) KOSPI considering positive returns, Example 8.15.

returns (both positive and negative), we calculate the mean and standard deviation values for the four indices, which are: (i) $E(X_{NIKKEI}) = -0.022066$, $Var(X_{NIKKEI}) = 0.0153958$; (ii) $E(X_{NIFTY}) = -0.022211493$, $Var(X_{NIFTY}) = 0.000330646$; (iii) $E(X_{FTSE}) = -0.01670011$, $Var(X_{FTSE}) = 0.000155784$; and (iv) $E(X_{KOSPI}) = -0.021793$, $Var(X_{KOSPI}) = 0.0002269$.

8.13 MLE Estimates of Parameters (Related to MND Only)

For

$$\boldsymbol{X}_{n\times p} = \begin{pmatrix} X_{1,1} & \cdots & X_{1,p} \\ \vdots & \ddots & \vdots \\ X_{n,1} & \cdots & X_{n,p} \end{pmatrix}$$

if one needs to estimate the parameters, then the total number of estimates required is $(1/2)p(p+1)$. Let us consider the case of MND, such that its log likelihood equation may be written as $log_eL(\mu, \boldsymbol{\Sigma}) = -(np/2)log_e 2\pi - (n/2)log_e|\boldsymbol{\Sigma}| - (1/2)\sum_{i=1}^{n}(\boldsymbol{X} - \mu)'\boldsymbol{\Sigma}^{-1}(\boldsymbol{X} - \mu)$. Solving $\partial log_eL(\mu, \boldsymbol{\Sigma})/\partial\mu_j$ and $\partial log_eL(\mu, \boldsymbol{\Sigma})/\partial\sigma_{j1,j2}$, $j_1 < j_2$, where $j = 1, \ldots, p$ and $j_1, j_2 = 1, \ldots, p$ we obtain

$$\hat{\mu} = \begin{pmatrix} \hat{\mu}_1 \\ \vdots \\ \hat{\mu}_p \end{pmatrix} = \bar{\boldsymbol{X}} = \begin{pmatrix} \bar{X}_1 \\ \vdots \\ \bar{X}_p \end{pmatrix} \quad \text{and} \quad \hat{\boldsymbol{\Sigma}} = \begin{pmatrix} \hat{\sigma}_{1,1} & \cdots & \hat{\sigma}_{1,p} \\ \vdots & \ddots & \vdots \\ \hat{\sigma}_{p,1} & \cdots & \hat{\sigma}_{p,p} \end{pmatrix} = \mathbf{S} = \begin{pmatrix} s_{1,1} & \cdots & s_{1,p} \\ \vdots & \ddots & \vdots \\ s_{p,1} & \cdots & s_{p,p} \end{pmatrix},$$

where $\bar{x}_j = (1/n)\sum_{i=1}^{n} x_{ij}$ and $s_{j_1,j_2} = (1/(n-1))\sum_{i=1}^{n}\left(x_{i,j_1} - \bar{x}_{j_1}\right)\left(x_{i,j_2} - \bar{x}_{j_2}\right)$, for $i = 1, \ldots, n$, $j = 1, \ldots, p$ and $j_1, j_2 = 1, \ldots, p$.

Let us now suppose the case of hypothesis testing. A few relevant results without proofs for the same can be stated as follows:

1. The test statistics for $H_o : \boldsymbol{a}'\mu = \boldsymbol{a}'\mu_o$ against $H_A : \boldsymbol{a}'\mu >$ or $\neq$ or $< \boldsymbol{a}'\mu_o$, given $\boldsymbol{\Sigma}$ is *known*, is $n(\bar{\boldsymbol{X}} - \mu_o)'\boldsymbol{\Sigma}^{-1}(\bar{\boldsymbol{X}} - \mu_o)$. The distribution $n(\bar{\boldsymbol{X}} - \mu_o)'\boldsymbol{\Sigma}^{-1}(\bar{\boldsymbol{X}} - \mu_o) \sim \chi^2_{\text{p}}$ and the value of α, that is, the level of confidence, is assumed, based on the problem formulation and practical requirements.
2. The test statistics for $H_o : \boldsymbol{a}'\mu = \boldsymbol{a}'\mu_o$ against $H_A : \boldsymbol{a}'\mu >$ or $\neq$ or $< \boldsymbol{a}'\mu_o$, given $\boldsymbol{\Sigma}$ is *unknown*, is $n(\bar{\boldsymbol{X}} - \mu_o)'\mathbf{S}^{-1}(\bar{\boldsymbol{X}} - \mu_o)$. The distribution $n(\bar{\boldsymbol{X}} - \mu_o)'\mathbf{S}^{-1}(\boldsymbol{X} - \mu_o) \sim T^2_p$ and the value of α, that is, the level of confidence, is assumed, based on the problem formulation and practical requirements.
3. The characteristics form of T^2 statistic is $T^2 = (\bar{\boldsymbol{X}} - \mu_o)'(\mathbf{S}/n)^{-1}(\bar{\boldsymbol{X}} - \mu_o)$. A few important points about the statistics are: (i) $\boldsymbol{S}/n$ is the sample covariance matrix of $\bar{\boldsymbol{X}}$, (ii) $\bar{\boldsymbol{X}} \sim N_p(\mu, (1/n)\boldsymbol{\Sigma})$, (iii) $(n-1)\boldsymbol{S} \sim W(n-1, \boldsymbol{\Sigma})$, and (iv) $\bar{\boldsymbol{X}}$ and $\boldsymbol{S}$ are independent.
4. One should always have $n - 1 > p$, otherwise, $\boldsymbol{S}$ is singular and hence T^2 cannot be calculated.

Using the property of sufficiency and the concept of factorization, we obtain the following results, which we state, again without any proofs:

1. If $\boldsymbol{x}_1, \ldots, \boldsymbol{x}_n$ are the observations from $N_p(\mu, \boldsymbol{\Sigma})$, then $\bar{\boldsymbol{x}}$ and $\boldsymbol{S}$ are sufficient for μ and $\boldsymbol{\Sigma}$.

2. The sufficient set of statistics $\bar{x}$ and S is complete for μ and Σ, where the sample is drawn from $N_p(\mu, \Sigma)$.
3. Let the mth component $Y_1, Y_2, \cdot s$ be $i.i.d.$, with means $E(Y_i) = \upsilon$ (do not confuse with the degree of freedom) and covariance matrices $E(Y_i - \upsilon)(Y_i - \upsilon)' = T$, then $(1/\sqrt{n})\sum_{i=1}^{n}(Y_i - \upsilon) \rightarrow N(\mathbf{0}, T)$ as $n \rightarrow \infty$, for $i = 1, \ldots, n$.

Without going into the detailed discussion and proofs, we would like to mention that for different loss functions (even considering the ubiquitous squared error loss), one should be careful to understand what are the best estimates for the mean and the standard deviation in the multivariate case, as it may not always be the sample means or the sample variances we all know so well in the univariate case. Some seminal work in this respect has been done by James and Stein (1961).

8.14 Copula Theory

When we talk about correlation coefficient, $\rho(X, Y)$, we generally refer to one of the following: Pearson product–moment correlation coefficient, intraclass correlation, rank correlation, Spearman's rank correlation coefficient, Kendall tau rank correlation coefficient, and Goodman and Kruskal's gamma. For the above definitions, the idea of linear correlation coefficient between two vectors of random variables X and Y is always assumed to be true. Furthermore, the following properties for $\rho(X, Y)$ are also important:

1. $-1 < \rho(X, Y) < 1$, for any range of X and Y.
2. If X and Y are independent, then $\rho(X, Y) = 0$.
3. $\rho(\alpha X + \beta, \gamma Y + \delta) = sgn(\alpha\gamma)\rho(X, Y)$, for any range of X and Y.

But in general, most random variables are not jointly elliptically distributed (normal is a class of elliptical distributions) and using linear correlation as a measure of dependence in such situations might prove very misleading. An example to illustrate how linear correlation is misused is as follows. Let $X \sim N(0, \sigma^2)$ and let $Y = X^2$. Then it is expected that both X and Y should be correlated, though on calculation we find that $Cov(X, Y) = 0$. Hence, from the discussion, it is obvious that linear correlation coefficient has some shortcomings and here is where copula comes into play. Before we define a copula function, we state its few properties which we think will benefit the readers so that he/she is in a much better position to appreciate the relevance of copula function later on. The properties are:

1. The variances of X and Y need to be finite.
2. Independence of two random variables implies that they are uncorrelated. The reverse is true only in case of MND.
3. Linear correlation is not invariant under nonlinear strictly increasing transformations, $T: \mathbb{R} \rightarrow \mathbb{R}$, since in general, for two real-valued random variables $\rho(T(X), T(Y)) \neq \rho(X, Y)$. Also, the value of linear correlation may change due to the presence of outliers.

One should remember that the linear relationship between two random variables breaks down at the tail. Hence, the concept of tail dependence is very important to understand why $\rho(X, Y)$ does not work at the extremes. This is where the concept of copula comes into play.

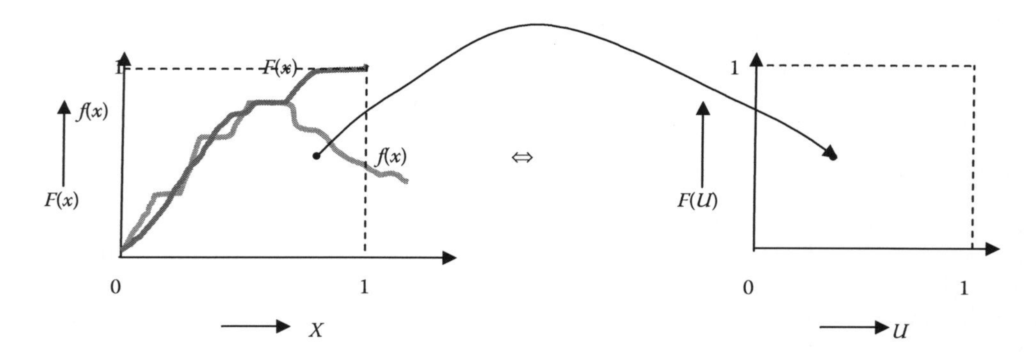

FIGURE 8.14
Copula concept using the mapping idea from $\boldsymbol{X}$ space to $\boldsymbol{U}$ space.

Example 8.13 has already given us the basic definition of copula, so rather than repeat the same, we proceed further to give an overview of copula theory and its use in general. The readers should understand that when we use a copula function, we are in a way trying to map from $F_{\boldsymbol{X}}(\boldsymbol{x})$ to $F_{\boldsymbol{U}}(\boldsymbol{u})$, that is, we are mapping from the $\boldsymbol{X}$ space to the unit vector, $\boldsymbol{U}$, space (which is a hypercube of unit dimension on all sides). This may be illustrated for the case when $p = 2$ and is shown in Figure 8.14.

Thus, in general, a copula $C : [0, 1]^p \to [0, 1]$ has the following properties:

1. $C(U_1, \ldots, U_p)$ is a nondecreasing distribution function in u_j, $j = 1, \ldots, p$.
2. $C(1, 1, \ldots, u_j, 1, \ldots, 1) = u_j$ for $j = 1, \ldots, p$ and $u_j \in [0, 1]$ since all marginal distributions of copula are uniformly distributed.
3. For all $(a_1, a_2), (b_1, b_2) \in [0, 1]^2$ with $a_1 \leq b_1$ and $a_2 \leq b_2$, we have $Pr(0 \leq x_1 \leq a_1, 0 \leq x_2 \leq a_2) - Pr(0 \leq x_1 \leq a_1, 0 \leq x_2 \leq b_2) - Pr(0 \leq x_1 \leq b_1, 0 \leq x_2 \leq a_2) + Pr(0 \leq x_1 \leq b_1, 0 \leq x_2 \leq b_2) \geq 0$.

One important concept in copula theory is the Sklar's theorem, which states that if $H_{X_1,\ldots,X_p}(x_1, \ldots, x_p)$ be the joint distribution of $(X_1, \ldots, X_p)$, and $F_{X_1}(x_1), \ldots, F_{X_p}(x_p)$ be the continuous marginal distributions of $(X_1, \ldots, X_p)$, then there exists a copula function $C\left\{F_{X_1}(x_1), \ldots, F_{X_p}(x_p)\right\}$ such that $H_{X_1,\ldots,X_p}\left(x_1, \ldots, x_p\right) = C\left\{F_{X_1}\left(x_1\right), \ldots, F_{X_p}\left(x_p\right)\right\}$. Thus, the distribution function $C(.)$ is in a way a mapping between the marginals and the joint distributions. We must remember that if $F_{X_1}(x_1), \ldots, F_{X_p}(x_p)$ are all continuous, then $C\left\{F_{X_1}(x_1), \ldots, F_{X_p}(x_p)\right\}$ is unique, else it may not be so.

Without going into the detailed concepts, we state a few important properties of copula, which are (i) invariance; (ii) comonotonicity and countermonotonicity; (iii) tail dependence; (iv) upper tail dependence; and (v) lower tail dependence.

To end this discussion about copula theory, we give a few examples of multivariate copula, which are the Gaussian and Student t-copula.

Gaussian copula: The copula of the p-variate normal distribution with linear correlation matrix $\boldsymbol{R}$ is of the following form: $C_R^{Ga}(\boldsymbol{U}) = \phi_R^p\{\phi^{-1}(u_1), \ldots, \phi^{-1}(u_p)\}$, where, ϕ_R^p denotes the joint distribution function of p-variate standard normal distribution function with linear correlation matrix $\boldsymbol{R}$, while ϕ^{-1} denotes the inverse of the distribution function of the univariate standard normal distribution. Another way of writing the Gaussian copula is $C_G(\boldsymbol{U}) = |\boldsymbol{\Sigma}|^{-(1/2)}\, exp(-(1/2)\boldsymbol{q}'\boldsymbol{\Sigma}^{-1}\boldsymbol{q} +$

$(1/2)\boldsymbol{q}'\boldsymbol{q})$. In the bivariate case, the expression takes the form of

$$C_R^{Ga}(\boldsymbol{U}) = \int_{-\infty}^{\phi^{-1}(u)} \int_{-\infty}^{\phi^{-1}(v)} \frac{1}{2\pi\sqrt{1-R_{12}^2}} exp\left\{-\frac{s^2 - 2R_{12}st + t^2}{2\left(1-R_{12}^2\right)}\right\} ds\, dt.$$

Here, R_{12} is a linear correlation coefficient of the corresponding bivariate normal distribution. Since elliptical distribution is radially symmetric, Gaussian copula does not have either upper or lower tail dependence.

Student t-copula: A p-dimensional t-copula is generally of the form

$$C_t(\boldsymbol{U}) = \int_{-\infty}^{t_\upsilon^{-1}(u_1)} \cdots \int_{-\infty}^{t_\upsilon^{-1}(u_p)} \frac{\Gamma((\upsilon+p)/2)}{\Gamma(\upsilon/2)\sqrt{(\pi\upsilon)^p|\boldsymbol{R}|}} \left(1 + \frac{\boldsymbol{x}'\boldsymbol{R}^{-1}\bar{\boldsymbol{x}}}{\upsilon}\right)^{-((\upsilon+p)/2)} d\boldsymbol{x}.$$

The bivariate t-copula is characterized by univariate Student t-distribution and is given as

$$C_{\upsilon,R}^t(u_1, u_2) = \int_{-\infty}^{t_\upsilon^{-1}(u_1)} \int_{-\infty}^{t_\upsilon^{-1}(u_2)} \frac{1}{2\pi\sqrt{1-R_{12}^2}} exp\left\{1 + \frac{s^2 - 2R_{12}st + t^2}{\upsilon\left(1-R_{12}^2\right)}\right\}^{-((\upsilon+2)/2)} ds\, dt,$$

where R_{12} is a linear correlation coefficient of the corresponding bivariate t_υ distribution if $\upsilon > 2$. A t-copula has both upper and lower tail dependences. Before we wind up this section, we mention the names of Cherubini et al. (2004) and Nelsen (2006), which are a few of the good references one may refer to understand copula theory.

8.15 Principal Component Analysis

Principal component analysis (PCA) is a multivariate ordination technique used to display patterns in multivariate data. It aims to graphically display the relative positions of data points in fewer dimensions while retaining as much information as possible, and also explore relationships between dependent variables. It is a hypothesis-generating technique that is intended to describe patterns in a data table, rather than test formal statistical hypotheses. PCA assumes linear responses of variables and has a range of applications other than data display, including multiple regression and variable reduction.

As mentioned, the main purpose of PCA is to reduce the dimensionality of multivariate data to make its structure clearer. It does this by looking for the linear combination of the variables, which accounts for as much as possible of the total variation in the data. It then goes on to look for a second combination, uncorrelated with the first, which accounts for as much of the remaining variation as possible and so on. If the greater part of the variation is accounted for by a small number of components, then they may be used in place of the original variables.

The principal idea of PCA is to reduce the dimension of $\boldsymbol{X}_{n\times p} = (\boldsymbol{X}_1, \ldots, \boldsymbol{X}_p)$, in order to find the best combination of $(X_1, \ldots, X_p)$, which is able to give us the maximum information as required. This reduction in dimension may be achieved using linear combinations. Thus, in PCA, one looks for linear combination aimed at creating the so-called largest spread among the variables, $X_1, \ldots, X_p$. This concept of largest spread invariably leads us to look into linear combinations, which have the largest variances. As the reader may be aware that PCA is performed on the covariance matrix, it is not scale invariant, as the units of measurement of X_1 or X_2 or, ..., or X_p may be different. Hence, we generally try to use the normalized version of PCA.

The main objective of PCA as mentioned above is to reduce the dimension of the observations, and the simplest way to do that would be to retain one of the variable, say X_j, and discard the rest, that is, $X_1, \ldots, X_{j-1}, X_{j+1}, \ldots, X_p$. Though the idea may seem plausible, but it is definitely not a reasonable approach as the strength or the ability of explanation is definitely not possible using any arbitrary X_j. An alternative plan may be to consider the simple average, that is, $(1/p)\sum_{j=1}^{p} X_j$ of all the elements of $\boldsymbol{X}_{n\times p} = (\boldsymbol{X}_1, \ldots, \boldsymbol{X}_p)$, but this again is not without its drawback as all the elements of $\boldsymbol{X}_{n\times p}$ are considered of equal importance. A more logical intuitive method would be to consider the weighted average $\sum_{j=1}^{p} \delta_j X_j$, given $\sum_{j=1}^{p} \delta_j^2 = 1$, where $\delta = (\delta_1, \ldots, \delta_p)$ is the weighting vector, which needs to be optimized.

Thus, the standard linear combination (SLC), that is, $\sum_{j=1}^{p} \delta_j X_j$, so that $\sum_{j=1}^{p} \delta_j^2 = 1$ should be chosen to *maximize* the variance of the projection of $\sum_{j=1}^{p} \delta_j X_j$.

Hence, we consider the following:

$$\max \left\{ Var\left(\sum_{j=1}^{p} \delta_j X_j\right) \right\}$$

$$\text{s.t.}: \sum_{j=1}^{p} \delta_j^2 = 1$$

$$-1 \le \delta_j \le 1, \quad \forall j = 1, \ldots, p.$$

Here, one may easily deduce that the *required direction* of δ may be found using spectral decomposition of the covariance matrix of $\boldsymbol{X}_{n\times p}$, that is, $\boldsymbol{\Sigma}$. Using basic rules of matrix algebra, we know that the first direction of δ is given by the eigen vector, γ_1, corresponding to the largest eigen vector value λ_1 of the covariance matrix, $\boldsymbol{\Sigma}$. Hence, the first SLC is the one with the highest variance, obtained from the optimization model and is termed as the first *principal component* (PC), that is, $Y_1 = \gamma_1' \boldsymbol{X}$. Once Y_1 is found, we proceed to find the second SLC with the second highest variance, that is, the second PC, which is given by $Y_2 = \gamma_2' \boldsymbol{X}$. Diagrammatically, it may be represented as shown in Figure 8.15. In Figure 8.15, let us consider three variables, X_1, X_2, and X_3. Thus, if one uses PCA, then the method would choose the first PCA axis as that line (marked here as PC # 1) that goes through the centroid, but at the same time it also minimizes the square of the distance of each point to that line. PC # 1 thus goes through the maximum variation in the data. If now one finds the second PCA axis (i.e., PC # 2), then this will also pass through the centroid, and would also go through the maximum variation in the data, but with a certain constraint, that it must be completely *uncorrelated* (i.e., at right angles, or *orthogonal*) to PC # 1. Hence, as shown, the angle between PC # 1 and PC # 2 is 90°. In a similar way, one can proceed and obtain PC # 3, such that the angle between PC # 2 and PC # 3 is also 90°.

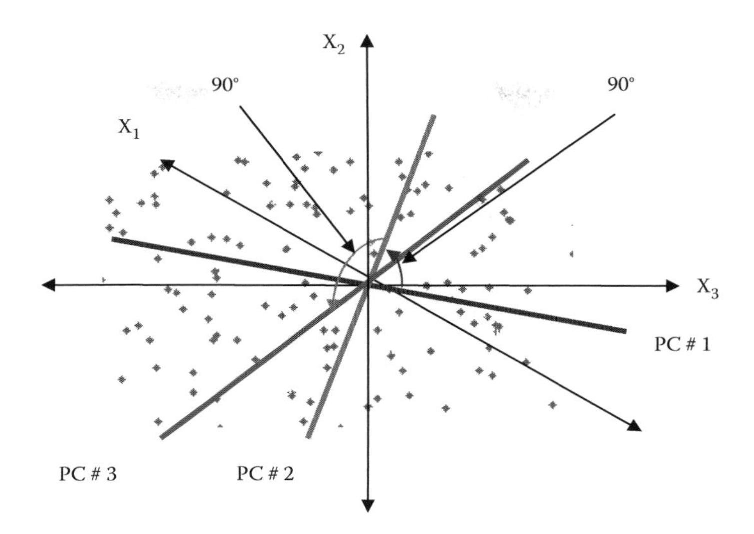

FIGURE 8.15
A hypothetical example illustrating the concept of PCA using orthogonality.

Example 8.16

Consider the MND $\boldsymbol{X} \sim N(\boldsymbol{\mu}, \boldsymbol{\Sigma})$, where $\boldsymbol{\mu} = (2.0, 3.0, 2.5)$ and

$$\boldsymbol{\Sigma} = \begin{pmatrix} 4.00 & -2.00 & 4.00 \\ -2.00 & 9.00 & 3.00 \\ 4.00 & 3.00 & 16.00 \end{pmatrix}.$$

Then the eigen values are $\lambda_1 = 1.6793$, $\lambda_2 = 9.4789$, and $\lambda_3 = 17.8418$, while the corresponding eigen vectors are $\gamma_1 = (0.8719, 0.3697, -0.3210)'$, $\gamma_2 = (0.4311, -0.8905, 0.1452)'$, and $\gamma_3 = (0.2322, 0.2650, 0.9359)'$, respectively. Thus, the PC transformation is given by

$$\boldsymbol{Y} = \begin{pmatrix} Y_1 \\ Y_2 \\ Y_3 \end{pmatrix} = \begin{pmatrix} 0.8719 & 0.3697 & -0.3210 \\ 0.4311 & -0.8905 & 0.1452 \\ 0.2322 & 0.2650 & 0.9359 \end{pmatrix} \begin{pmatrix} X_1 - 2.0 \\ X_2 - 3.0 \\ X_3 - 2.5 \end{pmatrix}.$$

Hence, the PCA axes are:

$$Y_1 = 0.8719X_1 + 0.3697X_2 - 0.3210X_3 - 2.0 \times 0.8719 - 3.0 \times 0.3697 - 2.5 \times (-0.3210)$$

$$Y_2 = 0.4311X_1 - 0.8905X_2 + 0.1452X_3 - 2.0 \times 0.4311 - 3.0 \times (-0.8905) - 2.5 \times 0.1452$$

$$Y_3 = 0.2322X_1 + 0.2650X_2 + 0.9359X_3 - 2.0 \times 0.2322 - 3.0 \times 0.2650 - 2.5 \times 0.9359$$

A way of double checking whether the PC transformations are correct is to calculate $\sum_{j=1}^{3} \delta_j^2$ for each of the eigen values. It is very intuitive to note that each of these values, that is, $\{0.8719^2 + 0.3697^2 + (-0.3210)^2\}$, $\{0.4311^2 + (-0.8905)^2 + 0.1452^2\}$, and $\{0.2322^2 + 0.2650^2 + 0.9359^2\}$ are equal to 1 as the case should be. Another method to double check is to find the variances of $\boldsymbol{Y}$.

Thus, we have $Var(Y_1) = 0.8719^2 \times Var(X_1) + 0.36976^2\ Var(X_2) + (-0.3210)^2\ Var(X_3) + 2 \times 0.8719 \times 0.36976 \times Covar(X_1, X_2) + 2 \times 0.36976 \times (-0.3210) \times Covar(X_2, X_3) + 2 \times$

$0.8719 \times (-0.3210) \times Covar(X_1, X_3) = 1.6792$, which is the value of the first eigen vector as calculated above. Similarly, $Var(Y_2) = 9.4789$ and $Var(Y_3) = 17.8418$.

Though not exhaustive, yet we state a few important results for PCA:

1. For a given $\boldsymbol{X} \sim N(\boldsymbol{\mu}, \boldsymbol{\Sigma})$, let $\boldsymbol{Y} = \Gamma'(\boldsymbol{X} - \boldsymbol{\mu})$ be the PC transformation, then
   a. $E(Y_j) = 0, j = 1, \ldots, p$.
   b. $Var(Y_j) = \lambda_j, j = 1, \ldots, p$.
   c. $Covar\left(Y_{j_1}, Y_{j_2}\right) = 0, j_1 \neq j_2 = 1, \ldots, p$.
   d. $\sum_{j=1}^{p} Var(Y_j) = tr(\boldsymbol{\Sigma})$.
   e. $\prod_{j=1}^{p} Var(Y_j) = |\boldsymbol{\Sigma}|$.
2. There exists no SLC which has larger variance than $\lambda_1 = Var(Y_1)$.
3. For the practical implementation of PCA, we replace $\boldsymbol{\mu}$ by $\bar{\boldsymbol{x}}$ and $\boldsymbol{\Sigma}$ by $\boldsymbol{S}$, and we evaluate the eigen values and the eigen vectors of $\boldsymbol{S}$.
4. The components of the eigen vectors are the weights of the original variables in the PC.
5. PCs are not scale invariant.

A few good references for PCA are Hastie et al. (2011), Jackson (2003), and Jolliffe (2002).

8.16 Factor Analysis

The origins of FA may be traced back to the work of Pearson (1901) and Spearman (1904). The term "FA" as we know today was first introduced by Thurstone (1931). It is a multivariate statistical method based on a model when the observed vector is partitioned into an *unobserved systematic* part and an *unobserved error* part. The components of the error vector are considered as uncorrelated or independent, while the systematic part is taken as a linear combination of a relatively smaller number of unobserved factor variables. Using FA, one can separate the effects of the factors (which are of primary interest to us) from the errors. Stated explicitly using FA, we intend to partition variables into particular groups such that within a particular group they are highly correlated among themselves. Moreover, these variables have relatively small correlations with variables in a different group. Thus, each group of variables represents a single underlying construct/factor that is responsible to provide information about the observed correlations. Before we go in to the mathematical discussion of FA and how it is used, we state here a few good references for FA: Anderson (2003), Basilevsky (1994), Child (2006), Gorsuch (1983), Harman (1976), Johnson and Wichern (2002), Lawley and Maxwell (1971), Mulaik (2009), Thompson (2004), and Thurstone (1931, 1947).

8.16.1 Mathematical Formulation of Factor Analysis

Suppose $\boldsymbol{X}_{(p\times 1)}$ be p number of variables, such that $E(\boldsymbol{X}) = \boldsymbol{\mu}_{(p\times 1)}$ and $Covar(\boldsymbol{X}) = \boldsymbol{\Sigma}_{(p\times p)}$, then using FA, one can express $\boldsymbol{X}_{(p\times 1)}$ as being dependent on

$$\boldsymbol{F}_{(m\times 1)} = \begin{pmatrix} F_1 \\ \vdots \\ F_m \end{pmatrix}$$

common factors and p additional specific factors denoted by $\varepsilon_{(p\times 1)}$. Mathematically, it can be expressed as $\boldsymbol{X} - \boldsymbol{\mu} = \boldsymbol{LF} + \boldsymbol{\varepsilon}$, that is,

$$\begin{bmatrix} X_1 - \mu_1 \\ \cdot \\ \cdot \\ X_p - \mu_p \end{bmatrix} = \begin{bmatrix} l_{1,1}F_1 + \cdots + l_{1,m}F_m + \varepsilon_1 \\ \cdot \\ \cdot \\ l_{p,1}F_1 + \cdots + l_{p,m}F_m + \varepsilon_p \end{bmatrix},$$

where

$$\boldsymbol{L}_{(p\times m)} = \begin{pmatrix} l_{1,1} & \cdots & l_{1,m} \\ \vdots & \ddots & \vdots \\ l_{p,1} & \cdots & l_{p,m} \end{pmatrix}$$

is the matrix of *factor* loading.

A careful look at this mathematical formulation, $\boldsymbol{X} - \boldsymbol{\mu} = \boldsymbol{LF} + \boldsymbol{\varepsilon}$, would distinguish this from MLR due to the fact that in MLR the independent variables can be observed while in FA it is not so. One also needs to make a distinction between FA and PCA. In the PCA method, the PCs are just the linear transformation arranged in the sense that the variance corresponding to the PCs *decreases* as one goes from the first PC to the second and so on. In doing so, the dimension of the data set is reduced, and this as we know is the main idea of PCA. On the other hand, in FA, one aims to model the variations using a linear transformation of a *fixed* number of variables, called the *factor* or the *latent* variables.

A few important properties/assumptions for FA, ($m < p$), are as follows:

1.

$$E(\boldsymbol{F}) = \boldsymbol{0}_{(m\times 1)} = \begin{bmatrix} 0 \\ \vdots \\ \vdots \\ 0 \end{bmatrix}_{(m\times 1)}.$$

2.

$$Covar(\boldsymbol{F}) = \boldsymbol{I}_{(m\times m)} = \begin{bmatrix} 1 & \cdots & 0 \\ \vdots & \ddots & \vdots \\ 0 & \cdots & 1 \end{bmatrix}_{(\boldsymbol{m}\times m)}.$$

3.

$$E(\boldsymbol{\varepsilon}) = \boldsymbol{0}_{(p\times 1)} = \begin{bmatrix} 0 \\ \vdots \\ \vdots \\ 0 \end{bmatrix}_{(p\times 1)}.$$

4.

$$Covar(\boldsymbol{\varepsilon}) = \boldsymbol{\Psi}_{(p\times p)} = \begin{bmatrix} \Psi_1 & \cdots & 0 \\ \vdots & \ddots & \vdots \\ 0 & \cdots & \Psi_p \end{bmatrix}_{(p\times p)}.$$

5.

$$Covar(\varepsilon, \boldsymbol{F}) = \boldsymbol{0}_{(p\times m)} = \begin{bmatrix} 0 & \cdots & 0 \\ \vdots & \ddots & \vdots \\ 0 & \cdots & 0 \end{bmatrix}_{(p\times m)}.$$

6.

$$\boldsymbol{\Sigma} = \boldsymbol{L}_{(p\times m)}\boldsymbol{L}'_{(m\times p)} + \boldsymbol{\Psi}_{(p\times p)} = \begin{bmatrix} \sigma_{11} & \sigma_{12} & \cdots & \sigma_{p1} \\ \sigma_{21} & \sigma_{22} & \cdots & \sigma_{p2} \\ \vdots & \vdots & \ddots & \vdots \\ \sigma_{1p} & \sigma_{2p} & \cdots & \sigma_{pp} \end{bmatrix}_{(p\times p)},$$

where $\sigma_{jj} = \left(l_{j,1}^2 + \cdots + l_{j,m}^2\right) + \Psi_j$, $j = 1, \ldots, p$. Here, $\left(l_{j,1}^2 + \cdots + l_{j,m}^2\right)$ is called the communality, while Ψ_j is the specific variance.

7. The eigen value and eigen vector for $\sum$ are $(\lambda_j, \boldsymbol{e}_j)$, $j = 1, \ldots, p$ such that $\lambda_1 \geq \cdots \geq \lambda_p \geq 0$.

8.

$$Covar(\boldsymbol{X}, \boldsymbol{F}) = \boldsymbol{L}_{(p\times m)} = \begin{pmatrix} l_{1,1} & \cdots & l_{1,m} \\ \vdots & \ddots & \vdots \\ l_{p,1} & \cdots & l_{p,m} \end{pmatrix}_{(p\times m)}.$$

9. Factor loadings, that is, $\boldsymbol{L}_{(p\times m)}$, are determined only up to an orthogonal matrix $\boldsymbol{T}_{(m\times m)}$. Thus, the loadings $\boldsymbol{L}^*_{(p\times m)} = \boldsymbol{L}_{(p\times m)}\boldsymbol{T}_{(m\times m)}$ and $\boldsymbol{L}$ both give the same representations. Furthermore, the communalities given by the diagonal elements of $\boldsymbol{L}_{(p\times m)}\boldsymbol{L}'_{(m\times p)}$ and $\boldsymbol{L}^*_{(p\times m)}\boldsymbol{L}^{*\prime}_{(m\times p)}$ are also unaffected by the choice of $\boldsymbol{T}_{(m\times m)}$.

The properties/assumptions stated above constitute the orthogonal factor model. When $m \ll p$, then FA as a method is very useful. On the other hand, if the off-diagonal elements of $\boldsymbol{S}(\boldsymbol{R})$ are small (zero), then the variables are not related and FA as a multivariate statistical technique is not useful. If we allow the $\boldsymbol{F}_{(m\times 1)}$ common factors to be correlated such that $Covar(\boldsymbol{F}) \neq \boldsymbol{I}_{(m\times m)}$, then we obtain the oblique factor model.

8.16.2 Estimation in Factor Analysis

In statistical literature, we have two methods for estimating the parameters in FA: (i) principal component method and (ii) maximum likelihood method.

8.16.3 Principal Component Method

1. For the population, we know

$$\boldsymbol{\Sigma} = \begin{bmatrix} \sigma_{11} & 0 & 0 & 0 \\ 0 & \sigma_{22} & 0 & 0 \\ \vdots & \vdots & \ddots & \vdots \\ 0 & 0 & 0 & \sigma_{pp} \end{bmatrix}_{(p\times p)} = \boldsymbol{L}_{(p\times m)}\boldsymbol{L}'_{(m\times p)} + \boldsymbol{\Psi}_{(p\times p)}$$

$$= \left[\sqrt{\lambda_1}\boldsymbol{e}_1\sqrt{\lambda_2}\boldsymbol{e}_2\cdots\sqrt{\lambda_m}\boldsymbol{e}_m\right]_{(p\times m)} \begin{bmatrix} \sqrt{\lambda_1}\boldsymbol{e}'_1 \\ \sqrt{\lambda_2}\boldsymbol{e}'_2 \\ \vdots \\ \sqrt{\lambda_m}\boldsymbol{e}'_m \end{bmatrix}_{(m\times p)} + \begin{bmatrix} \Psi_1 & 0 & 0 & 0 \\ 0 & \Psi_2 & 0 & 0 \\ \vdots & \vdots & \ddots & \vdots \\ 0 & 0 & 0 & \Psi_p \end{bmatrix}_{(p\times p)}$$

$$= \begin{bmatrix} \sqrt{\lambda_1} & 0 & \cdots & 0 \\ 0 & \sqrt{\lambda_2} & \cdots & 0 \\ \vdots & \vdots & \cdots & \vdots \\ 0 & 0 & \cdots & \sqrt{\lambda_m} \\ \vdots & \vdots & \cdots & \cdots \\ 0 & 0 & 0 & 0 \end{bmatrix}_{(p\times m)} \times \begin{bmatrix} \sqrt{\lambda_1} & 0 & \cdots & 0 & \cdots & 0 \\ 0 & \sqrt{\lambda_2} & \cdots & 0 & \cdots & 0 \\ \vdots & \vdots & \cdots & \vdots & \vdots & \vdots \\ 0 & 0 & 0 & \sqrt{\lambda_m} & 0 & 0 \end{bmatrix}_{(m\times p)}$$

$$+ \begin{bmatrix} \Psi_1 & 0 & 0 & 0 \\ 0 & \Psi_2 & 0 & 0 \\ \vdots & \vdots & \ddots & \vdots \\ 0 & 0 & 0 & \Psi_p \end{bmatrix}_{(p\times p)}$$

$$= \begin{bmatrix} (\lambda_1+\Psi_1) & 0 & 0 & 0 & 0 & \cdots & 0 \\ 0 & (\lambda_2+\Psi_2) & 0 & 0 & 0 & \cdots & 0 \\ \vdots & \vdots & \ddots & \vdots & \vdots & \vdots & 0 \\ 0 & 0 & 0 & (\lambda_m+\Psi_m) & 0 & \cdots & 0 \\ \vdots & \vdots & \vdots & \vdots & \vdots & \vdots & \vdots \\ 0 & 0 & 0 & 0 & 0 & 0 & \Psi_p \end{bmatrix}_{(p\times p)},$$

such that $\sigma_{jj} = \sum_{i=1}^{m} l_{ji}^2 + \Psi_j$, $j = 1, \ldots, p$. We assume the contribution of $(\lambda_{m+1}\boldsymbol{e}_{m+1}\boldsymbol{e}'_{m+1} + \cdots + \lambda_p\,\boldsymbol{e}_p\boldsymbol{e}'_p)$ is negligible.

2. We can also use

$$\boldsymbol{x}_i - \bar{\boldsymbol{x}} = \begin{bmatrix} x_{i,1} - \bar{x}_1 \\ \vdots \\ x_{i,p} - \bar{x}_p \end{bmatrix}, \quad \text{or} \quad z_i = \begin{bmatrix} \dfrac{(x_{i,1} - \bar{x}_1)}{\sqrt{s_{11}}} \\ \vdots \\ \dfrac{(x_{i,p} - \bar{x}_p)}{\sqrt{s_{pp}}} \end{bmatrix},$$

for $i = 1, \ldots, n$, the latter being used for the case to avoid problems of having one variable with large variance unduly affecting the factor loading.

3. In case we have a sample, then the eigen vector and eigen pair for **S** will be $(\hat{\lambda}_j, \hat{\boldsymbol{e}}_j), \ldots, p$ such that $\hat{\lambda}_1 \geq \cdots \geq \hat{\lambda}_p \geq 0$. In case $m < p$, then

$$\mathbf{S} = \begin{bmatrix} s_{11} & 0 & 0 & 0 \\ 0 & s_{22} & 0 & 0 \\ \vdots & \vdots & \ddots & \vdots \\ 0 & 0 & 0 & s_{pp} \end{bmatrix}_{(p\times p)} = \hat{\boldsymbol{L}}_{(p\times m)}\hat{\boldsymbol{L}}'_{(m\times p)} + \hat{\boldsymbol{\Psi}}_{(p\times p)}$$

such that $s_{jj} = \sum_{i=1}^{m} \hat{l}_{ji}^2 + \hat{\Psi}_j$, $j = 1, \ldots, p$.

4. If the number of common factors is not known beforehand or is not determined *a priori*, one can use the knowledge of previous researchers. A thumb rule is to find the value of the residual matrix, that is, $\boldsymbol{S} - \left(\hat{\boldsymbol{L}}_{(p\times m)}\hat{\boldsymbol{L}}'_{(m\times p)} + \hat{\boldsymbol{\Psi}}_{(p\times p)}\right) \leq \hat{\lambda}_{m+1}^2 + \cdots + \hat{\lambda}_p^2$. A small value of the sum of squares of the neglected eigen values means a small value for the sum of square errors of approximation.
5. The contribution to the total sample variance, that is, $s_{11} + \cdots + s_{pp}$ from the ith common factor is given by $\hat{l}_{i,1}^2 + \cdots + \hat{l}_{i,p}^2 = \left(\sqrt{\hat{\lambda}_i}\hat{\boldsymbol{e}}_i\right)'\left(\sqrt{\hat{\lambda}_i}\hat{\boldsymbol{e}}_i\right) = \hat{\lambda}_i$. Thus, the proportion of the total sample variance due to the jth factor is given by $\hat{\lambda}_j/(s_{11} + \cdots + s_{pp})$ or $(\sum_{j=1}^{i} \hat{\lambda}_j)/p$ depending on whether it is to do with $\boldsymbol{S}$ or $\boldsymbol{R}$.

   A modified approach called the principal factor method works in a similar method as stated above.

8.16.4 Maximum Likelihood Method

For the maximum likelihood methodology to work, a few important assumptions should hold:

1. $\boldsymbol{F}$ and ε are normally distributed
2. If the first holds true, then $\boldsymbol{X}_j - \mu = \boldsymbol{LF}_j + \varepsilon_j$ is also normally distributed
3. The likelihood function is of the form

$$L(\mu, \boldsymbol{\Sigma}) = (2\pi)^{-((n-1)p)/2}|\boldsymbol{\Sigma}|^{-(n-1)/2}e^{-(1/2)tr\left[\boldsymbol{\Sigma}^{-1}\{\sum_{i=1}^{n}(\mathbf{x}_i-\bar{\mathbf{x}})(\mathbf{x}_i-\bar{\mathbf{x}})'\}\right]}.$$

8.16.5 General Working Principle for FA

The general plan based on which one can use FA is as follows:

1. Generate a variance–covariance matrix of the observed variables, that is, $\boldsymbol{S}$, which is an estimate of $\boldsymbol{\Sigma}$.
2. Select the number of factors, that is, m, by first finding $\hat{\boldsymbol{\lambda}}$ and $\hat{\boldsymbol{e}}$, which are the respective estimates from the sample of size n. In general, find $\hat{\lambda}_j/(s_{11} + \cdots + s_{pp})$ or $(\sum_{j=1}^{i} \hat{\lambda}_j)/p$ for those $\hat{\lambda}_j s$ which are greater than 1.
3. Extract your initial set of factors, that is, find $F_1, \ldots, F_m, \hat{h}_i^2, \hat{\Psi}_i^2$.
4. Perform factor rotation to a terminal solution.
5. Interpret the factor structure, that is, $\boldsymbol{S} - (\hat{\boldsymbol{L}}_{(p\times m)}\hat{\boldsymbol{L}}_{(m\times p)} + \hat{\boldsymbol{\Psi}}_{(p\times p)})$.

```
1:  DEFINE: m, λ̂, ê, F(m×1) = (F_1 ... F_m)^T, L(p×m) = (l_{1,1} ... l_{1,m}; ... ; l_{p,1} ... l_{p,m})
2:  INPUT: X_{n×p}
3:  CALCULATE: E(X) = μ̂(p×1), Covar(X) = Σ̂(p×p) = S(p×p), λ̂, ê, λ̂_j/(s_11+...+s_pp), (Σ_{j=1}^{i} λ̂_j)/p and also max value of
    (Σ_{j=1}^{i} λ̂_j)/p for some fixed j
4:  START if: i = 1, ..., m
5:    CALCULATE: (i) λ̂_i which is the i^th eigen value of S(p×p) and ê_i which is its
      corresponding i^th eigen vector, (ii) in calculating these i^th values, also find
      out l̂_{i,j} = sqrt(λ̂_{i,j}) ê_{i,j}, j = 1,...,p, i = 1,...,m, ĥ_i^2 = l̂_{i,1}^2 + ... + l̂_{i,m}^2 and Ψ̂_i^2 = 1 − ĥ_i^2, i = 1,...,p
6:  END if
7:  REPORT: m, F(m×1), L(p×m) based on max value of (Σ_{j=1}^{i} λ̂_j)/p
8:  END
```

FIGURE 8.16
Pseudo-code to implement FA method.

6. Construct factor scores to use it in further analyses, that is,

$$\begin{bmatrix} l_{1,1}F_1 + \cdots + l_{1,m}F_m + \varepsilon_1 \\ \cdot \\ \cdot \\ l_{p,1}F_1 + \cdots + l_{p,m}F_m + \varepsilon_p \end{bmatrix}.$$

For the convenience of the reader, the general pseudo-code based on which one can work on a data set from the point of view of analyzing its use using FA is given in Figure 8.16.

Example 8.17

As an example, let us consider the data set taken from Holzinger and Swineford (1939). The brief background of the study is as follows. Twenty-six tests, intended to measure a general factor and five specific factors, were administered to seventh and eighth grade students in two schools, namely, Grant-White School ($n = 145$) and Pasteur School ($n = 156$). Students from the Grant-White School came from homes where the parents were American-born, while those from the Pasteur School were from homes where the parents were foreign-born. Data for the analysis include 19 tests intended to measure four domains, namely, (i) spatial ability (visual perception test, cubes, paper form board, lozenges), (ii) verbal ability (general information, paragraph comprehension, sentence completion, word classification, word meaning), (iii) speed (add, code, counting groups of dots, straight and curved capitals), and (iv) memory (word recognition, number recognition, figure recognition, object-number, number-figure, figure-word). For the FA study, consider 24, that is, $p = 24$, psychological tests are administered to the first group of students, that is, $n = 145$. Let us start with $m = 5$, such that we obtain Table 8.2, which has all the relevant information such as estimated factor loading, communalities, specific variances, etc. The method used is the principal component method.

Now, $S - (\hat{L}_{(p\times m)}\hat{L}_{(m\times p)} + \hat{\Psi}_{(p\times p)})$ will give the variability in the sample variance one is not able to explain using $m = 5$. In case $m = 15$, then the cumulative proportion goes up to 0.8879 from a value of 0.6021. One can also use the maximum likelihood method to get the solution, and we request the readers to solve this problem on their own to get a good understanding of the maximum likelihood method used in FA.

TABLE 8.2

FA Solution Using Data from Holzinger and Swineford (1939), Considering $m = 5$, Example 8.17

	Estimated Factor Loading $\hat{l}_{i,j} = \sqrt{\hat{\lambda}_{i,j}} e_{i,j}$ for						
Variables	F_1	F_2	F_3	F_4	F_5	$\hat{h}_i^2$	$\hat{\Psi}_i^2 = 1 - \hat{h}_i^2$
Visual perception	0.616	0.005	0.428	−0.205	0.009	0.6048	0.3952
Cubes	0.400	0.079	0.400	−0.202	−0.348	0.4881	0.5119
Paper form board	0.445	0.191	0.476	−0.106	0.375	0.6129	0.3871
Flags	0.511	0.178	0.335	−0.216	0.010	0.4518	0.5482
General information	0.695	0.321	−0.335	−0.053	−0.079	0.7073	0.2927
Paragraph comprehension	0.690	0.418	−0.265	0.081	0.008	0.7277	0.2723
Sentence completion	0.677	0.425	−0.355	−0.073	0.041	0.7720	0.228
Word classification	0.694	0.243	−0.144	−0.116	0.141	0.5948	0.4052
Word meaning	0.694	0.451	−0.291	0.080	0.005	0.7761	0.2239
Addition	0.474	−0.542	−0.446	−0.202	−0.079	0.7644	0.2356
Code	0.576	−0.434	−0.210	0.034	−0.003	0.5654	0.4346
Counting dots	0.482	−0.549	−0.127	−0.340	−0.099	0.6753	0.3247
Straight curved capitals	0.618	−0.279	0.035	−0.366	0.075	0.6006	0.3994
Word recognition	0.448	−0.093	−0.055	0.555	−0.156	0.5447	0.4553
Number recognition	0.416	−0.142	0.078	0.526	−0.306	0.5696	0.4304
Figure recognition	0.534	−0.091	0.392	0.327	−0.171	0.5833	0.4167
Object-number	0.488	−0.276	−0.052	0.469	0.255	0.6020	0.3980
Number-figure	0.544	−0.386	0.198	0.152	0.104	0.5181	0.4819
Figure-word	0.476	−0.138	0.122	0.193	0.605	0.6638	0.3362
Deduction	0.643	0.186	0.132	0.070	−0.285	0.5516	0.4484
Numerical puzzles	0.622	−0.232	0.100	−0.202	−0.174	0.5218	0.4782
Problem reasoning	0.640	0.146	0.110	0.056	0.023	0.4467	0.5533
Series completion	0.712	0.105	0.150	−0.103	−0.064	0.5552	0.4448
Arithmetic problems	0.673	−0.196	−0.233	−0.062	0.097	0.5589	0.4411
Eigen values	8.1354	2.096	1.6926	1.5018	1.0252		
Cumulative proportion of total sample variance	0.3390	0.4263	0.4968	0.5594	0.6021		

Data set # 1 for FA: Consider the data from the stock market, which one can obtain from http://in.finance.yahoo.com/. The data is related to the daily closing prices of Hang Seng Index of Hong Kong stock market (http://www.hkex.com.hk/eng/ index.htm). The number of stocks in Hang Seng is 48 and the time frame of our analysis is 2013–2015 or any appropriate time frame as appropriate. Our aim is to study the effect of a few fundamental financial ratios such as current ratio, cash ratio, return of assets, debt ratio, sales to revenue per employee, dividend payout ratio, price-to-book-value ratio, and price-to-sales ratio on the performance of the company and hence on the stock market index of that particular company. If we consider the returns, $r = log_e(P_t/P_{t-1})$, then one can use the concept of FA to study the effects of these ratios of a particular company on the stock market index of that company itself in more details.

Data set # 2 for FA: Let us consider the study performed by Linden (1977), which consists of the performance of $n = 160$ athletes. The $p = 10$ variables are: 100-m run, long jump, shot put, high jump, 400-m run, 110-m run, discus, pole vault, javelin, and 1500-m run.

One can access the data and study the analysis, which can be found in Basilevsky (1994). We urge the readers to work with this data set to gain a better appreciation of FA, which is an interesting multivariate statistical method.

Data set # 3 for FA: The information and background for the third data set can be found in Davis et al. (1997). The study deals with eating disorders and pertains to $n = 191$ individuals with respect to $p = 7$ variables. The data can be found in http://www.unt.edu/rss/class/mike/data/DavisThin.txt. The reader can study the analysis to appreciate the use of FA in a variety of fields, be it sociology or finance.

8.17 Multiple Analysis of Variance and Multiple Analysis of Covariance

8.17.1 Introduction to Analysis of Variance

Consider as a doctor you are interested to analyze the effect of both food habits as well as exercise regime/physical activity on a group of people who are your patients. The weights of these patients are known before the start of this experiment. Your emphasis is to study the reduction of individual weights, which is considered as the response variable. Each of the group of patients undergoes a particular food habit as well as an exercise regime/physical activity. Consider you have $I (i = 1, \ldots, I)$ number of patients in each group, $J (j = 1, \ldots, J)$ as the number of different food habits such as vegetarian, nonvegetarian, vegan, etc., and $K (k = 1, \ldots, K)$ as the number of different exercise regime/physical activities, such as weightlifting, yoga, aerobics, etc. In order to study the effects of food habit, exercise regime/physical activity on the reduction of individual weights, we may use the concept of analysis of variance (ANOVA). Formally, this method studies the total variation present in a set of observations, which is measured by the sum of squares of deviations of the observations from the mean, that is, *SS*. This deviation is partitioned into components associated with assignable effect due to fixed and/or random effects/unassignable effect due to residual random effect. The technique of ANOVA also provides the means for the systematic study of regression analysis and correlation coefficients. A few assumptions/properties that are relevant to ANOVA are:

1. The responses are independently and normally distributed, that is, $X_{i,j,k} \sim N(\mu_{j,k}, \sigma^2), i = 1, \ldots, I, j = 1, \ldots, J, k = 1, \ldots, K$, with *constant* variances (property of homoscedasticity) so that the only difference between the distributions of observations is the means, which is denoted by μ_{ij}.
2. If the number of observations in each group is *equal*, then the ANOVA model is termed as the *balanced* model, else it is an *unbalanced* model.
3. If the *assignable* effects are all fixed, then we have the *fixed-effects model*; otherwise, it is the *random-effects model* where the effects are random except for the additive constants. Note that a *mixed-effects model* contains effects of both fixed- and random-effects types.
4. We say we have a *two-way crossed* ANOVA model if we can categorize the observations in two ways, that is, categorizing in every possible food habit and exercise regime/physical activity pair. In case we have only I and J, then it is the *one-way* ANOVA model.
5. An ANOVA model is *additive* if one can express $\mu_{ij} = \gamma_i + \tau_j$, that is, the effect due to particular food habit and exercise regime/physical activity is the sum of the effect due to the food habit and an effect due to exercise regime/physical activity taken separately. In the additive model, the general hypotheses one is interested to study are that μ_{ij}sare neither

dependent on i nor j. In a *nonadditive* model, apart from the two hypotheses, we have a third one, which is that the model is additive (i.e., $\mu_{ij} = \gamma_i + \tau_j$).

6. When $\mu_{ij} = \mu_{ji}$, then we have the *symmetric* model; else it is the *asymmetric* model.
7. In case any exercise regime/physical activity occurs in only one type of exercise regime/physical activity, then we have the *twofold nested* ANOVA model; else it is the *twofold nonnested* model, an example of which is the one described above.

To continue with our discussion further, consider the general linear model $\boldsymbol{Y} = \boldsymbol{X}\beta + \epsilon$, such that $E(\epsilon) = \boldsymbol{0}$ and $Covar(\epsilon) = \sigma^2\boldsymbol{I}$, which implies that $E(\boldsymbol{Y}) = \boldsymbol{X}\beta$ and $Covar(\boldsymbol{Y}) = \sigma^2\boldsymbol{I}$. In case this is true, then a few relevant results for ANOVA are as stated below:

1. A linear estimator $\boldsymbol{c} + \boldsymbol{a}'\boldsymbol{Y}$ is unbiased for $\lambda'\beta$ iff $E(\boldsymbol{c} + \boldsymbol{a}'\boldsymbol{Y}) = \lambda'\beta$ for all β.
2. When $\lambda'\beta$ is estimable, then it is possible to find several estimators that are unbiased for $\lambda'\beta$ and the OLS estimator $\lambda'\hat{\beta}$ is also the best linear unbiased estimator (BLUE).
3. When $\hat{\beta}$ is any solution to the normal equations $\boldsymbol{X}'\boldsymbol{X}\beta = \boldsymbol{X}\boldsymbol{Y}$, then it is unbiased for $\lambda'\beta$.
4. Consider $\boldsymbol{\Lambda}'\beta$ is any d-dimensional estimable vector and $\boldsymbol{c} = \boldsymbol{A}'\boldsymbol{Y}$ is any vector of linear unbiased estimators of $\boldsymbol{\Lambda}'\beta$, then if $\hat{\beta}$ denotes any solution to the normal equations, then the matrix $Covar(\boldsymbol{c} + \boldsymbol{A}'\boldsymbol{Y}) - Covar(\lambda'\hat{\beta})$ is nonnegative definite.

It is generally acknowledged that Fisher (1921) developed the technique of ANOVA. A few relevant references in this area (along with scope of applications in social sciences and other areas) are: Hoaglin et al. (1991), Iversen and Norpoth (1987), Krishnaiah (1984), Lewis (1971), Rutherford (2001), Sahai and Ageel (2000), Scheffé (1999), Searle et al. (2009), Stuart et al. (1999), and Turner and Thayer (2001).

8.17.2 Multiple Analysis of Variance

With this short background about ANOVA, we come to the area of multianalysis of variance (MANOVA), which as a technique determines the effects of independent categorical variables on multiple continuous-dependent variables. It is usually used to compare several groups with respect to multiple continuous variables, as it tests for differences between centroids, that is, the vectors of the mean values of the dependent variables. On the other hand, one should remember that ANOVA tests for intergroup differences between the mean values of dependent variables. Before we discuss the methodology, we would like to stress the advantages of MANOVA over its univariate counterpart, which is ANOVA:

1. MANOVA can protect against Type I error that occurs if multiple ANOVAs are carried.
2. By measuring several dependent variables in a single experiment, there is a better chance of finding out which variables are important.
3. MANOVA is sensitive not only to mean difference but also to the direction and size of correlations among the dependent variables.

On the other hand, there are also some disadvantages of MANOVA when compared with ANOVA:

1. To do with loss of degrees of freedom, as we know that one degree of freedom is lost for each dependent variable. Hence, the gain of power obtained from the decrease in the sum of squares may be offset due to this loss in the degrees of freedom.

2. High level of dependence of variables does not give us a good picture of the data, so it may be prudent to use the ANOVA model instead.

Before we go into the methodology, we state the general assumptions for MANOVA, which though intuitive is important to understand as it sets the tone about the efficacy of MANOVA as a statistical method when used as a data analysis tool. The assumptions are:

1. The dependent variable should be normally distributed within groups.
2. There should be linear relationships among all pairs (i) of dependent variables, (ii) of covariates, and (iii) of dependent variables–covariables in each cell.
3. The dependent variables should exhibit equal levels of variance across the range of predictor variables. This property is termed as homogeneity.
4. Intercorrelations (covariance) should be homogeneous across the cells of the design.

MANOVA model: Let us consider the two-way MANOVA model of the form

$$\begin{pmatrix} Y_{ij1} \\ \vdots \\ Y_{ijp} \end{pmatrix}_{p\times 1} = \begin{pmatrix} v_1 \\ \vdots \\ v_p \end{pmatrix}_{p\times 1} + \begin{pmatrix} \alpha_{i1} \\ \vdots \\ \alpha_{ip} \end{pmatrix}_{p\times 1} + \begin{pmatrix} \beta_{j1} \\ \vdots \\ \beta_{jp} \end{pmatrix}_{p\times 1} + \begin{pmatrix} \varepsilon_{ij1} \\ \vdots \\ \varepsilon_{ijp} \end{pmatrix}_{p\times 1}.$$

Here, Y_{ijk} is the observation corresponding to the ith treatment, jth block, and kth variable; v_k is the overall mean for the kth variable; α_{ik} is the effect of the ith treatment on the kth variable; β_{jk} is the effect of the jth block on the kth variable; and finally, ε_{ijk} is the experimental error for the ith treatment, jth block, and kth variable. The relevant assumptions for the MANOVA model are the ones which one can refer in any good book in multivariate statistics.

Since this model assumes no interaction, we use this error to test the block and treatment effects. Thus, one can define the mean vector for a treatment i as $\mu_i = \boldsymbol{v} + \boldsymbol{\alpha}_i$. In case the null hypothesis states that all of the treatment mean vectors are identical, then we have H_O:$\mu_1 = \cdots = \mu_g$, or equivalently $\boldsymbol{\alpha}_1 = \cdots = \boldsymbol{\alpha}_g$, where $i = 1, \ldots, g$ are the number of treatments. The alternative hypothesis is H_A:$\mu_{ik} \neq \mu_{jk}$ for at least one $i \neq j$ and at least one variable k.

We now define the sample mean vector for treatment i and block j as

$$\begin{pmatrix} Y_{i\cdot 1} = \frac{1}{b}\sum_{j=1}^{b} Y_{ij1} \\ \vdots \\ Y_{i\cdot p} = \frac{1}{b}\sum_{j=1}^{b} Y_{ijp} \end{pmatrix} \text{ and } \begin{pmatrix} Y_{\cdot j1} = \frac{1}{a}\sum_{i=1}^{a} Y_{ij1} \\ \vdots \\ Y_{\cdot jp} = \frac{1}{a}\sum_{i=1}^{a} Y_{ijp} \end{pmatrix},$$

respectively, while the grand mean vector is

$$\begin{pmatrix} Y_{i\cdot 1} = \frac{1}{a\times b}\sum_{j=1}^{b}\sum_{i=1}^{a} Y_{ij1} \\ \vdots \\ Y_{i\cdot p} = \frac{1}{a\times b}\sum_{j=1}^{b}\sum_{i=1}^{a} Y_{ijp} \end{pmatrix}.$$

Furthermore, let us also define the total sum of squares and cross products matrix as $T = b\sum_{i=1}^{a}(\bar{y}_{i\cdot} - \bar{y}_{\cdot\cdot})(\bar{y}_{i\cdot} - \bar{y}_{\cdot\cdot})' + a\sum_{j=1}^{b}(\bar{y}_{\cdot j} - \bar{y}_{\cdot\cdot})(\bar{y}_{\cdot j} - \bar{y}_{\cdot\cdot})' + \sum_{i=1}^{a}\sum_{j=1}^{b}(Y_{ij} - \bar{y}_{i\cdot} - \bar{y}_{\cdot j} + \bar{y}_{\cdot\cdot})$ $(Y_{ij} - \bar{y}_{i\cdot} - \bar{y}_{\cdot j} + \bar{y}_{\cdot\cdot})'$, where the first, second, and the third terms are the *treatment* sum of squares and cross products matrix, *block* sum of squares and cross products matrix, and finally *error* sum of squares and cross products matrix, respectively. The individual element of the *treatment/block/error* sum of squares and cross products matrices can be expressed accordingly.

Finally, to close this section, we give a few of the statistics that are used in the study of MANOVA, but before that, let us define $A = SS_{Hypothesis}/SS_{Error}$, where SS means the sum of square. Now, based on $\boldsymbol{A}$, and the fact that λ_i denotes the ith eigen value of the matrix $\boldsymbol{A}$, the relevant statistics for MANOVA are as follows:

1. Wilk's lambda: $\prod_{i=1}^{q} 1/(1+\lambda_i)$, where q denotes the dependent variables in MANOVA study.
2. Pillai's trace: $\sum_{i=1}^{q} \lambda_i/(1+\lambda_i)$, where q denotes the dependent variables in MANOVA study.
3. Lawley–Hotelling trace: $\sum_{i=1}^{q} \lambda_i$, where q denotes the dependent variables in MANOVA study.
4. Roy's largest root: $\max_{i=1,\ldots,q}\lambda_i$, where q denotes the dependent variables in MANOVA study.

Example 8.18

A researcher randomly assigns 33 subjects to one of three groups. The first group receives technical dietary information interactively from an online website. Group 2 receives the same information from a nurse practitioner, while group 3 receives the information from a video tape made by the same nurse practitioner. The researcher looks at three different ratings of the presentation, which are to do with *difficulty*, *usefulness*, and *importance*, to determine if there is a difference in the modes of presentation. In particular, the researcher is interested to know whether the interactive website is superior because that is the most cost-effective way of delivering the information. Furthermore, the reader should note that (i) level 1 of the group variable is the treatment group; (ii) level 2 is control group 1, and finally (iii) level 3 is control group 2. The information about the data can be accessed at http://www.ats.ucla.edu/stat/stata/ado/analysis/. Without going through each and every step of the MANOVA calculation, we give the MANOVA test criteria and exact F statistics for the hypothesis of no overall group effect and the results are shown in Table 8.3.

Note: Multivariate analysis of covariance (MANCOVA) is an extension of analysis of covariance (ANCOVA) methods to cover cases where there is more than one dependent variable and where the

TABLE 8.3

MANOVA Test Criteria and Exact F Statistics for the Hypothesis of No Overall Group Effect, Example 8.18

Statistic	Value	F Value	Num dof	Den dof	$Pr > F$
Wilks' lambda	0.53598494	12.99	2	30	<0.0001
Pillai's trace	0.46401506	12.99	2	30	<0.0001
Hotelling–Lawley trace	0.86572405	12.99	2	30	<0.0001
Roy's greatest root	0.86572405	12.99	2	30	<0.0001

control of concomitant continuous independent variables—covariates—is required. The significant benefit of MANCOVA over MANOVA is the *factoring out* of noise or error that has been introduced by the covariant. The analysis one uses to solve problem using ANOVA versus MANOVA can be extended to the case when one is solving ANCOVA versus MANCOVA. We leave this for the reader to study and close this section with a few good references in this area such as Cooley and Lohnes (1971), Huberty and Olejnik (2006), and Morrison (1990).

8.18 Conjoint Analysis

When confronted with any decision process with different alternatives, human beings accept alternative(s) or reject alternative(s) or are ambivalent/indifferent (to different levels of degree) to alternative(s). Their choices are influenced by their likings, experiences, habits, role of advertisements, peer pressures, environmental effects, societal or family constraints, etc. Here is where conjoint analysis (CA) and discrete choice experimentation (DCE) may be used as tools for understanding how individuals develop preferences for alternatives. To study CA, discrete choice models, multiattribute utility theory (MAUT) and random utility theory (RUT) are used, but for the sake of brevity, we skip these discussions and concentrate on the general formulation, models, and applications of CA. Before we start our discussion, we state a few good texts in the area of CA, such as Louviere (1988), Orme and King (2006), Raghavarao et al. (2011), etc.

As a method, CA simultaneously finds a monotonic scoring of the dependent variable and numerical value for each level of each independent variable. This method is based on the main effects of ANOVA model. We state here the conjoint model in its simplest form for the ease of understanding of the readers. Consider $y_{i_1 \cdots i_p} = \mu + \beta_{1i_1} + \cdots + \beta_{pi_p} + \varepsilon_{i_1 \cdots i_p}$, such that $\sum \beta_{1i_1} = \cdots = \sum \beta_{pi_p} = 0$. Consider an example where you are an executive for a marketing firm and you are analyzing the factors that affect the decision of your target customer to buy a car. You want to investigate the preferences for the cars based on p attributes say, for example, mileage, price, safety, resale value, style, passenger space, luggage space, etc. Thus, $y_{i_1 \cdots i_p}$ denotes the a buyer's stated preference for a car with respect to i_1^{th} level of mileage, i_2^{th} level of price, and so on. The nonmetric CA model for the above model can be expressed as $\Phi\left(y_{i_1 \cdots i_p}\right) = \mu + \beta_{1i_1} + \cdots + \beta_{pi_p} + \varepsilon_{i_1 \cdots i_p}$, where $\Phi(\cdot)$ implies a monotonic transformation of the variable y. CA can be solved by the method of ANOVA. An important assumption is that the distance between any two adjacent preference ordering corresponds to the same difference in utility. Thus, we treat the ranking, which is a cardinal variable as if it were metric variable.

Example 8.19: (Härdle and Simar, 2007)

A manufacturer of food items intends to make a new margarine and varies the *product characteristics* as well as its *packaging*. The four different products made by the food manufacturer are ordered as shown in Table 8.4. The information about the data set can be found in Härdle and Simar (2007).

Let us consider the part worth X_1 as usage, and suppose a person ranks the six different products as shown in Table 8.5.

Solving, one obtains $\beta_{11} = -2, \beta_{12} = 0$, and $\beta_{13} = 2$, while on the other hand, we get $\beta_{21} = 0.16$ and $\beta_{22} = -0.16$, and $\mu = 3.5$. Using these values, we can easily obtain $\hat{Y}_1 = \beta_{11} + \beta_{21} + \mu = 1.66$, $\hat{Y}_2 = \beta_{11} + \beta_{22} + \mu = 1.34$, $\hat{Y}_3 = \beta_{12} + \beta_{21} + \mu = 3.66$, $\hat{Y}_4 = \beta_{12} + \beta_{22} + \mu = 3.34$, $\hat{Y}_5 = \beta_{13} + \beta_{21} + \mu = 5.66$, and $\hat{Y}_6 = \beta_{13} + \beta_{22} + \mu = 5.34$.

TABLE 8.4

Ranking of the Four Products, Example 8.19

Product Type	Product Characteristics	Packaging	Ranking
1	Low calories	Plastic pack	3
2	Low calories	Paper pack	4
3	High calories	Plastic pack	1
4	High calories	Paper pack	2

TABLE 8.5

Ranked Products, Example 8.19

		X_2 Calories	
		Low	**High**
X_1 **Usage**		**1**	**2**
Bread	1	2	1
Cooking	2	3	4
Universal	3	6	5

8.19 Canonical Correlation Analysis

Hotelling (1935, 1936) may be credited for developing the technique of canonical correlation analysis (CCA), where the author studied how arithmetic speed and arithmetic power are related to reading speed and reading power. Mathematically, CCA may be stated as follows. Suppose we are given $\boldsymbol{X} \in \mathbb{R}^p$ and $\boldsymbol{Y} \in \mathbb{R}^q$, $(p \leq q)$, then the idea is to find an index describing a possible link between $\boldsymbol{X}$ and $\boldsymbol{Y}$. Using CCA, we are interested to find vectors $\boldsymbol{a}$ of size $(p \times 1)$ and $\boldsymbol{b}$ of size $(q \times 1)$, such that the correlation coefficient between $U = \boldsymbol{a}'\boldsymbol{X}$ and $V = \boldsymbol{b}'\boldsymbol{Y}$, given by $\rho(U, V) = \rho(\boldsymbol{a}, \boldsymbol{b})$, is *maximized*. Remember, CCA is based on linear indices or linear combination as both $U = \boldsymbol{a}'\boldsymbol{X}$ and $V = \boldsymbol{b}^T\boldsymbol{Y}$ are linear combinations and may be expressed as $a_1X_1 + \cdots + a_pX_p$ and $b_1Y_1 + \cdots + b_qY_q$, respectively.

The idea of CCA is to go stepwise, where in the *first* step we determine the first pair of linear combinations, (U_1, V_1), which results in the largest value of correlation, ρ_1^{*2}. In the next step, one then determines the *second* pair of linear combinations, (U_2, V_2), which has the largest correlation, given by ρ_2^{*2}, among all the pairs such that they are *uncorrelated* with the initially selected pairs. We continue doing this till p stages, so that we obtain $(U_1, V_1), \ldots, (U_p, V_p)$ and $\left(\rho_1^{*2}, \rho_2^{*2}, \ldots, \rho_p^{*2}\right)$. In case the condition $(p \leq q)$ does not hold true, then we continue doing this till $\min(p, q)$. CCA is a simple and useful method to describe the correlation structure between two sets of variables. It is a generalization of the concept of multiple correlation and successively maximizes the correlation between appropriate pairs of linear combinations of the variables of the two sets. The method can be viewed as a dimension reduction technique in that it represents the correlation structure between two sets of variables in terms of a smaller number of canonical correlations. The pairs of linear combinations $\{(U_1, V_1), (U_2, V_2), \ldots, (U_p, V_p)\}$ thus obtained are termed as the *canonical variables*, while the corresponding correlations $\left(\rho_1^{*2}, \rho_2^{*2}, \ldots, \rho_p^{*2}\right)$ are known as the *canonical correlations* values.

Some interesting application areas of CCA are:

1. Study of how government policy variables such as interest rates, expenditures in various economic sectors, etc. are related to other economic goal variables such as foreign currency rates, inflation rates, etc.
2. Study of college performance variables of students with respect to their scholastic achievements before joining the college.

8.19.1 Formulation of Canonical Correlation Analysis

Suppose $\boldsymbol{X}$ is distributed with $E(\boldsymbol{X}) = \boldsymbol{\mu}_X$, $Covar(\boldsymbol{X}) = \boldsymbol{\Sigma}_{XX}$, while $\boldsymbol{Y}$ is distributed with $E(\boldsymbol{Y}) = \boldsymbol{\mu}_Y$, $Covar(\boldsymbol{Y}) = \boldsymbol{\Sigma}_{YY}$. Moreover, $Covar(\boldsymbol{X}, \boldsymbol{Y}) = \boldsymbol{\Sigma}_{XY} = \boldsymbol{\Sigma}'_{YX}$. Then one can easily verify that

$$\rho(U, V) = \rho(\boldsymbol{a}, \boldsymbol{b}) = \left\{ \frac{\boldsymbol{a}'\boldsymbol{\Sigma}_{XY}\mathbf{b}}{(\boldsymbol{a}'\boldsymbol{\Sigma}_{XX}\boldsymbol{a})^{1/2}(\boldsymbol{b}'\boldsymbol{\Sigma}_{YY}\mathbf{b})^{1/2}} \right\}.$$

One may also note that $\rho(\mathbf{c}\boldsymbol{a}, \boldsymbol{b}) = \rho(\boldsymbol{a}, \boldsymbol{b}) = \rho(\boldsymbol{a}, c\boldsymbol{b})$, where $c \in \mathbb{R}^+$.

From the optimization point of view, CCA can be stated simply as follows:

$$\max\left(\boldsymbol{a}'\boldsymbol{\Sigma}_{XY}\mathbf{b}\right)$$
$$\text{s.t.}: \boldsymbol{a}'\boldsymbol{\Sigma}_{XX}\boldsymbol{a} = 1$$
$$\boldsymbol{b}'\boldsymbol{\Sigma}_{YY}\mathbf{b} = 1$$

A closer look at the above optimization makes it obvious that in maximizing the ratio, which is the correlation coefficient, $\rho(U, V)$, we maximize the numerate, that is, $Covar(\boldsymbol{X}, \boldsymbol{Y})$ with the restrictions that both $Covar(\boldsymbol{a}'\boldsymbol{X})$ and $Covar(\boldsymbol{b}'\boldsymbol{Y})$ are equal to 1. Standard nonlinear optimization algorithms are available, which solves this problem. For a better explanation of how we go about in achieving this, one can refer to the algorithm described as a pseudo-code in Figure 8.17.

Before discussing the standardized form of CCA, we state a few important results relevant to CCA:

1: **DEFINE:** $\boldsymbol{a}$, $\boldsymbol{b}$, $\boldsymbol{X}_{n\times p}$, $\boldsymbol{Y}_{n\times q}$ where $p < q, U = \boldsymbol{a}'\boldsymbol{X}$, $V = \boldsymbol{b}'\boldsymbol{Y}$

2: **INPUT:** $\boldsymbol{X}_{n\times p}$, $\boldsymbol{Y}_{n\times q}$ where $p < q$

3: **CALCULATE:** $E(\boldsymbol{X}) = \boldsymbol{\mu}_X$, $E(\boldsymbol{Y}) = \boldsymbol{\mu}_Y$, $Covar(\boldsymbol{X}) = \boldsymbol{\Sigma}_{XX}$, $Covar(\boldsymbol{Y}) = \boldsymbol{\Sigma}_{YY}$, $Covar(\boldsymbol{X},\boldsymbol{Y}) = \boldsymbol{\Sigma}_{XY}$, $\boldsymbol{a}$, $\boldsymbol{b}$ such that $corr(U,V) = \frac{\boldsymbol{a}'\boldsymbol{\Sigma}_{XY}\mathbf{b}}{(\boldsymbol{a}'\boldsymbol{\Sigma}_{XX}\boldsymbol{a})^{\frac{1}{2}}(\boldsymbol{b}'\boldsymbol{\Sigma}_{YY}\mathbf{b})^{\frac{1}{2}}}$ is maximized

4: **START if:** $i = 1{:}p$

  Maximize: $Covar(U_i, V_i)$

  s.t.: $Covar(U_i) = 1$ and $Covar(V_i) = 1$

5: **CALCULATE:** (i) ρ_i^{*2} which is the ith eigen value of $\boldsymbol{\Sigma}_{XX}^{-\frac{1}{2}}\boldsymbol{\Sigma}_{XY}\boldsymbol{\Sigma}_{YY}^{-1}\boldsymbol{\Sigma}_{YX}\boldsymbol{\Sigma}_{XX}^{-\frac{1}{2}}$ and e_i which is its corresponding i^{th} eigen vector, (ii) ρ_i^{*2} which is the ith *largest*eigen value of $\boldsymbol{\Sigma}_{YY}^{-\frac{1}{2}}\boldsymbol{\Sigma}_{YX}\boldsymbol{\Sigma}_{XX}^{-1}\boldsymbol{\Sigma}_{XY}\boldsymbol{\Sigma}_{YY}^{-\frac{1}{2}}$ and $f_i$which is its corresponding i^{th} eigen vector, (iii) in calculating these i^{th} values, ensure that we find those linear combinations which are uncorrelated with the preceding $1,2,..,i-1$ number of canonical variables, (iv) $\boldsymbol{a}_i = \boldsymbol{e}_i'\boldsymbol{\Sigma}_{XX}^{-\frac{1}{2}}$ and $\boldsymbol{b}_i = \boldsymbol{f}_i'\boldsymbol{\Sigma}_{XX}^{-\frac{1}{2}}$, (v) $U_i = \boldsymbol{a}_i'\boldsymbol{X}$ and $V_i = \boldsymbol{b}_i'\boldsymbol{Y}$

6: **END if**

7: **REPORT:** $(\rho_1^{*2}, \rho_2^{*2}, \ldots, \rho_p^{*2})$ and $\{(U_1, V_1), \ldots\ldots, (U_p, V_p)\}$

8: **END**

FIGURE 8.17
Pseudo-code to implement CCA method.

1. For any given $r, 1 \leq j \leq p$, the maximum value of $\boldsymbol{a}'\boldsymbol{\Sigma}_{XY}\mathbf{b}$ subject to (i) $\boldsymbol{a}'\boldsymbol{\Sigma}_{XX}\boldsymbol{a} = 1$, (ii) $\boldsymbol{b}'\boldsymbol{\Sigma}_{YY}\mathbf{b} = 1$, and (iii) $\boldsymbol{a}'_j\ \boldsymbol{\Sigma}_{XX}\boldsymbol{a} = 0$ for $j = 1, \ldots, r-1$ is given by $\rho_j = \sqrt{\lambda_j}$, where e_j is the jth eigen value of $\left(\boldsymbol{\Sigma}_{\text{XX}}^{-1/2}\boldsymbol{\Sigma}_{\text{XY}}\boldsymbol{\Sigma}_{\text{YY}}^{-1/2}\right)\left(\boldsymbol{\Sigma}_{\text{XX}}^{-1/2}\boldsymbol{\Sigma}_{\text{XY}}\boldsymbol{\Sigma}_{\text{YY}}^{-1/2}\right)'$, and the maximum value is obtained when $\boldsymbol{a} = \boldsymbol{a}_j$ and $\boldsymbol{b} = \boldsymbol{b}_j$.
2. Let $U_j = \boldsymbol{a}'_j\boldsymbol{X}$ and $V_j = \boldsymbol{b}'_j\boldsymbol{Y}$ be the jth canonical correlation variables, $j = 1, \ldots, p$. Then

$$Var\begin{pmatrix} U \\ V \end{pmatrix} = \begin{pmatrix} \boldsymbol{I}_p & \Lambda \\ \Lambda & \boldsymbol{I}_p \end{pmatrix},$$

where $\boldsymbol{U} = (U_1, \ldots, U_p), \boldsymbol{V} = (V_1, \ldots, V_p)$ and $\Lambda = \text{diag}\left(\sqrt{\lambda_1}, \ldots, \sqrt{\lambda_p}\right)$.

Note: Thus, the canonical correlation coefficients, $\rho_j = \sqrt{\lambda_j}$, are the covariances between U_j and V_j, where $j = 1, \ldots, p$. Moreover, $\boldsymbol{a}'_1\boldsymbol{X}$ and $\boldsymbol{b}'_1\boldsymbol{Y}$ have the maximum covariance of value $\rho_1 = \sqrt{\lambda_1}$.

One should also remember a few important things (considering, $p \leq q$):

1. For the matrix $\left(\boldsymbol{\Sigma}_{XX}^{-1/2}\boldsymbol{\Sigma}_{XY}\boldsymbol{\Sigma}_{YY}^{-1/2}\right)\left(\boldsymbol{\Sigma}_{XX}^{-1/2}\boldsymbol{\Sigma}_{XY}\boldsymbol{\Sigma}_{YY}^{-1/2}\right)'$ which is of size $(p \times p)$, we have $\rho_1^2 \geq \cdots \geq \rho_p^2$ as the eigen values for which the associated eigen vectors are $\boldsymbol{e}_1, \ldots, \boldsymbol{e}_p$.
2. For the matrix $\left(\boldsymbol{\Sigma}_{YY}^{-1/2}\boldsymbol{\Sigma}_{YX}\boldsymbol{\Sigma}_{XX}^{-1/2}\right)\left(\boldsymbol{\Sigma}_{YY}^{-1/2}\boldsymbol{\Sigma}_{YX}\boldsymbol{\Sigma}_{XX}^{-1/2}\right)'$, which is of size $(q \times q)$, we have $\rho_1^2 \geq \cdots \geq \rho_p^2$ as the *largest* p eigen values for which the associated eigen vectors are $\boldsymbol{f}_1, \ldots, \boldsymbol{f}_p$.
3. $f_j \propto \left(\boldsymbol{\Sigma}_{YY}^{-1/2}\boldsymbol{\Sigma}_{YX}\boldsymbol{\Sigma}_{XX}^{-1/2}\right)\boldsymbol{e}_i, \quad j = 1, \ldots, p.$
4. $\left(U_1 = \boldsymbol{a}'_1\boldsymbol{X} = \boldsymbol{e}'_1\boldsymbol{\Sigma}_{XX}^{-1/2}\boldsymbol{X}, V_1 = \boldsymbol{b}'_1\boldsymbol{Y} = \boldsymbol{f}'_1\boldsymbol{\Sigma}_{YY}^{-1/2}\boldsymbol{Y}\right), \ldots,$ $\left(U_p = \boldsymbol{b}'_p\boldsymbol{X} = \boldsymbol{e}'_p\boldsymbol{\Sigma}_{XX}^{-1/2}\boldsymbol{X}, V_p = \boldsymbol{b}'_p\boldsymbol{Y} = \boldsymbol{f}'_p\boldsymbol{\Sigma}_{YY}^{-1/2}\boldsymbol{Y}\right).$
5. $Var(U_j) = Var(V_j) = 1, j = 1, \ldots, p.$
6. $Covar\left(U_{j_1}, U_{j_2}\right) = corr\left(U_{j_1}, U_{j_2}\right) = 0, \quad j_1 \neq j_2 = 1, \ldots, p.$
7. $Covar\left(V_{j_1}, V_{j_2}\right) = corr\left(V_{j_1}, V_{j_2}\right) = 0, \quad j_1 \neq j_2 = 1, \ldots, p.$
8. $Covar\left(U_{j_1}, V_{j_2}\right) = corr\left(U_{j_1}, V_{j_2}\right) = 0, \quad j_1 \neq j_2 = 1, \ldots, p.$

8.19.2 Standardized Form of CCA

In case

$$Z_{X_i} = \frac{X_i - \mu_{Y_i}}{\sqrt{\sigma_{X_iX_i}}},$$

where

$$\boldsymbol{\rho}_{XX} = \begin{pmatrix} \rho_{X_1X_1} & \cdots & \rho_{X_pX_1} \\ \vdots & \ddots & \vdots \\ \rho_{X_1X_p} & \cdots & \rho_{X_pX_p} \end{pmatrix}, \quad \boldsymbol{\Sigma}_{XX} = \begin{pmatrix} \sigma_{X_1X_1} & \cdots & \sigma_{X_pX_1} \\ \vdots & \ddots & \vdots \\ \sigma_{X_1X_p} & \cdots & \sigma_{X_pX_p} \end{pmatrix} \quad \text{and} \quad Z_{Y_j} = \frac{Y_j - \mu_{Y_j}}{\sqrt{\sigma_{Y_jY_j}}},$$

where

$$\boldsymbol{\rho}_{YY} = \begin{pmatrix} \rho_{Y_1Y_1} & \cdots & \rho_{Y_qY_1} \\ \vdots & \ddots & \vdots \\ \rho_{Y_1Y_q} & \cdots & \rho_{Y_qY_q} \end{pmatrix}, \quad \boldsymbol{\Sigma}_{YY} = \begin{pmatrix} \sigma_{Y_1Y_1} & \cdots & \sigma_{Y_qY_1} \\ \vdots & \ddots & \vdots \\ \sigma_{Y_1Y_q} & \cdots & \sigma_{Y_qY_q} \end{pmatrix},$$

then $U_k = \boldsymbol{a}'_k\boldsymbol{Z}_X = \boldsymbol{e}'_k\boldsymbol{\rho}_{XX}^{-1/2}\boldsymbol{Z}_X$ and $V_k = \boldsymbol{b}'_k\boldsymbol{Z}_Y = \boldsymbol{f}'_k\boldsymbol{\rho}_{YY}^{-1/2}\boldsymbol{Z}_Y, k = 1, \ldots, p$.

One should also remember the following:

1. $Covar(Z_X) = \boldsymbol{\rho}_{XX}$.
2. $Covar(Z_Y) = \boldsymbol{\rho}_{YY}$.
3. $Covar(Z_X, Z_Y) = \boldsymbol{\rho}_{XY} = \boldsymbol{\rho}'_{YX}$.
4. $\rho_1^{*2} \geq \cdots \geq \rho_p^{*2}$ are the nonzero eigen values of $\boldsymbol{\rho}_{XX}^{-1/2}\boldsymbol{\rho}_{XY}\boldsymbol{\rho}_{YY}^{-1}\boldsymbol{\rho}_{YX}\boldsymbol{\rho}_{XX}^{-1/2}$ for which $(\boldsymbol{e}_1, \ldots, \boldsymbol{e}_p)$ is the set of corresponding eigen vectors.
5. $\rho_1^{*2} \geq \cdots \geq \rho_p^{*2}$ are the nonzero *largest* p set of eigen values, from among q of them, of $\boldsymbol{\rho}_{YY}^{-1/2}\boldsymbol{\rho}_{YX}\boldsymbol{\rho}_{XX}^{-1}\boldsymbol{\rho}_{XY}\boldsymbol{\rho}_{YY}^{-1/2}$ for which $(\boldsymbol{f}_1, \ldots, \boldsymbol{f}_p)$ is the set of corresponding eigen vectors.

8.19.3 Correlation between Canonical Variates and Their Component Variables

In case one is interested to find the correlation between the original variables and their respective transformed variables, then a few interesting results can be stated. But before that, consider $\boldsymbol{A}_{p\times p} = [\boldsymbol{a}_1, \ldots, \boldsymbol{a}_p]'$, $\boldsymbol{B}_{q\times q} = [\boldsymbol{b}_1, \ldots, \boldsymbol{b}_q]'$, so that $\boldsymbol{U}_p \times 1 = \boldsymbol{A}_{p\times p}\boldsymbol{X}_{p\times 1}$ and $\boldsymbol{V}_{q\times 1} = \boldsymbol{B}_{q\times q}\boldsymbol{Y}_{q\times 1}$ and $p \leq q$. With these, the following results stated below hold:

1. $Covar(\boldsymbol{U}, \boldsymbol{X}) = \boldsymbol{A}'\boldsymbol{\Sigma}_{XX}$.
2. $Covar(\boldsymbol{V}, \boldsymbol{Y}) = \boldsymbol{B}'\boldsymbol{\Sigma}_{YY}$.
3. $$\boldsymbol{\rho}_{\boldsymbol{U},\boldsymbol{X}_{p\times p}} = \boldsymbol{A}'_{p\times p}\boldsymbol{\Sigma}_{XX\,p\times p}\begin{bmatrix} \sqrt{Var(X_1)} & \cdots & 0 \\ \vdots & \ddots & \vdots \\ 0 & \cdots & \sqrt{Var(X_p)} \end{bmatrix}_{p\times p}.$$
4. $$\boldsymbol{\rho}_{\boldsymbol{U},\boldsymbol{Y}_{p\times q}} = \boldsymbol{A}'_{p\times p}\boldsymbol{\Sigma}_{XY\,p\times q}\begin{bmatrix} \sqrt{Var(Y_1)} & \cdots & 0 \\ \vdots & \ddots & \vdots \\ 0 & \cdots & \sqrt{Var(Y_q)} \end{bmatrix}_{q\times q}.$$
5. $$\boldsymbol{\rho}_{\boldsymbol{V},\boldsymbol{X}_{q\times p}} = \boldsymbol{B}_{q\times q}\boldsymbol{\Sigma}_{YX\,q\times p}\begin{bmatrix} \sqrt{Var(X_1)} & \cdots & 0 \\ \vdots & \ddots & \vdots \\ 0 & \cdots & \sqrt{Var(X_p)} \end{bmatrix}_{p\times p}.$$
6. $$\boldsymbol{\rho}_{\boldsymbol{V},\boldsymbol{Y}_{q\times q}} = \boldsymbol{B}_{q\times q}\boldsymbol{\Sigma}_{YY\,q\times q}\begin{bmatrix} \sqrt{Var(Y_1)} & \cdots & 0 \\ \vdots & \ddots & \vdots \\ 0 & \cdots & \sqrt{Var(Y_q)} \end{bmatrix}_{q\times q}.$$

While in the standardized variable case, we have

1. $Covar(\boldsymbol{U}, \boldsymbol{Z}_X) = \boldsymbol{A}\rho_{XX}$.
2. $Covar(\boldsymbol{V}, \boldsymbol{Z}_Y) = \boldsymbol{B}\rho_{YY}$.
3.

$$\rho_{U,Z_{X_{p\times p}}} = \boldsymbol{A}_{Z_{X_{p\times p}}} \rho_{XX_{p\times p}} \begin{bmatrix} \sqrt{Var\left(Z_{X_1}\right)} & \cdots & 0 \\ \vdots & \ddots & \vdots \\ 0 & \cdots & \sqrt{Var\left(Z_{X_p}\right)} \end{bmatrix}_{p\times p}.$$

4.

$$\rho_{U,Z_{Y_{p\times q}}} = \boldsymbol{A}_{Z_{X_{p\times p}}} \rho_{XY_{p\times q}} \begin{bmatrix} \sqrt{Var\left(Z_{Y_1}\right)} & \cdots & 0 \\ \vdots & \ddots & \vdots \\ 0 & \cdots & \sqrt{Var\left(Z_{Y_q}\right)} \end{bmatrix}_{q\times q}.$$

5.

$$\rho_{V,Z_{X_{q\times p}}} = \boldsymbol{B}_{Z_{Y_{q\times q}}} \rho_{YX_{q\times p}} \begin{bmatrix} \sqrt{Var\left(Z_{X_1}\right)} & \cdots & 0 \\ \vdots & \ddots & \vdots \\ 0 & \cdots & \sqrt{Var\left(Z_{X_p}\right)} \end{bmatrix}_{p\times p}.$$

6.

$$\rho_{V,Z_{Y_{q\times q}}} = \boldsymbol{B}_{Z_{Y_{q\times q}}} \rho_{YY_{q\times q}} \begin{bmatrix} \sqrt{Var\left(Z_{Y_1}\right)} & \cdots & 0 \\ \vdots & \ddots & \vdots \\ 0 & \cdots & \sqrt{Var\left(Z_{Y_q}\right)} \end{bmatrix}_{q\times q}.$$

Note: A different perspective when analyzing CCA from the optimization point of view is stated here, with the idea that it acts as a good motivation for the interested readers. Let us consider the optimization problem stated before. In its Lagrangian form, considering the Lagrangian multipliers as λ_1 and λ_2, the expression to be differentiated is $g(\lambda_1, \lambda_2) = \boldsymbol{a}'\boldsymbol{\Sigma}_{XY}\mathbf{b} - \lambda_1\boldsymbol{a}'\boldsymbol{\Sigma}_{XX}\boldsymbol{a} - \lambda_2\boldsymbol{b}'\boldsymbol{\Sigma}_{YY}\mathbf{b}$. Now, putting $\partial g(\lambda_1, \lambda_2)/\partial\lambda_1 = \partial g(\lambda_1, \lambda_2)/\partial\lambda_2 = 0$, one obtains $\lambda_1 = \boldsymbol{a}'\boldsymbol{\Sigma}_{XX}\boldsymbol{a}$. With a few relevant mathematical changes, we obtain λ^2 (here, $\lambda_1 = \lambda_1 = \lambda$) and $\boldsymbol{b}$ as the eigen root and eigen vector corresponding to the determinantal equation $|\boldsymbol{\Sigma}_{YX}\boldsymbol{\Sigma}_{XX}^{-1}\boldsymbol{\Sigma}_{XY} - \lambda^2\boldsymbol{\Sigma}_{YY}| = 0$. On a similar line, one can also solve and get the other set of eigen root and eigen vector.

8.19.4 Testing the Test Statistics in CCA

A main concern in CCA is to find whether there is some significant relationship or dependence between the variables $\boldsymbol{X}$ and $\boldsymbol{Y}$. Gittins (1985) suggested the following test to verify the dependence between the variables $\boldsymbol{X}$ and $\boldsymbol{Y}$ using Wilk's likelihood ratio statistics, which is given by $T^{2/n} = |\boldsymbol{I} - \boldsymbol{S}_{YY}^{-1}\boldsymbol{S}_{YX}\boldsymbol{S}_{XX}^{-1}\boldsymbol{S}_{XY}| = \prod_{i=1}^{k}(1 - l_i)$. As this statistic has a complicated distribution, hence it is denoted by $-\{n - (p + q + 3)/2\}log\prod_{i=1}^{k}(1 - l_i) \sim \chi^2_{p\times q}$, provided $n \to \infty$ (Barlett, 1954). In case one is interested to find if only s of the total number of canonical correlations are nonzero, then the statistic is of the form $-\{n - (p + q + 3)/2\}log\prod_{i=s+1}^{k}(1 - l_i) \sim \chi^2_{(p-s)\times(q-s)}$, and this holds true in the approximate sense as $n \to \infty$.

Example 8.20

Consider we have a theoretical set of data where the following is given: $\boldsymbol{X} = (X_1, X_2, X_3)$, that is, $p = 3$, $\boldsymbol{Y} = (Y_1, Y_2, Y_3, Y_4, Y_5)$, that is, $q = 5$, such that $E(\boldsymbol{X}) = (2, 3, 6)$, $E(\boldsymbol{Y}) = (45, 44, 34, 32, 40)$,

$$Covar(\boldsymbol{X}) = \boldsymbol{\Sigma}_{XX} = \begin{pmatrix} 0.4 & 0.2449 & 0.45 \\ 0.2449 & 0.6 & 0.1837 \\ 0.45 & 0.1837 & 0.9 \end{pmatrix},$$

$$Covar(\boldsymbol{Y}) = \boldsymbol{\Sigma}_{YY} = \begin{pmatrix} 4 & 1.3416 & 1.4697 & 0 & 2.2627 \\ 1.3416 & 5 & 2.1909 & 5.3666 & 0.6325 \\ 1.4697 & 2.1909 & 6 & 0.7348 & 4.1569 \\ 0 & 5.3666 & 0.7348 & 9 & 5.9397 \\ 2.2627 & 0.6325 & 4.1569 & 5.9397 & 8 \end{pmatrix},$$

and

$$Covar(\boldsymbol{X}, \boldsymbol{Y}) = \boldsymbol{\Sigma}_{XY} = \begin{pmatrix} 0.1265 & 0.4243 & 1.3943 & 1.8974 & 0.7155 \\ 1.2394 & 0 & 0.3795 & 2.3238 & 0.6573 \\ 0.3795 & 0.8485 & 1.1619 & 1.7076 & 2.1466 \end{pmatrix}.$$

Given this set of information, let us calculate the following, the values of which are given alongside the formulae. We urge the reader to recalculate the values to get a good idea about the steps involved in CCA calculations:

1.

$$\boldsymbol{\Sigma}_{XX}^{-1/2} \boldsymbol{\Sigma}_{XY} \boldsymbol{\Sigma}_{YY}^{-1} \boldsymbol{\Sigma}_{YX} \boldsymbol{\Sigma}_{XX}^{-1/2} = \begin{pmatrix} -5.3437 & 13.4544 & -7.6743 \\ 13.4544 & -32.9776 & 19.9303 \\ -7.6743 & 19.9303 & -13.6489 \end{pmatrix}.$$

2.

$$\boldsymbol{e} = \begin{pmatrix} -50.7400 & 0 & 0 \\ 0 & 0.2008 & 0 \\ 0 & 0 & -1.4311 \end{pmatrix}.$$

3.

$$\boldsymbol{\rho}^* = \begin{pmatrix} 0.3227 & -0.8657 & -0.3827 \\ -0.8043 & -0.4640 & 0.3713 \\ 0.4990 & -0.1880 & 0.8460 \end{pmatrix}.$$

4.

$$\boldsymbol{\Sigma}_{YY}^{-1/2} \boldsymbol{\Sigma}_{YX} \boldsymbol{\Sigma}_{XX}^{-1} \boldsymbol{\Sigma}_{XY} \boldsymbol{\Sigma}_{YY}^{-1/2}$$

$$= \begin{pmatrix} -0.2219 - 0.1508i & 0.9157 - 2.1750i & -0.7086 + 0.4421i & 0.9504 - 2.1756i & 0.4862 + 1.8497i \\ 0.9157 - 2.1750i & -22.7845 + 1.1740i & 2.2736 - 0.9611i & -22.2198 - 2.8159i & 12.9425 - 4.9868i \\ -0.7086 + 0.4421i & 2.2736 - 0.9611i & -0.3291 + 0.8951i & 1.8779 - 0.3799i & -0.2866 + 0.6717i \\ 0.9504 - 2.1756i & -22.2198 - 2.8159i & 1.8779 - 0.3799i & -21.2278 - 6.7355i & 13.8364 - 2.2843i \\ 0.4862 + 1.8497i & 12.9425 - 4.9868i & -0.2866 + 0.6717i & 13.8364 - 2.2843i & -7.4071 + 4.8171i \end{pmatrix}.$$

5.

$$\boldsymbol{f} = \begin{pmatrix} -50.7400 & 0 & 0 \\ 0 & 0.2008 & 0 \\ 0 & 0 & -1.4311 \end{pmatrix}.$$

6.

$$\rho^* = \begin{pmatrix} 0.3227 & -0.8657 & -0.3827 \\ -0.8043 & -0.4640 & 0.3713 \\ 0.4990 & -0.1880 & 0.8460 \end{pmatrix}.$$

Example 8.21

As the next example, consider the data given in http://www.ats.ucla.edu/stat/r/dae/canonical.htm, which consists of 600 observations on eight variables. The psychological variables are (i) locus of control, X_1, (ii) self-concept, X_2, and (iii) motivation, X_3, such that $\boldsymbol{X}_{600\times 3}$ is the first set of variable, while the academic variables are standardized tests in (i) reading, Y_1 (ii) writing, Y_2, (iii) mathematics, Y_3, and (iv) science, Y_4 are such that $\boldsymbol{Y}_{600\times 4}$ is the second set of variables. Additionally, the variable female is a zero-one indicator variable with one indicating a female student, while a zero denotes a male student.

Solving the CCA problems yields

$$\boldsymbol{A} = (\boldsymbol{a}_1\boldsymbol{a}_2\boldsymbol{a}_3) = \begin{pmatrix} -1.2501 & 0.7660 & -0.4967 \\ 0.2367 & 0.8421 & 1.2051 \\ -1.2491 & -2.6360 & 1.0935 \end{pmatrix},$$

$$\boldsymbol{B} = (\boldsymbol{b}_1\boldsymbol{b}_2\boldsymbol{b}_3\boldsymbol{b}_4) = \begin{pmatrix} -0.0440 & -0.0016 & 0.0883 \\ -0.0551 & -0.0904 & -0.0961 \\ -0.0194 & -0.0030 & 0.0878 \\ 0.0038 & 0.1242 & -0.0885 \end{pmatrix},$$

$$\rho^{*^2} = (0.4464 \quad 0.1534 \quad 0.0225).$$

This means that the set of linear combinations of the variables are:

1.

$$U_1 = \boldsymbol{a}_1'\boldsymbol{X} = (-1.2501\ 0.2367\ -1.2491)\begin{pmatrix} X_1 \\ X_2 \\ X_3 \end{pmatrix} = -1.2501X_1 + 0.2367X_2 - 1.2491X_3,$$

$$V_1 = \boldsymbol{b}_1'\boldsymbol{Y} = (-0.0440\ -0.0551\ -0.0194\ 0.0038)\begin{pmatrix} Y_1 \\ Y_2 \\ Y_3 \\ Y_4 \end{pmatrix} = -0.0440Y_1 - 0.0551Y_2$$

$$-0.0194Y_3 + 0.0038Y_4,$$

2.

$$U_2 = \boldsymbol{a}_2'\boldsymbol{X} = (0.7660\ 0.8421\ -2.6360)\begin{pmatrix} X_1 \\ X_2 \\ X_3 \end{pmatrix} = 0.7660X_1 + 0.8421X_2 - 2.6360X_3,$$

$$V_2 = \boldsymbol{b}_2'\boldsymbol{Y} = (-0.0016 - 0.0904 - 0.0030\ 0.1242)\begin{pmatrix} Y_1 \\ Y_2 \\ Y_3 \\ Y_4 \end{pmatrix} = -0.0016Y_1 - 0.0904Y_2$$

$$-0.0030Y_3 + 0.1242Y_4,$$

3.

$$U_3 = \boldsymbol{a}_3'\boldsymbol{X} = (-0.4967 \; 1.2051 \; 1.0935)\begin{pmatrix} X_1 \\ X_2 \\ X_3 \end{pmatrix} = -0.4967X_1 + 1.2051X_2 + 1.0935X_3,$$

$$V_3 = \boldsymbol{b}_3'\boldsymbol{Y} = (0.0883 \; -0.0961 \; -0.0878 \; 0.0885)\begin{pmatrix} Y_1 \\ Y_2 \\ Y_3 \\ Y_4 \end{pmatrix} = 0.0883Y_1 \; -0.0961Y_2$$

$$-0.0878Y_3 \; -0.0885Y_4,$$

respectively. The corresponding linear combination graphs for (U_1, V_1), (U_2, V_2) and (U_3, V_3) are shown in Figure 8.18. Though not apparent but one can easily discern that the value of correlation coefficient or the slope of the set of (U_1, V_1) are the maximum, followed by (U_2, V_2), and then (U_3, V_3). This fact is also corroborated by the values of $\rho_1^{*^2} = 0.4464$, $\rho_2^{*^2} = 0.1534$, and $\rho_3^{*^2} = 0.0225$. Another way of verifying the values of $\rho_1^{*^2}, \rho_2^{*^2}, \rho_3^{*^2}$ is to have a look at

$$Covar(\boldsymbol{U}, \boldsymbol{V}) = \begin{bmatrix} 1 & 0.4464 & 0 & 0 & 0 & 0 \\ 0.4464 & 1 & 0 & 0 & 0 & 0 \\ 0 & 0 & 1 & 0.1543 & 0 & 0 \\ 0 & 0 & 0.1534 & 1 & 0 & 0 \\ 0 & 0 & 0 & 0 & 1 & 0.0225 \\ 0 & 0 & 0 & 0 & 0.0225 & 1 \end{bmatrix}.$$

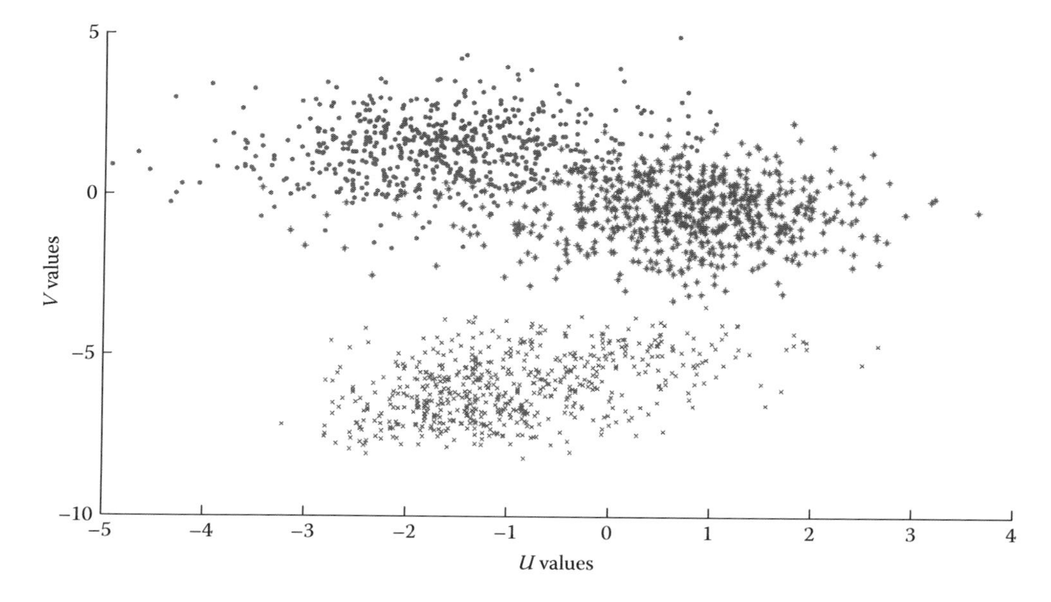

FIGURE 8.18
Graphs showing linear relationship between the set of variables using CCA method, Example 8.20.

For this problem, let us also find the following detailed calculations, which are shown for ease of understanding:

- $$Covar(\boldsymbol{U}, \boldsymbol{X}) = \boldsymbol{A}'\boldsymbol{\Sigma}_{XX} = \begin{pmatrix} -0.6128 & -0.0705 & -0.2006 \\ 0.2639 & 0.2972 & -0.2077 \\ -0.0640 & 0.6359 & 0.1846 \end{pmatrix}$$

- $$Covar(\boldsymbol{V}, \boldsymbol{Y}) = \boldsymbol{B}'\boldsymbol{\Sigma}_{YY} = \begin{pmatrix} -8.8950 & -8.8523 & -7.5317 & -6.7371 \\ 2.4743 & -2.1492 & 1.7693 & 6.5603 \\ 2.7587 & -3.3050 & 2.6697 & -2.3069 \end{pmatrix}$$

- $$\boldsymbol{\rho}_{\boldsymbol{U},\boldsymbol{X}_{p\times p}} = \boldsymbol{A}'_{p\times p}\boldsymbol{\Sigma}_{XX_{p\times p}} \begin{bmatrix} \sqrt{Var(X_1)} & \cdots & 0 \\ \vdots & \ddots & \vdots \\ 0 & \cdots & \sqrt{Var(X_p)} \end{bmatrix}_{p\times p}$$
$$= \begin{pmatrix} 0.6703 & 0 & 0 \\ 0 & 0.7055 & 0 \\ 0 & 0 & 0.3427 \end{pmatrix}.$$

- $$\boldsymbol{\rho}_{\boldsymbol{U},\boldsymbol{Y}_{p\times q}} = \boldsymbol{A}'_{p\times p}\boldsymbol{\Sigma}_{XY_{p\times q}} \begin{bmatrix} \sqrt{Var(Y_1)} & \cdots & 0 \\ \vdots & \ddots & \vdots \\ 0 & \cdots & \sqrt{Var(Y_q)} \end{bmatrix}_{q\times q}$$
$$= \begin{pmatrix} -5.6368 & -5.4272 & -5.2782 & -5.4888 \\ 4.3703 & 4.2584 & 4.0730 & 4.2295 \\ 5.0434 & 4.9145 & 4.7151 & 4.8921 \end{pmatrix}.$$

Furthermore, the value of

1. $$\boldsymbol{\rho}_{\boldsymbol{V},\boldsymbol{X}_{q\times p}} = \boldsymbol{B}'_{q\times q}\boldsymbol{B}_{q\times q}\boldsymbol{\Sigma}_{YX_{q\times p}} \begin{bmatrix} \sqrt{Var(X_1)} & \cdots & 0 \\ \vdots & \ddots & \vdots \\ 0 & \cdots & \sqrt{Var(X_p)} \end{bmatrix}_{p\times p}$$

   and

2. $$\boldsymbol{\rho}_{\boldsymbol{V},\boldsymbol{Y}_{q\times q}} = \boldsymbol{B}_{q\times q}\boldsymbol{\Sigma}_{YY_{q\times q}} \begin{bmatrix} \sqrt{Var(Y_1)} & \cdots & 0 \\ \vdots & \ddots & \vdots \\ 0 & \cdots & \sqrt{Var(Y_q)} \end{bmatrix}_{q\times q}$$

are left for the readers to calculate, which we are sure will give them a certain level of confidence to appreciate this multivariate statistical method in a better manner. Similarly, one can also find the standardized variables:

1. $Covar(\boldsymbol{U}, \boldsymbol{Z}_X) = \boldsymbol{A}\boldsymbol{\rho}_{XX}$,
2. $Covar(\boldsymbol{V}, \boldsymbol{Z}_Y) = \boldsymbol{B}\boldsymbol{\rho}_{YY}$,
3. $$\boldsymbol{\rho}_{U,Z_X}{}_{p\times p} = \boldsymbol{A}_{Z_{X_{p\times p}}} \boldsymbol{\rho}_{XX_{p\times p}} \begin{bmatrix} \sqrt{Var\left(Z_{X_1}\right)} & \cdots & 0 \\ \vdots & \ddots & \vdots \\ 0 & \cdots & \sqrt{Var\left(Z_{X_p}\right)} \end{bmatrix}_{p\times p},$$
4. $$\boldsymbol{\rho}_{U,Z_Y}{}_{p\times q} = \boldsymbol{A}_{Z_{X_{p\times p}}} \boldsymbol{\rho}_{XY_{p\times q}} \begin{bmatrix} \sqrt{Var\left(Z_{Y_1}\right)} & \cdots & 0 \\ \vdots & \ddots & \vdots \\ 0 & \cdots & \sqrt{Var\left(Z_{Y_q}\right)} \end{bmatrix}_{q\times q},$$
5. $$\boldsymbol{\rho}_{V,Z_{X_{q\times p}}} = \boldsymbol{B}_{Z_{Y_{q\times q}}} \boldsymbol{\rho}_{YX_{q\times p}} \begin{bmatrix} \sqrt{Var\left(Z_{X_1}\right)} & \cdots & 0 \\ \vdots & \ddots & \vdots \\ 0 & \cdots & \sqrt{Var\left(Z_{X_p}\right)} \end{bmatrix}_{p\times p},$$
6. $$\boldsymbol{\rho}_{V,Z_{Y_{q\times q}}} = \boldsymbol{B}_{Z_{Y_{q\times q}}} \boldsymbol{\rho}_{YY_{q\times q}} \begin{bmatrix} \sqrt{Var\left(Z_{Y_1}\right)} & \cdots & 0 \\ \vdots & \ddots & \vdots \\ 0 & \cdots & \sqrt{Var\left(Z_{Y_q}\right)} \end{bmatrix}_{q\times q}.$$

A few other sources for CCA are those provided by Karamouz et al. (2010) and Szakács et al. (2004). The interested readers can definitely have a look at the data sets in these references and work on them to hone their skills and understanding in CCA.

8.19.5 Geometric and Graphical Interpretation of CCA

It would not be out of context to say that a better appreciation of the CCA method can be obtained if one looks at the geometrical as well as graphical interpretation of CCA. A relevant reference for this is González et al. (2008). It is interesting to note that neural network (Hsieh, 2000) and kernel-based methods (Bach and Jordan, 2002, Lai and Fyfe, 2000, Melzer et al., 2001) have been used in the area of nonlinear CCA.

8.19.6 Conclusions about CCA

Before we wind up this topic, there are a few important points that should be remembered:

1. The classical CCA method may be used only for the condition when $n \geq (p + q + 1)$ (Eaton and Perlman, 1973).

2. In case X and Y are highly correlated, then the matrices Σ_{XX} and Σ_{XX} are ill conditioned and their respective inverses are unreliable.
3. In case $n < (p + q)$, then CCA cannot be utilized and for such cases, that is, $n < (p + q)$, partial least square (PLS) regression may be used. The advantage of PLS over CCA is the fact that in the former the asymmetry relationship between the predictors and dependent variables is preserved, while CCA treats them symmetrically.
4. Vinod (1976) and Leurgans et al. (1993) have shown the extension of ridge regression in the area of CCA.

8.20 Cluster Analysis

Cluster analysis (CA) is a statistical technique whereby we form clusters/groups of *similar* individuals/objects using data/information from individuals/objects. This statistical method develops tools and methods, where given a data matrix, $X_{(n \times p)}$, consisting of n number of individuals/objects where each of these n individuals/objectives are of dimension p, our aim is to build some natural subgroups or clusters of these individuals/objects. Using CA, we try to find some similarity or patterns in the data, for example, classification of plants/animals using taxonomy, diseases using epidemiology, etc. From a historical perspective, the origin of CA may be traced back to the work of Driver and Kroeber (1932) in anthropology. Later on, it was used in psychology (Cattell, 1943, Tryon, 1939, Zubin, 1938). Cluster analysis has been used in a variety of fields ranging from anthropology, agriculture, economics, psychology, geophysics, psychiatry, sociology, marketing, finance, behavioral sciences, different fields of engineering, etc. Even though old, good references with interesting applications can be found in Gordon (1981) and Hartigan (1975). Other good references from a theoretical points of view are Anderberg (1973), Duda et al. (2001), Duran and Odell (1974), Everitt and Dunn (2001), Gordon (1981), Hartigan (1975), Jain and Dubes (1988), Kaufman and Rousseeuw (2005), Späth (1980), etc. Another good book in the area of CA is by Xu and Wunsch (2008). Some other mathematical technique methods similar to cluster analysis are pattern recognition, numerical taxonomy, morphometrics, etc.

For a better understanding of clustering analysis as a technique, one should understand the basic four steps involved in cluster analysis:

- *Feature selection or extraction*: In the feature selection, step/stage one chooses the distinguishing features from a set of candidates, while on the other hand, in feature extraction step/stage, we utilize some transformations to generate useful and novel features from the original ones.
- *Clustering algorithm design and selection*: Depending on the proximity measure $d(P, Q)$, one constructs clustering criterion function so that the clustering algorithms may be developed. The main focus of the clustering algorithms is to cluster the objectives in groups based on some predefined criterion.
- *Cluster validation*: Effective validation standards and criteria are important to provide the degree of confidence for the clustering results derived from the used algorithms. This is what is done in the third stage step, which is the clustering validation step/stage.
- *Result interpretation*: The ultimate goal of clustering analysis step/stage is to provide the user with meaningful insights from the original data, so that they can effectively

solve the problems encountered, and this is what the result interpretation step/stage does.

The two fundamental steps in CA, which would be discussed by us here are: (i) choice of proximity (closeness) measure and (ii) choice of group-building algorithm.

Consider

$$\boldsymbol{X}_{n\times p} = \begin{pmatrix} x_{1,1} & \cdots & x_{1,p} \\ \vdots & \ddots & \vdots \\ x_{n,1} & \cdots & x_{n,p} \end{pmatrix}_{n\times p},$$

such that the proximity/distance matrix is given by

$$\boldsymbol{D}_{n\times n} = \begin{pmatrix} d_{1,1} & \cdots & d_{1,n} \\ \vdots & \ddots & \vdots \\ d_{n,1} & \cdots & d_{n,n} \end{pmatrix}_{n\times n},$$

where $d_{i,j}$ gives the measure of proximity/distance and is denoted by $\|x_i - x_j\|_2$ or $\{\max_{i,j}(d_{i,j}) - d_{i,j}\}$ as the case may be. In case, we have a *binary* structure pertaining to $\boldsymbol{X}$, that is, $x_{i,k} \in \{0, 1\}$, $i = 1,\ldots,n$ and $k = 1,\ldots,p$, then $d_{i,j} = (a_1 + \delta a_4)/(a_1 + \delta a_4 + \lambda(a_2 + a_3))$, where δ and λ are the weighting factors. Here, $a_1 = \sum_{k=1}^{p} I(x_{i,k} = x_{j,k} = 1)$, $a_2 = \sum_{k=1}^{p} I(x_{i,k} = 0, x_{j,k} = 1)$, $a_3 = \sum_{k=1}^{p} I(x_{i,k} = 1, x_{j,k} = 0)$, and $a_4 = \sum_{k=1}^{p} I(x_{i,k} = x_{j,k} = 0)$. A few examples of weighting factors are $(\delta = 0, \lambda = 1)$ (Jaccard, 1901), $(\delta = 1, \lambda = 2)$ (Tanimoto, 1957), and $(\delta = 0, \lambda = 0.5)$ (Dice,* 1945). On the other hand, when we have the *continuous* variable, then $d_{i,j} = \|x_i - x_j\|_r = \{\sum_{k=1}^{p} |x_{i,k} - x_{j,k}|^r\}^{1/r}$ (Minkowski metric), when expressed in the nonstandardized form, while $d_{i,j}^2 = \sum_{k=1}^{p}((x_{i,k} - x_{j,k})^2/s_{X_k,X_k})$ is the standardized version of this distance measure. A few other distance measures that have found use in CA are: Hamming distance, Euclidean distance: $d_{i,j}^2 = \sum_{k=1}^{p}(x_{i,k} - x_{j,k})^2$, Soergel distance: $d_{i,j} = \sum_{k=1}^{p} |x_{i,k} - x_{j,k}| / \sum_{k=1}^{p} \max(x_{i,k}, x_{j,k})$, Canberra metric: $d_{i,j} = \sum_{k=1}^{p}\{|x_{i,k} - x_{j,k}|/(|x_{i,k}| + |x_{j,k}|)\}$, Czekanowski metric:

$$d_{i,j} = \left\{1 - \frac{2\sum_{k=1}^{p} \min(x_{i,k}, x_{j,k})}{\sum_{k=1}^{p}(x_{i,k} + x_{j,k})}\right\},$$

etc. One should remember that both Canberra and Czekanowski measures are defined for nonnegative variables only. Even without the precise notion of a natural grouping, one is often able to cluster/group individuals/objects in 2-D or 3-D plots using eye, stars, and Chernoff faces.

When items (units or cases) are clustered, proximity is usually indicated by some sort of distance. For variables, the grouping is on the basis of correlation coefficients or such similar measure. Central to the goal of cluster analysis is the idea of the degree of similarity ($S(.,.)$) (or dissimilarity, $d(.,.)$) between the individual objects that are being clustered. It is important that the following properties are satisfied for the *distance* or *dissimilarity function* being used in CA.

1. $d(x_{i,k}, x_{j,k}) = d(x_{j,k}, x_{i,k})$, which is the property of symmetry.
2. $d(x_{i,k}, x_{j,k}) > 0$, if $x_{i,k} \neq x_{j,k}$, which is the property of positivity.

---

* Also independently developed by Sørensen, T. A method of establishing groups of equal amplitude in plant sociology based on similarity of species and its application to analyses of the vegetation on Danish commons, *Kongelige Danske Videnskabernes Selskab*, **5**, 1–34, 1957.

3. $d(x_{i,k}, x_{j,k}) = 0$, if $x_{i,k} = x_{j,k}$, which is the property of reflexivity.
4. $d(x_{i,k}, x_{j,k}) \leq d(x_{i,k}, x_{l,k}) + d(x_{l,k}, x_{j,k})$, which is generally called the triangle law.

Where $x_{i,k}$, $x_{j,k}$, and $x_{l,k}$ are some points in space. If along with the first two, the third and fourth property also holds for $d(.,.)$, then $d(.,.)$ is a *metric*. In line with *distance* or *dissimilarity* function, a *similarity* function, $S(.,.)$, can also be defined with the following properties, which are on similar lines as mentioned for $d(.,.)$:

1. $S(x_{i,k}, x_{j,k}) = S(x_{j,k}, x_{i,k})$, which is the property of symmetry.
2. $0 \leq S(x_{i,k}, x_{j,k}) \leq 1$, which is the property of positivity.
3. $S(x_{i,k}, x_{j,k})S(x_{j,k}, x_{l,k}) \leq \{S(x_{i,k}, x_{j,k}) + S(x_{j,k}, x_{l,k})\}S(x_{i,k}, x_{l,k})$, which is the property of reflexivity.
4. $S(x_{i,k}, x_{j,k}) = 1$, iff $x_{i,k} = x_{j,k}$.

Remember, $S(.,.)$ is called *similarity metric* if all the above four properties hold. If the original data was collected as similarities, then a suitable monotone decreasing function may be used to convert them to dissimilarities.

Typically, distance/dissimilarity functions are used to measure continuous features, whereas similarity functions are more appropriate for qualitative variables. Table 8.6 gives the similarity as well as dissimilarity measure for quantitative features/characteristics.

8.20.1 Clustering Algorithms

A widely agreed framework is to classify clustering as hierarchical clustering and partitioning clustering, based on the properties of the clusters generated. While generating the clusters, the concept of distance as a measure which groups objects into clusters with certain properties with respect to the idea of distance and its functional form comes in play. Most of the algorithms assume symmetric dissimilarity matrices. In case the original matrix $\boldsymbol{D}_{n\times n}$ is not symmetric, then one can replace the matrix by $(1/2)\left(\boldsymbol{D}_{n\times n} + \boldsymbol{D}^T_{n\times n}\right)$. The reader should remember that clustering algorithms may be classified as (i) exclusive clustering, (ii) overlapping clustering, (iii) hierarchical clustering, and (iv) probabilistic clustering. Without going into detailed analysis, we give here the pseudo-codes of a few of the clustering algorithms, so that it motivates the reader to understand them and do a thorough search of such algorithms which may be found in good references, a few of which have already been stated in due course of our discussion of CA.

Basic K-mean algorithm: The K-mean clustering algorithm works on the premise that centroids of a group of objects best depict the characteristics of that group/cluster. The pseudo-code for the K-mean clustering algorithm is as follows and is shown in Figure 8.19.

Bisecting K-mean algorithm: The bisecting K-mean algorithm is a simple extension of the basis K-mean algorithm. The pseudo-code for this algorithm is illustrated in Figure 8.20. The idea is to obtain K clusters and split the set of points into two clusters and then select one of them to split it again. We continue doing this until K clusters are obtained.

Basic agglomerative hierarchical clustering algorithm: It is a hierarchical clustering algorithm, whereby we start with points as individual clusters and at each step merge the closet pairs of clusters. Hence, a cluster proximity function is important, which needs to be defined before one ventures to use this clustering algorithm. For the benefit of the reader, the pseudo-code for this third algorithm is given in Figure 8.21.

Similarity as Well as Dissimilarity Measure for Quantitative Features/Characteristics

Measure	Mathematical Expression	Comments	Examples/ Applications	Pictorial Representation ($p = 2$ Where Vertical Axis Is $x_{.,1}$ and Horizontal Axis Is $x_{.,2}$)
Manhattan distance/taxi cab norm/sum of absolute difference (SAD) of the difference	$d(x_i, x_j) = \sum_{k=1}^{p} \lvert x_{i,k} - x_{j,k} \rvert$	• Equivalent to the L_1 norm Tends to form hyperrectangular cluster • Special case of Minkowski distance when $r = 1$	Fuzzy adaptive resonance theory	
Sum of square distance (SSD)	$d(x_i, x_j) = \sum_{k=1}^{p} \lvert x_{i,k} - x_{j,k} \rvert^2$	• Equivalent to the L_2 norm Tends to form hyperspherical cluster • Special case of Minkowski distance when $r = 2$	K-mean algorithm	
Mean absolute error (MAE), that is, normalized version of SAD	$d(x_i, x_j) = \frac{1}{p} \sum_{k=1}^{p} \lvert x_{i,k} - x_{j,k} \rvert$	• Tends to form hyperrectangular cluster	–	
Mean squared error (MSE), that is, normalized version of SSD	$d(x_i, x_j) = \frac{1}{p} \sum_{k=1}^{p} \lvert x_{i,k} - x_{j,k} \rvert^2$	• Tends to form hyperspherical cluster	–	
Euclidean distance	$d(x_i, x_j) = \left(\sum_{k=1}^{p} \lvert x_{i,k} - x_{j,k} \rvert^2 \right)^{1/2}$	• Equivalent to the L_2 norm Tends to form hyperspherical cluster	K-mean algorithm	

(Continued)

TABLE 8.6 (*Continued*)

Similarity as Well as Dissimilarity Measure for Quantitative Features/Characteristics

Measure	Mathematical Expression	Comments	Examples/ Applications	Pictorial Representation ($p = 2$ Where Vertical Axis Is $x_{.,1}$ and Horizontal Axis Is $x_{.,2}$)
Minkowski distance	$d(x_i, x_j) = \left(\sum_{k=1}^{p} \lvert x_{i,k} - x_{j,k}\rvert^r\right)^{1/r}$	• Invariant to translation/rotation • Features with large values/variances tend to dominate	Fuzzy c-means	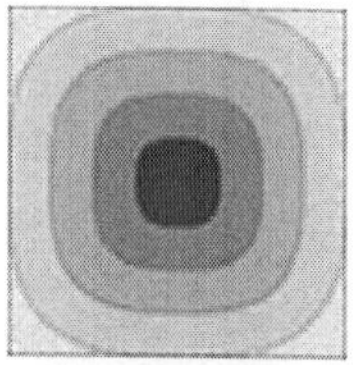
Canberra distance	$d(x_i, x_j) = \sum_{k=1}^{p} \frac{\lvert x_{i,k} - x_{j,k}\rvert}{\lvert x_{i,k}\rvert + \lvert x_{j,k}\rvert}$	–	–	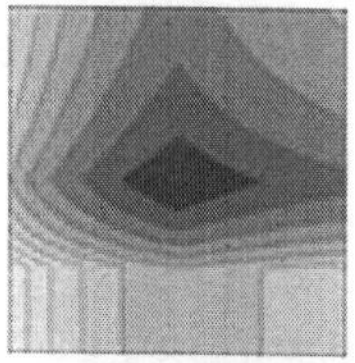
Cosine distance	$S(x_i, x_j) = \frac{\mathbf{x}_i^T \mathbf{x}_j}{\lVert\mathbf{x}_i\rVert \lVert\mathbf{x}_j\rVert}$	• Independent of vector length • Invariant to rotation • Noninvariant to linear transformation	Document clustering	
Maximum distance/ chessboard distance/ Chebyshev distance	$d(x_i, x_j) = \max_{1 \le k \le d} \lvert x_{i,k} - x_{j,k}\rvert$	• Equivalent to the L_∞ norm • Special case of Minkowski distance when $r = \infty$	Fuzzy c-means with sup norm	

(*Continued*)

TABLE 8.6 (*Continued*)

Similarity as Well as Dissimilarity Measure for Quantitative Features/Characteristics

Measure	Mathematical Expression	Comments	Examples/ Applications	Pictorial Representation ($p = 2$ Where Vertical Axis Is $x_{.,1}$ and Horizontal Axis Is $x_{.,2}$)
Pearson's correlation coefficient	$d(x_i, x_j) = \frac{1}{2}\left\{1 - \frac{\sum_{k=1}^{p}(x_{i,k} - \bar{x}_i)(x_{j,k} - \bar{x}_j)}{\sqrt{\sum_{k=1}^{p}(x_{i,k} - \bar{x}_i)^2}\sqrt{\sum_{k=1}^{p}(x_{j,k} - \bar{x}_j)^2}}\right\}$	• Not a metric • Derived from correlation coefficient • Magnitude of differences of variables not considered	Analysis of gene expression data	
Spearman's correlation coefficient	$d(x_i, x_j) = 1 - 6\left(\sum_{k=1}^{p}(x_{i,k} - x_{j,k})^2 / n(n^2 - 1)\right)$	• Not a metric • Derived from correlation coefficient • Magnitude of differences of variables not considered	–	–
Mahalanobis distance	$d(x_i, x_j) = (\boldsymbol{x}_i - \boldsymbol{x}_j)^T \boldsymbol{S}(\boldsymbol{x}_j - \boldsymbol{x}_i)$, where $\boldsymbol{S}$ is within covariance matrix	• Invariant to any nonsingular linear transformation • $\boldsymbol{S}$ is based on all the objects • Tend to form hyperellipsoid clusters • When correlation is zero, then squared Mahalanobis distance is equal to squared Euclidean distance	Ellipsoidal adaptive resonance theory Hyperellipsoidal clustering algorithm	–

```
1:  DEFINE: K points
2:  INPUT: K points as initial centroids
3:  REPEAT
4:     Form K clusters by assigning each point to its closest centroid
5:     Recompute the centroid of each cluster
6:  UNTIL:Cenroids do not change
```

FIGURE 8.19
Pseudo-code for K-mean algorithm.

```
1:  DEFINE: K points
2:  INPUT:  K  points  and  initialize  the  list  of  clusters  so  that  they  consists  of
    cluster containing all points
3:  REPEAT:
4:    Remove a cluster from the list of clusters and perform several bisection of the cluster
      choosen
5:    FOR: i:number of trials do
6:      Bisect the selected cluster using basic K-means
7:    END FOR
8:    Select two clusters from the bisection based on some criteria (e.g., SSE, etc.)
9:    Add these two clusters to the list of clusters
10: UNTIL:List of clusters contains K clusters
```

FIGURE 8.20
Pseudo-code for bisecting K-mean algorithm.

```
1:  DEFINE: K points, D proximity matrix
2:  INPUT: K points and D proximity matrix
3:  REPEAT
4:    Merge the closest two clusters based on the proximity property defined
5:    Update the proximity matrix
6:    Bisect the selected cluster using basic K-means
7:  UNTIL: Only one cluster remains
```

FIGURE 8.21
Pseudo-code for basic agglomerative hierarchical clustering algorithm.

This chapter is neither an exclusive discussion about clustering algorithms, nor about cluster analysis; hence, we desist ourselves from analyzing other algorithms along with their merits and demerits. We request the readers to check any good book in the area of clustering algorithm, a few examples of which are: Abonyi and Feil (2007), Everitt et al. (2011), and Höppner et al. (1999), to name a few.

Data set # 1: One can consider the data set given in Chapter 13 of Hartigan (1975), which relates to 13 different Indo-European languages equivalent of the names associated with certain common words. The data set has 13 rows and 17 columns (16 different words, e.g., black, eat, drink, fish, five, etc.). The interested reader can use any clustering technique to find the groups accordingly to which the languages can be clustered depending on these 16 different words, their pronunciation, diction, speech, etc.).

Data set # 2: Another interesting application can be the study of mutation sequence of amino acids in different species such as man, monkey, chicken, duck, kangaroo, and rattlesnake. Again, the data for the same can be referred to in Hartigan (1975). For this problem, we would try to group the species depending on the different characteristics one thinks are important and are found in the gene sequences related to amino acid.

Thus, in a nutshell, CA is ideally suited for defining groups of individuals/objects with maximal homogeneity within the groups, while also having maximum heterogeneity between groups, that is, determining the most similar groups that are also most different from each other. One difference between discriminant analysis and logistic regression and classification tree analysis is the fact that discriminant analysis and logistic regression make a number of assumptions about the underlying data, whereas classification tree analysis is a nonparametric technique. Another difference is that discriminant analysis and logistic regression can be used to derive probabilities of group membership for individuals, whereas classification tree analysis only produces average probabilities for the different groups.

8.21 Multiple Discriminant and Classification Analysis

Before discussing what multiple discriminant and classification analysis (MDCA) is all about, let us consider a few relevant examples to aid a better understanding of this multivariate statistical method, which consists of (i) a separation/discrimination rule along with (ii) an allocation/classification rule.

For a good motivation, let us consider a few practical applications of the MCDA technique. First, consider a pediatrician has with her the data of *height*, *age*, *sex*, and *age* of children (in the age group of 2–5 years) from the past. Based on this information, a child is categorized as being obese, normal, or malnourished. When a new patient (in this case, a child) visits the pediatrician, then she, that is, the pediatrician, has to categorize the child as being in any one of the above-mentioned *three* categories, such that she can suggest medical care for the child as deemed appropriate. As a second example of application, take into account a musicologist who is studying the composition written by composers between 1750 and 1820 AD (which is termed as the classical period of Western music). The musicologist is aware that a classification of the composer in that period may be made depending on the *melody*, *rhythm*, *dynamics*, *mood*, *timbre*, etc. When the musicologist is given a new piece or an unknown piece, then, depending on the characteristics of the musical piece, he/she may classify the musical piece as that belonging to Joseph Hayden, Wolfgang Amadeus Mozart, Ludwing van Beethoven, etc. Continuing our discussion further, next, think of a geologist who classifies rock as igneous, sedimentary, or metamorphic, depending on its *chemical composition*, *physical properties*, *texture* of its constituent particles, and *permeability*. Suppose the geologist is given a new sample of rock, and is told to classify the given sample, then he/she may do so as desired using MDCA. Finally, let us illustrate a fourth application where you as a credit risk analyst are first interested to discriminate a company as *good*, *average*, or *bankrupt*, depending on *price to earning ration* $\left(\frac{P}{E}\right)$, *amount of liability*, and *price of stock*. After having done that, you study the credentials of different new companies and classify them as belonging to any one of the categories as already decided.

In all these four application areas of MCDA, the essence of what one intends to achieve may be summarized as follows. Given observations/objects, as a *first* step, we separate/discriminate the observations/objects into clusters/groups, which are known *a priori*. Afterward, when a new set of observations/objects arrive, we intend to classify them into these known groups. Hence, the immediate goal of MCDA is as follows:

Step # 1: Separation/discrimination rule: In this step/rule, describe an algebraic or graphical rule such that one is able to differentiate observations/objects into different classes, depending on different characteristics/features, which are inherent in the observations/objects.

Step # 2: Allocation/classification rule: Once the first step is over, our next job is to sort out the new observations/objects into one of these classes depending on some logic/rule.

Thus, in MDCA, we first define Π_j, $j = 1, \ldots, J$, number of clusters/classes each with probability density function $f_j(x)$ such that all the observations *a priori* may be classified in any one of the J cluster/classes. Later, when a new set of observation/object arrive, we categorize that observations/objects, say x, into any one of the cluster/group. One should remember that the allocation/classification rules are developed based on the fact that the measured characteristics/features of randomly selected observations/objects are such that the possible sample space ($\Omega = \mathbb{R}^p$) is divided into R_j *disjoint* sets/regions, such that $\cup_{j=1}^{J} R_j = \mathbb{R}^p$ holds (Figure 8.22).

Now, if the new observation/object falls in R_j, then it is characterized as belonging to population, Π_j. For the classification/allocation rule, there are several situations that make it difficult to complete Step # 2 with minimum error: (i) incomplete information or knowledge of future performance; (ii) perfect information, and (iii) unavailable or expensive information. However one may try, there are always some misclassification/misallocation errors, and hence our main job is to have a good classification/allocation rule that results in a few of these errors. Apart from that, an optimal classification/allocation rule should also take into account both (i) *prior probabilities of occurrence* as well as (ii) *cost of classification/allocation*. Thus, the conditional probability of allocating/classifying an object/observation as Π_i when in fact it belongs to Π_j is $p_{i,j} = Pr(X \in R_i|\Pi_j) = \int I_{x \in R_i} f_j(x)dx, i \neq j = 1, \ldots, J$. Moreover, if the prior distributions are $\pi_j(x)$, then we have different cases which would dictate how the probability values would be calculated. One should remember that the concept of prior is practical and also logical as it gives the most likelihood case, that when a new observation is chosen, what is the probability that it will belong to a particular cluster/group? Considering there are J number of clusters/groups, classifying x in cluster/group i, although it is from cluster/group population Π_j, may be given by the conditional distribution $p_{i,j} = Pr(X \in R_i|\Pi_j) = \int I_{x \in R_i} f_j(x)dx, i \neq j = 1, \ldots, J$. Now, these R_j correspond to the Π_j population and one is interested to categorize an observation, say x, into any one of the group. In doing so, we entail a cost that is written as $C(i|j), i, j = 1, \ldots, J$ such that it signifies the cost of assigning the observation in the ith population, that is, Π_i, when actually it should belong to the jth, that is, Π_j population. Hence, the cost structure matrix may be given as shown in Figure 8.23, where $C(j|j) = 0, \forall j = 1, \ldots, J$.

Hence, the total gain/benefit given that the observation chosen belongs to population Π_j may be expressed as $TG(R_j) = -C(j|1)\pi_1 \int I_{x \in R_j} f_1(x)dx - \cdots - C(j|j-1)$ $\pi_{j-1} \int I_{x \in R_j} f_{j-1}(x)dx - C(j|j+1)\pi_{j+1} \int I_{x \in R_j} f_{j+1}(x)dx - \cdots - C(j|J)\pi_J \int I_{x \in R_j} f_J(x)dx$. Let us further define the Π_j population $\left(\mathbf{x}_{j1}, \ldots, \mathbf{x}_{jn_j}\right) \sim f_j(x)$, where $f_j(x)$ is the probability density function of the jth cluster/group. Then, the maximum likelihood discriminate rule would intuitively assign x to Π_j such that the likelihood function $L_j(x) = f_j(x)$ is maximized. Mathematically, the sets/regions, R_j, would be given by $R_j = \{x : L_j(x) > L_i(x), i = 1, \ldots, J, i \neq j\}$. Obviously, there is a *misclassification penalty* for assigning an observation in the wrong cluster/group, and hence in the same spirit there would be a *nonnegative penalty* for right classification of

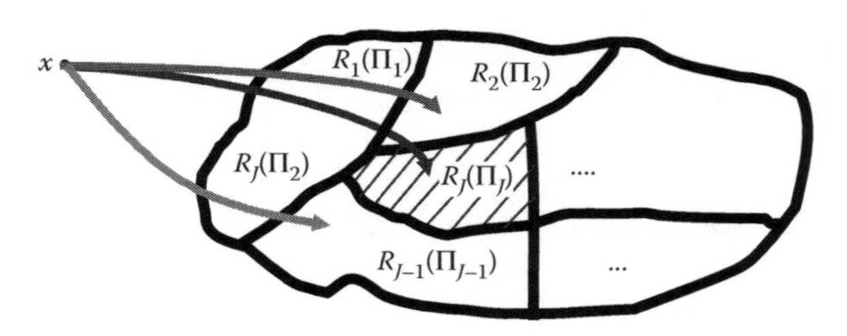

FIGURE 8.22
Pictorial illustration of the concept of MDCA.

	Classified population				
True population	C(1|1)	C(2|1)	···	C(J−1|1)	C(J|1)
	C(1|2)	C(2|2)	···	C(J−1|2)	C(J|2)
	⋮	···	⋱	···	⋮
	C(1|J−1)	C(2|J−1)	···	C(J−1|J−1)	C(J|J−1)
	C(1|J)	C(2|J)	···	C(J−1|J)	C(J|J)

FIGURE 8.23
Cost structure matrix for MDCA.

the same. To design a rule, let us assume γ_j and δ_j as the gain or loss in correct or erroneous classification of the observation, x, into the jth cluster/group. Furthermore, let the cost function be denoted as $C(i|j), i, j = 1, \ldots, J$. Intuitively one can easily comment that $C(j|j) = 0$ as we classify the observation in its right cluster/group and hence there is no error. Thus, the total gain/loss is given by the following equation: $TC(R_j) = -C(j|1)\pi_1 \int I_{x \in R_j} \times f_1(x)dx$. If one wants to diagrammatically represent the classification/misclassification probabilities, then one can refer to Figure 8.24 (refer Johnson and Wichern, 2002).

Before we discuss an example, let us highlight a few important points for the MCDA method, which may be useful for the readers to appreciate this multivariate statistical method in a much better way:

1. The MDCA method is appropriate when the dependent variable is categorical, while the independent variables are metric.
2. MDCA derives the variate that best distinguishes between *a priori* groups.
3. MDCA sets variate's weights to maximize between-group variance relative to within-group variance.

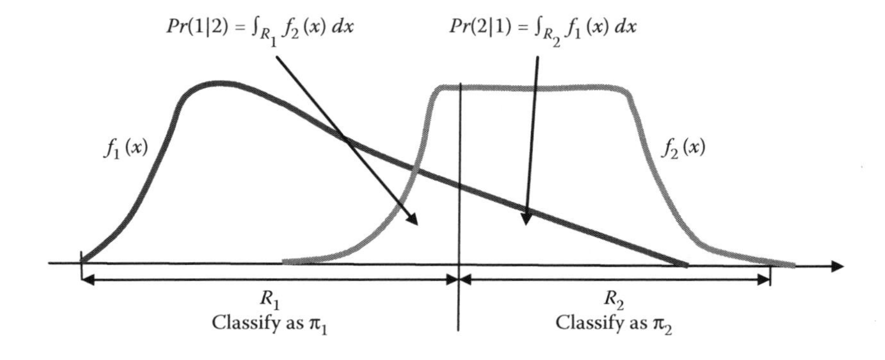

FIGURE 8.24
Misclassification probability for two regions marked as π_1 and π_2. (Adapted from Johnson, R.A. and Wichern, D.W. 2002. *Applied Multivariate Statistical Analysis*, Pearson Education, ISBN: 8178086867.)

Without being repetitive, we would like to mention the following important points for the MDCA method. It should be remembered that for each observation, we can obtain a discriminant Z-score, such that the average Z-score for a group gives the centroid for the group. Furthermore, the classification done using cutting scores which are derived from group centroids and finally the statistical significance of discriminant function is accomplished using distance between group centroids. Finally, to conclude this section, before discussing an example, we state the general steps one should remember for the MDCA method:

1. *Step # 1: Research problem/objectives*: In the research problem and objective formulation stage, one needs to do the following: (i) evaluate the differences between average scores for *a priori* groups on a set of variables; (ii) determine which set of independent variables account for most of the differences between groups; and finally (iii) classify the observations into groups.
2. *Step # 2: Research design*: The second stage is the research design stage in which the important things to remember are (i) there should be a proper selection of dependent as well as independent variables; (ii) the sample size considerations should be done appropriately, and finally (iii) the division of sample into analysis and holdout sample should be done rightly.
3. *Step # 3: Assumptions of MDCA*: As for any statistical method, MCDA also has some inherent assumptions, which are (i) multivariate normality for the independent variables; (ii) equal covariance matrices for the groups; (iii) low correlation among independent variables; and finally (iv) linear nature of the discriminant function.
4. *Step # 4: Estimation of MDCA and assessing fit*: In the fourth stage, that is, the estimation stage, one should remember that the estimation process can be either simultaneous or stepwise and to test the statistical significance of the discriminant function, one should use existing statistics such as Wilk's lambda, Hotelling's trace, Pillai's criterion, Roy's greatest root, and Mahalanobis distance function to test the efficacy of the data set as well as the MCDA method. It is important to note that the test statistic signifies the overall discrimination between groups and of each discriminant function. Moreover, to assess the overall fit, one should calculate the discriminant Z-score for each observation and then evaluate the group differences on Z-scores and predict the group membership accurately. To do this, we need to address the following rationale for classification matrices: (i) cutting score determination; (ii) considering costs of misclassification; (iii) constructing classification matrices; (iv) assessing classification accuracy; and (v) proper casewise diagnosis.
5. *Step # 5: Interpretation of results*: For this penultimate stage, remember the following and they are related to the interpretation of results, which can be further broken into those related to (i) methods for single discriminant function; (ii) discriminant weights; (iii) discriminant loadings; and (iv) partial F-values. Additional methods for more than two functions may be required to be used, and they are: (i) rotation of discriminant functions; (ii) potency index; and (iii) stretched attribute vectors.
6. *Step # 6: Validation of results*: Finally, for the validation stage, one needs to analyze results with what is practical/feasible, and then based on the feedback, conduct further tests if required.

A few relevant references are: Duda et al. (2001), Härdle and Simar (2007), and Lachenbruch (1975).

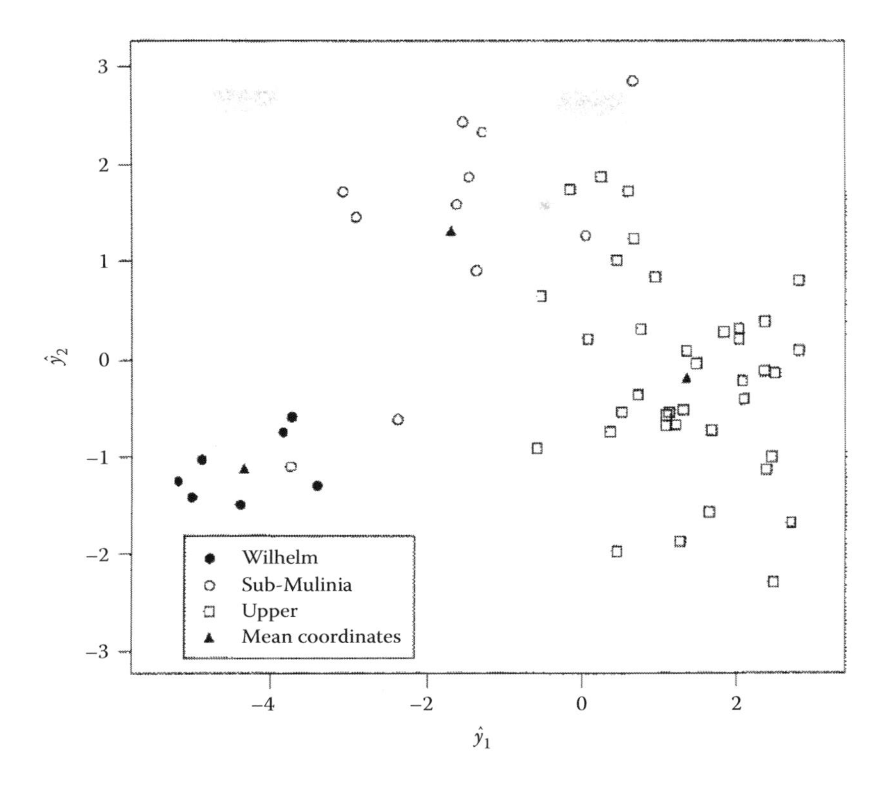

FIGURE 8.25
Crude oil sample in the discriminant space, Example 8.22. (Adapted from Johnson, R.A. and Wichern, D.W. 2002. *Applied Multivariate Statistical Analysis*, Pearson Education, ISBN: 8178086867.)

Example 8.22: (Johnson and Wichern, 2002)

As an example for the MDCA, let us consider the data set related to crude oil samples study by Gerrild and Lantz (1969). The crude obtained may be assigned to any one of the three populations, namely, π_1, π_2, and π_3, which are Wilhelm stone, sub-Mulinia sandstone, and upper sandstone, respectively. We are interested to study the characteristics, namely, vanadium, iron, beryllium, saturated hydrocarbon, and aromatic hydrocarbon, which may be denoted by X_1, X_2, X_3, X_4, and X_5. One can calculate the eigen values as 4.354 and 0.559. Finally, the Fisher linear discriminants are given by (i) $\hat{y}_1 = 0.312(x_1 - 6.180) - 0.710(x_2 - 5.081) + 2.764(x_3 - 0.511) + 11.809(x_4 - 0.201) - 0.235(x_5 - 6.434)$ and (ii) $\hat{y}_2 = 0.169(x_1 - 6.180) - 0.245(x_2 - 5.081) - 2.046(x_3 - 0.511) - 24.453(x_4 - 0.201) - 0.378(x_5 - 6.434)$. The crude oil sample in the discriminant space is illustrated in Figure 8.25. One can also find the Fisher values of discriminants which we omit and request the readers to study and do the necessary calculation as required.

8.22 Multidimensional Scaling

Like PCA, multidimensional scaling (MDS) is also a dimension reduction technique. Classic Torgerson metric MDS is actually done by transforming distances into similarities and performing PCA. Thus, PCA might be called the algorithm of the simplest MDS. MDS and PCA are not at the same level to be in line or opposite to each other. PCA is just a method while MDS is a class of analysis. As mapping, PCA is a particular case of MDS. On the other hand, PCA is a particular case of FA, which, being a data reduction, is more than only mapping, while MDS is only a mapping. Furthermore, PCA as a technique projects a multidimensional space onto direction of maximum variability, whereas, in MDS, the multidimensional space is projected while at the same time maintaining the interpoint distances.

In PCA, we use the concept of covariance matrix to study the correlation between design variables, and this is summarized using *dot* products, while, in MDS, one uses distance and loss function in order to study the similarity/dissimilarity, which is summarized using *cross* product.

Consider $\boldsymbol{X}_{n\times p} = (X_{i,j})$, with $i = 1, 2, \ldots, n$ and $j = 1, 2, \ldots, p$ such that $\boldsymbol{X}_{n\times p}$ signify a matrix corresponding to n number of readings where each of the readings are of dimension p. Our aim using MDS is to calculate interpoint distances $\delta(X)_{i_1,i_2} = \|\boldsymbol{X}_{i_1} - \boldsymbol{X}_{i_2}\|$, i_1, $i_2 = 1, 2, \ldots, n$ and then try to find $k(k \leq p)$-dimensional vector $\boldsymbol{Y}_i$, $i = 1, 2, \ldots, n$ with $\delta(Y)_{i_1,i_2} = \|\boldsymbol{Y}_{i_1} - \boldsymbol{Y}_{i_2}\|$, such that $\delta(X)_{i_1,i_2} \approx \delta(Y)_{i_1,i_2}$ for all i_1, $i_2 = 1, 2, \ldots, n$. The proximity measure, $\delta(X)_{i_1,i_2}$, need not be Euclidean or any distance measure as such. It can also be an error and in the general sense, the proximity measure is described as similarity or dissimilarity.

Before we solve a simple problem, one should be aware that the classical scaling concepts used in the literature are either ordinal or metrical. Furthermore, there can be many dimensions based on which distance measure has to be calculated and here is where the weighting function can be utilized to scale the distances and also to draw them in a 2-D scale.

Example 8.23: (Johnson and Wichern, 2002)

As an example for MDS, consider the distance matrix between 10 cities, namely, London, Berlin, Oslo, Moscow, Paris, Rome, Beijing, Istanbul, Gibraltar, and Reykjavik. The 2-D plot of the cities based on the distances is shown in Figure 8.26. There are a few important points that one needs to mention here related to multidimensional scaling: (i) the points can be reflected without changing the interpoint distances; (ii) the interpoint distances are not affected if one changes the origin by adding/subtracting a constant from the rows/columns; and (iii) the set of points can be rotated without affecting the interpoint distances.

8.23 Structural Equation Modeling

Many of the multivariate statistical methods such as multiple regression, FA, MANOVA, etc. suffer from one common limitation, which is to do with the fact that they can examine only a single relationship at a time. One method that is able to overcome this lacuna is structural equation modeling (SEM). SEM is an extension of several multivariate techniques notably multiple regression and FA, and is distinguished by two characteristics:

1. Estimation of multiple and interrelated dependence relationship.
2. Ability to represent unobserved concepts in these relationships and also the amount of measurement error in the estimation process.

Another interesting fact is that this topic under multivariate statistical analysis is the only one to have an exclusive peer-reviewed journal by the same name, which is *Structural Equation Modeling: A Multidisciplinary Journal*, Taylor & Francis, ISSN: 1070-5511 (Print) and 1532-8007 (Online). The origins of modern SEM is usually traced to biologist Sewall Wright's development of path analysis

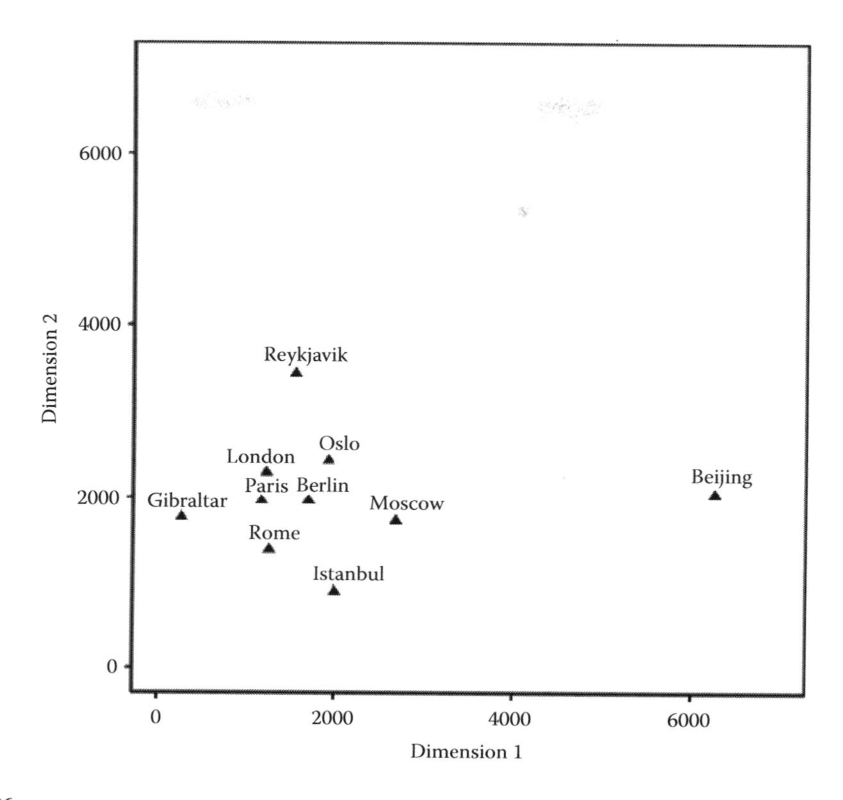

FIGURE 8.26
2-D plot of distances for the 10 cities, Example 8.23. (Adapted from Johnson, R.A. and Wichern, D.W. 2002. *Applied Multivariate Statistical Analysis*, Pearson Education, ISBN: 8178086867.)

(e.g., Wright, 1921, 1934). Another precursor of SEM is the path model (Duncan, 1966). If one follows the history of SEM, then a few good references that may be cited to trace the development of this statistical method are Bentler and Chou (1987), Bielby and Hauser (1977), Bollen (1989), and Bollen and Long (1993).

SEM as a statistical technique refers to the body of a comprehensive statistical methodology used to test and estimate the causal relations. The method uses a combination of cross-sectional statistical data and qualitative causal assumptions. It is different from another statistical technique, namely, multivariate linear regression model in the sense that the response variable in one regression equation in SEM may appear as a predictor in another equation. The variables in SEM may influence another reciprocally, either directly or through other variables. The proliferation of the use of SEM in social sciences, psychology, and related areas is due to the fact that implementation and the thinking process needed to actually theorize and practically apply this technique is akin to the informal thinking about causal relation that is common in the areas just mentioned.

The common aspects/concepts in SEM are: (i) model specification, (ii) estimation of free parameters, (iii) assessment of model and model fit, (iv) model modification, (v) sample size and power, and (vi) interpretation and communication.

We state a simple algorithm used for SEM, whereby we assume that latent variables are not present. First, let us consider x_i, $i = 1, \ldots, n$ points. Also assume the dependent variables are x_j, $j \in D$, while the predictor variables are x_k, $k \in P$ and $D \cap P \neq \emptyset$. Furthermore, consider there is a set T such that $p = P - T$ are the actual predictor variables that are used. Our task is to find the

```
1:  DEFINE: x_i, i = 1,···,n, variables, D as predictees set, P as predictors set, Q as queue for
    predictees and M as model
2:  INPUT: x_i, i = 1,···,n, set of variables; x_j, j ∈ D, set of predictees and x_k; k ∈ P, set of
    predictors
3:  SELECT: x_0 (x_0 ∈ D) and put it in set Q
4:  CREATE: Empty model, M, of n nodes with no links, Q as queue for predictees which is
    initially empty
5:  REPEAT
6:    Select x_i ∈ D
7:    Find x_j ∈ P, such that P = {x_j : x_j ≠ x_i and x_j → x_i}, i.e., x_j is a predictor of x_i
8:    For all x_j ∈ P, add x_j → x_i into set M
9:    For all x_j ∈ P, add x_j into set Q
10: UNTIL: Set Q is empty
```

FIGURE 8.27
Pseudo-code for structural equation modeling.

set of variables in p and T for x_i, $i \in D$. The algorithms (Figure 8.27) use covariance information, in the form of estimated standardized regression coefficients, to direct the construction of SEMs and to estimate the parameters of the models. Latent variables can result in biased estimates; hence, the algorithm might give erroneous results when latent variables are considered.

In the literature, one finds differing approaches to assess the best fit. Traditional approaches to modeling start from a null hypothesis, rewarding more parsimonious models, to others such as AIC that focus on how little the fitted values deviate from a comparison model, taking into account the number of free parameters used. Because different measures of fit capture different elements of the fit of the model, it is appropriate to report a selection of different fit measures. A few commonly used measures of fit are: (i) chi-squared, (ii) AIC; (iii) root mean square error (RMSE); (iv) standardized root mean residual (SRMR); and (v) comparative fit index. Though not exhaustive, but a few useful references for SEM are: Bagozzi (1982), Grace (2006), Hancock and Mueller (2006), Kaplan (2000), Kline (1998), Muthén (1983), Rabe-Hesketh et al. (2004), Raykov and Marcoulides (2006), McDonald and Ho (2002), and Skrondal and Rabe-Hesketh (2004).

Data set # 1: SEM process can be thought of as a four-stage process, namely, (i) model specification, (ii) model estimation, (iii) model evaluation, and finally (iv) model modification. One can use the data from Stein et al. (2003). The main idea of the study is to model two separate paths for alcohol and drugs and then test them in which psychosocial, environmental, and sociodemographic variables is the prediction the best. The behavioral and substance abuse-related factors as well as the key outcome of positive attitudes about quitting drugs ($N = 620$) or alcohol ($N = 526$) in a sample of 709 homeless women are used. A positive attitude about quitting alcohol was predicted by more addiction symptoms, fewer positive effects from using alcohol, and not having a partner who uses alcohol. A positive attitude about quitting drugs was predicted by more drug problems, greater drug use in the past 6 months, more active coping, more education, less emotional distress, not having a partner who uses drugs, and fewer addiction symptoms. As one understands that the primary goal of the study is to determine if a set of items that query both alcohol and drug problems are adequate indicators for the two underlying constructs, namely, alcohol use problem and drug use problem. The reader is urged to study the data set and utilize the concepts of SEM to solve this problem.

To end the discussion of MVA, we give a few of the necessary information related to journals, data sets, etc., which we think would make this chapter more interesting both from the point of view of theory as well as application. Finally, a few references that we definitely think are important with respect to this chapter and related topics are: Agresti (2007), Aitchison (1986), Atchley and Bryant

(1975), Bishop et al. (1975), Carroll and Green (1997), Cheung (2015), Gifi (1990), Harris (1975), Kachigan (1991), Karson (1982), Krzanowski (1988, 1995), McCullagh and Nelder (1989), Van de Geer (1993), Whittaker (1990), and others. This section has been an honest attempt to look into MVA and multivariate statistics from a fresh perspective and try to deal with the theoretical as well as practical aspects of different methods from the point of view of decision analysis. To end this chapter, we discuss very briefly the future trends in statistics and big data analysis, which we think would be an ever-burgeoning field considering the ever-increasing applications of statistics and its tools in our everyday life.

8.24 Future Areas of Research

Today, science is passing through an era of transformation. Any decision-making process today is based on the efficient analysis of data available at hand. Science is driven by the data and it is being termed as data science. In the coming few decades, the most important areas of research will be on the analysis of big data. Big data usually indicate data sets whose sizes are beyond the ability of commonly used software tools to manage and analyze within a tolerable time limit. The term "big data" is a constantly moving target. As of today, the size of "big data set" ranges from few dozen terabytes to many petabytes of data. Interestingly, big data are available today in every sphere of life. Starting from industry, environment, health care, and government security, big data are being collected and stored everyday. This large complex, structured or unstructured, and heterogeneous data in the form of big data has gained significant attention. The velocity of the expansion of the amount of data gives rise to a complete paradigm shift in how new-age data are processed.

The age of data science is in its infancy and is experiencing a tactical evolution by leaps and bounds in all dimensions of science. Even though over the past few years a few robust big data models have come into existence, there is still a need for the pool to expand at a faster pace to meet the challenges of data proliferation. The concept of big data is relatively new and needs further research. Big data sets cannot be practically analyzed on a single commodity computer because their sizes are too large to fit in memory or it is too time consuming to process when the current statistical methods are used. To circumvent this obstacle, one may have to resort to parallel and distributed architectures, with multicore and cloud computing platforms providing access to hundreds or thousands of processors. While the parallel and distributed architectures present new capabilities for storage and manipulation of data, from an inferential point of view, it is unclear how the current statistical methodologies can be transported to the paradigm of big data. Big data have put a great challenge on the current statistical methodology.

There are several algorithms that are recently developed and feasible for statistical inference of big data and workable on parallel machines, including the bag of little bootstraps by Kleiner et al. (2014), aggregated estimating equation of Xi et al. (2009), split and conquer algorithms of Chen and Xie (2014), and the subsampling-based stochastic approximation algorithm by Liang et al. (2013). On the other hand, iterative algorithms have been widely used in current society of scientific computing, and it mainly includes Markov chain Monte Carlo (MCMC) algorithms and the EM algorithm, which typically requires a large number of iterations and a complete scan of the full data set for each iterations. Given the success of the iterative algorithms in modern scientific computing, it would be of great interest to develop some innovative iterative algorithms that are feasible for big data.

There have been significant advances made by the statistical community on big data research in the past few years. One of the open problems is how to generalize and scale up such proposed techniques to the true big data setting. One of the key features of big data is that the statistical methods, which work well on small-scale data set, usually perform poorly in big data setting. Some of the other open problems include (i) to have better understanding of big data and associated statistical issues, (ii) to think more carefully about how to solve big data issues, and (iii) to have a more concrete focus on big data problems. There are lots of challenges in this exciting field of research. More people need to come and join this active area of research for further development.

References

Abonyi, J. and Feil, B. 2007. Cluster analysis for data mining and system identification, *Birkhäuser*, 978-3764379872.

Agresti, A. 2007. *An Introduction to Categorical Data Analysis*, Wiley, ISBN: 9780471226185.

Aitchison, J. 1986. *The Statistical Analysis of Compositional Data*, Chapman & Hall, ISBN: 978-94-010-8324-9.

Anderberg, M. R. 1973. *Cluster Analysis for Applications*, Academic Press, New York, ISBN: 0120576503.

Anderson, T. W. 2003. *An Introduction to Multivariate Statistical Analysis*, John Wiley & Sons, ISBN: 9780471360919.

Arnold, S. F. 1981. *The Theory of Linear Models and Multivariate Analysis*, John Wiley & Sons, ISBN: 9780471050650.

Atchley, W. R. and Bryant, E. H. (Editors). 1975. *Multivariate Statistical Methods: Within Group Covariation*, Dowden Hutchinson & Ross, ISBN: 9780470035955.

Athreya, K. B. and Lahiri, S. N. 2006. *Measure Theory*, Hindustan Book Agency, New Delhi.

Bach, F. R. and Jordan, M. I. 2002. Kernel independent component analysis, *Journal of Machine Learning Research*, **3**, 1–48.

Bagozzi, R. P. (Editor). 1982. Special issue on causal modeling, *Journal of Marketing*, **19**, 403–584.

Bandyopadhyay, S., Rao, A. R., and Sinha, B. K. 2011. *Models for Social Networks with Statistical Applications*, SAGE Publications, Thousand Oaks, CA.

Barlett, M. S. 1954. A note on multiplying factors for various chi-squared approximations, *Journal of the Royal Statistical Society (Series B)*, **16**, 296–298.

Basilevsky, A. T. 1994. *Statistical Factor Analysis and Related Methods: Theory and Applications*, John Wiley & Sons, ISBN: 978-0-471-57082-0.

Ben-Tal, A., El Ghaoui, L., and Nemirovski, A. 2009. *Robust Optimization*, Princeton University Press, ISBN: 9781400831050.

Ben-Tal, A. and Nemirovski, A. 1998. Robust convex optimization, *Mathematics of Operations Research*, **23**, 769–805.

Ben-Tal, A. and Nemirovski, A. 1999. Robust solutions to uncertain linear programs, *Operations Research Letters*, **25**, 1–13.

Ben-Tal, A. and Nemirovski, A. 2002. Robust optimization—Methodology and applications, *Mathematical Programming, Series B*, **92**, 453–480.

Bentler, P. M. and Chou, C. P. 1987. Practical issues in structural modeling, *Sociological Methods & Research*, **16**, 78–117.

Bertsimas, D. and Sim, M. 2003. Robust discrete optimization and network flows, *Mathematical Programming*, **98**, 49–71.

Bertsimas, D. and Sim, M. 2004. The price of robustness, *Operations Research*, **52**, 35–53.

Bertsimas, D. and Sim, M. 2006. Tractable approximations to robust conic optimization problems, *Mathematical Programming*, **107**, 5–36.

Bielby, W. T. and Hauser, R. M. 1977. Structural equation models, *Annual Review of Sociology*, **3**, 137–161.

Billingsley, P. 1995. *Probability and Measures*, 3rd edition, John Wiley, New York.

Bishop, Y. M. M., Fienberg, S. E., and Holland, P. W. 1975. *Discrete Multivariate Analysis: Theory and Practice*, MIT Press, ISBN: 9780262520409.

Bock, R. D. 1975. *Multivariate Statistical Methods in Behavioral Research*, McGraw-Hill, ISBN: 9780894980145.

Bollen, K. A. 1989. *Structural Equations with Latent Variables*, Wiley, ISBN: 978-0-471-01171-2.

Bollen, K. A. and Long, J. S. (Editors). 1993. *Testing Structural Equation Models*, Sage Publications, ISBN: 978-0803945074.

Carroll, J. D. and Green, P. E. 1997. *Mathematical Tools for Applied Multivariate Analysis*, Academic Press, ISBN: 978-0121609559.

Cattell, R. B. 1943. The description of personality: Basic traits resolved into clusters, *Journal of Abnormal and Social Psychology*, **38**, 476–506.

Chatfield, C. and Collins, A. J. 1980. *Introduction to Multivariate Analysis*, Chapman & Hall, ISBN: 978-0-412-16030-1.

Chen, X. and Xie, M. 2014. A split and conquer approach for extraordinary large data analysis, *Statistica Sinica*, **24**, 1655–1684.

Chen, Z., Aiyi, L., Qu, Y., Tang, L., Ting, N., and Tsong, Y. 2015. *Applied Statistics in Biomedicine and Clinical Trials Design*, Springer, New York.

Cherubini, U., Luciano, E., and Vecchiato, W. 2004. *Copula Methods in Finance*, John Wiley & Sons, ISBN 0470863447.

Cheung, M. W.-L. 2015. *Meta-Analysis: A Structural Equation Modeling Approach*, Wiley, ISBN: 978-1-119-99343-8.

Child, D. 2006. *The Essentials of Factor Analysis*, Continuum International, ISBN: 978-0826480002.

Chung, K. L. 1974. *A Course in Probability Theory*, Academic Press, New York.

Coles, S. G. 2001. *An Introduction to Statistical Modeling of Extreme Values*, Springer, ISBN: 978-1849968744.

Cooley, W. W. and Lohnes, P. R. 1971. *Multivariate Data Analysis*, John Wiley & Sons, ISBN: 978-0471170600.

Cuadras, C. M. and Rao, C. R. (Editors). 1993. *Multivariate Analysis: Future Directions 2*, North-Holland, ISBN: 978-0-444-81531-6.

Davis, C., Claridge, G., and Cerullo, D. 1997. Personality factors predisposing to weight preoccupation: A continuum approach to the association between eating disorders and personality disorders, *Journal of Psychiatric Research*, **31**, 467–480.

Davison, A. C. and Hinkley, D. V. 1997. *Bootstrap Methods and Their Applications*, Cambridge University Press, Cambridge, MA.

de Haan, L. and Resnick, S. I. 1977. Limit theory for multidimensional sample extremes, *Zeitschrift für Wahrscheinlichkeitstheorie und Verwandte Gebiete*, **40**, 317–337.

Dempster, A. P. 1969. *Elements of Continuous Multivariate Analysis*, Addison-Wesley, ISBN: 978-0201014853.

Dempster, A. P., Laird, M. N., and Rubin, D. B. 1977. Maximum likelihood estimation from incomplete data via EM algorithm (with discussion), *Journal of the Royal Statistical Society*, Series B, **39**, 1–38.

Dice, L. R. 1945. Measures of the amount of ecologic association between species, *Ecology*, **26**, 297–302.

Dillon, W. R. and Goldstein, M. 1984. *Multivariate Analysis, Methods and Applications*, Wiley, ISBN: 0-471-08317-8.

Doob, J. L. 1953. *Stochastic Processes*, John Wiley, New York.

Driver, H. E. and Kroeber, A. L. 1932. Quantitative expression of cultural relationships, *University of California Publications in American Archeology and Ethnology*, **31**, 211–256.

Duda, R. O., Hart, P. E., and Stork, D. G. 2001, *Pattern Classification*, Wiley, New York, ISBN: 978-0-471-05669-0.

Duncan, O. D. 1966. Path analysis: Sociological examples, *The American Journal of Sociology*, **72**, 1–16.

Duran, B. S. and Odell, P. L. 1974. *Cluster Analysis: A Survey*, Springer-Verlag, New York, ISBN 978-3-642-46309-9.

Eaton, M. L. 1983. *Multivariate Statistics: A Vector Space Approach*, John Wiley & Sons, ISBN: 9780471027768.

Eaton, M. L. and Perlman, M. D. 1973. The non-singularity of generalized sample covariance matrices, *Annals of Statistics*, **1**, 710–717.

Everitt, B. S. and Dunn, G. 2001. *Applied Multivariate Data Analysis*, Hodder Arnold Publication, ISBN: 9780340741221.

Everitt, B. S., Landau, S., Leese, M., and Stahl, D. 2011. *Cluster Analysis*, John Wiley & Sons, ISBN: 978-0-470-74991-3.

Fisher, R. A. 1921. Studies in crop variation. I. An examination of the yield of dressed grain from Broadbalk, *Journal of Agricultural Science*, **11**, 107–135.

Galton, F. 1886, Regression towards mediocrity in hereditary stature, *Journal of the Anthropological Institute*, **15**, 246–263

Galton, F. 1889. *Natural Inheritance*, MacMillan and Company, London and New York. https://archive.org/details/naturalinherita03galtgoog>

Gerrild, P. M. and Lantz, R. J. 1969. *Chemical Analysis of 75 Crude Oil Samples from Pliocence Sand Units, Elk Hills Oil Field, California*, U.S. Geological Survey Open File Report.

Ghosh, J. K., Delampady, M., and Samanta, T. 2006. *An Introduction to Bayesian Analysis: Theory and Methods*, Springer, New Delhi.

Gifi, A. 1990. *Nonlinear Multivariate Analysis*, Wiley, ISBN: 0-471-92620-5.

Gilks, W. R., Richardson, S., and Spiegelhalter, D. J. 1996. *Markov Chain Monte Carlo in Practice*, Chapman & Hall/CRC Press, New York.

Giri, N. C. 2004. *Multivariate Statistical Analysis*, Marcel Dekker, ISBN: 0824747135.

Gittins, R. 1985. *Canonical Analysis: A Review with Application in Ecology*, Springer-Verlag, ISBN: 3-540-13617-7.

Gnanadesikan, R. 2011. *Methods for Statistical Data Analysis of Multivariate Observations*, John Wiley & Sons, ISBN: 0471161195.

Goldberg, D. 1989. *Genetic Algorithm in Search, Optimization and Machine Learning*, Addison-Wesley Professional, Reading, MA.

González, I., Déjean, S., Martin, P. G. P., and Baccini, A. 2008. CCA: An R package to extend canonical correlation analysis, *Journal of Statistical Software*, **23**, 1–14.

Gordon, A. E. 1981. *Classification: Methods for the Exploratory Analysis of Multivariate Data*, Chapman & Hall, New York, ISBN: 9780412228506.

Gorsuch, R. L. 1983. *Factor Analysis*, Lawrence Erlbaum Associates, ISBN: 978-0898592023.

Grace, J. B. 2006. *Structural Equation Modeling and Natural Systems*, Cambridge University Press, ISBN: 9780521546539.

Hair, J. F., Black, W. C., Babin, B. J., Anderson, R. E., and Tatham, R. L. 2005. *Multivariate Data Analysis*, Pearson Education, ISBN: 9780130329295.

Hancock, G. R. and Mueller, R. O. 2006. *Structural Equation Modeling: A Second Course*, Information Age Publishing, ISBN: 978-1593110154.

Hanley, J. A. 2004, Transmuting women into men: Galton's family data on human stature, *The American Statistician*, **58**, 237–243

Härdle, W. K. and Simar, L. 2007. *Applied Multivariate Statistical Analysis*, Springer-Verlag, ISBN: 9783540722434.

Harman, H. H. 1976. *Modern Factor Analysis*, University of Chicago Press, ISBN: 978-0226316529.

Harris, R. J. 1975. *A Primer of Multivariate Statistics*, Academic Press, ISBN: 9780123272508.

Hartigan, J. A. 1975. *Clustering Algorithms*, Wiley, ISBN: 047135645X.

Hastie, T., Tibshirani, R., and Friedman, J. 2011. *The Elements of Statistical Learning: Data Mining, Inference, and Prediction*, Springer-Verlag, ISBN: 978-0387848570.

Hoaglin, D. C., Mosteller, F., and Tukey, J. W. (Editors). 1991, *Fundamentals of Exploratory Analysis of Variance*, John Wiley & Sons, ISBN: 0-471-52735-1.

Holzinger, K. J. and Swineford, F. 1939. *A Study in Factor Analysis: The Stability of Bi-Factor Solution*, University of Chicago: Supplementary Educational Monographs, **48**.

Höppner, F., Klawonn, F., Kruse, R., and Runkler T. 1999. *Fuzzy Cluster Analysis: Methods for Classification, Data Analysis and Image Recognition*, Wiley-Blackwell, 978-0471988649.

Hotelling, H. 1935. The most predictable criterion, *Journal of Educational Psychology*, **26**, 139–142.

Hotelling, H. 1936. Relationship between two set of variables, *Biometrika*, **28**, 321–377.

Huber, P. J. 1981. *Robust Statistics*, JohnWiley and Sons, New York.

Hsieh, W. W. 2000. Nonlinear canonical correlation analysis by neural networks, *Neural Networks*, **13**, 1095–1105.

Huberty, C. J. and Olejnik, S. 2006. *Applied MANOVA and Discriminant Analysis*, Wiley, ISBN: 978-0-471-46815-8.

Iversen, G. R. and Norpoth, H. 1987. *Analysis of Variance*, Sage Publications, ISBN: 0-8039-3001-1.

Jaccard, P. 1901. Étude comparative de la distribution florale dans une portion des Alpes et des Jura, *Bulletin de la Société Vaudoise des Sciences Naturelles*, **37**, 547–579.

Jain, A. K. and Dubes, R. C. 1988. *Algorithms for Clustering Data*, Prentice-Hall, ISBN: 0-13-022278-X.

James, W. and Stein, C. 1961. Estimation with Quadratic Loss, Berkeley Symposium on Mathematical Statistics and Probability, *Proceeding of the fourth Berkeley Symposium on Mathematical Statistics and Probability*, University of California Press, **1**, 361–379.

Jobson, J. D. 1991. *Applied Multivariate Data Analysis*, Springer-Verlag, ISBN: 978-1-4612-0955-3.

Jackson, J. E. 2003. *User's Guide to Principal Components*, Wiley-Interscience, ISBN: 978-0471471349.

Johnson, N. L. and Kotz, S. 1972. *Distributions in Statistics: Continuous Multivariate Distributions*, John Wiley & Sons, ISBN: 0471521620.

Johnson, R. A. and Wichern, D. W. 2002. *Applied Multivariate Statistical Analysis*, Pearson Education, ISBN: 8178086867.

Jolliffe, I. T. 2002. *Principal Component Analysis*, Springer-Verlag, ISBN: 978-0-387-22440-4.

Kachigan, S. K. 1991. *Multivariate Statistical Analysis: A Conceptual Introduction*, Radiu Press, ISBN: 978-0942154917.

Kaplan, D. 2000. *Structural Equation Modeling: Foundations and Extensions*, Sage Publications, ISBN: 978-1412916240.

Karamouz, M., Nazif, S., and Fallahi, M. 2010. Rainfall downscaling using statistical downscaling model and canonical correlation analysis: A case study. In: Palmer, R.N. (Ed.), *World Environmental and Water Resources Congress*, American Society of Civil Engineers, Reston, pp. 4579–4587.

Karson, M. J. 1982. *Multivariate Statistical Methods: An Introduction*, Iowa State University Press, ISBN: 9780813818450.

Kaufman, L. and Rousseeuw, P. J. 2005. *Finding Groups in Data: An Introduction to Cluster Analysis*, Wiley, ISBN: 978-0-471-73578-6.

Kelley, P. R. and Barry, R. 1997. Sparse spatial autoregressions. *Statistics and Probability Letters*, **33**, 291–297.

Kempton, R. A. and Fox, P. N. 2012. *Statistical Methods for Plant Variety Evaluation*, Chapman & Hall, London.

Kendall, M. G. 1980. *Multivariate Analysis*, Hodder Arnold, ISBN: 978-085264264.

Kennedy, W. J. and Gentle, J. E. 1980. *Statistical Computing*, Marcel Dekker, New York.

Kirkpatrick, S., Gelatt, Jr, C. D., and Vecchi, M. P. 1983. Optimization by simulated annealing, *Science*, **220**, 671–680.

Kleiner, A., Talwalkar, A., Sarkar, P., and Jordan, M. I. 2014. A scalable bootstrap for massive data, *Journal of the Royal Statistical Society*, Series B, doi:10.1111.rssb.12050.

Kline, R. B. 1998. *Principles and Practice of Structural Equation Modeling*, The Guilford Press, ISBN: 978-1606238769.

Kotz, S., Balakrishnan, N., and Johnson, N. L. 2000. *Continuous Multivariate Distributions: Volume 1—Models & Applications*, Wiley-Blackwell, ISBN: 978-0471183877.

Kotz, S. and Nadarajah, S. 2000. *Extreme Vale Distributions*, Imperial College Press, ISBN: 978-1-86094-224-2.

Kotz, S. and Nadarajah, S. 2004. *Multivariate Distributions and Their Applications*, Cambridge University Press, ISBN: 0521826543.

Krishnaiah, P. R. (Editor). 1984. *Handbook of Statistics 1: Analysis of Variance*, Elsevier, ISBN: 978-0-444-85335-6.

Krzanowski, W. J. 1988. *Principles of Multivariate Analysis: A User's Perspective*, Oxford University Press, ISBN: 0-198-52211-8.

Krzanowski, W. J. (Editor). 1995. *Recent Advances in Descriptive Multivariate Analysis*, Oxford University Press, ISBN: 9780198522850.

Kshirsagar, A. M. 1972. *Multivariate Analysis*, Mercel Dekker, ISBN: 9780824713867.

Lachenbruch, P. A. 1975. *Discriminant Analysis*, Macmillan Publishers, ISBN: 9780028482507.

Lai, P. L. and Fyfe, C. 2000. Kernel and nonlinear canonical correlation analysis, *International Journal of Neural Systems*, **10**, 365–377.

Lawley, D. N. and Maxwell, A. 1971. *Factor Analysis as a Statistical Method*, Macmillan, ISBN: 978-0408701525.

Lehmann, E. H. and Casella, G. 1998. *Theory of Point Estimation*, 2nd edition, Springer, New York.

Leurgans, S. E., Moyeed, R. A., and Silverman, B. W. 1993. Canonical correlation analysis when the data are curves, *Journal of Royal Statistical Society: Series B*, **55**, 725–740.

Lewis, D. G. 1971. *The Analysis of Variance*, Manchester University Press, ISBN: 0-7190-0467-5.

Liang, F., Cheng, Y., Song, Q., Park, J., and Yang, P. 2013. Aresampling-based stochastic approximation method for analysis of large geostatistical data, *Journal of the American Statistical Association*, **108**, 325Ű339.

Linden, M. 1977. A factor analytic study of Olympic decathlon data, *Research Quarterly*, **48**, 562–568.

Louviere, J. 1988. *Analyzing Decision Making: Metric Conjoint Analysis*, SAGE Publications, ISBN: 9780803927575.

Mardia, K. V., Kent, J. T., and Bibby, J. M. 1979. *Multivariate Analysis*, Academic Press, ISBN: 012471252.

Marshall, A. W. and Olkin I. 1967. A multivariate exponential distribution, *Journal of American Statistical Association*, **61**, 30–44.

Marshall, A. W. and Olkin I. 1983. Domains of attraction of multivariate extreme value distributions, *Annals of Probability*, **11**, 168–177.

McCullagh, P. and Nelder, J. A. 1989. *Generalized Linear Models*, Chapman & Hall, ISBN: 978-0412317606.

McDonald, R. P. and Ringo Ho, M.-H. 2002. Principles and practice in reporting structural equation analyses, *Psychological Methods*, **7**, 64–82.

McLachlan, G. J. and Krishnan, T. 1997. *The EM Algorithm and Extensions*, John Wiley and Sons, New York.

Melzer, T., Reiter, M., and Bischof, H. 2001. Nonlinear feature extraction using generalized canonical correlation analysis canonical correlation analysis, *Artificial Neural Networks-ICANN 2001* (Editors G. Dorner, H. Bischof, and K. Hornik), **2130** of LNCS, 353–360, Springer.

Mendenhall, W. and Sincich, T. 2006. *Statistics for Engineering and the Sciences*, Pearson Publication, ISBN: 978-0131877061.

Morrison, D. F. 1990. *Multivariate Statistical Methods*, McGraw-Hill Ryerson, ISBN: 9780070431867.

Mosteller, F. and Tukey, J. W. 1977. *Data Analysis and Regression*, Addison-Wesley, Reading, MA.

Muirhead. R. J. 2005. *Aspects of Multivariate Statistical Theory*, John Wiley & Sons, ISBN: 978-0471769859.

Mulaik, S. A. 2009. *The Foundations of Factor Analysis*, Chapman & Hall, ISBN: 978-1420099614.

Muthén, B. 1983. Latent variable structural equation modeling with categorical data, *Journal of Econometrics*, **22**, 48–65.

Nelder, J. A. and Mead, R. 1965. A simplex method for function minimization, *Computer Journal*, **7**, 308–313.

Nelsen, R. B. 2006. *An Introduction to Copulas*, Springer, ISBN 978-0-387-28678-5.

Orme, B. K. and King, W. C. 2006. *Getting Started with Conjoint Analysis: Strategies for Product Design and Pricing Research*. Research Publishers, Madison, WI, ISBN: 978-0972729741.

Osborne, M. R. 1972. Some aspects of non-linear least squares calculations, *Numerical Methods for Nonlinear Optimization* (Editor F. A. Lootsma), Academic Press, New York.

Pearson, K. 1901. On lines and planes of closest fit to systems of points in space, *Philosophical Magazine*, **2**, 559–572.

Pickards, J. 1981. *Multivariate Extreme Value Distributions*, *Proceedings 43rd Session International Statistical Institute*, Buenos Aires, 859–878.

Press, S. J. 1982. *Applied Multivariate Analysis: Using Bayesian and Frequentist Methods of Inference*, R. E. Krieger Publishing Company, ISBN: 9780882759760.

Rabe-Hesketh, S., Skrondal, A., and Pickles, A. 2004. Generalized multilevel structural equation modeling, *Psychometrika*, **69**, 167–190.

Raghavarao, D., Wiley, J. B., and Chitturi, P. 2011. *Choice Based Conjoint Analysis: Models and Designs*, CRC Press, Taylor & Francis, ISBN: 978-1-4200-9996-6.

Rao, C. R. 2008. *Linear Statistical Inference and Its Applications*, John Wiley & Sons, ISBN: 9780470316436.

Raykov, T. and Marcoulides, G. A. 2006. *A First Course in Structural Equation Modeling*, Lawrence Earlbaum Associates, ISBN: 0-8058-3569-5.

Roy, S. N. 1957. *Some Aspects of Multivariate Analysis*, Wiley, ISBN: 978-1124038551.

Roy, S. N., Gnanadesikan, R., and Srivastava, J. N. 1971. *Analysis and Design of Certain Quantitative Multiresponse Experiments*, Pergamon Press, ISBN: 9780080069173.

Rutherford, A. 2001. *Introducing ANOVA and ANCOVA: A GLM Approach*, Sage Publications, ISBN: 0-7619-5160-1.

Sahai, H. and Ageel, M. I. 2000. *The Analysis of Variance: Fixed, Random, and Mixed Models*, Springer, ISBN 978-1-4612-1344-4.

Scheffé, H. 1999. *The Analysis of Variance*, John Wiley & Sons, ISBN: 0-471-75834-5.

Searle, S. R., Casella, G., and McCulloch, C. E. 2009. *Variance Components*, John Wiley & Sons, ISBN: 0-470-00959-4.

Seber, G. A. F. 2004. *Multivariate Observations*, John Wiley & Sons, ISBN: 9780471691211.

Sen, P. K. 1986. Contemporary textbooks on multivariate statistical analysis: A panoramic appraisal and critique, *Journal of the American Statistical Association*, **81** (394), 560–564.

Serfling, R. J. 1980. *Approximation Theorems of Mathematical Statistics*, JohnWiley and Sons, New York.

Sibuya, M. 1960. Bivariate extreme statistics, *Annals of Institute of Mathematical Statistics*, **11**, 195–210.

Skrondal, A. and Rabe-Hesketh, S. 2004. *Generalized Latent Variable Modeling: Multilevel, Longitudinal, and Structural Equation Models*, Chapman & Hall, ISBN: 978-1584880004.

Späth, H. 1980. *Cluster Analysis Algorithms for Data Reduction and Classification of Objects*, Ellis Horwood, ISBN: 978-0853121411.

Spearman, C. 1904. General intelligence, objectively determined and measured, *The American Journal of Psychology*, **15**, 201–292.

Srivastava, M. S. and Khatri, C. G. 1979. *An Introduction to Multivariate Statistics*, North Holland, ISBN: 9780444003027.

Stein, J. A., Dixon, E., Longshore, D., and Galaif, E. 2003. Predicting positive attitudes about quitting drug- and alcohol-use among homeless women, *Psychology of Addictive Behaviors*, **17**, 32–41.

Stuart, A., Ord, J. K., and Arnold, S. 1999. *Kendall's Advanced Theory of Statistics, Volume 2A: Classical Inference and the Linear Model*, Wiley Publications, ISBN: 978-0-470-68924-0.

Szakács, G., Annereau, J. P., Lababidi, S., Shankavaram, U., Arciello, A., Bussey, K. J., Reinhold, W. et al. 2004. Predicting drug sensitivity and resistance: Profiling ABC transporter genes in cancer cells, *Cancer Cell*, **6**, 129–137.

Takeuchi, K., Yanai, H., and Mukherjee, B. N. 1982. *The Foundations of Multivariate Analysis*, Halstead Press, ISBN: 0852269641.

Tanimoto, T. 1957. *An Elementary Mathematical Theory of Classification and Prediction*, IBM Technical Report.

Tatsuoka, M. M. 1988. *Multivariate Analysis: Techniques for Educational and Psychological Research*, Macmillan, ISBN: 978-0024191205.

Theus, M. and Urbanek, S. 2008. *Interactive Graphics for Data Analysis: Principles and Examples*, CRC Press, Boca Raton, FL.

Thompson, B. 2004. *Exploratory and Confirmatory Factor Analysis: Understanding Concepts and Applications*, American Psychological Association, Washington DC.

Thurstone, L. L. 1931. Multiple factor analysis, *Psychological Review*, **38**, 406–427.

Thurstone, L. L. 1947. *Multiple-Factor Analysis*, The University of Chicago Press, Chicago, ISBN: 978-0226801094.

Tiago de Oliveira, J. 1958. Extremal distributions, *Revista der Faculdade de Ciencias de Lisboa: Series A*, **7**, 219–227.

Tiago de Oliveira, J. 1975. Bivariate and multivariate extreme distributions. In: G. P. Patil, S. Kotz, and J. K. Ord (eds), *A Modern Course on Statistical Distributions in Scientific Work, NATO Advanced Study Institutes Series*, Springer, Netherlands, **17**, 355–361.

Tibshirani, R. 1996. Regression shrinkage and selection via the lasso, *Journal of the Royal Statistical Society*, Series B, **58**, 267–288.

Tryon, R. C. 1939. *Cluster Analysis: Correlation Profile and Orthometric (Factor) Analysis for the Isolation of Unities in Mind and Personality*, Edwards Brothers.

Tukey, J.W. 1977. *Exploratory Data Analysis*, Addison-Wesley, Reading, MA.

Turner, J. R. and Thayer, J. F. 2001. *Introduction to Analysis of Variance: Design, Analysis & Interpretation*, Sage Publications, ISBN: 0-8039-7074-9.

Vinod, H. D. 1976. Canonical ridge and econometrics of joint production, *Journal of Econometrics*, **4**, 147–166.

Van de Geer, J. P. 1993. *Multivariate Analysis of Categorical Data: Applications*, SAGE Publications, ISBN: 978-0803945647.

Whittaker, J. 1990. *Graphical Models in Applied Mathematical Multivariate Statistics*, Wiley, ISBN: 978-0-471-91750-2.

Wishart, J. 1928. The generalized product moment distribution in samples from a normal multivariate population, *Biometrika*, **20**, 35–52.

Wright, S. 1921. Correlation and causation, *Journal of Agricultural Research*, **20**, 557–585.

Wright, S. 1934. The method of path coefficients, *Annals of Mathematical Statistics*, **5**, 161–215.

Xi, R., Lin, N., and Chen, Y. 2009. Compression and aggregation for logistic regression analysis in data cubes, *IEEE Transactions on Knowledge and Data Engineering*, **21**, 479–492.

Xu, R. and Wunsch, D. 2008. *Clustering*, Wiley, ISBN: 978-0-470-27680-8.

Zubin, T. 1938. A technique for measuring like-mindedness, *Journal of Abnormal Social Psychology*, **33**, 508–516.

Few Relevant URLs

S. No.	Details of Relevant URLs	URL
1	ACM Special Interest Group for Genetic and Evolutionary Computation	http://www.sigevo.org/
2	Ant Colony Optimization	http://www.aco-metaheuristic.org/
3	Artificial Immune System Web	http://www.artificial-immune-systems.org/
4	AURORA—Advanced Models, Applications and Software for High-Performance Computing in Finance, University of Vienna, Switzerland	http://www.univie.ac.at/sor/aurora6/index.html
5	Center of Banking and Financial Research, University of Cyprus, Cyprus	https://www.ucy.ac.cy/hermes/en/
6	Centre for Computational Finance and Economic Agents, University of Essex, UK	http://www.essex.ac.uk/ccfea/
7	Center for Machine Learning and Intelligent Systems, University of California Irwin, USA	https://archive.ics.uci.edu/ml/datasets.html
8	Centre for Optimization and Statistical Learning, Northwestern University, USA	http://www.mccormick.northwestern.edu/ research/optimization-machine-learning-center/ index.html
9	CiteSeerX (beta version)	http://citeseerx.ist.psu.edu/index
10	Convex Optimization	http://www.convexoptimization.com/
11	Computational Infrastructure for Operations Research	http://www.coin-or.org/
12	Data Publisher for Earth & Environmental Science (Pangaea)	http://www.pangaea.de/
13	Data Sets	http://www.statsci.org/datasets.html
14	Decision Tree for Optimization Software	http://plato.asu.edu/guide.html
15	Economic Papers	http://econpapers.repec.org/

Continued

S. No.	Details of Relevant URLs	URL
16	EMOO Repository	http://delta.cs.cinvestav.mx/~ccoello/EMOO
17	Evolutionary Algorithms for Solving Multi-Objective Problems	http://www.cs.cinvestav.mx/~emoobook/
18	IBM Research	http://researchweb.watson.ibm.com/
19	ICER—International Centre for Economic Studies	http://nf.vse.cz/english/science-and-research/icer-international-centre-for-economic-studies/
20	International Neural Network Society	http://www.inns.org/
21	International Society for Genetic and Evolutionary Computation	http://www.isgec.org/
22	International Society on Multiple Criteria Decision Making	http://www.mcdmsociety.org
23	International Statistical Institute	http://www.isi-web.org/
24	Kanpur Genetic Algorithm Lab	http://www.iitk.ac.in/kangal/index.shtml
25	Metaheuristic Network	http://www.metaheuristics.org/
26	Modeling & Optimization Research & Education, The University of Arizona	http://www.sie.arizona.edu/MORE/index.html
27	NEOS Guide: Companion Site to the NEOS Serve	http://www.neos-guide.org/
28	NEOS Server: State-of-the-Art Solvers for Numerical Optimization	http://www.neos-server.org/ neos/
29	Netlib Repository	http://www.netlib.org/
30	Network for Artificial Immune Systems	http://www.elec.york.ac.uk/ARTIST/
31	NIST, Information Technology Library	http://www.itl.nist.gov
32	Numerical Algorithm Group	http://www.nag.com/
33	Oak Ridge National Laboratory	http://www.ornl.gov/
34	Optimization Online	http://optimization-online.org/
35	Particle Swarm Intelligence	http://www.swarmintelligence.org/
36	Risk Management and Financial Engineering Lab, University of Florida, USA	http://www.ise.ufl.edu/rmfe/
37	Social Science Research Network	http://www.ssrn.com/
38	Stochastic Programming Society	http://stoprog.org/
39	Stochastic Optimization Research Group, Georgia Institute of Technology, USA	http://www2.isye.gatech.edu/so/
40	The Probability Web	http://probweb.berkeley.edu/

Few Relevant Softwares

S. No.	Details of Softwares	URL
1	A Quadratic Assignment Problem Library (QAPLIB)	http://www.seas.upenn.edu/qaplib/
2	AMPL Modeling Language for Mathematical Programming	http://www.ampl.com/
3	IBM ILOG CPLEX Optimizer	http://www-01.ibm.com/software/integration/optimization/cplex-optimizer/
4	Data Envelopment Analysis	http://www.dea-analysis.com/ and http://www.deazone.com/

Continued

S. No.	Details of Softwares	URL
5	Data Envelopment Analysis Online Software	https://www.deaos.com
6	EVIEWS	http://www.eviews.com/
7	GAUSS	http://www.aptech.com/products/gauss-mathematical-and-statistical-system/
8	GNU Octave	http://www.gnu.org/software/octave/
9	Graph Visualization Software	http://gephi.org/
10	High Performance 3D Visualization Software	http://www.vsg3d.com/
11	Indirect Optimization on the Basis of Self-Organization (IOSO)	http://iosotech.com/
12	jMetal (Metaheuristic Algorithms in Java)	http://jmetal.sourceforge.net/index.html
13	Julia Software	http://julialang.org/
14	LINDO	http://www.lindo.com/
15	Machine Learning Open Source Software	http://mloss.org/software/
16	Maple Soft	http://www.maplesoft.com/
17	MATLAB (MathWorks)	http://www.mathworks.com/
18	Mathematica (Wolfram Research)	http://www.wolfram.com/
19	MINITAB	http://www.minitab.com/en-us/
20	ModeFrontier	http://www.esteco.com/home.html
21	MOSEK	http://www.mosek.com/
22	MTC Software	http://www.hyperthermcam.com/en-us/
23	NetLearn (Interactive Demonstrations of Network Concepts)	http://www.ladamic.com/netlearn/
24	Neural Network Software	http://www.alyuda.com/ and http://www.alyuda.com/neural-networks-software.htm
25	NeuroSolutions	http://www.neurosolutions.com/
26	Numerical Algorithms Group (NAG)	http://www.nag.co.uk
27	Optimization Software	http://www.optimalon.com/
28	OR Softwares	http://www.mccormick.northwestern.edu/research/optimization-machine-learning-center/software-downloads/index.html
29	Paradiseo (A Software Framework for Metaheuristics)	http://paradiseo.gforge.inria.fr/
30	PISA (A Platform and Programming Language Independent Interface for Search Algorithms)	http://www.tik.ee.ethz.ch/pisa/
31	Robust Optimization Made Easy (ROME)	http://robustopt.com/
32	The Comprehensive R Archive Network	http://cran.r-project.org/
33	R-Studio	http://www.rstudio.com/
34	R Software	http://www.r-project.org
35	Reactive System, Inc.	http://www.reactive-systems.com/
36	Reflector CAD	http://www.breault.com/software/reflectorCAD.php
37	SAS	http://www.sas.com/
38	Scientific Software International, Inc.	http://www.ssicentral.com/
39	Scilab	http://www.scilab.org/

Continued

S. No.	Details of Softwares	URL
40	Simulation	http://www.simulations-plus.com/
41	Spotfire Analytical Tool	http://spotfire.tibco.com/discover-spotfire
42	SPSS Data Collection Data Model	http://www-01.ibm.com/software/analytics/spss/products/data-collection/data-model/
43	SPSS Modeler	http://www-01.ibm.com/ software/analytics/spss/products/modeler/
44	STATA Data Analysis and Statistical Software	http://www.stata.com/
45	Stuttgart Neural Network Simulator	http://www.ra.cs.uni-tuebingen.de/SNNS/
46	Statistica	http://www.statsoft.com/#
47	Systat	http://www.systat.com/

Few Relevant Societies/Department/Schools/Institutes, etc.

S. No.	Details of Societies/Department/ Schools/Institutes, etc.	URL
1	American Statistical Association	http://www.amstat.org/
2	The Royal Statistical Society	http://www.rss.org.uk/
3	Royal Economic Society	http://www.res.org.uk/
4	Institute of Operations Research and Management Science (INFORMS)	http://www.informs.org/
5	Society for Industrial and Applied Mathematics (SIAM)	http://www.siam.org/
6	Mathematical Optimization Society	http://www.mathopt.org/
7	The Association of European Operational Research Societies (EURO)	http://www.euro-online.org/web/pages/ 1/home
8	Canadian Operations Research Society	http://www.cors.ca/
9	French Society for Operations Research and Decisions	http://www.roadef.org/content/index.htm
10	The Operational Research Society	http://www.theorsociety.com/
11	Operational Research Society, Turkey	http://www.yad.org.tr/
12	Operations Research Society of China	http://www.orsc.org.cn/ and http://www.orsc.org.cn/engindex.html
13	The Operations Research Society of Japan	http://www.orsj.or.jp/
14	Operational Research Society of India	http://www.orsi.in/
15	Brazilian Society of Operations Research	http://www.sobrapo.org.br/
16	Stochastic Programming Community Home Page	http://www.stoprog.org/
17	International Society on Multiple Criteria Decision Making	http://www.mcdmsociety.org/
18	Society for Judgment and Decision Making	http://www.sjdm.org/
19	International Institute for Applied Systems Analysis	http://www.iiasa.ac.at/
20	The European Association for Decision Making	http://www.eadm.eu/
21	ESIGMA (European Summer Institute Group on Multicriteria Analysis): European Working Group on Multiple Criteria Decision Aiding	http://www.cs.put.poznan.pl/ewgmcda/
22	European Mathematical Information Service	http://www.emis.de/
23	Algorithms, Combinatorics, and Optimization	http://www.aco.gatech.edu/

Continued

S. No.	Details of Societies/Department/ Schools/Institutes, etc.	URL
24	University of Waterloo (Combinatorics & Optimization)	http://www.math.uwaterloo.ca/co/
25	Mathematical Science Research Institute	http://www.msri.org/web/msri
26	Institute of Quantum Information	http://www.iqi.caltech.edu/
27	Perimeter Institute of Theoretical Physics	http://www.perimeterinstitute.ca/
28	Centre for Discrete Mathematics & Theoretical Computer Science	http://www.dimacs.rutgers.edu/
29	Canadian Institute for Advanced Research (CIFAR)	http://www.cifar.ca/
30	QuantumWorks	http://www.quantumworks.ca/section/view
31	Center for the Mathematics of Information (CMI)	http://www.cmi.caltech.edu/index.shtml
32	Optimization Online	http://www.optimization-online.org/
33	The Fields Institute for Research in Mathematical Sciences	http://www.fields.utoronto.ca/
34	American Mathematical Society	http://www.ams.org/home/page
35	Canadian Mathematical Society	http://cms.math.ca/
36	Good e-books	http://sites.stat.psu.edu/~zuz13/resources.html
37	European Centre for Advanced Research in Economics and Statistics	http://www.ecares.org/
38	Cornell University Library	http://arxiv.org/
39	DEA Data Repository	http://www.etm.pdx.edu/dea/dataset/default.htm
40	Virtual Library For Economics and Business Studies	http://www.econbiz.de/en/
41	Data Envelopment Analysis: Applications for Measuring Efficiency	http://www.dea-analysis.com/
42	Decision Sciences Institute	http://www.decisionsciences.org/
43	DSpace MIT, USA	http://dspace.mit.edu/
44	Mathematical Programming Glossary	http://glossary.computing.society.informs.org/
45	Indian Statistical Institute	http://www.isical.ac.in/
46	Department of Statistics, Stanford University, USA	https://statistics.stanford.edu/
47	Department of Statistical Sciences (DSS), Cornell University, USA	http://stat.cornell.edu/
48	Department of Statistics, Harvard University, USA	http://statistics.fas.harvard.edu/
49	Department of Statistics, LSE, UK	http://www.lse.ac.uk/statistics/home.aspx
50	Department of Statistical Sciences, Duke University, USA	https://stat.duke.edu/
51	Department of Statistics, Oxford University, UK	https://www.stats.ox.ac.uk/
52	Department of Statistics, University of Washington, USA	https://www.stat.washington.edu/
53	Department of Statistics, University of California Berkeley, USA	http://statistics.berkeley.edu/
54	Department of Statistics, Columbia University, USA	http://stat.columbia.edu/
55	Department of Statistics, North Carolina State University, USA	http://www.stat.ncsu.edu/
56	Statistics Department, Wharton, University of Pennsylvania, USA	https://statistics.wharton.upenn.edu/

Continued

S. No.	Details of Societies/Department/ Schools/Institutes, etc.	URL
57	Department of Statistics, Yale University, USA	http://statistics.yale.edu/
58	Department of Statistics, University of Michigan Ann Arbor, USA	http://lsa.umich.edu/stats/
59	Department of Statistics, University of California Los Angeles, USA	http://statistics.ucla.edu/
60	Department of Statistics, Carnegie Mellon University, USA	http://www.stat.cmu.edu/
61	Department of Statistics, University of Wisconsin Madison, USA	https://www.stat.wisc.edu/
62	Department of Statistics, University of Florida, USA	http://www.stat.ufl.edu/
63	Department of Mathematics and Statistics, Indian Institute of Technology Kanpur, India	http://www.iitk.ac.in/math/
64	Department of Mathematics, Indian Institute of Technology Bombay, India	http://www.math.iitb.ac.in/
65	Department of Mathematics, Indian Institute of Technology Kharagpur, India	http://www.iitkgp.ac.in/academics/?page=acadunits&&dept=MM
66	Department of Mathematics, Indian Institute of Technology Madras, India	https://mat.iitm.ac.in/
67	Department of Statistics, Pune University, India	http://stats.unipune.ernet.in/
68	Faculty of Mathematics, Informatics and Mechanics, University of Warsaw, Poland	http://www.mimuw.edu.pl/
69	Department of Statistics and Applied Probability, National University of Singapore, Singapore	http://www.stat.nus.edu.sg/opencms/
70	Department of Mathematics, Hong Kong University of Science and Technology, China	http://www.math.ust.hk/welcome.php
71	List of Department in Statistics and Mathematics, Academia Sinica, Taiwan	http://www.math.sinica.edu.tw/addbook/default_e.jsp
72	Department of Mathematics, ETH Zurich, Switzerland	https://www.math.ethz.ch/
73	Institute for Operations Research, Department of Mathematics ETH Zurich, Switzerland	http://www.ifor.math.ethz.ch/
74	Graduate School of Mathematical Sciences, University of Tokyo, Japan	http://www.ms.u-tokyo.ac.jp/
75	Department of Mathematical Sciences, Tsinghua University, China	http://www.tsinghua.edu.cn/publish/mathen/ 2780/
76	Department of Mathematics, Katholieke Universiteit Leuven, Belgium	https://wis.kuleuven.be/english
77	Doctoral Program in Mathematics, École Normale Superieure, Paris (ENS Paris), France	http://www.math.u-psud.fr/~ecdoct/ecdoct/index.php?l=ANG
78	Department of Pure Mathematics, École Polytechnique (ParisTech), France	https://www.polytechnique.edu/en/department-of-pure-mathematics
79	Master de Sciences et Technologies, Université Pierre et Marie Curie (UPMC), France	http://www.master.ufrmath.upmc.fr/
80	Department of Statistics, Purdue University, USA	http://www.stat.purdue.edu/
81	Department of Statistics and Operations Research, University of North Carolina, USA	http://stat-or.unc.edu/

Continued

S. No.	Details of Societies/Department/ Schools/Institutes, etc.	URL
82	Department of Statistics, University of Chicago, USA	https://galton.uchicago.edu/
83	Department of Probability and Statistics, Peking University, China	http://www.stat.pku.edu.cn/en/
84	Institute of Mathematics and Statistics, University of Sao Paulo, Brazil	https://www.ime.usp.br/en
85	Leuven Statistics Research Centre, UCL Leuven	https://lstat.kuleuven.be/
86	Department of Statistics, Seoul National University, South Korea	http://gsis.snu.ac.kr/career/statistics
87	Institute of Operations Research and Statistics, National Tsing Hua University, China	http://www.ie.tsinghua.edu.cn/eng/content.php?pid=171&ty=173#
88	Department of Statistics and Operations Research, University of Vienna, Switzerland	https://isor.univie.ac.at/
89	Institute of Stochastics, University of Ulm, Germany	http://www.uni-ulm.de/index.php?id=3835&L=1
90	Department of Statistics, Ludwig-Maximilians-University Munich, Germany	http://www.statistik.lmu.de/index_e.html
91	Department of Mathematics, Humboldt-Universität zu Berlin, Germany	https://www.mathematik.hu-berlin.de/en
92	Department of Mathematics, Karlsruhe Institute of Technology, Germany	http://www.math.kit.edu/en
93	Department of Mathematics, RWTH Aachen University, Germany	http://www.mathematik.rwth-aachen.de/cms/~mxy/Mathematik/?lidx=1
94	Department of Mathematics, Technische Universität Dresden, Germany	http://tu-dresden.de/die_tu_dresden fakultaeten/fakultaet_mathematik_und_ naturwissenschaften/fachrichtung_mathematik
95	National Institute of Statistical Sciences	http://www.niss.org/
96	Statistical and Applied Mathematical Sciences Institute	http://www.samsi.info
97	List of Statistics Departments	http://www.stat.ufl.edu/vlib/statistics.html

Few Relevant Publication House/Publishers/Book and Journal Providers

S. No.	Details of Publication House/ Publishers/Book and Journal Providers	URL
1	ACM Publications	http://www.acm.org/publications
2	Addison-Wesley	http://www.pearsoned.co.uk/imprints/addison-wesley/
3	American Mathematical Society (AMS)	http://www.ams.org/home/page
4	Association of American University Presses (AAUP)	http://www.aaupnet.org/
5	Baltzer Science Publishers	http://www.baltzersciencepublishers.com/en/
6	Birkhäuser Basel	http://www.springer.com/birkhauser?SGWID=0-40290-0-0-0
7	Blackwell	http://as.wiley.com/WileyCDA/Section/ index.html

Continued

S. No.	Details of Publication House/ Publishers/Book and Journal Providers	URL
8	Cambridge University Press	http://www.cambridge.org
9	Centre de Recherches Mathématiques (University of Montréal, PQ)	http://www.crm.umontreal.ca/en/index.shtml
10	CRC Press	http://www.crcpress.com/
11	Dover Publications, Inc.	http://store.doverpublications.com/
12	Duke University Press	https://www.dukeupress.edu/
13	Elsevier	https://www.elsevier.com/
14	Gale Research, Cengage	http://www.cengage.com/search/showresults.do?N=197
15	Hindawi Publishing Corporation	http://www.hindawi.com/
16	IEEE	https://www.ieee.org/index.html
17	Institute of Mathematical Statistics Publications	http://imstat.org/publications/
18	Indian Institute of Science (IISc) Press	http://www.iiscpress.iisc.in/
19	Indiana University Press	http://www.iupress.indiana.edu/
20	JSTOR	http://www.jstor.org/
21	Mathematical Association of America	http://www.maa.org/
22	McGraw-Hill	http://www.mheducation.com/
23	Marcel Dekker, Inc.	http://www.dekker.com
24	MIT Press	https://mitpress.mit.edu/
25	Oxford University Press	http://global.oup.com
26	Pearson	http://www.pearsoned.co.uk/
27	Prentice-Hall	http://www.prenticehall.com/
28	Princeton University Press	http://press.princeton.edu/
29	Scientific & Academic Publishing	http://www.sapub.org/journal/index.aspx
30	SIAM	http://www.siam.org/
31	Springer	http://www.springer.com/gp/
32	Taylor & Francis	http://www.taylorandfrancis.com/
33	University of Chicago Press	http://www.press.uchicago.edu/index.html
34	Walter de Gruyter, Inc.	http://www.degruyter.com/
35	Wiley	http://www.wiley.com
36	Wolters Kluwer Group	http://wolterskluwer.com/

List of Few Relevant Journals

S. No.	Details of Relevant List of Few Journals	URL
1	*American Review of Mathematics and Statistics*	http://armsnet.info/
2	*Applied Econometrics and International Development*	http://www.usc.es/economet/eaa.htm
3	*The Annals of Probability*	http://www.imstat.org/aop/
4	*The Annals of Applied Probability*	http://www.imstat.org/aap/
5	*The Annals of Statistics*	http://www.imstat.org/aos/
6	*The Annals of Applied Statistics*	http://imstat.org/aoas/

Continued

S. No.	Details of Relevant List of Few Journals	URL
7	*Applied Stochastic Models in Business and Industry*	http://onlinelibrary.wiley.com/journal/10.1002/%28ISSN%291526-4025
8	*Australian & New Zealand Journal of Statistics*	http://onlinelibrary.wiley.com/journal/10.1111/%28ISSN%291467-842X
9	*Bayesian Analysis*	http://projecteuclid.org/euclid.ba
10	*Biometrika*	http://biomet.oxfordjournals.org/
11	*Biometrical Journal*	http://onlinelibrary.wiley.com/journal/10.1002/%28ISSN%291521-4036
12	*Biometrics*	http://onlinelibrary.wiley.com/journal/10.1111/%28ISSN%291541-0420
13	*Biostatistics*	http://biostatistics.oxfordjournals.org/
14	*Canadian Journal of Statistics*	http://onlinelibrary.wiley.com/journal/10.1002/%28ISSN%291708-945X
15	*Communications in Statistics—Simulation and Computation*	http://www.tandfonline.com/loi/lssp20#. VipG_iv2Qnk
16	*Communications in Statistics—Theory and Methods*	http://www.tandfonline.com/loi/lsta20#. VipG2yv2Qnk
17	*Computational Statistics*	http://link.springer.com/journal/180
18	*Computational Statistics and Data Analysis*	http://www.journals.elsevier.com/computational-statistics-and-data-analysis/
19	*Econometrica*	http://onlinelibrary.wiley.com/journal/10.1111/%28ISSN%291468-0262
20	*Econometrics*	http://onlinelibrary.wiley.com/journal/10.1111/%28ISSN%291368-423X
21	*Econometric Reviews*	http://www.tandfonline.com/action/journalInformation?journalCode=lecr20#. Vi3_YyusOnk
22	*Environmetrics*	http://onlinelibrary.wiley.com/journal/10.1002/%28ISSN%291099-095X
23	*International Journal of Forecasting*	http://www.journals.elsevier.com/international-journal-of-forecasting/
24	*International Statistical Review*	http://onlinelibrary.wiley.com/journal/10.1111/%28ISSN%291751-5823
25	*Journal of Agricultural, Biological, and Environmental Statistics*	http://link.springer.com/journal/13253
26	*Journal of the American Statistical Association*	http://www.tandfonline.com/loi/uasa20#. Vii_Qiv2Qnk
27	*Journal of Applied Econometrics*	http://onlinelibrary.wiley.com/journal/10.1002/%28ISSN%291099-1255
28	*Journal of Applied Statistics*	http://www.tandfonline.com/action/journalInformation?journalCode=cjas20#. Vi4CQCusOnk
29	*Journal of Business & Economic Statistics*	http://amstat.tandfonline.com/action/journalInformation?journalCode=ubes20#.Vi4CbCusOnk
30	*Journal of Chemometrics*	http://onlinelibrary.wiley.com/journal/10.1002/%28ISSN%291099-128X/issues
31	*Journal of Computational and Graphical Statistics*	http://amstat.tandfonline.com/action/journalInformation?journalCode=ucgs20#.Vi4CoyusOnk
32	*Journal of Econometrics*	http://www.journals.elsevier.com/journal-of-econometrics/

Continued

S. No.	Details of Relevant List of Few Journals	URL
33	*Journal of Economic and Social Measurement*	http://www.iospress.nl/journal/journal-of-economic-and-social-measurement/
34	*Journal of Environmental Statistics*	http://www.jenvstat.org/
35	*Journal of Japanese Society of Computational Statistics*	http://jscs.jp/oubun/
36	*Journal of the Japanese Statistical Association*	http://www.jss.gr.jp/en/journal/index.html
37	*Journal of Machine Learning Research*	http://www.jmlr.org/
38	*Journal of Modern Applied Statistical Methods*	http://www.jmasm.com/
39	*Journal of Multivariate Analysis*	http://www.journals.elsevier.com/journal-of-multivariate-analysis
40	*Journal of the Royal Statistical Society—Series A: Statistics in Society*	http://onlinelibrary.wiley.com/journal/10.1111/%28ISSN%291467-985X/
41	*Journal of the Royal Statistical Society—Series B: Statistical Methodology*	http://onlinelibrary.wiley.com/journal/10.1111/%28ISSN%291467-9868/
42	*Journal of the Royal Statistical Society—Series C: Applied Statistics*	http://onlinelibrary.wiley.com/journal/10.1111/%28ISSN%291467-9876
43	*Journal of Statistical Computation and Simulation*	http://www.tandfonline.com/action/journalInformation?journalCode=gscs20#.Vi4EICus0nk
44	*Journal of Statistics Education*	https://www.amstat.org/publications/jse/
45	*Journal of Statistical Physics*	http://www.springer.com/physics/complexity/ journal/10955
46	*Journal of Statistical Software*	http://www.jstatsoft.org/index
47	*Journal of Time Series Analysis*	http://onlinelibrary.wiley.com/journal/10.1111/%28ISSN%291467-9892
48	*Pharmaceutical Statistics*	http://onlinelibrary.wiley.com/journal/10.1002/%28ISSN%291539-1612
49	*Physica A: Statistical Mechanics and Its Applications*	http://www.journals.elsevier.com/physica-a-statistical-mechanics-and-its-applications/
50	*Psychometrika*	http://www.springer.com/psychology/journal/11336
51	*Quality and Reliability Engineering International*	http://onlinelibrary.wiley.com/journal/10.1002/%28ISSN%291099-1638/
52	*Sankhyā A: The Indian Journal of Statistics*	http://www.springer.com/statistics/journal/13171
53	*Sankhyā B: The Indian Journal of Statistics*	http://www.springer.com/statistics/journal/13571
54	*Scandinavian Journal of Statistics*	http://onlinelibrary.wiley.com/journal/10.1111/%28ISSN%291467-9469
55	*Statistics and Computing*	http://www.springer.com/statistics/computational+statistics/journal/11222
56	*Statistical Methods in Medical Research*	http://smm.sagepub.com/
57	*Statistics in Biopharmaceutical Research*	http://www.tandfonline.com/action/journalInformation?journalCode=usbr20#. Vi4G4Cus0nk
58	*Statistics in Medicine*	http://onlinelibrary.wiley.com/journal/10.1002/%28ISSN%291097-0258

Continued

S. No.	Details of Relevant List of Few Journals	URL
59	*Statistics and Probability Letters*	http://www.journals.elsevier.com/statistics-and-probability-letters/
60	*Statistical Analysis and Data Mining*	http://onlinelibrary.wiley.com/journal/10.1002/%28ISSN%291932-1872/
61	*Statistical Applications in Genetics and Molecular Biology*	http://www.degruyter.com/view/j/sagmb
62	*Statistical Modelling*	http://smj.sagepub.com/
63	*Statistica Neerlandica*	http://onlinelibrary.wiley.com/journal/10.1111/%28ISSN%291467-9574
64	*Statistica Sinica*	http://www3.stat.sinica.edu.tw/statistica/
65	*Statistical Science*	http://www.imstat.org/sts/
66	*Statistics and Risk Modeling*	http://www.degruyter.com/view/j/strm
67	*Statistics Surveys*	http://imstat.org/ss/
68	*Stochastic Environmental Research and Risk Assessment*	http://www.springer.com/environment/journal/477
69	*Stochastic Processes and Their Applications*	http://www.journals.elsevier.com/stochastic-processes-and-their-applications/
70	*Statistics and Probability Letters*	http://www.journals.elsevier.com/statistics-and-probability-letters/
71	*Structural Equation Modeling: A Multidisciplinary Journal*	http://www.tandfonline.com/toc/hsem20/current
72	*Technometrics*	http://amstat.tandfonline.com/action/journalInformation?show=aimsScope&journalCode=utch20#.Vi4I3yus0nk
73	*The Review of Economics and Statistics*	http://www.mitpressjournals.org/loi/rest

Few Relevant Data Sets

S. No.	Details of Data Sets	URL
1	Amazon Public Data	http://aws.amazon.com/public-data-sets/
2	Bernoulli Society	http://www.bernoulli-society.org/
3	Biologic Specimen and Data Repository Information Coordinating Center	https://biolincc.nhlbi.nih.gov/home/
4	Biostatistics Data, Vandebilt University, USA	http://biostat.mc.vanderbilt.edu/twiki/bin/view/Main/DataSets
5	Chemical Informatics	http://www.cheminformatics.org/datasets/index.shtml
6	Critical Assessment of Microarray Data Analysis	http://www.camda.duke.edu/camda03/datasets/
7	DataCite	http://www.datacite.org/
8	Data Sets	http://www.grappa.univ-lille3.fr/~torre/Recherche/Experiments/Datasets/
9	Data Sets for "The Elements of Statistical Learning"	http://statweb.stanford.edu/~tibs/ElemStatLearn/ data.html
10	Earthquake Data	http://earthquake.usgs.gov/data/

Continued

S. No.	Details of Data Sets	URL
11	European Cities 1M Data Sets	http://image.ntua.gr/iva/datasets/ec1 m/
12	Geo Data Centre, Arizona State University, USA	http://geodacenter.asu.edu/sdata
13	Hubble Space Data	http://hla.stsci.edu/
14	International Statistical Institute	http://isi-web.org/
15	MIT Airline Data Project	http://web.mit.edu/airlinedata/www/Revenue&Related.html
16	NCBI GenBank	http://www.ncbi.nlm.nih.gov/genbank/
17	NHLBI Data Repository and Biospecimen Repository Information Coordination Center	https://biolincc.nhlbi.nih.gov/redirect/
18	National Flight Data Center (NFDC)	https://nfdc.faa.gov/xwiki/bin/view/NFDC/WebHome
19	Network Data	http://www-personal.umich.edu/~mejn/netdata/
20	NOAA Data	http://www.ncdc.noaa.gov/
21	Open Sports Data/API	http://www.openligadb.de/
22	Online Glossary of Research Economics	http://www.econterms.com/
23	Online for Time Series Data	http://datamarket.com/
24	Online Statistical Software for Astronomy and Related Physical Sciences	http://astrostatistics.psu.edu/statcodes/
25	Precipitation Measurement Data	http://pmm.nasa.gov/data-access/google-earth
26	Public Data, University of Utah, USA	http://www.utah.gov/data/
27	Public Government Data Sets	http://catalog.data.gov/dataset
28	Quandl—Intelligent Search for Numerical Data	http://www.quandl.com/
29	Real-Time Space Weather Data Sources	http://space.rice.edu/ISTP/#RT
30	Stat Lib, Statistical Library, Carnegie Mellon University, USA	http://lib.stat.cmu.edu/
31	Statistical Data Set, University of Massachusetts Amherst, USA	http://www.umass.edu/statdata/statdata/
32	Statistical Data Set, University of Vienna, Switzerland	http://www.mat.univie.ac.at/~neum/statdat.html
33	Statistical Reference Data Set	http://www.itl.nist.gov/div898/strd/
34	Statistical Science Web (Data Sets)	http://www.statsci.org/datasets.html
35	Survival Analysis, Including Penalised Likelihood	http://sites.stat.psu.edu/~dhunter/R/html/survival/html/00Index.html
36	The Center for Innovation in Engineering and Science Education Real-Time Data Sites	http://www.k12science.org/materials/resources/realtimedata/
37	The Data and Story Library, Carnegie Mellon University, USA	http://lib.stat.cmu.edu/DASL/
38	The World Wide Web Virtual Library: Statistics	http://www.stat.ufl.edu/vlib/statistics.html
39	U.S. Department of Homeland Security Data	http://www.dhs.gov/topic/data

Continued

S. No.	Details of Data Sets	URL
40	University of Edinburgh School of Informatics Data Sets for Data Mining	http://www.inf.ed.ac.uk/teaching/courses/dme/html/datasets0405.html
41	Weibull.com, Reliability Engineering Resource Website	http://www.weibull.com/
42	U.S. Government Data	https://www.usa.gov/statistics
43	U.S. Web Traffic	https://analytics.usa.gov/
44	Ministry of Statistics and Programme Implementation (GoI)	http://mospi.nic.in/
45	U.K. Government Data	https://www.data.gov.uk/
46	German Government data	https://www.govdata.de/
47	National Institute of Statistics and Economic Studies (France)	http://www.insee.fr/en/
48	National Bureau of Statistics of China	http://chinadatacenter.org/AboutCDC/PartnersContent.aspx?id=23
49	Canadian Government Data	http://open.canada.ca/en
50	Brazilian Government Data	http://www.ibge.gov.br/english/
51	Statistics Bureau of Japan	http://www.stat.go.jp/english/
52	Statistics Sweden	http://www.scb.se/
53	Statistics Korea	http://kostat.go.kr/portal/english/index.action
54	World Bank Data	http://data.worldbank.org/

9

Univariate Time-Series Analysis and Forecasting: Theory and Practice

R. Krishnan

CONTENTS

ABSTRACT In this chapter, the aim is to analyze the basics of a univariate time-series model building, estimation, and forecasting in detail. We start with the nonseasonal time series and introduce the stationary models based on Wold's decomposition, explain the salient features of Box–Jenkins methodology to a time-series model identification, and the estimation of such models using the exact maximum likelihood procedure. The practical aspects in estimating such models are also discussed with the help of examples. The section on forecasting discusses the principles and the practical aspects of forecasting a time series, using symmetric and asymmetric loss functions, besides introducing the idea of forecast combination. Relaxing the assumption of stationarity, we explain the differences between the stationary and nonstationary processes and also outline the various econometric tests of nonstationarity. Next, we show how the frequency-domain techniques or spectral-analytic tools can be used to determine the periodicity of the cyclical component of a time series. An associated section explains the properties of filtering in general and highlights the distortion that such filtering may induce. The chapter concludes with a short note on the seasonality and the issues involved in testing for the deterministic and stochastic seasonality. We also include a brief discussion on the testing for unit roots at seasonal frequencies. A number of examples will highlight the important practical issues involved.

9.1 Preface

The data, used in applied research, can be broadly classified into cross-section, time-series, and panel data. A fundamental assumption, which the majority of the statistical techniques use to analyze such data, is that they have been *independently* generated. The resulting data sample, z_i, where $i = 1, \ldots, N$, is then taken to be *representative* of some population. The statistical analysis that follows is largely concerned with making inferences about the properties of the population from the sample. In such an analysis, the *order* in which the sample is presented is irrelevant, because of the independence assumption. However, in a real application, researchers often find that the data exhibit a lot of dependency, which is typical of a time-series data—that are data normally collected at prespecified fixed intervals of time—and this dependency was generally considered to be a nuisance. With such a dependency, one cannot ignore the order in which the data are presented. However, with the development of a body of techniques called the *time-series analysis*, such a dependency has been used to derive models that explain the main forces with a maximum simplicity and a minimum number of parameters, which drive the underlying system. More importantly, researchers using such time-series techniques provided much-improved forecasts of the underlying series, compared with the future values obtained through conventional methods.

9.2 Linear Filter

Time series can be collected either by sampling a continuous process at regular intervals or accumulating data over a period of time. Examples of such time series of the former can be temperature recorded every hour over several months and the latter is the gross domestic product (GDP) of an economy, etc. Such time series consists of many components, such as the trend, cyclical, seasonal, and the random components. Given the inherent possibility of interaction among such components, filters are often used to isolate the components of interest to the analyst. Filters are used in a variety of fields such as economics, finance engineering, astronomy, physics, etc. In communications engineering, filters are used to remove the unwanted signals or noise; in economics, they are often used to isolate trends and the seasonality. The necessity to filter the data does not really come either from the statistical necessity imposed by the estimation, testing, or forecasting of time-series models (although admittedly, stationary models are more tractable). Rather, it reflects the preference of applied researchers to focus on certain features of the data-generating mechanism that occur with a specific regularity.*

* Traditionally, economists use some popular filters such as the Beveridge–Nelson decomposition (Beveridge and Nelson 1981) or the Hodrick–Prescott filter (Hodrick and Prescott 1997) to isolate features such as the trend and the cycle. Filters

Time-series models that we encounter in practice often take the form of a filter, presented by Yule and popularized by Box–Jenkins (BJ) (1976). For example, if the time series z_t is defined as

$$z_t = \sum_{i=1}^{\infty} \psi_i e_{t-i} + e_t, \quad t = 1, 2, \ldots,$$

then, we say that z is a filtered version of e with ψ_i's as the weights or filter coefficients. The kind of filter that we are interested in results in a type of models called the autoregressive moving average (ARMA) models.

9.3 Types of Stochastic Models

The types of stochastic models that we are going to analyze can be classified as stationary and nonstationary models. Let us start with stationary models, and defer a discussion on the nonstationary models to a later section.

Stationary models can be either strictly stationary models or weakly stationary. A stochastic process is a *strictly stationary* process if

1. The probability distribution associated with any set of observations is unaffected by shifting the time origin either forward or backward.
2. That is, the probability distribution associated with m observations $z_1, z_2, \ldots, z_m$, made at any time points $1, 2, \ldots, m$, is the same as that of the m observations $z_{1+k}, z_{2+k}, \ldots, z_{m+k}$, made at times $1 + k, 2 + k, \ldots, m + k$.
3. The joint distribution of any two finite sets of random variables depends only on the distance between the time period, and not on time itself. Naturally, this does not mean that two realizations will look identical.

Intuitively, this assumption implies that a plot of a realization of the time series, over two equidistant time intervals, should exhibit similar statistical characteristics. Thus, for example, if $m = 2$, then, z_t and z_{t+k} have the same joint distribution and hence the same covariance for all k. Thus, a strictly stationary process with finite second moments is stationary. But the converse of the above statement is not true. Note, however, that the indices are not necessarily consecutive. If a time series is strictly stationary, then, we see that the distribution function of the random variable is the same at every point.

But this has its disadvantages. It is too difficult to test such a property and in most applications, it will be too ambitious even to claim that we know the form of the distribution function completely. However, a great deal can be accomplished, by dealing with only the first two moments of the time series. In line with this approach, we define a time series to be *weakly stationary* if

1. The expected values of z_t are the same for all t
2. The variance of $z_t = \sigma_z^2 < \infty$

in the frequency domain called the spectral filters are also used to remove powers from seasonal and any other frequencies that contribute more to explaining the variance, with a particular view to studying the contribution of other frequencies. In economics, it is normally used to remove the features such as seasonal and other frequencies other than the business cycle frequencies.

3. The covariance matrices of $(z_{t_1}, z_{t_2}, \ldots, z_{t_m})$ and that of $(z_{t_1+k}, z_{t_2+k}, \ldots, z_{t_m+k})$ are the same for all finite sets of indices and for all k
4. The covariance between any two time points will depend only on the time differences and not on time itself—that is, if we denote the covariance as γ, then, $\text{cov}(z_t z_{t+k}) = \gamma_k$

   And it is usually defined as

$$\gamma_k = \text{Cov}\{z_t z_{t+k}\} = E\{(z_t - \mu)(z_{t+k} - \mu)\}.$$

   The function γ_k is called the *autocovariance* of z_t. In other words, $\text{cov}(z_1 z_{1+k}) = \text{cov}(z_2 z_{2+k}) = \cdots \text{cov}(z_{t-k} z_t)$ for all k. We shall learn more about its properties as well as that of another useful function called the *autocorrelation* function (ACF) below.

In Figure 9.1, we display some time series that exhibit a stationary behavior.

It is also useful to note that the terms *stationary in the wide sense, covariance stationary, second-order stationary*, and *stationary* are also used to describe a weakly stationary time series. It follows from the definitions that a strict stationarity implies a weak stationarity. However, a covariance-stationary time series may possess higher moments that are not independent of time, and hence may not be strictly stationary. For us, however, the term stationarity will always refer to covariance stationarity, unless it is further qualified.

It is interesting to note that we have not said anything about the distribution of any process that we have considered so far. Now, suppose we assume that the probability distribution associated with any set of time is multivariate *normal,* then, the underlying process is called a *Gaussian* or *normal* process. The immediate implication of this assumption is that the existence of a finite mean and an autocovariance matrix will be sufficient to ensure the stationarity of a Gaussian process.

We shall explain weak stationarity with some examples.

- Let $z_t = A\cos(\theta t) + B\sin(\theta t)$, where A and B are uncorrelated random variables with zero means and unit variances, with $\theta \in [-\pi, \pi]$. The time series is stationary since

$$\begin{aligned}\text{Cov}(z_t, z_{t+k}) &= \text{Cov}(A\cos(\theta t) + B\text{sin}(\theta t), A\cos(\theta(t+k)) + B\text{sin}(\theta(t+k)))\\ &= \cos(\theta t)\cos(\theta(t+k)) + \sin(\theta t)\sin(\theta(t+k))\\ &= \cos(\theta k), \quad \text{which is independent of } t.\end{aligned}$$

- Let

$$z_t = \begin{cases} y_t & \text{if } t \text{ is even} \\ y_t + 1 & \text{if } t \text{ is odd,} \end{cases}$$

  where y_t is a stationary time series. $\text{Cov}(z_t, z_{t+k}) = \gamma_k$, z_t is not stationary because it does not have a constant mean.

9.3.1 Tools for the Analysis of Time-Series Data

Since we shall be exploiting the dependence of data points in analyzing a time series, we use two primary tools, *namely*, autocorrelations (ACs) and partial ACs to quantify such a dependence. We shall first take up the ACs. ACs are derived from autocovariances.

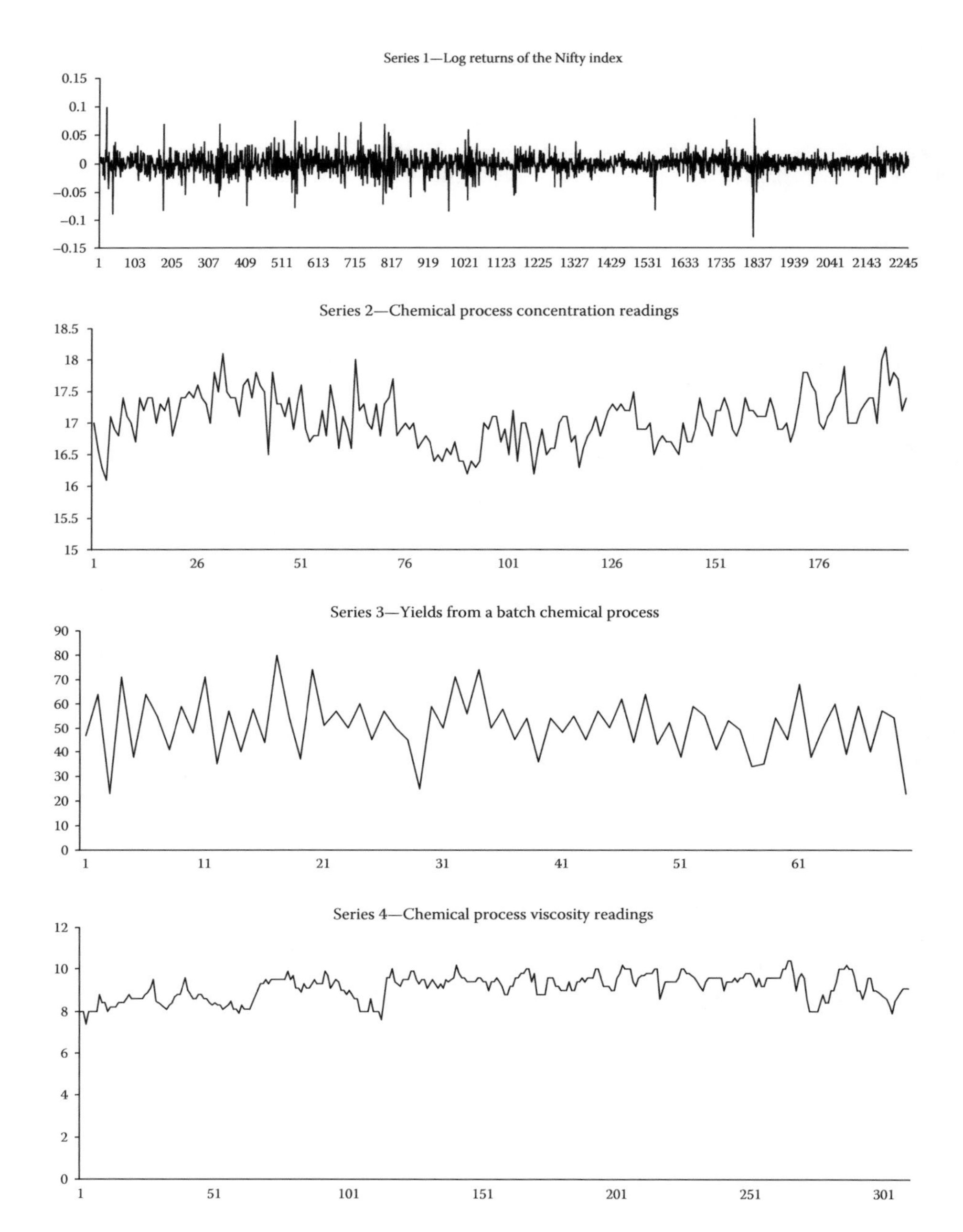

FIGURE 9.1
Some stationary time series. (Data for series 2, 3, and 4 were obtained from http://www.stat.wisc.edu/reinsel/ bjr-data/)

AC, at lag k is given by

$$\rho_k = \frac{E[(z_t - \mu)(z_{t+k} - \mu)]}{\sqrt{E[(z_t - \mu)^2 E(z_{t+k} - \mu)^2]}}$$

$$= \frac{E[(z_t - \mu)(z_{t+k} - \mu)]}{\sigma_z^2},$$

since, for a stationary process, the variance $\sigma_z^2 = \gamma_0$ is the same at time $t + k$ as well as at time t. Thus, the AC at lag k is

$$\rho_k = \frac{\gamma_k}{\gamma_0},$$

which implies that $\rho_0 = 1$.

9.3.2 Properties and Features of Autocovariances and ACs

Since ACs are fundamental to modeling any linear-stationary models, it will be interesting to spell out the properties that AC values have to satisfy, along with the main features.

1. Autocovariances are symmetric, that is, γ_k and γ_{-k} will represent the same magnitude. Notationally, $\gamma_k = \gamma_{-k}$. This can be established by observing from the definition of autocovariance that

$$\gamma_{-k} = \text{Cov}(z_t, z_{t-k}) = \text{Cov}(z_{t+k}, z_t) = \gamma_k.$$

2. An autocovariance matrix—and hence an AC matrix—is positive definite. Consider a linear function of the random variables $(z_1, \ldots, z_n)$, $L_t = l_1 z_1 + \cdots + l_n z_n$. If $\mathbf{a}$ is a $(T \times 1)$ vector, and $\mathbf{\Gamma}_n$ is the covariance matrix, then

$$\text{var}[L_t] = \mathbf{a}'\mathbf{\Gamma}_n\mathbf{a},$$

which is necessarily greater than zero.

3. The positive-definiteness condition imposes certain conditions that the AC values can take. The conditions are that the determinants implied by an AC matrix as well as that of all the principal minors must be greater than zero. So, it is not true that *any* choice of AC values can explain the given model. For instance, if we have three points, then, we should have

$$\begin{vmatrix} 1 & \rho_1 \\ \rho_1 & 1 \end{vmatrix} > 0, \quad \begin{vmatrix} 1 & \rho_2 \\ \rho_2 & 1 \end{vmatrix} > 0, \quad \begin{vmatrix} 1 & \rho_1 & \rho_2 \\ \rho_1 & 1 & \rho_1 \\ \rho_2 & \rho_1 & 1 \end{vmatrix} > 0,$$

which implies that

$$-1 < \rho_1 < 1, \quad -1 < \rho_2 < 1, \quad -1 < \frac{\rho_2 - \rho_1^2}{1 - \rho_1^2} < 1,$$

and so on.

Hence, if $\rho_1 = 0.8$, then, $0.28 < \rho_2 < 1$. So, ρ_2 cannot take any arbitrary value. This explanation shows that though the set of AC coefficients, $(1, \rho_1, \rho_2, \ldots)$, can be obtained for a given time series, there are some admissible conditions that the ACs have to satisfy. It also turns out that the obvious condition that AC values have to be less than one in the absolute value is only a necessary but not a sufficient condition. However, establishing such conditions that an AC matrix should satisfy is virtually impossible even for moderate samples. However, as it will be shown later, these conditions can be brought together under a single, simple definition called the *stationarity* condition.

4. The plot of γ_k against the lag k is called the *autocovariance function* $\{\gamma_k\}$ of the stochastic process. Similarly, a plot of the AC coefficient ρ_k as a function of the lag k is called the *ACF* $\{\rho_k\}$ of the process.

9.3.2.1 Estimation of Autocovariance and ACFs

We need an estimate of the above theoretical autocovariances and ACs. One estimate that is popularly used to find the kth-lag AC is

$$r_k = \frac{c_k}{c_0},$$

where

$$c_k = \frac{1}{T}\sum_{t=1}^{T-k}(z_t - \bar{z})(z_{t+k} - \bar{z}), \quad k = 0, 1, 2, \ldots, K$$

is the estimate of autocovariance γ_k; $\bar{z}$ is the mean of the time series; and K is the number of lags, for which one wishes to calculate ACs. There is no set rule for selecting this K. A rule of thumb that is normally used is that K may not be larger than, say, $T/4$, where T is the sample size.

9.3.2.2 Standard Errors of AC Estimates

Since AC estimates are one of the main tools of trade to identify a model for the given time series, we need to have a crude check on the statistical significance of such ACs. For example, if we want to check if the AC estimates, r_k's, are effectively zero from any hypothesized lag, then, we can use a formula given by Bartlett:

$$\text{var}[r_k] \simeq \frac{1}{T}\left\{1 + 2\sum_{v=1}^{q} r_v^2\right\}, \quad k > q.$$

We shall use the square root of the variance, which is the standard error, to pass conclusions about the significance of the AC. In this formula, k is the lag from which we expect the ACs to be effectively zero.

9.3.2.3 Example

Suppose we have calculated the ACs for a time series of length $T = 200$, which we report in Table 9.1.

1. Let us suppose we want to check that the series is completely random. Then, the series should not exhibit correlation at any lag. In effect, we are hypothesizing that ACs from lag 1 onward are zero. So, we have $q = 0$. Then, for all lags, from the formula, we have

$$\text{var}[r_k] \simeq \frac{1}{200} = 0.005,$$

the square root of which is 0.07. Since the very first value is more than five times the standard error, we can say that the series is not absolutely random. The implication here

TABLE 9.1

Estimated ACs

k	1	2	3	4	5	6	7	8	9	10
r_k	−0.39	0.30	−0.17	0.07	−0.10	−0.05	0.04	−0.04	−0.01	0.01

is, if we suspect that the data were generated by a Gaussian white noise, then, ρ_k for any $k \neq 0$ should lie between $\pm 2/\sqrt{T}$ at about 95% of the time.

2. Suppose we hypothesize that the first two ACs are significantly different from zero, then, the rest of the ACs are zero. Here, we expect ACs from lag 3 onward to be zero. Then, we have from the formula

$$\text{var}[r_k] \simeq \frac{1}{200}[1 + (2 * 0.2421)] = 0.0074, \quad k > 2,$$

which yields a standard error of 0.09. Since the estimated ACs for lags greater than 2 are small compared with twice the standard error, there is no reason to doubt our belief that r_1 and r_2 are not equal to zero and that $r_k = 0, \ (k \geq 3)$.

9.3.2.4 Linear-Stationary Models

A general linear-stationary stochastic model supposes that a time series is generated by a linear aggregation of random shocks. We shall discuss a number of stochastic models such as autoregressive (*AR*), moving average (*MA*), and *ARMA* models. Following the Box, Jenkins, and Reinsel (BJR) (1994) methodology, we show how to identify such models. More details can be found in the latest edition of their book, BJR (2008).

9.3.2.4.1 White Noise

A *white noise* is the basic *building block* for all stochastic processes. What is it?

A white-noise series is always identified by the *statistical properties* that it possesses. It is nothing but a series of random drawings or shocks. For example, $\{e_t\}_{t=-\infty}^{\infty}$ can be considered as a series of *shocks*. Such shocks or random drawings, naturally, are from a fixed distribution and it is these shocks that drive the system. The elements in e_t satisfy some statistical properties, such as

$$E(e_t|e_{t-1}, e_{t-2}, \ldots) = 0.$$
$$E(e_t^2) = \text{var}(e_t|e_{t-1}, e_{t-2}, \ldots) = \sigma^2.$$

And these are uncorrelated over time, meaning that $E(e_t, e_s) = 0, \ t \neq s$. A process satisfying these conditions is called an uncorrelated *white-noise* series. If we impose a stronger condition that the e_t's are independent across time and further assume that $e_t \sim N(0, \sigma^2)$, then, the white-noise series is called an independent Gaussian white-noise process. Note that independence implies uncorrelatedness but the converse is not true.

The models that we are going to see are based on an idea put forward by Yule and developed in detail by Wold. Wold pointed out that a time series in which successive values are highly dependent could be regarded as having been generated from an uncorrelated white-noise process. Such a time series can be generated by transforming the independent white-noise process by a *linear filter*. What is this linear filter?

This simply takes the form of a *weighted sum* of the current and past $e_t's$. That is,

$$\begin{aligned} z_t &= \kappa_t + e_t + \psi_1 e_{t-1} + \psi_2 e_{t-2} + \cdots \\ &= \mu + \psi(B) e_t, \end{aligned}$$

where κ_t is the linearly deterministic component of $z_t, \psi(B)e_t = e_t + \psi_1 e_{t-1} + \psi_2 e_{t-2} + \cdots$, where B is the backward-shift operator, is called the linearly indeterministic component. Generally, κ_t is set equal to μ, which is the mean of the process. Since the Wold representation requires an infinite number of parameters, which is never possible in practice, we express, as an alternative, $\psi(B)$ as the ratio of two finite-order polynomials:

$$\sum_{s=0}^{\infty} \psi_s B^s = \frac{\theta(B)}{\phi(B)} = \frac{(1 + \theta_1 B + \theta_2 B^2 + \cdots + \theta_q B^q)}{(1 + \phi_1 B + \phi_2 B^2 + \cdots + \phi_p B^p)}.$$

On the basis of this idea, BJ developed the popular *ARMA* models, which we shall see in detail later.

Wold representation of the process z_t has an equivalent alternate form that can be written, under suitable conditions, as a weighted sum of its own past deviations, $\tilde{z}_{t-1}, \tilde{z}_{t-2}, \ldots$ plus an added shock. That is,

$$\begin{aligned}\tilde{z}_t &= \pi_1 \tilde{z}_{t-1} + \pi_2 \tilde{z}_{t-2} + \cdots + e_t \\ &= \sum_{s=1}^{\infty} \pi_s \tilde{z}_{t-s} + e_t,\end{aligned}$$

where $\tilde{z} = z_t - \mu$. Since $\psi(B)\pi(B)\tilde{z}_t = \psi(B)e_t = \tilde{z}_t$, we have $\psi(B)\pi(B) = 1$. Hence, the relationship between the two weights can be easily found to be $\pi(B) = \psi(B)^{-1}$.

This only explains that π weights can be obtained given ψ weights and vice versa. We can use these two weights to define the two important concepts, namely, *stationarity* and *invertibility* attached to a general linear process. A linear process is *stationary* iff

$$\sum_{s=0}^{\infty} |\psi_s| < \infty,$$

and is *invertible* iff

$$\sum_{s=0}^{\infty} |\pi_s| < \infty.$$

These conditions are called as the *absolute summability* conditions. We can make a linear stochastic model satisfy these two conditions by simply restricting the parameters of the model. This issue will become clearer when we discuss some finite-order models, which can be inverted to obtain such infinite-order models.

Next, we shall consider some important linear models that are widely used in practice namely

1. AR processes
2. MA processes
3. ARMA processes, also called mixed processes

9.3.2.5 AR Processes

An AR model of order p, $AR(p)$, may be expressed as

$$\tilde{z}_t = \phi_1 \tilde{z}_{t-1} + \phi_2 \tilde{z}_{t-2} + \cdots + \phi_p \tilde{z}_{t-p} + e_t,$$

or

$$(1 - \phi_1 B - \phi_2 B^2 - \cdots - \phi_p B^p)\tilde{z}_t = e_t,$$

or

$$\phi(B)\tilde{z}_t = e_t.$$

9.3.2.6 Stationarity and Invertibility

The given $AR(p)$ model can be rewritten as

$$\tilde{z}_t = \phi^{-1}(B)e_t,$$

but writing out the restrictions on the parameters that satisfy the stationarity conditions is tedious for $p > 2$. An equivalent condition is that the roots of the equation $\phi(B) = 0$—also called the zeros of the polynomial—must lie outside the unit circle. The equation $\phi(B) = 0$ is called the characteristic equation. What about invertibility? Note that since the series $\pi(B) = \phi(B) = 1 - \phi_1 B - \phi_2 B^2 - \cdots - \phi_p B^p$ is finite, we do not need any condition on the parameters of an AR process to ensure invertibility. The implication here is that any finite-order AR model is invertible.

9.3.2.7 ACF of an AR(p) Process

To find the ACF of an $AR(p)$, premultiply the given model by $\tilde{z}_{t-s}$, take expectations, and divide by γ_0. Noting that $E(\tilde{z}_{t-s}e_t)$ will vanish when $s > 0$ and since $\tilde{z}_{t-s}$ can involve shocks only up to time $t - s$, and which are uncorrelated with e_t, we have the ACF

$$\rho_s = \phi_1\rho_{s-1} + \phi_2\rho_{s-2} + \cdots + \phi_p\rho_{s-p}, \quad s > 0.$$

For $s = 0$, we obtain the variance of the process. Since $E(z_t e_t) = \sigma_e^2$ for $s = 0$, we obtain

$$\gamma_0 = \phi_1\gamma_{-1} + \phi_2\gamma_{-2} + \cdots + \phi_p\gamma_{-p} + \sigma_e^2.$$

And hence

$$\sigma_z^2 = \frac{\sigma_e^2}{1 - \rho_1\phi_1 - \rho_2\phi_2 - \cdots - \rho_p\phi_p}.$$

One can consider $AR(1)$ and $AR(2)$ processes as special cases.

9.3.2.7.1 AR(1) Process

The first-order AR process can be written as

$$\begin{aligned}\tilde{z}_t &= \phi_1\tilde{z}_{t-1} + e_t\\ &= (1 - \phi_1 B)^{-1}e_t\\ &= e_t + \phi_1 e_{t-1} + \phi_1^2 e_{t-2} + \cdots.\end{aligned}$$

For stationarity, we have to impose conditions on the ψ weights or equivalently on the generating function $\psi(B)$. Here,

$$\psi(B) = (1 - \phi_1 B)^{-1} = \sum_{s=0}^{\infty}\phi_1^s B^s.$$

Thus, for stationarity, $\psi(B)$ must converge for $|B| \leq 1$. This means that for a given process, the parameter $|\phi_1| < 1$. In terms of the characteristic equation $(1 - \phi_1 B) = 0$, the only root lies outside the unit circle, since $B = \phi_1^{-1}$.

The ACF of an $AR(1)$ process can be easily derived from the *ACF* of an $AR(p)$ process. It simply becomes

$$\rho_s = \phi_1 \rho_{s-1}, \quad s > 0,$$

with a solution, given the initial condition that $\rho_0 = 1$,

$$\rho_s = \phi_1^s, \quad s \geq 0.$$

This says that the ACF of an $AR(1)$ process will exponentially decay to zero, when ϕ_1 is positive, but will oscillate if ϕ_1 is negative in sign. The variance of an $AR(1)$ process is also easily derived from the variance formula for a general $AR(p)$ process, as $\sigma_z^2 = \sigma_e^2/(1 - \phi_1^2)$.

Figure 9.2 demonstrates the behavior of some $AR(1)$ processes and their AC structures.

9.3.2.7.2 *AR(2) Process*

An $AR(2)$ process is written as

$$\tilde{z}_t = \phi_1 \tilde{z}_{t-1} + \phi_2 \tilde{z}_{t-2} + e_t.$$

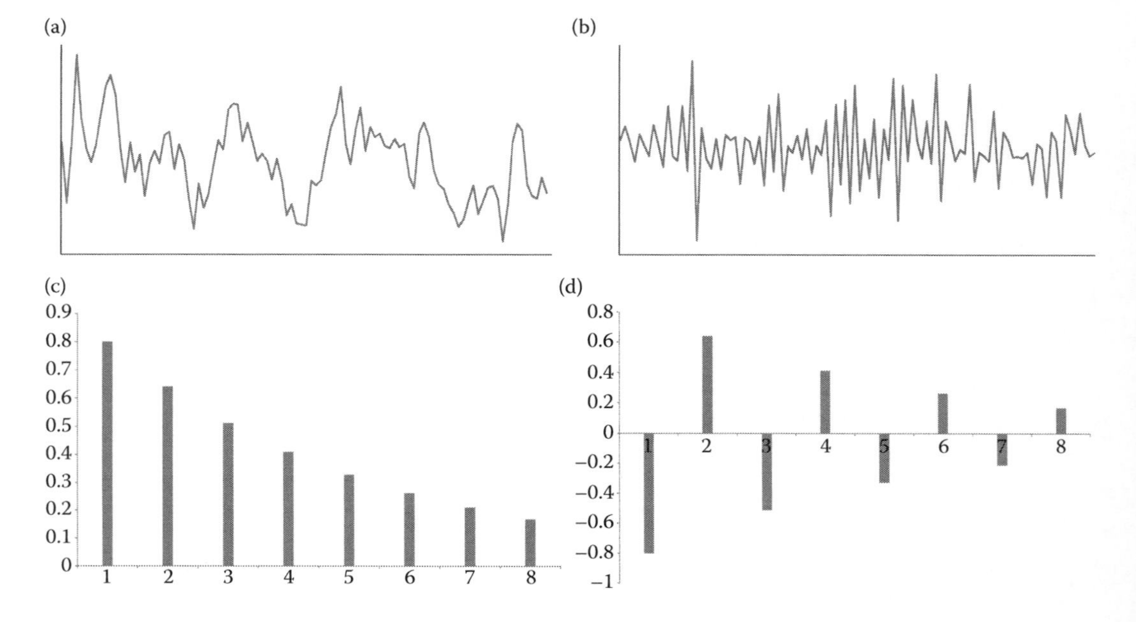

FIGURE 9.2
$AR(1)$ processes and their ACFs. Data generated from A. $\tilde{z}_t = 0.8\tilde{z}_{t-1} + e_t$, B. $\tilde{z}_t = -0.8\tilde{z}_{t-1} + e_t$, C. ACF for $AR(1)$ in A. and D. ACF for $AR(1)$ in B.

For stationarity, the roots of the characteristic equation $(1 - \phi_1 B - \phi_2 B^2) = 0$ should lie outside the unit circle. In terms of the parameters, this implies that

$$\begin{aligned} \phi_1 + \phi_2 &< 1, \\ \phi_2 - \phi_1 &< 1, \\ |\phi_2| &< 1. \end{aligned}$$

Writing again from the *ACF* of the *AR*(p) process, we obtain the *ACF* of an *AR*(2) process as

$$\rho_s = \phi_1 \rho_{s-1} + \phi_2 \rho_{s-2}, \quad s > 0, \quad \text{with } \rho_0 = 1, \quad \text{and} \quad \rho_1 = \phi_1/(1 - \phi_2).$$

The behavior of *ACF* for an *AR*(2) process can be explained as below:

1. If $\phi_1^2 + 4\phi_2 \geq 0$, then, the ACF consists of a mixture of damped exponentials. Depending on whether the roots are positive or negative, the *ACF*s will decay exponentially or oscillate.
2. Since a complex root is possible if the term $\phi_1^2 + 4\phi_2 < 0$, then, *ACF* exhibits a damped sine wave in such a case. The presence of complex roots in the data indicates cycles in the data. In an economic time series, finding such complex-valued roots is common, giving rise to business cycles. For an *AR*(2) model with a pair of complex roots, the average length of the cycle is given by the formula

$$k = \frac{2\pi}{\cos^{-1}\left[\phi_1/(2\sqrt{-\phi_2})\right]},$$

with the cosine inverse in radians. In terms of the complex roots $a \pm bi$, where $i = \sqrt{-1}$, we have $\phi_1 = 2a$, $\phi_2 = -(a^2 + b^2)$, and

$$k = \frac{2\pi}{\cos^{-1}\left(a/\sqrt{a^2 + b^2}\right)}.$$

In Figure 9.3, we display the data generated from *AR*(2) processes and their correlation structures. The data in A are from an *AR*(2) process with complex roots and the ACF plotted in D clearly displays the pseudocyclic behavior. The fundamental period of the ACF is 4.88.

From the formula for the variance of an *AR*(p) process, one can write the variance of the *AR*(2) process as

$$\sigma_z^2 = \frac{\sigma_e^2}{1 - \rho_1 \phi_1 - \rho_2 \phi_2}.$$

Substituting for ρ_1 and ρ_2, we obtain

$$\sigma_z^2 = \frac{(1 - \phi_2)\sigma_e^2}{(1 + \phi_2)\left[(1 - \phi_2)^2 - \phi_1^2\right]}.$$

With this, we shall now move onto another important class of linear-stochastic models, *namely*, the MA models.

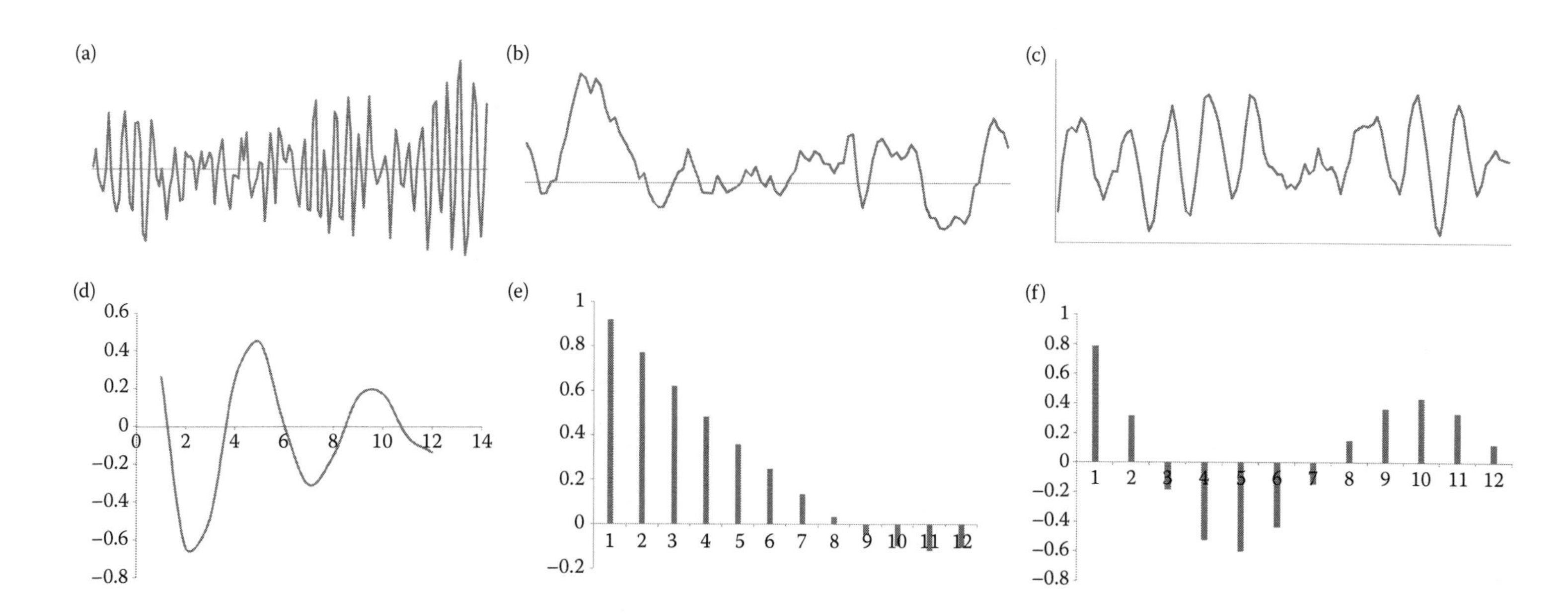

FIGURE 9.3
AR(2) processes and their ACFs. Data generated from A. $\tilde{z}_t = 0.5\tilde{z}_{t-1} - 0.8\tilde{z}_{t-2} + e_t$, B. $\tilde{z}_t = 1.3\tilde{z}_{t-1} - 0.4\tilde{z}_{t-2} + e_t$, C. $\tilde{z}_t = 1.4\tilde{z}_{t-1} - 0.8\tilde{z}_{t-2} + e_t$, D. ACF for $AR(2)$ in A, E. ACF for $AR(2)$ in B, and F. ACF for $AR(2)$ in C.

9.3.2.8 MA Models

MA models arise when the process $\tilde{z}_t$ is expressed in terms of the current and past shocks

$$z_t = \sum_{k=0}^{q} \theta_k e_{t-k},$$

where $\theta_0 \neq 0$, and $\theta_q \neq 0$ is called a *one-sided MA of order q*. Note that the number of terms on the right-hand side of the above equation is $q + 1$. In particular, the process is called a *left-MA*, because nonzero weights are applied only to e_t and to e's with indexes less than t. We shall most frequently use only such one-sided representation. Thus, the above representation for z_t is called an $MA(q)$ model. We can express the model in terms of the backward-shift operator as

$$\begin{aligned}\tilde{z}_t &= (1 + \theta_1 B + \theta_2 B^2 + \cdots + \theta_q B^q)e_t \\ &= \theta(B)e_t,\end{aligned}$$

where $\theta(B) = (1 + \theta_1 B + \theta_2 B^2 + \cdots + \theta_q B^q)$.

An $MA(q)$ model is invertible if the roots of the characteristic equation, $\theta(B) = 0$, lie outside the unit circle. What about stationarity? Since the $\psi(B)$ series, that is, $\theta(B) = 1 + \theta_1 B + \theta_2 B^2 + \cdots + \theta_q B^q$ is finite, no restrictions are needed on the parameters of the MA model to make them stationary.

9.3.2.8.1 ACFs of an MA(q) Process

Multiplying both the sides of

$$\tilde{z}_t = e_t + \theta_1 e_{t-1} + \theta_2 e_{t-2} + \cdots + \theta_q e_{t-q},$$

by $\tilde{z}_{t-s}$ and taking the expectations, the autocovariances of an $MA(q)$ process can be written as

$$\gamma_s = E\left[(e_t + \theta_1 e_{t-1} + \cdots + \theta_q e_{t-q})(e_{t-s} + \theta_1 e_{t-(s+1)} + \cdots + \theta_q e_{t-(s+q)})\right].$$

From this equation, we can see that the variance of the process is

$$\gamma_0 = (1 + \theta_1^2 + \theta_2^2 + \cdots + \theta_q^2)\sigma_e^2,$$

and the autocovariances for the various lags are

$$\gamma_s = \begin{cases} (\theta_s + \theta_1\theta_{s+1} + \cdots + \theta_{q-s}\theta_q)\sigma_e^2 & \text{when } s = 1, \ldots, q \\ 0 & \text{when } s > q. \end{cases}$$

Accordingly, the ACs are defined by

$$\rho_s = \begin{cases} \dfrac{\theta_s + \theta_1\theta_{s+1} + \theta_2\theta_{s+2} + \cdots + \theta_{q-s}\theta_q}{1 + \theta_1^2 + \theta_2^2 + \cdots + \theta_q^2}, & \text{for } s = 1, 2, \ldots, q \\ 0 & \text{for } s > q. \end{cases}$$

Note that the behavior of *ACF* of a qth-order *MA* process is distinctive. Here, unlike the ACF of an *AR* process, the *ACF* of an *MA* process abruptly becomes zero for all s satisfying $|s| > q$. In other words, the *ACF* of an $MA(q)$ process has a cutoff at lag q.

Lower-order *MA* models and the behavior of *ACF* can be easily derived from the discussion of an $MA(q)$ model. For example, for $q = 1$, the model simply is

$$\tilde{z}_t = e_t + \theta_1 e_{t-1} = (1 + \theta_1 B)e_t.$$

For invertibility, $|\theta_1| < 1$. The *ACF* of an $MA(1)$ process can also be found from the *ACF* of an $MA(q)$ process as

$$\theta_1^2 - \frac{\theta_1}{\rho_1} + 1 = 0,$$

with $\rho_s = 0$ for $s \geq 2$, and the variance is given by

$$\gamma_0 = (1 + \theta_1^2)\sigma_e^2.$$

The quadratic equation for ρ_1 has two solutions and we have to take that which gives an invertible solution. For example, if $\rho_1 = 0.4$, then, θ_1 has two solutions, 0.5 and 2.0 and also note that one is the reciprocal of the other. Here, we can take only 0.5, because it corresponds to an invertible process, though the *ACF* remains the same for both the values. In fact, as one can check, for an $MA(1)$ process, the AC values will lie between -0.5 and 0.5. If $\rho_1 = -0.5$, then, $\theta_1 = -1$ and θ_1 will be 1 if $\rho_1 = 0.5$.

This highlights a point that for every invertible *MA* representation, there is a noninvertible representation and vice versa. But only those innovations that are attached to the invertible process are called as *fundamental innovations.* And we get the invertible representations by simply taking the reciprocals of noninvertible factors. Generalizing, suppose we have a noninvertible *MA* process, which we write after factorizing the *MA* operator as follows:

$$\tilde{z}_t = \prod_{i=1}^{q}(1 + \lambda_i B)\,\tilde{e}_t,$$

$$E(\tilde{e}_t\tilde{e}_\tau) = \begin{cases} \tilde{\sigma}_e^2 & \text{for } t = \tau \\ 0 & \text{otherwise,} \end{cases}$$

where we have assumed that $|\lambda_i| < 1$ for $i = 1, 2, \ldots, n$ and $|\lambda_i| > 1$ for $i = n + 1, n + 2, \ldots, q$.

We obtain an invertible representation by simply rewriting the model for $\tilde{z}_t$ as

$$\tilde{z}_t = \left[\left\{\prod_{i=1}^{n}(1 + \lambda_i B)\right\}\left\{\prod_{i=n+1}^{q}\left(1 + \lambda_i^{-1}B\right)\right\}\right]e_t,$$

where

$$E(e_t e_\tau) = \begin{cases} \tilde{\sigma}_e^2\lambda_{n+1}^2\lambda_{n+2}^2\cdots\lambda_q^2 & \text{for } t = \tau \\ 0 & \text{otherwise.} \end{cases}$$

Note that we are not discussing the case where one or more roots are *equal* to one.

The main features of an $MA(2)$ process can be written likewise. The restrictions on the parameters are the same as for an $AR(2)$ process.

9.3.2.9 Mixed AR and MA Process: ARMA Process

ARMA models contain both *AR* and *MA* elements and these are recommended by Box and Jenkins for a parsimonious representation of time-series models, when the process is expected to contain a large number of either *AR* or *MA* elements alone. For example, an $ARMA(p,q)$ model looks like

$$\tilde{z}_t = \phi_1\tilde{z}_{t-1} + \phi_2\tilde{z}_{t-2} + \cdots + \phi_p\tilde{z}_{t-p} + \theta_1 e_{t-1} + \theta_2 e_{t-2} + \cdots + \theta_q e_{t-q} + e_t.$$

Using the backward-shift operator, we can write the model compactly as

$$\phi(B)\tilde{z}_t = \theta(B)e_t,$$

where

$$\phi(B) = 1 - \phi_1 B - \phi_2 B^2 - \cdots - \phi_p B^p, \quad \text{and}$$
$$\theta(B) = 1 + \theta_1 B + \theta_2 B^2 + \cdots + \theta_q B^q.$$

An *ARMA* model has to satisfy both the stationarity and invertibility conditions. The process is stationary if the roots of the characteristic equation $\phi(B) = 0$ lie outside the unit circle and is invertible if the roots of the characteristic equation $\theta(B) = 0$ lie outside the unit circle. Thus, it follows that a stationary *ARMA* process can always be represented by a high-order *MA* model, and if the process obeys the invertibility conditions, then, it can also be represented by a high-order *AR* process.

The ACF of a mixed process is found exactly the same way as before. However, there will be cross-covariance terms. On premultiplying the given *ARMA* model by $\tilde{z}_{t-s}$ and taking expectations, we obtain

$$\begin{aligned}\gamma_s = {} & \phi_1\gamma_{s-1} + \phi_2\gamma_{s-2} + \cdots + \phi_p\gamma_{s-p} + \theta_1\gamma_{ze}(s-1) \\ & + \theta_2\gamma_{ze}(s-2) + \cdots + \theta_q\gamma_{ze}(s-q) + \gamma_{ze}(s).\end{aligned}$$

Here, $\gamma_{ze}(s)$ is the cross-covariance between z and e and is defined by $\gamma_{ze} = E[\tilde{z}_{t-s}e_t]$. Since $\tilde{z}_{t-s}$ depends only on shocks that have occurred up to $t-s$, it follows that $E(\tilde{z}_{t-s}e_t) = \gamma_{ze}(s)$ exists only for $s \le 0$. That is, with $\psi_0 = 1$, we have

$$E(\tilde{z}_{t-s}e_t) = \gamma_{ze}(s) = \begin{cases} 0 & \text{for } s > 0 \\ \psi_{-s}\sigma_e^2 & \text{for } s \le 0. \end{cases}$$

Hence, the expression for γ_s can be rewritten for $s = 0, \ldots, q$ as

$$\gamma_s = \phi_1\gamma_{s-1} + \cdots + \phi_p\gamma_{s-p} + \sigma_e^2(\theta_s\psi_0 + \theta_{s+1}\psi_1 + \cdots + \theta_q\psi_{q-s}),$$

with the convention that $\theta_0 = 1$ and $\psi_0 = 1$.

From these discussions, one can easily write the features of any finite-order $ARMA(p,q)$ model. We shall concentrate on the most-encountered mixed process, the $ARMA(1,1)$ process.

9.3.2.9.1 ARMA(1,1)

From the general $ARMA(p,q)$ model setup, we can easily write an $ARMA(1,1)$ model as follows:

$$(1-\phi_1 B)\tilde{z}_t = (1+\theta_1 B)e_t.$$

First, we note that the process is stationary if $|\phi| < 1$, and invertible, if $|\theta| < 1$. From the formula for the ACF for the general $ARMA(p,q)$ process, we obtain the ACs as

$$\rho_1 = \frac{(1+\phi_1\theta_1)(\phi_1+\theta_1)}{1+\theta_1^2+2\phi_1\theta_1}$$

$$\rho_s = \phi_1\rho_{s-1}, \quad s \geq 2, \quad \text{or}$$

$$\rho_s = \phi_1^{s-1}\rho_1, \quad s \geq 2.$$

Thus, the ACF decays exponentially from the starting value of ρ_1, which depends on θ_1 as well as on ϕ_1. In contrast, the ACF for the $AR(1)$ process decays exponentially from the starting value of $\rho_0 = 1$. The exponential decay for the mixed model will be smooth if ϕ_1 is positive and will alternate if ϕ_1 is negative. Furthermore, the sign of ρ_1 is determined by the sign of $(\phi_1+\theta_1)$ and dictates from which side of zero the exponential decay takes place.

It is interesting to ask the question as to why mixed processes arise in practice. Granger and Newbold (GN) (1986) point out that a mixed model arises when the time series of various types are *aggregated.* The important result here is that even when the component series may be pure *AR* processes, the "aggregated" series will be a mixed process. Suppose that X_t and Y_t are two independent stationary-stochastic processes. If $X_t \sim ARMA(p_1,q_1)$, $Y_t \sim ARMA(p_2,q_2)$, and if $Z_t = X_t + Y_t$, then, $Z_t \sim ARMA(p_3,q_3)$, where $p_3 = p_1 + p_2$ and $q_3 = \max(p_1+q_2, p_2+q_1)$. This can be easily shown.

X_t and Y_t follow *ARMA* processes implying that

$$X_t = \frac{\theta_1(B)}{\phi_1(B)}e_t \quad \text{and} \quad Y_t = \frac{\theta_2(B)}{\phi_2(B)}\eta_t.$$

Hence, if $Z_t = X_t + Y_t$, we can write

$$\begin{aligned} Z_t &= \frac{\theta_1(B)}{\phi_1(B)}e_t + \frac{\theta_2(B)}{\phi_2(B)}\eta_t, \\ &= \frac{\phi_2(B)\theta_1(B)e_t + \phi_1(B)\theta_2(B)\eta_t}{\phi_1(B)\phi_2(B)} \\ \phi_1(B)\phi_2(B)Z_t &= \phi_2(B)\theta_1(B)e_t + \phi_1(B)\theta_2(B)\eta_t. \end{aligned}$$

Here, we have used a result that if X_t and Y_t are two independent *MA* processes of order q_1 and q_2, respectively, then, $Z_t = X_t + Y_t$ will also follow an *MA* process of order $\max(q_1,q_2)$. GN (1986) have highlighted some more interesting features of such an aggregation.

9.3.2.10 Partial ACs

We shall now introduce another tool that helps one to identify the type of model that the underlying process of interest follows, *namely,* the partial ACs, which also exhibit some distinctive features. The idea of partial ACs is based on the same principle of the partial correlations between two random variables. The correlation between two random variables is often due to the fact that both the variables are correlated with the same third variable. If we extend this concept to the analysis of a covariance-stationary time series, $(z_1, z_2, \ldots, z_t)$, then, a large portion of correlation between z_t and z_{t-k} can be due to the correlation that these variables have with the intervening variables, $(z_{t-1}, \ldots, z_{t-k+1})$ and to adjust for these correlations, we have to calculate the *partial ACs*. The partial AC of lag k can be thought of as the partial regression coefficient ϕ_{kk} in the representation

$$z_t = \phi_{11} z_{t-1} + \phi_{22} z_{t-2} + \cdots + \phi_{kk} z_{t-k} + e_t.$$

This representation measures the correlation between z_t and z_{t-k} after adjustments have been made for the intermediate variables $z_{t-1}, z_{t-2}, \ldots, z_{t-k+1}$. Extending this definition, a partial correlation to derive partial ACs of AR processes of various orders follows naturally.

9.3.2.11 Partial ACs of an AR Process

Following our definition of a partial AC, lag 1 partial AC is given by the partial regression coefficient ϕ_{11} in the representation

$$z_t = \phi_{11} z_{t-1} + e_t.$$

This is nothing but an $AR(1)$ model. Similarly, lag 2 partial ACs is the partial regression coefficient in the representation

$$z_t = \phi_{11} z_{t-1} + \phi_{22} z_{t-2} + e_t.$$

Again, this is an $AR(2)$ model.

From these interpretations, it is clear that the partial ACs follow a particular form for AR processes. If the true model is an $AR(1)$ process, then, it is clear that the true value of ϕ_{kk} for $k > 1$ is zero. This is because terms other than ϕ_{11} do not enter the model at all. Similarly, if $p = 2$, fitting a third-order model should make the coefficient of z_{t-3} statistically insignificant from zero. Thus, if the true model is an $AR(p)$ process, then, the true value of ϕ_{kk} is zero, for all $k > p$.

Now, one can discern a distinct pattern in the behavior of partial ACs of an AR process. If the true model is an $AR(1)$ process, then, we should have only one significant partial AC; that is, only ϕ_{11} will be significant and all the other coefficients will be zero. Thus, the partial AC for an $AR(1)$ process has a *cutoff* after lag 1. Similarly, the partial AC of an $AR(2)$ process will have a *cutoff* after lag 2. This fact, coupled with our observations about the structure of the ACs of an AR process, makes it easy to recognize an AR process.

In summary, the $AR(p)$ process is described as

1. An ACF that is infinite in extent and that is a combination of damped exponentials or damped oscillations.
2. Partial ACs that are zero for lags larger than the AR order; that is, the partial AC has a cut off after lag p.

9.3.2.12 Alternative Derivation of the Partial ACs

An alternative way of obtaining partial ACs for any stochastic process is through a set of simultaneous linear equations in ϕ's called the Yule–Walker equations. These are given by

$$\rho_s = \sum_{i=1}^{p} \phi_{ii} \rho_{s-i}, \quad s = 1, \ldots, p.$$

Written out explicitly, for the various s, the linear equations are given by

$$\begin{array}{ccccccccc}
\rho_1 & = & \phi_{11} & + & \phi_{22}\rho_1 & + & \cdots & + & \phi_{pp}\rho_{p-1}, \\
\rho_2 & = & \phi_{11}\rho_1 & + & \phi_{22} & + & \cdots & + & \phi_{pp}\rho_{p-2}, \\
\vdots & & \vdots & \vdots & & & \cdots & \vdots & \vdots \\
\rho_p & = & \phi_{11}\rho_{p-1} & + & \phi_{22}\rho_{p-2} & + & \cdots & + & \phi_{pp}.
\end{array}$$

These may be written in matrix form as

$$\begin{bmatrix} 1 & \rho_1 & \cdots & \rho_{p-1} \\ \rho_1 & 1 & \cdots & \rho_{p-2} \\ \vdots & \vdots & \ddots & \vdots \\ \rho_{p-1} & \rho_{p-2} & \cdots & 1 \end{bmatrix} \begin{bmatrix} \phi_{11} \\ \phi_{22} \\ \vdots \\ \phi_{pp} \end{bmatrix} = \begin{bmatrix} \rho_1 \\ \rho_2 \\ \vdots \\ \rho_p \end{bmatrix},$$

or

$$\mathbf{P}_p \boldsymbol{\phi}_p = \boldsymbol{\rho}_p,$$

or

$$\boldsymbol{\phi}_p = \mathbf{P}_p^{-1} \boldsymbol{\rho}_p.$$

Solving the equations for $p = 1, 2, \ldots$ successively, we obtain

$$\phi_{11} = \rho_1,$$

$$\phi_{22} = \frac{\begin{vmatrix} 1 & \rho_1 \\ \rho_1 & \rho_2 \end{vmatrix}}{\begin{vmatrix} 1 & \rho_1 \\ \rho_1 & 1 \end{vmatrix}} = \frac{\rho_2 - \rho_1^2}{1 - \rho_1^2},$$

$$\phi_{33} = \frac{\begin{vmatrix} 1 & \rho_1 & \rho_1 \\ \rho_1 & 1 & \rho_2 \\ \rho_2 & \rho_1 & \rho_3 \end{vmatrix}}{\begin{vmatrix} 1 & \rho_1 & \rho_2 \\ \rho_1 & 1 & \rho_1 \\ \rho_2 & \rho_1 & 1 \end{vmatrix}}.$$

9.3.2.13 Estimation of the Partial ACF

The partial ACs can be estimated by running the successive AR processes of orders $1, 2, 3, \ldots$ by least squares and picking out the estimates $\hat{\phi}_{11}, \hat{\phi}_{22}, \hat{\phi}_{33}, \ldots$ of the last coefficient fitted at each stage. Alternatively, the approximate Yule–Walker estimates of the successive AR processes may be used. Thus, we can substitute estimates, r_p, of the theoretical ACs, ρ_p, to yield $\hat{\phi}_{pp}$. A plot of $\hat{\phi}_{pp}$ against p gives the *partial correlogram.* For an *AR*(p) process, a feature of a partial correlogram will be that $\hat{\phi}_{pp}$ will be nonzero for any lag less than or equal to p and zero for all lags greater than p. One may use the measure

$$\text{SE}[\hat{\phi}_{pp}] \approx \frac{1}{\sqrt{T}} \quad p \geq p + 1,$$

to check if the partial ACs beyond the hypothesized lag are statistically insignificant.

9.3.2.14 Partial AC of an MA Process

Unlike the *AR* process where the partial ACs derived from the Yule–Walker equations were linear, the Yule–Walker equations of an *MA* process are nonlinear. Except for the simple case of an $MA(1)$ process, we have to solve these equations iteratively to get the partial ACs. Hence, we shall restrict our attention to the $MA(1)$ process.

Here, we recall the ACF for the $MA(1)$ process as

$$\rho_1 = \frac{\theta_1}{1+\theta_1^2}, \quad \rho_s = 0, \quad s \geq 2.$$

Again, we make use of the Yule–Walker equations setup to write

$$\phi_{11} = \rho_1 = \frac{\theta_1}{1+\theta_1^2} = \frac{\theta_1(1-\theta_1^2)}{(1-\theta_1^4)}$$

$$\phi_{22} = \frac{\rho_2 - \rho_1^2}{1-\rho_1^2} = -\frac{\rho_1^2}{1-\rho_1^2} = \frac{-\theta_1^2(1-\theta_1^2)}{(1-\theta_1^6)}$$

$$\phi_{33} = \frac{\rho_1^3}{(1-2\rho_1^2)} = \frac{\theta_1^3(1-\theta_1^2)}{(1-\theta_1^8)},$$

and so on.

In general,

$$\phi_{ss} = \frac{(-1)^{(s-1)} \cdot \theta_1^s(1-\theta_1^2)}{\left(1-\theta_1^{2(s+1)}\right)}, \quad s > 0.$$

Thus, with the increasing lags, the partial ACs ϕ_{ss} are dominated by damped exponentials or a damped sine wave (alternates in sign). If θ_1 is positive, then, the partial ACs alternate in sign and will decay exponentially if θ_1 is negative.

We now note a duality between the $AR(1)$ and $MA(1)$ processes. While the ACF of an $MA(1)$ process has a cutoff after lag 1, the ACF of an $AR(1)$ process tails off exponentially. Conversely, while the partial ACF of an $MA(1)$ process tails off and is dominated by a damped exponential, the partial ACF of an $AR(1)$ process has a cutoff after lag 1.

As mentioned before, the partial ACs of higher-order *MA* processes have to be solved iteratively. For example, suffice it to say that the partial ACs of an $MA(2)$ process will decay exponentially, if the roots of the characteristic equation are real and will exhibit a *pseudocyclic behavior*, if the roots are complex. Thus, it behaves more or less like the ACF of an $AR(2)$ process.

9.3.2.15 Autocovariance-Generating Function

Since *ACF* is an important tool to identify a possible model, another way to obtain the associated autocovariances of a linear process is through the *autocovariance-generating function* defined below:

$$\gamma(B) = \sum_{k=-\infty}^{\infty} \gamma_k B^k.$$

It can be shown that

$$\gamma(B) = \sigma^2\psi(B)\psi(B^{-1}) = \sigma^2\psi(B)\psi(F).$$

In this expansion, the coefficient of B_0, which is γ_0, will give the variance of the process whereas the coefficient, γ_k, is the coefficient of both B^k and B^{-k}.

For example, if z_t follows an $MA(1)$ process, then $\psi(B) = (1 + \theta B)$. From the autocovariance-generating function, we have

$$\begin{aligned}\gamma(B) &= \sigma^2(1 + \theta B)(1 + \theta B^{-1}) \\ &= \sigma^2 \left\{\theta B^{-1} + (1 + \theta^2) + \theta B\right\}.\end{aligned}$$

Compared with the expression $\gamma(B) = \sum_{s=-\infty}^{\infty} \gamma_s B^s$, the autocovariances are $\gamma_0 = (1 + \theta^2)\sigma^2$, $\gamma_1 = \theta\sigma^2$, and $\gamma_s = 0$, where $s \geq 2$. Similarly, for an $AR(1)$ process, $\psi(B) = (1 - \phi B)^{-1}$.

And hence, the autocovariance-generating function becomes

$$\gamma(B) = \sigma^2(1 + \phi B + \phi^2 B^2 + \cdots)(1 + \phi B^{-1} + \phi^2 B^{-2} + \cdots).$$

From this, it is very straightforward to get the autocovariances for the various s. For example, collecting the coefficients of B^0 from this expansion, we get γ_0, which is the variance of the process. The coefficients of B^0 are given by

$$\sigma^2(1 + \phi^2 + \phi^4 + \cdots) = \frac{\sigma^2}{(1 - \phi^2)}.$$

Similarly, for γ_1, we collect the coefficients of B^1 to give

$$\begin{aligned}\gamma_1 &= \sigma^2(\phi + \phi^3 + \phi^5 + \cdots) \\ &= \sigma^2 \cdot \phi(1 + \phi^2 + \phi^4 + \cdots) \\ &= \sigma^2 \frac{\phi}{(1 - \phi^2)}.\end{aligned}$$

We can generalize this for any B^k and write

$$\gamma_k = \sigma^2 \frac{\phi^k}{(1 - \phi^2)}.$$

Now, one can easily check that these expressions of $AR(1)$ are the same as those obtained earlier.

9.3.3 Forecasting: Principles and Practical Aspects

By forecasting, we mean to say something about the likely course of the future values of our sample, based on some information, that can be a formal model using past data or simply a hunch. Many techniques of varying degrees of difficulty exist to forecast the future values—ad hoc methods, surveys, regression techniques, time-series techniques, and Bayesian methods. And, sometimes, forecasts can be judgmental or an expert opinion. Such forecasts can be obtained for a long or a short horizon. Normally, one forecasts one period into the future, which is called the one-step-ahead forecasts, because it has some appealing statistical properties. Forecasts are of a great importance and figure prominently in many questions in macroeconomics and in finance, such as testing the rationality of expectations, efficiency of stock markets, or verifying if government projections are systematically overoptimistic. Such economic and financial forecasts may be point forecasts, interval forecasts, or density forecasts. Understanding the type of forecasts is essential to some of the issues such as constructing, evaluating, comparing, and combining forecasts. Since point forecast is the most-common type of an economic forecast, we cover only point forecasts and the issues related to such forecasts in the discussion below.

9.3.3.1 Principles of Forecasting

Let z_t be some discrete-time covariance stationary-stochastic process. We wish to forecast z_t, s time units ahead, so that attention is directed to the random variable z_{t+s}. Thus, for $s = 1$, a one-step-ahead forecasting situation arises. Any forecasting procedure will have to be based on some *information set*, I_t, consisting of sample values of the series at time t. To obtain any kind of the best value for our forecasts, one requires a criterion against which various alternatives can be judged. One way to proceed is to introduce the idea of a *cost or loss function.* A popular loss function is a *quadratic* function. A quadratic loss function means choosing the forecast $z^*_{t+1|t}$ so as to minimize

$$E(z_{t+1} - z^*_{t+1|t})^2.$$

This expression is known as the *mean-squared error* associated with the forecast $z^*_{t+1|t}$ and is given by

$$MSE(z^*_{t+1|t}) \equiv E(z_{t+1} - z^*_{t+1|t})^2.$$

The forecast with the smallest mean-squared error turns out to be the expectation of z_{t+1} that is conditional upon I_t:

$$z^*_{t+1|t} = E(z_{t+1} \mid I_t).$$

That is, an optimal predictor under a squared or a quadratic loss function is just the conditional mean of z_{t+1} given I_t. If we assume further that the forecast function is a *linear function* of the data available, I_t, then, only *linear forecasts* are made. Thus, the forecast $z^*_{t+1|t}$ is restricted to be a linear function of I_t:

$$z^*_{t+1|t} = \boldsymbol{\alpha}' I_t.$$

Suppose we find such a value for $\boldsymbol{\alpha}$ such that the forecast error $(z_{t+1} - \boldsymbol{\alpha}' I_t)$ is uncorrelated with I_t. That is,

$$E[(z_{t+1} - \boldsymbol{\alpha}' I_t) I_t'] = \mathbf{0}'.$$

If this holds, then, the forecast $\boldsymbol{\alpha}' I_t$ is called the *linear projection* of z_{t+1} on I_t. The linear projection produces the smallest mean-squared error among the class of linear-forecasting rules. If we use $\hat{P}(z_{t+1} \mid I_t) = \boldsymbol{\alpha}' I_t$ to denote such a forecast, then, $MSE[\hat{P}(z_{t+1} \mid I_t) \geq MSE[E(z_{t+1} \mid I_t)]$, because one has to remember that the conditional expectation offers the best-possible forecast among *all* classes of estimators.

9.3.3.2 Forecasting ARMA Models

Let $\tilde{z}_t$ follow the stationary, invertible, $ARMA(p, q)$ process. If we want to forecast $\tilde{z}_t$, s lead times ahead standing at time t, then, we want to find

$$\phi(B)\tilde{z}_{t+s} = \theta(B)e_{t+s}.$$

We denote the s-step-ahead forecast made at time origin t as $\hat{z}_{t+s|t}$ or simply $\hat{z}_t(s)$, and this has the smallest-expected squared errors among the set of all possible forecasts that are linear in $\tilde{z}_{t-s},\ s \geq 0$. A recurrence relation for the forecasts is obtained by replacing each element in the forecast expression according to the following rules:

1. Replace the unknown values $\tilde{z}_{t+s}$ by their forecasts $\hat{z}_t(s)$, for $s > 0$
2. The values of $\tilde{z}_{t+s},\ s \leq 0$, are simply the known past values of $\tilde{z}_t$
3. Since e_t is white noise, the optimal forecast of $e_{t+s},\ s > 0$ is simply zero
4. The values of $e_{t+s},\ s \leq 0$ are just the known values of e_t

9.3.3.3 Forecasting Particular Models

We can use these steps to forecast *ARMA* models. Suppose $\tilde{z}_t$ follows an $MA(1)$ model. Hence, $e_t = (1 + \theta B)^{-1}\tilde{z}_t$. Following the steps above, the optimal one-step-ahead linear forecast of an $MA(1)$ model will simply be

$$\hat{z}_{t+1|t} = \theta \hat{e}_t,$$

where $\hat{e}_t$ can be calculated from the recursion $\hat{e}_t = \tilde{z}_t - \theta\hat{e}_{t-1}$. It is easy to show that the forecast of an $MA(1)$ process for $s = 2, 3, \ldots$ periods is simply the unconditional mean, μ of the process. And, the *MSE* of the forecast or, equivalently, the conditional variance of the forecasts of an $MA(1)$ process, for $s = 2, 3\ldots,$ is simply the unconditional variance of the process, $(1 + \theta^2)\sigma^2$. One can generalize this to an $MA(q)$ process. The optimal linear forecast of an $MA(q)$ is

$$\hat{E}[z_{t+s} \mid e_t, e_{t+1}, \ldots] = \begin{cases} \mu + (\theta_s + \theta_{s+1}B + \cdots + \theta_q B^{q-s})\hat{e}_t & \text{for } s = 1, 2, \ldots, q \\ \mu & \text{for } s = q+1, q+2, \ldots \end{cases}$$

where $\hat{e}_t$ can be calculated from the recursion

$$\hat{e}_t = \tilde{z}_t - \theta_1\hat{e}_{t-1} - \cdots - \theta_q\hat{e}_{t-q}.$$

The *MSE* is

$$\begin{aligned} \sigma_e^2 \quad & \text{for } s = 1 \\ (1 + \theta_1^2 + \theta_2^2 + \cdots + \theta_{s-1}^2)\sigma_e^2 \quad & \text{for } s = 2, 3, \ldots, q \\ (1 + \theta_1^2 + \theta_2^2 + \cdots + \theta_q^2)\sigma_e^2 \quad & \text{for } s = q+1, q+2, \ldots \end{aligned}$$

Thus, for $MA(q)$ processes, forecasts beyond the order of the process are simply the process mean and the variance of the forecasts is the unconditional variance of the process.

Forecasting an *AR* process is quite straightforward. For example, the s-period-ahead forecast of the covariance-stationary $AR(1)$ process, $z_t = \mu + \phi z_{t-1} + e_t$, is

$$\hat{z}_t(s) = \mu + \phi^s(z_t - \mu).$$

As the forecast horizon increases, the forecast converges to the unconditional mean. The *MSE* of the s-period-ahead forecasts is

$$[1 + \phi^2 + \phi^4 + \cdots + \phi^{2(s-1)}]\sigma_e^2 = \sigma_e^2\frac{(1 - \phi^{2s})}{(1 - \phi^2)}.$$

Note that this error grows with s and asymptotically approaches $\sigma_e^2/(1-\phi^2)$, which is the unconditional variance of z.

9.3.3.4 Confidence Intervals for the Forecasts

Our job of forecasting does not end with getting point forecasts for the desired lead periods. Since forecasts are random variables, we have to provide an indication of their likely reliability. And, for this purpose, we shall construct confidence intervals for the forecasts. We shall do this first by calculating the ψ weights. We will need these ψ weights to also update the old forecast, the moment new information is available. The calculation of the ψ weights is accomplished by representing a given $ARMA(p,q)$ process as an infinite MA process $\tilde{z}_t = \psi(B)e_t$ so that the weights are calculated from the expression

$$\phi(B)\psi(B) = \theta(B).$$

That is, the elements of $\psi(B) = 1 + \psi_1(B) + \psi_2 B^2 + \cdots$ can be obtained by equating the coefficients of the same powers of B in the expansion

$$(1 - \phi_1 B - \phi_2 B^2 - \cdots - \phi_p B^p)(1 + \psi_1 B + \psi_2 B^2 + \cdots) = (1 + \theta_1 B + \theta_2 B^2 + \cdots + \theta_q B^q).$$

For example, for the model $(1-B)(1-0.5B)z_t = e_t$, we have

$$(1 - 1.5B + 0.5B^2)(1 + \psi_1 B + \psi_2 B^2 + \cdots) = 1.$$

We can obtain the ψ weights with the help of the recursive relations that are just outlined, and by using $\phi_1 = 1.5$ and $\phi_2 = -0.5$, as follows:

$$\begin{aligned} \psi_0 &= 1, \\ \psi_1 &= 1.5, \\ \psi_j &= 1.5\psi_{j-1} - 0.5\psi_{j-2}, \quad j = 2, 3, 4, \ldots \end{aligned}$$

The confidence limits can be calculated by noting that the expression for the forecast error for any origin is

$$e_t(s) = z_{t+s} - \hat{z}_{t+s} = e_{t+s} + \psi_1 e_{t+s-1} + \cdots + \psi_{s-1} e_{t+1}.$$

From this, we can derive the expression for the variance of the forecast error as

$$E(e_t^2(s)) = \left\{1 + \sum_{j=1}^{s-1} \psi_j^2\right\} \sigma_e^2.$$

We can replace σ_e^2 with an estimate, say, s_e^2, and assuming that the errors of e's are normally distributed, we can derive the confidence intervals for the forecasts. Thus, an approximate 95% confidence interval is given by

$$\hat{z}_{t+s} \pm 1.96 \left\{1 + \sum_{j=1}^{s-1} \psi_j^2\right\}^{1/2} s_e.$$

9.3.3.5 Evaluating Forecasts, Forecast Accuracy, and Combination

Evaluating forecasts for their optimality or accuracy varies depending on how forecasts were obtained. The methods to evaluate forecasts that are explicitly model based differ from those that are obtained from model-free methods. Let us begin by listing out the main issues behind evaluating forecasts and comparing the accuracy of rival forecasts:

1. How does one evaluate a single set of forecasts? Here, we have a record of point forecasts $\hat{z}_t(s)$ and the corresponding realizations or outturns of z_{t+s}. Thus, an investigator's information set contains only a set of forecasts and the actual values of the predictand. We naturally want to evaluate, monitor, and improve this set of forecasts.
2. How do we evaluate and compare the accuracy of the competing set of forecasts? Here, we have forecasts for the same random variable from several models or several sources, and we need a metric to select the best among them.

9.3.3.6 Evaluating a Single Forecast

Forecast evaluation essentially amounts to checking some properties that optimal forecasts and the errors are expected to follow. These properties are the following (see Diebold and Lopez 1996):

1. Optimal forecasts are unbiased
2. Optimal forecasts have one-step-ahead forecast errors that are of white noise
3. Optimal forecasts have s-step-ahead errors that are at most $MA(s-1)$
4. Optimal forecasts have s-step-ahead errors that have variances that are nondecreasing in s but converge to the unconditional variance of the process

Even though tests are available to ascertain that every set of forecasts satisfies each of these key properties (see Diebold and Lopez 1996 for a list of such tests), such single-optimal forecasts are unlikely in practice. In fact, many times, we may have many competing forecasts for the same random variable, all suboptimal in some sense, that are compared and are often combined to achieve better forecasts. Hence, a test of unbiasedness and the efficiency of a single set of forecasts is often presented as minimum requirements for optimal forecasts. Forecasts are *rational* when they are unbiased and efficient. If unbiasedness is not rejected, then, it can be retained as a maintained hypothesis and the next basic property of optimal-forecast errors—from which all the others (including the above four properties) follows—*namely*, the unforecastability on the basis of information available at the time forecasts were made, can be tested. This is the test for a full optimality of forecasts. And, it is also sometimes called the orthogonality tests.

Suppose we have a sequence of pairs of forecasts and outturns, $(z_{t+s|t}, z_{t+s})$, where, for example, $t = 1, 2, \ldots, T$ and s is a fixed integer. An obvious property is that the forecasts are unbiased:

$$E_t(z_{t+s|t} - z_{t+s}) = 0, \quad \text{or} \quad E_t(z_{t+s|t}) = E_t(z_{t+s}).$$

This equation implies that the forecasts are unconditionally unbiased, that is, the forecasts have not been conditioned on any information set. Unbiasedness is tested by checking if the sample mean of the forecast errors, $e_{t+s|t} = z_{t+s} - z_{t+s|t}$, is significantly different from zero.

The two main concepts of rationality that are tested in the literature are a weak and a strong rationality (see Clements 2005). Weak rationality or a consistency of forecasts means that the forecasts are conditionally unbiased and often refers to the property that forecasters are not systematically wrong in making their forecasts. A strong rationality—also called an efficiency—means the forecasts, besides being unbiased, are uncorrelated with any other information or series available at the time the forecasts were made. If so, then, it will be possible to exploit this information and produce better forecasts. There have been many studies that test the rationality of forecasts based on these notions, with the Mincer–Zarnowitz (1969) regression being the most popular. Forecasts can be efficient and unbiased, but still be highly inaccurate. Since unbiasedness can result if large positive and negative errors offset each other, making the mean of the forecast errors close to zero, we will also need to pay attention to the variance of the observed sample of forecast errors about the mean.

We shall first discuss the tests of rationality of the forecasts.

The tests of rationality can be conducted based on the regression equation of the form

$$z_{t+1} = \alpha + \beta z_{t+1|t} + e_{t+1}, \quad t = 1, \ldots, T.$$

We have assumed that $s = 1$, since the forecasts for $s > 1$ have their own set of issues. Moreover, e_{t+1}'s are independent and identically distributed (iid) with zero mean.

If the joint null hypothesis of $\alpha = 0$ and $\beta = 1$ is accepted, then, the forecasts are unbiased. From the above regression equation, we can see that

$$E_t(z_{t+1}) = \alpha + \beta E_t(z_{t+1|t}),$$

with the assumption that $E_t(e_{t+1}) = 0$. However, it has been pointed out that $\alpha = 0$ and $\beta = 1$ is only a necessary but not a sufficient condition. Note that $E_t(z_{t+1|t}) = E_t(z_{t+1})$ can also be satisfied if $\alpha = (1 - \beta)E_t(z_{t+1|t})$. With this condition, unbiasedness is still possible even if $\beta \neq 1$. So, a necessary and sufficient condition for unbiasedness is $\alpha = (1 - \beta)E_t(z_{t+1|t})$. An even more satisfactory inference about unbiasedness can be obtained by testing if the forecast error has a mean of zero. That is, use a t-test for the hypothesis $\gamma = 0$ in

$$\epsilon_{t+1|t} = z_{t+1} - z_{t+1|t} = \gamma + e_{t+1},$$

where e_{t+1} is white noise.

If unbiasedness is not rejected, then, this is typically retained as the maintained hypothesis, and various other tests of the forecast errors being uncorrelated with the past values of the process, past errors of the process, or in fact with any other variables known at time t can be conducted. Tests that check the uncorrelatedness of the errors with other variables are often termed as the orthogonality tests. An example will be $H_0 : \boldsymbol{\gamma} = \mathbf{0}$ in the regression

$$\epsilon_{t+1|t} = \boldsymbol{\gamma}'\mathbf{z}_t + e_{t+1},$$

where $\mathbf{z}_t$ is the designated vector of the variables known at period t.

But a joint test of $\alpha = 0$ and $\beta = 1$ can also be viewed as the test of efficiency, in the sense of checking that forecasts and forecast errors are uncorrelated (see Clements and Hendry 2000). If there is any correlation, then, this can be used to help predict future errors, which in turn can be used to adjust the forecast-generating mechanism. One way to check this when $s = 1$ is to check whether the residual variance in the regression is equal to the forecast-error variance. To note this, reparameterize the expression for $\epsilon_{t+1|t}$ as

$$\epsilon_{t+1|t} = z_{t+1} - z_{t+1|t} = \alpha + (\beta - 1)z_{t+1|t} + e_{t+1},$$

then,

$$\text{Var}(\epsilon_{t+1|t}) = (\beta - 1)^2\text{Var}(z_{t+1|t}) + \text{Var}(e_{t+1}) + 2(\beta - 1)\text{Cov}(z_{t+1|t}, e_{t+1}),$$

so that $\beta = 1$ implies $\text{Var}(\epsilon_{t+1|t}) = \text{Var}(e_{t+1})$, whatever the value of α would be. Additionally, if $\alpha = 0$ and $\beta = 1$, then, we have

$$\text{MSE} = E(\epsilon^2_{t+1|t}) = \text{Var}(\epsilon_{t+1|t}) = \text{Var}(e_{t+1})$$

Under these conditions, the forecast error and the forecast are uncorrelated:

$$E(\epsilon_{t+1|t} z_{t+1|t}) = \alpha E(z_{t+1|t}) + (\beta - 1)E(z^2_{t+1|t}) + E(e_{t+1} z_{t+1|t}) = 0.$$

9.3.3.7 Testing the Rationality of Multistep Forecasts

Let $s > 1$ in the regression

$$z_{t+s} = \alpha + \beta z_{t+s|t} + e_{t+s}.$$

In this case, forecasts will overlap and the forecast errors will be serially correlated. Suffice it to say that, because of this serial correlation, the ordinary least-square (OLS) estimates still remain unbiased; but the standard errors of the parameter estimates will turn to be inconsistent and have to be corrected. This is typically done by using the Newey–West standard errors, which correct the variance–covariance matrix for the unknown form of AC–heteroscedasticity and is implemented as follows:

If $\mathbf{x}_{t+s} = (1 \quad z_{t+s|t})'$ and $\boldsymbol{\gamma} = (\alpha \quad \beta)'$, then, we can write the above regression as

$$z_{t+s} = \mathbf{x}'_{t+s}\boldsymbol{\gamma} + e_{t+s}.$$

The OLS estimator of the covariance matrix of $\hat{\boldsymbol{\gamma}}$, $\hat{V}(\hat{\boldsymbol{\gamma}})$, is given by

$$\hat{V}(\hat{\boldsymbol{\gamma}}) = \hat{\sigma}^2 \left(\sum_{t=1}^{T} \mathbf{x}_{t+s}\mathbf{x}'_{t+s} \right)^{-1},$$

where $\hat{\sigma}^2$ is the usual OLS estimator of the residual error variance, σ^2. However, the OLS estimator assumes that the errors are serially uncorrelated. If this is violated, then, we have to use the Newey–West covariance matrix given by

$$\hat{V}^*(\hat{\boldsymbol{\gamma}}) = \left(\sum_{t=1}^{T} \mathbf{x}_{t+s}\mathbf{x}'_{t+s} \right)^{-1} T\mathbf{S}^* \left(\sum_{t=1}^{T} \mathbf{x}_{t+s}\mathbf{x}'_{t+s} \right)^{-1},$$

where

$$\mathbf{S}^* = \frac{1}{T}\sum_{t=1}^{T} \hat{e}^2_{t+s}\mathbf{x}_{t+s}\mathbf{x}'_{t+s} + \frac{1}{T}\sum_{j=1}^{s-1} w_j \sum_{i=j+1}^{T} \hat{e}_{i+s}\hat{e}_{i+s-j}(\mathbf{x}_{i+s}\mathbf{x}'_{i+s-j} + \mathbf{x}_{i+s-j}\mathbf{x}'_{i+s}).$$

The square roots of the diagonal elements of $\hat{V}^*(\hat{\boldsymbol{\gamma}})$ are called the heteroscedasticity-consistent standard errors (HCSEs) and are robust. These standard errors can also be used for a valid inference in place of the square root of the diagonal elements of $\hat{V}(\hat{\boldsymbol{\gamma}})$, in the event that e_{t+s} are heteroscedastic but are uncorrelated. The w_j's are the weights and they should be selected in such a way that the resulting covariance matrix is positive definite. Bartlett weights of $w_j = 1 - j/s$ ensure that.

9.3.3.8 Accuracy Measures and Tests of a Comparative Forecast Accuracy

Testing for unbiasedness is obviously of interest. However, as pointed out earlier, systematically unbiased forecasts could be wildly inaccurate, with large errors of opposite signs canceling, but resulting in a large variance of forecast errors. So, analyzing the variance of forecast errors, given by

$$E(e_{t+s|t}^2) = E(z_{t+s} - z_{t+s|t})^2,$$

may be more useful. Choosing the forecast with the smallest-error variance will amount to choosing the forecast with the smallest-expected squared error. The strategy is, when forecasts are unbiased, to choose forecasts with the smallest mean-squared error. Thus, the sample counterpart of $E(e_{t+s|t}^2)$, for a sample T, s-step-ahead forecast errors is simply

$$\text{MSE} = \frac{1}{T}\sum_{t=1}^{T} e_{t+s|t}^2.$$

And the square root of this quantity is called the *root mean-squared error (RMSE).* And this is one of the most popular stylized measures of accuracy.

It is pertinent to point out here that the evaluation of forecasts in economics and finance has been mostly conducted under a symmetric loss function, such as the squared-error loss function. The optimal properties that we have listed above are valid only under such a symmetric loss function. They can be invalid under an asymmetric loss function or if the underlying data-generating process (DGP) is nonlinear. Not only does the shape of the loss function matter, but also the forecast horizon. Ranking the forecasts may differ across the loss functions and the forecast horizon. This has led some to argue that we should have some universally applicable accuracy measures. However, the general practice is to adopt the appropriate loss function depending on a particular decision environment. So, this has led the researchers to mainly use some stylized loss function such as the RMSE. Once a loss function has been specified, next, we have to check which of the competing forecasts has the smallest RMSE.

Suppose we have two competing models, M_1 and M_2. And let $z_{1,t}(1), z_{1,t}(2), \ldots, z_{1,t}(s)$ and $z_{2,t}(1), z_{2,t}(2), \ldots, z_{2,t}(s)$ be the one- to s-step-ahead forecasts from M_1 and M_2, respectively. Let the corresponding errors be $e_{1,t}(1)$, $e_{1,t}(2)$, $\ldots, e_{1,t}(s)$ and $e_{2,t}(1)$, $e_{2,t}(2)$, $\ldots, e_{2,t}(s)$. We then compute the root mean-squared forecast errors for the various forecast horizons for both the models as follows:

$$RMSE_j(l) = \sqrt{\sum_{i=t}^{T-l} e_{j,i}^2(l)},$$

where $l = 1, 2, \ldots, s$ and $j = 1, 2$. For one-step-ahead forecasts, we select model M_1 over model M_2 if $RMSE_1(1) < RMSE_2(1)$; otherwise, model M_2 is preferred. The same rule applies to other forecast horizons.

However, it is possible that there can be RMSEs that are very close to each other. This leads us to the question if there are statistical measures to test if two RMSEs are statistically different. This is equivalent to testing if two competing forecasts are identical. Several tests have been recommended and these are called the "tests for equal forecast accuracy."

9.3.3.9 Morgan–Granger–Newbold Test

GN (1986) extend the idea of Morgan and suggest a test that is popularly called the Morgan–Granger–Newbold (MGN) test. This test tells us how to test if two one-step-ahead forecasts are equally accurate after allowing for a contemporaneous correlation. Other assumptions are however retained. That is, the loss function is quadratic and the forecast errors are zero mean, Gaussian, and serially uncorrelated, implying that we are considering only one-step-ahead forecasts.

Let us suppose that the two competing forecasts are given as $\hat{z}_t(1)$ obtained from model 1 and $\tilde{z}_t(1)$ obtained from model 2 and the corresponding forecast errors are collected in a column vector as $\hat{\mathbf{e}}$ and $\tilde{\mathbf{e}}$, where $\hat{\mathbf{e}} = (\hat{e}_{2|1}, \hat{e}_{3|2}, \ldots, \hat{e}_{T+1|T})'$ and $\tilde{\mathbf{e}} = (\tilde{e}_{2|1}, \tilde{e}_{3|2}, \ldots, \tilde{e}_{T+1|T})'$. Under unbiasedness, the equality of MSE of $\hat{\mathbf{e}}$, and $\tilde{\mathbf{e}}$ mean equality of variances, we can test the equality of variances by constructing

$$u_{1,t+1|t} = \hat{e}_{t+1|t} - \tilde{e}_{t+1|t},$$

$$u_{2,t+1|t} = \hat{e}_{t+1|t} + \tilde{e}_{t+1|t},$$

and testing for zero correlations between $u_{1,t+1|t}$ and $u_{2,t+1|t}$. Note that

$$E(u_{1,t+1|t}u_{2,t+1|t}) = E(\hat{e}^2_{t+1|t}) - E(\tilde{e}^2_{t+1|t}) \quad \text{so that}$$

$$E(u_{1,t+1|t}u_{2,t+1|t}) = 0 \implies E(\hat{e}^2_{t+1|t}) = E(\tilde{e}^2_{t+1|t}).$$

Alternatively, this implies that the variances will be equal only if the covariance is zero. One can calculate the test statistic

$$\frac{r}{\sqrt{(T-1)^{-1}(1-r^2)}} \sim t_{T-1}, \quad \text{where}$$

$$r = \frac{\mathbf{u}_1'\mathbf{u}_2}{\sqrt{\mathbf{u}_1'\mathbf{u}_1\mathbf{u}_2'\mathbf{u}_2}}, \quad \text{and}$$

$$\mathbf{u}_i' = (u_{i,2|1}, \ldots, u_{i,T+1|T}), \quad i = 1, 2.$$

If the t-statistic is significant, then, r is statistically different from zero, with the implication that model 1 has a larger MSE if r is positive and model 2 has a larger MSE if r is negative. Harvey, Leybourne, and Newbold (HLN) (1997) set up the MGN framework in a regression setup and a nonparametric approach, but the inherent limitations of a quadratic loss function and the applicability to only a one-step prediction error still remain even in these extensions.

9.3.3.10 Meese–Rogoff Test

Meese–Rogoff (MR) (1988) developed a test of an equal forecast accuracy that allows the forecast errors to be serially and contemporaneously correlated, implying that this test can be used to compare forecasts with horizon s greater than 1. And they base their test directly on the sample covariance between $u_{1,t+s|t}$ and $u_{2,t+s|t}$. Thus, the MR test allows for covariance between the transformed error vectors, $\hat{u}_t$ and $\tilde{u}_t$. Under the null of an equal-forecast accuracy and allowing for the fact that $\hat{u}_t$ and $\tilde{u}_t$ are themselves $MA(s-1)$, the following asymptotic result holds:

$$T^{1/2}(\hat{\gamma}_{\hat{u},\tilde{u}}(0) - \gamma_{\hat{u},\tilde{u}}(0)) \xrightarrow{D} N(0, \Omega),$$

where

$$\hat{\gamma}_{\hat{u},\tilde{u}}(0) = \hat{\mathbf{u}}'\tilde{\mathbf{u}}/T$$

and, Ω, which consists of cross- and autocovariances, can be consistently estimated by their sample counterparts as

$$\hat{\Omega} = \sum_{k=-m}^{m} (1 - |k|/T) \left[\hat{\gamma}_{\hat{u}\hat{u}}(k)\hat{\gamma}_{\tilde{u}\tilde{u}}(k) + \hat{\gamma}_{\hat{u}\tilde{u}}(k)\hat{\gamma}_{\tilde{u}\hat{u}}(k)\right].$$

MR test is then

$$MR = \frac{\hat{\gamma}_{\hat{u}\tilde{u}}(0)}{\sqrt{(\hat{\Omega}/T)}} \sim N(0,1) \quad \text{for a large } T.$$

9.3.3.11 Diebold–Mariano Test

Diebold–Mariano (DM) (1995) introduced a test that does not need any of the assumptions that the previous two tests needed. In fact, the loss function also need not be quadratic; it can be any arbitrary loss function. Let $g(x)$ be any such arbitrary loss function. x can be either $\hat{\mathbf{e}}$ or $\tilde{\mathbf{e}}$. We shall define the loss differential $d_{t+s|t} = [g(\hat{e}_{t+s|t}) - g(\tilde{e}_{t+s|t})]$ so that the equal-forecast accuracy entails that $E(d_{t+s|t}) = 0$. Given a covariance-stationary sample realization, $\{d_{t+s|t}\}$, the asymptotic distribution of the sample mean loss differential, $\bar{d}$,

$$\bar{d} = \frac{1}{T}\sum_{t=1}^{T} d_{t+s|t}$$

is given by

$$\sqrt{T}\,(\bar{d} - \mu) \xrightarrow{D} N(0, 2\pi f_d(0)),$$

where

$$f_d(0) = \frac{1}{2\pi}\sum_{\tau=-\infty}^{\infty} \gamma_d(\tau)$$

is the spectral density of the loss differential at frequency zero and $\gamma_d(\tau)$ is the autocovariance function of the loss differential at displacement τ.*

DM statistic is

$$DM = \frac{\bar{d}}{\sqrt{\frac{1}{T}2\pi\hat{f}_d(0)}} \xrightarrow{D} N(0,1),$$

where $\hat{f}_d(0)$ is a consistent estimate of $f_d(0)$. The null that $E(d_{t+s|t}) = 0$ is rejected in favor of the alternative that $E(d_{t+s|t}) \neq 0$, when DM exceeds the critical value. A consistent estimator of $f_d(0)$, can be of the form

$$\hat{f}_d(0) = \frac{1}{2\pi}\sum_{\tau=-m}^{m} \omega(\tau/m)\hat{\gamma}_d(\tau),$$

where $\omega(\cdot)$ is a weighting scheme or kernel for the autocovariances.

---

* We discuss the concept of a spectral density in a separate section on frequency-domain analysis, later in this chapter.

9.3.3.12 Modified DM Test

HLN (1997) suggest a small sample correction to the above DM test as follows.

The modified DM statistic is

$$DM^* = \frac{DM}{\sqrt{[\{T+1-2s+s(s-1)/T\}/T]}}.$$

However, HLN recommend that DM^* should be compared with critical values from a t distribution with $T-1$ degrees of freedom instead of the standard normal. Mariano (2004) outlines some nonparametric tests of the forecast accuracy.

9.3.3.12.1 Forecast Combination

A typical economic forecast can generally be thought to be suboptimal in some sense. So, a forecaster may try to improve her forecasts by comparing a set of forecasts and select the best set based on some statistical measure. A more-profitable way may be to incorporate all of them into an overall combined forecast rather than selecting the best set of forecasts and discarding the rest. This idea of combining the forecasts was first suggested by Bates and Granger (BG) (1969). They suggest to let the combined forecast to be a weighted average, with appropriately chosen weights, of the individual forecasts.

The main idea is, when two (or more) forecasting models are available, then, taking a weighted combination of the available forecasts may generate a better forecast. Even though many methods have been suggested, they can be classified as *regression-based methods* or the *variance–covariance methods.* Let us discuss the variance–covariance method of BG (1969), which is the case of how to combine two, one-step-ahead unbiased forecasts. An extension to combining several forecasts is standard and some details are available in GN (1986).

We have two sets of forecasts, $\hat{z}_t(1)$ and $\tilde{z}_t(1)$. Let us simply call them $\hat{z}_t$ and $\tilde{z}_t$, respectively. We combine the forecasts as

$$\hat{z}_{ct} = (1-\lambda)\hat{z}_t + \lambda\tilde{z}_t,$$

with the weights summing up to unity and $0 \leq \lambda \leq 1$. Continuing with the squared-error loss function, λ has to be chosen so as to minimize the MSE of the combined forecast $\hat{z}_{ct}$. Let the error of the combined forecast be $z_t - z_{ct} = e_{ct}$, so that

$$e_{ct} = (1-\lambda)\hat{e}_t + \lambda\tilde{e}_t.$$

Hence, the variance of the combined-forecast error can be easily seen to be

$$V(e_{ct}) = (1-\lambda)^2 V(\hat{e}_t) + \lambda^2 V(\tilde{e}_t) + 2\lambda(1-\lambda)\text{cov}(\hat{e}_t, \tilde{e}_t).$$

$V(e_{ct})$ is minimized for the value of λ given by

$$\lambda^* = \frac{V(\hat{e}_t) - \text{cov}(\hat{e}_t, \tilde{e}_t)}{V(\hat{e}_t) + V(\tilde{e}_t) - 2\text{cov}(\hat{e}_t, \tilde{e}_t)}.$$

Substituting λ^* in the expression for $V(e_{ct})$ gives the minimum-achievable error variance:

$$V(e^*_{ct}) = \frac{(1-\rho^2)V(\hat{e}_t)V(\tilde{e}_t)}{V(\hat{e}_t) + V(\tilde{e}_t) - 2\rho\sqrt{V(\hat{e}_t)V(\tilde{e}_t)}},$$

where

$$\rho = \text{cov}(\hat{e}_t, \tilde{e}_t)/\sqrt{V(\hat{e}_t)V(\tilde{e}_t)}.$$

Note that $V(e_{ct}^*) < \min(V(\hat{e}_t), V(\tilde{e}_t))$ unless ρ is exactly equal to either $\sqrt{V(\hat{e}_t)/V(\tilde{e}_t)}$ or $\sqrt{V(\tilde{e}_t)/V(\hat{e}_t)}$. In case such an equality holds, the variance of the combined forecast is equal to the smaller of the two error variances. This highlights the fact that the combined forecast may outperform the individual forecasts—it cannot, in any case, be worse. In practice, one can obtain the weights by replacing the population moments with sample moments in the expression for λ^* so that we have

$$\hat{\lambda} = \frac{\sum_{t=1}^{T}(\hat{e}_t - \tilde{e}_t)\hat{e}_t}{\sum_{t=1}^{T}(\hat{e}_t - \tilde{e}_t)^2}.$$

Note that this is nothing but the OLS estimator of λ in the simple regression

$$\hat{e}_t = \lambda(\hat{e}_t - \tilde{e}_t) + e_{ct},$$

which is easily seen to be the rearranged expression for e_{ct}. Thus, the optimal weight can be obtained by a simple OLS. Note that if a test of the null of $\lambda = 0$, against an alternative of $\lambda > 0$—that is, $\tilde{z}_t$ has a positive weight in the combination—is accepted, then, $\hat{z}_t$ completely encompasses $\tilde{z}_t$. This means that forecast $\hat{z}_t$ is conditionally efficient and the forecast $\tilde{z}_t$ has no useful information that is not already present in $\hat{z}_t$.

Many improvements over this simple method have been suggested. It can be easily generalized to a combination of an arbitrary number of forecasts. As an example of the regression method, consider the suggestion by Granger and Ramanathan (1984). They suggest to generalize the combination by regressing the actual values on the forecasts, without any intercept, as

$$z_t = \beta_1\hat{z}_t + \beta_2\tilde{z}_t + e_{ct},$$

and estimate it using least squares. Since the weights have to sum up to one and be nonnegative, we have to use a constrained least-squares approach. The above regression can be extended to include many forecasts and it can also include biased forecasts. The general problem can be solved as a quadratic-programming problem, with the possibility of corner solutions, where many forecasts will receive zero weights. Weights can be constrained to be equal or can be allowed to vary over time. For example, in the variance–covariance context, Granger and Newbold (1973) suggested time-varying combining weights and in the regression context, Diebold and Pauly (1987) recommended the same. Rapach, Strauss, and Zhou (RSZ) (2010) provide an empirical evidence that the U.S. market risk premium is consistently predictable out of sample with macroeconomic variables. RSZ (2010) show how combining forecasts from individual predictive regressions generate consistent and significant out-of-sample forecasts, and reduce the model instability risk arising out of relying on a single model. Timmermann (2006) outlines the reasons as to why we may want to combine the forecasts, their importance in the forecasting literature, and also lists out numerous points against the forecast combination.

9.3.3.12.2 Asymmetric Loss Functions

Squared-error loss functions assume that positive- and negative-forecast errors attract the same penalty. Suppose positive-forecast errors are penalized more heavily than negative errors, then, it is optimal to make a negative-expected forecast error, so that the realized errors will be predominantly negative. Granger (1969), who formalizes these ideas, shows that the conditional mean predictor will not be optimal but will be biased, if the cost or the loss function is asymmetric. However, for Gaussian processes, a simple adjustment to the conditional mean yields the optimal predictor, where the adjustment depends only on the loss function and the forecast variance. As Zellner (1986) points out, this implies that the conditional expectations, not being optimal, are consistent with a rational behavior. Christoffersen and Diebold (1997) study the optimal prediction problem under general loss structures and compute the optimal predictor in two tractable cases, namely, linex and linlin loss functions and suggest ways to numerically obtain the optimal predictor in some intractable cases. They show that adjustments to the conditional mean will be time varying and depend on higher-order conditional moments. For example, we have to take into account the generalized autoregressive conditional heteroscedasticity (GARCH) effects for an optimal point prediction of Gaussian processes, under an asymmetric loss function.

One of the most-popular asymmetric loss function is the "linex" loss function of Varian (1974). Varian discusses the losses faced by property valuers, which rise linearly on one side and exponentially on the other side, implying that underassessment leads to linear revenue losses, while overassessment results in appeals, litigation, and other costs. This observation led to the formulation of a linear–exponential loss function, which was popularly called as the "linex" loss function. It is explained as follows:

$$L(e_{t+s|t}) = \alpha \exp(\gamma e_{t+s|t}) - \beta e_{t+s|t} - \alpha, \quad \text{with } \gamma, \beta \neq 0, \ \alpha > 0.$$

$e_{t+s|t}$ is the s-step error in forecasts. Since $L(0) = 0$, for a minimum to exist at $e_{t+s|t} = 0$, we must have $\gamma\alpha = \beta$. So, we can write the loss function as

$$L(e_{t+s|t}) = \alpha \left[\exp(\gamma e_{t+s|t}) - \gamma e_{t+s|t} - 1\right], \quad \text{with } \gamma \neq 0, \quad \text{and} \quad \alpha > 0.$$

In this loss function, it is γ that determines the extent of asymmetry and α measures the scale of the loss. For $\gamma > 0$, the loss function is approximately linear for $e_{t+s|t} < 0$ (overpredictions) and exponential for $e_{t+s|t} > 0$ (underpredictions). For a small γ, the loss function is approximately quadratic, which we obtain as the first two terms of a Taylor-series expansion of $L(e_{t+s|t})$ about zero:

$$L(e_{t+s|t}) \simeq \frac{\alpha\gamma^2}{2} e_{t+s|t}^2.$$

Under linex loss, the optimal predictor solves

$$\min_{\hat{z}_{t+s}} E_t \left[\alpha \left(\exp(\gamma e_{t+s|t}) - \gamma e_{t+s|t} - 1\right)\right],$$

where $e_{t+s|t} = z_{t+s} - \hat{z}_{t+s}$. $\hat{z}_{t+s}$ is the optimal predictor that is assumed to have the form $\hat{z}_{t+s} = z_{t+s|t} + \alpha_{t+s}$, where the process is conditionally Gaussian, $z_{t+s}|I_{t-1} \sim N(z_{t+s|t}, \sigma_{t+s|t}^2)$. It can be shown that the optimal predictor becomes

$$\tilde{z}_{t+s|t} = z_{t+s|t} + \frac{\gamma}{2}\sigma_{t+s|t}^2.$$

As the asymmetry lessens, that is, $\gamma \to 0$, the optimal predictor tends to the conditional expectation. GARCH implies that the conditional variance $\sigma_{t+s|t}^2$ will be time dependent so that the optimal degree of overprediction or underprediction will depend on the perceived volatility when the prediction is made, and will change over time. Thus, in this case, a knowledge of the optimal predictor requires a knowledge of the joint specification of the conditional mean and variance.

When the forecast loss is symmetric, the optimal predictor is, $z_{t+s|t}$, whether or not there is the GARCH effect.

Linlin loss function is defined as

$$L(e_{t+s|t}) = \begin{cases} a|e_{t+s|t}| & \text{if } e_{t+s|t} > 0 \\ b|e_{t+s|t}| & \text{if } e_{t+s|t} \leq 0. \end{cases}$$

If the process is conditionally Gaussian, $z_{t+s}|I_t \sim N(z_{t+s|t}, \sigma^2_{t+s|t})$, then, Christoffersen and Diebold (1997) show that the optimal predictor under linlin loss is given by

$$\hat{z}_t(s) = z_{t+s|t}|I_t + \sigma_{t+s|t}\Phi^{-1}(a/(a+b)),$$

where Φ denotes the cumulative distribution function (cdf) of a standard normal. For $a > b$, $\Phi^{-1}(a/(a+b)) > 0$ and so, the optimal predictor is upward biased. As an example, Christoffersen and Diebold (1996) show that under the linlin loss function, the optimal predictor for a conditionally heteroscedastic process is simply

$$\hat{z}_t(s) = \Phi^{-1}(a/(a+b)|I_t),$$

so that $\hat{z}_t(s)$ is simply the $(a/(a+b))$th conditional quantile and when $a = b$ it is the median.

However, such a closed-form analytical solution for optimal predictors remains to be generally impossible, except under a conditional normality assumption. Christoffersen and Diebold (1997) and Patton and Timmermann (2007) have more details on how to obtain optimal predictors under general loss structures.

9.3.4 MLE and Prediction-Error Decomposition

Our discussion on forecasting assumed that the unknown population parameter values were known to the user. However, in practice, one has only an estimate of the population parameters. The primary method of estimating the unknown parameters of the *ARMA* model will be the maximum likelihood method. Though the classical maximum likelihood method was developed for iid variables, they could still be adapted to estimate the unknown parameters of time-series models, where the data would be dependent. A convenient approach is usually through the *prediction-error decomposition.* (see Harvey 1981, 1984). Suppose we have a set of T-dependent observations drawn from a normal distribution with mean μ and the variance–covariance matrix, Ω—that is, $z \sim N(\mu, \Omega)$. On the basis of the notion of conditional density, the joint-density function of the given observations in logs, $\log L(z)$, may be factored into two parts, by writing

$$\log L(z) = \log L(z_1, \ldots, z_{T-1}) + \log \ell(z_T \mid z_{T-1}, \ldots, z_1).$$

Now, consider the problem of estimating z_T, given that $z_{T-1}, \ldots, z_1$ are known. Suppose $\hat{z}_{T/T-1}$ is an estimator of z_T, based on an information set consisting of past observations, then, the minimum mean-squared estimator of z_T conditional on $z_{T-1}, \ldots, z_1$ is

$$\hat{z}_{T|T-1} = E(z_T \mid z_{T-1}, \ldots, z_1).$$

Let the prediction-error variance associated with this estimator be denoted as $\sigma^2 f_T$. Hence, we can write its distribution as

$$\begin{aligned} \log \ell(z_T \mid z_{T-1}, \ldots, z_1) = {} & (1/2)\log 2\pi - (1/2)\log\sigma^2 \\ & - (1/2)\log f_T - (1/2)\sigma^{-2}(z_T - \hat{z}_{T|T-1})^2/f_T, \end{aligned}$$

which can be equally interpreted as the distribution of the prediction error, $z_T - \hat{z}_{T|T-1}$. Such a prediction error is independent of past observations by definition and so, such a decomposition may be repeated with respect to the likelihood of the first $T - 1$ observations.

$\ell(z_1)$ is the unconditional distribution of z_1. If μ_1 is the unconditional mean of z_1, and the corresponding variance is σ_1^2, we can decompose the likelihood function into a joint distribution of T-independent prediction errors

$$\nu_t = z_t - \hat{z}_{t|t-1}, \quad t = 1, \ldots, T,$$

where $\hat{z}_{1|0} = \mu_1$. And hence the name *prediction-error decomposition.* Each prediction error has a mean of zero and a variance $\sigma^2 f_t$ and so the expression

$$\log L(z) = -\frac{T}{2}\log 2\pi - \frac{T}{2}\log\sigma^2 - \frac{1}{2}\sum_{t=1}^{T}\log f_t - \frac{1}{2}\sigma^{-2}\sum_{t=1}^{T}\nu_t^2/f_t$$

may be regarded as the general expression for the log-likelihood of any given $ARMA(p, q)$ model. For example, for the Gaussian $AR(1)$ model,

$$\tilde{z}_t = \phi\tilde{z}_{t-1} + e_t, \quad \tilde{z}_t = z_t - \mu,$$

where $e_t \sim N(0, \sigma^2)$ and the vector of the population parameter is $\theta = (\phi, \sigma^2)'$, and the log-likelihood can be seen to be

$$\log L(\theta) = -\frac{1}{2}\log(2\pi) - \frac{1}{2}\log[\sigma^2/(1-\phi^2)] - \frac{\tilde{z}_1^2}{2\sigma^2/(1-\phi^2)} - [(T-1)/2]\log(2\pi)$$
$$- [(T-1)/2]\log(\sigma^2) - \sum_{t=2}^{T}\left[\frac{(\tilde{z}_t - \phi\tilde{z}_{t-1})^2}{2\sigma^2}\right].$$

We obtain the estimates by differentiating the log-likelihood. In practice, however, such an attempt leads to a set of nonlinear equations in the unknown parameter vector θ, and thus involves some numerical procedures. The estimates obtained in this way are called the *exact maximum likelihood estimates.* See Hamilton (1994) for a second way to obtain the log-likelihood for the same model.

An alternative to avoid such numerical procedures is to condition on the initial values. To estimate the parameters of an $AR(p)$ model, we can condition on the p-initial values, that is, we treat p-initial values as given. Thus, the estimates obtained are called the *conditional maximum likelihood estimates.* For example, in the $AR(1)$ case, suppose we condition on the first observation, that is, treat $\tilde{z}_1$ as known, we can maximize the log-likelihood conditioned on the first observation. It can be shown that maximizing the resulting log-likelihood with respect to ϕ is equivalent to minimizing the sum of squares, $\sum_{t=2}^{T}(\tilde{z}_t - \phi\tilde{z}_{t-1})^2$ which can be obtained by simply running the least-squares regression of z_t on z_{t-1}.

However, for MA models, a similar conditioning information on the initial values does not help. One has to necessarily maximize the log-likelihood using some numerical maximization techniques. For example, in the case of a Gaussian $MA(1)$ model,

$$\tilde{z}_t = e_t + \theta_1 e_{t-1}, \quad e_t \sim N(0, \sigma_e^2).$$

Suppose we condition on the fact that $e_0 = 0$. It is clear that $(\tilde{z}_1 \mid e_0) \sim N(0, \sigma_e^2)$. Hence, the first prediction error is $\tilde{z}_1$ itself—that is, $e_1 = \tilde{z}_1$. It can be shown that, given $e_0 = 0$, the e_t's for the entire sample is given as

$$e_t = \tilde{z}_t - \theta_1 e_{t-1}, \quad t = 1, \ldots, T, \quad \text{with } e_0 = 0.$$

The conditional log-likelihood is

$$\log L(\theta) = -\frac{T}{2}\log(2\pi) - \frac{T}{2}\log\sigma_e^2 - \sum_{t=1}^{T} \frac{e_t^2}{2\sigma_e^2},$$

which can be maximized using numerical optimization procedures.

Conditioning on presample values may simplify the estimation, but in typically small samples, the conditioning assumption may have a significant effect on the likelihood. For example, in an $MA(1)$ process, the first prediction error is the same—it is z_1—in both conditional and exact ML estimation procedures. But the assumption that $e_0 = 0$ means that $\text{var}(z_1) = \text{var}(e_1) = \sigma^2$, whereas, if we allow e_0 to be random, then, $\text{var}(z_1) = \sigma^2(1 + \theta^2)$. And this will be reflected in f_1 in the exact log-likelihood function. And in typically small sample sizes, such an assumption may matter a lot for the estimates. Besides, if θ happens to be close to unity, then, the differences will be even more significant. Hence, the exact or unconditional maximum likelihood method is generally recommended. This brings us to the necessity of discussing the methods to obtain the exact finite-sample forecasts, which will enable us to obtain the exact ML estimates of the parameter vector. Even though many algorithms are available to build the components of the log-likelihood, two such methods are popularly discussed in the literature: one through a triangular factorization of any positive-definite matrix and the other through Kalman recursions. While an extensive discussion on both these methods is available in Hamilton (1994) and Harvey (1981, 1984, 1989), below, we summarize the important steps in calculating the components of the log-likelihood function using Kalman recursions, which is normally preferred. Though the state-space form and the associated Kalman filter (KF) can be explained in different ways, we closely follow the notations and steps set out in Harvey (1984), which are easy to understand. We shall explain the steps to calculate the prediction error and its variance, using the simplest $MA(1)$ model. Other useful algorithms are by Newbold (1974) and the *innovations algorithm* suggested by Ansley (1979).

9.3.4.1 State-Space Form and ARMA Models

To apply the KF approach, we have to cast the ARMA model in a state-space form. State- space models were originally developed by control engineers to represent a dynamic system or dynamic linear models. The interest normally centers on an $(m \times 1)$ vector of variables, called a *state vector,* that may be signals from a satellite or the actual position of a missile or a rocket. A state vector represents the dynamics of the process. More precisely, it retains all the memory in the process. All the dependence between the past and the future must funnel through the state vector. The elements of the state vector may not have any specific economic meaning, but the state-space approach is popular in economic applications involving the modeling of unobserved or latent variables, such as permanent income, NAIRU (nonaccelerating inflation rate of unemployment), expected inflation, state of the business cycle, etc. In most cases, such signals are not observable directly, but such a vector of variables is related to another $(n \times 1)$ vector z_t of variables that are actually observed, through an equation called the *measurement equation* or the *observation equation*, given by

$$z_t = A_t x_t + Y_t \alpha_t + N_t,$$

where Y_t and A_t are the parameter matrices of order $(n \times m)$ and $(n \times k)$, respectively; x_t is the $(k \times 1)$ vector of exogeneous or predetermined variables; and N_t is the $(n \times 1)$ vector of disturbances that has a zero mean and the covariance matrix H_t.

Although the state vector α_t is not directly observable, its movements are assumed to be governed by a well-defined process called the *transition equation* or *state equation* given by

$$\alpha_t = T_t \alpha_{t-1} + R_t \eta_t, \quad t = 1, \ldots, T,$$

where T_t and R_t are matrices of order $(m \times m)$ and $(m \times g)$, respectively, and η_t is the $(g \times 1)$ vector of disturbances with a mean of zero and the covariance matrix Q_t. Moreover, the dimension of the state vector has to be large enough so that the dynamics of the systems can be captured by the simple first-order Markov structure of the state equation. From a technical point of view, the aim of the state-space form is to set up α_t such that it has as few elements as possible. Such a state-space setup is called a *minimal realization* and it is the basic criterion for a good state-space form. A popular choice is $m = \max(p, q + 1)$ where, in an *ARMA* model setup, p and q refer, respectively, to the order of the *AR* and the *MA* polynomial.

But in many cases of interest, only one observation is available in each time period, that is, z_t is now a scalar in the measurement equation. Moreover, the transition matrix is much simpler than that given before, in the sense that the parameters in most cases, including the variance, are assumed to be *time invariant.* For many applications using KF, the vector of exogenous variables is simply not necessary. One may also assume that the variance of the noise term is time invariant. In some of the state-space applications—especially those that use *ARMA* models—the measurement error in the observation equation, that is, N_t, is assumed to be zero. This means that N_t in such applications will be absent. Therefore, the general system now boils down to

$$z_t = y_t' \alpha_t + N_t, \quad t = 1, \ldots, T,$$
$$\alpha_t = T \alpha_{t-1} + R \eta_t, \quad t = 1, \ldots, T.$$

Here, y_t' is a $(1 \times m)$ vector. There are many ways to write a given system in the state-space form. But written in any way, if our primary interest is on forecasting, then, we will get identical forecasts, no matter which form we use. Also note that we can write any state-space form as an *ARMA* model. In this way, there is an equivalence between the two forms.

For the first-order MA model, assuming that the model has a zero mean

$$z_t = e_t + \theta_1 e_{t-1}.$$

For an $MA(1)$ model, $m = 2$, so that the state and the measurement equations are given as follows:

$$\text{State equation:} \quad \alpha_t = \begin{bmatrix} 0 & 1 \\ 0 & 0 \end{bmatrix} \alpha_{t-1} + \begin{bmatrix} 1 \\ \theta_1 \end{bmatrix} e_t \quad \text{and}$$

$$\text{Observation equation:} \quad z_t = \begin{bmatrix} 1 & 0 \end{bmatrix} \alpha_t.$$

If we define $\alpha_t = (\alpha_{1t}\ \alpha_{2t})'$, then, $\alpha_{2t} = \theta_1 e_t$ and $\alpha_{1t} = \alpha_{2,t-1} + e_t = e_t + \theta_1 e_{t-1}$ and this is precisely the original model.

Next, consider the mixed $ARMA(1,1)$ model. For $ARMA(1,1)$, $m = 2$. So, the state and the observation equations are as follows:

$$\text{State equation:} \quad \epsilon_{t+1} = \begin{bmatrix} \phi_1 & 0 \\ 1 & 0 \end{bmatrix} \epsilon_t + \begin{bmatrix} 1 \\ 0 \end{bmatrix} e_{t+1},$$

$$\text{Observation equation:} \quad z_t = \mu + \begin{bmatrix} 1 & \theta_1 \end{bmatrix} \epsilon_t.$$

From the state equation, we have $\epsilon_{2,t+1} = \epsilon_{1,t}$ and $(1 - \phi_1 B)\epsilon_{1,t+1} = e_{t+1}$. And from the observation equation, $z_t = \mu + (1 + \theta_1 B)\epsilon_{1,t}$. And multiplying by $(1 - \phi_1 B)$ gives $(1 - \phi_1 B)(z_t - \mu) = (1 - \phi_1 B)(1 + \theta_1 B)\epsilon_{1,t}$ or $(1 - \phi_1 B)\tilde{z}_t = (1 + \theta_1 B)e_t$ that is the given model.

9.3.4.2 KF Recursions and the Estimation of ARMA Models

KF has many uses. Given our objectives, KF can be used either to obtain the values of unknown parameters or, given the parameter vectors, we may obtain the linear least-squares forecasts of the state vector on the basis of observed data. We are utilizing it primarily as an algorithm to evaluate the components of the likelihood function. And, later, we shall outline two algorithms to obtain smoothed estimates of the state vector.

Kalman filtering follows a two-step procedure. In the first step, the optimal predictor for the *next* observation is formed, based on all the information *currently* available. This is done by the *prediction equation.* In the second step, the moment the observation becomes available, it is then incorporated into the estimator of the state vector using the *updating equation.* These two equations collectively form the KF equations. Applied recursively, KF provides an optimal solution to the twin problems of prediction and updating. Assuming that the observations are normally distributed and also assuming that the current estimator of the state vector is the *best* available, the prediction and the updating estimators are the best, by which we mean that the estimators have the minimum mean-squared error (MMSE). It is very evident that the process of predicting the next observation and updating it as soon as the actual value becomes available has an interesting by-product—the *prediction error.* And we have seen in the section on estimation, how a set of dependent observations can be decomposed in terms of prediction errors. KF gives us a natural mechanism to carry out this decomposition.

9.3.4.3 KF Recursions: Main Equations

Kalman recursions consist of two equations, namely the *prediction equation* and the *updating equation*.

9.3.4.3.1 Prediction Equation

Let a_t denote the *MMSE* estimator of α_t based on all information up to and including the current observation z_t. Similarly, let $a_{t/t-1}$ be the *MMSE* estimator of α_t at time $t-1$. That is, $a_{t|t-1} = E(\alpha_t|I_{t-1})$. At time $t-1$, all the available information, including z_{t-1}, is incorporated into a_{t-1}, which is the *MMSE* estimator of α_{t-1}. The prediction error has a covariance matrix, $\sigma_e^2 P_{t-1} = E\left[(\alpha_{t-1} - a_{t-1})(\alpha_{t-1} - a_{t-1})'\right]$. The introduction of σ_e^2 is for convenience. Given this, we can write the prediction equation as

$$a_{t/t-1} = T a_{t-1},$$

and the covariance matrix of the estimation error as

$$P_{t/t-1} = T P_{t-1} T' + R Q R'.$$

With this set of prediction equations, we can write the *MMSE* estimator of z_t, at time $t-1$, which is clearly $\hat{z}_{t/t-1} = y_t' a_{t/t-1}$. The associated prediction error is $(z_t - \hat{z}_{t/t-1}) = \nu_t = y_t'(\alpha_t - a_{t/t-1}) + N_t$, the expectation of which is zero. And, the variance of the prediction error is $\text{var}(\nu_t) = \sigma^2 y_t' P_{t|t-1} y_t + \sigma^2 h = \sigma^2 f_t$.

9.3.4.3.2 Updating Equation

Once z_t is available, we can update the estimator, $a_{t-1|t}$ as

$$a_t = a_{t|t-1} + P_{t|t-1} y_t \left(z_t - y_t' a_{t|t-1}\right) / f_t.$$

And, the covariance matrix of the estimation error is given as

$$P_t = P_{t|t-1} - P_{t|t-1} y_t y_t' P_{t|t-1} / f_t \quad \text{where } f_t = y_t' P_{t|t-1} y_t + h.$$

Here, the $(m \times 1)$ vector, $K_t = (P_{t|t-1} y_t / f_t)$, is the Kalman gain.

9.3.4.3.3 ML Estimation of ARMA Models Using KF Recursions

In our explanation so far, we have motivated the discussion on KF in terms of a linear projection of the state vector, α_t and the observed times series, z_t. These are linear forecasts and are optimal among any function, if we assume that the state vector and the disturbances are multivariate Gaussian. Our main aim is to see how KF recursions calculate these forecasts recursively, generating $a_{1|0}, a_{2|1}, \ldots, a_{T|T-1}$, and $P_{1|0}, P_{2|1}, \ldots, P_{t|t-1}$ in succession.

9.3.4.3.4 How Do We Start the Recursions?

To start the recursions, we need to obtain $a_{1|0}$. This means that we should get the first-period forecast of α based on an information set. Since we don't have information on the zeroth period, we take the unconditional expectation as

$$a_{1|0} = E\left(\alpha_1\right),$$

where the associated estimation error has a zero mean and the covariance matrix $\sigma^2 P_{1|0}$. For pure MA models, it is easy to obtain these quantities. For the simplest $MA(1)$ model, $a_{1|0} = E(\tilde{z}_t \quad \theta_1 e_1)' = (0 \quad 0)'$ and the associated variance matrix of the estimation error, $\sigma^2 P_0$ or $\sigma^2 P_{1/0}$, is simply $E(\alpha_1 \alpha_1')$, so that we have

$$\begin{aligned} P_{1|0} &= \sigma^{-2} E\left(\alpha_1 \alpha_1'\right) \\ &= \sigma^{-2} E\left\{\begin{bmatrix} \tilde{z}_1 \\ \theta_1 e_1 \end{bmatrix} \left[\tilde{z}_1 \theta_1 e_1\right]\right\} = \begin{bmatrix} 1+\theta_1^2 & \theta_1 \\ \theta_1 & \theta_1^2 \end{bmatrix}. \end{aligned}$$

However, obtaining the covariance matrix for the initial state vector is too tedious for pure AR and mixed models. So, we need a closed-form solution to calculate this matrix. A simple solution is available, if we can make prior assumptions about the distribution of the state vector. If the state vector is covariance stationary, then, one can easily check from the state equation that the unconditional mean of the state vector is zero. That is, from the state equation, one can easily see that $E(\alpha_t) = 0$ and the unconditional variance of α_t is easily seen to be

$$E\left(\alpha_t \alpha_t'\right) = E\left[\left(T\alpha_{t-1} + R\eta_t\right)\left(T\alpha_{t-1} + R\eta_t\right)'\right].$$

If we denote the LHS of the above expression as Σ and note that the state vector depends on shocks only up to $t-1$, then, we obtain

$$\Sigma = T\Sigma T' + RQR'.$$

A direct closed-form solution is given by

$$\text{vec}\,(\Sigma) = \left[I_{m^2} - (T \otimes T) \right]^{-1} \text{vec}\left(RQR'\right).$$

What this implies is that, provided the state vector is covariance stationary, KF recursions can be started with $(a_{1\,|\,0}) = 0$, and the $(m \times m)$ matrix $P_{1\,|\,0}$, whose elements can be expressed as a column vector, is obtained from

$$\text{vec}\left(P_{1|0}\right) = \left[I_{m^2} - (T \otimes T) \right]^{-1} \text{vec}\left(RQR'\right).$$

For example, for the $MA(1)$ model, the prediction error and its variance can be obtained using Kalman recursions, as follows. Assuming that the process has a zero mean, we start the recursions by noting that $a_{1|0} = Ta_0 = 0$ and

$$P_{1|0} = \begin{bmatrix} (1+\theta_1^2) & \theta_1 \\ \theta_1 & \theta_1^2 \end{bmatrix}.$$

Let us calculate the prediction error for z_1. One can easily see that $\hat{z}_{1|0} = 0$, and hence, the associated prediction error $\nu_1 = z_1$ itself and the prediction-error variance is given as $(\sigma^2(1+\theta_1^2))$, with $f_1 = (1+\theta_1^2)$.

Applying the updating formula gives us

$$a_1 = \begin{bmatrix} z_1 \\ z_1\theta_1/(1+\theta_1^2) \end{bmatrix} \quad \text{and} \quad P_1 = \begin{bmatrix} 0 & 0 \\ 0 & \theta_1^4/(1+\theta_1^2) \end{bmatrix}.$$

Continuing in this manner, we see that

$$a_{2|1} = Ta_1 = \begin{bmatrix} z_1\theta_1/(1+\theta_1^2) \\ 0 \end{bmatrix} \quad \text{and} \quad P_{2|1} = \begin{bmatrix} \dfrac{(1+\theta_1^2+\theta_1^4)}{(1+\theta_1^2)} & \theta_1 \\ \theta_1 & \theta_1^2 \end{bmatrix}.$$

Predicting z_2 gives us

$$z_2 = [1 \quad 0] \begin{bmatrix} z_1\theta_1/(1+\theta_1^2) \\ 0 \end{bmatrix} = z_1\theta_1/(1+\theta_1^2),$$

and so, the prediction error is $\nu_2 = z_2 - z_1\theta_1/(1+\theta_1^2)$ and its variance f_2 can be easily calculated as $(1+\theta_1^2+\theta_1^4)/(1+\theta_1^2)$.

These steps show that, for the $MA(1)$ model, one can calculate the prediction error and its variance using the following recursions:

$$\nu_t = z_t - \frac{\theta_1 \nu_{t-1}}{f_{t-1}}, \quad t = 1, 2, \ldots, T, \quad \text{where } \nu_0 = 0, \quad \text{and}$$

$$f_t = 1 + \frac{\theta_1^{2t}}{1+\theta_1^2+\cdots+\theta_1^{2(t-1)}}.$$

As a final step toward finalizing the likelihood function, we shall note the following further simplification. Recall that we had decomposed the likelihood for a set of dependent observations, into a likelihood for the independent errors, using the concept of the prediction-error decomposition, as

$$\log L(z) = -\frac{T}{2}\log 2\pi - \frac{T}{2}\log\sigma^2 - \frac{1}{2}\sum_{t=1}^{T}\log f_t - \frac{1}{2}\sigma^{-2}\sum_{t=1}^{T}\nu_t^2/f_t.$$

From our derivation, we can see that ν_t and f_t do not depend on σ^2 and hence we can "concentrate" σ^2 out. This means that we have to differentiate the log-likelihood with respect to σ^2 and get an estimator for σ^2, say, $\hat{\sigma}^2$. So, we get

$$\hat{\sigma}^2 = \frac{1}{T}\sum_{t=1}^{T}\frac{\nu_t^2}{f_t}.$$

Evaluating the log-likelihood in terms of $\sigma^2 = \hat{\sigma}^2$ and simplifying, we obtain

$$\text{Log } L\,(z)_c = \frac{T}{2}\left(\log 2\pi + 1\right) - \frac{1}{2}\sum_{t=1}^{T}\log f_t - \frac{T}{2}\log\hat{\sigma}^2.$$

We either maximize this log-likelihood or minimize

$$\text{Log } L\,(z)_c = \sum_{t=1}^{T}\log f_t + T\log\hat{\sigma}^2.$$

One can make an initial guess about the underlying parameters and either apply the numerical estimation procedures to calculate the derivatives or analytically calculate the derivatives by differentiating the Kalman recursions. In either case, one has to keep in mind the restrictions to be imposed on the parameters, especially on the *MA* parameters, to take care of the identification problem.

9.3.4.4 Kalman Smoothing

We have motivated the discussion on KF so far as an algorithm for predicting the state vector, and obtaining the exact finite-sample forecasts as a linear function of past observations. We have also shown how the resulting prediction error and the prediction-error variance can be used to evaluate the log-likelihood. This is suboptimal if we are interested in estimating the sequence of states. In many cases, KF is used to obtain an estimate of the state vector itself. For example, in their model of the business cycle, Stock and Watson (1991) show how one may be interested in knowing the *state of the economy* or the phase of the business cycle the economy is in, which is unobservable at any given historical point. Stock and Watson suggest that comovements in many macroaggregates have a common element, which may be called the state of the economy and this is unobservable. They motivate the use of KF to obtain an estimate of this unobserved state of the economy.

Sometimes, elements of the state vector are even interpreted as estimates of missing observations, which can be higher-frequency data points from an observable lower-frequency one or simply an estimate of a missing data point. For example, if we have data on a macroaggregate from 1955 to 2014, we may be interested in obtaining an estimate of 1970 that may be missing. Or, we may be interested in extracting monthly data from quarterly data. Such estimates of the unobserved state of the economy or missing observations can be obtained from *smoothed estimates* of the state vector, α_t.

There are basically three forms of smoothing for a linear model. *Fixed-point* smoothing computes smoothed estimates of the state vector at some fixed point in time. That is, we obtain $a_{\tau|t}$ for any particular value of τ at all time periods of $t > \tau$. *Fixed-lag* smoothing computes smoothed estimates for a fixed delay, that is, $a_{t-j|t}$ for $j = 1, \ldots, M$ where M is some lag. *Fixed-interval* smoothing computes smoothed estimates of the entire state vector, for a fixed span of data. Of these, fixed-point and fixed-interval smoothing techniques are used more often. All the three techniques are closely linked to KF. Below, we state the important recursions of these two techniques. The details are available in Anderson and Moore (1979) and Harvey (1989).

9.3.4.4.1 Fixed-Point Smoother

A fixed-point smoother is run along with the regular KF. Below, we specify the smoothing recursions, in a univariate framework with time-invariant parameters

$$
\begin{aligned}
a_{\tau|t} &= a_{t|t-1} + K_t^* \nu_t, \quad t = \tau, \ldots, T \\
P_{t+1|t}^* &= P_{t|t-1}\left[T - K_t y_t'\right]' \\
P_{\tau|t} &= P_{t|t-1} - P_{t|t-1} y_t + K_t^{*'} \quad t = \tau, \ldots, T \\
K_t^* &= P_{t|t-1} y_t \quad t = \tau, \ldots, T,
\end{aligned}
$$

where K_t^* is the Kalman gain of the smoothing recursions and K_t is the Kalman gain obtained from the regular KF. ν_t's are the innovations obtained by running the regular filter, where we obtain the recursions for $a_{t|t-1}$ and $P_{t|t-1}$. $P_{\tau|t}$ yields the MSE matrix of the smoothed estimators.

9.3.4.4.2 Fixed-Interval Smoother

Each step of the Kalman recursions gives an estimate of the state vector, α_t, given all the current and past observations. However, an econometrician should use all the available information to estimate the sequence of states. Kalman smoother provides these estimates. The only smoothed estimator that utilizes all the sample observations is given by $a_{t|T} = \hat{E}(\alpha_t | I_T)$ and the *MSE* of this smoothed estimate is denoted as

$$
P_{t|T} = \hat{E}\left[(\alpha_t - a_{t|T})(\alpha_t - a_{t|T})'\right].
$$

The smoothing equations start from $a_{t|T}$ to $P_{t|T}$ and work backward.

The expressions for $a_{t|T}$ and $P_{t|T}$, which may be called the *smoothing algorithm,* are given below without a proof (see Hamilton 1994 for a detailed derivation):

$$
\begin{aligned}
a_{t|T} &= a_t + P_t^*\left(a_{t+1|T} - T_{t+1} a_t\right) \\
P_{t|T} &= P_t + P_t^*\left(P_{t+1|T} - P_{t+1|t}\right) P_t^{*'}
\end{aligned}
$$

where

$$
P_t^* = P_t T_{t+1}' P_{t+1|t}^{-1}, \quad t = T - 1, \ldots, 1
$$

with

$$
a_{T|T} = a_t \quad \text{and} \quad P_{T|T} = P_t.
$$

TABLE 9.2

Updated and Smoothed Estimates

t	1	2	3	4
z_t	4.4	4.0	3.5	4.6
a_t	0.306	0.029	−0.199	0.223
P_t	0.235	0.202	0.202	0.202
$a_{t\|T}$	0.290	0.019	−0.179	0.223
$P_{t\|T}$	0.223	0.193	0.193	0.202

A set of direct residuals can also be obtained from the smoothed estimators:

$$e_t = z_t - y_t' a_{t|T}, \quad t = 1, \ldots, T.$$

This is not to be confused with the prediction residuals, ν_t, defined earlier.

We shall explain the fixed-interval smoothing algorithm with an exercise from Harvey (1984, p. 119) as an example. Consider the exercise

$$z_t = 4 + \alpha_t + \epsilon_t, \quad \epsilon_t \sim WN(0, \sigma^2)$$
$$\alpha_t = 0.5\alpha_{t-1} + \eta_t, \quad \eta_t \sim WN(0, \sigma^2 q),$$

where the state, α_t, and the observation, z_t, are scalars. The state follows a stationary $AR(1)$ process but is contaminated by noise. We assume that q is known. Also note that in this example, we have allowed the observation z_t to be measured with error, ϵ_t. Here, we are given that $a_0 = 4$, $P_0 = 12$, and $q = 4$. Note that here $T = 0.5$, $R = 1$, and $y_t' = 1$.

Table 9.2 displays the updated and smoothed values with four observations: $z_1 = 4.4$, $z_2 = 4$, $z_3 = 3.5$, and $z_4 = 4.6$. The initial state vector has the property, $\alpha_0 \sim N(a_0, \sigma^2 P_0)$. a_t and $a_{t|T}$ refer to the estimated and smoothed state vectors, respectively; P_t and $P_{t|T}$ refer to the associated covariances.

9.3.5 Time-Series Modeling: Practical Aspects

BJ suggest the following three-step iterative procedure to determine the order of the AR or MA models and estimate them:

- Identification
- Estimation
- Diagnostic testing

Identification basically means that we tentatively fix the order of the model, that is, to obtain an initial estimate of p and q. This itself proceeds in two stages. First, we have to make sure that the data we analyze are *stationary*. Prior to that, a simple plot of the data can help to determine if any preliminary transformation of the data—such as taking logs of the data, to stabilize the variance—is necessary. Then, BJ suggest to plot the sample ACs of such transformed data for the first 20 lags or so. And, the tendency of the ACs not to *die* out rapidly is a definite indication of nonstationarity. And, the remedy suggested is to difference such transformed data as many times as possible to produce the stationarity. Mostly differencing once makes the data stationary. But, later, we are going to see that the decision to difference the data can be placed on a rigorous statistical footing by doing the so-called *unit-root testing* where we are going to use rigorous statistical tests that use a different distributional theory for hypothesis testing, to decide on the degree of differencing.

Second, after we have transformed the data, we again study the appearance of the sample ACs and partial ACs to tentatively identify the order of p and q. It is to be recalled here how we have already discussed how theoretical ACs and partial ACs behave for AR, MA, and mixed processes. In general, AR (MA) behavior, as indicated by the ACF, tends to mimic MA (AR) behavior as measured by the partial ACF. Besides, in general, the ACF of a mixed process containing a pth-order AR component and a qth-order MA component is a mixture of exponentials and damped sine waves after the first $q - p$ lags. Conversely, the partial ACF is dominated by a mixture of exponentials and damped sine waves after the first $p - q$ lags. One can use the previously referred Bartlett formula to adjudge if the ACs are statistically insignificant beyond any hypothesized lag. For partial ACs, one can use the fact that partial ACs of a order higher than p are approximately independent and normally distributed with a zero mean.

In the estimation stage, we obtain efficient estimates of the unknown parameters of the selected model. However, in practice, an examination of the plots of ACs and partial ACs does not conclusively point to a single model. So, one ends up fitting different models, by estimating the unknown parameters either by the least-squares method or by the maximum likelihood method, the important issues behind, which we have already discussed.

At the third stage, we need to subject the estimated model to diagnostic checks, by analyzing the residuals for any residual dependency and the goodness of fit, by using some information criterion, which we shall discuss shortly. If no lack of fit is indicated, then, the model is ready to use. But, if the lack of fit is indicated, then, the iterative cycle of identification, estimation, and diagnostic checking is repeated until a suitable model is found.

A disadvantage of the above iterative procedure is that there is a strong dependence on individual judgment in such an identification procedure. For the same data set, two analysts may arrive at very different models. This procedure may become even more difficult for seasonal data. Proceeding in this way may involve fitting several possible models and a final selection may be made on the basis of some summary statistics such as the t-statistic, correlation behavior of the residuals, and the predictive success of the model.

As an alternative method, one can estimate a battery of models in the range of say $p = 4$ and $q = 4$ and then use some information criterion to automatically select a suitable model. The information criterion involves using a function of the residual sum of squares combined with a penalty for a large number of parameters. Two such popular criteria are the Akaike information criterion (AIC) and the Schwarz information criterion (SIC)—that is also known as Schwarz Bayesian criterion (SBC) or Bayesian information criterion (BIC). These are defined as follows:

$$\text{Akaike information criterion: } \log(RSS/T) + 2K/T,$$

$$\text{Schwarz information criterion: } \log(RSS/T) + \log(T)K/T.$$

In the above criteria, *RSS* denotes the residual sum of squares; $K = p + q + 1$ denotes the number of parameters, including a constant; and T is the sample size. In the above formula, the second term corresponds to a “penalty factor” for the inclusion of additional parameters in the model. In the information-criteria approach, we select the model that yields the *minimum* value for the criterion used. And, we compare the various *AIC* and *BIC* values obtained for the various models as the basis for the selection of an appropriate model. Hence, since the *SIC* criterion imposes a greater penalty for the number of estimated model parameters than does *AIC*, the use of a minimum *SIC* for model selection will always result in a chosen model whose number of parameters is not greater than that chosen by *AIC*.

Though this is a popular procedure, an immediate disadvantage is that several models will have to be built, which may be computationally time consuming and expensive. Moreover, it may not work well in case we want to fit models that skip lags. Understanding the way that *ARMA* models are estimated needs a lot of practice. Though analysts generally rely on *SIC* to select an appropriate model, the theoretical results are conflicting about which of these is better. If our combination contains the true model, then *SIC*, given enough data, will definitely choose it, while *AIC* may choose a higher-order model—that is, *SIC* is consistent and *AIC* is not. But if our selected space of models does not contain the true model—such as in the case of infinite lags—then *AIC* is shown to be better in picking out an approximate model—that is, *AIC* is efficient while *SIC* is not.

Our job does not end with identifying and estimating the selected models. We have to apply some diagnostic checks on the estimated model. In a stationary-model setup, our aim is to exploit the dependence of the sample observations and build *ARMA* models. If we are confident that we have identified a model that represents the given data well, then, ideally, there should not be *any* indication of dependency after we have estimated the model. This essentially consists of checking the residuals for any more residual dependency and this residual series should be a white-noise series. This is popularly called the *diagnostic checking of the residuals.* We follow these steps:

1. After estimating the model, either by the *OLS* or *ML* method, we obtain the residuals.
2. A check of AC of the residuals should offer a valuable evidence concerning any lack of fit and a possible model of model inadequacy. This essentially means that the residual series should not display any dependency and should be a white-noise series, meaning that the AC values, $r_k(\hat{e})$ of the residuals, $\hat{e}$'s, should not be significantly different from their theoretical values of zero. Recalling from our earlier discussion of Bartlett formula, this means that one can use the standard error $T^{-1/2}$ to evaluate $r_k(\hat{e})$.
3. Subsequently, however, it has been proved that using this statistical measure to assess the significance of individual ACs may lead to dangerous and misleading conclusions. Hence, a better measure was suggested by Box–Pierce called the Box–Pierce Q statistic, which assessed the adequacy of an estimated model by considering, say the first 25 ACs *collectively* rather than individually. Box–Pierce show that if the estimated model is adequate, then, the statistic

$$Q = T \sum_{k=1}^{K} r_k^2(\hat{e})$$

should follow the χ^2 distribution with $(K - p - q)$ degrees of freedom, where K is the number of ACs that we select; p and q are the number of *AR* and *MA* parameters in the model; and T is the sample size. A very large value of Q will indicate an inadequate model.
4. However, later on, it was pointed out by Ljung–Box that the χ^2 distribution does not provide a sufficiently accurate approximation to the distribution of the Q statistic under the null hypothesis and suggest, the following modified statistic:

$$Q^* = T(T+2) \sum_{k=1}^{K} (T-k)^{-1} r_k^2(\hat{e}),$$

such that the modified Q^* statistic is more-approximately distributed as a χ^2 distribution with $(K - p - q)$ degrees of freedom. And, this modified statistic has been recommended for use in practice and is better suited to check the residuals, especially from *ARMA* models.
5. Examining the squared residuals may often suggest departures of data from the fitted model, which cannot be detected from the *levels* of the residuals. Checking the squared residuals

can be useful if one is checking for time-varying conditional variances, which is the so-called *ARCH* effect. To check this, McLeod and Li (1983) suggest to apply the Ljung–Box statistic, say Q_{ML}, to the *squared* residuals. McLeod and Li show that Q_{ML} has an approximate $\chi^2(K)$ distribution under the null hypothesis that the first K ACs of squared residuals, $\hat{e}^2$, are zero. This is also called testing for *ARCH* effects. But this test is normally done if the underlying time series is a financial time series, such as the returns from a financial asset.

9.3.5.1 Application

We build univariate models for all the four stationary series displayed in Figure 9.1. We start by estimating the ACs and partial ACs and these are plotted in Figure 9.4.

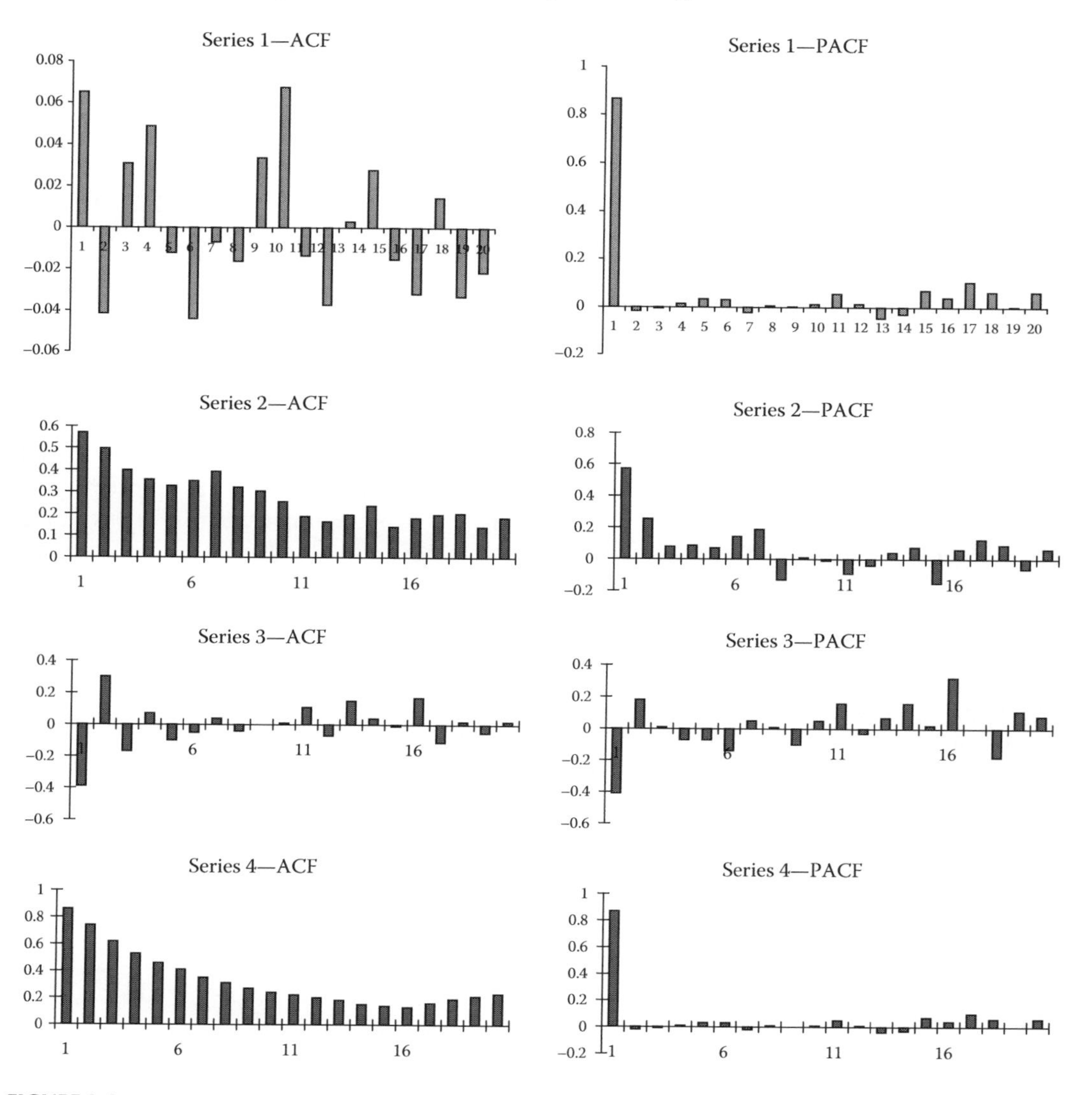

FIGURE 9.4
Correlations of a stationary time series.

TABLE 9.3

AIC and SIC Values

AR Models		MA Models 0		MA Models 1		MA Models 2		MA Models 3
0			1	−8.242/−8.240	1	−8.250 / −8.241	1	−8.246/−8.238
			2	−2.066/−2.045	2	−2.171/−2.138	2	−2.213/−2.163
			3	4.875/4.940	3	4.826/4.922	3	4.841/4.969
			4	−1.738/−1.713	4	−2.045/−2.009	4	−2.193/−2.145
1	1	−7.976/−7.974	1	−8.225/−8.220	1	−8.245/−8.237	1	−8.244/−8.234
	2	−2.226/−2.210	2	−2.296/−2.246	2	−2.282/−2.232	2	−2.302/−2.235
	3	4.816/ 4.880	3	4.818/4.914	3	4.921/5.049	3	4.862/5.023
	4	−2.392/ −2.368	4	−2.386/−2.350	4	−2.380/−2.332	4	−2.374/−2.313
2	1	−8.246/ −8.241	1	−8.243/−8.235	1	−8.243/8.233	1	−8.243/−8.231
	2	−2.160/−2.127	2	−2.306/−2.256	2	−2.338/−2.231	2	−2.292/−2.208
	3	4.811 /4.907	3	4.839/4.968	3	4.862/5.022	3	4.877/5.070
	4	−2.322/−2.286	4	−2.380/2.331	4	−2.362/2.302	4	−2.374/−2.302
3	1	−8.246/−8.239	1	−8.246/−8.236	1	−8.243/−8.231	1	−8.248/−8.232
	2	−2.186/−2.136	2	−2.298/−2.233	2	−2.358 / −2.275	2	−2.309/−2.209
	3	4.839/4.968	3	4.850/5.011	3	4.878/5.071	3	4.904/5.129
	4	−2.380/2.332	4	−2.372/−2.313	4	−2.393 /−2.321	4	−2.381/−2.300

Note: The above table shows the AIC/SIC values for the various models in combinations of $p = 0, 1, 2, 3$ and $q = 0, 1, 2, 3$. The model with the minimum AIC/SIC value will be selected as the representative one for the given sample.

A visual inspection of these figures says that for series 1 and 4, the AR(1) model may be a good fit with positive parameters. Series 2 may have a long memory or may follow a mixed process while series 3 may be a second-order AR process or may be a mixed process. But identifying in this way is tentative, time consuming, and a lot of judgement goes into it. So, we estimated all the models in the combination $p = 0, 1, 2$, and 3 and $q = 0, 1, 2$, and 3, which is a total of 15 models for each series. And we have tabulated the estimated AIC and SIC values in Table 9.3. In the table, the shaded figures refer to AIC values and the boxed values show the SIC values. And it also displays the models selected for the various series by these two criteria. Since SIC penalizes more for additional parameters, in Table 9.4, we have displayed the estimated coefficients of the models selected by the SIC and the results of the diagnostic analysis in Table 9.5. For these diagnostic tests, we used $K = 25$ for the first two series and $K = 20$ lags for the last two series. Referring to the χ^2 table with relevant degrees of freedom, the diagnostics clearly say that, at the 5% level, except for series 1, the models are quite representative of the series. Since series 1 contains data on stock returns, we show elsewhere that we have to account for the heteroscedasticity to obtain a representative model.

In Table 9.3, the values within boxes specify the minimum SIC values among all the calculated SIC values for all models. Similarly, the shaded values specify the minimum AIC values. One can see that the following models have been selected for the various series:

	AIC	SIC
1.	ARMA(0,2)	ARMA(0,2) (or) ARMA(2,0)
2.	ARMA(3,2)	ARMA(3,2)
3.	ARMA(2,0)	ARMA(1,0)
4.	ARMA(3,2)	ARMA(1,0)

TABLE 9.4

Estimates of Fitted Models Using SIC Criterion

Series	NOBS	Fitted Models	Residual Variance
1	2252	$z_t = 0.003\ (\pm 0.001) + e_t + 0.071\ (\pm 0.021)e_{t-1} - 0.041\ (\pm 0.020)e_{t-2}$	0.016
		$z_t - 0.068\ (\pm 0.021)z_{t-1} - 0.046\ (\pm 0.021)z_{t-2} = 0.004\ (\pm 0.001) + e_t$	0.016
2	197	$z_t - 2.326\ (\pm 0.064)z_{t-1} + 1.657\ (\pm 0.127)z_{t-2} - 0.330\ (\pm 0.064)z_{t-3}$	0.304
		$= 16.99\ (\pm 0.012) + e_t - 1.999\ (\pm 0.028)e_{t-1} 0.999\ (\pm 0.028)e_{t-2}$	
3	70	$z_t - 0.419\ (\pm 0.114)z_{t-1} = 51.266\ (\pm 0.925) + e_t$	10.955
4	310	$z_t - 0.869\ (\pm 0.028)z_{t-1} = 9.108\ (\pm 0.128) + e_t$	0.301

Note: Values in parentheses denote the standard errors.

TABLE 9.5

Summary of Diagnostic Tests on Residuals of Fitted Models

Series	NOBS	Fitted Models	$\hat{Q}$	DF
1	2252	ARMA(0,2) [ARMA(2,0)]	43.04 [43.30]	23 [23]
2	197	ARMA(3,2)	8.82	20
3	70	ARMA(1,0)	18.01	19
4	310	ARMA(1,0)	11.88	19

9.3.6 Modeling Nonstationary Processes: Unit Roots, Testing, and Implications

Any attempt at modeling the time-series data cannot confine to analyzing a stationary time series alone. In fact, most of the time series that we encounter in practice are not stationary. A time series can exhibit nonstationarity in many different ways. In particular, most economic time series exhibit nonstationarity either through their time-changing levels—that is, the mean changes through time—and/or through changes in variances over time.

To understand this phenomenon, we have to note that a time series can be decomposed into three components, namely, the trend, cyclical, and the random component. When we talk about a model for the given time-series data, we implicitly mean to construct a model for each component. What is a trend? A trend is the slow, long-run evolution in the variables that we want to model. In business, economics, and finance time series, a trend is usually produced by slowly evolving preferences, technologies, and demographics and this definition tells us that a trend is a slowly evolving component of a time series. And, this is also sometimes called the secular component of the time series. The cyclical component, on the other hand, is believed to be caused, mostly, by nominal factors such as the monetary factors and on a few occasions, by real factors such as those mentioned above. And this cyclical component, as the very name suggests, is generally believed to not show any tendency to grow over time, but is expected to be fluctuating around some level. And the endeavor of any time-series analyst will be to model these two components in the best-possible manner so that the only remaining random component is shown to be truly so, empirically.

The problem of modeling these components is simple in the case of a stationary series, which by nature fluctuates around some value and shows no tendency to grow over time. And this value is the *mean* of the series. Since this does not change over time, we obtain the cyclical component from such a stationary data by simply subtracting the mean and building an ARMA model with the demeaned data. Moreover, since the mean is not time varying, the second moment, which is the variance, will also not be time varying. We have already seen that such time series are nothing but realizations from weakly stationary processes.

But the problem is not so simple in modeling a nonstationary series. As observed before, though there are many ways by which a time-series data may cease to be a stationary series, literature concentrates mainly on the nonstationarity arising due to the time-varying behavior of the first two moments, *namely,* the mean and the variance of the process. Time-varying mean essentially implies that the series under question is growing and one has to model this growing component called a trend and remove it. Such a trend or a secular component could be either deterministic or stochastic and the methods used to remove it are distinct and have profound policy implications. We shall begin by briefly outlining the classical methods to remove such trends.

9.3.6.1 Classical Trend Estimation

Traditionally, the trend is approximated by a polynomial in time, of degree p, such as

$$z_t = \alpha_0 + \alpha_1 t + \cdots + \alpha_p t^p + e_t,$$

where e_t is the noise component that has a mean of zero and is uncorrelated.

The simplest linear time trend results when $p = 1$. The parameters can be estimated by OLS, if the errors are serially uncorrelated. This is unlikely to be the case. So, for example, if e_t follows a stationary $AR(1)$ model, $e_t = \phi e_{t-1} + u_t$, $|\phi| < 1$, $u_t \sim IID(0, \sigma^2)$, then, we have

$$z_t = \beta_0 + \beta_1 t + \phi z_{t-1} + u_t, \quad \text{where } \beta_0 = (1-\phi)\alpha_0 + \phi\alpha_1,\ \beta_1 = (1-\phi)\alpha_1.$$

The statistician has to select p though it must be kept low, since a high-degree polynomial results in a large variability in the trend estimates. Anderson (1994) suggests a way to decide on the appropriate degree within some range of possibilities. The unknown parameters can then be estimated either by OLS or feasible-generalized least squares (FGLS), though FGLS is more efficient. Ng and Vogelsang (2002) explore the conditions under which the FGLS trend estimation can lead to a forecast-error reduction. Besides polynomials in time, other popular functions used to capture a deterministic trend are the Gompertz curve, the modified exponential, and the logistic curve.

Sometimes, a simple function of time may not fit the entire range of data, but may fit only a small interval. This leads to the idea of smoothing. By smoothing, we mean to represent the trend at a point by considering a weighted average of a set of points around it, compared with a polynomial trend where all the observations are used to estimate the polynomial. Given the observations, $z_1, \ldots, z_T$, we shall estimate the trend at t by

$$z_t^* = \sum_{s=-m}^{m} \alpha_s z_{t+s}, \quad t = m+1, \ldots, T-m,$$

which is a weighted average of the observed z_t's with

$$\sum_{s=-m}^{m} \alpha_s = 1.$$

This is sometimes called the two-sided MA process and the sequence $z_{m+1}^*, \ldots, z_{T-m}^*$ is called the MA of the original z_t sequence. Note that we lose m observations both at the start and at the end. A particular case is the arithmetic average, when $\alpha_s = 1/(2m+1)$. The resulting MA is called the equal-weighted MA filter. And, the MA with equal weights is a special case of polynomial smoothing, when the underlying degree of the polynomial is one (see Anderson 1994 for a list of weights to higher-degree polynomials). The MA with unequal weights can be approximated by a succession of smoothing procedures using equal weights. Spencer's 15-point formula is an example of this and is given by

$$z_t^* = \frac{1}{4}\left(-3z_{t-2} + 3z_{t-1} + 4z_t + 3z_{t-1} - 3z_{t+2}\right),$$

then, average z_t^* using an equal-weighted $MA(5)$ and then, apply to this average, an equal-weighted $MA(4)$ filter successively twice. This produces an $MA(15)$ filter with the weights of $\alpha_0 = 74/320$, $\alpha_{\pm 1} = 67/320$, $\alpha_{\pm 2} = 46/320$, $\alpha_{\pm 3} = 21/320$, $\alpha_{\pm 4} = 3/320$, $\alpha_{\pm 5} = -5/320$, $\alpha_{\pm 6} = -6/320$, and $\alpha_{\pm 7} = -3/320$. Another popular procedure is Spencer's 21-point formula. Both formulas reproduce the series perfectly, if it is generated by a cubic polynomial. Kendall (1973) explains through an example, how to derive these weights.

An alternative approximation to higher-order polynomials is to approximate segments of functions by lower-order polynomials and then join the functions at the segments to form a continuous function. One such popular piecewise approximation is called the cubic-spline approximation.

Suppose a curve with cubic-polynomial segments is appropriate. Let the set $\Delta = \{\overline{x}_0 < \overline{x}_1 < \cdots < \overline{x}_k\}$ be a mesh of $[\overline{x}_0, \overline{x}_k]$ and the $k+1 \geq 3$ be the individual points; $\overline{x}_j (j = 0, 1, \ldots, k)$ in the interval are called the knots. The mesh can represent time and the knots can represent the individual time points. Let, $y = (y_0, \ldots, y_k)$, be the corresponding set of ordinates. Suppose $S_\Delta(x)$ is a polynomial of degree at most three on each of the subinterval $(\overline{x}_{j-1}, \overline{x}_j)$ in Δ and has continuous derivatives up to the second order over $[x_0, x_k]$, then, it is called a cubic spline, which is also called the grafted polynomials (see Fuller 1996). We can use the following parameterization to represent the cubic spline, $S_\Delta(x)$, as:

$$S_\Delta(x) = \sum_{j=0}^{3} w_j x^j + \sum_{j=1}^{k-1} w_j^* (x - \overline{x}_j)_+^3$$

where

$$(x - \overline{x}_j)_+ = \max(x - \overline{x}_j, 0) = \begin{bmatrix} (x - \overline{x}_j), & x > \overline{x}_j \\ 0, & x \leq \overline{x}_j \end{bmatrix}.$$

Then, following Pollock (1999), the problem can be reduced to

$$S_\Delta(x) = \sum_{j=-3}^{k-1} w_j^* (x - \overline{x}_j)_+^3.$$

We can estimate the unknown coefficients, w_j^* by regression. Given observations $(x_0, \ldots, x_k)$, we can estimate the coefficients by minimizing the function

$$\sum_{i=0}^{k} \{y_i - S_\Delta(x_i)\}^2 = \sum_{i=0}^{k} \left\{ y_i - \sum_{j=-3}^{k-1} w_j^* (x_i - \overline{x}_j)_+^3 \right\}^2.$$

An accessible introduction to the application of spline functions to a structural change estimation, which may be of a particular interest to economists, is available in Poirier (1976), which also discusses the hypothesis testing for linear and quadratic segments.

A trend-cycle decomposition is implicit in any detrending procedure and the various detrending procedures produce cyclical components with different features. Such classical detrending procedures, especially MAs, though appealing and simple, distort the cyclical and the random effects in a series. The apparent realization that systematic fluctuations can be generated merely by averaging a white-noise series, called the Slutsky–Yule effect, highlights the difficulty in the choice of an MA.*

A different approach to eliminate the trend component, if a series consists of polynomial trends with a superimposed random element, is to simply the difference of the series and get rid of the trend. That is, in the simple case of a linear trend, we have

$$\phi(B)\nabla(z_t - \alpha_0 - \alpha_1 t) = \theta(B)e_t,$$

which is equivalent to $\phi(B)\nabla z_t = c + \theta(B)e_t$ with $c = \phi(1)\alpha_1$ and $\nabla = (1 - B)$, which is the difference operator.

We can generalize this and show that the dth differencing annihilates a polynomial trend of degree $d - 1$. BJR (2008) call such a process as a homogeneous, linear nonstationary process, which can be modeled as the autoregressive-integrated moving-average (*ARIMA*) models. In such models, only the trends are supposed to be stochastic and exhibit the nonstationarity in levels, whereas the nonstationarity in both the levels and the slope will necessitate a second differencing of the series.

But a closer look at the above discussion raises an important question. The issue is not whether the underlying time series follows a polynomial trend, but if the deviations from the trend require differencing. If no differencing is required, then, the underlying trend is supposed to be deterministic and can be removed using any of the classical trend-removal methods. However, Nelson and Plosser (1982) prove that many U.S. macroeconomic time series have stochastic trends. They also point out that, if the secular components in a time series are stochastic, then, the models based on deterministic time-trend residuals are misspecified. Hence, they advocate differencing of the series to remove the trend. Moreover, the question of whether or not to difference the deviations has important implications for forecasting, economic growth, and business-cycle studies. This makes it imperative that next, we outline the differences between processes with a deterministic trend and a stochastic trend and suggest rigorous econometric testing procedures to distinguish the two processes. We shall start by explaining the two types of models in more detail.

9.3.6.2 Trend-Stationary and Difference-Stationary Models

Suppose we remove the trend using the classical method of a *deterministic time trend,* such as

$$z_t = \alpha_0 + \alpha_1 t + \psi(B)e_t.$$

Thus, the mean μ of the stationary process is replaced by a linear function of the date t. It is clear that $E[z_t] = \alpha_0 + \alpha_1 t$, which, by being time dependent, violates one of the conditions of second-order or covariance stationarity. However, it is also evident that the deviations are stationary. Thus, this specification satisfies the traditional view that, if both cyclical (or transitory) and trend (permanent) movements are observed in an economic time series, then, the former component results, after removing the permanent or secular component by the time trend. Hence, detrending the series yields the cyclical component that the business-cycle theory explains. Such a process is sometimes described as a *trend stationary,* or a TS, process.

* The distorting effects of filters are explained more fully in the section on spectral analysis, discussed below.

But there is an alternative view to look at nonstationarity. The second specification is the *unit-root* process, also called a *difference-stationary* (DS) process

$$(1-B)z_t = \alpha_1 + \psi(B)e_t,$$

where $\psi(1) \neq 0$. This condition rules out the possibility that the original series is stationary. If we set $\psi(B) = 1$ in the above unit-root process, then, we obtain

$$z_t - z_{t-1} = \alpha_1 + e_t.$$

This process is called a *random walk (RW) with drift* with α_1 being the drift parameter and this is one of the simplest models in this class of nonstationary time-series models. The model without the drift term is called a *pure RW* model. That is,

$$z_t - z_{t-1} = e_t.$$

By repeated substitution in the pure RW and RW-with-drift model, we obtain

$$z_t = z_0 + \sum_{i=1}^{t} e_i$$

$$z_t = z_0 + \alpha_1 t + \sum_{i=1}^{t} e_i,$$

where it is clear that z_t in a pure RW is just an accumulation of past shocks, while in the model with a drift, deviations from a time trend are *still* not stationary, since they are accumulations of past shocks. However, note that if one *first* differences these models, then, ∇z_t is indeed stationary. Hence, the name *DS* process. Such processes are also sometimes called as processes integrated of order *one*. That is indicated as $z_t \sim I(1)$. The term "integrated" has the same meaning as in calculus. Such processes along with the *AR* and the *MA* operators are called an *ARIMA*(p,d,q) model. Here, p refers to the *AR* operator, unit roots that are not included; d refers to the order of integration; and q refers to the *MA* operator. After taking d differences, the *ARIMA*(p,d,q) model reduces to a stationary *ARMA*(p,q) model.

9.3.6.2.1 *Differences between TS and DS Models*

How do we distinguish one process from the other? The two processes differ in terms of forecasts, roots of the polynomials, persistence of shocks, and, most importantly the asymptotics concerning these two processes. It is the asymptotic issues that make differentiating between the two processes compelling.

Forecasts from a *TS* process are obtained by just adding the known deterministic component to the stationary-stochastic component:

$$\hat{z}_{t+s|t} = \alpha_0 + \alpha_1(t+s) + \psi_s e_t + \psi_{s+1}e_{t-1} + \psi_{s+2}e_{t-2} + \cdots.$$

Note that as the forecast horizon increases, the absolute summability of the ψ weights implies that this forecast converges to the time trend, $\alpha_0 + \alpha_1 t$. Thus, the process is said to be *trend reverting.*

On the other hand, a *DS* process will behave differently. Recall that the level of a variable at time $t+s$, given the changes, is just the sum of changes between time t and $t+s$; that is,

$$\begin{aligned} z_{t+s} &= (z_{t+s} - z_{t+s-1}) + (z_{t+s-1} - z_{t+s-2}) + (z_{t+1} - z_t) + z_t \\ &= \nabla z_{t+s} + \nabla z_{t+s-1} + \cdots + \nabla z_{t+1} + z_t. \end{aligned}$$

Taking the linear projection of the above equation and adjusting, we obtain

$$\hat{z}_{t+s|t} = s\alpha_1 + z_t + (\psi_s + \psi_{s-1} + \cdots + \psi_1)e_t + (\psi_{s+1} + \psi_s + \cdots + \psi_2)e_{t-1} + \cdots.$$

Thus, forecasts from a *DS* process keep diverging and *do not* revert to the trend.

One can distinguish the TS process from the DS process by testing for a root equal to unity in the *MA* and the *AR* polynomials, respectively. To make this clear, let us start with a TS process:

$$z_t = \alpha_0 + \alpha_1 t + \psi(B)e_t.$$

Since $\psi(B) = \theta(B)/\phi(B)$, we rewrite the model as

$$\phi(B)z_t = \phi(B)\alpha_0 + \phi(B)\alpha_1 t + \theta(B)e_t.$$

Suppose this were the true model, and we first difference it by mistake. Upon first differencing the TS process and adjusting, we obtain

$$\phi(B)(1-B)z_t = \alpha_1\phi(B=1) + \theta(B)(1-B)e_t,$$

where $\phi(B=1)$ indicates the constant obtained by evaluating the polynomial $\phi(B)$ at $B=1$. Note that one of the roots of the *MA* polynomial is equal to one. This indicates that if the true model were a TS process, and we differenced it by mistake, then, we should end up obtaining a unit root in the *MA* polynomial, making the resultant model *noninvertible*.

On the other hand, what if we just fit a time trend to a *DS* process? This mistake should manifest in the form of a unit root in the *AR* polynomial. If we estimate the *TS* model by mistake,

$$z_t = \alpha_0 + \alpha_1 t + \psi(B)e_t,$$

then, $\psi(B)$ will have a unit root in the *AR* polynomial. To understand this, factorize the *AR* polynomial and allow one factor to be equal to one so that

$$\psi(B) = \frac{(1 + \theta_1 B + \theta_2 B^2 + \cdots + \theta_q B^q)}{(1-B)\phi^*(B)}.$$

Rewriting the "wrong" model, we obtain, after adjustment

$$\phi^*(B)(1-B)z_t = \alpha_1\phi(B=1) + \theta(B)e_t,$$

which clearly shows the presence of a unit root in the *AR* polynomial. To sum up, if the true process is a *TS* process, then, we should fail to reject a unit *MA* root in first the *differences* and if the process is a *DS* process, then, we should fail to reject an *AR* unit root in *levels*. Thus, checking if the series has a unit root in the *AR* polynomial is the so-called *unit-root* testing procedure. If it is proved, we say that $z_t \sim I(1)$. And, we have to *first* difference the series before any univariate modeling.

Chan, Hayya, and Ord (CHO) (1977) using spectral-analytic techniques show that first differencing a TS process does induce the stationarity, but produces a spurious first-order negative AC in the residuals. Moreover, the high-frequency portion of the spectral density will be exaggerated and the lower-frequency portion will be attenuated, resulting in high-frequency cycles. On the other hand, if the true model is an RW, and if we detrend it linearly, then, we will obtain a large spurious-positive AC in the first few lags and the residuals will be dominated by low-frequency cycles. Nelson and Kang (1981) qualify these results further. Taken on the whole, these results seem to suggest that an inappropriate detrending of the time series will produce an evidence of periodicity that is not part of the dynamics of the original time series. Nevertheless, such practices are still widespread.

Yet another difference between the TS and unit-root processes is the persistence of innovations or shocks. Here, we consider the consequences for z_{t+s}, if e_t were to increase by one unit with e's for all the other dates unaffected. It is natural to answer this question using the *MA* representation. In a TS process, the shocks will have only a temporary effect. That is, $\partial z_{t+s}/\partial e_t = \psi_s$. Because of an absolute summability, the effect of any stochastic disturbance eventually wears off. That is, $\lim_{s\to\infty}(\partial z_{t+s}/\partial e_t) = 0$. In contrast, for a unit root or a DS process, it can be easily seen that

$$\frac{\partial z_{t+s}}{\partial e_t} = \frac{\partial z_t}{\partial e_t} + \psi_s + \psi_{s+1} + \cdots + \psi_1 = 1 + \psi_1 + \psi_2 + \cdots + \psi_s.$$

Thus, an innovation e_t has a permanent effect on the level of z_t that is captured by

$$\lim_{s\to\infty} \frac{\partial z_{t+s}}{e_t} = 1 + \psi_1 + \psi_2 + \cdots = \psi(1).$$

$\psi(1)$ measures the permanent effect of a shock e_t on the level of z_t with $\psi(1) = \theta(1)/\phi(1)$. Sometimes, $\psi(1)$ is interpreted as the *long-run effect* of a shock *relative* to the immediate effect of a shock. So, if $\psi(1)$ is also equal to one, then, the long-run effect of a shock is equal to the immediate effect.

The most-important reason why identifying if a time series is a *DS* process is important is the fact that classical asymptotic results are invalid on coefficients of a *DS* process. This leads us to the so-called unit-root analysis. And the parameters of the test regressions have nonstandard asymptotic distributions.

9.3.6.3 Processes with Unit Roots

After Nelson and Plosser (1982) highlighted the importance of removing the trend by differencing, it has become a virtual necessity to check for the presence of unit roots in any time-series dataset before embarking on any further modeling. Just what is this unit-root problem? Unless stated otherwise, it refers to the presence of a root equal to +1 in the AR polynomial of an ARMA model. If proved, the series needs to be first differenced for any further analysis in a univariate setting. Can there be two roots equal to unity? It is possible but very rare and we shall not deal with such a case here. But often in practice there may be some doubt about the decision to first difference the data.

Informal tools are often used to help an analyst decide if he needs to first difference and such tools include the following:

1. Examination of plots of the series and its differences and check for the wandering behavior of the series, so characteristic of a nonstationary process. An indication of this will be the tendency for the given series to wander away from its mean μ for long stretches of time.
2. Inspection of the ACF of the series and its differences, for failure to damp out quickly.

3. In case the first two steps are not helpful, one can fit an ARMA model to the raw data, ignoring any nonstationarity for the time being. And a formal inspection of the AR polynomial of the fitted model for the presence of a root equal to or very near unity at this stage shall be very useful.

But we run into trouble when we have to decide about a root that is very close to unity. If $\hat{\phi}(B)$, the AR polynomial of the fitted model, has a root exactly equal to one, then unambiguously we can say that we need to first difference the data. But if $\hat{\phi}(B)$ has a factor $(1-\hat{\rho}B)$ that is close to $(1-B)$, a question arises if $(1-\hat{\rho}B)$ is significantly different from $(1-B)$; this is equivalent to checking if $\hat{\rho}$ is significantly different from one. We may need formal statistical inference tools. We shall see below the use of these tests in detecting roots equal to unity. As pointed out earlier, we shall confine our discussion to the case of a root being equal to $+1$ only.

Consider the *OLS* estimation of a Gaussian $AR(1)$ process,

$$z_t = \rho z_{t-1} + e_t,$$

where $e_t \sim N(0, \sigma_e^2)$ and $z_0 = 0$. The *OLS* estimate of ρ is given by

$$\hat{\rho} = \frac{\sum z_{t-1} z_t}{\sum z_{t-1}^2}.$$

Here the distribution property tells us that if the true value of ρ is less than 1 in absolute value, then

$$\sqrt{T}(\hat{\rho}-\rho) \xrightarrow{D} (0, (1-\rho^2)).$$

However, when the null is $\rho = 1$ this reduces to a degenerate distribution when $\sqrt{T}\big(\hat{\rho}-1\big)$ has a *zero* variance or that the distribution collapses to a point mass at zero. In such a scenario, to avoid getting a degenerate distribution for $\hat{\rho}$ in the unit-root case, we have to multiply $\big(\hat{\rho}-1\big)$ by T rather than by $\sqrt{T}$. The distributional properties are normally explained by rewriting the parameter as deviations from the true value, so that

$$T\big(\hat{\rho}-1\big) = \frac{1/T \sum z_{t-1} e_t}{1/T^2 \sum z_{t-1}^2}.$$

It can be shown that, as $T \to \infty$,

$$2(T\sigma^2)^{-1} \sum_{t=2}^{T} z_{t-1} e_t \xrightarrow{D} \chi^2(1)$$

$$T^{-2} \sum_{t=2}^{T} z_{t-1}^2 \xrightarrow{D} \sigma^2 \int_0^1 [W(r)]^2\, dr,$$

such that

$$\frac{T^{-1}\sum z_{t-1} e_t}{T^{-2}\sum z_{t-1}^2} \xrightarrow{D} \frac{1/2\big\{[W(1)]^2 - 1\big\}}{\int_0^1 [W(r)]^2 dr},$$

where $[W(1)]^2$ is $\chi^2(1)$ variable and $[W(r)]^2$ is the standard Brownian motion, also called a Wiener process (see Fuller 1996; Hamilton 1994 for a detailed derivation). This is sometimes called the normalized bias test (NBT). Thus, NBT has a nonstandard distribution and the critical values have to be obtained through simulation, though some exact numerical procedures are available. The popular t-statistic to test the null of $\rho = 1$ can also be computed the usual way, but the resulting limiting distribution is not the usual Gaussian distribution; hence the critical values have to be simulated. That is,

$$\hat{t}_\rho = \frac{\hat{\rho} - 1}{\hat{\sigma}_\rho} \quad \text{and} \quad \hat{t}_\rho \xrightarrow{D} \frac{1/2\{[W(1)]^2 - 1\}}{\left\{\int_0^1 [W(r)]^2 dr\right\}^{1/2}},$$

where $\hat{\sigma}_\rho$ is the standard error of $\hat{\rho}$.

Various ways have been suggested in the literature to check if the data need differencing. It can be classified as follows:

1. Those that test for the presence of unit roots by taking nonstationarity as the null. These are called nonstationarity tests. That is, in a first-order autoregression, we test the null, $\rho = 1$ against the alternative of $\rho < 1$. If accepted, then we conclude that the data have a root equal to unity. Popular tests in this category are the Dickey–Fuller (DF), the augmented DF (ADF), and the Phillips–Perron (PP) tests. Note that we normally conduct a left-tail test for the presence of unit roots. This is because $\rho > 1$ implies that the underlying series is an explosive series and such series are a rarity especially in economics, though stock prices may exhibit an explosive behavior. We discuss tests for such an explosive time series below, where we test the null of $\rho = 1$ against the alternative of $\rho > 1$. These tests are called the right-tail unit-root tests.
2. Those tests that have the stationary case as the null as against the alternative of nonstationarity. These are called stationarity tests. That is, we have here as the null, $|\rho| < 1$ as against the alternative $\rho = 1$. Popular tests in this category are the Kwiatkowski–Phillips–Schmidt–Shin (KPSS) (1992) test and Leybourne and McCabe (1994) test.
3. Those that exploit the idea that any overdifferencing will result in a root equal to unity in the *MA* polynomial. These may be called as tests of overdifferencing. Test procedures suggested by Breitung (1994) and Choi and Yu (1997) fall under this category. However, this procedure is not popular among applied researchers.

We shall however confine ourselves only to outlining those testing procedures that have nonstationarity as the null, for the simple reason that these are the most-popular procedures of testing for a unit root among researchers. We proceed as follows:

1. We check for unit roots in the *OLS* estimation of an $AR(1)$ process, where we assume that the errors are white noise. That is, in the model,

$$z_t = \rho z_{t-1} + e_t$$

   we shall test if the null $\rho = 1$ can be accepted. Testing this null under the assumption of white-noise errors is called the *DF* test for unit roots.
2. As a next step, we shall extend this testing procedure to cover processes, whose first differences exhibit serial correlations. That is, e_t in the above specification is not a white-noise series; but it is a stationary process that can follow an $AR(p)$ process. Testing for the null of $\rho = 1$ under this condition leads us to two popular tests in the literature: (1) *PP* tests and (2) the *augmented Dickey–Fuller (ADF)* tests. While the ADF test corrects for the serial correlation with a parametric *AR* structure, the PP test corrects the same in a nonparametric way.

We shall discuss below some popular models used to test for unit roots, under different assumptions about the trend term. While testing for unit roots, we have to remember the important distinction between the maintained null hypothesis (or the true DGP) and the test (or estimating) regression. We shall consider below various cases that allow z_t to follow different *DGPs*:

Suppose we are interested in testing if our data are a pure RW. That is, we posit our true model to be

$$z_t = z_{t-1} + e_t.$$

Test of unit roots will depend on the test regression that we use. Since we test the null of a pure RW, the best way to check our claim will be to use the following two test regressions:

$$\begin{aligned} \textit{Case 1: } z_t &= \rho z_{t-1} + e_t \\ \textit{Case 2: } z_t &= \alpha + \rho z_{t-1} + e_t \end{aligned}$$

It has to be emphasized here that the asymptotic distribution of ρ given by $\hat{\rho}$ will not be the same for these test regressions.

Case 1: Here we are interested in the asymptotic properties of the *OLS* estimate, $\hat{\rho}$, given by

$$\hat{\rho} = \frac{\sum_{t=1}^{T} z_{t-1} z_t}{\sum_{t=1}^{T} z_{t-1}^2}$$

The null and the alternative hypotheses to be tested are as follows:

$$\begin{aligned} H_0 : \rho = 1 \Rightarrow z_t \sim I(1) \quad & \text{a pure random walk,} \\ H_1 : |\rho| < 1 \Rightarrow z_t \sim I(0) \quad & \text{a stationary } AR(1) \text{ model with zero mean.} \end{aligned}$$

We have already shown that the asymptotic distribution of the OLS estimate of ρ is nonstandard and critical values are to be calculated either through simulation or exact numerical procedures. Critical values calculated through simulation by Fuller under various trend specifications are often used by applied researchers. The critical values for the NBT, with the test regression as in *Case 1*, are available in Fuller (1996, Table 10.A.1) under $\hat{\rho}$. One may also use the popular t-statistic to test the null of $\rho = 1$. Again, this statistic is calculated in the usual way; that is

$$t = \frac{\hat{\rho} - 1}{\hat{\sigma}_\rho},$$

where $\hat{\sigma}_\rho$ is the standard error of $\hat{\rho}$. But appropriate distribution is again a function of the standard Brownian motion. The relevant critical values are available in Fuller (1996, Table 10.A.2) under $\hat{\tau}$. It is pertinent to point out here that if we continue to use Gaussian distribution for inference, we will end up rejecting the null of unit roots more often, implying we will be committing a type-1 error. This is easily understood by noting that $P\{(\hat{\rho} - 1)/\hat{\sigma} < -1.95\} = 5\%$ according to DF critical values as $P\{(\hat{\rho} - 1)/\hat{\sigma} < -1.65\} = 5\%$ by the standard-normal distribution.

Case 2: Here we are still interested in checking if the DGP is a pure RW model, but with a nonzero mean under the alternative

$$z_t = \alpha + \rho z_{t-1} + e_t.$$

The null and the alternative hypotheses to be tested are as follows:

$$H_0: \rho = 1,\ \alpha = 0 \Rightarrow z_t \sim I(1) \quad \text{a pure random walk,}$$

$$H_1: |\rho| < 1 \Rightarrow z_t \sim I(0) \quad \text{a stationary } AR(1) \text{ model with nonzero mean.}$$

As already pointed out, the presence of deterministic terms such as a constant affects the asymptotic distributions of the estimated ρ and hence we need a different set of critical values to test the null for this case. The relevant critical values for the NBT, with the test regression as in *Case 2,* are given in Fuller (1996, Table 10.A.1) under $\hat{\rho}_{\mu}$ and the critical values to conduct the standard t-statistic are available in Fuller (1996, Table 10.A.2) under $\hat{\tau}_{\mu}$. The t-test of the constant term also follows a nonstandard distribution, a functional of the Wiener process. The relevant critical values are available in Dickey and Fuller (1981, p. 1062). However, the maintained assumption of $\alpha = 0$ followed while deriving the asymptotic distribution of $\hat{\rho}$ needs to be ascertained as well. So it makes sense to check the combined null that $\alpha = 0$, $\rho = 1$ through an F-test. Again these F-values are not the standard F-values but are those that have been calculated by Dickey and Fuller (1981, Table IV, p. 1063). It will be useful to point out here that the models discussed above are suitable to test nontrending economic series such as interest series, exchange rates, and rate spreads.

Case 3: How do we deal with series that exhibit a clear trend? Are such trends captured better by an RW model with a drift only or by the simple deterministic time-trend model? This question arises while we try to model the trending time series such as the asset price or the levels of macroeconomic aggregates such as a real GDP. To test this, consider the more-general test regression with which one can test both the specifications:

$$z_t = \alpha + \alpha_1 t + \rho z_{t-1} + e_t.$$

In this specification, neither the presence nor the value of the term α affects the asymptotic distribution of $\hat{\rho}$. That is, it does not matter if the true value of α is zero or not. Thus, the maintained hypothesis will be that $\alpha = \alpha_0$. The null and the alternative hypotheses to be tested are

$$H_0: \rho = 1,\ \alpha_1 = 0 \Rightarrow z_t \sim I(1) \quad \text{a random walk with drift,}$$

$$H_1: |\rho| < 1\ \alpha_1 \neq 0 \Rightarrow z_t \sim I(0) \quad \text{a stationary } AR(1) \text{ model with a deterministic time trend.}$$

The NBT, $T(\hat{\rho} - 1)$, and the t-statistic computed from the test regression converge asymptotically to a nonstandard distribution and the critical values to evaluate these statistics are available, respectively, in Fuller (1996, Table 10.A.1) under $\hat{\rho}_{\tau}$ and in Fuller (1996, Table 10.A.2) under $\hat{\tau}_{\tau}$. Some of the relevant joint tests are as follows. These are the usual-regression F-tests and are calculated the usual way.

1. Since the true value of α_1 was assumed to be zero, we need to conduct a joint test of ($\alpha = \alpha_0$, $\alpha_1 = 0$, $\rho = 1$) to ascertain that the asymptotic distribution of the hypotheses depends on the assumption. We calculate the F-test the usual way, but the resulting distribution however is a nonstandard one and the calculated F-statistic should be compared against the critical values calculated by DF through simulation and are available in Dickey and Fuller (1981, Table VI, p. 1063).
2. Suppose at this stage an analyst is interested in testing if the true process is a pure RW. That is, he is interested in testing the joint null of ($\alpha = \alpha_1 = 0$, $\rho = 1$). The relevant critical values are available in Dickey and Fuller (1981, Table-V, p. 1063). If this null is accepted, one can see that the underlying process boils down to a pure RW.

3. Continuing with the same null of a pure RW, suppose he is interested in the individual significance of the estimated t-statistic of α and α_1. These can also be evaluated; but again, the distributions of these statistics are nonstandard and can be compared against the critical values in Tables II and III of Dickey and Fuller (1981, p. 1062).

9.3.6.4 Unit-Root Tests for a Wider Class of Errors

The *DF* procedures discussed so far have assumed *iid* errors and $z_0 = 0$. These assumptions may not be tenable with the real-world data. So we shall discuss how to modify these test procedures for unit roots under the assumption of a wider class of errors where, say, the serial correlation in errors is allowed. When the errors are correlated, we may have to either change the test regression or modify the statistics obtained earlier to get consistent estimators and statistics. Dickey and Fuller adopt the first strategy of changing the test regression, using a parametric approach, resulting in what is widely known as the ADF test. On the other hand, Phillips and Perron follow the second approach of modifying the existing statistics using a nonparametric approach, leading to the PP test. We shall start with the ADF test.

9.3.6.4.1 Adjusting the Estimating Equation: ADF Test

With *ADF* tests, we check the null hypothesis that a time series z_t is $I(1)$ against the alternative that it is $I(0)$ assuming that the dynamics in the data have an ARMA structure, which will be approximated by a long-order *AR* model. But before we outline the *ADF* test procedure, we shall show that by transforming an *AR* model in levels, we can isolate a unit root as a testable coefficient in models of order higher than one. The transformed $AR(p)$ model can be written as

$$\{(1 - \rho B) - (\phi_1^* B + \phi_2^* B^2 + \cdots + \phi_{p-1}^* B^{p-1})(1 - B)\} z_t = e_t, \quad \text{or}$$

$$z_t = \rho z_{t-1} + \phi_1^* \nabla z_{t-1} + \phi_2^* \nabla z_{t-2} + \cdots + \phi_{p-1}^* \nabla z_{t-p+1} + e_t,$$

where $\rho = \phi_1 + \phi_2 + \cdots + \phi_p$. This transformation implies that to check for unit roots in higher-order *AR* models, we have to simply check if the coefficient ρ is equal to one in the transformed model. If the null of $\rho = 1$ is accepted, then, the transformed model implies that, apart from the advantage of isolating the unit root as a separate coefficient, such transformation makes it easy to estimate the unknowns, because under the null, all regressors are stationary. With this transformed model, we can test for unit roots in all three cases outlined before. We shall discuss only *Case 2,* since the very same adjustment that we discuss here is applicable to other cases also. Generalizing *Case 2* discussed before, we are interested in the following model setup and we want to check for unit roots:

$$z_t = \alpha + \rho z_{t-1} + \phi_1^* \nabla z_{t-1} + \phi_2^* \nabla z_{t-2} + \cdots + \phi_{p-1}^* \nabla z_{t-p+1} + e_t.$$

The null for this model will be $\alpha = 0$ and $\rho = 1$. If the null is accepted, then we have to run the $AR(p-1)$ model on the first differences of the data. If the null is not accepted, we can retrieve the original $AR(p)$ model with a nonzero mean model in levels. In this model, under the null, the coefficients on the stationary regressors, $(\nabla z_{t-1}, \nabla z_{t-2}, \ldots, \nabla z_{t-p+1})$, tend to a limiting normal distribution, with mean zero and covariance matrix $\mathbf{V}$. With this result, one can apply the regular Student t- and F-test statistics for hypothesis testing. However, the coefficients on the constant term and ρ have nonstandard asymptotic distributions. But it has been established that in spite of the presence of lagged changes, a generalization of the *DF* test given earlier for a similar model setup called the augmented Dickey–Fuller (*ADF*) test can be obtained. *ADF* coefficient test or the NBT $T(\hat{\rho}-1)$ divided by $(1-\phi_1^*-\phi_2^*-\cdots-\phi_{p-1}^*)$ has the same distribution as the *DF* test for the same case conducted with serially uncorrelated errors. That is, the *ADF* test for the current setup is given as

$$\frac{T(\hat{\rho}-1)}{1-\phi_1^*-\phi_2^*-\cdots-\phi_{p-1}^*} \xrightarrow{D} \left[\text{the same as given earlier for the DF test for } \textit{Case 2}\right]$$

However, the distribution of the *OLS* t test has the same distribution as before and does not need *any* correction even in the present case of serially correlated errors. Similarly, the joint *DF* F-test of $\alpha = 0$ and $\rho = 1$ is also not affected and is the same as the *DF* distribution tabulated with white-noise errors. It is also interesting to note that any hypothesis tests involving the coefficients of ∇z_{t-j}'s and ρ asymptotically follow a Gaussian distribution and hence regular testing procedures are valid.

One can generalize the above observations to the other two cases also. In these models too, only the NBT needs to be adjusted the way shown above. And the t-test and the F-test, if necessary, need *no* adjustment.

9.3.6.4.2 *Choice of Lag Length for ADF Test*

It has been reported in the literature that *ADF* test is sensitive to the number of lagged terms used. Too few lagged terms will bias the test while too many will reduce the power of unit-root tests. So analysts normally include a large number of lags. Several guidelines have also been suggested for the choice of p:

1. The number of lagged terms can be arbitrarily fixed based on the following rule suggested by Schwert:

$$p = \text{Int}\left\{c(T/100)^{1/d}\right\} \quad \text{where } c = 12 \quad \text{and} \quad d = 4.$$

   But this rule is ad hoc and it need not be optimal for all p and q in an $ARMA(p,q)$ model.

2. Information-based rules such as the AIC and BIC can also be used.
3. Ng and Perron suggest setting an upper-bound *pmax* for p and estimate the *ADF* test regression with $p = pmax$. If the absolute value of the t-statistic for testing the significance of the last lagged difference is greater than 1.6, then, set $p = pmax$ and perform the unit-root test. Otherwise, one can reduce the lag length by one and repeat the process.
4. One can use the sequential-testing procedure of Hall (1994). He suggests two rules to select p in the context of a pure autoregression. The first rule, called the *general-to-specific rule,* is to start with a large value of p, say *pmax*, and test the significance of the last coefficient. If found insignificant, reduce p iteratively till a significant test statistic is obtained. The lag, by the addition of which a significant F-test is obtained, may be selected as the lag order to perform the *ADF* unit-root test. The other rule, called the *specific-to-general rule,* is to start from the smallest number of lags and increase p successively till an insignificant joint statistic is obtained. Hall has shown that the specific-to-general method is not generally valid asymptotically. Several Monte Carlo studies also support using Hall's general-to-specific rule only.

9.3.6.4.3 *Modified Test Statistic: PP Test Statistic*

Unlike the *ADF* test, which modifies the estimating regression, PP suggest modifying *both* the NBT statistic and the t-test statistic obtained under the assumption of serially uncorrelated errors. *PP* suggest the following nonparametric correction of the test statistics for the unit-root null, by using consistent estimates of variances. They call the NBT statistic as Z_ρ statistic and the t-statistic as Z_t statistic. And most importantly, these statistics follow the same distribution as those obtained under the assumption of serially uncorrelated errors by Dickey and Fuller. And they are as follows:

$$Z_\rho = T(\hat{\rho} - 1) - \frac{1}{2}\left(T^2 \cdot \hat{\sigma}^2_{\hat{\rho}} \div s_T^2\right)\left(\hat{\lambda}^2 - \hat{\gamma}_0\right)$$

and

$$Z_t = \left(\frac{\hat{\gamma}_0}{\hat{\lambda}^2}\right)^{1/2} \cdot t_{\hat{\rho}} - \left\{\frac{1}{2}(\hat{\lambda}^2 - \hat{\gamma}_0)/\hat{\lambda}\right\} \times \left\{T \cdot \hat{\sigma}_{\hat{\rho}} \div s_T\right\}.$$

Here

$$\hat{\lambda}^2 = \hat{\gamma}_0 + 2\sum_{j=1}^{q}\left[1 - \frac{j}{q+1}\right]\hat{\gamma}_j,$$

$\hat{\gamma}_j$ are the autocovariances of the residual series $\hat{e}_t$, obtained after running the desired model

$\hat{\sigma}^2_{\hat{\rho}}$ = variance of $\hat{\rho}$

s_T^2 = variance of the residual series, $\hat{e}_t$

The question of selecting the lag order q, for which we calculate the autocovariances of the residual series, is essentially an empirical matter.

If, for example, we are interested in the model setup, defined as *case 2* earlier, we conduct the *PP* test as follows: Run the test regression

$$z_t = \alpha + \rho z_{t-1} + e_t.$$

Calculate Z_ρ and the Z_t statistics using the above formulas, with autocovariances calculated from the residual series, $\hat{e}_t$, obtained from the regression above. Both Z_ρ and the Z_t converge to the same asymptotic distribution as before for *Case 2*.

One can generalize these steps for the other cases also.

9.3.6.5 Efficient Unit-Root Tests and GLS Detrending

Even though ADF and PP tests are asymptotically equivalent, they differ substantially in finite samples. Moreover, subsequent studies on these two tests have found numerous econometric issues—the most important of them being the size distortion and the low power—associated with these two tests. Size distortion—rejecting the $I(1)$ null much too often when it is true—arises in these tests essentially because the associated distributions ignore the presence of an MA component and the low power of these two tests is due to the fact that they cannot distinguish the null of the unit root from alternatives that are highly persistent, but nevertheless stationary processes. Even though some simulation studies show that, with a high-order AR regression, the ADF test may be more useful in practice than the PP test in tackling size distortion, the necessity to find DF-type unit-root tests with more power against a range of stationary alternatives, led to what is called *efficient unit-root tests*, with the contribution of Elliott, Rothenberg, and Stock (ERS) (1996) being the most popular one.

We shall discuss this test below, along with a more-efficient PP test procedure, suggested by Ng and Perron, that has a substantially higher power and does not display any serious size distortion. And it is generally recommended in the literature that one uses such unit-roots tests with better power properties and with the least size distortions.

9.3.6.5.1 Efficient Unit-Root Tests

Suppose we have T observations that obey the model,

$$\begin{aligned} z_t &= \boldsymbol{\beta}'\mathbf{D}_t + e_t \\ e_t &= \rho e_{t-1} + u_t, \quad u_t \sim (0, \sigma^{*2}) \quad (ERS - 1). \end{aligned}$$

Here, $\mathbf{D}_t$ represents a vector of deterministic terms. Typically $\mathbf{D}_t = 1$ or $\mathbf{D}_t = (1, t)'$. Consider testing the null of $H_0 : \rho_1 = 1$ against the alternative $H_1 : |\rho| < 1$. If the true distribution of the data is known, an optimal test with maximum power can be obtained for any point alternative and one can plot the power of this test as a function of $\bar{\rho}$ to get an upper bound or power envelope. But this idea depends on the specific value of $\bar{\rho}$. There is no uniformly most-powerful test that can be used against all alternatives, $|\rho| < 1$. Using asymptotic concepts involving the *local-to-unity* alternative, $\rho = 1 + c/T$, for $c < 0$; ERS derive a class of test statistics that comes very close to the power envelope for a wide range of alternatives. These tests are called as the *efficient unit-root tests* and they have substantially higher power than the ADF and the PP unit-root tests when ρ is close to unity.

9.3.6.5.2 Local-to-Unity Tests

Local-to-unity tests use a sequence of models indexed by the sample size T:

$$\begin{aligned} z_t &= \rho z_{t-1} + e_t, \quad t = 1, \ldots, T \\ \rho &= 1 - \frac{c}{T}, \end{aligned}$$

where the AR parameter is now a function of sample size. Obviously, the true population value is *fixed* and does not change with sample size; but local to unity specifies a model where the parameter changes according to the sample size. This is done as a statistical device to ensure that the same asymptotics apply to both the stationary and nonstationary cases.

9.3.6.5.3 Point-Optimal Tests

Point-optimal tests are tests whose power is quite higher than that of other tests with the same size. That is, it maximizes the power of a test at a predetermined point under the alternative, in contrast to the uniformly most powerful test. We start by noting that a feasible test statistic that is asymptotically optimal for the point alternative can be $\bar{\rho} = 1 + \bar{c}/T, \bar{c} < 0$.

The test is constructed as follows:

1. From $(ERS - 1)$ we obtain on rearrangement, the following T-dimensional vector, $\mathbf{z}_\rho$ and $(T \times q)$ matrix $\mathbf{D}_\rho$:

$$\begin{aligned} \mathbf{z}_\rho &= (z_1, z_2 - \rho z_1, \ldots, z_T - \rho z_{T-1})' \\ \mathbf{D}_\rho &= \left(\mathbf{D}_1', \mathbf{D}_2' - \rho \mathbf{D}_1', \ldots, \mathbf{D}_T' - \rho \mathbf{D}_{T-1}'\right). \end{aligned}$$

   Note here that all variables are quasi-differenced using the operator $(1 - \rho B)$.

2. Next for any value of ρ define $S(\rho)$ as the sum of squared residuals from a least-squares regression of $\mathbf{z}_\rho$ on $\mathbf{D}_\rho$.
3. ERS show that the feasible point-optimal unit-root test against the alternative, $\bar{\rho} = 1 - \bar{c}/T$ has the form

$$P_T = \left[S(\bar{\rho}) - \bar{\rho}S(1)\right]/\hat{\sigma}^{*2},$$

   where $\hat{\sigma}^{*2}$ is a consistent estimate of σ^{*2}.
4. ERS derive the asymptotic distribution of P_T for $\mathbf{D}_t = 1$ and $\mathbf{D}_t = (1, t)'$ and provide asymptotic and finite sample critical values for various sizes. Rejection occurs for small values of P_T. More precisely, we reject the null if the sample or calculated statistic lies to the left of $100\alpha\%$ critical value.
5. Through a series of simulation experiments, ERS discover that if $\bar{\rho} = 1 + \bar{c}/T$ is chosen such that, with T given, if $\bar{c} = -7$ for $\mathbf{D}_t = 1$ and $\bar{c} = -13.5$ for $\mathbf{D}_t = (1, t)$, then the overall power of P_T will be closer to the power envelope for a wide range of alternatives.

9.3.6.5.4 DF–GLS Tests

ERS (1996) also suggest an efficient version of the *ADF* t-statistic using the same local-to-unity alternative.

1. Using the trend parameters, $\hat{\boldsymbol{\beta}}_\rho$, we detrend the data as follows:

$$z_t^d = z_t - \hat{\boldsymbol{\beta}}_\rho' \mathbf{D}_t.$$

   ERS call this procedure as GLS detrending. It is called so based on a theorem by Grenander–Rosenblatt, which says that for deterministically trended data with stationary deviations—which is what (*ERS* − 1) implies—least-square estimates of the trend parameters ignoring the serial correlation are asymptotically equivalent to GLS estimates that incorporate serial correlation.
2. Next using the GLS detrended data, z_t^d, estimate by OLS, the ADF test regression, omitting any deterministic terms:

$$\nabla z_t^d = \tilde{\rho} z_{t-1} + \sum_{i=1}^{p-1} \phi_i^* \nabla z_{t-i}^d + e_t, \quad \text{(DF–GLS reg.)}$$

   and compute the t-stat for the $\tilde{\rho}$ coefficient.
3. For $\mathbf{D}_t = 1$, ERS show that the asymptotic distribution of the DF–GLS t-test of the null of $\tilde{\rho} = 0$ in (DF–GLS reg.) is the same as that of the ADF t-test, but has a higher asymptotic power against the ADF t-test than against local alternatives.
4. When $\mathbf{D}_t = (1, t)'$ however, the distribution of DF–GLS t-test is different from the ADF t-test. ERS provide the critical values for the DF–GLS t-test in this case.
5. Note that the above discussion assumed $z_0 = 0$ which implies $z_1 = e_1$. Elliott (1999) also considers efficient unit-root tests under the alternative assumptions, $z_0 = 0$ when $\rho = 1$ but $z_1 \sim (0, \sigma^2/(1 - \rho^2))$ when $\rho < 1$. This will be labeled as the unconditional initial assumption and the test statistic will be referred to as the QT statistic and it is calculated in exactly the same way as the PT statistic was calculated.

9.3.6.5.5 Modified PP Tests

Perron and Ng (1996) develop a class of unit-root tests based on a weakly dependent error structure. The motivation for these tests is aimed at reducing the size distortions present in the traditional PP test when the errors are dependent. The earlier Z_ρ and the Z_t PP statistics are modified as MZ_ρ and MZ_t statistics and also introduce a new statistic, *MSB* to test the null:

$$MZ_\rho = Z_\rho + \frac{T}{2}(\hat{\rho} - 1)^2$$

$$MZ_t = Z_t + \frac{1}{2}\left(\frac{\sum_{t=2}^{T} z_{t-1}^2}{\hat{\sigma}^{*2}}\right)^{1/2} (\hat{\rho} - 1)^2$$

$$MSB = \left[\frac{T^{-2}\sum_{t=2}^{T} z_{t-1}^2}{\hat{\sigma}^{*2}}\right]^{1/2}$$

Under the null of $\rho = 1$, MZ_ρ and MZ_t will be the same as Z_ρ and Z_t calculated earlier using the traditional PP test. The modified PP statistics can be calculated using Z_ρ and the Z_t statistics obtained after estimating the regressions in any of the three cases discussed before under the DF test procedure. But note that if we are using *Cases 2* and *3*, then we have to use demeaned or the detrended z_t's, respectively, while calculating the above-modified MZ_t and *MSB* statistics.

9.3.6.5.6 Modified Efficient PP Tests

Using the GLS detrending procedure of ERS, Ng and Perron (2001) suggest efficient versions of the modified PP tests of Perron and Ng (1996). These modified efficient PP tests do not exhibit severe size distortions of the traditional PP tests, for errors with a large negative MA or AR roots. And they can have a substantially higher power than the traditional PP tests especially when ρ is close to unity.

Using the GLS detrended data z_t^d, the modified efficient PP tests are defined as

$$MZ_\rho^d = \left(T^{-1}(z_t^d)^2 - \hat{\sigma}^{*2}\right)\left(2\,T^{-2}\sum_{t=2}^{T}(z_{t-1}^d)^2\right)^{-1}$$

$$MSB^d = \left(\frac{k}{\hat{\sigma}^{*2}}\right)^{1/2}, \quad \text{where } k = \frac{\sum_{t=2}^{T}(z_{t-1}^d)^2}{T^2}$$

$$MZ_t^d = MZ_\rho^d \times MSB^d.$$

Ng and Perron have derived the asymptotic distributions of these statistics under the local alternative $\rho = 1 - c/T$ for $\mathbf{D}_t = 1$ and $\mathbf{D}_t = (1, t)'$. Specifically, they show that the asymptotic distribution of MZ_t^d is the same as the DF–GLS t-test. Since the estimation of σ^{*2} has important implications for the finite-sample behavior of the ERS point-optimal test and the modified efficient PP tests, Ng and Perron stress that an AR estimate of σ^{*2} should be used to achieve a stable finite-sample size. So note the following:

- They recommend estimating σ^{*2} from the ADF test regression based on the GLS detrended data in ERS tests, DF–ERS tests, and modified efficient PP tests.
- Similarly, for the modified PP tests it should be estimated from the traditional ADF regression.

The estimator is given below:

$$\hat{\sigma}^{*2} = \frac{\hat{\sigma}^2}{\left(1 - \hat{\phi}(1)\right)^2},$$

where $\hat{\phi}(1) = \sum_{i=1}^{p-1} \hat{\phi}_i^*$ and $\hat{\sigma}^2 = (T - p - 1)^{-1} \sum_{t=p}^{T} \hat{e}_t^2$ are obtained from the (DF–GLS reg.).

9.3.6.5.7 Modified Information Criteria

Ng and Perron also stress that good size and power properties of all the efficient unit-root tests rely on the proper choice of the lag length p used for specifying the (DF–GLS reg.) test regression. They argue, however, that the traditional model selection criteria such as AIC and BIC are not well suited for determining p with integrated data. Ng and Perron suggest the modified information criteria (MIC) that selects p by minimizing $MIC(p)$

$$MIC(p) = \ln(\hat{\sigma}^2) + \frac{C_T(\tau_T(p) + p)}{T - p_{max}},$$

$$\tau_T(p) = \frac{\hat{\beta}^2 \sum_{t=p_{max}+1}^{T} z_{t-1}^d}{\hat{\sigma}^2},$$

$$\hat{\sigma}^2 = \frac{1}{T - p_{max}} \sum_{t=p_{max}+1}^{T} \hat{e}_t^2,$$

with $C_T > 0$ and $C_T/T \to 0$ as $T \to \infty$. The maximum lag, p_{max}, may be set using Schwert's rule. The modified AIC (*MAIC*) results when $C_T = 2$, and the modified BIC (*MBIC*) results when $C_T = \ln(T - p_{max})$. Through a series of simulation experiments, Ng and Perron recommend selecting the lag length p by minimizing the MAIC.

All commercial software packages routinely calculate all the statistics discussed so far and therefore, applied researchers are encouraged to use these improved statistics for a better inference.

9.3.6.5.8 ADF Test and Long-Run Variance as a Nuisance Parameter

One may argue that correlated errors can equally imply a general *ARMA* process of an unknown order. *ADF* test accommodates such a wider-error structure for the e_t's by a long-order autoregression. Said and Dickey (1984) show that ADF t-test from a long-order autoregression remains asymptotically valid for any general *ARMA* process, provided the lag length in the ADF regression increases with the sample size at a rate less than $T^{1/3}$, where T is the sample size. On the other hand, Xiao and Phillips (1998) provide a coefficient test or NBT for any general *ARMA* process. They say that as the coefficient length goes to infinity, their limit distribution depends on some nuisance parameters. They suggest a generalization of the coefficient test based on their Z_ρ test, where the nuisance parameters can be estimated and the coefficient estimate transformed, so that the coefficient test will have the same limit distribution as the DF coefficient test or the Z_ρ test.

Consider the ADF test regression estimated before

$$z_t = \rho z_{t-1} + \phi_1^* \nabla z_{t-1} + \phi_2^* \nabla z_{t-2} + \cdots + \phi_{p-1}^* \nabla z_{t-p+1} + e_t.$$

Letting $\hat{\delta} = T(\hat{\rho} - 1)$ where $\hat{\rho}$ is the OLS estimator of ρ, Xiao and Phillips show that under the null of $\rho = 1$, the limiting distribution of $\hat{\delta}$ is

$$\hat{\delta} \xrightarrow{D} \frac{\sigma}{\sigma_{e,lr}} \left(\frac{\int_0^1 W(r) dW(r)}{\int_0^1 W(r)^2 dr} \right).$$

So clearly the ratio $\sigma/\sigma_{e,lr}$ is a nuisance parameter. But Xiao and Phillips also show that the corrected statistics

$$\frac{\sigma_{e,lr}}{\sigma}\hat{\delta} \xrightarrow{D} \left(\frac{\int_0^1 W(r) dW(r)}{\int_0^1 W(r)^2 dr} \right).$$

Suppose we replace $\sigma_{e,lr}$ and σ with some consistent estimators given by

$$\sigma_{e,lr}^2 = \left(\frac{\theta(1)}{\phi(1)} \right)^2 \sigma^2 = \psi(1)^2 \sigma^2.$$

The parameters can be estimated from the finite-order ADF regression and the quantities $\sigma_{e,lr}$ and σ can be obtained in the following way:

$$\hat{\sigma}_{e,lr}^2 = \left(\frac{1}{(1 - \hat{\Phi}(1))} \right)^2 \hat{\sigma}_p^2,$$

$$\hat{\sigma}_p^2 = T^{-1} \sum_{t=p}^{T} \hat{e}_{t,p}^2,$$

where $\hat{\Phi}(1) = \sum_{i=1}^{p} \hat{\phi}_i$ with $\hat{\phi}_i$ being the coefficients and $\hat{e}_{t,p}$ are the residuals, respectively, of the ADF regression of ∇y_t above. With these quantities it is easy to see that

$$\frac{T(\hat{\rho} - 1)}{(1 - \hat{\Phi}(1))} \xrightarrow{D} \left(\frac{\int_0^1 W(r) dW(r)}{\int_0^1 W(r)^2 dr} \right).$$

9.3.6.6 Unit Roots versus Trend-Break Alternative

TS models originally studied by Nelson and Plosser (1982) tested the TS alternatives against the null of DS models where the model setup tested the presence of the unit root as the null and deterministic nonstationarity as the alternative hypothesis. But a deterministic nonstationarity needs, by no means, to be the "only" alternative. There can be other alternatives. For example, one can witness in many economic time series, a "heterogeneous" behavior in trends. This behavior has been called "variable trends," "segmented trends," or "breaking trends." However the name implies that there are "structural breaks" in the data. It has been reported in the literature that unit-root tests, with misspecified alternatives, can lead to a "spurious" acceptance of a "stochastic" nonstationarity. One such alternative is the breaking-trend stationarity (*BTS*). Standard tests for unit roots—as have been discussed so far—will not reject the null of a unit root if the alternative happens to be *BTS*. This alternative way to test for the presence of unit roots has been suggested by Perron (1989). Perron bases his arguments on the most-important implication of the unit root finding that random shocks have *permanent* effects on the system. Fluctuations are not transitory. Perron seeks to counter this with his *BTS* alternative. Perron's claims are based on the fact that if random shocks have been driving the system, such shocks should have arisen from "within" the system or the economy and, not as a result of events, an economy or the forces that drive a variable having little control over. Basically such shocks should be termed as "exogenous" to the system and are not part of the true DGP. And, hence, Perron argues, such exogenous shocks have to be factored in properly, while developing tests for unit roots. Perron's results, using Nelson and Plosser (1982) data set, are as follows:

1. Most of the series are not characterized by unit roots and fluctuations are indeed transitory.
2. Only two events (or shocks) have had a permanent effect on the various economic time series analyzed by Nelson and Plosser: (1) the Great Crash of 1929 and (2) the oil price shock of 1973. But these two shocks were completely "exogenous" and had no role in the DGP of the macroseries.

The effects of these two exogenous shocks are the following:

1. The 1929 crash resulted in a dramatic "drop" in the mean of most of the series under consideration and this resulted in the level of most variables falling down.
2. The 1973 oil shock resulted in a change in the slope of the trend of most aggregates, that is, a slowdown in the growth of many series without any sudden change in the level of the variables.

So Perron says that most macrovariables are *TS* if one allows a "single" change in the intercept of the trend function after 1929 and a "single" change in the slope of the trend function after 1973. So he proposes the following unit-root testing procedure: the null will still test for a unit root—with or without a drift—but the added new feature now is that we allow a one-time change at a time T_B $(1 < T_B < T)$.

The following three models are considered under the null hypothesis:

1. One that permits an exogenous change in the level of the series—called a "crash model"
2. One that permits an exogenous change in the slope—referred to as a "changing growth" model
3. One that allows both types of change

The null and alternatives have been parameterized as follows:

Null hypotheses

$$\text{Model (A)} \quad z_t = \mu + dD(TB)_t + z_{t-1} + e_t$$

$$\text{Model (B)} \quad z_t = \mu_1 + z_{t-1} + (\mu_2 - \mu_1)DU_t + e_t$$

$$\text{Model (C)} \quad z_t = \mu_1 + z_{t-1} + dD(TB)_t + (\mu_2 - \mu_1)DU_t + e_t$$

$$\begin{aligned} dD(TB)_t &= 1 \quad \text{if } t = T_B + 1, \quad 0 \quad \text{otherwise} \\ DU_t &= 1 \quad \text{if } t > T_B \qquad\quad\; 0 \quad \text{otherwise} \end{aligned}$$

and e_t can be either an uncorrelated innovation sequence or a stationary process. Instead of considering the alternative hypothesis that z_t is a stationary series around a deterministic linear trend with time-invariant parameters, Perron considers the following alternative models:

$$\text{Model (A)} \quad z_t = \mu_1 + \beta t + (\mu_2 - \mu_1)DU_t + e_t$$

$$\text{Model (B)} \quad z_t = \mu + \beta_1 t + (\beta_2 - \beta_1)DT_t^* + e_t$$

$$\text{Model (C)} \quad z_t = \mu_1 + \beta_1 t + (\mu_2 - \mu_1)DU_t + (\beta_2 - \beta_1)DT_t + e_t,$$

where

$$DT_t^* = t - T_B, \quad \text{and} \quad DT_t = t \quad \text{if } t > T_B \quad \text{and} \quad 0 \quad \text{otherwise.}$$

Here T_B refers to the time of the break, that is, the period at which the change in the parameters of the trend function occurs.

Model (A) is referred to as the crash model, representing, under the null, a unit-root process that is characterized by a dummy variable that takes the value of 1 at the time of the break. The alternative *TS* system allows for a one-time change in the intercept of the trend function.

Model (B) is called the *changing-growth model* under the null. Moreover, the drift parameter changes from μ_1 to μ_2 at T_B. The alternative hypothesis allows a change in the slope of the trend function without any sudden change in the level at the time of the break.

Model (C) allows for both effects to take place simultaneously, that is, a sudden change in the "level" followed by a different growth path.

Perron motivates his analysis by fitting trends and through Monte Carlo simulation exercises. Figure 9.5 captures the behavior of some of the series that Nelson–Plosser had used in their paper, along with the trend breaks. Perron showed that the ACs of the detrended series, declined quite rapidly, which is clearly a feature of the *TS* process. His Monte Carlo simulations show how one can end up with a spurious evidence unit roots, if one does not account for breaks.

However, the models shown above are valid only for the case of uncorrelated errors. Perron suggests the following models for a wider class of errors, where he nests the null and the alternatives:

$$z_t = \hat{\mu}^A + \hat{\theta}^A DU_t + \hat{\beta}^A t + \hat{d}^A D(TB)_t + \hat{\rho}^A z_{t-1} + \sum_{i=1}^{k} \hat{\phi}_i \nabla z_{t-i} + \hat{e}_t$$

$$z_t = \hat{\mu}^B + \hat{\theta}^B DU_t + \hat{\beta}^B t + \hat{\gamma}^B DT_t^* + \hat{\rho}^B z_{t-1} + \sum_{i=1}^{k} \hat{\phi}_i \nabla z_{t-i} + \hat{e}_t$$

$$z_t = \hat{\mu}^C + \hat{\theta}^C DU_t + \hat{\beta}^C t + \hat{\gamma}^C DT_t + \hat{d}^C D(TB)_t + \hat{\rho}^C z_{t-1} + \sum_{i=1}^{k} \hat{\phi}_i \nabla z_{t-i} + \hat{e}_t.$$

Under the alternative of a *TS* process, the parameters should satisfy some restrictions. For example, for Model (A) to hold under the null, the parameter restrictions are $\hat{\rho}^A = 1$, $\hat{\beta}^A \neq 0$, $\hat{\theta}^A = 0$, and $\hat{d}^A \simeq 0$. The asymptotic distributions of the t-statistics of $t_{\hat{\rho}}^A$ and $t_{\hat{\rho}}^C$ are the same as for the case of uncorrelated errors, but Perron suggests a correction of dropping the term DU_t from Model (B) to improve the power of the unit-root test. With this correction, $t_{\hat{\rho}}^B$ also follow the same distribution as the uncorrelated case. For all the cases, however, one has to refer to the set of critical values calculated by Perron.

9.3.6.7 Extensions to Perron's Test

Zivot and Andrews (ZA) (1992) questioned Perron's suggestion of fixing the break points as *exogenously* given and suggested to *estimating* the break points, if any, in the data. They suggested a

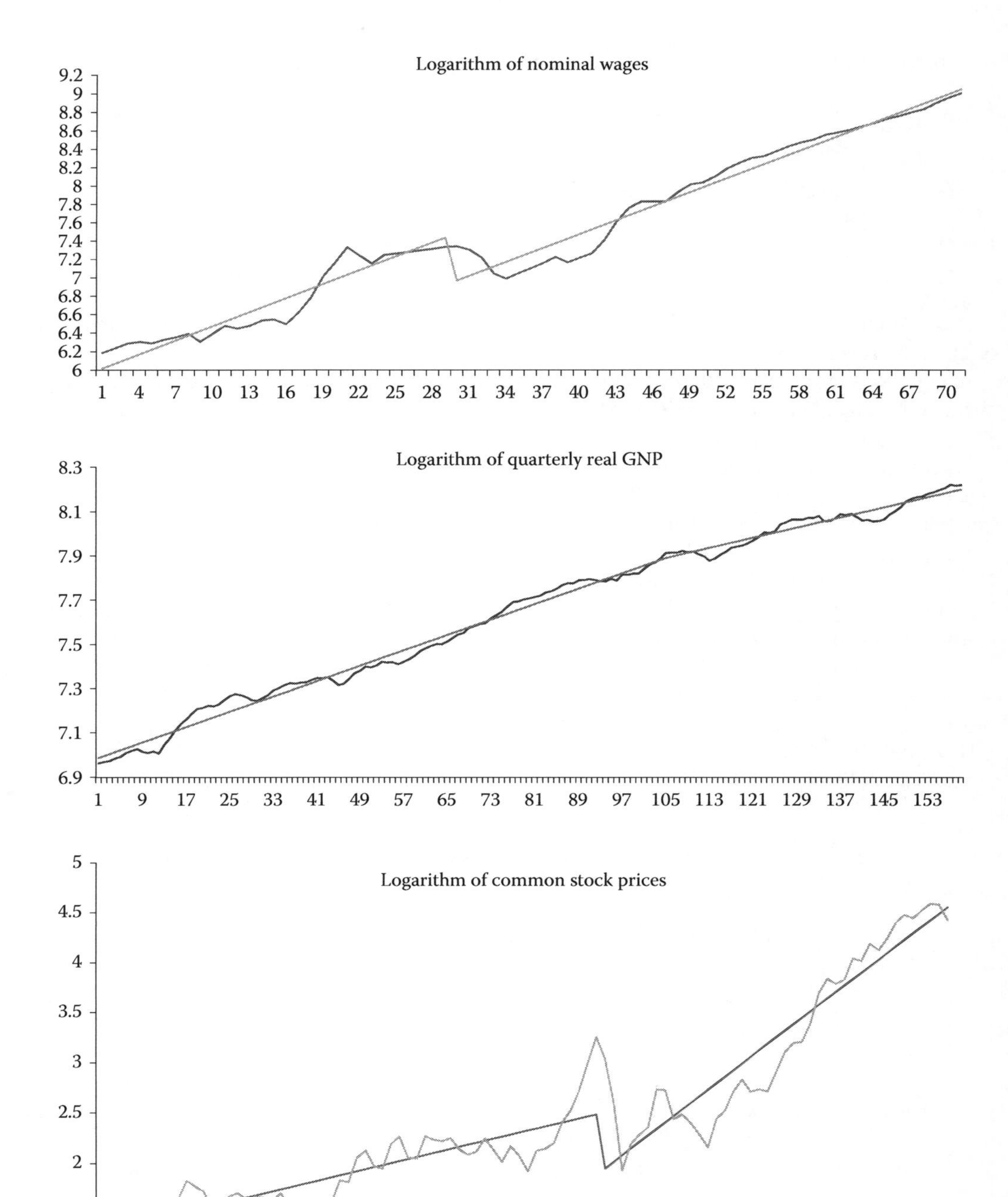

FIGURE 9.5
Nelson–Plosser data and the trend for models A, B, and C.

data-dependent algorithm to estimate them and call such a test an *unconditional* unit-root test as compared with the Perron test, which is conditional on the break point. Accordingly, to ZA, the null will be the standard unit-root test with a drift, but one that excludes any structural change. However, the alternative hypotheses will still be the same as in Perron's model.

Consistent with their arguments, ZA propose as their null, for all the three models used by Perron to be

$$z_t = \mu + z_{t-1} + e_t.$$

Since the alternative considered is a *TS* process with a one-time break, the goal is to estimate the breakpoint that gives the most weight to the alternative of a *TS* process but a least-favorable result to the null. A plausible estimation procedure consistent with this view is to choose that λ that minimizes the one-sided t-statistic for testing $\rho_i = 1, (i = A, B, C)$, when small values of the statistic lead to the rejection of the null. Let $\hat{\lambda}^i_{inf}$ denote such a minimizing value for model i. For each series, ZA propose to use the same ADF-testing procedure of Perron that nests both the null and the alternatives and estimates the regression equations by OLS with a break fraction $\lambda = T_B/T$, ranging from points $2/T$ to $(T-1)/T$. For each value of λ, the number of lagged regressors, k, can be determined by using the procedure used by Perron and the t-statistic for testing $\rho^i = 1$. The minimum t-statistics reported are the minimum over all $T-2$ regressions and the break years are the years corresponding to the minimum t-statistics. Also to be noted is the fact that the number of lagged regressors can be allowed to vary for each tentative choice of λ. Since ZA treat the selection of breakpoints as endogenous, Perron's tables of critical values are no longer valid; hence, they determine the asymptotic distribution of their minimum t-statistics and tabulate the critical values through simulation. They suggest to reject the null of the unit root if

$$\inf t_{\hat{\rho}}(\lambda) < k^i_{\text{inf},\alpha}, \quad i = A, B, C,$$

where $k^i_{\text{inf},\alpha}$ denotes the size α critical value from the asymptotic distribution of the minimum t-statistics. The critical values for ZA test as well as Perron's tests are available in Panels A and B of Tables 9.2 through 9.4 respectively of their paper.

9.3.6.8 Application of Unit-Root Test Procedures

We apply some of the unit-root testing procedures discussed above on some major Indian macro time series on money, prices, trade, and credit. Monthly data ranging from 1955 to 2003 were collected from various publications of the Reserve Bank of India (RBI), such as the various issues of the monthly bulletins of RBI, the Report on Currency and Finance (RCF), and the *Handbook of Statistics on the Indian Economy*. Trade data were collected from the CD-ROM version of the International Financial Statistics (IFS). Price indices were spliced to bring them to a common base year using symmetric means as recommended by Hill and Fox (1997). All data are in logs.

Tables 9.6 through 9.10 display the results of ADF, Perron, Zivot–Andrews, and efficient unit-roots test procedures. Starting with ADF test results in Table 9.6, the statistic of interest is $t_{\hat{\rho}}$. Recall that we are checking for the null of the unit root, $\rho = 1$ against the alternative $\rho < 1$. Since we conduct a left-tail test, we accept the null of unit roots, if the calculated value of the t-statistic is greater than the tabulated critical value. Now, consider the first row below the name of the series, which gives the ADF test result for the entire sample. For example, for M1, consider the period, 1955:1–2002:12. For this full sample, $t_{\hat{\rho}} = -2.62$, which exceeds the 5% critical value of -3.42, so that the null of the unit root is accepted. Extending this to the other series, we see that the null of the unit is not accepted in WPI and food-credit series.

TABLE 9.6

Results of ADF Unit-Root Test Procedure

				Regression: $z_t = \mu + \beta t + \rho z_{t-1} + \sum_{i=1}^{k} \phi_i^* \nabla z_{t-1} + e_t$				
Series	k	$\hat{\mu}$	$t_{\hat{\mu}}$	$\hat{\beta}$	$t_{\hat{\beta}}$	$\hat{\rho}$	$t_{\hat{\rho}}$	**S(e)**
M1								
1955:1–2002:12	13	0.087	2.82	0.1×10^{-3}	2.87	0.989	−2.62	0.3×10^{-3}
1955:1–1980:12	13	0.211	2.73	0.2×10^{-3}	2.68	0.972	−2.67	0.3×10^{-3}
1981:1–2002:12	14	0.363	1.73	0.4×10^{-3}	1.60	0.965	−1.66	0.2×10^{-3}
M3	7	0.250	3.64	0.4×10^{-3}	3.60	0.970	−3.16	0.2×10^{-2}
CPI								
1955:1–2002:12	12	0.046	3.46	0.8×10^{-4}	3.32	0.987	−3.27	0.7×10^{-4}
1955:1–1974:12	12	0.089	2.11	0.1×10^{-3}	2.26	0.975	−2.09	0.9×10^{-4}
1975:1–2002:12	12	0.123	2.48	0.1×10^{-3}	2.36	0.975	−2.41	0.5×10^{-4}
WPI								
1970:1–2000:12	12	0.151	5.09	0.3×10^{-3}	4.95	0.957	−5.01	0.8×10^{-4}
1970:1–1974:8	8	−0.010	−0.069	0.1×10^{-2}	3.12	1.001	0.04	0.99×10^{-2}
1974:9–2000:12	12	0.153	3.76	0.25×10^{-3}	3.64	0.962	−3.68	0.68×10^{-4}
Exports								
1970:1–2000:12	13	0.039	2.95	0.6×10^{-3}	1.51	0.959	−1.43	0.011
1970:1–1985:2	12	0.084	2.69	0.12×10^{-2}	1.31	0.896	−1.40	0.016
1985:3–2000:12	12	0.147	1.41	0.305×10^{-3}	0.38	0.969	−0.65	0.55×10^{-2}
Imports	12	0.069	3.96	0.1×10^{-2}	2.39	0.893	−2.43	0.0149
Food Credit								
1970:4–2003:3	12	0.227	3.92	0.4×10^{-3}	3.54	0.962	−3.89	0.0108
1970:4–1987:7	13	0.300	2.50	0.75×10^{-3}	1.90	0.950	−2.35	0.012
1987:8–2003:3	13	0.871	4.79	0.21×10^{-2}	4.18	0.89	−4.66	0.85×10^{-2}
Nonfood credit	12	0.319	3.10	0.5×10^{-3}	3.00	0.962	−3.00	0.2×10^{-3}

Note: k for most models was fixed at 12 considering the monthly nature of the data. All data are in logs.

Since the Indian economy had undergone many changes during the long period under consideration, we followed Perron and tested for the unit root under the possibility of a structural change. Following a visual plot of the data, we decided to split some of the series and fix the breakpoints, in the way shown in Table 9.6 and detrend according to Model A. The ACs of the detrended series are shown in Table 9.7 and these do not decay as rapidly as one expects, suggesting that the series may have unit roots even after accounting for any possible structural change. So, we tested for unit roots in split samples and the results are reported in the second and third rows below the respective series.

TABLE 9.7

Sample ACs of the "Detrended" Series

Series	T	**Variance**	r_1	r_2	r_3	r_4	r_5	r_6	r_7	r_8	r_9	r_{10}
M1	576	0.031	0.99	0.97	0.96	0.95	0.93	0.92	0.91	0.91	0.90	0.89
CPI	576	0.009	0.99	0.97	0.95	0.93	0.91	0.89	0.87	0.86	0.85	0.83
WPI	372	0.004	0.98	0.95	0.91	0.86	0.82	0.78	0.74	0.70	0.66	0.62
Exports	372	0.042	0.74	0.72	0.69	0.70	0.68	0.70	0.70	0.68	0.65	0.64
Food credit	396	0.164	0.93	0.83	0.76	0.71	0.68	0.64	0.59	0.54	0.49	0.46

Note: All series were detrended using Model A under the alternative hypothesis.

TABLE 9.8

Results of Perron Unit-Root Test Procedure

Series	Regression, Model A: $z_t = \mu + \theta DU_t + \beta t + dD(TB)_t + \rho z_{t-1} + \sum_{i=1}^{k} \phi_i^* \nabla z_{t-1} + e_t$													
	T	T_B	λ	$\hat{\mu}$	$t_{\hat{\mu}}$	$\hat{\theta}$	$t_{\hat{\theta}}$	$\hat{\beta}$	$t_{\hat{\beta}}$	$\hat{d}$	$t_{\hat{d}}$	$\hat{\rho}$	$t_{\hat{\rho}}$	S(e)
M1	576	312	0.54	0.093	3.05	−0.006	−1.88	0.0001	3.59	−0.003	−0.54	0.988	−2.77	0.0003
CPI	576	240	0.42	0.045	3.51	−0.001	−0.83	0.0001	3.47	0.004	0.79	0.986	−3.39	0.0001
WPI	372	57	0.15	0.131	4.73	−0.006	−2.81	0.0002	4.72	0.001	0.19	0.962	−4.67	0.0001
Exports	372	183	0.49	0.020	1.05	−0.021	−0.76	0.001	1.81	−0.015	−0.35	0.935	−1.78	0.012
Food credit	396	209	0.53	0.44	5.51	−0.12	−3.82	0.001	4.97	−0.05	−0.43	0.920	−5.45	0.0104

Note: $\lambda = T_B/T$ and k for all the models were fixed at 12. All data are in logs. S(e) is variance of the residuals.

TABLE 9.9

Results of ZA Unit-Root Test Procedure

	Model A		Model B		Model C	
Series	**t-Statistics**	**Break Year**	**t-Statistics**	**Break Year**	**t-Statistics**	**Break Year**
M1	−3.75	1989 : 2 (410)	−3.35	1980 : 5 (305)	−3.57	1980 : 4 (304)
CPI	−4.64	1958 : 10 (58)	−5.02	1970 : 11 (191)	−5.12	1968 : 1 (157)
WPI	−5.68	1972 : 5 (29)	−5.67	1973 : 5 (41)	−5.94	1977 : 2 (86)
Exports	−2.81	1979 : 5 (113)	−2.20	1985 : 6 (186)	−3.12	1977 : 4 (88)
Imports	−3.83	1984 : 2 (170)	−3.12	1974 : 2 (50)	−3.89	1972 : 11 (35)
Food credit	−5.72	1987 : 3 (204)	−4.38	1975 : 10 (67)	−5.71	1987 : 3 (204)
Nonfood credit	−5.25	1978 : 9 (102)	−5.14	1983 : 8 (161)	−5.52	1980 : 6 (123)

None of the results obtained is overturned even with split samples. Next, we confirm this result using Perron's method with a full sample and Table 9.8 displays the results of Perron's unit-root testing procedure and the point of break, T_B, fixed by us. For convenience, we report the results of Model A only. Here also, the statistic of interest is $t_{\hat{\rho}}$ but the critical values depend on the value of λ fixed by the user. Perron has tabulated these values for all possible λ values and for Model A these are available in Table IV.B in Perron (1989, p. 1376) and also in ZA (1992, p. 256). Again, none of the evidence obtained with the ADF test procedure has been overturned. For example, the $t_{\hat{\rho}}$ value of -2.77 for M1 clearly exceeds the 5% critical value of -3.76 for a fixed $\lambda = 0.54$ from Perron's table, which clearly says that the null of the unit root can be accepted, even after accounting for any possible structural breaks. The rest of the results can be similarly interpreted.

Next, we extend Perron's test to accommodate the ZA criticism that the so-called structural breaks should not be fixed by the user; but should be estimated by the data. For this test, we include imports and nonfood credit too. Though visually the plots of these series do not clearly show any break, it will be a good idea to allow the data to decide the break, if any. Recall here that we had accepted the presence of unit roots in these series. The results of ZA test for the unit root are displayed in Table 9.9. Here also, we fixed the lag order to be 12 for simplicity. (We highlight that the correct lag order has to be estimated, and not fixed.) We estimate all the three models for the ZA test. If we allow the break to be decided by the data, then, the unit root is clearly rejected by ZA test for nonfood credit. We had accepted the null of the unit root using the standard ADF test and hence, this is a reversal. Our "guesstimate" of the breakpoints, based on the visual inspection of the data, has been proved to be a way off the mark for almost all the series, except FC, for which, incidentally, we had rejected the null of the unit root using the ADF test. In general, there are no major surprises in store for us here too, though CPI poses an interesting question. Model A for CPI accepts the unit root, but with a break at 1958 : 10, which is way down the sample. The other two models have decisively rejected the unit roots. So, it will be interesting to check for the presence of unit roots from 1959 : 1 to 2002 : 12. For the other series, the null of the unit root is more or less accepted. For some series, all three models accept the unit roots.

TABLE 9.10

Results of ADF–GLS Unit-Root Test Procedure

ADF–GLS regression: $\nabla z_t^d = \tilde{\rho} z_{t-1} + \sum_{i=1}^{p-1} \phi_i^* \nabla z_{t-i}^d + e_t$, $D_t = (1\ t)$.								
	Nondeseasonalized		**Nondeseasonalized**		**Deseasonalized**		**Deseasonalized**	
Series	***PT* Statistics**	**DF–GLS *t*-Statistics**	***QT* Statistics**	**DF–GLS(U) *t*-Statistics**	***PT* Statistics**	**DF–GLS *t*-Statistics**	***QT* Statistics**	**DF–GLS(U) *t*-Statistics**
M1	41.39	−0.84	22.72	−2.07	135.08	−0.12	73.99	−1.64
M3	65.21	−0.56	34.78	−1.89	115.99	−0.22	61.21	−1.73
CPI	20.17	−1.52	11.28	−2.37	33.82	−1.16	19.49	−1.87
WPI	6.94	−2.48	6.22	−2.68	10.29	−1.99	9.01	−2.20
Exports	6.79	−2.74	4.83	−3.34	13.16	−1.86	11.01	−2.08
Imports	7.82	−2.43	7.82	−2.44	3.59	−2.94	3.51	−2.99
Food credit	4.19	−2.73	1.56	−3.67	16.45	−1.71	10.74	−2.02
Nonfood credit	2.76	−2.95	2.55	−3.00	12.02	−1.75	11.80	−1.73

To improve the power of the unit-root tests, we also conducted the efficient unit-root tests recommended by ERS (1996). The results are tabulated in Table 9.10. Two features of this table need highlighting: first, we have highlighted the difference between using seasonalized data and non-deseasonalized data. Second, we have also reported the test results based on the unconditional distribution for e_1 when $\rho < 1$. So, we have reported the relevant QT statistics also along with PT statistics for the conditional case. The relevant 5% asymptotic critical values for PT statistic and *DF–GLS* t-statistic for $\tilde{\rho}$ in the ADF–GLS regression are, respectively, 5.62 and −2.89. And the critical values for the unconditional distribution case are, respectively, 2.85 and −3.17. Values of PT and QT smaller than the reported critical values imply rejection of the null of the unit root. According to these two statistics, credit series with nondeseasonalized data rejects the null of the unit root. But credit offtake in India is strongly seasonal and deseasonality is strongly recommended. So with deseasonalized data, however, the unit root null is accepted. However, the null of the unit root is largely accepted, if we consider the t-statistics, in both these cases. The null of unit roots is accepted in all series if we consider deseasonalized data. These results also indicate that an unconditional distribution may matter, going by the large differences in the statistics of interest in the two cases.

So, what we are not sure is if there is a way to select the best among these three competing procedures. In sum, these test procedures have not thrown up any majorly conflicting evidence on the presence of unit roots in major Indian macro time series. One can safely summarize these findings as, some of the macroseries such as M1, M3, CPI and the trade data, exhibiting stochastic nonstationarity.

9.3.6.8.1 Unit Root with a Nonlinear Trend

A standard TS alternative assumes that the deterministic trend is linear (or trivially a constant). Suppose the trend is not linear but a higher-order polynomial, the ideas discussed in the DF test extend naturally. In DF and PP tests, the maintained hypothesis is that of an integrated series along with a drift but no trend. Ouliaris et al. (1989) extended this analysis to include the polynomial trend of an arbitrary order, as well as a general deterministic trend as part of the maintained hypothesis.

We assume that the time series y_t is generated according to

$$(y_t - \mu_t) = u_t$$
$$(1 - \rho B)u_t = e_t$$

where

$$\mu_t = \sum_{p=0}^{P} \beta_p t^p$$

so that

$$(1 - \rho B)\left(y_t - \sum_{p=0}^{P} \beta_p t^p\right) = e_t$$

$$\nabla y_t = (\rho - 1)y_{t-1} + (1 - \rho)\beta_0 + (1 - \rho B)\sum_{p=1}^{P} \beta_j t^p + e_t.$$

If the null of $\rho = 1$ is accepted, then $\nabla y_t = e_t$. For a more-general process, it will be $\nabla y_t = u_t$, where u_t will be a stationary process. The DF and ADF test procedures can still be used, but the limiting distributions will depend on the order of the fitted polynomial trend. Ouliaris et al. (1989)

set $\beta_P = 0$ when $\rho = 1$. Under the alternative, $\rho < 1$, however, β_P may not be zero. The underlying statistics to test the null hypotheses is invariant to the presence of β_p, $p = 1, \ldots, P-1$. Thus, we maintain a polynomial trend of order P under both the null and an alternative. Ouliaris et al. (1989) show that the limiting distribution of the test statistic, $h_p(\hat{\alpha}) = T(\hat{\rho} - 1)$, and the Wald F-statistic to test the joint null of $\rho = 1$ and $\beta_P = 0$, $F_p(\hat{\rho}, \hat{\beta}_p)$ are hindered by nuisance parameters. So, they redefine these statistics as follows:

$$K_p(\hat{\rho}) = T(\hat{\rho} - 1) - \frac{T(\hat{\omega}^2 - \hat{\sigma}^2)}{2\hat{s}_0^2}$$

$$S_p(\hat{\rho}) = \frac{\hat{\sigma}}{\hat{\omega}} t(\hat{\rho}) - \frac{T(\hat{\omega}^2 - \hat{\sigma}^2)}{2\hat{\omega} s_0}$$

$$F_p^*(\hat{\rho}, \hat{\beta}_p) = \frac{\hat{\sigma}^2}{\hat{\omega}^2} F_1(\hat{\rho}, \hat{\beta}_p) + \frac{T^2(\hat{\omega}^2 - \hat{\sigma}^2)}{4\hat{\omega}^2 s_0^2} - T(\hat{\rho} - 1)(1 - [\hat{\sigma}/\hat{\omega}]^2)$$

where

$$s_0^2 = \text{residual sum of squares from the regression of } y_{t-1} \text{ on } 1, t, \ldots, t^p,$$

$$\hat{\omega}^2 = \text{any consistent estimator of } \omega^2 \text{obtained from the estimated residuals, } e.$$

To complete the computational requirement, Ouliaris et al. (1989) provide a consistent estimator for the long-run variance, ω^2. Ouliaris et al. (1989) outline the asymptotic distributions of these statistics and provide simulated critical values for the purpose of hypothesis testing. They provide asymptotic distributions for the test statistics for $p = 2, 3, 4, 5$. They also suggest a nonparametric unit-root test, when the trend term is an arbitrary deterministic function of time. This procedure exploits the fact that differencing a TS series will result in a negative unit root in the MA representation. Testing uses the univariate-bounds procedure for no cointegration developed by Phillips and Ouliaris (1988).

To explain this approach for $P = 0$,

$$e_t = \nabla y_t = y_t - y_{t-1}$$

Under the null hypothesis of a unit root, e_t is a stationary process with a finite asymptotic variance. If, however, y_t is stationary, then, e_t will have an $MA(1)$ representation with a unit root and its spectrum will be zero at zero frequency. This suggests a test to check if the estimated spectrum at the zero frequency is negligible and thus supports the hypothesis that the long-run variance ω^2 is zero.

The test statistic is formed as a ratio given as

$$\rho^2 = \left(\frac{\omega}{\sigma}\right)^2$$

and $\hat{r}^2 = (\hat{\omega}/\hat{\sigma})^2$, and a consistent estimator of ρ^2, is used in practical applications. The alternative hypothesis is then

$$H_a = \rho^2 = \frac{\omega^2}{\sigma^2} = 0.$$

The alternative of trend stationarity is accepted if the true ρ^2 is sufficiently small. On the basis of the bounds test, we may say that if r^2 is smaller than the upper bound, then, we may accept the alternative hypothesis of trend stationarity and if it exceeds the upper bound, then, the null may be accepted. Phillips and Ouliaris (1988) recommend using 0.10 as the rejection point for the upper and lower bounds.

To make these tests operational, we require a consistent estimator for the long-run variance, ω^2. Ouliaris et al. (1989) suggest using a weighted-periodogram ordinates around the zero frequency. Other methods that have better econometric properties may also be used. For example, Ng and Perron (2001) suggest using an AR estimator for the long-run variance.

9.3.6.9 Right-Tail Unit-Root Test and Test for Bubbles

In an interesting new development in the field of unit-root testing, a series of papers by Phillips, Jun Yu, and their coauthors focuses on empirical tests for bubbles and rational exuberance. With an unusual surge or fall in the prices of a stock or any asset, macroeconomists debate if such a behavior can be termed as bubbles and if they are rational or behavioral. Since standard unit-root tests fail to detect such an explosiveness (see Evans 1991), Phillips, Wu, and Yu (PWY) (2011) suggest a unit-root test based on the right tail of distribution to check for the presence of bubbles. They call this test sup ADF (SADF) test (or a forward-recursive right-tailed ADF test).

For each time series z_t, we apply the *ADF* test for a unit root against the alternative that the series has an explosive unit root (right tailed). So, we estimate the following regression:

$$z_t = \mu + \rho z_{t-1} + \sum_{i=1}^{p} \phi_i \nabla z_{t-i} + e_t, \quad e_t \sim NID(0, \sigma^2).$$

The null is $H_0 : \rho = 1$ and the right-tailed alternative is $H_1 : \rho > 1$. The above model is estimated repeatedly, in a forward-recursive manner, using subsets of the sample data, and adding one observation at each pass. If we start the first regression with $T_0 = \text{Int}[Tr_0]$ observations, for some fraction r_0 of the total sample, subsequent regressions use this originating data set, adding one successive observation, and giving a sample of size $T^* = [Tr]$ for $r_0 \leq r \leq 1$. Denote the corresponding t-statistic by ADF_r and hence ADF_1 corresponds to the full sample. Under the null we have

$$ADF_r \xrightarrow{D} \frac{\left(\int_0^1 W(r)dW(r)\right)}{\left\{\int_0^1 W(r)^2 dr\right\}^{1/2}} \quad \text{and}$$

$$\sup_{r\in[r_0,1]} ADF_r = \sup_{r\in[r_0,1]} \frac{\left(\int_0^1 W(r)dW(r)\right)}{\left\{\int_0^1 W(r)^2 dr\right\}^{1/2}},$$

where W is the standard Brownian motion. Note that this is a stochastic process that evolves with r. To date the beginning and the end of exuberance, one can compare the time series of ADF_r statistics against the right-tailed critical values of the standard D–F t-statistic. The starting date of the bubble is the smallest value of r for which the estimated ADF_r is larger than the critical value. In particular, if r_e is the beginning and r_f is the collapse dates of the explosive behavior in the data, then, we construct estimates of these dates as follows:

$$\hat{r}_e = \inf_{s\geq r_0} \left\{s : ADF_s > cv_{\alpha_n}^{ADF}(s)\right\}, \quad \hat{r}_f = \inf_{s\geq \hat{r}_e} \left\{s : ADF_s < cv_{\alpha_n}^{ADF}(s)\right\},$$

where $cv_{\alpha_n}^{ADF}(s)$ is the right-side critical value of ADF_s corresponding to the significance level, α_n.

In a later paper, Phillips, Shi, and Yu (PSY) (2012) demonstrate that when the sample period includes multiple episodes of exuberance and collapse, the SADF test may suffer from reduced power and can be inconsistent, failing to reveal the existence of bubbles. To overcome this weakness, they propose an alternative approach named the generalized SADF (GSADF) test. The GSADF test is also based on the idea of repeatedly implementing a right-tailed ADF test, but the new test extends the sample sequence to a broader and more-flexible range. Instead of fixing the starting point of the sample (viz., on the first observation of the sample), the GSADF test extends the sample sequence by changing both the starting point and the ending point of the sample over a feasible range of flexible windows. The test involves running the ADF test regression on a sample sequence. However, unlike in the SADF test, where we changed only the endpoint of the regression, r_2 from r_0 to 1, we also change the starting points r_1 to change within a feasible range, which is from 0 to $r_2 - r_0$. The largest ADF statistic over the feasible ranges r_1 and r_2 is the GSADF statistic. The GSADF statistic is utilized to conduct an inference of the existence of an explosive behavior within the whole sample period. Suppose there is an evidence of an explosive behavior, then, one can date stamp the occurrence periods using the backward SADF statistic. PSY derive the limit distribution of the GSADF test statistic and the critical values are displayed in their paper.

The above discussion suggests a possible two-step procedure to test for the existence of bubbles. As a first step, we may use the SADF test to test for the null of a unit root or no bubbles. If the null is rejected, as a second step, we can use the GSADF test to date stamp the start and the end of these bubbles. This step will be useful if we suspect the existence of multiple bubbles in our data. As an application of these tests, we seek to identify periods of bubbles, if any, in the Indian stock market. We use the closing Nifty-index data for the period from January 31st 1997 to September 28th 2012.* Figure 9.6 displays these data.

One can identify two distinct periods during which the Nifty index rose sharply and declined steeply. It rose sharply over December 1999–2007 and fell steeply over January 2008 to March 2009 and started rising again and remained steady over the remaining part of the sample. The first two graphs display this split clearly. Initially employing the SADF test, we find no evidence of any bubble episode in the Nifty index. The result of the SADF test for the entire sample of 3912 observations is summarized in Table 9.11 along with the critical values.

A similar analysis of the split samples also did not yield any evidence of bubbles in the Indian context. Figure 9.7 displays the entire set of SADF test statistics along with the critical values.

9.3.6.10 Other Approaches to Unit-Root Tests: Variance-Ratio Test

Variance ratio (VR) tests and the concept of RW are closely connected with the concept of the efficiency of the stock market. It is now a part of financial economics folklore that if a stock market

TABLE 9.11

SADF Statistic for Nifty Index, 1997–2012

	Critical Level	*t*-Statistic	*p*-Value
SADF statistic		1.0948	0.208
Test the critical values	99% level	2.1011	
	95% level	1.6233	
	90% level	1.4149	

* Historical data were collected from National Stock Exchange of India website, http://nseindia.com/products/content/equities/indices/historical_index_data.htm

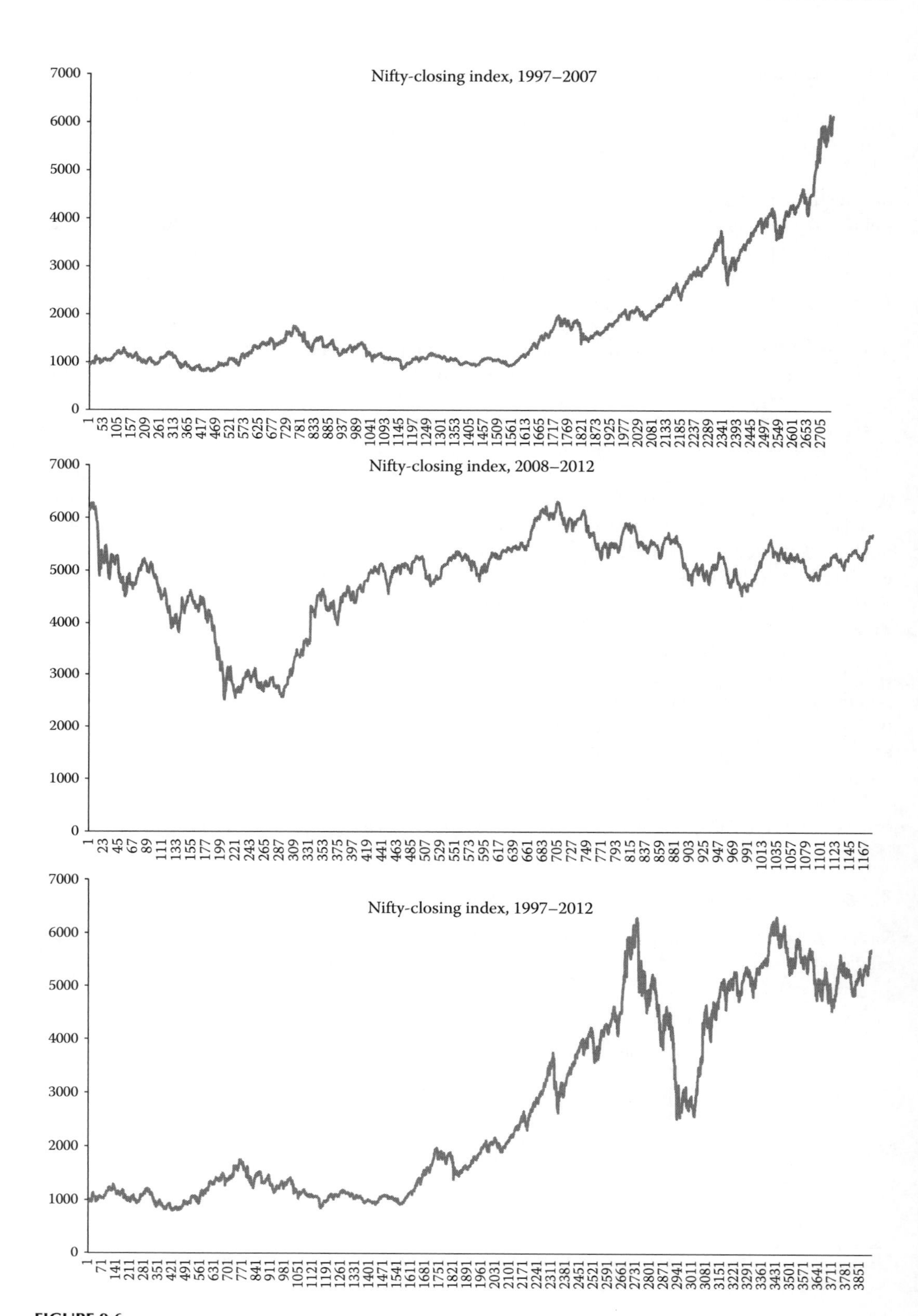

FIGURE 9.6
Nifty-closing price index.

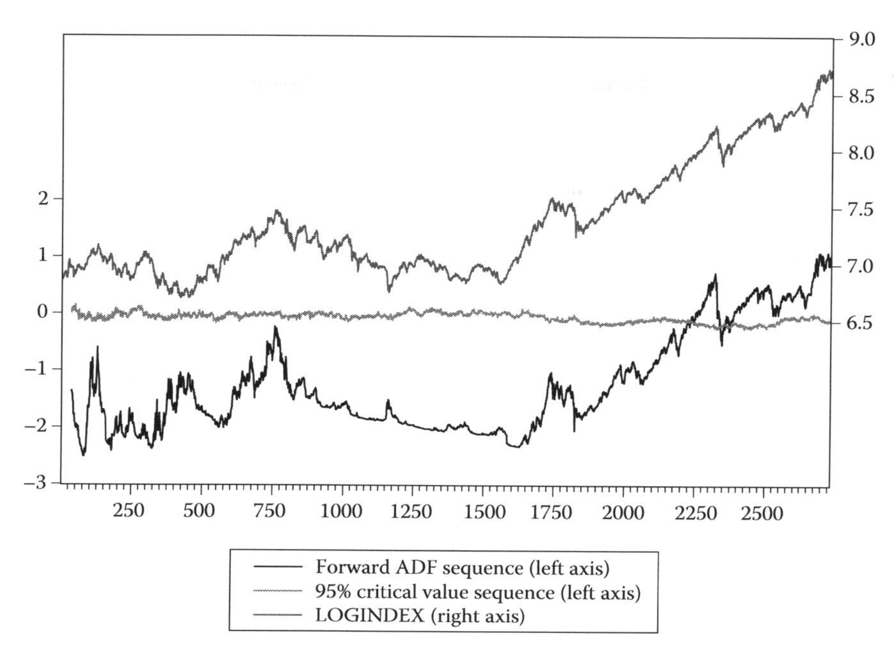

FIGURE 9.7
SADF test statistics, Nifty index 1997–2012.

is efficient, then, the stock prices are supposed to follow a RW. But our primary aim is to introduce the concept of VR and explain how it can be used to test for RW. This implies that the VR test can also be used as a test for the presence of a unit root in the data. The VR test makes use of the different behavior of variance of the series under $I(1)$ and $I(0)$ possibilities. It is based on a property of the RW that the variance of RW increments must be a linear function in the time interval.

Suppose, for example, we rewrite a pure RW as

$$z_t = z_0 + \sum_{i=1}^{t} e_i \quad e_t \sim IID(0, \sigma^2).$$

Let z_0 be nonstochastic. If we define, ∇_2, to be the two-period difference operator—that is, $\nabla_2 z_t = z_t - z_{t-2}$, then, $\text{var}(\nabla_2 z_t) = \text{var}(e_t + e_{t-1}) = 2\sigma^2$. Let us call this two-period difference, $VR(2)$. Therefore, the plausibility of the RW may be checked by comparing the variance of $e_t + e_{t-1}$ to twice the variance of e_t. In practice, this equality may not hold exactly, but a ratio involving these two quantities must be statistically indistinguishable from one. Thus, to conduct a statistical test of VR, we need to understand the statistical distribution of the VR under the null of the RW hypothesis. Upon generalizing it to any qth-period difference, we obtain the ratio

$$VR(q) = \frac{\text{Var}[e_t(q)]}{q\ \text{Var}[e_t]}.$$

The quantity $VR(q)$ can be calculated for various q's and for each q, we have to test if the ratio is a unit under the null that z_t is RW. If it is not accepted, then, e_t can be serially correlated and z_t is not an uncorrelated RW process. In such a case, $VR(2)$ will be equal to $1 + \rho_1$ and $VR(q)$ will be equal to

$$\begin{aligned} VR(q) &= 1 + \frac{2}{q}\sum_{j=1}^{q-1}(q-j)\,\rho_j \\ &= 1 + 2\sum_{j=1}^{q-1}\left(1-\frac{j}{q}\right)\rho_j. \end{aligned}$$

The second equality shows that $VR(q)$ is a particular linear combination of the first $q-1$ ACs of e_t with linearly declining weights. Under RW, however, $VR(q) = 1$, because $\rho_j = 0$ for $j \geq 1$.

Though many versions of VR tests are discussed in the literature, below, we shall discuss the tests proposed by Lo and MacKinlay (LM) and Cochrane's measure. A joint test of *VR*, called the multiple VR test, has also been recommended in the literature. A useful survey of some of these test procedures is available in the survey paper by Charles and Darne (2009).

A possible test of this assertion can be based on the empirical counterpart of the ratio. LM (1988) suggest the following two unbiased estimators, using overlapping data points. (They also suggest tests using nonoverlapping data sets, but they admit that the tests are clearly inefficient, because they don't use all the information). Let the sample consist of $Tq+1$ observations, $(z_0, z_1, \ldots, z_{Tq})$. Then,

$$\begin{aligned} \hat{\mu} &= \frac{1}{Tq}\sum_{k=1}^{Nq}(p_k - p_{k-1}) = \frac{1}{Tq}\left(p_{Tq} - p_0\right) \\ \hat{\sigma}_a^2 &= \frac{1}{Tq-1}\sum_{k=1}^{Tq}\left(z_k - z_{k-1} - \hat{\mu}\right)^2 \\ \hat{\sigma}_c^2(q) &= \frac{1}{M}\sum_{k=q}^{Tq}\left(z_k - z_{k-q} - q\hat{\mu}\right)^2 \\ M &= q(Tq+1-q)\left(1-\frac{q}{Tq}\right). \end{aligned}$$

With this, the empirical counterpart of the VR can be written as

$$\widehat{VR}(q) = \frac{\hat{\sigma}_c^2(q)}{\hat{\sigma}_a^2}.$$

Under the null, we can then state the following asymptotic distributions:

$$\sqrt{Tq}\,(\widehat{VR}(q) - 1) \xrightarrow{D} N\left(0, \frac{2(2q-1)(q-1)}{3q}\right)$$

These statistics can be standardized in the usual way to yield asymptotically standard-normal test statistics. Since VR tests are often used with stock price data or continuously compounded returns data, both of which exhibit time-varying volatility, LM (1988) recommend a robust version of the above VR test, so that RW is not rejected because of heteroscedasticity. Details are available in their paper.

9.3.6.10.1 Multiple VR Tests

While many such tests for measuring VR were suggested in the literature, all such tests were considered as individual tests. That is, VR for each q was calculated and checked against the null of RW separately. The question of whether a time series is mean reverting or not requires that the null holds true for all values of q. In view of this, it is necessary to conduct a joint test where a multiple comparison of VRs over a set of different time horizons is made. Many tests for a joint testing of VRs have been suggested in the literature but we will discuss only one such test, which is called the Chow–Denning (CD) (1993) test. The multiple VR tests consider the joint null hypothesis $H_0 : VR(q_i) = 1$ for all $i = 1, 2, \ldots, m$, against the alternative $H_1 : VR(q_i) \neq 1$ for some i. An F-test may sound appropriate here in testing multiple VRs. But in the VR context, if H_0 is rejected, then, further information on which of the individual VRs are different from one is desirable. CD suggest a test that may be useful from this context, as shown below.

9.3.6.10.2 C–D Multiple VR Test

CD (1993) extend the LM (1988) VR methodology and propose a simple joint-testing procedure for the multiple comparison of the set of VR estimates with unity, which allows us to examine a vector of individual test statistics while controlling for the overall test size. For a set of m-test statistics, the RW hypothesis is rejected, even if any one estimated VRs is significantly different from one.

To test the null hypothesis, CD consider the standardized VR statistic for *iid* increments:

$$\psi(q) = \sqrt{Tq}\,(\widehat{VR}(q) - 1)\,[2(2q - 1)(q - 1)/3q]^{-1/2}\,.$$

CD consider a set of VR statistics, $\{(\widehat{VR}(q_i) - 1),\ i = 1, 2, \ldots, m\}$. Under the RW hypothesis, CD test a set of hypotheses implied by the VRs by considering the largest-absolute statistic:

$$Z_1^*(q) = \max_{1 \leq i \leq m} |\psi(q_i)|.$$

CD however point out that assuming that ψ's are independent, $Z_1^*(q)$ can still be tested against the standard-normal distribution but after adjusting the size of the test—also called the significance level—α, to guard against an inappropriately large probability of Type-I error that arises in multiple testing. (Briefly stating, when we conduct two independent tests of a null hypothesis at 5% level, then, the probability that neither will be significant is given by $0.95 \times 0.95 = 0.90$. That is, the significance level is now actually 10% rather than the original 5% level. In general, if we conduct κ-independent tests at the α significance level, then, the overall significance level will now be $1 - (1 - \alpha)^{\kappa}$. Thus if $\kappa = 7$, then, α will be nearly 0.30!! leading to Type-I error. To keep the significance level at α, we make use of α/κ as the correct test size at the separate test level. This is generally referred to as the Bonferroni correction). But CD suggest to use a sharper Bonferroni correction, $\alpha^* = 1 - (1 - \alpha)^{1/m}$ as the test size. However, the approach by CD is valid only if the vector of the individual VRs is multivariate normal. This implies that there should be a little overlap while calculating the various ratios; that is, q/T is small. CD opine that the same procedure can be adopted to jointly test, $\psi^*(q)$, with heteroscedastic increments.

CD also suggest a test if the ψ's are correlated. In such a case, CD suggest to check the test statistic, $Z_1^*(q)$, against what is called the studentized maximum modulus (*SMM*) distribution. That is, $SMM(\alpha, m, T)$, where α is the test size; m is the number of q values; and T is the sample size. Asymptotically, however, when $T = \infty$, SMM converges to a standard normal with the test size, α^*.

As an example of this test, we estimated the VR for various periods, using continuously compounded weekly returns over May 1997 and January 2006, calculated from the daily Nifty index,

TABLE 9.12
Results of VR Tests

Individual Tests				
Period	**Variance Ratio**	**Standard Error**	***Z*-Statistic**	***P*-Value**
2	0.4975	0.0474	−10.5994	0
4	0.2532	0.0887	−8.4207	0
5	0.1921	0.1039	−7.7792	0
8	0.1292	0.1402	−6.2098	0
10	0.0988	0.1601	−5.6305	0
16	0.0602	0.2087	−4.5038	0
30	0.0340	0.2923	−3.3048	0.001

Joint Tests		
Joint Tests	**Value**	***P*-Value**
Max $\lvert Z\text{-stat}\rvert$ at period 2	10.59942	0

to verify if the Indian stock market can be termed as efficient. Table 9.12 shows the results and it is evident that the null of prices being a RW is strongly rejected, implying correlations in weekly returns.

9.3.6.10.3 Cochrane's Measure of VR

Cochrane (1988) applies the concept of VR through what he calls as the variance of long differences and measures the persistence of shocks. Recall that if a series belongs to the DS class—that is, it has a unit root—the effect of a shock to innovations—say, a decline in log gross national product (GNP)—leaves a permanent effect on the level of the series. That is, after the shock, the series never "reverts" to the existing trend. So, we say that shocks are persistent. On the other hand, if the series belongs to a TS process, then, a decline in log GNP today has no effect on the forecasts of the level of GNP in the future. This implies that the GNP growth rates must rise in the short run until it reverts to the trend line. Cochrane (1988) contests this difference as an extreme and claims that a trending series can be better characterized as falling somewhere between the two extreme views. In his paper, Cochrane shows that a series such as U.S. GNP does in fact revert to the trend, but after a long time, which may be several years later. Cochrane's claim is tantamount to asking the question as to how much does the long-run forecast of GNP respond to shocks?

- If by one unit, then it finds an RW.
- If by zero, then, it finds a TS process.
- It can also find numbers between one and zero, which characterizes a series that returns toward the trend in the long run but does not get all the way there.
- If it exceeds one, then, of course, we have a series that will diverge forever, following a shock.

The possibility of a value lying between 0 and 1, perhaps, prompts Cochrane to hypothesize that a trending series can be modeled as a combination of a stationary series and an RW. Since it is the RW component that carries the permanent part of a shock, one can ask the question of how important this component is to the series. Cochrane tries to answer this question by asking another related question: how big is the variance of shocks to the RW or the permanent component compared with the variance of yearly growth rates? If the variance of shocks to the RW component is zero, then, the series is TS. If the variance of shocks to the RW component is equal to the variance of the first differences, then, the series is an RW. As before, there can be values between 0 and 1 and greater than 1.

Cochrane attempts to measure this using the VR concept.

- Assume that log GNP, z_t, follows a pure RW. Then, the variance of its k-period difference will be k times the variance of the first difference: $\text{var}(z_t - z_{t-k}) = k\sigma^2$.
- Suppose it is TS, the variance of its k differences approaches twice the unconditional variance of the series: $\text{var}(z_t - z_{t-k}) = 2\sigma_z^2$.
- And if we plot $(1/k)\text{var}(z_t - z_{t-k})$ as a function of k, then, the plot will be a constant at σ^2 if z_t is an RW. If z_t is TS, then, the plot should decline toward zero.
- Next, suppose that the fluctuations in log GNP are partly temporary and partly permanent, which we can model as a combination of a stationary series and a RW. A plot of $(1/k)\text{var}(z_t - z_{t-k})$ as a function of k will settle down to the variance of the RW component. And according to Cochrane, the variance of k differences can explain the fluctuations in such a structure, better than many other competing models.

Cochrane sets about proving his claims about the variances of k differences as a better statistic to explain the fluctuations in z_t by showing that the first differenced stationary time series is equivalent to a time series that is composed of a stationary and a RW component. To show that such a representation exists, he makes use of a decomposition called the Beveridge–Nelson decomposition.

The idea that cyclical or transitory movements can be observed in an economic time series and separated from the trend or the permanent components is a very old one and the traditional application of this concept is of course the business-cycle analysis. A number of approaches have been suggested in the literature to measure the business cycle. Beveridge and Nelson (BN) (1981) propose a measure based on the fact that any DS time series, with or without a drift, can be decomposed into two additive components, a stationary series and a stochastic part. The stationary part, which BN call as the cyclical component, is defined to be the forecastable momentum in the series at each point in time. The stochastic part traces out the predictive distribution for the future path of the original series. The application of this technique begins with an investigation of the stochastic structure of each series and then exploits the particular structure of each to arrive at the appropriate filters.

Let z_t be an $I(1)$ process with or without a drift. Then, we can decompose it into a stochastic (random) component and an $I(0)$ stationary or the cyclical component, $z_t = TS_t + C_t$, where TS_t is the stochastic trend that captures the effects of shocks that have a permanent effect on the level of z_t and c_t captures the effects of shocks that have only temporary effects. BN in their paper show what this stochastic trend comprises of and how to quantify the permanent part of this trend component in a DS time series that may or may not have a drift and that allows for a Wold decomposition. Suppose

$$\nabla z_t = \alpha_1 + u_t, \quad u_t = \psi(B)e_t.$$

BN define the permanent component as follows: the permanent component of an $I(1)$ process with a drift is the limiting forecast of z_t, as the forecast horizon goes to infinity, which is adjusted for the mean rate of growth over the forecast horizon. In other words, they show that

$$TD_t + BN_t = \lim_{s \longrightarrow \infty} z_{t+s|t} - \alpha_1 s,$$

where TD_t is the deterministic part of the trend and BN_t is the stochastic part of the trend, so that $C_t = z_t - TD_t - BN_t$. Thus, according to BN decomposition, the stochastic-trend part in any DS process can be broken up into a deterministic part, TD_t and a stochastic one, BN_t. In this sense, it is BN_t that is the permanent component of z_t. (Note that BN call BN_t as the stochastic trend.) And BN show that BN_t follows a pure RW. That is,

$$\begin{aligned} BN_t &= BN_{t-1} + \psi(1)e_t \\ &= BN_0 + \psi(1)\sum_{j=1}^{t} e_j. \end{aligned}$$

Since BN_0 is the initial value, we can say that $\psi(1)\sum_{j=1}^{t} e_j$ essentially *quantifies* the stochastic trend. And the innovation variance of this RW component, BN_t, $\sigma^2_{\nabla BN_t}$, is a measure of the importance of the RW component of the series, z_t.

Given the BN decomposition, Cochrane next establishes that the variance of k differences can be used to estimate the innovation variance of an RW component. To document that claim and provide asymptotic distributions, Cochrane outlines the following steps:

- Let σ_k^2 denote $(1/k)$ times the population variance of k differences of z_t, $\sigma_k^2 = k^{-1}\text{var}(z_t - z_{t-k})$; σ_k^2 is related to the AC coefficients of ∇z_t by

$$\sigma_k^2 = \left(1 + 2\sum_{j=1}^{k-1} \frac{k-j}{k}\rho_j\right)\sigma^2_{\nabla z},$$

where $\sigma^2_{\nabla z} = \text{var}(z_t - z_{t-1})$ and $\rho_j = \text{cov}(\nabla z_t \nabla z_{t-j})/\sigma^2_{\nabla z}$.
A proof of this claim is available in Cochrane (1988, pp. 916–917).

- Cochrane sets out next to prove that the limit of σ_k^2 is indeed the innovation variance of the RW component. Noting that the RW component according to BN decomposition discussed above is BN_t, Cochrane shows that

$$\lim_{k\longrightarrow\infty} \sigma_k^2 = \left(1 + 2\sum_{j=1}^{\infty} \rho_j\right)\sigma^2_{\nabla z} = \sigma^2_{\nabla BN_t}.$$

The proof of this follows from the definition of spectral density.

- It is clearly evident that under the null of RW, the VR between the variance of k differences and variance of the RW component goes to one in the limit, as all ACs vanish under the null. Thus, Cochrane also uses the VR concept but conceptualizes it in a different way.
- Intuitively, $\sigma^2_{\nabla BN_t}$ is called the *long-run variance*; meaning it is the variance of the process when all the transitory fluctuations have died out.
- How does one estimate σ_k^2? Just replace the population quantities with their sampling counterparts:

$$\hat{\sigma}_k^2 = \left(1 + 2\sum_{j=1}^{k-1} \frac{k-j}{k}\hat{\rho}_j\right)\hat{\sigma}^2_{\nabla z}.$$

- The asymptotic variance of $\hat{\sigma}_k^2$ is

$$\text{Asy.Var}\,(\hat{\sigma}_k^2) = \frac{4k\hat{\sigma}_k^4}{3T},$$

where T is the sample size. Experiments show that $k = 30$ is good enough to estimate $\sigma^2_{\nabla BN_t}$ but since Cochrane uses annual data, he believes that after 30 years, the temporary effects of business cycles are over.

9.4 Spectral or Frequency-Domain Analysis: An Overview

Given an economic time series, we are basically interested in the following three points: the way the time-series data have been collected, trends and seasonal variations (which are mostly periodic), and business-cycle and economic fluctuations. Trends and annual components (seasonal) are frequently the most-obvious features of an economic time series. But if one prepares to "filter" out these components, then, what remains is a long series of irregular fluctuations. Some economists have suggested that these could have arisen as the sum of a number of more or less regular fluctuations. But such fluctuations were mistakenly labeled as cycles and many such cycles have been discussed in the literature. Some of the most-important ones are the following:

1. The Kondratieff long wave of 40–60-years duration
2. Kuznets long wave—20 years in duration, found mostly in GNP, population variables, and migration
3. Building cycle—a cycle of 15–20-year duration in building industries of various countries
4. Major and minor cycle—major cycles have a duration of 6–11 years and inventories have been found to have cycles of such duration, and minor cycles have 2–4-years duration

But it was left to the National Bureau of Economic Research (NBER) to clarify and define the business cycle. In none of these "cycles," there is any sense of regularity implied. They should aptly be described as "fluctuations." There is also an evidence to suggest that business cycles tend to recur over and over again in virtually the same form, with the same duration and some amplitude of movement. There have been mild cycles and severe ones. Upswings have been longer than downswings. It is this mixture of regularity but yet nonregularity that has posed statistical difficulties to econometricians (see Granger and Hatanaka 1964 for more on this, which is a classic reference on spectral analysis, especially for economists).

One of the advantages of spectral methods is that they provide us with a way to tackle exactly such a mixture of regularity and nonregularity. Such methods basically help the analyst to

1. Explain fluctuations similar to the business cycles, as a sum of a number of sine series with suitably chosen amplitudes and phases and all of them having nearly equal periods.
2. Describe the essential fluctuations of the business-cycle type by generalizing such a sum of sine terms, to an integral of a sine function over a band of periods.

9.4.1 Uses of Spectral Analysis

Spectral-analytic techniques are mainly used to study the cyclical characteristics of the data. That is, we use such techniques to understand how important the cycles of different frequencies are in explaining the behavior of the given time series. Hence, the name frequency-domain analysis. However, frequency-domain and time-domain techniques are complementary and not competitive. It is just that they explain different features of the same data set. And in this section, we shall motivate the discussion on spectral analysis with the specific intention of demonstrating how this technique can be used to unearth the hidden cycles in any time-series data. Why can't a correlogram detect such cycles? It is useful only if a time series consists of one cycle. Since a time series may consist of many cycles, spectral analysis will identify the cycles of all frequencies and will let the analyst select the one that he needs.

9.4.2 Search for Periodicity

This is the problem of describing or finding periodicities, if any, present in the time-series data. Here, we have to note the following:

1. In some cases, we may already know that a collection of periodicities are present; but we may still need to know their amplitudes and the phase.
2. Many times, we may not know the periods also; so, we may have to find them too. In these two cases, least squares can be profitably used to estimate the unknowns.

But what shall we do when we want to analyze an arbitrary set of data into periodic components, *whether or not* the data appear to be periodic. By this, we mean that a visual inspection of the plots of such data may not reveal the presence of cycles; but such cycles or periodicities may be *hidden*. It is in detecting such hidden periodicities that a harmonic or Fourier analysis is very useful. In Fourier analysis, the periodic functions used are the sine and the cosine functions. They have the important properties that an approximation consisting of a given number of such trigonometric terms achieves the MMSE between the signal and the approximation, and also that they are orthogonal; so, the coefficients can be determined independently of one another. In other words, by summing up a lot of sine and cosine terms of different amplitudes and varying periods, one can construct an artificial time series that resembles a real one. Thus, the goal of spectral analysis is, given a real-time series, how to construct it using sines and cosines—that is, how to write it, for example, as

$$z_t = a_0 + \sum_j a_j \sin(2\pi\omega_j t) + b_j \cos(2\pi\omega_j t).$$

This is called the Fourier representation of a time series. Different time series will need different a_j's and b_j's. Here, ω is the angular frequency defined as $\omega = 2\pi f$, which is essentially f, the frequency, normalized to the frequency of the harmonic (sin,cos) functions, that is equal to $1/2\pi$. (Harmonic frequencies are multiples of the fundamental frequency.) An accessible introduction to Fourier analysis of a time series and many related concepts is available in Bloomfield (2000).

But such classical Fourier methods fail when they are applied to a time series that are stochastic in nature. The reasons are that the classical Fourier methods are applied on the assumption of fixed amplitudes, phases, and frequencies. But a stochastic time series, on the other hand, are characterized by random changes of these quantities. So, the classical Fourier methods have to be adapted to describe such a stochastic time series. The appropriate way then is to apply Fourier transform to the autocovariance function of a stochastic time series, which will give us the *theoretical or power spectrum* or simply the *spectrum*. Hamilton (1994), Harvey (1984), and Sargent (1987) provide an easy introduction to the underlying issues, including the properties of filters, which we explain below.

9.4.3 Power Spectrum and Its Properties

The theoretical spectrum can be derived by exploiting the link between the autocovariances and the population spectrum. Since the autocovariance-generating function of a stochastic process can be written as $\gamma(z) = \sum_{\tau=-\infty}^{\infty} \gamma_\tau z^\tau = \sigma_e^2 \psi(z)\psi(z^{-1})$, the theoretical spectrum can be written in two different forms, for some complex scalar $z = e^{-i\omega}$, as follows:

$$f(\omega) = \frac{1}{2\pi} \sum_{\tau=-\infty}^{\infty} \gamma_\tau e^{-i\omega\tau},$$
$$= \frac{1}{2\pi} \sigma_e^2 \psi(e^{-i\omega})\psi(e^{i\omega}).$$

The resulting $f(\omega)$ is the *power spectrum* or the *theoretical spectrum.* And the spectrum is the Fourier transform of the autocovariance function. This can be calculated given a sequence of autocovariances using the following relation: $e^{-i\omega\tau} = \cos(\omega\tau) - i\sin(\omega\tau)$. With this, we can write the second form as

$$f(\omega) = \frac{1}{2\pi} \sum_{\tau=-\infty}^{\infty} \gamma_\tau [\cos(\omega\tau) - i \sin(\omega\tau)].$$

Since $\gamma_\tau = \gamma_{-\tau}$, this can be simplified to be

$$f(\omega) = \frac{1}{2\pi}\left[\gamma_0 + 2\sum_{\tau=1}^{\infty} \gamma_\tau \cos(\omega\tau)\right] \quad 0 \le \omega \le \pi.$$

Since $\cos(\omega + 2\pi\tau) = \cos(\omega)$, $\tau = 0, \pm 1, \pm 2, \ldots$, it follows that the spectrum is a periodic function with a period of 2π. Since the spectrum is symmetric around $\omega = 0$, we can confine ourself to the interval $[0, \pi]$. And the spectrum exists, because the autocovariances are summable. If the underlying series, z_t, is of white noise, then, we can easily see that the theoretical spectrum is $(2\pi)^{-1}\sigma^2$, where σ^2 is the variance of z_t and the spectrum will be flat. Similarly, for an $MA(1)$ process, we have $\gamma(0) = (1 + \theta^2)\sigma^2$ and $\gamma(1) = \theta^2\sigma^2$, and $\gamma(s) = 0$, for $s \ge 2$. The theoretical spectrum can be calculated as $f(\omega) = (\sigma^2/2\pi)(1 + \theta^2 + 2\theta\cos\omega)$.

It is pertinent to ask if one can calculate the autocovariances given the spectrum. This is possible using the inverse Fourier transform. If we know the value of $f(\omega)$, for all $\omega \in (0, \pi)$, then, we can calculate the autocovariances for any lag τ. The formula for doing this is

$$\int_{-\pi}^{\pi} f(\omega)e^{i\omega\tau} = \gamma_\tau$$

or

$$\int_{-\pi}^{\pi} f(\omega)\cos(\omega\tau)d\omega = \gamma_\tau.$$

These expressions along with those of the theoretical spectrum are called as the *Fourier-transform pairs.* They are theoretically equivalent to one another. As a special case, consider the case of $\tau = 0$. Then we have

$$\int_{-\pi}^{\pi} f(\omega)d\omega = \gamma_0.$$

This has the interpretation that the area under the population spectrum between $\pm\pi$ gives the variance of the process. Since $f(\omega)$ is symmetric around $\omega = 0$, we can write it as

$$2\int_0^{\omega_1} f(\omega)d\omega = \gamma_0.$$

9.4.4 Estimation of the Spectrum

We need to get an estimate of $f(\omega_j)$ for various j's. And this can be obtained using both parametric and nonparametric methods. Among the parametric methods, it is the *ARMA* methods that are most commonly used. But it is the nonparametric method that is more popular and we shall also concentrate on this method. Nonparametric estimates are also called *kernel* estimates. Here, the suggestion is to estimate the population spectrum, at any given frequency, $\omega_j = (2\pi j/T)$, $j = 1, \ldots, n$, where $n = T/2$ if T is even and $n = (t-1)/2$ if T is odd. The reason for the choice of the selected frequencies is purely computational, since we will be making use of some orthogonality relations. We obtain the spectral values by averaging over the adjacent frequencies. Since the unsmoothed sample spectrum behaves very violently around the theoretical spectrum, providing a meaningful interpretation of the spectrum is difficult. Hence, it is normal to get a smoothed estimate of the spectrum called a smoothed spectrum. We can obtain such a smoothed spectrum at a particular frequency either by weighting the spectral values around the spectrum of interest or by weighting the autocovariances. The former weighting pattern is called the spectral windows and the latter is called the lag windows. Obtaining smoothed spectral estimates using lag windows, or alternatively called as weighting in the time domain, is more popular.

A sample analog of the theoretical spectrum can be written as

$$\hat{f}(\omega) = (2\pi)^{-1}\left[\hat{\gamma}_0 + 2\sum_{\tau=1}^{T-1}\hat{\gamma}(\tau)\cos(\omega\tau)\right].$$

This is called as the sample periodogram. Note that a sample of size T can have $T-1$ autocovariances. What we do here is to weight the autocovariances by a weighting pattern, named K^* in the sample periodogram. Such a weighting pattern is called the lag window. Then, the spectral estimate $\hat{f}(\omega)$ may look like

$$\hat{f}(\omega) = (2\pi)^{-1}\left\{\hat{\gamma}_0 + 2\sum_{\tau=1}^{T-1} K^*\hat{\gamma}_\tau \cos(\omega\tau)\right\}.$$

Smaller values of K^*_τ result in a smoother spectrum. Some smoothing schemes have been suggested in the literature and the most popular among them are *Bartlett kernel or window* and the *Parzen kernel or window.* The Bartlett window, for example, is given by

$$K^*_\tau = \begin{cases} 1 - \dfrac{\tau}{q+1}, & \text{for } \tau = 1, \ldots, q \\ 0, & \text{for } \tau > q \end{cases}$$

If we use the Bartlett window, then, the smoothed sample spectrum estimate will be

$$\hat{f}(\omega) = \frac{1}{2\pi}\left[\hat{\gamma}_0 + 2\sum_{\tau=1}^{q}\left[1 - \tau/(q+1)\right]\hat{\gamma}_\tau \cos(\omega\tau)\right], \quad 0 \le \omega \le \pi.$$

This is the formula used if one wishes to calculate the smoothed sample spectral estimates with the Bartlett-weighting pattern. The parameter q is called the bandwidth parameter and how to select it is a question of practical importance. Since we estimate the spectral value at a frequency by averaging over different frequencies, the variance is reduced but this increases the bias. The severity of the bias depends on the steepness of the population spectrum and the size of the bandwidth. One practical guide is to estimate the spectrum for many bandwidth parameters and use judgment to choose that q that gives us a satisfactory estimate.

9.4.5 Application of Spectral Analysis

We illustrate the use of spectral-analytic techniques to uncover the hidden periodicities in the industrial production of India. Figure 9.8 displays plots of the spectrum estimates for the Indian IIP data. Unseasonalized monthly data on industrial production over the period from 1971 to 2002, with a sample of 384 observations, were used. We spliced the index to a common base year using the symmetric mean method of Hill and Fox (1997). The first column of Figure 9.8a–c shows the spectral estimates calculated according to the Parzen window and Figure 9.8d–f in the second column shows the plots of spectral estimates calculated using the Bartlett-weighting pattern. Since the spectra obtained through the two weighting patterns look similar, we shall explain the spectral estimates obtained using the Bartlett window.

Sample spectral estimates of logs of raw data plotted in Figure 9.8d display $\hat{f}(\omega)$ against the values of j, implying that the period of the cycle is T/j. One can easily see that almost all the values are concentrated around the lowest frequencies, that is, j near zero. This can be easily attributed to the upward trend in the raw data. Moreover, since the frequency is near zero, we will obtain a period of near infinity. We encounter such a result, if there is a stochastic trend in the series, which may be removed by first differencing the data.

So, next, we estimated the spectrum after taking the first differences of the logs of raw data. The estimates are in Figure 9.8e. Since the sample size is 383, the first peak occurring at $j = 32$ corresponds to a period of about 12 months. The second peak occurring at $j = 64$ corresponds roughly to a period of 6 months and this clearly refers to the existence of a seasonal cycle and it is natural to attribute the finding of three subsequent peaks occurring at $j = 95, 128$ and $j = 160$ to be associated with cycles of periods $4, 3$, and 2.4, respectively. This confirms the presence of seasonal cycles in the data. This is an important result because this finding justifies using the filter $(1 - B^{12})$ as a whole to filter out seasonal cycles. In the time domain, this is an issue related to the problem of testing for unit roots at the seasonal frequencies, along with the standard unit-root testing at a frequency of zero. The details on testing for seasonal unit roots are covered in the section on seasonality later in this chapter.

If one wants to spike the seasonal effects, so that we can isolate the business-cycle frequencies alone, one can use the seasonal filter, $(1 - B^{12})$. Figure 9.8f plots the spectral estimates obtained after applying the seasonal filter on raw data. The only substantial peak occurs at $j = 8$ that corresponds to a period of 46.5 months or roughly 4 years, which is generally believed to be the business-cycle frequency for India. So, filtered in this way, one can easily isolate the business-cycle frequency. Specific frequencies, other than seasonal, can also be removed; but we need to use different filters.

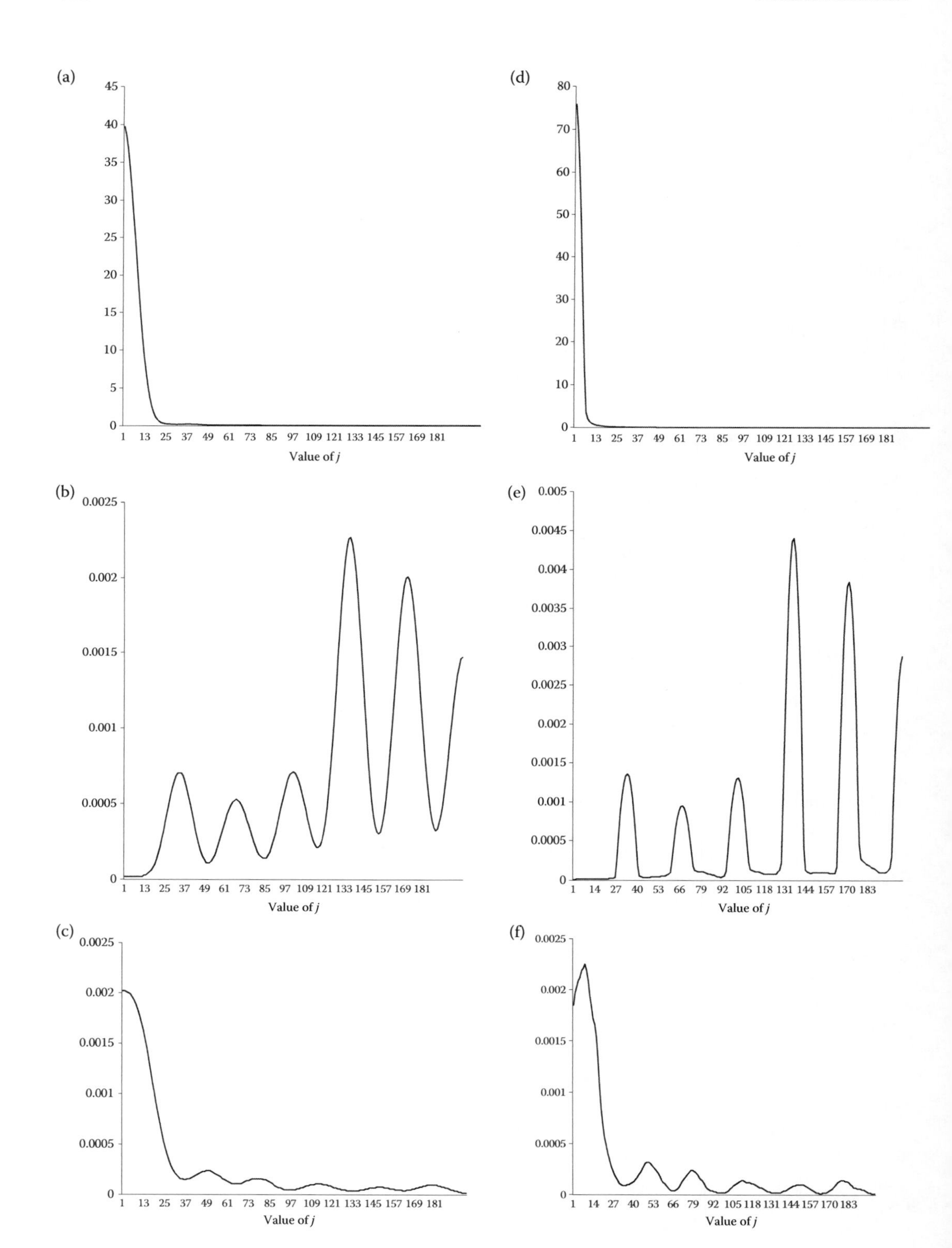

FIGURE 9.8
Sample periodogram with Parzen and Bartlett windows.

9.4.6 A Note on Filters

Sometimes, data are filtered or transformed in a particular way. The AR models, MA models, or the first differences are basically some filtered models only. For example,

1. $z_t = (1 - B)x_t$ is a first-difference filter.
2. $z_t = \sum_{j=-r}^{s} w_j x_{t-j}$ is an MA filter, with weights summing up to one. This is *not* an MA process.
3. $z_t = \sum_{j=0}^{s} x_{t-j}$ is an MA with constant weights.

Combined with spectral analysis, such filters are useful in checking the relevance of filtering operations and the effect of inappropriate filters on the spectrum of the original series. Filters remove whatever power that is in the frequencies not wanted by us. By implication, this also warns the users of applying filtering operations arbitrarily, since any incorrect filtering may distort the spectrum of the original series and may introduce spurious cycles. For example, it can be shown that filtering a white noise produces a series that have spurious cycles—the so-called *Slutsky effect*. It is very difficult to investigate this possibility in the time domain; spectral methods are more suited to tackle such issues.

9.4.6.1 *Complex Spectral Representation*

For a purely indeterministic process,

$$y_t = \int_{-\pi}^{\pi} e^{i\omega t} z_y(\omega) d\omega,$$

where $d\omega$ is any small interval in the range $-\pi \leq \omega \leq \pi$ and i is a number such that $i^2 = -1$.

$z_y(\omega)$ is an orthogonal increment process, with the properties

$$E[z_y(\omega)d\omega] = 0$$

$$E[z_y(\omega)d\omega \cdot \overline{z_y(\omega)d\omega}] = f(\omega)d\omega$$

$$E[z_y(\omega_1)d\omega_1 \cdot \overline{z_y(\omega_2)d\omega_2}] = 0$$

$$E[z_y(\omega_1) \cdot z_y(\omega_2)] = 0,$$

where $d\omega_1$ and $d\omega_2$ are two nonoverlapping intervals. It follows that $E(y_t) = 0$ and

$$\gamma_\tau = \int_{-\pi}^{\pi} e^{i\omega\tau} f(\omega) d\omega.$$

The inverse Fourier transform of the above expression is the power spectrum or the theoretical spectrum, $f(\omega)$, defined earlier.

9.4.6.2 *ARMA in the Frequency Domain*

We can reproduce some of the results arrived earlier using the complex representation also. Let x_t be a stationary time series.

Suppose we have

$$z_t = \sum_{j=-r}^{s} w_j x_{t-j}.$$

This operation uses a time-invariant filter. When the weights sum up to one, this is called the MA filter, not to be confused with the MA process.

Let $f_x(\omega)$ be the spectrum of x_t and define

$$W(\omega) = \sum_{j=-r}^{s} w_j e^{-i\omega j}.$$

This is called as the *frequency-response function.* It tells us which frequency components the filter captures from the original series. One may see that it is basically the Fourier transform of the linear filter. The spectrum of z_t is then

$$f_z(\omega) = |W(\omega)|^2 f_x(\omega),$$

and the factor $|W(\omega)|^2$ is called as the transfer function. And $|W(\omega)|^2 = W(\omega)\overline{W(\omega)}$. Hence, the spectral representation of z_t is

$$z_t = \int_{-\pi}^{\pi} e^{i\omega t} W(\omega) z_x(\omega) d\omega,$$

where the spectral representation of x_t in a complex form is

$$x_t = \int_{-\pi}^{\pi} e^{i\omega t} z_x(\omega) d\omega,$$

and z_x is an orthogonal increment process with the properties as explained before. Suppose z_t follows an $MA(1)$ model, in terms of the filter, $\omega_0 = 1$, $\omega_1 = \theta$, and other weights being zero. So, $|W(\omega)|^2 = (1 + \theta^2 + 2\theta\cos\omega)$ and $f_x(\omega) = \sigma^2/2\pi$ since x_t is a white-noise process. The spectrum of z_t, $f_z(\omega)$ is therefore

$$f_z(\omega) = \left(\frac{\sigma^2}{2\pi}\right)(1 + \theta^2 + 2\theta \cos \omega).$$

And this is the same as was derived earlier using the theoretical spectrum. Similarly, if z_t follows a stationary $AR(1)$ process, then, we can express it as an infinite-order MA process so that

$$W(\omega) = \frac{1}{(1 - \phi e^{-i\omega})}$$

$$f_z(\omega) = \left(\frac{\sigma^2}{2\pi}\right)\left[\frac{1}{|1 - \phi e^{-i\omega}|^2}\right] = \left(\frac{\sigma^2}{2\pi}\right)\frac{1}{(1 + \phi^2 - 2\phi \cos \omega)}.$$

Using the techniques described above, it is not too difficult to write the spectrum of the general $ARMA(p, q)$ model as

$$f_z(\omega) = \left(\frac{\sigma^2}{2\pi}\right) \cdot \frac{\left|1 + \sum_{j=1}^{q} \theta_j e^{-i\omega j}\right|^2}{\left|1 - \sum_{j=1}^{p} \phi_j e^{-i\omega j}\right|^2}.$$

This is referred to as the rational spectrum. In terms of the associated polynomials, we can write the same as

$$f_z(\omega) = \left(\frac{\sigma^2}{2\pi}\right) \cdot \frac{|\theta(e^{-i\omega})|^2}{|\phi(e^{-i\omega})|^2}.$$

9.4.7 Properties of Linear Filters

It is with the help of the frequency-response function, along with the gain function and phase function, that we can assess the efficacy of filters. Time-invariant linear filters are commonly used on an economic time series to remove a cyclical component. Suppose there is a 5-year cycle in the data that needs to be removed. We may use the filter

$$z_t = x_t - x_{t-5}, \quad t = 6, \ldots, T$$

We can assess the effect of using the method discussed below, only if the filter used is linear, and not the nonlinear filters such as those used by government agencies.

9.4.7.1 Gain and Phase

A filter changes the relative importance of the various cyclical components that will be captured by the *gain* function. Another property of the filter is to induce a shift in the time period of the series under consideration and this shift is captured by the *phase* function. If an application of the filter, x_t, on z_t results in a phase shift, then, it means that the peaks and lows of x_t and z_t have a different timing. In business-cycle literature, if an indicator series establish a sizable phase lead, at low business-cycle frequencies, then, the series is considered as a leading indicator of movements in some important reference series such as the GNP. But Sargent (1987) illustrates with examples that a phase lead is neither a necessary nor a sufficient condition for one series to be of use in predicting another. Nevertheless, these two statistics are widely reported.

The gain function, denoted as $G(\omega)$, is nothing but the square root of the transfer function, $|W(\omega)|$. If $W(\omega) = W^*(\omega) + iW^\dagger(\omega)$, then, the phase effect is defined as

$$\text{Ph}(\omega) = \tan^{-1}\left[-W^\dagger(\omega)/W^*(\omega)\right], \quad 0 \leq \omega \leq \pi.$$

We can rewrite the frequency-response function in terms of gain and phase as follows:

$$W(\omega) = G(\omega)e^{-i\text{Ph}(\omega)},$$

so that the spectral representation for y_t is now

$$y_t = \int_{-\pi}^{\pi} e^{i\omega(t-\text{Ph}(\omega)/\omega)} G(\omega) z_x(\omega) d\omega.$$

The implication of the filter in terms of gain is that the amplitude, z_x, is multiplied by $G(\omega)$ and the phase lag tells us that y lags x by $\text{Ph}(\omega)/\omega$ periods.

For example, for the linear filter that we saw earlier,

$$z_t = x_t - x_{t-5},$$

the frequency-response function is given as

$$W(\omega) = 1 - e^{-i5\omega},$$

and so,

$$G(\omega)^2 = 2(1 - \cos 5\omega),$$

and the phase is given by

$$\text{Ph}(\omega) = \tan^{-1}\left[(-\sin 5\omega)/(1 - \cos 5\omega)\right].$$

If there was a 5-year cycle in the data that needs to be removed, then, the spectral power at the related fundamental frequency $2\pi/5$ and its first harmonic $4\pi/5$ must be removed completely, that is, the gains at these frequencies should be zero. One can easily see from the gain function that the gain at these frequencies is indeed zero. Additionally, the gain at $\omega = 0$ is also zero, so that the differencing filter essentially removes the trend as well.

On the basis of the gain function, we can classify filters as high-pass filters, low-pass filters, band-pass filters, and all-pass filters. High-pass filters pass the high-frequency components of a signal and attenuate lower frequencies and involve differentiation of the input. Low-pass filters allow low frequencies with different gains but high frequencies are rejected and such low-pass filters are associated with some form of integration of the output. Band-pass filters capture signals over a specified frequency range and shut all powers outside this range and all-pass filters capture all the frequency components of the signal. The gain function of this filter is simply one and it leaves the original signal unaltered. Band-pass filters are frequently applied in business-cycle analysis and filters such as the Hodrick–Prescott filter and the Baxter–King filter are widely used to extract the cyclical components. We explain how such a band-pass filter works below using Sargent's (1987) example.

Let x_t be a covariance-stationary random process. Suppose we define z_t as

$$z_t = \sum_{j=-\infty}^{\infty} w_j x_{t-j} = W(B)x_t.$$

The spectrum of z_t is simply

$$f_z(\omega) = |W(\omega)|^2 f_x(\omega).$$

Suppose we choose $W(\omega)$ to be such that

$$W(\omega) = \begin{cases} 1 & \text{for } \omega \in [a,b] \cup [-b,-a], \quad 0 < a < b < \pi \\ 0 & \text{otherwise.} \end{cases}$$

This filter eliminates the spectral powers completely, at frequencies outside the region $[a, b]$ or $[-b, -a]$. We can obtain the set of w_j that satisfies the filter by using the "inverse" Fourier transform

$$w_j = \frac{1}{2\pi}\int_{-\pi}^{\pi} W(\omega)e^{i\omega j}d\omega = \frac{1}{2\pi}\int_{-b}^{-a} e^{i\omega j}d\omega + \int_{a}^{b} e^{i\omega j}d\omega$$

$$= \frac{1}{2\pi}\int_{a}^{b}\left(e^{i\omega j} + e^{-i\omega j}\right)d\omega$$

$$= \frac{1}{\pi}\left(\frac{\sin jb - \sin ja}{j}\right), \quad \text{for all integers } j.$$

Given the w_j coefficients, process z_t has all its variance defined in the frequency bands, $\omega \in [a, b]$, $\omega \in [-b, -a]$. And hence, the spectrum of z_t can be defined as follows:

$$\frac{1}{2\pi}\int_{-\pi}^{\pi} f_z(\omega)d\omega = \frac{1}{2\pi}\int_{-b}^{-a} f_x(\omega)d\omega + \frac{1}{2\pi}\int_{a}^{b} f_x(\omega)d\omega.$$

In general, the first term on the right-hand side of this formation also tells that a decomposition of the variance of series x by frequency, over the frequency band $[-b, -a]$, can be obtained by integrating the spectrum over that band and dividing by 2π.

9.4.7.2 Spurious Cyclical Behavior

Filters distort the features of a time series. Some filters can induce spurious filters in the data. Especially, it was found that applying a set of summing operations on a white noise or a serially uncorrelated process, e_t, could induce a spurious cycle in the data, if a certain amount of differencing operations had already taken place. This was called the *Yule–Slutsky effect.* If we denote the frequency-response function of the first filter as $W_1(\omega)$ and that of the second filter as $W_2(\omega)$, the combined effect of the filtering operation is given by

$$W(\omega) = W_1(\omega)W_2(\omega),$$

and the corresponding transfer function is

$$|W(\omega)|^2 = |W_1(\omega)|^2|W_2(\omega)|^2.$$

An interesting example of this is the Kuznets cycle that postulates a cycle of around 20 years in a certain economic time series.

Kuznets removed the cyclical components at higher frequencies by applying two filters. First, a simple 5-year MA filter of the form

$$z_t = \frac{1}{5}\sum_{j=-2}^{2} x_{t-j}, \quad t = 3, \ldots, T-2$$

was applied with an aim to attenuate the 5-year trade cycle. This filter has a frequency-response function

$$W(\omega) = \frac{\sin(5\omega/2)}{5\sin(\omega/2)}.$$

Then, he performed another operation of the form

$$y_t = z_{t+5} - z_{t-5},$$

which has a simple frequency-response function, $2i\sin(5\omega)$. Combining the two filters, we obtain an overall transfer function

$$|W(\omega)|^2 = \left[\frac{2\sin(5\omega)\sin(5\omega/2)}{5\sin(\omega/2)}\right]^2.$$

This has a very high peak centered at a frequency corresponding to a period of 20.3 years. So, Kuznets' conclusion of a 20-year cycle could have been a spurious finding, induced by the two filtering operations!

9.4.8 A Note on Analyzing Seasonality

The discussion so far has dealt with only a nonseasonal time series or has implicitly assumed that the given time series has been seasonally adjusted. Most macroeconomic time series are released every month or quarterly. Financial time series are available at higher frequencies such as weekly or daily. Seasonality is the systematic and the mostly predictable interyear movement. Many economic time series exhibit seasonality, but it is uncommon in engineering and scientific data. Since a substantial part of the seasonal components is predictable, the common practice, even among economists, is to remove such seasonality using seasonal-adjusting procedures, such as the Census $X - 11$ or the much-improved $X - 12$ method, developed by the U.S. Census Bureau (see Hylleberg 1992 for a discussion of these methods).*

But in many applications, the interest may lie in focusing on the seasonal components separately to enable decision makers to understand the preferences of economic agents and frame policies. In this section, we shall discuss some issues involved in modeling such components, if any, in a monthly or quarterly data only. Seasonality could occur in the mean or in the higher moments such as variance. The class of models used to model seasonality in the mean includes deterministic seasonal processes, linear-stochastic stationary and nonstationary processes, and unobserved component models. Seasonality at higher moments is modeled using nonlinear models such as the stochastic-seasonal unit-root processes and seasonal GARCH processes. However, our focus will be on modeling changes in the seasonal mean, especially, the deterministic seasonal processes and the stationary and nonstationary seasonal processes.

9.4.8.1 Deterministic Seasonality

Deterministic seasonality describes the behavior in which the unconditional mean of the process changes with the season of the year. It has two equivalent representations, *namely,* the dummy-variable representation and the trigonometric representation.

* The current version of the Census method $X - 13$, named *ARIMA − SEATS* can be downloaded from the Bureau of Census website, https://www.census.gov/srd/www/x13as/. See also the TRAMO/SEATS developed by Gomez and Maravall (1996), which is used by the EU statistical office, *EUROSTAT*.

9.4.8.1.1 Dummy-Variable Representation

We can write the conventional representation of the dummy-variable model as follows:

$$z_t = \sum_{s=1}^{S} \gamma_s \delta_{st} + e_t,$$

where e_t is the weakly stationary stochastic process with a mean of zero; δ_{st} is a seasonal dummy variable, which takes the value of 1 in season s and 0 otherwise; and S is the number of observations in a year. Thus, the seasonal mean for year t is

$$E(z_{st}) = \gamma_s, \quad s = 1, \ldots, S$$

and the above expression implies a seasonally shifting mean. Owing to this shifting mean, the process is actually nonstationary; but this nonstationarity can be removed by simply subtracting the mean for each season. Thus, the deviations, $z_t - E(z_t) = e_t$, are weakly stationary. If the overall mean is given by

$$E(z_t) = \mu = \frac{1}{S} \sum_{s=1}^{S} \gamma_s,$$

and it is also subtracted, then, the deterministic seasonal effect for the season s is $m_s = \gamma_s - \mu$. So, this definition imposes the restriction $\sum_{s=1}^{S} m_s = 0$ with the natural interpretation that there is no deterministic seasonality when the observations are summed over a year. If the overall level of the series, μ is separated from the seasonal components, then, we have

$$z_t = \mu + \sum_{s=1}^{S} m_s \delta_{st} + e_t.$$

When μ is replaced by $\mu_0 + \mu_1 t$, the last equation can be generalized to include a trend component that is constant over the seasons.

9.4.8.1.2 Trigonometric Representation

A deterministic function with period S can be equivalently expressed as the sum of sine and cosine terms. Correspondingly, we have

$$\sum_{s=1}^{S} \delta_{st} m_s = \mu + \sum_{k=1}^{S/2} \left[\alpha_k \cos\left(\frac{2\pi kt}{S}\right) + \beta_k \sin\left(\frac{2\pi kt}{S}\right) \right], \quad t = 1, \ldots, T,$$

where

$$\alpha_k = \frac{2}{S} \sum_{s=1}^{S} m_s \cos\left(\frac{2\pi ks}{S}\right), \quad k = 1, 2, \ldots, \frac{S}{2} - 1,$$

$$\alpha_{S/2} = \frac{1}{S} \sum_{s=1}^{S} m_s \cos(\pi s),$$

$$\beta_k = \frac{2}{S} \sum_{s=1}^{S} m_s \sin\left(\frac{2\pi ks}{S}\right), \quad k = 1, 2, \ldots, \frac{S}{2}$$

Note that $\beta_{S/2} \sin(2\pi kt/S) = 0$. This representation is the basis of spectral analysis and seasonal adjustment. For quarterly data, the relation between the seasonal dummy-variable coefficients in the conventional representation and the deterministic trigonometric components is given by

$$\begin{aligned}
\gamma_1 &= \mu + \beta_1 - \alpha_2 \\
\gamma_2 &= \mu - \alpha_1 + \alpha_2 \\
\gamma_3 &= \mu - \beta_1 - \alpha_2 \\
\gamma_4 &= \mu + \alpha_1 + \alpha_2.
\end{aligned}$$

We can write the relation more compactly as

$$\Gamma = RB,$$

where $\Gamma = (\gamma_1, \gamma_2, \gamma_3, \gamma_4)'$, $B = (\mu, \alpha_1, \beta_1, \alpha_2)'$, and

$$R = \begin{bmatrix} 1 & 0 & 1 & -1 \\ 1 & -1 & 0 & 1 \\ 1 & 0 & -1 & -1 \\ 1 & 1 & 0 & 1 \end{bmatrix}.$$

Note that, here, the coefficients α_1 and β_1 are associated with the spectral frequency $\pi/2$ and hence have a full cycle for every four periods. Similarly, the coefficient α_2 is associated with the spectral frequency π and hence contributes a cycle for every two periods. With quarterly data, the frequencies $\pi/2$ and π are referred to as the spectral frequencies. Similarly, in the case of monthly data, the seasonal frequencies with $S = 12$ are $\pi/6, \pi/3, \pi/2, 2\pi/3, 5\pi/6$, and π. Any monthly data can also be written in terms of sine and cosine functions at these frequencies.

9.4.8.2 Stochastic Seasonality

Seasonality can also be modeled as the sum of a deterministic component and a stochastic process. *ARMA* processes, as discussed earlier, can be easily adapted to build such stochastic seasonal models. A simple example is the first-order seasonal AR process

$$z_t = \Phi_S z_{t-S} + e_t,$$

where $e_t \sim$ i.i.d.$(0, \sigma^2)$ and $|\Phi_S| < 1$. The AC values take the form

$$\rho(kS) = \Phi_S^k, \quad k = 1, 2, \ldots$$

and $\rho(k)$ is zero for all other k. An alternative spectral representation for any seasonal *ARMA* model can be written in terms of the associated polynomials as

$$f_z(\omega) = \left(\frac{\sigma^2}{2\pi}\right) \cdot \frac{|\Theta_S(e^{-i\omega S})|^2}{|\Phi_S(e^{-i\omega S})|^2}.$$

Just as the economic time series exhibit a unit root at zero frequency, such seasonal stochastic models may also exhibit unit roots in seasonal frequencies. If we set $\Phi_S = 1$, then, we obtain the simplest nonstationary seasonal process, which is the seasonal RW, defined as

$$\nabla_S z_t = e_t,$$

where $\nabla_S = (1 - B^S)$ and this is called as the seasonal unit-root process. For the quarterly seasonal RW process, we can factorize the AR operator as $(1 - B^4) = (1 - B)(1 + B)(1 + B^2)$. From the expression for the spectrum of a seasonal *ARMA* model, one can write the spectral density of the quarterly seasonal RW as

$$f_z(\omega) = \left(\frac{\sigma^2}{2\pi}\right) \cdot \frac{1}{|(1 - e^{-4i\omega})|^2}.$$

This implies that the spectral density of z_t has infinite power at zero frequency, which represents the conventional unit root, and at seasonal frequencies of π and $\pi/2$ that correspond to the unit roots equal to -1 and $\pm i$ arising from the factors $(1 + B)$ and $(1 + B^2)$. Similarly, one can write the frequencies that give rise to infinite power at zero and seasonal frequencies for monthly data also.

An issue that an applied researcher often confronts is whether the observed time series should be modeled as a deterministic seasonal process or as a seasonal unit process. For a finite sample, the two processes can exhibit similar empirical properties, with a strong empirical power at seasonal frequencies. Despite this apparent complementarity, applied researchers routinely start by assuming a seasonal unit-root process, even though Bell (1987) shows that in samples that are normally encountered in practice, it may be difficult to distinguish between a deterministic and a seasonal unit-root process. Canova and Hansen (1995) propose a test procedure to verify if seasonal variations in any observed economic time series can be captured by a simple deterministic seasonal representation, explained by time dummies, against the alternative that there are unit roots at the seasonal frequencies. Alternatively, the most popular HEGY (1990) procedure tests for the null of seasonal unit roots at zero frequency and at one or more seasonal frequencies for quarterly data, while the BM test of Beaulieu and Miron (1993) extends the HEGY test for monthly data.

9.4.8.3 Canova–Hansen Test

Canova–Hansen (CH) test procedure is based on the trigonometric representation of a deterministic seasonality, based on Lagrange-multiplier statistic. We follow Ghysels and Osborn (2001) in presenting this test. We start by rewriting the representation of z_t as follows:

$$z_t = \sum_{s=1}^{S} F_s' B_t \delta_{st} + e_t, \quad \text{(CH 1)}$$

where F_s' is the $1 \times S$ vector, which is the sth row of matrix R and $B_t = (\mu_t, \alpha_{1t}, \beta_{1t}, \ldots, \alpha_{S/2,t})'$. e_t's are assumed to be normally distributed and stationary. CH test proposes to test if the B vector evolves according to an RW. That is,

$$B_t = B_{t-1} + V_t,$$

where V_t is i.i.d. with $\sigma^2 H$, where H is a known positive-definite matrix. And V_t is independent of e_t. The goal of CH test is to test the null that σ^2 is zero, implying that B is constant that will support the fact that seasonality is deterministic, against the alternative that $\sigma^2 > 0$ implying that the elements of the vector contain unit roots, meaning that the seasonality is stochastic and nonstationary. The entire model can be written in vector notation as

$$Z_\tau = \Gamma + E_\tau = RB + E_\tau,$$

where $Z_\tau = (z_{1\tau}, \ldots, z_{S\tau})'$ is the vector of observations for year τ and E_τ is the corresponding zero-mean stationary-disturbance process with a covariance matrix, Ω_E. The test can be constructed as follows:

- Estimate (CH-1) by OLS and collect the residuals, $\hat{e}_t, t = 1, \ldots, T$.
- From the OLS residuals, form the $S \times 1$ vectors, $\hat{E}_t^a$, where $\hat{E}_t^{a'} = \left(\sum_{j=1}^{t} \hat{e}_j \delta_{1j}, \ldots, \sum_{j=1}^{t} \hat{e}_j \delta_{Sj}\right)$ for $t = 1, \ldots, T$.
- From the Lagrange-multiplier statistic,

$$L = \frac{S}{T^2} \sum_{t=1}^{T} \left(R'\hat{E}_t^a\right)' \hat{\Omega}_{RE}^{-1} \left(R'\hat{E}_t^a\right),$$

$$= \frac{S}{T^2} \sum_{t=1}^{T} \hat{E}_t^a \hat{\Omega}_E^{-1} \hat{E}_t^a,$$

  where $\hat{\Omega}_{RE} = R'\hat{\Omega}_E R$ and can be consistently estimated.
- The distribution of the statistic, however, is nonstandard and sometimes it is called as the von Mises distribution with S degrees of freedom, $VM(S)$. The asymptotic critical values have been tabulated by Canova and Hansen (1995). The null is rejected for large values of L.
- The above test procedure conducts the test on the entire vector, B_t. Analogous tests can be considered if we want to test for unit roots in any particular frequency.*

9.4.8.4 HEGY Test Procedure

HEGY procedure was actually proposed for quarterly data and was extended for monthly data first by Franses (1991) and later by BM (1993). We explain the extended procedure of BM and the linear transformations proposed by them. According to BM, their version has the added advantage of making the set of regressors to be mutually orthogonal.

HEGY procedure starts by noting that the seasonal difference filter recommended by BJR (1994) is $(1 - B^S)$—where S is the number of observations per year, and the typical values being $S = 2, 4, 12, 52$—can be factorized and rewritten as

$$\left(1 - B^S\right) = (1 - B)\, S(B), \quad \text{where } S(B) = \left(1 + B + B^2 + \cdots + B^{S-1}\right).$$

The first factor denotes the long-run or the zero-frequency unit root and the seasonal unit roots are in the seasonal summation filter $S(B)$. The seasonal summation filter has the real root of -1 if $S = 2$, the real root -1, and the two complex conjugate roots of $\pm i$ if $S = 4$, and one real and five pairs of complex conjugate roots if $S = 12$, etc. What HEGY tests imply is the simple fact that *unless* we test and show that there are roots equal to one at *all* the above-mentioned frequencies, any routine application of the seasonal filters $(1 - B^4)$ or $(1 - B^{12})$, which implicitly assume and impose unit roots in all the frequencies, may lead to serious misspecifications. HEGY propose tests to check the hypothesis that a series is *seasonally* integrated by checking that the roots are exactly

* The computer code to calculate the CH statistic is available in R language at this URL: http://www.jalobe.com:8080/blog/testing-for-seasonal-stability-canova-and-hansen-test-statistic/ that we have also used to obtain the results of CH test displayed in Table 9.13.

on the unit circle at these frequencies as against the stationary alternative. It also facilitates checking about any particular unit root independently of the presence of unit roots at the zero frequency or any other seasonal frequency. BM (1993) extend it to test if monthly data are seasonally integrated. It involves running the following regression:

$$\phi(B)\nabla_{12}z_t = \sum_{i=1}^{12} \pi_i x_{i,t-1} + e_t,$$

where $x_{i,t}$'s are the linear transformations involving lagged *levels* of z_t that are mutually orthogonal (see BM 1993, p. 308 for these transformations). The aim is to check for the presence of a unit root at both zero frequency and at seasonal frequencies. One way to interpret the results is as follows:

1. If $\pi_i = 0,\ i = 1, \ldots, 12$, then, the series is seasonally integrated—that is, the filter $(1 - B^{12})$ is applicable.
2. If $\pi_k = 0$ is accepted against the alternative $\pi_k < 0$, for $k = 1, 2$, then, the series has a unit root *only* at zero frequency and at frequency π, respectively.
3. To accept unit roots at any other frequency, first, we should test the null $\pi_k = 0$ for k even with a two-sided test, since under the alternative, the even coefficients may be negative or positive. If one fails to reject the null, then, test for $\pi_{k-1} = 0$ with a one-sided test that $\pi_{k-1} < 0$.
4. Another way to show that no unit root exists in any seasonal frequency is to show that π_k is not equal to zero for $k = 2$ and at least one element of each of the sets $(\pi_k,\ \pi_{k+1},\ k = 3, 5, \ldots, 11)$ is shown to be equal to nonzero.

The tests for unit roots at the zero as well as the seasonal frequencies are done based on the t-statistics and the F-values as displayed in Table A.1 of BM (1993, pp. 325 and 326). This test can be extended to include a trend as well as seasonal dummies; but, of course, the distributions of the statistics of importance are different and nonstandard. So, the estimating regression becomes

$$\phi(B)\nabla_{12}z_t = \sum_{i=1}^{12} \pi_i x_{i,t-1} + \mu + \boldsymbol{\alpha}'\mathbf{D}_t + \delta t + e_t,$$

where μ is a constant; δ is the slope coefficient of the trend term; $\mathbf{D}_t$ is the vector of 11 dummy variables with the coefficient vector $\boldsymbol{\alpha}$; and the *AR* operator $\phi(B)$ has enough terms to reduce the error to be of white noise.

GLN (1994) extend the above HEGY procedure and propose two overall F-statistics for quarterly data: (1) to test for the overall null and (2) to check for unit roots at all seasonal frequencies simultaneously. This examines the unit roots implied by the seasonal summation filter $(1 + B + \cdots + B^{S-1})$. These test statistics are extended to the monthly data by Taylor (1998). In terms of the null, these two tests are (1) $\pi_i, = 0,\ i = 1, \ldots, 12$ and (2) $\pi_i, = 0,\ i = 2, \ldots, 12$. We refer to these overall tests as $F_{1,\ldots,12}$ and $F_{2,\ldots,12}$, respectively. But any finding of the so-called seasonal unit roots should be interpreted with care, since they imply varying seasonal patterns that may not be feasible in most situations.

We applied these two tests on some major Indian macro time series, *namely,* the money series (M1 and M3), price series (CPI and WPI), and industrial production (IIP). Monthly data on these series have been collected from various publications of RBI. Index numbers on price series and production data were spliced to form the series with a common base. Table 9.13 displays the results of CH test, where we have used the trigonometric representation of seasonal dummies.

TABLE 9.13

CH Test Results for Stability of Seasonal Dummies (Asymptotic p-Values in Parentheses)

	Series				
	M1	M3	CPI	IIP	WPI
$\pi/6$	0.799(0.037)	0.332(0.398)	0.521(0.163)	1.025(0.010)	1.067(0.008)
$2\pi/6$	3.698(0.000)	0.252(0.580)	0.184(0.773)	1.290(0.002)	1.274(0.002)
$3\pi/6$	0.355(0.356)	0.176(0.796)	0.245(0.599)	0.702(0.061)	0.129(0.926)
$4\pi/6$	0.384(0.309)	0.186(0.767)	0.168(0.819)	0.646(0.082)	0.230(0.647)
$5\pi/6$	0.136(0.904)	0.179(0.787)	0.191(0.751)	0.596(0.107)	0.647(0.081)
π	0.284(0.154)	0.110(0.553)	0.644(0.017)	0.242(0.207)	0.262(0.181)

Seasonal stability is accepted only for M3 and the stability of CPI is accepted for all the frequencies, except at the Nyquist frequency of π that implies cycles for every 2 months, which one may ignore. IIP and WPI are quite unstable, as one may expect. So, by this test, seasonality is deterministic in M3 and almost so in CPI while it may be stochastic and nonstationary in the rest of the series, implying unit roots only in *some* seasonal frequencies. The important implication here is that one may not need the seasonal summation filter as a whole to remove the seasonality. This is further confirmed by HEGY test results, conducted in the time domain and explained below.

The results of HEGY test for monthly data and that of GLN extension to HEGY test are shown in Tables 9.14 and 9.15. (More detailed results of a similar analysis are available in Krishnan (2007)).

TABLE 9.14

HEGY Test Results for Seasonality

	M1	M3	CPI	IIP	WPI
Lags	**6**	**0**	**4**	**12**	**12**
π_1	−2.30	−2.75	−3.29	−2.66	−3.36
π_2	−2.79	−3.73**	−4.60**	−1.77	−4.38**
π_3	−6.74**	−9.28**	−4.10**	−3.12	−3.75*
π_4	−2.08*	1.48	−7.18**	−0.45	−4.71**
π_5	−5.80**	−9.37**	−6.50**	−2.85	−3.25
π_6	−0.81	−2.62*	4.51**	−0.22	3.41**
π_7	−4.79**	−10.58**	−3.85*	−4.11**	−4.00**
π_8	−5.41**	−1.58	−7.90**	−3.22**	−3.65**
π_9	−6.19**	−6.68**	−5.98**	−3.34	−3.42*
π_{10}	1.26	−1.98*	4.08**	−1.55	1.26
π_{11}	−2.39	−9.77**	−2.65	−2.68	−2.09
π_{12}	−7.86**	−4.48**	−10.06**	−3.23**	−4.54**
$F_{3,4}$	29.96**	43.17**	44.45**	5.49	24.38**
$F_{5,6}$	17.22**	47.11**	32.41**	4.08	11.60**
$F_{7,8}$	25.68**	57.86**	36.70**	14.23**	15.52**
$F_{9,10}$	19.80**	25.19**	24.12**	7.16*	6.52
$F_{11,12}$	33.82**	61.17**	51.66**	8.87**	12.13**

Note: *indicates the statistic significant at the 5% level, **indicates the statistic significant at the 1% level. All statistics in this table have been compared against the asymptotic critical values from BM (1993), Table-A1.

TABLE 9.15

Extended Test Results for Seasonal Unit Roots

	GLN Extension of HEGY Test	
	$F_{1,...,12}$	$F_{2,...,12}$
M1	28.86**	30.65**
M3	52.07**	52.90**
CPI	46.84**	48.80**
IIP	8.54**	8.63**
WPI	24.31**	25.38**

Note: *indicates the statistic significant at the 5% level, **indicates the statistic significant at the 1% level. Estimated statistics have been compared against the asymptotic critical values from Table-II of Taylor (1998). At the 5% and 1% significance levels, respectively, these are 4.60 and 5.31 for $F_{1,...,12}$ and 4.45 and 5.19 for $F_{2,...,12}$.

The striking result from HEGY test in Table 9.14 is that all series have a root at zero frequency (lag π_1), justifying the first-difference filter. So, we have to first difference all the series. However, none of the series needs the seasonal summation filter $S(B)$ as a whole, though this does not rule out the possibility of unit roots at *some* of the seasonal harmonics. For example, M1 and IIP show the presence of a unit root at the Nyquist frequency π, lag π_2 referring to six cycles per year and hence the filter $(1 - B)$ needs to be augmented with $(1 + B)$. There is also an evidence of roots at the seasonal harmonics of other series. For example, for *WPI*, the insignificant $F_{9,10}$ says that an additional filter of $(1 + 3^{1/2}B + B^2)$ is needed. Similarly, for *IIP*, an insignificant $F_{3,4}$ and $F_{5,6}$ mean, taking into account the zero- and Nyquist frequency filters, we have to transform the series using the "combined" filter of $(1 - B^4)(1 + B + B^2)$. As a further check, we filtered the *IIP* series according to this finding and estimated the AC values of the filtered series for the first 24 lags. A visual inspection of the *ACF* has shown that there was still some strong dependence left in the filtered series, but there was no evidence of any nonstationarity left in the series. Both the *ACF* and the *PACF* strongly suggested a mixed process for the filtered series.

The results of the extended test of GLN are shown in Table 9.15. Checking the estimated F-values against Table-II asymptotic critical values—that are 4.60 and 4.45 for $F_{1,...,12}$ and $F_{2,...,12}$, respectively—calculated by Taylor (1998) for monthly data—does not throw any surprise. It confirms our earlier findings of HEGY test that we don't need the seasonal summation filter as a whole. These results confirm that any blind application of the seasonal filter may distort the characteristics of the original time series.

9.5 Recent Developments and Research Trends

With the advent of massive and ultra-high-frequency time-series data sets on diverse fields, the traditional BJ type of linear models was found to be inadequate and this recognition led researchers to develop novel methodologies such as the nonstationary and nonlinear models, models with a local stationarity, time-varying coefficients, long-range dependence, clustering and classification, and quantile estimation, among others. While the theoretical structure of some of the above methodologies, especially on unit-root models, has been reasonably well established, a statistical inference on several of the above topics and the extension to the multivariate counterpart has not yet been well established.

Many new open-research questions on the existing topics such as the forecast combination and forecast intervals, predicting the missing observations, have also been raised. For example, Stock and Watson (2004) discussed a puzzle in forecast combination, where repeated empirical findings show that simple combinations such as averages outperform more-sophisticated combinations that the theory suggests should do better. This is an important practical issue that will no doubt receive further research attention in the future. A good reference paper that raises new research questions on the topic of combining forecasts is that of Terasvirta (2006).

An application of quantile information to time-series issues has witnessed a lot of research attention in recent times. In the univariate context, quantile autoregression (QAR) models, which can be viewed as a special case of random-coefficient regression models, capture the effect of conditioning variables on the location, scale, and shape of the conditioning distribution of the response variables. Koenker and Xiao (2006) show how QAR models can be used to capture a systematic asymmetries in the dynamics of a time series. They apply QAR models on U.S. unemployment rate and show that it reacts asymmetrically to different types of shocks and such an asymmetric response has important policy implications. Xiao (2012) considers a wide range of quantile-regression models to show how to conduct unit-root tests and estimate the ARCH family of volatility models. Several software packages exist for quantile-regression applications. For example, both parametric and nonparametric quantile-regression estimations can be implemented by the function rq() and rqss() in the package quantreg in the free computing language R. SAS has a suite of procedures that are closely modeled on the functionality of the R package quantreg.

9.6 Information on Software Packages

All empirical exercises in this chapter have been conducted with the help of the commercial software packages named as the regression analysis of time series (RATS) and Eviews and the free-to-use R software. RATS is developed and sold by estima http://www.estima.com and Eviews is developed and marketed by IHS Global Inc. SAS is another hugely popular commercial software package suited for time-series applications. Among the free software, one can use GRETL (Gnu regression, econometrics, and time-series library) that is available at http://gretl.sourceforge.net/. It is free and very easy to use, and all the graphs are generated by GNUPlot (http://www.gnuplot.info/). To estimate more-complicated models, one can use the E4 toolbox (http://www.ucm.es/info/icae/e4/). It is also free, and runs in both MATLAB® and GNU Octave (http://www.gnu.org/software/octave/). The Octave language is quite similar to MATLAB so that most programs are easily portable, and it is free.

Another software, which is considered the best, is the R software for statistical analysis and graphical display. There are several packages to such problems, such as, for example

- http://cran.r-project.org/web/packages/hydroTSM/
- http://cran.r-project.org/web/packages/EcoHydRology/index.html
- "http://rwiki.sciviews.org/doku.php?id=guides:tutorials:hydrological_data_analysis" are the examples for hydrology. Moreover, R also has other packages for a powerful spatial analysis, geographic information system (GIS) integration, etc. A big advantage of using R is, besides being free, it has a huge user community, so that finding an advice is easy. One can obtain general and technical information on a number of proprietary and free statistical packages using the link.
- http://en.wikipedia.org/wiki/Comparison_of_statistical_packages

Renfro (2004) provides a comprehensive account of the existing econometric software packages. A very useful source on software is the *Journal of Statistical Software* (http://www.jstatsoft.org/) that publishes peer-reviewed articles on statistical software, along with a source code. It also publishes software reviews and its articles are freely available online.

9.7 Conclusion

In this chapter, we have attempted to give a broad overview of the issues involved in modeling a univariate time series both in the time domain and in the frequency domain. This is by no means a comprehensive overview, but it does cover some topics that are not yet a part of the time-series folklore, such as the right-tail unit-root test or the extended discussion on forecasting topics such as an asymmetric loss function or the forecast combination. Empirical applications using major macro time series of the Indian economy point out that most of them are nonstationary. An overview of filtering explains the importance of correct filters, because an inappropriate filtering will result in spurious cycles in the filtered data. A detailed discussion on topics such as fractional differencing, and long-memory procedures, would make this chapter even more broad based.

References

Anderson, B.D.O., and J.B. Moore, 1979, *Optimal Filtering*, Prentice-Hall, Englewood Cliffs.

Anderson, T.W., 1994, *The Statistical Analysis of Time Series*, Wiley Classics Library Edition, New York.

Ansley, C.F., 1979, An algorithm for the exact likelihood of a mixed autoregressive moving average process, *Biometrica*, 66, 59–65.

Bates, J.M., and C.W.J. Granger, 1969, The combination of forecasts, *Operations Research Quarterly*, 20, 451–468.

Beaulieu, J.J., and Miron, J.A., 1993, Seasonal unit roots in aggregate U.S. data, *Journal of Econometrics*, 55, 305–328.

Bell, W.R., 1987, A note on overdifferencing and the equivalence of seasonal time series models with monthly means and models with $(0, 1, 1)_{12}$ seasonal parts when $\Theta = 1$, *Journal of Business and Economic Statistics*, 5, 383–387.

Beveridge, S., and C.R. Nelson, 1981, A new approach to decomposition of economic time series into permanent and transitory components with particular attention to measurement of "business cycle," *Journal of Monetary Economics*, 7, 151–74.

Bloomfield, P., 2000, *Fourier Analysis of Time Series—An Introduction*, John Wiley & Sons, New York.

Box, G.E.P., and G.M. Jenkins, 1976, *Time Series Analysis: Forecasting and Control*, Revised edition, Holden-Day, San Francisco.

Box, G.E.P., G.M. Jenkins, and G.C. Reinsel, 2008, *Time Series Analysis: Forecasting and Control*, John Wiley & Sons, New York.

Breitung, J., 1994, Some simple tests of the moving average unit root hypothesis, *Journal of Time Series Analysis*, 15, 351–370.

Canova, F., and B.E. Hansen, 1995, Are seasonal patterns constant over time? A test for seasonal stability, *Journal of Business and Economic Statistics*, 13, 237–252.

Chan, K.H., J.C. Hayya, and J.K. Ord, 1977, A note on trend removal methods: The case of polynomial regression versus variate differencing, *Econometrica*, 45, 737–744.

Charles, A., and O. Darne, 2009, Variance-ratio tests of random walk: An overview, *Journal of Economic Surveys*, 23, 503–527.

Choi, I., and B.C. Yu, 1997, A general framework for testing $I(m)$ against $I(m+k)$, *Journal of Economic Theory and Econometrics*, 3, 103–138.

Chow, K.V., and K.C. Denning, 1993, A simple multiple variance ratio test, *Journal of Econometrics*, 58, 385–401.

Christoffersen, P.F., and F.X. Diebold, 1996, Further results on forecasting and model selection under asymmetric loss, *Journal of Applied Econometrics,* 11, 561–572.

Christoffersen, P.F., and F.X. Diebold, 1997, Optimal prediction under asymmetric loss, *Econometric Theory*, 13, 808–817.

Clements, M.P., 2005, *Evaluating Econometric Forecasts of Economic and Financial Variables*, Palgrave Macmillan, New York.

Clements, M.P., and D.F. Hendry, 2000, *Forecasting Economic Time Series*, Cambridge University Press, Cambridge.

Cochrane, J.H., 1988, How big is the random walk in GNP? *Journal of Political Economy*, 96, 893–920.

Dickey, D.A., and W.A. Fuller, 1981, Likelihood ratio statistics for autoregressive time series with a unit root, *Econometrica*, 49, 1057–1072.

Diebold, F.X., and J.A. Lopez, 1996, Forecast evaluation and combination, in G.S. Maddala, and C.R. Rao (Editors), *Handbook of Statistics, Volume 14, Statistical Methods in Finance*, North-Holland, Amsterdam.

Diebold, F.X., and R.S. Mariano, 1995, Comparing predictive accuracy, *Journal of Business and Economic Statistics*, 13, 253–63.

Diebold, F.X., and P. Pauly, 1987, Structural change and the combination of forecasts, *Journal of Forecasting,* 6, 21–40.

Elliott, G., 1999, Efficient tests for a unit root when the initial observation is drawn from its unconditional distribution, *International Economic Review*, 40, 767–783.

Elliott, G., T.J. Rothenberg, and J.H. Stock, 1996, Efficient tests for an autoregressive root, *Econometrica*, 64, 813–836.

Evans, G.W., 1991, Pitfalls in testing for explosive bubbles in asset prices, *American Economic Review*, 81, 922–930.

Franses, P.H., 1991, Seasonality, nonstationarity and the forecasting of monthly time series, *International Journal of Forecasting,* 7, 199–208.

Fuller, W., 1996, *Introduction to Statistical Time Series*, John Wiley & Sons, New York.

Ghysels, E., H.S. Lee, and J. Noh, 1994, Testing for unit roots in time series: Some theoretical extensions and a Monte Carlo investigation, *Journal of Econometrics*, 62, 415–442.

Ghysels, E., and D.R. Osborn, 2001, *The Econometric Analysis of Seasonal Time Series*, Cambridge University Press, UK.

Gomez, V., and A. Maravall, 1996, *Programs TRAMO and SEATS*, Banco de Espana, Madrid.

Granger, C.W.J., 1969, Prediction with a generalized cost of error function, *Operational Research Quarterly*, 20, 199–207.

Granger, C.W.J., and M. Hatanaka, 1964, *Spectral Analysis of Economic Time Series*, Princeton University Press, Princeton, New Jersey.

Granger, C.W.J., and P. Newbold, 1973, Some comments on the evaluation of economic forecasts, *Applied Economics*, 5, 35–47.

Granger, C.W.J., and P. Newbold, 1986, *Forecasting Economic Time Series*, Academic Press, London.

Granger, C.W.J., and R. Ramanathan, 1984, Improved methods of combining forecasts, *Journal of Forecasting*, 3, 197–204.

Hall, A., 1994, Testing for unit root in time series with pretest data based model selection, *Journal of Business and Economic Statistics*, 12, 461–470.

Hamilton, J., 1994, *Time Series Analysis*, Princeton University Press, New Jersey.

Harvey, A.C., 1981, *The Econometric Analysis of Time Series*, Philip Allan, Oxford.

Harvey, A.C., 1984, *Time Series Models*, Heritage Publishers, New Delhi.

Harvey, A.C., 1989, *Forecasting Structural Time Series Models and Kalman Filter*, Cambridge University Press, New York.

Harvey, D., S. Leybourne, and P. Newbold, 1997, Testing the equality of prediction mean squared errors, *International Journal of Forecasting*, 13, 281–291.

Hill, R.J., and K.J. Fox, 1997, Splicing index numbers, *Journal of Business and Economic Statistics*, 15, 387–389.

Hodrick, R.J., and E.C. Prescott, 1997, Postwar U.S. business cycles: An empirical investigation, *Journal of Money, Credit and Banking*, 29, 1–16.

Hylleberg, S., R. Engle, C.W.J. Granger, and S. Yoo, 1990, Seasonal integration and cointegration, *Journal of Econometrics*, 44, 215–238.

Hylleberg, S., 1992, *Modelling Seasonality*, Oxford University Press, UK.

Kendall, M.G., 1973, *Time Series*, Charles Griffin, London.

Koenker, R., and Z. Xiao, 2006, Quantile autoregression, *Journal of the American Statistical Association*, 101, 980–1006.

Krishnan, R., 2007, Seasonal characteristics of Indian Time Series, *Indian Economic Review*, 42, 191–210.

Kwiatkowski, D., P.C.P. Phillips, P. Schmidt, and Y. Shin, 1992, Testing the null hypothesis of stationarity against the alternative of a unit root, *Journal of Econometrics*, 54, 159–178.

Leybourne, S.J., and B.P.M. McCabe, 1994, A consistent test for unit root, *Journal of Business and Economic Statistics*, 12, 157–166.

Lo, A., and A.C. MacKinlay, 1988, Stock prices do not follow a random walk: Evidence from a simple specification test, *Review of Financial Studies*, 1, 41–66.

Mariano, R.S., 2004, Testing forecast accuracy, in M.P. Clements and D.F. Hendry (Editors), *A Companion to Economic Forecasting*, Blackwell Publishing, Cornwall, UK, pp. 284–298.

McLeod, A.I., and W.K. Li, 1983, Diagnostic checking ARMA time series models using squared residual autocorrelations, *Journal of Time Series Analysis*, 4, 269–273.

Meese, R.A., and K. Rogoff, 1988, Was it real? The exchange rate-interest differential relation over the modern floating-rate period, *Journal of Finance*, 43, 933–948.

Mincer, J., and V. Zarnowitz, 1969, The evaluation of economic forecasts, in J. Mincer (Editor), *Economic Forecasts and Expectations,* National Bureau of Economic Research, New York, pp. 3–46.

Nelson, C.R., and H. Kang, 1981, Spurious periodicity in inappropriately detrended time series, *Econometrica*, 49, 741–751.

Nelson, C.R., and C. Plosser, 1982, Trends and random walks in macro economic time series: Some evidence and implications, *Journal of Monetary Economics*, 10, 139–162.

Newbold, P., 1974, The exact likelihood function for a mixed autoregressive moving average process, *Biometrica*, 61, 423–426.

Ng, S., and P. Perron, 2001, Lag length selection and the construction of unit root tests with good size and power, *Econometrica*, 69, 1519–1554.

Ng, S., and T. Vogelsang, 2002, Forecasting autoregressive time series in the presence of deterministic components, *Econometrics Journal*, 5, 196–224.

Ouliaris, S., J.Y. Park, and P.C.B. Phillips, 1989, Testing for unit root in the presence of maintained trend, in B. Raj (Editor), *Advances in Econometrics and Modelling*, Springer, New York, pp. 7–28.

Patton, A.J., and A. Timmermann, 2007, Testing forecast optimality under unknown loss, *Journal of the American Statistical Association*, 102, 1172–1184.

Perron, P., 1989, The Great Crash, the oil price shock and the unit root hypothesis, *Econometrica*, 57, 1361–1401.

Perron, P., and S. Ng, 1996, Useful modifications to some unit root tests with dependent errors and their local asymptotic properties, *Review of Economic Studies*, 63, 435–463.

Phillips, P.C.B., and S. Ouliaris, 1988, Testing for cointegration using principal component methods, *Journal of Economic Dynamics and Control*, 12, 205–230.

Phillips, P.C.B., S. Shi, and J. Yu, 2012, Testing for multiple bubbles, *Cowles Foundation Discussion Paper, No. 1843*, Yale University, New Haven, Connecticut, USA.

Phillips, P.C.B., Y. Wu, and J. Yu, 2011, Explosive behavior in the 1990s NASDAQ: When did exuberance escalate asset values? *International Economic Review*, 52, 201–226.

Poirier, D.J., 1976, *The Econometrics of Structural Change*, North-Holland Publishing Company, Amsterdam.

Pollock, D.S.G., 1999, *A Handbook of Time-Series Analysis, Signal Processing and Dynamics*, Academic Press, UK.

Rapach, D.E., J.K. Strauss, and G. Zhou, 2010, Out-of-sample equity premium prediction: Combination forecasts and links to the real economy, *Review of Financial Studies*, 23, 822–862.

Renfro, C.G., 2004, A compendium of existing econometric software packages, *Journal of Economics and Social Measurement*, 29, 359–409.

Said, S.E., and D.A. Dickey, 1984, Testing for unit roots in autoregressive-moving average model of unknown order, *Biometrika*, 71, 599–608.

Sargent, T.J., 1987, *Macroeconomic Theory*, Academic Press, London.

Stock, J., and M.W. Watson, 1991, A probability model of the coincident economic indicators, in K. Lahiri and G.H. Moore (Editors), *Leading Economic Indicators: New Approaches and Forecasting Records,* Cambridge University Press, England, pp. 63–90.

Stock, J.H., and M.W. Watson, 2004, Combination forecasts of output growth in a seven-country data set, *Journal of Forecasting*, 23, 405–430.

Taylor, A.M.R., 1998, Testing for unit roots in monthly time series, *Journal of Time Series Analysis*, 19, 349–368.

Terasvirta, T., 2006, Forecasting economic variables with nonlinear models, in G. Elliot, C.W.J. Granger, and A. Timmermann (Editors), *Handbook of Economic Forecasting*, Elsevier Science, Amsterdam, pp. 413–457.

Timmermann, A., 2006, Forecast combinations, in G. Elliot, C.W.J. Granger, and A. Timmer-mann (Editors), *Handbook of Economic Forecasting*, Elsevier Science, Amsterdam, 135–196.

Varian, H., 1974, A Bayesian approach to real estate assessment, in S.E. Feinberg and A. Zellner (Editors), *Studies in Bayesian Econometrics and Statistics in Honor of L.J. Savage*, North-Holland, Amsterdam, pp. 195–208.

Xiao, Z., 2012, Time series quantile regressions, in T. Subba Rao, S. Subba Rao, and C.R. Rao (Editors), *Handbook of Statistics*, Elsevier Science, Amsterdam, pp. 213–257.

Xiao, Z., and P.C.B. Phillips, 1998, An ADF coefficient test for a unit root in ARMA models of unknown order with empirical applications to the U.S. economy, *Econometrics Journal*, 1, 27–43.

Zellner, A., 1986, Bayesian estimation and prediction using asymmetric loss functions, *Journal of the American Statistical Association*, 81, 446–451.

Zivot, E., and D.W.K. Andrews, 1992, Further evidence on the Great Crash, the oil price shock and the unit root hypothesis, *Journal of Business and Economic Statistics*, 10, 251–270.

10

Univariate Volatility Modeling: Theory and Practice

R. Krishnan

CONTENTS

ABSTRACT Accounting for and predicting volatility, especially in asset prices, occupy a central role in finance. The importance of correctly specifying conditional volatility has deep implications for quantifying risk. Naturally, this led to a plethora of volatility models, both deterministic and stochastic. Such volatility models have been estimated both in univariate and multivariate contexts. This chapter focuses on univariate volatility models. In the first part, we begin listing the various notions of volatility and concentrate on some of the pioneering statistical models that have been discussed in the literature to quantify conditional volatility. This part is a tour to understand both the theoretical underpinnings and the practical aspects in the estimation of deterministic volatility models such as the linear ARCH and GARCH family of models and their nonlinear counterparts. Our discussion will revolve around models that explain stylized facts such as high kurtosis and fat tails in financial returns, the presence of positive autocorrelations in both the absolute and squared returns, indicating substantially more linear dependence than autocorrelations of returns. Along the way, we outline the practical issues involved in model building and forecasting, from a practitioner's viewpoint. In the second part, we describe stochastic volatility models, where conditional volatility is described as a latent variable that depends not on past observations but on some unobserved past latent structure. Steps involved in estimating such models using state space and Markov chain Monte Carlo methods are outlined. A number of examples will highlight the important practical issues involved.

10.1 Preface

Accounting for and predicting volatility, especially in asset prices, have been a source of research since Mandelbrot's (1963) observations about volatility clusters. Volatility occupies a central role in finance, and being an unobservable variable, it must be predicted, say, from historical prices or option prices. A popular method, among many that analysts employ to obtain such forecasts, is to estimate historical volatility and project it; but such a procedure is based on the critical assumption

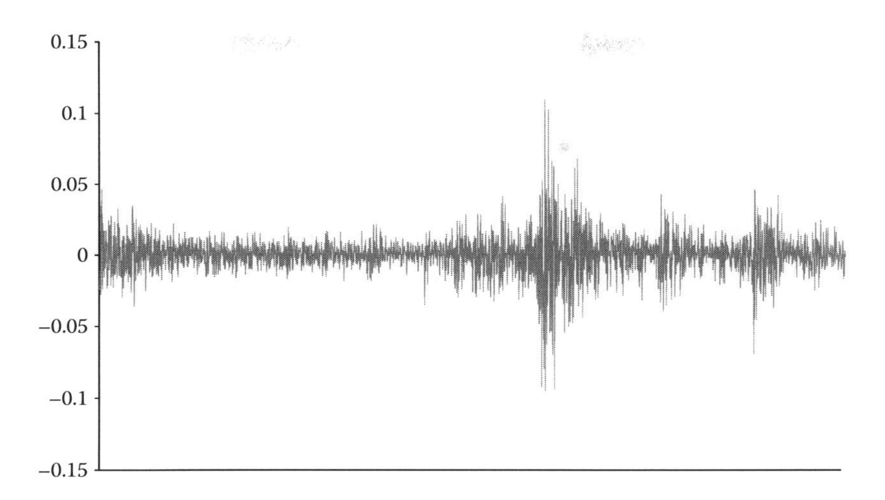

FIGURE 10.1
Daily returns of S&P 500 index.

that volatility remains constant over time, where evidence suggests that such an assumption is untenable. Hence, one may say that the procedures used to predict volatility, assuming it to remain time invariant, are essentially ad hoc. Plots of various time series displayed below clearly exhibit this fact. Figures 10.1 through 10.3 present daily returns, respectively, from two popular stock indices, namely, the Standard & Poor's 500 consisting of stocks from 500 blue chip companies and Nifty index, another popular stock index consisting of stocks from 50 blue chip Indian companies and from Coca-Cola stock. While there is no visible trend in any of the series indicating stationarity, there are periods in which returns cluster together, indicating that during those periods, returns are

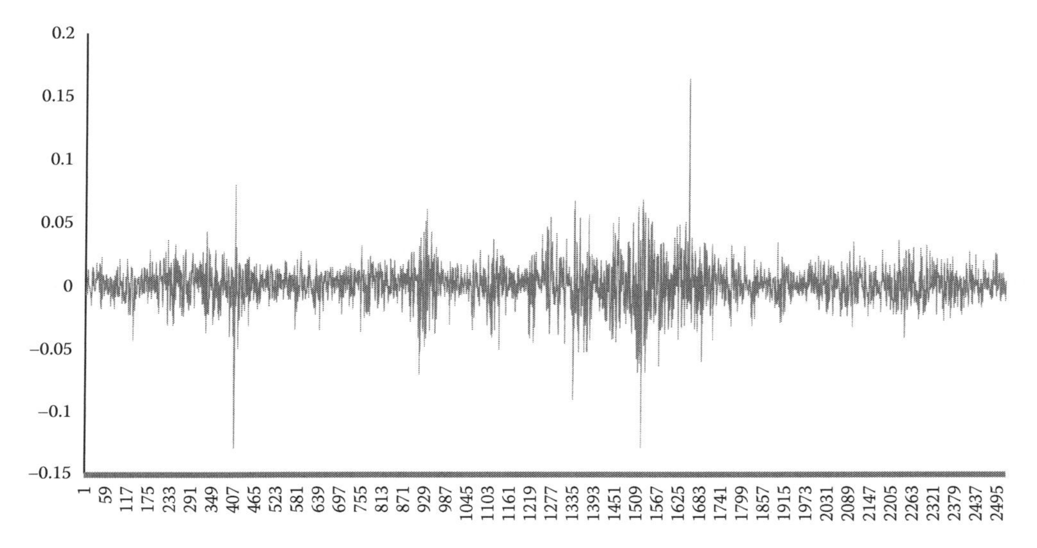

FIGURE 10.2
Daily returns of Nifty stocks.

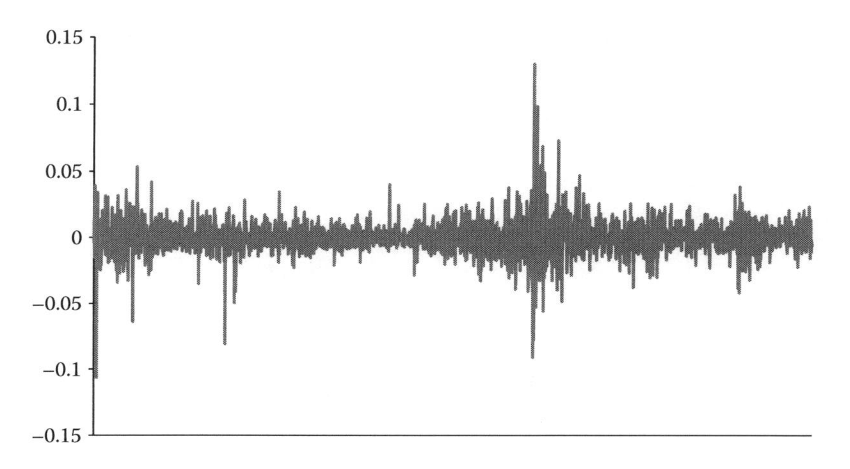

FIGURE 10.3
Daily returns of Coca-Cola stock.

more volatile than other periods, implying volatility may be time varying. One can note the dramatic fluctuations in such volatility over the financial crisis period of 2003–2007.

What is volatility? It refers to the sudden and unpredictable variability in the level of the series under question. It is often referred to as time-varying volatility. Normally, such time-varying volatility is encountered while employing cross-section data, where heteroscedasticity is explained as a function of some independent variables. But heteroscedasticity is pervasive in financial and macro time series also, and volatility clusters is one of the stylized facts of a time series of asset returns.

How does one measure volatility? Alternatively, how does one define volatility? It is defined in many ways. In science, it refers to the sudden evaporation at normal temperatures. Until Engle (1982), who pioneered a rigorous statistical model called the autoregressive conditional heteroscedasticity (ARCH) model that captured such a time-varying volatility, time series analysts had tried many ad hoc procedures such as rolling standard deviations or five-year moving averages to describe volatility in their data. Engle showed that such time-varying volatility could be modeled using conditional variance or standard deviation. Such has been the success and popularity gathered by ARCH models ever since that it has mutated into several forms over the years and we now have a veritable family of ARCH models. It is about such a family of models that this chapter deals with. Such models may be called *time series models of heteroscedasticity.* And the literature on such models of volatility is so vast that many surveys on the theoretical developments and applications have been published periodically. To name a few, one can quote surveys by Bollerslev et al. (1992), Bera and Higgins (1993), Palm (1996), and Li et al. (2002).

It is a very pertinent question to ask here, why one should know what the variances will be at a future date. Why is it important?

1. Why is the variance of inflation higher in some periods than others? Here, the importance of separating predictable movements from the unpredictable residuals is important. One may need to predict the variance of the residuals if we want to have a better inflation forecast.
2. In an asset market setup, knowing variances is important because changing variances influence investor behavior. Mandelbrot (1963) observed that “random variables with an infinite population variance are indispensable for a workable description of price changes.” His observation that variances change over time and large(small) changes are followed by large(small) changes of either sign are “stylized facts” for many economic and financial

variables. But even such facts need to be captured by some appropriate modeling. The simplest case of variance influencing financial market behavior is where the risk of an asset can be equated to its variance and its expected return in equilibrium must depend upon this risk. Thus, the conditional variance of an asset can influence its conditional mean.

3. At the methodological level, a variance that changes over time also has implications for the efficiency of statistical inference about the parameters that describe the dynamics of the level of the series.
4. Such models are useful, not only to capture some stylized facts about stock market, but also in diverse areas. For example, ARCH models have been used in testing capital asset pricing models (CAPM) and in developing tests for market efficiency by accounting for volatility clusters. It has been used in the macro context to model inflationary uncertainty, to model the exchange rate fluctuations, and to check for the efficacy of central bank interventions in foreign exchange market. In fact, Engle employed such a model to explain time-varying variance in U.K. inflation data.

Despite such a wide applicability of volatility models, there is no gainsaying the point that it is the stylized facts observed in financial markets all over the world that have constantly resulted in refinements to the standard ARCH models. Apart from the foreign exchange market, it is in the financial variables, especially in stock prices or returns from stocks, that one witnesses very volatile patterns. Hence, it should not be surprising to note that such ARCH models and the subsequent improvements have been geared to explaining changing variance in some financial variables, mostly continuously compounded daily returns. Cont (2001) lists the stylized facts that characterize many financial markets. Though it is extremely difficult to find a stochastic process that will capture all these stylized facts, the literature observes that any satisfactory model for daily returns must be consistent with at least three stylized facts that are of particular importance:

- The distribution of returns is approximately symmetric and has high kurtosis, fat tails, and a peaked center compared to a normal distribution.
- Autocorrelations of returns are all close to zero.
- The autocorrelations of both the absolute and squared returns are positive for many lags and they indicate substantially more linear dependence than the autocorrelation of returns.

We confirm these stylized facts by analyzing the correlation properties of returns series from financial markets in India and abroad. We calculated the continuously compounded returns of these assets and estimated the correlation coefficients of the returns, absolute returns and squared returns. These are displayed in Tables 10.1 through 10.3. Even the autocorrelations of returns display some correlations and the absolute and squared returns show strong correlations.

Our discussion in this chapter will also revolve around models using financial returns that explain the above three important stylized facts, though such models can be used to explain heteroscedasticity in *any* time series. While by no account is this chapter an exhaustive survey, the main aim is to explain the theoretical and empirical features of some important pioneering volatility models that explain the above-mentioned three stylized facts, from a practitioner's viewpoint. The handbook edited by Bauwens et al. (2012) touches upon several topics on volatility not covered by this chapter.

Such volatility models have been estimated both in univariate and multivariate contexts. This chapter focuses on univariate volatility models. Univariate volatility models can be either deterministic or stochastic. The plan of this chapter is as follows. In the first part, we begin listing the various notions of volatility and concentrate on some of the pioneering statistical models that have

TABLE 10.1

Autocorrelations for Returns

	Lags 1–5					Q-Statistic				
	Autocorrelations									
Series	**1**	**2**	**3**	**4**	**5**	**Q4**	**Q8**	**Q12**	**Q16**	**Q20**
S&P 500	−0.115	−0.044	0.034	−0.011	−0.051	42.04	55.62	65.59	87.31	102.94
Nifty	0.066	−0.038	−0.004	0.007	−0.026	14.74	29.99	33.64	43.69	56.95
Coca-Cola	−0.061	−0.022	0.005	−0.016	−0.034	**11.35****	23.86	34.48	38.21	44.36
Reliance Industries	0.007	−0019	−0.057	0.027	−0.018	**11.14****	**14.89****	50.87	62.70	74.54
SAIL	0.017	−0.034	−0.008	0.004	−0.023	**4.00***	**7.56***	**19.41***	**28.84****	**34.33****
Shell	−0.020	−0.052	−0.056	0.116	−0.045	36.72	43.47	60.21	81.21	93.20
TISCO	−0.226	−0.007	0.003	−0.026	0.004	132.04	139.01	147.26	148.02	155.10
ICICI	0.062	−0.019	−0.035	−0.008	−0.057	13.78	43.77	45.06	45.91	52.50
General Electric	−0.017	0.035	−0.015	0.040	−0.060	**8.41***	24.88	63.19	70.30	104.15
Glaxo	−0.069	−0.044	−0.019	0.042	−0.069	22.67	41.16	44.40	48.28	50.50
Hindalco	0.078	0.00	−0.027	−0.017	0.008	17.86	**19.66****	**22.26****	**23.64***	**26.10***
Nikkei	−0.026	−0.012	−0.013	−0.020	−0.045	**3.38***	**9.74***	**17.94***	**26.09***	**29.99***
Re$xrate	0.367	−0.009	−0.065	0.005	0.180	16.66	29.54	38.80	52.96	80.44
Ster$xrate	0.009	−0.003	−0.028	0.017	−0.046	**3.93***	**18.05****	33.00	50.29	55.39
Yen$xrate	−0.037	−0.012	−0.003	−0.000	−0.031	**4.94***	**13.94***	**18.87***	**25.52***	**36.98****
Euro$xrate	0.011	−0.020	0.005	0.036	−0.031	**6.10***	**17.66****	**19.48***	**22.69***	**25.19***

Note: Autocorrelations have been calculated using the daily log returns of the respective series, except rupee–dollar exchange rate, for which monthly returns have been used. * and **, and bold numbers denote insignificant Ljung–Box statistics at 5% and 1% levels, respectively, signifying that the null of uncorrelatedness is accepted.

TABLE 10.2

Autocorrelations for Absolute Returns

	Lags 1–5					Q-Statistic				
	Autocorrelations									
Series	**1**	**2**	**3**	**4**	**5**	**Q4**	**Q8**	**Q12**	**Q16**	**Q20**
S&P 500	0.249	0.379	0.305	0.337	0.394	1049.70	2385.53	3506.57	4349.46	5216.32
Nifty	0.265	0.248	0.244	0.268	0.229	662.39	1144.37	1522.42	1844.31	2090.92
Coca-Cola	0.267	0.306	0.252	0.234	0.229	703.47	1257.60	1723.99	2179.02	2438.45
Reliance Industries	0.222	0.191	0.182	0.176	0.202	378.87	649.28	834.58	1006.242	1153.88
SAIL	0.259	0.169	0.170	0.152	0.185	387.19	631.38	837.84	969.20	1068.08
Shell	0.335	0.276	0.321	0.313	0.309	724.10	1238.28	1760.08	2202.08	2562.94
TISCO	0.388	0.113	0.107	0.085	0.103	456.66	554.65	629.82	703.954	788.97
ICICI	0.239	0.196	0.164	0.205	0.191	411.30	850.82	1230.12	1634.94	1898.15
General Electric	0.346	0.352	0.268	0.300	0.342	1030.88	2344.77	3269.94	4109.82	5031.14
Glaxo	0.254	0.161	0.133	0.106	0.117	306.97	463.86	568.24	662.42	763.27
Hindalco	0.270	0.192	0.216	0.186	0.173	483.65	793.33	1117.60	1251.11	1384.83
Nikkei	0.243	0.324	0.319	0.277	0.298	847.54	1535.69	2103.85	2549.63	2891.74
Re$xrate	0.390	0.044	0.059	0.131	0.314	20.95	44.24	54.21	62.99	85.49
Ster$xrate	0.120	0.138	0.155	0.177	0.186	289.11	686.19	1090.25	1454.51	1740.68
Yen$xrate	0.081	0.117	0.062	0.052	0.086	87.35	168.45	252.26	341.93	413.55
Euro$xrate	0.018	0.103	0.063	0.102	0.101	82.00	213.51	337.02	458.42	564.37

Note: Autocorrelations have been calculated using the daily log returns of the respective series, except rupee–dollar exchange rate, for which monthly returns have been used. The calculated Q statistics for all series clearly reject the null of uncorrelatedness.

TABLE 10.3

Autocorrelations for Squared Returns

	Lags 1–5					Q-Statistic				
	Autocorrelations									
Series	1	2	3	4	5	Q4	Q8	Q12	Q16	Q20
S&P 500	0.206	0.401	0.187	0.317	0.346	865.48	1842.59	2912.56	3403.56	3987.78
Nifty	0.214	0.168	0.110	0.161	0.129	283.29	398.72	541.03	620.76	669.38
Coca-Cola	0.236	0.264	0.194	0.125	0.132	444.14	616.33	912.64	1038.08	1108.45
Reliance Industries	0.129	0.101	0.099	0.084	0.133	110.33	216.31	257.87	304.55	335.88
SAIL	0.316	0.162	0.119	0.127	0.233	407.99	674.84	753.90	803.72	830.41
Shell	0.355	0.294	0.327	0.342	0.342	813.48	1356.32	1950.36	2367.07	2709.64
TISCO	0.485	0.004	0.003	0.000	0.001	590.92	590.94	591.00	591.03	591.28
ICICI	0.222	0.142	0.112	0.174	0.124	280.85	589.91	861.87	1144.30	1267.10
General Electric	0.277	0.288	0.148	0.167	0.260	531.31	1438.37	1905.59	2318.54	2860.59
Glaxo	0.429	0.069	0.051	0.052	0.052	498.86	516.99	525.8	534.79	544.68
Hindalco	0.318	0.162	0.275	0.224	0.171	640.85	899.51	1238.62	1298.03	1342.51
Nikkei	0.320	0.444	0.389	0.269	0.276	1296.07	2003.50	2714.25	3117.45	3321.83
Re$xrate	0.334	−0.002	0.021	0.128	0.334	15.43	33.04	38.21	43.06	51.17
Ster$xrate	0.101	0.105	0.214	0.170	0.233	311.50	787.35	1242.08	1546.00	1766.16
Yen$xrate	0.081	0.142	0.049	0.019	0.074	95.89	155.76	246.17	373.36	399.28
Euro$xrate	0.026	0.152	0.054	0.118	0.064	131.66	249.09	395.98	531.50	645.85

Note: Autocorrelations have been calculated using the daily log returns of the respective series, except rupee–dollar exchange rate, for which monthly returns have been used. The calculated Q statistics for all series clearly reject the null of uncorrelatedness.

been discussed in the literature to quantify conditional volatility. The basic focus of this part is to describe the salient features of some deterministic conditional volatility models that capture the important stylized facts, such as the leverage effects, excess kurtosis observed especially in financial time series. In part two, we describe stochastic volatility (SV) models, where conditional volatility is described as a latent variable that depends not on past observations but on some unobserved past latent structure.

Part 1: Deterministic Conditional Volatility Models

10.2 Notions of Volatility

Volatility in general refers to the sudden and unpredictable variability of the series under consideration. In finance, volatility is the price variability during certain periods and is simply denoted as the conditional variance or standard deviation of the series under consideration, which will be mostly the continuously compounded one period returns in case of asset returns. There are many types of volatility, which we explain below.

10.2.1 Realized Volatility

Realized volatility—also called historical volatility—is simply calculated as the standard deviations of the historical returns, and an estimate of the standard deviation is, if we have returns for N trading periods, $r_{t-1}, \ldots, r_{t-N}$, and $\bar{r}$ is the average returns over the period

$$s = \sqrt{\frac{1}{N-1} \sum_{i=1}^{N} (r_{t-i} - \bar{r})^2}$$

The trading period over which the returns are calculated has been normally one day. But with the availability of intraday data, research considers trading period measured in minutes.

10.2.2 Conditional Volatility

This is measured through a statistical model such as the models from the ARCH/GARCH family. Specifically, one uses time series models where conditional variance is specified as a function of past values. Forecasting volatility using such models is straightforward. SV models recognize that volatility is a latent variable and modeled explicitly as a stochastic process.

10.2.3 Implied Volatility

Implied volatility is calculated from option prices and it is denoted as σ. The famous Black–Scholes formula provides us with a rule to price European call options and suffice it to say that the time-varying volatility parameter σ plays a vital role in deciding the price.

We confine ourselves only to conditional volatility models in this chapter. An important purpose in building a model for a univariate time series is to predict the future values of the series. Forecasting principles tell us that best linear forecasts are obtained from expectations conditioned on past information of the series. If we allow past information to affect the conditional variance of such a forecast, one may obtain better confidence intervals for the forecasts. Such a model was suggested by Engle, which was different from the standard model for heteroscedasticity.

The importance of correctly specifying conditional volatility stems from the wide range of applications using the conditional variance as a measure of risk. In finance, for example, Merton (1980) showed that the expected returns and conditional volatility are related. Bekaert and Wu (2000) use conditional volatility to find support for the volatility feedback story. Li (1998) uses conditional volatility to explain the predictability of risk premia for stock and bond returns. Lobo and Tufte (1998) check if conditional volatility is impacted by political factors. Kho (1996) exploits conditional volatility to assess the efficiency of currency futures market and concludes that profit opportunities arising out of exploiting volatility are not substantial. By analyzing the behavior of conditional volatility during various regimes, using post-war U.S. data, Vazquez (2004) uses conditional volatility to evaluate the utility of Lucas proof (LP) equilibrium performance against the fundamental equilibrium. Engle (2002) appraises the users of conditional volatility models, of some new areas where the concept can be gainfully employed.

10.3 Autoregressive Conditional Heteroscedasticity

We shall introduce a model first proposed by Engle (1982) to model changing variances in time series data. It is called the ARCH model.

Definition

An ARCH process can arise under many circumstances. Let $\tilde{z}$ be the time series of interest, where $\tilde{z}_t = z_t - \mu$ and μ is the mean of the series. Consider initially the stationary first-order autoregression for $\tilde{z}_t$ as the mean equation:

$$\tilde{z}_t = \phi_1 \tilde{z}_{t-1} + e_t,$$

where e_t is white noise with $\text{var}(e_t) = \sigma^2$. Here, the conditional mean of $\tilde{z}_t$ is equal to $\phi_1 \tilde{z}_{t-1}$, while the unconditional mean is zero. From our discussion on forecast of $\tilde{z}_t$, we know that the best forecasts are obtained by using conditional means. One can expect even better forecast *intervals* if additional information from the past is allowed to affect the forecast variance. The conditional variance $\tilde{z}_t$ is σ^2, while the unconditional variance is $\sigma^2/(1 - \phi_1^2)$. Though the unconditional variance is time-independent, conditional variance *could* change with time. So, modeling fluctuations in variance could be fruitful and we have already outlined some possible reasons for the same.

Engle suggests a model where mean and variance vary jointly:

$$\begin{aligned} e_t &= \upsilon_t h_t^{1/2}, \\ h_t &= \alpha_0 + \alpha_1 e_{t-1}^2 \end{aligned}$$

with $V(\upsilon_t) = 1$ and h_t being the variance equation. Adding the normality assumption and using conditional densities, the same could be expressed in terms of the information set, ψ_{t-1}:

$$\begin{aligned} e_t | \psi_{t-1} &\sim N(0, h_t) \\ h_t &= \alpha_0 + \alpha_1 e_{t-1}^2. \end{aligned}$$

One can extend and generalize the function to "p" lags, where "p" will be the order of the ARCH process and $\boldsymbol{\alpha}$ is a vector of unknown parameters.

We can describe the ARCH model in a regression context also. The mean equation expresses the dependent variable z_t as being generated by

$$z_t = \mathbf{x}_t'\boldsymbol{\beta} + e_t, \quad t = 1, \ldots, T, \quad \text{where}$$
$$\mathbf{x}_t = (k \times 1)\text{vector of lagged exogenous and endogeneous variables, and}$$
$$\boldsymbol{\beta} = (k \times 1)\text{vector of coefficients.}$$

The ARCH model characterizes the distribution of the stochastic error, e_t, conditional on the information set ψ_t. Formally, Engle's original model assumes that adding the assumption of normality, and using conditional densities,

$$z_t|\psi_{t-1} \sim N(\mathbf{x}_t'\boldsymbol{\beta}, h_t), \quad \text{where}$$
$$e_t|\psi_{t-1} \sim N(0, h_t), \quad \text{and}$$
$$h_t = \alpha_0 + \alpha_1 e_{t-1}^2 + \cdots + \alpha_p e_{t-p}^2.$$

Volatility clusters often occur due to large shocks to the dependent variable. In the regression model, for example, a large shock is represented by a large deviation of z_t from its conditional mean $\mathbf{x}'\boldsymbol{\beta}$ or equivalently a large positive or negative value of the error e_t. To make conditional variance positive, Engle imposes the restrictions that $\alpha_0 > 0$, and $\alpha_i \geq 0$, $i = 1, \ldots, p$, so that the variance function is an increasing function of the magnitude of the lagged errors, irrespective of their signs. Hence, large(small) errors of either sign tend to be followed by a large(small) errors of either sign. The order of the lag, p, determines the length of time for which the shock persists in conditioning the variance of subsequent errors. The larger the value of p, the longer the volatility period.

Before we proceed further, it would be ideal to illustrate through graphs, how a data exhibiting a typical ARCH behavior appears to be. For comparison purposes, we first generate a white noise series, which would enable us to compare it with the series drawn with ARCH disturbances.

Suppose we specify the following model for the ARCH process:

$$e_t = \upsilon_t\sqrt{h_t}, \quad \upsilon_t \sim NID(0, 1),$$
$$h_t = \alpha_0 + \alpha_1 e_{t-1}^2 + \alpha_2 e_{t-2}^2 + \cdots + \alpha_p e_{t-p}^2.$$

We impose the conditions that $\alpha_0 = 1$ and $\alpha_i = 0$, $i = 1, \ldots, p$, thus making $h_t = 1$ and hence, $e_t = \upsilon_t$, which is a white noise series.

We present below one such series.

The displayed data in Figure 10.4 is simply a simulated Gaussian white noise series, the process usually assumed for errors in a linear model. As expected, we do not see any clustering of data points. Next, we graph in Figures 10.5 and 10.6 simulated data from ARCH(1) and ARCH(4) models.

Figures 10.5 and 10.6 clearly show data points clustered over certain ranges. Also note that in Figure 10.6 such clusters are thicker, denoting longer episodes of volatility, and hence a model with longer lags. Figures 10.7 and 10.8 represent returns from the popular company Intel Corporation and the popular stock index S&P 500 index. These data clearly exhibit volatility clusters, and returns from S&P 500 index may follow an ARCH(4) model.

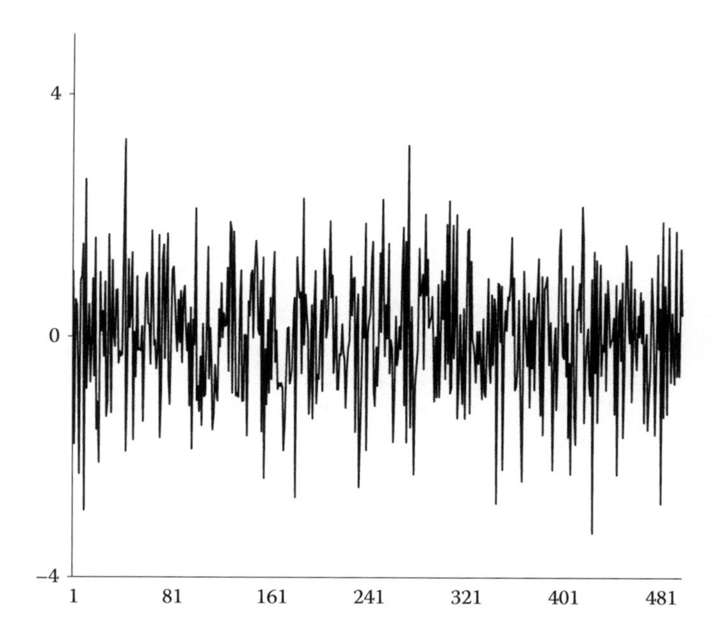

FIGURE 10.4
Simulated white noise.

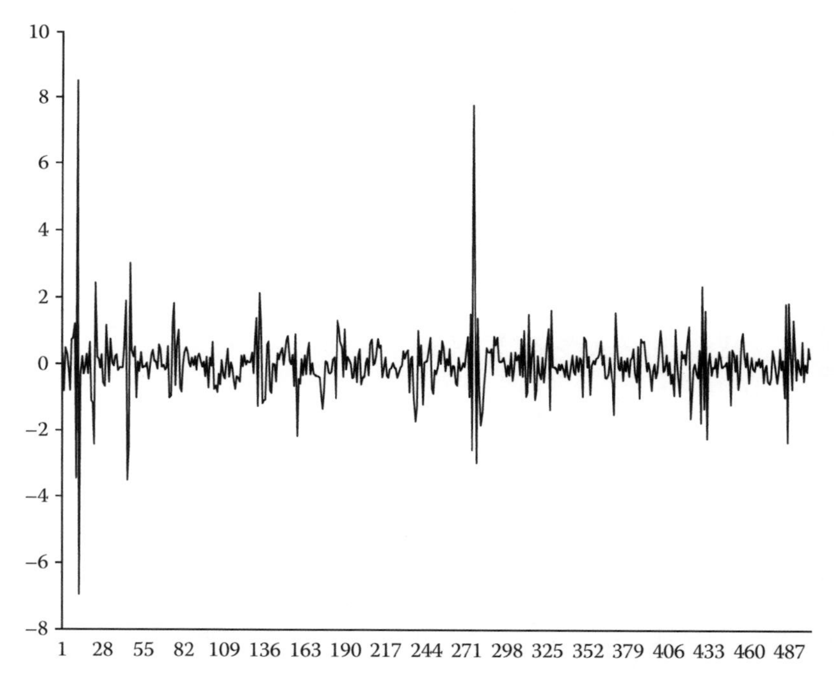

FIGURE 10.5
Simulated ARCH(1) data.

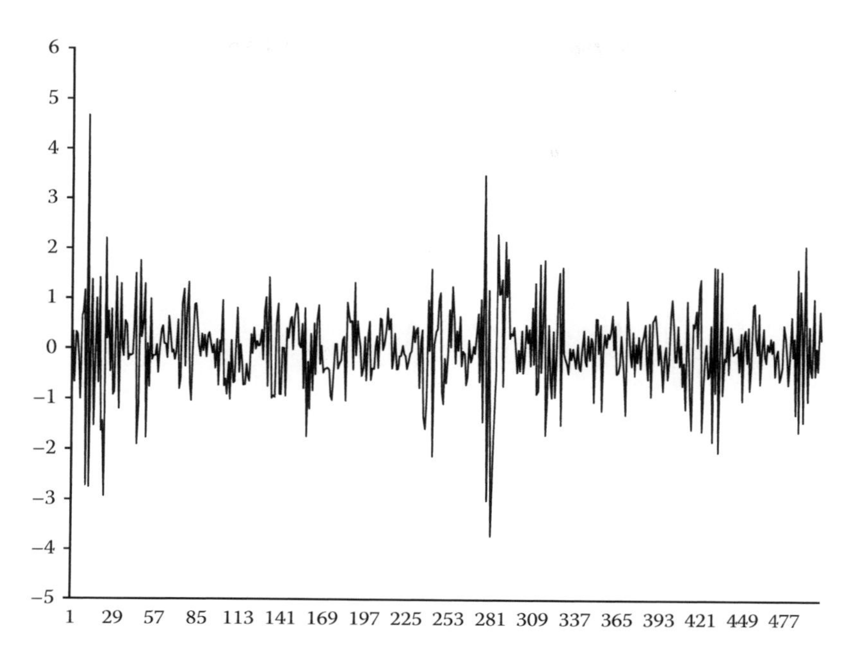

FIGURE 10.6
Simulated ARCH(4) data.

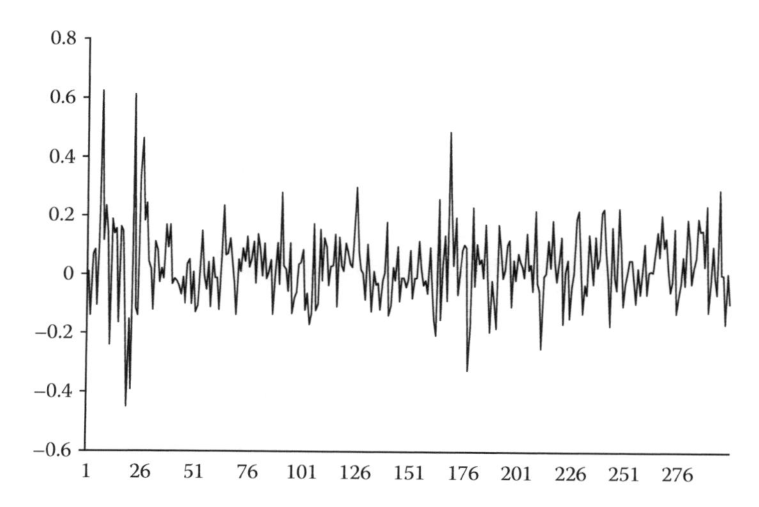

FIGURE 10.7
Monthly log returns of Intel Corporation.

10.3.1 Moments of a General ARCH Process

It can be easily shown that the simplest ARCH model explains the three stylized facts of daily returns in any financial market, under the assumption that the underlying series has a constant mean. But we shall lay down conditions for the stationarity of the process and other moment conditions for a general ARCH process, where z_t itself follows a low-order autoregressive (AR) process, but the volatility changes still explain the stylized facts. For many functions of h and values of α,

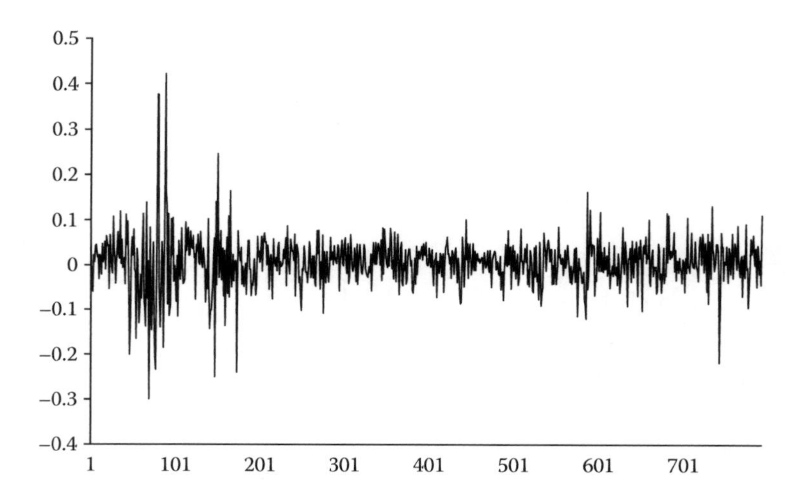

FIGURE 10.8
Monthly excess returns of SP 500 index.

the variance is independent of t and finite—an important condition for covariance stationarity of the process. We shall make explicit below sufficient conditions for covariance stationarity and the conditions for some higher-order moments to exist. For instance, we shall lay down the condition for the fourth moment to exist to study the tail behavior.

The innovation sequence could be from a time series model or a regression model. Take the AR(1)–ARCH(1) model for illustration. Suppose we have

$$\begin{aligned}\tilde{z}_t &= \phi_1\tilde{z}_{t-1} + e_t,\\ e_t &= \upsilon_t\sqrt{h_t},\\ h_t &= \alpha_0 + \alpha_1 e_{t-1}^2, \quad \text{with } V(\upsilon_t) = 1.\end{aligned}$$

In terms of the conditional densities, with the information set ψ_{t-1},

$$e_t|\psi_{t-1} \sim N\big(0, h_t\big), \quad \text{where } h_t = \alpha_0 + \alpha_1 e_{t-1}^2.$$

Besides the nonnegativity conditions on the α parameters, for covariance stationarity, we further need that $\alpha_1 < 1$. But this is not all. If we need higher-order moments also to exist, α_1 must satisfy some additional restriction. In some cases, we may need kurtosis to exist to explain the tail behavior. For example, we may want to check if the excess kurtosis is positive. If it is so, the tail behavior of e_t is heavier than the normal distribution. One can verify these by spelling out the moment conditions using the above ARCH specification.

The unconditional mean of e_t is zero because

$$E(e_t) = E[E(e_t|I_{t-1})] = E[\sqrt{h_t}E(e_t)] = 0.$$

The unconditional variance is given by

$$\begin{aligned}\text{Var}(e_t) &= E(e_t^2) = E[E(e_t^2|I_{t-1})]\\ &= E(\alpha_0 + \alpha_1 e_{t-1}^2) = \alpha_0 + \alpha_1 E(e_{t-1}^2).\end{aligned}$$

Since e_t is a stationary process, we have

$$\text{Var}(e_t) = \frac{\alpha_0}{1-\alpha_1}.$$

Because the variance of e_t must be positive, we need that $0 \leq \alpha_1 < 1$. Under the assumption of fourth-order stationarity of e_t, we have, after some algebra

$$E(e_t^4) = \frac{3\alpha_0^2(1+\alpha_1)}{(1-\alpha_1)(1-3\alpha_1^2)}.$$

This has the following important implications. Since the fourth moment of e_t is positive, α_1 must also satisfy the condition that $0 \leq \alpha_1^2 < 1/3$ or $1 - 3\alpha_1^2 > 0$. The unconditional kurtosis of e_t is given by

$$\frac{E(e_t^4)}{[\text{Var}(e_t)]^2} = 3\frac{(1-\alpha_1^2)}{(1-3\alpha_1^2)} > 3.$$

Thus the excess kurtosis of e_t is positive. This simply means that the given model setup can capture outliers that occur in asset returns better than iid errors. Similarly, for the ARCH(2) process, the necessary and sufficient condition for the existence of the fourth moment is given by

$$\alpha_2 + 3\alpha_1^2 + 3\alpha_2^2 + 3\alpha_1^2\alpha_2 - 3\alpha_1^2 < 1.$$

Given this, Bollerslev (1998) lists out the autocorrelations and partial autocorrelations of an ARCH(2) process. Given this condition, the autocorrelations for $\{e_t^2\}$ are found to be

$$\begin{aligned}
\rho_1 &= \alpha_1/(1-\alpha_2),\\
\rho_2 &= (\alpha_2 + \alpha_1^2 - \alpha_2^2)/(1-\alpha_2),\\
\rho_k &= \alpha_1\rho_{k-1} + \alpha_2\rho_{k-2} \quad (k > 2),
\end{aligned}$$

whereas the partial autocorrelations cut-off after lag 2:

$$\phi_{22} = \alpha_2, \quad \phi_{kk} = 0 \quad (k > 2).$$

After some algebra, one can show that the first two autocorrelations must satisfy

$$\begin{aligned}
&0 \leq \rho_1 \leq (1/3)^{1/2},\\
&\rho_2 \geq \rho_1^2,\\
&\rho_2 < \rho_1^2 + (1/3)^{1/2}(1-3\rho_1^2)^{1/2}(1-\rho_1^2)^{1/2}.
\end{aligned}$$

However, establishing such moment conditions for higher-order models is difficult. The minimum condition of covariance stationarity must be satisfied. Milhoj (1985) however relaxed Engle's assumption on the presence of e_t's from infinite past but with finite variance and showed that an ARCH(q) model is covariance stationary iff $\alpha_1 + \alpha_2 + \cdots + \alpha_q < 1$. This condition is also sufficient to prove the strict stationarity and ergodicity of the process. Besides, Milhoj also showed that Engle's condition for the existence of the $2m$th moment for ARCH(1) model can be proved without the assumption that the errors started infinitely in the past.

10.4 Estimation of ARCH Models

In this section, we shall see how to estimate the parameters of a model describing ARCH errors or innovations. Consistent with our discussion so far, we have to realize that such errors themselves normally come either from a time series application or from a regression model. We shall see the important steps involved in estimating a regression model with ARCH errors. The preferable method of estimation is the maximum likelihood approach, since there is efficiency gain using maximum likelihood methods, rather than least squares. Maximum likelihood (ML) estimator is nonlinear but more efficient. One can either maximize the likelihood function by direct methods, though most applied work on ARCH models uses the Bernt, Hall, Hall, Hausman (BHHH) algorithm. A simple alternative numerical optimization routine is the method of scoring algorithm. The main advantage of this algorithm is that only the first derivatives are needed to evaluate the likelihood and we have the first derivatives readily available for our case. Engle demonstrates how an ordinary least square (OLS) program is enough to run these iterative steps and obtain the estimates. We outline the important steps in estimation below.

10.4.1 ML Estimation of an ARCH Regression Model

For the pth-order linear case, we can straightaway write the model specification and the likelihood, apart from a constant, as

$$\begin{aligned} z_t|\psi_{t-1} &\sim N(\mathbf{x}_t\beta, h_t), \quad \text{where} \\ h_t &= \alpha_0 + \alpha_1 e_{t-1}^2 + \cdots + \alpha_p e_{t-p}^2, \\ e_t &= z_t - \mathbf{x}_t\beta, \\ l_t &= -\frac{1}{2}\log h_t - \frac{e_t^2}{2h_t}. \end{aligned}$$

The unknown parameter β has to be estimated by maximizing the likelihood. As usual, the ML estimator is found by solving the first-order conditions. The derivative with respect to β is

$$\begin{aligned} \frac{\partial l_t}{\partial \beta} &= -\frac{1}{2h_t}\frac{\partial h_t}{\partial \beta} + \frac{e_t\mathbf{x}_t'}{h_t} + \frac{e_t^2}{2h_t^2}\frac{\partial h_t}{\partial \beta} \\ &= \frac{e_t\mathbf{x}_t'}{h_t} + \frac{1}{2h_t}\frac{\partial h_t}{\partial \beta}\left[\frac{e_t^2}{h_t} - 1\right]. \end{aligned}$$

This is the familiar first-order condition for exogenous heteroscedastic correction. Substituting the linear variance function gives

$$\frac{\partial l_t}{\partial \beta} = \frac{e_t\mathbf{x}_t'}{h_t} - \frac{1}{h_t}\left(\frac{e_t^2}{h_t} - 1\right)\sum_j \alpha_j e_{t-j}\mathbf{x}_{t-j}'.$$

The Hessian is given by

$$\frac{\partial^2 l_t}{\partial \boldsymbol{\beta} \partial \boldsymbol{\beta}'} = -\frac{\mathbf{x}_t' \mathbf{x}_t}{h_t} - \frac{1}{2h_t^2} \frac{\partial h_t}{\partial \boldsymbol{\beta}} \frac{\partial h_t}{\partial \boldsymbol{\beta}'} \left(\frac{e_t^2}{h_t} \right) - \frac{\partial h_t}{\partial \boldsymbol{\beta}} \frac{2e_t \mathbf{x}_t'}{h_t^2}$$
$$+ \left[\frac{e_t^2}{h_t} - 1 \right] \frac{\partial}{\partial \boldsymbol{\beta}'} \left[\frac{1}{2h_t} \frac{\partial h_t}{\partial \boldsymbol{\beta}} \right].$$

If we take conditional expectations, the last two terms will vanish and the conditional expectation of $\left[e_t^2 / h_t \right]$ is one. The diagonal blocks of the information matrix—which is the negative expectation of the Hessian—is given by

$$\mathbf{I}_{\boldsymbol{\beta}\boldsymbol{\beta}} = \sum E \left[\frac{\mathbf{x}_t' \mathbf{x}_t}{h_t} - \frac{1}{2h_t^2} \frac{\partial h_t}{\partial \boldsymbol{\beta}} \frac{\partial h_t}{\partial \boldsymbol{\beta}'} \right].$$

and this can be consistently estimated, after substituting for the linear variance function.

One can express in a similar fashion, the off-diagonal blocks of the information matrix as

$$\mathbf{I}_{\boldsymbol{\alpha}\boldsymbol{\beta}} = \sum E \left[\frac{1}{2h_t^2} \frac{\partial h_t}{\partial \boldsymbol{\alpha}} \frac{\partial h_t}{\partial \boldsymbol{\beta}'} \right].$$

Here, Engle proves an important theorem, where it is shown that this off-diagonal block is zero. The theorem states that if an ARCH regression model is symmetric and regular, then $\mathbf{I}_{\boldsymbol{\alpha}\boldsymbol{\beta}} = \mathbf{0}$. The implications are far reaching; now the estimation of $\boldsymbol{\alpha}$ and $\boldsymbol{\beta}$ can be undertaken *separately* without asymptotic loss of efficiency and their variances can be calculated separately. It can be shown that $\boldsymbol{\beta}$ and $\boldsymbol{\alpha}$ are $\sqrt{T}$ consistent with asymptotic covariance matrices given, respectively, as $\hat{\mathbf{I}}_{\boldsymbol{\beta}\boldsymbol{\beta}}$ and $\hat{\mathbf{I}}_{\boldsymbol{\alpha}\boldsymbol{\alpha}}$. With this vital result, one can proceed with some of the practical aspects of the estimation of an ARCH model.

10.4.2 Estimation of ARCH Models: Practical Aspects

Let us suppose we are interested in the estimation of an ARCH regression model. That is, we have estimated a regression model, errors from which are heteroscedastic. How do we proceed? Given the block diagonality of the information matrix, one can estimate $\boldsymbol{\alpha}$ and $\boldsymbol{\beta}$ separately without loss of asymptotic efficiency. Furthermore, one can be estimated efficiently with only a consistent estimate of the other.

1. Obtain an initial estimate of $\boldsymbol{\beta}$ using OLS, which is still consistent but linear. Calculate the residuals.
2. From these residuals, construct an efficient estimate of $\boldsymbol{\alpha}$ using these residuals and based on these estimates $\hat{\boldsymbol{\alpha}}$, we can update our initial estimate of $\boldsymbol{\beta}$ and iterate until convergence.
3. If one uses the method of scoring algorithm, this iteration involves a four-step procedure, which is demonstrated by Greene (2000, Chapter 18).
4. In practice, only one-step iteration is sufficient, because no efficiency is gained after the first updating. But the resulting estimates are only asymptotically equivalent to ML estimates.

10.4.3 Testing for ARCH Disturbances

Since the ARCH model requires iterative procedures, it may be desirable to test whether it is appropriate before going through the effort of estimating it. We shall be using the Lagrange multiplier (LM) test to test for the presence of ARCH disturbances. The null hypothesis to be tested is $\alpha_1 = \alpha_2 = \cdots = \alpha_p = 0$. The LM test is based upon the slope vector and the information matrix under the null. Shorn of technical details, the steps involved in the test procedure are as follows:

1. Estimate the mean equation using OLS and save the residuals.
2. Regress the squared residuals on a constant and p lags of the squared residuals, and calculate the R^2 of this regression. If, for example, one wants to test if ARCH(1) effect is present, one would run the regression with $p = 1$.
3. Test $\hat{\lambda} = T \cdot R^2$ as a $\psi^2_{(p)}$ distribution. The test confirms the presence of heteroscedasticity if $\hat{\lambda}$ is greater than the critical value. Other tests also have been suggested in the literature.

10.4.4 An Example

We applied the above procedure to estimate a simple ARCH model on the continuously compounded returns, r_t, calculated using daily Nifty index of the NSE (National Stock Exchange) of India, for the period February 1997 to January 2006, containing 2252 observations. We used the commercial software package, RATS. Our autocorrelation function (ACF) and partial autocorrelation function (PACF) structures of the returns suggested some low-order AR model for the returns. Summary statistics displayed in Table 10.4, for the residual series e_t, clearly show that the series is skewed to the left and has a significantly fatter tail than normal. The Ljung–Box Q statistics for squared residuals calculated for 12 lags clearly confirm the presence of time-varying volatilities.

TABLE 10.4

Summary Statistics for the Residual Series

	Sample Statistics
Number of observations	2252
Variance	0.000263
Skewness	−0.429661
	(0.000)
Excess kurtosis	5.568470
	(0.000)
Ljung–Box (12) for the residuals in levels	33.8554
	(0.001)
Ljung–Box (12) for the squared residuals	325.4121
	(0.000)

Note: The summary statistics displayed in this table are for the residual series e_t obtained after running an AR(1) model on the returns data. The Ljung–Box Q statistics follow a χ^2 distribution with 12 degrees of freedom. Skewness and excess kurtosis have been checked against the null of zero. Asymptotic probability values are in parentheses.

Assuming that $\upsilon_t \sim N(0,1)$, we estimated the following AR(1)–ARCH(6) model:

$$r_t = \mu + \phi_1 r_{t-1}, + e_t, \quad e_t = \upsilon_t \sqrt{h_t},$$
$$h_t = \alpha_0 + \alpha_1 e_{t-1}^2 + \alpha_2 e_{t-2}^2 + \alpha_3 e_{t-3}^2 + \alpha_4 e_{t-4}^2 + \alpha_5 e_{t-5}^2 + \alpha_6 e_{t-6}^2$$

and the model with the estimated parameters is

$$r_t = 1.1307 \times 10^{-3} + 0.1131 r_{t-1} + e_t$$
$$h_t = 9.8946 \times 10^{-5} + 0.1776 e_{t-1}^2 + 0.1816 e_{t-2}^2 + 0.1099 e_{t-3}^2 - 0.00767 e_{t-4}^2 + 0.1291 e_{t-5}^2 + 0.0849 e_{t-6}^2$$

All the t-statistics, except for the coefficient of e_{t-4}^2, were significant. However, the Ljung–Box statistic calculated with the standardized residuals was estimated to be $Q(10) = 26.033$ with a p-value of 0.004, which clearly indicated that the model was not adequate. The expected return on the Nifty index is not very good with only about 0.11%. But establishing if higher-order moments exist for this model is very difficult and hence we shall return to this series after we discuss other popular models of conditional volatility.

10.5 Generalized Autoregressive Conditional Heteroscedasticity

GARCH belongs to a general class of process that allows a much more flexible lag structure. As we develop this model, we shall see that it bears much resemblance to the extension of the standard time series AR process to an autoregressive-moving average (ARMA) process, and like ARMA processes, GARCH also permits a more parsimonious description in many situations. GARCH model was developed basically to correct some of the shortfalls of the ARCH type of models. We can note the following limitations of ARCH models:

1. A practical problem with the ARCH type of model proposed by Engle is that sometimes a long lag structure in the conditional variance equation is called for. Engle overcame this problem by adopting an arbitrary weighting pattern as given below:

   $$h_t = \alpha_0 + \alpha_1 \sum_{i=1}^{q} w_i e_{t-i}^2,$$

   where the weights given by

   $$w_i = \frac{(q+1) - i}{(1/2)\, q(q+1)}$$

   decline linearly, and are constructed so that $\sum_i w_i = 1$. For Engle, $q = 4$ so that the weights were 0.4, 0.3, 0.2, and 0.1, respectively. This way one can accommodate long lags and yet estimate only two parameters. The lag selection is of course ad hoc. But the GARCH process extends the problem naturally by allowing for long lags to be modeled.
2. The ARCH model assumes that positive and negative shocks have the same effect on volatility because in ARCH models volatility depends on the square of the previous shocks.

But in practice it is well known that the price of a financial asset responds differently to positive and negative shocks.

3. The nonnegativity restrictions on the parameters are very often violated in practice. The moment structure of an ARCH model is found to be too restrictive and the constraint becomes too complicated for higher-order models.

In an attempt to answer the first limitation pointed above and to allow for a longer memory and a more flexible lag structure, the ARCH class of models were extended and a generalized ARCH model—GARCH—structure was proposed by Bollerslev (1986).

As in the previous model specification, let e_t denote a real-valued discrete-time stochastic process and ψ_t be the information set containing all information through time t. The GARCH(p, q) process—generalized autoregressive heteroscedasticity—is then given by

$$\begin{aligned} e_t|\psi_t &\sim N(0, h_t) \\ h_t &= \alpha_0 + \sum_{i=1}^{q} \alpha_i e_{t-i}^2 + \sum_{i=1}^{p} \beta_i h_{t-i} \\ &= \alpha_0 + A(L) e_t^2 + B(L)h_t, \end{aligned}$$

where

$$\begin{aligned} & p \geq 0, \quad q > 0, \\ & \alpha_0 > 0, \quad \alpha_i \geq 0, \quad i = 1, \ldots, q, \\ & \beta_i \geq 0, \quad i = 1, \ldots, p. \end{aligned}$$

In an interesting study, Nelson and Cao (1992) show that the set of restrictions that the parameters of a GARCH model have to satisfy need not be imposed in estimation, as violation of these inequalities does not necessarily mean that the conditional variance function is misspecified. They pointed out that, in the inverted representation—that is, the model expressed in infinite e's—the conditions

$$\alpha_0^* > 0, \quad \delta_i \geq 0, \quad i = 1, \ldots, \infty, \quad \text{and} \quad \alpha_0^* = \frac{\alpha_0}{\big(1 - B(1)\big)}$$

are sufficient to ensure that the conditional variance is strictly positive. For example, Nelson and Kao show that in a GARCH(1,2) process, the conditions

$$\alpha_0 > 0, \quad \alpha_1 \geq 0, \quad \beta_1 \geq 0, \quad \text{and} \quad \beta_1\alpha_1 + \alpha_2 \geq 0$$

are sufficient to guarantee that $h_t > 0$. Therefore in this model, α_2 *may* be negative. They present general results for GARCH$(1, q)$ and GARCH$(2, q)$; but proving beyond $p \geq 3$ is pretty difficult.

A GARCH(p,q) is stationary if $A(1) + B(1) < 1$. But the same condition can be understood better if we can rewrite the GARCH model as an ARMA(p,q) model as follows:

$$\begin{aligned} e_t^2 = \alpha_0 &+ (\alpha_1 + \beta_1) e_{t-1}^2 + \cdots + (\alpha_s + \beta_s) e_{t-s}^2 \\ &+ \omega_t - \beta_1\omega_{t-1} - \beta_2\omega_{t-2} - \cdots - \beta_p\omega_{t-p}, \end{aligned}$$

where h_t is the forecast of e_t^2, $\omega_t = e_t^2 - h_t$, and $s = \max(p,q)$. We have defined by convention, $\beta_j = 0$, for $j > p$ and $\alpha_j = 0$, for $j > q$. With this representation, it is easy to see that for covariance stationarity, a GARCH(p,q) model has to satisfy the condition $A(1) + B(1) < 1$. However, for strict stationarity, this condition is only a sufficient but not necessary condition. For example, for GARCH(1,1) to be strictly stationary, we need

$$E\left[\log(\beta_1 + \alpha_1 \upsilon_{t-1}^2)\right] < 0$$

so that, following Jensen's inequality,

$$E\left[\log(\beta_1 + \alpha_1 \upsilon_{t-1}^2)\right] \le \log\left(E\left[\beta_1 + \alpha_1 \upsilon_{t-1}^2\right]\right) = \log(\alpha_1 + \beta_1)$$

and the process is strictly stationary even if $\alpha_1 + \beta_1 = 1$.

Bollerslev shows that the order of a GARCH model can be identified with the help of the autocorrelation and the partial autocorrelation functions of e_t^2 also, which has a pattern similar to an ARMA(p,q) process. However, for most purposes, the simplest GARCH(1, 1) model is sufficient. We describe the main features of the model below:

$$\begin{aligned} e_t|\psi_{t-1} &\sim N(0, h_t) \\ h_t &= \alpha_0 + \alpha_1 e_{t-1}^2 + \beta_1 h_{t-1}, \\ \alpha_0 &> 0, \quad \alpha_1 \ge 0, \quad \beta_1 \ge 0. \end{aligned}$$

For stationarity, $\alpha_1 + \beta_1 < 1$ is sufficient to ensure wide-sense stationarity. A general expression for the existence of the $2r$th moment of a GARCH(1,1) process is provided by Bollerslev. In particular, the fourth moment exists if

$$3\alpha_1^2 + 2\alpha_1\beta_1 + \beta_1^2 < 1,$$

and the coefficient of kurtosis is given by

$$\kappa = \frac{6\alpha_1^2}{\left[1 - 2\alpha_1^2 - (\alpha_1 + \beta_1)^2\right]},$$

which is greater than zero by assumption. Hence, the GARCH(1,1) process is leptokurtic (heavy tailed). The general conditions for the existence of higher moments of a GARCH(p,q) process are derived in He and Terasvirta (1999) and Bollerslev states the conditions for finite fourth moment to exist for GARCH(1,2) and GARCH(2,1) processes. Though GARCH(1,1) model suffices for most purposes, Bollerslev also shows how one can use the autocorrelation function and the partial autocorrelation function of e_t^2 to decide on a possible model. From the ARMA representation of e_t^2, we can easily see that

$$\begin{aligned} \gamma_k &= \sum_{i=1}^{q} \alpha_i \gamma_{k-i} + \sum_{i=1}^{p} \beta_i \gamma_{k-i} \\ &= \sum_{i=1}^{m} \varphi_i \gamma_{k-1}, \quad k \ge p+1, \end{aligned}$$

where

$$\begin{aligned} m &= \max\{p, q\} \\ \varphi_i &= \alpha_i + \beta_i \\ i &= 1, \ldots, q, \end{aligned}$$

$\alpha_i = 0$, for $i > q$ and $\beta_i = 0$ for $i > p$. And we get the following analog to the Yule–Walker equations:

$$\rho_k = \gamma_k \gamma_0^{-1} = \sum_{i=1}^{m} \varphi_i \rho_{n-i}, \quad k \geq p+1.$$

Generally, the autocorrelation function of e_t^2 is dominated by exponential decays. Partial autocorrelations of GARCH processes behave very much like that of an autoregressive process. That is, the partial autocorrelation function has a *cut-off* after the true order of the ARCH model.

Specifically for a GARCH(1,1) model, Bollerslev establishes that the autocorrelations are, given the fourth moment,

$$\rho_1 = \alpha_1 \left(1 - \alpha_1 \beta_1 - \beta_1^2\right) / \left(1 - 2\alpha_1 \beta_1 - \beta_1^2\right),$$

$$\rho_k = (\alpha_1 + \beta_1)^{k-1} \rho_1 \quad (k > 1).$$

The partial autocorrelation function will be dominated by a damped exponential.

10.5.1 Estimation of GARCH(p, q) Models

GARCH models are normally estimated either by maximum likelihood method or quasi-maximum likelihood method (QML). For simplicity, we discuss the ML estimation of a GARCH model. Steps leading to the estimation of GARCH(p, q) models are very similar to ARCH(p) model. For the GARCH(p, q) regression model with z_t as the dependent variable, $\mathbf{x}_t$ the vector of independent variables, and $\mathbf{b}$ the vector of unknown parameters, let $\mathbf{y}_t' = (1, e_{t-1}^2, \ldots, e_{t-q}^2, h_{t-1}, \ldots, h_{t-p})$ and $\omega' = (\alpha_0, \alpha_1, \ldots, \alpha_q, \beta_1, \ldots, \beta_p)$ so that

$$e_t = z_t - \mathbf{x}_t' \mathbf{b}$$

$$e_t | \psi_{t-1} \sim N(0, h_t), \quad \text{where}$$

$$h_t = \alpha_0 + \sum_{i=1}^{q} \alpha_i e_{t-i}^2 + \sum_{i=1}^{p} \beta_i h_{t-i},$$

$$l_t = -\frac{1}{2} \log h_t - \frac{e_t^2}{2h_t}.$$

Differentiating with respect to the variance parameters gives

$$\frac{\partial l_t}{\partial \omega} = -\frac{1}{2h_t} \frac{\partial h_t}{\partial \omega} + \frac{e_t^2}{2h_t^2} \frac{\partial h_t}{\partial \omega}$$

$$= \frac{1}{2h_t} \frac{\partial h_t}{\partial \omega} \left(\frac{e_t^2}{h_t} - 1 \right),$$

where

$$\frac{\partial h_t}{\partial \omega} = \mathbf{y}_t + \sum_{i=1}^{p} \beta_i \frac{\partial h_{t-i}}{\partial \omega}.$$

One can similarly write the first-order conditions with respect to the mean equation parameters. Since GARCH also belongs to the symmetric function in the sense of Engle (1982), one can show that the off-diagonal blocks of the information matrix vanish. But the presence of the recursive terms in the first-order conditions makes estimation difficult. One may have to use BHHH algorithm to maximize the log-likelihood. And a consistent estimate of the asymptotic variance matrix can be obtained from the last iteration of the BHHH algorithm. Lumsdaine (1996) provides a proof of the consistency and asymptotic normality of GARCH(1,1) model under the condition $E\ln(\alpha_1 \upsilon_t^2 + \beta_1) < 0$.

When the true distribution of e_t is not normal, then a pseudo- or a quasi-maximum likelihood estimator is still obtained, by using the above likelihood for a GARCH process. Asymptotic properties of QML are however unclear. So, for inference, many authors compute the asymptotic standard errors using the technique suggested by Bollerslev and Wooldridge (1992), which is available in almost all commercial software packages.

10.5.2 Testing for GARCH Effects

Tests for the presence of GARCH largely follows the same steps as ARCH tests. Here also, we use the Lagrange multiplier test $\hat{\lambda} = T \cdot R^2$. But there is a difficulty in constructing a test. When the null is that the innovations are white noise, a test against a GARCH alternative is not feasible. Lee (1991) shows that the tests for GARCH(r, q) and ARCH$(q + r)$ asymptotically coincide or the two tests are asymptotically equivalent, under an ARCH(q) null. Consequently, to test for GARCH effects, one can run the regression, under the null of an ARCH(r) model:

$$e_t^2 = \alpha_0 + \alpha_1 e_{t-1}^2 + \cdots + \alpha_r e_{t-r}^2 + \alpha_{r+1} e_{t-r-1}^2 + \cdots + \alpha_{r+s} e_{t-r-s}^2 + \eta_t$$

and test for the significance of

$$\alpha_{r+1} = \cdots = \alpha_{r+s} = 0$$

10.6 Stationary ARMA–GARCH Models

The ARCH–GARCH family of models are predominantly used to model the volatilities arising in, mainly, financial data, though not much attention has been given to the specification or the estimation of the conditional mean equation. Conditional mean may have a richer specification than a simple mean corrected uncorrelated white noise. For example, conditional mean could follow an AR or ARMA model, but we have to allow for the conditional variances to be time varying.

One can define an ARMA–GARCH the following way:

$$z_t = \sum_{i=1}^{p} \phi_i z_{t-i} + \sum_{i=1}^{q} \theta_i e_{t-i} + e_t,$$

$$e_t = \upsilon_t \sqrt{h_t} \quad h_t = \alpha_0 + \sum_{i=1}^{q} \alpha e_{t-i}^2 + \sum_{i=1}^{p} \beta_i h_{t-i}.$$

10.6.1 Estimation

There are only few papers that have analyzed the asymptotic properties of the LS estimators. Weiss (1984) considers the class of ARMA models with ARCH errors and reports that many U.S. macroeconomic time series can be successfully explained by such models. Setting the unknown parameters of the mean equation equal to θ, Weiss (1986) showed that, when the parameter vector δ of the variance equation is assumed known, the least square estimator $\hat{\theta}$ is consistent and asymptotically normal. Weiss's result however requires that z_t has finite fourth moment.

But still, least square estimate (LSE) is inefficient for ARMA–GARCH models. Here again, Weiss shows that QMLE is consistent and asymptotically normal under a finite fourth-order moment condition. Engle (1982) demonstrates that QMLE is more efficient than LSE for ARMA–GARCH models. Bougerol and Picard (1992) present a necessary and sufficient condition ensuring the existence of a strictly stationary solution.

10.7 Nonstationary ARMA–GARCH Models

The last few decades saw seminal research undertaken to analyze nonstationarity in time series, but only under the assumption of a constant variance. But in recent times, the problem of nonstationarity with time-varying conditional variance has also attracted much attention.

The ARMA–GARCH model is called nonstationary if the characteristic polynomial $\phi(B)$ has a root equal to unity. For example,

$$y_t = \phi y_{t-1} + e_t$$

where $\phi = 1$, and e_t follows the GARCH(1,1) process

$$e_t = \upsilon_t \sqrt{h_t}, \quad h_t = \alpha_0 + \alpha_1 e_{t-1}^2 + \beta_1 h_{t-1}.$$

Under the condition that $\beta_1 = 0$ and the existence of a finite fourth moment, Pantula (1989) shows that, with the resulting first-order ARCH process, the traditional Dickey–Fuller test can still be applied. Ling et al. (2003) prove the same result for the GARCH(1,1) model, under the second moment condition, $\alpha_1 + \beta_1 < 1$.

However, it should be noted that QMLE-based unit root tests offer a more powerful alternative to least squared-based tests (refer to Ling and Li (1998), Ling et al. (2003), and Seo (1999) for more details).

10.8 Forecasting ARCH and GARCH Models

10.8.1 Forecasting an ARCH Model

Forecasting an ARCH model is the same as forecasting an AR model. We have to find a recursive formula to obtain the forecasts. Recall that we have to forecast both the mean and the variance equations. Suppose we have a time series of returns, r_t, which is uncorrelated, so that

$$r_t = c + e_t$$

then the minimum mean squared error s-step ahead forecast at time origin t is just c. Recall that if we confine to the linear forecasting rules, the best forecast that minimizes the MSE is the conditional expectation. The corresponding forecast error is then

$$e_{t+s} = r_{t+s} - c.$$

The conditional variance of this forecast error is then

$$\text{Var}_t(e_{t+s}) = E_t(h_{t+s}).$$

Next, we can write the forecasts of the variance equation. At the forecast origin t, the one-step ahead forecast of h_{t+1} is

$$\hat{h}_t(1) = \alpha_0 + \alpha_1 e_t^2 + \cdots + \alpha_p e_{t-p+1}^2.$$

Similarly, the two-step ahead forecast is

$$\hat{h}_t(2) = \alpha_0 + \alpha_1 \hat{h}_t(1) + \cdots + \alpha_p e_{t-p+2}^2.$$

Upon generalizing, the s-step ahead forecast is

$$\hat{h}_t(s) = \alpha_0 + \sum_{i=1}^{p} \alpha_i \hat{h}_t(s-i),$$

where $\hat{h}_t(s-i) = e_{t+s-i}^2$, if $s - i \leq 0$.

10.8.2 Forecasting a GARCH Model

Forecasts of a GARCH model can be obtained using methods similar to those of an ARMA model. Let us illustrate the steps with a GARCH(1,1) model. We continue to assume that the conditional mean is just a constant and hence we concentrate on getting the forecasts of the variance equation. For one-step ahead, we have the GARCH(1,1) model

$$h_{t+1} = \alpha_0 + \alpha_1 e_t^2 + \beta_1 h_t.$$

Since e_t^2 and h_t are known, the forecast is simply

$$\hat{h}_t(1) = \alpha_0 + \alpha_1 e_t^2 + \beta_1 h_t.$$

For multistep ahead forecasts, we use the fact that $e_t^2 = h_t \upsilon_t^2$, and rewrite the volatility equation for two-step ahead as

$$h_{t+2} = \alpha_0 + (\alpha_1 + \beta_1) h_{t+1} + \alpha_1 h_{t+1}(\upsilon_{t+1}^2 - 1).$$

Since $E(\upsilon_{t+1}^2 - 1 | I_t) = 0$, the two-step ahead forecast is given by

$$\hat{h}_t(2) = \alpha_0 + (\alpha_1 + \beta_1)\hat{h}_t(1).$$

In general, we have

$$\hat{h}_t(s) = \alpha_0 + (\alpha_1 + \beta_1)\hat{h}_t(s-1), \quad s > 1.$$

With repeated substitutions, we rewrite the s-step ahead forecast as

$$\hat{h}_t(s) = \frac{\alpha_0[1 - (\alpha_1 + \beta_1)^{s-1}]}{1 - \alpha_1 - \beta_1} + (\alpha_1 + \beta_1)^{s-1}\hat{h}_t(1).$$

Therefore

$$\hat{h}_t(s) \to \frac{\alpha_0}{1 - \alpha_1 - \beta_1}, \quad \text{as} \quad s \to \infty,$$

provided $\alpha_1 + \beta_1 < 1$. Consequently, the multistep ahead forecasts of a GARCH(1,1) model converge to the unconditional variance of e_t as forecast horizon goes to infinity, provided it exists.

10.9 Integrated GARCH Models

A class of models propounded by Engle and Bollerslev (1986), with the property *persistence variance*, in which the current information remains important for forecasting conditional variances for all horizons, is the IGARCH(p, q) model.

We obtain the integrated GARCH model when

$$\sum_{i=1}^{p} \alpha_i + \sum_{i=1}^{p} \beta_i = 1.$$

In finance, a question often asked is, how long does a shock to conditional variance persist? A long persistence implies that the entire structure of risk premia may be affected. z_t, however, is not covariance stationary anymore. To illustrate, consider an IGARCH(1,1) model given by

$$h_t = \alpha_0 + \alpha_1 e_{t-1}^2 + \beta_1 h_{t-1},$$

where $\alpha_1 + \beta_1 = 1$. Given that $e_t^2 = \upsilon_t^2 h_t$, we can rewrite the IGARCH(1,1) model:

$$h_t = \alpha_0 + h_{t-1}\left[(1 - \alpha_1) + \alpha_1 \upsilon_{t-1}^2\right].$$

We can write the s-step forecast for h_t as

$$h_t(s) = (s-1)\alpha_0 + h_{t-1}, \quad s \geq 1.$$

One may recall that this model closely resembles that of the forecast of a random walk model with drift. Thus, the effect of h_{t-1} is persistent, even with increasing horizon.

RiskMetrics (see Longerstaey and Spencer (1996)) developed by J.P. Morgan assumes an IGARCH (1,1) for its value at risk (VaR) calculations. VaR is a measure of prospective loss. What does it say? VaR is a single number that summarizes an institution's exposure to risk. For a time horizon s and confidence level p, the value at risk is the loss in market value over the time horizon s that is exceeded with probability $(1 - p)$. Normally, p is either 99% or 95%, though there is no uniform practice. For example, if a bank says that VaR on its daily trading portfolio is USD35 millions at the 99% confidence level, it means that there is a 1% chance that the loss exceeds USD35 millions. But we have to keep in mind that VaR is a prediction concerning possible loss of a portfolio in a given time horizon. So it should be computed using the predictive distribution of future returns of financial returns. But predictive distribution is hard to get. An IGARCH(1,1) specification provides us with a convenient way to calculate the predictive distribution.

RiskMetrics's main assumption is that the continuously compounded returns of a portfolio follow a conditional normal distribution. Let r_t be the daily log return of a financial asset. Information set at time $t - 1$ is I_{t-1}. So RiskMetrics assumes that

$$r_t | I_{t-1} \sim N(\mu_t, h_t).$$

Also, RiskMetrics assumes that

$$\mu_t = 0, \quad h_t = \alpha h_{t-1} + (1 - \alpha) r_{t-1}^2, \quad 1 > \alpha > 0.$$

So, if $log P_t = p_t$ then $r_t = p_t - p_{t-1}$. And for the current setup, $r_t = e_t$. With $e_t = \sqrt{h_t}\upsilon_t$, it follows that h_t follows an IGARCH(1,1) process, without drift. The value of α is often in the interval (0.9,1) with a typical value being 0.94.

10.9.1 Advantage of IGARCH

The main advantage is that the conditional distribution of a multiperiod return is easily available. Suppose we have a k-period horizon. The log return from time $t + 1$ and time $t + k$ is

$$r_t(k) = r_{t+1} + r_{t+2} + \cdots + r_{t+k}.$$

For this model

$$r_t(k) | I_t \sim N\left(0, h_t(k)\right).$$

It is easy to show that the one-step ahead forecast of the variance equation is

$$h_{t+1} = \alpha h_t + (1 - \alpha) r_t^2$$

and

$$\text{var}(r_{t+i} | I_t) = h_{t+1}, \quad i \geq 1,$$

and hence

$$h_t(k) = k\, h_{t+1}.$$

Thus, the results show that

$$r_t(k) | I_t \sim N(0, k\, h_{t+1}),$$

so that, under IGARCH(1,1) model, the conditional variance of $r_t(k)$ is proportional to the time horizon k and the conditional standard deviation is $\sqrt{k\, h_{t+1}}$.

If the probability is set to 5%, RiskMetrics uses $1.65\sqrt{h_{t+1}}$. It uses only one-sided 5% quantile of a normal distribution with mean zero and standard deviation $\sqrt{h_{t+1}}$. (The actual 5% quantile is $-1.645\sqrt{h_{t+1}}$; but the negative sign is ignored with the understanding that it signifies a loss.) Thus, if the standard deviation is measured in percentage, then the daily VaR of the portfolio under the RiskMetrics is

$$\text{VaR} = \text{amount of exposure} \times 1.65\sqrt{h_{t+1}}.$$

Similarly, that of k-day horizon is

$$\text{VaR} = \text{amount of exposure} \times 1.65\sqrt{k\, h_{t+1}}.$$

This is referred to as the square root of time rule in the VaR calculation under RiskMetrics.

For example, suppose the sample standard deviation of the continuously compounded daily return of the exchange rate between Euro and USD is 0.25%. For a long exposure of, with 10 million USD worth of Euro/USD exchange contract, the 5% VaR for a 1-day horizon is $10{,}000{,}000 \times (1.65 \times 0.0025) = \$41{,}250$. This actually means that 95% of the times, one will not lose more than \$41,250 in the next 24 hours. And the corresponding VaR for a 1-month (30-day) horizon is $10{,}000{,}000 \times \left(\sqrt{30} \times 1.65 \times 0.0025\right) \approx \$225{,}935$.

10.10 Asymmetric GARCH Models

While GARCH addressed the problem of long lag structure, the important problem of asymmetry and the nonnegativity restrictions on the parameters needed to be addressed. Asymmetry refers to larger increases in volatility to bad news than to positive news of the same magnitude. Many parameterizations have been proposed in the literature to capture this important empirical regularity, chief among them being the EGARCH model of Nelson and the GJR (Glosten, Jagannathan, Runkle)–GARCH model or the threshold GARCH (TGARCH) model.

10.10.1 EGARCH Model

Nelson (1991) proposes a different functional form to model conditional heteroscedasticity, which, he says, will take care of some of the limitations of GARCH process of Bollerslev. He proposes an *exponential* ARCH model, also called an EGARCH model. Since h_t clearly must be nonnegative, Nelson makes log of h_t linear in some function of time and lagged υ's. That is, for some suitable function, g,

$$\ln h_t = \alpha_t + \sum_{k=1}^{\infty} \beta_k\, g\left(\upsilon_{t-k}\right) \quad \text{with } \beta_1 \equiv 1. \qquad \text{(DBN-1)}$$

To accommodate the asymmetric relation between stock returns and volatility changes, Nelson makes the value of $g(\upsilon_t)$ as a function both of the sign and magnitude of υ_t by selecting the following specification:

$$g\left(\upsilon_t\right) = \theta\upsilon_t + \gamma\left[\mid \upsilon_t \mid - E \mid \upsilon_t \mid\right]. \qquad \text{(DBN-2)}$$

(DBN-1) and (DBN-2) together are called the *exponential* GARCH model. Here, θ captures the effect of the sign of the error and γ captures the magnitude of the error. That this functional form allows the conditional variance to respond asymmetrically to rises and falls in stock prices can easily be seen by rewriting:

$$g(\upsilon_t) = \begin{cases} (\theta+\gamma)\upsilon_t - \gamma E(|\ \upsilon_t\ |) & \text{if } \upsilon_t \geq 0, \\ (\theta-\gamma)\upsilon_t - \gamma E(|\ \upsilon_t\ |) & \text{if } \upsilon_t < 0. \end{cases} \qquad \text{(DBN-3)}$$

In this setup, an apparent finding of $\theta < 0$ is called the *leverage* effect according to finance literature, implying that negative shocks or bad news have a larger impact on volatility than positive shocks or good news. If $\theta = 0$, then there is no asymmetry, while $\theta > 0$ will lead to the strange result of positive news implying a stronger impact on volatility than negative news. In practice, a parsimonious version of (DBN-1) is often employed:

$$\ln h_t = \alpha_t + \frac{(1+\theta_1 L + \cdots + \theta_q L^q)}{(1-\phi_1 L - \cdots - \phi_p L^p)} g(\upsilon_{t-1}).$$

Here, we assume that $(1+\theta_1 L + \cdots + \theta_q L^q)$ and $(1-\phi_1 L - \cdots - \phi_p L^p)$ have no common roots. So, for $\ln h_t$ to be stationary, the roots of the AR parameter should lie outside the unit circle. To ensure the existence of finite unconditional moments of an arbitrary order, Nelson recommends a commonly used family of distributions called the generalized error distribution (GED) for the innovations υ_t. GED includes as a special case, the normal distribution, along with many other distributions. Nelson proposed using the GED, normalized to have zero mean and unit variance, and the density of such a normalized GED variable is given as follows:

$$f(\upsilon) = \frac{\nu \exp[-(1/2)\ |\ \upsilon/\lambda\ |^{\nu}]}{\lambda\, 2^{(1+1/\nu)}\, \Gamma(1/\nu)}, \qquad \text{(DBN-4)}$$

where $-\infty < \upsilon < \infty$, and $0 < \nu < \infty$ and where $\Gamma(\cdot)$ is the gamma function and

$$\lambda = \left[2^{(-2/\nu)} \Gamma(1/\nu)/\Gamma(3/\nu)\right]^{1/2},$$

where ν is the tail thickness parameter. When $\nu = 2$, then υ has a standard normal distribution. For $\nu < 2$, the distribution of υ has thicker tails than the normal.

One can note that the expected absolute value of a variable drawn from a distribution given in (DBN-4) is given by

$$E\,|\upsilon_t| = \frac{\lambda\, 2^{1/\nu}\ \Gamma\,(2/\nu)}{\Gamma\,(1/\nu)}.$$

For the standard normal case, ($\nu = 2$), this becomes

$$E(|\ \upsilon_t\ |) = \sqrt{2/\pi}.$$

The exponential form of the EGARCH can be explained with an example. Let us take the (1,0) model for illustration. And let $\alpha_t = \alpha_0$, for simplicity.

$$e_t = \upsilon_t \sqrt{h_t},$$

$$(1-\phi_1 B)\ln(h_t) = (1-\phi_1)\,\alpha_0 + g\,(\upsilon_{t-1}),$$

where we assume that the υ_t's are standard normal, so that, $E(|\upsilon_t|) = \sqrt{2/\pi}$. With this, the model for $\ln(h_t)$ becomes

$$h_t = h_{t-1}^{\phi_1} \exp(\alpha_*) \exp(\theta + \gamma) \frac{e_{t-1}}{\sqrt{h_{t-1}}}, \quad \text{if } \upsilon_{t-1} \geq 0,$$

$$h_t = h_{t-1}^{\phi_1} \exp(\alpha_*) \exp(\theta - \gamma) \frac{e_{t-1}}{\sqrt{h_{t-1}}}, \quad \text{if } \upsilon_{t-1} < 0,$$

where $\alpha_* = (1 - \phi_1)\,\alpha_0 - \sqrt{2/\pi}\gamma$.

The coefficients $(\theta + \delta)$ and $(\theta - \delta)$ show the asymmetry in response to positive and negative υ_{t-1}. Rewriting higher-order models this way, however, is too difficult.

As an illustration, we shall show how to model EGARCH using Nelson's idea. Nelson also models the contribution of nontrading days to market variance, where he assumes that each nontrading day contributes as much to the variance as some fixed fraction of a trading day. He expects the coefficient to be between 0 and 1. He also allows the conditional variance to be a factor in deciding the returns, so that he essentially builds an EGARCH–M model. Excess returns—defined as the difference between the returns on a particular asset of interest and a risk less asset, usually T-bills—are modeled by him as follows:

$$r_t = a + br_{t-1} + \delta h_t + e_t.$$

The residual is modeled as $e_t = \upsilon_t\sqrt{h_t}$, where υ_t is an i.i.d. random variable from a GED distribution whose density is given in (DBN-4) and h_t evolves according to the following ARMA(1,1) model:

$$(\ln h_t - \alpha_t) = \phi_1(\ln h_{t-1} - \alpha_{t-1}) + g(\upsilon_{t-1}) + \theta_1 g(\upsilon_{t-2}).$$

Nelson allows α_t, the unconditional mean of $\ln(h_t)$, to be time-dependent:

$$\alpha_t = \alpha + \ln(1 + \rho N_t),$$

where N_t is the number of nontrading days between t and $t-1$. α and ρ are parameters that are to be estimated along with the other parameters, by maximum likelihood estimation. The sample log-likelihood is given by

$$L(\cdot) = \sum_{t=1}^{T} \ln\left(\frac{\nu}{\lambda}\right) - \frac{1}{2}\left|(r_t - a - br_{t-1} - \delta h_t)/\lambda \cdot \sqrt{h_t}\right|^{\nu} - (1 + \nu^{-1})\ln(2)$$
$$- \ln\left[\Gamma\left(\frac{1}{\nu}\right)\right] - \frac{1}{2}\ln h_t.$$

The sequence of h_t, $t = 1, \ldots, T$, is calculated from the above expression for $\ln(h_t)$ with the presample values of $\ln(h_t)$ set to their unconditional mean, α_t, and setting, υ_t, as

$$\upsilon_t = (r_t - a - br_{t-1} - \delta h_t)/\sqrt{h_t}.$$

10.10.2 GJR–GARCH Model

Another GARCH model recommended by Glosten et al. (1993) also captures the asymmetry effect and is popular among researchers. It is called the GJR–GARCH model or the threshold GARCH (TGARCH) model. A simple GJR–GARCH(1,1) is defined as follows:

$$h_t = \alpha_0 + \beta h_{t-1} + \alpha_1 e_{t-1}^2 + \beta^* e_{t-1}^2 \cdot S_{t-1},$$
$$S_{t-1} = 1, \quad \text{if } e_{t-1} \geq 0,$$
$$S_{t-1} = 0, \quad \text{if } e_{t-1} < 0.$$

In this specification, β^* should be negative for the leverage effect. (Note that alternatively, β^* will be positive, if we define the indicator function as $S_{t-1} = 1$, if $e_{t-1} \leq 0$, and $S_{t-1} = 0$, if $e_{t-1} > 0$). Additionally, α_1, β are nonnegative parameters, like in the GARCH model. From the model, it is seen that a positive e_{t-1} contributes $(\alpha_1 + \beta^*)$ to h_t, whereas a negative e_{t-1} contributes α_1 to h_t. Note that the model uses zero as the *threshold* value to separate the impacts of past shocks. Other threshold values can also be used.

This process is covariance stationary if

$$\alpha_1 + \beta^*/2 + \beta < 1,$$

and hence the unconditional variance of e_t is

$$\text{var}(e_t) = \alpha_0 / \left[1 - \alpha_1 + \beta + \beta^*/2\right].$$

10.10.3 An Alternative Form of the EGARCH Model

EGARCH model can be written in many forms. One such form is given below (see Tsay (2005)).

Assuming that $\upsilon_t \sim N(0, 1)$, let

$$g^*(\upsilon_{t-i}) = \left(|\upsilon_{t-i}| - \sqrt{\frac{2}{\pi}}\right) + \theta_i \upsilon_{t-i}, \quad i = 1, 2, \ldots.$$

The volatility model is given as

$$\ln h_t = \alpha_0 + \sum_{i=1}^{s} \alpha_i \frac{|e_{t-i}| + \theta_i e_{t-i}}{\sqrt{h_{t-i}}} + \sum_{j=1}^{n} \phi_j \ln h_{t-j}.$$

What will a positive e_{t-i} mean? A positive e_{t-i} contributes $\alpha_i(1 + \theta_i) \cdot |\upsilon_{t-i}|$ to $\ln(h_t,)$ whereas a negative e_{t-i} contributes $\alpha_i(1 - \theta_i) \cdot |\upsilon_{t-i}|$ to $\ln(h_t,)$ where $\upsilon_t = e_t/\sqrt{h_t}$. And the parameter θ_i signifies the leverage or asymmetry effect of e_{t-i} and is expected to be negative.

10.10.4 An Application

We next demonstrate how to interpret the GARCH, EGARCH, and GJR–GARCH models with the returns calculated from the Nifty index described earlier. We estimate the EGARCH model assuming a GED for the innovations υ_t and normal distribution for the other two models. Table 10.5 displays the results.

TABLE 10.5

Parameter Estimates of the Variance Equations—Standard GARCH Models

	Estimation Results
GARCH(1,1)	$h_t = \underset{(3.153\times 10^{-6})}{1.425\times 10^{-5}} + \underset{(0.021)}{0.158}\, e_{t-1}^2 + \underset{(0.023)}{0.799}\, h_{t-1}$
EGARCH(1,1)	$\ln h_t = \left[\underset{(0.55)}{-8.578} + \ln(1 + \underset{(0.005)}{0.238}\, N_t)\right] + \left[\dfrac{1 + \underset{(0.197)}{0.220}\, B}{1 - \underset{(0.046)}{0.815}\, B}\right] g(\upsilon_{t-1})$ $g(\upsilon_t) = \underset{(0.026)}{-0.141}\, \upsilon_t + \underset{(0.038)}{0.288}\left[\lvert\upsilon_t\rvert - E\lvert\upsilon_t\rvert\right]$
GJR–GARCH	$h_t = \underset{(4.089\times 10^{-6})}{2.196\times 10^{-5}} + \underset{(0.027)}{0.755}\, h_{t-1} + \underset{(0.016)}{0.053}\, e_{t-1}^2 + \underset{(0.038)}{0.232}\, S_{t-1}^{-} e_{t-1}^2$

Note: The table displays estimates of the simple GARCH(1,1) models. Residual series $\mathbf{e}_t$ was obtained after running an AR(1) model on the returns data. Asymptotic, robust standard errors are given in parentheses.

In Table 10.5, the coefficient estimate of the variable, υ_t, the variable that captures asymmetry or the leverage in the EGARCH model, is highly significant and *negative.* Similarly, the parameter corresponding to the $S_{t-1}^{-} e_{t-1}^2$ term is positive and highly significant in the GJR model. These results are consistent with the notion that negative shocks make the conditional variance more volatile than positive shocks of the same magnitude, the leverage effects. Another well-known fact that stock returns have fatter tails was also confirmed with the tail thickness parameter, ν, in the EGARCH model estimated to be 1.417 with a standard error of 0.0471. A random variable drawn from a GED has a distribution with thicker tails if $\nu < 2$; and hence this result from EGARCH clearly confirms that returns have significantly fatter tails. Another feature that has been captured by the EGARCH specifically is the role of nontrading days. This feature is very much relevant in the Indian context and our estimate corresponding to the variable N_t clearly shows that nontrading days account for a significant effect on the return variance. The estimate of 0.238 clearly shows that a nontrading day contributes nearly one-fourth as much to the volatility as any trading day. We reckon this to be an important result hitherto being ignored in Indian studies. Results in this table confirm that standard GARCH models may not be adequate enough to capture the returns volatility, because of the presence of asymmetry and fatter tails. And the role of nontrading days cannot be ignored.

10.11 News Impact Curve

A handy tool employed by Pagan and Schwert (1990) and Engle and Ng (1993) to compare the various asymmetric models is the *news impact curve*, which plots the relation between the conditional volatility and the shocks. Using the news impact curve, Engle and Ng compare the various asymmetric volatility models and conclude that the models differ in the way they accommodate asymmetry. To explain this phenomenon, they categorize the return shocks or error term of the mean equation as collectively measuring *news*. So a positive e_t—an unexpected increase in price—suggests the arrival of some good news, while a negative e_t—an unexpected decrease in price—denotes the arrival of some bad news. Similarly, a large value of $|e_t|$ implies that the news is "significant" or "big" in the sense that it produces a large unexpected change in price.

Recall Engle's ARCH(p) model:

$$h_t = \alpha_0 + \sum_{i=1}^{p} \alpha_i e_{t-i}^2,$$

where the parameter α_i measures the impact of the shock i period ago, $(i \leq p)$. Generally, $\alpha_i < \alpha_j$, $i > j$. That is, the older the news, the less will be the impact on current volatility. This means, in an ARCH(p) model, news that arrived at the market more than p periods ago has no effect at all on current volatility. Similarly, Bollerslev recommended the GARCH(p,q) specification to tackle the long lag order structure of an ARCH model. In most cases, GARCH(1,1) is the preferred model. Despite the popularity of these two models, they do not address an important stylized fact of financial returns, namely, the asymmetry effect. To recall, asymmetry means unexpected bad news increases predictable (conditional) volatility more than an unexpected good news of similar magnitude. So, Nelson proposed the EGARCH model to accommodate such asymmetric effects. We state below, assuming that υ_t follows a normal distribution, the simplest EGARCH(1,0) model:

$$\ln(h_t) = \alpha_0 + \phi \ln(h_{t-1}) + \theta \frac{e_{t-1}}{\sqrt{h_{t-1}}} + \gamma \left[\frac{|e_{t-1}|}{\sqrt{h_{t-1}}} - \sqrt{\frac{2}{\pi}} \right].$$

Engle and Ng suggest a diagnostic tool to evaluate how these models capture the effect of news on volatility. The diagnostic tool suggested by them is called the *news impact curve*, NIC. Evaluating all lagged conditional variances at the level of the unconditional variance (say, σ^2) of the return equation or equally, the mean equation, and holding constant all information dated earlier than and from e_{t-2}, NIC examines the implied relation between e_{t-1} and h_t. This curve measures how new information is incorporated into the volatility estimates. In particular, for a GARCH(1,1) model, the NIC is a quadratic function and symmetric, centered on $e_{t-1} = 0$, or equivalently, has a minimum at $e_{t-1} = 0$, which can be easily seen by writing the equation for the GARCH(1,1) model news impact curve as follows, under the current assumptions:

$$h_t = A + \alpha_1 \cdot e_{t-1}^2, \quad \text{where } A = \alpha_0 + \beta_1 \sigma^2.$$

Similarly for the EGARCH, it has its minimum at $e_{t-1} = 0$, and is exponentially increasing in both directions, but with different parameters. For example, for the EGARCH(1,0) model, assuming that υ_t is from a normal distribution, we have the equation for the news impact curve given as follows:

$$h_t = A \cdot \exp\left[\frac{(\theta + \gamma)}{\sigma} \cdot e_{t-1}\right] \quad \text{for } e_{t-1} > 0,$$

$$h_t = A \cdot \exp\left[\frac{(\theta - \gamma)}{\sigma} \cdot e_{t-1}\right] \quad \text{for } e_{t-1} < 0,$$

$$A = (\sigma^2)^{\phi} \cdot \exp\left[\alpha_0 - \gamma \cdot \sqrt{\frac{2}{\pi}}\right].$$

We have plotted below NIC for GARCH and EGARCH models and we can see that for the GARCH model it is a quadratic curve centered at $e_{t-1} = 0$, whereas for the EGARCH model, it has a minimum at $e_{t-1} = 0$ and is exponentially increasing in both directions but with different parameters.

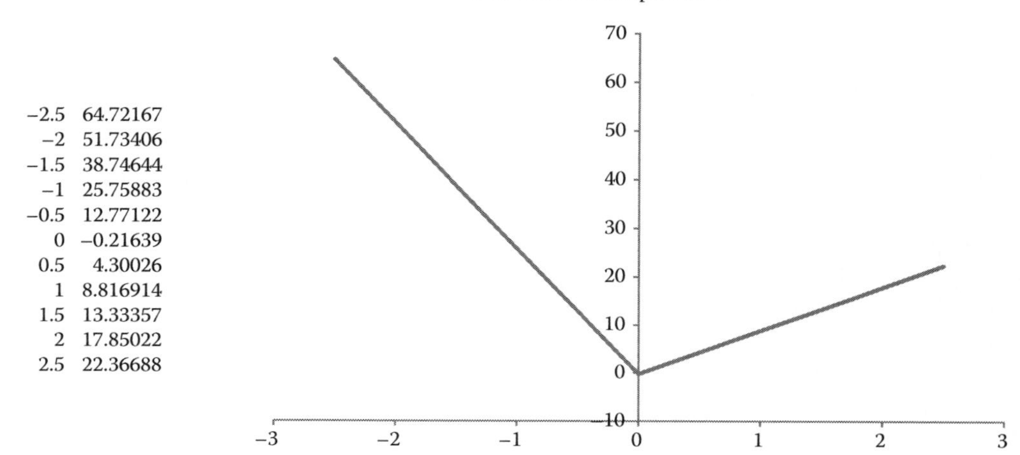

−2.5	64.72167
−2	51.73406
−1.5	38.74644
−1	25.75883
−0.5	12.77122
0	−0.21639
0.5	4.30026
1	8.816914
1.5	13.33357
2	17.85022
2.5	22.36688

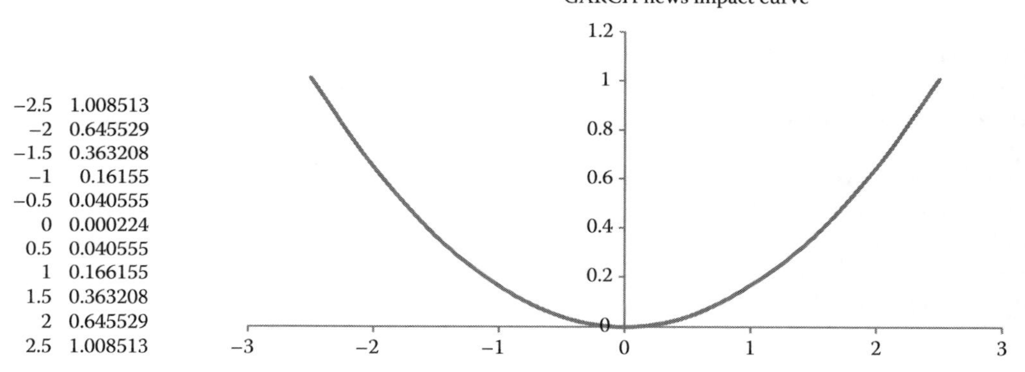

−2.5	1.008513
−2	0.645529
−1.5	0.363208
−1	0.16155
−0.5	0.040555
0	0.000224
0.5	0.040555
1	0.166155
1.5	0.363208
2	0.645529
2.5	1.008513

Similarly, the equation for news impact curve for the AGARCH(1,1) model is

$$h_t = A + \alpha_1 \cdot (e_{t-1} + \gamma)^2,$$

where $A = \alpha_0 + \beta\sigma^2$. And the equations for the GJR news impact curve are

$$h_t = A + \alpha_1 e_{t-1}^2, \qquad \text{for } e_{t-1} < 0$$

$$h_t = A + (\alpha_1 + \beta^*) e_{t-1}^2, \qquad \text{for } e_{t-1} > 0$$

where $A = \alpha_0 + \beta\sigma^2$.

Hence, implicit in any volatility model is a particular news impact curve. We have listed below some alternative predictable volatility models recommended in the literature, and listed in Engle and Ng (1993), to capture time-varying heteroscedasticity. The shape of the news impact curve is also given.

10.12 Some Alternative Predictable Volatility Models

Nonlinear ARCH model

$$h_t = \alpha_0 + \alpha_1 |e_{t-1}|^{\gamma} + \beta h_{t-1}.$$

Multiplicative ARCH model

$$\log(h_t) = \alpha_0 + \alpha_1 \log(e_{t-1}^2).$$

GJR model

$$h_t = \alpha_0 + \alpha_1 e_{t-1}^2 + \beta h_{t-1} + \beta^* S_{t-1}^- e_{t-1}^2, \quad \text{where}$$

$$S_{t-1}^- = 1 \text{ if } e_{t-1} \text{ is } \geq 0, \quad S_{t-1}^- = 0 \quad \text{if} \quad e_{t-1} \text{ is } < 0.$$

EGARCH model

$$\ln(h_t) = \alpha_0 + \phi \ln(h_{t-1}) + \theta \frac{e_{t-1}}{\sqrt{h_{t-1}}} + \gamma \left[\frac{|e_{t-1}|}{\sqrt{h_{t-1}}} - \sqrt{\frac{2}{\pi}} \right].$$

Autoregressive standard deviation model

$$h_t = [\alpha_0 + \alpha_1 |e_{t-1}|]^2.$$

Asymmetric GARCH model

$$h_t = \alpha_0 + \alpha_1 (e_{t-1} + \gamma)^2 + \beta h_{t-1}.$$

Nonlinear asymmetric GARCH model

$$h_t = \alpha_0 + \beta h_{t-1} + \alpha_1 (e_{t-1} + \gamma \sqrt{h_{t-1}})^2.$$

VGARCH model

$$h_t = \alpha_0 + \beta h_{t-1} + \alpha_1 (e_{t-1}/\sqrt{h_{t-1}} + \gamma)^2.$$

- The news impact curve of the nonlinear model of Engle and Bollerslev is symmetric.
- The news impact curve of the multiplicative model of Milhoj, Geweke, and Pantula is symmetric and passes through the origin. Depending on the value of α_i's, the two sides of the news impact curve can be either steeper or less steep than the GARCH(1,1) news impact curve.
- The autoregressive standard deviation model of Schwert has a news impact curve that is symmetric and centered at $e_{t-1} = 0$.
- The news impact curve of asymmetric GARCH model of Engle, AGARCH is centered at $e_{t-1} = -\gamma$, which is to the right of the origin for when $\gamma < 0$.
- The news impact curve of GJR model is centered at $e_{t-1} = 0$, but has different slopes for its positive and negative sides.

- The news impact curve of both the nonlinear asymmetric GARCH model (NGARCH) and the VGARCH model is symmetric and centered at $e_{t-1} = (-\gamma) \cdot \sqrt{h_{t-1}}$. However, the slope of the two upward sloping portions of the VGARCH is steeper than that of the NGARCH model.

What is to be noted is that the news impact curves of some asymmetric models capture asymmetry or leverage, either by allowing the news impact curves to be *tilted* or *rotated*, that is, allow the slope of the two sides of the news impact curve to differ, or by shifting the center of the news impact curve to locate at a point, where e_{t-1} is positive. Such differences have important implications for portfolio selection and asset pricing. Since an unexpected drop in prices will now give rise to two different predictable volatilities given by, say, a GARCH model and an EGARCH model, we will obtain two very different market premiums, and hence two different risk premiums for individual stocks under a conditional version of the CAPM.

To summarize, this discussion clearly underlines the importance of understanding the impact of news on volatility and Engle and Ng in their paper have proposed some simple diagnostic tests to test if the proposed conditional volatility models have accounted properly for the asymmetric effect of news.

10.13 Diagnostic Tests Based on News Impact Curves: Test of Asymmetry

To test if our conditional volatility models have properly accounted for news, Engle and Ng recommend the following three diagnostic tests:

- Sign bias test
- Negative size bias test
- Positive size bias test

These tests enable us to check if any variable not included in the volatility model can help predict squared normalized residuals. If so, then the variance equation or model is misspecified.

Sign bias test: This test considers a dummy variable S_{t-1}^{-} that takes a value of one when e_{t-1} is negative and zero if it is positive. This test just examines the impact of positive and negative return shocks on the volatility not predicted by the model under consideration.

Negative size bias test: This test considers the variable, $S_{t-1}^{-}e_{t-1}$, that focuses on the different effects that large and small negative return shocks have on volatility, but not accounted for by the volatility model under consideration.

Positive size bias test: This test considers the variable, $S_{t-1}^{+}e_{t-1}$, where S_{t-1}^{+} is defined as 1 minus S_{t-1}^{-}. It focuses on the different impacts of large and small positive return shocks on volatility, but not accounted for by the volatility model under consideration.

10.13.1 How to Conduct These Tests?

- After estimating the desired conditional volatility model, calculate the standardized shocks, $e_t^* = e_t/\sqrt{h_t}$.

- Run the following three regressions:

$$(1) \quad (e_t^*)^2 = a + b\, S_{t-1}^- + \epsilon_t$$

$$(2) \quad (e_t^*)^2 = a + b\, S_{t-1}^- e_{t-1} + \epsilon_t$$

$$(3) \quad (e_t^*)^2 = a + b\, S_{t-1}^+ e_{t-1} + \epsilon_t$$

- The t ratio for the coefficient b, t_b, in regressions (1), (2), and (3) above defines, respectively, the *sign bias test*, *negative sign bias test*, and *positive sign bias test.*
- Engle and Ng also suggest to conduct these tests jointly through the following regression:

$$(4) \quad (e_t^*)^2 = a + b_1\, S_{t-1}^- + b_2\, S_{t-1}^- e_{t-1} + b_3\, S_{t-1}^+ e_{t-1} + \epsilon_t$$

- The joint test is an LM test and it follows a chi-square distribution with three degrees of freedom. And this test statistic is calculated as T times the R^2 of regression (4). If the volatility model being used is correct, then $b_1 = b_2 = b_3 = 0$.

Table 10.6 displays the diagnostic test results conducted on some predictable volatility models. In the table, the Ljung–Box test was done on 12 lags of squared normalized residuals. None is significant, pointing out that the lower-order models are reliable. We also conducted the sign bias test as recommended by Engle and Ng based on models identified under restrictions $b = \theta = 0$. We see that most of the models capture the news effect reasonably well except GARCH. The joint test statistic, distributed as $\chi^2(3)$, also indicates that many of the models can reasonably capture the impact of news on volatility. The only asymmetric model that fails the joint test is the nonlinear ARCH model.

10.14 Nesting the GARCH Family of Models

Looking at the functional forms used by the various asymmetrical volatility models listed out earlier and the discussion about the news impact curve, the two aspects that stand out are:

1. There are models that quantify conditional volatility by tracing the evolution of conditional variance and there are models that use conditional standard deviation for the same purpose.
2. There are models that also differ in the way they capture asymmetry. Some do so by allowing the news impact curve to be shifted and the rest by rotating it.

All these suggest the need for a more rigorous approach than available presently, to narrow down the correct model or family of models that the stock returns obey. One possible approach is to *nest* the family of GARCH models listed above within a *general* model and use the classical log-likelihood ratio (LR) test to select a proper model. Hentschel (1995) suggests using the Box–Cox rule to generalize the *absolute value GARCH model* and allow powers of $f(\upsilon_t)$ and $\sqrt{h_t}$ to describe the conditional variance equation.

An absolute value GARCH model is also recommended sometimes as an alternative to the traditional GARCH models. It can be stated as

$$e_t = \upsilon_t \sqrt{h_t},$$

$$\sqrt{h_t} = \alpha_0 + \alpha_1 \sqrt{h_{t-1}} |\upsilon_{t-1}| + \beta \sqrt{h_{t-1}}.$$

TABLE 10.6

Diagnostic Test Results

	Diagnostic Statistics				
Model	**Ljung–Box (12)**	**Sign Bias**	**Negative Bias**	**Positive Bias**	**Joint Test**
GARCH	8.924	0.573	−2.034*	−1.387	16.952
	(0.709)				(0.001)
EGARCH	16.016	0.418	0.123	−0.914	2.291
	(0.191)				(0.514)
GJR	13.925	0.946	−0.054	−0.279	2.666
	(0.301)				(0.446)
NAGARCH	13.178	−0.201	−1.224	−0.696	3.200
	(0.356)				(0.362)
TGARCH	18.238	−0.229	−0.125	−0.946	1.132
	(0.109)				(0.769)
AGARCH	15.301	0.873	0.531	−0.443	1.946
	(0.225)				(0.584)
NARCH	8.785	0.785	−0.898	−1.950	15.014
	(0.721)				(0.002)
APARCH	15.383	0.470	0.416	−0.977	2.365
	(0.221)				(0.500)

Note: *indicates significance at the five percent level. Column 2 of the table gives the estimated Ljung–Box Q statistic for 12 lags, along with the asymptotic probability values in parentheses. These were calculated with the squared normalized residuals obtained after estimating the models listed in column 1 and are distributed as $\chi^2(12)$ random variables. Columns 3, 4, and 5 give the calculated Student t ratios for the coefficient b, in regressions (1), (2), and (3) explained in the text. The coefficient b from these regressions indicates if the models in column 1 have captured successfully the asymmetric effects of news on volatility. The last column gives the calculated LM statistic, obtained after estimating equation 4 in the text, for a joint test of the three biases listed in columns 3, 4, and 5. It is distributed as a χ^2 (3) random variable. Asymptotic probability values are displayed within parentheses.

But to handle asymmetry, the absolute value GARCH can be generalized as follows:

$$\sqrt{h_t} = \alpha_0 + \alpha_1\sqrt{h_{t-1}}f(\upsilon_{t-1}) + \beta\sqrt{h_{t-1}},$$

$$f(\upsilon_t) = |\upsilon_t - b| - c(\upsilon_t - b).$$

Here, b is called the *shift* parameter and c the tilt parameter. These two parameters capture the two different ways asymmetry is measured. For example, if we restrict $b = c = 0$, we get the absolute value GARCH model. If we leave b free but restrict $c = 0$, we get a news impact curve that is shifted and if we restrict b but allow c to be free, we get impact curves that are tilted. We give below figures of a shifted and a tilted news impact curve.

The generalized version of the absolute value GARCH model recommended by Hentschel is as follows:

$$\frac{(h_t)^{\lambda/2} - 1}{\lambda} = \alpha_0' + \alpha_1(h_{t-1})^{\lambda/2}f^{\nu}(\upsilon_{t-1}) + \beta\frac{(h_{t-1})^{\lambda/2} - 1}{\lambda}.$$

The above model may be called the *general* or the *unrestricted* model and it nests many popular GARCH models. It is by restricting the values of λ, ν, b, and c in the general model that we are going to retrieve the various GARCH models that differ in the two aspects pointed above and hence this specification is advantageous. We shall give two examples:

1. Restricting $\lambda = \nu = 1$ will fetch us the absolute value GARCH model (4), with $\alpha_0 = \alpha_0' + 1 - \beta$.
2. Similarly, restricting $\lambda = \nu = 2$ and $b = c = 0$ will give us the standard GARCH specification given by

$$h_t = \alpha_0'' + 2\alpha_1 h_{t-1} f^2(\upsilon_{t-1}) + \beta h_{t-1}, \quad \text{where } \alpha'' = (2\alpha_0' + 1 - \beta). \tag{10.1}$$

Likewise, if we let b alone to be free, we get the nonlinear asymmetric GARCH model of Engle and Ng (1993). On the other hand, if we let c alone to be free, we can retrieve the GJR model. Similarly, if we restrict $\lambda = 0$, $\nu = 1$, and $b = 0$ but allow c to be free, then, using l'Hôpital's rule, we get the exponential GARCH model of Nelson. Thus, the idea is to estimate the general or the unrestricted model and the various nested GARCH models and compare the log-likelihood of the general and the restricted models to select a model.

We tested Hentschel's idea on the continuously compounded returns calculated using daily Nifty index of the NSE of India, for the period February 1997 to January 2006, containing 2252 observations. We use the LR test to pick among the asymmetric models that best describe the data. Owing to space constraints, we present here only the LR statistics. (More detailed results of a similar analysis are available in Krishnan and Mukherjee [2010].) We have calculated two sets of LR statistics. The first set displayed in Table 10.7 will help us decide (1) if asymmetry is definitely a feature of Indian stock returns; (2) whether such an asymmetry, if present, is captured better by a shift or rotation of the news impact curve or both affect the news impact curve. If the presence of asymmetry has been established, then the second set of LR statistics displayed in Table 10.8 will help us select a possible model from among the asymmetric family of GARCH models.

- Comparing the LR statistics in the first row of each panel of Table 10.4, against the chi-squared critical values with two degrees of freedom, the null of $b = c = 0$ is decisively rejected in favor of the alternative that some type of asymmetry is

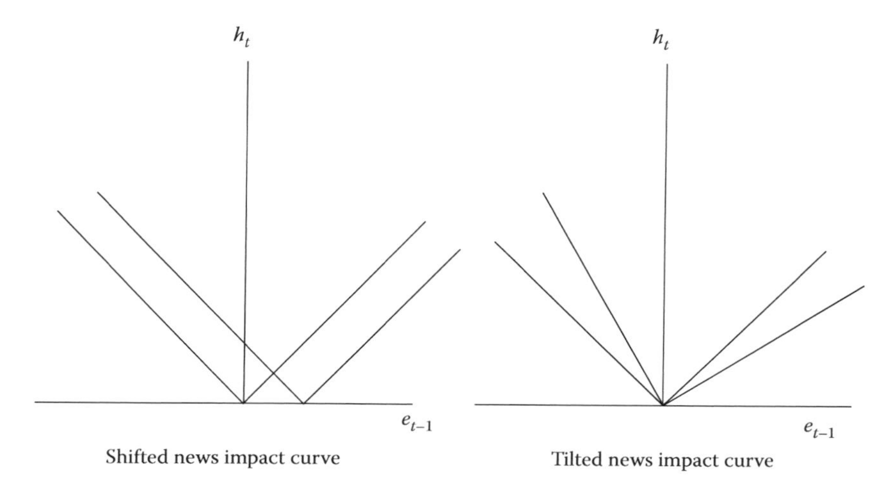

Shifted news impact curve Tilted news impact curve

present, confirming the presence of leverage effects in returns. So the Bollerslev type of symmetric GARCH model may not explain fully the volatility in the Indian stock market.

TABLE 10.7

Likelihood Ratio Statistics–I

		H_A		
Maintained hypothesis	H_0	$b = 0;\ c$ **free**	b **free**, $c = 0$	b, c **free**
$\lambda = \nu$	$b = c = 0$	30.993 (0.000)	39.367 (0.000)	40.206 (0.000)
	$b = 0$, c free			9.213 (0.002)
	$c = 0$, b free			0.839 (0.360)
$\lambda = \nu = 1$	$b = c = 0$	38.284 (0.000)	43.102 (0.000)	43.251 (0.000)
	$b = 0$, c free			4.967 (0.026)
	$c = 0$, b free			0.149 (0.700)
$\lambda = \nu = 2$	$b = c = 0$	32.363 (0.000)	33.117 (0.000)	35.293 (0.000)
	$b = 0$, c free			2.930 (0.087)
	$c = 0$, b free			2.176 (0.140)
$\lambda = 0$, $\nu = 1$	$b = c = 0$	29.978 (0.000)	38.608 (0.000)	39.463 (0.000)
	$b = 0$, c free			9.485 (0.002)
	$c = 0$, b free			0.855 (0.355)
λ free, ν free	$b = c = 0$	34.604 (0.000)	40.158 (0.000)	40.398 (0.000)
	$b = 0$, c free			5.794 (0.016)
	$c = 0$, b free			0.240 (0.624)

Note: The estimated likelihood ratio statistics, displayed in the table, test restrictions within each family of models used for estimation and are distributed as χ^2 random variables with either one or two degrees of freedom. For example, statistics in the third row of each panel reveal that we may reject those models that accommodate asymmetry by rotating the news impact curve. Associated asymptotic probability values are given in parentheses.

- We turn next to deciding if such an asymmetry can be accommodated by shifting or rotating the news impact curve and this will help us to select a model among the asymmetric family of models. The second and third rows in each panel display, respectively, LR statistics to test if the asymmetry can be fixed by either a shift or a rotation in the news impact curve, as against the alternative that it is caused both by rotation and shift. The insignificant statistics in the third row in all the panels clearly reject rotating the news impact

TABLE 10.8

Likelihood Ratio Statistics–II

	H_A	
H_0	$\lambda = \nu$	λ, ν free
EGARCH		1.65
($\lambda = 0$, $\nu = 1$)		(0.438)
AGARCH	9.620	9.98
($\lambda = \nu = 1$)	(0.008)	(0.006)
GARCH	17.370	17.730
($\lambda = \nu = 2$)	(0.000)	(0.000)
NARCH		0.360
($\lambda = \nu$)		(0.549)

Note: The estimated likelihood ratio statistics, displayed in the table, test restrictions among the four family of models used for estimation and are distributed as χ^2 random variables with either one or two degrees of freedom. Asymptotic probability values are in parentheses. This table helps us in deciding the family of models the data may belong to. For example, the statistics in column two and three clearly reject the null that the data conform to the GARCH or the AGARCH family.

curve to tackle asymmetry. This result rejects the popular GJR type of asymmetric models or the asymmetric power ARCH type of models recommended by Ding et al. (1993) for the Indian stock market data. But the results in the second row of each panel show that *shifts* in the news impact curve alone are enough to explain the asymmetry at the 5% level, except the GARCH family, which accepts the role of shift only at the 10% level. These findings support the observation that it is small shocks that contribute significantly to the asymmetry, contrary to the belief that only very large shocks matter in explaining asymmetry.

- Turning to the second set of LR statistics displayed in Table 10.5, we exploit the results and the LR statistics arising out of the models where the parameters b and c are freely estimated, to identify a probable family, to which the asymmetrical model belongs. First, comparing the results of the second column, we see that the data does not conform to the AGARCH and GARCH family. This is because, while it is possible that λ is equal to ν, they are neither equal to one nor two. Lastly, comparing the results in the last column, we see that the test results are not conclusive. AGARCH and GARCH are again rejected in favor a model that freely estimates the exponents; but choosing between the two remaining options is not clear. Models belonging both in the completely different EGARCH and the nonlinear GARCH family models seem to be possible candidate models, since the respective nulls are clearly accepted. Because of the inherent model specification, we cannot nest one in the other. But since our previous LR result clearly rejects models such as EGARCH and GJR that rotates the news impact curve, one may add here that a nonlinear asymmetric model where the exponents λ and ν are equal but somewhere between one and two may describe the data well. Though our exercise has not given us a unique solution, we have narrowed down our selection.

10.15 Forecasting an EGARCH Model

Recall the closed form expressions we wrote for the ARCH and GARCH models. Unfortunately, such closed form expressions are not available for the EGARCH model. However, forecasts of many such nonlinear models may be obtained by Monte Carlo methods or bootstrapping. If one is convinced that the distribution for υ_t is standard normal, then Cao and Tsay (1992) and Tsay (2005) demonstrate how to forecast an EGARCH model. We shall write a recursive expression for forecasting the variance EGARCH(1,0) model, the following way:

Rewrite an EGARCH(1,0) model in the exponential form:

$$h_t = h_{t-1}^{\phi_1}\exp[(1-\phi_1)\alpha_0]\exp[g(\upsilon_{t-1})],$$

$$g(\upsilon_{t-1}) = \theta\upsilon_{t-1} + \gamma\left[|\,\upsilon_{t-1}\,| - \sqrt{2/\pi}\right].$$

Let the forecast origin be t. For the one-step ahead forecast, the variance equation becomes

$$h_{t+1} = h_t^{\phi_1}\exp[(1-\phi_1)\alpha_0]\exp[g(\upsilon_t)],$$

where all quantities on the right-hand side are known. So the one-step ahead forecast $\hat{h}_t(1)$ can be easily calculated. For the two-step ahead forecast, note that the model becomes

$$h_{t+2} = h_{t+1}^{\phi_1}\exp[(1-\phi_1)\alpha_0]\exp[g(\upsilon_{t+1})].$$

Taking conditional expectation, we have

$$\hat{h}_t(2) = \hat{h}_t^{\phi_1}(1)\exp[(1-\phi_1)\alpha_0]E_t\{\exp[g(\upsilon_{t+1})]\},$$

where E_t denotes the conditional expectation taken at time origin t.

A closed form expression for the prior expectations is

$$E\{\exp[g(\epsilon_t)]\} = \exp\left(-\gamma\sqrt{2/\pi}\right) \times \left[e^{(\theta+\gamma)^2/2}\Phi(\theta+\gamma) + e^{(\theta-\gamma)^2/2}\Phi(\gamma-\theta)\right],$$

where $\Phi(x)$ is the cumulative distribution function of a standard normal distribution. Accordingly, the two-step ahead volatility forecast is written as

$$\hat{h}_t(2) = \hat{h}_t^{\phi_1}(1)\exp\left[(1-\phi_1)\alpha_0 - \gamma\sqrt{2/\pi}\right] \times \left[\exp\{(\theta+\gamma)^2/2\}\Phi(\theta+\gamma) + \exp\{(\theta-\gamma)^2/2\}\Phi(\gamma-\theta)\right].$$

A recursive form for any s-step ahead forecast therefore becomes

$$\hat{h}_t(s) = \hat{h}_t^{\phi_1}(s-1)\exp\left[(1-\phi_1)\alpha_0 - \gamma\sqrt{2/\pi}\right] \times \left[\exp\{(\theta+\gamma)^2/2\}\Phi(\theta+\gamma) + \exp\{(\theta-\gamma)^2/2\}\Phi(\gamma-\theta)\right].$$

Note that the values of $\Phi(\theta + \gamma)$ and $\Phi(\gamma - \theta)$ can be obtained from any standard statistical package. If unavailable, accurate approximations for the same are available in the statistics literature. Such a recursive form, however, is not available for other distributions such as the GED. Such forecasted volatilities can be used with forecasted series values to generate forecast confidence intervals for the forecasted series. Many times, it is the forecasted volatilities that are of prime interest. Hence, confidence intervals for them have to be obtained. But analytical expressions for such confidence intervals are available only for some special cases. Mostly, such intervals are obtained using the simulation-based method.

The models that we have described above are only a small selection of models that have been recommended in the literature to describe volatility. We now turn to explaining another important class of volatility models called the *SV models*. Though these models are definitely more sophisticated than the GARCH type of models, there are no clear answers to the question if one should prefer stochastic conditional volatility models to the deterministic ones, at least, to explain the volatility in daily returns.

Part 2: Stochastic Volatility Models

10.16 Stochastic Volatility Models

In ARCH models, we modeled volatility as a deterministic variable as

$$r_t - \mu = \upsilon_t \sigma_t.$$

But the recognition that volatility can be better captured if treated explicitly as an unobserved latent variable led to an alternative way to explain the time-varying volatility of a financial time series. And that is to attach an *innovation* to the conditional variance equation σ_t^2 and this innovation captures the unpredictable events of the same day. The resulting model is called an *SV model*. Though not quite as popular as its deterministic counterpart, the ARCH family of models, SV models arise quite naturally in models for pricing options and in models involving trading volume or transaction counts, where the latent σ_t^2 is expected to capture the random flow of new information. The presence of such unpredictable component means we have to redefine our volatility model as follows:

$$r_t = \mu + e_t, \quad e_t = \upsilon_t \sigma_t$$
$$(1 - \alpha_1 B - \cdots - \alpha_p B^p)\sigma_t^2 = \alpha_0 + \eta_t.$$

Here, η_t is assumed to be independent and identically distributed with zero mean and unit variance. The main difference between ARCH and SV models is that the presence of the unpredictable volatility component η_t in σ_t^2, which makes $\text{var}(\sigma_t^2 | r_{t-1}, r_{t-2}, \ldots) > 0$. And since true volatility is unobservable, SV processes estimate h_t along with other unknown parameters. This improvement of course comes at the cost of ease of computation.

Since σ_t^2 is also stochastic now, in SV models, it is normal to assume that the two processes σ_t^2 and υ_t are independent. We shall call the SV model *independent*, if the two error processes are stochastically independent. Dependency between the two processes is also discussed in the literature. For example, Ghysels et al. (1996) discuss dependent processes where one process does not Granger cause the other.

10.16.1 Alternative Forms of the SV Model

There are two different version of the SV model used in the literature. It is needless to say that both specifications are equivalent. The standard SV model, first proposed by Taylor (2008) and which we call Model 1, is given as follows:

$$r_t = \mu + \upsilon_t \sigma_t,$$

$$\ln(\sigma_t) = \alpha_0 + \alpha_1 \ln(\sigma_{t-1}) + \eta_t, \quad |\alpha_1| < 1, \quad \upsilon_t \sim N(0,1), \quad \eta_t \sim N\left(0, \sigma_\eta^2\right).$$

An alternative version of the above model, which is sometimes used in the literature and which we label Model 2 for our purpose, is given below:

$$r_t - \mu = e_t, \quad e_t = \upsilon_t \exp(h_t/2), \quad h_t = \alpha_0 + \alpha_1 h_{t-1} + \eta_t, \quad \eta_t \sim N\left(0, \sigma_\eta^2\right), \quad \upsilon_t \sim N(0,1).$$

In such a parameterization, normally the mean μ is set to zero for ease of exposition.

10.16.2 Properties of Returns in SV Model

We shall first check if the SV model captures the important stylized facts of returns using Model 1 of Taylor (Taylor 2008, 2005), who has also listed the important moments. Under strict stationarity of σ_t^2 and the assumption that σ_t^2 is stochastically independent of υ_t and all the autocorrelations of σ_t^2 are positive, we shall first derive the moments of the independent SV processes. One can easily see that the returns are uncorrelated. Since $\text{cov}(XY) = E(XY) - E(X)E(Y)$, we can write

$$\begin{aligned}
\text{cov}(r_t, r_{t+k}) &= \text{cov}(\sigma_t \upsilon_t, \sigma_{t+k}\upsilon_{t+k}) \\
&= E[\sigma_t \upsilon_t \sigma_{t+k}\upsilon_{t+k}] - E[\sigma_t \upsilon_t] E[\sigma_{t+k}\upsilon_{t+k}] \\
&= 0.
\end{aligned}$$

Similarly, we can show that kurtosis of returns under the present model is

$$\text{kurtosis}(r_t) = \frac{3E[\sigma_t^4]}{E[\sigma_t^2]^2} = 3\left(1 + \frac{\text{var}(\sigma_t^2)}{E[\sigma_t^2]^2}\right) > 3.$$

Functions of returns can have substantial autocorrelations, even though returns have very small autocorrelations. For instance, stylized facts of returns clearly show that absolute returns and squared returns are positively correlated. Any volatility model is expected to capture such a fact. So we can check if an SV also accounts for such a fact. Let $s_t = (r_t - \mu)^2$. Then

$$\text{cov}(s_t, s_{t+k}) = \left[\frac{\text{var}(\sigma_t^2)}{\text{var}(s_t)}\right] \text{cov}(\sigma_t^2, \sigma_{t+k}^2) > 0.$$

Hence, we can also express the autocorrelations of s_t, $\rho_{k,s}$ to be a multiple of the autocorrelations of σ_t^2, ρ_{k,σ^2}. Since $\text{cov}(s_t, s_{t+k}) = \text{cov}\left(\sigma_t^2, \sigma_{t+k}^2\right)$, we can write

$$\begin{aligned}
\frac{\text{cov}(s_t, s_{t+k})}{\text{var}(s_t)} &= \left[\frac{\text{var}(\sigma_t^2)}{\text{var}(s_t)}\right] \frac{\text{cov}(\sigma_t^2, \sigma_{t+k}^2)}{\text{var}(\sigma_t^2)} \\
\rho_{k,s} &= \left[\frac{\text{var}(\sigma_t^2)}{\text{var}(s_t)}\right] \rho_{k,\sigma^2}, \quad k > 0.
\end{aligned}$$

We can rewrite further as follows, since $\text{var}(X) = E(X^2) - [E(X)]^2$,

$$\rho_{k,s} = \left[\frac{E[\sigma_t^4] - E[\sigma_t^2]^2}{k_\upsilon E[\sigma_t^4] - E[\sigma_t^2]^2}\right]\rho_{k,\sigma^2} = \frac{\left[\dfrac{E[\sigma_t^4] - E[\sigma_t^2]^2}{E[\sigma_t^2]^2}\right]}{\left[\dfrac{k_\upsilon E[\sigma_t^4] - E[\sigma_t^2]^2}{E[\sigma_t^2]^2}\right]}\rho_{k,\sigma^2}$$

$$= \frac{\left[\dfrac{E[\sigma_t^4]}{E[\sigma_t^2]^2} - 1\right] \times k_\upsilon}{(k_r - 1) \times k_\upsilon}\rho_{k,\sigma^2} = \frac{k_r - k_\upsilon}{k_\upsilon(k_r - 1)}\rho_{k,\sigma^2}, \; k > 0,$$

whenever the kurtosis of returns is finite.

Taylor (2005) gives the closed form expression for autocorrelations of the absolute excess returns $a_t = |r_t - \mu| = \sigma_t|\upsilon_t|$ or any power of a_t. These returns are positively correlated.

10.17 Standard SV Model

Having established that an SV model also captures some of the salient stylized facts of returns, we shall now turn to discussing some specific SV models suggested in the literature. The simplest continuous distribution for the SV σ_t is log-normal for daily returns and this specification was recommended by Taylor (2008), and which we have labeled Model 1. Then $\ln \sigma_t \sim N\left(\mu_{\sigma^2}, \sigma_{h,SV}^2\right)$. This choice assures positive conditional variances, permits calculation of moments, and accounts for excess kurtosis in returns. This simple model has also strong empirical support, though this model *does not* allow for returns to react asymmetrically to price falls and rises, σ_t is independent of the signs of all previous returns. We shall then outline a state space representation of the model, its moments, and its autocorrelations.

We shall recall the standard SV model:

$$r_t = \mu + \upsilon_t \sigma_t$$
$$\ln(\sigma_t) = \alpha_0 + \alpha_1 \ln(\sigma_{t-1}) + \eta_t, \quad |\alpha_1| < 1, \quad \upsilon_t \sim N(0,1), \quad \eta_t \sim N\left(0, \sigma_\eta^2\right).$$

We continue to assume that the errors υ_t and η_t are independent, Gaussian white noise series. Owing to the Gaussianity of η_t, this model is also called *log-normal AR(1) SV model.* Thus the returns process is strictly stationary, since it is the product of independent strictly stationary processes. It is also covariance stationary because the returns have finite variance. Since $E(\sigma_t) = \exp(\mu_{\sigma^2} + 1/2\sigma_{h,SV}^2)$, and $E(\sigma^r) = \exp\left(r\mu_{\sigma^2} + (1/2)r^2\sigma_{h,SV}^2\right)$, we can show that

$$\text{var}(r_t) = \exp\left(2\mu_{\sigma^2} + 2\sigma_{h,SV}^2\right)$$
$$k_{rt} = \text{kurtosis}(r_t) = 3\exp\left(4\sigma_{h,SV}^2\right).$$

And we can derive the autocorrelations of squared excess returns s_t also. Note that

$$\begin{aligned}\text{var}(\sigma_t^2) &= E(\sigma_t^4) - [E(\sigma_t^2)]^2 \\ &= \exp(4\mu_{\sigma^2} + 8\sigma_{h,SV}^2) - \exp(4\mu_{\sigma^2} + 4\sigma_{h,SV}^2) \\ &= \exp(4\mu_{\sigma^2} + 4\sigma_{h,SV}^2)[\exp(4\sigma_{h,SV}^2) - 1].\end{aligned}$$

Similar arguments will give us

$$\text{var}(s_t) = \exp(4\mu_{\sigma^2} + 4\sigma_{h,SV}^2)[(3\exp(4\sigma_{h,SV}^2) - 1)].$$

Now, from the expression for $\rho_{k,s}$ before, we have

$$\frac{\rho_{k,s}}{\rho_{k,\sigma^2}} = \frac{\left[\exp(4\sigma_{h,SV}^2) - 1\right]}{\left[3\exp(4\sigma_{h,SV}^2) - 1\right]}.$$

Similarly, one can derive the moments and autocorrelations of the absolute returns, a_t also.

10.17.1 An Alternative Parameterization of the Log-Normal SV Model

Some authors prefer the following parameterization of the log-normal SV model, which we have labeled Model 2:

$$r_t - \mu_t = e_t, \quad e_t = \upsilon_t \exp(h_t/2), \quad h_t = \alpha_0 + \alpha_1 h_{t-1} + \eta_t, \quad \eta_t \sim N(0, \sigma_\eta^2), \quad \upsilon_t \sim N(0, 1).$$

In such a parameterization, normally the mean μ_t is set to zero for ease of exposition.

10.17.2 Basic Properties of the Alternative Log-Normal SV Model

As η_t is Gaussian, h_t is also standard Gaussian autoregression, which will be strictly covariance stationary if $|\alpha_1| < 1$ with

$$E(h_t) = \mu_h = \frac{\alpha_0}{1 - \alpha_1}, \quad \sigma_h^2 = \text{var}(h_t) = \frac{\sigma_\eta^2}{1 - \alpha_1^2}.$$

Using the properties of the log-normal distribution, we can write the moment conditions, for the alternative parameterization of the log-normal SV model, as follows:

$$\begin{aligned}E(e_t^2) &= \exp(\mu_h + \sigma_h^2/2), \\ E(e_t^4) &= 3\exp(2\mu_h + 2\sigma_h^2).\end{aligned}$$

Upon generalizing, we can write the moment generating expression for any even rth moment, following Shephard (1996):

$$E(e_t^r) = E(\upsilon_t^r)E\left\{\exp\left(\frac{r}{2}h_t\right)\right\}$$
$$= r!\exp\left(\frac{r}{2}\mu_h + r^2\sigma_h^2/8\right) / \left(2^{r/2}(r/2)!\right),$$

$$\text{so that kurtosis} = \left[E(e_t^4)/(\sigma_{e^2}^2)^2\right] = \frac{3\exp(2\mu_h + 2\sigma_h^2)}{\exp(2\mu_h + \sigma_h^2)} = 3\exp(\sigma_h^2) \geq 3.$$

Note here that $3\exp(\sigma_h^2) = 3\exp\left(\sigma_\eta^2/(1-\alpha_1^2)\right)$ will be greater than 3, so long as $\sigma_\eta^2 > 0$. Hence, the kurtosis generated by this specification of the SV process will be greater than 3 and will increase with σ_η^2 and $|\alpha_1|$, given that $|\alpha_1| < 1$.

Note that $E(e_t^2 e_{t-k}^2) = E\{\exp(h_t + h_{t-k})\}$, and h_t is Gaussian AR(1).

So, we can write

$$\text{cov}(e_t^2 e_{t-k}^2) = \exp\{2\mu_h + \sigma_h^2(1+\alpha_1^k)\} - \{E(e_t^2)\}^2$$
$$= \exp\left(2\mu_h + \sigma_h^2\right)\left\{\exp(\sigma_h^2\alpha_1^k) - 1\right\}.$$

So,

$$\rho_{e_t^2}(k) = \text{cov}(e_t^2 e_{t-k}^2)/\text{var}(e_t^2) = \frac{\exp\left(2\mu_h + \sigma_h^2\right)\left\{\exp(\sigma_h^2\alpha_1^k) - 1\right\}}{3\exp(2\mu_h + 2\sigma_h^2) - \exp(2\mu_h + \sigma_h^2)}$$
$$= \frac{\exp\left(2\mu_h + \sigma_h^2\right)\left\{\exp(\sigma_h^2\alpha_1^k) - 1\right\}}{(3\exp(\sigma_h^2) - 1)(\exp(2\mu_h + \sigma_h^2))} = \frac{\exp(\sigma_h^2\alpha_1^k) - 1}{3\exp(\sigma_h^2) - 1} \backsimeq \frac{\exp(\sigma_h^2) - 1}{3\exp(\sigma_h^2) - 1}\alpha_1^k.$$

Note that if $\alpha_1 < 0$, $\rho_{e_t^2}(k)$ can be negative. Note that this is the autocorrelation function of an ARMA(1,1) process.

We can also write the model in logs. Clearly,

$$\ln e_t^2 = h_t + \ln\upsilon_t^2, \quad h_t = \alpha_0 + \alpha_1 h_{t-1} + \eta_t.$$

Since the i.i.d. term $\ln\upsilon_t^2$ has been added to h_t, the model for $\ln e_t^2$ is now an ARMA(1,1) model. If υ_t is normal, then $\ln\upsilon_t^2$ is not normally distributed; but it follows a highly skewed log-chi squared distribution with one degree of freedom. It can be approximated by a normal density with mean and variance given as

$$E(\ln\upsilon_t^2) = \left(\frac{\Gamma'(1/2)}{\Gamma(1/2)} - \ln\left(\frac{1}{2}\right)\right) \approx -1.2704,$$

$$\text{var}(\ln\upsilon_t^2) = \left.\frac{\partial^2\ln\Gamma(\nu)}{\partial\nu^2}\right|_{\nu=1/2} = \frac{\pi^2}{2},$$

where the first term within parentheses on the right side of the expression for the mean is the digamma function, which is the logarithmic derivative of the Gamma function:

$$\psi(x) = \frac{\partial}{\partial x}\ln(\Gamma(x)) = \frac{\Gamma'(x)}{\Gamma(x)}.$$

But from hereon, we shall continue with the standard SV model using the first parameterization, that is, Model 1. We have already listed out its basic properties.

One can describe several properties of the standard SV model, cast in a state space representation, its moments, and its autocorrelations. Let us define, for the standard SV model, the following:

$$l_t = \ln(|r_t - \mu|), \quad L_t = \ln(\sigma_t), \quad \text{and} \quad \epsilon_t = \ln(|\upsilon_t|).$$

With these, there exists a linear state space representation for the process $\{l_t\}$. The measurement equation is

$$l_t = L_t + \epsilon_t,$$

and the transition equation is

$$L_t = \alpha_0 + \alpha_1 L_{t-1} + \eta_t.$$

But the state space model is not a Gaussian model because ϵ_t is not a normal variable. It is skewed with a long left tail with mean $\mu_\epsilon = -0.635$ and variance $\sigma_\epsilon^2 = \pi^2/8$. But the moments and the autocorrelations can be obtained.

10.17.3 Autocorrelations of l_t, σ_t, a_t, s_t

Excess returns of $(r_t - \mu)$ are uncorrelated and hence its autocorrelations are zero. So we shall concentrate on calculating the autocorrelations of $l_t = \ln(|r_t - \mu|)$, which can be easily derived from the state space representation.

Since $l_t = \ln a_t = \ln(\sigma_t) + \ln(|\upsilon_t|)$ (recall that $(|r_t - \mu|) = a_t$)

$$\begin{aligned}\text{cov}(l_t, l_{t+k}) &= \text{cov}(\ln(\sigma_t) + \ln(|\upsilon_t|), \ln(\sigma_{t+k}) + \ln(|\upsilon_{t+k}|)) \\ &= \text{cov}(\ln(\sigma_t), \ln(\sigma_{t+k})) \ (\because \upsilon_t \text{ is iid}) \\ &= \alpha_1^k \sigma_{h,SV}^2 \ (\because \ln \sigma_t \text{ follows an } AR(1) \text{ process}).\end{aligned}$$

Then,

$$\rho_{k,l} = \text{corr}(l_t, l_{t+k}) = \frac{\text{cov}(l_t, l_{t+k})}{\text{var}(l_t)} = \frac{\alpha_1^k \sigma_{h,SV}^2}{\text{var}(\ln(\sigma_t) + \ln(|\upsilon_t|))}.$$

But we know that

$$\text{var}(\ln(\sigma_t)) = \sigma_{h,SV}^2, \quad \text{var}(\ln(|\upsilon_t|)) = \frac{\pi^2}{8}.$$

Hence,

$$\rho_{k,l} = \frac{\alpha_1^k \sigma_{h,SV}^2}{\sigma_{h,SV}^2 + (\pi^2/8)} = \frac{8\alpha_1^k \sigma_{h,SV}^2}{8\sigma_{h,SV}^2 + \pi^2}.$$

The above algebra extends to the autocorrelations of any power of $a_t = (|r_t - \mu|)$ for the standard SV model also, once we calculate the autocorrelations of σ_t^p.

We can sum up the main results of the discussion on SV model so far:

- All moments of returns are finite.
- Note that, like the EGARCH model, in SV models, we normally define the variance model for $\ln(\sigma_t^2)$ to ensure positiveness of conditional variance.

- SV models definitely bring in a lot of flexibility into volatility modeling but also make parameter estimation difficult.
  - Difficulty is understandable because note that we now have two innovations, υ_t and ϵ_t for each shock e_t.
  - Estimation cannot be done using ML methods, which the ARCH models allow. To estimate the parameters of an SV model, we need a quasi-likelihood method, whose components are built through Kalman filter or by applying Markov chain Monte Carlo (MCMC) methods or we can use the GMM estimation procedures. All these methods are involved.

10.18 Estimation of SV Models: State Space Form

To recall, the standard SV in its state space form in the first parameterization can also be written as follows:

$$\text{Let} \quad r_t - \mu = e_t = \sigma_t \upsilon_t \quad \text{Measurement equation}$$
$$\ln(\sigma_t) = \alpha_0 + \alpha_1 \ln(\sigma_{t-1}) + \eta_t \quad \text{Transition equation}$$

Then,

$$\ln|e_t| = \ln\sigma_t + \ln\upsilon_t,$$
$$E[\ln|\upsilon_t| = -0.63518, \text{var}(\ln|\upsilon_t|) = \frac{\pi^2}{8}.$$

The measurement equation can be rewritten as

$$\ln|e_t| = -0.63518 + \ln\sigma_t + \tilde{\epsilon}_t, \quad \text{where } \tilde{\epsilon}_t = \ln|\upsilon_t| + 0.63518, \ \tilde{\epsilon}_t \sim iid\,(0, \pi^2/8).$$

This adjustment allows us to show that the disturbances $\tilde{\epsilon}_t$ and η_t are uncorrelated.

State space parameters are: $\ln\sigma_t$, which is a state variable, $y_t = 1$, $T = \phi$, and $R_t = 1$. State space formulation will be complete if we can specify the behavior of the initial state—that is, $\boldsymbol{\alpha}_0 \sim N(\mathbf{a}_0, \mathbf{P}_0)$. For the problem on hand

$$a_0 = \frac{\alpha_0}{(1-\alpha_1)}, \qquad P_0 = \frac{\sigma_\eta^2}{(1-\alpha_1^2)}.$$

The same model can be written under the alternative parameterization also. Recall that under the alternate parameterization, setting the mean μ to zero for simplicity, we have, from Harvey et al. (1994),

$$r_t = \upsilon_t \exp(\sigma_t^2/2),$$
$$\sigma_t^2 = \alpha_0 + \alpha_1 \sigma_{t-1}^2 + \eta_t.$$

Then, we can write

$$\ln r_t^2 = \sigma_t^2 + \ln \upsilon_t^2, \quad \ln \upsilon_t^2 \sim iid\,(-1.27, \pi^2/2).$$

The measurement equation is

$$\ln r_t^2 = -1.27 + \sigma_t^2 + \epsilon_t^*, \quad \text{where } \epsilon_t^* = \ln \upsilon_t^2 + 1.27, \quad \epsilon_t^* \sim iid\,(0, \pi^2/2).$$

and the transition equation is

$$\sigma_t^2 = \alpha_0 + \alpha_1 \sigma_{t-1}^2 + \eta_t, \quad \eta \sim N(0, 1).$$

Though the transition equation is a perfectly valid part of a state space representation, measurement equation is not, since $\ln(\upsilon_t^2)$, and hence, ϵ_t^*, is not Gaussian. But it is a distribution with a known mean and variance, which we have already described. Though the model is not normal, estimation can proceed by pretending as though it is normal; that is, we can proceed assuming $\epsilon_t^* \sim iid\, N(0, \pi^2/2)$. The advantage is that the model's parameters can be estimated by QMLs. Harvey et al. (1994) apply Kalman filter effectively to estimate such SV models.

The goal is to maximize the corresponding log-likelihood:

$$\ln L = -\frac{1}{2} \sum_{t=1}^{T} \left(\ln(2\pi) + \ln(F_t) + \frac{\tilde{e}_t^2}{F_t} \right).$$

Note that the form of the log-likelihood is exactly the same as that of the ARCH model. The resulting quasi-ML estimates are asymptotically normal and consistent.

In this chapter, we shall concentrate on the Monte Carlo-based estimation of the parameters of the standard SV model.

10.19 MCMC-Based Estimation of the Standard SV Model

Among the methods suggested to estimate the unknowns of the standard SV model, the quasi-maximum likelihood estimation method of Harvey et al. (1994) and the MCMC methods using the Metropolis–Hastings and Gibbs algorithms are widely used. We shall explain below the MCMC methods.

In the estimation of the standard SV model, the likelihood function

$$f(\mathbf{y}|\theta) = \int f(\mathbf{y}|\mathbf{h}, \theta) f(\mathbf{h}|\theta) d\mathbf{h}$$

is intractable. Among the MCMC methods suggested to estimate the unknowns of the standard SV model, the approaches of Jacquier et al. (1994) and Kim et al. (1998) are popular. Bayesian methods are an alternative approach to the quasi-likelihood method of Harvey et al. (1994) to estimate the unknowns. And MCMC methods are well suited for Bayesian inference and we outline below the Kim, Shephard, Chib (KSC) approach for estimating the standard SV model using the Bayesian method.

10.20 Bayesian Inference and MCMC Algorithm

We shall give a simple introduction to the concept of Bayesian inference and the MCMC algorithm.

Let θ be the unknown vector of parameters of the model under consideration and $\mathbf{y}$ be the data. Considering that θ is a random vector, inference is based on its posterior distribution. Bayes' theorem states that

$$p(\theta|\mathbf{y}) \propto p(\mathbf{y}|\theta)p(\theta),$$

where $p(\mathbf{y}|\theta)$ is the likelihood function and $p(\theta)$ is our prior belief on the distribution of θ and $p(\theta|\mathbf{y})$ is called the posterior distribution of θ. The principle behind the MCMC-based Bayesian inference is to simulate $p(\theta|\mathbf{y})$ based on repeated samples of a Markov chain, whose invariant distribution is the target density of interest.

A discrete stochastic process, X_t, is said to be a *Markov chain* if it has the following *Markov* property:

$$P(X_t = x_t|X_{t-1} = x_{t-1}, X_{t-2} = x_{t-2}, \ldots,) = P(X_t = x_t|X_{t-1} = x_{t-1}).$$

This definition states that, given the entire past history, the process depends only on the immediate past. The distribution of a Markov chain is fully specified by its transition probabilities, $P(X_t = x_t|X_{t-1} = x_{t-1})$ and the initial condition $P(X_0 = x_0)$.

In an inference problem involving the parameter vector θ, where $\theta \in \boldsymbol{\Theta}$, and the data vector, $\mathbf{y}$, we need to know the distribution of $p(\theta|\mathbf{y})$. The idea of a Markov chain is to simulate a Markov process on $\boldsymbol{\Theta}$, which will converge to the invariant transition distribution that is $p(\theta|\mathbf{y})$. The key is to run the simulation sufficiently long so that the distribution of the current values is close enough to the target density of interest. Methods used to arrive at the distribution, $p(\theta|\mathbf{y})$, are chiefly called the MCMC methods. There are different ways to construct such a Markov chain. Gibbs sampling, or Gibbs sampler, and the Metropolis–Hastings algorithm are the two most used algorithms to simulate the draws. Several resources of varying degrees of complexity are available on the subject of MCMC. While Tsay (2005) provides an accessible introduction to the topic, we refer to Robert and Casella (2010), Shephard and Pitt (1997), Hautsch and Ou (2009), Jacquier et al. (1994), and Tsay (2005) among others.

10.20.1 Gibbs Sampling

Suppose we have a joint distribution of $p(\theta_1, \ldots, \theta_k)$ that we want to sample from. It could be the posterior distribution in a Bayesian context. We can use the Gibbs sampler to draw from this joint distribution, provided we know the full conditional distributions for each parameter. That is, in the parameter vector $\theta = (\theta_1, \ldots, \theta_n)$, the full conditional distribution for every single parameter given the others, is known: $p(\theta_j|\theta_{-j}, \mathbf{y})$, where θ_{-j} denotes all elements of θ except θ_j. The basis of this assertion is the important Hammersley–Clifford (HC) theorem, which proves that the joint distribution, $p(\theta|\mathbf{y})$ is completely determined by the conditional distributions, $p(\theta_j|\theta_{-j})$, under a positivity condition.

The Gibbs algorithm can be summarized in the following steps:

1. Let $\theta^0_{-1} = (\theta^0_2, \ldots, \theta^0_n)$ be the arbitrary starting values for $\theta_2, \ldots, \theta_n$. One can start with any θ. Order does not matter. We are starting with θ_1 and proceeding in a sequence for convenience.

2. Generate:
   a. θ_1^1 from $p(\theta_1|\theta_2^0, \theta_3^0, \ldots, \theta_n^0, \mathbf{y})$
   b. θ_2^1 from $p(\theta_2|\theta_2^0, \theta_3^0, \ldots, \theta_1^1, \mathbf{y})$

      $\vdots$

   c. θ_n^1 from $p(\theta_n|\theta_1^1, \theta_2^1, \ldots, \theta_{n-1}^1, \mathbf{y})$
3. Using the new parameter vector, $\boldsymbol{\theta}^1 = (\theta_1^1, \ldots, \theta_n^1)$ as the starting values, repeat step (2) to get an updated parameter vector $\boldsymbol{\theta}^2 = (\theta_1^2, \ldots, \theta_n^2)$.
4. Repeat these iterations m times to get the following sequence of random draws:
   $\boldsymbol{\theta}^1 = (\theta_1^1, \ldots, \theta_n^1), \boldsymbol{\theta}^2 = (\theta_1^2, \ldots, \theta_n^2), \ldots, \boldsymbol{\theta}^m = (\theta_1^m, \ldots, \theta_n^m)$.

Using some weak regularity conditions and convergence theorem results from Markov chain theory, we can show that, for a sufficiently large $m, (\theta_1^m, \ldots, \theta_n^m)$ will be approximately equal to a random draw from the joint distribution, $p(\boldsymbol{\theta}|\mathbf{y})$. But in practice, we draw a sufficiently large number of samples, say k samples, but discard the first m draws so that we finally use $(\theta_1^{m+1}, \ldots, \theta_n^{m+1}), \ldots, (\theta_1^k, \ldots, \theta_n^k)$. The discarded samples are called the *burn-in* samples.

10.20.2 Metropolis–Hastings Algorithm

Obtaining the posterior or the conditional posterior distributions is often difficult. If a conjugate prior distribution is available, then closed form expressions of conditional posterior distributions can be easily written. This implies that random draws of Gibbs sampler can be obtained using computer routines commonly available in standard computer packages. If we do not know some or all of the conditional posterior distributions, then Gibbs sampler cannot be used. However, many alternative algorithms have been suggested in the literature and the Metropolis–Hastings algorithm, developed first by Metropolis et al. (1953) and later generalized by Hastings (1970), is a popular one. Chib and Greenberg (1995) provide an easy introduction to this algorithm.

It consists of the following steps:

1. Choose any arbitrary starting value for the parameter vector, $\boldsymbol{\theta}_0$, such that $p(\boldsymbol{\theta}_0|\mathbf{y}) > 0$.
2. For $t = 1, 2, \ldots$
   a. Draw a candidate sample $\boldsymbol{\theta}_*$ from a known *jumping distribution* or a *proposal distribution*, $J_t(\boldsymbol{\theta}_t|\boldsymbol{\theta}_{t-1})$. The original suggestion of Metropolis was that the jumping distribution be a symmetric distribution, such as a normal distribution, which implies $J_t(\boldsymbol{\theta}_*|\boldsymbol{\theta}_{t-1}) = J_t(\boldsymbol{\theta}_{t-1}|\boldsymbol{\theta}_*)$. This was relaxed later by Hastings who suggested that the jumping distribution need not be symmetric.
   b. Calculate the acceptance ratio r, the probability of acceptance:

$$r = \frac{p(\boldsymbol{\theta}_*|\mathbf{y})/J_t(\boldsymbol{\theta}_*|\boldsymbol{\theta}_{t-1})}{p(\boldsymbol{\theta}_{t-1}|\mathbf{y})/J_t(\boldsymbol{\theta}_{t-1}|\boldsymbol{\theta}_*)}.$$

   In case our jumping distribution is symmetric, then

$$r = \frac{p(\boldsymbol{\theta}_*|\mathbf{y})}{p(\boldsymbol{\theta}_{t-1}|\mathbf{y})}.$$

c. Set

$$\theta_t = \begin{cases} \theta_* & \text{with probability } \min[r, 1] \\ \theta_{t-1} & \text{otherwise} \end{cases}$$

The acceptance–rejection rule implies that the jump to θ_* from θ_{t-1} is always accepted if our candidate draw increases the conditional posterior density compared to our current draw. In that case, set $\theta_* = \theta_t$. If it decreases, acceptance or rejection obeys the following rule:

i. For each θ_*, draw a value u from a uniform distribution, $U(0, 1)$. If $u \leq r$, then accept $\theta_* = \theta_t$ with a probability equal to the ratio of the probabilities, r.

ii. Set $\theta_t = \theta_{t-1}$ otherwise.

By this rule, each iteration always results in a draw, either θ_t or θ_{t-1}. Needless to say that the acceptance–rejection rule has to be monitored closely. The proposal distribution should be such that we have a good acceptance rate with good mixing—that is, the chain moves on the parameter space quickly enough. Too high acceptance rates may indicate that the chain is moving slowly on the space. Low acceptance rates, though less of an issue, nevertheless indicates a poor algorithm, resulting in slow convergence and a large number of simulations.

10.21 MCMC-Based Estimation of the Standard SV Model (*Continued*)

We shall continue our discussion on the estimation of the SV model by viewing it as a hierarchical structure of conditional distributions.

For our convenience we shall restate, a version of the original SV model of Taylor that we shall be using, as follows:

$$r_t = \upsilon_t \exp(h_t/2), \quad h_t = \mu + \phi h_{t-1} + \eta_t, \quad \eta_t \sim N(0, \sigma_\eta^2), \quad \upsilon_t \sim N(0, 1).$$

Let $\theta = (\mu, \phi, \sigma_\eta^2)'$, $\mathbf{h} = (h_1, \ldots, h_T)'$, and $\mathbf{r} = (r_1, \ldots, r_T)'$ be the observed data. Applying Bayes' theorem

$$p(\mathbf{h}, \theta|\mathbf{r}) \propto p(\mathbf{r}|\mathbf{h}, \theta)p(\mathbf{h}|\theta)p(\theta).$$

Bayesian inference for model parameters θ and the unobserved volatility states $\mathbf{h}$ is based on the posterior distribution, $p(\mathbf{h}, \theta|\mathbf{r})$, which is proportional to the above product. One can see that the posterior distribution is a product of the likelihood function for the observed data $\mathbf{r}$ conditional on θ and the volatility states $\mathbf{h}$, conditional distribution of the volatility states, $\mathbf{h}$ and our prior distribution for the parameter vector θ. The above idea of expressing the joint posterior distribution of $\mathbf{h}$ and θ is termed data augmenting technique in Bayesian inference. From this joint posterior, we use the marginal distributions $p(\theta|\mathbf{r}, \mathbf{h})$ and $p(\mathbf{h}|\mathbf{r}, \theta)$ to infer about the parameters of the variance equation and the unobserved volatilities. We compute these marginals by constructing the Markov chain and use the Gibbs sampler to alternate between the posterior densities $p(\theta|\mathbf{r}, \mathbf{h})$ and $p(\mathbf{h}|\mathbf{r}, \theta)$.

The steps of a single-move Gibbs sampler for the standard SV model are:

1. Start with any arbitrary starting values for $\theta^0, \mathbf{h}^0$.
2. For $g = 1, \ldots, G$,
   a. for $t = 1, \ldots, T$:
      sample $h_t^{(g)}$ from $p\left(h_t|\mathbf{r}, \mathbf{h}_{-t}^{(g-1)}, \theta^{(g-1)}\right)$, where
      $$\mathbf{h}_{-t}^{(g-1)} = \left(h_1^{(g)}, \ldots, h_{t-1}^{(g)}, h_{t+1}^{(g-1)}, \ldots, h_T^{(g-1)}\right).$$
   b. Sample $\sigma_\eta^{2(g)}$ from $p\left(\sigma_\eta^2|\mathbf{r}, \mathbf{h}^{(g)}, \mu^{(g-1)}, \phi^{(g-1)}\right)$
   c. Sample $\phi^{(g)}$ from $p\left(\phi|\mathbf{r}, \mathbf{h}^{(g)}, \sigma_\eta^{2(g)}, \mu^{(g-1)}\right)$
   d. Sample $\mu^{(g)}$ from $p\left(\phi|\mathbf{r}, \mathbf{h}^{(g)}, \phi^{(g)}, \sigma_\eta^{2(g)}\right)$

Our next task is to see how to implement steps 2(a)–2(d). We shall first concentrate on steps 2(b)–2(d). Once we know the conditional posteriors of μ, ϕ, and σ_η^2, then sampling from these distributions is easily accomplished.

10.21.1 Sampling of μ, ϕ, and σ_η^2

We begin by specifying a prior distribution for the elements of θ. Following Jacquier et al. (1994), we specify a standard natural conjugate priors for the elements as follows: $p(\mu) \sim N(\alpha_\mu, \beta_\mu^2)$ and $p(\phi) \sim N(\alpha_\phi, \beta_\phi^2)$. And for the variance, we can specify an inverted gamma prior distribution, $p(\sigma_\eta^2) \sim IG(\alpha_\sigma, \beta_\sigma)$. Given the prior distributions, we can write specifically the conditional posteriors as

$$p(\mu|\mathbf{r}, \mathbf{h}, \phi, \sigma_\eta^2) \propto p(\mathbf{r}|\mathbf{h}, \mu, \phi, \sigma_\eta^2)p(\mathbf{h}|\mu, \phi, \sigma_\eta^2)p(\mu)$$
$$p(\phi|\mathbf{r}, \mathbf{h}, \mu, \sigma_\eta^2) \propto p(\mathbf{r}|\mathbf{h}, \mu, \phi, \sigma_\eta^2)p(\mathbf{h}|\mu, \phi, \sigma_\eta^2)p(\phi)$$
$$p(\sigma_\eta^2|\mathbf{r}, \mathbf{h}, \phi, \mu) \propto p(\mathbf{r}|\mathbf{h}, \mu, \phi, \sigma_\eta^2)p(\mathbf{h}|\mu, \phi, \sigma_\eta^2)p(\sigma_\eta^2)$$

In all these conditional posteriors, the likelihood function $p(\mathbf{r}|\mathbf{h}, \mu, \phi, \sigma_\eta^2)$ is a constant with respect to the model parameters and can be omitted. And closed form expressions for the full conditional posteriors can be easily derived using standard theory on linear models. All these are available in Hautsch and Ou (2009). Since we can sample from the conditional posterior distributions, we can use the single-move Gibbs sampler described before. But we still have to find a way to sample h_t, the latent states, in step $2(a)$ in the Gibbs algorithm explained above, from the full conditional posterior.

10.21.2 Sampling of the Latent Volatility

To sample the latent volatilities from their full conditional posterior is difficult. Several suggestions exist in the literature on how to sample these latent states from $p\left(h_t|\mathbf{r}, \mathbf{h}_{-t}^{(g-1)}, \theta^{(g-1)}\right)$.

These include the methods suggested by Shephard (1993), Jacquier et al. (1994), Shephard and Kim (1994), Shephard and Pitt (1997), and Kim et al. (1998).

A common feature of all these suggestions is to exploit the Markovian structure of the volatility process and write (where we write $\mathbf{h}_{-t}^{(g-1)}$ as $\mathbf{h}_{-t}$ and $\theta^{(g-1)}$ as θ):

$$\begin{aligned} p(h_t|\mathbf{h}_{-t}, \theta, \mathbf{r}) &= p(h_t|h_{t-1}, h_{t+1}, r_t, \theta) \\ &\propto p(r_t|h_t)p(h_t|h_{t-1}, \theta)p(h_{t+1}|h_t, \theta)(\text{note that } r_t \text{ does not depend on } \theta) \\ &\propto \frac{1}{\sqrt{2\pi\exp(h_t)}}\exp\left[-\frac{r_t^2}{2\exp(h_t)}\right] p(h_t|h_{t-1}, \theta)p(h_{t+1}|h_t, \theta) \\ &= p^*(r_t|h_t, \theta)\, p(h_t|h_{t-1}, \theta)p(h_{t+1}|h_t, \theta). \end{aligned}$$

Our next step is to show that

$$p(h_t|h_{t-1}, h_{t+1}, \theta) = p(h_t|h_{t-1}, \theta)p(h_{t+1}|h_t, \theta) = p_N(h_t|\alpha_t, \beta^2),$$

where $p_N(h_t|\alpha_t, \beta^2)$ denotes a normal density function with mean α_t and β^2 as variance, which we show below.

$$\begin{aligned} & p(h_t|h_{t-1}, \theta)p(h_{t+1}|h_t, \theta) \\ &\propto \exp\left\{-\frac{1}{2\sigma_\eta^2}\left[(h_t - \mu - \phi h_{t-1})^2 + (h_{t+1} - \mu - \phi h_t)^2\right]\right\} \\ &\propto \exp\left\{-\frac{1}{2\sigma_\eta^2}\left[h_t^2 - 2h_t(\mu + \phi h_{t-1}) + (\mu - \phi h_{t-1})^2 + (h_{t+1} - \mu)^2\right.\right. \\ &\quad \left.\left. - 2\phi h_t(h_{t+1} - \mu) + \phi^2 h_t^2\right]\right\} \\ &\propto \exp\left\{-\frac{1}{2\sigma_\eta^2}\left\{(1 + \phi^2)h_t^2 - 2h_t[\mu(1 - \phi) + \phi(h_{t-1} + h_{t+1})] + (h_{t+1} - \mu)^2\right\}\right\} \\ &\propto \exp\left[-\frac{1}{2\beta^2}\left\{(h_t - \alpha_t)^2 - \alpha_t^2 + (h_{t+1} - \mu)^2\right\}\right] \\ & \text{where } \alpha_t = \frac{\mu(1 - \phi) + \phi(h_{t-1} + h_{t+1})}{1 + \phi^2} \quad \text{and} \quad \beta^2 = \frac{\sigma_\eta^2}{1 + \phi^2} \end{aligned}$$

and terms, unrelated to h_t, h_{t+1} have been ignored as constants. It is still not easy to sample from this density. Because we have to find a way to linearize $\exp(h_t)$ so that an accept–reject algorithm can be implemented. We have to find a bound density, say $g^*(.)$, which is easier to simulate than $p^*(.)$. Two suggestions as to how to proceed with sampling from the above posterior density are to use a Metropolis–Hastings accept–reject algorithm as used by Jacquier et al. (1994) or use a simple reject algorithm as recommended by Kim et al. (1998). We shall outline the simple accept–rejection method of the latter.

10.21.3 Simple Rejection Sampling

Kim et al. (1998) consider the same model specification as Shephard and Pitt (1997); but they employ a first-order Taylor series expansion to obtain the pseudo-dominating density. They use the fact that, since $\exp(-h_t)$ is bounded below by its tangent at any point, we expand $\exp(-h_t)$ around it's mean α_t to obtain:

$$\begin{aligned}\log p^*(r_t|h_t,\theta) &\le -\frac{1}{2}\log 2\pi - \frac{1}{2}h_t - \frac{r_t^2}{2}\left[\exp(-\alpha_t)\left\{(1+\alpha_t)-h_t\right\}\right] \\ &\le \log g^*(r_t|h_t,\theta).\end{aligned}$$

Since we have shown that $p(h_t|h_{t-1},h_{t+1},\theta) = p_N(\alpha_t,\beta^2)$, we have

$$p^*(r_t|h_t,\theta)p(h_t|h_{t-1},h_{t+1},\theta) \le g^*(r_t|h_t,\theta)\, p_N(h_t|\alpha_t,\beta^2).$$

We can further demonstrate that the right-hand side of the above expression, ignoring constants and terms unrelated to h_t, is dominated by a normal density with mean α_t^* and variance β^2, as shown below:

$$\begin{aligned}p^*(r_t|h_t,\theta)p(h_t|h_{t-1},h_{t+1},\theta) &\propto \exp\left(-\frac{h_t}{2}\right) - \frac{r_t^2}{2}\exp(-\alpha_t)(1+\alpha_t) + \frac{r_t^2}{2}\exp(-\alpha_t)h_t \\ &\quad + \exp\left[-\frac{(h_t-\alpha_t)^2}{2\beta^2}\right] \\ &\propto \exp\left\{-\frac{1}{2\beta^2}\left[(h_t-\alpha_t)^2 - h_t\left(r_t^2\exp(-\alpha_t)-1\right)\beta^2\right]\right\} \\ &\propto \exp\left\{-\frac{(h_t-\alpha_t^*)^2}{2\beta^2}\right\} = g_N(\alpha_t^*,\beta^2), \\ &\quad \text{where } \alpha_t^* = \alpha_t + \frac{\beta^2}{2}\left[r_t^2\exp(-\alpha_t)-1\right]\end{aligned}$$

and $g_N(h_t|\alpha_t^*,\beta^2)$ denotes a normal density function with mean α_t^* and β^2 as variance. Since the target distribution, $p(h_t|\mathbf{h}_{-t},\mathbf{r},\theta)$, is bounded by $g_N(h_t|\alpha_t^*,\beta^2)$, simple rejection sampling can be used. Thus, we update log-volatility states for each time period by drawing proposals, h_t^*, from an $N(\alpha_t^*,\beta^2)$ density, with acceptance probability given by the rule $\min\left\{1,\ p\left(h_t|\mathbf{h}_{-t},\mathbf{r},\theta\right)/g_N\left(x_t|\alpha_t^*,\beta^2\right)\right\}$. If accepted, we set $h_t = h_t^*$ and proceed to update h_{t+1}. Kim et al. (1998) show that draws from their proposal density have a very high acceptance rate.

The important steps can be outlined as follows:

For $t = 1,\ldots,T$:

1. Draw h_t^* from $g_N(h_t|\alpha_t^*,\beta^2)$.
2. Draw U from $U[0,1]$.

3. Calculate $r = \dfrac{p^*(r_t|h_t^*, \theta)}{g^*(r_t|h_t^*, \theta)}$.
4. If $U < r$, then set $h_t = h_t^*$.
   Else
   go to step 1

10.21.4 Application

We consider the Nifty returns with 2252 observations, the same data set we had used before. We estimate the following version of original SV model of Taylor's SV model:

$$r_t = \upsilon_t \exp(h_t/2), \quad h_t = \mu + \phi h_{t-1} + \eta_t, \quad \eta_t \sim N(0, \sigma_\eta^2), \quad \upsilon_t \sim N(0, 1).$$

We implemented the Jacquier et al. (1994) accept–rejection method with the refinement suggested by Kim et al. (1998) to the rejection technique. We used the codes available in RATS. We ran an initial GARCH(1,1) model and used the estimated volatilities for h_t, as starting values. We used a total of 20,000 draws, with the burn-in sample being 10,000.

We have plotted below the posterior densities of all the three parameters of the variance equation. In the plots, density for Alpha refers to the density of constant in the variance equation, density for Delta refers to the density of the GARCH coefficient, and density for Sigma refers to variance of the error in the variance equation. The plots reveal that none of the three parameters are concentrated and are wide.

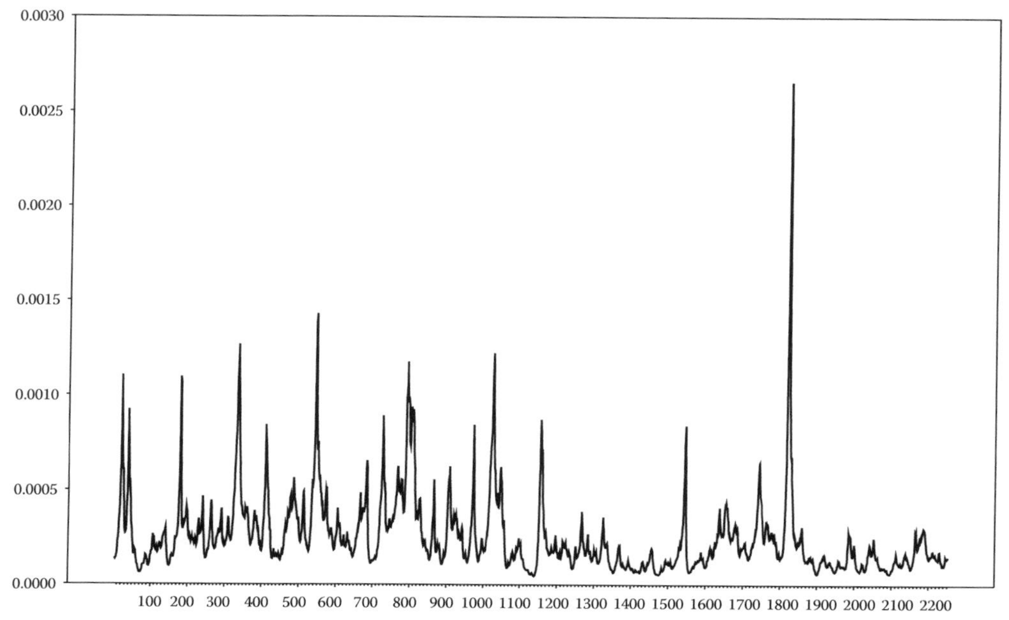

Mean estimate of volatility

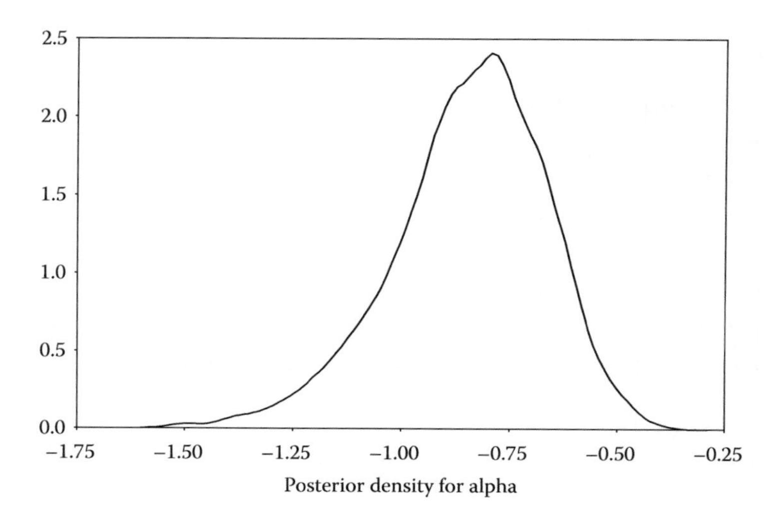
2.5
2.0
1.5
1.0
0.5
0.0
−1.75
−1.50
−1.25
−1.00
−0.75
−0.50
−0.25
Posterior density for alpha

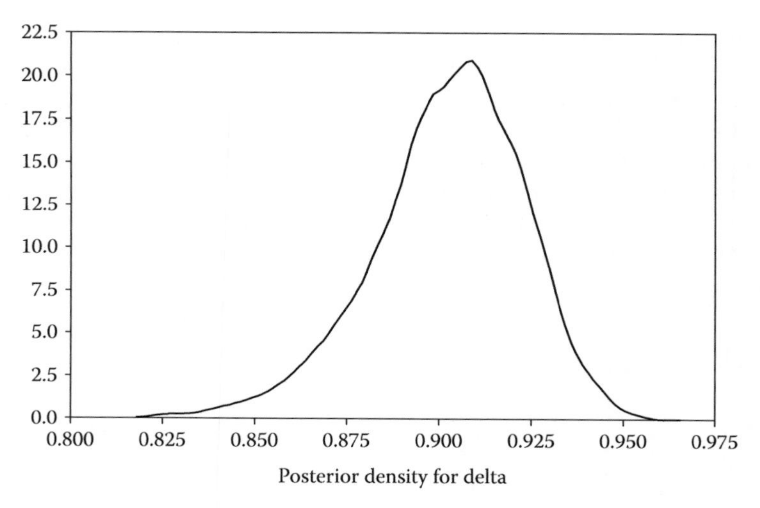
22.5
20.0
17.5
15.0
12.5
10.0
7.5
5.0
2.5
0.0
0.800
0.825
0.850
0.875
0.900
0.925
0.950
0.975
Posterior density for delta

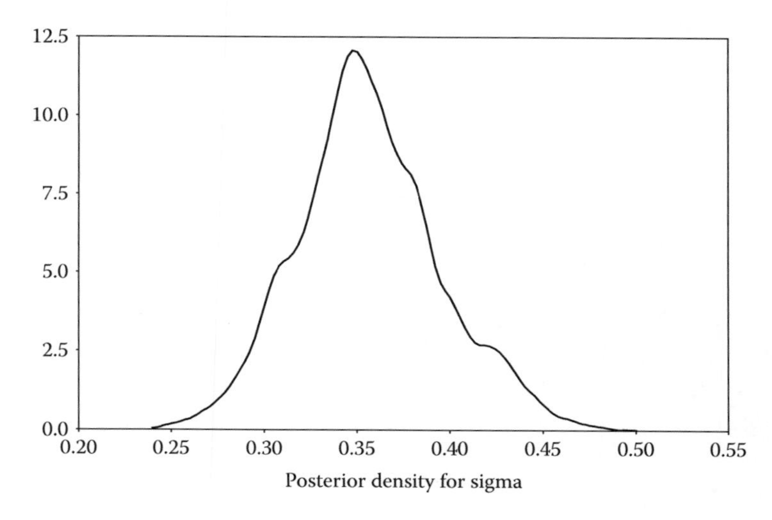
12.5
10.0
7.5
5.0
2.5
0.0
0.20
0.25
0.30
0.35
0.40
0.45
0.50
0.55
Posterior density for sigma

The posterior mean and standard error of the three coefficients of the variance equation are given below:

Parameter	α	δ	σ_η^2
Mean	−0.792	0.908	0.342
Standard error	0.165	0.019	0.038

It is interesting to note that the posterior mean of the GARCH coefficient is not too far off that obtained in the GARCH(1,1) model for the same data set.

10.22 GARCH versus SV Models

Both GARCH and SV models aim to explain the stylized facts of returns and forecast volatility. Since SV models allow for volatility to be expressed as an unobserved variable to be estimated explicitly, it implies that such models may provide a better in-sample fit than GARCH models. But the parameters of an SV model are difficult to estimate. Several comparisons between these two models have been made in the literature. Kobayashi and Shi (2005) compare the two models, with the EGARCH model as the null against the SV model, where Franses et al. (2008) compare a GARCH model against an SV model. Kim et al. (1998) propose a nonnested likelihood ratio test between SV and GARCH models and show that for financial series, SV models are a better fit than the GARCH models. Hansen et al. (2003) suggest a model confidence set procedure to select the best volatility model based on superior forecasting abilities. The model that minimizes the expected loss, for any loss function specified by the user, is the best model. However, as shown by Nelson (1990a,b), in the limit, both GARCH and SV models display similar characteristics and EGARCH, in the limit, is the diffusion process that motivates the SV models which prompt Taylor (2005) to conclude that comparing the two models may not be a fruitful exercise.

10.23 Conclusions

The importance of correctly specifying conditional volatility has deep implications for quantifying risk. Naturally, this led to a plethora of conditional volatility models, both deterministic and stochastic. In this chapter, we have outlined the theoretical steps of some important models, which are generally considered to be pioneering ones. And we have also applied these models on the Nifty returns of NSE of India. Our results convincingly support the presence of asymmetric effects, but there is no clear evidence about the type of asymmetric models. Also, in the Indian market, small shocks contribute to innovation as much large shocks. While this chapter is not an exhaustive survey of volatility models, including topics such as GARCH with jumps will definitely make it more complete.

References

Bauwens, L., C. Hafner, and S. Laurent (Eds.). 2012. *Handbook of Volatility Models and Their Applications*, John Wiley, New Jersey.

Bekaert, G. and G. Wu. 2000. Asymmetric volatility and risk in equity markets, *Review of Financial Studies,* 13, 1–42.

Bera, A. K. and M.L. Higgins. 1993. ARCH models: Properties, estimation and testing, *Journal of Economic Surveys*, 7, 305–362.

Bollerslev, T. 1986. Generalized autoregressive conditional heteroscedasticity, *Journal of Econometrics*, 31, 307–327.

Bollerslev, T. 1998. On the correlation structure for the generalized autoregressive conditional heteroscedastic process, *Journal of Time Series Analysis*, 9, 121–131.

Bollerslev, T., R.Y. Chou, and K.F. Kroner. 1992. Arch modelling in finance: A review of the theory and empirical evidence, *Journal of Econometrics*, 52, 5–59.

Bollerslev, T. and J.M. Wooldridge. 1992. Quasi maximum likelihood estimation and inference in dynamic models with time varying covariances, *Econometric Review*, 11, 143–172.

Bougerol, P. and N. Picard. 1992. Stationarity of GARCH processes and of some nonnegative time series, *Journal of Econometrics*, 52, 115–127.

Cao, C.Q. and R. Tsay. 1992. Nonlinear time-series analysis of stock volatilities, *Journal of Applied Econometrics*, 7, S165–S185.

Chib, S. and E. Greenberg. 1995. Understanding the Metropolis–Hastings algorithm, *The American Statistician*, 49, 327–335.

Cont, R. 2001. Empirical properties of asset returns: Stylized facts and statistical issues, *Quantitative Finance*, 1, 223–236.

Ding, Z., C.W.J. Granger, and R.F. Engle. 1993. A long memory property of stock market returns and a new model, *Journal of Empirical Finance*, 1, 83–106.

Engle, R.F. 1982. Autoregressive conditional heteroscedasticity with estimates of the variance of U.K. inflation, *Econometrica*, 50, 267–288.

Engle, R.F. 2002. New frontiers for ARCH models, *Journal of Applied Econometrics*, 17, 425–446.

Engle, R.F. and T. Bollerslev. 1986. Modelling the persistence of conditional variances, *Econometric Reviews*, 5, 1–50.

Engle, R.F. and V.K. Ng. 1993. Measuring and testing the impact of news on volatility, *Journal of Finance*, 48, 1749–1778.

Franses, P. H., M.V.D. Leij, and R. Paap. 2008. A simple test for GARCH against a stochastic volatility model, *Journal of Financial Econometrics*, 6, 291–306.

Ghysels, E., A.C. Harvey, and E. Renault. 1996. Stochastic volatility. In *Handbook of Statistics, Volume 14, Statistical Methods in Finance*, Eds. C.R. Rao and G.S. Maddala, 119–191, North-Holland, Amsterdam.

Glosten, L.R., R. Jagannathan, and D.E. Runkle. 1993. On the relation between expected value and the volatility of the nominal excess returns on stocks, *Journal of Finance*, 48, 1779–1801.

Greene, W.H. 2000. *Econometric Analysis*, Fourth Edition, Prentice Hall International, Upper Saddle River, New Jersey.

Hansen, P.R., A. Lunde, and J.M. Nason. 2003. Choosing the best volatility models: The model confidence set approach, *Oxford Bulletin of Economics and Statistics*, 65, 839–861.

Harvey, A.C., E. Ruiz, and N. Shephard. 1994. Multivariate stochastic variance models, *Review of Economic Studies*, 61, 247–264.

Hastings, W.K. 1970. Monte Carlo sampling methods using Markov chains and their applications, *Biometrika*, 57, 97–109.

Hautsch, N. and Y. Ou. 2009. Stochastic volatility estimation using Markov chain simulation, in *Applied Quantitative Finance*, Eds. W. Hardle, N. Hautsch, and L. Overbeck, Springer, Berlin.

He, C. and T. Terasvirta. 1999. Properties of moments of a family of GARCH processes, *Journal of Econometrics*, 92, 173–192.

Hentschel, L. 1995. All in the family: Nesting symmetric and asymmetric GARCH models, *Journal of Financial Economics*, 39, 71–104.

Jacquier, E., N.G. Polson, and P.E. Rossi. 1994. Bayesian analysis of stochastic volatility models, *Journal of Business and Economic Statistics*, 12, 371–389.

Kim, S., N. Shephard, and S. Chib. 1998. Stochastic volatility: Likelihood inference and comparison with ARCH models, *Review of Economic Studies*, 65, 361–393.

Kho, B.C. 1996. Time-varying risk premia, volatility, and technical trading rule profits: Evidence from foreign currency futures markets, *Journal of Financial Economics*, 41, 249–290.

Kobayashi, M. and X. Shi. 2005. Testing for EGARCH against stochastic volatility models, *Journal of Time Series Analysis*, 26, 135–150.

Krishnan, R. and C. Mukherjee. 2010. Volatility in Indian stock markets—A conditional variance tale re-told, *Journal of Emerging Market Finance*, 9, 71–93.

Lee, J.H.H. 1991. A Lagrange multiplier test for GARCH models, *Economics Letters*, 37, 265–271.

Li, W.K., S. Ling, and M. McAleer. 2002. Recent theoretical results for time series models with GARCH errors, *Journal of Economic Surveys*, 16, 245–269.

Li, Y. 1998. Time variations in risk premia, volatility, and reward-to-volatility, *Journal of Financial Research*, 21, 431–446.

Ling, S. and W.K. Li. 1998. Limiting distributions of maximum likelihood estimators for unstable ARMA models with GARCH errors, *Annals of Statistics*, 26, 84–125.

Ling, S., W.K. Li, and M. McAleer. 2003. Estimation and testing for unit root processes with GARCH (1,1) errors: Theory and Monte Carlo evidence, *Econometric Reviews*, 22, 179–202.

Lobo, B.J. and D. Tufte. 1998. Exchange rate volatility: Does politics matter? *Journal of Macro Economics*, 20, 351–365.

Longerstaey, J. and M. Spencer. 1996. *RiskMterics™—Technical Document*, Fourth Edition, Morgan Guarantee Trust Company, New York.

Lumsdaine, R.L. 1996. Consistence and asymptotic normality of the quasi-maximum likelihood estimator in IGARCH(1,1) and covariance stationary GARCH(1,1) models, *Econometrica*, 64, 575–596.

Mandelbrot, B. 1963. New methods in statistical economics, *Journal of Political Economy*, 71, 441–440.

Milhoj, A. 1985. Moment structure of ARCH processes, *Scandinavian Journal of Statistics*, 12, 281–292.

Merton, R.C. 1980. On estimating the expected return on the market: An exploratory investigation, *Journal of Financial Economics*, 8, 323–361.

Metropolis, N., A. Rosenbluth, M. Rosenbluth, A. Teller, and E. Teller. 1953. Equations of state calculations, by fast computing machines, *Journal of Chemical Physics*, 21, 1087–1092.

Nelson, D.B. 1990a. ARCH models as diffusion approximations, *Journal of Econometrics*, 45, 7–38.

Nelson, D.B. 1990b. Stationarity and persistence in the GARCH(1,1) model, *Econometric Theory*, 6, 318–334.

Nelson, D.B. 1991. Conditional heteroscedasticity in asset returns: A new approach, *Econometrica*, 62, 347–370.

Nelson, D.B. and C.Q. Cao. 1992. Inequality constraints in the univariate GARCH model, *Journal of Business and Economic Statistics*, 10, 229–235.

Pagan, A.R. and W.G. Schwert. 1990. Alternative models for conditional stock volatility, *Journal of Econometrics*, 45, 267–290.

Palm, F.C. 1996. GARCH models of volatility, in *Handbook of Statistics*, Eds. G.S. Maddala and C.R. Rao, Vol. 14, North-Holland, Elsevier Science, Amsterdam.

Pantula, S.G. 1989. Estimation of autoregressive models with ARCH errors, *Sankhya*, 50, 119–138.

Robert, C.P. and G. Casella. 2010. *Introducing Monte Carlo Methods with R*, Springer, USA.

Seo, B. 1999. Distribution theory for unit roots with conditional heteroscedasticity, *Journal of Econometrics*, 91, 113–144.

Shephard, N. 1993. Fitting non-linear time series models with applications to stochastic variance models, *Journal of Applied Econometrics*, 8, S135–S152.

Shephard, N. 1996. Statistical aspects of ARCH and stochastic volatility, In *Time Series Models: In Econometrics, Finance and Other Fields*, Eds. D.R. Cox, D.V. Hinkley, and O.E. Bandorff-Nielsen, Chapman-Hall, London.

Shephard, N. and S. Kim. 1994. Comment on 'Bayesian analysis of stochastic volatility models' by Jacquier, Polson and Rossi, *Journal of Business and Economic Statistics*, 12, 406–410.

Shephard, N. and M.K. Pitt. 1997. Likelihood analysis of non-Gaussian measurement time series, *Biometrika*, 84, 653–667.

Taylor, S.J. 2008, *Modelling Financial Time Series*, World Scientific Publishing, Singapore.

Taylor, S.J. 2005. *Asset Prices, Dynamics, Volatility and Prediction*, Princeton University Press, New Jersey, USA.

Tsay, R. 2005. *Analysis of Financial Time Series*, John Wiley, USA.

Vazquez, J. 2004. Switching regimes in the term structure of interest rates during U.S post-war: A case for Lucas proof equilibrium? *Studies-in-Nonlinear-Dynamics-and-Econometrics*, 8, 1–39.

Weiss, A.A. 1984. ARMA models with ARCH errors, *Journal of Time Series Analysis*, 5, 129–143.

Weiss, A.A. 1986. Asymptotic theory for ARCH models: Estimation and testing, *Econometric Theory*, 2, 107–131.

11

Metaheuristic Techniques

Sunith Bandaru and Kalyanmoy Deb

CONTENTS

ABSTRACT Most real-world search and optimization problems involve complexities such as nonconvexity, nonlinearities, discontinuities, mixed nature of variables, multiple disciplines and large dimensionality, a combination of which renders classical provable algorithms to be either ineffective, impractical or inapplicable. There do not exist any known mathematically motivated algorithms for finding the optimal solution for all such problems in a limited computational time. Thus, in order to solve such problems to practicality, search and optimization algorithms are usually developed using certain heuristics that, though lacking in strong mathematical foundations, are nevertheless good at reaching an approximate solution in a reasonable amount of time. These so-called metaheuristic methods do not guarantee finding the exact optimal solution, but can lead to a near-optimal solution in a computationally efficient manner. Owing to this practical appeal combined with their ease of implementation, metaheuristic methodologies are gaining popularity in several application domains. Most metaheuristic methods are stochastic in nature and mimic a natural, physical, or biological principle resembling a search or an optimization process. In this chapter, we discuss a number of such methodologies, specifically evolutionary algorithms, such as genetic algorithms and evolution strategy, particle swarm optimization, ant colony optimization, bee colony optimization, simulated annealing, and a host of other methods. Many metaheuristic methodologies are being proposed by researchers all over the world on a regular basis. It therefore becomes important to unify them to understand common features of different metaheuristic methods and simultaneously to study fundamental differences between them. Hopefully, such endeavors will eventually allow a user to choose the most appropriate metaheuristic method for the problem at hand.

11.1 Introduction

Most real-world search and optimization problems involve complexities such as nonconvexity, nonlinearities, discontinuities, mixed nature of variables, multiple disciplines, and large dimensionality, a combination of which renders classical provable algorithms to be ineffective, impractical, or inapplicable. There do not exist any known mathematically motivated algorithms for finding the optimal solution for all such problems in a limited computational time. Thus, in order to solve such problems to practicality, search and optimization algorithms are usually developed using certain *heuristics* that, though lacking in strong mathematical foundations, are nevertheless good at reaching an approximate solution in a reasonable amount of time. These so-called metaheuristic methods do not guarantee finding the exact optimal solution, but can lead to a near-optimal solution in a computationally efficient manner. Owing to this practical appeal combined with their ease of implementation, metaheuristic methodologies are gaining popularity in several application domains. Most metaheuristic methods are stochastic in nature and mimic a natural, physical, or biological principle resembling a search or an optimization process. In this chapter, we discuss a number of such methodologies, specifically evolutionary algorithms, such as genetic algorithms (GAs) and evolution strategy (ES), particle swarm optimization (PSO), ant colony optimization (ACO), bee colony optimization, simulated annealing (SA), and a host of other methods. Many metaheuristic methodologies are being proposed by researchers all over the world on a regular basis. It therefore becomes important to unify them to understand common features of different metaheuristic methods and simultaneously to study fundamental differences between them. Hopefully, such endeavors will eventually allow a user to choose the most appropriate metaheuristic method for the problem at hand.

This chapter does not claim to be a thorough survey of all existing metaheuristic methods and their related aspects. Instead, the purpose is to describe some of the most popular methods and provide pseudocodes to enable beginners to easily implement them. Therefore, this chapter should be seen more as a quick-start guide to popular metaheuristics than as a survey of methods and applications. Readers interested in the history, variants, empirical analysis, specific applications, tuning, and parallelization of metaheuristic methods should refer to the following texts:

1. Books: [1–13]
2. Surveys: [14–24]

3. Combinatorial problems: [25–34]
4. Analysis: [35–43]
5. Parallelization: [44–52]

The development of new metaheuristic methods has picked up pace over the last 20 years. Nowadays, many conferences, workshops, symposiums, and journals accept submissions related to this topic. Some of them (in no particular order) are

Conferences/Symposiums:

1. Genetic and Evolutionary Computation Conference (GECCO)
2. IEEE Congress on Evolutionary Computation (CEC)
3. Evolutionary Multi-criterion Optimization (EMO)
4. Parallel Problem Solving from Nature (PPSN)
5. Foundations of Genetic Algorithms (FOGA)
6. Simulated Evolution And Learning (SEAL)
7. Learning and Intelligent OptimizatioN (LION)
8. International Joint Conference on Computational Intelligence (IJCCI)
9. International Conference on Artificial Intelligence and Soft Computing (ICAISC)
10. IEEE Symposium Series on Computational Intelligence (SSCI)
11. IEEE Swarm Intelligence Symposium (SIS)

Journals:

1. *IEEE Transactions on Evolutionary Computation* (IEEE Press)
2. *Applied Soft Computing* (Elsevier)
3. *Computers & Operations Research* (Elsevier)
4. *European Journal of Operational Research* (Elsevier)
5. *Information Sciences* (Elsevier)
6. *Evolutionary Computation* (MIT Press)
7. *Computational Optimization and Applications* (Springer)
8. *Soft Computing* (Springer)
9. *Engineering Optimization* (Taylor & Francis)
10. *IEEE Transactions on Systems, Mans and Cybernetics* (IEEE Press)
11. *Engineering Applications of Artificial Intelligence* (Elsevier)
12. *International Transactions in Operational Research* (Wiley)
13. *Intelligent Automation & Soft Computing* (Taylor & Francis)
14. *Applied Computational Intelligence and Soft Computing* (Hindawi)
15. *Journal of Multi-Criteria Decision Analysis* (Wiley)
16. *Artificial Life* (MIT Press)
17. *Journal of Mathematical Modelling and Algorithms in Operations Research* (Springer, discontinued)
18. *International Journal of Artificial Intelligence & Applications* (AIRCC)

Some publications that are entirely dedicated to metaheuristics also exist, namely,

1. *Journal of Heuristics* (Springer)
2. *Swarm and Evolutionary Computation* (Elsevier)
3. *Swarm Intelligence* (Springer)
4. *Natural Computing* (Springer)
5. *Genetic Programming and Evolvable Machines* (Springer)
6. *International Journal of Metaheuristics* (Interscience)
7. *International Journal of Bio-Inspired Computation* (Interscience)
8. *Memetic Computing* (Springer)
9. *International Journal of Applied Metaheuristic Computing* (IGI Global)
10. *Computational Intelligence and Metaheuristic Algorithms with Applications* (Hindawi)
11. *European Event on Bio-Inspired Computation* (EvoStar Conference)
12. *International Conference on Adaptive & Natural Computing Algorithms* (ICANNGA)
13. *Ant Colony Optimization and Swarm Intelligence* (ANTS Conference)
14. *Swarm, Evolutionary and Memetic Computing Conference* (SEMCCO)
15. *Bio-Inspired Computing: Theories and Applications* (BICTA)
16. *Nature and Biologically Inspired Computing* (NaBIC)
17. *International Conference on Soft Computing for Problem Solving* (SocProS)
18. *Metaheuristics International Conference* (MIC)
19. *International Conference on Metaheuristics and Nature Inspired Computing* (META Conference)

Implementation of metaheuristic methods, though mostly straightforward, can be a tedious task. Luckily, several software frameworks are freely available on the Internet, which can be used by beginners to get started with solving their optimization problems. Notable examples are

1. PISA: A platform and programming language-independent interface for search algorithms (`http://www.tik.ee.ethz.ch/pisa/`)
2. ParadisEO: A C++ software framework for metaheuristics (`http://paradiseo.gforge.inria.fr/`)
3. Open BEAGLE: A C++ evolutionary computation framework (`https://code.google.com/p/beagle/`)
4. Evolving Objects: An evolutionary computation framework (`http://eodev.sourceforge.net/`)
5. GAlib: A C++ library of genetic algorithm components (`http://lancet.mit.edu/ga/`)
6. METSlib: A metaheuristic modeling framework and optimization toolkit in C++ (`https://projects.coin-or.org/metslib`)
7. ECF: A C++ evolutionary computation framework (`http://gp.zemris.fer.hr/ecf/`)
8. HeuristicLab: A framework for heuristic and evolutionary algorithms (`http://dev.heuristiclab.com/`)
9. ECJ: A Java-based evolutionary computation research system (`https://cs.gmu.edu/~eclab/projects/ecj/`)

10. jMetal: Metaheuristic algorithms in Java (`http://jmetal.sourceforge.net/`)
11. MOEA Framework: A free and open source Java framework for multiobjective optimization (`http://www.moeaframework.org/`)
12. JAMES: A Java metaheuristics search framework (`http://www.jamesframework.org/`)
13. Watchmaker Framework: An object-oriented framework for evolutionary/genetic algorithms in Java (`http://watchmaker.uncommons.org/`)
14. Jenetics: An evolutionary algorithm library written in Java (`http://jenetics.io/`)
15. Pyevolve: A complete genetic algorithm framework in Python (`http://pyevolve.sourceforge.net/`)
16. DEAP: Distributed evolutionary algorithms in Python (`https://github.com/DEAP/deap`)

These implementations provide the basic framework required to run any of the several available metaheuristics. Software implementations specific to individual methods are also available in plenty. For example, genetic programming (GP; discussed later in Section 11.10), which requires special solution representation schemes, has several variants in different programming languages, each handling the representation in a unique way.

This chapter is arranged as follows. In Section 11.2, we discuss basic concepts and classification of metaheuristics. We also lay down a generic metaheuristic framework which covers most of the available methods. From Section 11.3 to Section 11.14, we cover several popular metaheuristic techniques with complete pseudocodes. We discuss the methods in alphabetical order in order to avoid preconceptions about performance, generality, and applicability, thus respecting the No Free Lunch theorem [53] of optimization. In Section 11.15, we enumerate several less popular methods with a brief description of their origins. These methodologies are ordered by their number of Google Scholar citations. We conclude this chapter in Section 11.16 with a few pointers to future research directions.

11.2 Concepts of Metaheuristic Techniques

The word "heuristic" is defined in the context of computing as a method of denoting a rule-of-thumb for solving a problem without the exhaustive application of a procedure. In other words, a heuristic method is one that (i) looks for an approximate solution, (ii) need not particularly have a mathematical convergence proof, and (iii) does not explore each and every possible solution in the search space before arriving at the final solution, and hence is computationally efficient.

A metaheuristic method is particularly relevant in the context of solving search and optimization problems. It describes a method that uses one or more heuristics and therefore inherits all the three properties mentioned above. Thus, a metaheuristic method (i) seeks to find a near-optimal solution, instead of specifically trying to find the exact optimal solution, (ii) usually has no rigorous proof of convergence to the optimal solution, and (iii) is usually computationally faster than exhaustive search. These methods are iterative in nature and often use stochastic operations in their search process to modify one or more initial candidate solutions (usually generated by random sampling of the search space). Since many real-world optimization problems are complex due to their inherent practicalities, classical optimization algorithms may not always be applicable and may not fare well in solving such problems in a pragmatic manner. Realizing this fact and without disregarding the importance of classical algorithms in the development of the field of search and optimization, researchers and practitioners sought for metaheuristic methods so that a near-optimal solution can be obtained in a computationally tractable manner, instead of waiting for a provable optimization algorithm to be developed before attempting to solve such problems. The ability of the metaheuristic methods to handle different complexities associated with practical problems and arrive at a reasonably acceptable solution is the main reason for the popularity of metaheuristic methods in the recent past.

Most metaheuristic methods are motivated by natural, physical, or biological principles and try to mimic them at a fundamental level through various operators. A common theme seen across all metaheuristics is the balance between *exploration* and *exploitation*. Exploration refers to how well the operators diversify solutions in the search space. This aspect gives the metaheuristic a global search behavior. Exploitation refers to how well the operators are able to utilize the information available from solutions from previous iterations to intensify search. Such intensification gives the metaheuristic a local search characteristic. Some metaheuristics tend to be more explorative than exploitative, while some others do the opposite. For example, the primitive method of randomly picking solutions for a certain number of iterations represents a completely explorative search. On the other hand, hill climbing where the current solution is incrementally modified until it improves is an example of completely exploitative search. More commonly, metaheuristics allow this balance between diversification and intensification to be adjusted by the user through operator parameters.

The characteristics described above give metaheuristics certain advantages over the classical optimization methods, namely,

1. Metaheuristics can lead to *good enough* solutions for computationally easy (technically, P class) problems with large input complexity, which can be a hurdle for classical methods.
2. Metaheuristics can lead to *good enough* solutions for the NP-hard problems, that is, problems for which no known exact algorithm exists that can solve them in a reasonable amount of time.
3. Unlike most classical methods, metaheuristics require no gradient information and therefore can be used with nonanalytic, black-box, or simulation-based objective functions.
4. Most metaheuristics have the ability to recover from local optima due to inherent stochasticity or deterministic heuristics specifically meant for this purpose.
5. Because of the ability to recover from local optima, metaheuristics can better handle uncertainties in objectives.
6. Most metaheuristics can handle multiple objectives with only a few algorithmic changes.

11.2.1 Optimization Problems

A nonlinear programming (NLP) problem involving n real or discrete (integer, Boolean, or otherwise) variables or a combination thereof is stated as follows:

$$\begin{aligned} &\text{Minimize} f(\mathbf{x}), && \\ &\text{subject to} \quad g_j(\mathbf{x}) \geq 0, && j = 1, 2, \ldots, J, \\ &\qquad\qquad h_k(\mathbf{x}) = 0, && k = 1, 2, \ldots, K, \\ &\qquad\qquad x_i^{(L)} \leq x_i \leq x_i^{(U)}, && i = 1, 2, \ldots, n, \end{aligned} \tag{11.1}$$

where $f(\mathbf{x})$ is the objective function to be optimized, $g_j(\mathbf{x})$ represent J inequality constraints, and $h_k(\mathbf{x})$ represent K equality constraints. Typically, equality constraints are handled by converting them into soft inequality constraints. A survey of constraint handling methods can be found in References 54 and 55. A solution $\mathbf{x}$ is feasible if all $J + K$ constraints and variable bounds $[x_i^{(L)}, x_i^{(U)}]$ are satisfied.

The nature of the optimization problem plays an important role in determining the optimization methodology to be used. Classical optimization methods should always be the first choice for solving convex problems. An NLP problem is said to be convex if and only if (i) f is convex, (ii) all $g_j(\mathbf{x})$ are concave, and (iii) all $h_k(\mathbf{x})$ are linear. Unfortunately, most real-world problems are nonconvex and NP-hard, which makes metaheuristic techniques a popular choice.

Metaheuristics are especially popular for solving combinatorial optimization problems because not many classical methods can handle the kind of variables that they involve. A typical combinatorial optimization problem involves an n-dimensional permutation $\mathbf{p}$ in which every entity appears only once:

$$\begin{aligned} &\text{Minimize} f(\mathbf{p}), \\ &\text{subject to} \quad g_j(\mathbf{p}) \geq 0, \quad j = 1, 2, \ldots, J \\ &\qquad\qquad\quad h_k(\mathbf{p}) \geq 0, \quad k = 1, 2, \ldots, K. \end{aligned} \tag{11.2}$$

Again, a candidate permutation $\mathbf{p}$ is feasible only when all J constraints are satisfied. Many practical problems involve combinatorial optimization. Examples include knapsack problems, bin-packing, network design, traveling salesman, vehicle routing, facility location, and scheduling.

When multiple objectives are involved that conflict with each other, no single solution exists that can simultaneously optimize all the objectives. Rather, a multitude of optimal solutions are possible which provide a trade-off between the objectives. These solutions are known as the Pareto-optimal solutions and they lie on what is known as the Pareto-optimal front. A candidate solution $\mathbf{x_1}$ is said to *dominate* $\mathbf{x_2}$ and denoted as $\mathbf{x_1} \prec \mathbf{x_2}$ if and only if the following conditions are satisfied:

1. $f_i(\mathbf{x_1}) \leq f_i(\mathbf{x_2}) \; \forall i \in \{1, 2, \ldots, m\}$
2. $\exists j \in \{1, 2, \ldots, m\}$ such that $f_j(\mathbf{x_1}) < f_j(\mathbf{x_2})$

If only the first of these conditions is satisfied, then $\mathbf{x_1}$ is said to *weakly dominate* $\mathbf{x_2}$ and is denoted as $\mathbf{x_1} \preceq \mathbf{x_2}$. If neither $\mathbf{x_1} \preceq \mathbf{x_2}$ nor $\mathbf{x_2} \preceq \mathbf{x_1}$, then $\mathbf{x_1}$ and $\mathbf{x_2}$ are said to be nondominated with respect to each other and denoted as $\mathbf{x_1} || \mathbf{x_2}$. A feasible solution $\mathbf{x}^*$ is said to be Pareto-optimal if there does not exist any other feasible $\mathbf{x}$ such that $\mathbf{x} \prec \mathbf{x}^*$. The set of all such $\mathbf{x}^*$ (which are nondominated with respect to each other) is referred to as the Pareto-optimal set. Multiobjective optimization poses additional challenges to metaheuristics due to this concept of nondominance.

In this chapter, we will restrict ourselves to describing metaheuristics in the context of single-objective optimization. Most metaheuristics can be readily used for multiobjective optimization by simply considering Pareto-dominance or other suitable selection mechanisms when comparing different solutions and by taking extra measures for preserving solution diversity. However, it is important to note that since multiobjective optimization problems with conflicting objectives have multiple optimal solutions, metaheuristics that use a "population" of solutions are preferred so that the entire Pareto-optimal front can be represented simultaneously.

11.2.2 Classification of Metaheuristic Techniques

The most common way of classifying metaheuristic techniques is based on the number of initial solutions that are modified in subsequent iterations. Single-solution metaheuristics start with one initial solution which gets modified iteratively. Note that the modification process itself may involve more than one solution, but only a single solution is used in each following iteration. Population-based metaheuristics use more than one initial solution to start optimization. In each iteration, multiple solutions get modified, and some of them make it to the next iteration. Modification of solutions is done through operators that often use special statistical properties of the population. The additional parameter for size of the population is set by the user and usually remains constant across iterations.

Another way of classifying metaheuristics is through the domain that they mimic. Umbrella terms such as bio-inspired and nature-inspired are often used for metaheuristics. However, they can be further subcategorized as evolutionary algorithms, swarm-intelligence-based algorithms, and physical-phenomenon-based algorithms. Evolutionary algorithms (such as GAs, ESs, differential evolution [DE], GP, evolutionary programming [EP], etc.) mimic various aspects of evolution in nature such as survival of the fittest, reproduction, and genetic mutation. Swarm-intelligence algorithms mimic the group behavior and/or interactions of living organisms (such as ants, bees, birds, fireflies, glowworms, fishes, white blood cells, bacteria, etc.) and nonliving things (such as water drops, river systems, masses under gravity, etc.). The rest of the metaheuristics mimic various physical phenomena such as annealing of metals, musical aesthetics (harmony), etc. A fourth subcategory may be used to classify metaheuristics whose source of inspiration is unclear (such as tabu search [TS] and scatter search [SS]) or those that are too few in number to have a category for themselves.

Other popular ways of classifying metaheuristics are [1,35]

1. Deterministic versus stochastic methods: Deterministic methods follow a definite trajectory from the random initial solution(s). Therefore, they are sometimes referred to as trajectory methods. Stochastic methods (also discontinuous methods) allow probabilistic *jumps* from the current solution(s) to the next.
2. Greedy versus nongreedy methods: Greedy algorithms usually search in the neighborhood of the current solution and immediately move to a better solution when it is found. This behavior often leads to a local optimum. Nongreedy methods either hold out for some iterations before updating the solution(s) or have a mechanism to backtrack from a local optimum. However, for convex problems, a greedy behavior is the optimum strategy.
3. Memory usage versus memoryless methods: Memory-based methods maintain a record of past solutions and their trajectories and use them to direct search. A popular memory method called tabu search will be discussed in Section 11.14.
4. One versus various neighborhood methods: Some metaheuristics such as SA and TS only allow a limited set of moves from the current solution. But many metaheuristics employ operators and parameters to allow multiple neighborhoods. For example, PSO achieves this through various swarm topologies.
5. Dynamic versus static objective function: Metaheuristics that update the objective function depending on the current requirements of search are classified as dynamic. Other metaheuristics simply use their operators to control search.

Most metaheuristics are population-based, stochastic, nongreedy, and use a static objective function.

In this chapter, we avoid making any attempt at classifying individual metaheuristic techniques because of two reasons:

1. Single-solution metaheuristics can often be transformed into population-based ones (and vice versa) either by hybridization with other methods or, sometimes, simply by consideration of multiple initial solutions.

2. The large number of metaheuristics that have been derived off late from existing techniques use various modifications in their operators. Sometimes, these variants blur the extent of mimicry so much that referring to them by their domain of inspiration becomes unreasonable.

Therefore, for each metaheuristic, we discuss the most popular and generic variant.

11.2.3 A Generic Metaheuristic Framework

Most metaheuristics follow a similar sequence of operations and therefore can be defined within a common generic framework as shown in Algorithm 11.1.

The `REQUIRE` statement indicates that in order to prescribe a metaheuristic algorithm, a minimum of five plans (SP, GP, RP, UP, TP) and four parameters (N, μ, λ, ρ) are needed. The individual plans may involve one or more parameters of their own.

Step 1 initializes the iteration counter (t) to zero. It is clear from the repeat-until loop of the algorithm that a metaheuristic algorithm is by nature an iterative procedure.

Step 2 creates an initial set of N solutions in the set S_t. Mostly, the solutions are created at random in between the prescribed lower and upper bounds of variables. For combinatorial optimization problems, random permutations of entities can be created. When problem-specific knowledge of good initial solutions is available, they may be included along with a few random solutions to allow exploration.

Step 3 evaluates each of N set members by using the supplied objective and constraint functions. Constraint violation, if any, must be accounted for here to provide an evaluation scheme which will be used in subsequent steps of the algorithm. One way to define constraint violation $CV(\mathbf{x}/\mathbf{p})$ is given by

$$CV(\mathbf{x}/\mathbf{p}) = \sum_{j=1}^{J} \langle g_j(\mathbf{x}/\mathbf{p}) \rangle, \tag{11.3}$$

Algorithm 11.1: A Generic Metaheuristic Framework

Require: SP, GP, RP, UP, TP, N (≥ 1), μ ($\leq N$), λ, ρ ($\leq N$)
1: Set iteration counter $t = 0$
2: Initialize N solutions S_t randomly
3: Evaluate each member of S_t
4: Mark the best solution of S_t as $\mathbf{x}_t^*$
5: **repeat**
6: Choose μ solutions (set P_t) from S_t using a *selection plan (SP)*
7: Generate λ new solutions (set C_t) from P_t using a *generation plan (GP)*
8: Choose ρ solutions (set R_t) from S_t using a *replacement plan (RP)*
9: Update S_t by replacing R_t using ρ solutions from a pool of any combination of at most three sets P_t, C_t, and R_t using an *update plan (UP)*
10: Evaluate each member of S_t
11: Identify the best solution of S_t and update $\mathbf{x}_t^*$
12: $t \leftarrow t + 1$
13: **until** (a *termination plan (TP)* is satisfied)
14: Declare the near-optimal solution as $\mathbf{x}_t^*$

where $\langle\alpha\rangle$ is $|\alpha|$ if $\alpha < 0$, and is zero, otherwise. Before adding the constraint values of different constraints, they need to be normalized [56]. The objective function value $f(\mathbf{x})$ and constraint violation $CV(\mathbf{x}/\mathbf{p})$ value can both be sent to subsequent steps for a true evaluation of the solutions.

Step 4 chooses the best member of the set S_t and saves it as $\mathbf{x}_t^*$. This step requires a pairwise comparison of set members. One way to compare two solutions (A and B) for constrained optimization problems is to use the following parameterless strategy to select a winner:

1. If A and B are feasible, choose the one having smaller objective (f) value.
2. If A is feasible and B is not, choose A and vice versa.
3. If A and B are infeasible, choose the one having smaller CV value.

Step 5 puts the algorithm into a loop until a TP in Step 13 is satisfied. The loop involves Steps 6–12. TP may involve achievement of a prespecified target objective value (f_T) and will be satisfied when $f(\mathbf{x}_t^*) \leq f_T$. TP may simply involve a prespecified number of iterations T, thereby setting the termination condition $t \geq T$. Other TPs are also possible and this is one of the five plans that is to be chosen by the user.

Step 6 chooses μ solutions from the set S_t by using an SP. SP must carefully analyze S_t and select better solutions (using f and CV) of S_t. A metaheuristic method will vary largely based on what SP is used. The μ solutions form a new set P_t.

In Step 7, solutions from P_t are used to create a set of λ new solutions using a GP. This plan provides the main search operation of the metaheuristic algorithm. The plans will be different for function optimization and combinatorial optimization problems. The created solutions are saved in set C_t.

Step 8 chooses ρ worse or random solutions of S_t to be replaced by an RP. The solutions to be replaced are saved in R_t. In some metaheuristic algorithms, R_t can simply be P_t (requiring $\rho = \mu$), thereby replacing the very solutions used to create the new solutions. To make the algorithm more greedy, R_t can be ρ worst members of S_t.

In Step 9, ρ selected solutions from S_t are to be replaced by ρ other solutions. Here, a pool of at most $(\mu + \lambda + \rho)$ solutions of the combined pool $P_t \cup C_t \cup R_t$ is used to choose ρ solutions using an UP. If $R_t = P_t$, then the combined pool need not have both R_t and P_t in order to avoid duplication of solutions. UP can simply be choosing the best ρ solutions of the pool. Note that a combined pool of $R_t \cup C_t$ or $P_t \cup C_t$ or simply C_t is also allowed. If members of R_t are included in the combined pool, then the metaheuristic algorithm possesses an elite-preserving property, which is desired in an efficient optimization algorithm.

RP and UP play an important role in determining the robustness of the metaheuristic. A greedy approach of replacing the worst solutions with the best solutions may lead to faster yet premature convergence. Maintaining diversity in candidate solutions is an essential aspect to be considered when designing RPs and UPs.

Step 10 evaluates each of the new members of S_t and Step 11 updates the set-best member $\mathbf{x}_t^*$ using the best members of updated S_t. When the algorithm satisfies TP, the current-best solution is declared as the near-optimal solution.

The above metaheuristic algorithm can also represent a single solution optimization procedure by setting $N = 1$. This will force $\mu = \rho = 1$. If a single new solution is to be created in Step 7, $\lambda = 1$. In this case, SP and RP are straightforward procedures of choosing the singleton solution in S_t. Thus, $P_t = R_t = S_t$. The GP can choose a solution in the neighborhood of singleton solution in S_t using a Gaussian or other distribution. The UP will involve choosing a single solution from two solutions of $R_t \cup C_t$ (elite preservation) or choosing the single new solution from C_t alone and replacing P_t.

By understanding the role of each of the five plans and choosing their sizes appropriately, many different metaheuristic optimization algorithms can be created. All five plans may involve stochastic operations, thereby making the resulting algorithm stochastic.

With this generic framework in mind, we are now ready to describe a number of existing metaheuristic algorithms. Readers are encouraged to draw parallels between different operations used in these algorithms with the five plans discussed above. It should be noted that this may not always be possible since some operations may be spread across or combine multiple plans.

11.3 Ant Colony Optimization

ACO algorithms fall into the broader area of swarm intelligence in which problem-solving algorithms are inspired by the behavior of swarms. Swarm intelligence algorithms work with the concept of self-organization, which relies on any form of central control over the swarm members. A detailed overview of the self-organization principles exploited by these algorithms, as well as examples from biology, can be found in Reference 57. Here, we specifically discuss the ACO procedure.

As observed in nature, ants often follow each other on a particular path to and from their nest and a food source. An inquisitive mind might ask the question: "How do ants determine such a path and importantly, is the path followed by the ants optimal in any way?" The answer lies in the fact that when ants walk, they leave a trail of chemical substance called *pheromone*. The pheromone has the property of depleting with time. Thus, unless further deposit of pheromone takes place on the same pheromone trail, the ants will not be able to follow one another. This can be observed in nature by wiping away a portion of the trail with a finger. The next ant to reach the wiped portion can be observed to stop, turn back, or wander away. Eventually, an ant will "risk" crossing the wiped portion and finds the pheromone trail again. Thus, in the beginning, when the ants have no specific food source, they wander almost randomly on the lookout for one. While they are on the lookout, they deposit a pheromone trail which is followed by other ants if the pheromone content is strong enough. The stronger the pheromone trail, the higher the number of ants that are likely to follow the trail. When a particular trail of ants locates a food source, the act of repeatedly carrying small amounts of food back to the nest increases the pheromone content on that trail, thereby attracting more and more ants to follow it. However, if an ant ends up just wandering and eventually becomes unsuccessful in locating a food source, despite attracting some ants to follow its trail will only frustrate the following ants and eventually the pheromone deposited on that trail will decrease. This phenomenon is also true for suboptimal trails, that is, trails to locations with limited food source.

These ideas can be used to design an optimization algorithm for finding the shortest path. Let us say that two different ants are able to reach a food source using two different pheromone trails—one shorter and the other somewhat longer from source to destination. Since the strength of a pheromone trail is related to the amount of pheromone deposit, for an identical number of ants in each trail, the pheromone content will get depleted quickly for the longer trail, and by the same reason the pheromone content on the shorter trail will get increasingly strengthened. Thus, eventually there will be so few ants following the longer trail that the pheromone content will go below a limit that is not enough to attract any further ants. This fact of a strengthening shorter trails with pheromone allows the entire ant colony to discover an eventual minimal trail for a given source–destination combination. Interestingly, such a task is also capable of identifying the shortest distance configuration in a scenario having multiple food source and multiple nests.

Although observed in 1944 [58], the ants' ability to self-organize to solve a problem was only demonstrated in 1989 [59]. Dorigo and his coauthors suggested an ACO method in 1996 [60] that made the approach popular in solving combinatorial optimization problems. Here, we briefly describe two variants of the ACO procedure—the ant system (AS) and the ant colony system (ACS)—in the context of a traveling salesman problem (TSP).

In a TSP, a salesman has to start from a city and travel to the remaining cities one at a time, visiting each city exactly once so that the overall distanced traveled is minimum. Although it sounds simple, such a problem is shown to be NP-hard [61] with increasing number of cities. The first task in using an ACO is to represent a solution in a manner amenable by an ACO. Since pheromone can dictate the presence, absence, and strength of an path, a particular graph connecting cities in a TSP problem is represented in an ACO by the pheromone value of each edge. While at a particular node (i), an ant (say k-th ant) can then choose one of the neighboring edges (say from node i to neighboring node j) based on the existing pheromone value (τ_{ij}) of each edge (ij edge) probabilistically as follows:

$$p_{ij}^k = \begin{cases} \dfrac{\tau_{ij}^{\alpha}\eta_{ij}^{\beta}}{\sum_{l \in C_k} \tau_{ij}^{\alpha}\eta_{ij}^{\beta}}, & \text{if } j \in C_k, \\ 0, & \text{otherwise,} \end{cases} \tag{11.4}$$

where τ_{ij} is the existing pheromone value of the ij-th edge, heuristic information η_{ij} is usually the part of the objective value associated with the ij-th edge, C_k is the set of cities that are not visited by the k-th ant yet, and α and β are problem-specific parameters indicating the relative importance of existing pheromone content and heuristic information in deciding the probability event, respectively. For the AS approach, η_{ij} is chosen as the inverse of the distance between nodes i and j. Thus, among all allowable node from C_k, the probability is highest for the node (j) that is nearest to node i.

After the complete path is identified, the pheromone content of each edge is updated as follows:

$$\tau_{ij} = (1 - \rho)\tau_{ij} + \rho \sum_{k=1}^{m} \Delta\tau_{ij}^k, \tag{11.5}$$

where $\rho \in (0, 1)$ is the evaporation rate or pheromone decay rate, m is the number of ants in the AS, $\Delta\tau_{ij}^k$ is the quantity of pheromone laid by the k-th ant on edge ij given by

$$\Delta\tau_{ij}^k = \begin{cases} \dfrac{Q}{L_k}, & \text{if ant } k \text{ used the edge } ij \text{ in its tour,} \\ 0, & \text{otherwise.} \end{cases} \tag{11.6}$$

The parameter Q is a constant and L_k is the overall length of the tour by ant k. Thus, if the k-th ant finds a shorter tour (having smaller L_k), the pheromone content τ_{ij}^k gets increased by a larger amount. The first term in Equation 11.5 causes evaporation or decay of pheromone and the second term strengthens the pheromone content if an overall shorter tour is obtained.

The AS approach works as follows. Initially, random values of pheromone content within a specified range are assumed for each edge. Each ant in the system starts with a random city, and visits allowable neighboring cities selected probabilistically using Equation 11.4. For m ants, m such trails are constructed. Thereafter, pheromone levels are updated for each edge using Equation 11.5. The whole process is repeated until either all ants take the same trail or the maximum number of iterations is reached, in which case the shortest path found becomes the solutions. The AS algorithm has been used to solve quadratic assignment problems [62], job-shop scheduling problems [63], vehicle routing problem [64], etc.

In the ACS approach, in addition to a global pheromone update rule described later, a local pheromone update rule is also applied [65]. Each ant modifies the pheromone content of each edge as soon as it traverses it, as follows:

$$\tau_{ij} = (1 - \psi)\tau_{ij} + \psi\tau_0, \tag{11.7}$$

where $\psi \in (0, 1)$ and τ_0 is the initial value of the pheromone. The effect of this local pheromone update is to diversify the search procedure by allowing subsequent ants to be aware of the edges chosen by previous ants. The global pheromone update rule in ACS only effects the best tour of the k-ants in the current iteration (having a tour length L_b). The rest of the edges are only subject to evaporation. That is,

$$\tau_{ij} = \begin{cases} (1 - \rho)\tau_{ij} + \rho\Delta\tau_{ij}, & \text{if edge } ij \text{ belongs to tour,} \\ (1 - \rho)\tau_{ij}, & \text{otherwise,} \end{cases} \tag{11.8}$$

where $\Delta\tau_{ij} = 1/L_b$ is used. Another modification adopted in ACS concerns the edge probability. For an ant k, the choice of city j depends on a parameter q_0 (user-defined within [0,1]). If a random number $q \in [0, 1]$ is such that $q \le q_0$, then the next city is

$$j = \text{argmax}_{l \in C_k} \left\{ \tau_{il} \eta_{il}^{\beta} \right\}, \tag{11.9}$$

otherwise Equation 11.4 is used.

Both AS and ACS can be represented by the pseudocode in Algorithm 11.2.

ACO for solving continuous optimization problems is not straightforward, yet attempts have been made [66–68] to change the pheromone trail model, which is by definition a discrete probability distribution, to a continuous probability density function. Moreover, presently, there is no clear understanding of which algorithms may be called ant based. Stigmergy optimization is the generic name given to algorithms that use multiple agents exchanging information in an indirect manner [69].

11.4 Artificial Bee Colony Optimization

Karaboga [70,71] suggested a global-cum-local search-based optimization procedure based on the bee's foraging for nectar (food) in a multidimensional space. Food sources span the entire variable space and the location of a food source is a point in the variable space. The nectar content at a food source is considered as the objective function of that point in variable space. Thus, internally, artificial bee colony optimization (ABCO) method maximizes the objective function similar to bee's locating the food source having the highest nectar content. In other words, the task of finding the optimal solution of an objective function $f(\mathbf{x})$ is converted to an ABCO problem in which artificial bees will wander in an artificial multidimensional space to locate the highest-producing nectar source.

Algorithm 11.2: Ant Colony Optimization

Require: $m, n, \alpha, \beta, \rho, \psi, q_0, \tau_0, Q$
1: Set iteration counter $t = 0$
2: Set best route length L_{best} to ∞
3: **if** AS is desired **then**
4: Randomly initialize edge pheromone levels within a range
5: **end if**
6: **if** ACS is desired **then**
7: Initialize all edge pheromone levels to τ_0
8: **end if**
9: **repeat**
10: **for** $k = 1$ **to** m **do**
11: Randomly place ant k at one of the n cities
12: **repeat**
13: **if** ACS is desired **and** $rand(0, 1) \leq q_0$ **then**
14: Choose next city to be visited using Equation 11.9
15: **else**
16: Choose next city to be visited using Equation 11.4
17: **end if**
18: **if** ACS is desired **then**
19: Update current edge using local pheromone update rule in Equation 11.7
20: **end if**
21: **until** unvisited cities exist
22: Find best route having length L_b
23: **if** ACS is desired **then**
24: Update edges of best route using global pheromone update rule in Equation 11.8
25: **end if**
26: **if** AS is desired **then**
27: **for all** edges **do**
28: Update current edge using pheromone update rule in Equation 11.5
29: **end for**
30: **end if**
31: **end for**
32: **if** $L_b < L_{best}$ **then**
33: $L_{best} = L_b$
34: **end if**
35: $t \leftarrow t + 1$
36: **until** TP is satisfied
37: Declare route length L_{best} as near-optimal solution

In order to achieve the search task, the essential concept of bee colony's food foraging procedure is simulated in an artificial computer environment. A population of initial food sources $\mathbf{x}^{(k)}$ ($k = 1, 2, \ldots, N$, where N is the colony size) is randomly created in the entire variable space:

$$x_i^{(k)} = x_i^{(L)} + u_i \left(x_i^{(U)} - x_i^{(L)} \right), \tag{11.10}$$

where u_i is a random number in $[0, 1]$. Three different types of bees are considered, each having to do a different task. The *employed* bees consider a food source from their respective *memories* and find a new food source $\mathbf{v}_e$ in its neighborhood. Any neighborhood operator can be used for this purpose. Simply, a uniformly distributed food source within $\pm\mathbf{a}$ of the current memory location $\mathbf{x}_m$ can be used:

$$v_{e,i} = x_{m,i} + \phi_i \left(x_{m,i} - x_i^{(k)} \right), \tag{11.11}$$

where $\mathbf{x}^{(k)}$ is a randomly selected food source and ϕ_i is a random number in $[-a_i, a_i]$. The newly created food source $\mathbf{v}_e$ is then compared with $\mathbf{x}_m$ and the better food source is kept in the memory of the employed bee. The number of employed bees are usually set as 50% of the food sources (N).

Employed bees share their food source information (that are stored in their memories) with *onlooker* bees, that stay in the bee hive and observe the foraging act of employed bees. An onlooker bee chooses the food source location $\mathbf{v}_e$ found by an employed bee with a probability which is proportional to the nectar content of the food source $\mathbf{v}_e$. The higher the nectar content, the higher is the probability of choosing the food source. Once a food source is selected, it is again modified to find $\mathbf{v}_o$ in its neighborhood by using a similar procedure with the selected $\mathbf{v}_e$, as depicted in Equation 11.11. The better of the two food sources are kept in the memory of the onlooker bee. The number of onlooker bees is usually set as 50% of the food sources (N).

Finally, the third kind of bees—the *scout* bees—randomly choose a food source location using Equation 11.10 and act more like global overseers. If any employed bee cannot improve their memory location by a predefined number of trials (a limit parameter L), it becomes a scout bee and reinitializes its memory location randomly in the variable space. The number of scout bee is usually chosen as one. The algorithm is run for a maximum of T generations.

Each food source is associated with either an employed or an onlooker bee, thereby having a single food source in each of them. The overall ABCO algorithm has the typical structure as shown in Algorithm 11.3.

ABCO method has been applied to solve integer programming [72–74], combinatorial optimization problems [75,76], and multiobjective optimization problems [77].

Algorithm 11.3: Artificial Bee Colony Optimization Algorithm

Require: N, L
1: Set iteration counter $t = 0$
2: Initialize $N/2$ employer bees (solutions) to $N/2$ food sources using Equation 11.10
3: **repeat**
4: Find nectar content at employer bee locations (i.e., evaluate solutions $\mathbf{x}_m$)
5: Move each employer bee to a new food source in its neighborhood using Equation 11.11
6: Find nectar content at new employer bee locations (i.e., evaluate solutions $\mathbf{v}_e$)
7: Record employer bees that return to $\mathbf{x}_m$ due to lower nectar content at $\mathbf{v}_e$
8: Recruit $N/2$ onlooker bees to employer bee locations with probabilities proportional to their nectar content
9: Move each onlooker bee to a new food source in its neighborhood (similar to Equation 11.11)
10: Find nectar content at new onlooker bee locations (i.e., evaluate solutions $\mathbf{v}_o$)
11: Move each employer bee to the best location found by onlooker bees assigned to it
12: Record the best food source among all $\mathbf{x}_m$, $\mathbf{v}_e$, and $\mathbf{v}_o$
13: Convert employer bees that cannot find better food sources in L trials into scout bees
14: Initialize scout bees, if any, using Equation 11.10
15: $t \leftarrow t + 1$
16: **until** TP is satisfied
17: Declare the best food source (near-optimal solution)

11.5 Artificial Immune System

Immune systems in biological organisms protect living cells from an attack from foreign bodies. When attacked by *antigens*, different antibodies are produced to defend the organism. In a macro sense, the natural immune system works as follows. Antibodies that are efficient in fighting antigens are cloned and hypermutated, and more such antibodies are produced to block the action of antigens. Once successful, the organism learns to create efficient antibodies. Thus, if similar antigens attack later, the organism can quickly produce the right antibodies to mitigate their effect. The process of developing efficient antibodies can be simulated to solve optimization problems, in which antibodies can represent decision variables and the process of cloning, hypermutation, and selection can be thought as a search process for finding *optimal* antibodies that tend to match or reach the extremum values of antigens representing objectives. There are mainly two different ways such an immunization task is simulated, which we discuss next.

11.5.1 Clonal Selection Algorithm

Proposed by Burnet [78] in 1959, the clonal selection procedure works with the concept of cloning and *affinity maturation*. When exposed to antigens, B-cells (so called because they are produced in the bone marrow) that best bind with the attacking antigens get emphasized through cloning. The ability of binding depends on how well the *paratope* of an antibody matches with the *epitope* of an antigen. The closer the match, the stronger is the affinity of binding. In the context of a minimization problem, the following analogy can drive the search toward the minimum. The smaller the value of the objective function for an antibody representing a solution, the higher can be the affinity level. The cloned antibodies can then undergo hypermutation according to an inverse relationship to their affinity level. Receptor editing and introduction of random antibodies can also be implemented to maintain a diverse set of antibodies (solutions) in the population. These operations sustain the affinity maturation process.

The clonal selection algorithm CLONALG [79,80] works as follows. First, a population of N antibodies is created at random. An antibody can be a vector of real, discrete, or Boolean variables, or it can be a permutation or a combination of the above, depending on the optimization problem being solved. At every iteration, a small fraction (say, n out of N) of antibodies is selected from the population based on their affinity level. These high-performing antibodies are then cloned and then hypermutated to construct N_c new antibodies. The cloning is directly proportional to the ranking of the antibodies in terms of their affinity level. The antibody with the highest affinity level in a population has the largest pool of clones. Each clone antibody is then mutated with a probability that is higher for antibodies having a lower affinity level. This allows a larger mutated change to occur in antibodies that are worse in terms of antigen affinity (or worse in objective values). Since a large emphasis of affinity level is placed on the creation mechanism of new antibodies, some randomly created antibodies are also added to the population to maintain adequate diversity. After all newly created antibodies are evaluated to obtain their affinity level, $R\%$ worst antibodies are replaced by the best newly created antibodies. This process continues till a prespecified termination criterion is satisfied. Algorithm 11.4 shows the basic steps.

On careful observation, the clonal selection algorithm described above is similar to a GA, except that no recombination-like operator is used in the clonal selection algorithm. The clonal selection algorithm is a population-based optimization algorithm that uses overlapping populations and heavily relies on a mutation operator whose strength directly relates to the value of the objective function of the solution being mutated.

Algorithm 11.4: Artificial Immune Systems—Clonal Selection Algorithm

Require: N, n, N_c, R
1: Set iteration counter $t = 0$
2: Randomly initialize N antibodies to form S_t
3: **repeat**
4: Evaluate affinity (fitness) of all antibodies in S_t
5: Select n antibodies with highest affinities to form P_t
6: Clone the n antibodies in proportion to their affinities to generate N_c antibodies
7: Hypermutate the N_c antibodies to form C_t
8: Evaluate affinity of all antibodies in C_t
9: Replace $R\%$ worst antibodies in S_t with the best antibodies from C_t
10: Add new random antibodies to S_t
11: $t \leftarrow t + 1$
12: **until** TP is satisfied
13: Declare population best as near-optimal solution

11.5.2 Immune Network Algorithm

A natural immune system, which although becomes active when attacked by external antigen, sometimes can work against its own cells, thus stimulating, activating, or suppressing each other. To reduce the chance of events, the clonal selection algorithm has been modified to recognize antibodies (solutions) that are similar to one another in an evolving antibody population. Once identified, similar antibodies (in terms of a distance measure) are eliminated from the population through *clonal suppression*. Moreover, antibodies having very small affinity level toward attacking antigens (meaning worse than a threshold objective function value) are also eliminated from the population. In the artificial immune network (aiNET [81]) algorithm, a given number of affinity clones are selected to form an immune network. These clones are thought to form a *clonal memory*. The clonal suppression operation is performed on the antibodies of the clonal memory set. The aiNET algorithm works as follows. A population of N antibodies are created at random. A set of n antibodies are selected using their affinity level. Thereafter, each is cloned based on their affinity level in a manner similar to that in the clonal selection algorithm. Each clone is then hypermutated as before and is saved within a clonal memory set (M). After evaluation, all memory clones that have objective value worse than a given threshold ϵ are eliminated. Antibodies that are similar (e.g., solutions clustered together) to each other are also eliminated through the clonal suppression process using a suppression threshold σ_s. A few random antibodies are injected in the antibody population as before, and the process is continued till a termination criterion is satisfied. The pseudocode for a typical immune network algorithm is shown in Algorithm 11.5.

In the context of optimization, artificial immune system (AIS) algorithms are found to be useful in following problems. Dynamic optimization problems in which the optimum changes with time is a potential application domain [82] simply because AIS is capable of quickly discovering a previously found optimum if it appears again at a later time. The immune network algorithm removes similar solutions from the population, thereby allowing multiple disparate solutions to coexist in the population. This allowed AIS algorithms to solve multimodal optimization problems [83] in which the goal is to find multiple local or global optima in a single run. AIS methods have also been successfully used for solving TSPs [84]. Several hybrid AIS methods coupled with other evolutionary approaches are proposed [85]. AIS has also been hybridized with ACO algorithms [86]. More details can be found elsewhere [87].

Algorithm 11.5: Artificial Immune Systems—Immune Network Algorithm

Require: $N, n, N_c, \epsilon, \sigma_s, R$
1: Set iteration counter $t = 0$
2: Set clonal memory set $M = \Phi$
3: Randomly initialize N antibodies to form S_t
4: **repeat**
5: Evaluate affinity (fitness) of all antibodies in S_t
6: Select n antibodies with highest affinities to form P_t
7: Clone the n antibodies in proportion to their affinities to generate N_c antibodies
8: Mutate the N_c antibodies in inverse proportion to their affinities to form C_t
9: Evaluate affinity of all antibodies in C_t
10: $M = M \cup C_t$
11: Eliminate memory clones from M with affinities $< \epsilon$
12: Calculate similarity measure s for all clonal pairs in M
13: Eliminate memory clones from M with $s < \sigma_s$
14: Replace $R\%$ worst antibodies in S_t with the best antibodies from M
15: Add new random antibodies to S_t
16: $t \leftarrow t + 1$
17: **until** TP is satisfied
18: Declare the best antibody in M as near-optimal solution

11.6 Differential Evolution

DE was proposed by Storn and Price [88] as a robust and easily parallelizable alternative to global optimization algorithms. DE borrows the idea of self-organizing from Nelder–Mead's simplex search algorithm [89], which is also a popular heuristic search method. Like other population-based metaheuristics, DE also starts with a population of randomly initialized solution vectors. However, instead of using variation operators with predetermined probability distributions, DE uses the difference vector of two randomly chosen members to modify an existing solution in the population. The difference vector is weighted with a user-defined parameter $F > 0$ and to a third (and different) randomly chosen vector as shown below:

$$\mathbf{v}_{i,t+1} = \mathbf{x}_{r_1,t} + F \cdot (\mathbf{x}_{r_2,t} - \mathbf{x}_{r_3,t}). \tag{11.12}$$

At each generation t, it is ensured that r_1, r_2, and r_3 are different from each other and also from i. The resultant vector $\mathbf{v}_i$ is called a mutant vector of $\mathbf{x}_i$. F remains constant with generations and is fixed in $[0, 2]$ (recommended $[0.4, 1]$ [90]).

DE also uses a variation of crossover in which instead of recombining different members of the population, each member recombines with its own mutant vector. This recombination is similar to the discrete recombination of evolution strategies (ES), where each component (j) of the offspring is randomly chosen from one of the parents. Of course, in DE, the parents are simply $\mathbf{x}_i$ and its corresponding mutant vector $\mathbf{v}_i$. The recombinant is called a trial vector and denoted by $\mathbf{u}_i$. The crossover can be given by

$$u_{ji,t+1} = \begin{cases} v_{ji,t+1} & \text{if } (rand_j[0,1] \leq CR \text{ or } j = rand_i), \\ x_{ji,t} & \text{otherwise,} \end{cases} \tag{11.13}$$

where $rand_i$ is a dimension randomly selected once for each i. Thus, $j = rand_i$ ensures that the first condition is true at least once, that is, at least one of the components of the trial vector $\mathbf{u}_i$ comes from $\mathbf{v}_i$. $CR \in [0, 1]$ is called the crossover constant.

Different DE strategies are generally represented using the notation $DE/x/y/z$ where

1. x specifies the vector to be mutated.
2. y specifies the number of difference vectors to be used.
3. And z denotes the crossover scheme. The one described above is called independent binomial and is represented by "bin."

The most popular variants are $DE/rand/1/bin$ described above and $DE/best/2/bin$ [91] which uses the mutation,

$$\mathbf{v}_{i,t+1} = \mathbf{x}_{best,t} + F \cdot (\mathbf{x}_{r_1,t} - \mathbf{x}_{r_2,t}) + F \cdot (\mathbf{x}_{r_3,t} - \mathbf{x}_{r_4,t}). \tag{11.14}$$

However, as many as 10 different variants were proposed by Price et al. [92] by combining binomial ("bin") and exponential ("exp") crossovers with $DE/rand/1/$, $DE/rand/2/$, $DE/best/1/$, $DE/best/2/$, and $DE/target-to-best/2$. Other variants which use objective function information $DE/rand/2/dir$ [93], trigonometric mutation [94], arithmetic recombination [95], and pure mutants and pure recombinants [92] have also been proposed. A summary of these and many other DE variants can be found in Reference 90. Here we lay out the pseudocode for the classic $DE/rand/1/bin$ variant in Algorithm 11.6.

Algorithm 11.6: Differential Evolution

Require: N, F, CR

1: Set iteration counter $t = 0$
2: Randomly initialize N population members to form S_t
3: **repeat**
4:   Evaluate fitness of all members in S_t
5:   **for** $i = 1$ **to** N **do**
6:     Generate random integer $rand_i$ between 1 and search space dimension D
7:     **for** $j = 1$ **to** D **do**
8:       Generate random number $rand_j \in [0, 1]$
9:       **if** $rand_j \leq CR$ **or** $j = rand_i$ **then**
10:         $u_{ji,t+1} = v_{ji,t+1} = x_{jr_1,t} + F \cdot (x_{jr_2,t} - x_{jr_3,t})$
11:       **else**
12:         $u_{ji,t+1} = x_{ji,t}$
13:       **end if**
14:     **end for**
15:     Evaluate trial vector $\mathbf{u}_{i,t+1}$
16:     **if** $f(\mathbf{u}_{i,t+1}) \leq f(\mathbf{x}_{i,t})$ **then**
17:       Replace $\mathbf{x}_{i,t}$ with $\mathbf{u}_{i,t+1}$
18:     **end if**
19:   **end for**
20:   $t \leftarrow t + 1$
21: **until** TP is satisfied
22: Declare population best as near-optimal solution

11.7 Evolution Strategies

ESs initially employed the most basic form of adaptation among all of evolutionary algorithms. Developed in the 1960s [96], the simplistic version called the (1+1)-ES uses one parent to produce an offspring through binomially distributed mutation. The better of the two is selected for the next iteration. An approximate analysis of the (1+1)-ES using Gaussian mutation was presented by Rechenberg in Reference 97. Over the years, much theory in metaheuristics has been developed with regard to the convergence of the (1+1)-ES on different problem types. The first multimembered ES [98], called the (μ+1)-ES or steady-state ES, uses a population of μ parents and introduced recombination of parents. Two randomly selected parents are crossed over to produce an offspring and after it undergoes mutation, the worst of the μ+1 individuals is eliminated for the next iteration. (μ+1)-ES showed the effectiveness of recombination in speeding up evolution [99]. Schwefel introduced two new versions of the multimembered ES [100,101], which generate $\lambda \geq 1$ offspring in each generation. The first version, denoted as $(\mu + \lambda)$-ES, eliminates λ worst individuals from the combined parent–offspring population of $\mu + \lambda$ individuals, thus performing elitism. The second version, denoted as (μ, λ)-ES, simply replaces the μ parents with μ individuals from λ offspring, thus requiring $\lambda > \mu$. Note that $\lambda = \mu$ will not work since all offspring will have to be selected, thus providing no selection pressure. The two versions are also sometimes referred to as plus-selection and comma-selection ES, respectively. It is generally recommended that plus-selection be used with discrete search spaces and comma-selection be used for unbounded search spaces [99].

The canonical form of state-of-the-art ES algorithms is $(\mu/\rho \overset{+}{,} \lambda)$-ES. Here, ρ (mixing number) represents the number of parents (out of μ) that are used for recombination to produce one offspring. Naturally, an ES with $\rho = 1$ uses no recombination. A distinguishing feature of ES is the use of strategy parameters **s** that are unique for each individual. These so-called *endogenous* strategy parameters can evolve with generations. Most self-adaptive ES use these evolvable parameters to tune the mutation strength. These strategy parameters can also be recombined and mutated like the solution vectors. On the other hand, *exogenous* strategy parameters μ, ρ, and λ remain unchanged with generations. Algorithm 11.7 shows the generic ES procedure.

Recombination of the ρ members can either be discrete (dominant) or intermediate. In the former, each dimension of the offspring is randomly chosen from the corresponding dimension of the parents, whereas in the latter, all dimensions of the offspring are arithmetically averaged values of corresponding dimensions over all parents. For strategy parameters, intermediate (μ/μ) recombination has been recommended in References 99 and 102 to avoid overadaptation or large fluctuations.

Traditionally, ES variants have only used mutation as the variation operator. As a result, many studies exist which discuss, both theoretically [103,104] and empirically, the effect of mutation strength on the convergence of ES and the self-adaptation of strategy parameters in mutation operators [105], most notably that of the standard deviation in Gaussian distribution mutation. Self-adaptive ES have emerged to be one of the most competitive metaheuristic methods in recent time. We describe them next.

11.7.1 Self-Adaptive ES

A self-adaptive ES is one in which the endogenous strategy parameters (whether belonging to recombination or mutation) can tune (adapt) themselves based on the current distribution of solutions in the search space. The basic idea is to try bigger steps if past steps have been successful and smaller steps when they have been unsuccessful. However, as mentioned before, most self-adaptive ES are

Algorithm 11.7: ($\mu/\rho \overset{+}{,} \lambda$) **Evolution Strategy**

Require: μ, ρ, λ
1: Set iteration counter $t = 0$
2: Randomly initialize μ population members in S_t and their strategy parameters $\mathbf{s_t}$
3: **repeat**
4: Evaluate fitness of all members in S_t
5: Set $C_t = \Phi$
6: **for** $i = 1$ **to** λ **do**
7: Randomly select ρ members from S_t to form P_t
8: Recombine the strategy parameters of the ρ members to get $\mathbf{s}'^{(i)}_t$
9: Recombine the members in P_t to form an offspring $\mathbf{x}'^{(i)}_t$
10: Mutate the strategy parameters $\mathbf{s}'^{(i)}_t$ to get $\mathbf{s}^{(i)}_{t+1}$
11: Mutate offspring $\mathbf{x}'^{(i)}_t$ using $\mathbf{s}^{(i)}_t$ to get $\mathbf{x}^{(i)}_{t+1}$
12: $C_t = C_t \cup \mathbf{x}^{(i)}_{t+1}$
13: **end for**
14: Evaluate fitness of all members in C_t
15: **if** comma-selection is desired **then**
16: Select best μ members from C_t and update S_t
17: **end if**
18: **if** plus-selection is desired **then**
19: Select best μ members from $P_t \cup C_t$ and update S_t
20: **end if**
21: $t \leftarrow t + 1$
22: **until** TP is satisfied
23: Declare population best as near-optimal solution

concerned with the strategy parameter of mutation and this is what we discuss here. Discussions on other strategy parameters can be found Reference 106.

For unconstrained real-valued search spaces, the maximum entropy principle says that the variation operator should use a normal distribution so as to not introduce any bias during exploration [99]. The mutation operation on a solution (or "recombinant" if crossover is used) $\mathbf{x}'$ is therefore given by

$$\mathbf{x} = \mathbf{x}' + \sigma\mathbf{N}(\mathbf{0}, \mathbf{I}) = \sigma\mathbf{N}(\mathbf{x}', \mathbf{I}), \tag{11.15}$$

where σ ($= \mathbf{s}$ in Algorithm 11.7) is the only strategy parameter and $\mathbf{N}(\mathbf{0}, \mathbf{I})$ is a zero-mean unit-variance normally distributed random vector. Thus, each component i of the mutant $\mathbf{x}$ obeys the following density function:

$$prob(x'_i) = \frac{1}{\sqrt{2\pi\sigma}} \exp\left(-\frac{(x'_i - x_i)^2}{2\sigma^2}\right). \tag{11.16}$$

The mutation strength σ is thus the standard deviation of the mutant distribution, which implies $\sigma > 0$. It is also sometimes called the step-size when referring to Equation 11.15. Note that for σ being an endogenous strategy parameter, also undergoes recombination and mutation. As discussed above, recombination of strategy parameters is mainly intended to reduce fluctuations in values and therefore intermediate recombination (arithmetic averaging of standard deviations of parents) is often used. Mutation of the mutation strength is multiplicative since an additive mutation (like that in Equation 11.15) cannot guarantee $\sigma > 0$. For a strategy parameter value (or "recombinant" if crossover is used) σ', the "log-normal" mutation operator [100] is given by

$$\sigma = \sigma' \exp(\tau N(0, 1)), \tag{11.17}$$

where τ is called the learning parameter and is proportional to $1/\sqrt{N}$ (usually $\tau = 1/\sqrt{2N}$), N being the search space dimension. A self-adaptive ES using Equation 11.17 in Step 10 and Equation 11.15 in Step 11 of Algorithm 11.7 is called $(\mu/\rho \overset{+}{,} \lambda)$-$\sigma$SA-ES in the literature.

When multiple strategy parameters are used, that is, when $\mathbf{s} = \boldsymbol{\sigma}$, Equation 11.15 becomes

$$\mathbf{x} = \mathbf{x}' + \operatorname{diag}(\sigma_1, \sigma_2, \ldots, \sigma_N)\mathbf{N}(\mathbf{0}, \mathbf{I}). \tag{11.18}$$

Here, diag() is a diagonal matrix with elements σ_1, σ_2, etc. representing the standard deviations for mutation along different dimensions. This allows for axes-parallel ellipsoidal mutant distributions. Self-adaptation in this case is given by the extended log-normal rule [101]

$$\boldsymbol{\sigma} = \exp(\tau_0 N(0, 1)) \begin{pmatrix} \sigma_1' \exp(\tau N_1(0, 1)) \\ \sigma_2' \exp(\tau N_1(0, 1)) \\ \vdots \\ \sigma_N' \exp(\tau N_N(0, 1)) \end{pmatrix}. \tag{11.19}$$

τ_0 acts as a global learning parameter and τ acts coordinatewise on each normal distribution $N_i(0, 1)$. Recommended values for them are $\tau_0 \propto 1/\sqrt{2N}$ and $\tau \propto 1/\sqrt{2\sqrt{N}}$.

11.7.2 Covariance Matrix Adaptation-ES

Covariance matrix adaptation-ES (CMA-ES) [107] and its variants [108–110] represent the state of the art of self-adaptive ES algorithms for continuous search space optimization. The main difference in CMA-ES from self-adaptive ES is that it uses a generic multivariate normal distribution for mutation which is given by

$$\mathbf{x} = \mathbf{x_m} + \sigma\mathbf{N}(\mathbf{0}, \mathbf{C}) = \sigma\mathbf{N}(\mathbf{x_m}, \mathbf{C}), \tag{11.20}$$

where $\mathbf{C}$ is the covariance matrix which enables arbitrarily rotated ellipsoidal mutant distributions. By contrast, the mutant distribution obtained from Equation 11.15 is always spherical while that obtained from Equation 11.18 is always an axes-parallel ellipsoid.

Another important difference is that while self-adaptive ES starts with a population of randomly initialized solutions, CMA-ES starts with a random population mean $\mathbf{x}_m$ and generates λ offspring using Equation 11.20. $\mathbf{C}$ is initialized to an identity matrix so that the initial offspring distribution is spherical. After evaluating the λ offspring, μ best solutions are selected and recombined using a weighted intermediate recombination given by,

$$\mathbf{x}_\mathbf{m}' = \sum_{i=1}^{\mu} w_i \mathbf{x}_{i:\lambda} \tag{11.21}$$

to produce a single recombinant representing the new population mean $\mathbf{x}'_{\mathbf{m}}$. The best offspring gets the largest weight and the μ-th best offspring gets the least weight. Additionally, the weights satisfy $w_1 \geq w_2 \geq w_3 \cdots \geq w_\mu > 0$, $\sum w_i = 1$ and $\mu_w = 1/\sum w_i^2 \approx \lambda/4$. Owing to this selection scheme, CMA-ES is also known as $(\mu/\mu_w, \lambda)$-CMA-ES.

Rewriting $\mathbf{x}'_{\mathbf{m}}$ as below, we can define $\mathbf{y}$ and $\mathbf{y_w}$.

$$\mathbf{x}'_{\mathbf{m}} = \mathbf{x_m} + \sigma \sum_{i=1}^{\mu} w_i \left(\frac{\mathbf{x}_{i:\lambda} - \mathbf{x_m}}{\sigma} \right) = \mathbf{x_m} + \sigma \sum_{i=1}^{\mu} w_i \mathbf{y}_{i:\lambda} = \mathbf{x_m} + \sigma \mathbf{y_w}. \tag{11.22}$$

Next, the step-size σ and the covariance matrix $\mathbf{C}$ are updated using the following formulae:

$$\mathbf{p}'_\sigma = (1 - c_\sigma)\mathbf{p}_\sigma + \sqrt{1 - (1 - c_\sigma)^2}\sqrt{\mu_w}\sqrt{\mathbf{C}^{-1}}\mathbf{y_w}, \tag{11.23}$$

$$\mathbf{p}'_{\mathbf{c}} = (1 - c_c)\mathbf{p_c} + \mathbf{1}_{(||\mathbf{p}'_\sigma|| < 1.5\sqrt{N})}\sqrt{1 - (1 - c_c)^2}\sqrt{\mu_w}\mathbf{y_w}, \tag{11.24}$$

$$\mathbf{C}' = (1 - c_1 - c_\mu)\mathbf{C} + c_1 \mathbf{p}'_{\mathbf{c}}\mathbf{p}'^{T}_{\mathbf{c}} + c_\mu \sum_{i=1}^{\mu} w_i \mathbf{y}_{i:\lambda}\mathbf{y}^T_{i:\lambda}, \tag{11.25}$$

$$\sigma' = \sigma \exp\left(\frac{c_\sigma}{d_\sigma}\left(\frac{||\mathbf{p}'_\sigma||}{E\left[||\mathbf{N}(\mathbf{0},\mathbf{I})||\right]} - 1\right)\right). \tag{11.26}$$

Here, the values $c_c, c_\sigma \approx 4/N$, $c_1 \approx 2/N^2$, $c_\mu \approx \mu_w/N^2$, and $d_\sigma = 1 + \sqrt{\mu_w/N}$ are often used. $\mathbf{1}_{(||\mathbf{p}'_\sigma|| < 1.5\sqrt{N})}$ represents an indicator function which is equal to one when its subscript is true; else it is equal to zero. $E\left[||\mathbf{N}(\mathbf{0},\mathbf{I})||\right]$ represents the expected norm of a normally distributed random vector. $\mathbf{p_c}$ and $\mathbf{p}_\sigma$ are called the evolution paths of the covariance matrix and step-size, respectively. Initializations $\mathbf{C} = \mathbf{I}$ and $\mathbf{p_c}, \mathbf{p}_\sigma = \mathbf{0}$ are used at the start of the algorithm. The above updates ensure that $\mathbf{C}$ is symmetric and positive semidefinite, as required for covariance matrices, at each iteration. The derivation of the above equations is mathematically involved and readers are referred to the original CMA-ES paper [111] and References 107–109 for details and explanation of parameter settings. The pseudocode for CMA-ES is shown in Algorithm 11.8 for an N-dimensional search space.

11.8 Evolutionary Programming

EP together with GAs and evolutionary strategies are the earliest paradigms of evolutionary algorithms. It was developed by Lawrence J. Fogel [112] in 1960. The main difference from GA and ES is that EP does not use a recombination or crossover operator. The reason for this is that EP was originally designed as an abstraction of evolution at the macro level of species, rather than at the genetic level. Thus, instead of individuals of a population which can mate and reproduce, EP considers each individual as a reproducing population and the aim is to mimic the linkage between parent and its offspring [113,114].

EP, like any other evolutionary algorithm, starts with a population of μ randomly initialized population members. Each member undergoes mutation to produce an offspring. Different EP variants use

Algorithm 11.8: $(\mu/\mu_w, \lambda)$ **Covariance Matrix Adaptation—Evolution Strategy**

Require: λ, μ ($= \lambda/2$ typically)
1: Set recombination weights w_i $\forall\, i = 1, \ldots, \mu$, Recommended $w_i = \ln(\frac{\lambda+1}{2}) - \ln(i)$
2: Renormalize weights such that $\sum w_i = 1$
3: Set $\mu_w = 1/\sum w_i^2$, $c_c, c_\sigma \approx 4/N$, $c_1 \approx 2/N^2$, $c_\mu \approx \mu_w/N^2$ and $d_\sigma = 1 + \sqrt{\mu_w/N}$
4: Set iteration counter $t = 0$
5: Set $\mathbf{C}_t = \mathbf{I}$, $\mathbf{p_c}_t, \mathbf{p}_{\sigma t} = \mathbf{0}$
6: Randomly initialize mean point $\mathbf{x_m}_t$ and step-size $\sigma_t > 0$, Recommended $\sigma_t = 0.3$
7: **repeat**
8: Generate λ offspring from $\mathbf{x_m}_t$ using Equation 11.20 and form C_t
9: Evaluate fitness of $f(\mathbf{x})$ all members in C_t
10: Sort members of C_t according to their fitness
11: Choose μ best individuals from C_t and form P_t
12: Generate updated mean $\mathbf{x}'_\mathbf{m}$ from P_t using Equation 11.21
13: Generate updated step-size evolution path $\mathbf{p}'_\sigma$ using Equation 11.23
14: Generate updated covariance evolution path $\mathbf{p}'_\mathbf{c}$ using Equation 11.24
15: Generate updated covariance matrix $\mathbf{C}'$ using Equation 11.25
16: Generate updated step-size σ' using Equation 11.26
17: $\mathbf{x_m}_{t+1} = \mathbf{x}'_\mathbf{m}$
18: $\mathbf{p_c}_{t+1} = \mathbf{p}'_\mathbf{c}$ and $\mathbf{p}_{\sigma t+1} = \mathbf{p}'_\sigma$
19: $\mathbf{C}_{t+1} = \mathbf{C}'$ and $\sigma_{t+1} = \sigma'$
20: $t \leftarrow t + 1$
21: **until** TP is satisfied
22: Declare current $\mathbf{x_m}$ as near-optimal solution

different mutation schemes. The original EP [112] described the evolution of finite-state machines using uniform random mutations on discrete variables. David B. Fogel [115] introduced normally distributed mutations for real-valued variables where the standard deviation for an individual's mutation is a function of its fitness value, that is,

$$\begin{aligned} \mathbf{x} &= \mathbf{x}' + \sigma\mathbf{N}(\mathbf{0}, \mathbf{I}), \\ \sigma &= \sqrt{G(f(\mathbf{x}'), \kappa)}, \end{aligned} \tag{11.27}$$

where G is a fitness function that scales the objective function values to positive values and also sometimes uses random alteration κ. Fogel later introduced meta-EP [116] that uses self-adaptation in a manner somewhat similar to that used in ES for multiple strategy parameters, that is,

$$\begin{aligned} \mathbf{x} &= \mathbf{x}' + \text{diag}(\sigma_1, \sigma_2, \ldots, \sigma_N)\mathbf{N}(\mathbf{0}, \mathbf{I}), \\ \sigma_i^2 &= \sigma_i'^2 + \sqrt{\zeta}\sigma_i' N_i(0, 1)\ \forall i = 1, \ldots, N, \end{aligned} \tag{11.28}$$

where ζ is an exogenous parameter. Meta-EP self-adaptation differs with respect to the underlying stochastic process. Unlike the log-normal mutation in ES which guarantees positiveness of σ, negative variances can occur in EP. When this happens, $\sigma_i = \sqrt{\epsilon_c} > 0$ is used. Later versions of EP [117] use ES's self-adaptation as shown in Equation 11.19. Fogel also proposed Rmeta-EP [118] for $N = 2$ dimensions which uses correlation coefficients similar to CMA-ES. However, its generalization for $N > 2$ is not obvious [119].

Once μ offspring are created using mutation, EP uses a kind of $(\mu + \mu)$ stochastic selection. For each solution in the combined parent–offspring population of size 2μ, q other solutions are randomly selected from the population. The score $w \in \{0, \ldots, q\}$ of each solution is the number of those q solutions that are worse than it, that is,

$$w_i = \sum_{j=1}^{q} \begin{cases} 1 & \text{if } f(\mathbf{x}_i) \leq f(\mathbf{x}_{j:2\mu}) \\ 0 & \text{otherwise} \end{cases} \quad \forall i = \{1, \ldots, 2\mu\}. \tag{11.29}$$

Thereafter, μ individuals with the best scores are chosen for the next iteration. Note that this selection strategy becomes more and more deterministic, approaching that in ES, as q increases. A little thought also reveals that the selection is elitist. The pseudocode for EP is shown in Algorithm 11.9.*

Algorithm 11.9: Evolutionary Programming

Require: μ, q, Mutation parameters ($\{G, \kappa\}$ **or** $\{\zeta, \epsilon_c\}$ **or** $\{\tau_0, \tau\}$)
1: Set iteration counter $t = 0$
2: Randomly initialize μ population members and σ_i $\forall i = \{1, \ldots, N\}$ (if used) to form S_t
3: **repeat**
4: Evaluate fitness $f(\mathbf{x}')$ of all members in $P_t = S_t$
5: **if** standard EP is used **then**
6: Mutate individuals in P_t using Equation 11.27 to form C_t
7: **end if**
8: **if** meta-EP is used **then**
9: Mutate individuals in P_t using Equation 11.28 to form C_t
10: **end if**
11: **if** EP with ES's self-adaptation is used **then**
12: Mutate individuals in P_t using Equations 11.18 and 11.19 to form C_t
13: **end if**
14: **for** $i = 1$ **to** 2μ **do**
15: Select $\mathbf{x}_i$ from $P_t \cup C_t$
16: Set $w_i = 0$
17: Select q individuals randomly from $P_t \cup C_t$ to form Q_t
18: **for** $j = 1$ **to** q **do**
19: Select $\mathbf{x}_j$ from Q_t
20: **if** $f(\mathbf{x}_i) \leq f(\mathbf{x}_j)$ **then**
21: $w_i \leftarrow w_i + 1$
22: **end if**
23: **end for**
24: **end for**
25: Sort members of $P_t \cup C_t$ according to their scores w
26: Replace S_t with μ best score members of $P_t \cup C_t$
27: $t \leftarrow t + 1$
28: **until** TP is satisfied
29: Declare population best as near-optimal solution

* Note that this section used "primed" symbols, $\mathbf{x}'$ and σ'_i, even though they are not recombinants, in order to facilitate comparison with ES.

11.9 Genetic Algorithms

GAs are among the most popular of metaheuristic techniques. Introduced in 1975 through the seminal work of John Holland [120], its framework for search and optimization, now popularly known as binary-coded genetic algorithm (BGA), was developed by Goldberg [121]. In a BGA, solutions are represented using binary-bit strings which are iteratively modified through recombination and random bit-flips, much like the evolution of genes in nature, which also undergo natural selection, genetic crossovers, and mutations. The ease of implementation and generality of BGA made it a popular choice of metaheuristic in the following years. However, inherent problems were also identified along the way, most notably Hamming cliffs, unnecessary search space discretization, and ineffective schema propagation in case of more than two virtual alphabets. These paved the way for real-parameter genetic algorithm (RGA). In an RGA, a solution is directly represented as a vector of real-parameter decision variables. Starting with a population of such solutions (usually randomly created), a set of genetic operations (such as selection, recombination, mutation, and elite preservation) are performed to create a new population in an iterative manner. Clearly, the SP must implement the selection operator of RGAs, the GP must implement all variation operators (such as recombination and mutation operators), and the UP must implement any elite-preservation operator. Although most RGAs differ from each other mainly in terms of their recombination and mutation operators, they mostly follow one of the two algorithmic models discussed later. First, we shall discuss some commonly used GA operators which can be explained easily with the help of the five plans outlined before.

Among GA's many selection operators, the proportionate selection scheme (or roulette wheel selection), the ranking scheme and the tournament selection operators are most popular. With the proposed metaheuristic algorithm, these operators must be represented in the SP. For example, the proportionate selection operator can be implemented by choosing the i-th solution for its inclusion to P_t with a probability $f_i / \sum_i f_i$. This probability distribution around S_t is used to choose μ members of P_t. The ranking selection operator uses *sorted ranks* instead of actual function values to obtain probabilities of solutions to be selected. The tournament selection operator with a size s can be implemented by using an SP in which s solutions are chosen and the best is placed in P_t. The above procedure needs to be repeated μ times to create the parent population P_t.

Traditionally, BGAs have used variants of point crossover and bit-flip mutation operators. RGAs, on the other hand, use blending operators [122] for recombination of solutions. BLX-α [123], simulated binary crossover (SBX) [124], and fuzzy recombination [125] are often used in the literature. SBX emulates a single-point binary crossover in that the average decoded values of parents and their offspring are equal. The offspring are thus symmetrically located with respect to the parents. Moreover, it uses a polynomial probability distribution function which ensures that the probability of creating offspring closer to the parents is greater. The distribution can also be easily modified when the variables are bounded [124]. Mutation operators for RGAs such as random mutation [126], uniform mutation [127], nonuniform mutation [126], normally distributed mutation [128], and polynomial mutation [129] have also been developed.

Elite preservation in a GA is an important task [130]. This is enforced by allowing best of parent and offspring populations to be propagated in two consecutive iterations. The UP of the proposed algorithm generator achieves elite preservation in a simple way. As long as the previous population S_t or the chosen parent population P_t is included in the UP for choosing better solutions, elite preservation is guaranteed. Most GAs achieve this by choosing the best μ solutions from a combined population of P_t and C_t.

A niche-preservation operator is often used to maintain a diverse set of solutions in the population. Only the SP gets affected for implementing a niche-preservation operator. While choosing the parent population P_t, care should be taken to lower the selection probability of population-best solutions in P_t. Solutions with a wider diversity in their decision variables must be given priorities. The standard sharing function approach [131], clearing approaches [132], and others can be designed using an appropriate SP.

A mating restriction operator, on the other hand, is used to reduce the chance of creating *lethal* solutions arising from mating of two dissimilar yet good solutions. This requires an SP in which the procedures of choosing of each of the μ parents become dependent on each other. Once the first parent is chosen, the procedure of choosing other parents must consider a similarity measure with respect to the first parent.

11.9.1 Generational versus Steady-State GAs

Evolutionary algorithms, particularly GAs, are often used with a generational or with a steady-state concept. In the case of the former, a complete population of λ solutions are first created before making any further decision. The proposed metaheuristic algorithm can be used to develop a generational GA by repeatedly using the SP–GP plans to create λ new offspring solutions. Thereafter, the RP simply chooses the whole parent population to be replaced, or $R_t = S_t$ (and $\rho = \mu$). With an elite-preservation operator, the UP chooses the best μ solutions from the combined population $S_t \cup C_t$. In a GA without elitism, the UP only chooses the complete offspring population C_t.

On the other extreme, a steady-state GA can be designed by using a complete SP–GP–RP–UP cycle for creating and replacing only one solution ($\lambda = r = 1$) in each iteration. It is interesting to note that the SP can use a multiparent ($\mu > 1$) population, but the GP creates only one offspring solution from it. The generational gap GAs can be designed with a nonsingleton C_t and R_t (or having $\lambda > 1$ and $r > 1$).

The generational model is a direct extension of the canonical binary GAs to real-parameter optimization. In each iteration of this model, a complete set of N new offspring solutions are created. For preserving elite solutions, both the parent and offspring populations are compared and the best N solutions are retained. In most such generational models, the tournament selection (SP) is used to choose two parent solutions and a recombination and a mutation operator are applied to the parent solutions to create two offspring solutions. Algorithm 11.10 shows the pseudocode for the above-described GA.

Another commonly used RGA is called the CHC (Cross-generational selection, Heterogeneous recombination, and Cataclysmic mutation) [133] in which both parent and offspring population (of the same size N) are combined and the best N members are chosen. Such an algorithm can be realized by using an UP, which chooses the best N members from a combined $B \cup C$ population. The CHC algorithm also uses a mating restriction scheme.

Besides the classical and evolutionary algorithms, there exist a number of hybrid search and optimization methods in which each population member of a GA undergoes a local search operation (mostly using one of the classical optimization principles). In the so-called Lamarckian approach, the resulting solution vector replaces the starting GA population member, whereas in the Baldwinian approach, the solution is unchanged but simply the modified objective function value is used in the subsequent search operations. Recent studies [134,135] showed that instead of using a complete Baldwin or a complete Lamarckian approach to all population members, the use of Lamarckian approach to about 20%–40% of the population members and the Baldwin approach to the remaining solutions is a better strategy. In any case, the use a local search strategy to update a solution can be considered as a special-purpose *mutation* operator associated with the generational plan of the proposed metaheuristic algorithm. Whether the mutated solution is accepted in the population (the Lamarckian approach) or simply the objective function value of the mutated solution is used

Algorithm 11.10: Genetic Algorithm

Require: N, p_c, p_m
1: Set iteration counter $t = 0$
2: Randomly initialize N population members to form S_t
3: **repeat**
4: Evaluate fitness of all members in S_t
5: **if** niching is desired **then**
6: calculate sharing function
7: **end if**
8: Perform N tournament selections on randomly chosen pairs from S_t to fill P_t
9: **if** mating restriction is desired **then**
10: calculate similarity measures of members in P_t
11: **end if**
12: Perform $N/2$ crossover operations on randomly chosen pairs from P_t with probability p_c to form P_t'
13: Perform mutation operation on each member of P_t' with probability p_m to form C_t
14: Evaluate fitness of all members in C_t
15: **if** elite preservation is desired **then**
16: Set S_{t+1} with the best N solutions of $P_t \cup C_t$
17: **else**
18: Set $S_{t+1} = C_t$
19: **end if**
20: $t \leftarrow t + 1$
21: **until** TP is satisfied
22: Declare population best as near-optimal solution

(the Baldwin approach) or they are used partially (the partial Lamarckian approach) is a matter of implementation.

11.10 Genetic Programming

GP extends the concept of GAs from the evolution of binary strings and real-valued variables to computer programs. It is defined as a "weak" search algorithm for automatically generating computer programs to perform specified tasks. Weak search methods do not require specification of the form or structure of the solution in advance. A classic case study for such methods is the Santa Fe Trail problem [136]. An irregular twisting trail of food pellets is placed on a 32×32 grid. The trail consists of single and double gaps along straight regions and single, double, and triple gaps at corners. An artificial ant starting at a given cell can move from cell to cell and consume these food pellets as it encounters them. The task is to generate a computer program for an ant which maximizes the food intake in a given number of steps (or minimize the number of steps required to consume all food pellets). Another common application of GP is seen in symbolic regression, where the regression function has no specified mathematical form and need to be derived from data.

GP systems evolve programs in a domain-specific language specified by primitives called *functions* and *terminals*. The terminal set ($\mathcal{T}$) may consist of the program's external inputs, random (ephemeral) constants, and nullary (zero-argument) functions/operators/actions, whereas the function set ($\mathcal{F}$) may contain arithmetic operators ($+$, $-$, $\times$, etc.), mathematical functions (sin, cos, etc.), Boolean operators (AND, OR, etc.), conditional operators (`IF-THEN-ELSE`, etc.), programming constructs (`FOR`, `DO-WHILE`, etc.), or any other functions that are defined in the language being used. In symbolic regression problems, the function set contains mathematical and arithmetic operators and the terminal set consists of the independent variables and ephemeral constants. The Santa Fe Trail problem discussed above has been successfully solved by Koza using actions `MOVE FORWARD`, `TURN RIGHT`, and `TURN LEFT` as the terminals and `FOOD AHEAD?`, `PROGN2`, and `PROGN3` as the functions. The `FOOD AHEAD?` function allows the ant to sense whether the cell directly in front of it contains food, based on which the program can make a decision. `PROGN2` and `PROGN3` are functions that take two and three subprograms, respectively, and execute them in order.

The search space for such programs (or mathematical expressions) is immense and as such a metaheuristic is required to find a near-optimal solution. More importantly, the structure and length of the program are unknown, making it difficult to efficiently solve the problem with any common metaheuristic. The representation in GP, however, allows the evolution of programs. The programs are most often represented as trees, just as most compilers use parse trees internally to represent programs. Other common ways of expressing programs include linear [137] and graph-based representations, parallel distributed GP [138], and Cartesian GP [139].

A typical GP algorithm starts with a population of randomly created individuals or programs or trees. Two initialization methods are very common in GP, the `FULL` method and the `GROW` method [140]. The `FULL` method always generates trees in which all leaves (end nodes) are at the same user-specified depth value. This is achieved by randomly selecting nodes only from the $\mathcal{F}$ set until the depth limit is reached, at which point nodes are selected from the $\mathcal{T}$ set. On the other hand, the `GROW` method creates trees of varied sizes and shapes by randomly selecting nodes from the full primitive set ($\mathcal{F} \cup \mathcal{T}$) for all nodes until the depth limit is reached, at which point nodes are chosen from $\mathcal{T}$ as in the case of `FULL` method. Next, the fitness for each individual is determined by "running" the program. For example, in case of Santa Fe Trail problem, GP executes the trail that each individual represents and calculates the number of food pellets consumed by the ant. High-fitness individuals are selected to form the mating pool, on which primary genetic operations, namely, crossover and mutation, are applied to create a new population of programs. In GP, subtree crossover refers to exchange of randomly selected (but compliant) parts of the parent programs. Mutation usually involves randomly replacing one or more functions or terminals with other compliant primitives. The process is repeated until some stopping criterion (such as maximum number of generations) is met.

Algorithm 11.11 shows the pseudocode for a generic GP implementation. Note its structural similarity to Algorithm 11.10. Niching, mating restriction, and elite-preservation operations are also relevant to GP. Additionally, GP also uses automatically defined functions (ADFs) [141] and architecture-altering operations [142], both introduced by Koza. ADFs are subprograms that are dynamically evolved during GP and can be called by the program being evolved. ADFs contain a set of reusable program statements which occur together frequently in the main program. For example, an ADF for calculating the scalar product of two vectors can greatly simplify the evolution of a program for calculating the product of two matrices. Automatically defined iterations (ADIs), automatically defined loops (ADLs), and automatically defined recursions (ADRs) have also been proposed in Reference 142. ADFs can either be attached to specific individuals in the population, as suggested by Koza, or be provided as a dynamic library to the population [143,144].

Architectural-altering operators, such as argument and subroutine creation/deletion, act upon the program and its automatically defined components to modify their architecture, something that is not possible through crossover or mutation. Modern GP implementations also use bloat control mechanisms. Bloat refers to a nonfunctional or redundant part of the program being evolved, which grows with generations [145]. Bloating is undesirable because it increases GP's computational time.

Algorithm 11.11: Genetic Programming

Require: N, p_c, p_m, n, d
1: Set iteration counter $t = 0$
2: Randomly initialize N population members (trees) to form S_t
3: **if** ADFs are used **then**
4: Attach randomly initialized ADFs to the population members
5: **end if**
6: **repeat**
7: Evaluate fitness of all members in S_t
8: **if** niching is desired **then**
9: calculate sharing function
10: **end if**
11: Perform N tournament selections on randomly chosen pairs from S_t to fill P_t
12: **if** mating restriction is desired **then**
13: calculate similarity measures of members in P_t
14: **end if**
15: Set $C_t = \Phi$
16: **for** $i = 1$ **to** $N/2$ **do**
17: Choose two random individuals $I_{1,t}$ and $I_{2,t}$ from P_t
18: **if** $rand(0, 1) \leq p_c$ **then**
19: **if** ADFs are used **then**
20: Perform subtree crossover on ADF parts of $I_{1,t}$ and $I_{2,t}$ to yield offspring $I'_{1,ADF}$ and $I'_{2,ADF}$
21: Perform subtree crossover on non-ADF parts of $I_{1,t}$ and $I_{2,t}$ to yield offspring $I'_{1,NADF}$ and $I'_{2,NADF}$
22: Combine $I'_{1,ADF}$ and $I'_{1,NADF}$ to yield $I'_{1,t}$
23: Combine $I'_{2,ADF}$ and $I'_{2,NADF}$ to yield $I'_{2,t}$
24: **else**
25: Perform subtree crossover on $I_{1,t}$ and $I_{2,t}$ to yield $I'_{1,t}$ and $I'_{2,t}$
26: **end if**
27: **else**
28: Set $I'_{1,t} = I_{1,t}$ and $I'_{2,t} = I_{2,t}$
29: **end if**
30: **if** $rand(0, 1) \leq p_m$ **then**
31: Perform mutation on $I'_{1,t}$ to yield $I_{1,t+1}$
32: Perform mutation on $I'_{1,t}$ to yield $I_{1,t+1}$
33: **else**
34: Set $I_{1,t+1} = I'_{1,t}$ and $I_{2,t+1} = I'_{2,t}$
35: **end if**
36: **if** architectural-altering operations are used **then**
37: Apply randomly chosen operations to ADF parts of $I_{1,t+1}$ and $I_{2,t+1}$
38: **end if**
39: $C_t = C_t \cup \{I_{1,t+1}, I_{2,t+1}\}$
40: **end for**
41: **if** bloat control is desired **then**
42: Apply size constraints $\{n, d\}$ or parsimony pressure-based measures
43: **end if**
44: Evaluate fitness of all members in C_t
45: **if** elite preservation is desired **then**
46: Set S_{t+1} with the best N solutions of $P_t \cup C_t$
47: **else**
48: Set $S_{t+1} = C_t$
49: **end if**
50: $t \leftarrow t + 1$
51: **until** TP is satisfied
52: Declare population best as near-optimal solution

The most common way of reducing bloat is to constrain program size (number of tree nodes n and its depth d). Other methods use some measure of parsimony or program complexity [146]. A comparison of these methods can be found in Reference 147.

GP has been applied in several areas other than automatic programming, such as automatic synthesis, machine learning of functions, scientific discovery, pattern recognition, symbolic regression (also known as model induction or system identification), and art. Readers are referred to Reference 148 for a thorough review.

11.11 Particle Swarm Optimization

The canonical form of PSO consists of a population of particles, known as a swarm, with each member of the swarm being associated with a position vector $\mathbf{x}_t$ and a velocity vector $\mathbf{v}_t$. The size of these vectors is equal to the dimension of the search space. The term velocity ($\mathbf{v}_t$) at any iteration t indicates the directional distance that the particle has covered in the $(t-1)$-th iteration. Hence, this term can also be called as a *history* term. The velocity of the population members moving through the search space is calculated by assigning stochastic weights to $\mathbf{v}_t$ and the attractions from a particle's personal-best or "pbest" ($\mathbf{p}_l$) and swarm's best or "gbest" ($\mathbf{p}_g$), and computing their resultant vector. The "pbest" indicates the best position attained by a particle so far, whereas "gbest" indicates the best location found so far in the entire swarm (this study implements popularly used fully informed swarm topology, that is, each particle knows the *best* location in the entire swarm rather than in any defined neighborhood).

The following equations describe the velocity and position update for i-th particle at any iteration t (we refer to this as the *child creation rule*):

$$\mathbf{v}_{i,t+1} = w\mathbf{v}_{i,t} + c_1\mathbf{r}_1 .* (\mathbf{p}_{l,i} - \mathbf{x}_{i,t}) + c_2\mathbf{r}_2 .* (\mathbf{p}_g - \mathbf{x}_{i,t}), \tag{11.30}$$

$$\mathbf{x}_{i,t+1} = \mathbf{x}_{i,t} + \mathbf{v}_{i,t+1}. \tag{11.31}$$

Here, $\mathbf{r}_1$ and $\mathbf{r}_2$ are random vectors (with each component in $[0, 1]$), and w, c_1, and c_2 are prespecified constants. The $.*$ signifies componentwise multiplication of two vectors. In each iteration, every particle in the population is updated serially according to the above position and velocity rules.

A little thought will reveal that the above child creation rule involves a population member $\mathbf{x}_i$ at the current generation, its position vector in the previous generation, and its best position so far. We argue that these vectors are all *individualistic* entities of a population member. The only nonindividualistic entity used in the above child creation is the position of the globally best solution $\mathbf{p}_g$, which also justifies the population aspect of the PSO. The standard PSO is shown in Algorithm 11.12.

The parametric studies on coefficients (w, c_1, and c_2) for terms $\mathbf{v}_{i,t}$, $(\mathbf{p}_{l,i} - \mathbf{x}_{i,t})$ and $(\mathbf{p}_g - \mathbf{x}_{i,t})$, respectively, were conducted in References 149 and 150 and empirical studies revealed that in Equation 11.31, the w value should be about 0.7–0.8, and c_1 and c_2 around 1.5–1.7. Irrespective of the choice of w, c_1, and c_2, while working with the bounded spaces, the velocity expression often causes particles to "fly-out" of the search space. To control this problem, a velocity clamping mechanism was suggested in Reference 151. The clamping mechanism restricted the velocity component in $[-v_{\max,j}\ v_{\max,j}]$ along each dimension (j). Usually, $v_{\max,j}$ along the j-th dimension is taken as 0.1–1.0 times the maximum value of x_i along the i-th dimension. Such a clamping mechanism does not necessarily ensure that particles shall remain in the search space.

The velocity term ($\mathbf{v}_{i,t}$) indicates a particle's ability to explore the search space while it moves under the attraction of "pbest" and "gbest." In initial phases of the search, wide explorations are

Algorithm 11.12: Particle Swarm Optimization Algorithm

Require: N, w, c_1, c_2
1: Set iteration counter $t = 0$
2: Randomly initialize positions $\mathbf{x}_t$ and velocities $\mathbf{v}_t$ of N particles to form S_t
3: **for** $i = 1$ **to** N **do**
4: Set $\mathbf{p}_{l,i} = \mathbf{x}_{i,t}$
5: **end for**
6: Evaluate fitness $f(\mathbf{x}_t)$ of all particles in S_t
7: Assign best fitness particle as $\mathbf{p}_g$
8: **repeat**
9: **for** $i = 1$ **to** N **do**
10: Calculate new velocity for particle i using Equation 11.30 or 11.32
11: Calculate new position for particle i using Equation 11.31
12: **if** $f(\mathbf{x}_{i,t+1}) < f(\mathbf{p}_{l,i})$ **then**
13: Set $\mathbf{p}_{l,i} = \mathbf{x}_{i,t+1}$
14: **end if**
15: **if** $f(\mathbf{x}_{i,t+1}) < f(\mathbf{p}_g)$ **then**
16: Set $\mathbf{p}_g = \mathbf{x}_{i,t+1}$
17: **end if**
18: Update particle i's position and velocity to $\mathbf{x}_{i,t+1}$ and $\mathbf{v}_{i,t+1}$
19: **end for**
20: $t \leftarrow t + 1$
21: **until** TP is satisfied
22: Declare swarm best as near-optimal solution

favored, whereas toward the end more focused search is desired. To achieve this, the concept of decreasing inertia weight (w) was introduced in Reference 152. The strategy has gained popularity in promoting convergence. The idea of varying coefficients was also extended successfully to dynamically update the parameters c_1 and c_2 in References 153 and 154.

Premature convergence has been a major issue in PSO. The particles often accelerate toward the swarm's best location and the population collapses. A study classified under swarm stability and explosion [155] proposed a velocity update rule based on "constriction factor" (χ). In the presence of χ, the velocity update Equation 11.30 becomes

$$\mathbf{v}_{i,t+1} = \chi \left(\mathbf{v}_{i,t} + c_1\mathbf{r}_1 .* \ (\mathbf{p}_{l,i} - \mathbf{x}_{i,t}) + c_2\mathbf{r}_2 .* \ (\mathbf{p}_g - \mathbf{x}_{i,t})\right). \tag{11.32}$$

This is one of the most popular versions of velocity update rule in PSO studies.

11.12 Scatter Search

SS was first introduced by Fred Glover in 1977 [156] to solve integer programming problems. Later, the basic SS method was enhanced to solve other optimization problems. The original SS method was developed using five different operators, as described below:

Diversification operator: In this operation, a set of diverse solutions is generated using a seed solution.

Improvement operator: In this operation, a solution is transformed into one or more enhanced solutions. This is similar to a local search heuristic. If the operator is unable to produce a better solution, the original solution is considered as the enhanced solution.

Reference set update operator: A set of b best solutions are collected through this operator. This set can be considered as an "elite set." Solutions gain membership to the reference set based on their quality and/or their diversity within the reference set.

Subset generation operator: This operator takes the reference set and chooses a subset for performing the solution combination operator described next.

Solution combination operator: This operator transforms a given subset of solutions produced by the subset generation operator into one or more combined solutions.

The SS methodology works as follows. Initially, a set S_0 of N solutions is created using a combination of diversification operator and the improvement operator. Then, reference set update operator is used to select the best b solutions in terms of objective values and their diversity from each other and saved into a reference set RS. Usually, $|RS| = N/10$. One way to use both objective function value and diversity measure is to select b_1 solutions (set B_1) from the top of the list ranked according to objective values (of S_0) and then select b_2 solutions (set B_2) using the maximum of minimum distances of unselected solutions in S_0 from the chosen reference set B_1. In this case, $b = b_1 + b_2$, where b_1 and b_2 can be chosen appropriately to emphasize convergence and diversity maintenance, respectively, of the resulting algorithm. They can also be made adaptive by monitoring different performance measures during iterations.

New subsets of solutions are then created using the subset generation operator. In its simplest form, every pair of solutions from RS is grouped as a subset, thereby making a total of $\binom{b}{2}$ or $b(b-1)/2$ subsets. Then, the solution combination operator is applied to each subset one at a time to obtain one or more combined solutions. Each of these new solutions is then modified using the improvement operator. Again the reference set update operator is used to choose b best solutions from a combined set of RS and improved solutions. If the new reference set is identical to the previous reference set after all subsets are considered, this means that no improvement in the reference set is obtained and the algorithm is terminated. On the other hand, if at least one subset generates a new solution that makes the reference set better than before, the algorithm is continued. The subset combination operator can use a recombination-like operator (which may introduce user parameters) by using two solution vectors to create a single or multiple solutions. The above generic SS procedure is shown in Algorithm 11.13.

More sophisticated methodologies for the above five operators exist, including path relinking [157], dynamic reference set updating, reference set rebuilding, etc. [158]. The primary reference for the SS approach is the book by Laguna and Marti [159].

11.13 Simulated Annealing

The SA method resembles the cooling process of molten metals through a structured annealing process. At a high temperature, the atoms in the molten metal can move freely with respect to each another, but as the temperature is reduced, the movement of the atoms gets restricted. The atoms start to get ordered and finally form crystals having the minimum possible energy. However, the formation of the crystal mostly depends on the cooling process and cooling rate. If the temperature is reduced at a very fast rate, the crystalline state may not be achieved at all, instead, the system may end up in a polycrystalline state, which may have a higher energy state than that in the crystalline

Algorithm 11.13: Scatter Search

Require: N, b_1, b_2
1: Set iteration counter $t = 0$
2: Set $RS = \Phi$
3: Apply diversification operator to generate N solutions to form S_t
4: Apply improvement operator on all solutions in S_t
5: **repeat**
6: Evaluate fitness of all members in S_t
7: Set $B_1 = \Phi$, $B_2 = \Phi$
8: Find b_1 best fitness solutions in $RS \cup S_t$ to form B_1
9: Evaluate minimum distance d for each solution in $(RS \cup S_t) - B_1$ from set B_1
10: Find b_2 solutions with highest d values to form B_2
11: Set $RS = B_1 \cup B_2$
12: Apply subset generation operator on RS to form new solution set C'_t
13: Apply improvement operator on all solutions in C'_t to form C_t
14: Set $S_t = C_t$
15: $t \leftarrow t + 1$
16: **until** TP is satisfied
17: Declare best solution in RS as near-optimal solution

state. Therefore, in order to achieve the absolute minimum energy state, the temperature needs to be reduced at a slow rate. The process of slow cooling is known as *annealing* in metallurgical parlance.

The SA procedure simulates this process of slow cooling of molten metal to achieve the minimum function value in a minimization problem. The cooling phenomenon is simulated by controlling a temperature-like parameter introduced with the concept of the Boltzmann probability distribution. According to the Boltzmann probability distribution, a system in thermal equilibrium at a temperature T has its energy distributed probabilistically according to $P(E) = \exp(-E/kT)$, where k is the Boltzmann constant. This expression suggests that a system at a high temperature has almost uniform probability of being at any energy state, but at a low temperature, it has a small probability of being at a high energy state. Therefore, by controlling the temperature T and assuming that the search process follows the Boltzmann probability distribution, the convergence of an algorithm can be controlled. Metropolis et al. [160] suggested one way to implement the Boltzmann probability distribution in simulated thermodynamic systems. The same can also be used in the function minimization context. Let us say, at any instant, the current point is $s^{(t)}$ and the function value at that point is $E(t) = f(s^{(t)})$. Using the Metropolis algorithm, we can say that the probability of the next point being at $s^{(t+1)}$ depends on the difference in the function values at these two points or on $\Delta E = f(s^{(t+1)}) - f(s^{(t)})$ and is calculated using the Boltzmann probability distribution:

$$P(E(t+1)) = \min\,[1, \exp(-\Delta E/kT)].$$

If $\Delta E \leq 0$, this probability is one and the point $s^{(t+1)}$ is always accepted. In the function minimization context, this makes sense because if the function value at $s^{(t+1)}$ is better than that at $s^{(t)}$, the point $s^{(t+1)}$ must be accepted. The interesting situation happens when $\Delta E > 0$, which implies that the function value at $s^{(t+1)}$ is worse than that at $s^{(t)}$. According to many traditional algorithms, the point $s^{(t+1)}$ must not be chosen in this situation. But according to the Metropolis algorithm, there is some finite probability of selecting the point $s^{(t+1)}$ even though it is a worse than the point $s^{(t)}$. However, this probability is not the same in all situations. This probability depends on relative magnitude of ΔE and T values. If the parameter T is large, this probability is more or less high for points with largely disparate function values. Thus, any point is almost acceptable for a large value of T. On the other hand, if the parameter T is small, the probability of accepting an arbitrary point is small. Thus, for small values of T, the points with only small deviation in function value are accepted.

SA is a point-by-point method. The algorithm begins with an initial point and a high temperature T. A second point is created at random in the vicinity of the initial point and the difference in the function values (ΔE) at these two points is calculated. If the second point has a smaller function value, the point is accepted; otherwise the point is accepted with a probability $\exp(-\Delta E/T)$. This completes one iteration of the SA procedure. In the next generation, another point is created at random in the neighborhood of the current point and the Metropolis algorithm is used to accept or reject the point. In order to simulate the thermal equilibrium at every temperature, a number of points (n) is usually tested at a particular temperature, before reducing the temperature. The algorithm is terminated when a sufficiently small temperature is obtained or a small enough change in function values is found. Algorithm 11.14 shows the pseudocode for SA.

The initial temperature (T) and the number of iterations (n) performed at a particular temperature are two important parameters which govern the successful working of the SA procedure. If a large initial T is chosen, it takes a number of iterations for convergence. On the other hand, if a small initial T is chosen, the search is not adequate to thoroughly investigate the search space before converging to the true optimum. A large value of n is recommended in order to achieve quasi-equilibrium state at each temperature, but the computation time is more. Unfortunately, there are no unique values of the initial temperature and n that work for every problem. However, an estimate of the initial temperature can be obtained by calculating the average of the function values at a number of random points in the search space. A suitable value of n can be chosen (usually between 20 and 100) depending on the available computing resource and the solution time. Nevertheless, the choice of the initial temperature and subsequent cooling schedule still remain an art and usually require some trial-and-error efforts.

Algorithm 11.14: Simulated Annealing

Require: Cooling schedule $g(T)$, n
1: Set iteration counter $t = 0$
2: Initialize random solution s_0
3: Initialize temperature T to a very high value
4: Set $s_{best} = s_0$
5: **repeat**
6: Evaluate fitness of s_t
7: **repeat**
8: Create s_{t+1} randomly in the neighborhood of s_t
9: Set $\Delta E = f(s_{t+1}) - f(s_t)$
10: **until** $\Delta E < 0$ **or** $rand(0,1) \leq \exp(-\Delta E/T)$
11: **if** $(t \mod n) = 0$ **then**
12: Update $T = g(T)$
13: **end if**
14: **if** $f(s_{t+1}) < f(s_t)$ **then**
15: $s_{best} = s_{t+1}$
16: **end if**
17: $t \leftarrow t + 1$
18: **until** TP is satisfied
19: Declare s_{best} as near-optimal solution

11.14 Tabu Search

TS algorithm proposed by Glover [161] is a popular single-solution metaheuristic. It is a sophisticated version of the SA algorithm, in which a memory of past solutions is kept and used to prevent the algorithm from revisiting those solutions (cycling) when a "nonimproving" move is made to escape local optima. For computational tractability, this memory is basically a "tabu list" of past moves (instead of the solutions themselves) which should be prevented in the following iterations. Each move in the tabu list is forbidden for a given number of iterations, called the tabu tenure (τ). Tabu solutions may be accepted if they meet certain conditions known as *aspiration criteria*. A commonly used criterion it to go ahead with a forbidden move if it generates a solution better than the best found solution.

A TS algorithm starts with a single solution like in SA and N new solutions are sampled from its neighborhood. Solutions that require a forbidden move are replaced by other samples. The best of these neighborhood solutions is chosen and if its fitness is better than the current solution, then the former replaces the latter, just like in any local search algorithm. However, if none of the neighborhood solutions are better than the current solution, a nonimproving move is made. In SA, this move is a controlled randomization. In TS, a move that is not forbidden by the tabu list is performed instead to prevent cycles. The size of the tabu list k puts an upper bound on the size of the cycle that can be prevented. The size of k will determine the computational complexity of the algorithm. Sometimes, multiple tabu lists may also be used to capture other attributes of visited solutions such as variables that change values (continuous search space) or elements that change positions (discrete search space) [162].

In addition to the tabu list, which is categorized as short-term memory, a TS can also have a medium-term memory which controls search intensification and a long-term memory which enables search diversification. The purpose of the medium-term memory is to utilize the information of the recorded elite solutions to guide the search process toward promising regions of the search space. One way to achieve this is to find features that are common to recorded elite solutions and force the common features to exist in the newly created solution. This process is called the intensification. The purpose of the long-term memory is to introduce diversity in the search space by creating solutions in the relatively unexplored regions of the search space. The frequency of occurrence of different regions of the search space is recorded in a memory and a new solution is created in the region having the least frequency of occurrence thus far. This is called diversification.

TS has been found to be efficient in solving combinatorial optimization problems having discrete search spaces [163,164] and vehicle routing [165]. The bare framework of TS is shown in Algorithm 11.15. Actual implementations of the three types of memory can be found in Reference 166.

11.15 Other Metaheuristic Techniques

A host of other metaheuristic optimization methods inspired by nature, biological, and physical phenomena have been suggested in the recent past. In this section, we have attempted to give an exhaustive list of these metaheuristics without going into implementation specifics and instead very briefly describe their basis. We also specify the number of citations as of June 2015 of the oldest paper that we could find on Google Scholar which describes the methodology. This number can be thought of as an indicator of the popularity and impact of the method in the field of metaheuristic search and optimization.

Algorithm 11.15: Tabu Search

```
Require: N,τ, k
 1: Set iteration counter t = 0
 2: Initialize random solution s_0
 3: Initialize tabu list T, medium-term memory M and long-term memory L
 4: Set s_best = s_0
 5: repeat
 6:    Evaluate fitness of s_t
 7:    Set N(s_t) = Φ
 8:    repeat
 9:       Create a neighbor to s_t
10:       if move to neighbor occurs in T and aspiration criteria is false then
11:          Discard the neighbor
12:       else
13:          Add neighbor to N(s_t)
14:       end if
15:    until |N(s_t)| = N or no more neighbors exist
16:    Find best solution s'_t in N(s_t)
17:    if f(s'_t) < f(s_t) then
18:       s_best = s'_t
19:       Delete moves older than τ iterations from T
20:       Add move s_t ⇒ s'_t to T
21:       while size(T) > k do
22:          Delete oldest move in T
23:       end while
24:    end if
25:    s_t = s'_t
26:    if intensification is desired then
27:       Modify s_t
28:       Update M
29:    end if
30:    if diversification is desired then
31:       Modify s_t
32:       Update L
33:    end if
34:    t ← t + 1
35: until TP is satisfied
36: Declare s_best as near-optimal solution
```

1. *Harmony search [167,168]* (2119): The analogy with jazz musicians adjusting pitches to attain harmony (musical aesthetics) is used to generate offspring, first by discrete recombination and then through mutation.
2. *Estimation of distribution algorithms [169,170]* (1912): This method uses a probabilistic model of the current population to generate offspring instead of using crossover or mutation operators. The best of the offspring population are selected to update the probability distribution, which is usually either a Bayesian network or a multivariate normal distribution.

3. *Bacterial foraging [171]* (1662): The biology and physics underlying the chemotactic (foraging) behavior of *Escherichia coli* bacteria are modeled through operators in optimization that mimic chemotaxis (guided movement), swarming (co-operation), reproduction, elimination (death), and dispersal (random movement).
4. *Memetic algorithms [172]* (1445): A hybridization of population-based metaheuristics (usually GA), local search heuristics, problem-specific information, and sometimes machine learning is used to evolve a population of solutions.
5. *Gene expression programming [173]* (1435): It is a variant of linear GP where fixed length genes are encoded using left to right and top to bottom traversal of parse trees.
6. *Coevolutionary algorithms [174,175]* (1204): The analogy with species (subpopulations) in an ecosystem is used so that each subpopulation represents a component of the solution, evolved independently by a standard GA.
7. *Firefly algorithm [176,177]* (1182): The analogy with fireflies using light to communicate with their neighbors is used, where each firefly (solution) moves toward brighter fireflies (brightness proportional to the fitness function) based on the distance between them, and the brightest firefly moves randomly.
8. *Gravitational search algorithm [178]* (1052): It uses a gravitationally isolated system of objects (solutions) in the search space with masses proportional to the fitness function and with accelerations and velocities determined by formulae loosely based on the law of gravitation.
9. *Cuckoo search [179]* (985): The analogy with *brood parasitism* and *Lévy flight* behavior seen in cuckoo birds is used. Each cuckoo performs a Lévy flight to reach a new location and lays an egg(s) (solution) which replaces a randomly chosen egg in a randomly chosen nest depending on its fitness. Host birds can either throw away (discard) an egg or abandon their nests to build new ones.
10. *Biogeography-based optimization [180]* (907): The analogy with migration of species is used to assign emigration and immigration probabilities for each habitat (solution) based on its habitat suitability index (fitness), and these probabilities decide the movement of species between habitats thus modifying them.
11. *Shuffled frog leaping algorithm [181,182]* (875): It is a variant of memetic algorithm, where a population of frogs (solutions) is organized into different groups (memeplexes) within which a PSO-type search is performed locally, following which the memeplexes are shuffled.
12. *Bees algorithm [183]* (787): It is similar to ABCO with the difference that the former uses elitist strategy to assign more neighborhood-searching bees to good fitness sites, following which the best bee from each path is selected to form new bees.
13. *Cultural algorithms [184]* (747): The analogy with human social evolution is used to evolve a population using a *belief space* which itself gets adjusted by high-performing solutions in the population.
14. *Imperialist competitive algorithm [185,186]* (727): The analogy with imperialism and colonization is used, where powerful imperialist countries (high-fitness solutions) are assigned colonies (other solutions) which can "move" toward their center of power and can even exchange places with it if a better location is found. Powerful empires (imperialists and their colonies) try to capture the weaker ones by competing with each other (proportionate probabilistic selection) in terms of their combined power (weighted fitness).

15. *Bat algorithm [187]* (532): It mimics the echolocation ability of microbats. Each microbat (solution) is associated with a velocity and a frequency of emitted sound, which it uses to move through the search space. When a prey (better fitness location) is located, the sound emitted by the bat are modified by decreasing the amplitude and increasing the pulse emission rate.
16. *Grammatical evolution [188,189]* (463): It uses the Backus–Naur form of context-free grammar to encode programs as integer (and hence binary) strings which can be evolved by most metaheuristics, thus making automatic programming language independent.
17. *Invasive weed optimization [190]* (425): It mimics robustness, adaptation, and randomness of colonizing weeds. A fixed number of seeds (solutions) are dispersed in the search space. Each seed grows into a flowering plant and disperses its own seeds in the neighborhood based on its fitness. After a maximum number of plants is reached, weaker ones are eliminated in favor of stronger ones.
18. *Charged system search [191]* (291): Solutions are represented by charged particles associated with positions and velocities, the amount of charge representing the fitness value. A highly charged particle can attract any other particle with a lower charge. However, only some low charge particles can attract particles with higher charge. The Coulomb's law is used to calculate the forces of attraction between particles and their positions and velocities are adjusted.
19. *Big bang–big crunch optimization [192]* (290): This algorithm is designed to simulate big bang and big crunch occurring one after the other in each iteration. The first big bang generates random solutions. The following big crunch converges all solutions to the center of mass where mass is proportional to the fitness value. Thereafter, each big bang generates new solutions around this center of mass using a normal distribution. The solutions eventually collapse to the optimum.
20. *Marriage in honey bees optimization [193]* (271): It models the mating flight of queen bees as seen in bee colonies. The queen bee moves based on its speed, energy, and position and collects genetic material (solution coordinates) from good fitness drones. Upon reaching the colony, it generates broods by combining its genome with a randomly selected drone genome (crossover). The best broods replace the worst queens and the remaining broods are killed.
21. *Bee colony optimization [194]* (261): It uses a similar analogy to ABCO described in Section 11.4. In the *forward pass*, employer bees fly out of the colony and perform a series of moves to construct partial solutions. In the *backward pass*, they return to the colony and share information with other bees. Based on the collective information from all bees, each bee can either abandon its partial solution, continue to construct its partial solution, or recruit bees to help further explore its partial solution.
22. *Group search optimizer [195]* (257): It uses the analogy with the producer–scrounger model of animal foraging. The producer (best fitness member) performs *scanning* to find a better resource location. The scanning strategy mimics the *white crappie* fish, which uses a series of cone-shaped visual scans to locate food. Other members (scroungers) perform random walk toward the producer and can become a producer if they find a better fitness location along the way.
23. *Honey-bee mating optimization [196]* (204): Like marriage in honey bees optimization, this metaheuristic also mimics the mating flight of queen bees of a honey bee colony. However, it can also handle continuous variables.

24. *Krill herd algorithm [197]* (196): It is based on the herding behavior of krill individuals. The movement of each krill individual depends on the minimum of its distance from food (foraging) and from highest density of the herd (herding). Additionally, random movement is also used to enhance exploration.
25. *Bee hive algorithm [198]* (182): The metaheuristic was developed to solve routing problems using *short distance bee agents* for neighborhood search and *long distance bee agents* for global exploration.
26. *Glowworm swarm optimization [199]* (170): Similar to ACO, this metaheuristic uses glowworms instead of ants as agents. The glowworms emit light whose intensity is proportional to the amount of *luciferin* (fitness value) they have. Brighter glowworms act as attractor to other glowworms given that the former are within the sensor range of the latter. This approach allows multimodal function optimization.
27. *Intelligent water droplet algorithm [200]* (163): It is a swarm-based metaheuristic inspired from water drops carrying soil particles in rivers. In combinatorial problems, each water drop gradually builds a solution as it moves and increases its velocity according to the soil content of the path. More soil corresponds to a lower increase in velocity. Over iterations, paths with lesser soil content are preferred probabilistically.
28. *Bee system [201]* (133): It adopts the honey bee colony analogy to solve combinatorial problems. For example, in the TSP, the nectar quantity represents the length of the path. The bees are allowed to fly to any node with a probability that decreases with increasing distance from their current location. The influence of this distance increases with iterations, making search more random at the beginning and more restricted toward the end.
29. *Stochastic diffusion search [202]* (120): It uses agents with hypotheses communicating directly with each other. In the *test phase*, each agent evaluates its hypothesis and finds it to be either successful or unsuccessful. In the *diffusion phase*, agents with unsuccessful hypotheses either select other agents to follow or generate new hypotheses.
30. *Cat swarm optimization [203]* (108): As the name suggests, the algorithm attempts to model the behavior of cats which represent solutions having position and velocity. Each cat has two modes of movement in the search space: the seeking mode and the tracking. The former plays the role of the local search component of the metaheuristic while the latter tends to global search.
31. *Central force optimization [204]* (96): It also uses the analogy with particle motion under gravitational force such as gravitational search algorithm, although with different formulations. Central force optimization is inherently deterministic unlike most other metaheuristics.
32. *Queen-bee evolution [205]* (95): It is basically an integration of queen bee's mating flight behavior with GAs. Instead of randomly selecting parents for crossover, each individual mates with the queen bee (best fitness individual of previous generation). The mutation operation is also modified.
33. *Cooperative bees swarm optimization [206]* (94): Unlike other bee algorithms, this method starts with a single bee which provides the best location based on neighborhood search. Other bees are then recruited to a specified region around the best-known location. After performing neighborhood searches, all bees communicate their individual best locations to the swarm. A tabu list of visited locations is also maintained to avoid cycles.
34. *Artificial fish swarm optimization [207]* (91): It tries to mimic the social behavior of fishes such as preying for food, swarming, and following their neighbors. Each fish performs a local search by *visually* examining its surroundings and moving toward a better food source

(objective value). It may also either swarm another fish's food source or simply follow one of its neighbors.

35. *Eagle strategy [208]* (91): It uses the foraging behavior of eagles as a metaheuristic. Eagles representing candidate solutions initially perform a slow search where they search a larger area using DE. Once promising regions (prey) are identified, the chasing phase begins which uses a fast and local search algorithms such as hill-climbing methods.
36. *Differential search algorithm [209]* (89): This algorithm simulates the Brownian-like random-walk movement used by various organisms to migrate as a superorganism. Each superorganism represents a candidate solution and each of its members corresponds to one variable dimension. As superorganisms move in the search space, they exchange information about their best location with other superorganisms.
37. *Backtracking search optimization [210]* (74): It is a simplified form of evolutionary algorithm which uses the following operations: initialization, selection-1, mutation, crossover, and selection-2. Selection-1 uses the best of the historical population members to generate a population. Mutation mutates the current population using the output of selection-1. Crossover recombines the best individuals of the mutant population and finally selection-2 uses greedy selection to update the current population with better offspring if found.
38. *Flower pollination algorithm [211]* (61): Each solution is represented by a pollen gamete from a flowering plant. Flower pollen are carried by pollinators such as insects which move according to Lévy flights. All gametes either combine with the fittest pollen (global pollination) with a given probability or randomly combine with other pollen gametes (local pollination).
39. *Analytic programming [212]* (58): This metaheuristic is capable of generating programs and can be used for symbolic regression. Unlike GP, it can be used with any evolutionary algorithm since it acts as a superstructure requiring no changes to the variation operators.
40. *Fish school search [213,214]* (54): It uses the analogy with feeding, swimming, and breeding behavior found in fish schools. Feeding of a fish is analogous to a candidate solutions' fitness value. Thus, the fish can gain or lose weight depending on the regions it swims in. Swimming and breeding refer to the actual search process where solutions swarm potential regions.
41. *Black hole optimization [215]* (54): It mimics the gravitation interaction of a black hole with nearby stars. The best solution in the current population becomes the black hole and attracts other solutions. Other stars may become a black hole if they reach a point with better fitness than the current black hole. When a solution passes the event horizon of the current black hole, it vanishes and a new star (solution) is randomly initialized.
42. *Monkey search [216]* (51): The solutions are analogous to monkeys and undergo climb, watch-jump, and somersault steps. The climb step acts like a local optimizer for each monkey. The watch-jump step represents exchange of fitness landscape information and the somersault step enables exploration of new regions in the search space.
43. *Bacterial swarming algorithm [217]* (50): This metaheuristic uses the tumble-and-run behavior of a bacterium's chemotactic process. The tumble operation acts like the global search component, where each bacterium (solution) moves independently to find good regions. Thereafter, the run operation causes local search.
44. *Water cycle algorithm [218]* (49): It is loosely based on the hydrologic cycle which consists of precipitation (random initialization of solutions), percolation into the sea (attraction toward the best solution), evaporation, and again precipitation (reinitialization of some solutions).

45. *Gaussian adaptation [219]* (49): Similar in principle to CMA-ES, this method samples single candidate solutions from a multivariate normal distribution and iteratively updates its mean and covariance so as to maximize the entropy of the search distribution. Strategy parameters are either fixed or adapted as in CMA-ES.
46. *River formation dynamics [220]* (45): It is used to solve the TSP using the analogy with a flowing river eroding soil from high altitudes and depositing it at lower altitudes. Initially, all nodes are at the same altitude. But as the water drops move from origin to destination, they erode various edges until all drops follow the same path.
47. *Brain storm optimization [221]* (40): The solutions are considered as individuals coming up with different ideas during a brainstorming session. The best ideas are selected and new ideas are created based on their combinations. Individuals with rejected ideas try to come up with new ideas (reinitialization).
48. *Roach infestation optimization [222]* (39): It mimics the collective behavior of cockroaches in finding dark regions in an enclosed space. The level of darkness is proportional to the fitness value. Initially each roach (solution) acts individually to find a dark region. However, when it comes in close proximity to another roach, they exchange information about their respective regions. The information of the darkest known location will thus spread through the roach population.
49. *League championship algorithm [223]* (35): It is based on the metaphor of sporting competitions in sport leagues. Each candidate solution is a team and its position in the search space defines the team's strategy. Each week (iteration) the teams participate in pairwise competitions and based on their ranking modify their strategies.
50. *Electromagnetism optimization [224]* (31): It is similar in principle to charged system search. Charge particles represent solutions which move under the influence of electrostatic forces. The nature of the force on a particle (attractive or repulsive) due to another particle depends on the charge (fitness) of the other particle. Therefore, the motion does not strictly conform to physical electrostatic laws.
51. *Galaxy-based search [225]* (25): It uses the metaphor of the spiraling arms of a galaxy sweeping the solution space to find better regions. This step is referred to as the spiral chaotic move. Thereafter, local search is performed within these regions to fine-tune the obtained solutions.
52. *Dolphin echolocation [226]* (24): It mimics the echolocation ability of dolphins to find prey. Solutions (echoes) are initially randomly distributed in the search space. The fitness (echo quality) at each location is evaluated and more or less echoes are allocated in the neighborhood of each location based on the fitness values.
53. *Wolf search algorithm [227]* (20): The analogy with wolfs hunting for prey in packs is used in this algorithms. Each wolf (solution) has a fixed visual area and can sense companions within it. The quality of a wolf's position is represented by the fitness value of the solutions. A wolf which has a better-placed companion in its sight moves toward it. At times, a wolf may sense an enemy and sprint to a random location (reinitialization).
54. *Artificial cooperative search [228]* (20): It uses two populations referred to as superorganisms. Each superorganism is a collection of sub-superorganisms (solutions) and each sub-superorganism is a collection of variable dimensions (i.e., a solution). The sub-superorganisms have a predator–prey relationship based on the relative fitness values. Whether a superorganism is the prey or the predator is determined randomly in each generation.

55. *Rubber band technique-based GA [229]* (20): This method is specifically targeted to solve multiple sequence alignment problems encountered in bioinformatics. It is inspired by the behavior of an elastic rubber band on a plate with several poles. The poles can either be primary (through which the band ought to pass) or secondary. GA is used to arrive at the rubber band configuration which best addresses the sequence alignment problem.
56. *Bacterial–GA foraging [230]* (19): It combines bacterial foraging algorithm with GA. In addition to tumble and swim, the bacteria (solutions) also undergo crossover and mutation as in a GA. When individuals are eliminated, new individuals are formed by mutating the eliminated ones instead of random reinitialization.
57. *Spiral dynamics-inspired optimization [231]* (19): The population consists of candidate solutions, each forming the center of a spiral. The shape and size of the spirals is determined by two parameters, the rotation angle, and the convergence rate to the center. These parameters control the diversification and intensification of the spiral around the solution. Spirals corresponding to high fitness centers tend toward intensification.
58. *Great salmon run algorithm [232]* (16): The algorithm takes its inspiration from the salmon run phenomenon where millions of salmons migrate upstream for spawning, even risking becoming prey. Each solution represents a salmon subgroup that selects a migration pathway. Different pathways pose different threats to the subgroup, such as either being hunted by humans or by bears. At the end, salmons from safer pathways spawn to produce offspring which are more likely to take safer pathways.
59. *Paddy field algorithm [233]* (15): It uses the analogy with paddy fields where the seeds laid in the most fertile (high fitness) region of the field (search space) grow to produce more seeds. The seeds represent solutions and therefore eventually the optimum gets crowded.
60. *Japanese tree frogs algorithm [234]* (13): The algorithm simulates the mating behavior of male tree frogs which intentionally desynchronize their mating calls to allow female frogs to correctly localize them. The analogy is used to solve the graph coloring problem, whose optimization version aims to minimize the number of colors to be used.
61. *Consultant-guided search [235]* (13): Each solution in the population is a virtual person who can act both as a client and a consultant. The reputation of a person as a consultant depends on the number of successes achieved by its clients following the person's advice. Reputation can either increase or decrease, but when it falls below a certain value, the person takes a sabbatical leave (reinitialization) and stops giving advice.
62. *Hierarchical swarm model [236]* (13): It generalizes the standard flat framework of swarm intelligence methods. Instead of all particles interacting with each other, this method introduces a multiple-level hierarchy. At each level, the interaction is limited to the particles within the agents. Different agents interact at the next higher level. The topmost level consists of a single agent.
63. *Social emotional optimization [237]* (11): This algorithm is loosely based on the way humans modify their behavior to improve their social status. Based on an emotion index, individuals (solutions) behave in a certain way. If this behavior is approved by the society (i.e., the solution has high fitness), the emotion index is increased so that this behavior becomes stronger.
64. *Anarchic society optimization [238]* (10): It is inspired by a hypothetical society whose members behave anarchically. The movement of solutions is based on the member's fickleness index, internal irregularity index, and external irregularity index. Each of these

represents the member's dissatisfaction with the current position, comparison with previous positions, and affinity to others' positions.

65. *Eco-inspired evolutionary algorithm [239]* (10): It consists of multiple populations of individuals dispersed in the search space. Populations belonging to the same habitat are allowed to mate but those from different habitats are reproductively isolated. However, individuals may migrate to different habitats. While intrahabitat reproduction leads to intensification of solutions, interhabitat migration leads to diversification. Both operations depend on the fitness landscape.
66. *OptBees algorithm [240]* (9): Like the many other bee-based algorithms, it uses the concept of employer and onlooker bees which recruit more bees by dancing in relation to the best found food source (fitness). Scout bees are also used to search the decision space randomly.
67. *Termite colony optimization [241]* (9): The algorithm uses termites as a representation of solutions. A part of the termite population moves randomly and is responsible for providing diversity to the algorithm. The rest of the termites scan their neighborhood for pheromone content left by other termites and move toward the best location. If none is found, they perform a random walk.
68. *Egyptian vulture optimization [242]* (8): It mimics the nature, behavior, and skills of Egyptian vultures for acquiring food. The algorithm primarily favors combinatorial optimization problems.
69. *Virtual ant algorithm [243]* (7): This is similar to ACO but is used to solve optimization problems other than TSP. The virtual ants move through the search space and deposit pheromone whose levels are determined by the fitness of the region traversed by them.
70. *Bumblebees multiagent combinatorial optimization [244]* (5): The bumblebees algorithm was also developed to solve the graph coloring problem. A toroidal grid is used to represent solutions to this problem. The bumblebees' nest is randomly placed in one of the cells and the bees are allowed to move to any other cell in search for food. Bumblebees that find food increase their lifespan by two units. After each generation, the bumblebee's life is decreased by one unit and those which reach zero are removed.
71. *Atmosphere clouds model optimization [245]* (3): In this method, the search space is first divided into many disjoint regions. The humidity and air pressure in these regions is defined by the best-known fitness value and the number of evaluated points in that region. Clouds form over a region when the humidity value is above a specified threshold. The movement of clouds to low-pressure regions is equivalent to assigning more function evaluation to the lease explored regions. Water droplets from the clouds represent new solutions to be evaluated.
72. *Human-inspired algorithm [246]* (3): It is based on the way human climbers locate the highest peak by scanning the surroundings and communicating with fellow climbers. It divides the whole search space into equal subspaces and evenly assigns population members. The best fitness subspaces are allocated more members in the following generations by further dividing them into finer subspaces.
73. *Weightless swarm algorithm [247]* (2): It is a simplification of the conventional PSO and has been used to solve dynamic optimization problems. The inertia weights are discarded and it also uses a modified update strategy.
74. *Good lattice swarm algorithm [248]* (1): It replaces the random initialization of particles and their velocities in PSO with a good lattice uniform distribution, which according to number theory has the least discrepancy.

75. *Photosynthetic algorithm [249]* (1): This algorithm uses the analogy with carbon molecules transforming from one substance to another during the Calvin–Benson cycle and mimics the reaction taking place in the chloroplast during photorespiration for recombining different solutions.
76. *Eurygaster algorithm [250]* (0): It mimics the behavior of eurygasters attacking grain farms in different groups, always settling for the closest unattacked regions of the farm. The method has been applied to graph partitioning problem.

Many such methods are yet to be discovered and practiced, but an interesting future research would be to identify essential differences among these algorithms and specific features that make each of them unique and efficient in solving a specific class of optimization problems. Recent approaches [251–253] in this direction have provided useful insights about similarities of a few metaheuristic methodologies.

Despite their success and popularity, metaheuristic techniques have received some criticism over the years. The main criticism concerns their nonmathematical nature and lack of a proof of convergence. Owing to the stochasticity involved, it is difficult to derive an exact theory of how these algorithms work. Few theoretical analyses do exist for some of the methods. However, they often relate to a simplified and basic version of the algorithm. While there is no immediate remedy for this, the best approach for practitioners is to provide empirical results on a wide variety of benchmark test problems available in the literature. Reproducibility of results is an essential part of dissemination in this regard. In order to achieve this under stochasticity, uniformly seeded pseudorandom number generators are used and the algorithm is run several times. A common practice is to use statistical analysis to draw conclusions from such empirical studies.

Recent criticism has been with respect to the ever-growing number of "novel" metaheuristics as is evident from the list above. Sörensen [254] argues that many of these recently proposed metaheuristics are similar in almost all respects except the metaphor that is used to describe them. For example, in References 255 and 256, the author shows that one of the recent metaheuristics, harmony search, is a special case of evolutionary strategy. Sörensen also points out that the terminology used to convey and justify the metaphor sometimes completely obfuscates the similarities with existing methods. Indeed, as described in Section 11.2, all metaheuristics have two major characteristics, exploration and exploitation. There may be several different ways of accomplishing these tasks in an algorithm, but calling each such method "new," "novel," and "innovative" would be pushing the metaphor too far. The major disadvantage of the undeserved attention that many of these recent techniques get is the distraction it causes away from key developments in the field of metaheuristics.

11.16 Conclusions

This chapter began with a discussion on the main concepts concerning metaheuristics such as their classification and characteristics. We laid down a generic framework into which most metaheuristic techniques can fit. The key components essential to the working of any metaheuristic were identified under five plans, namely, selection, generation, replacement, update, and termination. Some important parameters were also introduced. Thereafter, we reviewed a number of popular metaheuristic methods for solving optimization problems. Pseudocodes have been provided for 14 of the most commonly used methods for ease of understanding and implementation. Key variants of some of the algorithms were also discussed. Except SA and TS, all other methods are population-based.

Moreover, we mainly concentrated on evolutionary and swarm-based methods, which are older, well known, and frequently used. For completeness, we have included an extensive list of other metaheuristics and ordered them by their Google Scholar citations. Though this metric may not be a true count of citations, it serves as a measure of popularity of these methods. Finally, we briefly discussed the criticism that the field of metaheuristics currently faces.

A recent trend in the field of metaheuristics is hybridization. This not only refers to the combination of different available metaheuristics but also to their integration with classical search methods. Classical algorithms are efficient local optimizers and therefore can aid the intensification process. Not surprisingly, such hybrid methods have been shown to outperform standalone methods. Hybrid-metaheuristics also refer to the use of data-mining and machine-learning techniques to enhance search. With the increased accessibility to computing power, resource, and memory-intensive tasks such as data-mining can now be performed alongside metaheuristic optimization to learn patterns, predict outcomes, and assist the search behavior.

Parallel metaheuristics can refer to different levels of parallelization, algorithmic level, iteration level, and solution level. Algorithmic level involves multiple metaheuristics running simultaneously, communicating and collaborating with each other to solve a complex search problem. Iteration level involves parallelizing the operations within an iteration. This is especially useful in population-based metaheuristics. Solution-level parallelization pertains to individual solutions. For example, if the evaluation of the objective function involves a finite element analysis, then this analysis itself can be parallelized. All these are very active and relevant areas of research within metaheuristics.

Another common research problem that needs to be addressed is that of setting the best parameters for the algorithm at hand. It is well known that for most metaheuristics the performance can be drastically affected through their parameters. Parameter tuning refers to the selection of the best configuration of the algorithm for solving a particular problem. Both offline and online parameter tuning methods have been proposed in the literature. However, as with optimization algorithms, there is no one best approach.

Metamodeling is an entire research area in itself. As metamodels are increasingly being used within optimization, more empirical studies need to be performed on the compatibility of different metamodels with different metaheuristics. Again, it is improbable that one unique combination will emerge as the winner. However, these studies can give a lot of insight about which methods work best on specific use-cases.

Metaheuristics have been and will continue to be used to solve complex optimization problems. Instead of developing more and more new algorithms, the focus should now shift toward improving current algorithms by unifying different ideas into a common and generic framework as the one described in this chapter. There is also a growing need for classifying different metaheuristic methods according to their functional or algorithmic similarities. The algorithmic connection between two such methods provides a way to introduce one algorithm's key feature into another algorithm, a matter that will help unify different metaheuristic methods under one umbrella of a meta-algorithm for solving complex real-world optimization problems.

Acknowledgment

The first author acknowledges the financial support received from The Knowledge Foundation (KK-stiftelsen, Stockholm, Sweden) for the ProSpekt project KDISCO.

References

1. E.-G. Talbi, *Metaheuristics: From Design to Implementation*. John Wiley & Sons, Hoboken, New Jersey, USA, 2009.
2. G. Zäpfel, R. Braune, and M. Bögl, *Metaheuristic Search Concepts: A Tutorial with Applications to Production and Logistics*. Springer Science & Business Media, Heidelberg, 2010.
3. M. Gendreau and J.-Y. Potvin, *Handbook of Metaheuristics*. Springer, New York, USA, 2010.
4. S. Luke, *Essentials of Metaheuristics*. Lulu, 2013. Available for free at http://cs.gmu.edu/~sean/book/metaheuristics/.
5. C. C. Ribeiro and P. Hansen, *Essays and Surveys in Metaheuristics*. Springer Science & Business Media, New York, USA, 2012.
6. F. Glover and G. A. Kochenberger, *Handbook of Metaheuristics*. Kluwer Academic Publishers, Dordrecht, 2003.
7. I. H. Osman and J. P. Kelly, *Meta-Heuristics: Theory and Applications*. Kluwer Academic Publishers, Norwell, Massachusetts, USA, 2012.
8. S. Voß, S. Martello, I. H. Osman, and C. Roucairol, *Meta-Heuristics: Advances and Trends in Local Search Paradigms for Optimization*. Springer Science & Business Media, New York, USA, 2012.
9. T. F. Gonzalez, *Handbook of Approximation Algorithms and Metaheuristics*. CRC Press, Boca Raton, FL, USA, 2007.
10. J. Dréo, A. Petrowski, P. Siarry, and E. Taillard, *Metaheuristics for Hard Optimization: Methods and Case Studies*. Springer Science & Business Media, Berlin Heidelberg, 2006.
11. P. Siarry and Z. Michalewicz, *Advances in Metaheuristics for Hard Optimization*. Springer Science & Business Media, Berlin Heidelberg, 2007.
12. K. F. Doerner, M. Gendreau, P. Greistorfer, W. Gutjahr, R. F. Hartl, and M. Reimann, *Metaheuristics: Progress in Complex Systems Optimization*. Springer Science & Business Media, New York, USA, 2007.
13. X.-S. Yang, *Nature-Inspired Optimization Algorithms*. Elsevier, London, UK, 2014.
14. D. F. Jones, S. K. Mirrazavi, and M. Tamiz, Multi-objective meta-heuristics: An overview of the current state-of-the-art, *European Journal of Operational Research*, vol. 137, no. 1, pp. 1–9, 2002.
15. I. H. Osman and G. Laporte, Metaheuristics: A bibliography, *Annals of Operations Research*, vol. 63, no. 5, pp. 511–623, 1996.
16. E.-G. Talbi, A taxonomy of hybrid metaheuristics, *Journal of Heuristics*, vol. 8, no. 5, pp. 541–564, 2002.
17. L. Jourdan, M. Basseur, and E.-G. Talbi, Hybridizing exact methods and metaheuristics: A taxonomy, *European Journal of Operational Research*, vol. 199, no. 3, pp. 620–629, 2009.
18. G. R. Raidl, A unified view on hybrid metaheuristics, in *Hybrid Metaheuristics*, F. Almeida, M. J. B. Aguilera, C. Blum, J. M. M. Vega, M. P. Pérez, A. Roli, and M. Sampels (Eds.), pp. 1–12, Springer, Berlin Heidelberg, 2006.
19. I. Boussaïd, J. Lepagnot, and P. Siarry, A survey on optimization metaheuristics, *Information Sciences*, vol. 237, pp. 82–117, 2013.
20. S. Olafsson, Metaheuristics, in *Handbook in Operations Research and Management Science*, S. Henderson and B. Nelson (Eds.), vol. 13, pp. 633–654, Elsevier, Amsterdam, 2006.
21. B. Melián, J. A. M. Pérez, and J. M. M. Vega, Metaheuristics: A global view, *Inteligencia Artificial*, vol. 7, no. 19, pp. 7–28, 2003.
22. C. Blum and A. Roli, Hybrid metaheuristics: An introduction, in *Hybrid Metaheuristics*, C. Blum, M. J. B. Aguilera, A. Roli, and M. Sampels (Eds.), pp. 1–30, Springer, Berlin Heidelberg, 2008.
23. C. Rego and F. Glover, Local search and metaheuristics, in *The Traveling Salesman Problem and Its Variations*, G. Gutin and A. P. Punnen (Eds.), pp. 309–368, Springer, USA, 2007.
24. L. M. Rios and N. V. Sahinidis, Derivative-free optimization: A review of algorithms and comparison of software implementations, *Journal of Global Optimization*, vol. 56, no. 3, pp. 1247–1293, 2013.
25. C. Blum and A. Roli, Metaheuristics in combinatorial optimization: Overview and conceptual comparison, *ACM Computing Surveys*, vol. 35, no. 3, pp. 268–308, 2003.

26. L. Bianchi, M. Dorigo, L. M. Gambardella, and W. J. Gutjahr, A survey on metaheuristics for stochastic combinatorial optimization, *Natural Computing*, vol. 8, no. 2, pp. 239–287, 2009.
27. M. Gendreau and J.-Y. Potvin, Metaheuristics in combinatorial optimization, *Annals of Operations Research*, vol. 140, no. 1, pp. 189–213, 2005.
28. C. Blum, J. Puchinger, G. R. Raidl, and A. Roli, Hybrid metaheuristics in combinatorial optimization: A survey, *Applied Soft Computing*, vol. 11, no. 6, pp. 4135–4151, 2011.
29. T. G. Stützle, *Local Search Algorithms for Combinatorial Problems: Analysis, Improvements, and New Applications*. IOS Press, Amsterdam, 1999.
30. L. Bianchi, M. Birattari, M. Chiarandini, M. Manfrin, M. Mastrolilli, L. Paquete, O. Rossi-Doria, and T. Schiavinotto, Hybrid metaheuristics for the vehicle routing problem with stochastic demands, *Journal of Mathematical Modelling and Algorithms*, vol. 5, no. 1, pp. 91–110, 2006.
31. O. Bräysy and M. Gendreau, Vehicle routing problem with time windows, Part II: Metaheuristics, *Transportation Science*, vol. 39, no. 1, pp. 119–139, 2005.
32. B. L. Golden, E. A. Wasil, J. P. Kelly, and I.-M. Chao, The impact of metaheuristics on solving the vehicle routing problem: Algorithms, problem sets, and computational results, in *Fleet Management and Logistics*, T. G. Crainic and G. Laporte (Eds.), pp. 33–56, Springer, Heidelberg, 1998.
33. W.-C. Chiang and R. A. Russell, Simulated annealing metaheuristics for the vehicle routing problem with time windows, *Annals of Operations Research*, vol. 63, no. 1, pp. 3–27, 1996.
34. J. Puchinger and G. R. Raidl, *Combining Metaheuristics and Exact Algorithms in Combinatorial Optimization: A Survey and Classification*. Springer, Berlin Heidelberg, 2005.
35. M. Birattari, L. Paquete, T. Stützle, and K. Varrentrapp, *Classification of Metaheuristics and Design of Experiments for the Analysis of Components*, tech. rep., Darmstadt University of Technology, Germany, 2001.
36. D. J. Rosenkrantz, R. E. Stearns, and P. M. Lewis, II, An analysis of several heuristics for the traveling salesman problem, *SIAM Journal on Computing*, vol. 6, no. 3, pp. 563–581, 1977.
37. W. J. Gutjahr, Convergence analysis of metaheuristics, in *Matheuristics*, V. Maniezzo, T. Stützle, and S. Voß (Eds.), pp. 159–187, Springer, USA, 2010.
38. X.-S. Yang, Metaheuristic optimization: Algorithm analysis and open problems, in *Experimental Algorithms*, P. M. Pardalos and S. Rebennack (Eds.), pp. 21–32, Springer, Berlin Heidelberg, 2011.
39. T. Stützle and S. Fernandes, New benchmark instances for the qap and the experimental analysis of algorithms, in *Evolutionary Computation in Combinatorial Optimization*, J. Gottlieb and G. R. Raidl (Eds.), pp. 199–209, Springer, Berlin Heidelberg, 2004.
40. M. Birattari, T. Stützle, L. Paquete, and K. Varrentrapp, A racing algorithm for configuring metaheuristics, in *Proceedings of the Genetic and Evolutionary Computation Conference*, vol. 2, pp. 11–18, 2002.
41. M. Birattari and M. Dorigo, *The problem of tuning metaheuristics as seen from a machine learning perspective*. PhD thesis, Universite Libre de Bruxelles, 2004.
42. O. Rossi-Doria, M. Sampels, M. Birattari, M. Chiarandini, M. Dorigo, L. M. Gambardella, J. Knowles et al., A comparison of the performance of different metaheuristics on the timetabling problem, in *Practice and Theory of Automated Timetabling IV*, E. Burke and P. De Causmaecker (Eds.), pp. 329–351, Springer, Berlin, 2003.
43. E. Alba, G. Luque, and E. Alba, Measuring the performance of parallel metaheuristics, in *Parallel Metaheuristics: A New Class of Algorithms*, E. Alba (Ed.), vol. 47, pp. 43–62, John Wiley & Sons, Hoboken, New Jersey, USA, 2005.
44. T. G. Crainic and M. Toulouse, *Parallel Strategies for Meta-Heuristics*. Springer, New York, USA, 2003.
45. E. Alba, *Parallel Metaheuristics: A New Class of Algorithms*. John Wiley & Sons, USA, 2005.
46. V.-D. Cung, S. L. Martins, C. C. Ribeiro, and C. Roucairol, Strategies for the parallel implementation of metaheuristics, in *Essays and Surveys in Metaheuristics*, C. C. Ribeiro and P. Hansen (Eds.), pp. 263–308, Springer, USA, 2002.
47. T. G. Crainic and M. Toulouse, Parallel meta-heuristics, in *Handbook of Metaheuristics*, M. Gendreau and J.-Y. Potvin (Eds.), pp. 497–541, Springer, USA, 2010.
48. S. Cahon, N. Melab, and E.-G. Talbi, Paradiseo: A framework for the reusable design of parallel and distributed metaheuristics, *Journal of Heuristics*, vol. 10, no. 3, pp. 357–380, 2004.

49. E.-G. Talbi, *Parallel Combinatorial Optimization*. John Wiley & Sons, Hoboken, New Jersey, USA, 2006.
50. E. Alba, G. Luque, and S. Nesmachnow, Parallel metaheuristics: Recent advances and new trends, *International Transactions in Operational Research*, vol. 20, no. 1, pp. 1–48, 2013.
51. E. Alba and G. Luque, Evaluation of parallel metaheuristics, in *Parallel Problem Solving from Nature*, pp. 9–14, Springer, Berlin, 2006.
52. S. D. Eksioglu, P. Pardalos, and M. Resende, Parallel metaheuristics for combinatorial optimization, in *Models for Parallel and Distributed Computation*. R. Corrêa, I. Dutra, M. Fiallos, and F. Gomes (Eds.), Springer, USA, 2002.
53. D. Wolpert and W. Macready, No free lunch theorems for optimization, *IEEE Transactions on Evolutionary Computation*, vol. 1, no. 1, pp. 67–82, 1997.
54. Z. Michalewicz, A survey of constraint handling techniques in evolutionary computation methods, *Evolutionary Programming*, vol. 4, pp. 135–155, 1995.
55. C. A. C. Coello, Theoretical and numerical constraint-handling techniques used with evolutionary algorithms: A survey of the state of the art, *Computer Methods in Applied Mechanics and Engineering*, vol. 191, no. 11, pp. 1245–1287, 2002.
56. K. Deb and R. Datta, A fast and accurate solution of constrained optimization problems using a hybrid bi-objective and penalty function approach, in *IEEE Congress on Evolutionary Computation*, pp. 165–172, IEEE, 2010.
57. S. Camazine, J.-L. Deneubourg, N. R. Franks, J. Sneyd, G. Theraulaz and E. Bonabeau, *Self-Organization in Biological Systems*. Princeton University Press, Princeton, New Jersey, USA, 2003.
58. P. Grassé, Recherches sur la biologie des termites champignonnistes (Macrotermitinae), *Annales des Sciences Naturelles (Zoologie)*, vol. 6, pp. 97–171, 1944.
59. J.-L. Deneubourg, S. Aron, S. Goss, and J. M. Pasteels, The self-organizing exploratory pattern of the argentine ant, *Journal of Insect Behavior*, vol. 3, no. 2, pp. 159–168, 1990.
60. M. Dorigo, V. Maniezzo, and A. Colorni, Ant system: Optimization by a colony of cooperating agents, *IEEE Transactions on Systems, Man, and Cybernetics, Part B: Cybernetics*, vol. 26, no. 1, pp. 29–41, 1996.
61. E. L. Lawler, J. K. Lenstra, A. H. G. R. Kan, and D. B. Shmoys, *The Traveling Salesman Problem: A Guided Tour of Combinatorial Optimization*. Wiley, Chichester, UK, 1985.
62. V. Maniezzo and A. Colorni, The ant system applied to the quadratic assignment problem, *IEEE Transactions on Knowledge and Data Engineering*, vol. 11, no. 5, pp. 769–778, 1999.
63. A. Colorni, M. Dorigo, V. Maniezzo, and M. Trubian, Ant system for job-shop scheduling, *Belgian Journal of Operations Research, Statistics and Computer Science*, vol. 34, no. 1, pp. 39–53, 1994.
64. B. Bullnheimer, R. F. Hartl, and C. Strauss, Applying the ant system to the vehicle routing problem, in *Meta-Heuristics*, S. Voß, S. Martello, I. H. Osman, and C. Roucairol (Eds.), pp. 285–296, Springer, USA, 1999.
65. M. Dorigo and L. M. Gambardella, Ant colony system: A cooperative learning approach to the traveling salesman problem, *IEEE Transactions on Evolutionary Computation*, vol. 1, no. 1, pp. 53–66, 1997.
66. G. Bilchev and I. C. Parmee, The ant colony metaphor for searching continuous design spaces, in *Evolutionary Computing*, T. C. Fogarty (Ed.), pp. 25–39, Springer, Berlin Heidelberg, 1995.
67. K. Socha and M. Dorigo, Ant colony optimization for continuous domains, *European Journal of Operational Research*, vol. 185, no. 3, pp. 1155–1173, 2008.
68. M. Mathur, S. B. Karale, S. Priye, V. Jayaraman, and B. Kulkarni, Ant colony approach to continuous function optimization, *Industrial & Engineering Chemistry Research*, vol. 39, no. 10, pp. 3814–3822, 2000.
69. G. Crina and A. Ajith, Stigmergic optimization: Inspiration, technologies and perspectives, in *Stigmergic Optimization*, A. Ajith, G. Crina, and R. Vitorino (Eds.), pp. 1–24, Springer, Berlin Heidelberg, 2006.
70. D. Karaboga, *An Idea Based on Honey Bee Swarm for Numerical Optimization*, tech. rep., Erciyes University, Turkey, 2005.
71. D. Karaboga and B. Basturk, A powerful and efficient algorithm for numerical function optimization: Artificial bee colony (ABC) algorithm, *Journal of Global Optimization*, vol. 39, no. 3, pp. 459–471, 2007.

72. R. S. Rao, S. Narasimham, and M. Ramalingaraju, Optimization of distribution network configuration for loss reduction using artificial bee colony algorithm, *International Journal of Electrical Power and Energy Systems Engineering*, vol. 1, no. 2, pp. 116–122, 2008.
73. A. Singh, An artificial bee colony algorithm for the leaf-constrained minimum spanning tree problem, *Applied Soft Computing*, vol. 9, no. 2, pp. 625–631, 2009.
74. N. Karaboga, A new design method based on artificial bee colony algorithm for digital iir filters, *Journal of the Franklin Institute*, vol. 346, no. 4, pp. 328–348, 2009.
75. Q.-K. Pan, M. F. Tasgetiren, P. N. Suganthan, and T. J. Chua, A discrete artificial bee colony algorithm for the lot-streaming flow shop scheduling problem, *Information Sciences*, vol. 181, no. 12, pp. 2455–2468, 2011.
76. D. Teodorović and M. Dell'Orco, Bee colony optimization—A cooperative learning approach to complex transportation problems, in *Proceedings of the 16th Mini-EURO Conference on Advanced OR and AI Methods in Transportation*, pp. 51–60, 2005.
77. S. Omkar, J. Senthilnath, R. Khandelwal, G. N. Naik, and S. Gopalakrishnan, Artificial bee colony (ABC) for multi-objective design optimization of composite structures, *Applied Soft Computing*, vol. 11, no. 1, pp. 489–499, 2011.
78. S. F. M. Burnet, *The Clonal Selection Theory of Acquired Immunity*. Cambridge University Press, Cambridge, UK, 1959.
79. L. N. De Castro and F. J. Von Zuben, The clonal selection algorithm with engineering applications, in *Proceedings of the Genetic and Evolutionary Computation Conference*, vol. 2000, pp. 36–39, 2000.
80. L. N. De Castro and F. J. Von Zuben, Learning and optimization using the clonal selection principle, *IEEE Transactions on Evolutionary Computation*, vol. 6, no. 3, pp. 239–251, 2002.
81. J. Timmis, C. Edmonds, and J. Kelsey, Assessing the performance of two immune inspired algorithms and a hybrid genetic algorithm for function optimisation, in *IEEE Congress on Evolutionary Computation*, vol. 1, pp. 1044–1051, IEEE, 2004.
82. A. Gasper and P. Collard, From GAs to artificial immune systems: Improving adaptation in time dependent optimization, in *IEEE Congress on Evolutionary Computation*, vol. 3, IEEE, 1999.
83. L. N. De Castro and F. J. Von Zuben, Learning and optimization using the clonal selection principle, *IEEE Transactions on Evolutionary Computation*, vol. 6, no. 3, pp. 239–251, 2002.
84. L. N. De Castro and F. J. Von Zuben, The clonal selection algorithm with engineering applications, in *Proceedings of the Genetic and Evolutionary Computation Conference*, vol. 2000, pp. 36–39, 2000.
85. P. Hajela and J. Lee, Constrained genetic search via schema adaptation: An immune network solution, *Structural Optimization*, vol. 12, no. 1, pp. 11–15, 1996.
86. X. Wang, X. Z. Gao, and S. J. Ovaska, An immune-based ant colony algorithm for static and dynamic optimization, in *International Conference on Systems, Man and Cybernetics*, pp. 1249–1255, IEEE, 2007.
87. D. DasGupta, *An Overview of Artificial Immune Systems and Their Applications*. Springer, Berlin Heidelberg, 1999.
88. R. Storn and K. Price, Differential evolution—a simple and efficient heuristic for global optimization over continuous spaces, *Journal of Global Optimization*, vol. 11, no. 4, pp. 341–359, 1997.
89. J. A. Nelder and R. Mead, A simplex method for function minimization, *The Computer Journal*, vol. 7, no. 4, pp. 308–313, 1965.
90. S. Das and P. N. Suganthan, Differential evolution: A survey of the state-of-the-art, *IEEE Transactions on Evolutionary Computation*, vol. 15, no. 1, pp. 4–31, 2011.
91. R. Storn and K. Price, Minimizing the real functions of the ICEC'96 contest by differential evolution, in *IEEE International Conference on Evolutionary Computation*, pp. 842–844, IEEE, 1996.
92. K. V. Price, R. M. Storn, and J. A. Lampinen, *Differential Evolution—A Practical Approach to Global Optimization*. Springer-Verlag, Berlin Heidelberg, 2005.
93. V. Feoktistov and S. Janaqi, Generalization of the strategies in differential evolution, in *Proceedings of the 18th International Parallel and Distributed Processing Symposium*, pp. 2341–2346, IEEE, 2004.
94. H.-Y. Fan and J. Lampinen, A trigonometric mutation operation to differential evolution, *Journal of Global Optimization*, vol. 27, no. 1, pp. 105–129, 2003.

95. K. V. Price, An introduction to differential evolution, in *New Ideas in Optimization*, D. Corne, M. Dorigo, F. Glover, D. Dasgupta, P. Moscato, R. Poli, and K. V. Price (Eds.), pp. 79–108, McGraw-Hill, London, UK, 1999.
96. H.-P. Schwefel, *Kybernetische evolution als strategie der experimentellen forschung in der strömungstechnik*, Master's thesis, Technical University of Berlin, 1965.
97. I. Rechenberg, *Evolutionsstrategien*. Springer, Berlin Heidelberg, 1978.
98. I. Rechenberg, *Evolutionsstrategie Optimierung Technischer Systeme Nach Prinzipien der Biologishen Evolution*. Friedrich Frommann Verlag, Struttgart-Bad Cannstatt, 1973.
99. H.-G. Beyer and H.-P. Schwefel, Evolution strategies—A comprehensive introduction, *Natural Computing*, vol. 1, no. 1, pp. 3–52, 2002.
100. H.-P. Schwefel, *Evolutionsstrategie und numerische Optimierung*. PhD thesis, Technical University of Berlin, 1975.
101. H.-P. Schwefel, *Numerische Optimierung von Computer-Modellen Mittels der Evolutionsstrategie: Mit einer Vergleichenden Einführung in die Hill-Climbing-und Zufallsstrategie*. Birkhäuser Verlag, Basel, 1977.
102. T. Back, F. Hoffmeister, and H.-P. Schwefel, A survey of evolution strategies, in *Proceedings of the Fourth International Conference on Genetic Algorithms*, pp. 2–9, Morgan Kaufmann, 1991.
103. H.-G. Beyer, *The Theory of Evolution Strategies*. Springer, Berlin, 2001.
104. A. Auger and N. Hansen, Theory of evolution strategies: A new perspective, *Theory of Randomized Search Heuristics: Foundations and Recent Developments*, vol. 1, pp. 289–325, 2011.
105. H.-G. Beyer, Toward a theory of evolution strategies: Self-adaptation, *Evolutionary Computation*, vol. 3, no. 3, pp. 311–347, 1995.
106. H.-G. Beyer and K. Deb, On self-adaptive features in real-parameter evolutionary algorithms, *IEEE Transactions on Evolutionary Computation*, vol. 5, no. 3, pp. 250–270, 2001.
107. N. Hansen and A. Ostermeier, Completely derandomized self-adaptation in evolution strategies, *Evolutionary Computation*, vol. 9, no. 2, pp. 159–195, 2001.
108. N. Hansen, S. D. Müller, and P. Koumoutsakos, Reducing the time complexity of the derandomized evolution strategy with covariance matrix adaptation (CMA-ES), *Evolutionary Computation*, vol. 11, no. 1, pp. 1–18, 2003.
109. N. Hansen and S. Kern, Evaluating the CMA evolution strategy on multimodal test functions, in *Parallel Problem Solving from Nature*, X. Yao, E. K. Burke, J. A. Lozano, J. Smith, J. J. Merelo-Guervos, J. A. Bullinaria, J. E. Rowe, P. Tiňo, A. Kaban, and H.-P. Schwefel (Eds.), pp. 282–291, Springer, Berlin, 2004.
110. A. Auger and N. Hansen, A restart CMA evolution strategy with increasing population size, in *IEEE Congress on Evolutionary Computation*, vol. 2, pp. 1769–1776, IEEE, 2005.
111. N. Hansen and A. Ostermeier, Adapting arbitrary normal mutation distributions in evolution strategies: The covariance matrix adaptation, in *IEEE International Conference on Evolutionary Computation*, pp. 312–317, IEEE, 1996.
112. L. J. Fogel, A. J. Owens, and M. J. Walsh, *Artificial Intelligence through Simulated Evolution*. John Wiley, New York, USA, 1966.
113. D. B. Fogel, An introduction to simulated evolutionary optimization, *IEEE Transactions on Neural Networks*, vol. 5, no. 1, pp. 3–14, 1994.
114. T. Bäck and H.-P. Schwefel, An overview of evolutionary algorithms for parameter optimization, *Evolutionary Computation*, vol. 1, no. 1, pp. 1–23, 1993.
115. D. B. Fogel, An analysis of evolutionary programming, in *Proceedings of the First Annual Conference on Evolutionary Programming*, pp. 43–51, 1992.
116. D. B. Fogel, L. J. Fogel, and J. W. Atmar, Meta-evolutionary programming, in *Conference on Signals, Systems and Computers*, pp. 540–545, IEEE, 1991.
117. X. Yao, Y. Liu, and G. Lin, Evolutionary programming made faster, *IEEE Transactions on Evolutionary Computation*, vol. 3, no. 2, pp. 82–102, 1999.
118. D. B. Fogel, *Evolving artificial intelligence*. PhD thesis, University of California, San Diego, 1992.

119. T. Bäck, G. Rudolph, and H.-P. Schwefel, Evolutionary programming and evolution strategies: Similarities and differences, in *Proceedings of the Second Annual Conference on Evolutionary Programming*, pp. 11–22, Evolutionary Programming Society, 1993.
120. J. Holland, *Adaptation in Natural and Artificial Systems*. MIT Press, Cambridge, Massachusetts, USA, 1992.
121. D. Goldberg, *Genetic Algorithms in Search, Optimization, and Machine Learning*. Addison-Wesley, Reading, Massachusetts, USA, 1989.
122. F. Herrera, M. Lozano, and J. Verdegay, Tackling real-coded genetic algorithms: Operators and tools for behavioural analysis, *Artificial Intelligence Review*, vol. 12, no. 4, pp. 265–319, 1998.
123. L. Eshelman and J. Schaffer, Real-coded genetic algorithms and interval-schemata, in *Proceedings of the Second Workshop on Foundations of Genetic Algorithms*, vol. 2, pp. 187–202, Morgan Kaufmann, 1992.
124. K. Deb and R. B. Agrawal, Simulated binary crossover for continuous search space, *Complex Systems*, vol. 9, no. 3, pp. 1–15, 1994.
125. H.-M. Voigt, H. Mühlenbein, and D. Cvetkovic, Fuzzy recombination for the breeder genetic algorithm, in *Proceedings of the Sixth International Conference on Genetic Algorithms*, pp. 104–113, Morgan Kaufmann, 1995.
126. Z. Michalewicz, *Genetic Algorithms + Data Structures = Evolution Programs*. Springer, Berlin Heidelberg, 1996.
127. H.-P. Schwefel, Collective phenomena in evolutionary systems, in *Problems of Constancy and Change—The Complementarity of Systems Approaches to Complexity*, pp. 1025–1033, International Society for General Systems Research, 1987.
128. D. Fogel, *Evolutionary Computation: Toward a New Philosophy of Machine Intelligence*. Wiley-IEEE Press, Hoboken, New Jersey, 2006.
129. K. Deb and M. Goyal, A combined genetic adaptive search (GeneAS) for engineering design, *Computer Science and Informatics*, vol. 26, no. 4, pp. 30–45, 1996.
130. G. Rudolph, Convergence analysis of canonical genetic algorithms, *IEEE Transactions on Neural Networks*, vol. 5, no. 1, pp. 96–101, 1994.
131. D. E. Goldberg and J. Richardson, Genetic algorithms with sharing for multimodal function optimization, in *Proceedings of the Second International Conference on Genetic Algorithms*, pp. 41–49, Hillsdale, 1987.
132. A. Pétrowski, A clearing procedure as a niching method for genetic algorithms, in *IEEE International Conference on Evolutionary Computation*, pp. 798–803, 1996.
133. L. J. Eshelman, The CHC adaptive search algorithm: How to have safe search when engaging in nontraditional genetic recombination, in *Foundations of Genetic Algorithms*, Morgan Kaufmann, San Mateo, CA, 1990.
134. J. Joines and C. R. Houck, On the use of non-stationary penalty functions to solve nonlinear constrained optimization problems with GA's, in *IEEE Congress on Evolutionary Computation*, pp. 579–584, IEEE, 1994.
135. D. Whitley, V. S. Gordon, and K. Mathias, Lamarckian evolution, the Baldwin effect and function optimization, in *Parallel Problem Solving from Nature*, Y. Davidor, H.-P. Schwefel, and R. Männer (Eds.), pp. 5–15, Springer, Berlin Heidelberg, 1994.
136. J. R. Koza, *Genetic Programming: On the programming of Computers by Means of Natural Selection*. MIT Press, Cambridge, Massachusetts, USA, 1992.
137. M. F. Brameier and W. Banzhaf, *Linear Genetic Programming*. Springer, New York, USA, 2007.
138. R. Poli, Evolution of graph-like programs with parallel distributed genetic programming, in *Proceedings of the Seventh International Conference on Genetic Algorithms*, pp. 346–353, Morgan Kaufmann, 1997.
139. J. F. Miller and P. Thomson, Cartesian genetic programming, in *Genetic Programming*, R. Poli, W. Banzhaf, W. B. Langdon, J. Miller, P. Nordin, and T. C. Fogarty (Eds.), pp. 121–132, Springer, Berlin Heidelberg, 2000.
140. R. Poli, W. Langdon, N. McPhee, and J. Koza, *Genetic Programming: An Introductory Tutorial and a Survey of Techniques and Applications*, tech. rep., University of Essex, UK, 2007.
141. J. R. Koza, *Genetic Programming II: Automatic Discovery of Reusable Programs*. MIT Press, Cambridge, Massachusetts, USA, 1994.

142. J. R. Koza, F. H. Bennett III, and O. Stiffelman, *Genetic Programming as a Darwinian Invention Machine*. Springer, Berlin Heidelberg, 1999.
143. P. J. Angeline and J. B. Pollack, The evolutionary induction of subroutines, in *Proceedings of the Fourteenth Annual Conference of the Cognitive Science Society*, pp. 236–241, 1992.
144. J. P. Rosca and D. H. Ballard, Discovery of subroutines in genetic programming, in *Advances in Genetic Programming 2*, P. J. Angeline and K. E. Kinnear, Jr. (Eds.), pp. 177–202, MIT Press, Cambridge, MA, 1996.
145. P. Nordin and W. Banzhaf, Complexity compression and evolution, in *Proceedings of the Sixth International Conference on Genetic Algorithms*, pp. 310–317, 1995.
146. B.-T. Zhang and H. Mühlenbein, Balancing accuracy and parsimony in genetic programming, *Evolutionary Computation*, vol. 3, no. 1, pp. 17–38, 1995.
147. S. Luke and L. Panait, A comparison of bloat control methods for genetic programming, *Evolutionary Computation*, vol. 14, no. 3, pp. 309–344, 2006.
148. W. Banzhaf, P. Nordin, R. E. Keller, and F. D. Francone, *Genetic Programming: An Introduction*. Morgan Kaufmann, San Francisco, CA, USA, 1997.
149. M. Clerc, *Particle Swarm Optimization*. ISTE Ltd, London, UK, 2010.
150. Y. Shi and R. C. Eberhart, Parameter selection in particle swarm optimization, in *Proceedings of the Seventh International Conference on Evolutionary Programming*, pp. 591–600, Springer, 1998.
151. R. Eberhart, P. Simpson, and R. Dobbins, *Computational Intelligence PC Tools*. Academic Press Professional, San Diego, CA, USA, 1996.
152. Y. Shi and R. Eberhart, A modified particle swarm optimizer, in *IEEE Congress on Evolutionary Computation*, pp. 69–73, IEEE, 1998.
153. A. Ratnaweera, S. K. Halgamuge, and H. C. Watson, Self-organizing hierarchical particle swarm optimizer with time-varying acceleration coefficients, *IEEE Transactions on Evolutionary Computation*, vol. 8, no. 3, pp. 240–255, 2004.
154. P. Tawdross and A. König, Local parameters particle swarm optimization, in *International Conference on Hybrid Intelligent Systems*, pp. 52–52, IEEE, 2006.
155. M. Clerc and J. Kennedy, The particle swarm-explosion, stability, and convergence in a multidimensional complex space, *IEEE Transactions on Evolutionary Computation*, vol. 6, no. 1, pp. 58–73, 2002.
156. F. Glover, Heuristics for integer programming using surrogate constraints, *Decision Sciences*, vol. 8, no. 1, pp. 156–166, 1977.
157. F. Glover, Scatter search and path relinking, in *New Ideas in Optimization*, D. Corne, M. Dorigo, F. Glover, D. Dasgupta, P. Moscato, R. Poli, and K. V. Price (Eds.), McGraw-Hill, New York, 1999.
158. F. Glover, M. Laguna, and R. Martí, Fundamentals of scatter search and path relinking, *Control and Cybernetics*, vol. 29, no. 3, pp. 653–684, 2000.
159. M. Laguna and R. Marti, *Scatter Search: Methodology and Implementations in C*. Springer Science & Business Media, New York, USA, 2012.
160. N. Metropolis, A. W. Rosenbluth, M. N. Rosenbluth, A. H. Teller, and E. Teller, Equation of state calculations by fast computing machines, *The Journal of Chemical Physics*, vol. 21, no. 6, pp. 1087–1092, 1953.
161. F. Glover, Future paths for integer programming and links to artificial intelligence, *Computers & Operations Research*, vol. 13, no. 5, pp. 533–549, 1986.
162. F. Glover and E. Taillard, A user's guide to tabu search, *Annals of Operations Research*, vol. 41, no. 1, pp. 1–28, 1993.
163. F. Glover, Tabu search—Part I, *ORSA Journal on Computing*, vol. 1, no. 3, pp. 190–206, 1989.
164. F. Glover, Tabu search—Part II, *ORSA Journal on Computing*, vol. 2, no. 1, pp. 4–32, 1990.
165. M. Gendreau, A. Hertz, and G. Laporte, A tabu search heuristic for the vehicle routing problem, *Management Science*, vol. 40, no. 10, pp. 1276–1290, 1994.
166. F. Glover and M. Laguna, *Tabu Search*. Springer, New York, USA, 2013.
167. W. G. Zong, *Music-Inspired Harmony Search Algorithm: Theory and Applications*. Springer, Berlin Heidelberg, 2009.
168. Z. W. Geem, J. H. Kim, and G. Loganathan, A new heuristic optimization algorithm: Harmony search, *Simulation*, vol. 76, no. 2, pp. 60–68, 2001.

169. P. Larrañaga and J. A. Lozano, *Estimation of Distribution Algorithms: A New Tool for Evolutionary Computation*. Springer, New York, USA, 2002.
170. S. Baluja, *Population-Based Incremental Learning: A Method for Integrating Genetic Search Based Function Optimization and Competitive Learning*, tech. rep., Carnegie Mellon University, USA, 1994.
171. K. M. Passino, Biomimicry of bacterial foraging for distributed optimization and control, *IEEE Control Systems Magazine*, vol. 22, no. 3, pp. 52–67, 2002.
172. P. Moscato, *On Evolution, Search, Optimization, Genetic Algorithms and Martial Arts: Towards Memetic Algorithms*, tech. rep., Caltech Concurrent Computation Program, California Institute of Technology, USA, 1989.
173. C. Ferreira, Gene expression programming: A new adaptive algorithm for solving problems, *Complex Systems*, vol. 13, no. 2, pp. 87–129, 2001.
174. W. D. Hillis, Co-evolving parasites improve simulated evolution as an optimization procedure, *Physica D: Nonlinear Phenomena*, vol. 42, no. 1, pp. 228–234, 1990.
175. M. A. Potter and K. A. De Jong, A cooperative coevolutionary approach to function optimization, in *Parallel Problem Solving from Nature*, Y. Davidor, H.-P. Schwefel, and R. Männer (Eds.), pp. 249–257, Springer, Berlin Heidelberg, 1994.
176. X.-S. Yang, *Nature-Inspired Metaheuristic Algorithms*. Luniver Press, Frome, UK, 2010.
177. X.-S. Yang, Firefly algorithms for multimodal optimization, in *Stochastic Algorithms: Foundations and Applications*, O. Watanabe and T. Zeugmann (Eds.), pp. 169–178, Springer, Berlin Heidelberg, 2009.
178. E. Rashedi, H. Nezamabadi-Pour, and S. Saryazdi, GSA: A gravitational search algorithm, *Information Sciences*, vol. 179, no. 13, pp. 2232–2248, 2009.
179. X.-S. Yang and S. Deb, Cuckoo search via Lévy flights, in *World Congress on Nature & Biologically Inspired Computing*, pp. 210–214, IEEE, 2009.
180. D. Simon, Biogeography-based optimization, *IEEE Transactions on Evolutionary Computation*, vol. 12, no. 6, pp. 702–713, 2008.
181. M. Eusuff, K. Lansey, and F. Pasha, Shuffled frog-leaping algorithm: A memetic meta-heuristic for discrete optimization, *Engineering Optimization*, vol. 38, no. 2, pp. 129–154, 2006.
182. M. M. Eusuff and K. E. Lansey, Optimization of water distribution network design using the shuffled frog leaping algorithm, *Journal of Water Resources Planning and Management*, vol. 129, no. 3, pp. 210–225, 2003.
183. D. Pham, A. Ghanbarzadeh, E. Koc, S. Otri, S. Rahim, and M. Zaidi, The bees algorithm—A novel tool for complex optimisation problems, in *International Conference on Intelligent Production Machines and Systems*, pp. 454–459, 2006.
184. R. G. Reynolds, An introduction to cultural algorithms, in *Proceedings of the Third Annual Conference on Evolutionary Programming*, pp. 131–139, World Scientific, 1994.
185. E. Atashpaz-Gargari and C. Lucas, Imperialist competitive algorithm: An algorithm for optimization inspired by imperialistic competition, in *IEEE Congress on Evolutionary Computation*, pp. 4661–4667, IEEE, 2007.
186. B. Xing and W.-J. Gao, Imperialist competitive algorithm, in *Innovative Computational Intelligence: A Rough Guide to 134 Clever Algorithms*, B. Xing and W.-J. Gao (Eds.), pp. 203–209, Springer, International Publishing, Switzerland, 2014.
187. X.-S. Yang, A new metaheuristic bat-inspired algorithm, in *Nature Inspired Cooperative Strategies for Optimization*, J. R. González, D. A. Pelta, C. Cruz, G. Terrazas, and N. Krasnogor (Eds.), pp. 65–74, Springer, Berlin Heidelberg, 2010.
188. C. Ryan, J. Collins, and M. O. Neill, Grammatical evolution: Evolving programs for an arbitrary language, in *Genetic Programming*, W. Banzhaf, R. Poli, M. Schoenauer, and T. C. Fogarty (Eds.), pp. 83–96, Springer-Verlag, Berlin Heidelberg, 1998.
189. M. O'Neill and C. Ryan, Grammatical evolution, *IEEE Transactions on Evolutionary Computation*, vol. 5, no. 4, pp. 349–358, 2001.
190. A. R. Mehrabian and C. Lucas, A novel numerical optimization algorithm inspired from weed colonization, *Ecological Informatics*, vol. 1, no. 4, pp. 355–366, 2006.
191. A. Kaveh and S. Talatahari, A novel heuristic optimization method: Charged system search, *Acta Mechanica*, vol. 213, no. 3–4, pp. 267–289, 2010.

192. O. K. Erol and I. Eksin, A new optimization method: Big bang–big crunch, *Advances in Engineering Software*, vol. 37, no. 2, pp. 106–111, 2006.
193. H. Abbass, MBO: Marriage in honey bees optimization—A haplometrosis polygynous swarming approach, in *IEEE Congress on Evolutionary Computation*, vol. 1, pp. 207–214, IEEE, 2001.
194. D. Teodorović and M. Dell'Orco, Bee colony optimization—A cooperative learning approach to complex transportation problems, in *Advanced OR and AI Methods in Transportation*, pp. 51–60, 2005.
195. S. He, Q. H. Wu, and J. Saunders, Group search optimizer: An optimization algorithm inspired by animal searching behavior, *IEEE Transactions on Evolutionary Computation*, vol. 13, no. 5, pp. 973–990, 2009.
196. A. Afshar, O. Bozorg Haddad, M. A. Mariño, and B. Adams, Honey-bee mating optimization (HBMO) algorithm for optimal reservoir operation, *Journal of the Franklin Institute*, vol. 344, no. 5, pp. 452–462, 2007.
197. A. H. Gandomi and A. H. Alavi, Krill herd: A new bio-inspired optimization algorithm, *Communications in Nonlinear Science and Numerical Simulation*, vol. 17, no. 12, pp. 4831–4845, 2012.
198. H. F. Wedde, M. Farooq, and Y. Zhang, Beehive: An efficient fault-tolerant routing algorithm inspired by honey bee behavior, in *Ant Colony Optimization and Swarm Intelligence*, M. Dorigo, M. Birattari, C. Blum, L. M. Gambardella, F. Mondada, and T. Stützle (Eds.), pp. 83–94, Springer, 2004.
199. K. Krishnanand and D. Ghose, Detection of multiple source locations using a glowworm metaphor with applications to collective robotics, in *IEEE Swarm Intelligence Symposium*, pp. 84–91, IEEE, 2005.
200. H. Shah-Hosseini, The intelligent water drops algorithm: A nature-inspired swarm-based optimization algorithm, *International Journal of Bio-Inspired Computation*, vol. 1, no. 1–2, pp. 71–79, 2009.
201. P. Lucic and D. Teodorovic, Bee system: Modeling combinatorial optimization transportation engineering problems by swarm intelligence, in *Preprints of the TRISTAN IV Triennial Symposium on Transportation Analysis*, pp. 441–445, 2001.
202. J. Bishop, Stochastic searching networks, in *First IEE International Conference on Artificial Neural Networks*, pp. 329–331, IET, 1989.
203. S.-C. Chu, P.-W. Tsai, and J.-S. Pan, Cat swarm optimization, in *PRICAI 2006: Trends in Artificial Intelligence*, Q. Yang and G. Webb (Eds.), pp. 854–858, Springer, Berlin Heidelberg, 2006.
204. R. A. Formato, Central force optimization: A new metaheuristic with applications in applied electromagnetics, *Progress in Electromagnetics Research*, vol. 77, pp. 425–491, 2007.
205. S. H. Jung, Queen-bee evolution for genetic algorithms, *Electronics Letters*, vol. 39, no. 6, pp. 575–576, 2003.
206. H. Drias, S. Sadeg, and S. Yahi, Cooperative bees swarm for solving the maximum weighted satisfiability problem, in *Computational Intelligence and Bioinspired Systems*, J. Cabestany, A. Prieto, and F. Sandoval (Eds.), pp. 318–325, Springer, Berlin Heidelberg, 2005.
207. X.-L. Li and J.-X. Qian, Studies on artificial fish swarm optimization algorithm based on decomposition and coordination techniques, *Journal of Circuits and Systems*, vol. 1, pp. 1–6, 2003.
208. X.-S. Yang and S. Deb, Eagle strategy using Lévy walk and firefly algorithms for stochastic optimization, in *Nature Inspired Cooperative Strategies for Optimization*, J. R. González, D. A. Pelta, C. Cruz, G. Terrazas, and N. Krasnogor (Eds.), pp. 101–111, Springer-Verlag, Berlin Heidelberg, 2010.
209. P. Civicioglu, Transforming geocentric Cartesian coordinates to geodetic coordinates by using differential search algorithm, *Computers & Geosciences*, vol. 46, pp. 229–247, 2012.
210. P. Civicioglu, Backtracking search optimization algorithm for numerical optimization problems, *Applied Mathematics and Computation*, vol. 219, no. 15, pp. 8121–8144, 2013.
211. X.-S. Yang, Flower pollination algorithm for global optimization, in *Unconventional Computation and Natural Computation*, J. Durand-Lose and N. Jonoska (Eds.), pp. 240–249, Springer, Berlin Heidelberg, 2012.
212. I. Zelinka, Analytic programming by means of SOMA algorithm, in *Proceedings of the Eighth International Conference on Soft Computing*, vol. 2, pp. 93–101, 2002.
213. J. Carmelo Filho, F. B. De Lima Neto, A. J. Lins, A. I. Nascimento, and M. P. Lima, A novel search algorithm based on fish school behavior, in *IEEE International Conference on Systems, Man and Cybernetics*, pp. 2646–2651, IEEE, 2008.

214. C. J. Bastos Filho, F. B. de Lima Neto, A. J. Lins, A. I. Nascimento, and M. P. Lima, Fish school search, in *Nature-Inspired Algorithms for Optimisation*, R. Chiong (Ed.), pp. 261–277, Springer, Berlin Heidelberg, 2009.
215. A. Hatamlou, Black hole: A new heuristic optimization approach for data clustering, *Information Sciences*, vol. 222, pp. 175–184, 2013.
216. A. Mucherino and O. Seref, Monkey search: A novel metaheuristic search for global optimization, in *Data Mining, Systems Analysis and Optimization in Biomedicine*, O. Seref , O. E. Kundakcioglu, and P. Pardalos (Eds.), vol. 953, pp. 162–173, AIP Publishing, Florida, USA, 2007.
217. Y. Chu, H. Mi, H. Liao, Z. Ji, and Q. Wu, A fast bacterial swarming algorithm for high-dimensional function optimization, in *IEEE Congress on Evolutionary Computation*, pp. 3135–3140, IEEE, 2008.
218. H. Eskandar, A. Sadollah, A. Bahreininejad, and M. Hamdi, Water cycle algorithm—A novel metaheuristic optimization method for solving constrained engineering optimization problems, *Computers & Structures*, vol. 110-111, pp. 151–166, 2012.
219. G. Kjellstrom and L. Taxen, Stochastic optimization in system design, *IEEE Transactions on Circuits and Systems*, vol. 28, no. 7, pp. 702–715, 1981.
220. P. Rabanal, I. Rodríguez, and F. Rubio, Using river formation dynamics to design heuristic algorithms, in *Unconventional Computation*, S. G. Akl, C. S. Calude, M. J. Dinneen, G. Rozenberg, and H. T. Wareham (Eds.), pp. 163–177, Springer, 2007.
221. Y. Shi, An optimization algorithm based on brainstorming process, *International Journal of Swarm Intelligence Research*, vol. 2, no. 4, pp. 35–62, 2011.
222. T. C. Havens, C. J. Spain, N. G. Salmon, and J. M. Keller, Roach infestation optimization, in *Swarm Intelligence Symposium*, pp. 1–7, IEEE, 2008.
223. A. H. Kashan, League championship algorithm: A new algorithm for numerical function optimization, in *International Conference of Soft Computing and Pattern Recognition*, pp. 43–48, IEEE, 2009.
224. E. Cuevas, D. Oliva, D. Zaldivar, M. Pérez-Cisneros, and H. Sossa, Circle detection using electromagnetism optimization, *Information Sciences*, vol. 182, no. 1, pp. 40–55, 2012.
225. H. Shah-Hosseini, Principal components analysis by the galaxy-based search algorithm: A novel metaheuristic for continuous optimisation, *International Journal of Computational Science and Engineering*, vol. 6, no. 1, pp. 132–140, 2011.
226. A. Kaveh and N. Farhoudi, A new optimization method: Dolphin echolocation, *Advances in Engineering Software*, vol. 59, pp. 53–70, 2013.
227. R. Tang, S. Fong, X.-S. Yang, and S. Deb, Wolf search algorithm with ephemeral memory, in *IEEE Seventh International Conference on Digital Information Management*, 2012.
228. P. Civicioglu, Artificial cooperative search algorithm for numerical optimization problems, *Information Sciences*, vol. 229, pp. 58–76, 2012.
229. J. Taheri and A. Y. Zomaya, RBT-GA: A novel metaheuristic for solving the multiple sequence alignment problem, *BMC Genomics*, vol. 10, no. 1, p. Article ID S10, 2009.
230. T.-C. Chen, P.-W. Tsai, S.-C. Chu, and J.-S. Pan, A novel optimization approach: Bacterial-GA foraging, in *International Conference on Innovative Computing, Information and Control*, pp. 391–391, IEEE, 2007.
231. K. Tamura and K. Yasuda, Spiral dynamics inspired optimization, *Journal of Advanced Computational Intelligence and Intelligent Informatics*, vol. 15, no. 8, pp. 1116–1122, 2011.
232. A. Mozaffari, A. Fathi, and S. Behzadipour, The great salmon run: A novel bio-inspired algorithm for artificial system design and optimisation, *International Journal of Bio-Inspired Computation*, vol. 4, no. 5, pp. 286–301, 2012.
233. U. Premaratne, J. Samarabandu, and T. Sidhu, A new biologically inspired optimization algorithm, in *International Conference on Industrial and Information Systems*, pp. 279–284, IEEE, 2009.
234. H. Hernández and C. Blum, Distributed graph coloring: An approach based on the calling behavior of Japanese tree frogs, *Swarm Intelligence*, vol. 6, no. 2, pp. 117–150, 2012.
235. S. Iordache, Consultant-guided search: A new metaheuristic for combinatorial optimization problems, in *Proceedings of the Genetic and Evolutionary Computation Conference*, pp. 225–232, ACM, 2010.
236. H. Chen, Y. Zhu, K. Hu, and X. He, Hierarchical swarm model: A new approach to optimization, *Discrete Dynamics in Nature and Society*, vol. 2010, p. Article ID 379649, 2010.

237. Y. Xu, Z. Cui, and J. Zeng, Social emotional optimization algorithm for nonlinear constrained optimization problems, in *Swarm, Evolutionary, and Memetic Computing*, B. K. Panigrahi, S. Das, P. N. Suganthan, and S. S. Dash (Eds.), pp. 583–590, Springer, Berlin Heidelberg, 2010.
238. H. Shayeghi and J. Dadashpour, Anarchic society optimization based PID control of an automatic voltage regulator (AVR) system, *Electrical and Electronic Engineering*, vol. 2, no. 4, pp. 199–207, 2012.
239. R. S. Parpinelli and H. S. Lopes, An eco-inspired evolutionary algorithm applied to numerical optimization, in *World Congress on Nature & Biologically Inspired Computing*, pp. 466–471, IEEE, 2011.
240. R. D. Maia, L. N. de Castro, and W. M. Caminhas, Bee colonies as model for multimodal continuous optimization: The OptBees algorithm, in *IEEE Congress on Evolutionary Computation*, pp. 1–8, IEEE, 2012.
241. R. Hedayatzadeh, F. Akhavan Salmassi, M. Keshtgari, R. Akbari, and K. Ziarati, Termite colony optimization: A novel approach for optimizing continuous problems, in *18th Iranian Conference on Electrical Engineering*, pp. 553–558, IEEE, 2010.
242. C. Sur, S. Sharma, and A. Shukla, Egyptian vulture optimization algorithm—A new nature inspired metaheuristics for knapsack problem, in *International Conference on Computing and Information Technology*, pp. 227–237, Springer, 2013.
243. X.-S. Yang, J. M. Lees, and C. T. Morley, Application of virtual ant algorithms in the optimization of CFRP shear strengthened precracked structures, in *Proceedings of the Sixth International Conference on Computational Science*, pp. 834–837, Springer, 2006.
244. F. Comellas and J. Martinez-Navarro, Bumblebees: A multiagent combinatorial optimization algorithm inspired by social insect behaviour, in *Proceedings of the First ACM/SIGEVO Summit on Genetic and Evolutionary Computation*, pp. 811–814, ACM, 2009.
245. G.-W. Yan and Z.-J. Hao, A novel optimization algorithm based on atmosphere clouds model, *International Journal of Computational Intelligence and Applications*, vol. 12, no. 1, p. Article ID 1350002, 2013.
246. L. M. Zhang, C. Dahlmann, and Y. Zhang, Human-inspired algorithms for continuous function optimization, in *International Conference on Intelligent Computing and Intelligent Systems*, vol. 1, pp. 318–321, IEEE, 2009.
247. T. Ting, K. L. Man, S.-U. Guan, M. Nayel, and K. Wan, Weightless swarm algorithm (WSA) for dynamic optimization problems, in *Network and Parallel Computing*, J. J. Park, A. Zomaya, S.-S. Yeo, and S. Sahni (Eds.), pp. 508–515, Springer, Berlin Heidelberg, 2012.
248. S. Su, J. Wang, W. Fan, and X. Yin, Good lattice swarm algorithm for constrained engineering design optimization, in *International Conference on Wireless Communications, Networking and Mobile Computing*, pp. 6421–6424, IEEE, 2007.
249. B. Alatas, Photosynthetic algorithm approaches for bioinformatics, *Expert Systems with Applications*, vol. 38, no. 8, pp. 10541–10546, 2011.
250. F. Ahmadi, H. Salehi, and K. Karimi, Eurygaster algorithm: A new approach to optimization, *International Journal of Computer Applications*, vol. 57, no. 2, pp. 9–13, 2012.
251. K. A. DeJong, *Evolutionary Computation: A Unified Approach*. MIT Press, Cambridge, Massachusetts, USA, 2006.
252. K. Deb and N. Padhye, Development of efficient particle swarm optimizers by using concepts from evolutionary algorithms, in *Proceedings of the Genetic and Evolutionary Computation Conference*, pp. 55–62, ACM, 2010.
253. N. Padhye, P. Bhardawaj, and K. Deb, Improving differential evolution through a unified approach, *Journal of Global Optimization*, vol. 55, no. 4, pp. 771–799, 2013.
254. K. Sörensen, Metaheuristics—The metaphor exposed, *International Transactions in Operational Research*, vol. 22, no. 1, pp. 3–18, 2015.
255. D. Weyland, A rigorous analysis of the harmony search algorithm: How the research community can be misled by a "novel" methodology, *International Journal of Applied Metaheuristic Computing*, vol. 1, no. 2, pp. 50–60, 2010.
256. D. Weyland, A critical analysis of the harmony search algorithm—How not to solve Sudoku, *Operations Research Perspectives*, vol. 2, pp. 97–105, 2015.

12

Neural Networks

Amit Bhaya

CONTENTS

ABSTRACT This chapter contains a brief history of the origins of artificial neural networks (ANNs), subsequently discussing the main learning algorithms and paradigms. These include perceptrons and multilayer perceptrons and a presentation of the backpropagation algorithm for the problem of classifying data. Support vector machines are then discussed for the cases of both linearly separable as well as nonseparable data, followed by a discussion on radial basis functions networks and interpolation using such networks. A discussion of universal approximations serves to contextualize the mathematical properties of neural networks. Neural networks for optimization are presented from the point of view of gradient dynamical systems. The application of these dynamical neural networks to the real-time solution of linear systems is presented. The chapter ends with an annotated bibliography, a list of neural network software suites, a list of journals that publish ANN-related research, and a list of websites that make data sets publicly available.

12.1 Introduction

What are neural networks?
Wikipedia's definition of artificial neural networks (ANNs) is as follows:

> Artificial neural networks are composed of interconnecting artificial neurons (programming constructs that mimic the properties of biological neurons). Artificial neural networks may either be used to gain an understanding of biological neural networks, or for solving artificial intelligence problems without necessarily creating a model of a real biological system. The real, biological nervous system is highly complex: artificial neural network algorithms attempt to abstract this complexity and focus on what may hypothetically matter most from an information processing point of view. Good performance (e.g. as measured by good predictive ability, low generalization error), or performance mimicking animal or human error patterns, can then be used as one source of evidence towards supporting the hypothesis that the abstraction really captured something important from the point of view of information processing in the brain. Another incentive for these abstractions is to reduce the amount of computation required to simulate artificial neural networks, so as to allow one to experiment with larger networks and train them on larger data sets.

This is a good definition that captures many of the important aspects of ANNs. In particular, this chapter is exclusively devoted to describing ANNs that are designed as problem-solving tools, rather than as tools that allow an understanding of biological systems. Since the field of ANNs has grown tremendously in the last two decades, the focus will be on those aspects that are more relevant to decision making.

12.1.1 Brief History of the Origins of ANNs

Biologists have been studying the nervous system and, more specifically, the brain for more than a century, experimenting and speculating on how the brain works and what its components are. This activity led to an interest in developing mathematical models for the brain, as well as for the more macroscopic tasks of learning, giving birth to what are now called artificial neural networks. Early seminal work by McCulloch and Pitts [59] and Hebb [39] proposed independent processors, later called artificial neurons, that generate an output that is a simple function of the total input and, the outputs of these processors, in turn, serve as the inputs to other processors. One of the first words that arose to describe this situation was connectionism, because it was felt that the processing power of the brain emerges from the interconnections of its neurons, even though each of the latter may have rather limited capability.

We give a brief and rather sketchy timeline of the early history of neural networks, emphasizing the mathematical contributions which had a great impact on the field of ANNs. For a more detailed history, including all the fundamental work done by biologists, neuroscientists, psychologists, and so on, we refer the reader to the annotated bibliography (Section 12.9).

- 1943. McCulloch and Pitts [59] put forth the first mathematical model, described as a binary threshold function, of a biological neuron and noted that many arithmetic and logical operations can be implemented using such binary threshold neurons.
- 1948. Wiener [79] published a seminal book *Cybernetics, or Communication and Control in the Animal and the Machine* establishing abstract principles of organization in complex systems and proposing a mathematical approach to neurodynamics, extending original work initiated by the Russian-American mathematical biophysicist Nicholas Rashevsky.
- 1949. Hebb [39] published his extremely influential book *The Organization of Behavior* in which he introduced the learning rule that repeated excitatory activation of one neuron by another at a particular synapse increases its conductance.
- 1954. Gabor [33] described his learning filter which used the steepest descent algorithm to minimize the mean squared error between the observed output signal and a signal generated based upon the past information, in order to find optimal filter coefficients or weights. This paper also established a strong connection between signal processing and neural networks.
- 1960. Widrow and Hoff [78] published their Adaline network that is also trained by a gradient descent rule to minimize mean squared error.
- 1962. Rosenblatt [67] published his book *Principles of Neurodynamics: Perceptrons and the Theory of Brain Mechanisms* in which perceptrons as well as a learning algorithm for McCulloch–Pitts neurons were presented and the perceptron convergence theorem was conjectured.
- 1969. Minsky and Papert [62] published *Perceptrons: An Introduction to Computational Geometry*, which showed both the strengths of the perceptron model through mathematical proofs as well as some limitations. Both were exploited by different communities with different agenda, but there is no doubt that this book, despite, or perhaps because of, the controversy that it engendered, stimulated research both in artificial intelligence and, later, in ANNs as well.

In the 1970s and 1980s, several researchers explored the fundamental ideas mentioned in the timeline. We highlight the development of learning rules applicable to multilayer networks which were discovered and rediscovered, more or less independently, in the dynamic programming and optimal control community by Bryson [15], Kelley [52], and Dreyfus [30], and then brought to the attention of the ANN community by McClelland and Rumelhart [68] several years later. Important contributions were also made by Cybenko (universal approximation properties) [27], Amari (information geometry) [4], Grossberg (adaptive resonance theory) [34,35], Hopfield (optimization and associative memories) [42], and Kohonen (self-organizing maps) [53], to name just a few. In addition, advances were made in understanding neural networks from the point of view of statistics and regression by Bishop [12], Ripley [66], and others, to the extent that we can now say that the field is now in a mature state of development, with its strong connections to approximation theory, nonlinear regression, signal processing, and statistical and machine learning theory well understood and documented. Given the extremely broad scope of the subject of ANNs, some decisions to limit the topics considered in this chapter had to be made and, in general, preference was given to those aspects of neural networks accessible to a broad audience and directly relevant to decision making.

12.2 Basic Concepts and Terminology

The mathematical model of an artificial neuron is arrived at as an abstraction of a real biological neuron, which typically integrates a large number of electrical signals through structures called dendrites. In a real neuron, these electrical signals are modified in a complex manner, involving both chemical and electrical processes, as well as a threshold, before appearing at the output of the neuron. Since we wish to deal exclusively with artificial neurons, we will pass immediately from this extremely simplified description to the mathematical model of an artificial neuron. In words, an artificial neuron is an object that receives input signals from various sources with different gains, commonly referred to as connection weights. These signals are combined into a weighted sum plus a threshold, generating a quantity known as an input activation signal. In terms of the biological metaphor which inspires this mathematical model, the connection weights model the synaptic efficiencies; positive weights correspond to excitatory synapses, while negative ones correspond to inhibitory synapses. The threshold represents the internal firing threshold of a biological neuron and the activation signal can be thought of as the aggregated cell potential. This input activation signal is then transformed through the so-called activation function, generating the output signal of the neuron. For artificial neurons, activation functions are of several standard types: binary threshold, linear threshold, sigmoidal, Gaussian, and represent approximations of the sigmoidal or logistic functions often encountered in real neurons.

Thus the four items that constitute an artificial neuron are

1. A set of links, also referred to as synapses, each with an associated weight (also called gain or strength). In particular, a signal x_i at the input of synapse j connected to neuron i is multiplied by the (synaptic) weight or gain w_{ij}. We follow the convention that the first subscript refers to the neuron, while the second refers to the input synapse, noting that some authors follow the opposite convention.
2. A summer or adder that carries out the weighted sum or linear combination of the input signals.
3. An activation function that limits the amplitude of the neuron output—as mentioned above, this function often takes the form of a threshold function of some specified type and which we will denote as ϕ.
4. An exogenous or externally applied bias, denoted b_i for neuron i, which has the effect of increasing or decreasing the overall input to the activation function: that is, a term that is added to the linear combination of the input signals, before passing through the activation function.

With this notation, the ith neuron can be represented mathematically as follows:

$$u_i = \sum_{j=1} w_{ij} x_j \tag{12.1}$$

$$v_i = u_i + b_i \tag{12.2}$$

$$y_i = \phi(v_i) \tag{12.3}$$

For notational and mathematical convenience, the bias or threshold b_i is often considered as an additional weight, denoted w_{i0}, for instance, which always receives a constant input $x_0 = 1$. With this convention, $v_i = \sum_{j=0} w_{ij} x_j$ and $y_i \phi(v_i)$, as before. An important observation at this point is that if the weights associated to the ith neuron are stacked into a vector denoted as $\mathbf{w}_i := (w_{i0}, w_{i1}, \ldots)$ and the corresponding input vector denoted as $\mathbf{x} = (x_0, x_1, \ldots)$, then the so-called internal activation v_i is simply the inner product of these two vectors, that is, $v_i = \mathbf{w}_i^\mathsf{T}\mathbf{x}$. Note that this inner product is naturally interpreted as a measure of similarity between the input signal $\mathbf{x}$ and the weight vector $\mathbf{w}_i$, in the sense that it is largest when the two vectors are aligned and equal to zero when they are orthogonal. In other words, input vectors that are aligned and point in the same (respectively, opposite) direction as the weight vector of the neuron result in its maximal positive (respectively, negative) activation, while orthogonal input vectors result in zero activation. This mechanism, in increasingly sophisticated forms, lies at the heart of the discriminative property of a large class of ANNs. It is also worth mentioning that the formation of an inner product between an input vector and a given weight vector is also referred to as a linear filter. Thus, a neuron with fixed weights can be thought of as a linear filter and one with weights that are modified or learned in some way as an adaptive filter. Finally, note that an input vector is often referred to as a *pattern*—this terminology comes from the field of pattern recognition, in which a pattern is essentially a vector whose components are *features* (of the pattern). We will use the terms "vector," "data point," and "pattern" interchangeably.

As far as the activation function ϕ is concerned, there are many types, but two basic ones that are ubiquitous are the threshold, McCulloch–Pitts or Heaviside function

$$\phi(v) = \begin{cases} 1 & \text{if } v \geq 0 \\ 0 & \text{if } v < 0 \end{cases} \tag{12.4}$$

and the class of sigmoid or logistic functions, with an S-shaped graph, described mathematically as

$$\phi(v) = \frac{1}{1 + \exp(-\alpha v)} \tag{12.5}$$

Note that the parameter α in Equation 12.5 determines the slope of the sigmoid: specifically, the slope is $\alpha/4$ at the origin and the threshold function can be seen as the limit of a sigmoidal function as the slope parameter α tends to infinity. The most important difference, which is fundamental to learning algorithms that we will present below, is that sigmoidal activation functions are differentiable, while threshold functions present a discontinuity at the origin and are not differentiable. In some applications, it is necessary to use an odd activation function, symmetrical about the origin and assuming both positive and negative values. These symmetric versions of Equations 12.4 and 12.5 are as follows:

$$\phi(v) = \begin{cases} 1 & \text{if } v > 0 \\ 0 & \text{if } v = 0 \\ -1 & \text{if } v < 0 \end{cases} \tag{12.6}$$

$$\phi(v) = \tanh(v) \tag{12.7}$$

Another popular activation function that we will have occasion to discuss and use is the Gaussian function with center μ and spread σ:

$$\phi(v) = \exp\left(-\frac{(v - \mu)^2}{2\sigma^2}\right) \tag{12.8}$$

In contrast with sigmoidal functions, which are monotonic, the Gaussian function, also referred to as a representative of the class of radial basis functions (RBFs), is nonmonotonic, reaching its peak value at the center μ and decreasing smoothly to zero as the argument v moves away from the center. The intuition, which has obvious biological inspiration, is that activations close to the center elicit strong responses, whereas those farther away yield weaker responses. Thus, tuning of the "receptive" center of a neuron can make it capable of recognizing (= responding to) a specific input range.

With these preliminaries, we can informally describe an ANN as a set of interconnected artificial neurons. The architecture of the neural network describes the organization of these neurons into layers or fields. A very general description of an ANN is a massively parallel adaptive network of simple computing elements called artificial neurons. Evidently, the model of an artificial neuron is an abstract and vastly simplified version of its biological counterpart, but one that attempts to capture some of computational properties of the latter. The important point is that, despite the simplicity of the individual computing element, the massively parallel interaction between these elements should be designed in such a way as to lead to the capability to solve interesting and complex problems—in this sense, it may be said that complex behavior emerges from the interaction, described by the network (interconnection) of simple neurons. More formally, a neural network is specified by the following items:

- Neurons—These can be of three types: input, hidden, and output. As the names suggest, input neurons receive the external inputs to the network, while output neurons are responsible for generating the output of the network. Hidden neurons, not accessible to the outside of the network, are connected, possibly through other layers of neurons, either to the input or output neurons and are involved in intermediate stages of the overall computation.
- Model of each neuron—The vector of inputs, the summing rule, which is usually of the inner product form, in which case, it is enough to specify the weights associated to the neuron and the activation function.
- Architecture—Referring to the connectivity graph of the neurons. By specifying the input neurons, the output neurons, as well as the manner in which they are connected to the so-called hidden layer neurons, a complete description of the flow of signals from the input to the output of the network is obtained.
- Learning rule—The main feature of a neural network is its ability to learn from examples. The learning rule specifies how to use the example inputs and feedback from the corresponding outputs in order to adjust neuron weights in such a manner that the resulting network is able, in principle, to perform well on inputs that have never been presented to it.

In order to specify the architecture of a neural network, it can be viewed as a weighted directed graph in which each neuron is a node and each directed edge represents the connection (and its direction) between one node and another, while the weight represents the strength of the connection. In broad terms, architectures can be classified as

- Feedforward: The directed graph representing the network has no loops.
- Feedback: The directed graph contains loops, which represent feedback connections. The term "recurrent" is often used instead of "feedback."

In either case (feedforward or feedback), from a mathematical point of view, a neural network is a mapping from the space of inputs to the space of outputs (which could coincide) and, in this context, the term "heteroassociative" refers to the case when the input and output spaces are different. The word "associative" emphasizes the fact that the neural network associates or maps an input vector to a certain output vector.

The term "autoassociative," on the other hand, refers to the case in which there are feedback connections between the outputs and the inputs and the network associates an element of the input space with another element of the same space, since the output space is the same as the input space.

Another classification of neural networks has to do with the manner in which the input is related to the output. If the output depends only on the current input, then the network is said to be static or memoryless. The latter term refers to the fact that the response of the network to an input depends only on the current input and is independent of the previous network state. This is usually the case with feedforward architectures. In this case, from a mathematical point of view, a network that learns a certain mapping between the input space and the output space after being presented, during the learning process, with associated pairs of inputs and outputs has approximated or interpolated the function between the input and output spaces. In statistics, the term "regression" is used to refer to this function approximation process.

Feedback or recurrent networks, on the other hand, when presented with a new input vector, compute an output vector in accordance with the neuron activations and interconnections as usual; however, since the outputs are fed back to the inputs, there is a change in the input activations. As a result, there is a change in the output of the network. A concise mathematical description of this situation is a dynamical system and it can be affirmed that if a vector called the state, made up of all the neuron activations, is known, together with the current input, then the next state and, indeed, all subsequent states can be determined. For such a dynamical system, a crucial question is that of stability—does the changing sequence of states eventually converge to an equilibrium state? In a typical application of such a network, stability can be interpreted as recall or associative memory, meaning that if the input presented to a network is a distorted version of one of the stable stored vectors, then the network state converges to this stable state and its output indicates this recall or recuperation of a remembered (=stored) state. Another application of recurrent networks is to problems in optimization in which it is often the case that there is a unique equilibrium state that corresponds to the unique optimal point to be found.

In summary, neural networks can be designed for a variety of tasks that can be loosely classified into two groups: the first denominated as function approximation, regression or pattern classification and the second as associative memory or optimization. The main features of a neural network solution to these tasks are robustness and capacity for generalization. The term "robustness" refers to the ability of neural networks to continue to perform their designated tasks even if some of the neurons or connections between them get damaged or inoperative. This behavior stems from the massively parallel nature of these networks.

12.3 Learning Algorithms and Paradigms

In the context of neural networks, learning algorithms are procedures that encode training information about labeled or classified patterns by suitably modifying the weights in the architecture chosen for the network. Broadly speaking, there are two types of learning: supervised and unsupervised. In supervised learning, data is often in the form of input vectors associated to output vectors. The components of the input vectors are also commonly referred to as *features* in the pattern recognition literature.

The input space will be referred to as $\mathcal{X}$ and will usually be a subset of $\mathbb{R}^n$, while the output space will be denoted $\mathcal{D}$ and generally will be a subset of $\mathbb{R}$.

In supervised learning, although there appear to exist two categories of rules, namely, error correction rules and gradient-descent-based rules, it will be seen below that both these categories can be viewed more profitably as arising from the minimization of an appropriate cost function.

The defining feature of a supervised learning system or machine is that it uses a so-called *training data set* $\mathcal{T} = \{\mathbf{x}(k), \mathbf{d}(k)\}_{k=1}^{N}$, where $\mathbf{x}_k \in \mathbb{R}^n$ is an input, also called an *n-feature vector*, for which the desired output $d_k \in \mathbb{R}$ is known. Training refers to the process of finding a function $f : \mathbb{R}^n \rightarrow \mathbb{R} : \mathbf{x} \mapsto y$ such that, for a vector $\mathbf{x}_{\text{new}} \neq \mathbf{x}(k)$ for any k, the corresponding output, which is commonly referred to as the predicted output and denoted $\hat{y} = f(\mathbf{x}_{\text{new}})$, is, in the ideal case, equal to the expected output, denoted d_{new}. When this happens, the machine is said to have *generalized* successfully. Of course, this is too simplistic a view and a more rigorous approach takes into account the probability distribution of the training data as well as the new or unseen data for which the trained machine is expected to perform well. In short, if the machine is provided with a set of data $\mathcal{T}$ that has a given probability distribution, then training should be interpreted to mean that the mean square error on the training data set is reduced to sufficiently low levels, and generalization is then interpreted to mean that if unseen data are drawn from the same probability distribution as the training data, then the network should attain low mean square error once again. As we will see in the sequel, for most neural networks, training involves adjusting a set of weights in order that the mean square error be minimized to within some tolerance. Choosing this tolerance adequately is quite critical, since a low mean square error on the training data can be achieved by choosing this tolerance to be low, but this usually means that the weights have been adjusted to learn the particular features of the training data, such as the noise contained in the training samples and, when the trained network is presented with new, unseen inputs, the predicted outputs are very inaccurate. Figure 12.1 puts this in perspective, showing the result of fitting a curve with a multilayer perceptron (MLP) with many degrees of freedom (weights in the layers) to a relatively small data set for a simple one-dimensional problem. Specifically, consider the target function, that is, the unknown function which ideally generates the data, to be a straight line. Suppose that the observed data correspond to points on this line with a little added noise. Then, as we know from the theory of interpolation, it is possible to find a polynomial of suitably high order (in this case, since there are only five data points, a fourth-order polynomial) that passes exactly through all the given data points. In fact, choosing an MLP with many weights or degrees of freedom results in this type of behavior. However, a glance at Figure 12.1

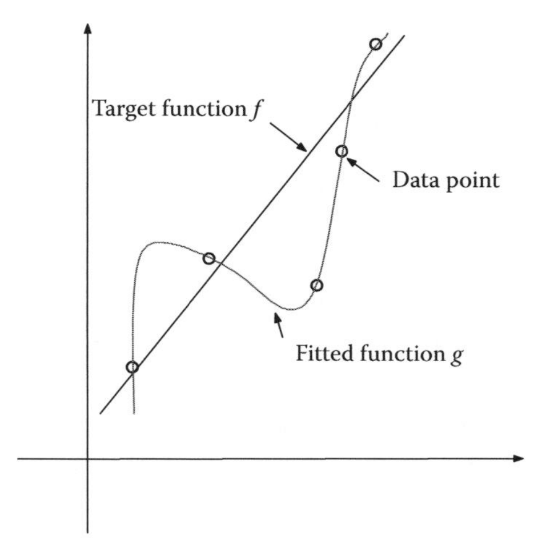

FIGURE 12.1
Illustrating the overfitting problem for a small one-dimensional data set.

shows that the fitted function g looks nothing like the target function f. Another way to say this is to say that overfitting has occurred, meaning that the learning algorithm has been misled by the noise and ended up "learning" the noise as well. The undesirable result of this overfitting is that, although the fitted curve has zero in-sample error, it has very large out-of-sample error and can be said to generalize very badly. Any practical learning algorithm must therefore take steps to avoid overfitting and this is usually achieved by tools such as regularization or validation, and we will give some pointers to the literature on this subject below.

12.4 Perceptrons and Multilayer Perceptrons

In order to understand the basic geometry of perceptrons and also recall the recent history of neural networks, we briefly discuss the pattern recognition or classification problem for networks of McCulloch–Pitts or binary threshold neurons. Examples of classification problems that are performed effortlessly by humans are ubiquitous and some of the simpler ones, such as postal code reading, are now considered to be benchmark examples for machine learning algorithms, including neural networks. Typically, a classification task is solved by collecting a large amount of data which has already been classified or labeled and using this data to "teach" or train a human or a machine to recognize classes, in order that, when presented with data that has not been seen before or used in the training process, correct classification, or classification with a low error rate, results.

In the simplest and most frequent case, a typical data point is a vector of measurements of different quantities called *features* and it is assumed that each labeled data point can be put into one of two categories, depending on whether it has or does not have a particular property. In this situation, the design of a classifier involves the determination of a boundary that separates the two categories and, as soon as such a boundary or decision frontier has been found, the classification task for a given feature vector is reduced to task of computing which side of the decision frontier it lies on. A decision frontier or boundary could be a hyperplane or a more complicated hypersurface: in either case, the feature space is divided into two decision regions. In the case that a hyperplane suffices to separate the data into two classes, the data are said to be *linearly separable*. It is often the case that data that are not linearly separable, after a suitable transformation or embedding into a higher-dimensional space, can become linearly separable and this is very desirable since linear separation is well understood and computationally tractable. This will be explained through an example in the sequel.

12.4.1 Classification of Data into Two Classes: Basic Geometry

We start with a discussion of the simplest case: separation of data into two classes by a separating hyperplane. For this case of linearly separable data, the problem can be stated as follows:

LINEAR CLASSIFICATION PROBLEM: Given a set $X = \{x_i \in \mathbb{R}^n, i = 1, \dots, M\}$ that are labeled as belonging to one of two disjoint classes, C_1 or C_2, find a hyperplane that separates the two classes, meaning that all points in class C_1 lie on one side of the hyperplane, while all points in class C_2 lie on the other side.

A more practical version of this problem that will be discussed later will remove the assumption of perfect linear separability and require the computation of a separating hyperplane, using only a subset of the labeled sample points, in such a way that an appropriate measure of misclassification is minimized.

The geometric notions required to characterize perfect linear separability come from convex analysis and the concepts needed to enunciate the fundamental separating hyperplane theorem are introduced below.

Definition 12.1

A set $\mathcal{S} \subset \mathbb{R}^n$ is *convex* if and only if, for all $\lambda \in [0, 1]$ and for all points $\mathbf{x}, \mathbf{y} \in \mathcal{S}$, the *convex combination* $\lambda\mathbf{x} + (1 - \lambda)\mathbf{y}$ belongs to $\mathcal{S}$. Stated in words, a set $\mathcal{S}$ is convex if and only if it contains all points on all line segments with end points in $\mathcal{S}$.

Figure 12.2 shows examples of convex and nonconvex sets in the plane.

Definition 12.2

The *convex hull* $\text{conv}(X)$ of an arbitrary set $X \subset \mathbb{R}^n$ is the smallest convex set in $\mathbb{R}^n$ which contains the set X. Equivalently, it is the set of all convex combinations of points in X.

Definition 12.3

Sets X_1 and X_2 are said to be *linearly separable* if their convex hulls are disjoint.

With these preliminaries, the separating hyperplane theorem can be stated as follows:

Theorem 12.1

Suppose X_1 and X_2 are two disjoint convex subsets of $\mathbb{R}^n$ ($X_1 \cap X_2 = \emptyset$). Then there exist $\mathbf{a} \neq \mathbf{0}$ in $\mathbb{R}^n$, and b in $\mathbb{R}$ such that $\mathbf{a}^\mathsf{T}\mathbf{x} \leq b$ for all $\mathbf{x} \in X_1$ and $\mathbf{a}^\mathsf{T}\mathbf{x} \geq b$ for all $\mathbf{x} \in X_2$. The hyperplane $\{\mathbf{x} : \mathbf{a}^\mathsf{T}\mathbf{x} = b\}$ is called a *separating hyperplane* for the sets X_1 and X_2, or is simply said to separate the sets X_1 and X_2 (see Figure 12.3).

To illustrate the concept of linear separability and motivate the introduction of perceptrons and MLP, two classical examples of the AND and XOR (exclusive OR) Boolean functions are presented, the first being linearly separable and the second linearly inseparable. From the truth table of

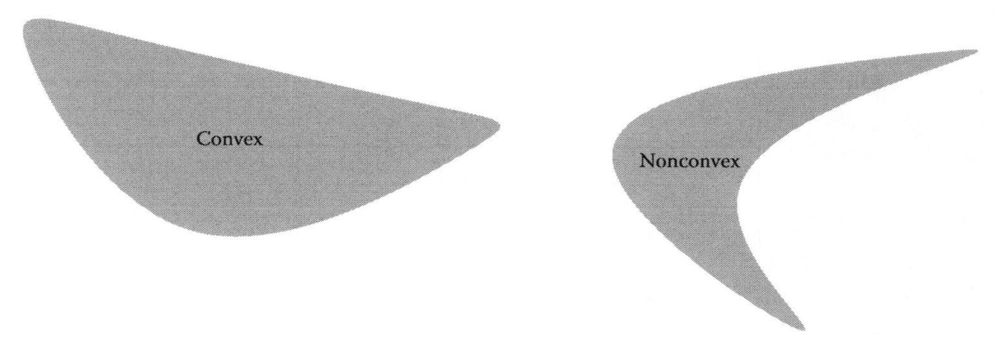

FIGURE 12.2
Examples of a convex and a nonconvex set in $\mathbb{R}^2$.

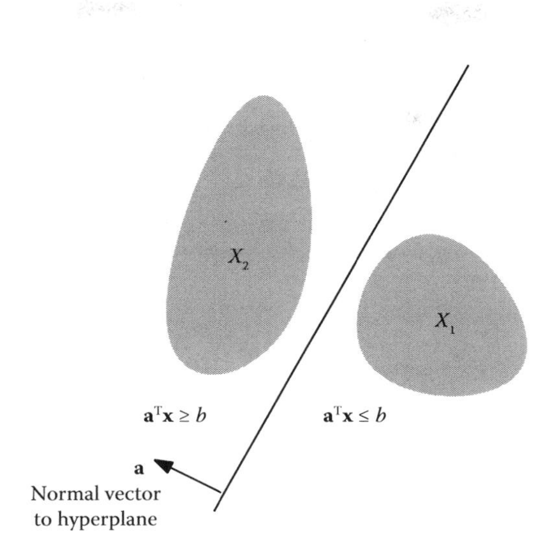

FIGURE 12.3
The hyperplane $\{\mathbf{x} : \mathbf{a}^{\mathrm{T}}\mathbf{x} = \mathbf{b}\}$ separates the disjoint convex sets X_1, X_2.

the Boolean AND function, we know that the three points $(0,0), (0,1), (1,0)$ are mapped to 0 and the points $(1,1)$ mapped to 1, so that a classifier that performs the Boolean AND function should classify the first three points into one class (X_1) and the remaining point $(1,1)$ into a second class (X_2). Similarly, the truth table of the Boolean XOR function tells us that $(0,0), (1,1)$ should be mapped to the class X_1 (representing output 0) and the points $(0,1), (1,0)$ to class X_2 (representing output 1). Figure 12.4 shows the geometry of these two functions and it is clear that the AND function corresponds to a linearly separable problem, while the XOR function does not. A McCulloch–Pitts or binary threshold neuron, shown in Figure 12.5, can classify the linearly separable classes corresponding to the AND function. For this two-dimensional problem, a single McCulloch–Pitts neuron has two inputs x_1, x_2 and the corresponding weights w_1, w_2, as well as a bias, which we denote w_0, in order to associate it to a constant input $+1$. Then, defining $\mathbf{x} := (1, x_1, x_2)$ and $\mathbf{w} := (w_0, w_1, w_2)$,

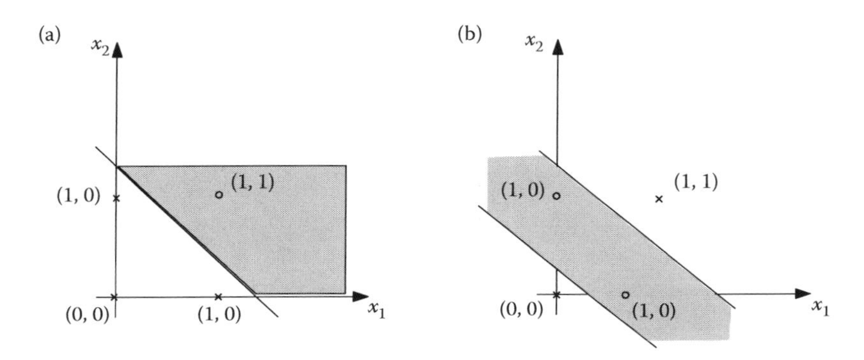

FIGURE 12.4
(a): The Boolean AND function, which is linearly separable. (b): The Boolean XOR function, which is not linearly separable, since two lines are required in order to separate the classes. Crosses (×) denote points mapped to zero, while circles (○) denote points mapped to one.

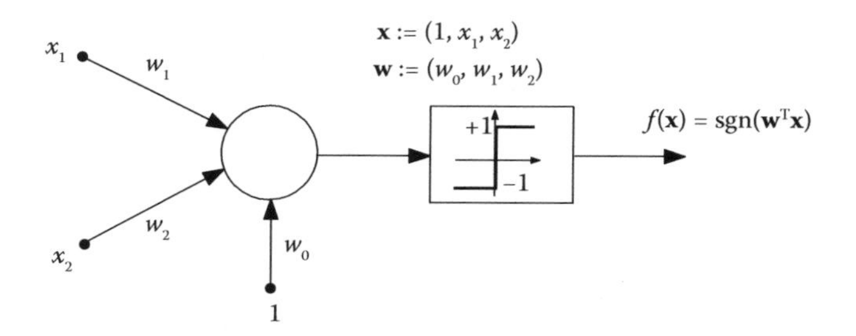

FIGURE 12.5
A McCulloch–Pitts or binary threshold neuron that realizes an inner product between the input and weight vectors, which is then mapped to its output through a threshold or signum function.

this is easily seen by writing the *discriminant function*:

$$x_2 = -\frac{w_1}{w_2}x_1 - \frac{w_0}{w_2} \tag{12.9}$$

which is the equation of a straight line with slope $-w_1/w_2$ and intercept $-w_0/w_2$ which are functions of the weights, the appropriate choice of which determines the classes that are distinguished by the associated McCulloch–Pitts neuron. These are the two half planes above and below the line (12.9). This makes it clear that if a set of data points is linearly separable, then there exists a set of weights for a McCulloch–Pitts or binary threshold neuron that is capable of classifying this data set. A glance at Figure 12.4 for the XOR function suggests immediately that, even though the XOR data set is not linearly separable with one straight line, it is certainly separable with two straight lines. Specifically, since the two points $(1, 0)$ and $(0, 1)$ should be in the same class (output of 1), and the other two points $(0, 0)$ and $(1, 1)$ in the other class (output 0), it follows that the two straight lines correspond to the OR function (lower line) and the negation of the AND function, called the NAND function (upper line). This leads immediately to the idea of a combination, through an AND, of the two McCulloch–Pitts neurons, one corresponding to an OR and the other to a NAND, as shown in Figure 12.6. We may summarize the discussion above by saying that a single McCulloch–Pitts neuron implements a function from $\mathbb{R}^n$ into the set $\{0, 1\}$ that performs linear separation: all points belong to one class are mapped to zero and all points belonging to the other class mapped to one. Furthermore, it appears that, for sets of points that are not linearly separable, some interconnection of McCulloch–Pitts neurons might be able to do the job of separating them. The most general combinatorial question that can be asked in this context is as follows. If k points in $\mathbb{R}^n$ are arbitrarily assigned to two classes, labeled 0 and 1, how many are linearly separable?

At this point, a natural question would be to ask how far one can go with interconnections of multiple levels or layers of binary threshold neurons This is a purely geometrical question, which we shall set aside here, instead referring the reader to References 20 and 24 for further details. A more practical question will take us forward. Given a set of labeled data points that is known to be linearly separable but is accessed only sequentially (meaning that the data points become available one at a time, sequentially), is it possible to use this information to find the weights (i.e., the normal vector to the separating hyperplane)?

12.4.2 Perceptron Learning Algorithm

In order to arrive at a geometric understanding of this algorithm, we consider the case of a single McCulloch–Pitts neuron, with weight vector **w**. Recall that the weight vector is the normal to the

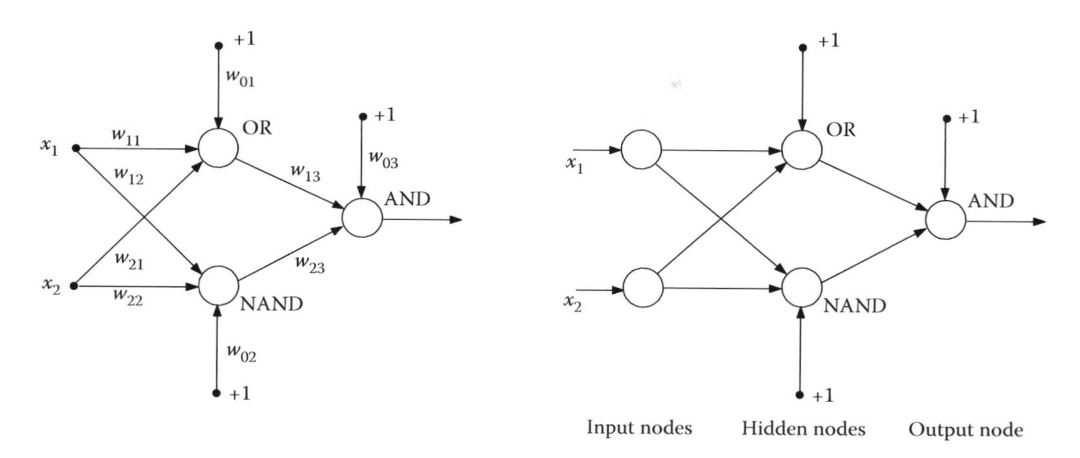

FIGURE 12.6
A two-layer McCulloch–Pitts or binary threshold neuron network that implements the XOR function. The figure on the left depicts each McCulloch–Pitts neuron as a circle, with the arrows representing inputs, which undergo a weighted sum (the w_{ij}s being the weights) and then pass through a threshold function. The figure on the right is a common way to redraw the one on the left: the input nodes now correspond to linear neurons which transmit the inputs without any modification. The other nodes, referred to as hidden or output nodes, are McCulloch–Pitts neurons.

(pattern) separating hyperplane corresponding to this neuron. This means that the activation $y = \mathbf{w}^{\mathsf{T}}\mathbf{x}$ resulting from an input pattern $\mathbf{x}$ is positive whenever the angle between the vectors $\mathbf{x}$ and $\mathbf{w}$ is less than $\pi/2$ and negative otherwise, resulting in the separation of input vectors into two classes on either side of the hyperplane. It is useful to introduce the terminology of weight space to refer to the space to which the weight vectors belong and pattern space to refer to the space to which the input vectors belong. Using these terms, the separating hyperplane $\mathbf{w}^{\mathsf{T}}\mathbf{x} = 0$ can be interpreted in two ways. The first interpretation is the usual one in pattern space, in which a useful weight vector is one that is normal to a hyperplane that separates the labeled patterns into two classes according to their labels. On the other hand, fixing attention on a specific pattern $\mathbf{x}_k$ and its corresponding label, 1 say, setting the inner product $\mathbf{x}_k^{\mathsf{T}}\mathbf{w}$ to 0 can be thought of as defining a hyperplane in weight space, called the pattern hyperplane (corresponding to $\mathbf{x}_k$). The vector $\mathbf{x}_k$ is the normal to its pattern hyperplane in weight space. Figure 12.7 shows how this geometry leads to a straightforward identification of all vectors in weight space that classify a given set of linearly separable patterns: it is enough to find the convex cone that is the intersection of all the positive regions of the pattern hyperplanes and any weight vector in this region solves the classification problem.

This geometrical approach also leads directly to the perceptron learning algorithm. Suppose that the labeled patterns are denoted $\mathbf{x}_i$ and that the weights are updated by an iterative procedure, so that the weight vector at the kth iteration is $\mathbf{w}_k$. Suppose that the pattern presented to the neuron is $\mathbf{x}_k$ and that it belongs to class 1, but that $\mathbf{w}_k^{\mathsf{T}}\mathbf{x}$ is negative, instead of assuming a positive value, which would be correct. The iterative procedure should be designed in such a way that, given this information, it modifies the weight vector to a new one such that the inner product becomes positive. This situation is depicted in Figure 12.8, from which it is clear how to proceed in the general case.

Thus, if the misclassification is such that corrective action requires increasing the inner product, then the weight vector should be incremented in the direction of the current pattern vector; otherwise, it should be decremented. This recipe for updating weights for a single McCulloch–Pitts neuron is

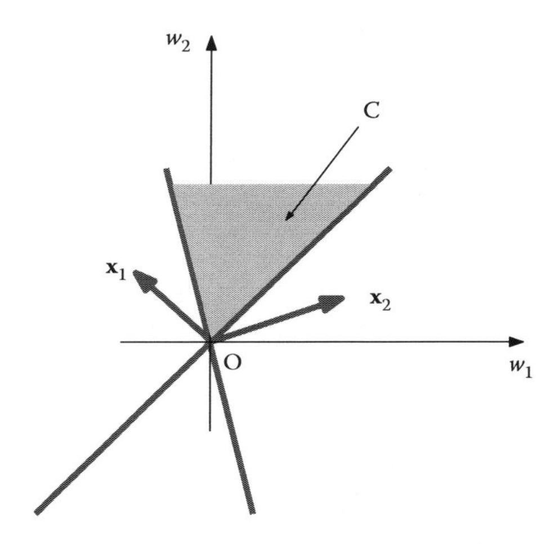

FIGURE 12.7
Two patterns $\mathbf{x}_1, \mathbf{x}_2$ depicted in weight space $\mathbb{R}^2$. The conical shaded region C contains all weight vectors that classify both patterns in class 1 (positive inner product).

known as the *perceptron learning algorithm* and can be written as follows:

$$\mathbf{w}_{k+1} = \begin{cases} \mathbf{w}_k + \eta_k \mathbf{x}_k, & \text{if } \mathbf{x}_k \in X_1 \text{ and } \mathbf{w}_k^\top \mathbf{x}_k \leq 0 \\ \mathbf{w}_k - \eta_k \mathbf{x}_k, & \text{if } \mathbf{x}_k \in X_2 \text{ and } \mathbf{w}_k^\top \mathbf{x}_k \leq 0 \end{cases} \tag{12.10}$$

where η_k is the step size in the direction of the current pattern vector, usually referred to as the *learning rate*.

Since we will not really use the perceptron learning algorithm, which is of interest more from the point of view of insight and history, we state the perceptron convergence theorem informally as follows: the perceptron learning algorithm (12.10) converges in a finite number of steps for a linearly separable set of patterns and for any initial weight vector.

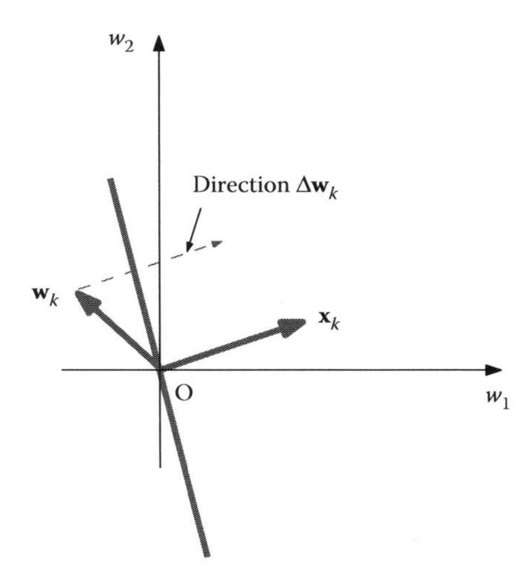

FIGURE 12.8
A misclassified pattern $\mathbf{x}_k$ depicted in weight space $\mathbb{R}^2$ in which the weight at the kth iteration is $\mathbf{w}_k$. The smallest perturbation $\Delta\mathbf{w}_k$ that can be applied to the vector $\mathbf{w}_k$ in order that its inner product with $\mathbf{x}_k$ become positive is evidently in the direction of the normal vector to the pattern hyperplane, which is $\mathbf{x}_k$.

12.4.3 Multilayer Perceptrons and the Backpropagation Algorithm

The perceptron and least mean squares (LMS) learning algorithms which use, respectively, a single threshold and a single linear neurons are successful in finding weights that solve a linear classification problem and, in fact, this is the major application for these types of neurons. This subsection explores the enhanced capabilities of multilayer networks that were touched on in the discussion of the XOR problem. We will start by considering a simple multilayer architecture, depicted in Figure 12.9 in which the input layer has n linear neurons, plus an additional input to all the bias connections of the next layer, known as the hidden layer. The hidden layer, in turn, is assumed to have q neurons, each with a sigmoidal activation function, plus an additional neuron that generates the bias signal for the output layer. The hidden layer neurons receive their signals from the input layer. Finally, the output layer consists of q sigmoidal neurons, receiving signals from the hidden layer. Note that the flow of signals in the MLP network is unidirectional, from the inputs through the hidden layer to the outputs and, to emphasize this characteristic, MLPs are often also referred to as feedforward neural networks. Note that the neural network is a composition of a map from $\mathbb{R}^n$ to $\mathbb{R}^q$ (the linear map from inputs to hidden layer) and a map from $\mathbb{R}^q$ to $\mathbb{R}^p$ (the nonlinear map from hidden layer to outputs), so that, overall, it is a map from $\mathbb{R}^n$ to $\mathbb{R}^p$. The problem that we wish to solve with an MLP can be stated as follows. Given a set of input–output pairs, we wish to find an algorithm to adjust the weights of an MLP so that, after adjustment, the MLP approximates or "learns" the functional relationship between inputs and outputs. In other words, if the inputs are presented to an MLP, with some initially chosen weights, the resulting outputs do not match the given ones and the desired algorithm should use this error information to adjust the weights in such a way that, eventually, the MLP produces the correct input–output pairs. More specifically, we are interested in two major questions:

- Does the MLP architecture described above permit the approximation of an arbitrary mapping from $\mathbb{R}^n$ to $\mathbb{R}^p$ by the suitable choice of its parameters (number of hidden layer neurons, weights, bias values, activation functions)? In fact, we are asking if an MLP network possesses the *universal approximation property*.
- Assuming that the answer to the previous question is affirmative and the MLP network is trained on a certain input–output data set, how well does it predict outputs for inputs that do not belong to the data set used for training? This is a question about the so-called *generalization ability* of the MLP network.

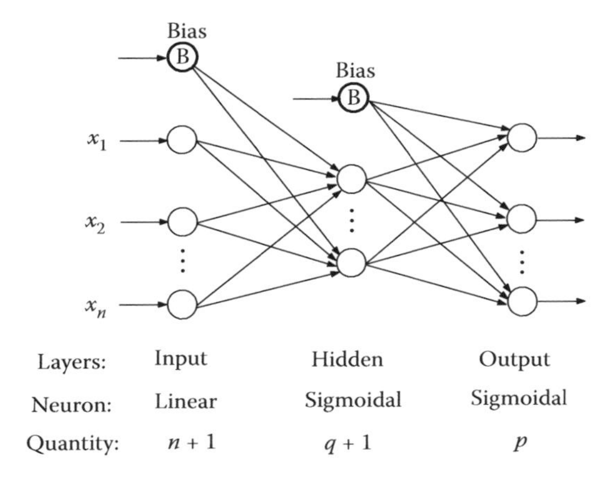

FIGURE 12.9
A multilayer perceptron, introducing the notation that will be used in the derivation of the backpropagation algorithm.

Before discussing the mathematical answers to these questions, which belong formally to the area of mathematics known as approximation theory, we state an affirmative answer informally. Roughly speaking, it is possible to prove that if an MLP network has a sufficiently large number of neurons (interconnections), then it has the universal approximation property, meaning that it can approximate any input–output function belonging to a sufficiently smooth class to any desired degree of accuracy. This is most often phrased as a nonconstructive existence result and, as such, not of great interest to practitioners, except insofar as it guarantees the existence of a solution to the universal approximation problem using MLPs. In fact, the great explosion in the field of neural networks took place after the discovery that, for differentiable activation functions (of the sigmoidal type, for instance), the use of the classical gradient descent on the output error results, after some manipulation, in a viable algorithm that performs weight updates in such a way as to solve the function approximation problem. This is the famous backpropagation (BP) algorithm, arguably the most popular algorithm for the training of MLP networks and we will derive this algorithm now, deferring the discussion of theoretical matters to Section 12.7.

12.4.4 Backpropagation Algorithm

We will derive the BP algorithm for the three-layer MLP network shown in Figure 12.9, noting that the generalization to networks with more layers is just a matter of bookkeeping a larger number of indices and more complex notation, but no fundamentally new idea.

Suppose that the given training data of N input–output pairs is denoted as follows:

$$\mathcal{T} = \{\mathbf{x}(k), \mathbf{d}(k)\}_{k=1}^{N} \tag{12.11}$$

where the $\mathbf{x}(k)$ are input vectors in $\mathbb{R}^n$ and $\mathbf{d}(k)$ the corresponding output vectors in $\mathbb{R}^p$. The remaining notation is displayed in Table 12.1. In Table 12.1, as the subscripts suggest, the notation w_{ih} denotes weights between the input and hidden layers, while w_{hj} denotes weights between the hidden and output layers: in both cases, the indices $i = 0, h = 0$ denote the bias neuron.

For the sake of concreteness, we will assume that, for the linear neurons in the input layer:

$$\ell(x) = x \tag{12.12}$$

while for the hidden and output layers which have sigmoidal neurons

$$\phi(x) = \frac{1}{1 + e^{\alpha x}} \tag{12.13}$$

TABLE 12.1

Notation for BP Algorithm

Layer → **Neuron ↓**	**Input**	**Hidden**	**Output**
Number	$n+1$	$q+1$	p
Type	Linear	Sigmoidal	Sigmoidal
Index	i	h	j
Input signal	x_i	z_h	y_j
Output signal	$\ell(x_i)$	$\phi(z_h)$	$\phi(y_j)$
Weights	w_{ih}	w_{hj}	

For simplicity, we will assume $\alpha = 1$. Then, assuming that k is an iteration counter and denoting the weights at the kth iteration as $w_{ih}(k)$, $w_{hj}(k)$, on presenting the input $\mathbf{x}(k)$ to the ANN, the calculated output will not, in general, be equal to the specified output $\mathbf{d}(k)$ and we may define the error at the kth iteration as

$$\mathbf{e}(k) = \mathbf{d}(k) - \Phi(\mathbf{y}(k)) \tag{12.14}$$

where

$$\Phi(\mathbf{y}(k)) := (\phi(y_1(k)), \ldots, \phi(y_p(k))) \tag{12.15}$$

The standard procedure is to work with the square of the two-norm of the error vector for the kth input. Specifically, we define

$$e(k) = \frac{1}{2}\mathbf{e}(k)^{\mathsf{T}}\mathbf{e}(k) \tag{12.16}$$

With these preliminaries, we sketch the main steps of the BP algorithm before detailing how each step is actually performed. Each presentation of a pattern is identified with an iteration and the import of step 3 is that we may have to cycle more than once through the given data set of N input–ouput pairs before convergence occurs. Each such cycle or presentation of all N samples of the data set is referred to as an *epoch*, in the standard jargon of neural networks.

An important observation in regard to the overview of the BP Algorithm 12.1 is that an important simplification has already been made. Namely, that it is valid to perform gradient descent on the error function at the kth iteration, sometimes referred to as the instantaneous error, rather than the conventional route of using the total error or its mean, referred to as the mean square error, defined as

$$e_{\mathrm{MSE}} = \frac{1}{N}\sum_{k=1}^{N} e(k) \tag{12.17}$$

where $e(k)$ is instantaneous error as defined in Equation 12.16. In fact, these two approaches to iteratively updating the weights, based respectively on the mean square error and the instantaneous error, are referred to as *batch updating* and *pattern updating*. It can be shown [8] that if the step size in the direction of weight change, also referred to as learning rate, is sufficiently small, then

Algorithm 12.1: Overview of the Backpropagation Algorithm

1: $k = 1$
2: **while** Stopping criterion based on error is not met **do**
3: $k \leftarrow k(\mathrm{mod}\, N)$
4: Present input $\mathbf{x}(k)$ to the network, calculating, successively, the activation of the hidden layer neurons, their outputs, thence the activations of the output layer neurons and, finally, the outputs.
5: Calculate the errors at the outputs in accordance with (12.16).
6: Regarding the errors as functions of the weights, use a descent algorithm to calculate a descent direction (= changes in weights) in weight space that reduces the squared error (12.16).
7: Update the weights in accordance with the changes computed in the previous step.
8: $k = k + 1$
9: **end while**

the weight changes computed by pattern updating are good estimates of those computed by batch updating. The standard BP algorithm uses pattern updating. The boxes below show the main steps in detail: they are the forward phase which consists of propagating the input signals through the hidden layer to the outputs; the backward phase in which the errors at the output layer are propagated backward to the input layer, using the error gradients calculated by the chain rule; and, finally, the weight update phase in which the information of the preceding phase is used to calculate the increments to the weights and apply them. It is the crucial backward phase that gives the algorithm its name.

Forward phase of the backpropagation algorithm
Propagating the input signals through the MLP to the outputs

1. Input layer:

$$\ell(x_i(k)) = x_i(k), i = 1, \ldots, n \tag{12.18}$$

$$\ell(x_0(k)) = x_0(k) = 1 \quad \text{(bias neuron)} \tag{12.19}$$

2. Hidden layer:

$$z_h(k) = \sum_{i=0}^{n} w_{ih}(k)\phi(x_i(k)), h = 1, \ldots, q \tag{12.20}$$

$$\phi(z_h(k)) = \frac{1}{1+e^{-z_h(k)}}, h = 1, \ldots, q \tag{12.21}$$

$$\phi(z_0(k)) = 1, \quad \text{for all } k \tag{12.22}$$

where, once again, the subscript zero is reserved for the biases of the hidden neurons.

3. Output layer:

$$y_j(k) = \sum_{h=0}^{q} w_{hj}(k)\phi(z_h(k)), h = 1, \ldots, p \tag{12.23}$$

$$\phi(y_j(k)) = \frac{1}{1+e^{-y_j(k)}}, h = 1, \ldots, p \tag{12.24}$$

The second phase, which concerns the computation of error gradients, consists of two steps:

Step 1: Computation of error gradients for hidden to output layer
Observe that the signal flow from hidden to output layer can be described as follows. A signal $\phi(z_h(k))$ activates an output layer neuron with $y_j(k)$ after passing through a weight $w_{hj}(k)$. The resulting final output $\phi(y_j(k))$ is compared with the given desired output $d_j(k)$ to calculate the error component $e_j(k)$ of $\mathbf{e}(k)$ which is used in the

calculation of the error $e(k)$ in Equation 12.16. Given this sequence of dependencies, from the chain rule we can write:

$$\frac{\partial e(k)}{\partial w_{hj}(k)} = \frac{\partial e(k)}{\partial \phi(y_j(k))} \frac{\partial \phi(y_j(k))}{\partial y_j(k)} \frac{\partial y_j(k)}{\partial w_{hj}(k)} \tag{12.25}$$

Each of the three partial derivatives on the right-hand side of Equation 12.25 can be calculated directly as

$$\frac{\partial e(k)}{\partial \phi(y_j(k))} = -(d_j(k) - \phi(y_j(k))) = -e_j(k) \tag{12.26}$$

$$\frac{\partial \phi(y_j(k))}{\partial y_j(k)} = \phi'(y_j(k)) = \phi(y_j(k))(1 - \phi(y_j(k))) \tag{12.27}$$

$$\frac{\partial y_j(k)}{\partial w_{hj}(k)} = \phi(z_h(k)) \tag{12.28}$$

Plugging these expressions into Equation 12.25 gives

$$\frac{\partial e(k)}{\partial w_{hj}(k)} = -e_j(k)\phi'(y_j(k))\phi(z_h(k)) \tag{12.29}$$

$$= -\delta_j(k)\phi(z_h(k)) \tag{12.30}$$

where

$$\delta_j(k) := e_j(k)\phi'(y_j(k)) \tag{12.31}$$

In words, $\delta_j(k)$ is the product of the error and the slope of the sigmoidal function ϕ. Given the nature of the sigmoidal function (namely, small slope for very small or very large arguments and largest slope for zero arguments), we conclude that $\delta_j(k)$ has the same characteristics.

Step 2: Computation of error gradients for input to hidden layer
Once again, following the signal flow and using the chain rule, we can write

$$\frac{\partial e(k)}{\partial w_{ih}(k)} = \frac{\partial e(k)}{\partial \phi(z_h(k))} \frac{\partial \phi(z_h(k))}{\partial z_h(k)} \frac{\partial z_h(k)}{\partial w_{ih}(k)} \tag{12.32}$$

Since the error $e(k)$ is a function of the errors at each of the p output neurons, from the chain rule:

$$\frac{\partial e(k)}{\partial \phi(z_h(k))} = \sum_{j=1}^{p} \left\{ \frac{\partial e_k}{\partial y_j(k)} \frac{\partial y_j(k)}{\partial \phi(z_h(k))} \right\} \tag{12.33}$$

Substituting this expression in Equation 12.32 gives

$$\frac{\partial e(k)}{\partial w_{ih}(k)} = \sum_{j=1}^{p} \left\{ \frac{\partial e(k)}{\partial y_j(k)} \frac{\partial y_j(k)}{\partial \phi(z_h(k))} \right\} \phi'(z_h(k))\phi(x_i(k)) \tag{12.34}$$

$$= \sum_{j=1}^{p} \left\{ \frac{\partial e(k)}{\partial \phi(y_j(k))} \frac{\partial \phi(y_j(k))}{\partial y_j(k)} \frac{\partial y_j(k)}{\partial \phi(z_h(k))} \right\} \phi'(z_h(k))\phi(x_i(k)) \tag{12.35}$$

$$= \sum_{j=1}^{p} \left\{ -e_j(k)\phi'(y_j(k))w_{hj}(k) \right\} \phi'(z_h(k))x_i(k) \tag{12.36}$$

$$= -\sum_{j=1}^{p} \left(\delta_j(k)w_{hj}(k) \right) \phi'(z_h(k))x_i(k) \tag{12.37}$$

In order to understand how this calculation leads to the well-known BP procedure, we examine Equation 12.37 more closely. In particular, in the first summation term $\sum_{j=1}^{p} \left(\delta_j(k)w_{hj}(k) \right)$, the $\delta_j(k)$s are known, since they have already been computed for the output neurons (12.31). Thus, if we define the quantities

$$e_h(k) := \sum_{j=1}^{p} \delta_j(k)w_{hj}(k) \tag{12.38}$$

and

$$\delta_h(k) := e_h(k)\phi'(z_h(k)) \tag{12.39}$$

then we can rewrite Equation 12.37 as

$$\frac{\partial e_k}{\partial w_{ih}(k)} = -\delta_h(k)x_i(k) \tag{12.40}$$

in close analogy with Equation 12.30. The important points to note are that

- The expression (12.38) can be interpreted as propagating the scaled errors $\delta_j(k)$ backward through the weights $w_{hj}(k)$, to compute the contribution of node h to the error.
- If we are dealing with an architecture that has more hidden layers, this process can be repeated for each of them, propagating errors backward from the output, successively through each of the hidden layers in the direction of the input layer.

These observations lead to the name backpropagation associated to this algorithm. The equations for weight update can now be written down.

Step 3: Calculation of the weight updates

1. Hidden to output layer weight updates:

$$w_{hj}(k+1) = w_{hj}(k) + \Delta w_{hj}(k) \tag{12.41}$$

$$= w_{hj}(k) - \eta \frac{\partial e(k)}{\partial w_{hj}(k)} \tag{12.42}$$

$$= w_{hj}(k) + \eta \delta_j(k) \phi(z_h(k)) \tag{12.43}$$

2. Input to hidden layer weight updates:

$$w_{ih}(k+1) = w_{ih}(k) + \Delta w_{ih}(k) \tag{12.44}$$

$$= w_{ih}(k) - \eta \frac{\partial e(k)}{\partial w_{ih}(k)} \tag{12.45}$$

$$= w_{ih}(k) + \eta \delta_h(k) x_i(k) \tag{12.46}$$

12.4.4.1 Backpropagation with Momentum

As the derivation makes clear, the BP algorithm is a gradient-based algorithm and can be viewed as a variant of the steepest descent algorithm. In fact, it inherits two well-known properties of the steepest descent algorithm. Namely, in general, it converges to a local minimum of the error function and, moreover, the convergence can be very slow. Furthermore, the learning rate usually has to be kept small in order to avoid oscillations and follow a relatively smooth trajectory in weight space. In order to get around these limitations, a popular method, shown in Algorithm 12.2, is to use a so-called momentum term in the weight update equations:

$$\Delta w_{hj}(k) = \eta \delta_j(k) \phi(z_h(k)) + \alpha \Delta w_{hj}(k-1) \tag{12.47}$$

$$\Delta w_{ih}(k) = \eta \delta_h(k) x_i(k) + \alpha \Delta w_{ih}(k-1) \tag{12.48}$$

The intuitive justification for the introduction of the momentum term is that, if it is large enough, it allows the trajectories to overshoot local minima that are located in shallow valleys and find other local minima in deeper neighboring valleys. Mathematically, it can be shown that, for a quadratic error function, backpropagation with momentum is exactly equivalent to the conjugate gradient algorithm [9]. In general, the introduction of the momentum term transforms the error dynamics from first order to second order, allowing extra degrees of freedom.

12.5 Support Vector Machines

Support vector machines (SVMs) embody a state-of-the-art learning technique that is firmly grounded in theory—the Vapnik–Chervonenkis theory of structural risk minimization discussed in Reference 75. In addition, SVMs are easily implemented and have been successful in a wide variety of applications. They belong to the general class of kernel methods. Two key ideas underlie the

Algorithm 12.2: Backpropagation with Momentum Algorithm

1: Data: $\mathcal{T} = \{\mathbf{x}(k), \mathbf{d}(k)\}_{k=1}^{N}$.
2: Objective: Adjust the weights of a neural network with n input nodes, q hidden nodes and p output nodes to $\mathcal{T}$ so as to minimize a given error criterion.
3: $k = 1$
4: **while** Stopping criterion based on error is not met **do**
5: $k \leftarrow k(\text{mod } N)$
6: Present input $\mathbf{x}(k)$ to the network, calculating, successively, the activation of the hidden layer neurons, their outputs, thence the activations of the output layer neurons and, finally, the outputs.
7: Calculate the errors at the outputs in accordance with (12.16).
8: Regarding the errors as functions of the weights, use a descent algorithm to calculate a descent direction (= changes in weights) in weight space that reduces the squared error (12.16).
9: Update the weights in accordance with the changes computed in the previous step.
10: $k = k + 1$
11: **end while**

SVM. The first idea, already discussed above, is to map the feature vector, using a nonlinear map, into a higher-dimensional space, in which linear classification becomes possible. This clearly results in nonlinear classifiers in the original feature space, so that the well-known limitations of linear classifiers can be overcome. The second idea is referred to informally as *large margin classification* and is best explained with reference to Figure 12.10. This figure shows the case of linearly separable data in two dimensions, with an infinity of separating lines, each of which is able to classify the training data shown without error. The natural question to ask is whether the generalization performance on new unseen data is the same for each of these lines. In fact, in the worst case, one can even expect that all the separating lines or hyperplanes will have similar performances and, in general, this is true. This does not, however, exclude the possibility that data distributions that lead to

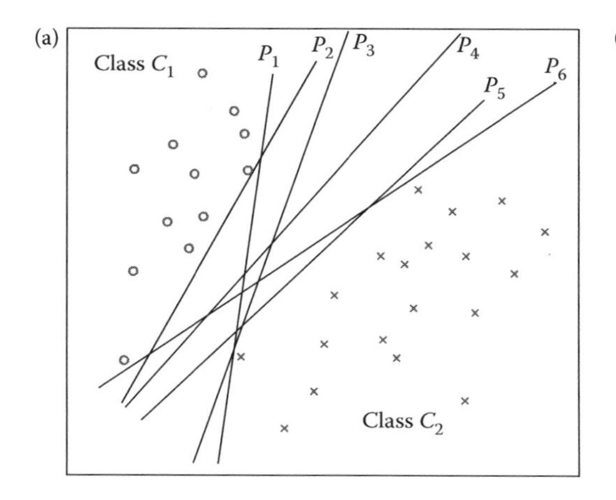

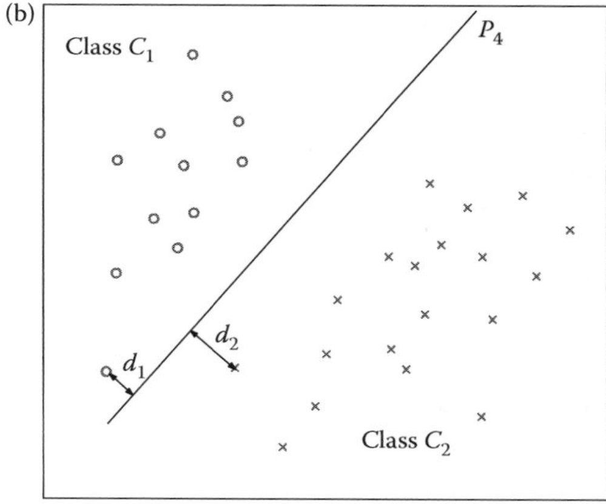

FIGURE 12.10
(a) Data clouds belonging to two different classes C_1, C_2 can be separated by many hyperplanes, P_i. (b) Each of these hyperplanes separates the two clouds with a certain margin, $M = d_1 + d_2$, where d_i is the distance from the separating hyperplane to the closest point in C_i, illustrated for P_4.

worst-case performance are uncommon, so that for most "common" or typical distributions, it may make sense to try to optimize the choice of separating hyperplane. In fact, it turns out that choosing the hyperplane that maximizes the margin (distance to the closest point of each class) actually helps to improve generalization performance. This makes intuitive sense, because if the margin is large, then the separation of the training points should be robust to small changes in the location of the hyperplane, leading to more reliable predictive performance. Statistical learning theory, pioneered by Vapnik and coauthors [75], formalizes the intuition that large margin classifiers tend to have good generalization performance and practical observations bear this out as well.

12.5.1 SVMs for Linearly Separable Classes

By way of introduction, we consider a two class linearly separable classification problem: the training data set $\mathcal{T} = \{\mathbf{x}_k, d_k\}_{k=1}^N$, $\mathbf{x}_k \in \mathbb{R}^n$, $d_k \in \{-1, +1\}$. Since the assumption of linear separability is being made, it follows that there exists a weight vector $\mathbf{w} \in \mathbb{R}^n$ and a bias w_0, such that $\mathbf{w}^\mathsf{T}\mathbf{x} + w_0 = 0$ defines a separating hyperplane, also called a *decision hyperplane*. In fact, there are infinitely many separating hyperplanes, as can be seen in Figure 12.10a. A glance at Figure 12.10b bolsters the intuition that a decision boundary drawn in the middle of the gap between the two data class clouds should be less prone to errors, either in the calculation of the boundary itself, or in the location of points in the two classes. Some learning methods that we have seen, such as the perceptron algorithm, stop when any linear separating hyperplane is found, while others seek to optimize, in the class of all linear separators, some appropriate criterion. In the case of the SVM, this criterion is to maximize the distance between the separating hyperplane and the data points. This maximal distance from the separating hyperplane to the closest data point is called the *margin* of the SVM. From this it follows that the separating hyperplane is completely specified by the subset of closest points in the data, and these are termed the *support vectors*, making the usual linguistic identification between data points in $\mathbb{R}^n$ and vectors. Figure 12.11 illustrates all the terms introduced in this paragraph and will be used to formalize the SVM mathematically. In order to calculate the distance from an arbitrary point $\mathbf{x} \in C_1$ to the closest point $\mathbf{x}_p$ on the separating hyperplane, observe that the line joining $\mathbf{x}$ to $\mathbf{x}_p$ is perpendicular to the hyperplane and therefore parallel to the normal vector $\mathbf{w}$ to the hyperplane and can be written as a scalar multiple α of the unit vector $\mathbf{w}/\|\mathbf{w}\|$ in the direction of $\mathbf{w}$. From Figure 12.11, we can write that

$$\mathbf{x} - \mathbf{x}_p = \alpha \frac{\mathbf{w}}{\|\mathbf{w}\|} \tag{12.49}$$

and, since $\mathbf{x}_p$ lies on the separating hyperplane:

$$\mathbf{w}^\mathsf{T}\mathbf{x}_p + w_0 = 0 \tag{12.50}$$

$$\mathbf{w}^\mathsf{T}\left(\mathbf{x} - \alpha \frac{\mathbf{w}}{\|\mathbf{w}\|}\right) + w_0 = 0 \tag{12.51}$$

which yields

$$\alpha = \frac{\mathbf{w}^\mathsf{T}\mathbf{x} + w_0}{\|\mathbf{w}\|} \tag{12.52}$$

The term *geometric margin* of the SVM is defined as the maximum width of the band that can be drawn separating the support vectors of the two classes. This is twice the minimum value of α (12.52), where the minimum is taken over all the data points. From the formula (12.52), it is also

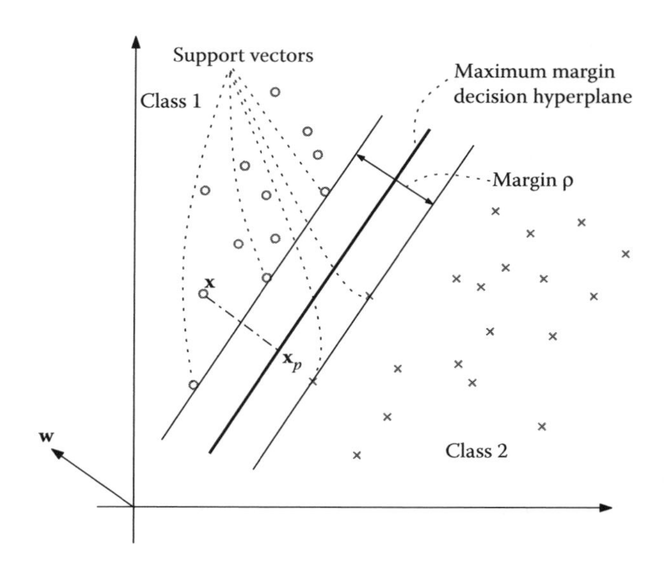

FIGURE 12.11
For the linearly separable classes 1 and 2, the distance vector from an arbitrary point $\mathbf{x}$ in class 1 to the closest point $\mathbf{x}_p$ on the separating (or decision) hyperplane is parallel to its normal vector $\mathbf{w}$.

immediate that the geometric margin is invariant to the scaling of $\mathbf{w}$, w_0. A convenient choice for the scaling is to require that, for all data points

$$d_k(\mathbf{w}^\top\mathbf{x}_k + w_0) \geq 1, k = 1, \ldots, N \tag{12.53}$$

with equality being attained for the support vectors, there being at least one in each class C_i. Thus, from Equation 12.52, it follows that the distance of a support vector in either class from the separating hyperplane is $1/\|\mathbf{w}\|$, so that the margin is $\rho = 2/\|\mathbf{w}\|$. We wish to maximize this margin, which is formulated as the following constrained optimization problem:

$$\max_{\mathbf{w},w_0} \frac{2}{\|\mathbf{w}\|}$$

$$\text{s.t. } d_k(\mathbf{w}^\top\mathbf{x}_k + w_0) \geq 1, k = 1, \ldots, N$$

Since maximizing $2/\|\mathbf{w}\|$ is equivalent to minimizing $\|\mathbf{w}\|$, or, more conveniently, $(1/2)\mathbf{w}^\top\mathbf{w}$, we arrive at the standard formulation of the SVM as an optimization problem, namely:

$$\begin{aligned} &\min_{\mathbf{w},w_0} \frac{1}{2}\mathbf{w}^\top\mathbf{w} \\ &\text{s.t. } d_k(\mathbf{w}^\top\mathbf{x}_k + w_0) \geq 1, k = 1, \ldots, N \end{aligned} \tag{12.54}$$

The optimization problem described in Equation 12.54 belongs to the class of quadratic programming (QP) problems, in which a quadratic objective function is minimized subject to linear constraints. The class of QP problems has been extensively studied and many algorithms have been proposed for it. However, since data sets of interest in practical problems, now often generated and accessed online, have exploded in size, this implies that scalability of any algorithm to find an SVM is now of great importance. Recent research has concentrated on the special structure of QP problems arising from SVM problems and has led to libraries that are tailored to be much faster and more scalable. The details of these algorithms are best left to the literature, to which pointers are given in Section 12.9. We concentrate on understanding the kind of solutions that arise, from a theoretical point of view.

12.5.2 Construction of the SVM Solution in the Linearly Separable Case

To solve the QP problem, the standard first step is to use the Lagrange multiplier method for constrained optimization problems. Two important advantages result: it becomes easier to handle the constraints (in the dual problem) and the training data appear as a scalar or inner product, which is crucial in the generalization of the linear SVM to an SVM for data that are not linearly separable (allowing the use of the so-called kernel trick). For Equation 12.54, the Lagrangian is

$$\mathcal{L}(\mathbf{w}, w_0, \Lambda) := \frac{1}{2}\mathbf{w}^{\mathrm{T}}\mathbf{w} - \sum_{i=1}^{N} \lambda_i (d_i(\mathbf{w}^{\mathrm{T}}\mathbf{x}_i + w_0) - 1) \tag{12.55}$$

where $\Lambda := (\lambda_1, \ldots, \lambda_N), \lambda_i \geq 0$ is a nonnegative vector of Lagrange multipliers (also called dual variables), one for each constraint. This is the primal problem. In brief, the strategy for transforming the primal problem into its dual is to equate the partial derivatives of the Lagrangian with respect to the primal variables $(\mathbf{w}, w_0)$ to zero, and use the resulting relations to eliminate the primal variables from the Lagrangian, which, in fact leads to the dual function which is to be maximized, under new constraints (in the dual variables) that turn out to be much simpler. The partial derivatives of $\mathcal{L}$, which must vanish at an optimal point, are

$$\frac{\partial \mathcal{L}(\mathbf{w}, w_0, \Lambda)}{\partial \mathbf{w}} = \mathbf{w} - \sum_{i=1}^{N} \lambda_i d_i \mathbf{x}_i = \mathbf{0} \tag{12.56}$$

$$\frac{\partial \mathcal{L}(\mathbf{w}, w_0, \Lambda)}{\partial w_0} = \sum_{i=1}^{N} \lambda_i d_i = \Lambda^{\mathrm{T}}\mathbf{d} = 0 \tag{12.57}$$

where $\mathbf{d} = (d_1, \ldots, d_N) \in \mathbb{R}^N$ is the vector of data labels. The Karush–Kuhn–Tucker (KKT) complementarity conditions must also be satisfied:

$$\lambda_i (d_i(\mathbf{w}^{\mathrm{T}}\mathbf{x}_i + w_0) - 1) = 0, i = 1, \ldots, N \tag{12.58}$$

The interpretation of the KKT complementarity condition is that λ_j can only be positive for vectors that satisfy the constraint with equality, that is, are support vectors. Following the strategy outlined above and substituting Equations 12.56 and 12.57 into Equation 12.55 leads to the dual formulation:

$$\mathcal{L}'(\Lambda) = \sum_{i=1}^{N} \lambda_i - \frac{1}{2}\sum_{i=1}^{N}\sum_{i=1}^{N} \lambda_i \lambda_j d_i d_j \left(\mathbf{x}_i^{\mathrm{T}}\mathbf{x}_j\right) \tag{12.59}$$

Introducing the notation $\mathbf{1} := (1, 1, \ldots, 1) \in \mathbb{R}^N$ and $\mathbf{H} := (h_{ij})$, where $h_{ij} = d_i d_j (\mathbf{x}_i^{\mathrm{T}}\mathbf{x}_j)$, we can rewrite $\mathcal{L}'$ compactly as

$$\mathcal{L}' = \mathbf{1}^{\mathrm{T}}\Lambda - \frac{1}{2}\Lambda^{\mathrm{T}}\mathbf{H}\Lambda \tag{12.60}$$

The dual optimization problem is

$$\max_{\Lambda} \mathbf{1}^{\mathsf{T}}\Lambda - \frac{1}{2}\Lambda^{\mathsf{T}}\mathbf{H}\Lambda \tag{12.61}$$

$$\text{s.t. } \mathbf{d}^{\mathsf{T}}\Lambda = 0 \tag{12.62}$$

$$\text{and } \Lambda \geq \mathbf{0} \tag{12.63}$$

where $\mathbf{0}$ denotes the zero vector in $\mathbb{R}^N$ and the last inequality is to be interpreted componentwise. Note that the dual problem is again a QP problem, but with constraints in a very convenient and simple form. Also observe that w_0 has disappeared from the dual formulation, but it is easy to calculate its optimal value from the optimal values of the Lagrange multipliers and weights. Specifically, denoting optimal quantities with asterisk superscripts, from Equation 12.56, the optimal weight vector can be written as

$$\mathbf{w}^* = \sum_{i=1}^{N} \lambda_i^* d_i \mathbf{x}_i \tag{12.64}$$

As observed above, only the multipliers λ_k corresponding to support vectors will be positive (all other multipliers are zero), so that the summation in Equation 12.64 needs to be performed only over the support vectors. Finally, from the KKT complementarity conditions, the optimal bias w_0^* is computed as

$$w_0^* = d_{\text{sv}}^{-1} - (\mathbf{w}^*)^{\mathsf{T}}\mathbf{x}_{\text{sv}} \tag{12.65}$$

where the subscript sv refers to any support vector (for which $\lambda_{\text{sv}} > 0$). In closing, note that, from Equation 12.64, the SVM decision function that classifies a new data point $\mathbf{x}$ can be written as

$$f_{\text{svm}}(\mathbf{x}) = \text{sign}\left(\sum_{i=1}^{n_{\text{sv}}} d_i \lambda_i^* \mathbf{x}_i^{\mathsf{T}}\mathbf{x} + w_0^*\right) \tag{12.66}$$

where n_{sv} denotes the number of support vectors and sign denotes the signum function, equal to $+1$ for positive arguments, and to -1 for negative arguments, so that patterns in class C_1 lie "above" the hyperplane, producing an output of $+1$, whereas patterns in class C_2 produce an output of -1 and lie "below" the hyperplane.

12.5.3 SVM with Soft Margins

Even when the data are known to be linearly separable, noise and classification errors may transform the real data set into one that is not exactly linearly separable. It may also be desirable to seek a solution that is robust to bad data points. The natural approach is to allow some misclassification, meaning that some data points are on the wrong side of the margin. In order to control the number of such misclassified points, also called mistakes, each incurs a cost or penalty proportional to its distance from the correct side of the margin. The standard way of implementing this idea is to introduce the so-called slack variables, illustrated in Figure 12.12, in which a nonzero value of ξ_i allows $\mathbf{x}_i$ to violate the margin constraint at a cost proportional to ξ_i.

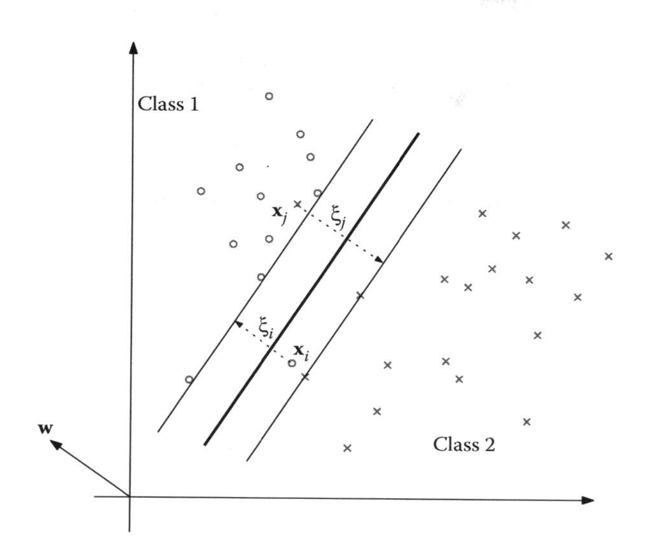

FIGURE 12.12
A soft margin classifier allows misclassification of data points and the figure shows two such points $\mathbf{x}_i$, with its corresponding violation (which is the distance or slack variable ξ_i) as well as $\mathbf{x}_j$ and ξ_j.

The formulation of the SVM optimization problem with slack variables, also called the *soft margin SVM* problem, is

$$\min_{\mathbf{w},w_0,\xi_i} \frac{1}{2}\mathbf{w}^{\mathsf{T}}\mathbf{w} + C\sum_{i=1} \xi_i \tag{12.67}$$

$$\text{s.t. } d_i(\mathbf{w}^{\mathsf{T}}\mathbf{x}_i + w_0) \geq 1 - \xi_i \tag{12.68}$$

$$\xi_i \geq 0 \tag{12.69}$$

We can interpret this optimization problem as follows. The width of the margin is being traded off against the number of misclassifications, in the sense that the margin can be less than one for a point $\mathbf{x}_i$ if the corresponding slack variable ξ_i is chosen to be positive; however, this entails an additional cost term $C\xi_i$ in the objective function that is being minimized. Evidently, the sum of the slack variables ξ_i provides an upper bound on the number of training errors. Thus, an alternative verbal interpretation of the soft margin SVM is that it minimizes training error traded-off versus margin. The parameter C that controls overfitting is thus seen to be a *regularization* term: when its value is large, this provides a disincentive to allow training errors in order to obtain a better margin. Conversely, with a small value of C, a wider margin can be obtained by accounting for some bad data with slack variables.

Using the Lagrangian, similar steps and the same notation as in the previous case, the dual problem for soft margin classification is

$$\max_{\Lambda} \mathbf{1}^{\mathsf{T}}\Lambda - \frac{1}{2}\Lambda^{\mathsf{T}}\mathbf{H}\Lambda \tag{12.70}$$

$$\text{s.t. } \mathbf{d}^{\mathsf{T}}\Lambda = 0 \tag{12.71}$$

$$\text{and } \Lambda \geq \mathbf{0} \tag{12.72}$$

$$\Lambda \leq C\mathbf{1} \tag{12.73}$$

In fact, it turns out that the dual Lagrangian for this case is the same as the one for the separable case (12.60), since the slack variables and their Lagrange multipliers do not appear. The only difference is in the new upper bound on the Lagrange multipliers (12.73). As in the separable case, the optimal weight vector ($\mathbf{w}^*$) can be computed from the optimal Lagrange multipliers (λ_i^*), superscripted with asterisks:

$$\mathbf{w}^* = \sum_{i=1}^{N} \lambda_i^* d_i \mathbf{x}_i \tag{12.74}$$

$$= \sum_{i=1}^{n_{\text{sv}}} \lambda_i^* d_i \mathbf{x}_i \tag{12.75}$$

where n_{sv} denotes the number of support vectors, which, as before, are those which have nonzero Lagrange multipliers. The computation of the optimal bias w_0^* is now more involved and requires the manipulation of the KKT complementarity conditions and the first-order optimality conditions—we refer the reader to Reference 18 for a very accessible and detailed discussion. Once again, the SVM decision function that classifies a new data point $\mathbf{x}$ can be written as

$$f_{\text{svm}}(\mathbf{x}) = \text{sign}\left(\sum_{i=1}^{n_{\text{sv}}} d_i \lambda_i^* \mathbf{x}_i^{\mathrm{T}} \mathbf{x} + w_0^*\right) \tag{12.76}$$

We make some brief comments about computational aspects. In many real problems, the support vectors form a small proportion of the training data. However, for problems which have small margins or are nonseparable, for which the set of misclassified points is potentially large, every misclassified point corresponds to a nonzero Lagrange multiplier. This can lead to poor performance of the SVM algorithm, particularly in the nonlinear case, which we shall discuss later.

The major chunk of time required to train an SVM is that used to solve the associated QP problem. Thus, both the theoretical and empirical complexity depend on the QP solution method. In order to escape from the classical result that the computational time is proportional to the cube of the size of the data set [55], recent research has focused on various approaches to reduce complexity, one of them being the use of approximate solutions. Nevertheless, empirical complexity remains superlinear [48], making it impractical to use SVMs on very large data sets and even more so for nonseparable problems, for which the complexity is higher by a factor equal to the size of the data set.

12.5.4 SVM for Linearly Nonseparable Data

This subsection deals with the case where the data set is not merely a linearly separable data set contaminated with a few exceptions or noise, but rather one that does not allow linear separability. In the one-dimensional case, it is easy to exemplify linear nonseparability as well as a straightforward fix to this problem.

Example 12.1

Consider the set of 10 points in $\mathbb{R}$ given by

$$X = [-2, -1, 0, 0.5, 2.5, 3, 4, 6, 6.5, 8]$$

where the points $2.5, 3, 4$ belong to class 2 and all the other points to class 1. These points and the two classes to which they belong are depicted in Figure 12.13, with the points in class 1 denoted

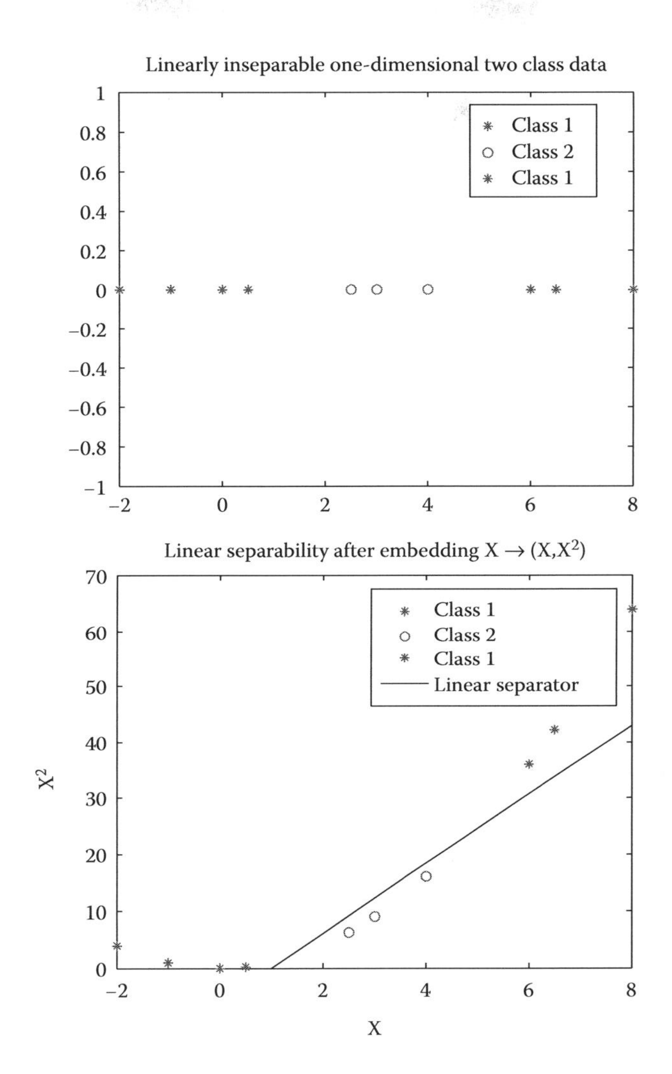

FIGURE 12.13
The plot on the left shows points in $\mathbb{R}$ that are not separable into two classes by any point on the real line. The figure on the right shows that, if these points are embedded into $\mathbb{R}^2$, using the mapping $X_k \mapsto (X_k, X_k^2)$, they become linearly separable, since all points of class 1 lie above the straight line and all points of class 2 below it.

by "*" and the points in class 2 by "o." From the figure, it is clear that it is not possible to choose a point on the real line such that all points in class 1 lie on one side of it and all points of class 2 on the other, which shows that, in this simple one-dimensional example, the set of points X is not "linearly" separable. However, a simple embedding into the plane results in a two-dimensional data set of points, which is linearly separable (Figure 12.13). This is the essential idea of mapping a given data set by some possibly nonlinear transformation into a higher-dimensional space in such a way that the mapped points are linearly separable in the image space.

In addition to achieving linear separability, it is also desirable to maintain the original topology in order to preserve nearness relations between data points in order that the classifier in the higher-dimensional space continues to present good generalization performance. A standard way of doing this for the SVM, which also works for several other linear classifiers, has come to be known as the *kernel trick*. Let us denote the mapping of input vectors belonging to X into a higher-dimensional feature space $\mathcal{F}$ by

$$\Phi : \mathbb{R}^n \rightarrow \mathcal{F} : \mathbf{x} \mapsto \Phi(\mathbf{x}) \tag{12.77}$$

where the space $\mathcal{F}$ has a larger dimension than n and can even be infinite dimensional. Suppose that the mapping into the feature space is successful and results in linearly separable data in $\mathcal{F}$. Then, following Equation 12.76, we know that the linear SVM classifier in feature space can be written as

$$f_{\text{svm}}(\mathbf{x}) = \text{sign}\left(\sum_{i=1}^{n_{\text{sv}}} d_i \lambda_i^* \Phi(\mathbf{x}_i)^{\mathsf{T}} \Phi(\mathbf{x}) + w_0^*\right) \tag{12.78}$$

Observe that in the decision function (12.78) and, in fact, in all the intermediate steps such as the formation of the dual Lagrangian, calculation of the optimal bias w_0, etc., the training data appear only in the form of inner products of the type $\mathbf{x}_i^{\mathsf{T}}\mathbf{x}_j$. Now suppose that there exists a kernel function $K(\mathbf{x}_i, \mathbf{x}_j)$ such that

$$K(\mathbf{x}_i, \mathbf{x}_j) = \Phi(\mathbf{x}_i)^{\mathsf{T}} \Phi(\mathbf{x}_j) \tag{12.79}$$

then we could use $K(\mathbf{x}_i, \mathbf{x}_j)$ directly in the training procedure, without explicitly computing $\Phi(\mathbf{x})$ and, in fact, even without knowledge of the mapping Φ. This is what was referred to above as the kernel trick. Another way to express this is to say that a kernel function corresponds to an inner product in some expanded feature space. Some examples clarify the kernel concept.

Example 12.2

Consider a two-dimensional input space ($X = \mathbb{R}^2$) and suppose that $K(\mathbf{x}_i, \mathbf{x}_j) = (\mathbf{x}_i^{\mathsf{T}}\mathbf{x}_j)^2$. We wish to find a feature space $\mathcal{F}$ and a mapping Φ from X to $\mathcal{F}$ such that $(\mathbf{x}^{\mathsf{T}}\mathbf{y})^2 = \Phi(\mathbf{x})^{\mathsf{T}}\Phi(\mathbf{y})$. One possible choice is $\mathcal{F} = \mathbb{R}^3$ and

$$\Phi(\mathbf{x}) = \left(x_1^2, \sqrt{2}x_1x_2, x_2^2\right) \tag{12.80}$$

We can check that Equation 12.79 is satisfied for this choice of the pair $(\mathcal{F}, \Phi)$:

$$\Phi(\mathbf{x})^{\mathsf{T}}\Phi(\mathbf{y}) = \left(x_1^2y_1^2 + 2x_1x_2y_1y_2 + x_2^2y_2^2\right) \tag{12.81}$$

$$= (x_1y_1 + x_2y_2)^2 \tag{12.82}$$

$$= (\mathbf{x}^{\mathsf{T}}\mathbf{y})^2 \tag{12.83}$$

$$= K(\mathbf{x}, \mathbf{y}) \tag{12.84}$$

This choice is far from unique, even if the dimension of the feature space is kept at 3. Some alternative valid kernel choices that can be easily checked by the reader follow.

1. $\mathcal{F} = \mathbb{R}^3$ and $\Phi(\mathbf{x}) = (1/\sqrt{2})\left(\left(x_1^2 - x_2^2\right), 2x_1x_2, \left(x_1^2 + x_2^2\right)\right)$
2. $\mathcal{F} = \mathbb{R}^4$ and $\Phi(\mathbf{x}) = \left(x_1^2, x_1x_2, x_1x_2, x_2^2\right)$
3. $\mathcal{F} = \mathbb{R}^6$ and $\Phi(\mathbf{x}) = \left(1, \sqrt{2}x_1, \sqrt{2}x_2, x_1^2, x_2^2, \sqrt{2}x_1x_2\right)$

The fundamental question that arises in regard to examples like the one above is as follows. For which kernel does there exist a mapping Φ into a higher-dimensional feature space $\mathcal{F}^*$ such that the kernel admits an inner product representation (12.79)? A basic result from functional analysis known as Mercer's condition characterizes such kernel functions and they are then known as *Mercer kernels* or as *inner product kernels*. In brief, Mercer's condition can be stated as follows.

Theorem 12.2

There exists a mapping Φ from X to a Hilbert space $\mathcal{F}$ such that the kernel function has an inner product representation

$$K(\mathbf{x}, \mathbf{y}) = \Phi(\mathbf{x})^{\mathsf{T}}\Phi(\mathbf{y})$$

if and only if, for any $g(\mathbf{x})$ that is square integrable (i.e., $\int g^2(\mathbf{x})d\mathbf{x} < \infty$),

$$\int\int K(\mathbf{x}, \mathbf{y})g(\mathbf{x})g(\mathbf{y})d\mathbf{x}d\mathbf{y} \geq 0$$

It is easy to see that such a kernel function must be continuous and symmetric and possess a positive definite Gramian matrix. From a practical point of view, the great majority of applications use either polynomial kernels (of the type illustrated in Example 12.2), or the class of RBFs, which are essentially Gaussians. Three of the most frequently used inner product kernels are

1. Polynomial kernels:

$$K(\mathbf{x}, \mathbf{y}) = (1 + \mathbf{x}^{\mathsf{T}}\mathbf{y})^n$$

2. RBF kernels:

$$K(\mathbf{x}, \mathbf{y}) = \exp\left(-\frac{(\mathbf{x} - \mathbf{y})^{\mathsf{T}}(\mathbf{x} - \mathbf{y})}{\sigma^2}\right)$$

3. Logistic kernels:

$$K(\mathbf{x}, \mathbf{y}) = \tanh(2\mathbf{x}^{\mathsf{T}}\mathbf{y} + 1)$$

   where the particular choice of the constants (2 and 1) is made in order to satisfy the Mercer condition.

We can summarize the SVM learning procedure as follows:

Main steps of the SVM learning procedure

1. Data: Training set $\mathcal{T} = \{\mathbf{x}_i, d_i\}_{i=1}^N$, where $\mathbf{x}_i \in \mathbb{R}^n$, for all i and $d_i \in \{-1, +1\}$.
2. Initialization: Choice of kernel function K, regularization parameter C, followed by calculation of matrix $\mathbf{H} = (h_{ij})$, where $h_{ij} = d_i d_j K(\mathbf{x}_i, \mathbf{x}_j)$.

* Technically, this higher-dimensional space must be an inner product space, which is complete with respect to the norm induced by the inner product: such a space is commonly known as a Hilbert space, or, more specifically, a reproducing kernel Hilbert space.

3. Solution of the (dual) quadratic optimization problem:

$$\max_{\Lambda} \mathbf{1}^{\mathsf{T}}\Lambda - \frac{1}{2}\Lambda^{\mathsf{T}}\mathbf{H}\Lambda \tag{12.85}$$

$$\text{s.t. } D^{\mathsf{T}}\Lambda = 0 \tag{12.86}$$

$$\Lambda \geq \mathbf{0} \tag{12.87}$$

$$\Lambda \leq C\mathbf{1} \tag{12.88}$$

4. The decision function is then given by

$$f_{\text{svm}}(\mathbf{x}) = \text{sign}\left(\sum_{i=1}^{n_{\text{sv}}} d_i \lambda_i^* K(\mathbf{x}, \mathbf{x}_i) + w_0^*\right)$$

where the $*$ superscripts denote the optimal values of the respective variables, found in the previous step.

Finally, Figure 12.14 depicts how the SVM can be represented in the style of a neural network in which the nodes are designated to perform kernel evaluations, while the optimal Lagrange multipliers and corresponding class labels define the weighted sum that generates the final output after passing through a signum threshold function. In a very rough analogy, the support vectors can be seen as "bias" inputs to the kernel evaluation nodes.

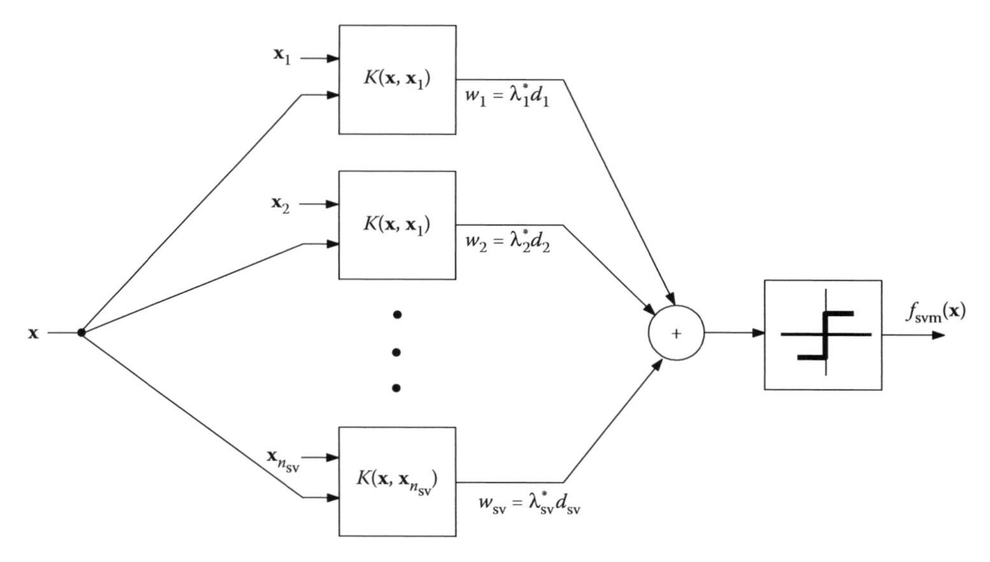

FIGURE 12.14
An SVM depicted as a neural network in which the nodes perform kernel evaluation, with their outputs weighted by the product of the optimal Lagrange multipliers and corresponding class labels. The weighted sum passes through a signum threshold function to produce the final output (class label) corresponding to the test input.

12.6 RBF Networks

In terms of biological inspiration for ANNs and, in particular, for RBF networks, it should be noted that the nervous system contains many examples of neurons with so-called local receptive fields, which are tuned to respond to stimuli: examples are the orientation-selective cells in visual cortex and somatosensory cells responsive to specific body regions. On the other hand, the RBF method for multivariate approximation is one of the most often applied approaches in modern approximation theory when the task is to approximate scattered data in several dimensions. Thus, although RBF kernels were briefly mentioned in the previous section as part of the SVM methodology, they were actually introduced to the neural network community somewhat earlier than SVMs in their capacity to perform universal approximation of functions, or, more specifically, for the *exact interpolation* of a given set of data points $\mathcal{T} = \{\mathbf{x}_i, d_i\}_{i=1}^{N}$, where $\mathbf{x}_i \in \mathbb{R}^n$ and $d_i \in \mathbb{R}$, which means that

$$f(\mathbf{x}_i) = d_i, \quad i = 1, \ldots, N \tag{12.89}$$

We will show how an RBF network solves this interpolation problem and then relax the requirement of exactness. We first define an RBF formally.

Definition 12.4

An RBF is a real-valued function, conventionally denoted ϕ, whose value at a point $\mathbf{x} \in \mathbb{R}^n$ depends only on the distance from some other point $\mathbf{c}$, called a center, so that $\phi(\mathbf{x}, \mathbf{c}) = \phi(\|\mathbf{x} - \mathbf{c}\|)$.

The norm is usually the Euclidean distance or two-norm, although other norms may also be used. We will see that sums of RBFs are typically used to approximate given functions, and, moreover, that this approximation process can also be interpreted as a simple kind of neural network. Using the notation $r := \|\mathbf{x} - \mathbf{x}_i\|$, the commonly used types of RBFs, with $\epsilon > 0$ denoting an adjustable parameter are the Gaussian, with $\phi(r) = e^{-\epsilon r^2}$ and the *inverse multiquadric*, with $\phi(r) = 1/\sqrt{1 + (\epsilon r)^2}$. The Gaussian RBF is often expressed in the following form:

$$\phi(\mathbf{x}) = \exp\left(\frac{\|\mathbf{x} - \boldsymbol{\mu}\|^2}{2\sigma^2}\right) \tag{12.90}$$

where $\boldsymbol{\mu}$ denotes the center and the parameter σ is referred to as the spread, which influences the smoothness of the interpolation.

In the simplest case of interpolation by a superposition of RBFs, we suppose that the number of RBFs and the number of data points are exactly the same, so that the interpolating map f can be written as

$$f_{\text{rbf}}(\mathbf{x}) := \sum_{i=1}^{N} w_i \phi(\|\mathbf{x} - \mathbf{x}_i\|) \tag{12.91}$$

Exact interpolation means that we must have

$$\sum_{i=1}^{N} w_i \phi(\|\mathbf{x}_j - \mathbf{x}_i\|) = d_j, \quad j = 1, \ldots, N \tag{12.92}$$

We introduce some notation for the specified output and weight vector, as well as the interpolation matrix

$$\mathbf{d} := (d_1, \ldots, d_N) \tag{12.93}$$

$$\mathbf{w} := (w_1, \ldots, w_N) \tag{12.94}$$

$$\Phi := \begin{bmatrix} \phi(\|\mathbf{x}_1 - \mathbf{x}_1\|) & \cdots & \phi(\|\mathbf{x}_1 - \mathbf{x}_N\|) \\ \vdots & \ddots & \vdots \\ \phi(\|\mathbf{x}_N - \mathbf{x}_1\|) & \cdots & \phi(\|\mathbf{x}_N - \mathbf{x}_N\|) \end{bmatrix} \in \mathbb{R}^{N \times N} \tag{12.95}$$

With this notation, noting that Φ is a symmetric matrix, Equation 12.92 becomes

$$\Phi \mathbf{x} = \mathbf{d} \tag{12.96}$$

which admits a unique solution for the weight vector $\mathbf{w}$ iff Φ is nonsingular. For the choices of RBFs given above (Gaussian and inverse multiquadric), making the assumption that the data points are distinct ensures the nonsingularity of Φ and this result is known as Micchelli's theorem.

To extend the simple case above, the natural modifications to be considered are to choose less basis functions than the number of data points and, as a consequence, open up the possibility of not necessarily placing all the centers of the basis functions at data points. In addition, in the Gaussian case, for example, the basis functions themselves, instead of being identical, can be permitted to have different spreads. In other words, the center locations and spreads can be made into parameters that need to be learned by some appropriate technique.

12.6.1 Interpolation with Fewer RBFs than Data Points

We first examine the simplest case, in which the centers and spreads are fixed and only the weights need to be determined. As usual in the undetermined case of k equations in N unknowns ($k < N$), we will use the least squares method. The squared error function is expressed as

$$E = \frac{1}{2} \sum_{m=1}^{N} \left(d_m - \sum_{i=1}^{k} w_i \phi(\|\mathbf{x}_m - \mathbf{x}_i\|) \right)^2 \tag{12.97}$$

In matrix notation, defining $\Phi = (\phi_{ij})$, where $\phi_{ij} := \phi(\|\mathbf{x}_i - \mathbf{x}_j\|)$ and $\mathbf{d}, \mathbf{w}$ are as before, we can write the normal expression for the least squares solution $\mathbf{w}_{\text{ls}}$ as

$$\Phi^\top \Phi \mathbf{w} = \Phi^\top \mathbf{d}$$

12.7 Universal Approximation Using Neural Networks

In 1900, in his famous lecture to the International Congress of Mathematicians in Paris, Hilbert formulated 23 challenging problems which inspired many important discoveries in the last century, in the attempt to solve them. Of these, the 13th problem was concerned with the solution of higher-order algebraic equations. Hilbert conjectured that such equations are not solvable by functions which can be constructed as finite compositions of continuous functions of only two variables.

Hilbert's conjecture was refuted by Kolmogorov in 1957 [54] who proved the superposition theorem which now bears his name and states that any multivariate continuous real-valued function can be represented as the superposition and composition of continuous functions of only one variable. Some researchers suggested that Kolmogorov's theorem provides theoretical support for a universal approximation property for neural networks, since they implement functions in a similar fashion, using finite combinations and compositions of a simple (sigmoidal) univariate function. However, it was soon pointed out that the basis functions used in Kolmogorov's theorem were highly nonsmooth and, if approximated by smooth functions, the resulting superposition would not, in general, lead to an approximate implementation of the original function.

Cybenko [27] demonstrated that finite linear combinations of compositions of a fixed univariate function and a set of affine functionals can uniformly approximate any continuous function of n real variables supported on the unit hypercube, with only mild conditions imposed on the univariate function. In particular, he showed that arbitrary decision regions can be arbitrarily well approximated by continuous feedforward neural networks (MLPs), with only a single hidden layer and any continuous sigmoidal nonlinearity. This result, although proved in a nonconstructive manner, provided a firm theoretical basis for universal approximations by MLPs. Later, similar results were derived for RBF networks [65,69].

12.8 Neural Networks for Optimization

As we have seen, the term "artificial neural network" refers to a large class of dynamical systems. In the taxonomy or "functional classification scheme" of Reference 61, from a theoretical point of view, ANNs are divided into two classes: *feedforward* and *feedback* networks; the latter class is also referred to as *recurrent*. The feature common to both classes and, indeed, to all ANNs is that they are used because of their "learning" capability. This means that ANNs are dynamical systems that depend on some control parameters, also known as weights or gains and therefore they can be described from the viewpoint of control systems [46]. When a suitable adjustment of these parameters is achieved, the ANN is said to have learnt some functional relationship between a class of inputs and a class of outputs.

Three broad classes of applications of ANNs can be identified. In the first, described in detail above, a set of inputs and corresponding desired outputs is given, and the problem is to adjust the parameters in such a way that the system fits this data in the sense that if any of the given inputs is presented to it, the corresponding desired output is in fact generated. This class of problem goes by many names: mathematicians call it a *function approximation problem* and the parameters are adjusted according to time-honored criteria such as least squared error, etc.; for statisticians, it is a *regression problem* and, once again, there is an array of techniques available; finally, in the ANN community, this is referred to as a *learning problem* and, on the completion of the process, the ANN is said to have *learned* a functional relationship between inputs and outputs.

In the second class, it is desired to design a dynamical system whose equilibria correspond to the solution set of an optimization problem: this used to be referred to as *analog computation*, since the idea is to start from some arbitrary initial condition and follow the trajectories of the dynamical system which converge to its equilibria, which are the optimum points that it is required to find.

For the third class, loosely related to the second, the objective is to design a dynamical system with multiple stable equilibria. In this case, if for each initial condition, the corresponding trajectory converges to one of the stable equilibria, then the network is referred to as an *associative memory* or *content-addressed memory*. The reason behind this picturesque terminology is that an initial condition in the basin of attraction of one of the stable equilibria may be regarded as a corrupted or noisy version of this equilibrium; when the ANN is "presented" with this input, it "associates" it with or "remembers" the corresponding "uncorrupted" version, which is the attractor of the basin (to which nearby trajectories converge). According to the statistical physicists Herz, Krogh, and Palmer who wrote an influential book on neural computation [40], the associative memory problem is the "fruit fly" or "Bohr atom" problem of the ANN field, since it "illustrates in about the simplest possible manner the way that collective computation can work." The basic problem, for an n-unit McCulloch–Pitts network, is stated informally as follows:

> ASSOCIATIVE MEMORY PROBLEM: Store a set of p patterns $\mathbf{x}_i$ in such a way that, when presented with a new pattern $\mathbf{x}$, the network responds by producing, at its output, whichever pattern $\mathbf{x}_i$ most closely resembles the pattern $\mathbf{x}$.

However, despite the historical and ideological importance of the associative memory problem, it is fair to say that, in recent times, especially from the point of view of applications, there has not been much activity in this area and thus we will focus mainly on the second class of problems in this section, since the first class has already been discussed in the previous sections.

In the ANN community, a distinction is often made between the parameter adjustment processes in the first and second classes of problems. For instance, if a certain type of dynamical system, referred to as a *feedforward* ANN or *multilayer perceptron*, is used, then a popular technique for adjustment of the parameters (weights) is referred to as *backpropagation* and usually there is no explicit mention of feedback, so that control aspects of the weight adjustment process are not effectively taken advantage of. In the second class of problems, the parameters to be chosen are referred to as *gains* and are usually chosen once and for all in the design process, so that the desired convergence occurs. However, in this case, the ANN community refers to the feedback aspect by using the name *recurrent ANNs*. This recurrent aspect of ANNs used for optimization and associative memory applications was first put forward in the pioneering paper of Hopfield [41,42].

While it is well known that Hopfield neural networks are gradient dynamical systems [10], in the reverse direction, it is not equally well appreciated that, with suitable assumptions, a gradient dynamical system can be interpreted as a neural network. In this section, we explore this connection in a systematic manner.

12.8.1 Gradient Dynamical Systems for Unconstrained Optimization

The simplest class of unconstrained optimization problem is that of finding $\mathbf{x} \in \mathbb{R}^n$ that minimizes the real scalar function $E : \mathbb{R}^n \rightarrow \mathbb{R} : \mathbf{x} \mapsto E(\mathbf{x})$.

A point $\mathbf{x}^* \in \mathbb{R}^n$ is a *global minimizer* for $E(\mathbf{x})$ if and only if

$$E(\mathbf{x}^*) \leq E(\mathbf{x}), \quad \text{for all } \mathbf{x} \in \mathbb{R}^n$$

and a *local minimizer* for $E(\mathbf{x})$ if

$$E(\mathbf{x}^*) \leq E(\mathbf{x}), \quad \text{for all } \mathbf{x} \text{ in a ball } B(\mathbf{x}^*, \epsilon)$$

Assuming here, for simplicity, that the first and second derivatives of $E(\mathbf{x})$ with respect to $\mathbf{x}$ exist, then necessary and sufficient conditions for the existence of a local minimizer are

$$\nabla E(\mathbf{x}^*) = 0 \quad \text{and} \quad \nabla^2 E(\mathbf{x}^*) > 0$$

It is quite natural to design procedures that seek a minimum of a given function based on its gradient. More specifically, the class of methods known as steepest descent methods uses the gradient direction as the one in which the largest decrease (or descent) in the function value is achieved.

The idea of moving along the direction of the gradient of a function leads naturally to a dynamical system of the following form:

$$\frac{\mathrm{d}\mathbf{x}(t)}{\mathrm{d}t} = -\mathbf{M}(\mathbf{x}, t)\nabla E(\mathbf{x}) \tag{12.98}$$

where $\mathbf{M}(\mathbf{x}, t)$ is, in general, a positive definite matrix called the *learning matrix* in the neural network literature. Integrating the differential Equation 12.98, for a given arbitrary initial condition $\mathbf{x}(0) = \mathbf{x}_0$, corresponds to following a trajectory that leads to a vector $\mathbf{x}^*$ that minimizes $E(\mathbf{x})$. More precisely, provided that some appropriate conditions (given below) are met, the solution $\mathbf{x}(t)$ along time, of the system (12.98), will be such that

$$\lim_{t\to\infty} \mathbf{x}(t) = \mathbf{x}^* \tag{12.99}$$

In order to ensure that the above limit is attained, and consequently that a desired solution $\mathbf{x}^*$ to the minimization problem is obtained, we recall another connection which is to associate to function $E(\mathbf{x})$ the status, or the interpretation, of an energy function. In the present context, $E(\mathbf{x})$ is also frequently called a *computational energy function*. The notation $E(\mathbf{x})$ is chosen to help in recalling this connection.

12.8.2 Liapunov Stability of Gradient Dynamical Systems

The natural Liapunov function associated to a gradient dynamical system is now discussed in the present context.

Consider an unconstrained optimization problem and, to aid intuition, rename the objective function that is to be minimized as an *energy function* $E(\mathbf{x})$, where $\mathbf{x}$ is the vector of variables to be chosen such that $E(\cdot)$ attains a minimum. Thinking of $\mathbf{x}$ as a state vector, consider the gradient dynamical system that corresponds to steepest descent or descent along the negative of the gradient of $E(\cdot)$, where the matrix $\mathbf{M}(\mathbf{x}, t)$, as in Equation 12.98, can be thought of as a "gain" matrix that is positive definite for all $\mathbf{x}$ and t and whose role will become clear in the sequel. Then the time derivative of E decreases along the trajectories of Equation 12.98, because, for all $\mathbf{x}$ such that $\nabla E(\mathbf{x}) \neq \mathbf{0}$:

$$\frac{\mathrm{d}E}{\mathrm{d}t} = \nabla E(\mathbf{x})^T \dot{\mathbf{x}} = -\nabla E(\mathbf{x})^T \mathbf{M}(\mathbf{x}, t)\nabla E(\mathbf{x}) < 0 \tag{12.100}$$

Under the assumption that the matrix $\mathbf{M}(\mathbf{x}, t)$ is positive definite for all $\mathbf{x}$ and t, equilibria of Equation 12.98 are points at which $\nabla E = \mathbf{0}$, that is, are local minima of $E(\cdot)$. By Liapunov theory, the conclusion is that all isolated equilibria of Equation 12.98, which are isolated local minima of $E(\cdot)$, are locally asymptotically stable.

One of the simplest choices for $\mathbf{M}(\mathbf{x}, t)$ is clearly $\mathbf{M}(\mathbf{x}, t) = \mu\mathbf{I}$, where $\mu > 0$ is a scalar, $\mathbf{I}$ is the $n \times n$ identity matrix, and for this choice, system, Equation 12.98 becomes

$$\frac{\mathrm{d}\mathbf{x}}{\mathrm{d}t} = -\mu\nabla E(\mathbf{x}); \quad \mathbf{x}(0) = \mathbf{x}_0 \tag{12.101}$$

Clearly, choice of the parameter μ affects the rate at which trajectories converge to an equilibrium. In general, the choice of the positive definite matrix $\mathbf{M}(\mathbf{x}, t)$ in Equation 12.98 is guided by the desired rate of convergence to an equilibrium. In fact, the choice of $\mathbf{M}(\mathbf{x}, t)$, as a function of $\mathbf{x}$ (the state of the system) and time t, amounts to the use of time-varying state feedback control.

As a concrete example of the connection between gradient dynamical systems and neural networks, we will now briefly discuss the well-known class of Hopfield–Tank neural networks from this perspective.

12.8.3 Hopfield–Tank Neural Networks Written as Gradient Dynamical Systems

The basic mathematical model of a Hopfield neural network (see, for example, [23,41]) in normalized form can be written as

$$\mathbf{C}\frac{\mathrm{d}\mathbf{u}}{\mathrm{d}t} = -\mathbf{K}\mathbf{u} + \mathbf{W}\boldsymbol{\phi}(\mathbf{u}) + \boldsymbol{\theta} \tag{12.102}$$

where $\mathbf{u} = (u_1, u_2, \ldots, u_n)$, $\mathbf{C} = \mathrm{diag}(\tau_1, \tau_2, \ldots, \tau_n)$, $\mathbf{K} = \mathrm{diag}(\alpha_1, \alpha_2, \ldots, \alpha_n)$, $\boldsymbol{\phi}(\mathbf{u}) = (\phi_1(u_1), \phi_2(u_2), \ldots, \phi_n(u_n))$, $\boldsymbol{\theta} = (\theta_1, \theta_2, \ldots, \theta_n)$, and, finally, $\mathbf{W} = (w_{ij})$ is the symmetric interconnection matrix, where $w_{ji} = r_j G_{ji}$; $\theta_j = r_j I_j$; $\tau_j = r_j C_j > 0$; $\alpha_j = r_j / R_j > 0$, $\phi_i(\cdot)$ are the nonlinear activation functions that are assumed to be differentiable and monotonically increasing; r_j are the so-called scaling resistances, and G_{ji} are the conductances. The notation in the above equation originated from the circuit implementation of ANNs where, in addition, C_j, R_j correspond, respectively to capacitors and resistors and I_j to constant input currents.

The recurrent Hopfield–Tank neural net can be viewed from a feedback control perspective in different ways and a block diagram representation is depicted in Figure 12.15, in which the interconnection or weight matrix has been shown as part of the plant.

If the Hopfield–Tank neural network is to be used as an associative memory, then it must have multiple equilibria, each one corresponding to a state that is "remembered" as well as a local minimum of an appropriate energy function. In fact, the gradient of this energy function defines a gradient dynamical system that is referred to as a Hopfield–Tank associative memory neural network.

On the other hand, if the Hopfield–Tank neural network is to be used to carry out a global optimization task, then the energy function, which corresponds to the objective function of the optimization problem to be solved, should admit a unique global minimum.

These statements are now made precise.

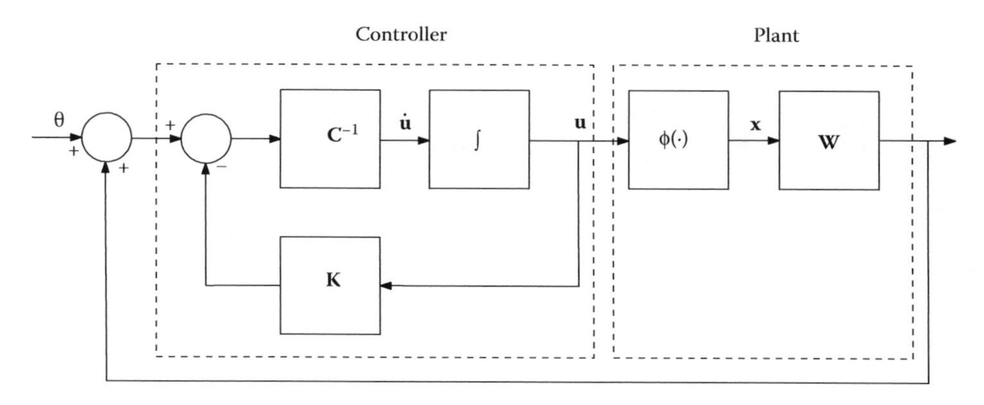

FIGURE 12.15
Dynamical feedback system representation of the Hopfield–Tank network (12.102).

12.8.3.1 Hopfield–Tank Network as Associative Memory

Defining

$$\mathbf{x} = \boldsymbol{\phi}(\mathbf{u}) \tag{12.103}$$

$\dot{\mathbf{x}} = \mathbf{D}_{\phi}\dot{\mathbf{u}}$ and thus Equation 12.102 can be written as

$$\dot{\mathbf{x}} = \mathbf{D}_{\phi}\mathbf{C}^{-1}(-\mathbf{K}\mathbf{u} + \mathbf{W}\mathbf{x} + \boldsymbol{\theta}) \tag{12.104}$$

Since $\boldsymbol{\phi}(\mathbf{u})$ is a diagonal function, with components $\phi_i(u_i)$ continuously differentiable and monotonically increasing, bounded and belonging to the first and third quadrants, two conclusions follow: $\mathbf{D}_{\phi}$ is a positive diagonal matrix and $\boldsymbol{\phi}$ is invertible [64], so that

$$\mathbf{u} = \boldsymbol{\phi}^{-1}(\mathbf{x}) \tag{12.105}$$

and Equation 12.104 can be written in $\mathbf{x}$-coordinates as

$$\dot{\mathbf{x}} = \mathbf{D}_{\phi}\mathbf{C}^{-1}(-\mathbf{K}\boldsymbol{\phi}^{-1}(\mathbf{x}) + \mathbf{W}\mathbf{x} + \boldsymbol{\theta}) \tag{12.106}$$

Now, defining $\mathbf{k}$ as the vector containing the diagonal elements of the diagonal matrix $\mathbf{K}$, that is

$$\mathbf{k} = (\alpha_1, \ldots, \alpha_n) \tag{12.107}$$

and the function $\boldsymbol{\phi}_I : \mathbb{R}^n \to \mathbb{R}^n$ as

$$\boldsymbol{\phi}_I(\mathbf{x}) := \left(\int_0^{x_1} \phi_1^{-1}(\tau)d\tau, \ldots, \int_0^{x_n} \phi_n^{-1}(\tau)d\tau \right) \tag{12.108}$$

an energy function, $E : \mathbb{R}^n \to \mathbb{R}$ can be defined as follows:

$$E(\mathbf{x}) := -\frac{1}{2}\mathbf{x}^T\mathbf{W}\mathbf{x} - \boldsymbol{\theta}^T\mathbf{x} + \mathbf{k}^T\boldsymbol{\phi}_I(\mathbf{x}) \tag{12.109}$$

Note that this energy function is not necessarily positive definite; it is, however, continuously differentiable. Calculating the gradient of $E(\mathbf{x})$ gives

$$\nabla E(\mathbf{x}) = -\mathbf{W}\mathbf{x} - \boldsymbol{\theta} + \mathbf{K}\boldsymbol{\phi}^{-1}(\mathbf{x}) \tag{12.110}$$

From Equations 12.106 and 12.110, it follows that we can write

$$\dot{\mathbf{x}} = -\mathbf{D}_{\phi}\mathbf{C}^{-1}\nabla E(\mathbf{x}) \tag{12.111}$$

showing, as claimed above, that the Hopfield–Tank network is a gradient dynamical system, recalling that the matrix $\mathbf{D}_{\phi}\mathbf{C}^{-1}$ is positive diagonal. This means that results on gradient systems can be used. In particular, it can be concluded that all trajectories of the Hopfield–Tank network tend to equilibria, which are extrema of the energy function, and, furthermore, that all isolated local minima of the energy function $E(\cdot)$ are locally asymptotically stable. Thus, the Hopfield–Tank network functions as an associative memory that "remembers" its equilibrium states, in the sense that if it is presented with (= initialized with) a state that is a slightly "corrupted" version of an equilibrium state, then it converges to this uncorrupted state.

12.8.3.2 Hopfield–Tank Net as Global Optimizer

Under different hypotheses and a different choice of energy function, it is possible to find conditions for convergence of the trajectories of Equation 12.102 to a unique equilibrium state.

First, Equation 12.102 is rewritten as

$$\frac{d\mathbf{u}}{dt} = -\mathbf{L}\mathbf{u} + \mathbf{T}\phi(\mathbf{u}) + \eta \tag{12.112}$$

where $\mathbf{L} := \mathbf{C}^{-1}\mathbf{K}$ is a diagonal matrix, $\mathbf{T} := \mathbf{C}^{-1}\mathbf{W}$ is the diagonally scaled interconnection matrix, and $\eta := \mathbf{C}^{-1}\theta$ is a constant vector. Suppose that it is desired to analyze the stability of an equilibrium point $\mathbf{u}^*$. For Liapunov analysis, it is necessary to shift the equilibrium to the origin, using the coordinate change $\mathbf{z} = \mathbf{u} - \mathbf{u}^*$. In $\mathbf{z}$-coordinates, Equation 12.112 becomes

$$\dot{\mathbf{z}} = -\mathbf{L}\mathbf{z} + \mathbf{T}\psi(\mathbf{z}) \tag{12.113}$$

where $\psi(\mathbf{z}) := (\psi_1(z_1), \ldots, \psi_n(z_n))$ and $\psi_i(z_i) := \phi_i(z_i + u_i^*) - \phi_i(u_i^*)$. It is easy to see that, under the smoothness, monotonicity, boundedness, and first-quadrant–third-quadrant assumptions on the ϕ_i, the functions ψ_i inherit these properties, and, in fact, the function ψ has bounded components, i.e., for all i, $|\psi_i(z_i)| \leq b_i|z_i|$, so that, defining $\mathbf{B} = \text{diag}(b_1, \ldots, b_n)$, the following vector inequality holds (componentwise):

$$\mathbf{B}^{-1}\psi(\mathbf{z}) \leq \mathbf{z} \tag{12.114}$$

In order to define a computational energy function, we introduce the following notation: $\mathbf{p}$ is a vector with all components positive, $\mathbf{P}$ is a diagonal matrix with components of the diagonal equal to that of the vector $\mathbf{p}$, and

$$\psi_S(\mathbf{z}) := \left(\int_0^{z_1} \psi_1(\tau)d\tau, \ldots, \int_0^{z_n} \psi_n(\tau)d\tau \right) \tag{12.115}$$

A computational energy function is then defined as follows:

$$\overline{E}(\mathbf{z}) = 2\mathbf{p}^T \psi_S(\mathbf{z}) \tag{12.116}$$

From Equations 12.115 and 12.116, it is clear that the computational energy function is of the Persidskii type.

Calculating the time derivative of $\overline{E}(\cdot)$ along the trajectories of Equation 12.113 gives

$$\dot{\overline{E}}(\mathbf{z}) = -2\psi(\mathbf{z})^T\mathbf{P}\mathbf{L}\mathbf{z} + 2\psi(\mathbf{z})^T\mathbf{P}\mathbf{T}\psi(\mathbf{z}) \tag{12.117}$$

From Equation 12.114, noting that $\mathbf{PL}$ is a positive diagonal matrix, the following componentwise majorization is immediate:

$$\psi(\mathbf{z})^T\mathbf{P}\mathbf{L}\mathbf{z} \geq \psi(\mathbf{z})^T\mathbf{P}\mathbf{L}\mathbf{B}^{-1}\psi(\mathbf{z}) \tag{12.118}$$

Thus, defining the positive diagonal matrix $\overline{\mathbf{L}} := \mathbf{L}\mathbf{B}^{-1}$, the following holds:

$$\begin{aligned} \dot{\overline{E}}(\mathbf{z}) &\leq -2\psi(\mathbf{z})^T\mathbf{P}\overline{\mathbf{L}}\psi(\mathbf{z}) + 2\psi(\mathbf{z})^T\mathbf{P}\mathbf{T}\psi(\mathbf{z}) \\ &= \psi(\mathbf{z})^T[(\mathbf{T} - \overline{\mathbf{L}})^T\mathbf{P}^T + \mathbf{P}(\mathbf{T} - \overline{\mathbf{L}})]\psi(\mathbf{z}) \\ &< 0 \end{aligned} \tag{12.119}$$

provided that

$$(\mathbf{T} - \overline{\mathbf{L}})^T \mathbf{P}^T + \mathbf{P}(\mathbf{T} - \overline{\mathbf{L}}) < 0 \qquad (12.120)$$

An interconnection matrix $\mathbf{T}$ is said to be *additively diagonally stable* if there exist positive diagonal matrices $\mathbf{P}$ and $\overline{\mathbf{L}}$ such that Equation 12.120 holds [51]. It turns out that if the interconnection matrix $\mathbf{T}$ is additively diagonally stable, then it is also true that Equation 12.112 has a unique equilibrium, and so it has just been proved, by the use of the energy function (12.116), which in this case is also a Liapunov function, that this unique equilibrium is globally asymptotically stable. Finally, in view of the definition of the computational energy function (12.116), it is easy to verify that Equation 12.113 can be written as

$$\dot{\mathbf{z}} = -\mathbf{L}\mathbf{z} + \mathbf{T}\mathbf{P}^{-1}\nabla \overline{E}(\mathbf{z}) \qquad (12.121)$$

The right-hand side of Equation 12.121 contains two terms: a dissipative linear term $\mathbf{Lz}$ and a gradient term. The additive diagonal stability condition, which relates the matrix $\mathbf{TP}^{-1}$ that multiplies the gradient of the energy function and the matrix $\mathbf{L}$ of the linear term, ensures that the energy function works as a Liapunov function for the system. Note that, in this case, unlike the associative memory case, symmetry of the interconnection matrix is not required.

For more details on global stability of neural networks and on the matrix classes involved in this stability analysis, see Reference 51.

12.8.4 Gradient Dynamical Systems That Solve Linear Systems of Equations

Another perspective on the connection between the iterative methods considered as dynamical systems with control and the neural networks studied in this chapter is obtained by recalling that a zero finding problem can be recast as an optimization problem.

Given a positive definite matrix $\mathbf{A} \in \mathbb{R}^{n \times n}$ and a vector $\mathbf{b} \in \mathbb{R}^n$, consider the following quadratic energy function:

$$E(\mathbf{x}) = \frac{1}{2}\mathbf{x}^T \mathbf{A}\mathbf{x} - \mathbf{b}^T \mathbf{x} \qquad (12.122)$$

The corresponding GDS that minimizes $E(\cdot)$ is

$$\dot{\mathbf{x}} = -\mathbf{M}\nabla E(\mathbf{x}) = -\mathbf{M}(\mathbf{A}\mathbf{x} - \mathbf{b}) \qquad (12.123)$$

Since trajectories of this GDS converge to the unique equilibrium, which is the solution $\mathbf{x}^*$ of the linear system

$$\mathbf{A}\mathbf{x} = \mathbf{b} \qquad (12.124)$$

this means that Equation 12.123 can be said to "implement" a continuous algorithm to solve the linear system with positive definite coefficient matrix (12.124).

Since the gradient of $E(\mathbf{x})$ is $\nabla E(\mathbf{x}) = \mathbf{A}\mathbf{x} - \mathbf{b}$ and the Hessian matrix $\nabla^2 E(\mathbf{x}) = \mathbf{A}$ (which is positive definite), thus $\mathbf{x}^* = \mathbf{A}^{-1}\mathbf{b}$ is the unique global minimizer of the energy function $E(\mathbf{x})$.

The function

$$V(\mathbf{x}) = E(\mathbf{x}) - E(\mathbf{x}^*) \qquad (12.125)$$

is a Liapunov function, since, assuming that $\mathbf{M}$ is positive definite, clearly $\dot{V}(\mathbf{x}) = -(\mathbf{A}\mathbf{x} - \mathbf{b})^T \mathbf{M}(\mathbf{A}\mathbf{x} - \mathbf{b})$ is negative definite. The matrix $\mathbf{M}$ is interpreted as a learning matrix or, equivalently, as a feedback gain matrix. The feedback gain or learning matrix can also be identified, in the appropriate context, as a preconditioner.

Minimizing the energy function $E(\mathbf{x})$ in Equation 12.122 corresponds to a two-norm minimization problem. If, instead, the one-norm is chosen to formulate the minimization problem, a least absolute deviation solution is found and the resulting dynamical system can be represented as a neural network with discontinuous activation functions. For rectangular matrices, minimizing the square of the two-norm of the residue leads to the normal equations and the solution in the least squares sense.

12.8.5 Discrete-Time Neural Networks

There are many different methods to derive discrete-time versions of the differential equations that describe the Hopfield model above [11,73]. For example, the simple forward Euler method applied to the Equation 12.102, after normalization of step size and time constants, yields the following difference equation:

$$\mathbf{u}_{k+1} = (\mathbf{I} - \mathbf{K})\mathbf{u}_k + \mathbf{W}\boldsymbol{\phi}(\mathbf{u}_k) + \boldsymbol{\theta} \tag{12.126}$$

Equation 12.126 describes a dynamical system that is also known as a discrete-time recurrent ANN.

Another simple discrete-time version of the Hopfield model [23] is arrived at by choosing $\mathbf{K} = \mathbf{I}$ and using the change of variables (12.103), that is, $\mathbf{x}_k = \boldsymbol{\phi}(\mathbf{u}_k)$, which yields, in the $\mathbf{x}$-variables:

$$\mathbf{x}_{k+1} = \boldsymbol{\phi}(\mathbf{W}\mathbf{x}_k + \boldsymbol{\theta}) \tag{12.127}$$

In digital implementations of Equation 12.127, the nonlinear activation function $\phi_j(\cdot)$ commonly used is the signum function, which, in addition to being easily implemented, allows useful theoretical manipulations. This leads to the form below:

$$\mathbf{x}_{k+1} = \text{sgn}(\mathbf{W}\mathbf{x}_k + \boldsymbol{\theta}) \tag{12.128}$$

Observe that the discrete-time model (12.128), in fact, is a nonlinear recurrence. Note also that the signum function is applied after the summation (in Equation 12.128) or addition of state variables. There are other situations in which the nonlinearity occurs before the addition (i.e., $\mathbf{x}_{k+1} = \mathbf{W}\text{sgn}(\mathbf{x}_k) + \boldsymbol{\theta}$, see [11,51]). The discrete-time model (12.128) is shown in block diagram control system representation in Figure 12.16.

System (12.128) can also be represented as a neural network. Figure 12.17 shows this representation for the general form Equation 12.127.

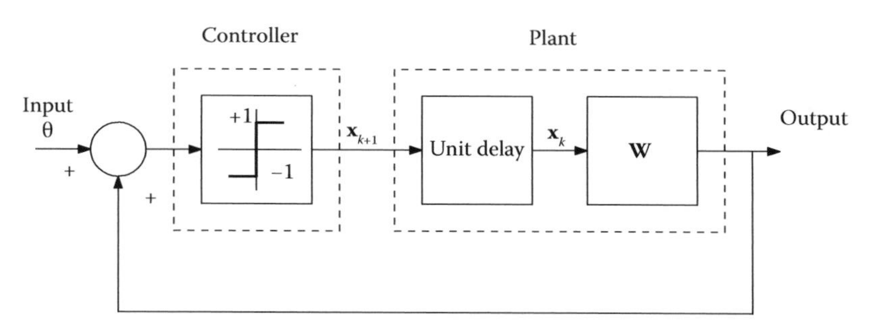

FIGURE 12.16
The discrete-time Hopfield network (12.128) represented as a feedback control system.

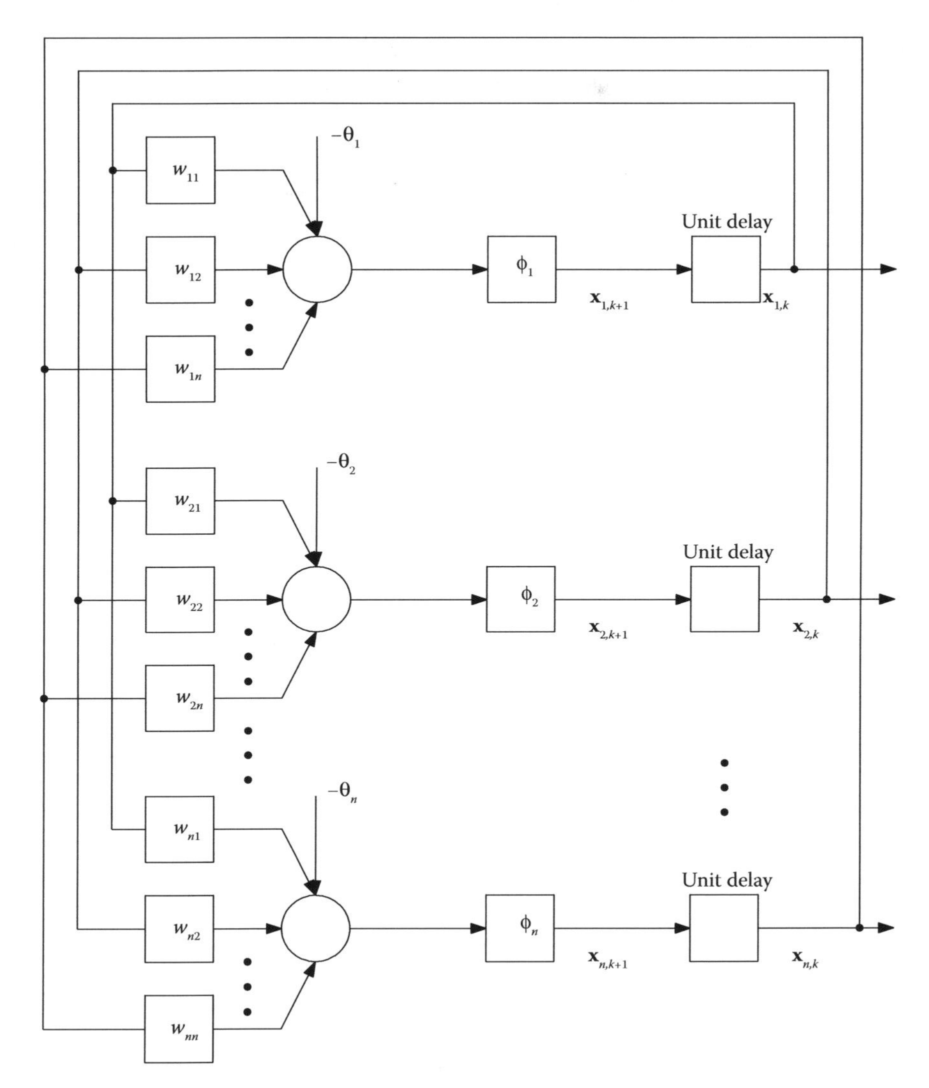

FIGURE 12.17
The discrete-time recurrence (12.127) represented as an artificial neural network.

12.9 Annotated Bibliography

As mentioned in the introduction, the field of ANNs is now in a mature enough state that there are several excellent textbooks and monographs on the details of every aspect mentioned in this chapter, as well as on several others that have been omitted. We will give pointers to some of these resources, noting that it is now possible to first consult a secondary source, such as a textbook, before consulting the vast literature for primary sources. This is the approach we have taken in order to keep the list of references to a small size.

The boom in ANN activity can probably be traced to the book by Rumelhart and McClelland [68]. A first-hand account of some of the early developments in neurodynamics can be found in Wiener's introduction to his book on cybernetics [79]. Sources for history are Anderson and Rosenfeld [5] and for the original, seminal papers, the IEEE collection [61].

There are several textbooks on the subject of neural networks, providing different perspectives on the subject. A clear and concise, but now rather old, treatment from the viewpoint of statistical physics can be found in *Introduction to the Theory of Neural Computation* by Hertz et al. [40]. The many interconnections of neural networks with statistics and, in particular, nonlinear regression and Bayesian methods are given a thorough and insightful treatment in Ripley [66] and, more recently in Reference 77. A comprehensive introduction, containing many topics not touched upon in this chapter, from an engineering (signal processing) viewpoint can be found in Haykin [38], now in its third edition.

Fairly detailed treatments of neural networks can also be found in textbooks on pattern recognition and machine learning, such as Bishop [12,13] and Duda et al. [31].

Many features that are today associated to SVMs, such as large margin classifiers and the use of optimization techniques, were already used in the 1960s, but the basic SVM technique described above was only introduced in 1995 and, since then, a large literature has emerged. A clear and succinct book length treatment is provided in Cristianini and Shawe-Taylor [25] and a comprehensive account in the monographs by Schölkopf and Smola [70] and Abe [1]. An important simplification of the SVM approach, useful in many applications, goes by the name of least squares SVM, described in detail in the monograph [72].

The statistical theory of learning is authoritatively described by one of its inventors, Vapnik, in References 75 and 76. Recent accessible accounts can be found in References 3 and 56.

The treatment of neural networks for optimization and from a gradient dynamical system viewpoint is based on References 10 and 50.

12.9.1 Software

Many neural network toolboxes are freely available on the Internet. Commonly used ANN simulators include the Stuttgart Neural Network Simulator (SNNS), Emergent, JavaNNS, Neural Lab, and NetMaker and the open source WEKA Data Mining Software, described in Witten and Frank [80]. The reader should search on the web for the most up-to-date information on these, since new networks, algorithms, and heuristics appear in the literature almost every day. In particular, MATLAB® routines for RBFs and MLPs are described in the book by Nabney [63].

Similarly, many software packages are available for implementing SVMs, such as the freely available LIBSVM C++ software package (see Chang and Lin [21] and Hsu et al. [43]).

More pointers can be found in http://www.signalprocessingsociety.org/technical-committees/list/mlsp-tc/mlsp-resources/#databases under the heading "Public Domain Neural Network Software Information."

12.9.2 Journals That Publish ANN-Related Research

A quantitative way to get updated information on ANN journals is to use the Journal Citation Report©, Thomson Reuters, which, in addition to evaluating journals based on impact factors, also has a "Related Journal" feature, which, when applied to a particular journal, ranks other journals which have cited and citing relationships with it, taking into account the number of citations from the citing journal to the cited journal, the total number of articles in the related journal, and the total number of citations from the citing journal. Applied to the journal *IEEE Transactions on Neural Networks and Learning Systems*, this produced a list of 180 journals, the first 20 of which are given below, including the *IEEE Transactions on Neural Networks and Learning Systems* itself in the list (so that there are a total of 21 journals). The large number of journals, covering topics such as control and automation, pattern recognition, chaos and nonlinear dynamics, statistics, chemometrics, signal processing, image and vision processing as well as neuroscience (in the biological sense), giving an idea of the extremely wide range of areas that use ANN techniques.

1. *IEEE Transactions on Neural Networks and Learning Systems* (known as *IEEE Transactions on Neural Networks* from 1990 to 2011, when the words "and Learning Systems" were added to its title.)
2. *Neural Processing Letters*
3. *Neurocomputing*
4. *Neural Networks*
5. *Machine Learning*
6. *Neural Computation*
7. *Journal of Machine Learning Research*
8. *Neural Computing and Applications*
9. *Journal of Intelligent and Fuzzy Systems*
10. *Neural Network World*
11. *IEEE Transactions on Systems Man and Cybernetics Part B—Cybernetics*
12. *Connection Science*
13. *International Journal on Neural Systems*
14. *Cognitive Neurodynamics*
15. *IEEE Transactions on Pattern Analysis*
16. *IEEE Transactions on Fuzzy Systems*
17. *IEEE Computational Intelligence Magazine*
18. *Journal of Chemometrics*
19. *Applied Intelligence*
20. *Applied Soft Computing*
21. *IEEE Transactions on Automatic Control*

12.9.3 Data Sets

At the current time, there are probably more real data sets available on the Internet than most of us will ever be able to analyze in one lifetime. Some of these, however, are used as benchmarks: namely, the UCI Machine Learning Repository, which maintains the classic data sets and also adds new ones all the time; the University of Waikato's repository, which contains pointers to other repositories (including the UCI one), the compilation of sites maintained by the IEEE Signal Processing Society and many individual compilations, such as the one built up by Rashidi, Torgo, etc.

- http://archive.ics.uci.edu/ml/datasets.html
- http://www.cs.waikato.ac.nz/ml/weka/datasets.html
- http://www.signalprocessingsociety.org/technical-committees/list/mlsp-tc/mlsp-resources/#databases
- https://sites.google.com/site/parisar/links/datasets/generaldatasets
- http://www.dcc.fc.up.pt/ ltorgo/Regression/DataSets.html

The UCI data set collection is popular because, in addition to providing the real raw data, it also contains documentation as well as published results of various analyses.

In addition to assessing methods on real data, which is an essential part of validating an algorithm, simulated data are very useful, since large data sets with known statistical properties can be generated at little cost, enabling estimation of asymptotic error rates and other statistics.

12.10 Discussion

This chapter has concentrated mainly on sketching the evolution of a basic linear discriminant model to models which are essentially linear, but with nonlinear decision boundaries (the RBF model) or models in which linear combinations of univariate nonlinear functions (of linear functions) are used, once again to build nonlinear decision boundaries (the MLP model). The basic idea behind neural network approximation methods is the combination of simple nonlinear functions, processed in a hierarchical manner, which allows for the creation of efficient algorithms for learning (= weight adjustment). It should be mentioned that neural networks are not confined to the feedforward or MLP nets for supervised pattern classification or regression, as presented in this chapter. More complex networks, suitable for specific tasks, which possess feedback from one layer to the previous layer, as well as unsupervised networks also have been developed and the interested reader should refer to the annotated bibliography in Section 12.9 to follow up on this.

The SVM, which can be thought of as defining basis functions implicitly through the definition of a kernel function in the data space, gives very good performance on many problems and today is often the method of choice. Reasons for this are that there are few parameters to set in an SVM, essentially just the kernel parameters and the regularization parameter, which can be adjusted quite easily to get optimum performance on a validation set. However, it should be kept in mind that the standard SVM, with its emphasis on the decision boundary, will not be suitable in situations in which its conditions of use involve noise or drift and differ from the training conditions.

12.10.1 Some Recommendations

Before applying the techniques described in this chapter, the reader should ask some questions of himself. Is there a reason to think that the decision boundary is nonlinear? Is the performance provided by linear techniques inadequate? If so, then the use of a neural network or an SVM may be justifiable, although one cannot be sure of this. Here are some general guidelines:

1. A simple pattern recognition technique (e.g., linear discriminant analysis, naive Bayes) should be tested, before considering more complex neural network methods.
2. A simple RBF (unsupervised selection of centers, weights optimized using a squared error criterion) should be tried to get a feel for whether nonlinear methods provide some gain for your problem.
3. Data preprocessing and scaling is always good practice, for numerical and algorithmic reasons.
4. Cross-validation or regularization should always be used for model selection and avoidance of overfitting.
5. In MLP training, preference should be given to the batch mode.

6. For classification problems in high-dimensional spaces and in which misclassification rate can be used to measure classifier performance, SVMs should be preferred.

12.11 Applications

Applications of neural networks are widespread and a famous statement attributed to John Denker explains this: "neural networks are the second best way of doing just about anything" by which he meant that they provide acceptable performance on a wide variety of problems that are difficult to solve well using other methods.

There are many surveys of neural network/SVM applications and comparative studies in different fields. In order to remain reasonably up to date, we will limit ourselves to citing a sampling of surveys and applications, published during or after the year 2000, grouped by a sample of broad application areas, but the reader would be well advised to do a literature search, in the specific application area of interest, using the many web search engines available.

- Classification by ANNs: Broad overview in Reference 81, with almost 2000 references
- Finance, including prediction: [2,6,19,47,57,60]
- Management: [37,71,74]
- Bioinformatics, including drug discovery: [17,22,45,49]
- Medicine: [7,16,32,36,44,58]
- Plagiarism detection: [26,29]
- Process engineering: [14,28]

References

1. S. Abe. *Support Vector Machines for Pattern Classification*. Springer-Verlag, London, 2nd edition, 2010.
2. Y. S. Abu-Mostafa, A. F. Atiya, M. Magdon-Ismail, and H. White. Special issue on "neural networks in financial engineering." *IEEE Trans. Neural Networks*, 12(4):653–656, 2001.
3. Y. S. Abu-Mostafa, M. Magdon-Ismail, and H.-T. Lin. *Learning from Data: A Short Course*. AML-book.com, 2012.
4. S.-I. Amari. Information geometry of neural networks. *Trans. IEICE*, E75-A:531–536, 1992.
5. J. Anderson and E. Rosenfeld, editors. *Neurocomputing: Foundations of Research*. MIT Press, Cambridge, MA, 1988.
6. E. Angelini, G. di Tollo, and A. Roli. A neural network approach for credit risk evaluation. *Q. Rev. Econ. Finan.*, 48(4):733–755, 2008.
7. D. Ansari, J. Nilsson, R. Andersson, S. Regnér, B. Tingstedt, and B. Andersson. Artificial neural networks predict survival from pancreatic cancer after radical surgery. *Am. J. Surg*, 205(1):1–7, 2013.
8. D. P. Bertsekas and J. N. Tsitsiklis. *Neuro-Dynamic Programming*. Athena Scientific, Belmont, MA, 1996.
9. A. Bhaya and E. Kaszkurewicz. Steepest descent with momentum for quadratic functions is a version of the conjugate gradient method. *Neural Networks*, 17(1):65–71, 2004.
10. A. Bhaya and E. Kaszkurewicz. *Control Perspectives on Numerical Algorithms and Matrix Problems, Volume DC10 of Advances in Design and Control*. SIAM, Philadelphia, 2006.
11. A. Bhaya, E. Kaszkurewicz, and V. S. Kozyakin. Existence and stability of equilibria in continuous-variable discrete-time neural networks. *IEEE Trans. Neural Networks*, 7(3):620–628, 1996.

12. C. M. Bishop. *Neural Networks for Pattern Recognition*. Oxford University Press, New York, 1995.
13. C. M. Bishop. *Pattern Recognition and Machine Learning*. Springer, New York, 2006.
14. G. Bloch and T. Denoeux. Neural networks for process control and optimization: Two industrial applications. *ISA Trans.*, 42(1):39–51, 2003.
15. A. E. Bryson. A gradient method for optimizing multi-stage allocation processes. In *Harvard University Symposium on Digital Computers and Their Applications*, Boston, MA, 1961.
16. A. Buchner, M. May, M. Burger, and 22 others. Prediction of outcome in patients with urothelial carcinoma of the bladder following radical cystectomy using artificial neural networks. *Eur. J. Surg. Oncol.*, 39(4):372–379, 2013.
17. R. Burbidge, M. Trotter, B. Buxton, and S. Holden. Drug design by machine learning: Support vector machines for pharmaceutical data analysis. *Comput. Chem.*, 26:5–14, 2001.
18. C. J. C. Burges. A tutorial on support vector machines for pattern recognition. *Data Mining Know. Discov.*, 2:121–167, 1998.
19. B. Cao, D. Zhan, and X. Wu. Application of SVM in financial research. In *2009 International Joint Conference on Computational Sciences and Optimization*, 507–511. Sanya, Hainan, IEEE, 2009.
20. W. Cao and G. Mirchandani. On hidden nodes for neural networks. *IEEE Trans. Circuits Syst.*, 36(5):661–664, 1987.
21. C. C. Chang and C. J. Lin. LIBSVM: A library for support vector machines. *ACM Trans. Intell. Syst. Technol.*, 2(3):1–27, 2011.
22. F. Cheng and S. Vijaykumar. Applications of artificial neural network modeling in drug discovery. *Clin. Exp. Pharmacol.*, 2:e113, 2012.
23. A. Cichocki and R. Unbehauen. *Neural Networks for Optimization and Signal Processing*. John Wiley, Chichester, 1993.
24. T. Cover. Geometrical and statistical properties of systems of linear inequalities with applications in pattern recognition. *IEEE Trans. Electron. Comput.*, 14:326–344, 1965.
25. N. Cristianini and J. Shawe-Taylor. *An Introduction to Support Vector Machines*. Cambridge University Press, Cambridge, UK, 2000.
26. D. Curran. An evolutionary neural network approach to intrinsic plagiarism detection. In L. Coyle and J. Freyne, editors, *Artificial Intelligence and Cognitive Science*, volume 6206 of *Lecture Notes in Computer Science*, 33–40. Springer, Berlin, Heidelberg, 2010.
27. G. Cybenko. Approximation by superposition of sigmoidal functions. *Math. Control Signals Syst.*, 2:303–314, 1989. Correction in vol. 5 [op.cit.], p. 455, 1992.
28. J. Fernández de Cañete and A. B. Bulsari. Special issue on "neural networks in process engineering." *Neural Comput. Appl.*, 9(3):163–164, 2000.
29. J. Diederich. Authorship attribution with support vector machines. *Appl. Intell.*, 19(1–2):109–123, 2006. Special Issue: Neural Networks and Machine Learning for Natural Language Processing.
30. S. E. Dreyfus. The numerical solution of variational problems. *J. Math. Anal. Appl.*, 5(1):30–45, 1962.
31. R. O. Duda, P. E. Hart, and D. G. Stork. *Pattern Classification*. John Wiley, New York, 2nd edition, 2001.
32. R. Dybowski and V. Gant, editors. *Clinical Applications of Artificial Neural Networks*. Cambridge University Press, New York, 2001.
33. D. Gabor. Communication theory and cybernetics. *IRE Trans. Circuit Theory*, CT-1:19–31, 1954.
34. S. Grossberg. Adaptive pattern classification and universal recoding, I: Parallel development and coding of neural feature detectors. *Biol. Cybernetics*, 23:121–134, 1976.
35. S. Grossberg. Adaptive pattern classification and universal recoding, II: Feedback, expectation, olfaction, and illusions. *Biol. Cybernetics*, 23:187–202, 1976.
36. I. Guyon, J. Weston, S. Barnhill, and V. Vapnik. Gene selection for cancer classification using support vector machines. *Machine Learn.*, 46:389–422, 2002.
37. H. Hakimpoor, K. A. B. Arshad, H. H. Tat, N. Khani, and M. Rahmandoust. Artificial neural networks applications in management. *World Appl. Sci. J.*, 14(7):1008–1019, 2011.
38. S. Haykin. *Neural Networks and Learning Machines*. Prentice-Hall, Pearson, Upper Saddle River, NJ, 3rd edition, 2009.
39. D. O. Hebb. *The Organization of Behavior: A Neurophysiological Theory*. Wiley, New York, 1949.
40. J. Hertz, A. Krogh, and R. G. Palmer. *Introduction to the Theory of Neural Computation*. Addison-Wesley, Reading, MA, 1991.

41. J. Hopfield and D. Tank. Computing with neural circuits: A model. *Science*, 233:625–633, 1986.
42. J. J. Hopfield. Neural networks and physical systems with emergent collective computational abilities. *Proc. Natl. Acad. Sci.*, 79:3088–3092, 1982.
43. C. W. Hsu, C. C. Chang, and C. J. Lin. A practical guide to support vector classification. Technical report, Department of Computer Science, National Taiwan University, Taipei, 2003.
44. X. Hu, H. Cammann, H. A. Meyer, K. Miller, K. Jung, and C. Stephan. Artificial neural networks and prostate cancer—Tools for diagnosis and management. *Nat. Rev. Urol.*, 10(3):174–182, 2013.
45. S. Hua and Z. Sun. A novel method of protein secondary structure prediction with high segment overlap measure: Support vector machine approach. *J. Mol. Biol.*, 308:397–407, 2001.
46. K. J. Hunt, D. Sbarbaro, R. Żbikowski, and P. J. Gawthrop. Neural networks for control systems—A survey. *Automatica*, 28(6):1083–1112, 1992.
47. T. Jiří. Application of neural networks in finance. *J. Appl. Math.*, 3(3):269–277, 2010.
48. T. Joachims. Training linear SVMs in linear time. In *Proceedings of the 12th ACM SIGKDD International Conference on Knowledge Discovery and Data Mining*, KDD '06, pages 217–226, New York, NY, USA, 2006. ACM.
49. R. Kakumani, V. Devabhaktuni, and M. Ahmad. A two-stage neural network based technique for protein secondary structure prediction. In *Conference Proceedings of IEEE Engineering in Medicine and Biology Society*, pages 1355–1358, Vancouver, British Columbia, Canada, 2008.
50. E. Kaszkurewicz and A. Bhaya. On a class of globally stable neural circuits. *IEEE Trans. Circuits Syst. I Fundamental Theory Appl.*, 41(2):171–174, 1994.
51. E. Kaszkurewicz and A. Bhaya. *Matrix Diagonal Stability in Systems and Computation*. Birkhäuser, Boston, 2000.
52. H. J. Kelley. Gradient theory of optimal flight paths. *Am. Rocket Soc. J.*, 30(10):941–954, 1960.
53. T. Kohonen. *Self-Organizing Maps*. Springer-Verlag, Heidelberg, 3rd, extended edition, 2001.
54. A. N. Kolmogorov. On the representation of continuous functions of several variables by superposition of continuous functions of one variable and addition. *Dokl. Akad. Nauk. SSSR*, 114:953–956, 1957.
55. M. K. Kozlov, S. P. Tarasov, and L. G. Khachiyan. Polynomial solvability of convex quadratic programming. *Soviet Math. Doklady*, 20:1108–1111, 1979. Russian original published in Doklady Akademiia Nauk SSR, 228 (1979).
56. S. Kulkarni and G. Harman. *An Elementary Introduction to Statistical Learning Theory*. John Wiley, Hoboken, NJ, 2011.
57. G.-Q. Li, S.-W. Xu, and Z.-M. Li. Short-term price forecasting for agro-products using artificial neural networks. *Agric. Agric. Sci. Proc.*, 1:278–287, 2010.
58. B. Liu and Y. Jiang. A multitarget training method for artificial neural network with application to computer-aided diagnosis. *Med. Phys.*, 40(1):011908–011916, 2013.
59. W. S. McCulloch and W. Pitts. A logical calculus of the ideas immanent in nervous activity. *Bull. Math. Biophys.*, 5:115–133, 1943.
60. P. D. McNelis. *Neural Networks in Finance*. Elsevier Academic Press, Amsterdam, 2005.
61. P. Mehra and B. W. Wah. *Artificial Neural Networks: Concepts and Theory*. IEEE Computer Society Press, Los Alamitos, CA, 1992.
62. M. Minsky and S. Papert. *Perceptrons*. MIT Press, Cambridge, MA, 1969. Expanded edition published in 1988.
63. I. Nabney. *NETLAB: Algorithms for Pattern Recognition*. Springer-Verlag, London, 2002.
64. J. M. Ortega and W. C. Rheinboldt. *Iterative Solutions of Nonlinear Equations in Several Variables*. Academic Press, New York, 1970.
65. J. Park and I. W. Sandberg. Approximation and radial-basis functions networks. *Neural Comput.*, 5:305–316, 1993.
66. B. D. Ripley. *Pattern Recognition and Neural Networks*. Cambridge University Press, Cambridge, UK, 1996.
67. F. Rosenblatt. *Principles of Neurodynamics: Perceptrons and the Theory of Brain Mechanisms*. Spartan, New York, 1962.
68. D. E. Rumelhart, G. E. Hinton, and R. J. Williams. Learning internal representations by error propagation. In D. E. Rumelhart and J. L. McClelland, editors, *Parallel Distributed Processing: Explorations in the*

Microstructure of Cognition, Volume I: Foundations, 318–362. MIT Press/Bradford Books, Cambridge, MA, 1986.
69. F. Scarselli and A.-C. Tsoi. Universal approximation using feedforward neural networks: A survey of some existing methods, and some new results. *Neural Networks*, 11(1):15–37, 1998.
70. B. Schölkopf and A. Smola. *Learning with Kernels*. The MIT Press, Cambridge, MA, 2002.
71. K. A. Smith and J. N. D. Gupta. Neural networks in business: Techniques and applications for the operations researcher. *Comput. Oper. Res.*, 27:1023–1044, 2000.
72. J. A. Suykens, T. V. Gestel, J. D. Brabanter, B. D. Moor, and J. Vandewalle. *Least Squares Support Vector Machines*. World Scientific, Singapore, 2002.
73. M. Takeda and J. W. Goodman. Neural networks for computation: Number representations and programming complexity. *Appl. Optics*, 25(18):3033–3046, 1986.
74. S. Thomassey and M. Happiette. A neural clustering and classification system for sales forecasting of new apparel items. *Appl. Soft Comput.*, 7(4):1177–1187, 2007.
75. V. N. Vapnik. *The Nature of Statistical Learning Theory*. Springer, New York, 1995.
76. V. N. Vapnik. An overview of statistical learning theory. *IEEE Trans. Neural Networks*, 10(5):988–999, 1999.
77. A. R. Webb and K. D. Copsey. *Statistical Pattern Recognition*. Wiley, Hoboken, NJ, 2011.
78. B. Widrow and M. E. Hoff. Adaptive switching circuits. In *WESCON Convention*, volume IV, 96–104, Los Angeles, CA, 1960.
79. N. Wiener. *Cybernetics or Control and Communication in the Animal and the Machine*. MIT Press, Cambridge, MA, 1948. Second edition published in 1961.
80. I. H. Witten and E. Frank. *Data Mining: Practical Machine Learning Tools and Techniques*. Morgan Kaufmann, San Mateo, CA, 2nd edition, 2005.
81. G. P. Zhang. Neural networks for classification: A survey. *IEEE Trans. Syst. Man Cybernetics C: Appl. Rev.*, 30(4):451–462, 2000.

13

Simulation Modeling and Analysis

Aparna Gupta

CONTENTS

ABSTRACT Simulation modeling and analysis is a powerful suite of techniques for addressing a broad array of decision analytical problems. In this chapter, we highlight the most important aspects of the topic, taking advantage of prior development of discrete and continuous stochastic models in this book. Beginning with discussing model building for simulation analysis relevant in various application areas, we review the fundamentals of simulation in terms of random variates generation, model validation, and simulation output analysis. We consider manufacturing and service systems, where we are concerned with the optimal layout and configuration of manufacturing operations or service delivery setup for best-performance characteristics, or sensitivity analysis from changes made in layout or changes appearing due to breakdowns. The second application domain we consider is that of communication and transportation networks, where flow of entities is on a fixed infrastructure designed to meet customer demand, while satisfying "physical" constraints of the system. Therefore, capability analysis is often the motivation of evaluating such systems for meeting demand as well as quality of service delivered to customers. We finally consider two example application domains in banking and finance, one in asset valuation and allocation and the other in risk management. In the second part of the chapter, we delve into specialized topics of interest for a simulation study, such as variance reduction techniques, rare-event simulation, and stochastic Kriging. For a more sophisticated use of simulation modeling and analysis, we develop modeling approaches for simulation of dynamic systems. Acknowledging that many decision analytics problems extend to assessing better or the best configurations of systems or ways of performing tasks, the chapter presents simulation-based optimization methodologies and techniques to address optimal decision-making problems. While we have initiated the chapter with several specific application domains where simulation modeling and analysis is extensively utilized, the entire chapter is interspersed with examples for making a stronger context for an analyst to benefit from the topics covered.

This chapter is devoted to developing simulation modeling and analysis concepts and techniques to address decision analysis problems. Simulation modeling and analysis is a vast topic, and a powerful suite of techniques, for solving a broad array of problems. This chapter will highlight in brevity the most important aspects of the topic. Taking advantage of prior discussions on discrete and continuous random variables and distributions, this chapter will begin with describing model building for simulation analysis of decision making. The next several major topics will develop the fundamentals of simulation modeling and analysis, through topics such as random variates generation, model validation, and simulation output analysis. This is followed by specialized topics on variance reduction techniques, rare-event simulation, stochastic Kriging, and simulation of dynamic systems for more sophisticated use of simulation modeling and analysis. The later part of this chapter focuses on simulation-based optimization methodologies and techniques to address optimal decision-making problems that may be solved using simulation in conjunction with optimization techniques. While the entire chapter is interspersed with examples, we initiate the chapter with several specific application domains where simulation modeling and analysis is extensively utilized. We will end the chapter by pointing to the future challenges facing simulation modeling and analysis for decision making today.

13.1 Introduction to Simulation Modeling

Simulation is a key tool in the tool-set to aid decision making in the presence of risk and uncertainty. Simulation modeling and analysis constitutes a set of quantitative tools and methodologies, and indeed software systems, to facilitate delving into the role risks can play in a variety of contexts of decision making. These contexts can be quite broad and varied, and hence simulation is considered a versatile tool, relevant and widely used in many sectors and areas of study, as the example domains we will present in this chapter will demonstrate. In fact, the modeling techniques underlying simulation are also quite broad, not all of which incorporate randomness. Some "simply" describe how a complex system evolves deterministically over time. Models built using complex systems of ordinary or partial differential equations and solved by numerical techniques fall in this category. For our purposes here, however, simulation will be used to model and analyze risk in order to address decision-making problems in the presence of risk and uncertainty.

Simulation often means different things to different people, with Pritsker [88] being the first to provide a compilation of definitions of simulation in 1979. More recently, Oren [82] came up with about 400 definitions of simulation. The word "simulation" is derived from the Latin "simulatus," past participle of "simulare," which simply means to copy, represent. Simulation is an attempt to duplicate the operation of a system or the behavior of a quantity of interest without incurring the expense and expending the effort to build or operate the system, or generate observations for the quantity of interest by natural means. More operationally, simulation involves creating an experimental setup to study the dynamics of the system or the quantity of interest. Instead of "solving" the system, a simulation analyst operates a "model" of the system in the experimental setup under different conditions to assess the system behavior and determine appropriate decisions to improve or respond to system behavior.

The model of a system almost always is an abstraction of the relatively complex reality, where a simplified description is constructed for the purpose of gaining insight into the system behavior. Modeling involves not only creating abstractions, but also coding it in an appropriate computing environment, which is used for running experiments. There are dedicated commercial products to facilitate building simulation models for different industry segments—manufacturing, communications network, civil infrastructure—roadways, railways, air-traffic, medical and healthcare services, financial services, etc. The objectives of a simulation study in any of these domains can range from performance analysis, capability analysis, comparison studies, sensitivity analysis, optimization study, constraint analysis, or asset valuation. Models are abstractions of the real system, where a system is viewed as a group of objects or quantities joined together in some regular interaction or interdependence. Models developed for facilitating simulation are constructed with a view of using them to conduct specific experiments, predict events, or determine a future course of action that would in some sense be best. In short, models are built with a definite use in mind; hence, models themselves are not the focus. Instead, the use they would be put to is the focus. Models must always be treated as a means to an end. Identifying the objective of a simulation study helps determine the level of scope and detail appropriate in the model of a system for the set goals. Abstraction in model building is justified since it brings perspective to the need for detail, introducing a level of detail that improves understanding of system behavior and the necessary modifications needed for the system. Controlling details allows the simulation analyst to have more control over variations to the system studied, and organize beliefs and empirical observations.

Conducting a simulation analysis for a decision problem can be organized in logical stages, laid out in Figure 13.1. Once the objectives of a study are stated, including the measures for evaluating the outcomes, appropriate data must be acquired, and equations and algorithms need to be defined to describe the model. It is generally useful to run a simplification of the model or just a portion of the model for testing a model's validity. After reviewing the usefulness and robustness of the model, translating the mathematical model into the desired computational environment can be initiated. Developing portions of the model separately and putting these modules together makes for a more robust model implementation strategy. A well-tested and documented computer model should be the

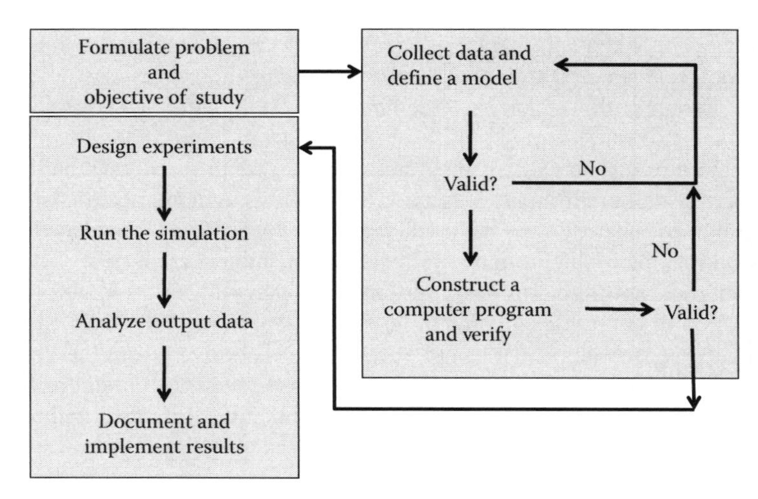

FIGURE 13.1
A guideline for the flow of activity to design a successful simulation study.

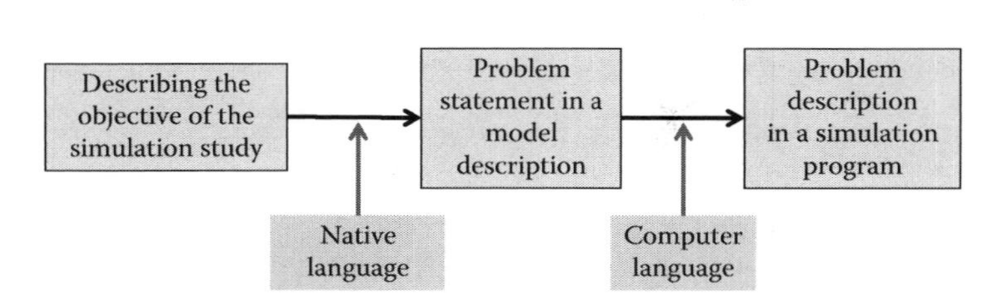

FIGURE 13.2
The stages to adopt in building a simulation model.

basis of running the necessary designed experiments, and the corresponding simulation runs. The output data needs to be analyzed to achieve the goals of the simulation study, as well as guide further development of the study.

Before moving forward, we offer one word of caution at the outset. When conducting a simulation study, it is often tempting to jump into writing up the computer model to perform the simulation runs and experiments, and generate the output results, cutting short the due diligence necessary for the conceptual model development. This can lead to erroneous assumptions or implementation for the intended problem and study objectives. Figure 13.2 emphasizes the two stages for a simulation study. The first stage is where the conceptual model building and assessment in natural language and mathematical formalism takes place. The second stage would then translate the conceptual model into a computer representation of the model in a chosen simulation software using appropriate programming structures.

We have briefly reviewed the concept of simulation and its purpose and guidelines for implementing a simulation study. In the next section, we will identify some sample problem domains and advance the model development process in these domains geared toward addressing specific problems. All subsequent sections will continue to advance the simulation modeling and analysis topics at greater depth, thus building on the conceptual and mathematical modeling proficiency required for solving problems using simulation.

13.2 Problems Domains and Model Building

As stated earlier, simulation modeling and analysis is a powerful suite of techniques, for solving a broad array of problems in a variety of application domains. These application domains range from manufacturing processes, service systems, supply-chain networks, communications networks, urban and civil infrastructure—roadways, railways, air-traffic, medical and healthcare services, electrical grids, financial services, risk management, among others. The objectives of a simulation study in these domains can range from system design, performance analysis, capability analysis, comparison studies, sensitivity analysis, optimization study, constraint analysis, risk assessment, or asset valuation.

In this section, picking some sample problem domains as the context, we will present some approaches for model building, as well as motivate some problem objectives to justify utilization of simulation modeling and analysis. Our first example application domain is that of manufacturing and service systems. Here, we will be concerned with the optimal layout and configuration of manufacturing operations or service delivery setup for best-performance characteristics, or sensitivity analysis from changes made in layout or changes appearing due to breakdowns. The second application domain we examine will be that of communication and transportation networks, which we have picked since in these domains flow of entities is on a fixed infrastructure, designed to meet customer demand, while satisfying "physical" constraints of the system. Here, the motivation behind the simulation study will be a capability analysis for meeting demand as well as quality of service delivered to the customers. The last two example application domains are in banking and finance, one in asset valuation and allocation and second in risk management. In general, these issues of asset allocation and valuation arise in other application domains; therefore, our study of them in the context of finance is generalizable.

13.2.1 Manufacturing and Service Systems

Today's high-tech manufacturing systems for automotive, aerospace, consumer electronics, semiconductor, biomedical industries, etc. can be extremely complex. Manufacturing steps running into hundreds, with thousands of part types, being utilized in many assembly and subassembly lines in batch processing mode to produce the complex products can by no means be designed and managed by intuition, using simple rules of thumb or even static queuing models. The high cost associated with establishing, running, and maintaining such complex manufacturing systems further necessitates utilizing formal dynamic models and analysis of such systems. The complexity of these systems reflected in complexity of models often entails that the models may not be solvable analytically, in which case simulation analysis is the method of choice [48,81]. This is often termed discrete-event simulation [8,66,91].

Discrete-event simulation refers to simulation of a system where unit objects, which can be of great many variety, move through the system as a unit. Therefore, the number of unit objects of a certain kind at a certain location of the system varies discretely over time. Simulation of these changes in counts of objects at different locations can be considered as events, hence discrete-event simulation. Figure 13.3 displays a simple caricature of a system where discrete-event simulation tracks movement of parts through an assembly line, as they wait for being processed, are processed, and then moved to the next station in the assembly line.

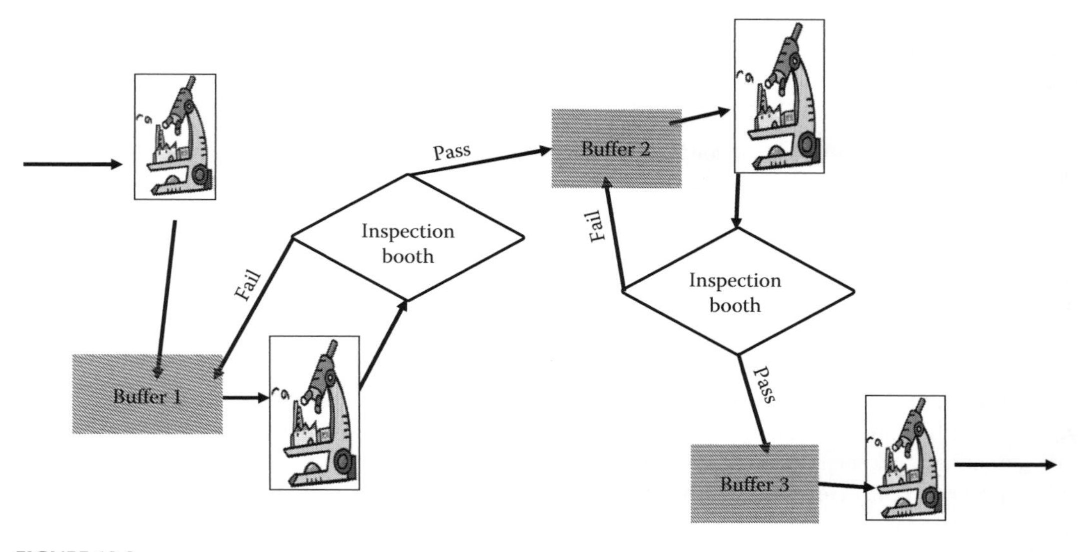

FIGURE 13.3
The layout of a manufacturing floor with waiting stations (buffers), processing units, inspection, and routing of entities.

Complex transformation and routing of objects may be necessitated in the dynamic model to reflect the processing of objects in the real system. The routing of objects through the assembly lines may also require critical logical steps. Examples of such logical steps are included in the work flow caricature of Figure 13.3, where after two main processing steps, the outputs are checked for defects and errors, routed forward only if they are found fit.

For modeling these systems, two variables types stand out as most significant. These are interarrival times of different objects at different locations or processing stations and the processing times at the processing stations. Interarrival times of objects may be an independent variable or may be dependent on the condition and status of the upstream process. Similarly, processing time at a station may be an independent variable or may be dependent on the condition or type of object being processed. Typically, plenty of data from the system will be necessary to correctly and appropriately characterize the model of the system in these terms.

A model for the arrival process for each object type at a processing station may be chosen to reasonably capture work flow in an assembly line. A simple and popular model for arrival process is the Poisson process, which assumes that the interarrival times of objects' arrival at a processing station are exponentially distributed. A homogeneous Poisson process assumes a constant arrival rate, λ, which implies that the number of objects, N_t, arriving at the processing station in t duration of time is Poisson distributed, with mean λt. If the arrival rate fluctuates with time, a nonhomogeneous Poisson process will better capture this behavior, where the arrival rate, λ_t, is time-variant. In some cases, the arrival rate may not only be time-variant, it may also vary stochastically. The Cox process is a generalization of a nonhomogeneous Poisson process with stochastic arrival rate for objects.

In the arrival processes discussed thus far, the interarrival times follow an exponential distribution, albeit with time-varying parameters. In some cases, the arrival pattern may not be appropriately represented by the exponential distribution. A fractional Poisson process is a stochastic process used to model long-memory dynamics, where the interarrival time is taken to follow the nonexponential power-law distribution. In general, such generalization of the Poisson process with the interarrival time for objects being modeled by different distributions than the exponential distribution is called renewal process [90].

The most important performance measures in manufacturing systems tend to be the throughput of acceptable quality products [3,18,49,56]. In some cases, additional performance measures of utilization of the processing stations or equipment, inventory levels being used to support the assembly line, etc. may also be considered. Upon defining the model and performance measures, a simulation model can be developed and implemented in a chosen computing environment depending on the study objectives. The objective could be to have a platform for a one-time use for design of production line or one for a continual use through design and future improvements of the production line.

Service systems have gained tremendous importance in the recent decades, as consumption in the developed countries, and increasingly in the emerging economies, has moved from physical goods to personal experiences. Design and analysis of service systems bears many similarities with manufacturing systems; therefore, the discrete-event simulation methods and models for manufacturing systems can be adapted for analysis of service systems. There are, however, some significant difference that the analyst must bear in mind. In service systems, the consideration of system performance shifts to the quality, timeliness, and efficiency of service, rather than the throughput or utilization of the servers.

Servers in service systems are often humans, instead of automated machines; therefore, this can inherently result in variability and unpredictability. Services, such as banking [84,104], insurance [87], entertainment, and medical services [1,38,89], can also be visualized to constitute rather complex stochastic processes, which must also often operate under many resource constraints. High degree of variation in customer needs and characteristics may necessitate significant customization in the service. The 1990s and 2000s saw a high growth in call center service systems, especially with increasing business use of IT&S solutions and growth in e-business and online commerce. Discrete-event simulation has been extensively utilized for design and improvement of call center systems [6,64,94].

13.2.2 Communication and Transportation Networks

Simulation modeling is valuable for addressing design and management issues for manufacturing and service enterprises, and as discussed in the previous section, has been widely investigated in the literature and amply utilized in practice. By changing the definition of entities tracked, jobs being processed, and objectives being served, simulation analysis can be quite useful for decision making in the management of critical infrastructural systems, such as communication network, transportation networks, logistics, and supply chain networks.

In communication networks, the core entity is data, whether it is the command and control data or content data, that is routed through "nodes" and "links" of the network for the purpose of computing and communications. The routers, servers, processors, memory and storage devices, and channels of data flow are resources that are utilized by the data for accomplishing the objectives of communications. The objectives or performance measures in communication networks are often measured in terms of throughput of data, say kilobytes per second (kbps), end-to-end or point-to-point latency in dataflow, congestion at a specific resource or set of resources, utilizations of nodes or links, calls dropped, or denial of service. Evaluation of communication networks using simulation modeling and analysis has been performed for over two decades [2,7], with dedicated software solutions, such as COMNET, OPNET, NETSIM, etc., developed for this purpose [67]. With an increasing popularity of simulation analysis utilized for communication systems, came the concern if these studies were being done credibly [86]. The use of appropriate pseudo-random number generators of independent uniformly distributed numbers and appropriately developed analysis of simulation output data is emphasized in these studies [33,62].

The transportation networks are critical for commerce and mobility of individuals. The developed economies have extensive and sophisticated transportation networks and infrastructure, while the emerging economies and developing countries recognize the significance of this infrastructure, and are making progress toward developing this infrastructure. Design and management of transportation networks and the traffic is significantly facilitated by simulation modeling and analysis [11,68,71,85]. Some models are developed for strategic planning, others are meant for tactical planning, while some modeling effort is done for network management and traffic control. As opposed to data being the entities flowing in communication networks, in transportation systems, entities are a variety of vehicles, trucks, rails, airplanes, helicopters, etc. Resources are all the equipment and human resources needed to design, build, run, and manage the transportation system.

Some of the simulation studies are focused at a micro level of a transportation system [73], such as the design of a traffic signal timer [22], or analyzing mixed traffic flow on a multilane road [27], while other studies may be designed for system-level analysis [42,46]. Transportation systems form a critical infrastructure and must be designed to continue to serve their necessary role under stressed and adverse circumstances, such as earthquakes, a terrorist attack, hurricanes, etc. Simulation study of transportation systems is also conducted under extreme-event conditions, such as evacuation under congested traffic [80], seismic response at seaports [77], strategies for response under extreme conditions [29], and development of capability assessment and emergency response for transportation systems.

In transportation systems, when the focus shifts from the carriers to what is being carried, numerous freight and logistics planning issues can be addressed using simulation modeling and analysis. Investigation can be an enterprise-wide integrated logistics and transportation analysis [108] or a more specific logistics optimization using simulation [63,78]. Similar to transportation systems, in logistics, planning the granularity may be reduced to a real-time micro-level issue [95] and emergency response [10]. Logistics coordination between firms creates the supply chains, where simulation models are utilized for improving design and operations [15,47,72,99] of supply chains.

Smart transportation systems interface with communication networks to create the transportation systems of the future. Intervehicle communication opens new possibilities for transportation systems, freight and logistics management, where simulation modeling and analysis is utilized and can continue to contribute [28,51]. Similarly, the convergence of communication networks with the power grid is creating smart grids of the future, as is the frontier of reliably integrating renewables for power generation on the grid. Simulation analysis is applied for addressing these issues as well [96]. These infrastructural domains provide rich contexts for application of simulation modeling and analysis for important decision making.

13.2.3 Asset Valuation and Allocation

Assets or investments that generate stochastic cash flows in the future must be valued to assess them as investment prospects. Such need for valuation may arise for equity valuation of a firm, project valuation for capital budgeting in a firm, or derivative instruments for the purpose of risk management. Additionally, when an investor needs to assess allocation of capital among potential investment options, simulation-based optimization can be useful should the problem characteristics require such an approach. Asset valuation, project valuation, derivative pricing, and portfolio selection are problems in finance that widely utilize simulation modeling and analysis for these financial decisions [16,23,31,36,41,45,75,76,97].

We consider pricing a financial derivative instrument as an example to illustrate the use of simulation analysis for valuation of such contracts. A plain-vanilla European call option, a specific derivative instrument, gives the buyer of the contract the right to buy the underlying asset at a future time, T, for a set price, K. The price, also called the option premium, is given by

$$c(t, S_t) = E_Q[e^{-r(T-t)}(S_T - K)_+ | S_t], \tag{13.1}$$

with the price of the underlying asset, S_t, evolving by the following equation:

$$dS_t = rS_t dt + \sigma S_t dB_t, \tag{13.2}$$

driven by the standard Brownian motion, B_t, in the risk-neutral world. Equivalently, the underlying asset price can be described as evolving by the following geometric Brownian motion process:

$$S_t = S_0 \exp\left(\left(r - \frac{1}{2}\sigma^2\right)t + \sigma B_t\right), \tag{13.3}$$

where B_t is the standard Brownian motion or the Wiener process. Once the exact solution of Equation 13.2 is substituted in Equation 13.8, after a few steps of derivation utilizing the lognormal distribution for the underlying asset price at T, we obtain the European call option price to be

$$c(t, S_t) = S_t \Phi(d_1) - Ke^{-r(T-t)}\Phi(d_2), \tag{13.4}$$

where $\Phi(x)$ is the cumulative distribution function of the standard normal distribution $(N(0, 1))$, and

$$d_1 = \frac{\ln(S_t/K) + (r + (\sigma^2/2))(T - t)}{\sigma\sqrt{T - t}}, \tag{13.5}$$

and

$$d_2 = \frac{\ln(S_t/K) + (r - (\sigma^2/2))(T - t)}{\sigma\sqrt{T - t}}, \tag{13.6}$$

$$= d_1 - \sigma\sqrt{T - t}. \tag{13.7}$$

This formula in Equations 13.4 through 13.6 is the famous Black–Scholes–Merton option pricing formula [44]. However, if the model describing the underlying asset price dynamics or the definition of the derivative contract is more complex, such as in a variety of exotic options, Equation 13.8 cannot be computed analytically. Therefore, simulation-based assessment becomes one of the promising alternatives.

One would generate a large sample of realizations $\{S_{iT}\}$ emanating from its value S_t at t, using Monte Carlo simulation based on the underlying asset price dynamics in the risk-neutral world. This sample would be used to estimate the expectation in Equation 13.8 by the sample mean estimator as follows:

$$\hat{c}(t, S_t) = e^{-r(T-t)} \sum_{i=1}^{N} \frac{(S_{iT} - K)_+}{N} \tag{13.8}$$

for the plain-vanilla European call option, and with other pay-off functions for other option types.

In a portfolio allocation strategy, if the total resources available for investment initially are W_0, the investor must allocate the wealth among N stocks. The dynamic investment decisions are $\{w_i(t)|t \in \mathcal{T}\}$, which must be made to achieve the investor's investment objectives. Let us say the initial price per share of the N stocks is given by $S_i(0); i = 1, \ldots, N$. We can model the price evolution of the N stocks by the following stochastic differential equation model:

$$dS_i(t) = \mu_i S_i(t)dt + \sigma_i S_i(t)dB_i(t), \tag{13.9}$$

where the N standard Brownian motion processes, $B_i(t)$, are correlated. The correlation between the N Brownian motion processes can be described by

$$E[dB_i(t)dB_j(t)] = \rho_{ij}dt, \text{ for } i = 1, \ldots, N; j = 1, \ldots, N. \tag{13.10}$$

The parameter ρ_{ij} is the correlation coefficient between increments of the Brownian motion processes $B_i(t)$ and $B_j(t)$, with $\rho_{ii} = 1$ for all $i = 1 \ldots N$. If the investment weight for the ith stock at a time, t, is maintained at $w_i(t)$, then we can derive that the wealth, $W(t)$, will accumulate or evolve by the following equation:

$$dW(t) = \sum_{i=1}^{N} w_i(t)W(t)(\mu_i dt + \sigma_i dB_i(t)). \tag{13.11}$$

The dynamic investment strategies, $\{w_i(t)|t \in \mathcal{T}\}$, can be evaluated by the investor's utility for accumulated wealth. A utility function-based performance measure can be constructed to evaluate the investment strategies as follows:

$$U(W) = \int_0^T E[u(W(t))]. \tag{13.12}$$

Here again, in some simpler cases of the utility function, analytical solutions are obtainable. In more complex cases, optimal investment strategy can be sought by simulation-based optimization approaches.

13.2.4 Risk Management

In banking and insurance industries, the business model of the firms provides a fundamental economic infrastructure for fund allocation and risk management, respectively, for a modern society. Therefore, a regulatory framework is put in place to ensure proper functioning of these sectors. Besides the regulatory requirements for risk management, for internal risk assessment and control, banks and insurance firms utilize extensive risk management framework, processes, and methodologies.

For investment assets in banking or insurance, appropriate market risk assessment and management is necessary to ensure that the firm's risk exposure is suitable for the overall riskiness of the firm. Depending on the firm's assets and investment strategy, interest rates, currency, equity, and commodity market risks may be relevant for inclusion in this analysis. Appropriate risk measures and corresponding risk limits are utilized for guidelines for risk monitoring and control across the firm [25,41].

Similarly, individual loans and loan portfolios must be assessed for their credit risk evolution for a bank's credit risk exposure and credit risk management strategy [30,41,55,101]. Risk aggregation across business lines and activities of a firm allows determining a firm's overall riskiness, which is utilized for judging the capital the firm should possess to ensure solvency in extreme stress conditions. The regulatory, or internally imposed, minimum capital requirement is the regulatory, or economic, capital a bank maintains. These risk assessments and quantifications is a rich consumer of simulation modeling and analysis [36,41]. An insurance firm must conduct detailed actuarial analysis for the loss distribution underlying its insurance products, in order to determine the appropriate pricing, underwriting, and loss reserves determination. Finally, both in banking and insurance, securitization is an important risk management strategy, which utilizes simulation methodologies [52,105].

13.2.5 Selecting Input Distributions and Processes

As discussed in the above sections, the problem domains addressed using simulation modeling and analysis could be manufacturing or service systems, communication or transportation networks, banking and insurance, or one of the other domains not discussed above, such as biological systems, environmental or ecological systems, behavioral or social sciences, and other engineering applications. In each domain, the problem must be framed in the context of a model for the system, where appropriate objects/entities or stochastic factors/variables must be defined, along with the processing stations/resources and other controlling factors.

Model development requires identification of the important sources of randomness, such as arrival rates, arrival times, service times, rates of change of stochastic factors, resource availability, downtime or breakdown, and interaction between the sources of randomness. The modeler must decide if the randomness requires a discrete or a continuous distribution representation. In some cases, this may be obvious by the nature of the variable in question, whereas in others, the choice may be made based on tractability of the model and/or familiarity of the modeler with the choice. Dynamic evolution of stochastic variables will need to be modeled using stochastic processes, such as in the examples discussed in Sections 13.2.1 through 13.2.4, or arrival processes for describing movement of entities/objects between resources or service stations.

Input distribution and process selection must be guided by some amount of data available from the system under study. In case of design studies, data must be available from trial runs, test beds, or best substitutes. Based on these data, tests must be conducted for their independence, as independent, identically distributed (i.i.d.) observations are a prerequisite for the calibration of model distributions. In case of stochastic processes, these independence tests are often applied to increments of the process, that is, $S_{t+\Delta} - S_t$. Autocorrelation plots and scatter plots are effective tools to visually and simply assess independence of data.

In selecting model distributions, as stated earlier, a decision is required for the representation of realizations of the random variables as discrete or continuous valued. Other guidance for selection of input distribution is in terms of range of values one expects for the random behavior in the real system, that is, are the values positive, lower or upper bounded, or infinite ranged? Plotting histograms and box plots of the data obtained from the real system for the sources of randomness is instructive to assess the shape of distribution suitable for the random variables, such as skewness and modality. Guided by these observations, one can proceed to select a standard probability distribution from the nonnegative, bounded, or unbounded class of distributions or opt to work with empirical (or historical) distribution.

For the dynamic evolution of randomness using stochastic processes, the model choice is for a discrete-time or a continuous-time process. The former represents a variability that is observed only at discrete time points, defined by the behavior of object/entities, resources/processing stations, or stochastic factors, while the latter is applicable when such logical discrete time points are not obvious and the variables evolve continuously. The behavior of the dynamic systems is captured by time increments and their properties. The time increments may have homogeneous distributional characteristics or display nonhomogeneity, which guides the selection of stochastic processes. Poisson process, which possesses independent increments, is a popular model for arrival processes, as discussed earlier. In many cases, such as in communication and transportation networks, the arrival rate for objects/entities is not homogeneous through times of the day. Therefore, nonhomogeneous Poisson process model is considered. In finance, the Brownian motion process is widely used to model log-return of prices, as seen in Equation 13.3. Stochastic calculus developed for Brownian motion and jump-diffusion processes allows exploring a vast variety of continuous-time stochastic processes, which we discuss in Section 13.8.

A static or a dynamic model for stochastic behavior of a system is characterized by some parameters for its complete representation in the model. This specification, which defines location, scale, shape, etc. characteristics of the randomness, is developed by calibrating the chosen distributional or stochastic process model to the data available for the variability they represent. Some of the methods utilized for calibration are discussed later in this chapter in Section 13.8.3. Once a calibrated model is constructed, the model must be implemented by constructing a mechanism for simulating the randomness inherent in the model. This is achieved by utilizing specialized methods and routines for generating appropriately distributed random variates. This is the topic of the next section. Moreover, the model must be validated for goodness of its representation of the real system. A variety of goodness-of-fit tests are available for this purpose, some of which are discussed later in Section 13.3.5, and the reader is referred to other sources for more details [8,9,65].

13.3 Random Variates Generation

Simulation is to copy; therefore, for simulating random variables, their inherent nature of randomness—generating random outcomes—needs to be copied. Before the advent of digital computers, random numbers were directly obtained from actual random processes, such as rolling a die or tossing a coin, electronically by the noisy output of a valve. Such numbers were not statistically reliable, and a particular sequence of random numbers could not be reproduced for comparative studies. Today digital computers allow the implementation of simple deterministic algorithms to generate sequences of random variables, quickly and reproducibly. Such numbers are, strictly speaking, not truly random, but with sufficient care they can be made to resemble random numbers by most of their properties. Hence, these numbers are called pseudo-random numbers.

Actually, pseudo-random numbers is a term reserved for random numbers corresponding to the simplest distribution, the uniform random variable, $U(0, 1)$. Random numbers generated for other distributions, such as normal, lognormal, Weibull, etc., are called random variates. We will first describe methods for generating pseudo-random numbers, followed by some methods for generating other random variates. This order of presentation is particularly meaningful, since pseudo-random numbers are the building block for all other random variates. As such to a decision analyst, it may not be essential to know the mechanics of the random number generator used; however, for an advanced use of simulation, this knowledge is important so that appropriate modifications to their use may be developed, as we will see later in this chapter. Two basic deterministic, recursive methods for creating sequence of pseudo-random numbers follow.

13.3.1 Linear Congruential Generator

The linear congruential generator (LCG), a basic iterative method for generating pseudo-random numbers, is based on the primary relation as follows:

$$X_{n+1} = aX_n + b \quad (mod\ c), \tag{13.13}$$

where a, c (>0) are positive integers, and b (≥0) is a nonnegative integer. The recursion needs to start from a number, X_0, which is called the "seed." The "mod" operator is short for "modulo," representing the remainder after division of a number by another number. For example, $7(mod\ 3) = 1$, $9(mod\ 5) = 4$, etc. Therefore, the LCG recursion generates a sequence of numbers taking integer values from 0 to $c - 1$, the remainders when $aX_n + b$ is divided by c. When a, b, and c are picked appropriately,

$$U_n = X_n/c \tag{13.14}$$

seem uniformly distributed in the unit interval [0, 1]. For example, $a = 16,807 = 7^5$, $b = 0$, $c = 2^{31} - 1$ (prime) is a relatively carefully selected set of values for these parameters.

The congruential algorithm is widely used; however, it displays a looping characteristic captured by the "period" of a random number generator. The period of a random number generator is defined as the smallest positive integer p which satisfies $X_{i+p} = X_i$ for all $i > k$, for some $k \geq 0$. In a large-scale simulation, this becomes a major weakness. For example, a 32-bit LCG will have a period less than 2^{32}, which will get exhausted in just a few minutes in a large-scale simulation. Therefore, a, b, and c need to be chosen with care, with not only long period in mind, but also good statistical properties, such as apparent independence, computational and storage efficiency, reproducibility, and facilities for separate streams.

13.3.2 Lagged Fibonacci Generator

The alternatives to LCG random number generator are also based on iterative procedures, developed with the goal of achieving the desirable properties listed above. One of them is the lagged Fibonacci generator, where the recurrence relation for the generator is

$$X_n = X_{n-r} \quad op \quad X_{n-s}, \tag{13.15}$$

where s and r are the lags with $0 < s < r$ and $n \geq r$ and op is a binary operator. For example, the binary operator could be "addition (mod c)" or "subtraction (mod c)." If op is "addition (mod c)," the lagged Fibonacci generator will become $X_n = X_{n-r} + X_{n-s}(mod\ c)$. It should be noted that the lagged Fibonacci generator iterations begin from $n \geq r$; therefore, the first r iterates, $\{X_k; k = 1, \ldots, r\}$, have to be obtained from some other scheme. The LCG can be used for initialization of the lagged Fibonacci generator.

A good property of the lagged Fibonacci generators is that extremely long periods are possible with these generators, and several have been shown to exhibit good global properties if the parameters are chosen carefully. Excellent references for more detailed information on random number generators are books by Law and Kelton [66], Glasserman [36], and Knuth [61]. These books offer rigorous discussion of properties and implementation issues for various random number generators.

As stated earlier, while performing a simulation-based analysis, it is good to know the random number generator being used. A particular random number generator may have some artifacts that can sometimes result in misleading conclusions. Hence, it may even be advisable to test run the simulation model with two different random number generators to make sure the results are genuine, and not artifacts of the properties of a generator.

13.3.3 Generation of Discrete Random Variates

Pseudo-random numbers are the building block for generating all other random variates. Therefore, the basis for generating random variates with good properties for all other distribution models is using good pseudo-random numbers. We will now describe methods for generating variates by other distribution models, starting with simple discrete distribution that has finite number of outcomes, 2 or more. The common aspect of these methods for generating random variates is an appropriately designed transformation of the random numbers.

13.3.3.1 n-Outcome Random Variate

Two-point random variable, X, is one that takes two values $x_1 < x_2$ with probabilities p_1 and $p_2(= 1 - p_1)$, respectively. Even though this is a very simple distribution, it is popularly used in describing time-dependent stochastic models, such as the random walk process. The transformation of pseudo-random numbers, $U(0, 1)$, that can be used to generate random variates for a two-point random variable is

$$X_n = x_1 \quad \text{if } 0 \leq U_n \leq p_1, \tag{13.16}$$

$$= x_2 \quad \text{if } p_1 < U_n \leq 1, \tag{13.17}$$

for $n \geq 1$. Therefore, for every pseudo-random number generated, a random variate for the two-point random variable model is produced. This is not always the case for other methods and for other distribution models.

The method for the two-outcome random variable can be easily extended to N-state random variable taking values $x_1 < x_2 < \cdots < x_N$, with nonzero probabilities $p_1, p_2, \ldots, p_N$, respectively. From the rules for probability distributions, we have that $\sum_{i=1}^{N} p_i = 1$. We define $s_0 = 0$,

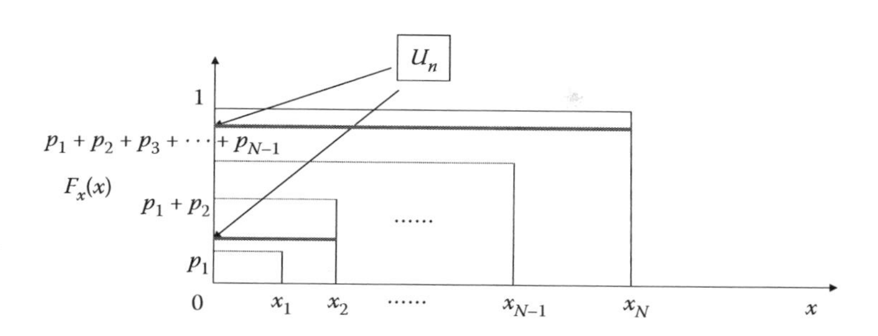

FIGURE 13.4
N-outcome discrete random variate generation.

$s_j = \sum_{i=1}^{j} p_i$ for $j = 1, \ldots, N$, and using these quantities generate the N-point random variates as follows:

$$X_n = x_{j+1} \quad \text{if } s_j \leq U_n \leq s_{j+1} \quad j = 0, \ldots, N, \tag{13.18}$$

for $n \geq 1$. Figure 13.4 is a pictorial depiction of the above algorithm, which shows that the above algorithm is effectively an inversion of the cumulative mass function.

Not all discrete random variables have a finite number of outcomes; some can have infinite number of outcomes. If a discrete random variable has infinite outcomes, the above transformation has to be modified into an iterative procedure. We describe this next in the context of the Poisson random variate.

13.3.3.2 Poisson Random Variate

Poisson random variate is a counting discrete random variate with an infinite number of outcomes starting from $n = 0$. Poisson model is widely used for arrival process in many manufacturing and service systems and in communication and transportation networks, as discussed earlier. The possible outcomes of a Poisson random variable are $0, 1, 2, \ldots$. Given the parameter λ, the probability that there will be j number of occurrences of the event of interest by this stochastic model is given by $p_j = e^{-\lambda}\lambda^j/j!$ (this is the probability mass function of the Poisson distribution). The following algorithm generalizes the transformation for finite outcome discrete random variable to generate random variates by the infinite outcome Poisson distribution.

Step 1: Initialize Set $s_1 = 0$, $s_2 = p_1$, $j = 1$ and generate a uniform random number U_n;

Step 2: Check If $s_1 \leq U_n \leq s_2$, then $X_n = j$.
And exit.

Step 3: Else Update $s_1 = s_2$, $s_2 = s_2 + p_{j+1}$, $j = j + 1$;
Go to Step 2.

Therefore, the method transforms every pseudo-random number U_n generated into a Poisson random variate, X_n, for all $n \geq 1$. However, the generation of random variates now needs more work than was needed in the n-outcomes random variate case.

13.3.4 Generation of Continuous Random Variates

We now move from discrete random variables to considering continuous random variables. We have, in fact, already seen one example of a continuous random variate generation, the uniform distribution, $U(0, 1)$. The uniform random variates generated by the random number generators were key to generating the discrete random variates. They will continue to be so for continuous random variates. Let X be a continuous random variable with a probability distribution function, $F_X(x)$. Therefore, by the property of probability distribution functions, $F_X : R \to [0, 1]$. This basic property of all probability distribution functions is exploited in the first method presented, the inverse transform method.

13.3.4.1 Inverse Transform Method

The principle behind the inverse transform method for generating random variates for a random variable, X, is as follows. Let U $(0 < U < 1)$ be a uniform random variate generated by a random number generator. If we can find an X such that $X(U) = F_X^{-1}(U)$ for every uniform random variate, U, generated, then X will be the desired random variate. Here, F^{-1} is the inverse of the probability distribution function of the random variable, X, assuming it exists and can be computed with sufficient ease. In general, if the cumulative distribution function of the random variable, X, is not continuous, we will define $X(U) = \inf\{x : U \leq F_X(x)\}$. Figure 13.5 shows the principle behind the inverse transform method pictorially.

The inverse transform method is very effective for the generation of random variates for the exponential random variable model. The probability distribution function for the exponential random variable is $F_X(x) = 1 - e^{-\lambda x}$, for the parameter $\lambda > 0$. Applying the inverse transform method gives

$$X_n(U_n) = F_X^{-1}(U_n) = -(ln(1 - U_n))/\lambda = -ln(U_n)/\lambda \quad \text{for } 0 < U_n < 1, \tag{13.19}$$

where $n \geq 1$. The last equality in Equation 13.19 is true since, if U_n is $U(0, 1)$, then so is $1 - U_n$. Therefore, given a sequence of random numbers, U_n, generated by a random number generator, $X_n = -ln(U_n)/\lambda$ is a sequence of exponentially distributed random variates. This is a very simple and efficient transformation.

In principle, the inverse transform method should be all one should need to know about continuous random variates generation. However, this does translate to reality since computing inverse of many cumulative distribution functions is not easy. Consider the normal distribution as an example, where the integral of the probability density function must be evaluated numerically. Therefore, inverting the cumulative distribution function is not an efficient option. In such cases, other random variate generating techniques are needed.

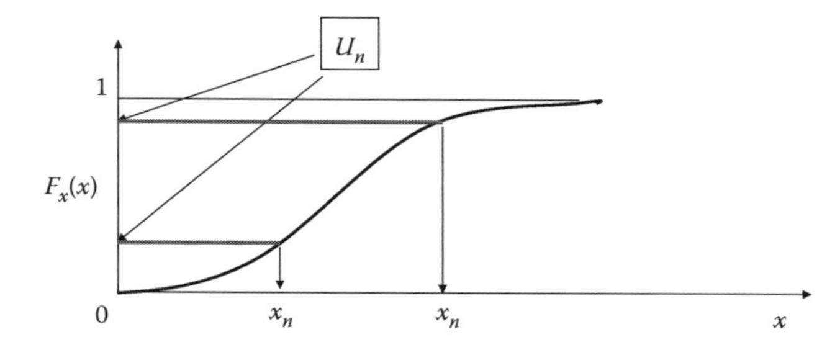

FIGURE 13.5
A pictorial depiction of the principle behind the inverse transform method.

13.3.4.2 Acceptance–Rejection Method

Acceptance–rejection method, as the name suggests, builds an iterative procedure by which in every iteration, the output is either accepted as a valid random variate outcome or is rejected. The method is useful when the inverse transform method or any other direct method is not available for a probability distribution model. For a random variable, X, with a probability density function, $f_X(x)$, the method relies on a second random variable, Y, with a probability density function, $g_Y(y)$. The key determinant for the selection of the second random variable, Y, is that a method for generating random variates for Y should be available. The acceptance–rejection method will use this fact as a basis for generating random variates for X by the probability density, $f_X(x)$.

The method requires picking a constant, c, such that $f_X(y)/g_Y(y) \leq c$, for all y. We have assumed for the definition of the constant c that the two random variables, X and Y, are defined on the same sample space, say Ω. The algorithm behind the acceptance–rejection method is as follows.

13.3.4.2.1 Acceptance–Rejection Algorithm

Step 1: Generate Y_n with probability density $g_Y(y)$.

Step 2: Generate a random number U_n.

Step 3: If $U_n \leq f_X(Y_n)/(cg_Y(Y_n))$, set $X_m = Y_n$ (Accept); $m \leftarrow m + 1$.
Otherwise Reject, and *return to Step 1*.

In Figure 13.6, we give a pictorial description of the acceptance–rejection method. Every time a Y_n is generated by its probability density, $g_Y(y)$, for which we have an easier method for generating random variates, we also generate a uniform random variate, U_n. The probability that $U_n \leq (f_X(Y_n)/(cg_Y(Y_n)))$ is $f_X(Y_n)/(cg_Y(Y_n))$, which is the probability that the Y_n generated could also be a realization of X generated by $f_X(x)$. Hence, Y_n is accepted as a realization of X_n in this scenario, but rejected otherwise. Therefore, for every random variate generated for density, $f_X(x)$, there are at the least two uniform random numbers needed, one to generate Y_n and the other for the acceptance–rejection decision.

While one may be concerned about the efficiency of this method on the grounds of frequency of rejection versus acceptance, it is reassuring that the following result can be proven for the acceptance–rejection method.

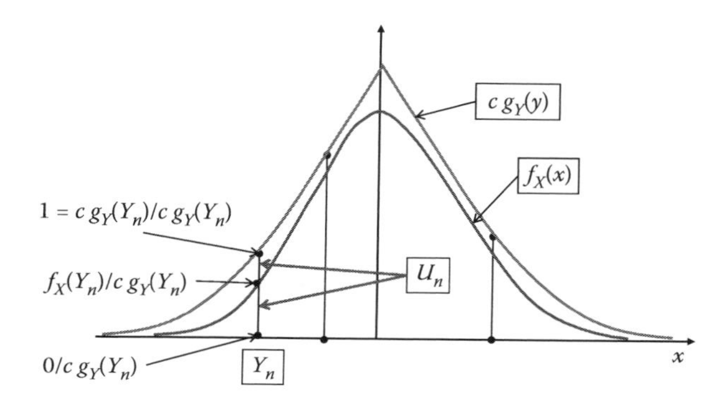

FIGURE 13.6
A pictorial depiction of the principle behind the inverse transform method.

Theorem 13.1

The random variates generated by the acceptance–rejection method, in fact, have the density $f_X(x)$. The number of iterations of the algorithm that are needed to create a desired random variate is a Geometric random variable with mean c. For the proof of this result, the reader may refer to Nelson [79].

Therefore, the acceptance–rejection is in contrast to the inverse transform method, where every uniform random number generated produces a random variate by the desired distribution. In the acceptance–rejection method, the number of iterations before obtaining the next random variate by the desired distribution is not deterministic. The lower the value of c, the fewer may be the iterations needed before successfully generating a random variate for X. The trick, therefore, is to find the smallest constant, c, such that $f_X(y)/g_Y(y) \leq c$.

Let us consider an example to demonstrate the application of acceptance–rejection method. Let $g_Y(y) = \exp(-|y|)/2$ defined on $(-\infty, \infty)$, which is the double-exponential density. The inverse transform method can be easily modified to generate random variates for the double-exponential distribution, as follows:

Step 1: Generate two random numbers, U_1 and U_2.

Step 2: Let $Y_n = -ln(U_1)$.

Step 3: If $U_2 \leq \frac{1}{2}$, then set $Y_n \leftarrow -Y_n$.
Otherwise return Y_n.

If $f_X(x) = (\exp(-x^2/2))/\sqrt{2\pi}$, which is the probability density function for the standard normal distribution. As stated earlier, for the normal distribution, the integrals of the probability density function to compute the distribution function must be evaluated numerically. Therefore, inverting the probability distribution function to apply the inverse transform method is not an efficient option. In order to apply the acceptance–rejection method, we need to find a constant c, such that $f_X(y)/g_Y(y) \leq c$ for all y. It can be shown that $f_X(y)/g_Y(y) \leq 1.3155$ in this case; hence, we can take $c = 1.3155$.

The acceptance–rejection method is a good method to generate random variates for the normal distribution; however, the normal distribution is such a popular distribution that specialized methods have been developed for it. We study some of these next.

13.3.4.3 Normal Random Variate

The normal distribution is a popular random variable model and will be used to construct stochastic processes in later sections. Owing to the frequent use of the normal distribution model, we first describe the Box–Muller specialized method for the generation of normal random variates.

13.3.4.3.1 Box–Muller Method

The Box–Muller method avoids using the probability distribution function for the generation of normal random variates. It instead uses the fact that if U_1 and U_2 are two independent $U(0, 1)$ uniformly distributed random variables, then G_1 and G_2 defined by the following transformation are two independent standard normal (Gaussian) random variates:

$$G_1 = \sqrt{-2\ln(U_1)}\cos(2\pi U_2), \tag{13.20}$$

$$G_2 = \sqrt{-2\ln(U_1)}\sin(2\pi U_2). \tag{13.21}$$

If the two uniforms random numbers, U_1 and U_2, are truly independent, we would have generated two independent standard normal variates. However, if the two uniform random numbers are two successive random numbers from a congruential random number generator, this will not be the case; the two normal random variates, (G_1, G_2), will make a spiral structure in the $R \times R$ space. Therefore, to be able to use both the normal random variates generated in one calculation, where independent normal random variates are necessary, the two uniform random numbers must come from different linear congruential streams—corresponding to different seeds.

The transformation underlying G_1 and G_2, that is, Equations 13.20 and 13.21, is simple, but there is one disadvantage of the Box–Muller method. The transformations require computing trigonometric functions (sin, cos), which are somewhat computationally demanding, thus resulting in inefficiency in normal random variates generation. An alternate method that bypasses this computational burden is the Polar–Marsaglia method.

13.3.4.3.2 Polar–Marsaglia Method

The method is based on the fact that if U is a uniformly distributed random variable, $U(0,1)$, then $V = 2U - 1$ is $U(-1,1)$, that is, V is uniformly distributed on the interval $(-1,1)$. If V_1 and V_2 are independent and distributed as $U(-1,1)$, obtained by transforming as above two independent standard uniform random variables, U_1 and U_2, then we define $W = V_1^2 + V_2^2$, when $V_1^2 + V_2^2 \leq 1$. Therefore, W lies in the unit circle.

It can be shown that W so defined is uniformly distributed, $U(0,1)$. This can be visualized by considering the pair (V_1, V_2) as points on the $R \times R$ plane. Since the points (V_1, V_2) are equally likely to fall anywhere in the unit circle inscribed within the box, $[-1,1] \times [-1,1]$, their distance from the origin, $(0,0)$, is equally likely to be between 0 and 1. Similarly, the angle made by (V_1, V_2) on the positive x-axis, θ, is uniformly distributed as $U(0, 2\pi)$, and is independent of W. See Figure 13.7 for a pictorial representation of the (W, θ) pair constructed.

The square defined by the values (V_1, V_2) has an area of 4, whereas the area of the circle inscribed by W is $\pi r^2 = \pi$. Therefore, the area of the circle is $\pi/4$ fraction of the area of the square in Figure 13.7. The point (V_1, V_2) will fall inside the circle with a probability of $\pi/4$. When that happens, $W = V_1^2 + V_2^2 \leq 1$, and we can use the random numbers, V_1, V_2, to make the following computations:

$$\cos\theta = V_1/\sqrt{W}, \tag{13.22}$$

$$\sin\theta = V_2/\sqrt{W}. \tag{13.23}$$

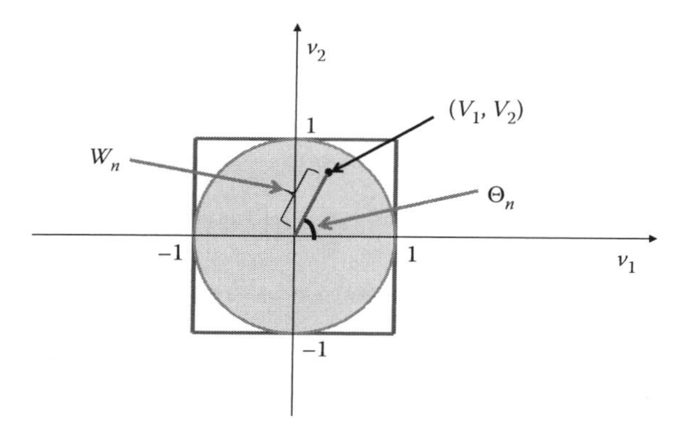

FIGURE 13.7
A pictorial depiction of the construction of the Polar–Marsaglia method.

This allows rewriting the Box–Muller method by replacing the sin and cos functions as follows:

$$G_1 = V_1\sqrt{-2\ln W/W}, \tag{13.24}$$

$$G_2 = V_2\sqrt{-2\ln W/W}. \tag{13.25}$$

In the above modification of the Box–Muller method, θ replaces $2\pi U_2$, since it has the same distribution as $2\pi U_2$. W plays the role of U_1 in the Box–Muller method, again due to distributional similarity.

The Polar–Marsaglia method can also be seen as an example of the acceptance–rejection method. As seen in the previous section, there are other acceptance–rejection methods for the generation of normal random variates, where acceptance/rejection is the central theme of the method. Here, the acceptance/rejection was implemented depending on whether the (V_1, V_2) pair fell inside the unit circle or not. If it fell inside the circle, the algorithm could continue and generate a pair of independent normal random variates; otherwise not. Therefore, although a proportion of the generated uniformly distributed random numbers are discarded, this method is often computationally more efficient than the Box–Muller method, especially when a large quantity of normal random variates are to be generated.

In simulation experiments, since often a very large number of random variates need to be generated to serve the purpose of the simulation study, the efficiency of the random variates generator algorithms is of significant concern [5].

13.3.4.3.3 Generation of Multivariate Normal

The Box–Muller and Polar–Marsaglia methods give independent normal (Gaussian) random variates with zero mean and standard deviation of 1, when uniform random numbers U_1 and U_2 used are independent uniform, $U(0, 1)$. In practice, often a pair, (X_1, X_2), or an n-dimensional set, $(X_1, \ldots, X_n)$, of correlated Gaussian random variates are required. Let X_1, X_2 be jointly Gaussian with mean μ_1, μ_2, and variance σ_1^2, σ_2^2, respectively, and covariance of $\rho\sigma_1\sigma_2$. We now describe how the independent standard normal random variates can be transformed to be a correlated pair of normal variates of a given correlation structure.

To begin with, assume μ_1, $\mu_2 = 0$ and σ_1, $\sigma_2 = 1$. We first generate Y_1, Y_2, and Y_3 random variates that are independent, standard normal, $N(0, 1)$. Using these three independent standard normal random variates, we define the following:

$$X_1 = \sqrt{1-|\rho|}Y_1 + \sqrt{|\rho|}Y_3, \tag{13.26}$$

$$X_2 = \sqrt{1-|\rho|}Y_2 \pm \sqrt{|\rho|}Y_3, \tag{13.27}$$

where in Equation 13.27, "+" is used when $\rho \geq 0$ and "−" for $\rho < 0$. The (X_1, X_2) pair thus obtained is jointly standard normal, with desired correlation structure. In order to get random variates with the general mean and standard deviation characteristics, the following transformations are applied: $X_1 \leftarrow \mu_1 + \sigma_1 X_1$ and $X_2 \leftarrow \mu_2 + \sigma_2 X_2$.

The method described above for bivariate normal random variates with general correlation structure can be extended for n-dimensional $(n > 2)$ correlated normal random variates. This requires using the Cholesky factorization of the correlation matrix, given as $Corr = RR^T$. Once n-dimensional standard, independent normal variates, Y, are produced, the desired correlation structure is introduced as $X = RY$. Following this, as done for the two-dimensional case, the required mean and variance can be introduced for each element of X.

13.3.4.4 Chi-Square and Other Random Variates

Several additional specific distribution models may be picked for describing the stochastic behavior of entities, objects, and resources in different application contexts. Some of these, such as the lognormal distribution and Weibull distribution, are extensions of more basic distributions. Therefore, if a method is known for generating random variates for the simpler model, the method can be extended to create random variates for the extended models. For instance, we discussed a few methods for generating normal random variates in Section 13.3.4.3. If a normal random variate, X_n, is produced by any of these methods, after appropriately choosing its mean and standard deviation, the desired lognormal random variate can be produced as $V_n = \exp(X_n)$.

The Weibull distribution is a generalization of the exponential distribution [41]. The inverse transform method is efficiently utilized for generating random variates for the exponential distribution. The same can be applied to generate Weibull random variates. Similarly, the gamma distribution, $\Gamma(x; \alpha, k)$, is a sum of exponential random variables. This relation can be utilized for generating gamma random variates.

Finally, we consider the chi-square, χ_d^2, distribution, where d is a positive integer denoting the degrees of freedom. If $Y_1, \ldots, Y_d$ are independent standard normal random variates ($N(0,1)$), then $Y_1^2 + \cdots + Y_d^2$ has a χ_d^2 distribution. Moreover, for constants $\alpha_1, \ldots, \alpha_d$, the distribution of $\sum_{i=1}^{d}(Y_i + \alpha_i)^2$ is noncentral chi-square, $\chi_d^2(\nu)$, with d degrees of freedom and noncentrality parameter, $\nu = \sum_{i=1}^{d} \alpha_i^2$. This relationship between chi-square and normal distributions can be utilized for generating chi-square random variates. In general, however, for more general values of parameters, such as degrees of freedom, d, in case of chi-square and scale-shape (α,k) parameters in case of the gamma model, more specialized methods for random variates generation would be needed [66].

We have so far seen some basic methods for random variates generation, both for discrete and continuous stochastic models. The intention of this overview was to give the reader a sense of the building blocks behind the sophisticated simulation software available today, and the underlying assumptions and issues. We next move to discussing testing of random variates for their quality and accuracy.

13.3.5 Testing Random Variates

Once appropriate methods for random variates generation are created, they may be used to perform the required experiments for the simulation study. The quality and reliability of the experiments, however, will rely on the correctness and quality of random variates generated. Therefore, testing the outputs of the random number generators and random variates generating algorithms is an important step for assuring the validity of a simulation analysis. Since the random numbers are the basic building block for all other random variates, it is necessary to have independence wherever this is a requirement. We begin with testing independence of the uniform random numbers.

13.3.5.1 Testing for Independence of Random Numbers

Statistical independence is usually a difficult property to test. No single test for it is totally satisfactory and fully reliable. Among the random number generators, a generator like the LCG is an important case to consider to test independence of its output, since in the linear congruential iterations, each random number is determined only by its immediate predecessor. If the generator fails one of the independence tests, a remedial strategy may be developed.

Different plotting techniques is one approach for testing independence. Plotting techniques, such as autocorrelation plots and scatter plots, are some of the visual tests that may be used. A simple test

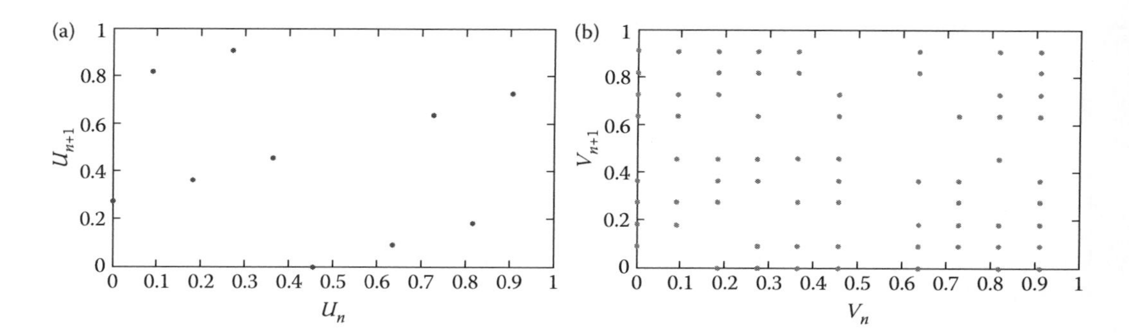

FIGURE 13.8
Display of output from a linear congruential generator. (a) The1000 numbers generated lie on three parallel lines. (b) The 1000 numbers after implementing shuffling.

for independence involves plotting the successive pairs (U_n, U_{n+1}) for $n = 1, 2, 3, \ldots$ as points in the unit square of the $R \times R$ plane, with the U_n on the x-coordinate and U_{n+1} on the y-coordinate. If the random numbers generated by the random number generator are independent, the plots will scatter about with no apparent patterns.

Applying this plotting technique to the output from an LCG reveals that the points lie on one of c different straight lines of slope a/c. We display this for a simple generator to enhance the pattern, with $a = 6, b = 3, c = 11$, in the left panel of Figure 13.8. For a more sophisticated LCG with parameters suggested in Section 13.3.1, a large number of random numbers generated should fairly evenly fill the unit square. The presence of patches without any of these points is an indication of bias in the generator. These patterns can be eliminated by applying additional remedial procedures to the numbers produced by a generator.

13.3.5.1.1 Shuffling Procedure

A shuffling procedure attempts to introduce greater "randomness" in the deterministic output from a random number generator. Sample steps of a shuffling procedure can be as follows:

1. Generate 20 or more random numbers from a random number stream: $\{U_1, \ldots, U_{20}\}$.
2. Pick one of these 20 numbers with equal probability, 1/20. This will require generating another random number from a different stream, W_1. The randomly generated index ranging from $1 \ldots 20$ is obtained as, $j = \text{floor}(20 * W_1) + 1$. Assign the random number from the $U(1:20)$ stream corresponding to the randomly picked index, say U_j, to the shuffled random number stream V, that is, $V_1 = U_j$.
3. Replace the random number picked from the U stream, say U_j, with the next random number in the U sequence, that is, U_{21}.
4. Repeat Steps 2–3. This results in a shuffled sequence of random numbers, V, of the initial sequence, U.

Note that this procedure requires two times more random numbers generated than an unshuffled sequence of the same length. This is because a random number stream also needs to be generated to determine the index of a randomly picked number from a set of (20 or so, in the above setup) numbers. Shuffling procedures have been found to be effective in reducing patchiness in poor generators. They also provide a possibility to lengthen the periods when using LCGs.

In the right panel of Figure 13.8, we apply a shuffling procedure to the output of the left panel from a simple LCG. The shuffling improves the output, although this does not make this generator attractive for practical use. Shuffling improves the independence properties of a random number generator, and maintains reproducibility as long as randomly picking index in the shuffling procedure can be reproduced. Experiments being reproducible is a desired property for a simulation environment.

Besides plotting strategies to test independence, other quantitative hypothesis tests can also be conducted to test the independence of the output of a random number generator. The most useful hypothesis tests for this purpose are the Runs tests, such as Runs above and below the Median, Runs Up, Runs Down, etc. Beyond independence, we will also need to test if the random variates generated are truly representative of the desired distribution.

13.3.5.2 Testing for Correctness of Distribution

Testing accuracy of the distribution of random variates generated is not any different from statistical inference for testing the distribution of data acquired from real-world experiments. We had addressed this in Section 13.2.5 in the context of input distribution selection. As in standard statistical inference for real-world data, one simple way to test correctness of distribution is to plot a histogram for the output of a random variates generator and compare it visually with the graph of the true density function they are supposed to simulate. This is the simplest assessment possible, and remains a subjective evaluation.

A less subjective evaluation is to produce a probability plot for the simulated data. For instance, if the random variates are supposed to represent a normal probability distribution, one can plot the data in a normal probability plot. If the points fall effectively on a straight line, the hypothesis that the data represent a normal probability model cannot be rejected. In Figure 13.9, we plot lognormal and Weibull probability plots in the left and right panel, respectively, for lognormal data generated using the Polar–Marsaglia method. The lognormal random variates are obtained by taking the exponential of the output of the Polar–Marsaglia method, that is, $\exp(G_1)$, where G_1 is defined in Equation 13.24. The lognormal probability plot is arguably a straight line, and would pass the so-called *fat-pencil* test, that is, if a fat-pencil were put on the points in the plot, it will hide all the points. However, the Weibull probability plot, which was chosen to create a contrast, is by no means a straight line.

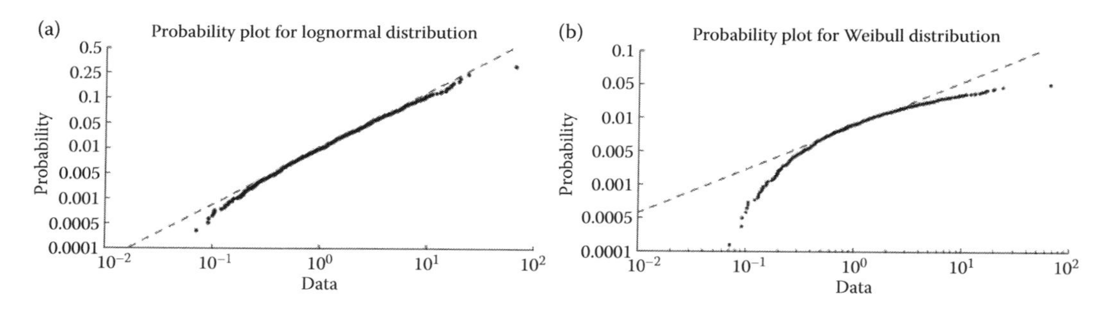

FIGURE 13.9
Display of probability plots. (a) Lognormal probability plot. (b) Weibull probability plot.

For more rigorous quantitative evaluation, we would set up standard statistical tests for testing validity of a hypothesized distribution for the simulated random variates data. The tests are developed based on quantifying the "distance" between the histogram of the simulated data and the true probability density. The first test is more suitable for discrete distribution models.

13.3.5.2.1 χ^2 Goodness-of-Fit Test

Using the designed random variates generator, generate a large number, N, of independent and identically distributed random variates. We form a cumulative frequency histogram $F_N(x)$ for these random variates, which we wish to compare with the true cumulative mass function, $F(x)$. The procedure requires subdividing the random variates data into $k+1$ mutually exclusive categories and count the numbers, $N_1, N_2, \ldots, N_{k+1}$, of them falling into each of these categories, with $N = N_1 + N_2 + \cdots + N_{k+1}$.

Breaking the data into $k+1$ categories works easily for discrete random variables, with possibly each discrete outcome of the random variable making a category. For continuous random variables, one needs to set arbitrarily chosen breakpoints. Following the categorization of the data, the *true* probability for each of the categories is computed, say $p_1, p_2, \ldots, p_{k+1}$. In the discrete case, this would essentially be the probability mass function of the random variable. Given the true probability for each category, the *true* expected number of values falling in each category should be $Np_1, Np_2, \ldots, Np_{k+1}$.

The *Pearson statistic* designed to test the hypothesis of whether the simulated data represents the hypothesized distribution, measures the "distance" between the true expected number of observations in each category and the observed number of observation in each category, as follows:

$$\chi^2 = \sum_{j=1}^{k+1} \frac{(N_j - Np_j)^2}{Np_j}. \tag{13.28}$$

If the hypothesis is supported, in other words, if the random variates are generated from the desired distribution, the statistics should have a small value. The Pearson statistic is asymptotically distributed according to the χ^2-distribution with k degrees of freedom, expected value $E[\chi^2] = k$, and $Var(\chi^2) = 2k$. Therefore, to complete the hypothesis test, we pick a significance level 100α % and determine a value of $\chi^2(1-\alpha, k)$, such that

$$P(\chi^2 \leq \chi^2(1-\alpha, k)) = 1 - \alpha. \tag{13.29}$$

Restating Equation 13.29 for the probability of the complementary event, $\chi^2 > \chi^2(1-\alpha, k)$, gives

$$P(\chi^2 > \chi^2(1-\alpha, k)) = \alpha. \tag{13.30}$$

The event $\chi^2 > \chi^2(1-\alpha, k)$ is the event of Type I error, when the null hypothesis is rejected when in fact it is true. The null hypothesis in our case is that the simulated data are accurate, that is, they represent the desired distribution model. However, the particular sample of size N is finite, and hence, there is a possibility that the distance measured in Equation 13.28 comes out to be large, leading to the erroneous conclusion of rejecting the null hypothesis when in fact it is true. The probability of type I error measures the probability of making this error, rejecting the null when the null is true, in Equation 13.30. Therefore, $0 < \chi^2 \leq \chi^2(1-\alpha, k)$ defines our acceptance region for the null hypothesis that $\chi^2 \sim 0$, or the "distance" between the observed number of observations and true expected observations in each category is essentially zero. If the χ^2 value computed by Equation 13.28 satisfies $\chi^2 \leq \chi^2(1-\alpha, k)$, we fail to reject the null hypothesis at the significance level of 100α %. If however, $\chi^2 > \chi^2(1-\alpha, k)$, then the null is rejected, which implies that the random variates are not acceptably generated according to the desired distribution.

13.3.5.2.2 Kolmogorov–Smirnov Test

For a continuous random variable, the discrete categories of the χ^2 goodness-of-fit test are (i) artificial, (ii) subjective, and (iii) do not fully take into account the variability in the data. These disadvantages are avoided in the Kolmogorov–Smirnov (KS) test, which is based on the Glivenko–Cantelli theorem [14,60]. If $\{X_i; i = 1, \ldots, N\}$ are the N random variates generated for a continuous distribution model with a cumulative distribution function, $F(x)$, then the cumulative frequency function, $F_N(x)$, for the random variates generated can be described as

$$F_N(x) = (\#\text{of} X_i's \leq x)/N. \tag{13.31}$$

Therefore, if the sample $\{X_i\}$ is sorted in increasing order to get $\{X_{(i)}\}$, then

$$F_N(X_{(i)}) = i/N, \quad i = 1, 2, 3, \ldots, N. \tag{13.32}$$

The Glivenko–Cantelli theorem states that the cumulative frequency function, $F_N(x)$, will converge to the true cumulative distribution function, $F(x)$, as the number of random variates generated, N, becomes large. Formally, it states the following:

$$D_N = \sup_{-\infty<x<\infty} |F_N(x) - F(x)| \rightarrow 0 \ \ a.s. \ \ \text{as } N \rightarrow \infty. \tag{13.33}$$

In order to apply the KS one-sided test at $100\alpha\%$ significance level, based on the above result, for the null hypothesis: $H_0 : \sqrt{N} D_N = 0$, we follow the following steps. The test utilizes the fact that the test statistic, $\sqrt{N} D_N$, follows a Kolmogorov distribution with cumulative density function, $H(x)$. The steps involve first computing the test statistic, followed by defining the acceptance region for the chosen significance level, and finally, checking if the computed test statistic falls in the acceptance region.

1. Compute the value of the test statistic, $\sqrt{N} D_N$, from the random variates as follows:
   a. Define $D_N^+ = \max_{1\leq i\leq N}\{\frac{i}{N} - F(x_{(i)})\}$.
   b. Define $D_N^- = \max_{1\leq i\leq N}\{F(x_{(i)}) - (i - 1/N)\}$.
   c. Compute $D_N = \max\{D_N^+, D_N^-\}$.
2. Find the acceptance region $(0, x_{1-\alpha})$, where $x_{1-\alpha}$ value is such that $H(x_{1-\alpha}) = 1 - \alpha$, where H is the Kolmogorov distribution function.
3. Finally, if $\sqrt{N} D_N < x_{1-\alpha}$, then we fail to reject the null hypothesis at $100\alpha\%$ significance level; otherwise, the null hypothesis is rejected and the random variates are concluded to not represent the desired distribution.

Being based on asymptotic results, the sample size or number of random variates generated to conduct the test, N, is important, both for the χ^2 goodness-of-fit test and the KS test. A guideline suggests that $N > 35$ suffices for these tests, but in a simulation analysis lot many random variates can be generated without much effort, hence sample size is not an issue.

We have spent considerable effort in describing the procedures for testing random variates' independence and accuracy. The primary motivation for this description was to understand the nuts and bolts behind the otherwise blackbox simulation software routines. Just as random variates generation can be conveniently accomplished by using the packaged routines in a simulation software, most statistical softwares come packaged with functions and routines to conduct the above tests.

13.4 Model Validation

The models developed for complex systems, even after considerable abstraction and simplification, can be quite complex. Each module or component of the model can have several variables, each described by various stochastic distribution or process models, with complex interaction between the variables, as well as the model components, to determine the overall system behavior or performance. We have so far discussed the principles behind building a model for a simulation study and the methods underlying capturing the randomness of various variables of the model. Besides capturing the randomness of the variables by generating random variates by various probability distributions, we also looked at methods to test the accuracy of the distribution models. We now need to move to the higher level of testing the accuracy of simulation models.

Testing the accuracy of the simulation model can be broken down into two major pieces, testing the accuracy of the computer representation of the model (right-most box in Figure 13.2) and testing the accuracy of the model for its ability to capture desired characteristics of the real system (the top loop in the right box in Figure 13.1). Testing of programming accuracy is often referred to as *model verification*, and includes programming error detection and debugging, while testing the model accuracy is *model validation*. Model validation can ask fundamental questions about whether the conceptual model correctly reflects the real system, or whether the conceptual model is really capable of addressing the necessary issues about the real system. The results of a validation analysis may result in going back to the drawing board, as suggested in Figure 13.1.

Clearly, model verification and validation is a very important exercise, since without a level of confidence on the model accuracy, its recommendations cannot be trusted. Despite its importance, often, a modeler can miss paying sufficient attention to this step. Moreover, a key point to note is that model validation should be a continuous, ongoing activity throughout the time the model is in use to make decisions. This is important, since no model is good for all times and under all conditions. As times and conditions change, the assumptions underlying a model may no longer hold, and hence, must be assessed. Reasons for overlooking the need for testing model accuracy can be multifold, ranging from good-old laziness, ignorance, to overconfidence in one's modeling capabilities, or pressures of time and budget.

The challenge behind testing models is, while model building is a fun and creative activity, model testing can be quite effortful and laborious. But, being skeptical of one's own work is a good rule to follow throughout model building and model usage. As in the planning for any other project, explicitly setting aside time and resources for model testing is a good practice to combat pressures of time and budget, or as a means for instilling the discipline to overcome laziness.

13.4.1 Techniques for Model Verification

Errors creeping in model building can be classified as syntactical or semantical. Syntactical errors are unintentional addition, omission, or misplacement of notation that either prevents the model from running or causes it to run incorrectly. Misplaced decimal point, parentheses can have a dramatic impact on the outcome. Semantical errors are errors in the meaning or intention of the modeler, such as a wrong condition inserted in an if-then-else statement. Semantical errors are harder to detect, but can have a very damaging effect on the usefulness of the model.

The best practice for developing good models is that the entire development of the simulation project should be done so that it facilitates testing its accuracy. Writing spaghetti software code, or other poor organization of code, such as without good descriptive variable names, descriptive comments, good flow of code logic, that makes the code hard to understand even by its own creator, is clearly not advisable. A stitch in time does, indeed, save nine, if not more!

As stated in Section 13.1, the code for a simulation model should be built modularly, starting simple and gradually growing to capture the complexity of the model, with staged verification and, preferably, validation. Stepwise refinement and progressively adding complexity to the model, verifying and validating the model in each step, develops the model in several passes, and ensures the model's accuracy, along with guaranteeing that the model possesses the right level of complexity needed for the purpose of the simulation study. Use of unstructured control, such as "goto"-type statements, should be avoided; instead logic control should be structured, such as using "if-then-else," "do-while," etc. statements. Not only model code and its logic, but also data supporting the model should be thoroughly and clearly documented. This helps detect and remove unintentional errors in model data, logic, and construction. It also facilitates communication and collaboration for efficient, error-free model building and usage.

Performing a top-down and bottom-up model code review helps in a thorough inspection of the code and its accuracy. A top-down review begins with looking at the major module and works its way to the lower-level modules, while a bottom-up review begins looking at the smallest modules and builds up the verification process toward the upper major modules. Running the model, provided it runs, to check for reasonable output can also help identifying semantical errors. Plotting outputs provides a visual aid for verification, where some errors can be detected visually that may otherwise go unnoticed. The plots of outputs provide help in identifying a problem, rather than discovering the cause of a problem. For aid in locating syntactical errors or tracing the code to identify semantical errors, software tools come with debuggers, which help uncover the source of the error.

To a seasoned programmer, most of the points made here would be trivialities, or second nature to account for when building the model, but for a newbie, learning the discipline can save many hours of neck-breaking, mind-numbing debugging work.

13.4.2 Techniques for Model Validation

Model validation tests if the model is a meaningful and accurate representation of the real system for the purpose of the simulation study. Validation of models should be a continuous activity throughout the time the model is in use to make decisions. As stated earlier, this is important, since, as times and conditions change, the assumptions underlying a model may no longer hold and, hence, must be assessed. Validation is not only of the model structure, but begins at the data-gathering stage to support building the model. As is said, "garbage-in, garbage-out." Validation looks for functional validity that the model's output behavior has sufficient accuracy for the model's intended purpose. Therefore, among other things, model validation helps develop a trust that simulation results may be used for real-world decisions.

Validation is a hard and painstaking process, but just as important, if not more, than any other activity in a simulation study. Therefore, the more the number of novel ideas generated for performing validation the better. We suggest some ideas here. It helps to compare the simulation results with other models, or simpler versions of the same model, for which analytical results may be available. Conducting the simulation under degenerate or extreme parameters or conditions should give anticipated degenerate or extreme output, for instance, should $X_0 = 0$ imply $X_T = 0$, or if investment weights $w_t^i = 0$ (for all i and t), what should be the anticipated portfolio performance in Section 13.2.3, similarly, if the strike, $K = 0$, what should the equity option discussed in Section 13.2.3 be worth, etc. On the other hand, reasonable values should result in reasonable outcomes. Not only that, if you show the results from the model to a knowledgeable expert, would he/she agree that the results seem reasonable and representative of the real system? In this regard, the simulation model can be subjected to a Turing test, where we ask the knowledgeable expert to discriminate between model and system outputs. If the expert cannot detect a difference, there is more evidence for model validity.

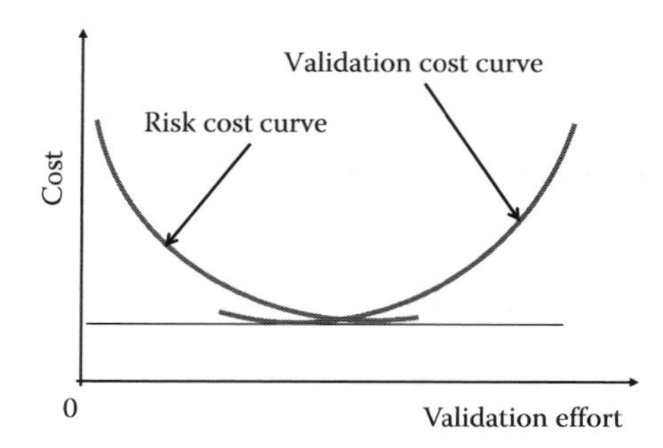

FIGURE 13.10
Display of validation cost versus risk cost curve.

As in model verification, plots and summary statistics of output variables are useful means for model assessment. Comparing these with similar information for the real system creates a context for judging if the model is capturing the necessary system characteristics. Conducting sensitivity analysis with the model is useful for validation, by changing some input parameters to determine their effect on model's behavior and output. These effects of sensitivity analysis on the model should be similar to how the real system would behave under these changes. Current and historical data for the system is useful to support the validation.

Although validation is a continuous process, and should be performed throughout the simulation study, there is still an optimum level of effort on validation, beyond which returns may be less valuable. Overdoing is not good, although this is hardly ever the problem in practice. A balance is needed between validation costs and cost incurred due to risk of making decisions based on an invalid model. This trade-off is illustrated in Figure 13.10, where the lowest total cost point between the validation cost curve and the risk cost curve is sought. Some models meant for ongoing use need continued validation, which should be performed only in a controlled modification mode to prevent unwieldy growth of the model that hurts its robustness.

13.5 Simulation Output Analysis

A simulation model is a computer-based statistical sampling experiment; therefore, appropriate statistical techniques must be used to design and analyze the simulation experiments, consistent with the goal of the study. Once the verification and validation steps are satisfactorily accomplished, in the final box of Figure 13.1, the simulation model is used for running the required simulation experiments and simulation output analysis. There is a risk of spending a great deal of time and money on model development and "programming," while making lesser effort for analyzing simulation output data appropriately. Care and attention is needed for designing and implementing the output analysis in a simulation study.

The level of effort or precision required in a simulation output analysis depends on several factors, such as the nature of the problem, importance of decision, and validity of the input data, and availability of a verified and valid model. In some cases, a rough analysis using judgmental procedures may suffice, while in others a detailed statistical analysis will be necessary. We will focus here on statistical analysis of simulation data.

As with statistical analysis of data obtained from the real system, statistical analysis of simulation output comprises of descriptive and inferential statistics. The goal of descriptive statistics, as the name suggests, is to *describe* the properties of the system based on the statistical properties of the simulated data obtained from running experiments using the model. The data generated by running experiments using the model is finite, depending on the design for sample size sought, and is hoped to be representative of the population. Inferential statistics tries to make conclusions or *infer* knowledge about the population based on the sample data produced by running simulation experiments, assuming the data are representative of the population. These concepts of sample and population, and descriptive or inferential statistics, are similar to those in statistical analysis of data obtained from the real world.

The simulation output variables of a model are in their turn random variables, potentially a complex function of the input variables and stochastic factors and their interactions in the model. Experimental samples, called replications, are intended to be independent observations of the output variables obtained after every "run" of the simulation model. The independence of observations of the output variables from different runs of the simulation model depends on the properties of the random number generator in use, such as long cycle period and ability to use different seeds. A reasonable number of replications (sample size) is a good indicator of what can be expected in any subsequent replication. Clearly, the sample size, amount of data generated from simulation experiments to make descriptive or inferential statistics, determines the accuracy and quality of inferences made from the data. Independence of observations in the simulated data is a requirement of most descriptive and inferential statistics procedure. The idea is to generate large enough sample to draw valid inferences about the population, where more is always better, but sample generation time and computing cost are the primary constraints.

13.5.1 Descriptive Output Analysis

Let the simulation output quantity of interest or the performance measure for the simulation study be θ. Using the replications, we want to estimate the value of the performance measure. We can seek two types of estimates under a descriptive output analysis.

Point estimator: A *formula* for a single value estimate of the performance measure, denoted by $\hat{\Theta}$.

Point estimate: The actual value a point estimator takes when specific data values are plugged into the formula, denoted by $\hat{\theta}$.

Interval estimate: This gives a range of values the performance measure will have with a degree of confidence. It is also called a *confidence interval*.

For example, the descriptive statistic of interest of an output variable could be its mean, μ. To create an estimate of the mean, we can use a point estimator or an interval estimator. Sample mean estimator defined by

$$\bar{X} = \frac{\sum_{i=1}^{N} X_i}{N} \tag{13.34}$$

is a point estimator, while a confidence interval based on sample mean is an interval estimator. The confidence interval estimator will be

$$\left(\bar{X} - z_{\alpha/2}\frac{\sigma}{\sqrt{N}}, \bar{X} + z_{\alpha/2}\frac{\sigma}{\sqrt{N}}\right), \tag{13.35}$$

if the standard deviation of the performance measure σ is known; otherwise, it will be

$$\left(\bar{X} - t_{N-1,\alpha/2}\frac{s}{\sqrt{N}}, \bar{X} + t_{N-1,\alpha/2}\frac{s}{\sqrt{N}}\right), \tag{13.36}$$

where s is a point estimator for the standard deviation of the performance measure. $z_{\alpha/2}$ is the $1-\alpha/2$-th percentile of a standard normal distribution and $t_{N-1,\alpha/2}$ is the $1-\alpha/2$-th percentile of a standard t-distribution with $N-1$ degrees of freedom. When we substitute the values for X_i's to be data obtained from N runs of the simulation model, say x_i, we obtain point estimates and interval estimates for mean. The interval estimates are created so that with $100(1-\alpha)\%$ confidence the true mean, μ, lies within the confidence interval.

The narrower the confidence interval, the better the accuracy of the interval estimate. However, there is an inverse relationship between confidence level, $100(1-\alpha)\%$, and the width of the confidence interval. If we desire a higher confidence level, the width of the confidence interval becomes bigger. On the other hand, if we require a tighter confidence interval, the confidence level will drop. For a given confidence level, getting a tighter confidence interval can be accomplished by increasing the number of observations generated. In particular, for a confidence interval with width $2e$ and a confidence level of $100(1-\alpha)\%$, the number of observations exceeding

$$N > \left(\frac{z_{\alpha/2}\sigma}{e}\right)^2 \tag{13.37}$$

will suffice. If the standard deviation of the output variable, σ, is not known, a point estimate for it is used to get an approximate number of observations needed for the desired interval estimate accuracy.

The mean of the output variable is one example of a performance measure. Often, other summary descriptive statistics are required for the output variables of a simulation study. A general functional of the output variable, X, that may be defined as, $\theta = E[f(X)]$, or can be a conditional expectation, $\theta = E[f(X)|g(X)]$. Functions $f(.)$ and $g(.)$ are appropriately well-defined functions. For instance, variance of the output variable can be computed by picking function, $f(x) = (x-\mu)^2$. Similarly, semivariance, conditional variance, percentiles, expected shortfall, value-at-risk (VaR), etc. can be estimated. For each functional, a point estimator needs to be used to create the point estimate or the interval estimate. The point estimators may be chosen specific to the context where the functionals are used.

13.5.1.1 Designing Simulation Run by Properties of Estimators

Every point estimator for a functional of a simulation output random variable, $f(X)$, has properties that should be understood. These properties would guide the development of design of simulation experiments. An estimator is said to be unbiased if in expectation it gets right what it is attempting to estimate, that is, in our earlier notation $E(\hat{\Theta}) = \theta$. Clearly, being unbiased is a good property for a point estimator to have. However, beyond bias, there is a second important property of an estimator to consider, which is the variance of the estimator, $V(\hat{\Theta})$. The higher the variance of an estimator, the poorer the estimator, since for any given simulated data, the estimate produced by the estimator can be quite off from the true population value for that functional of the output variables, $\theta = E[f(X)]$. In that, one seeks a minimum variance unbiased estimator (MVUE) for the functionals of interest.

The variance of an estimator, in the simulation approach to solving problems, depends heavily on the computational effort made. The greater the computational budget, the lower the level of achievable variance of an estimator. However, computational budget is never infinite, just as compute time is rarely unlimited. Therefore, the design of a simulation study must determine the trade-off between bias, variance, and compute time for various quantities being estimated.

The general guideline for the design of simulation runs decision process is as follows:

1. If the compute time for each replication to generate θ_i is fixed, say τ, and the estimator is unbiased, select number of runs to fit $V(\hat{\Theta})\tau$ within the computational budget.
2. If the compute time for each replication to generate θ_i is stochastic, say $\tilde{\tau}$, and the estimator is unbiased, select number of runs to fit $V(\hat{\Theta})E[\tilde{\tau}]$ within the computational budget.
3. If more than one estimators are available, select one with least mean-squared error ($MSE(\hat{\Theta}) = bias(\hat{\Theta})^2 + V(\hat{\Theta})$). Follow guideline for step 1 or 2, depending on the nature of replication compute time.

13.5.2 Inferential Output Analysis

Inferential analysis of simulation output variable is in essence identical to the inferential analysis of real-world data. The analyst postulates a hypothesis regarding the value of a functional of a simulation output variable and utilizes the data generated from simulation experiments to test if the hypothesis has support or not. The test is conducted on the basis of a test statistic, which is essentially a point estimator for the hypothesized functional. For instance, the sample mean, $\bar{X}$, estimator for the population mean functional, $\mu = E[X]$.

Developing the test statistic utilizes knowledge of the *sampling distribution* of the estimator. The sampling distribution of the sample mean estimator, $\bar{X}$, is asymptotically a normal distribution, by the central limit theorem. This fact is used to construct the test statistic

$$\frac{\bar{X} - \mu}{\sigma/\sqrt{N}}, \tag{13.38}$$

if σ is known. And

$$\frac{\bar{X} - \mu}{s/\sqrt{N}}, \tag{13.39}$$

when σ is not known, and is estimated using the point estimator, s. In the σ unknown setting, the test statistic (Equation 13.39) is approximately a t-distribution with $N - 1$ degrees of freedom.

Under the hypothesis that the true population mean of the output variable, $\mu = \mu_0$, an acceptance region is constructed as

$$\left(\mu_0 - z_{\alpha/2}\frac{\sigma}{\sqrt{N}}, \mu_0 + z_{\alpha/2}\frac{\sigma}{\sqrt{N}}\right), \tag{13.40}$$

in the case when the standard deviation of the performance measure, σ, is known. Otherwise, the acceptance region becomes

$$\left(\mu_0 - t_{N-1,\alpha/2}\frac{s}{\sqrt{N}}, \mu_0 + t_{N-1,\alpha/2}\frac{s}{\sqrt{N}}\right), \tag{13.41}$$

when σ is not known. If the computed sample mean, $\bar{X}$, falls in the acceptance region, the simulated data supports the hypothesis of the mean output variable level being μ_0. If the computed sample mean does not fall in the acceptance region, the hypothesized value of mean can be rejected. As before, $z_{\alpha/2}$ is the $1 - \alpha/2$-th percentile of a standard normal distribution and $t_{N-1,\alpha/2}$ is the $1 - \alpha/2$-th percentile of a standard t-distribution with $N - 1$ degrees of freedom. The value α is called the significance level, which is the probability of making an erroneous conclusion, namely, rejecting the hypothesis when it is in fact true (Type I error). Clearly, we would like to minimize the probability of making an erroneous conclusion, but we cannot indefinitely reduce this probability without creating another problem, that of not being able to reject the hypothesis when it is in fact false (Type II error).

The principle applied above to the mean functional, $\mu = E[X]$, can be applied to any other functional for which inferential analysis is needed. A point estimator, its sampling distribution, and a hypothesized value of the functional of the output variable will need to be defined. Using these and a chosen significance level, an acceptance region will be constructed and the test performed. Specifications of other inferential analysis may be developed in the context of specific problems. For instance, we introduced some examples of hypothesis tests for testing random variates (Section 13.3.5.2).

13.6 Variance Reduction Techniques

All tasks of decision making under uncertainty can benefit from simulation analysis, especially when analytical solutions are difficult to construct. However, every problem solved using simulation must deal with the fact that in simulation tasks are accomplished by generating samples of observations. Estimates for the quantities of interest are obtained by applying the chosen estimators to the samples generated. This is true for all the decision-making problems discussed in Sections 13.2.1 through 13.2.4, where simulation analysis is applied for the assessment for performance measures for the problem.

For pricing of derivatives, as discussed in Section 13.2.3, defined for equity, interest rate, commodities, exchange rates, and credit risks, when the choice of model for the underlying risk does not readily yield an analytical solution, simulation offers an alternative for price estimation. Simulation analysis is also useful when these variety of derivatives for different risk types are used to develop hedging strategies for those risk exposures. These derivatives may be defined for single instruments, but can also be defined and utilized for hedging portfolio of instruments, such as a portfolio of equities or debt instruments.

A key goal of portfolio optimization is to construct portfolios that achieve the desired risk–reward trade-off. These portfolios could consist of a variety of instruments affected by market risks, where each instrument in the portfolio may be affected by multiple risk factors and their interactions. Moreover, as sketched in Section 13.2.3, we can pose the asset allocation or portfolio optimization problems in static as well as dynamic settings. Simulation-based optimization is an area of simulation modeling and analysis that allows addressing these optimization problems using simulation analysis, especially when analytical methods are not available or efficient for solving the problems. This methodology is also applicable for determining hedging strategies, both static and dynamic ones, designed to transfer risk to achieve desired risk–reward profile.

For portfolio of instruments, whether it is market risk instruments, credit risk-sensitive instruments or pure risk instruments, simulation analysis is useful to assess the portfolio level risk, as discussed in Section 13.2.4. Risk measures, such as VaR and conditional VaR (CVaR), often require simulation analysis for their estimation, given the complexity of the task at hand. At the portfolio level, the models of individual risk factors and their interactions become complex enough that analytical solutions are rarely obtainable. Risk assessment and monitoring at the portfolio level, especially to address nonstationarity of risk factors, requires extensive scenario analysis and stress testing. The high dimensionality of these problems, due to large portfolio sizes and number of risk factors, poses significant challenge for these assessments.

In the design and management of manufacturing and service systems, as well as in routing and congestion management decisions in transportation and communication networks, discussed in Sections 13.2.1 and 13.2.2, a certain fidelity in estimates of performance measures is a necessity for them to be actionable. Furthermore, just as in risk assessment and portfolio optimization decisions, design optimizations for these systems must have the benefit of judging the impact of choice of decision variables on the performance measures reasonably accurately.

Using simulation to solve the above problems based on estimates of key quantities using finite generated samples implies there will always be some uncertainty regarding the theoretical value of these estimated quantities. We discussed in Section 13.5 that the uncertainty of estimates is summarized by developing confidence intervals of desired confidence level. Construction of confidence intervals relies on the distribution of the estimator for the quantity of interest, more specifically, on the variance of the estimator. For example, if the estimator is the sample mean estimator, $\bar{X}$, then we utilize the fact that $\bar{X} \approx N(\mu, \sigma/\sqrt{N})$, under certain conditions. From this fact, the confidence interval for the theoretical value of the quantity of interest is obtained as $(\bar{X} - z_{\alpha/2}(\sigma/\sqrt{N}), \bar{X} + z_{\alpha/2}(\sigma/\sqrt{N}))$. A tighter confidence interval will assure higher reliability of decisions made based on simulation analysis.

One way to make the confidence interval tighter, or more accurate, is to increase the sample size, N. We do this as the first response; however, there is a limitation to take this to extreme due to time and computational resource restriction, as discussed in Section 13.5.1.1. The other option available is to reduce σ^2, which is the variance of random sample element, X_i, used to construct the estimator. The latter response is the focus of this section, which will benefit all the tasks we have proposed to achieve using simulation analysis.

Once we have explored a variety of variance reduction techniques, we will move our attention to simulation optimization in Section 13.10, which is a direct beneficiary of improvements in accuracy of performance measure estimates. As mentioned, optimization problems show up in a variety of decision making under uncertainty contexts, such as in the design of manufacturing plant, routing packets to meet quality of service, portfolio optimization, developing hedging strategies, management of strategic, and business or operational risk. We will develop the principles of simulation optimization, and discuss several methods for implementing simulation optimization in the concluding sections of this chapter.

As the name suggests, variance reduction techniques are designed to reduce the variance of the estimator by means other than simply raising the number of simulation runs conducted to increase the sample size. Variance reduction methods are designed on a variety of themes, all geared toward the same goal of improving the accuracy of quantities estimated using simulation. We discuss this topic immediately after general discussion of simulation output analysis since in practice variance reduction is a basic need of almost all simulation analysis. Since improving efficiency for better decision making is often a necessity, not an option.

Variance reduction techniques are built around two broad strategies. The first set of strategies take advantage of tractable features of a model, interrelation between variables of the model, to adjust or correct simulation output. The other strategy adopted in developing variance reduction techniques is by reducing variability in simulation inputs. We will consider some methods of both categories.

We will discuss and illustrate the following methods in the coming sections:

Control variate: The control variate method utilizes information regarding correlation between variables of the model to develop a new estimator, which is designed to have a lower variance.

Antithetic variates: This method attempts to reduce variance by modifying how random variate inputs are used to generate a sample of observation for the quantity of interest.

Stratified sampling: In this technique, the input random variates are sampled in a controlled manner to reduce the variance.

Latin hypercube sampling: The Latin hypercube method is most advantageous for variance reduction as the dimension of the problem increases.

Importance sampling: This method utilizes the properties of the probability distribution of the quantity of interest to design a second probability distribution which emphasizes the "important" observations of the first probability distribution.

It is possible, where appropriate, to attempt to combine the application of more than one variance reduction method for estimation of a single quantity, or for estimation of different quantities, in a decision-making process.

In general, in implementing any variance reduction technique, attention is required for how the simulation study is designed. As discussed in Section 13.5.1.1, the important trade-off to construct in any simulation-based estimation is between bias, variance, and compute time for an estimator. In case of unbiased estimators, the focus narrows down to variance and compute time. We had defined τ as the compute time for each replication toward generating a sample for estimating the quantity of interest. Including a variance reduction technique can result in an increase in compute time for each replication; hence, care is needed regarding the computational burden implied by the variance reduction technique.

The guideline from Section 13.5.1.1 applies, regarding comparing the product of compute time and variance of estimators. Let us say Θ_2 and Θ_1 are two estimators for the quantity of interest, θ, when variance reduction is applied versus not, respectively. We compare the efficiency of applying variance reduction by comparing whether $\tau_1 V(\Theta_1)$ is greater or less than $\tau_2 V(\Theta_2)$, if the compute times τ_1 and τ_2 are deterministic. If the compute times are stochastic, we compare whether $E[\tau_1]V(\Theta_1)$ is greater than or less than $E[\tau_2]V(\Theta_2)$. If the application of variance reduction technique ends up reducing the efficiency of the estimation task, it may actually become counterproductive to use one. Therefore, attention is required for the design and implementation of the variance reduction technique, including in terms of efficiency of the code and data management for the implementation of the technique.

13.6.1 Control Variates

Control variate method is perhaps the most effective and broadly applicable technique in many applications of simulation modeling and analysis. The method exploits information about the errors in estimates of known quantities to reduce the error in an estimate of an unknown quantity; hence, the use of the word "control," where one variable helps control the variance in estimate of another.

Let $Y_1, Y_2, \ldots, Y_n$ be output of n replications of a simulation for a random variable, Y, where the quantity of interest is $\theta = E[Y]$. For example, $Y_i = C_i = e^{-rT}(S_i(T) - K)_+$, where $\theta = E[e^{-rT}(S(T) - K)_+]$ is the price of a plain-vanilla European call option defined on the underlying asset, $S(t)$, with strike price K and maturity T. Y_i are i.i.d. random variates, and in order to estimate $E[Y]$, we utilize the standard sample mean estimator, $\Theta_1 = \sum_{i=1}^{n} \frac{Y_i}{n}$.

Suppose for each replication, Y_i, we also calculate another output X_i, where (X_i, Y_i) are i.i.d., and the expected value of X, $E[X]$, is known analytically. We will call the variable X a *control variate*, which we will use to create a new estimator of $\theta = E[Y]$. We first define a modified replication, $Y_i(b)$, as follows:

$$Y_i(b) = Y_i - b(X_i - E[X]), \tag{13.42}$$

where b is a fixed number picked appropriately. We compute the sample mean of $Y_i(b)$, $\Theta_2 = \sum_{i=1}^{n} Y_i(b)/n$, which would be the control variate estimator. It is clear that this estimator is unbiased, from the way it is constructed:

$$E[\Theta_2] = \sum_{i=1}^{n} E\left[\frac{Y_i(b)}{n}\right], \tag{13.43}$$

$$= \sum_{i=1}^{n} \frac{1}{n}(E[Y_i] - bE[X_i - E[X]]), \tag{13.44}$$

$$= \sum_{i=1}^{n} \frac{1}{n}\theta. \tag{13.45}$$

It can also be shown that the estimator, Θ_2, is consistent, that is, $\Theta_2(n) \rightarrow \theta$ in probability, as $n \rightarrow \infty$.

We need to assess if the new estimator in fact results in variance reduction, and what makes this possible. For this purpose, we first compute the variance of the replications, $Y_i(b)$, as follows:

$$Var[Y_i(b)] = Var[Y_i - b(X_i - E[X])], \tag{13.46}$$

$$= \sigma_Y^2 - 2b\rho_{XY}\sigma_X\sigma_Y + b^2\sigma_X^2, \tag{13.47}$$

where σ_Y and σ_X are the standard deviations of Y and X, respectively, while ρ_{XY} is their correlation. The variance of replications, $Y_i(b)$, is lower than variance of original replications, Y_i, if we can construct the control variate replication such that

$$-2b\rho_{XY}\sigma_X\sigma_Y + b^2\sigma_X^2 < 0. \tag{13.48}$$

Let us examine when the condition in Equation 13.48 can be achieved. If $\rho_{XY} \sim -1$, that is, the variables X and Y are strongly negatively correlated, picking an appropriate negative value of b can satisfy Equation 13.48. Similarly, if $\rho_{XY} \sim +1$, picking an appropriate positive value of b can satisfy the above condition. We will be best served if we sought the best possible value of b.

The optimal choice of b is obtained by taking the first derivative of the expression on the left-hand-side of Equation 13.48 with respect to b, and solving for its zero. We obtain the solution to be $b = Cov[X, Y]/Var[X]$, which is confirmed to be the optimal choice by taking the second derivative of the expression on the left-hand-side of Equation 13.48 with respect to b. In general, the theoretical value of b may not be known; therefore, it must be estimated from a sample as follows:

$$\hat{b}_n = \frac{\sum_{i=1}^{n}(X_i - \bar{X})(Y_i - \bar{Y})}{\sum_{i=1}^{n}(X_i - \bar{X})^2}. \tag{13.49}$$

Using the estimated value of b, the control variate replicates become $Y_i(\hat{b}_n) = Y_i - \hat{b}_n(X_i - E[X])$, and the control variate estimator is $\Theta_2 = \sum_{i=1}^{n} Y_i(\hat{b}_n)/n$. With this choice of alternate replications, if the correlation between X and Y is high, irrespective of its sign, we would achieve variance reduction, since the variance of the control variate estimator would be lower. We consider some examples of control variate-based variance reduction applied to asset valuation next.

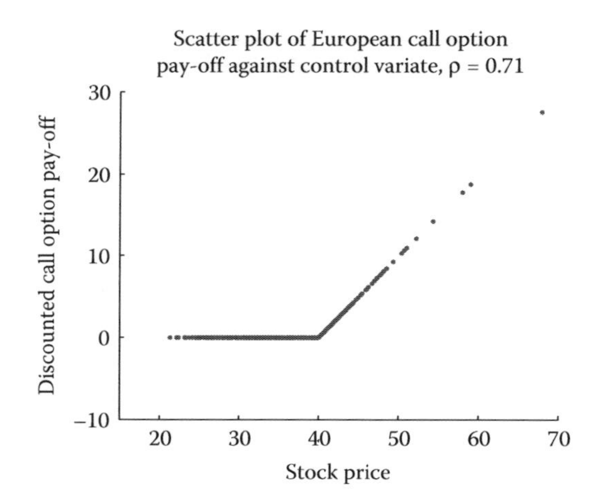

FIGURE 13.11
Scatter plot of replications of discounted European call option pay-off against replications of the stock price at option maturity. The dependence of these two quantities is as expected, which results in the strong positive correlation that the control variate method can utilize.

Example 13.1

Consider the case of pricing a plain-vanilla European call option. We utilize the underlying asset price as the control variate for pricing the option using simulation. This choice is justified since in the risk-neutral world, we note that $E[e^{-rT}S_T] = S(0)$; therefore the theoretical mean of the control variate is known. The control variate replications are set up as follows:

$$Y_i(b) = Y_i - b[S_i(T) - e^{rT}S(0)]. \tag{13.50}$$

To verify that this choice of control variate will be successful in variance reduction, we need to assess the correlation between $Y_i = e^{-rT}(S_i(T) - K)_+$ and $X_i = S_i(T)$. In Figure 13.11, we display a scatter plot of Y_i and X_i. As expected, the plot shows the dependence of discounted pay-off of European call option on terminal stock price realizations, which results in the desired positive correlation.

Example 13.2

In this example, we go beyond the simple case of pricing a plain-vanilla European option. Pricing path-dependent options is particularly challenging, since in case of a path-dependent option, the price of the option depends on the entire trajectory of the underlying asset price during the life of the option. Pathwise accuracy requires strong convergence for simulation approximations, as will be studied later in Section 13.8.2, which ends up being computationally more demanding for the same level of accuracy in estimates. Therefore, variance reduction can provide significant help in maintaining the accuracy and compute burden trade-off. Let us consider an arithmetic average Asian option with strike K and maturity T. The price of this option can be shown to be $\theta = E[e^{-rT}(\bar{S}_A - K)_+]$, where $\bar{S}_A$ is the arithmetic mean of the price of the underlying asset at periodic time points, $\{t_i\}$, through the life of the option.

For the purpose of pricing this Asian option using control variate variance reduction, we utilize a tractable option, that is, an option whose price is known analytically. We choose the pay-off of a plain-vanilla European call option with the same strike and maturity as our control variate; therefore $X_i = (S_i(T) - K)_+$. Under the Black–Scholes option pricing model, $C_{bls}(0, S_0; T, K, \sigma) =$

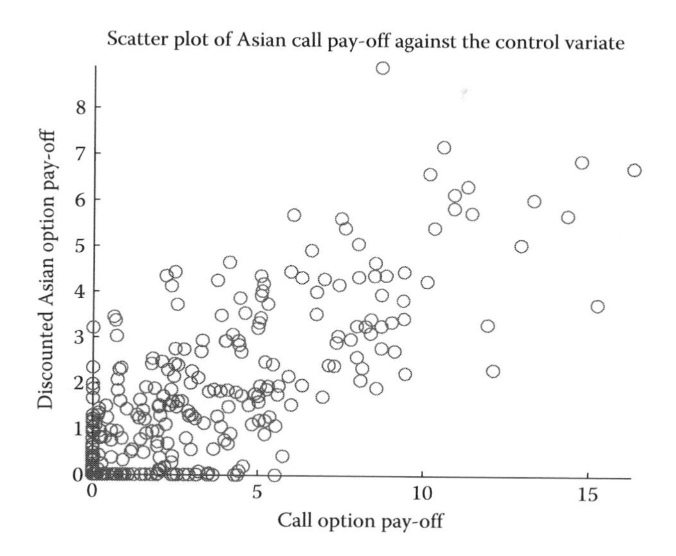

FIGURE 13.12
Scatter plot of replications of discounted Asian call option pay-off against replications of European call pay-off at option maturity. A strong positive correlation is visible, which is utilized in the design of the control variate method.

$e^{-rT}E[X_i]$, where C_{bls} is the Black–Scholes European call option price given in Equation 13.4. If the pay-off of the arithmetic average Asian option is given by

$$Y_i = e^{-rT}(\bar{S}_A - K)_+ = e^{-rT}\left(\frac{1}{m}\sum_{i=1}^{m} S(t_i) - K\right)_+, \tag{13.51}$$

then we create the control variate replication as $Y_i(b) = Y_i - b(X_i - E[X])$. As stated above, the vanilla call control variate is $X_i = (S_i(T) - K)_+$, where we know $E[X_i] = e^{rT}C_{bls}(0, S_0; T, K, \sigma)$. Similar to Example 13.1, we can examine the degree of variance reduction obtained in this case by evaluating $\rho_{XY} = corr(e^{-rT}(\bar{S}_A - K)_+, (S_i(T) - K)_+)$. The correlation is quite visible in Figure 13.12 in which we display a scatter plot of Y_i and X_i.

In general, there is no restriction regarding the number of control variates that may be applied simultaneously. For instance, in the above example of pricing an arithmetic average Asian option, one may simultaneously apply several European call options corresponding to a range of strike prices as control variates. As seen in Figure 13.12 for a single vanilla European call option, each of these options will have a similar correlation resulting in contribution to reduction in variance of the control variate estimator. A more detailed description and analysis of control variate-based variance reduction can be found in Glasserman [36], Law [65], and Asmussen [5].

13.6.2 Antithetic Variables

Instead of seeking a second variable, or set of variables, which promise to have a high correlation with the quantity of interest, as utilized in control variates, in the antithetic method of variance reduction, we create a negative correlation between replications of a single variable. This is done in a specific manner, and is successful in variance reduction under specific circumstances. Here, replicates are produced in pairs, where one replicate in the pair is negatively correlated with other; hence, the pair is called antithetic variates pair.

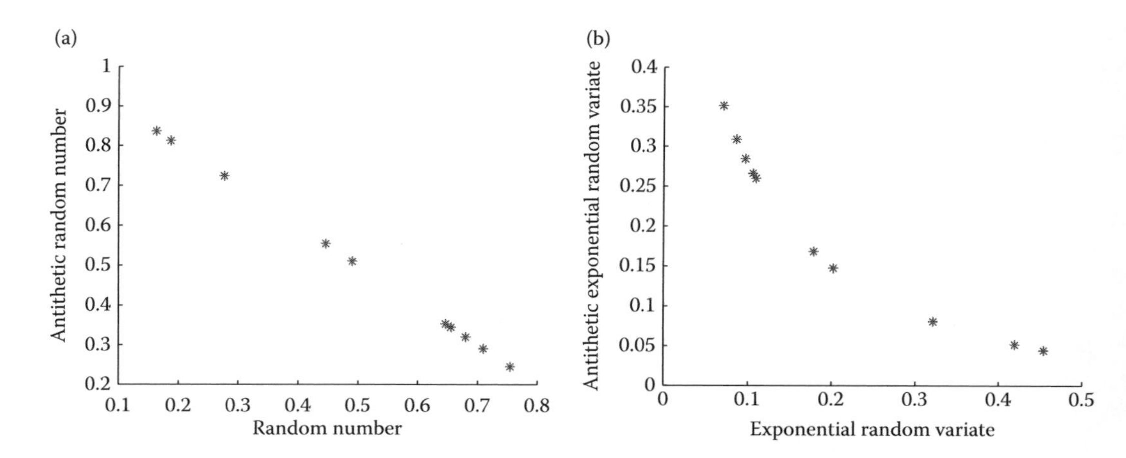

FIGURE 13.13
Scatter plot of (a) antithetic random numbers and (b) antithetic exponential random variates.

Antithetic variates can induce a negative correlation by using a few different themes. One theme is that of generating random numbers in pairs, where the complementary random numbers are produced noting the properties of continuous uniform distribution. If $U_k \sim U(0, 1)$, then automatically $1 - U_k \sim U(0, 1)$; moreover, if U_k is small, $1 - U_k$ is large, and vice versa. Therefore, every random number generated, U_k, is accompanied by its antithetic, $1 - U_k$, where the pair is used to generate random variates for the quantity of interest. The reader should be reminded that random numbers play a fundamental role in generating all other random variates, as presented at length in Section 13.3.

Antithetic variates method of variance reduction works when the method for generating random variates from the uniform random numbers maintains the monotonicity of the random numbers. For instance, this is achieved in the direct methods of random variates generation, such as the inverse transform method described in Section 13.3.4.1. An antithetic random variate pair, (X_{1k}, X_{2k}), is generated using the antithetic random number pair $(U_k, 1 - U_k)$ using the method intended to generate the random variates by the desired distribution. Figure 13.13 shows a scatter plot of antithetic random numbers, which show a perfect negative correlation. A strong negative correlation is maintained in the antithetic exponential random variates. Therefore, the use of U_k and $1 - U_k$ must be synchronized; otherwise, the variance reduction may backfire.

Variance reduction is achieved from antithetic variates construction due to the following reason. Say the quantity of interest is $\theta = E[X]$. We generate n replicates X_{1k}, along with their antithetic pairs X_{2k}, to estimate the quantity of interest as $\Theta_2 = \bar{X}(n) = (1/2)((\sum_{i=1}^{n} X_{1k}/n) + (\sum_{i=1}^{n} X_{2k}/n))$. The variance of this estimator can be computed as follows:

$$var(\bar{X}(n)) = var\left(\frac{1}{2}\left(\frac{\sum_{i=1}^{n} X_{1k}}{n} + \frac{\sum_{i=1}^{n} X_{2k}}{n}\right)\right), \tag{13.52}$$

$$= \frac{var(X_{1k})}{4n} + \frac{var(X_{2k})}{4n} + 2\frac{cov(X_{1k}, X_{2k})}{4n}, \tag{13.53}$$

$$= \frac{\sigma_X^2}{2n} + \frac{1}{2n} cov(X_{1k}, X_{2k}), \tag{13.54}$$

where σ_X is the standard deviation of each replicate, X. Therefore, the key for obtaining variance reduction is if $cov(X_{1k}, X_{2k}) < 0$. This was in fact the design for choice of antithetic pairs; therefore, the antithetic estimator, Θ_2, has lower variance than the usual sample mean estimator Θ_1, that is

$$var(\bar{X}(n)) = \frac{\sigma_X^2}{2n} + \frac{1}{2n}cov(X_{1k}, X_{2k}), \quad (13.55)$$

$$< \frac{\sigma_X^2}{2n}, \quad (13.56)$$

$$= var(\Theta_1). \quad (13.57)$$

The above design can also be applied if the quantity of interest is a function, $f(x)$, of the random variate, X, provided the function is monotonic. A monotonic function, whether it is nondecreasing or nonincreasing, maintains the relationship between antithetic pairs, that is, $cov(f(X_{1k}), f(X_{2k})) < 0$. Therefore, variance is reduced by applying antithetic variates.

The antithetic design of variance reduction need not be applied only to uniform random variates. In fact, it can be applied to any symmetric distribution by observing that once a random variate X_{1k} is generated by that distribution, we can generate an antithetic pair as $X_{2k} = 2\mu_X - X_{1k}$, where μ_X is the mean (and the median) of the random variable, X. It can be shown in this case also that $cov(X_{1k}, X_{2k}) < 0$, and for a monotonic function, $f(x)$, $cov(f(X_{1k}), f(X_{2k})) < 0$. Therefore, in the numerous contexts of simulation modeling and analysis for decision making, where distributions such as the normal distribution, lognormal distribution, t-distribution, and processes such as the Wiener process is utilized for developing the stochastic models, antithetic variate-based variance reduction can be applied based on this theme.

It is worth elaborating at this time that given increments of Wiener process, $\Delta W \sim N(0, \sqrt{\Delta t})$, we can generate antithetic variates pairs $(\Delta W_{1k}, \Delta W_{2k}) = (\Delta W_{1k}, -\Delta W_{1k})$. Synchronizing antithetic pairs thus constructed can help generate antithetic trajectories or sample paths of various Ito processes. We will examine the Ito processes, as an example of type of stochastic processes developed on the basis of the Gaussian distribution, later in Section 13.8. Figure 13.14 shows a couple of antithetic sample paths of an Ito process. Antithetic trajectories can be utilized in variance reduction where the directionality of a trajectory is of importance, such as high versus low asset price, or high versus low utilization of a resource. For instance, when pricing an up-and-in barrier options, where such an option becomes valuable only when the underlying asset price crosses an upper threshold

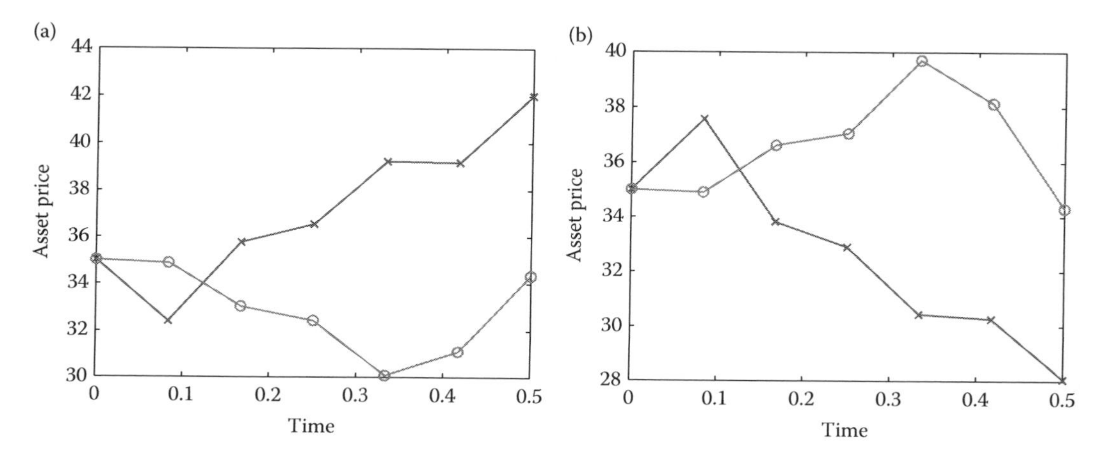

FIGURE 13.14
Sample paths for asset price evolution in antithetic pairs. (a) Sample paths. (b) Antithetic sample paths.

or barrier during the life of the option. Therefore, in pricing such as an option, if one trajectory is leading to a knock-in (or the barrier/threshold being reached) of the barrier option, the antithetic trajectory is likely to not lead to a knock-in, and hence have a zero pay-off. This can yield a negative correlation in replications of barrier option pay-offs, thus resulting in a variance reduction.

Feasibility and efficacy of antithetic variates is model dependent. The fundamental requirement that a model should satisfy for antithetic variates method to work is that the random number or variate generated is transformed monotonically in the model. Therefore, implementation of antithetic variate-based variance reduction can benefit from care in programming. Programming tricks that can help in the necessary synchronization for antithetic variates include random number stream dedication, using inverse-transform wherever possible, judicious wasting of random numbers, and pregeneration. In some cases, we can also seek to apply partial antithetic variates, which means only some quantities of interest in a large simulation utilize antithetically generated variates, while others are performed based on independent variates. This may be needed where full synchronization is difficult, and partial application is both feasible and effective in variance reduction.

13.6.3 Stratified Sampling

The stratified sampling method of variance reduction is built on developing a sampling mechanism that constraints the fraction of observations drawn from specific subset or "stratum" of the sample space. Therefore, the two basic steps of stratified sampling constitute

1. Break the sample space into several strata or subsets using an appropriately defined rule.
2. Generate a sample of constrained number of observations from each stratum.

The advantage of this approach for variance reduction stems from the greater uniformity by which it represents "subpopulations" of the entire sample space, rather than attempting to represent the entire population at once. We describe the method in detail in order to elaborate this point.

Let us say we want to estimate, $\theta = E[Y]$ for a stochastic factor, Y, where Y is defined on the probability space $(\Omega, \mathbf{P}, \mathcal{F})$. We identify, $A_1, A_2, \ldots, A_K$, as the disjoint subsets of the sample space, Ω, such that $\bigcup_{i=1}^{K} A_i = \Omega$. The stratified sampling-based estimator is designed based on the following relationship:

$$\theta = E[Y] = \sum_{i=1}^{K} p_i E[Y|Y \in A_i], \tag{13.58}$$

where $p_i = \mathbf{P}(Y \in A_i)$. Based on Equation 13.58, we define the new estimator Θ_2 for the quantity of interest, θ, as follows:

$$\Theta_2 = \sum_{i=1}^{K} p_i \bar{Y}_i, \tag{13.59}$$

where $\bar{Y}_i = \sum_{j=1}^{N_i} Y_{ji}/N_i$ is the sample mean of observations $\{Y_{ji}\}$ generated from the subset A_i. Figure 13.15 shows the strata, sampling within each stratum, and the conditional mean, $\bar{Y}_i$, for each stratum.

In using this variance reduction estimator, a few questions need to be addressed. How does one decide the specific disjoint subsets, $A_1, A_2, \ldots, A_K$, of the sample space, Ω, to use and how does one decide the weight p_i and number of replicates to draw from each stratum, N_i, in Equation 13.59? The

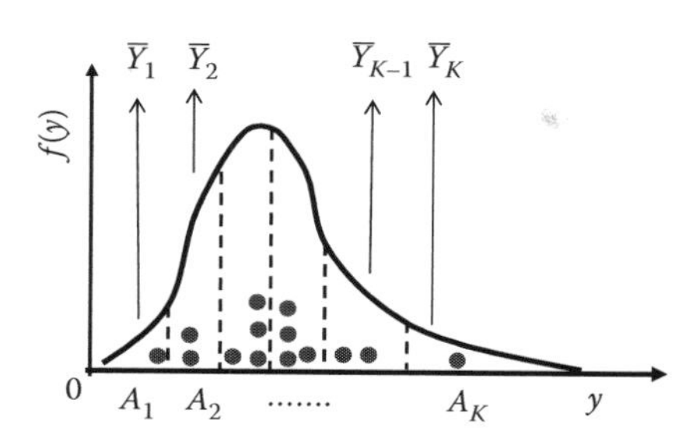

FIGURE 13.15
The sample space of the quantity of interest is broken down into strata, where samples are drawn from each stratum to create the estimator.

strata are often generally defined by using a second random variable, X, as follows: $A_i = \{X = x_i\}$. For instance, the random variable X may be simply defined as a K-outcome discrete random variable indicating when Y takes values in a subset A_i, that is, $\mathcal{I}_{\{X=x_i\}} = \mathcal{I}_{\{Y \in A_i\}}$. However, using a second random variable for defining the strata allows more general definitions of the strata.

The choice of weights, p_i, and number of replicates, N_i, is not too far removed from the stratum selection. For instance, if K subsets are sought, the weight allocation to each subset may be $p_i = 1/K$ and $N_i = Np_i$, if a total of N replicates are to be generated. This also helps define the stratum as $A_i = (y_i, y_{i+1}]$ such that $\mathbf{P}(Y \in A_i) = p_i$. More general definitions of both weights, p_i, and the number of replicates to draw from each stratum, N_i, may also be used. For a specific choice of design of the stratified sampling-based estimator of θ results in the following variance of the estimator:

$$var(\Theta_2) = \sum_{i=1}^{K} \frac{p_i^2}{N_i} \sigma_i^2, \tag{13.60}$$

where $\sigma_i^2 = var(Y|X = x_i)$, that is, it is the conditional variance of Y for values lying within a stratum, A_i. Just as in the case of control variates, the optimum variance reduction in stratified sampling can be shown to be obtained by the following choice of N_i:

$$N_i^* = N \frac{p_i \sigma_i}{\sum_{k=1}^{K} p_k \sigma_k}, \tag{13.61}$$

where $\sigma_k^2 = var(Y|X = x_k)$, as defined earlier. Therefore, the number of observations sampled from each stratum should be proportional to the amount of variability the quantity of interest, Y, has in that stratum.

In Figure 13.15, we apply stratified sampling to the normal distribution, $N(\mu, \sigma)$, where the strata are defined by σ deviations from the mean, μ. Therefore, we have $A_1 = (-\infty, \mu - 3\sigma]$, $A_2 = (\mu - 3\sigma, \mu - 2\sigma]$, $A_3 = (\mu - 2\sigma, \mu - \sigma]$, $A_4 = (\mu - \sigma, \mu]$, $A_5 = (\mu, \mu + \sigma]$, $A_6 = (\mu + \sigma, \mu + 2\sigma]$, $A_7 = (\mu + 2\sigma, \mu + 3\sigma]$, $A_8 = (\mu + 3\sigma, -\infty)$. We define the remaining parameters for stratified sampling by first defining $p_i = \mathbf{P}(Y \in A_i)$, and then utilize the optimal subsample size for each stratum as per Equation 13.61.

If Y is instead a stochastic process, such as an Ito process defined by a Wiener process-driven stochastic differential equation model defined later in Section 13.8, then stratified sampling can be applied if the distribution of the process is known at a specific time point, t, such as when $Y = S_t$

has the lognormal distribution. When applying stratified sampling to the simulation of sample paths of stochastic processes, where information regarding distribution of increments of the process alone may be utilized, the sample size, N, to draw in each increment of the process, Y, must be determined. This bears resemblance with simulating a multinomial tree of N branches at each time step, with the difference being the N outcomes are not fixed. They are instead randomly generated from the K strata, N_i observations from ith stratum.

For a general Ito process driven by the Wiener process, the above-described stratified sampling approach for the normal distribution can be applied to the Wiener increments, $\Delta W_t \sim N(0, \sqrt{\Delta t})$, in order to generate sample paths for the Ito process. When N outcomes are generated for the Wiener increment using stratified sampling, stratified sampling is applied again for each of the N outcomes in order to create the next N increments per outcome for advancing the sample paths forward. Hence, the resemblance with a multinomial tree. Similarly, other processes for which the distribution of the process at a specific time is known or when the distribution of the increments of the process is known, stratified sampling can be adapted for generating sample paths of the stochastic process. For instance, for a homogeneous discrete- or continuous-time Markov chain, such as a homogeneous Poisson process, $Y = \Delta N_t \sim Po(\lambda t)$, stratified sampling can help reduce the variance in the increments.

13.6.4 Latin Hypercube Sampling

Stratified sampling is an effective method for variance reduction; however, as the number of stochastic components in a model increases, that must be simultaneously simulated due to their interdependence and joint relevance to the problem objectives, stratifying a multidimensional state space becomes prohibitively cumbersome. Latin hypercube sampling is the extension of stratification for sampling to the case of multiple dimensional random variables. Multidimensional random variables are inescapable since stochastic factors and characteristics of a system do not appear in singletons in almost all practical application contexts. Therefore, Latin hypercube sampling becomes very crucial for interaction of multiple stochastic components of a model, such as when developing stress tests for asset-liability management or VaR calculations for market or credit risk, or for developing congestion scenarios for communication or transportation networks. As stated before, stratified sampling becomes infeasible in higher dimensions due to the rapid growth in the number of strata that must be sampled from in the higher-dimensional sample space.

Latin hypercube sampling surmounts the curse of dimensionality by treating all coordinates of the d-dimensional sample space equally, and sampling for each coordinate of the higher-dimensional space by stratified sampling of the marginal distribution. The exponential growth, K^d, of d-dimensional stratified sampling, where K is the number of strata for each dimension, is avoided by collating the outcomes generated for each coordinate. Instead of stratifying the d-dimensional space and generating outcomes from a d-dimensional hypercube, Latin hypercube method stratifies the one-dimensional marginal distributions of Y_i corresponding to the multidimensional joint distribution of $\mathbf{Y} = [Y_1, Y_2, \ldots, Y_d]$. It then permutes and collates each dimension to get the d-dimensional realizations. Furthermore, randomly permuting the coordinates of the d-dimensional realizations creates more realizations, thus more readily filling realizations in the K^d hypercubes.

In Figure 13.16, we display the two marginal distributions of two-dimensional stochastic components of a model, such as loss distribution from two loan portfolios. The approach developed in stratified sampling is applied to each dimension, generating the outcomes shown along the two axes. The two-dimensional outcomes are generated by randomly picking realizations from a stratum in each dimension and collating them in a two-dimensional vector (indicated by stars in the two-dimensional space in the figure). As an example, say the two-dimensional joint distribution is

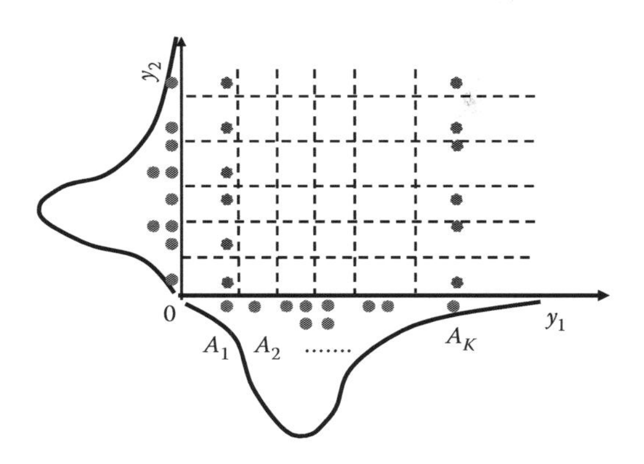

FIGURE 13.16
The two-dimensional sample space is broken down into strata constructed for marginal distribution of each coordinate. The two-dimensional realizations are constructed by randomly collating single-dimensional outcomes from single-dimensional stratified sampling.

given as

$$f(y_1, y_2) = 2y_2e^{-y_1}, \quad \text{for } 0 \le y_1, 0 \le y_2 \le 1 \tag{13.62}$$

$$= 0, \quad \text{otherwise,} \tag{13.63}$$

then the marginal distribution of Y_1 is obtained as $f_{Y_1}(y_1) = e^{-y_1}$, which is an exponential distribution, while the marginal distribution of Y_2 is $f_{Y_2}(y_2) = 2y_2$. It is trivial to sample from both these marginal distributions by applying stratified sampling, followed by randomly permuting the coordinates and collating to obtain the two-dimensional samples.

As is evident from the plot in Figure 13.16, Latin hypercube sampling is also well suited for generating scenarios for scenario analysis and stress testing, especially beneficial when high-dimensional stochastic inputs must be studied for their joint impact. Scenarios generated from single-dimensional marginal distributions that define stressed outcomes, or simply scenarios of interest for that dimension, can be picked and collated to create multidimensional scenarios.

13.6.5 Importance Sampling

Importance sampling method of variance reduction attempts to reduce variance of estimates of a quantity of interest by changing the probability measure from which samples are generated to create the estimate. For performing this change of measure, it recruits another random variable, much like the control variate method of variance reduction.

The name "importance" sampling is given to this method because in this method, we change the measure of the primary random variable(s) of interest to try to give more weight to the *important* outcomes of the random variable(s), thereby increasing the sampling efficiency. Let us say we are interested in estimating θ given below for a function, $h(.)$, of a set of random variables X.

$$\theta = E[h(X)] = \int_{-\infty}^{\infty} h(x)f(x)dx, \tag{13.64}$$

where X is the set of random variables with $f(x)$ being its joint probability density. We select another probability density, $g(x)$, defined on R^d that is positive wherever the original joint density function $f(x)$ is positive, that is, $g(x) > 0$ wherever $f(x) > 0$. We utilize this new probability density to make the following restatement for θ:

$$\theta = E[h(X)] = \int_{-\infty}^{\infty} h(x)[f(x)/g(x)]g(x)dx. \tag{13.65}$$

Alternatively, Equation 13.65 can be interpreted as follows:

$$\theta = \tilde{E}\left[h(X)\frac{f(X)}{g(X)}\right], \tag{13.66}$$

where the $\tilde{E}[]$ corresponds to taking expectation with respect to the new probability measure. The term $f(X)/g(X)$ is called the *likelihood ratio* or the *Radon–Nikodym derivative* for the change of probability measure.

Success of importance sampling in variance reduction lies in the selection of the new probability density, $g(X)$. For variance reduction to materialize, we need to have the following relations to hold:

$$\tilde{E}\left[\left(h(X)\frac{f(X)}{g(X)}\right)^2\right] = E\left[h(X)^2\frac{f(X)}{g(X)}\right], \tag{13.67}$$

$$< E[h(X)^2]. \tag{13.68}$$

This would definitely be accomplished if $f(x) < g(x)$, for all values of x. However, this would violate $g(x)$ remaining a probability density function. Therefore, one must investigate what values of x we can maintain $f(x) < g(x)$ for a more effective variance reduction while using importance sampling. One can intuitively surmise that should happen for values of random variable more likely to occur by the $f(x)$ distribution of random variable, X. Therefore, $g(x) > f(x)$ for values of x that more likely to occur, hence importance sampling.

Importance sampling can be quite beneficially applied in specific cases where estimation of likelihood of events must be done that occur with low probability. This can be accomplished by choosing the new probability measure that emphasizes these low likelihood values of the sample space. We investigate the topic of rare-event simulation next.

13.7 Rare-Event Simulation

Rare events refer to events of importance in the study of a system that occur, but not frequently. However, when they do occur, they can cause significant change in behavior of the system, thus, necessitating a careful study of the impact of such events. In transportation networks, when adverse conditions arise due to construction, weather conditions, or accidents, traffic patterns may change and these characteristics must be understood for better emergency response and traffic management. In air-traffic control, there is a nonzero probability of midair aircraft collisions, which, while it is hopefully very low, must be assessed in attempts to make air travel even more safe. In telecommunication networks or communication networks, the performance of the network can be significantly affected by high congestion due to a sudden hike in the number of calls or data being transmitted due, for instance, to an emergency, a terrorist attack. Massive failures in parts of communication network due to weather conditions, hurricane, tornado, etc., result in denial of service and can seriously hamper emergency response.

In manufacturing systems, equipment breakdowns can result in serious backups in production line, or increased number of defective items or drop in quality of items being produced. Simulating occurrence of such breakdowns is crucial to ensure quality assurance and minimizing rework or loss due to breakdowns. Human errors and fraudulent activities in service systems, such as in medical services, insurance claims, and banking, which hopefully occur rarely, can have quite damaging impact on the performance and reputation of a firm. Massively damaging weather events, such as the Indonesian tsunami, Japanese nuclear disaster, or hurricane Katrina's impact on New Orleans, can wreak havoc, running up insurance claims that threaten solvency of the private insurance providers. In banking, bank-runs or large losses due to clustered defaults in loan portfolio can threaten a bank from remaining a going concern. Therefore, while these events are less frequent, hence rare, it is important to develop an understanding of them and study their characteristics in order to improve the system performance and reliability.

As such, the different methods discussed in the previous section for variance reduction may be applied for rare-event simulation and estimation. Without a doubt, in rare-event estimation, the need for improvement in efficiency of estimation by variance reduction techniques is much needed. However, not all methods may be equally effective for the task at hand. We will focus here on importance sampling and splitting as the most effective techniques from the perspective of performance and broad applicability. Additionally, we will briefly describe some more methodologies that may be useful in specific contexts.

A rare event is an event that occurs with a small probability; however, how small is small depends on the context. In some cases, the probabilities being estimated may be as low as 10^{-6}, while in others, an event with a likelihood of 10^{-3} may be considered devastatingly rare enough. While one in a million microchip coming out defective could be a violation of a quality control target, a one in a thousand day, or roughly three years, severe daily trading loss may be bad enough to wipe out the performance of an investment fund manager.

In almost all the above example systems and rare-event estimation problems, the models required to depict reality of the system are complicated enough to not be solvable analytically. This could be because either the assumptions cannot be stringent enough for analytical solution to be obtainable, or the dimension of the problem could be too large. In such situations, simulation may be the only tool available at hand, so that, even when the state space is large, the results may be obtained with desired accuracy in reasonable time. This is necessary when dealing with critical systems, where the rare event is a catastrophic failure with possible human or monetary losses.

The first quantity of interest we may need to estimate is $\rho = P(E)$, where E is a rare event. A crude simulation approach would be to generate N i.i.d. random variates, $X_i = \mathcal{I}_E$, where $\mathcal{I}_E$ is the indicator function of the event E, and estimate $\hat{\rho}_N = \bar{X}$. Since X is a Bernoulli random variable, its variance is $\sigma_X^2 = P(E)(1 - P(E))$. Therefore, the confidence interval around this estimate, for any choice of sample size N, is $\left(\hat{\rho}_N \pm z_{\alpha/2}\sqrt{P(E)(1 - P(E))/N}\right)$. The relative accuracy of this estimate is $z_{\alpha/2}\sqrt{(1 - P(E))/P(E)N}$, which becomes progressively worse as $P(E)$ becomes smaller. This is the core challenge before the task of rare-event estimation.

13.7.1 Methodologies for Rare-Event Estimation

There are many techniques for addressing the rare-event characteristics of a problem [54,83]. These methodologies include:

Importance sampling: As seen in Section 13.6.5, importance sampling is a technique that, when applied for rare-event estimation, employs a change of measure for the variable of interest in the model so as to make the rare events happen often rather than rarely.

Splitting: Splitting is the other important rare-event estimation technique, where a nested sequence of events is constructed containing the rare event, each event itself not being rare. Say E is the rare event whose probability of occurrence must be estimated. We construct $E = E_m \subset E_{m-1} \subset \cdots \subset E_1$, such that $E_{m-1}, \ldots, E_1$ are not "rare." We use the following relation to estimate $P(E)$:

$$P(E) = P(E_m|E_{m-1}) \cdots P(E_2|E_1)P(E_1). \tag{13.69}$$

Each individual conditional probability is estimated using crude Monte Carlo without changing the measure of the original variable of interest in the model.

Stratified sampling: Stratified sampling was discussed in Section 13.6.3 as a method for variance reduction where the sample space is broken into subsets or strata. A designed approach to sampling is constructed from each stratum with the goal of improving efficiency of estimation. Customizing this method of rare events implies constructing the strata and sampling in a way that more observations are drawn from the desired regions.

Cross-entropy method: This is an adaptive importance sampling method, where instead of attempting a single shot at obtaining the best change of measure, an iterative process is employed in improving the change of measure to obtain the optimal asymptotic zero-variance estimator [83,92,93].

Heavy-tailed simulations: Rare events for the light-tailed distributions occur in a very different manner than those for the heavy-tailed distributions. Therefore, specialized methods are developed for the heavy-tailed settings. An approach based on hazard rate twisting, as opposed to exponential twisting in the light-tailed case, is used for constructing the importance sampling distribution [53].

We consider a couple of example contexts to further elaborate on some of the above methodologies for rare-event simulation. Queuing networks are relevant for many of the domains we discussed in Section 13.2 where simulation modeling and analysis is useful. Communication networks and transportation networks, as well as many of the manufacturing and service systems, are in effect queuing networks, where flow of entities through routes and processing units makes the fundamental purpose of the system served. The rare events may occur in the queuing networks of these systems from heavy traffic due to arrival rates being large or from heavy-tailed service times. The former may happen due to nonstationarity in arrival process characteristics, such as due to episodic or seasonally higher demand for product, services, and events of emergency. Heavy-tailed service times can arise from some job sizes being unusually large, such as large files being transferred over a communication network, requiring much longer to transmit or complete. This can clog the queuing network, causing congestion and delays.

As discussed in Section 13.2.4, in banking and insurance, severe losses due to many loan defaults or losses due to severely adverse events are rare events with likely very devastating consequences. Rare-event estimation for these scenarios is essential for sound risk management. Both in banking and insurance, capital reserves are maintained to ensure solvency of the firm should the rare but severely adverse events transpire up to some confidence level. Measures, such as VaR, CVaR, or expected shortfall, for evaluating the tail characteristics of the loss distribution are utilized for determining the capital reserve. We will apply stratified sampling and splitting techniques for the tail estimates of loss given defaults in loan portfolios to determine the capital reserves requirement.

Let us consider a portfolio of M loans that a bank has extended to a set of obligors. An obligor may be unable to make a timely repayment of the debt, and as a result may default causing losses to the bank. Usually, this likelihood of default is assessed for a period of time in the future, thus requiring the assessment of loss given default experienced by the loan portfolio for a set horizon, say one year, one quarter, or a couple of years. Corresponding to each obligor, we may assign a default indicator, X_i, such that, $X_i = 1$ if the ith obligor defaults in the given time horizon, and is zero otherwise. The probability of default $p_i = P(X_i = 1)$ for the ith obligor and l_i is the loss given default from the ith obligor's loan. The bank's loss exposure for this loan portfolio is given by $L = \sum_{i=1}^{M} l_i X_i$. We would wish to efficiently estimate the probability of the portfolio loss to exceed a large enough value, $P(L > x)$, or the conditional expectation of the loss being high, $E[L|L > x]$.

We apply stratified sampling and splitting to create an estimate for the probability, $P(L > x)$, as follows:

$$\theta = P(L > x) = \sum_{i=1}^{K} P(L > x|L \in A_i)P(L \in A_i), \tag{13.70}$$

$$= \sum_{i=2}^{K} P(L > x|L \in A_i)P(L \in A_i)$$
$$+ P(L > x|L \in A_1 \cap E_{m-1}) \cdots P(E_2|E_1)P(E_1)P(L \in A_1), \tag{13.71}$$

where, as shown in Figure 13.17, $\mathcal{I}_{\{L>x|L\in A_1\}} = E_m \subset E_{m-1} \subset \cdots \subset E_1$, such that $E_{m-1}, \ldots, E_1$ are not "rare." The stratification used in Equation 13.70 is in the M-dimensional space for the M obligors defaulting during the time horizon and the extent of loss the default may cause the bank. The rare-event estimator is defined in terms of the Equation 13.71.

We consider an application of rare-event simulation using importance sampling for heavy-tailed phenomena. Rare events occur in a very different manner in the heavy-tailed settings than in the light-tailed ones. In queuing networks arising in communication or transportation networks, or in service or production systems, rare events may occur in the system from heavy traffic due to arrival rates being large or from heavy-tailed service times. For heavy-tailed distribution, defined as one for which the moment generating function of the distribution is infinite for any positive value of the argument, as in Reference 54, the framework of exponential twisting-based importance sampling used of light-tailed cases may no longer be directly applicable. A heavy-tailed problem must first be transformed into a light-tailed problem and then the light-tailed exponential twisting importance sampling framework can be applied to them.

Let us the quantity of interest, $Y = g(X_1, X_2, \ldots, X_N)$, for some function $g : R^N \rightarrow N$ and independent random variables, $X_1, \ldots, X_N$. X_i's have a probability density function, $f_i(x)$, and cumulative distribution function, $F_i(x)$. These random variables could be quantities, arrival rates,

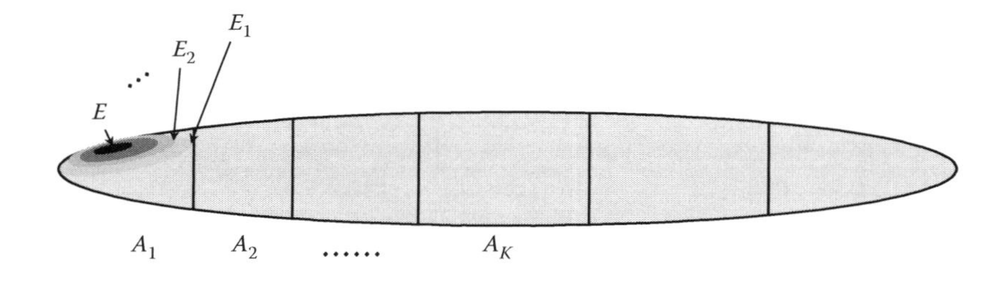

FIGURE 13.17
Combining stratified sampling and splitting in order to estimate rare-event loss due to default or adverse weather event.

services times, etc., for a queuing network, while Y is a performance measure for the queuing network. We are interested in computing, $P(Y > y)$ for some threshold value of performance y.

Importance sampling would seek a new probability density for X_i's, say $\tilde{f}_i(x)$, with the same support as $f_i(x)$, and result in the following estimation:

$$P(Y > y) = E[\mathcal{I}_{\{Y>y\}}] = \tilde{E}[\mathcal{I}_{\{Y>y\}} L(X_1, \ldots, X_N)], \quad (13.72)$$

where $\tilde{E}[]$ is expectation with respect to the new density and

$$L(x_1, \ldots, x_N) = \prod_{i=1}^{N} \frac{f_i(x_i)}{\tilde{f}_i(x_i)}. \quad (13.73)$$

For light-tailed random variables, the change of measure done as follows is called exponentially twisting the original distributions, and yields an asymptotically optimal change of measure for importance sampling. Exponentially twisting $f_i(x)$ by amount $\theta(> 0)$ gives the new density, $f_{i,\theta}(x) = f_i(x)e^{\theta x}/M_{X_i}(\theta)$, where $M_{X_i}(\theta)$ is a finite-valued moment generating function of X_i. The likelihood ratio becomes

$$L(X_1, \ldots, X_N) = \prod_{i=1}^{N} \left(M_{X_i}(\theta) e^{-\theta X_i} \right). \quad (13.74)$$

One seeks the optimal $\theta = \theta^*$ that minimizes the variance of the exponential twisting importance sampling estimator. In heavy-tailed phenomena, the exponential twisting will not directly apply, as the moment generating function is infinite in the case of heavy-tailed distributions. Therefore, transformations must be constructed to convert the heavy-tailed distributions into light-tailed ones before the above importance sampling procedure may be applied. For examples of transformations, refer to References 36, 37, 43, and 54.

13.8 Simulation for Dynamic Systems

In Section 13.2, several problem domains were discussed and an overview of modeling approaches that find use in these problem domains was provided. Continuous-time and continuous-space stochastic models, constructed in terms of stochastic processes, such as the Brownian motion, find broad application in finance, financial engineering, and risk management. Integrals defined in terms of the Brownian motion helps construct stochastic differential equation models for defining a variety of processes in these applications. Some of these models were discussed in Sections 13.2.3 and 13.2.5. For instance, Ito integral is defined very similarly to the Riemann integral, with increments in time being replaced by increments in the Brownian motion or the Wiener process, as follows:

$$\sum_{j=0}^{N} f(t_j^*)[W_{t_{j+1}} - W_{t_j}] \rightarrow \int_0^T f(t) dW_t, \quad \text{as } \Delta t_j \rightarrow 0, \forall j, \quad (13.75)$$

for measurable functions, $f(t, \omega)$ and $t_j^* = t_j$.

The Ito integral with respect to the Wiener process is utilized to define more general models driven by the Brownian motion. These models are written as follows:

$$dX_t = \mu(t, X_t)dt + \sigma(t, X_t)dW_t, \tag{13.76}$$

and imply that the process X_t evolves by the following integrals:

$$X_t = X_0 + \int_0^t \mu(s, X_s)ds + \int_0^t \sigma(s, X_s)dW_s. \tag{13.77}$$

In some cases, an analytical solution of the model constructed in Equation 13.77 is obtainable. However, when analytical solutions of stochastic differential equations describing time-dependent evolution of stochastic variables in a system are not readily available, we must resort to numerical techniques using simulation analysis. In this section, we develop these simulation techniques, as well as develop the terminology for assessing the accuracy of these numerical solutions. In order to develop the methodology for simulation-based solutions of stochastic differential equations, we will need to recall random variate generation topics covered in Section 13.3, as well as develop the basic principles of numerical solutions of deterministic differential equations.

The primary idea employed is to approximate the differential equation-based representation of the model to a difference equation-based model on a chosen time discretization of the interval on which the solution is required. The solution can then be obtained by an iterative procedure, once the initial condition of the stochastic process is known. As stated earlier, the advantage of modeling risks by continuous-time model, even though one is not able to solve these models analytically, is that an arbitrary choice of discretization can be picked to achieve the desired accuracy of the numerical solution. We begin with investigating the simplest of methods for simulation-based solutions of stochastic differential equations, which is the Euler method.

13.8.1 Euler Method for Solving Differential Equations

We first apply the Euler scheme to solve differential equations in the deterministic case. Let $x = x(t; t_0, x_0)$ be the solution of an initial value problem (IVP)

$$\frac{dx}{dt} = a(t, x), \tag{13.78}$$

with $x(0) = x_0$. An IVP consists of a deterministic differential equation-based model, where the initial value of the function satisfying the differential equation is known. Suppose we want to solve the system given in Equation 13.78 on a time interval, $[0, T]$. We will begin with creating a discretization of the interval into subintervals, $0 = t_0 \leq t_1 \leq \cdots \leq t_N = T$. We may keep the gap between these discretized time points equidistant or not. If we choose to keep them equidistant, although this is not essential, we have $t_n - t_{n-1} = \Delta = T/N$ for all n, such that $0 \leq n \leq N$. We will control the value of Δ to achieve the desired accuracy. In general, what is essential is that the maximum of all subinterval lengths gets smaller for improved accuracy.

In Equation 13.78, we first approximate the first derivative by a simple difference scheme. The difference scheme is $dx/dt \simeq (x(t_n) - x(t_{n-1}))/\Delta$. This allows us to write the following approximation of the model in Equation 13.78,

$$y_{n+1} = y_n + a(t_n, y_n)\Delta \quad \text{for } n = 1, \ldots, N, \tag{13.79}$$

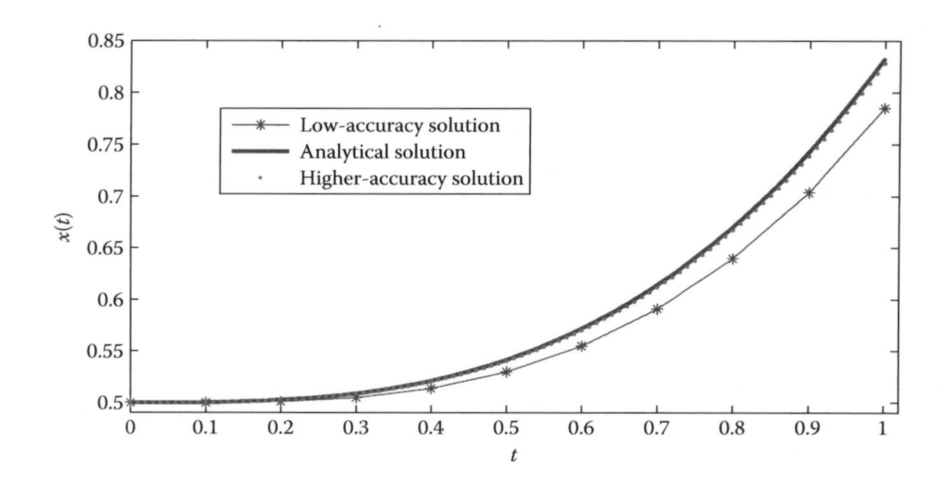

FIGURE 13.18
Comparison of exact and numerical solution for an example ordinary differential equation.

where we are careful to indicate the approximate solution by a new symbol, indicating $y_n \simeq x(t_n)$. Using this approximate equation, the approximate solution y_n, for $n = 1, \ldots, N$, can be obtained iteratively. Once we set $y_0 = x_0$, then y_1 can be obtained from Equation 13.79 by plugging in the value of y_0, and so on.

In Figure 13.18, we apply the Euler scheme to a problem for which we know the analytical solution. We also compute the simulation-based numerical solution to demonstrate how well the numerical solution approximates the analytical solution. In general, we do not have this luxury of comparing analytical solution with the numerical one, since the reason we seek numerical solution is that we do not know a way to obtain the analytical solution. In this simple case, we get to make this comparison, and see that the numerical solution gets very close to the analytical solution for reasonably coarse discretization (here, $N = 100$). Figure 13.18 displays the solution of the following problem and its discretized approximation:

$$\frac{dx}{dt} = t^2, \tag{13.80}$$

$$y_{n+1} = y_n + t_n^2 \Delta \quad \text{for } n = 1, \ldots, N, \tag{13.81}$$

where the initial condition is taken as $x(0) = y_0 = 1/2$.

We now consider a general stochastic differential equation model for a process, X_t, given as

$$dX_t = a(X_t)dt + b(X_t)dW_t, \quad t \in [t_0, T], \tag{13.82}$$

with the initial value of the process given as $X_0 = x_0$.

An Euler approximation of the original model in Equation 13.82 can be obtained and simulated to obtain an approximate solution. Once again, we construct the discretization, $t_0, t_1, t_2, t_3, \ldots, t_{N-1}, t_N = T$, of equidistant points with a time step of length $\Delta = (T - t_0)/N$. The Euler approximation in this case is a continuous-time stochastic process $Y = \{Y(t), t_0 \leq t \leq T\}$ satisfying the iterative scheme

$$Y_{n+1} = Y_n + a(Y_n)\Delta + b(Y_n)\Delta W_n. \tag{13.83}$$

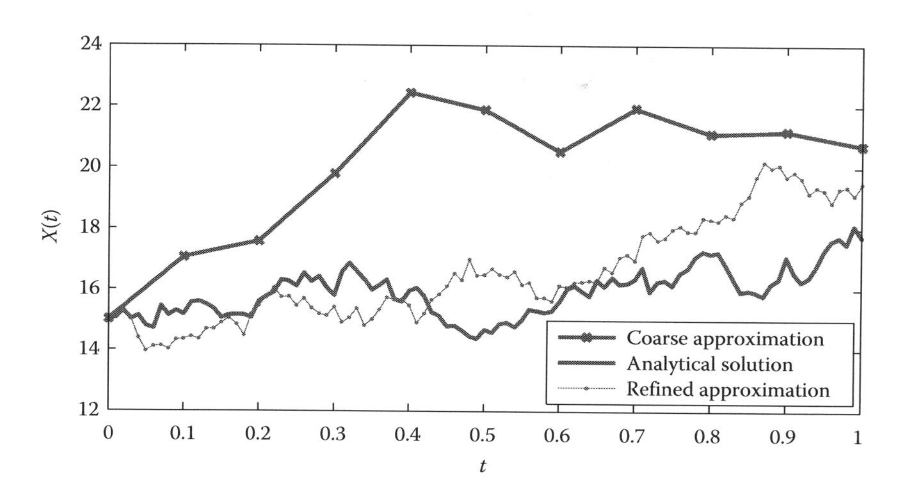

FIGURE 13.19
Comparison of exact and numerical solution for an example stochastic differential equation.

The main difference in the stochastic case is that here ΔW_n needs to be generated as a random variate with the appropriate distribution. Given the properties of the Wiener process, these Wiener increments are independent Gaussian random variables with mean, $E[\Delta W_n] = 0$, and variance, $E[(\Delta W_n)^2] = \Delta$.

In Figure 13.19, we apply the Euler scheme to a problem for which we know the analytical solution. We also compute the simulation-based solution to demonstrate how the numerical solution compares with the analytical one. Figure 13.19 displays the solution of the following model we have described before in Section 13.2.3, as well as its discretized approximation:

$$dX_t = \mu X_t dt + \sigma X_t dW_t, \tag{13.84}$$

$$Y_{n+1} = Y_n + \mu Y_n \Delta + \sigma Y_n \Delta W_n, \quad \text{for } n = 1, \ldots, N, \tag{13.85}$$

where the initial value of the variable is taken as $X(0) = Y_0 =\$15$, the drift is taken as $\mu = 14\%$ and volatility is $\sigma = 20\%$. As seen in Figure 13.19, the refined approximation, which corresponds to $N = 100$, performs much better than the coarse approximation, and is already picking up the characteristics of the analytical solution.

In the simulation approach, not only is the $\{Y_n\}$ sequence obtained from applying the Euler scheme in Equation 13.83 an approximation of the actual solution, X_t, of the original model in Equation 13.82, it is also only observed at discrete time points of the time discretization, $t_0, \ldots, t_{N-1}, t_N = T$. To make the $Y(t)$ process a continuous-time stochastic process, the value at the intermediate points must be determined by interpolation, such as by applying a linear interpolation, as is done for sample paths in Figure 13.19. The linear interpolation will not capture the fine structure of sample paths of a general diffusion processes that they inherit from the Wiener process. However, the finer the grid size, the finer will be the structure inherited. The question of quality of approximation in the approximate solution still remains.

In Figure 13.20, we plot the histogram of the approximate and analytical solution of the example problem in Equations 13.84 and 13.85 at the terminal time point, $t = 1$. Since we know the properties of the analytical solution, namely, that it is the geometric Brownian motion, we also plot a lognormal probability plot for both the analytical and simulation-based solution in Figure 13.20 panel (c) and

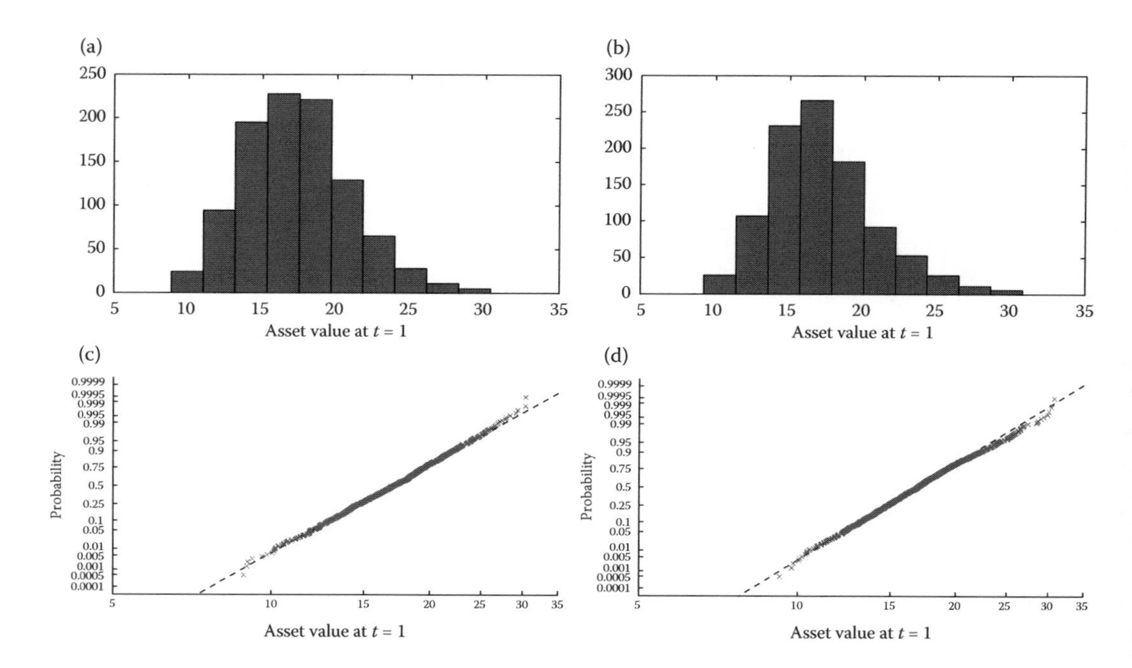

FIGURE 13.20
Comparison of distributional properties of the exact and numerical solution for the example stochastic differential equation. (a) Histogram for asset value at $t = 1$ using analytical soln. (b) Histogram for asset value at $t = 1$ using simulation-based soln. (c) Probability plot for analytical soln for lognormal distribution. (d) Probability plot for simulation-based soln for lognormal distribution.

(d). In both cases, the lognormal distribution is supported. The difference in mean and standard deviation of asset value at $t = 1$ by the two solutions is of order 10^{-2}.

We next turn our attention to creating precise measures for accuracy of solutions, and how they respond to improved accuracy as time discretization is refined.

13.8.2 Evaluating Simulation Solutions

When attempting to solve a differential equation using numerical techniques, two most important questions must be addressed. For any specific method of discrete approximation of the continuous problem and for a specific selected time discretization, how "far" will the approximate solution end up being from the actual solution? Second, for a specific method of discrete approximation of the continuous problem, as the time discretization is refined, how rapidly does the approximate solution become "closer" to the actual problem? The first enquiry is termed *error analysis*, while the latter is called *rate of convergence* of an approximate method. We begin with defining the latter.

13.8.2.1 Convergence Properties of Solutions

We have so far looked at only one method of approximating the continuous problem, the Euler scheme. The Euler method was utilized to define an approximate solution in the deterministic and stochastic examples in the previous section. We can measure the goodness of the approximate solution in two ways. There is the *local discretization error*, which is the error made in a single time step, or a single iteration of the numerical procedure to obtain y_n in Equation 13.79. This is obtained assuming that the exact solution at t_n, $x(t_n)$, is known, and matches y_n, then the difference between $x(t_{n+1})$ and y_{n+1} is observed. This error is usually not going to be zero.

Once several such steps of the iteration are made to obtain a sequence of y_n values, the (local discretization) errors accumulate. *Global discretization error* is all the errors accumulated up to a time point, t. Therefore, global discretization error can be seen as the sum of propagated truncation error and local discretization error at any point of the iterative procedure. The size of this error, and how rapidly it can go down as discretization is refined, provides us the definition of convergence rate.

Order of convergence: A method converges with order γ if there exists a constant K ($< \infty$) such that the $|x(t_{n+1}) - y_{n+1}| = |e_{n+1}|$ can be bounded from above by $K\Delta^{\gamma}$, for all $\Delta \in (0, \delta_0)$, for some $1 > \delta_0 > 0$.

By this definition of order of convergence, the Euler method can be shown to have an order, $\gamma = 1.0$. However, it should be noted that the above definition of order of convergence is stated assuming that there are no round-off errors. In practice, due to round-off errors, there is a minimum time-discretization level, Δ_m, below which we cannot hope to improve accuracy by taking finer time-discretization levels. Therefore, in order to get more accurate approximate solutions, we need to consider methods with higher orders of convergence. Different discretization schemes may be explored and adopted, each with a level of accuracy and computational burden.

13.8.2.2 Error Analysis: Absolute Error Criterion

The above definition of order of convergence was developed with a focus on deterministic problems. When we apply simulation to solve a stochastic differential equation, the solution is not a deterministic function. The solution is a realization of a stochastic process. From the point of view of accuracy of the solution, the question arises regarding the sense in which the approximate solution converges to the actual one.

In fact, the quality of a discrete-time approximate solution can only be judged on the basis of the main goal of simulation. There are two basic tasks connected with the simulation of solution of a stochastic differential equation: (1) when a good *pathwise approximation* is required, and (2) when an approximation of *expectation of a functional* of an Ito process is required, such as probability distribution or its moments. Depending on the requirement, an appropriate error criterion needs to be used. We first define the criterion for pathwise convergence of approximate solution.

Absolute error criterion is defined as

$$\epsilon = E[|X_T - Y(T)|], \tag{13.86}$$

which gives a measure of pathwise closeness at the end of the time interval $[0, T]$. In case of the deterministic IVP, where the function $b(.) = 0$ in Equation 13.82, the absolute error criterion coincides with the usual deterministic error criterion for the absolute global discretization error.

13.8.2.2.1 Method to Estimate the Absolute Error

Generate N sample paths for the Wiener process in $[0, T]$ and compute the discrete solution $Y(t)$ for each sample path of the Wiener process. Using the same sample paths, compute the exact solution (provided this is known). Denote the quantities obtained from the kth simulation by $Y(T, k)$ and corresponding exact solution by $X_{T,k}$. Then the absolute error can be estimated as

$$\hat{\epsilon} = \frac{1}{N}\sum_{k=1}^{N} |X_{T,k} - Y(T,k)|. \tag{13.87}$$

13.8.2.2.2 Confidence Interval for the Absolute Error Estimate

The estimated absolute error $\hat{\epsilon}$ is a random variable. It will be asymptotically (or in the limit) Gaussian if $|X_{T,k} - Y(T,k)|$ is believed to have the same distribution for all k. The estimated absolute error, $\hat{\epsilon}$, will converge in distribution to the nonrandom expectation ϵ as $N \to \infty$. These facts can be used to create a confidence interval estimate of the true absolute error, ϵ.

This requires getting an estimate of the standard deviation σ_ϵ of the estimated absolute error $\hat{\epsilon}$. This is accomplished by performing M batches of N simulations each. Now, $X_{T,k,j}$ will be the kth exact solution in the jth batch and $Y(T,k,j)$ will be the kth discrete approximate solution in the jth batch. Compute the absolute error for the jth batch as follows:

$$\hat{\epsilon}_j = \frac{1}{N}\sum_{k=1}^{N} |X_{T,k,j} - Y(T,k,j)|. \tag{13.88}$$

Then compute the overall absolute error across batches:

$$\hat{\epsilon} = \frac{1}{M}\sum_{j=1}^{M} \hat{\epsilon}_j = \frac{1}{MN}\sum_{j=1}^{M}\sum_{k=1}^{N} |X_{T,k,j} - Y(T,k,j)|. \tag{13.89}$$

Once the batch mean and grand mean across batches is computed, the estimated variance of the absolute error estimates can be computed as follows:

$$\hat{\sigma}_\epsilon^2 = \frac{1}{M-1}\sum_{j=1}^{M} (\hat{\epsilon}_j - \hat{\epsilon})^2. \tag{13.90}$$

This implies that

$$t = \frac{\hat{\epsilon} - \epsilon}{\hat{\sigma}_\epsilon/\sqrt{M}} \tag{13.91}$$

will have a t-distribution with $M - 1$ degrees of freedom, since, if we knew the variance of absolute error precisely, the ratio would be approximately normally distributed. We select a confidence level of $100(1-\alpha)\%$ for getting a confidence interval for ϵ, then the interval will be obtained as $(\hat{\epsilon} - \Delta\epsilon, \hat{\epsilon} + \Delta\epsilon)$, where $\Delta\epsilon = t_{1-\alpha/2,M-1}(\hat{\sigma}_\epsilon/\sqrt{M})$ and $P(-t_{1-\alpha/2,M-1} \le t \le t_{1-\alpha/2,M-1}) = 1 - \alpha$.

13.8.2.2.3 Convergence by Absolute Error Criterion

A discrete-time approximation Y with maximum time step size δ *converges strongly* to X at time T, if $\lim_{\delta \to 0} E[|X_T - Y(T)|] = 0$. In order to compare schemes for the quality of solutions they give, we develop the measure for their *order of strong convergence*. We say a discrete time approximation

Y converges strongly with order $\gamma > 0$ at time T, if there exists a positive constant C, which does not depend on δ and a $\delta_0 > 0$, such that

$$E[|X_T - Y(T)|] \leq C\delta^{\gamma} \tag{13.92}$$

for each $\delta \in (0, \delta_0)$. This is a direct extension of the definition in the deterministic case of order of convergence, and reduces to it when the diffusion term is zero. Euler scheme studied earlier was the simplest useful scheme, but in general is not particularly efficient. Euler scheme has an order of strong convergence, $\gamma = 0.5$.

Let us look at a new scheme that has a higher order of strong convergence, the *Milstein scheme*, which when applied to the model in Equation 13.82 gives

$$Y_{n+1} = Y_n + a(Y_n)\Delta_n + b(Y_n)\Delta W_n + \frac{1}{2}b(Y_n)b'(Y_n)(\Delta W_n^2 - \Delta_n). \tag{13.93}$$

This is obtained by a Taylor expansion of $X(t)$ about $X(t_0)$, and truncating after first three terms. The strong order of convergence for Milstein scheme is 1.0.

13.8.2.3 Error Analysis: Mean Error Criterion

At other times, we may be interested not in the paths of an Ito process, but instead might want to capture some distributional information about them, such as specific moments, probability of events, etc. Under such circumstances, the requirements are less stringent than for the pathwise case. Here, we are seeking a good approximation in moments of the solution, and will need to define order of weak convergence. Let us begin with estimating the error in approximate solution of a stochastic differential equation when the criterion is being able to match the mean of the Ito process. Therefore, we are interested in computing the $E[X_T]$ using the discrete-time approximation, $Y(T)$, and its mean $E[Y(T)]$.

Mean error criterion: We define the error by

$$\mu = E[Y(T)] - E[X_T]. \tag{13.94}$$

Note that the mean error can take both negative and positive values. The estimated values of *mean error* will be obtained by running the simulation N times and calculating the mean of these outcomes of the $Y(T)$, as follows:

$$\hat{\mu} = \frac{1}{N}\sum_{k=1}^{N} Y(T,k) - E[X_T]. \tag{13.95}$$

As in the case of estimating absolute error, if we want to construct a confidence interval for the mean error estimate, we will generate M batches with N simulations each and estimate the above mean error for each batch.

$$\hat{\mu}_j = \frac{1}{N}\sum_{k=1}^{N} Y(T,k,j) - E[X_T]. \tag{13.96}$$

Using this set of estimates, we compute the overall average across batches:

$$\hat{\mu} = \frac{1}{M}\sum_{j=1}^{M}\hat{\mu}_j = \frac{1}{MN}\sum_{j=1}^{M}\sum_{k=1}^{N} Y(T,k,j) - E[X_T]. \tag{13.97}$$

Combining the batch mean and overall mean, we compute the standard deviation of mean error estimate as follows:

$$\hat{\sigma}_{\mu}^2 = \frac{1}{M-1}\sum_{j=1}^{M}(\hat{\mu}_j - \hat{\mu})^2. \tag{13.98}$$

This implies that

$$t = \frac{\hat{\mu} - \mu}{\hat{\sigma}_{\mu}/\sqrt{M}} \tag{13.99}$$

follows t-distribution with $M - 1$ degrees of freedom. As before, if we knew the theoretical variance of absolute error, the ratio in Equation 13.99 would be approximately normally distributed. Now if we select a confidence level of $100(1 - \alpha)\%$ for constructing a confidence interval for μ, then the confidence interval will be obtained as

$$(\hat{\mu} - \Delta\mu, \hat{\mu} + \Delta\mu), \tag{13.100}$$

where $\Delta\mu = t_{1-\alpha/2,M-1}\frac{\hat{\sigma}_{\mu}}{\sqrt{M}}$ and $P(-t_{1-\alpha/2,M-1} \leq t \leq t_{1-\alpha/2,M-1}) = 1 - \alpha$.

13.8.2.3.1 Convergence by Mean Error Criterion

In general, there could be a general functional to be estimated for the Ito process, say, $E[g(X_T)]$. Then we would do a similar computation using the simulated value $Y(T)$ as we would do for the mean, $E[X_T]$. We will say that a general discrete-time approximation $Y(T)$ with maximum step-size, δ, *converges weakly* to X at time T as $\delta \to 0$ with respect to a class $\mathcal{C}$ of test functions $g : R \to R$, if we have

$$\lim_{\delta \to 0} |E[g(X_T)] - E[g(Y(T))]| = 0 \tag{13.101}$$

for all $g \in \mathcal{C}$. So, if $\mathcal{C}$ contains all polynomials, then all moments of $Y(T)$ converge to the moment of the true process.

We say a discrete-time approximation Y *converges weakly with order* $\beta > 0$ to X at time T as $\delta \to 0$, if for each polynomial $g(x)$ there exists a positive constant C, which does not depend on δ, and a finite $\delta_0 > 0$, such that

$$\mu(\delta) = |E[g(X_T)] - E[g(Y(T))]| \leq C\delta^{\beta}, \tag{13.102}$$

for each $\delta \in (0, \delta_0)$. Euler scheme has an order of weak convergence of 1.0. We will next look at another scheme with a higher order of weak convergence.

The important point to note is that weak and strong convergence criteria lead to the development of different discrete-time approximations, which are only efficient with respect to one of the two criteria. This fact makes it important to clarify the aim of a simulation before choosing the approximation scheme. The questions to ask are whether a good pathwise approximation of the Ito process is required or if the approximation of some functional of the Ito processes the real objective.

13.8.2.3.2 Weak Higher-Order Methods

The Euler scheme has a weak order of convergence of 1.0. We consider a discretization scheme with a higher-order truncation of Ito–Taylor expansion. The resulting scheme is as follows:

$$\begin{aligned} Y_{n+1} &= Y_n + a(Y_n)\Delta_n + b(Y_n)\Delta W_n + \frac{1}{2}b(Y_n)b'(Y_n)(\Delta W_n^2 - \Delta_n) \\ &\quad + a'(Y_n)b(Y_n)\Delta Z_n + \left(\frac{1}{2}a(Y_n)a'(Y_n) + \frac{1}{2}a''(Y_n)b^2(Y_n)\right)\Delta_n^2 \\ &\quad + \left(a(Y_n)b'(Y_n) + \frac{1}{2}b''(Y_n)b^2(Y_n)\right)(\Delta W_n \Delta_n - \Delta Z_n), \end{aligned} \tag{13.103}$$

where ΔZ_n is normally distributed with mean $E[\Delta Z_n] = 0$, $Var(\Delta Z_n) = 1/3\Delta_n^3$ and $cov(\Delta Z_n, \Delta W_n) = 1/2\Delta_n^2$. This method has an order of weak convergence of 2.0.

In this section, we have given an overview of a variety of improved methods for numerically solving stochastic differential equations, along with principle on which error analysis of these improved methods is conducted. For additional improved schemes for stochastic models, the reader is advised to refer to Kloeden and Platen [60].

13.8.3 Estimating Parameters

In the above section, we have constructed a framework for developing new models for dynamic evolution of stochastic variables utilizing the standard Wiener process. Similar constructions can be done using other processes, those discussed in Section 13.2.5. These will likely involve developing new definition of integrals, as we did with Ito integral, and the related calculus. For instance, we can explore utilizing the Poisson process for this purpose in order to capture sudden large changes along with diffusive changes in a random process. Developing these new integrals beyond the Ito integral is beyond the scope of this chapter.

Once a model is constructed, described by a stochastic differential equation, it can be solved analytically or numerically. The reader would have noticed that each model gets stated in terms of some crucial parameters. Unless some meaningful values of these parameters can be identified, the models cannot be used. In this section, we present some methods for calibrating the models developed in this chapter.

13.8.4 Geometric Brownian Motion

The simplest stochastic model we obtained in earlier discussions of this chapter was

$$dS_t = \mu S_t dt + \sigma S_t dW_t, \tag{13.104}$$

which was used to describe price dynamics, S_t, for a risky asset, along the initial value of the asset known to be S_0. Analytical solution is obtainable for this model as $S_t = S_0 \exp((\mu - \sigma^2/2)t + \sigma W_t)$. However, the two parameters in the model, namely, μ and σ, remain to be estimated in order to use the model.

Estimation of these parameters requires data, say M historical observations of the risky asset price, $\{S_{t_1}, S_{t_2}, \ldots, S_{t_M}\}$. It is noted that for these observations, given they are all realized by the model in Equation 13.104, we have

$$S_{t_n} = S_{t_{n-1}} \exp\left(\left(\mu - \frac{\sigma^2}{2}\right)(t_n - t_{n-1}) + \sigma\left(W_{t_n} - W_{t_{n-1}}\right)\right), \tag{13.105}$$

for all values of $n \in \{1, 2, \ldots, M\}$. Rearranging the terms in Equation 13.105 gives

$$\ln\left(\frac{S_{t_n}}{S_{t_{n-1}}}\right) = \left(\mu - \frac{\sigma^2}{2}\right)(t_n - t_{n-1}) + \sigma\left(W_{t_n} - W_{t_{n-1}}\right), \tag{13.106}$$

where $\ln\left(S_{t_n}/S_{t_{n-1}}\right) = r_{t_n}$ is called the log-return of the asset. If t_n's are equispaced, $(\mu - \sigma^2/2)(t_n - t_{n-1}) + \sigma\left(W_{t_n} - W_{t_{n-1}}\right)$ are i.i.d. by the normal distribution. The estimation of the parameters can be accomplished by first estimating the variance of the i.i.d. normal observations, followed by their mean. We obtain the following estimates:

$$\hat{\sigma}^2 = \frac{1}{(t_n - t_{n-1})} variance\left(\ln\left(\frac{S_{t_n}}{S_{t_{n-1}}}\right)\right), \tag{13.107}$$

$$\hat{\mu} = \frac{1}{(t_n - t_{n-1})} mean\left(\ln\left(\frac{S_{t_n}}{S_{t_{n-1}}}\right)\right) + \frac{\hat{\sigma}^2}{2}. \tag{13.108}$$

These estimates utilize the fact that the distribution of log-returns is known to be normal in this geometric Brownian motion model. Since the estimates are obtained utilizing the first and the second moments of the normal distribution, we can call this approach the method of moments. We discuss this method in more detail later. However, there is another method these estimates can be obtained by; we discuss this important method next.

13.8.4.1 Method of Maximum Likelihood

We describe the maximum likelihood method for calibrating models in general terms, which utilizes the knowledge of distributional properties of the variable being modeled. Assume that the model describes the risky asset dynamics, S_t, and we can describe the probability distribution of a function of these risky asset price dynamics. For example, the log-returns described in the previous section, that is, $r_t = \ln(S_t/S_{t-\Delta t})$. Consider a sample of observations for log-return of the risky asset are available, $\{r_{t_0}, r_{t_1}, \ldots, r_{t_N}\}$, where $t_0 = 0$ and $t_N = T$ are two time end-points of the sample. Desirably the observations are equi-spaced, that is, Δt apart, which will make the notation simpler.

If we know the distributional properties of r_{t_k}, even if they are conditional on precisely knowing the value of $r_{t_{k-1}}$, we can make this useful in this parameter estimation method. Let us say the conditional distribution for r_{t_k}, given $r_{t_{k-1}}$, is denoted by $f(r_k|r_{k-1}; \theta)$, for $k = 1, \ldots, N$ and with the set of parameters to be estimated in the model being, $\theta = [a; b; c; d]$. For instance, if the conditional distribution of r_{t_k}, given $r_{t_{k-1}}$, is normally distributed, we would have

$$f(r_k|r_{k-1}; \theta) = \frac{1}{\sqrt{2\pi}\sigma(r_{k-1}, \theta, \Delta t)} e^{-\frac{[r_k - \mu(r_{k-1}, \theta, \Delta t)]^2}{2\sigma^2(r_{k-1}, \theta, \Delta t)}}, \tag{13.109}$$

for $k = 1, \ldots, N$, and $\mu(r_{k-1}, \theta, \Delta t)$ and $\sigma(r_{k-1}, \theta, \Delta t)$, the mean and standard deviation expressed in terms of the parameters, $\theta = [a; b; c; d]$, and a given value of r_{k-1}. In particular, in the geometric Brownian motion case, we can describe $\sigma(r_{k-1}, \theta, \Delta t) = c\sqrt{\Delta t}$ and $\mu(r_{k-1}, \theta, \Delta t) = a\Delta t$.

We construct the likelihood function utilizing the conditional distribution of all the N observations as follows:

$$L(\theta) = \prod_{k=1}^{N} f(r_k|r_{k-1}; \theta) f(r_0). \tag{13.110}$$

Maximum likelihood estimate of parameters attempts to find those values of the parameters, θ, for which the likelihood that the observations resulted from the purported conditional distribution is the highest (maximized). Noting that the maximizers of a function also maximize the logarithm of that function, we define the log-likelihood function. Taking the logarithm of the likelihood function gives us the log-likelihood function as

$$\ln L(\theta) = \sum_{k=1}^{T} \ln f(r_k | r_{k-1}; \theta) + \ln f(r_0). \tag{13.111}$$

The advantage of a log-likelihood function over the likelihood function is that the former has less cumbersome summations of conditional density, while the latter has a product of conditional density. In particular, for the normal distribution case, we have

$$\ln L(\theta) = \sum_{k=1}^{T} \frac{-1}{2} \ln[2\pi\sigma(r_{k-1}, \theta, \Delta t)] - \frac{[r_k - \mu(r_{k-1}, \theta, \Delta t)]^2}{2\sigma(r_{k-1}, \theta, \Delta t)} + \ln f(r_0). \tag{13.112}$$

In order to maximize the log-likelihood function, which is the same as maximizing the likelihood function, we take the first derivative of the log-likelihood function with respect to all the four parameters, $\theta = [a; b; c; d]$, and equate each of them to zero to obtain

$$\frac{\partial \ln L(\theta)}{\partial a} = 0, \tag{13.113}$$

$$\frac{\partial \ln L(\theta)}{\partial b} = 0, \tag{13.114}$$

$$\frac{\partial \ln L(\theta)}{\partial c} = 0, \quad \text{and finally} \tag{13.115}$$

$$\frac{\partial \ln L(\theta)}{\partial d} = 0. \tag{13.116}$$

For the specific geometric Brownian motion case, these equations become

$$\frac{\partial \ln L(\theta)}{\partial a} = \sum_{k=1}^{N} \frac{2\Delta t[r_k - a\Delta t]}{2c^2\Delta t} = 0, \tag{13.117}$$

and

$$\frac{\partial \ln L(\theta)}{\partial c} = \sum_{k=1}^{N} \frac{-1}{c} - \frac{[r_k - a\Delta t]^2}{c^3\sqrt{\Delta t}} = 0. \tag{13.118}$$

Solving Equations 13.117 and 13.118 yields $\hat{a}\Delta t = \frac{1}{N}\sum_{k=1}^{N} r_k$, and $\hat{c}^2\Delta t = (1/N)\sum_{k=1}^{N}(r_k - a\Delta t)^2$. These estimates are not too different from those obtained in Equations 13.107 and 13.108 utilizing the method of moments, and are in fact identical if the variance is defined as a biased sample variance, $\sum_{i=1}^{N}(x_i - \bar{x})^2/N$. Therefore, in the case of normal distribution, the estimates obtained from method of moments and maximum likelihood method are the same.

13.8.4.2 Method of Quasi-Maximum Likelihood

The method of maximum likelihood is applicable only when the exact density of the stochastic factor is known. When this is the case, the method produces the most efficient way to determine the parameters that drive the risk. However, in many cases, the exact density for the stochastic factor, as suggested by the model, may not be determined or stated in the closed form. In such cases, one can benefit from making an approximation to the density, by picking a density which is tractable, and yet in some way not too far improved from the true density of the stochastic factor.

This approximated method of maximum likelihood is generally called the quasi-maximum likelihood method or the pseudo-maximum likelihood method. Once the true density is approximated by an approximate density, the actual procedure for determining the quasi-maximum likelihood estimates follows the steps of maximum likelihood method. Suppose the true density $f(r_k|r_{k-1};\theta)$ is approximated by the density

$$g(r_k|r_{k-1};\theta), \tag{13.119}$$

with $k = 1,\ldots,T$ and $\theta = [a;b;c;d]$. Using the approximate density, the quasi-maximum likelihood function is constructed in the same manner as for likelihood function, as follows:

$$L^Q(\theta) = \prod_{k=1}^{T} g(r_k|r_{k-1};\theta)g(r_0). \tag{13.120}$$

Taking the logarithm to simplify the product into a summation yields

$$\ln L^Q(\theta) = \sum_{k=1}^{T} \ln g(r_k|r_{k-1};\theta) + \ln g(r_0). \tag{13.121}$$

We then maximize the log-quasi likelihood function to determine the parameters that best describe the data coming from this approximate likelihood function. The reader may be rightly unconvinced about this method generating good estimates. The benefit of the method lies in making a good approximation of the density, and also in conducting an analysis of the consistency of the estimated parameters. Detailed analysis of consistency of the estimated parameters is beyond the scope of this chapter. The reader is referred to books on econometric analysis [39] and financial econometric analysis [17].

13.9 Stochastic Kriging

In all application problem domains, simulation modeling and analysis gets employed to develop a detailed understanding of the characteristics and behavior of a system. If the entities and servers have certain properties, what would the impact of these be on the performance of the system? Often, however, the investigation must go beyond a detailed understanding of the baseline characteristics and behavior of the system. Specifically, if any of the properties of the entities and servers should change, it is also important to assess the impact of these changes on the performance of the system. This assessment is often made using sensitivity analysis. Small or large perturbations made to the inputs of the simulation model that define the properties of the entities, servers, or risk factors would require the simulation experiments to be repeated to determine performance measures under the new configurations.

In many instances of simulation studies, the goal is to improve, or even design, the system for better or optimal performance. For such motivation, a more comprehensive view of the relationship between the inputs with the outputs of the simulation model must be developed. This is often called a response surface, or a metamodel, emulator, etc. For generating a response surface, as one would imagine, many simulation runs must be conducted with different configurations of the inputs. Being exhaustive with this exercise would be computationally prohibitive. Therefore, techniques must be constructed and utilized for developing the response surface with much fewer than exhaustive simulation runs corresponding to all combinations of inputs. This is the goal of Kriging [102,103]. Kriging is credited to South African mining engineer Mr. Danie G. Krige's master's thesis constituting pioneering distance-weighted average gold grades plotting at the Witwatersrand reef complex in South Africa. Kriging was originally developed for a deterministic application in geostatistics, or spatial statistics, and has more recently been extended to stochastic applications [57].

Optimization algorithms utilize the response surface to guide the search for the selection of optimal input parameters defining the properties of entities and servers toward the best system performance. Depending on the nature of relation between input and output of the system, a global construct of the response surface is beneficial for identifying the optimal input configuration of the system. We will investigate suitable optimization algorithms for simulation-based optimization in Section 13.10, where response surface developed using Kriging will be utilized.

In conventional numerical analysis, complete information for a response function is constructed using interpolation of observed input–output observations. Methods utilizing univariate, bivariate, or multivariate splines are utilized for this purpose, depending on the dimensionality of the input variables. If it is not necessary, or in fact is desired, that the response surface does not pass through the observed input–output observation points, perhaps due to inherent variability in the output observations, least-squares fits of linear regression models also serve this purpose. Appropriate generalizations of nonlinear functions may also be considered for least-squares fits. Kriging combines both these principles in constructing a response surface. It allows some variability in unobserved input–output pairs on the response surface, and forces the variability to be zero for the observed input–output points on the response surface. We begin with describing the basic formulation for Kriging.

13.9.1 Basic Kriging Formulation

To obtain a global response surface or metamodel, a classical fractional factorial design for the input combinations for simulation runs may be used to generate the observed input–output points or scenarios. However, a space-filling design may prove to be more useful to obtain observed input–output points or scenarios that cover the entire range of interest. In higher dimensions, Latin hypercube sampling, which we studied in Section 13.6.4 as a variance reduction technique, is a space-filling design for sampling global set of observations [57,103].

We describe the basic Kriging formulation, the so-called ordinary Kriging, using a set of space-filling inputs, $\{\mathbf{x_i} : i = 1, \ldots, n\}$, and corresponding observed outputs, $\{Y(\mathbf{x_i}) : i = 1, \ldots, n\}$. The input vector, $\mathbf{x_i}$, has k components. We seek the ordinary Kriging predictor, $\hat{Y}(\mathbf{x_{n+1}})$, for a "new," nonsimulated input, $\mathbf{x_{n+1}}$ as a weighted linear combination of the n observed outputs $\{Y(\mathbf{x_i}) : i = 1, \ldots, n\}$ as follows:

$$\hat{Y}(\mathbf{x_{n+1}}) = \sum_{i=1}^{n} \lambda_i Y(\mathbf{x_i}) = \lambda^T \mathbf{Y}, \tag{13.122}$$

where $\lambda = \{\lambda_1, \ldots, \lambda_n\}^T$ and $\mathbf{Y} = \{Y(\mathbf{x_1}), \ldots, Y(\mathbf{x_n})\}^T$, and we impose the constraint, $\sum_{i=1}^{n} \lambda_i = 1$. This formulation for Kriging assumes a single output variable, where in case of multiple outputs, the predictor may be applied to each output.

The Kriging weights, $\lambda = \{\lambda_1, \ldots, \lambda_n\}^T$, in Equation 13.122 are obtained as a minimal mean-squared error (MSE) defined as

$$\sigma_e^2 = E[(Y(\mathbf{x_{n+1}}) - \hat{Y}(\mathbf{x_{n+1}}))^2]. \tag{13.123}$$

The definition of this mean-squared error requires picking a spatial covariance structure between the observed input–output, $\{Y(\mathbf{x_1}), \ldots, Y(\mathbf{x_n})\}$, and the new input–output, $Y(\mathbf{x_{n+1}})$. Therefore, the optimal Kriging weights, λ_i^*, depend on the specific point, $\mathbf{x_{n+1}}$, to be predicted. This is a significant and notable difference from linear regression where a fixed estimated coefficient, β, is determined for all $\mathbf{x_{n+1}}$. The covariance structure among observed and new output points is picked so that the predictor is exact for the input points or scenarios already observed, which makes the metamodel estimates unbiased for the observed input–output points.

Covariance models in Kriging assume stronger positive correlation for input data that are closer. A covariance structure that is second-order stationary in spatial dimension serves this purpose. This implies that means and variances at an observed input–output point, $Y(\mathbf{x_i})$, namely, μ_Y and σ_Y^2 are constants, while covariances between outputs depend on the distance between their corresponding inputs, $h = \|\mathbf{x_i} - \mathbf{x_j}\|$. These covariances are taken to decrease as the distances between input points or scenarios increase.

Kriging can utilize different types of covariance functions. For instance, for a system with k-dimensional input and exponentially decaying covariance structure, we can consider the following covariance function:

$$cov[Y(\mathbf{x_i}), Y(\mathbf{x_j})] = \sigma_Y^2 \prod_{l=1}^{k} exp(-\theta_l |x_{i,l} - x_{j,l}|), \tag{13.124}$$

where the parameters θ_l specify the relative importance of the various inputs, $\{x_{i,l} : l = 1, \ldots, k\}$. Alternate functional forms are possible, such as a Gaussian function, $exp(-\theta_l (x_{i,l} - x_{j,l})^2)$, which lends smoothness to the objective. In all these forms, the structure is chosen so that covariance decreases to zero as the distance between input points increases, and at zero distance, the variance reduces to a constant, σ_Y^2.

Using the above forms of spatial covariance structure among observed input–output points or scenarios, $\{Y(\mathbf{x_1}), \ldots, Y(\mathbf{x_n})\}$, and new point, $Y(\mathbf{x_{n+1}})$, the optimal weights in Equation 13.122 can be derived. From the nature of the covariance structure, the observed points closer to the new input to be predicted get more weight. Furthermore, as stated earlier, the optimal weights λ_i^* vary with the specific input, $\mathbf{x_{n+1}}$, to be predicted.

Kriging is an active field of research and methodological development. We next consider some additional issues being investigated both in the methodology and application of Kriging.

13.9.2 Additional Issues and Applications of Kriging

It is worth noting that accuracy in the observed input–output points or scenarios is essential in the quality of the metamodel. In case of too few replicates, the estimated covariance function is very noisy, and as a result the Kriging weights and predictions are very inaccurate. It is demonstrated that as the number of replicates increases, the Kriging predictor's accuracy increases [59]. Therefore in stochastic Kriging for simulation metamodeling, it is important to explicitly capture inherent uncertainty in input–output and extrinsic uncertainty in the unobserved inputs [4,98].

Furthermore, the use of common random numbers as a variance reduction technique induces correlation between observed input–output points or scenarios. Studies have investigated the influence of common random numbers on the accuracy of stochastic Kriging. There is evidence that use of common random numbers leads to less precise predictions, but provides better gradient estimation and better estimation of the "slope" terms in any trend model [19]. Similarly, studying the role of the type of correlation function in stochastic Kriging metamodels in prediction accuracy has shown that the twice or higher-order continuously differentiable correlation functions demonstrate good capability to fit both differentiable and nondifferentiable multidimensional response surfaces [106].

Quality of parameter estimates is an issue in any kind of model, and is specifically important in stochastic Kriging where the intrinsic uncertainty of observed input–output points or scenarios affects the accuracy of metamodels. Studies have been conducted using different Kriging models on the effects of parameter estimation on prediction error of Kriging metamodel [109]. The study shows that the random noise in stochastic simulations can increase the parameter estimation uncertainties and the overall prediction error.

We developed a formulation for Kriging for a univariate output variable in Section 13.9.1, with a view that in the case of multiple necessary outputs of a simulation study, the predictor may be applied to each output. However, one may intuitively expect that multivariate Kriging would yield lower mean-squared error than univariate Kriging, as multivariate Kriging can take advantage of the cross-correlations between different output performance measures. Univariate Kriging does not obtain this opportunity, and must only utilize autocorrelations between outputs of the same performance measure for different input combinations. A recent study [58] has investigated this intuition with mixed results.

As mentioned earlier, in practice, the Kriging parameters are unknown and must be estimated, which increases the mean-squared error (MSE). In multivariate Kriging, additional parameters must be estimated related with the cross-correlations, which can end up increasing the MSE. The study [58] applies a nonseparable dependence model to ensure symmetric, positive-definiteness of covariance matrix of all the observed simulation outputs, along with satisfying all the assumptions of multivariate Kriging, and finds the simpler univariate Kriging to yield lower MSE than the more complicated multivariate Kriging. They utilize the true known multivariate Kriging in their study, and yet find that univariate Kriging performs better. Therefore, developing techniques for efficient multivariate Kriging remains a topic of future research.

While optimization based on the metamodel or response surface constructed by Kriging is the most prominent application of Kriging, the methodology has also been applied for risk assessments in finance and banking. For instance, for regulatory purposes and for internal risk assessment VaR and CVaR or expected shortfall [69] are computed for credit or investment portfolios. Kriging techniques have been developed for these assessments. For instance, stochastic Kriging metamodel-based method is developed for efficient estimation of risk measures and risk sensitivities by using gradient estimators of assets in a portfolio. The method yields a best linear unbiased predictor of the risk sensitivities with minimum mean-squared error [20].

Stochastic Kriging has mostly been developed and applied to simulation studies with continuous input variables or factors. There has been some effort to extend the approach to allow applying Kriging to systems with qualitative input factors [21]. This requires introducing basic steps of constructing valid spatial correlation functions for handling correlations across levels of qualitative factors. Furthermore, being able to apply Kriging for qualitative factors or discrete input variables leads to developing efficient discrete optimization algorithms via simulation [107].

Kriging is meant to aid optimization, and in this light, studies have examined the data enhancement, smoothing, and reconstruction capabilities of stochastic Kriging [13,40,50]. In a well-controlled setting, an assessment of Kriging method is done for its capability to enhance and smooth a noisy data set and reconstruct large missing regions of lost data. On appropriate selection of the correlation function and related parameters, the degree of smoothness can be robustly controlled, and is seen to be a viable ingredient for constructing effective global optimization algorithms. Therefore, it is not surprising that new algorithms are proposed to deal with global optimization challenges, for instance, for worst-case optimization of blackbox functions evaluated through costly computer simulations [74]. Worst-case optimization problems seek a minimax solution, that is, identify the values of the control variables that minimize the maximum of the objective function with respect to the input state variables, and are particularly difficult when the state and control variables are continuous. Finally, either by multivariate Kriging or by metamodeling of a weighted sum of output performance measures, stochastic Kriging can be applied to multiobjective simulation optimization [110].

In the context of stochastic Kriging, we have examined how the methodology can aid in a variety of optimization problems. We next take a detailed look at simulation-based optimization, or simply simulation optimization.

13.10 Simulation-Based Optimization

In the previous section, the need for optimization was motivated as the objective often encountered in simulation modeling and analysis projects. Numerous problems in service system design and optimization, manufacturing plant layout, inventory planning, communication networks, finance and risk management either directly involve solving an optimization problem, or indirectly precipitate into an optimization problem. For instance, in finance, portfolio optimization, optimal asset allocation, development of optimal hedging strategy, and asset-liability management are all essentially rife with optimization problems. Indirectly, calibration of simulation models, developing forecasting models, etc. involve solving some form of optimization problems. In general, optimization problems are classified by their key characteristics, which aids in developing methods for solving the problem.

A canonical optimization problem has an objective function, $f(x)$, where x's are decision variables, and there may be additional constraints that the decision variables must satisfy. An optimization problem with constraints is called a constrained optimization problem, while one without constraints is an unconstrained optimization problem. Therefore, a typical optimization problem may be stated as

$$\min_{x \in R^n} f(x), \tag{13.125}$$

$$\text{such that } Ax \leq b, \tag{13.126}$$

$$g(x) \leq 0, \tag{13.127}$$

$$h(x) = 0, \tag{13.128}$$

$$l \leq x \leq u, \tag{13.129}$$

where the matrix A of size $m \times n$ defines m linear constraints, $g(x)$ defines a set of nonlinear inequality constraints, and $h(x)$ are a set of nonlinear equality constraints. Additionally, l and u may be bounds on the decision variables. The most significant classification from the perspective of this presentation is that between deterministic and stochastic optimization problems. In the former, the objective function and constraints are obtainable as deterministic functions of the decision variables. In case of stochastic optimization, the presence of stochastic factors in the definition of the objective function and/or constraints makes them a functional of random variables. Therefore, these functions must be either computed probabilistically or must be estimated statistically.

Suppose r is a set of stochastic factors, and the objective or constraints of the optimization problem must be stated in terms of these stochastic factors, along with the decision variables, x. This is so because the stochastic factors in conjunction with the decision variables determine if the goals of the problem are achieved. The optimization problems must be modified as follows:

$$\min_{x \in R^n} E[f(x;r)], \tag{13.130}$$

$$\text{such that } E[A(r)]x \leq b, \tag{13.131}$$

$$E[g(x;r)] \leq 0, \tag{13.132}$$

$$E[h(x;r)] = 0, \tag{13.133}$$

$$l \leq x \leq u, \tag{13.134}$$

where the matrix $A(r)$ of size $m \times n$ is a function of the stochastic factors, r, defining m linear constraints for the decision variables, x. The set of nonlinear inequality and equality constraints are defined, respectively, in terms of $E[g(x;r)]$ and $E[h(x;r)]$. A stochastic optimization problem would essentially reduce to a deterministic optimization problem if the functions $E[f(x;r)]$, $E[g(x;r)]$, and $E[h(x;r)]$ can be computed probabilistically or analytically. For instance, this is the case for linear and quadratic functions of the stochastic factors, such as those appearing in the mean-variance portfolio optimization problem discussed in Section 13.2.3. Optimization problems can also be static versus dynamic, depending on whether the decision variable is independent or dependent on time, respectively. Therefore, these involve dynamic optimization to obtain decisions that adapt to evolving stochastic factors.

An additional complexity can arise in some optimization problems when the decision variables are not continuous or are not quantitative. In some contexts for simulation modeling and analysis project, variables such as "education," "geographical region," or "urban–suburban–rural" characterization of clients, or "advertising or marketing channels" or "system specifications" "training program choices" may all be nonordinal, qualitative choices. Similarly, in other cases, while decision variables are quantitative, they may not take values in a range of continuum.

In any optimization problem, one must conduct a preliminary assessment of the above characteristics of the optimization problem. The set of input factors that define the problem must be identified, along with specification of whether the factors are qualitative, discrete, or continuous-valued. The input factors may be static or evolving dynamically in response to the evolving stochastic behavior of the system. Some of the input factors are controllable, while others may be uncontrollable due to a variety of reasons. Optimization can be done with respect to all the input factors, but will be meaningful to do only for the controllable ones. Moreover, there may be constraints and bounds required for the controllable input factors or decision variables. Finally, a detailed analysis is needed whether the problem is deterministic or stochastic in nature. In other words, in the presence of stochastic factors, can the objective and constraints functions be determined analytically or probabilistically in terms of the controllable input factors?

When the objective function or constraints cannot be computed analytically is when simulation optimization is applicable and useful. We explore this in detail next.

13.10.1 Challenges of Simulation-Based Optimization

We will begin with thinking of the problem in terms of a classical mathematical optimization problem. However, the output performance measure or objective function, $R = E[f(x;r)]$, will be computed using simulation, since it cannot be computed analytically. The value R takes depends on the values of input factors, $x = [x_1, x_2, \ldots, x_k]$. As stated earlier, there may be bounds on the decision variables, $l_i \leq x_i \leq u_i$, as well as other constraints. Let us first consider the unconstrained stochastic optimization problem, where further investigation of the problem will only require considering properties of the objective function.

In computational optimization, the solution for the problem is sought iteratively, in each step attempting to make a better guess of the optimum. Seeking the optimal value of the objective function within the bounds for the decision variables through the iterative search is greatly facilitated by gaining some more information along the way of the objective function. If the function is continuous and differentiable, only continuous but not differentiable, or discontinuous, this can be utilized for guiding the search process. The slope or gradient of a differentiable function provides a direction of descent (or ascent, in case of a maximization problem). In case of continuous, but not differentiable, objective function, the direction of descent can be approximated using estimated slope or subgradients. When the objective function is discontinuous, such guidance is missing, and the search process would need to accommodate this fact.

Beyond continuity and differentiability of the problem, the objective function being linear versus nonlinear qualifies the difficulty of solving the optimization problem. A linear objective function is continuous and differentiable, but it is also both convex and concave. Therefore, the optimal solution lies at the boundary of the feasible region of the problem. This knowledge can be utilized in obtaining a solution for the problem. When the objective function is not linear, that is, it is nonlinear, solving the problem may be more challenging. Determining whether a nonlinear objective function is convex or concave is instructive for the existence and uniqueness of the optimal solution of the problem.

The expectation of a function is convex if for any two arbitrarily picked values for the decision variables, x_1 and x_2, and for any scalar $\lambda \in [0, 1]$, we can demonstrate that

$$E[f(\lambda x_1 + (1-\lambda)x_2; r)] \leq \lambda E[f(x_1; r)] + (1-\lambda)E[f(x_2; r)], \tag{13.135}$$

and the fact that expectation, $E[.]$, is a linear functional.

If the inequality holds in the reverse for any arbitrarily picked values for decision variables, the objective function is concave. Whether an objective function is convex or concave in the entire feasible region, assuming the feasible region is a convex set, this indicates the optimization problem has a unique solution. In case of a convex problem, the solution is an interior point, while for a concave problem, the solution lies in the boundary of the feasible region. A convex optimization problem is characterized by the gradient of the objective function, $\nabla E[f(x_0; r)]$, as follows:

$$E[f(x; r)] \geq E[f(x_0; r)] + \nabla E[f(x_0; r)]^T (x - x_0), \tag{13.136}$$

for all x, x_0. For a convex optimization problem, the Hessian of the objective function, $H(x) = [\partial^2 E[f(x; r)]/\partial x_i \partial x_j]$, is positive semidefinite. Both these properties can be utilized for constructing a method for solving the optimization problem. Local convexity of an objective function, even for a nonconvex problem, can be utilized for guidance in the search for the optimum.

When the objective function of a stochastic optimization problem cannot be computed analytically, there is good chance the function is not very well behaved. Since the objective function, $R = E[f(x; r)]$, does not have a closed-form, analytical formulation, it can only be estimated with a certain degree of precision, or at a chosen confidence level. This implies that even if theoretically the objective function is continuous and differentiable, the gradient or the Hessian will also need to be estimated with some degree of accuracy. Therefore, errors in these estimates essentially create additional challenge for the solution procedures.

Since the objective is estimated with significant computational effort, with or without utilizing stochastic Kriging, for the desired high accuracy, solving the optimization problem is a computationally intensive exercise. Moreover, in some cases, the objective function may be highly nonlinear, multimodal, and due to the estimation error, noisy. Therefore, methods constructed to solve these optimization problems must adapt to meet these challenges. Despite the challenges, seeking even an approximately optimal decision set is a worthwhile activity in most cases.

In a constrained optimization problem, attention is needed not just to the objective function, but also to the linear, nonlinear equality and inequality constraints in Equations 13.131 through 13.133, respectively. These constraints define a feasible region, and remaining within the feasible region in the process of search for optimal solution, or at the termination of the search process, is a must to satisfy the constraints. Various penalty function or barrier function approaches are developed to this end. For instance, if the problem is defined by nonlinear equality constraints, $E[h(x;r)] = 0$, we can define a new objective, as follows:

$$\min_{x \in R^n} E[f(x;r)] + \frac{\rho}{2} E[h(x;r)]^T E[h(x;r)], \tag{13.137}$$

$$\text{such that} \quad l \leq x \leq u, \tag{13.138}$$

where $\rho \geq 0$ is chosen as the penalty parameter. Searching for the optimal solution to this modified problem implies, depending on the size of the penalty parameter, ρ, the search may wander into infeasible region. However, if the penalty parameter is adjusted to a large enough value, the penalty term would dominate the objective function of the modified problem; hence the solution will be feasible by the original problem. A barrier function approach modifies the objective so that the search process is not allowed to wander in the infeasible region due to the barrier introduced in the objective. For instance, for nonlinear inequality constraints, $E[g(x;r)] \leq 0$, we can define a new objective, as follows:

$$\min_{x \in R^n} E[f(x;r)] + \eta \sum_{i=1}^{m_1} \ln(-E[g_i(x;r)]), \tag{13.139}$$

$$\text{such that} \quad l \leq x \leq u, \tag{13.140}$$

where m_1 is the number of nonlinear constraints and $\eta \geq 0$ is the chosen barrier parameter. In the barrier function approach, the barrier parameter is initially taken to have a large enough value, but must be gradually decreased to relax the imposition of a barrier. In the above description, the penalty and barrier function approach provide an intuitive guidance for constrained optimization; however, in their practical implementation, additional issues may need to be resolved for convergence of search.

The necessary optimality conditions for a constrained optimization problem are defined by the Karush–Kuhn–Tucker (KKT) conditions. These are defined in terms of a Lagrangian function for the constrained optimization problem, given as

$$L(x, \lambda, \mu) = E[f(x;r)] + \sum_{i=1}^{m_1} \lambda_i E[g_i(x;r)] + \sum_{i=1}^{m_2} \mu_i E[h_i(x;r)]. \tag{13.141}$$

If the objective function and nonlinear constraints are defined in terms of continuously differentiable functions, and if x^* is the local minimum of the constrained optimization problem, then there exist $\lambda^* \geq 0$ and μ^*, called KKT multipliers, such that

1. Feasibility: $E[g(x^*;r)] \leq 0$ and $E[h(x^*;r)] = 0$.
2. Stationarity: $\nabla E[f(x^*;r)] + \sum_{i=1}^{m_1} \lambda_i^* \nabla E[g_i(x^*;r)] + \sum_{i=1}^{m_2} \mu_i^* \nabla E[h_i(x^*;r)] = 0$.
3. Complementary slackness: $\lambda_i^* E[g_i(x^*;r)] = 0$ for all $i = 1, \ldots, m_1$.

These conditions can be utilized to develop search algorithms, where the search is done for both values of x and the KKT multipliers, λ and μ. We have provided a brief overview of challenges of constrained optimization, specifically as it is applicable for simulation-based constrained optimization. For additional discussion and development of these topics, the reader should refer to Gill et al. [34], Bertsekas [12], Fletcher [32], and Luenberger and Ye [70].

13.10.2 Simulation Optimization Methodologies

Simulation optimization must rely on simulation to compute the value of the objective function and constraints, and any related constructs developed for solving the optimization problem, such as Lagrangian function, barrier, or penalty function. Computational optimization or numerical optimization is an iterative process of obtaining a better solution for the problem, until the "best" is obtained. Therefore, for simulation optimization, an optimization routine has to work in tandem with a simulation routine that helps compute the required quantities that define the optimization problem. This can include using tools such as stochastic Kriging to aid in objective function or response surface evaluation. A schematic for this interaction is provided in Figure 13.21.

With this synchronized development in view, the general steps adopted for simulation optimization as depicted in Figure 13.21 are structured as follows:

1. The process begins at the "Start" with initializing the search process with a single or a sequence of decision choices.
2. Results from simulating earlier choices (or initial ones) are used to generate promising new directions to search in the space of possible input factor combinations.
3. Traditional or heuristic optimization search techniques may be employed. This stage will require interface with simulation routine to compute the quantities that define the problem.
4. As search progresses, the process should keep track of the best choices visited thus far. This is particularly important when objective or constraint function(s) are nonsmooth, multimodal functions.

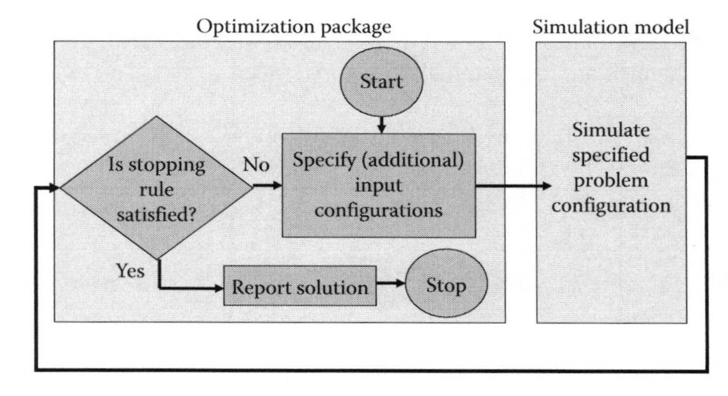

FIGURE 13.21
Typical interaction between optimization and simulation model.

5. Finally, search must be terminated, which is done either when good enough solution is obtained, significant improvements are not happening, or the process has run long enough.

Implementing a simulation optimization software from scratch is for most purposes not advisable, especially given the already available good options. In Section 13.13.3, we provide a list of software tools that come equipped with computational optimization, and specifically simulation optimization, depending on the user needs and problem characteristics. In general, features in a simulation optimization tool that one would seek for a reliable, effective, and useful implementation are as follows:

- The foremost important feature is that the quality of solution obtained in the amount of execution time required should be efficient. For this purpose, different algorithms designed for specific problem characteristics are developed.
- In many cases, as the search progresses, being able to keep track of certain number, say "m," of best decision choice configurations should be available to the user. The "m" being user-defined is an added benefit. This is particularly important in simulation optimization, where objective and constraints are computed to only certain degree of precision.
- For solving constrained optimization problem, as discussed in the previous section, accounting for constraints on decision variables besides the objective function of the optimization problem is required. For this purpose, specific optimization algorithm, function routines, and additional considerations must be incorporated.
- There should be several stopping rules for the termination of the search algorithm.
- Given the optimization routine is working in tandem with a simulation routine to estimate the objective and constraint functions, it is imperative that a confidence interval estimate for these performance measures corresponding to "m" best decision configurations is available.
- Incorporation of Kriging-based response surface generation should be available for inclusion in optimization search guidance.
- Moreover, in the decision configurations for which variability in the objective function or constraints is higher, it would be desirable to do more replications in order to obtain the same degree of precision.
- Finally, in cases where the problem size is large, either due to number of variables or problem complexity, having access to parallel implementations can greatly help with simultaneous simulation runs on networked computers.

With the general framework for simulation optimization laid out and with an understanding of the desirable features for a simulation optimization implementation, we now begin a discussion of several important search techniques utilized for simulation optimization in the following sections.

13.10.3 Continuous Simulation-Based Optimization

In cases where the objective function is smooth enough that its derivatives may be defined, as discussed in Section 13.10.1, this information may be gainfully utilized in the search for the optimal solution.

13.10.3.1 Gradient-Based Methods

For a continuous objective function, the gradient information, and for a continuously differentiable objective function, gradient as well as Hessian information may be employed to direct the search. In simulation optimization, however, the objective function, its gradient and Hessian must all be estimated using simulation. As opposed to analytical methods for implementing optimization algorithms, this poses additional challenge regarding accuracy of these estimates, and how well they may guide the search. In response to this challenge, being able to construct a confidence interval around the estimated objective function value, its gradient, and Hessian can prove helpful. Additionally, utilizing stochastic Kriging to create response surface information can help improve efficiency.

In these approaches, the flowchart of Figure 13.21 is primarily followed based on pursuing a single decision choice, x_k. The decision choice is iteratively improved to obtain a "better" iterate, x_{k+1}, by taking a step, s_k, of length h. Therefore, the iterations can be summarized as follows:

1. Pick a starting guess x_0.
2. Find a direction of descent for the objective function, s_k.
3. Determine an appropriate step length, $h_k \in R_+$, to move along the descent direction.
4. Update the guess, $x_{k+1} = x_k + h_k s_k$.
5. Check if termination criteria are met, else go to Step 2.

The iterations in the search are continued until either a convergence criterion is met or an iteration limit is reached. Different gradient-based algorithms adopt specific principles to determine the directions of descent and step length. As discussed in Section 13.10.1, the above scheme can be applied to constrained optimization problems by appropriately adjusting the objective function.

The *steepest descent or gradient method* uses just gradient, that is, first-order information in a Taylor expansion, for the direction of descent in each iteration. Therefore,

$$s_k = -\frac{\nabla E[f(x_k; r)]}{\|\nabla E[f(x_k; r)]\|}. \tag{13.142}$$

The step length h_k can be chosen by solving a one-dimensional problem $\min_{h \geq 0} l(h) = E[f(x_k + hs_k)]$. Alternatively, step length can be determined by creating a local quadratic approximation of the objective function and picking the minimum of this quadratic fit.

A first-order Taylor expansion-based method can be improved by incorporating second-order information in the Taylor expansion, provided this is available. This is the basis for the Newton's method or its variant, the quasi-Newton's method. In the *Newton's method*,

$$E[f(x_k + s; r)] = E[f(x_k; r)] + \nabla E[f(x_k; r)]^T s + \frac{1}{2} s^T H(x) s, \tag{13.143}$$

where $H(x)$ is the Hessian of the objective function. If the Hessian is positive definite, implying the function is locally strictly convex, we may find the minimizer of the second-order Taylor approximation of the objective as a solution of the system, $H(x)s = -\nabla E[f(x_k; r)]$. The solution of the linear system is taken as the descent direction, that is, $s^* = s_k$.

Indeed in simulation optimization, as stated earlier, objective function, its gradient, and Hessian must be estimated. When the gradient and Hessian information is approximated in each iteration, the method is called *quasi-Newton's method*. In fact, the gradient and/or Hessian may be adjusted iteratively as more information regarding the objective is accumulated. Moreover, in simulation optimization, these algorithms will also need to be adapted for the accuracy with which these estimates are made. For instance, at an iterate, x_k, if the variance of $f(x_k; r)$ is higher, the search algorithm will need to generate a higher number of replicates for the same desired level of accuracy.

Beyond the gradient (and Hessian)-based strategies for searching for the optimal decision, we also consider some heuristic strategies next. As stated earlier, gradient-based optimization algorithms are feasible for objective functions that are well behaved, that is, those that are continuously differentiable. For objective functions that are not well-behaved, these algorithms would have limited applicability. Moreover, when the problem has discrete decision variables, or when the problem is combinatorial, by definition convexity of the feasible region is lost. In these cases, heuristic approaches may be necessary.

Additional to the cases when gradient and Hessian information of the objective is not available, if the simulation-based estimates of these quantities are too noisy, or are too computationally demanding to compute at the appropriate accuracy, heuristic methods must be considered. One need not make an "either-or" decision in this regard, since heuristic algorithms can be combined with gradient-based information to construct hybrid algorithms, where such an approach holds merit. Since these hybrid methods will venture beyond the traditional realm of theoretical convergence results of optimization algorithms, such methods would still be considered heuristics.

13.10.4 Global Optimization Approach for Simulation-Based Optimization

The key to success of an optimization algorithm when gradient and Hessian information cannot be used is the quality of explorative search for the optimum decisions. We explore some methods in a few themes in this section.

13.10.4.1 Simulated Annealing

In gradient (and Hessian)-based methods, the exploration is guided by this information. In the absence of local slope and curvature information, the farthest extreme one might swing to is to pick random selection of candidate solutions in the feasible region. Repeatedly randomly picking candidate solutions from the entire feasible region may be counterproductive; therefore once an initial solution is guessed, random selection of subsequent guesses may be made from a predefined neighborhood of the previously guessed solutions. Moreover, a randomly selected decision in the neighborhood of a guessed solution may be allowed to graduate to be the next guessed solution only if it results in a significant improvement in the objective function value. This is the *random search* algorithm, which we summarize in Figure 13.22a.

The size of neighborhood to pick the next random solution from is a critical parameter in the design of random search. If chosen to be big, it will give the opportunity to explore broad and wide

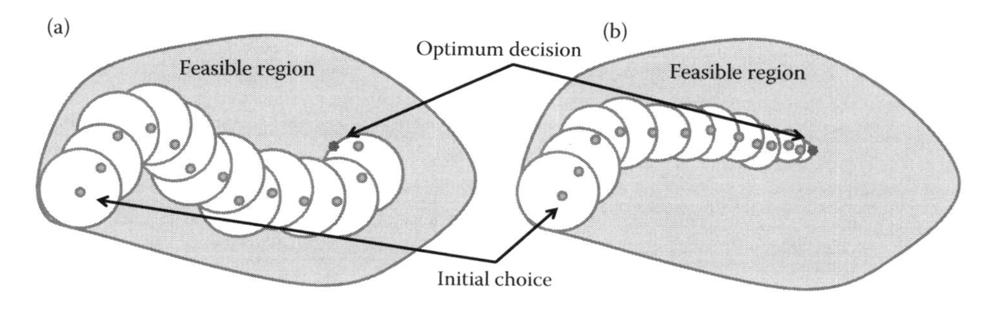

FIGURE 13.22
(a) A sample search trajectory for random search. (b) A sample search trajectory for simulated annealing.

from current guessed solution. However, once the guessed solution approaches an optimal solution, a large neighborhood to select the next solution from carries the risk of wandering away from the solution. This issue can be remedied by progressively reducing the size of the neighborhood from which to select the next candidate solution. This modification results in a new algorithm, called *simulated annealing*.

In simulated annealing, the logic applied is that as better candidate solutions are obtained, the search space is reduced to avoid wandering away in the progress toward the optimal solution. This is shown in Figure 13.22b. The method initializes as in random search, and progresses similarly with one important difference. Every time a new candidate solution is adopted, the admissible size of next neighborhood to select the next candidate solution from is simultaneously adjusted. The algorithm for simulated annealing would have the following structure.

> Pick an initial guess of a solution, x_0,
> initial size of neighborhood, d_0,
> desired reduction in function value in each iteration, ρ, and
> reduction in neighborhood size in each iteration, ϵ.
> While termination criteria are not fulfilled;
> Define a neighborhood, N_i, of size d_i such that $x_i \in N_i$;
> Generate a randomly selected candidate $x_c \in N_i$;
> Compute $E[f(x_c;r)]$ and $\Delta = E[f(x_c;r)] - E[f(x_i;r)]$;
> If $\Delta \leq \rho$ then $x_{i+1} = x_c$, and $d_{i+1} = \max(d_i - \epsilon, \epsilon)$
> Else go to Step "Generate a randomly selected candidate...."
> End

In the above algorithm, we have made sure that the neighborhood size does not become zero, or worse, negative. By not changing the neighborhood size in each iteration, the algorithm would mimic a random search. Moreover, the reduction in neighborhood size can be iteratively adjusted by making it depend on the iteration, namely, ϵ_i. Several variation of simulated annealing may be developed, where in one variation, the size of neighborhood is not always reduced. It is in the contrary expanded in cases where reduction in objective function is large. This variation can help faster convergence, since it is adaptive to when better progress is made in the search process.

13.10.4.2 Tabu Search

In random search or simulated annealing, even though the trajectories in Figure 13.22 paint a rather favorable scenario, in reality, there may be instances where the search process gets misdirected and delays the convergence to an optimum decision. Tabu search addresses this shortcoming by identifying a set of candidate solutions that have been marked either inferior or visited, and must not be revisited in future exploration. This set is called a tabu set, T, and hence the algorithm is called *tabu search*. The algorithmic description of tabu search, as a modification of the simulated annealing algorithm, is as follows:

> Pick an initial guess of a solution, x_0,
> define an initial size of neighborhood, d_0,
> desired reduction in function value in each iteration, ρ, reduction in neighborhood size in each iteration, ϵ, and
> a tabu set, $T = \emptyset$.

While termination criteria are not fulfilled;
  Define a neighborhood, N_i, of size d_i such that $x_i \in N_i$;
  Generate a randomly selected candidate $x_c \in N_i \backslash T$;
  Assign $T = T \bigcup \{x_c\}$;
  Compute $E[f(x_c; r)]$ and $\Delta = E[f(x_c; r)] - E[f(x_i; r)]$;
  If $\Delta \leq \rho$ then $x_{i+1} = x_c$, and $d_{i+1} = \max(d_i - \epsilon, \epsilon)$
  Else go to Step "Generate a randomly selected candidate. . . ."
End

An alternate version of tabu search may be developed that can help avoid converging to a local optimum. Each time search for an optimum is conducted, the local optimum obtained is included in the tabu set to avoid revisiting it in future exploration, thus forcing to explore alternatives. This version of tabu search can be combined with random search or simulated annealing toward the later iterations, once the search has reached a promising region, to avoid convergence to local optima. The algorithm can be structured as follows:

Pick an initial guess of a solution, x_0, and
Define a neighborhood, N_0, of size d_0 such that $x_0 \in N_0$;
  While termination criteria are not fulfilled;
    Obtain a local optimum, $x_c = argmin\ E[f(x; r)]$ for $x \in N_0 \backslash T$;
    Assign $T = T \bigcup \{x_c\}$;
  End
Pick the best solution visited throughout as the solution, x_{sol}.

The above heuristic methods have been examples of the general optimization process described in Figure 13.21. More specifically, they have been instances of solution algorithms where a single decision choice is considered in each iteration. We next develop solution algorithms that work with a set of decision choices in each iteration, and evolve the entire set in an iteration.

13.10.4.3 Scatter Search

There are many variants of scatter search possible, the common theme of them will involve creating a set of possible decision choices, selecting the better performing ones from this set, creating a new set of solutions based on the better ones, and continuing this process until satisfactory decision choices are obtained. This general theme is depicted in Figure 13.23. Specific implementations of scatter

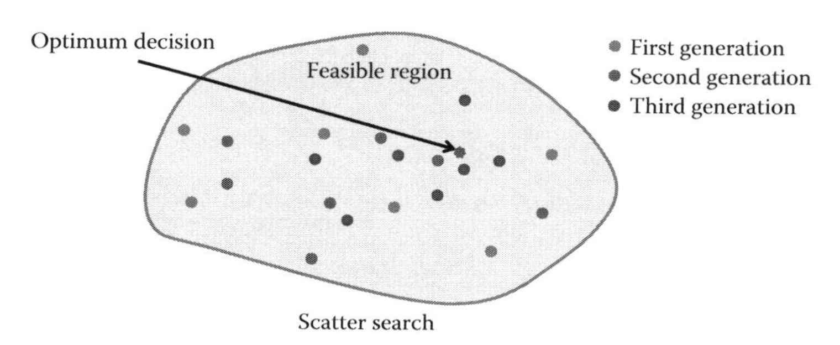

FIGURE 13.23
Three generations of a population-based scatter search algorithm.

search identify specific mechanisms for accomplishing each of the three tasks, until sufficient effort is made for the search process and good enough solutions are obtained, at which point the search is terminated. The best decision choice in the final set of decision choices is taken to be optimum solution.

It is totally conceivable and advisable to apply tabu search after scatter search, based on treating the solution of scatter search as the initial guess for tabu search. This allows combining the benefit of scatter search, of allowing a broad and wide search of the feasible region, with benefit of tabu search of not settling into a local optimum. We next present two biologically inspired population-based optimization algorithms that implement specific principles for accomplishing the three tasks of scatter search.

13.10.4.4 Evolutionary Strategies

Optimization algorithms that base search on objective function value alone, and not gradient or Hessian information, are often called class of direct methods. Evolutionary strategies are population-based direct methods, where concepts of biological evolution of living species form the thematic basis. In the Darwinian theory of evolution, a population of a living species adapts to its environment in order to survive. Survival favors the fittest, while the poor and weak members of the population fade away. The fittest survive, get to mate, and propagate their genetic material into new generations of population. Progressively, the population gets fitter and more adapted to its environment.

13.10.4.4.1 Genetic Algorithm

A genetic algorithm is a specific evolutionary strategy for solving optimization problems, where the above theme of evolution is utilized to obtain better solutions. As in any population-based method, a population of decision choices are evaluated in every iteration. Considering a population of solutions increases the chances of finding the global optimal. At any stage of the iterations of the algorithm, there is a population of solutions under consideration, which is in contrast with traditional optimization approaches, such as gradient-based methods or first two heuristic methods considered in this section.

The objective function value corresponding to each decision choice in the population indicates the fitness of that member of the population. The less fit, or higher objective function valued (in a minimization problem), decision choice is a poor solution, and is allowed to fade away. The more fit solutions are allowed to produce the next generation of set of decision choices. This production of a new generation, in genetic reproductive terminology, involves transfer, crossover, and mutation of genetic material of parents as it is transferred to offspring decision choices.

The main steps of an algorithmic implementation of genetic algorithm take the following form.

Generate an initial population of solutions of size N randomly scattered in the feasible region;

While termination criteria are not fulfilled;

Compute the objective value (or fitness) of each member of the population;

Based on fitness of the decision choices, let a fraction, ρ, of the population fade away;

Apply idealized-type genetic operators to the remaining ρN

members of the population to produce N "offspring" decision
choices;
End
Pick the best solution from the final population as the solution, x_{sol}.

A variation in the above algorithm may be made in terms of probabilistic fading of a member of a population, where the probability is inversely proportional to the fitness of the member.

The above properties of genetic algorithm make it suitable for simulation optimization. These population-based methods make no restrictive assumptions or require no prior knowledge of topology of response surface. They can work even if the objective function is high-dimensional, multimodal, discontinuous, nondifferentiable, and stochastic. Their popularity is also justified by their relative ease of use, as well as reasonable reliability, especially when combined with tabu search that hones in the final solution obtained from a genetic algorithm.

As is evident from the description of the algorithm, an evolutionary strategy, and for that matter all population-based strategies, requires a high amount of objective value evaluation. This can be somewhat tough if the simulation-based estimates of the objective function are computationally very intensive. Computation of population's fitness is, however, very amenable to parallelization; therefore, implementations of genetic algorithms are favored on parallel computing environments.

13.10.4.5 Particle Swarm Optimization

This population-based optimization algorithm is also biologically inspired, more specifically by certain zoological characteristics. If one has seen the collective movement of a flock of birds, a shoal of fish, or a swarm of flying insects, this approach can be visualized as attempting to mimic this movement to achieve the optimum solution. As in any population-based approach, the algorithm begins with an initial set of decision choices. Thereafter, for moving all the decision choices of the population in the improved direction, some "leaders" or best decision choices must be identified. The best decision choices of a single population and historically best decision choices are used to determine how each member of a population migrates to a new location. This guided movement continues until satisfactory improvement is achieved and further improvement ceases.

The general direction of improvement for each member of a population in each iteration can be summarized by a velocity vector, v_i. This is used to advance the progress to the optimum. In each iteration, the velocity vector is modified by the updated information for that iteration to obtain the change in velocity, Δv_i. The particle swarm optimization algorithm can be summarized as follows:

Generate an initial population of solutions of size N randomly scattered
in the feasible region;
Compute the objective value of each member of the initial population;
Define the universal best solution, x_u^*, as the best of initial population,
and a vector of movement for the population, $v = 0$;
While termination criteria are not fulfilled;
Compute the objective value of each member of the
population;

Based on objective function value for the decision choices, pick
the best solution, x_i^*;

If $E[f(x_i^*;r)] < E[f(x_u^*;r)]$, update the historically best solution, $x_u^* = x_i^*$;

Use a linear combination of x_u^* and x_i^* to obtain the change in velocity, Δv_i;

Update every member of the population by new velocity vector, $v_i = v_i + \Delta v_i$;

End

Pick the best solution from the final population as the solution, x_{sol}.

As stated earlier, the particle swarm optimization algorithm can be appended with a single decision choice-based method in the end, such as tabu search, to further fine-tune the solution. We have provided a set of example heuristic methods for simulation optimization, as well as gradient and Hessian-based methods for optimization. Using the given algorithms, as well as other heuristic algorithms, hybrid methods can also be constructed to take advantage of algorithm characteristics that suit the specific problem at hand. Interested readers should access more dedicated discussion of these algorithms in Gill et al. [34], Fletcher [32], and Gilli et al. [35].

13.11 Future Challenges for Simulation Modeling

Experts agree that simulation is increasingly the only method capable of analyzing, designing, evaluating, or controlling the large-scale, complex, uncertain systems we are interested in today and would be likely interested in the future [24,100]. However, there is also a very clear consensus that progress made in simulation modeling and analysis to date faces significant challenges in being prepared to meet the needs of operations research, management science, and decision analysis practitioners of the future. In this section, we discuss some of these challenges to motivate young researchers to begin tackling some of these grand challenges.

In modeling and simulation, one of the biggest challenges irrespective of the application domain is building trustable stochastic models of large complicated systems. In any real-world system, there are many sources of risks and uncertainties, which are often not independent. Moreover, the dependence structures are often very difficult to model, with temporal interrelations and time-varying characteristics, especially in settings that involve humans. For instance, the arrival rates in call centers are highly nonstationary, depending on the time of the day, day of the week, type of day, and other seasonal effects. Complex temporal interrelations could arise from the nature of the call affecting the call routing and service times needed. The arrival rates are also seen to be highly stochastic, making the deterministic time-varying arrival rates under the Poisson assumption inappropriate.

Similar challenges in arrival behavior and service time modeling occur in other systems that involve humans, such as customer arrivals at physical or online stores, incoming demands for products, arrivals at hospital emergency, etc. The challenge of simulation modeling and analysis adoption in healthcare systems is so significant that, even though there has been progress in the past decades, experts believe we need to invest in building generic models, or suite of generic building blocks, which are software platform independent, and are easily understood by clinical and managerial stakeholders in healthcare systems. This would allow hospital simulations to be rapidly built and reused, and would make creating customizable solutions to support wide-scale adoption and utilization of simulation models in such application domains.

Even while sophistications in modeling for real-world system characteristics are appropriately accomplished, the simulation analysis is insightful to the user if the user is adequately equipped with tools to visualize the results. In this regard, there are critical challenges to be addressed ahead by researchers and the practice community. In particular, we need to improve the state of the art in reporting of simulation results, as well as design tools that allow users to create their own reports. It is feared that failing to be equipped with these capabilities, simulation users are prone to misinterpret or misunderstand the simulation results, which could in fact be quite harmful.

Simulation modeling has focused on reporting means of quantities of interest, along with confidence intervals around them, as real-world systems would not exactly behave like the simulation models predict. This has so far been instructive with simulation results still helping guide users to make better choices. It may however need to be further improved, as the width of a confidence interval says nothing about the real-world system variability. It does not instruct about the risk of the real-world system and the actual values one would observe when implementing the solution. For this predicted value, a prediction interval (PI) is needed as a measure of risk in the real-world system.

As simulation outputs go, one is more often than not interested in multiple objectives and competing scenarios. For each output quantity of interest, there can be several properties one might care about, where trade-offs between costs, quality of service, etc. may be needed. For each of these quantities of interest, mean, variability, tail percentiles, PI, etc. may be relevant and necessary for decision making. Help in visualizing these multiple quantities and making necessary trade-offs across multiple scenarios or system designs are essential for improved data visualization under simulation modeling and analysis. Results reporting should also be cognizant of the fact that users prefer a user-guided sequence of experiments on different significant parameters and scenarios, as they explore better system configurations. Reporting tools should facilitate such exploration with ease.

Simulation modeling and analysis tools have to help the analysts step back from operational nitty gritty details of a system being modeled, and help link the insights obtained from the analysis to the high-level (big picture) or macro implications of the decisions. The value of this kind of information to the analyst and the managerial hierarchy would be tremendous, and truly begin to make the contribution of simulation analysis understood. This context provides opportunities to develop theory as to when simulation initiative for group decision making or decision making in organizational hierarchy does or does not create value. This linking from micro to macro also raises an interesting set of theoretical issues relating to sequential sampling and sequential decision making, where the many decision variables have complex interrelations and correlation structures. We also addressed this challenge above in regard to improving simulation modeling.

Paying greater attention to capturing human behavior, whether as system operators, in organizational structure, or in an increasingly networked world, seems to be a worthwhile goal. The agent-based simulation paradigm has fortunately made some progress in this direction. In agent-based simulation, agents have behaviors that depend on the agents' states. An agent-based simulation consists of a set of agents, a set of agent interactions, and the mechanisms that update the agent states as a result of the agents' behaviors and interactions. Thus, agent interaction algorithms form the basis for decentralized system optimization algorithms, for instance, ant colony optimization and particle swarm optimization studied earlier in this chapter. The interaction between agents approach of agent-based simulation allows connecting the micro to the macro of a system. Owing to these attributes, some researchers believe agent-based simulation is the future of simulation.

In summary, in recent years, leading researchers have reflected on the grand challenges for simulation modeling and analysis, in areas of ubiquitous simulation, high-performance computing, spatial simulation, human behavior, multidomain design, systems engineering, network simulation, and education. The themes have been diverse, from overall modeling and simulation methodology, interaction of models from different paradigms, coordinated modeling, parallel, and distributed simulation, agent-based modeling and simulation, ubiquitous computing, supercomputing, grid computing, cloud computing, big data and complex adaptive systems, "big" simulation applications, model abstractions, embedded simulation for real-time decision making, simulation on-demand, simulation-based acquisition, simulation interoperability, simulation composability, high-speed optimization, web simulation science, spatial simulation, cyber systems, democratization, and education. One thing is clear from these discussions: simulation modeling and analysis has a promising future for broad applicability and usefulness in decision sciences.

13.12 Conclusion

Simulation modeling and analysis is a powerful suite of techniques, for solving a broad array of problems in a variety of application domains. In this chapter, we started with a basic introduction of simulation modeling and methodologies, and developed the topics to the state-of-the-art considerations of solving problems using simulation modeling and analysis. The wide application domains for simulation modeling and analysis range from manufacturing processes, service systems, supply-chain networks, communications networks, urban and civil infrastructure—roadways, railways, air-traffic, medical and healthcare services, electrical grids, financial services, and risk management, among others. Many of these application domains were utilized to motivate the methodologies covered in this chapter.

The objectives of a simulation study in the application domains range from system design, performance analysis, capability analysis, comparison studies, sensitivity analysis, optimization study, constraint analysis, risk assessment, or asset valuation. A sequence of logical steps is necessary for conducting a simulation analysis, which should be followed to keep track of progress in a simulation study project. Today, the systems studied using simulation are increasingly complex and large. In light of this fact, all the steps of a simulation study need to be advanced to address the related challenges, specifically in aspects of rare-event simulation, stochastic Kriging, and simulation optimization.

A simulation modeler must be talented to make good abstractions of the system to depict in the model. The challenge could involve an ability to handle large, yet limited, data for defining and calibrating large and complex models. Software systems or programming needed to implement the model can be significant, and must be done with care to avoid costly errors. Beyond software implementation, the output analysis and design of simulation runs with variance reduction, stochastic Kriging, etc. requires sophistication in sight of the objectives of the study. Responding to all these challenges with necessary care for addressing important decision problems for large and complex systems is the future of simulation modeling and analysis.

13.13 Annotated Bibliography

Simulation modeling and analysis is an active area of research, both in its core methodology and technologies, as well as in the application domains. As the systems and the related challenges get more complex, simulation is increasingly seen as the promising tool to address these problems. Therefore, the research literature, venues of research outlets, and software tools available for simulation have been growing. We provide an overview of these here.

13.13.1 Conferences

We provide a list of simulation conferences. These conferences are either specialized simulation conferences, or have a significant participation and coverage in simulation.

1. WSC'14—Winter Simulation Conference, 2014, http://wintersim.org
2. INFORMS Annual Meeting, http://informs.org
3. IIE Conference, http://www.iienet2.org/
4. EURO-OR Annual Conference, http://www.euro-online.org/
5. SIMULTECH 2013—International Conference on Simulation and Modeling Methodologies; Proceedings of the 3rd International Conference on Simulation and Modeling Methodologies, Technologies and Applications
6. SCS Summer/Autumn/Spring Simulation Multi-Conference (SummerSim'14), http://www. scs.org/summersim, http://www.scs.org/autumnsim, http://www. scs.org/springsim
7. ESM'2014—The 28th Annual European Simulation and Modelling Conference, FEUP, http://sigarra.up.pt/feup/en/
8. 8th International Workshop on Multi-Agent Systems and Simulation (MAS&S'14), 2014, http://fedcsis.org/mass, e-mail: mass2014@fedcsis.org
9. SIMUtools 2014, 7th International ICST Conference on Simulation Tools and Techniques, http://simutools.org/2014/
10. ACM SIGSIM, Special Interest Group (SIG) on SImulation and Modeling (SIM), http://www.acm-sigsim-pads.org/
11. EUROSIS, The European Multidisciplinary Society for Modelling and Simulation Technology, http://www.eurosis.org/cms/
12. IEEE/ACM DS-RT 2014, The 18th IEEE/ACM International Symposium on Distributed Simulation and Real Time Applications, http://ds-rt.com/2014/
13. Modeling and Simulation Center (MSC-LES) Conferences, http://www.msc-les.org/new-conferences
14. SIMS—Scandinavian Simulation Society, http://www.scansims.org

13.13.2 Simulation Journals

There are many journals suitable for publishing original research in the area of simulation modeling and analysis. There are the core methodology journal, where the theory behind simulation approaches is developed, as well as new methodologies are developed. On the other hand, there is a long list of journals that publish research work based on application of simulation techniques for solving problems in specific application domains. Developing insight from the context of a specific application area requires great amount of domain knowledge as well as simulation modeling and analysis skills. The following is a list of both types of journals, organized in alphabetical order:

1. *ACM Transactions on Modeling and Computer Simulation*
2. *Annals of Operations Research*
3. *Business Process Management Journal*
4. *Computers & Industrial Engineering*
5. *Decision Support Systems*

6. *European Journal of Operational Research*
7. *IEEE Transactions on Communications*
8. *IEEE Transactions on Intelligent Transportation Systems*
9. *IEEE Vehicular Technology Magazine*
10. *IIE Transactions*
11. *INFORMS Journal on Computing*
12. *Insurance: Mathematics and Economics*
13. *Interfaces*
14. *International Journal of Emergency Management*
15. *International Journal of Physical Distribution & Logistics Management*
16. *International Journal of Production Research*
17. *Journal of Applied Finance*
18. *Journal of Global Optimization*
19. *Journal of Strategy and Management*
20. *Journal of Simulation*
21. *Journal of Transportation Engineering*
22. *Management Accounting Quarterly*
23. *Management Science*
24. *Mathematical Finance*
25. *Methodology and Computing in Applied Probability*
26. *Operations Research*
27. *Production Planning & Control*
28. *Reliability Engineering & System Safety*
29. *Simulation*
30. *The Journal of the Operational Research Society*
31. *Transportation Planning and Technology*
32. *Transportation Research. Part E, Logistics & Transportation Review*

13.13.3 Simulation Software

The following are general and popular discrete-event simulation softwares.

Arena is a discrete-event simulation and automation software developed by Rockwell Automation. The standard edition of Arena is designed for use throughout an enterprise. It models both simple and complex discrete processes and generates very detailed output statistics. It enables user-defined expressions and features and direct read/write access to external data files. The standard edition is used in all industries, most often in manufacturing, health care, customer service, supply chain, and transportation/logistics. The professional edition of Arena is designed to give additional model development flexibilities, as the professional edition models virtually any simple or complex process, including discrete, semicontinuous, batch, and high-speed/high-volume processing lines. It also has template development tools that enable users to create reusable, industry-specific modeling tools for their unique applications. Arena uses the SIMAN processor and simulation language. Access to the low-level SIMAN flowchart modules in the professional edition enables the most complex processes to be modeled and executing external programs within the model logic.

ProModel optimization suite is a discrete-event simulation tool that helps you to make better decisions faster, while also allows modeling of continuous processes. ProModel is used for evaluating, planning, or designing manufacturing, warehousing, logistics, and other operational and strategic applications. It can be used to plan, design, and improve new or existing manufacturing, logistics, and other tactical and operational systems. It empowers to accurately replicate complex real-world processes with their inherent variability and interdependencies, to conduct predictive performance analysis on potential changes, and then to optimize the system based on chosen key performance indicators.

ProModel views a facility as a collection of resources that are intended to function together cost-effectively. Each person and piece of equipment is related to every other component (by coincidence or convenience). Together, they define how the facility works. It can help address how one could dissemble all or part of the factory and reconfigure the pieces to find ways to run the entire system more efficiently. It lets you visualize which new configurations work best and which ones fail, by watching them for a week, month, or year on a trial basis. ProModel allows creating a dynamic, animated computer model of a business environment from CAD files, process or value stream maps, or Process Simulator models. It lets the modeler see and understand current processes and policies in action. Brainstorm using the model to identify potential changes and develop scenarios to test improvements which will achieve business objectives. Run scenarios independently of each other and compare their results in the Output Viewer. It lets the modeler test the impact of changes on current and future operations, risk free, with predictive scenario comparisons.

For continuous systems, any mathematical software is suitable as a generic simulation tool, in which a simulation project may be implemented. Specifically, MATLAB® mathematical software has a vast array of functions for working with random variates, simulation methodologies, and optimization algorithms. We list a few of these relevant functions here. The reader is advised to look up the extensive help documentation available with MATLAB to see the details of these and other related functions. At the bottom of each function description in MATLAB help documentation, look for "See Also" to explore other related functions. Resources such as MATLAB Primer [26] are also useful.

Random number generator: `rand`

Random variate distribution: `normrnd`, `unifrnd`, `unidrnd`, `binornd`, `poissrnd`, `exprnd`, `wblrnd`, `lognrnd`, `chi2rnd`, `ncx2rnd`, `gamrnd`

Debugging support: `assert`, `echo`, `error`, `keyboard`, `return`, `warning`

Distribution tests: `probplot`, `kstest`, `chi2gof`

Calibrating models: `polyfit`, `roots`, `fzero`, `fsolve`, `normlike`, `explike`, `wbllike`, `lognlike`

Variance reduction: `transprob`, `transprobbytotal`, `lhsample`

Optimization: `classify`, `fmincon`, `fminsearch`, `linprog`, `quadprog`

Global optimization: `patternsearch`, `ga`, `simulannealbnd`, `gamultiobj`, `GlobalSearch`, `MultiStart`, `run`

There are open source solutions available for developing programming tools for simulation analysis. Similar to MATLAB, R programming language is a power tool, which benefits from a community that develops functions for their and other's simulation projects. For more details, visit http://www.r-project.org/.

While there are numerous other simulation softwares that have a more or less broad applicability, such as VenSim, ExtendSim, GoldSim, etc., there are numerous application domain-specific simulation softwares, such as NetSim is the Network Simulation software with built-in development environment, Automod, developed by Applied Materials, is a 3D simulation software for manufacturing and distribution operations with a focus on large and complex manufacturing and automation models, etc. The Wikipedia page on simulation software is quite comprehensive in providing a list of simulation software and their relevance in types of application domains and simulation methodological themes. Please refer to http://en.wikipedia.org/wiki/Simulation_software and http://en.wikipedia.org/wiki/List_of_discrete_event_simulation_software.

References

1. W. Abo-Hamad and A. Arisha. Simulation-based framework to improve patient experience in an emergency department. *European Journal of Operational Research*, 224(1):154, 2013.
2. W.A. Al-Qaq, M. Devetsikiotis, and J.K. Townsend. Stochastic gradient optimization of importance sampling for the efficient simulation of digital communication systems. *IEEE Transactions on Communications*, 43(12):2975–2985, 1995.
3. M.M. AlDurgham and M.A. Barghash. A generalised framework for simulation-based decision support for manufacturing. *Production Planning & Control*, 19(5):518, 2008.
4. B. Ankenman, B.L. Nelson, and J. Staum. Stochastic Kriging for simulation metamodeling. In *Proceedings of the 2008 Winter Simulation Conference*, volume 1, Miami, December 2008.
5. S. Asmussen and P.W. Glynn. *Stochastic Simulation: Algorithms and Analysis*, 1st Edition. Springer, New York, 2010.
6. J. Atlason, M.A. Epelman, and S.G. Henderson. Optimizing call center staffing using simulation and analytic center cutting-plane methods. *Management Science*, 54(2):295–309, 2008.
7. R. Ayani, Y. Ismailov, M. Liljenstam, A. Popescu, and H. Rajaei. Modeling and simulation of a high speed LAN. *Simulation*, 64(1):7–14, 1995.
8. J. Banks. *Handbook of Simulation*. John Wiley & Sons, Inc., New York, 1998.
9. J. Banks, J.S. Carson, B.L. Nelson, and D.M. Nicol. *Discrete-Event System Simulation*. Prentice Hall, Englewood Cliffs, NJ, USA, 2000.
10. R. Banomyong and A. Sopadang. Using Monte Carlo simulation to refine emergency logistics response models: A case study. *International Journal of Physical Distribution & Logistics Management*, 40(8):709–721, 2010.
11. M.E. Ben-Akiva, H.N. Koutsopoulos, R.G. Mishalani, and Q. Yang. Simulation laboratory for evaluating dynamic traffic management systems. *Journal of Transportation Engineering*, 123(4):283–289, 1997.
12. D.P. Bertsekas. *Nonlinear Programming*, 2nd Edition. Athena Scientific, Belmont, MA, September 1999.
13. W.E. Biles, J.P.C. Kleijnen, W.C.M. van Beers, and I. van Nieuwenkuyse. Kriging metamodeling in constrained simulation optimization: An explorative study. In *Proceedings of the 2007 Winter Simulation Conference*, volume 1, Washington DC, December 2007.
14. P. Billingsley. *Probability and Measure*, 3rd Edition. Wiley-Interscience, Hoboken, NJ, 1995.
15. E. Bottani and R. Montanari. Supply chain design and cost analysis through simulation. *International Journal of Production Research*, 48(10):2859, 2010.
16. D.B. Brown and J.E. Smith. Dynamic portfolio optimization with transaction costs: Heuristics and dual bounds. *Management Science*, 57(10):1752–1770, 2011.
17. J.Y. Campbell, A.W. Lo, and A.C. MacKinlay. *The Econometrics of Financial Markets*. Princeton University Press, Princeton, 1996.
18. N. Chen and S. Zhou. Simulation-based estimation of cycle time using quantile regression. *IIE Transactions*, 43(3):176, 2011.

19. X. Chen, B. Ankenman, and B.L. Nelson. Common random numbers and stochastic Kriging. In *Proceedings of the 2010 Winter Simulation Conference*, volume 1, Baltimore, December 2010.
20. X. Chen, B.L. Nelson, and K.-K. Kim. Stochastic Kriging for conditional value-at-risk and its sensitivities. In *Proceedings of the 2012 Winter Simulation Conference*, volume 1, Berlin, December 2012.
21. X. Chen, K. Wang, and F. Yang. Stochastic Kriging with qualitative factors. In A. Tolk, R. Hill, R. Pasupathy, S.-H. Kim and M. E. Kuhl, editors, *Proceedings of the 2013 Winter Simulation Conference*, volume 1, page 790, Washington DC, December 2013.
22. C.-Y. Chou, C.-H. Chen, and M.-H.C. Li. Application of computer simulation to the design of a traffic signal timer. *Computers & Industrial Engineering*, 39(1):81–94, 2001.
23. V. Clark, M. Reed, and J. Stephan. Using Monte Carlo simulation for a capital budgeting project. *Management Accounting Quarterly*, 12(1):20–31, 2010.
24. R. Crosbie. Grand challenges in modeling and simulation. *SCS M&S Magazine*, 1(1):2010–01, 2010.
25. M. Crouhy, D. Galai, and R. Mark. *The Essentials of Risk Management*, 1st Edition. McGraw-Hill, New York, 2005.
26. T.A. Davis. *MATLAB Primer*, 8th Edition. CRC Press, Boca Raton, 2010.
27. P.P. Dey, S. Chandra, and S. Gangopadhyay. Simulation of mixed traffic flow on two-lane roads. *Journal of Transportation Engineering*, 134(9):361, 2008.
28. F. Dressler, C. Sommer, D. Eckhoff, and O.K. Tonguz. Toward realistic simulation of intervehicle communication. *IEEE Vehicular Technology Magazine*, 6(3):43–51, 2011.
29. J. Duanmu, K.M. Taaffe, M. Chowdhury, and R.M. Robinson. Simulation analysis for evacuation under congested traffic scenarios: A case study. *Simulation*, 88(11):1379–1389, 2012.
30. D. Duffie and K.J. Singleton. *Credit Risk, Pricing Measurement and Management*. Princeton University Press, Princeton, 2003.
31. J.D. Fink and K.E. Fink. Monte Carlo simulation for advanced option pricing: A simplifying tool. *Journal of Applied Finance*, 16(2):92–105, 2006.
32. R. Fletcher. *Practical Methods of Optimization*, 2nd Edition. Wiley, New York, May 2000.
33. T. Frantti. Reliable simulation of communication networks. *WSEAS Transactions on Communications*, 5(5):758–765, 2006.
34. P.E. Gill, W. Murray, and M.H. Wright. *Practical Optimization*. Emerald Group Publishing Limited, Bingley, UK, January 1982.
35. M. Gilli, D. Maringer, and E. Schumann. *Numerical Methods and Optimization in Finance*. Academic Press, Cambridge, January 2011.
36. P. Glasserman. *Monte Carlo Methods in Financial Engineering*. Springer, New York, 2004.
37. P. Glasserman, P. Heidelberger, and P. Shahabuddin. Portfolio value-at-risk with heavy-tailed risk factors. *Mathematical Finance*, 9:117–152, 2002.
38. C.J. Gonzalez, M. Gonzalez, and N.M. Rios. Improving the quality of service in an emergency room using simulation-animation and total quality management. *Computers & Industrial Engineering*, 33(1):97–100, 1997.
39. W.H. Greene. *Econometric Analysis*, 6th Edition. Pearson Prentice Hall, Upper Saddle River, 2011.
40. H. Gunes, E.C. Hakki, and U. Rist. Data enhancement, smoothing, reconstruction and optimization by Kriging interpolation. In *Proceedings of the 2008 Winter Simulation Conference*, volume 1, Miami, December 2008.
41. A. Gupta. *Risk Management and Simulation*. CRC Press, Boca Raton, 2013.
42. T.-Y. Hu and T.-Y. Liao. An empirical study of simulation-based dynamic traffic assignment procedures. *Transportation Planning and Technology*, 34(5):467, 2011.
43. Z. Huang and P. Shahabuddin. Rare-event, heavy-tailed simulations using hazard function transformations with applications to value-at-risk. *Proceedings of the 2003 Winter Simulation Conference*, 2003:276–284, New Orleans, 1987.
44. J.C. Hull. *Options, Futures, and Other Derivatives and DerivaGem CD Package*, 8th Edition. Prentice Hall, Upper Saddle River, 2011.
45. J.C. Hull and A. White. The pricing of options on assets with stochastic volatility. *Journal of Finance*, 42(2):281–300, June 1987.

46. M. Hunter, H.K. Kim, and W. Sun. Ad hoc distributed dynamic data-driven simulations of surface transportation systems. *Simulation*, 85(4):243–255, 2009.
47. A.P. Iannoni and R. Morabito. A discrete simulation analysis of a logistics supply system. *Transportation Research. Part E, Logistics & Transportation Review*, 42(3):191–210, 2006.
48. M. Jahangirian, T. Eldabi, A. Naseer, L.K. Stergioulas, and T. Young. Simulation in manufacturing and business: A review. *European Journal of Operational Research*, 203(1):1–13, 2010.
49. A. Jamalnia and A. Feili. A simulation testing and analysis of aggregate production planning strategies. *Production Planning & Control*, 24(6):423, 2013.
50. J. Janusevskis and R. Le Riche. Simultaneous Kriging-based estimation and optimization of mean response. *Journal of Global Optimization*, 55:313–336, 2013.
51. W.-L. Jin and W.W. Recker. Monte Carlo simulation model of intervehicle communication. *Transportation Research Record*, 2000:8–15, 2007.
52. A.A. Jobst. Tranche pricing in subordinated loan securitization. *Journal of Structured Finance*, 11(2):64–78, 2005.
53. S. Juneja and P. Shahabuddin. Simulating heavy-tailed processes using delayed hazard rate twisting. *ACM TOMACS*, 12:94–118, 2002.
54. S. Juneja and P. Shahabuddin. Rare-event simulation techniques: An introduction and recent advances. *Handbooks in Operations Research and Management Science*, 13:291–350, 2006.
55. Y.M. Kaniovski and G.Ch. Pflug. Risk assessment for credit portfolios: A coupled Markov chain model. *Journal of Banking & Finance*, 31:2303–2323, 2007.
56. C. Kim and T.-E. Lee. Modelling and simulation of automated manufacturing systems for evaluation of complex schedules. *International Journal of Production Research*, 51(12):3734, 2013.
57. J.P.C. Kleijnen. Kriging metamodeling in simulation: A review. *European Journal of Operational Research*, 192:707–716, 2013.
58. J.P.C. Kleijnen and E. Mehdad. Multivariate versus univariate Kriging metamodels for multi-response simulation models. *European Journal of Operational Research*, 236:573–582, 2014.
59. J.P.C. Kleijnen and W.C.M. van Beers. Robustness of Kriging when interpolating in random simulation with heterogeneous variances: Some experiments. *European Journal of Operational Research*, 165(3):826–834, 2004.
60. P.E. Kloeden and E. Platen. *Numerical Solution of Stochastic Differential Equations*. Springer, New York, 2000.
61. D. Knuth. *The Art of Computer Programming*, volume 2. Addison-Wesley, Boston, 2000.
62. S.S. Kolahi. Issues in simulation and modelling of communication systems. *WSEAS Transactions on Communications*, 6(2):372–376, 2007.
63. S. Kuhn and S. Wenzel. Information acquisition for modelling and simulation of logistics networks. *Journal of Simulation*, 4(2):109–115, 2010.
64. K. Lam and R.S.M. Lau. A simulation approach to restructuring call centers. *Business Process Management Journal*, 10(4):481–494, 2004.
65. A.M. Law. *Simulation Modeling and Analysis*, 4th Edition. McGraw-Hill Publishing Co, New York, 2006.
66. A.M. Law and W.D. Kelton. *Simulation Modeling and Analysis*, 3rd Edition. McGraw Hill, New York, 2000.
67. A.M. Law and M.G. McComas. Simulation of communications networks. In J.M. Charnes, D.J. Morrice, D.T. Brunner, and J.J. Swain, editors, *Proceedings of the 1996 Winter Simulation Conference*, volume 1, Coronado, CA, December 1996.
68. G. Liu, A.S. Lyrintzis, and P.G. Michalopoulos. Numerical simulation of freeway traffic flow. *Journal of Transportation Engineering*, 123(6):503–513, 1997.
69. M. Liu. Estimating expected shortfall with stochastic Kriging. In *Proceedings of the 2009 Winter Simulation Conference*, volume 1, pages 1249–1260, Austin, December 2009.
70. D.G. Luenberger and Y. Ye. *Linear and Nonlinear Programming (International Series in Operations Research & Management Science)*, 3rd Edition. Springer, November 2010.
71. M.E. Mahmoud and K. El-Araby. A robust dynamic highway traffic simulation model. *Computers & Industrial Engineering*, 37(1):189–193, 1999.

72. I. Manuj, J.T. Mentzer, and M.R. Bowers. Improving the rigor of discrete-event simulation in logistics and supply chain research. *International Journal of Physical Distribution & Logistics Management*, 39(3):172–201, 2009.
73. J. Maroto, E. Delso, J. Felez, and J.M. Cabanellas. Real-time traffic simulation with a microscopic model. *IEEE Transactions on Intelligent Transportation Systems*, 7(4):513–527, 2006.
74. J. Marzat, E. Walter, and H. Piet-Lahanier. Worst-case global optimization of black-box functions through Kriging and relaxation. *Journal of Global Optimization*, 55:707–727, 2013.
75. A. Melino and S.M. Turnbull. Pricing foreign currency options with stochastic volatility. *Journal of Econometrics*, 45:239–265, 1990.
76. K. Muthuraman and H. Zha. Simulation-based portfolio optimization for large portfolios with transaction costs. *Mathematical Finance*, 18(1):115–134, 2008.
77. U.J. Na and M. Shinozuka. Simulation-based seismic loss estimation of seaport transportation system. *Reliability Engineering & System Safety*, 94(3):722–731, 2009.
78. M. Narciso, M.A. Piera, and A. Guasch. A methodology for solving logistic optimization problems through simulation. *Simulation*, 86(5):369–389, 2010.
79. B.L. Nelson. *Stochastic Modeling, Analysis and Simulation*. McGraw Hill Inc., New York, 1995.
80. D. Ni. Challenges and strategies of transportation modelling and simulation under extreme conditions. *International Journal of Emergency Management*, 3(4):298–312, 2006.
81. S. Onggo. Introduction to discrete event simulation and agent-based modeling: Voting systems, health care, military, and manufacturing. *Interfaces*, 42(3):320–321, 2012.
82. T. Ören. A critical review of definitions and about 400 types of modeling and simulation. *SCS M&S Magazine*, 2(3):142–151, 2011.
83. D.P. Kroese, P.-T. de Boer, S. Mannor, and R.Y. Rubinstein. A tutorial on the cross-entropy method. *Annals of Operations Research*, 134(1):19, 2005.
84. S. Paisittanand and D.L. Olson. A simulation study of IT outsourcing in the credit card business. *European Journal of Operational Research*, 175(2):1248, 2006.
85. G.N. Papageorgiou. Computer simulation for transportation planning and traffic management: A case study systems approach. *WSEAS Transactions on Systems and Control*, 2(3):272–277, 2007.
86. K. Pawlikowski, H.D.J. Jeong, and J.S.R. Lee. On credibility of simulation studies of telecommunication networks. *IEEE Communications Magazine*, 40(1):132–139, 2002.
87. A. Prakash, S.K. Jha, and R.P. Mohanty. Scenario planning for service quality: A Monte Carlo simulation study. *Journal of Strategy and Management*, 5(3):331–352, 2012.
88. A.A.B. Pritsker. Compilation of definitions of simulation. *Simulation*, 33(2):61–63, 1979.
89. S. Robinson, Z.J. Radnor, N. Burgess, and C. Worthington. Simlean: Utilising simulation in the implementation of lean in healthcare. *European Journal of Operational Research*, 219(1):188, 2012.
90. S. Ross. *Stochastic Processes*. Wiley Publishers, Hoboken, NJ, 1995.
91. S.M. Ross. *Simulation*. Academic Press, Cambridge, 2006.
92. R.Y. Rubinstein. A stochastic minimum cross-entropy method for combinatorial optimization and rare-event estimation. *Methodology and Computing in Applied Probability*, 7(1):5, 2005.
93. R.Y. Rubinstein and D.P. Kroese. *The Cross-Entropy Method: A Unified Approach to Combinatorial Optimization, Monte Carlo Simulation and Machine Learning*. Springer-Verlag, New York, 2004.
94. R.M. Saltzman and V. Mehrotra. A call center uses simulation to drive strategic change. *Interfaces*, 31(3):87–101, 2001.
95. L. Schmidt and P. Gazmuri. Online simulation for a real-time route dispatching problem. *The Journal of the Operational Research Society*, 63(11):1492–1498, 2012.
96. J.M.S. Silva, J.R. Phillips, and L.M. Silveira. Efficient simulation of power grids. *IEEE Transactions on Computer-Aided Design of Integrated Circuits and Systems*, 29(10):1523, 2010.
97. T.K. Siu, H. Yang, and J.W. Lau. Pricing currency options under two-factor Markov-modulated stochastic volatility models. *Insurance: Mathematics and Economics*, 43:295–302, 2008.
98. J. Staum. Better simulation metamodeling: The why, what, and how of stochastic Kriging. In *Proceedings of the 2009 Winter Simulation Conference*, volume 1, pages 119–133, Austin, December 2009.
99. A. Tako and S. Robinson. The application of discrete event simulation and system dynamics in the logistics and supply chain context. *Decision Support Systems*, 52(4):802, 2012.

100. S.J.E. Taylor, S. Brailsford, S.E. Chick, P. L'Ecuyer, C.M. Macal, and B.L. Nelson. Modeling and simulation grand challenges: An or/ms perspective. In *Proceedings of the 2013 Winter Simulation Conference: Simulation: Making Decisions in a Complex World*, pages 1269–1282. IEEE Press, 2013.
101. L.C. Thomas. *Consumer Credit Models, Pricing, Profit and Portfolios*. Oxford University Press, Oxford, 2009.
102. W.C.M. Van Beers. Kriging metamodeling in discrete-event simulation: An overview. In *Proceedings of the 2005 Winter Simulation Conference*, volume 1, Orlando, December 2005.
103. W.C.M. Van Beers and J.P.C. Kleijnen. Kriging interpolation in simulation: A survey. In *Proceedings of the 2004 Winter Simulation Conference*, volume 1, Washington DC, December 2004.
104. R. Verma, G.D. Gibbs, and R.J. Gilgan. Redesigning check-processing operations using animated computer simulation. *Business Process Management Journal*, 6(1):54, 2000.
105. M.P. Wattman and K. Jones. Insurance risk securitization. *Journal of Structured Finance*, 12(4):49–54, 2007.
106. W. Xie, B. Nelson, and J. Staum. The influence of correlation functions on stochastic Kriging metamodels. In *Proceedings of the 2010 Winter Simulation Conference*, volume 1, Baltimore, December 2010.
107. J. Xu. Efficient discrete optimization via simulation using stochastic Kriging. In *Proceedings of the 2012 Winter Simulation Conference*, volume 1, page 12, Berlin, December 2012.
108. J. Xu and K.L. Hancock. Enterprise-wide freight simulation in an integrated logistics and transportation system. *IEEE Transactions on Intelligent Transportation Systems*, 5(4):342–346, 2004.
109. J. Yin, S.H. Ng, and K.M. Ng. A study on the effects of parameter estimation on Kriging model's prediction error in stochastic simulations. In *Proceedings of the 2009 Winter Simulation Conference*, volume 1, Austin, December 2009.
110. M. Zakerifar, W.E. Biles, and G.W. Evans. Kriging metamodeling in multi-objective simulation optimization. In *Proceedings of the 2009 Winter Simulation Conference*, volume 1, Austin, December 2009.

14

Web-Enabled Decision-Support Systems

Deepu Philip, Dmitri G. Markovitch, and Lois S. Peters

CONTENTS

ABSTRACT This chapter provides an overview of web-based decision-support systems (DSS) that are seeing a widespread use among decision makers, educators, and researchers in various disciplines and domains: business, natural resource management, medicine, manufacturing, engineering, and many others. The chapter opens with an introduction to DSS, complex systems, and decision theory. It then proceeds to discuss the design of web-enabled DSS and back-end database systems. Next, it addresses the front-end user interface and the intermediate application-layer concepts. In addition, the chapter also covers the topics related to interfacing the model base. The chapter concludes with one important application of DSS in higher education—DSS for management simulation games and a discussion of the future evolution of web-based DSS.

14.1 Introduction and Overview

The psychology literature defines decision making as the process of making a choice from a set of alternatives (Kahneman and Tversky, 2000). Decision making involves problem analysis, goal setting, alternatives evaluation, decision implementation, and feedback (Druzdzel and Flynn, 2002). In the real world, problems tend to be complex and ambiguous, and computers were initially employed to support the decision process by enabling a more rapid data analysis. Subsequent advancements in the fields of information technology, statistics, operations research, artificial intelligence (AI), and cognitive sciences were incorporated into computer programs to further facilitate the complex decision making. Such systems are broadly known as decision-support systems (DSS).

Although DSS were initially defined as "interactive computer based systems, which help decision makers utilize data and models to solve unstructured problems" (Mora et al., 2003, p. 102; Gorry and Scott-Morton, 1971), until recently, they were most commonly used in conjunction with structured problems. In such problems, the amount of information and complexity are oftentimes considerable, and the required domain knowledge may be beyond the unaided human cognition. Most DSS assist human decision makers by seamlessly integrating the inputs from multiple information sources and allowing a smart access to the necessary knowledge base, data, analyses, and recommendations.

In contrast, unstructured decision problems tend to be more ambiguous, where no standard approaches to obtain the optimal solution may exist or where there is no optimal solution. Human intuition is often used to make decisions involving unstructured problems (Turban and Aronson, 1998; Turban and Watkins, 1986; Turban et al., 2005). Semistructured decision problems fall between structured and unstructured decision problems, displaying the characteristics of both types of problems (Zhang et al., 2015). However, Sprague and Carlson (1982) and Zhang et al. (2015) further point out that by allowing DSS to support the structured part of any decision-making problem, decision makers can combine the resultant information or solution with their limited cognitive resources to tackle the unstructured element of the same problem. An example can be found in the area of capacity planning: the transportation method of linear programming (LP) is first used to create an initial aggregate production plan (the structured part). Next, decision makers use their intuition to plan for new products using the available capacity information (the unstructured part).

Before the Internet, DSS were implemented as stand-alone systems, and system access was conditional on access to the computers hosting the system. With recent advancements in Internet technologies, web-based DSS applications have become pervasive. There are at least two mechanisms by which technological advances in computers and telecommunications are driving the increased prevalence of web-based DSS. First, new technologies increase the functionality and accessibility of web-based DSS. Second, implementing and updating the models and data that underlie a web-based DSS is typically easier, quicker, and less costly than updating the same number of user sites where off-line DSS is installed on local computers.

The typical capabilities that are unique to web-based DSS can be summarized as follows:

- Global ease of access without the need for special software. A simple web browser is all that is required;
- Real-time collaborative decision making with decision makers is distributed globally. This contrasts with a stand-alone DSS that forces the decision makers to travel to the mainframe computer location where the system is made available;
- Support for choice from a set of alternatives enriched with a broader (multicultural) perspective, because DSS access is available to decision makers located in different countries;
- Distributed computing that pools DSS resources from multiple locations where the organization is present;
- Real-time analytics of big data from multiple sources in multiple global locations;
- Ease of use with mobile devices or thin clients (discussed below), because most of the computational load is taken up by the computer server that is hosting the DSS application.

Apart from providing more extensive computing power and the necessary analytical tools, web-based DSS can help to reduce expenses and the need for a costly specialized hardware and software at multiple locations (Bryan 2012; Watson, 2005). Computer terminals that interact with web-based DSS usually have minimal hardware requirements. Such computer terminals are called "thin-clients." Dedicated servers associated with web-enabled DSS can optimize information access, sharing, maintenance, and platform modification. All users can have access to the same software version but can run different scenarios and generate customized reports. Since data and applications are located on dedicated servers at a single location, input errors can be minimized and the maintenance and distribution of the software can be greatly simplified. Also, all the users can have access to the same versions of the decision model(s), which, all else equal, should require less analysis integration and interpretation effort across different locations (because the decision makers are familiar with the models), more streamlined communication involving DSS results, and fewer errors due to miscommunication.

An equally important advantage of web-based DSS is online access to a vast repository of relevant data that can be used in the analysis; this enables a faster response to changed situations and information flows (Paradice et al., 1987; Shim et al., 2002). Web-based systems also allow the diversity of participants in the location, time, expertise, and background. Web-based DSS have the potential to increase the productivity and profitability (Watson, 2005), as well as embrace more sophisticated models drawing on the advances in decision-making theory and neuroscience, making DSS more useful in highly complex decision situations.

The aforementioned benefits of web-based DSS also increase the effective competitive disadvantage incurred by firms that do not use such systems. Historically, simple models were used to represent the social, economic, and productive systems in which organizations operate. In their most primitive form, these models could be analyzed using paper and pencil. In eras characterized by exceedingly high computation costs, these models made simplifications to ensure the computational tractability. For example, nonlinear systems were often approximated by linear ones. Additionally, the naturally occurring heterogeneity of economic actors and their cognitive capabilities were often ignored in favor of simpler models focusing on homogeneous agents. When all rivals used equally poor models, the competitive disadvantage accruing to low-quality models was minimal. However, the rapidly decreasing cost of computation coupled with the accessibility of big data now enables superior models. DSS exploiting such benefits allow global businesses to be more productive and efficient (Boreisha and Myronovych, 2008). Firms that fail to use or rapidly update DSS based on such models risk are putting themselves at a competitive disadvantage.

DSS have been traditionally conceptualized as tools for top-level managers. However, this traditional conceptualization is misleading. Decisions are required at every level of an organization and not just at the managerial level. Different types of DSS can be used at different levels of the organization. As web-based DSS gain popularity, DSS are being deployed in various spheres of human activity, including business, manufacturing, design, medicine, defense, and even to facilitate consumer decision making, as discussed in the concluding section.

14.2 Grounding in Decision Theories

A review of the DSS literature by Bhargava et al. (2007) determined that DSS research was comparatively poorly grounded in theories of judgment and decision making. Furthermore, a review by Arnott and Pervan in 2014 demonstrated that grounding DSS in decision theories has not improved since their 2005 review. In fact, the use of decision-making theory in DSS research has declined. Karunakaran et al. (2015) also note in their review of business decisions that an emphasis on the theory behind DSS is generally weak and calls for future research that focuses on theory-driven models. With this in mind, we provide a brief review of decision theory to encourage the consideration of how to tie the decision theory into web-based DSS research and models.

The fields of psychology, cognitive science, organizational behavior, economics, mathematics, and philosophy have all contributed to our understanding of decision making by advancing the various perspectives, models, and techniques. Broadly speaking, the theoretical models can be grouped into those that view decision makers as engaging in deliberate and extensive reasoning and those that view human decisions as more or less extemporaneous, such as ad hoc decision making or opportunistic matching of solutions and problems as depicted in the garbage-can theory of decision making (Cohen et al., 1972). The classical model of decision making postulates that individuals are presented with a problem or opportunity and make an optimal decision based on perfect information about alternative choices. While decision models historically assume rationality and internal decision consistency, there is an increasing recognition in most fields that strong rationality assumptions are not supported by recent behavioral and empirical evidence (e.g., see Ariely, 2008). This suggests that the development of robust theoretical models for web-enabled DSS will require multidisciplinary collaboration and potentially more computationally intensive models.

The scholarly literature on bounded rationality (Simon, 1955, 1960; Tversky and Kahneman, 1974, 1981) supports the usefulness of DSS; humans only have the capacity for processing limited amounts of information. Thus, an important benefit of DSS systems is that they can help to relieve the cognitive load-facing decision makers by assuming the burden of processing large volumes of data (Deck and Jahedi, 2015). This ameliorates the issues associated with information overload and can enable decision makers to evaluate a greater number of alternatives.

Moreover, most decision making in organizations involves more than the linear and step-by-step approaches that the rational choice modeling entails (Dawes, 1988; Sprague, 1980). Although not purely chaotic, the process is often not as rational as even that which behavioral models assume. Decisions are usually made under ambiguity and uncertainty and can result in an idiosyncratic behavior and decision anomalies. This is particularly the case with those decisions that entail nonroutine problems and opportunities, and decisions made under time pressure, risk, and limited information (i.e., uncertainty).

Current extensions of the process-oriented decision theory investigate how an evaluative and creative behavior impacts decision steps and recursion in decision making. Recursive decision making is a process of moving back and forth between choice criteria and alternatives progressing as if going in a linear direction (Shim et al., 2002). It reflects the reality that the available alternatives influence the criteria applied and the established criteria influenced the alternatives considered. Franklin (2013) proposes an extending recursion between criteria and alternatives to all decision-making steps and discusses decision-analysis processes as a system interaction of the steps and processes. Because web-based DSS query models are hosted remotely, the models that underlie such DSS's are relatively unconstrained by computational power. As such, web-based DSS can easily go beyond a linear three-step (intelligence, design, and choice) approach to decision making by employing recursive engines (Clemen, 1990).

Recent trends in model building include incorporating cognitive processes involved in decision making. Identifying systematic failures in these processes allows the identification of corrective action, which has implications for practice and DSS. To the extent that web-based systems bring together diverse groups of people to make decisions, differences in biases and how they get resolved may need to be addressed. Heydenfeldt (2013) makes the case for training that strengthens a conscious information-processing capacity through being mindful about perceptive limitations and reviewing decision-making processes. However, neuroscience research also documents the importance of intuition and gut instinct in decision making, especially in complex situations with multiple variables under consideration (Brocas, 2012; Brocas and Carillo, 2014; Waytz and Mason, 2013).

Importantly, there is a growing acceptance in the scholarly community that group decision making, is generally becoming more common (Dennis, 1996; Hightower and Sayeed, 1996; Immelt et al., 2009), and is more prevalent in organizations than individual decision making (Csaszar and Enrione,

2015). Decision making in groups is characterized by dynamic interactions that may dampen or intensify individual tendencies. The research of Bazerman et al. (1984) indicates that groups, like individuals, are affected by cognitive dissonance and bias. Others (e.g., Seibert and Goltz, 2001; Whyte, 1991) have shown that group interactions and dynamics impact the outcomes. Therefore, the research on group decision making has gained a greater prominence in recent years. Accordingly, group decision-support systems (GDSS) have emerged as an important type of DSS that are also implemented on web-based platforms (Istudor and Duta, 2010; Lu et al., 2005). It is surprising, therefore, that some researchers document the decreased research intensity focused on GDSS from 30.3% of publications dealing with DSS in the 1997–2003 timeframe to 19.3% of such publications in the 2004–2010 timeframe (Arnott and Pervan, 2005, 2014).

An important subject within GDSS that deserves special mention is the study of virtual teams, or geographically distributed groups of individuals who rely on information and communication technologies, such as telephone, instant messaging, videoconferencing, and e-mail, to interact in the course of pursuing organizational goals. Managerial teams are increasingly virtual and multinational as a result of internationalization and advances in communication technology. This is particularly true in the field of technology management, because the emerging market growth has pushed many technology-oriented firms to adopt multinational research and development (R&D) strategies (Immelt et al., 2009). Web-based DSS should be a useful support tool for virtual teams (Paul et al., 2005), and the research on virtual teams will not only help to improve GDSS but also research through the use of web-based GDSS will provide an input into developing more robust GDSS models.

Recently, researchers interested in the impact of social interactions on effective decision-making strategies have demonstrated that the level of consensus required for getting the most out of each team member's expertise is context dependent. They show that the consensus is best when there is a need to minimize commission errors (i.e., pursuing a bad project, or type-1 statistical error). Situations where omission errors (i.e., missing a good project, or type-2 error) might be problematic (e.g., missing a disruptive technology) are probably best served by a low consensus with the best level of consensus dependent on the cost of errors (Csaszar and Enrione, 2015).

Finally, naturalistic decision making (NDM), which refers to the descriptive modeling of proficient (experienced) decision makers in natural contexts, in particular those characterized by considerable time pressures, ill-defined goals, uncertainty, and high personal stakes (Lipshitz et al., 2001) is reflected in a subtype of DSS called "expert systems." NDM theory takes on a process orientation by matching to the typical cases or mental models as opposed to a choice. The premise is that decision makers in natural (field) settings use situational–content-driven cognitive processes to solve domain-specific problems and take concrete actions (Klein 2008, p. 140). Therefore, NDM better reflects "real-world" decision making, whereas an emphasis on stylized contexts and assumptions has been characterized as a shortcoming of many analytical (e.g., game-theoretic) models.

We now proceed to discuss DSS in more detail and the added value of key web-based DSS technologies.

14.2.1 DSS Subsystems

As shown in Figure 14.1, both legacy and web-based DSS normally consist of three subsystems (Sauter, 2011; Turban and Aronson, 1998; Turban and Watkins, 1986) as follows:

1. *Dialog generation and management system (DGMS):* The DGMS is commonly referred to as the user interface (UI) (e.g., Sauter, 2011; Sprague, 1980). However, the DGMS may exceed the functionality of a typical UI. The other DSS subcomponents are routinely

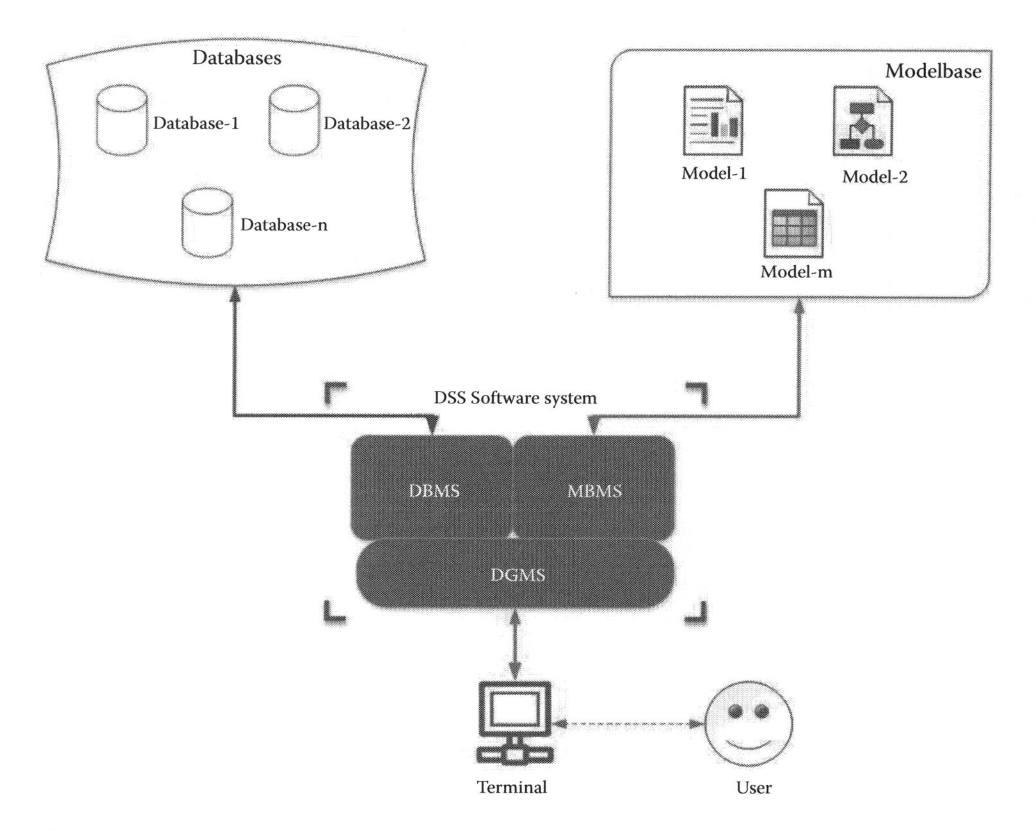

FIGURE 14.1
DSS subsystems overview.

accessed through the DGMS. Consider an example of a handheld Global Positioning System (GPS) device (i.e., a basic DSS). Using the touchscreen DGMS (or the UI), the user inputs the origin and destination to the device and clicks the button to initiate route planning. Internally, invisible to the user, the DGMS first triggers the appropriate queries to retrieve the necessary map data stored in a database management system (DBMS) (e.g., using geocodes, or an edge-weighted directed graph). Afterward, the DGMS interfaces with the model-based management system (MBMS) using the retrieved data to access the appropriate route-planning models (the shortest path, a scenic path, only highways, etc.). The results from the route-planning models are displayed on a map, allowing the user to decide the route from the available choices, while hiding from the user the entire complex operations of data retrieval (i.e., interactions with the DBMS) and model triggering and computation (i.e., interactions with the MBMS). Thus, the DGMS not only accepts user inputs, but also embeds the logic through which the other computer subcomponents can be instructed to provide an appropriate user feedback. In some configurations, the DGMS may, alternatively, trigger other applications that contain the logic.

2. *DBMS:* This DSS subsystem deals with the storage of data relevant to the decision problem using logical data structures (Elmasri and Navathe, 2003; Post, 1998). The user interacts with DBMS through the DGMS during DSS-assisted decision making. The DBMS separates the physical aspects of data storage and manipulation from the user and allows the

user to focus on data relevant to the decision problem at hand. The data are stored in a permanent-storage device, such as a hard-disk drive (HDD), solid-state drive (SSD), or optical media (e.g., a DVD or Blu-ray). It is not essential that data comprising a table in the database are stored contiguously. The storage is usually affected by using an unordered, ordered, or hashed organization within the storage device. Depending on how the DBMS is designed, auxiliary access structures called indexes (Elmasri and Navathe, 2003), which identify secondary access paths to retrieve records without affecting the physical placement of records on the disk, may be used to speedup data access. The DBMS decides the indexing based on its design. Approaches such as search trees, B-trees, or B+ trees are used by DBMS to quickly store or retrieve the data as per user requirements. Of note, the storage of numeric data may be done through one approach, whereas the storing of binary large-object (BLOB) data may be affected through the other approaches. All such aspects are hidden from the user by DBMS, so that the user can solely focus on appropriately developing and using the database. (See Chapters 14 and 15 of Elmasri and Navathe, 2003 for further details.)

The DBMS can manage multiple databases containing different types of data (e.g., financial data, sales data, an inventory, etc.). The DBMS manages the what, where, and how of this data storage, retrieval, and manipulation without exposing the user to such complexities. Most of the DBMS are Structured Query Language (SQL)-based systems (Hoffer et al., 2009). Recently, No-SQL (not-only SQL) DBMS systems have also seen an increased use (Leavitt, 2010; Sullivan, 2015). However, in all cases, the DBMS provides a platform for the user to develop an appropriate DSS database while hiding from the user the implementation details, including the storage, retrieval, optimization, backup, redundancy, etc.

3. *MBMS:* Decision models residing in a computer are usually referred to as the model base. It may include models preloaded by the DSS provider or custom models built specifically by or for the client organization. The storage of and access to specific models used in the decision-making process is the primary function of the MBMS (as shown in Figure 14.1). Typically, the data are retrieved using the DBMS and stored using appropriate data structures in the computer memory. The data then get transformed through various models in the MBMS into information that will aid the decision-making process (Sauter, 2011).

   The MBMS may be a stand-alone or an integrated software application that provides access to the model base, while ensuring that the requirements of individual models are fulfilled during the DSS run time. Also, the MBMS sometimes includes interfaces for building custom models, which is commonly known as model development environment (MDE) (see Turban and Watkins, 1986; Turban et al., 2005 for a more extensive treatment of the subject). The user normally interacts with the MBMS through the DGMS, because the technical aspects of data requirements, model storage, access to the appropriate data values, and the like, are hidden from the user within the MBMS (Paradice et al., 1987). It should also be noted that software applications such as MATLAB®, MATHCAD, SAS, R, SciLab, OPL Studio, or SPSS allow for the development of custom-decision models. These models require triggering of the run-time environment mandated by such applications. The MBMS should, ideally, facilitate this step.

14.2.2 Web-Based DSS: Key Technologies

Figure 14.2 presents a conceptual schematic of the web-based DSS that has all the three components of legacy DSS systems, however, integrated differently. The usage of hypertext transfer protocol

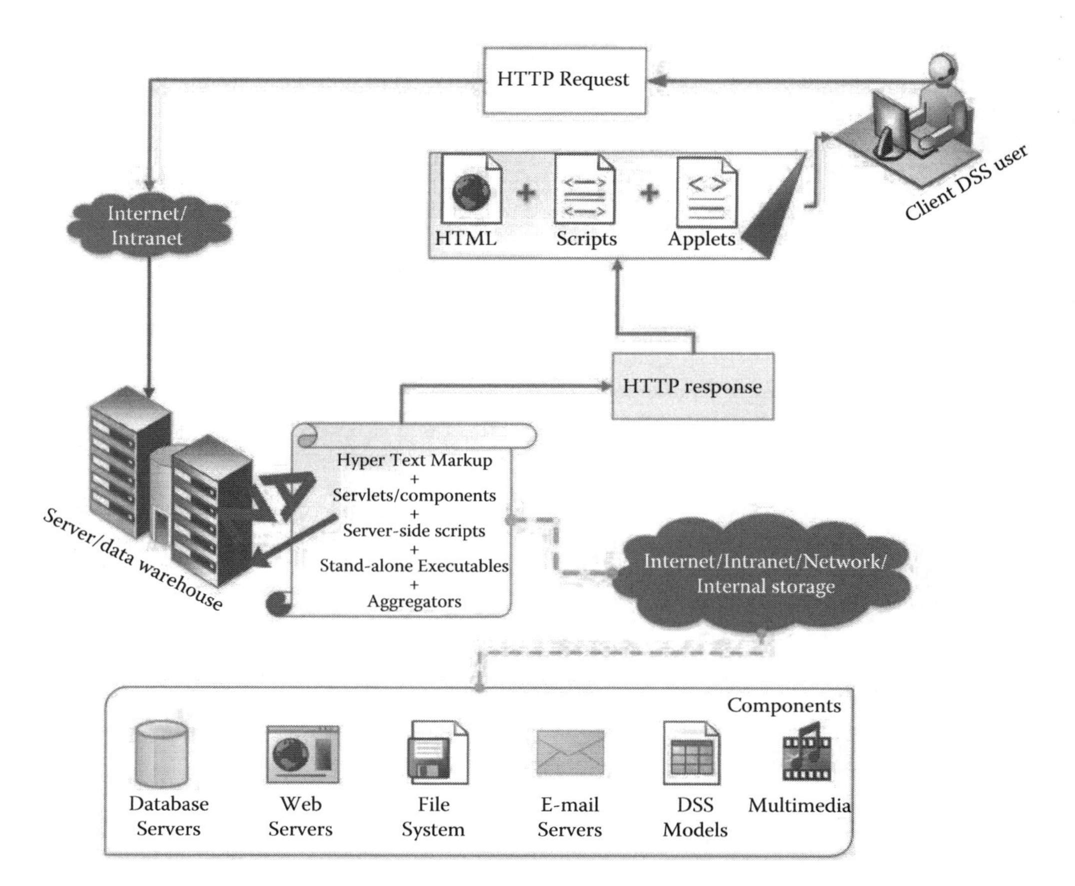

FIGURE 14.2
Web-based DSS schematic.

(HTTP) over the Internet/intranet allows the user to access the DSS from any remote location. Users connect to the computer server hosting the DSS through a network (the Internet or an intranet) using transmission control protocol/Internet protocol (TCP/IP). Thus, the "Web-based" aspect of DSS implies that the entire DSS is implemented using web technologies, where the application is accessed using a web browser and the web integrates the application and the database residing on a server or different servers. More recently, no-database web DSS have become available. These DSS allow the storage and retrieval of data using means other than database tables used in relational DBMS (RDBMS). The common storage forms include a key-value pair, a graph, or a document.

When implemented, most web-based DSS employ a three-tier architecture, although a limited number of systems utilize a four-tier architecture. The first tier is typically composed of the web browser (or a customized, Internet-enabled app), through which the user accesses the system, as shown in Figure 14.2. The browser transmits the user's requests using a transfer protocol (e.g., HTTP) to a server system over the Internet/intranet running a web-server application. Web browsers are software applications that are capable of retrieving, parsing, and presenting information resources (web pages) on the World Wide Web. Examples of web browsers include Firefox, Safari, and Chrome. The web pages may also contain appropriate scripts (i.e., a code) that trigger different events during run-time parsing of the web page. Such codes are written in interpretable scripting languages, such as PHP, Python, and Ruby.

The servers/data warehouse constitutes the second tier of the architecture. The web-server application can be an entire computer, appliance, or software capable of accepting and processing HTTP requests. Its primary function is to store, process, and deliver web pages to clients based on specific HTTP requests. Web-based DSS are designed to interact with users having appropriate credentials. The web server also ensures user authentication and the secure connection between the user's terminal and the server hardware. Examples of common web servers include Apache, Internet Information System (IIS), nginx, and Google Web Server.

The third tier of the architecture is the components of the web-based DSS in which the DBMS plays a key role. Other major components of the web-based DSS are mail server, file system, multimedia server, and DSS models. These web-based DSS components provide capabilities such as e-mail, file management, embed multimedia, etc. to the system. In web-based DSS, the DBMS either resides on the same computer as the web server or in another computer that is tightly integrated with the web-server computer. When the web server processes HTTP requests, appropriate calls (queries) to specific databases are passed on to the DBMS and the results are manipulated within the script during run time. Access to other inputs, including multimedia and models, is executed at run time when the web server parses the HTTP requests (refer to Figure 14.2).

In certain cases, sophisticated programs written using efficient programming languages, such as C++, Java, and ProLog, or sophisticated statistical models developed using applications such as R might be a part of the web-based DSS. Such stand-alone executables may form the fourth, and less-common, tier of the web-based DSS. Typically, web-based DSS that are more dependent on open-source scripting languages such as PHP or Python tend to include customized stand-alone executables (comprising the fourth tier). The reduction in computation time is the main reason for incorporating such compiled and optimized stand-alone executables. The alternative approach involving the use of an interpreted language takes significantly more time to complete the computations due to its inherent nonoptimized code.

Various tools are used to build web-based DSS systems. In addition to hypertext markup language (HTML), other popular tools include Common Gateway Interface (CGI) scripts, Extensible Markup Language (XML), Application-Programming Interface (API), Applets, Servlets, JavaScript, JQuery, ActiveX components/plugins, Asynchronous JavaScript and XML (AJAX), and Cascading Style Sheets (CSS). Scripting languages including PHP, Python, ASP, Ruby, or Perl are used for developing dynamic applications and UIs, whereas programming languages, such as C/C++, ProLog, and Java, are used for building complex models or algorithms that are accessed as per user requirements. Finally, the DBMS, such as MySQL, Oracle, and SQL Server, are used to facilitate data storage, retrieval, and manipulation.

14.3 Web-Based DBMSs

Business decisions are increasingly informed by rich and voluminous data from diverse sources comprising the firm's business environment. These data are stored in one or more databases. The DBMS, which serves as the backend of the DSS, performs the primary function of storing large quantities of diverse data using logical data structures on permanent-storage devices such as HDD, SSD, and optical disk, so that problem-relevant data can be retrieved and manipulated to generate information by the DSS upon the user's request.

The DBMS are computer programs that focus on the collection and management of large quantities of data on physical storage devices. Here, the term "management" refers to the functions such as creating, updating, and querying databases in an optimal manner. DSS capabilities for the retrieval, manipulation, and display of data items represent a typical functionality provided by the DBMS. The DBMS also provides critical information to the user on the types of data available (i.e., the data dictionary) and the data-access protocols (i.e., queries). Thus, the DBMS serves as a functional layer between the user and the actual data, where the physical aspects of the database structure and processing remain hidden from the user.

Since the data and models that use them are fundamentally interdependent, DSS designers usually focus on data and model management concurrently. In many cases, the database design for DSS entails decisions on the level of normalization that one should follow. Such decisions are mostly predicated on the type and complexity of the DSS models used by the system. Normalization refers to the process of organizing the attributes and relations in a relational database (Elmasri and Navathe, 2003; Turban et al., 2005). Here, "relations" refer to the tables of a database, and "attributes" refer to columns of the table.

Often, intentional redundancy is maintained in the DBMS to facilitate a quick turn-around time for complex decision models. Intentional redundancy can be achieved by duplicating information in different tables. Such a practice can reduce the time taken to perform the "JOIN" operation, which combines multiple large tables to obtain a temporary table, and then selects the required information from such a temporary table. Of note, doing this may violate the transitive dependency and thus make the database not in its third normal form (3NF). However, this may improve information-retrieval speeds in certain cases.

When compared with legacy DSS, the database designs of web-based DSS tend to prioritize the speed and availability over normality and storage-space optimization. This trade-off is dictated by the fact that a web-based DSS usually has many simultaneous users, and most of the usage requests are of a data-retrieval variety. This makes the read speed and availability critical (Hoffer et al., 2009; Turban and Watkins, 1986). In contrast, the traditional DBMS emphasized enforcing the security and integrity while avoiding the inconsistency. These goals were important to the success of transaction-processing systems (TPS), such as the bank teller, or a computerized inventory system, where the integrity of "data-write" operations is paramount.

Popular DBMS used in web-based DSS include MySQL, PostgreSQL, MariaDB, Microsoft Access/SQL Server, Oracle, DB2 Express-C, and SQLite. The DBMS choice usually depends on the type of data and the application program coding language. For example, either MySQL or MariaDB is frequently chosen where read speeds are critical and open-source scripting languages are used to develop the applications. These programs' popularity among web-based application developers and users can be attributed to their high-speed reads, a robust design, and their low cost (being open source). MySQL or MariaDB DBMS also allows a ready integration with popular programming languages, such as PHP and Python. Similarly, PostgreSQL is preferred where write speeds and capabilities are of a significant importance. For applications that mandate a high security, developers tend to prefer Oracle DBMS because of the many automatic security features, including Oracle security, label security, database vault, identity management, a transparent encryption, security-based data classification, internal realms, real-time access control, and a secure backup.

An additional noteworthy development in the DBMS space is the advent of No-SQL databases. No-SQL is increasingly used in web-based DSS. Such databases allow the storage and retrieval of data using means other than relations (i.e., database tables) that are typically used in RDBMS (Leavitt, 2010; McCreary and Kelly, 2013; Sullivan, 2015). In No-SQL, the typical forms of storage are a key-value pair, a graph, or a document. No-SQL-based DBMS tend to emphasize design simplicity (moving away from normalization), better horizontal scaling, and control. No-SQL-based DBMS place less emphasis on consistency, which is the key aspect of RDBMS (McCreary and Kelly, 2013). By sacrificing the consistency, No-SQL systems focus on the availability and partition tolerance (i.e., the capability of a system to continue to function even in the instances of an arbitrary partitioning due to network failures). Examples of popular No-SQL programs are Druid, Vertica,

CouchDB, MongoDB, Dynamo, MUMPS, Allegro, OrientDB, and CortexDB (for further details, see Chodorow, 2013; McCreary and Kelly, 2013; Sullivan, 2015).

14.4 Web-Based MBMSs

Models are used to analyze data in support of the decision-making process. Two distinct modeling approaches exist. Some models imitate the reasoning approach of a human expert (Kahneman et al., 1982), whereas others use a set of normatively sound principles of decision making (Druzdzel and Flynn, 2002). Applications that use models composed of rules elicited from human experts in the focal domain are broadly known as "expert systems."

Expert systems (ES) employ models that are typically stand-alone AI-type programs that use heuristic or algorithmic inference procedures comprising the knowledge about a particular problem domain. ES boast a considerable flexibility relative to other types of DSS in that ES are easier to modify during the execution. This allows for the management of variability during decision making, where changes in the inputs, environment, or system state are considered while proposing the decision options. For example, ES are better able to handle unanticipated inputs. Consider a scenario where one is developing an animal-classification system based on animal features. According to this perspective, fish may be classified as animals living in water that have a streamlined body, scales, gills, fins, and a tail. When the system is exposed to information about a whale, it will initially classify it as a fish. However, once the rules for marine mammals are developed for the ES, the whale is appropriately reclassified as a mammal rather than a fish. However, in a normative DSS, until the system obtains data containing whales classified as marine mammals, such an inference is usually not possible. This flexibility is due to the ES architecture, which separates the inference engine from the knowledge base that tends to grow exponentially. Typically, an inference engine applies logical rules to the database or knowledge base to deduce new knowledge. Inference engines are capable of starting with known facts and deduce new facts (forward chaining) or start with goals and work backward to determine factors that should be present to achieve the goal (backward chaining). Since the knowledge base resides outside the engine submodule (i.e., the code), it can be updated continuously without affecting the model's performance. ES models are usually programmed using ProLog-like languages, which allow an easy embedding of efficient heuristics. ProLog-like languages support logical programming, where the program can be stated as declarative formal-logic statements. Many aspects of inference engines are embedded in such languages, including forward or backward chaining, pattern matching, searches, unification, list processing, sequence, and tree representations, in addition to the ease of representing knowledge.

However, in many decision situations, the second DSS paradigm based on a normative-decision framework may be preferable. An example of such a DSS system is the control of an aircraft using an autopilot system. While it is computationally possible to design an autopilot using heuristics and ad hoc rules based on the relevant knowledge base, systems built using aerodynamic principles along with probability and reliability principles are more trusted and proven. In general, whenever the cost of an error is high, DSS designers tend to prefer MBMS based on normative models over others.

In a web-based DSS, the DGMS is usually developed using a scripting language that can be easily integrated with the web server and a customized MBMS. The integration of a customized MBMS in a web-based DSS involves a specialized handling of the requests by the scripting language. Such capabilities eliminate the need for developing DSS models in a particular programming language. This facilitates the rapid development, deployment, and maintenance of DSS.

Ideally, an MBMS should be able to support different types of models and trigger them as requested by the user or based on user queries of the DGMS. Most of the models associated with DSS can be broadly classified into four categories (Druzdzel, 1992; Sauter, 2011; Turban and Watkins, 1986) as follows:

1. *Domain models:* These are models that are built using an extensive domain knowledge (i.e., valid knowledge about the environment in which the specific system operates). Typically, a network or a tree-like representation is used in such models. The most common examples of such systems are decision trees (Lipshitz et al., 2001; Sauter, 2011; Turban and Aronson, 1998). Consider a medical diagnostic-support system, where a medical doctor's diagnostic path can be accurately captured using a decision-tree model. In such a model, symptoms and vital statistics form decision nodes and the course of action is chosen according to the most-likely scenario (based on probability); the path is followed until a confirmatory or disconfirmatory outcome is obtained. Then, the path is retraced and the next option is explored. Such models are typically built in customized programming languages, such as C, ProLog, and LISP. As mentioned previously, the scripting languages used in the development of web-based DSS are quite capable of accessing domain models during run time as per the user's request. Hence, web-based DSS are naturally capable of integrating a wide range of domain models, developed using different programing languages.
2. *Customized models:* These models are predominantly used in situations where each decision is unique and decision makers' extensive involvement is important in the decision-making process. Essentially, the computer model plays the role of a decision analyst in that it analyzes the decision maker's choice and presents back the results or projected outcomes. If the outcome is unsatisfactory, then, the decision maker uses the results to modify the decision and resubmit it to the model to reanalyze. The process continues until the decision maker is satisfied with the outcome. The major differentiating aspect of such models from ES models is in the formal and structured framework that uses proven probabilistic/quantitative approaches. Scripting languages including PHP, Python, Perl, and Ruby have considerable capabilities and dedicated libraries for the development of such customized models. For example, the pear library of PHP incorporates a vast collection of models that can be customized to suit individual analytic needs. A ready example of customized models that most readers will have been exposed to is the product recommender system provided to online shoppers by Amazon.com (Yiyang and Jiao, 2007).
3. *Learning models:* With the advancements in AI, learning models have gained a considerable popularity among DSS designers. These models may use a statistical guided or unguided learning (Russell and Norvig, 2003). At a high level of abstraction, learning processes may be viewed as comprising three steps: (i) observing a phenomenon, (ii) constructing a model of that phenomenon based on the observations, and (iii) using the model to do predictions (Bousequet et al., 2004). The goal of machine learning is to automate this learning process. The general idea behind the design of any learning algorithm is to look for regularities or commonalities in the observed phenomenon and relate them to future cases. Past observations on the focal phenomenon are thus used as "training data" for the learning model (Bousequet et al., 2004; Russell and Norvig, 2003). Statistical learning is the process of developing a predictive function based on the training data and establishing how well the model fits the data (Mohri et al., 2012). A guided statistical learning involves providing specific goals and cues to the system on how to conduct the learning, whereas unguided learning involves a self-exploratory learning by the system. However, both types of statistical learning assume that the phenomenon is stationary (future observations are related to

past observations), and the observations are independent and identically distributed (IID). The advantage of statistical learning models is that if the focal data-generating mechanism does not change over time, then, learning approaches can provide an excellent support for decisions. Other increasingly common tools include neural networks, support-vector machines, and evolutionary algorithms. One example is the PyBrain library for machine-learning AI algorithms based on Python scripting language, which can be easily integrated with a web-based DSS system.

4. *Equation-based models:* Oftentimes, interactions among the variables used in decision making can be accurately described by equations (e.g., Turban et al., 2005; Zhang et al., 2015). For example, the relationship between force, mass, and acceleration is generally well defined by the Newtonian equation $F = ma$, where "m" denotes the mass of the body in kilograms and "a" represents the acceleration of the body in meters per second squared. Hence, in most scenarios where force exerted by a body is to be studied, such equation-based models provide the most-parsimonious solution. (It must be noted that some authors do not consider such models as a separate class because they could also fit in as a subcategory of the other classes.) Perhaps, the simplest form of such models is the class of models used in LP problems where a decision can be modeled as a set of objective functions and constraints that can be solved using standard approaches. Many DSS systems provide such capabilities. For example, CPLEX by ILOG OPL Studio is a popular simplex algorithm implementation to solve the LP models quickly.

14.4.1 Web-Based Dialog Generation Management System

In any DSS, whether legacy or web based, the dialog generation management system (DGMS) is crucial. DSS with a cumbersome and unclear DGMS usually require more specialized skills. Many users believe that the DGMS *is* the DSS, because it is the medium through which they interact with DSS. In web-based DSS, the DGMS is delivered using web browsers or equivalent apps (e.g., smartphone apps) over the Internet, where the goal is to allow the user to interact with the system anywhere in the world with a minimal specialized knowledge.

In legacy DSS, the UIs were mostly text-oriented windowed applications (Sauter, 2011). In contrast, web-oriented programming languages, such as HTML (being a markup language), are aimed at developing interactive web pages. HTML has inbuilt capabilities to create intuitive, interactive, and informative UIs rapidly in comparison with stand-alone programming languages such as C++, FORTRAN, or JAVA. Extended capabilities, such as embedding pictures, voice, video, and tool tips, are readily available in HTML (currently HTML 5). Other HTML add-ons, such as JavaScript, CSS, and JQuery, generally make it a much-better tool for building DGMS than stand-alone compiled programming languages. Additionally, the ability of HTML interfaces to readily integrate with smart devices, such as smartphones and tablets, makes it a versatile cross-platform and cross-device choice.

Although the DGMS is sometimes referred to as the UI, the UI is just one component of the DGMS. Other components include connectivity components to the DBMS and MBMS, embedded calls to other software applications, session management, user authentication, and encryption. There are oftentimes different design philosophies underlying the UI. For example, consider the two popular search engines' home pages, Yahoo and Google. The side-by-side comparisons are shown in Figure 14.3.

Google maintains a simple and clean look with the search interface attaining the maximum prominence. It has a simple white base with no clutter. The UI suggests on how to proceed along with the available options. Yahoo, on the other hand, attempts to provide a potpourri of search, news, and

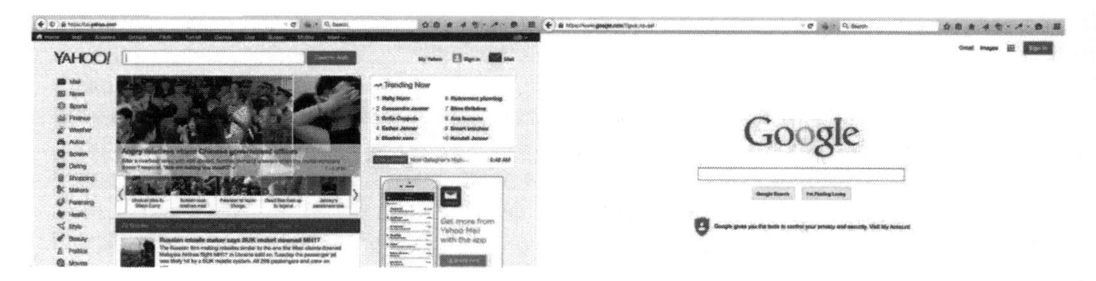

FIGURE 14.3
Comparison of Yahoo and Google search engine interface.

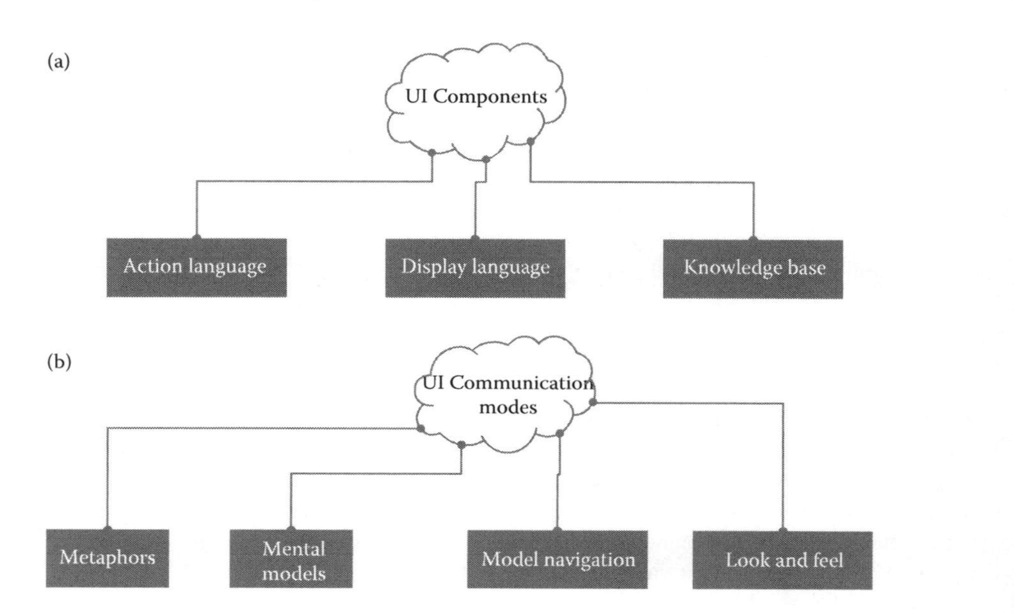

FIGURE 14.4
Components and communication modes of UI.

updates on social networks, aiming to serve as a single portal for multiple user needs. Some might believe that search engine users would prefer to have news, updates from friends, e-mails, and chat lists accessible on the side of the search page. However, this design is known to confuse many users (Krug, 2000). Similar considerations apply when implementing the UI of complex web-based DSS.

When describing the UI in terms of its components, it becomes apparent that the components commonly reflect the communication modes used (Sauter, 2011). Figure 14.4a depicts the various components of UI and Figure 14.4b shows the modes of communication with the UI. For a web-based DSS, the action language and display language are of the most importance, and hence, we elaborate on them here.

14.4.1.1 Action Language

The action language is the type of input used by decision makers to submit requests to the DSS (Sauter, 2011; Turban and Aronson, 1998). The action language includes the processes whereby the requests are submitted by the decision maker, which in turn invokes the decision models, data requisition, and conducting of the sensitivity analysis. Sauter (2011) classifies action languages into five subcategories: (i) menu format, (ii) question and answer (Q&A) format, (iii) command language format, (iv) input–output (IO) structured format, and (v) free-form natural-language format. For web-based DSS, we focus on the menu format and IO structured format as the main aspects of an action language.

1. *Menu format:* Menus present lists of commands, alternatives, or results from which the user can choose. Most web-based DSS employ a menu format for interaction. Menus are usually viewed as user-friendly for their ability to guide the users through the necessary steps (Krug, 2000). The usability of the menu system increases as the menus become more intuitive. The most important factor in menu design is that the menus use the same language as the language used by the decision maker (Sauter, 2011; Sprague and Carlson, 1982; Turban and Aronson, 1998). For example, if computer salespeople who are the primary users of a particular DSS favor the term "processor" in their professional activities, then, the DSS should use the same term, instead of the technically correct term "CPU."

   The menu system should, ideally, present the options in a logical order. For example, electronic address forms should preferably list countries in the address field in an alphabetical order, yet place the country in which the form is predominantly used at the top of the list. An example of a well-designed menu structure is shown in Figure 14.5 (courtesy: informationbuilders.com). It can be seen that the HTML frames are used to group information by the type and functionality. The topmost frame contains the UI to the DSS where

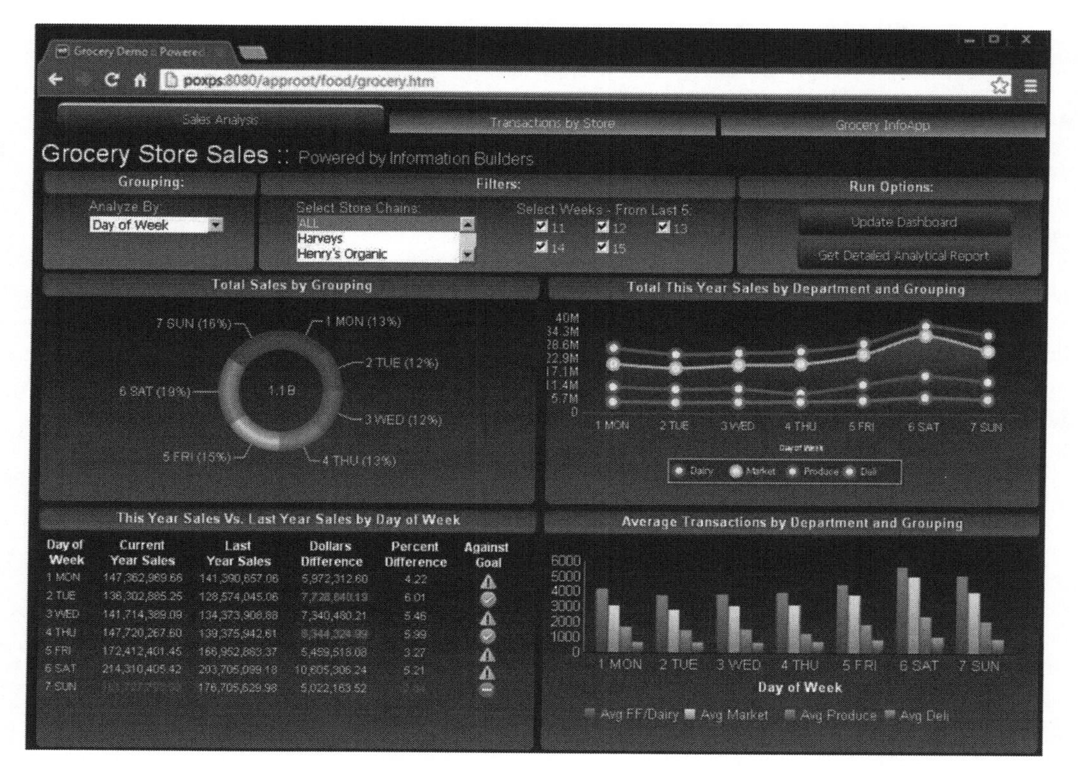

FIGURE 14.5
Action language usage in a DSS displaying grocery store sales.

different choices are presented as drop-down menus with different run options. Once the decision maker submits the inputs to the DSS, relevant data are retrieved from the DBMS and multiple charting models are activated. These charts are presented to the user in the frames below. The bottom two frames depict the outputs of comparisons with other similar stores, including simple icons to indicate whether the annual sales goals were achieved or not. This single-window approach allows the decision maker to stay on one web page and interact with the DSS with a few mouse clicks.

The menus are of primary benefit for inexperienced users, allowing for an immediate usage of the system with a minimal learning curve. This is particularly important in web-based DSS where users are located around the globe and providing personal training might be infeasible or costly. Additionally, the menus work well for decision makers who use the system sporadically. A notable disadvantage of the menu system is that experienced users (i.e., "power" users) of the DSS may get frustrated by the increased time and keystrokes needed to obtain important information through a menu-based display.

2. *IO-structured format:* This format is akin to paper forms in which fields are filled with relevant information. The user can navigate through the form and provide the required information, similar to a manual-form completion. Some experts consider this format as a part of the legacy system (e.g., Gorry and Scott-Morton, 1971; Sprague and Carlson, 1982). However, modern advancements in web technology, such as HTML forms, make the IO format interactive and easy to use, and are therefore quite relevant. For example, popular travel-planning software programs use HTML forms to obtain user information and also assist in the travel decision-making process. An example of a common IO HTML form is shown in Figure 14.6 (courtesy: hellobar.com).

14.4.2 Display Language

Just as the action language focuses on obtaining user input to the DSS, the display language is used to present information to the decision maker based on his/her queries of the DSS. The display language is also known as the presentation language. It should convey the necessary information to the user in an understandable and pleasing format without unnecessary clutter. While doing so, it should also include sufficient details on the intermediate steps in the process for the user to understand the DSS rationale. The display language has four major parts based on the display organization (Sauter, 2011; Sprague, 1980; Sprague and Carlson, 1982): (i) windowing, (ii) representation, (iii) graphs, and (iv) support of the decision-making phases. We discuss each of them in relation to the web-based DSS.

1. *Windowing:* represents the usage of separate graphical windows. Windowing helps to organize information for the decision maker. Typically, each window should represent a different kind of output from the DSS. Some DSS-display windows may show the relevant graphs and diagrams. Other windows may include spreadsheets or tabular information. Finally, some windows may flag cases that warrant additional analyses. In web-based DSS, windowing can be accomplished by tabbing, where similar information is grouped in tabs. Figure 14.7 shows how tabs (primary, social, and promotions) are used instead of windows to organize information on this web-based application.
2. *Representation:* shows the results of analyses based on user input. The purpose of the information determines its representation. For example, a multisite retailer may display store locations with increasing sales with upward-pointing arrows and stores with declining sales with downward-pointing arrows. Additionally, the size of the different arrows could reflect

SIGNUP FOR PRO

Hello Bar PRO
$4.95/month

CONTACT INFORMATION

First Name

Last Name

Email Address

ADDITIONAL INFO

Company

BILLING METHOD

VISA PayPal DISCOVER

Card Number (no dashes or spaces)

Card Expiration Date (mm/yyyy) Card Verification Code

BILLING ADDRESS

Street Address street address and apt/suite number

City

State Zip

Country
United States

ORDER OVERVIEW

Hello Bar PRO
$4.95/month

SUBMIT FORM

By submitting this form, you agree to Hello Bar's Terms of Use.

FIGURE 14.6
HTML form as an input mechanism.

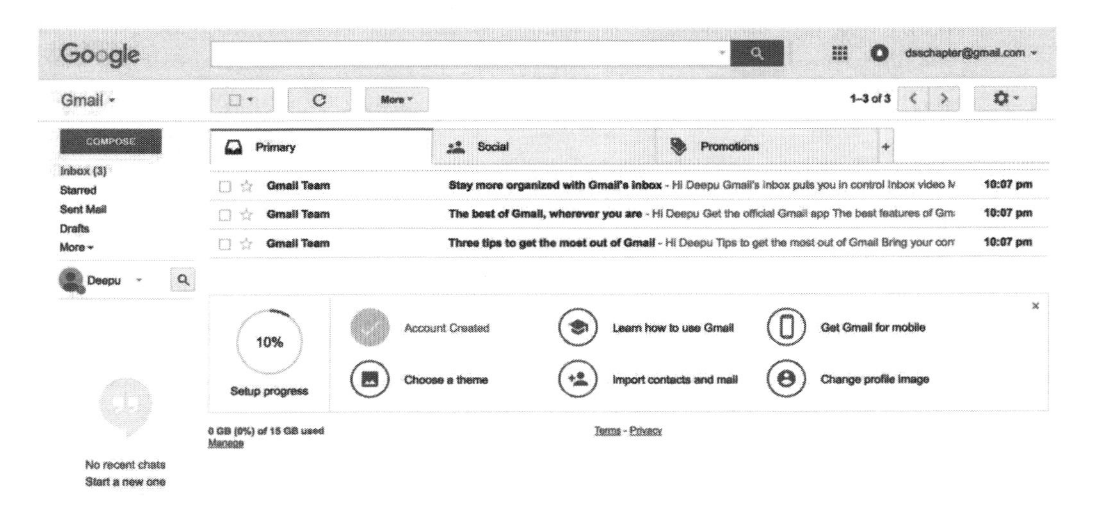

FIGURE 14.7
Organization of e-mails in specific tabs by Gmail.

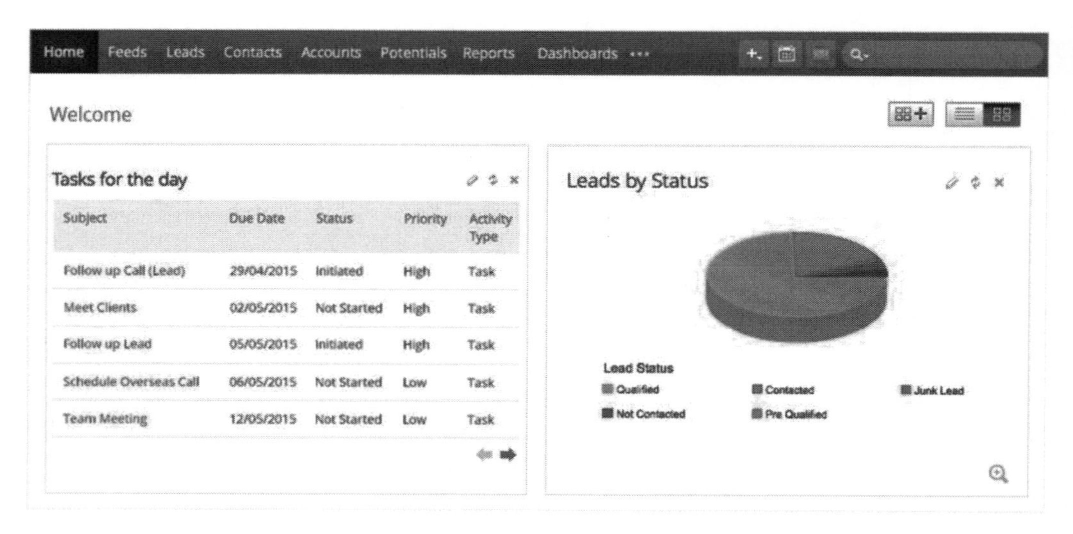

FIGURE 14.8
Web-based DSS information representation of sales leads.

the magnitude of the change. Such a representation allows the decision maker to quickly decide on the stores requiring special attention. Another example, shown in Figure 14.8, shows a web-based DSS capable of managing the sales leads and incorporating the analysis and reports. Figure 14.8 also demonstrates how the tabbed display of options and the augmentation of tabular data with a graphical representation facilitate the comprehension.

Besides providing the appropriate output for DSS results, the system should allow the user to control how the results are displayed. The facility to tweak and modify the results quickly ensures the creative ownership of the analysis, which leads to a better acceptance (Gorry and Scott-Morton, 1971; Sauter, 2011; Sprague and Carlson, 1982). Hence, an effective and flexible information presentation to the decision maker is the main objective of representation.

3. *Graphs:* The way information is presented is known to affect individuals' perceptions (Tversky and Kahneman, 1981). Graphs offer a quick way to visualize and comprehend the data and/or results. Hence, this aspect of the DSS aims at enabling the use of the DSS in a manner that increases the process effectiveness and minimizes the bias in modeling and output. For example, consider the case of mobile-phone sales to consumers, where the sales statistics are reported by a region or state and there is considerable variability within the regions. If the system reports only average sales, then, it might appear that the regions have different average sales. However, the conclusion drawn by a user might change dramatically if the additional information of the standard deviation in sales is made available. Hence, one can reduce the misperceptions and bias in decisions by simply providing the appropriate supplementary information.

   An additional common problem in displaying the results of analyses using graphs arises from scaling. By choosing different scales for the x- and y-axes, one can change the perception on the graph, as illustrated in Figure 14.9 that depicts the growth rate in house prices. It might appear from Figure 14.9 that the average house prices increased dramatically (even seemingly doubled) from 1990 to 1991, whereas the actual change is a more moderate 4%

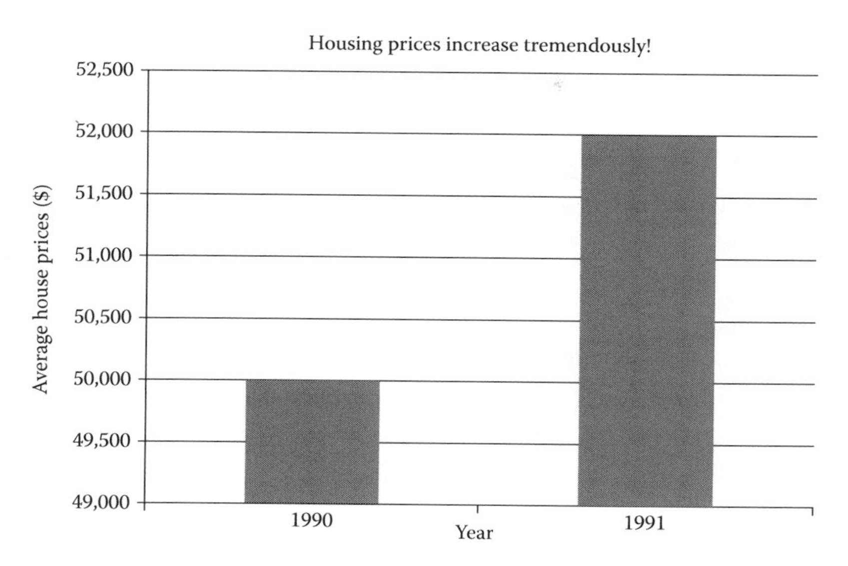

FIGURE 14.9
Graph-scaling bias.

increase. If the y-axis of Figure 14.9 was chosen from zero, then, both the vertical bars would show a small increase, consistent with the reality.

4. *Support of decision-making phases:* A DSS's presentation formats and displays must be designed to aid DSS users through all phases of the decision process, namely, the intelligence phase, design phase, and choice phase.

   In the intelligence phase, a DSS can help the decision maker to identify the issues or opportunities by making the relevant information to be more salient. For example, if a DSS user is a production manager, then, an automated report about the current productivity of the company's plants or production facilities may provide an opportunity to identify and solve production-related issues in a timely manner. When production facilities are geographically dispersed, integrating the data inputs requires a seamless connectivity, which a web-based DSS can provide. Siemens TeamCenter on the cloud is one such web-based system that facilitates a complex product lifecycle management (PLM) across different facilities.

   The design phase involves an analysis of multiple possible courses of action. In this phase, the decision makers focus their attention on developing, testing, and validating various models that can solve the problem. Sensitivity analysis of these models and assumptions is an important step during this phase. Since this is a highly involved decision phase, the system should make sharing and brainstorming ideas as easy as possible, especially in the era of a global business where geographically distributed global teams are the norm. Web-based DSS allows for a seamless user interaction in distributed decision-making environments, while being physically present in multiple locations (Istudor and Duta, 2010; Lu et al., 2005; Paul et al., 2004). In particular, web-based DSS allows for an easy incorporation of e-mails, notifications, blogs, shared notes, and similar tools for an extensive information sharing and cooperation among decision makers (Hinds and Mortensen, 2005; Mortensen and Hinds, 2001). In stand-alone DSS, such tools are not readily available and may be cumbersome to incorporate.

The choice phase is the stage in the decision-making process during which the decision maker makes his/her selections from the available courses of action. The most beneficial DSS feature needed to facilitate decisions in this phase is the comparison of various options based on the specified scenarios. Ideally, this information can be made available to the user in the most natural and understandable way throughout the decision process.

More generally, the wider access allowed by web-based DSS makes it more important that the system facilitates its use by the "average user." For example, allowing for context-sensitive help regarding the various options of the system is a beneficial feature in all the three phases of the decision process. Providing information and methodological knowledge, that is, both domain knowledge and decision methodology knowledge also plays an important role in "hand-holding" and improving the decision quality.

14.4.3 Knowledge Base

The DSS knowledge base represents the users' knowledge about the decision and the DSS, or what knowledge the user should have to use the DSS effectively. The DSS knowledge base also includes the information about how to trigger the model for various cases such as classification, clustering, anomaly detection, a structured prediction, etc., along with the details on data storage, and possibly results in interpretation. In web-based DSS, context-sensitive help and embedded tutorial videos capturing screen-by-screen suggestions are heavily used for providing user support. In legacy DSS, this was largely accomplished through personal training. HTML is ideally suited to create and support a knowledge base that is hyperlinked and capable of incorporating the information content in various formats.

In addition to the key aspects of the DGMS discussed so far, facilitating user communication with the DSS is essential for DSS-supported decision processes and outcomes. In any DSS, there exists a possibility of misinterpretation or miscommunication, because computers are not yet capable of discerning nuances, hints, or nonverbal communication cues. It is, therefore, up to the UI designer to develop interfaces that eliminate the ambiguity and error. Modern advancements in graphical user interface (GUI) technology have made UIs to be more interactive, intuitive, and user-friendly. According to various experts (e.g., Courtney, 2001; Druzdzel and Flynn, 2002; Sauter, 2011), DSS should be designed in a manner that supports and facilitates four aspects of user communication: metaphors, mental models, model navigation, and look and feel.

Metaphors refer to the notion that images and concepts familiar to the user may be employed by the system to streamline the communication process. Typical examples include the use of the question-mark symbol to designate the help function or a magnifying glass—for the search.

Mental models refer to the notion that the data and functions should capture the user's understanding of how the modeled variables and phenomena relate to each other in the real world. Reflecting upon the mental models in the DSS design will serve to reduce the mistakes due to a misunderstanding, promote a more-extensive use of the system, and better acceptance of the results. With the new graphical capabilities of HTML 5, web-based DSS facilitate the mental model communication. For example, web-based project management DSS allows the user to represent the activities in the form of a network graph with nodes and arcs, mimicking the mental representation of the precedence of the project activities and activity timings. When using mental models in DSS, specific attention should be given to ensure the accuracy and authenticity of such models.

Model navigation refers to the movement or interconnections between the data and the associated functions to provide the user with an easier access and understanding of the system, thus enabling a better or a more-efficient analysis. Moreover, depending on the data, algorithms, and the approaches used, the end result may vary. As an example, DSS may be used in situations where the user is trying to identify patterns in the data. When the user chooses the relevant data, the system should assist in the appropriate choice of the models, so that the analysis goals will be met. For example, if the marketing-department head is trying to forecast the weekly demand, then, line-balancing models associated with the production planning should not be in the possible-options list. Also, in situations where the decision maker triggers a whole suite of models, the results should be grouped into the appropriate tabs (e.g., regression models, forecasting models, time series, etc.), or a grouping based on data types (e.g., survey data models, sales data models, etc.). Such a grouping provides for a more-efficient way to compare the alternatives.

Look and feel refers to the appearance of the system and the manner in which its visible (UI) elements are standardized across tasks and contexts. Appearance assumes a particular importance in a global environment where cross-cultural interactions are a commonplace and DSS users may be from multiple cultures. Therefore, care should be taken when creating visual cues that could be perceived differently in different cultures. For example, the use of a green button to indicate the "proceed" is common across an array of cultures (Krug, 2000), but a flashing yellow button may signify different things in different cultures.

14.5 Application: DSS in Business Education

As discussed previously, DSS involve the use of computers in conjunction with models of specific application domains to (1) aid decision makers in decision processes involving complex tasks; (2) support, rather than supersede decision-maker judgment; and (3) improve both the effectiveness and efficiency of decision making. This indicates that DSS tools are particularly appropriate for use in simulation studies of various phenomena of interest and also educational simulations that emphasize the skill development through data-driven decision making.

Simulation studies are used in situations where the systems being modeled are complex (oftentimes, a system of systems) and the analysis is not possible through mathematical abstractions and the associated algorithms (Law and Kelton, 2000). Usually, complex simulations require dedicated simulation languages such as SIMAN, SimScript, MATLAB, or Simula. However, in web-based scripting languages, such as Python, SimPy libraries are available that allow a discrete next-event simulation programming. Consider an example where an automotive manufacturer wants to evaluate the impact of random-machine breakdowns on its overall productivity. Creating a simulation model of the production facility and subjecting it to a numerical analysis for a specified period of time repeatedly allows the decision maker to estimate with a high degree of confidence in the key performance measures, such as the queuing length, machine utilization, work in process, and the inventory. However, such studies tend to be quite time consuming because of the time required to build and validate such simulation models.

Although educational simulations are used in various domains—business, medicine, sciences, and engineering—the following application illustration concerns one popular class of educational simulations—software-based management simulation games (with which we are most familiar).

Many business schools use management simulations to inject experiential learning into their curriculum. These management simulations mimic the various aspects of the business context.

Simulation participants are typically charged with managing a virtual firm or division. To help encourage a more rigorous data-driven, rather than an intuition-driven, decision making, many simulations directly integrate DSS tools into game play. Some of the tools may enable a more usable data display and presentation. These may include tabular data-extraction utilities or graphing and visualization capabilities to allow the simulation participants a better use of voluminous data by isolating and focusing on the key relationships at a time. For example, a basic sorting and presentation tool may help the participants to identify unprofitable product configurations and regions, or project a customer segment evolution in future periods. Other built-in DSS tools may draw on models of the simulated industry to enable standard quantitative analyses, such as future demand, sales forecasts, and risk assessment.

DSS utilities used in simulations can, potentially, be of any complexity level. For example, some marketing simulations, such as Markstrat, enable a sophisticated conjoint analysis that can be used to assist "managers" with complex tasks involved in the new product design and strategy formulation, including brand positioning and marketing communications. In the same vein, the strategic innovation game (SIG) developed by the authors includes a "what-if" analysis submodule that is, conceptually, like a simulation within a simulation. It allows the participants to assess the profit impact of their current decisions, given the firm's cost structure and sales projections. In this way, DSS tools can represent an important component of experiential learning that features prominently in modern management education curricula.

14.6 Web-Based DSS: Future Developments

In the early days of DSS evolution, the major thrusts of DSS development included data structuring and storage (Gorry and Scott-Morton, 1971), UIs (Mason and Mitroff, 1973; Newell and Simon, 1972), and data visualization (Gopal et al., 2011; Pracht, 1986). Currently, issues such as real-time analysis of big data (Leavitt, 2010; Watson, 2005) and ad hoc question-based insights (Bousequet et al., 2011; Paradice et al., 1987) are receiving considerable attention.

Shim et al.'s (2002) work is widely regarded as an insightful review of the past and future of DSS. Many important achievements in the DSS field including GDSS, ES, executive information systems, data warehousing, online analytical processing, data mining, and web-based DSS are discussed. In addition to Shim et al.'s (2002) predictions, DSS appears to be fulfilling an increasing variety of roles. The explosion of information availability in the Internet domain has mandated the development of more specialized and widely applicable modular tools (e.g., Benyon et al., 2002; Bolloju et al., 2002; Courtney, 2001).

Scholars continue to debate the likely evolutionary paths of future DSS systems. Taylor et al. (2010) suggest that information systems research will evolve in six ways—inter-business system research, strategy, Internet applications, information systems thematic research, qualitative methods research, and decision-support research. Additionally, the building of DSS has migrated from configured systems delivered out of the box to the mash up of services put together to address specific problems (Arnott and Pervan, 2005). This mash-up approach is gaining popularity due to its rapid development and deployment time.

We envision that future DSS will be based, to a considerable extent, on smart-device platforms. The evolution of smartphones and other smart handheld devices has made computing to be much more real time, immediate, and full featured. In particular, compared with the simple microcontrollers that restricted the computational use of mobile phones and tablets in earlier years, the advent of powerful multicore processors, such as Tegra 3 and Qualcomm Snapdragon, has provided sufficient computational power to mobile devices to change the way that such devices are used. Concurrently, the evolution of sophisticated operating systems such as Android, iOS, and Windows 10 have allowed these devices to incorporate thin-client features, such as web browsers, light versions of DBMS (e.g., SQLite), and high-resolution display capabilities (e.g., Apple's Retina Display). These advances greatly facilitate the use of web-based DSS in conjunction with mobile-computing devices. However, historically, the most restrictive feature of mobile devices with respect to DSS use has been the cost and speed of telecommunications based on wireless technologies as compared with the relatively inexpensive and greater-throughput wireline (including optic fiber cables) telecommunications used by thin clients. A more-extensive use of mobile devices with web-based DSS capabilities has also been restricted until recently by the relatively limited energy-storage capabilities of smart mobile devices. As more complex applications (requiring a significant computational power) get integrated into smart handheld devices, the batteries have struggled to provide sufficient power and run time to these devices. However, the battery technology for mobile devices has also largely caught up with the requirements of an intensive mobile computing. Therefore, we predict that future DSS systems for mobile devices will be similar to the current web-based DSS in the communication, reach, and usability, but technologically different due to the architecture and network requirements. (Of note, a similar observation was made by Bryan (2012).)

Also, we anticipate that the advancements in data warehousing and knowledge management will drastically change the way that DSS are designed. For example, web-based DSS systems will move from RDBMS to No-SQL systems. Major corporations, including Google and Amazon, already use such systems (Leavitt, 2010). Such new DSS platforms challenge the traditional DSS approaches in terms of size, speed, and reach, along with the diversity of data that can be used and such systems' continuous availability (Alter, 2007).

Additionally, we predict that the advent and growth of social media as a set of business tools will usher in changes to the web-based DSS growth curve over the next decade. Specifically, for social media, web-based DSS requirements will emphasize group-support systems, where voting systems, negotiations, collective brainstorming, and the like will become more important (see also Griffin and deLeaster 2009; Scott, 2011). Furthermore, the area of virtual worlds and communities is now being explored by businesses, the military, and education for various purposes, including test beds, training tools, research, and recruiting. Developing DSS for such avenues, especially on the web platform, will be an important challenge (Condic, 2009).

We also subscribe to the view previously expressed by Bryan (2012) and Holsapple et al. (1998) that negotiation-support systems (NSS) will emerge as an important derivative of web-based DSS, designed to facilitate negotiations. More generally, many issues, including communication and coordination, group management, information ranking, and managing diverse information, will be areas where specialized tools and applications will have to be developed.

Finally, our discussion to this point has focused on web-based DSS used by organizational or industrial users (i.e., various decision makers in organizations). However, a distinct category of web-based DSS that emerged with the widespread popular adoption of the Internet includes consumer-centric DSS systems. This category comprises various systems vendors that incorporate into their storefronts and portals with the objective of facilitating consumer choice. The primary user of such systems is the consumer. Perhaps, the most-prominent manifestation of consumer-centric web-based DSS is the recommendation agents. These systems combine data on past consumer behavior with stated preferences (where available) to suggest new items for purchase or consumption. We anticipate that the advances in online big data analytics and more intuitive AI-type (predictive) algorithms will enable the creation of web-based recommendation agents that will be able to make better time- and place-specific product recommendations than consumers could themselves, because such web-based DSS will possess more information and be free of memory limitations and cognitive biases that impair consumer judgment.

References

Alter, S.L., Customer-centric systems: A multi-dimensional view, *Proceedings of WeB 2007 (Sixth Workshop on eBusiness)*, Montreal, Canada, pp. 130–141, 2007.

Ariely, D., *Predictably Irrational*, HarperCollins, New York, 2008.

Arnott, D., and Pervan, G., A critical analysis of decision support systems research, *Journal of Information Technology*, 20(2), pp. 67–87, 2005.

Arnott, D., and Pervan, G., A critical analysis of decision support systems research revisited: The rise of design science, *Journal of Information Technology*, 29, pp. 269–293, 2014.

Bazerman, M.H., Giuliano, T., and Appelman, A., Escalation of commitment in individual and group decision making, *Organizational Behavior and Human Performance*, 33, pp. 141–152, 1984.

Benyon, M., Rasmequan, S., and Russ, S., A new paradigm for computer-based decision support, *Decision Support Systems*, 33(2), pp. 127–142, 2002.

Bhargava, H., Power, D.J., and Sun, D., Progress in web-based decision support technologies, *Decision Support Systems*, 43, pp. 1083–1095, 2007.

Bolloju, N., Khalifa, M., and Turban, E., Integrating knowledge management into enterprise environments for the next generation of decision support, *Decision Support Systems*, 33(2), pp. 163–176, 2002.

Boreisha, Y., and Myronovych, O., Web-based decision support systems as knowledge repositories for knowledge management systems, *UbiCC Journal*, 3(2), pp. 22–29, 2008.

Bousequet, F., Fomin, V.V., and Drillion, D., Anticipatory standards development and competitive intelligence, *International Journal of Business Intelligence Research*, 2(1), pp. 16–30, 2011.

Bousequet, O., Boucheron, S., and Lugosi, G., Introduction to statistical learning theory, In: *Advanced Lectures on Machine Learning*, Vol. 3176 of Lecture Notes in Computer Science, pp. 175–213, Springer, Berlin/Heidelberg, 2004.

Brocas, I., Information processing and decision-making: Evidence from the brain science and implications for economics, *Journal of Economic Behavior and Organization*, 83, pp. 292–310, 2012.

Brocas, I., and Carillo, J.D., Dual-process theories of decision-making: A selective survey, *Journal of Economic Psychology*, 41, pp. 45–54, 2014.

Bryan, H., Dianne, H., David, P., and Courtney, J.F., A look toward the future: Decision support systems research is alive and well, *Journal of the Association for Information Systems*, 13(5), pp. 315–340, 2012.

Chodorow, K., *MongoDB: The Definitive Guide* (2nd Edition), O'Reilly Media, San Diego, California, 2013.

Clemen, R.T., *Making Hard Decisions: An Introduction to Decision Analysis*, PWS-Kent Publishing Company, Boston, Massachusetts, 1990.

Cohen, M.D., March, J.G., and Olsen, J.P., A garbage can model of organizational choice, *Administrative Science Quarterly*, 17(1), pp. 1–25, 1972.

Condic, K., Using second life as a training tool in an academic library, *The Reference Librarian*, 50(4), pp. 333–345, 2009.

Courtney, J.F., Decision making and knowledge management in inquiring organizations: Toward a new decision-making paradigm for DSS, *Decision Support Systems*, 31(1), pp. 17–38, 2001.

Csaszar, F.A., and Enrione, A., When consensus hurts the company, *Sloan Management Review*, 56(3), pp. 17–20, 2015.

Dawes, R.M., *Rational Choice in an Uncertain World*, Hartcourt Brace Jovanovich Publishers, San Diego, California, 1988.

Deck, C., and Jahedi, S., The effect of cognitive load on economic decision making: A survey and new experiments, *European Economic Review*, 78, pp. 97–119, 2015.

Dennis, A.R., Information exchange and use in group decision making: You can lead a group to information, but you can't make it think, *MIS Quarterly*, 20, pp. 433–457, 1996.

Druzdzel, M.J., Probabilistic reasoning in decision support systems: From computation to common sense, PhD thesis, Department of Engineering and Public Policy, Carnegie Mellon University, 1992.

Druzdzel, M.J., and Flynn, R.R., *Encyclopedia of Library and Information Science* (2nd Edition), Marcel Dekker Inc., New York, 2002.

Elmasri, R., and Navathe, S.B., *Fundamentals of Database Systems* (4th Edition), Pearson Education Inc., Boston, Massachusetts, 2003.

Franklin, C.L.-II, Developing expertise in management decision-making, *Academy of Strategic Management Journal*, 12(1), pp. 21–36, 2013.

Gopal, R., Marsden, J.R., and Vanthienen, J., Information mining—Reflections on recent advancements and the road ahead in data, text, and media mining, *Decision Support Systems*, 51(4), pp. 727–731, 2011.

Gorry, G.A., and Scott-Morton, M.S., A framework for management information systems, *Sloan Management Review*, 13(1), pp. 50–70, 1971.

Griffin, L., and deLeaster, E., Social networking healthcare, *IEEE 6th International Workshop on Wearable Micro and Nano Technologies for Personalized Health (pHealth)*, Oslo, pp. 75–78, 2009.

Heydenfeldt, J., Decision science and applied neuroscience: Emerging possibilities, *Performance Improvement*, 52(6), pp. 18–25, 2013.

Hightower, R., and Sayeed, L., Effects of communication mode and pre-discussion information distribution characteristics on information exchange in groups, *Information Systems Research*, 7, pp. 451–65, 1996.

Hinds, P.J., and Mortensen, M., Understanding conflict in geographically distributed teams: The moderating effects of shared identity, shared context, and spontaneous communication, *Organization Science*, 16(3), pp. 290–307, 2005.

Hoffer, J.A., Prescott, M., and Topi, H., *Modern Database Management Systems* (9th Edition), Pearson Education Inc., New Jersey, 2009.

Holsapple, C., Lai, H., and Whinston, A.B., A formal basis for negotiation support system research, *Group Decision and Negotiation*, 7, pp. 203–227, 1998.

Immelt, J.R., Govindarajan, V., and Trimble, C., How GE is disrupting itself, *Harvard Business Review*, 87(10), pp. 56–65, 2009.

Istudor, I., and Duta, L., Web-based group decision support system: An economic application, *Informatica Economică*, 14(1), pp. 191–200, 2010.

Kahneman, D., Slovic, P., and Tversky, A. (editors), *Judgment under Uncertainty: Heuristics and Biases*, Cambridge University Press, Cambridge, 1982.

Kahneman, D., and Tversky, A., *Choice, Values, Frames*, Cambridge University Press, New York, 2000.

Karunakaran, S., Venkataraghavan, K., and Sundarraj, R.P., Business view of cloud decisions, models and opportunities—A classification and review of research, *Management Research Review*, 38(6), pp. 582–604, 2015.

Klein, G., Naturalistic decision making, *Human Factors*, 50(3), pp. 456–460, 2008.

Krug, S., *Don't Make Me Think! A Commonsense Approach to Web Usability*, New Riders Publishing, Indianapolis, USA, 2000.

Law, A.M., and Kelton, W.D., *Simulation Modeling and Analysis* (3rd Edition), McGraw-Hill Higher Education, Boston, 2000.

Leavitt, N., Will NoSQL databases live up to their promise? *Computer*, 43(2), pp. 2–14, 2010.

Lipshitz, R., Klein, G., Orasanu, J., and Salas, E., Focus article: Taking stock of naturalistic decision making, *Journal of Behavioral Decision Making*, 14, pp. 331–352, 2001.

Lu, J., Zhang, G., and Wu, F., Web-based multi-criteria group decision support system with linguistic term processing function, *IEEE Intelligent Informatics Bulletin*, 5(1), pp. 35–43, 2005.

Mason, R., and Mitroff, I., A program on research for management information systems, *Management Science*, 19(5), pp. 475–487, 1973.

McCreary, D., and Kelly, A., *Making Sense of NoSQL: A Guide for Managers and the Rest of Us* (1st Edition), Manning Publications, Shelter Island, NY, 2013.

Mohri, M., Rostamizadeh, A., and Talwalkar, A., *Foundations of Machine Learning*, MIT Press, Cambridge, MA, 2012.

Mora, M., Forgionne, G.A., and Gupta, J.N.D., *Decision Making and Support Systems: Achievements, Trends, and Challenges for the New Decade*, Idea Group Publishing Inc., Hershey, Pennsylvania, 2003.

Mortensen, M., and Hinds, P.J., Conflict and shared identity in geo-graphically distributed teams, *International Journal of Conflict Management*, 12(3), pp. 212–238, 2001.

Newell, A., and Simon, H., *Human Problem Solving*, Prentice-Hall, Englewood Cliffs, NJ, 1972.

Paradice, D.B., Courtney, J., and James, F., Causal and non-causal relationships and dynamic model construction in a managerial advisory system, *Journal of Management Information Systems*, 3(4), pp. 40–53, 1987.

Paul, S., Seetharaman, P., Samarah, I., and Mykytyn, J.P., An empirical investigation of collaborative conflict management style in group support system-based global virtual teams, *Journal of Management Information Systems*, 21, pp. 185–222, 2004.

Paul, S., Seetharaman, P., Samarah, I., and Mykytyn, J.P., Understanding conflict in virtual teams: An experimental investigation using content analysis, *38th Hawaii International Conference of System Sciences*, Hawaii, 2005.

Post, G.V., *Database Management Systems: Designing and Building Business Applications*, McGraw-Hill Publishing Company, London, 1998.

Pracht, W.E., GISMO: A visual problem-structuring and knowledge-organization tool, *IEEE Transactions on Systems, Man, and Cybernetics*, 16(2), pp. 265–270, 1986.

Russell, S., and Norvig, P., *Artificial Intelligence: A Modern Approach* (2nd Edition), Pearson Education Inc., New Jersey, 2003.

Sauter, V.L., *Decision Support Systems for Business Intelligence* (2nd Edition), John Wiley & Sons, Inc., New Jersey, USA, 2011.

Scott, D.W., Using social media in engineering support and space flight operations control, *IEEE Automatic Control*, pp. 1–14, 2011, doi: 10.1109/AERO.2011.5747662.

Seibert, S.E., and Goltz, S.M., Comparison of allocations by individuals and interacting groups in an escalation of commitment situation, *Journal of Applied Social Psychology*, 31(1), pp. 134–156, 2001.

Shim, J.P., Courtney, J.F., Power, D.J., Warkentin, M.E., Sharda, R., and Carlsson, C., Past, present, and future of decision support technology, *Decision Support Systems*, 33(2), pp. 111–126, 2002.

Simon, H.A., A behavioral model of rational choice, *American Economic Review*, 69, pp. 99–118, 1955.

Simon, H.A., *The New Science of Management Decision*, Harper and Row, New York, 1960.

Sprague, R.H.-Jr., A framework for the development of decision support systems, *MIS Quarterly*, 4(4), pp. 1–26, 1980.

Sprague, R.H.-Jr., and Carlson, E.D., *Building Effective Decision Support Systems*, Prentice- Hall, Englewood Cliffs, New Jersey, 1982.

Sullivan, D., *NoSQL for Mere Mortals* (1st Edition), Addison-Wesley Professional, Hoboken, NJ, 2015.

Taylor, H., Dillon, S., and Van Wingen, M., Focus and diversity in information systems research: Meeting the dual demands of a healthy applied discipline, *MIS Quarterly*, 34(4), pp. 647–667, 2010.

Turban, E., and Aronson, J.E., *Decision Support Systems and Intelligent Systems*, Pearson Education, India, 1998.

Turban, E., Aronson, J.E., and Liang, T.P., *Decision Support Systems and Intelligent Systems* (7th Edition), Pearson, Prentice-Hall, NJ, 2005.

Turban, E., and Watkins, P.R., Integrating expert systems and decision support systems, *MIS Quarterly*, 10, pp. 121–136, 1986.

Tversky, A., and Kahneman, D., Judgment under uncertainty: Heuristics and biases, *Science*, 185.4157, pp. 1124–1131, 1974.

Tversky, A., and Kahneman, D., The framing of decisions and the psychology of choice, *Science*, 211.4481, pp. 453–458, 1981.

Watson, H., Sorting out what's new in decision support, *Business Intelligence Journal*, 10(1), pp. 4–8, 2005.

Waytz, A., and Mason, M., You're brain at work, *Harvard Business Review*, pp. 86–93, 2013.

Whyte, W.F., *Participatory Action Research*, Sage, Thousand Oaks, CA, 1991.

Yiyang, Z., and Jiao, J.R., A web-based product portfolio decision support system, *International Journal of Manufacturing Technology and Management*, 11(3/4), pp. 290–304, 2007.

Zhang, G., Lu, J., and Gao, Y., *Multi-Level Decision Making: Models, Methods and Applications*, Springer-Verlag, Berlin, Heidelberg, 2015.

15

Algorithms and Their Design

Surya Prakash, Phalguni Gupta, and Raghunath Tewari

CONTENTS

ABSTRACT An algorithm is a sequence of unambiguous instructions or steps to solve a particular problem. It is meant for obtaining a required output for any legitimate input in a finite amount of time. Algorithms are frequently expressed in the form of a pseudocode that can be converted into a program using a programming language. Hence, the concept of an algorithm is considered distinct from that of a program. This chapter discusses the basics about an algorithm. It defines the term "algorithm" formally and lists out various characteristics of an algorithm. It discusses the complexity analysis of an algorithm. It also presents various popular algorithms design paradigms such as divide and conquer, greedy algorithm, dynamic programming, and back tracking. Each of the algorithm design paradigm is discussed with several examples to illustrate the concept. This chapter briefly throws some light on the types of computational problems by presenting a discussion on the P-problem and NP-problem. These problems are very important in theoretical computer science and constitute the bulk of our practical computational problems. They have been central to the theory of computation for many years. A deterministic algorithm always gives the correct solution and the number of computational steps executed by the algorithm is the same for different executions of the algorithm with the same input data. However, it is practically realized many a times that for a given computational problem, it may be difficult to design a deterministic algorithm with a good execution time. Also, sometimes, it is observed that a deterministic algorithm performs reasonably well for a small size of input data; however, the execution time may go very high as the number of inputs increases. Approximation algorithms and randomized algorithms are the solutions to these problems. Approximation algorithms are the types of algorithms that are used to find approximate solutions to optimization problems whereas a randomized algorithm makes use of randomness in its logical steps to achieve a better performance in the average case. At the end of the chapter, a brief introduction of approximation and randomized algorithms is presented.

15.1 Introduction

15.1.1 What Is an Algorithm?

An algorithm is a sequence of unambiguous instructions or steps to solve a particular problem. This means that an algorithm is meant for obtaining a required output for any legitimate input in a finite amount of time. In the context of an algorithm, unambiguous instructions refer to instructions having only one possible interpretation. Algorithms are commonly expressed in the form of a pseudocode that can be converted into a program using a programming language such as C, Pascal, FORTRAN, Java, etc. Hence, the concept of an algorithm is thought to be distinct from that of a program.

15.1.2 Characteristics of an Algorithm

An algorithm should possess certain characteristics. Some of the important characteristics are listed below:

- *Uniqueness:* The result of each and every step of the algorithm is defined uniquely and it only depends on the input data and the outcome of the preceding instructions (steps).
- *Finiteness:* The algorithm must terminate after the execution of a finite number of instructions (steps). Also, each instruction must be executable in a finite amount of time.
- *Precision:* The steps of the algorithm are precisely defined. That is, there is no vagueness in the instructions (steps).
- *Input and output:* The algorithm has the provision to take zero or more but only a finite number of inputs and it produces at least one output. The requirement of at least one output is essential.
- *Generality:* An algorithm is not specific to a certain input and can be used on a set of inputs.

15.1.3 Complexity of an Algorithm

The complexity of an algorithm is to determine how fast or slow a particular algorithm runs. It is normally defined with the help of a function $T(n)$—time versus the input size n. $T(n)$ depends on the implementation and a given algorithm takes different amounts of time on different machines on the same inputs depending on various machine-dependent factors such as the processor speed, disk speed, instruction set, compiler, etc. However, it is required to define the time taken by an algorithm without depending on the implementation details specific to the machine. The solution to this problem is to represent the complexity (efficiency) of an algorithm asymptotically by measuring time $T(n)$ as the number of elementary "steps," theoretically assuming that each such step takes a constant amount of time.

Let us take an example of computing the addition of two vectors, each of size n. Assume that the algorithm adds two vectors element by element and the addition of two elements defines an elementary "step" in the algorithm. Clearly, the addition of two vectors would execute n steps. Further, assume that c is the time taken in adding two elements of vectors; then, the total computational time taken by the algorithm would be $T(n) = c^*n$. On two different machines, the addition of two vector elements may take different times, say c_1 and c_2. Hence, the addition of two vectors, each of n elements will take $T(n) = c_1{}^*n$ and $T(n) = c_2{}^*n$ time, respectively, in these two machines. This clearly shows that the time taken by the algorithm to add two vectors on different machines is different. If we plot $T(n)$ with respect to n, then, we get two lines with different slopes for two different machines. However, it can be noticed that there is a common factor in both these implementations and it is that time $T(n)$ grows linearly as the input size increases in both the cases. This technique of abstracting away the implementation details of an algorithm and representing the rate of usage of resources by the algorithm in terms of the input size n is an important idea in computer science with respect to the complexity analysis of algorithms.

There are two main characteristics that can be used to get an indication about the complexity of an algorithm. These characteristics are the running time of the algorithm and the space requirement of the algorithm. These characteristics are stated below:

- *Running time*: It is referred to as the time complexity of the algorithm. An algorithm is considered as efficient if it takes less running time as compared to the other algorithms for the same task. The running time of an algorithm is stated as a function that relates the input length to the number of steps.
- *Space requirement*: It is referred to as space complexity of the algorithm. An algorithm is considered as efficient if it uses a less amount of space for its execution.

15.1.4 Asymptotic Notations

The fundamental objective of the analysis of computational complexity is to characterize algorithms according to their performances. There are some standard ways to represent the run-time complexity of an algorithm.

15.1.4.1 Big-O Notation

It is also called as a "Big OH" or an "Order." Big-O notation is used to give an upper bound on a function, to within a constant factor and is written as $O(f(n))$. Figure 15.1 shows the intuition behind the O-notation. The definition of Big-O is formally given below.

Definition 15.1

$f(n) = O(g(n))$ means, $\exists$ a constant $c > 0$ and some number n_0 such that for all $n > n_0$,

$$f(n) \leq cg(n).$$

So, for example, if for an algorithm $T(n) = O(n^2)$, then, it says that the algorithm has a quadratic time complexity and is bounded by n^2 on top. The following are few more functions and their upper bound using Big-O:

- $n = O(n)$
- $n = O(n^2)$
- $n^2 = O(n^2)$
- $\log_2 n = O(\log_2 n)$

Exercise 15.1

Prove that $n^2 + 3n + 1 = O(n^2)$. For this, it is required to compute the values of constants c and n_0 such that $n^2 + 3n + 1 \leq c^*n^2$ holds. Let us consider $n_0 = 1$, then for $n \geq 1$

$$1 + 3n + n^2 \leq cn^2$$

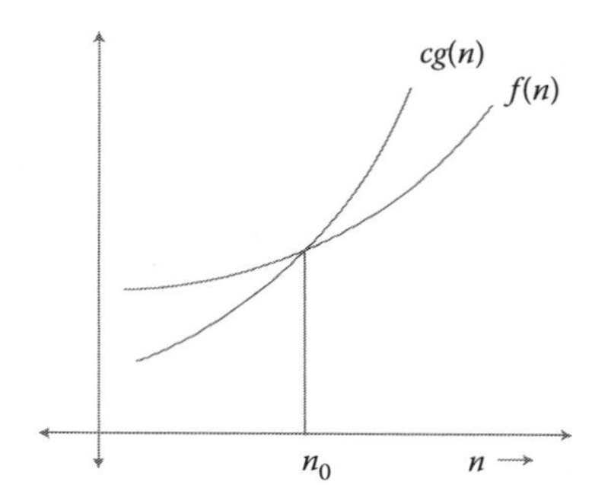

FIGURE 15.1
$f(n) = O(g(n))$ curve.

Therefore, for $c = 5$ and $n \geq n_0 = 1$, it can be seen that the above equation holds. Hence, it is proved that $n^2 + 3n + 1 = O(n^2)$.

On the basis of time complexity, algorithms can be classified into various classes. Few standard classes are constant-time algorithm, linear-time algorithm, quadratic-time algorithm, logarithmic-time algorithm, etc.

- *Constant-time algorithm*: An algorithm is said to run in constant time if it requires the same amount of time irrespective of the input size. For example, accessing any element in an array, push and pop methods in a fixed-size stack, and enqueue and dequeue methods in fixed-size queue require a constant-time algorithm.
- *Linear-time algorithm*: An algorithm is said to run in linear time if the execution time of the algorithm is directly proportional to the input size. In other words, the execution time grows linearly with respect to the input size. For example, an algorithm for linearly searching of an element in the array, and an algorithm for finding out the maximum and minimum in an array using linear traversing take linear time.
- *Quadratic-time algorithm*: An algorithm is said to run in quadratic time if the execution time of the algorithm is proportional to the square of the input size. Few examples of the algorithms that have a quadratic-time complexity are bubble sort, insertion sort, and selection sort.
- *Logarithmic-time algorithm*: An algorithm is said to run in logarithmic time if its execution time is proportional to the logarithm of the input size. Binary search is an example of an algorithm of a logarithmic-time complexity.

15.1.4.2 Big-Ω Notation

It is read as "Big Omega." It provides an asymptotic lower bound on a function, to within a constant factor and is written as $\Omega(f(n))$. Figure 15.2 shows the intuition behind the Ω-notation. The definition of Big-Ω is formally given below.

Definition 15.2

$f(n) = \Omega(g(n))$ means, $\exists$ constant $c > 0$ and some number n_0 such that for all $n > n_0$, $f(n) \geq cg(n)$.

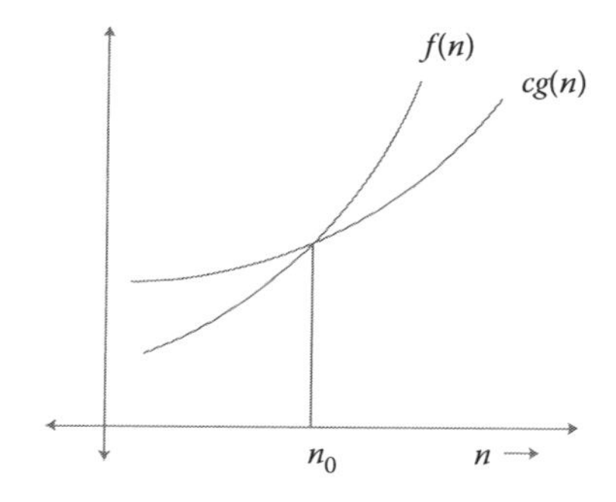

FIGURE 15.2
$f(n) = \Omega(g(n))$.

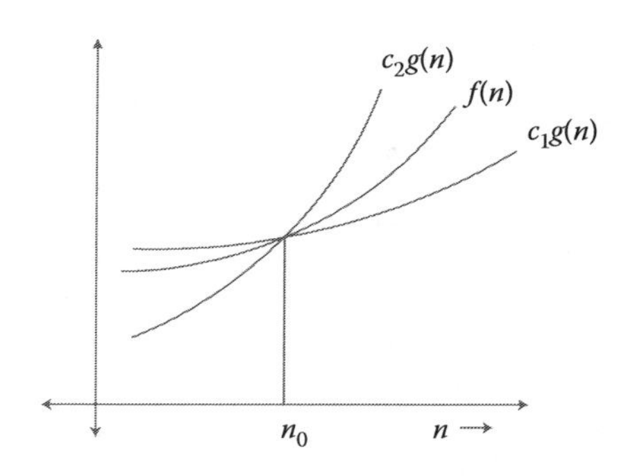

FIGURE 15.3
$f(n) = \Theta(g(n))$.

Example 15.1

1. $n^2 = \Omega(n)$
2. $n^2 = \Omega(n^2)$
3. $n^3 = \Omega(n^3)$
4. $n^c = \Omega(\log n)$ for any constant c

15.1.4.3 Big-Θ Notation

It is read as "Big Theta." It provides an asymptotic bound on a given function $f(n)$, to within constant factors above and below and is written as $\Theta(f(n))$. Figure 15.3 shows the intuition behind the Θ-notation. The definition of Big-Θ is formally given below.

Definition 15.3

$f(n) = \Theta(g(n))$ means that $\exists c_1, c_2 > 0$ and $n_0 > 0$ such that $\forall$ all $n > n_0$, and $c_1 g(n) \leq f(n) \leq c_2 g(n)$.

Example 15.2

1. $\log_2 n = \Theta(\log_2 n)$
2. $n^2 = \Theta(n^2)$
3. $n^3 = \Theta(n^3)$

15.1.5 Algorithm-Design Paradigms

15.1.5.1 Divide and Conquer

This is a technique of designing algorithms where a given instance of the problem to be solved is divided into several, smaller, subinstances that are solved independently and then, the solutions of the subinstances are combined to yield a solution for the original instance of the problem. Divide and conquer is a top-down technique for the designing of algorithms. It divides the problem into smaller subproblems assuming that the computation of the solutions of the subproblems and then composing the partial solutions into the solution of the original problem is easy to perform. One disadvantage of using the divide and conquer is that the process of recursively solving separate subinstances can result in the same computations being performed repeatedly since identical subinstances may arise.

15.1.5.2 Greedy Algorithm

The greedy strategy is often used to design algorithms for solving optimization problems, that is, the problems that require a minimization or maximization of some quantity. Greedy algorithms do not always yield an optimal solution; however, solutions obtained by the greedy strategy are often used as the basis of a heuristic approach. Many a times, it is seen that even for the problems that can be optimally solved by a greedy algorithm, establishing the correctness of the method may not be a trivial process.

15.1.5.3 Dynamic Programming

Like the greedy strategy, dynamic programming is also often applied in the construction of algorithms to solve optimization problems. In contrary to the greedy strategy that does not guarantee an optimal solution to a problem, dynamic programming always produces an optimal solution. In the divide and conquer strategy, a solution to a problem is obtained by recursively solving separate subinstances that may result into the same computations being performed repeatedly since identical subinstances may arise. The idea in dynamic programming is to avoid this by obviating the requirement to calculate the same quantity twice. The method usually accomplishes this by maintaining a table of subinstance results.

15.1.5.4 Backtracking

Backtracking is an algorithm-design strategy for finding out solutions to a problem. It systematically searches for a solution to a problem among all the available options. It incrementally computes the candidates to the solutions, and discards each partial candidate (i.e., backtracks) as soon as it finds that it cannot possibly lead to a valid solution.

15.2 Divide and Conquer

15.2.1 General Strategy

It is a strategy of designing algorithms where a given problem is solved by splitting it into several smaller subproblems that are solved independently. The solutions of the subproblems are then combined to obtain the solution of the original problem. The idea behind splitting of a problem into smaller subproblems is that it is easier to compute the solution of the smaller subproblems in comparison to directly computing the solution of the original problem. The divide-and-conquer strategy follows the top-down approach in the designing of an algorithm. Formally, the divide-and-conquer strategy consists of the following major three phases:

- *Division of a problem to subproblems*: In this phase, the problem is divided into several subproblems that are similar to the original problem however smaller the sizes are.
- *Computation of the solution of subproblems*: This phase involves the computation of the solution of the subproblems. The solution of each subproblem is computed in an independent manner.
- *Computation of the solution of the original problem*: After computing the solutions of all the subproblems, these solutions are finally combined to obtain the solution of the original problem.

15.2.1.1 Examples on the Divide-and-Conquer Strategy

In the following discussion, the solutions of a few problems are computed using the divide-and-conquer strategy.

```
Algorithm sequential-search(A[], key, low, high)
    Input: Array A, element key to be searched; and low and high are
           the minimum and maximum array indices between which search is
           to be performed
    Output: Index i such that A[i]=key or
                  0 if element not present in the array

begin
  for i = low to high do
     if A [i] == key then
        return i;
  end-for
  return 0;    /* 0 shows failure or non-existence
                  of searched element */
end-sequential-search
```

15.2.2 Binary Search

Before going into the details of a binary search, let us first briefly discuss the sequential search. Formally, the problem of searching can be stated as follows. Given an array A with n elements and the element x, find the index i such that $1 \leq i \leq n$ and $x = A[i]$ and return i. If no such i exists, then, declare the search as a *failure*. A sequential search compares the element x that is being searched with the elements of the array A in sequential manner until either end of an array A is reached or the element x that is being searched is found. An algorithm for a sequential search is presented above.

15.2.2.1 Analysis of the Algorithm

This algorithm clearly takes a $\Theta(m)$ time, where m is the index where element x is found. It is $\Omega(n)$ in the worst case (where n is the total number of elements present in the array) when the searched element is present at the end of the array or it is not in the array; and $O(1)$ in the best case when the first element in the array is the searched element.

It can be easily noted that if the elements of an array A are distinct and the query element x is present in the array, then, on an average, the for loop is executed $(n+1)/2$ times. Hence, the average-case complexity of a sequential search is $\Theta(n)$.

Now with this background, let us now discuss the binary search. Binary search is a good example of the divide-and-conquer strategy. It is a technique that is used to search an element in a given array. The basic idea of the binary search is as follows. Given an ordered array of n elements and an element x to search in the ordered array, the binary search checks the middle element of the array. If the middle element is equal to x, then, the search is terminated with success. If the middle element is greater than x, then, the binary search is invoked in the lower segment of the array else the binary search is invoked in the upper segment of the array. This process is continued until we reach the required element that is being searched or a complete array is exhausted.

Let us assume an array $A[1, \ldots, n]$ of a nondecreasing sorted order, that is, $A[i] \leq A[j]$ whenever $1 \leq i \leq j \leq n$ and let x be the element to be searched. Now, the problem is to search x in the array

A. If x is not in A, then, declare the search as a *failure* and return 0 else declare it as a *success* and return the index of the element. The binary search compares x to an element in the middle of the array, say at $k = \lceil n/2 \rceil$. If $x = A[k]$, then, the search is terminated with success and k is returned. If $x \leq A[k]$, then the element x is searched in $A[1, \ldots, k-1]$; otherwise the element x is searched in $T[k+1, \ldots, n]$. Algorithm for binary search is given below.

Let $T(n)$ be the time complexity of the above algorithm. In every call, the search space is reduced to a half; hence, the recurrence relation for computing the time complexity can be given as follows:

$$T(n) = T\left(\frac{n}{2}\right) + O(1)$$

The solution of the above recurrence relation is $T(n) = \Theta(\log n)$.

```
Algorithm binary-search(A[], key, low, high)
     Input: Array A[1..n] in non-decreasing sorted order, element key to be
            searched, low and high are the minimum and maximum indices
            between which search is to be performed in A
     Output: Index mid such that A[mid] = key or
             0 if element not present in the array
begin
   if (low > high)
         return 0; /* 0 shows failure or non-existence
                      of searched element */
   mid = ⌈(low + high)/2⌉;
   if (key == A[mid]) then
       return mid;
   if key ≤ A[mid] then
      return binary-search(A, key, low, mid-1);
   else
      return binary-search(A, key, mid+1, high);
end-binary-search
```

The above implementation uses recursion for implementation of the binary search. Like any other recursive routine, a recursive binary search can have an equivalent iterative implementation as given below.

```
Algorithm binary-search-iterative(A[], key, low, high)
     Input: Array A sorted in non-decreasing order, element key to be
            searched, low and high are the minimum and maximum indices
            between which search is to be performed in A
     Output: Index k such that A[mid] = key or
             0 if element not present in the array
begin
   if (key > A[high]) or (x < A[low]) then return 0;
   end-if
   i = low; j = high;
   while (i ≤ j) do
       mid = ⌈(i + j)/2⌉;
       if (x == A[mid] then return mid;
       end-if
```

```
        if (key ≤ A[mid]) then j = mid-1;
        else
           i = mid + 1;
        end-if
  end-while
  return 0; /*0 is returned when element key is not present in A*/
end-binary-search-iterative
```

The number of steps executed in a recursive version of the binary search and in the iterative implementation are the same. Hence, the time complexity of this iterative version of the binary search will also be $T(n) = \Theta(\log n)$.

15.2.3 Quick Sort

The quick sort was proposed by Tony Hoare (Hoare 1961) and is found to be a very fast sorting algorithm in practice. It follows the divide-and-conquer strategy that is very similar to the one used in binary search. The quick sort has two phases: the partition phase (divide phase) and the sort phase (conquer phase). In the partition phase, the given array is divided into two subarrays while the sort phase sorts out the two smaller subarrays that are generated in the partition phase. Once the two subarrays are sorted out, the resultant sorted subarrays are put back together again. The quick sort does more work in dividing the array into two subarrays, and a comparatively less work in the sort phase. The sort phase just sorts out the two smaller subarrays that are generated in the partition phase.

In short, the quick sort divides the array to be sorted into two partitions and then calls the quick sort procedure recursively to sort out the two partitions, that is, it divides the bigger problem into two smaller ones and conquers (sorts) them by solving the smaller ones recursively.

The quick sort uses a simple idea to partition the array. It picks one element called as a *pivot* element from the list. With the help of the pivot element, it splits the rest of the elements in the array into two subarrays, one subarray containing all the elements less than the *pivot* and another having elements greater than the *pivot*. Figure 15.4 demonstrates the partitioning and sorting in the quick sort. The partitioning strategy in the quick sort must make sure that the *pivot* is larger than all the elements in the lower part and less than all those elements in the upper part. A choice of the *pivot* element is important. In an ideal scenario, a *pivot* element should be chosen in such a way that it should divide the array into two equal halves after partitioning. In a general case, we do not have

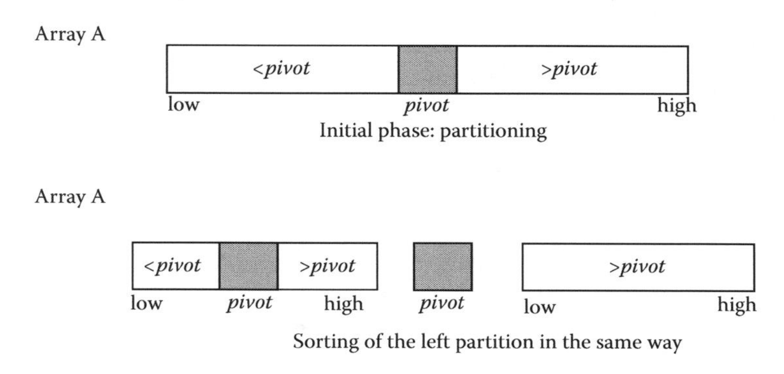

FIGURE 15.4
Partitioning and sorting in the quick sort.

any idea about the data items that are to be sorted; hence, any choice of the *pivot* element should work. Usually, it is convenient to consider the first or last element of the array as the *pivot* element.

The divide (partitioning) part of the quick sort is given below. The inputs to the algorithm are an array *A* containing the elements to be sorted; and the two indices of a *low* and *high*, respectively, to define the start and the end of the array. The algorithm considers the last element of the array as the *pivot* element.

```
Algorithm partition(A[], low, high)
    Input: A is the array to be sorted, low and high defines the minimum
           and maximum indices of the array between which sorting is
           to be performed.
    Output: position of the pivot element
    /*This algorithm considers last element of the array as pivot, puts the
    pivot element at its correct position in sorted array by placing all
    elements smaller than pivot to left of it and all greater elements
    to right of it. */
begin
    pivot = A[high]; /*pivot element*/
    i = low;
    for j = low:high - 1
        if (A[j] ≤ pivot)
              swap(A[i],A[j]); /*swap values of A[i] and A[j]*/
              i = i + 1;
        end-if
        swap(A[i],A[high]);
   end-for
   return i;
end-partition
```

The conquer (sorting) part of the quick sort is given below. Similar to the partitioning phase, the input to this phase is also an array *A* containing the elements to be sorted; and the two indices of a *low* and *high*, respectively, to define the start and the end of the array.

```
Algorithm quicksort(A[], low, high)
    Input: A is the array to be sorted, low and high defines the minimum and
           maximum indices of the array between which sorting is to be performed.
    Output: Sorted array A
begin
    if (high > low)
        pivot = partition(A, low, high); //call to partition algorithm
                                         //defined above
        quicksort(A, low, pivot-1);
        quicksort(A, pivot+1, high);
    end-if
    return A; /*Sorted array*/
end-quicksort
```

Let us now analyze the complexity of the quick sort algorithm. As discussed above, the quick sort picks an element (*pivot*) in each step and then arranges the elements in the array in such a way that all the elements less than the *pivot* element are placed to the left of the *pivot* element, and all the elements larger than the *pivot* element are placed on the right side of the *pivot* element. This creates two partitions of the array, one on the left side of the pivot element and another on the right side of the *pivot* element. In all subsequent iterations of the sorting, the position of the *pivot* element remains unchanged as it has been placed at its correct place. The partitioning step of the quick sort uses comparisons and exchange operations to arrange the array elements as stated above and to place the *pivot* element at its appropriate place. As it can be seen from the partitioning algorithm, it performs $(n - 1)$ comparisons and an exchange operation for every comparison if the comparison results are to be true. So, the number of exchanges is less than or equal to $(n - 1)$. Hence, for an array of size n, the partition step has a time complexity of $O(n)$ or αn where α is a constant. So, we can write the following recurrence relation to compute the total time taken by the quick sort algorithm to sort out n elements:

$$T(n) = T(a) + T(b) + \alpha n$$

where a and b are the sizes of the two partitions. In the worst case, we might pick a *pivot* element to be the smallest one among all the elements in A. In that case, $a = 0$, $b = n - 1$, and the recurrence relation simplifies to the following:

$$T(n) = T(n - 1) + \alpha n$$

The solution to this recurrence relation turns out to be $O(n^2)$ and gives the worst-case complexity of the quick sort. This shows that the worst case of the quick sort is not very efficient. This has happened due to the selection of a bad *pivot* element. As stated above, a *pivot* element should be chosen in such a way that it partitions the array into roughly two equal halves. If the chosen *pivot* partitions the array into two equal halves, then, we can rewrite the above recurrence relation as follows:

$$T(n) = T\left(\frac{n}{2}\right) + T\left(\frac{n}{2}\right) + \alpha n$$

or

$$T(n) = 2T\left(\frac{n}{2}\right) + \alpha n$$

The solution of this recurrence relation turns out to be $O(n \log n)$ and gives the complexity of the best case of the quick sort. In fact, this is the best time complexity that is achieved by any comparison-based sorting algorithm.

Now, the question is how to choose a *pivot* element that is not bad. The answer to this question lies in *randomization*. Suppose we pick $x = A[k]$ where k is chosen randomly between the *low* and *high* indices of array A. In that case, it is equally likely to choose an element of array A between the *low* and *high* indices. To do the average-case analysis, let us consider all the possibilities of the partitioning of array A between the indices of a *low* and *high*. Let *low* be 1 and *high* be n to consider the entire array. We can assume that each partitioning is equally likely; hence, we get the probability of each partitioning as $1/n$. To get the average-case complexity of the quick sort, we can take the probability-weighted average of the time complexities of all the possibilities. Hence, the average-case analysis of a randomized quick sort algorithm generates a randomized recurrence as follows:

$$T(n) = \frac{1}{n}\left[\sum_{i=0}^{n-1}[T(i) + T(n - i - 1) + \alpha n]\right]$$

The relation can be simplified as given below:

$$T(n) = \frac{1}{n}\left[2\sum_{i=0}^{n-1}[T(i) + \alpha n]\right]$$

$$nT(n) = 2\sum_{i=0}^{n-1}[T(i) + \alpha n]$$

The above equation can be written as follows by leaving $O(n)$ terms and only considering $O(n^2)$ terms from the later part of the expression:

$$nT(n) = 2\sum_{i=0}^{n-1} T(i) + cn^2 \tag{15.1}$$

where c is a positive constant. Now, if we put $(n - 1)$ in place of n, then, the expression turns out to be

$$(n - 1)T(n - 1) = 2\sum_{i=0}^{n-2} T(i) + c(n - 1)^2 \tag{15.2}$$

Subtracting Equation 15.2 from Equation 15.1

$$nT(n) - (n - 1)T(n - 1) = 2T(n - 1) + 2cn - c$$

Leaving the constant terms and rearranging the remaining terms, we can write

$$nT(n) = (n + 1)T(n - 1) + 2cn$$

Dividing both the sides by $n(n + 1)$

$$\frac{T(n)}{n + 1} = \frac{T(n - 1)}{n} + \frac{2c}{n + 1}$$

Now, we can telescope and write the above expression as follows:

$$\frac{T(n)}{n + 1} = \frac{T(n - 1)}{n} + \frac{2c}{n + 1}$$

$$\frac{T(n - 1)}{n} = \frac{T(n - 2)}{n - 1} + \frac{2c}{n}$$

$$\frac{T(n - 2)}{n - 1} = \frac{T(n - 3)}{n - 2} + \frac{2c}{n - 1}$$

$$\frac{T(2)}{3} = \frac{T(1)}{2} + \frac{2c}{3}$$

Adding all the above equations, we get

$$\frac{T(n)}{n + 1} = \frac{T(1)}{2} + 2c\sum_{i=3}^{n+1}\frac{1}{i}$$

Without losing the generality, we can add a constant $2c(1 + (1/2))$ on the right side of the above equation to start the series from $i = 1$. The equation now becomes as follows:

$$\frac{T(n)}{n+1} = \frac{T(1)}{2} + 2c\sum_{i=1}^{n+1}\frac{1}{i} \tag{15.3}$$

The second term in Equation 15.3 is the sum of a harmonic series and we use the following standard result to compute the summation of the series:

$$\sum_{i=1}^{n}\frac{1}{i} = \log_e(n) + 0.577$$

Hence, Equation 15.3 converts into the following:

$$\frac{T(n)}{n+1} = \frac{T(1)}{2} + 2c[\log_e(n) + 0.577] + \frac{2c}{n+1}$$

$T(1)$ is the time required to sort out an array of one element and can be considered as zero. Hence, the above expression can be rewritten as

$$T(n) = (n+1)[2c(\log_e(n) + 0.577)] + 2c$$

This gives

$$T(n) = O(n \log n)$$

This shows that even the average-case time complexity of the quick sort is $O(n \log n)$.

15.2.4 Merge Sort

The merge sort is one of the earliest sorting algorithms, invented by John von Neumann in 1945 (Knuth 1998). It is a divide-and-conquer-based sorting algorithm. In the *divide* phase, the n-elements array that is to be sorted is divided into two subarrays of $n/2$ elements each. These subarrays are sorted out recursively in the *conquer* phase. The two sorted subarrays are merged to produce one combined sorted array. To sort out the subarrays recursively, the division of subarrays keeps happening until the subarrays of size one are achieved. Since an array with one element is always sorted, merging starts with the merge of sorted arrays of size one and producing the merge of subarrays of size two. In the next level, the obtained subarrays of size two are merged. The whole procedure of the merge sort is summarized in the following two steps:

1. Divide the unsorted array into n subarrays, each containing one element. An array of one element is considered to be sorted.
2. Repeatedly merge the sorted subarrays, starting from subarrays of size one, to produce new sorted subarrays until there is only one array remaining. This final merged array will be the sorted array.

Let us first look at the merging step. Suppose we have two sorted arrays given. We need to merge these arrays to get one unified sorted array. The algorithm on the next page presents the merging of two sorted arrays.

Let us analyze the time complexity of the this algorithm. In the worst case, both arrays A and B get empty at about the same time. This happens when elements in both the arrays are in such a way that in the merged array, adjacent elements come from different arrays. There is an example given below where A and B are the arrays to be merged and C is the merged array.

$$A = \{1, 3, 5, 7, 9\}$$
$$B = \{2, 4, 6, 8\}$$
$$C = \{1, 2, 3, 4, 5, 6, 7, 8, 9\}$$

Each comparison adds one item to C; so, in the worst case, we need $(m + n - 1)$ comparisons. That means to merge $m + n$ elements of array A and B, the complexity would be $O(m + n)$. The best case of merging happens when all the elements of one array are smaller than all the elements of the other array. In this case, the number of comparisons required to merge two arrays would be at most $\max(m, n)$.

```
Algorithm merge-sorted-arrays(A[], B[], m, n)
    Input: Sorted arrays A and B which are to be merged, m and n are the
           number of elements in A and B respectively.
    Output: Sorted array C obtained after merging of arrays A and B
begin
    i = 1;
    j = 1;
    k = 1;
    while (i ≤ m and j ≤ n)
         if (A[i] ≤ B[j])
               C[k] = A[i];
               i = i + 1;
         else
               C[k] = B[j];
               j = j + 1;
         end-if
         k = k + 1;
    end-while

    if (i < m)
         for p = i:m
             C[k] = A[p];
             k = k + 1 ;
         end-for
    else
         for p = j:n
             C[k] = B[p];
             k = k + 1;
         end-for
    end-if
end-merge-sorted-arrays
```

Once we know how to combine two sorted arrays, we can construct a divide-and-conquer sorting algorithm that simply divides the array into two subarrays, sorts the two subarrays recursively, and merges the results. The algorithm for the merge sort is given below.

```
Algorithm merge-sort (A[], low, high)
    Input: Array A, low and high are the minimum and maximum indices of
    the array between which sorting is to be performed.
    Output: Sorted array A
begin
    if (low < high) then
        mid = ⌊(low + high)⌋/2;
        merge-sort(A, low, mid);
        merge-sort (A, mid+1, high);
        m = (mid - low) + 1; /*Number of elements in first half of the array*
        n = (high - mid); /*Number of elements in second half of the array*/
        merge(A[low..mid], A[mid+1..high],m,n)
    else
        return; /*Termination of recursive calls*/
    end-if
end merge-sort
```

Let us analyze the complexity of the merge sort. As we see in the above algorithm, every time, the array is divided into two halves and the two subarrays are recursively sorted. Let n be the number of elements to be sorted in the original array. The array is divided into two halves of $n/2$ elements each, and these subarrays are recursively sorted. Once the two subarrays are sorted, they are merged to get the final sorted array. As shown earlier, in the worst case, the merging of two sorted arrays having a total of n elements takes $O(n)$ time. Hence, the recurrence relation for the merge sort can be given as follows:

$$T(n) = 2T\left(\frac{n}{2}\right) + O(n)$$

The solution of this recurrence relation turns out to be $O(n \log n)$. Hence, the time complexity of the merge sort is $O(n \log n)$.

15.3 Greedy Strategy

15.3.1 General Strategy

The greedy strategy is used to design algorithms for solving optimization problems; however, it should be noted that it does not always yield an optimal solution. Often, the problems for which an optimal solution can be obtained using the greedy strategy, establishing the correctness of the solution is not trivial. Since the greedy strategy does not give a guarantee for the solution to be optimal, it frequently serves as the basis of a heuristic approach.

Greedy algorithms are relatively easy to design for optimization problems. Before discussing the greedy strategy in detail, let us first consider a more-detailed definition of the environment in which typical optimization problems occur. All greedy algorithms follow the same general form. A greedy algorithm for a particular problem is specified by describing three predicates: *solution*, *feasible*, and the selection function *select*. Let us define the environment in more details. It consists of following:

- A collection of **candidates** from which the solution set is obtained
- A set of candidates that have already been used in search of the solution
- A **predicate** function to test whether a given set of candidates give a solution (not necessarily an optimal one)
- A predicate function to test if a set of candidates can be **extended** to a solution (not necessarily an optimal solution)
- A **selection function** that chooses some candidate that has not yet been investigated
- An **objective function** that evaluates a solution and assigns a value to it

Let us define the following supportive functions for the above environment for a greedy algorithm:

- *Select*(...): It takes the candidate set and selects and returns some candidates from it that has not yet been explored.
- *IsSolution*(...): It considers the candidate set and checks if a given set of candidates give a solution or not.
- *IsFeasible*(...): It considers the candidate set and checks if a given set of candidates can be extended to give a solution or not.

We can briefly summarize the above environmental item used in the greedy strategy as follows. The use of a greedy strategy in the computation of a solution to an optimization problem involves finding a subset S from a collection of candidates C where the subset S must satisfy a specified criterion. This means the selected subset should be such that the objective function is optimized by S and it is a solution to the optimization problem. Optimization may be either minimization or maximization depending on the problem being solved. In the greedy strategy, the selection function assigns a numerical value to each candidate, x, and among the various candidates, chooses the one for which *select*(x) is the smallest or the largest.

Let us now define a general form of the greedy algorithm with the help of the above functions. The input to the algorithm is a candidate set "C" and the algorithm returns a set "S" of candidates that makes a solution set for the given problem.

To illustrate the concept of a greedy strategy, let us discuss the solution to the following two problems that require an optimization to achieve its solution:

- Minimum cost spanning tree
- Fractional knapsack

It is seen that the greedy strategy produces an optimal solution to a fractional knapsack and the minimum cost spanning-tree problems. We will discuss them here.

```
Algorithm greedy-algorithm(C)
    Input C : Candidate set
    Output S: Candidates which makes Solution set
begin
  S = { }; /*Initially, solution set is empty*/
  while (not isSolution(S)) and C ≠ { }
```

```
    x = select(C);
    C = C - {x};
    if isFeasible(S ∪ {x}) then
        S = S ∪ {x};
    end if;
  end-while;
  return S;
end greedy-algorithm;
```

15.3.2 Minimum Cost Spanning Tree

In graph theory, a graph is defined as $G = (V, E)$ where V is a set of vertices and E is the set of edges. A graph is said to be connected if any vertex in the graph can be reached from any other vertex of the graph, either directly or through some other vertices. A tree T is defined as a connected graph that contains all the vertices of graph G; however, it has exactly $|V| - 1$ edges, where $|V|$ is the number of vertices in the graph G.

A spanning tree T of an undirected connected graph $G = (V, E)$, where V is a set of vertices and E is the set of edges, is a subgraph that includes all the vertices of G and that is a tree. In general, a graph may have several spanning trees. If a weight is assigned to each edge of the graph, then, we can obtain a weight to a spanning tree by computing the sum of the weights of the edges currently in the spanning tree. A minimum cost spanning tree of a graph can then be defined as the tree that has a weight less than or equal to the weight of every other spanning tree of graph G.

The problem of the minimum cost spanning tree deals with the computation of a spanning tree of a graph having the least weight. The input for this problem is an undirected connected graph, $G = (V, E)$ in which each edge, $e \in E$, has an associated positive weight. The output of this problem is a spanning tree, $T = (V, F)$ of G such that the total edge weight is the least among all the possible spanning trees of G. It should be noted that T can be called the spanning tree of G if and only if T is a tree and $F \subseteq E$.

The general terms (discussed above) used in the greedy algorithm formulation for the minimum cost spanning tree can be expressed as given below:

- The candidates set here contains the edge set E of G.
- A subset S of edges E (i.e., $S \subseteq E$) is a solution if the graph $T = (V, S)$ is a spanning tree of G.
- A subset H of edges E (i.e., $H \subseteq E$) is feasible if there is a spanning tree $T = (V, S)$ of G for which H is the subset of S.
- The objective function that is to be minimized for the computation of the minimum cost spanning tree is the sum of the edge weights in a solution.
- The *select*(...) function chooses a candidate edge e from E whose weight is the smallest {} from the remaining edges.

The complete Kruskal algorithm (Kruskal 1956) for the computation of the minimum cost spanning tree using a greedy strategy is as follows:

```
Algorithm minimum-cost-spanning-tree(G)
    Input: connected graph G=(V,E) where V and E are the vertices and edges
           of the graph respectively.
    Output: tree T=(V,S), where S is the solution edge set
```

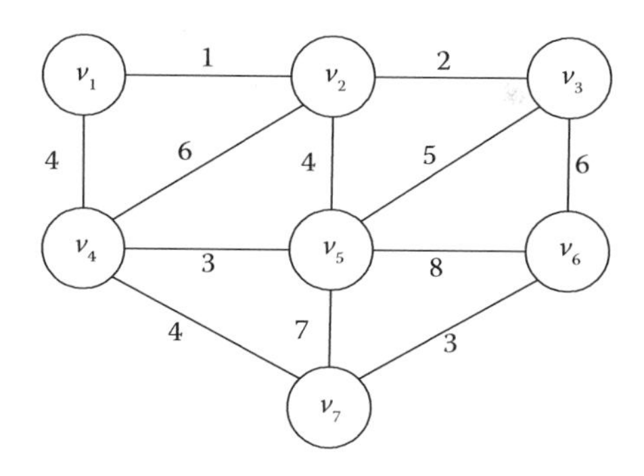

FIGURE 15.5
Example of a weighted graph with seven vertices.

```
begin
    S := { }
    while (T(V,S) not a tree) and E ≠ {}
    e := select(E) /*selects of minimum weight edge from E*/
    E := E - {e};
    if T(V, S ∪ {e}) is acyclic then
    S := S ∪ {e};
    end if;
  end while-loop;
  return T;
end minimum-cost-spanning-tree
```

The above algorithm can be viewed as dividing the set of vertices $V = \{v_1, v_2, v_3, \ldots, v_n\}$ of graph G into n-distinct subsets as follows:

$$\{v_1\}, \{v_2\}, \{v_3\} \cdots \{v_n\}$$

Initially set S is empty. Now, the algorithm adds an edge to set S if and only if it joins two vertices that belong to different subsets. Further, if an edge $e = (v_i, v_j)$ is added to set S, then, the two subsets containing the vertices v_i and v_j are merged into a single set. In this way, the algorithm terminates when only a single set is left as shown below:

$$\{v_1, v_2, v_3, \ldots, v_n\}$$

Let us consider an example of the execution of Kruskal algorithm on a graph shown in Figure 15.5. Table 15.1 shows the subsets arising during the execution of Kruskal algorithm on this graph whereas the final minimum cost spanning tree obtained using the Kruskal algorithm is shown in Figure 15.6.

15.3.3 Fractional Knapsack

The fractional knapsack problem can be stated as follows. There are n items available in a store where the weight of the ith item is $w_i > 0$ and the value of each item is $v_i > 0$, for $i = 1, 2, \ldots, n$. A thief has come to the store to steal items from the store. He can carry a maximum weight of W in a knapsack that he is carrying. The thief may decide to carry only a fraction x_i of item i, where $0 \leq x_i \leq 1$. In that case, item i contributes $x_i w_i$ to the total weight and $x_i v_i$ contributes to the total value of the load of the knapsack.

TABLE 15.1

Step-by-Step Execution of Kruskal Algorithm on a Graph Shown in Figure 15.1

Iteration	Edge Selected	Vertex Subsets	Remarks
0	–	$\{v_1\}, \{v_2\}, \{v_3\}, \{v_4\}, \{v_5\}, \{v_6\}$, and $\{v_7\}$	*Initially every vertex makes a distinct set*
1	$\langle v_1, v_2\rangle$	$\{v_1, v_2\}, \{v_3\}, \{v_4\}, \{v_5\}, \{v_6\}$, and $\{v_7\}$	
2	$\langle v_2, v_3\rangle$	$\{v_1, v_2, v_3\}, \{v_4\}, \{v_5\}, \{v_6\}$, and $\{v_7\}$	
3	$\langle v_4, v_5\rangle$	$\{v_1, v_2, v_3\}, \{v_4, v_5\}, \{v_6\}$, and $\{v_7\}$	
4	$\langle v_6, v_7\rangle$	$\{v_1, v_2, v_3\}, \{v_4, v_5\}$, and $\{v_6, v_7\}$	
5	$\langle v_1, v_4\rangle$	$\{v_1, v_2, v_3, v_4, v_5\}$ and $\{v_6, v_7\}$	
6	$\langle v_2, v_5\rangle$	$\{v_1, v_2, v_3, v_4, v_5\}$ and $\{v_6, v_7\}$	*Not included as it makes a cycle*
7	$\langle v_4, v_7\rangle$	$\{v_1, v_2, v_3, v_4, v_5, v_6, v_7\}$	*The algorithm terminates when all the vertices appear in one set*

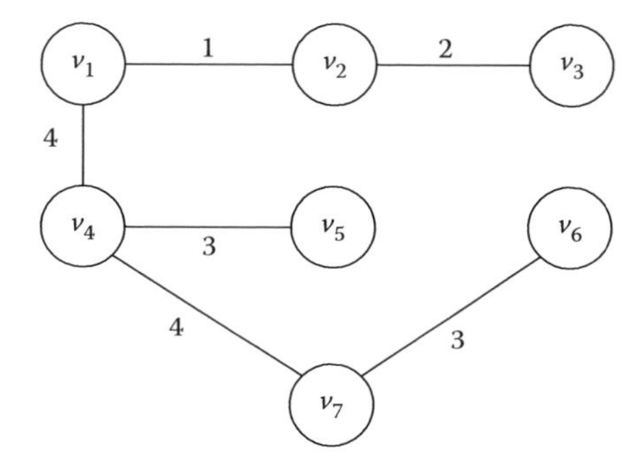

FIGURE 15.6
Minimum cost spanning tree of the graph shown in Figure 15.5 computed using Kruskal's algorithm.

The fractional knapsack problem can be stated as follows.

Maximize $\sum x_i v_i$ subject to constraint $\sum x_i w_i \leq W$, where $0 \leq i \leq n$. It is evident that an optimal solution must completely fill the knapsack. To fill the knapsack completely, a thief may need to take m items ($m \leq n$) where $m - 1$ items are taken completely and the mth item is either considered completely or he considers a fraction of it depending on the left space in the knapsack. Thus, in an optimal solution, $\sum_{i=1}^{m} x_i w_i = W$.

The fractional knapsack problem exhibits the greedy property. An algorithm for the solution of the fractional knapsack problem using a greedy strategy is presented here on the next page.

If the items are *a priori* sorted into the decreasing order of v_i/w_i, then, the while-loop in the above algorithm takes a time in $O(n)$; hence, the overall complexity of the above algorithm including the sort can be given as follows:

$$\begin{aligned}
\text{Total complexity } T(n) &= \text{Time required for sorting} \\
&\quad + \text{time required for the above knapsack algorithm} \\
&= O(n \log n) + O(n) \\
&= O(n \log n)
\end{aligned}$$

If we use the heap data structure and maintain a heap of all the items by keeping the item with the largest v_i/w_i at the root, then, the complexity of the knapsack algorithm would be as follows:

- Creating the heap will take $O(n)$ time
- The while-loop will now take $O(\log n)$ time (since the heap property is maintained after the removal of the root)

Note that the heap is a tree-based data structure that satisfies the heap property. The heap property is defined as follows. If X is a parent node of Y in a tree, then, the key of node X is ordered with respect to the key of node Y with the same ordering applying all over the heap. Heaps can be further divided into two types, namely, the *max heap* or *min heap*. In a max heap, keys of parent nodes are always greater than or equal to the keys of their children and the highest key is present in the root node. On the contrary, in a min heap, keys of parent nodes are less than or equal to the keys of their children and the lowest key is present in the root node.

Although the use of this data structure does not change the worst-case complexity, it may be faster in the case where only a small number of items are required to fill the knapsack. There is a variant of the knapsack problem that is called the integer knapsack or 0–1 knapsack problem. In the 0–1 knapsack problem, an item is fully considered or it is completely left. So, the elements of vector x are either 1 or 0. The solution provided by a greedy strategy for the integer knapsack problem is not optimum. That means the solution vector x does not necessarily provide the list of items for which the value of the items carried in the knapsack is maximum.

```
Algorithm fractional-knapsack
    Input: Set w={w1, w2,..., wn} containing weights for n items, set
           v={v1, v2, ..., vn} containing respective values of n
           items present in w, maximum weight W that can be
           carried in the knapsack
    Output: vector x={x1, x2, .... xn}, where xi is 1 if i-th item
            is completely taken in the knapsack, it is 0 if i-th item is
            not taken at all and it is a fractional value if i-th item is
            partially taken in the knapsack
begin
    for i =1:n
          xi =0;
    end-for

weight = 0
  while (weight < W)
      Consider item i (where vi/wi is maximum among remaining items)
        if (weight + wi  ≤ W);
              xi = 1
                      weight = weight + wi
              else
                      xi = (w - weight) / wi;
                      weight = W;
  end-while
return x;
end fractional-knapsack
```

15.4 Dynamic Programming

Dynamic programming is usually used in developing algorithms to solve a certain class of optimization problems. It is used to write algorithms for the problems that require a maximization or minimization of some measure. It is an algorithm-design paradigm for solving a large problem by breaking it down into a set of simpler subproblems. It provides an efficient way to solve problems whose subproblems are of an overlapping nature. Dynamic programming provides an efficient solution as compared to the other techniques that do not take advantage of the overlapping nature of the subproblems. For example, in the divide-and-conquer strategy, a solution to a problem is obtained by recursively solving separate subproblems where it may lead to the same computations being performed repeatedly due to the arising of identical subproblems. The idea behind dynamic programming is to avoid this problem by removing the calculation of the same quantity multiple times. Owing to this nature, dynamic programming also provides efficient solutions as compared to the depth-first search strategy.

In the dynamic programming strategy, different subproblems are solved and the obtained solutions of the subproblems are combined to reach the final solution. Often while using a simpler optimization technique, many of the subproblems are generated and solved multiple times. The dynamic programming approach ensures that each subproblem is solved only once. This reduces the number of computations. In dynamic programming, once the solution to a given subproblem is computed, it is stored for future use. In future when the solution to the same problem is required, then simply the stored solution is reused. Dynamic programming follows the bottom-up approach in which the smallest subproblems are explicitly solved first and then the solutions of these subproblems are used to construct solutions to the subsequent larger subproblems.

Dynamic programming examines the previously solved subproblems and uses their solutions to give the best solution for the given problem. The reuse of already-computed solutions in the dynamic programming strategy is usually accomplished by maintaining a table of results of the subproblems. This approach of reuse of the solution of subproblems is very useful when the number of repeating subproblems grows exponentially as a function of the size of the input.

The solutions to many optimization problems are efficiently computed using dynamic programming. Some of the problems where dynamic programming is successfully used are the shortest-path computation between two points, the fastest way to multiply many matrices, etc. There are many other algorithm-design strategies that can alternatively be used in place of dynamic programming. One such strategy is the greedy strategy that chooses the locally optimal choice at each branch while taking a decision in search of an optimum solution. The locally optimal choice may be a poor choice for the overall solution of the problem. Though a greedy algorithm does not guarantee an optimal solution for a problem, it is often found to be faster to compute. Fortunately, some greedy algorithms are proven to lead to the optimal solution. For example, Kruskal (Kruskal 1956) and Prim's (Prim 1957) algorithm to compute the minimum cost spanning trees is based on the greedy strategy and produces optimal solutions.

We can understand the difference between dynamic programming and the greedy strategy with the help of this example. Let us say that in a city, we have to go from point A to point B as quick as possible during peak rush hours. A dynamic programming approach will find the shortest paths to points close to A, and use these solutions to compute the shortest path to B. A greedy algorithm on the other hand will ask us to start driving immediately and suggest us to pick the next road that looks the fastest at every junction. As we can see, the strategy of choosing the best at every junction might not lead to the shortest journey time. This is due to the fact that we might take some easy roads at the beginning and then find ourselves stuck in a traffic jam.

In contrary to dynamic programming, divide and conquer is a top-down approach that logically proceeds from the initial original problem and progressively goes down to the smallest subproblems through intermediate subproblems.

To illustrate the dynamic programming paradigm, let us consider a few problems whose solutions can be efficiently achieved using dynamic programming. The first problem that we shall consider is computation of the binomial coefficient. The second problem is the Floyd–Warshall algorithm for the all-pair shortest path and the third is the 0–1 knapsack problem.

15.4.1 Binomial Coefficient

The binomial coefficient is defined as the number of ways of picking k-unordered outcomes from n possibilities. It is also known as a combinatorial number. It is represented as $\binom{n}{k}$ and mathematically defined as follows:

$$\binom{n}{k} = \frac{n!}{k!(n-k)!} \quad (0 \le k \le n)$$

It is trivial to verify that the above relationship can also be written as follows:

$$\binom{n}{k} = \begin{cases} \binom{n-1}{k-1} + \binom{n-1}{k} & 0 < k < n \\ 1 & k = 0 \quad \text{or} \quad k = n \end{cases}$$

Using this relationship, a simple divide-and-conquer-based solution to this problem of calculating the binomial coefficient can be written as follows:

```
Algorithm binomial-coeff(n, k)
      Input: n and k
      Output: (n k)
begin
     if k = 0 or k = n then
        return 1;
     else
        t = binomial-coeff(n-1, k-1) + binomial-coeff(n-1, k);
     return t;
     end-if
end binomial-coeff
```

The dynamic programming approach also uses the same formula to compute the binomial coefficient; however, it constructs a table of all $(n+1) \times (k+1)$ binomial coefficients, that is, $\binom{i}{j}$ for each value of $i \in [0, n]$, and $j \in [0, k]$. These coefficients are calculated in a particular order as follows:

- First, the table cells corresponding to the coefficients $\binom{i}{0}$ for all $i \in [0, n]$, and $\binom{1}{1}$ are initialized to a value of 1.
- The remaining cells of the table corresponding to the binomial coefficient $\binom{i}{j}$ are calculated in an increasing order of the value of $i + j$.

It should be noted that since the computation of the coefficient $\binom{i}{j}$ needs only the values of $\binom{i-1}{j-1}$ and $\binom{i-1}{j}$, the computation of entries of various table cells in the order of an increasing $i + j$ makes sure that the values needed to compute $\binom{i}{j}$ have already been computed. This can be verified by looking at the following relationship, that is,

$$(i-1) + (j-1) < (i-1) + j < i + j$$

The following is the algorithm for a dynamic programming based solution for the computation of a binomial coefficient.

```
Algorithm binomial-coeff-dynamic(n, k)
     Input: n and k
     Output: (n k)
begin
    for i=0:n
         BiCoef(i,0)  := 1;
    end-of
    BiCoef(1,1)  := 1;
    sum = 3;
      i = 2;
      j = 1;
    while (sum <= n+k)
      BiCoef(i,j)  := BiCoef(i-1,j-1)  + BiCoef(i-1,j);
      i = i-1;
      j = j+1;
      if (i < j) or (j > k)
                 sum = sum + 1;
                 if (sum <= n+1)
                        i = sum-1;
                        j = 1;
                 else
                        i = n;
                        j = sum-n;
                 end-if
      end-if
      end-while
    return BiCoef(n,k);
end binomial-coeff-dynamic
```

Now, let us discuss the differences between the divide-and-conquer approach and the dynamic programming approach with respect to the computation of a binomial coefficient. Divide and conquer computes the values, for example, $\binom{2}{1}$ many times to compute $\binom{n}{k}$. If the value of n is large and k is dependent on n (i.e., k is not a constant), then, $\binom{2}{1}$ may be computed a huge number of times. In the case of dynamic programming, $\binom{2}{1}$ is computed only once and reused in future whenever it is required. The time complexity of the computation of a binomial coefficient using divide and conquer can be shown to be $\Omega\binom{n}{k}$. Its worst case will happen when $k = n/2$, and then, in that case, it is asymptotically bounded by $\Omega(2^n/n)$. Though the algorithm for divide and conquer looks simple, it is practically infeasible.

Since the dynamic programming strategy computes $\binom{i}{j}$ exactly once and reuses it whenever required, it is very efficient as compared to divide and conquer approach. The time complexity of a dynamic programming based binomial computation is $O(nk)$. Even in the worst case when $k = n/2$, the complexity is just $O(n^2)$ that is polynomial on the input size.

15.4.2 All-Pair Shortest Path Using Floyd–Warshall Algorithm

Floyd–Warshall algorithm is a dynamic programming based algorithm used for solving the all-pairs shortest-path problem. The problem is to find the shortest paths (distances) between every pair of vertices in a weighted directed graph. Let us consider a directed graph, $G(V, E)$, with nodes $V = \{1, 2, \ldots, n\}$ and edges $E \subseteq V \times V$ and each edge in the graph has a weight associated with it. The Floyd–Warshall algorithm computes the shortest path between each pair of vertices and returns an $n \times n$ matrix D in which the cell (i, j) contains the length of the shortest path from node i to node j in graph G.

Let us consider following directed graph $G = (V, E)$ with four vertices $V = \{1, 2, 3, 4\}$ and four edges $E = \{\langle 1, 2\rangle, \langle 2, 3\rangle, \langle 3, 4\rangle, \langle 1, 4\rangle\}$.

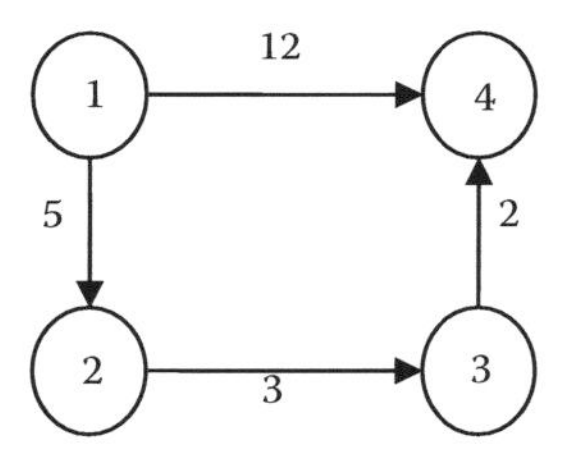

The adjacency matrix for this graph can be written as follows. In the matrix, value *INF* (infinite) at the cell (i, j) shows that there is no direct edge between node i and node j in the graph. The value at the cell (i, j) is 0 if i is equal to j.

0	5	*INF*	12
INF	0	3	*INF*
INF	*INF*	0	2
INF	*INF*	*INF*	0

The output for this graph is expected to be the following matrix where every cell (i, j) tells the distance of the shortest path between node i and node j. A value of *INF* in cell (i, j) shows that there is no path existing between node i and node j.

0	5	8	10
INF	0	3	5
INF	*INF*	0	2
INF	*INF*	*INF*	0

The algorithm constructs a sequence of matrices: $D_0, D_1, \ldots, D_n$ during the execution of the algorithm. Here, matrix D_k ($k \in [1, n]$) contains the length of the shortest path between every pair of vertices, when only k vertices, that is, nodes $\{1, 2, 3, \ldots, k\}$ can be used as intermediate nodes on the path. For example, entry $D_k(i, j)$ in matrix D_k denotes the length of the shortest path from node i to node j when only the k vertices are used. It is obvious to see that $D_n = D$.

Matrix D_0 (which is actually the adjacency matrix of the graph) is the matrix that corresponds to the smallest subproblem here. The matrix is initiated as follows:

$$D_0(i, j) = \begin{cases} 0 & \text{if } i = j \\ INF & \text{if } edge\ (i, j) \notin E \\ length(i, j) & \text{if } edge\ (i, j) \in E \end{cases}$$

Now, let us see how to construct D_{k+1}, for some $k < n$, assuming that we have already computed or constructed D_k. This means that we have to construct the shortest path from i to j **only** with considering vertices $\{1, 2, 3, \ldots, k, k + 1\}$ of the graph. There are two possibilities.

Case 1: The path from i to j does not contain the node $k + 1$. In this case, $D_{k+1}(i, j)$ is given as follows:

$$D_{k+1}(i, j) = D_k(i, j)$$

Case 2: The path from i to j contains the node $k + 1$. In this case, $D_{k+1}(i, j)$ is given as follows:

$$D_{k+1}(i, j) = \min[D_k(i, j), D_k(i, k + 1), D_k(k + 1, j)]$$

If we notice the expressions in the above relationships, it looks trivial to use a recursive algorithm to implement this algorithm. However, like the previous example, a recursive implementation will be very inefficient.

We would use an iterative algorithm to implement this. We do not require k matrices in the actual implementation as once the matrix $D_{(k+1)}$ has been computed, the matrix D_k is no longer required. So, we overwrite the values of D_k by D_{k+1}. Hence, in the algorithm given below, we shall only use one matrix of size $n \times n$ throughout this algorithm. Let us call this matrix as D.

The input to the algorithm is given in terms of the adjacency matrix L of the graph. The adjacency matrix is defined exactly in the same way as D_0 given above and hence it is used to initialize matrix D in the algorithm. An entry in cell (i, j) of matrix L states the length of the direct edge between node i and node j in graph G. So, matrix D with the initial values obtained from L is the same as matrix D_0 in the above discussion.

There are three nested loops in the algorithm; hence, the time complexity of the algorithm is $O(n^3)$. It can be seen that $O(n)$ steps are required to calculate each of the n^2 entries of matrix D.

```
Algorithm all-pair-shortest-path(L)
   Input: Adjacency Matrix L of graph G
   Output: Matrix D with shortest path lengths between every pair
           of vertices
begin
  for i=1:n
      for j=1:n
           D(i,j) = L(i,j);
      end-for
  end-for

  for k=1:n
      for i=1:n
          for j=1:n
                if (D(i,j) > D(i,k) + D(k,j))
```

```
                    D( i,j ) = D(i,k) + D(k,j);
                end-if
            end-for
        end-for
    end-for
    return D;
end all-pair-shortest-path
```

15.4.3 0–1 Knapsack Problem

The 0–1 knapsack problem can be stated as follows. Given n items with their respective weights and values, the objective is to put these items in a knapsack of capacity W so that we get the maximum total value in the knapsack. Let us say that we have a set X of n items: $X = \{x_0, x_1, \ldots, x_{n-1}\}$. The available qualities (or weights) of these items are $w_0, w_1, \ldots, w_{n-1}$ respectively and make set W, which means that $W = \{w_0, w_1, \ldots, w_{n-1}\}$. The values of these items are $v_0, v_1, \ldots, v_{n-1}$ and make set V, which means that $V = \{v_0, v_1, \ldots, v_{n-1}\}$. We have a knapsack with capacity W. Our objective is to find out the maximum value subset of V such that the sum of the weights of the items chosen is smaller than or equal to W. We cannot partially take an item; so, either we have to pick an item completely, or we have to completely leave it. Since to fill the knapsack, either we leave an item completely or take it completely, the problem is termed as 0–1 knapsack problem.

A brute force solution to this problem is as follows. From the available items, consider all the subsets of items and compute the total weight and value of each subset. Further, pick only those subsets whose total weight is less than or equal to W. Now, from all the chosen subsets, we can pick the subset with the maximum value.

Another solution is presented here using dynamic programming. While choosing the optimal subsets of items, there can be two possibilities for every item, first the item is chosen for the optimal subset, and second that item is not included in the optimal set. Hence, we can say that the maximum value that can be obtained for a knapsack from n items is the maximum of the following two values:

- The maximum value obtained by $n - 1$ items and W weight (excluding the nth item).
- The sum of the value of the nth item and the maximum value obtained by $n - 1$ items and W minus the weight of the nth item.

If the weight of the nth item is greater than W, then, the nth item cannot be included and case 1 is the only possibility. The following is a recursive naive implementation that simply follows the recursive structure mentioned above:

```
Algorithm knapSack01(W,w[1..n],v[1..n],n)
     Input: W: capacity of knapsack, w and v are respectively
              weight and value vector and n is the number of distinct items
     Output: Maximum value that a knapsack of capacity W can accommodate.
begin
 if (n == 0 || W == 0)
              return 0;
  /* If weight of the n-th item is more than W (Knapsack
   capacity), then it cannot be kept in the knapsack,
  Return the maximum of two cases: (1) n-th item included
                                   (2) n-th item not included
  */
```

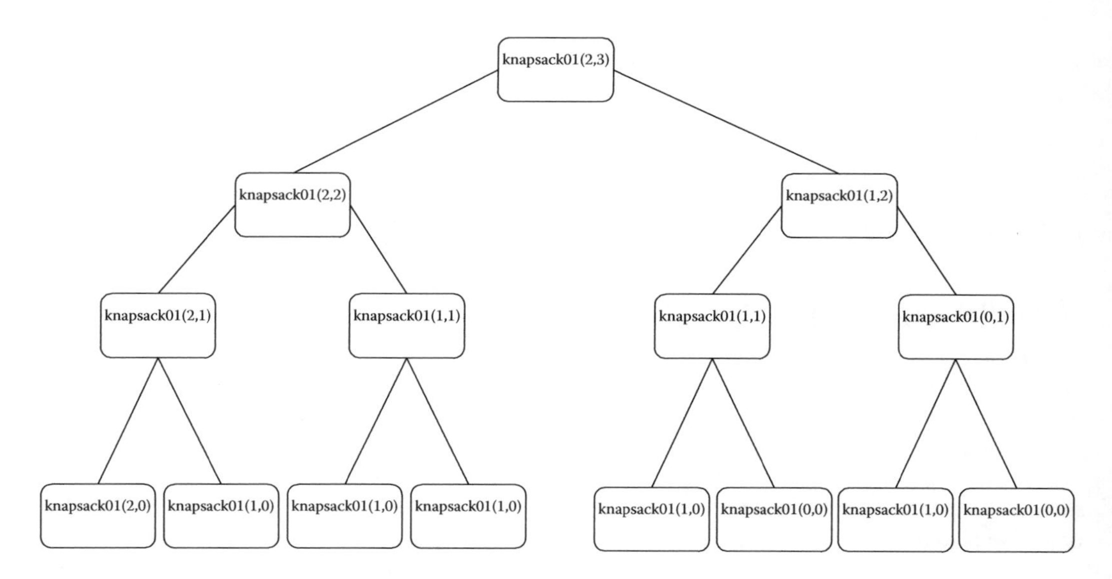

FIGURE 15.7
Recursion tree for a knapsack of capacity $W = 2$ units and 3 items, each having a unit weight.

```
if (w[n-1] > W)
      return knapSack01(W, w, v, n-1);
else
   return max(v[n-1]+knapSack01(W-w[n-1],w,v,n-1,knapSack01(W, w, v,n-1));
end-knapSack01
```

The time complexity of this naive recursive solution is exponential with ($O(2^n)$). Point to note in the above function is that the solution to the same subproblems is computed again and again. Let us see the recursion tree for the function call knapsack01(3,2) in the following example where there are three items with weight $w = \{1, 1, 1\}$ and value $v = \{10, 20, 30\}$, respectively. The size of the knapsack is $W = 2$. We can see that knapsack01(1,1) is evaluated twice while computing knapsack01(3,2).

The two parameters indicated in the recursion tree are only W and n for simplicity. The recursion tree for knapsack01(W, n) on the above sample inputs is given below in Figure 15.7.

Since the subproblems are evaluated again and again, this solution exhibits the overlapping subproblems property. To overcome this drawback of the solution, we can use dynamic programming to compute the solution where as seen earlier in other problems, the solution to a subproblem is computed only once and it is reused if required. Like other dynamic programming problems, a repeated computation of the same subproblems is avoided by maintaining a table to store the already-computed solutions in a bottom-up manner. The following is a dynamic programming based algorithm for the 0–1 knapsack problem.

```
Algorithm knapSack01(W, w[1..n], v[1..n], n)
     Input: W: capacity of knapsack, w and v are respectively weight and
            value vector and n is the number of distinct items
     Output: Maximum value that a knapsack of capacity W can accommodate.
             This value is stored in the bottom most right corner of the tabl
             i.e. B[n,W] is the maximal value that can be placed in the knaps
```

```
begin
1: for w = 0:W
2:      B[0,w] = 0;
3:    end-for
4: for i = 1:n
5:       B[i,0] = 0;
6: end-for
7: for i = 1:n
8:    for wt = 0:W
9:       if (w[i]<= wt) /* item i can be part of the solution */
10:               if (v[i] + B[i-1,wt-w[i]] > B[i-1,wt])
11:                    B[i,wt] = v[i] + B[i-1,wt- w[i]];
12:               else
13:                    B[i,wt] = B[i-1,wt];
14:               end-if
15:      else
16.            B[i,wt] = B[i-1,wt]; /*w[i]> wt */
17.      end-if
18:     end-for
19:   end-for
20. return B[1..n,1..W];
end knapSack01
```

The time complexity of the above solution is $O(nW)$ where n is the number of items and W is the capacity of the knapsack.

Let us consider that the number of elements n equals to 4 and the capacity of the knapsack W equals to 5. Let the weights of the four elements be 2, 3, 4, and 5, respectively, and the values (benefit) of these elements be 2, 4, 5, and 6, respectively. The given values are summarized below:

$$n = 4, \quad W = 5, \quad w = \{2, 3, 4, 5\}, \quad v = \{2, 4, 5, 6\}$$

The following is a step-by-step implementation of the above algorithm on these data. After the execution of steps 1–3 of the above algorithm, matrix B looks as follows:

i\ *wt*	**0**	**1**	**2**	**3**	**4**	**5**
0	0	0	0	0	0	0
1						
2						
3						
4						

After the execution of steps 4–6:

i\ *wt*	**0**	**1**	**2**	**3**	**4**	**5**
0	0	0	0	0	0	0
1	0					
2	0					
3	0					
4	0					

$i = 1$, $wt = 1$, $b[i] = 3$, $w[i] = 2$, and ($w[i] > wt$), Step 16 is executed: Now $wt - w[i] = -1$

$i\backslash wt$	**0**	**1**	**2**	**3**	**4**	**5**
0	0	0	0	0	0	0
1	0	0				
2	0					
3	0					
4	0					

$i = 1$, $wt = 2$, $b[i] = 3$, $w[i] = 2$, and $(w[i] = w])$, Step 11 is executed: $wt - w[i] = 0$

$i\backslash wt$	**0**	**1**	**2**	**3**	**4**	**5**
0	0	0	0	0	0	0
1	0	0	3			
2	0					
3	0					
4	0					

$i = 1$, $wt = 3$, $b[i] = 3$, $w[i] = 2$, and $(w[i] < w)$, Step 11 is executed: $wt - w[i] = 1$

$i\backslash wt$	**0**	**1**	**2**	**3**	**4**	**5**
0	0	0	0	0	0	0
1	0	0	3	3		
2	0					
3	0					
4	0					

$i = 1$, $wt = 4$, $b[i] = 3$, $w[i] = 2$, and $(w[i] < w)$, Step 11 is executed: $wt - w[i] = 2$

$i\backslash wt$	**0**	**1**	**2**	**3**	**4**	**5**
0	0	0	0	0	0	0
1	0	0	3	3	3	
2	0					
3	0					
4	0					

$i = 1$, $wt = 5$, $b[i] = 3$, $w[i] = 2$, and $(w[i] < w)$, Step 11 is executed: $wt - w[i] = 3$

$i\backslash wt$	**0**	**1**	**2**	**3**	**4**	**5**
0	0	0	0	0	0	0
1	0	0	3	3	3	3
2	0					
3	0					
4	0					

$i = 2$, $wt = 1$, $b[i] = 4$, $w[i] = 3$, and $(w[i] > w)$, Step 16 is executed: $wt - w[i] = -2$

$i\backslash wt$	**0**	**1**	**2**	**3**	**4**	**5**
0	0	0	0	0	0	0
1	0	0	3	3	3	3
2	0	0				
3	0					
4	0					

$i = 2$, $wt = 2$, $b[i] = 4$, $w[i] = 3$, and $(w[i] > w)$, Step 16 is executed: $wt - w[i] = -1$

$i\backslash wt$	**0**	**1**	**2**	**3**	**4**	**5**
0	0	0	0	0	0	0
1	0	0	3	3	3	3
2	0	0	3			
3	0					
4	0					

$i = 2$, $wt = 3$, $b[i] = 4$, $w[i] = 3$, and $(w[i] = w,\ b[i] + B[i-1, wt - w[i]] > B[i-1, w])$, Step 11 is executed: $wt - w[i] = 0$

$i\backslash wt$	**0**	**1**	**2**	**3**	**4**	**5**
0	0	0	0	0	0	0
1	0	0	3	3	3	3
2	0	0	3	4		
3	0					
4	0					

$i = 2$, $wt = 4$, $b[i] = 4$, $w[i] = 3$, and $(w[i] < w,\ b[i] + B[i-1, wt - w[i]] > B[i-1, w])$, Step 11 is executed: $wt - w[i] = 1$

$i\backslash wt$	**0**	**1**	**2**	**3**	**4**	**5**
0	0	0	0	0	0	0
1	0	0	3	3	3	3
2	0	0	3	4	4	
3	0					
4	0					

$i = 2$, $wt = 5$, $b[i] = 4$, $w[i] = 3$, and $(w[i] < w, b[i] + B[i-1, wt - w[i]] > B[i-1, w])$, Step 11 is executed: $wt - w[i] = 2$

$i\backslash wt$	**0**	**1**	**2**	**3**	**4**	**5**
0	0	0	0	0	0	0
1	0	0	3	3	3	3
2	0	0	3	4	4	7
3	0					
4	0					

$i = 3$, $wt = 1$ to 3, $b[i] = 5$, $w[i] = 4$, and $(w[i] > wt)$, Step 16 is executed: $wt - w[i] = -$ve

$i\backslash wt$	**0**	**1**	**2**	**3**	**4**	**5**
0	0	0	0	0	0	0
1	0	0	3	3	3	3
2	0	0	3	4	4	7
3	0	0	3	4		
4	0					

$i = 3$, $wt = 4$, $b[i] = 5$, $w[i] = 4$, and $(w[i] = w, b[i] + B[i-1, wt - w[i]] > B[i-1, w])$, Step 11 is executed: $wt - w[i] = 0$

$i\backslash wt$	**0**	**1**	**2**	**3**	**4**	**5**
0	0	0	0	0	0	0
1	0	0	3	3	3	3
2	0	0	3	4	4	7
3	0	0	3	4	5	
4	0					

$i = 3$, $wt = 5$, $b[i] = 5$, $w[i] = 4$, and $(w[i] < w, b[i] + B[i-1, wt - w[i]] < B[i-1, w])$, Step 13 is executed: $wt - w[i] = 1$

$i\backslash wt$	**0**	**1**	**2**	**3**	**4**	**5**
0	0	0	0	0	0	0
1	0	0	3	3	3	3
2	0	0	3	4	4	7
3	0	0	3	4	5	7
4	0					

$i = 4$, $wt = 1$ to 4, $b[i] = 6$, $w[i] = 5$, and $(w[i] > wt)$, Step 16 is executed: $wt - w[i] = -\text{ve}$

$i\backslash wt$	**0**	**1**	**2**	**3**	**4**	**5**
0	0	0	0	0	0	0
1	0	0	3	3	3	3
2	0	0	3	4	4	7
3	0	0	3	4	5	7
4	0	0	3	4	5	

$i = 4$, $wt = 5$, $b[i] = 6$, $w[i] = 5$, and $(w[i] = w,\ b[i] + B[i-1,\ wt - w[i]] < B[i-1, w])$, Step 13 is executed: $wt - w[i] = 0$

$i\backslash wt$	**0**	**1**	**2**	**3**	**4**	**5**
0	0	0	0	0	0	0
1	0	0	3	3	3	3
2	0	0	3	4	4	7
3	0	0	3	4	5	7
4	0	0	3	4	5	7

The $B(n, W)$ cell of the above table gives the maximal value of the items that can be placed in the knapsack. The following algorithm can be used to find out the actual knapsack items:

```
Algorithm get-knapsack-items(W,w[1..n],v[1..n],n,B[1..n,1..W])
     Input: W: capacity of knapsack, w and v are respectively weight
            and value vectors and n is the number of distinct items, B is the
            table returned by knapsack01 algorithm above

     Output: Binary vector x[1..n]. If x[i]=1, that means i-th item
             is kept in the knapsack.
begin
     i=n;
     k=W;
     for j=1:n
       x[i] = 0; /*Initially no item is taken in the knapsack*/
```

```
    end-for;

    while (i > 0 and k > 0)
      if B[i,k] ≠  B[i-1,k] then
         x[i] = 1; /*Take i-th item in the stack*/
         i = i - 1;
         k = k - w[i]
      else
         i = i - 1; /*Assumed that i-th item is not in the knapsack*/
      end-if
   end-while
   return x[1..n];
end get-knapsack-items
```

When this algorithm is executed on table B computed by the previous example, it returns vector x as [1, 1, 0, 0]. That means items 1 and 2 are considered for the knapsack. Other items are just neglected.

15.4.4 Challenges in Dynamic Programming

Though in most of the cases, it is possible to formulate the solution of a problem using the dynamic programming paradigm and solve it optimally, there are some practical difficulties in implementing the dynamic programming for some real-life problems. These problems have components that are very large as well as complex to make them computationally tractable. For example, as a dynamic programming based algorithm proceeds, it creates an exponential growth in computational and storage requirements. This problem is often referred to as the "curse of dimensionality." Occasionally, there are real-time solution constraints for solving a problem and the data of a problem being solved may be available only little in advance. Sometimes, the data may get changed and may need online replanning during the execution. All these problems motivated the researchers to go for an approximate dynamic programming (Powell 2011) that provides a faster and tractable solution in these scenarios.

15.5 Backtracking

15.5.1 General Strategy

Backtracking is an algorithm-design paradigm that systematically searches for the solution to a problem among all the available options. It examines various sequences of the decisions, until we find one that works and produces the solution. While solving a problem, many a times, we are faced with a number of options and are supposed to choose one of the available ones to go ahead in finding the solution. Once we make out the choice, we may get a new set of options from which we have to further choose the best option to go to the next stage to move toward the solution. At a particular stage, what set of options we get depends on what choice we made earlier. This procedure is repeated over and over again until we reach the final solution state (called the goal state) or a dead end from where there are no further options available to explore. If the goal state is achieved, then, the solution search is terminated with success; otherwise, we go back to the previous stage (backtrack) and explore the unexplored options. This is continued until we get the solution to the

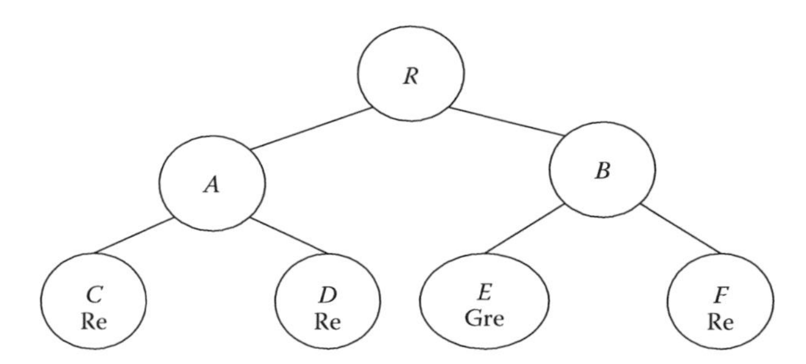

FIGURE 15.8
Example of backtracking.

problem or all the available options are explored. The following points can be noted regarding the backtracking:

- First, we do not have enough information to find out what to choose.
- Second, each decision takes us to a new set of choices.
- Third, some sequence of choices may lead to a solution of the problem.

Backtracking can be considered as one application of the divide-and-conquer approach that is used to compute the solution to a problem instance where the solution may not be straightforward and might consist of many small parts. To compute the solution, backtracking computes the solution of the first part. Further, it tries all the possibilities of solving the second part. This is continued until it finally achieves a complete solution. For example, let us say that we are searching for the best solution. To do so, we produce all the solutions part by part and remember the best solution that we find till that stage.

We can visualize backtracking with the help of an example. Consider the tree shown in Figure 15.8. The tree has some green leaves and some red leaves. In general, a tree may contain all green or all red leaves. Our objective is to reach a green leaf in the tree. To achieve this objective, we start at the root of the tree. Starting with the root, at each stage, we choose one of its children to move to the next level, and we keep going until we reach a leaf. Suppose the leaf where we land up is a red leaf, then, we *backtrack* to continue the search for a green leaf by revoking our *most recent* choice and trying out the next option in that set of options. There may be a situation when we run out of options, in that case, we revoke the choice that got us there, and try another choice at that node. In case there are no other options left, then, we can conclude that there are no green leaves in the tree. Another recursive way of stating the backtracking could be as follows. To search a tree, if the tree consists of a single node, then, test if it is a goal node. Otherwise, search the subtrees until we find one containing a goal node, or until we have searched for all the nodes unsuccessfully. A complete list of the steps to be followed in this example to search for the green leaf is as follows:

1. Starting at the root, here, our options are A and B. Let us say we choose A.
2. At A, our options are C and D, let us say we choose C.
3. C is red, hence go back to A.
4. At A, we have already examined C, and it has failed. Now examine D.
5. D is also found to be red. Hence go back to A.
6. At A, we have exhausted all the possible options. Hence we go back to the root.

7. At the root, we have already examined A, hence now try B.
8. At B, our available options are E and F. Now, let us say we try E.
9. E is green. Hence the solution is achieved. Stop.

15.5.1.1 Backtracking Algorithm: Recursive

Let us say that we are exploring at node n. The algorithm is expressed as a *Boolean* function: explore(n). If explore(n) returns true, then, it shows that node n is a part of the solution. This essentially means that node n is one of the nodes on a path from the root to some goal node. If *explore* (n) returns false, then, it shows that there is no path that contains n to any goal node. Now, we write the algorithm for backtracking.

```
Algorithm explore(n)
     Input: node n
     Output: Returns true if node 'n' present else returns false
begin
    if (n is a leaf node)
           if (leaf is our goal node)
               return true;
           else
               return false;
           end-if
    else
        for each child x of n
            if explore(x) succeeds
                  return true
            end-if
        end-for
        return false
   end-if
end explore
```

Let us understand the working of this algorithm. We notice that if *explore*(x) is true for any child of n, then, *explore*(n) will be true. And, if *explore*(x) is false for all children x of n, then *explore*(n) will also be false. Hence, to decide whether any nonleaf node n is a part of the path to a goal node, we have to check whether, for any child x of n, *explore*(x) is true. This is carried out recursively on each child of n. In the above algorithm, this is done in the last part of the algorithm.

Finally, the recursion will terminate at the leaf node, say l. If the leaf node l is a goal node, then, *explore*(l) returns true. If the leaf node l is not a goal node, then, *explore*(l) will return false. This is the base case or a terminating condition of the algorithm. In the algorithm, it is written at the beginning of the algorithm.

When the algorithm finds a goal node, the path information can be obtained following the sequence of recursive calls in the reverse order. As the recursion unrolls with the termination of the algorithm, the path can be recovered in one node at a time by making a note of the node at the current level.

15.5.1.2 Backtracking Algorithm: Nonrecursive Algorithm Using a Stack

Earlier, we have seen a recursive algorithm for backtracking. As we know, any recursive algorithm can be converted into a nonrecursive algorithm; here, we present an iterative version of the backtracking algorithm using a stack. In fact, while compiling, a recursive algorithm is translated into a machine or an assembly language using a stack.

```
Algorithm explore(n)
     Input: node n
     Output: Returns true if node 'n' present else returns false
begin
     push(S,n) /*Push node n in stack S*/
     while (stack is not empty)
     {
        x = top(S) /*top(S) returns the top element of the stack S
        without removing it from the stack. */
        if (x is a leaf node )
            if (x is the goal node)
                     return true
        else
                     pop(S)
        end-if
     else
         if there exists an unexplored child of x, say y
              push(S,y)
          else
              pop(S)
          end-if
    end-if
  return false
end explore
```

The algorithm starts from the root. The only nodes that can be pushed into the stack are the children of the node that are currently present on top of the stack. Also, only one child is pushed into the stack at a time. Hence, the nodes present in the stack at any time describe a valid path in the tree. A node is removed from the stack only when it is concluded that there is no goal node among its descendents. Hence, if it is discovered that the root node has been removed from the stack making it empty, then, it can be concluded that there is no goal node present in the tree or in other words, there is no solution that exists for the problem. If the algorithm terminates successfully, then, the reverse order of the nodes present in the stack forms a path from the root to a goal node.

15.5.2 Solving 4: Queen Problem by Backtracking

The n-queen problem can be stated as follows. We are given with an n-by-n board and n queens. The problem is to place the n queens $Q_1, Q_2, \ldots, Q_n$ on the board in such a way that they do not attack each other. When n is equal to 4, the problem is called the 4-queen problem. Let us formulate the solution to this problem. So, here, we are given with a 4-by-4 board and four queens, and we need to place all four queens in such a way that they do not attack on each other and every column gets at least one queen.

Let us say start by putting the first queen at A_1. On the left below is the actual board, whereas on the right is the board showing all the available and attacking positions with respect to the already-placed queens. The positions painted as dark gray positions are the ones where a new queen cannot be placed. These are the attacking positions for Q_1.

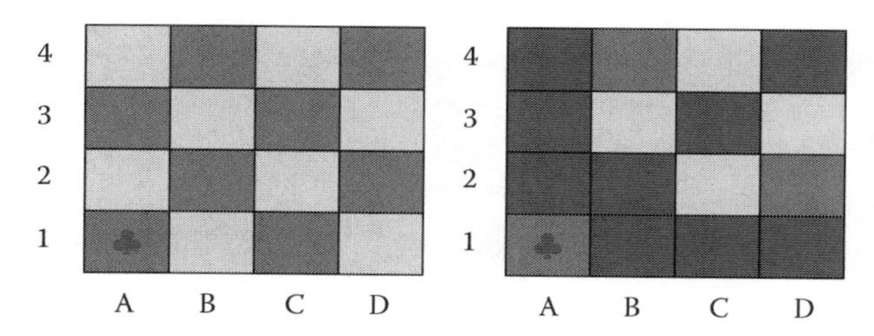

While putting the new queen on the board, we need to check the positions on which Q_1 is attacking. So, the next queen Q_2 has two options available and they are B_3 or B_4. Let us choose B_3 first.

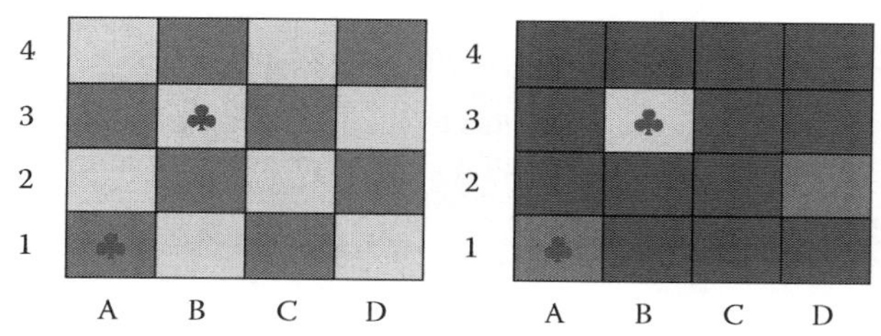

As mentioned earlier, dark gray color-painted cells depict the prohibited or attacking positions with respect to Q_1 and Q_2. Given this situation, it is clear that one cannot place the third queen on the third column. Since we need to have a queen for each column, this does not lead to a feasible solution.

At this situation, the constraints are in such a way that no longer we can satisfy them to reach the final solution. Therefore, we need to rearrange the placement of the queens on the board to move forward. Now, we need to find out what we need to change. As we have observed, the problem has happened after placing Q_2, let us first try out with this queen. Remember, there were two possible places for Q_2 to be placed and they were B_3 and B_4. We have checked by putting Q_2 at B_3 and found that this creates a problem for the third queen. So, now, there is only one position left for Q_2 to be placed, and hence, let us keep Q_2 at cell B_4. The left board below shows the two queens placed and the right board shows the free and attacking positions for the new queen with respect to the already-placed queens.

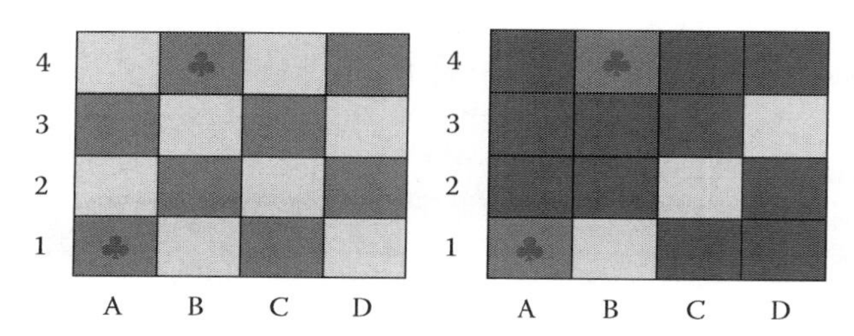

With this setup, we can see that we have a position at C_2 where Q_3 can be placed without any conflict. However, the placement of Q_3 at C_2 will make it impossible for Q_4 to be placed. This is because after the placement of Q_3 at C_2, there will not be any feasible position left in column 4 for Q_4 to be placed. This shows that just the placement of Q_2 at B_4 that was the only one position left does not help. Hence, backtracking only one step does not help. We need to backtrack one more step.

The second step of backtrack is necessary as there is no position for Q_2 to explore that will give any feasible positions for Q_3 or Q_4 to be placed without any conflict. This suggests that we need to change the position of Q_1 to look for the new possibilities. As can be recalled, we have started with placing Q_1 at A_1. Now, let us try continuing upward and placing Q_1 at A_2.

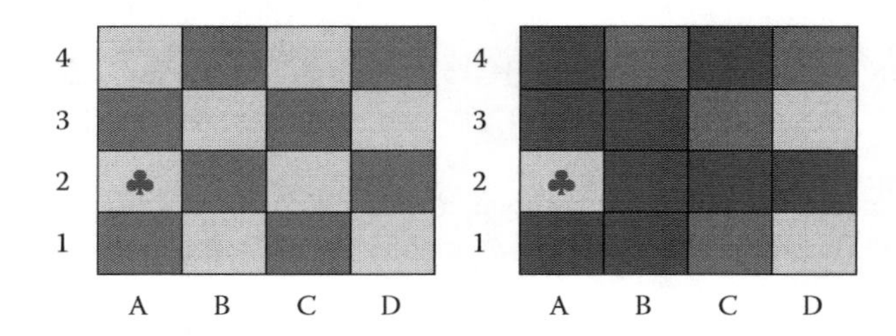

It is now trivial to place the other left-out queens. Now, Q_2 can be placed at B_4. Further, Q_3 can be safely placed at C_1 whereas Q_4 can be kept at D_3. The resultant board looks as follows. We can notice that to reach this solution, we had performed two backtracks.

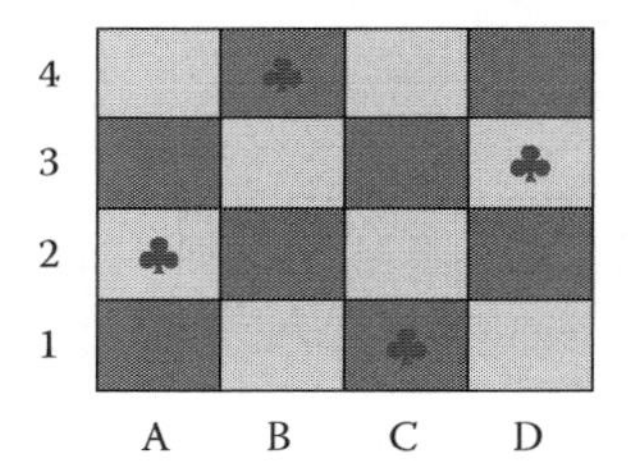

This obtained solution is one among the various possible solutions. Now, the question is what we should do to compute other solutions. And, again, the answer to this question lies in backtracking.

We need to restart in the reverse order in the above-obtained solution and consider placing Q_4 somewhere else. Let us say that we place Q_4 in some location up in column 4. We can notice that this is not possible as it comes to the attacking positions of Q_2. Hence, we backtrack to Q_3 and attempt to get an acceptable place for Q_3 that is different from C_1. We notice that it is not possible to allocate any other position to Q_3 other than C_1. If Q_3 is moved to any other location, then, it comes in the attacking position of some other queen. Hence, again, we need to backtrack to Q_2. For Q_2 also, we see that hence there is no other choice and finally we land up by replacing Q_1. We place Q_1 on A_3.

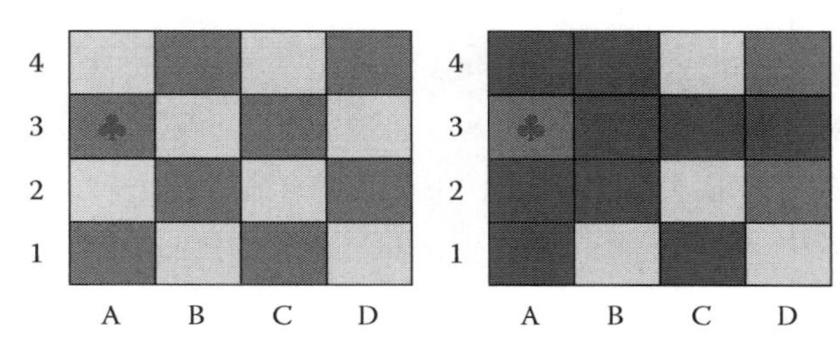

The board above on the right shows the cells shaded in dark gray color as the attacking positions after the placement of Q_1 at A_3. This leaves only position B_1 left for Q_2 to be placed in column 2.

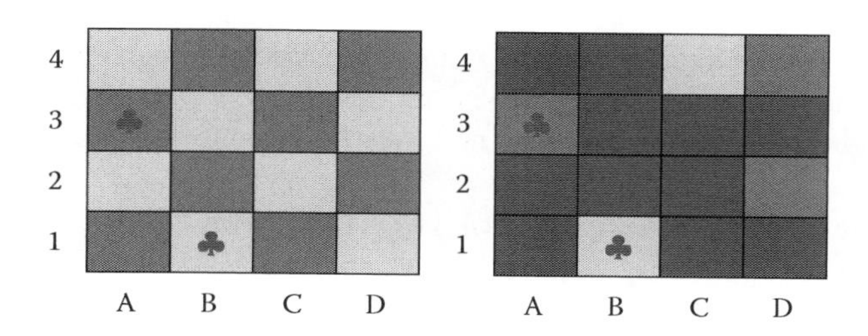

The board above on the right shows the attacking positions (painted as dark gray color) and free positions after placing Q_2 at B_1 in column 2. After the placement of two queens Q_1 and Q_2 in column 1 and column 2, respectively, we shall place the third queen Q_3 in column 3. It can be trivially seen that C_4 is the only place left where Q_3 can be placed. Hence, we put Q_3 at C_4 in column 3.

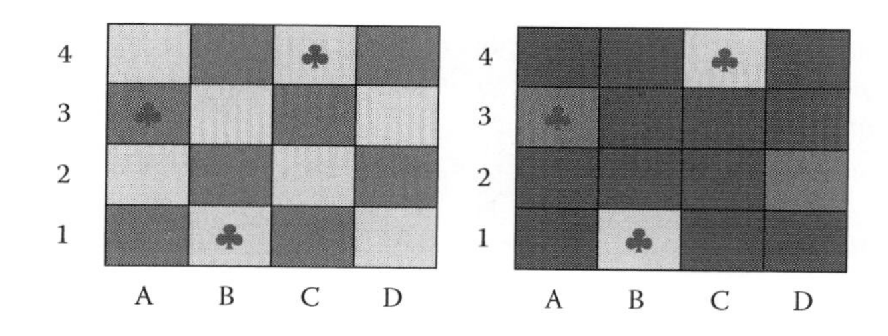

Here, the board above on the right shows the attacking positions (painted as dark gray color) and free positions after placing Q_3 at C_4 in column 3. After the placement of Q_3, D_2 is the only one position left in column 4 that is not an attacking position and hence we can finally place Q_4 at D_2. Hence, the final solution looks as follows.

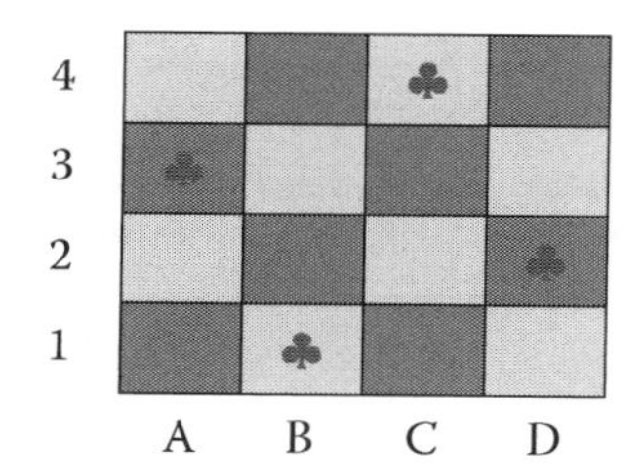

15.6 NP-Completeness

Now, we shall discuss two families of problems, which are very important in theoretical computer science. These two families that constitute the bulk of our practical computational problems have been central to the theory of computation for many years.

15.6.1 P-Problem

A computational problem is said to be a P-problem if there exists an algorithm that solves the problem in polynomial time using a deterministic Turing machine. In other words, the algorithm to solve that problem is such that the number of steps of the algorithm is bounded by a polynomial in n, where n is the length of the input, that is, the time complexity $T(n)$ of the algorithm is expressed as $T(n) = O(n^c)$, where c is a positive constant. Few examples of P-problems are given below:

- Sorting algorithms: Time complexity $O(n^2)$ or $O(n \log n)$
- All-pairs shortest path: Time complexity $O(n^3)$
- Minimum cost spanning tree: Time complexity $O(E \log E)$ or $O(E^2)$
- Searching of a number in a list of n numbers: Time complexity $O(n)$ or $O(\log n)$
- Multiplication of matrices: Time complexity $O(n^3)$

P-problems are the problems that can be efficiently solved using computers without any practical difficulty. If we think about the practical problems that are solved in reasonable time by a computer for a reasonably large input, then, they require not more than $O(n^3)$ or $O(n^4)$ time. In fact, it has been observed that the time complexity for most of these important algorithms lies somewhere in the range of $O(\log n)$ to $O(n^3)$. Thus, we can say that the practical computation resides within polynomial-time bounds.

15.6.2 NP-Problem

A problem is called nondeterministically polynomial (NP) if a nondeterministic Turing machine can be found that can solve the problem in a polynomial number of nondeterministic moves. In other words, a problem is called an NP-problem if it can be solved by a nondeterministic Turing machine in polynomial time. An alternate definition of an NP-problem can be given as follows without using the notion of Turing machine. A problem is called an NP-problem if a candidate solution of the problem comes from a finite set of possibilities, and it requires polynomial time to verify the correctness of the solution.

NP-problems include all nondeterministically polynomial problems whereas P-problems include all deterministically polynomial problems. Clearly, it can be concluded that P is a subset of NP. There exists a very famous and open question in theoretical computer science if P = NP?

NP-problems make an important class of problems that we know how to solve in exponential time using a deterministic Turing machine; however, we do not know how to solve them in polynomial time. Moreover, it is not even known if these problems can be solved in polynomial time at all.

Few examples of NP-problems are given below:

- *Integer factorization problem*: This problem involves determining given two integers n and k, whether there is a factor f with $1 < f < k$ and f dividing n?
- *Graph isomorphism problem*: This problem involves determining whether two graphs can be drawn identically.
- *Traveling-salesman problem*: This problem involves determining if there is a route of some length that goes through all the nodes in a certain network.
- *Boolean satisfiability problem*: This problem involves determining whether or not a certain formula in propositional logic with Boolean variables is true for some value of the variables.

15.6.3 NP-Complete Problem

A problem is said to be NP-hard if an algorithm for solving it can be translated into the algorithm for solving any other NP-problem. It is relatively easier to demonstrate that a problem is NP than to show that it is NP-hard. A problem that is both NP and NP-hard is called an NP-complete problem. There are several computational problems that have been shown to be NP-complete. Few examples of NP-complete problems are given below:

- Traveling-salesman problem
- Boolean satisfiability problem
- *Graph-coloring problem*: It is a special case of graph labeling that involves the assignment of labels popularly called "colors" to the elements of a graph subjected to certain constraints.
- *Clique problem*: It refers to the problem of finding out whether there exists a clique (a complete subgraph) of a given size in a graph.

15.7 Approximation Algorithms

A deterministic algorithm always produces the correct solution and the number of computational steps executed by the algorithm is the same for different executions of the algorithm with the same input data. Figure 15.9 shows the block diagram of a deterministic algorithm. Practically, it is realized many a times that for a given computational problem, it may be difficult to design a deterministic algorithm with a good running time. Also, sometimes, a deterministic algorithm may perform reasonably well for a small size of input data; however, the running time may go very high as the number of inputs increase. *Approximation algorithms* and *randomized algorithms* are the solutions to these problems. This section introduces the concept of an approximation algorithm whereas the concept of a randomized algorithm is discussed in the next section.

Approximation algorithms are the types of algorithms that are used to find approximate solutions to optimization problems. These algorithms are often associated with NP-complete problems. The running time required by the deterministic algorithms for the NP-complete problem may be very very high. Sometimes, these algorithms may take days and months to terminate and to produce the output for a given input. Though a deterministic algorithm always produces the correct solution, it becomes practically infeasible many a times due to its huge running time. An approximation algorithm is a way to deal with NP-completeness for optimization problems. The aim of an approximation algorithm is to reach the optimum solution as close as possible in a reasonable amount of time that is at most polynomial in terms of the number of inputs. It should be noted that an approximation algorithm does not guarantee the best solution for a problem. Since it is believed that there can never be efficient polynomial-time exact deterministic algorithms for solving NP-complete problems, it is often acceptable to have at least polynomial-time suboptimal solutions for an NP-complete problem.

Approximation algorithms should not be confused with heuristic-based algorithms. Heuristic-based algorithms often produce good solutions that are reasonably fast; however, sometimes, it may

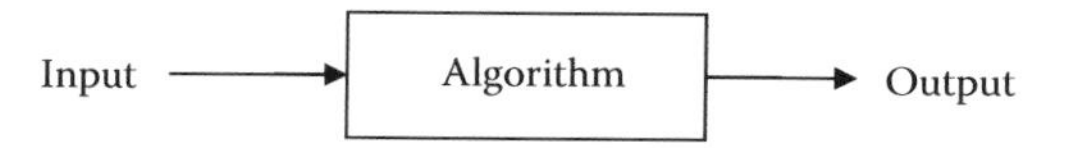

FIGURE 15.9
Block diagram of a deterministic algorithm.

be required to have a provable solution quality and provable run-time bounds for a problem. Approximation algorithms give the guarantee to be optimal up to a small constant factor. For example, an approximation algorithm may be proved to be capable of producing an optimal solution within a constant factor of 3%.

NP-complete problems vary quite a lot in their approximation ability. For example, there are some problems such as the *bin-packing problem* that can be approximated within any factor greater than 1. Approximation algorithms that can be approximated within any factor greater than 1 are often called polynomial-time approximation algorithms. However, there also exist some problems where there is no possibility to achieve an approximate solution within any constant factor, or even a polynomial factor unless P = NP. For example, the *maximum clique problem* is one such problem where achieving an approximate solution within a constant factor is not possible.

Though approximation algorithms provide workable solutions most of the time, it is important to note that all the approximation algorithms are not appropriate for practical applications. Also, there are some approximation algorithms that have very impractical running times in spite of being polynomial on the input size. For example, an approximation algorithm with $O(n^{2156})$ time complexity (Vazirani 2003) is impractical.

15.7.1 Example of an Approximation Algorithm

Approximation algorithms are mostly used for problems where the exact polynomial-time deterministic algorithms are known but are found to be very expensive for a reasonably large input size. A classical example of an approximation algorithm is finding out the *minimum vertex cover* in a graph (Dinur and Safra 2005). The vertex cover for a graph is defined as the set of vertices such that each edge of the graph is incident to at least one vertex of the set. Computation of the vertex cover for a graph deals with finding an edge in the graph that is not considered yet and adding both endpoints of the edge to the vertex cover, until no such edge remains. It is clear that the resulting cover will be at most two times as large as the optimal one (the minimum vertex cover). So, this algorithm is a constant-factor approximation algorithm with a factor of 2.

15.8 Randomized Algorithms

A randomized algorithm makes use of randomness in its logical steps. For this purpose, it uses random bits as an auxiliary input to guide its behavior. Randomness is introduced in the hope of obtaining a good performance of the algorithm in the *average case* over all the possible choices of random bits. The performance of a randomized algorithm can be considered to be a random variable determined with the help of random bits. A randomized algorithm may have either the running time or the output or both as random variables. Figure 15.10 shows the block diagram of a randomized algorithm.

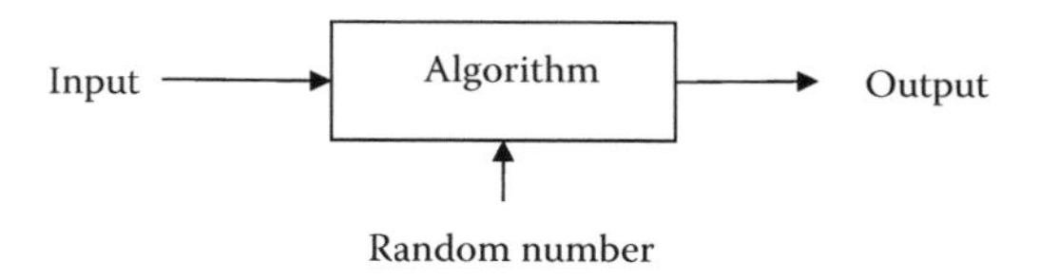

FIGURE 15.10
Block diagram of a randomized algorithm.

It is important to note that randomized algorithms are not the probabilistic analysis of the expected running time of a deterministic algorithm. The objective in the probabilistic analysis is very different from a randomized algorithm where the aim is to compute the expected running time of the algorithm. The input to this analysis is assumed to come from a probability distribution.

15.8.1 Example: Verification of Polynomial Identities

A very basic example of a randomized algorithm, described in Mitzenmacher and Upfal (2005), is presented here. The problem is to design a randomized algorithm for the following problem. We are given with two products of polynomials and our task is to determine if both the polynomials compute the same function. For example, we have two polynomials $p(x)$ and $q(x)$ defined as follows:

$$p(x) = (x-7)(x-3)(x-1)(x+2)(2x+5)$$

$$q(x) = 2x^5 - 13x^4 - 21x^3 + 127x^2 + 121x - 210$$

Both the polynomials $p(x)$ and $q(x)$ are of degree 5. Just by looking at the polynomials, it is not trivial to tell if both of them are the same or not without multiplying out the factors of $p(x)$. However, assuming that each integer multiplication takes a unit time, multiplying out all the factors of $p(x)$ may take $O(d^2)$ running time, d being the degree of the polynomial.

There can be an alternate approach to do this task using randomization. The algorithm is based on the following facts. The evaluation of $p(x)$ and $q(x)$ takes only $O(d)$ integer operations, and we can have $p(x) = q(x)$ only when either

- $p(x)$ or $q(x)$ are the same polynomial
- x is a root of polynomial $p(x) - q(x)$

As we know that at most, the polynomial $p(x) - q(x)$ has degree d, it cannot have more than d roots. So, if x is uniformly chosen at random from a larger range, then, it is probable that we will not get a root. For example, just the evaluation of $p(11) = 1{,}12{,}320$ and $q(11) = 1{,}20{,}306$ shows that p and q are not the same.

15.8.2 Advantage of a Randomized Algorithm

There are some practical challenges associated with the implementation of randomized algorithms. For example, there is always a finite probability of getting the wrong results in a randomized algorithm. In spite of the challenges, randomized algorithms offer several advantages. Some of them are listed below:

- Randomized algorithms are simple and easy to implement.
- Randomized algorithms are very fast with a very high probability.
- The output produced by randomized algorithms is optimum with a very high probability. The probability of obtaining a wrong output can be made arbitrarily small by the repeated use of randomness.

The quality of the result of a randomized algorithm is highly dependent on the quality of the random numbers. However, the computation of truly random numbers is practically impossible and one needs to be dependent on pseudorandom numbers. In practice, since randomized algorithms are approximated using pseudorandom numbers in place of true random bits, such an implementation of a randomized algorithm is expected to deviate from the ideal theoretical behavior.

References

Dinur, I. and S. Safra, On the hardness of approximating minimum vertex cover, *Annals of Mathematics*, 162(1), 439–485, 2005.

Hoare, C. A. R., Algorithm 64: Quicksort, *Communication of the ACM*, 4(7), 321, 1961.

Knuth, D., Section 5.2.4: Sorting by merging, sorting and searching, In Sloane, N. J. A., *The Art of Computer Programming*, 2nd edition, Addison-Wesley, Reading, MA, 158–168, 1998.

Kruskal, J. B., On the shortest spanning subtree of a graph and the traveling salesman problem, *Proceedings of the American Mathematical Society*, 7, 48–50, 1956.

Mitzenmacher, M. and E. Upfal, *Probability and Computing: Randomized Algorithms and Probabilistic Analysis*. Cambridge University Press, New York, NY, 2005.

Powell, W. B., *Approximate Dynamic Programming: Solving the Curses of Dimensionality*, 2nd edition, Wiley-Interscience, Hoboken, NJ, 2011, ISBN: 978-0-470-60445-8.

Prim, R. C., Shortest connection networks and some generalizations, *Bell System Technical Journal*, 36, 1389–1401, 1957.

Vazirani, V. V., *Approximation Algorithms*, Springer, Berlin, Germany, 2003, ISBN: 3-540-65367-8.

16

Data Envelopment Analysis: An Overview

Subhash C. Ray

CONTENTS

ABSTRACT Over the past few decades, data envelopment analysis (DEA) has emerged as an important nonparametric method of evaluating the performance of decision-making units through benchmarking. Although developed primarily for measuring the technical efficiency, DEA is now applied extensively for measuring the scale efficiency, cost efficiency, and profit efficiency as well.

This chapter integrates the different DEA models commonly applied in empirical research with their underlying theoretical foundations in neoclassical production economics. Under the assumptions of the free disposability of inputs and outputs and convexity of the production possibility set, the free-disposal convex hull of the observed input–output bundles provides an empirical "estimate" of the technology set. The graph of the technology provides an appropriate benchmark for the efficiency evaluation under alternative orientations. The radial models consider either a proportional upward scaling of all the outputs or a downward scaling of all the inputs and the efficiency measure does not reflect the presence of slacks. Nonradial models allow different inputs to decrease or individual outputs to increase at different rates, thereby ruling out input or output slacks at the optimal solution. Unless constant returns to scale are explicitly assumed, (locally) increasing, constant, or diminishing returns will hold at different points on the graph of the technology. A point where locally the constant returns to scale hold is the most-productive scale size. One can characterize a firm as too small, if it is located at a point on the frontier where increasing returns to scale holds or if both input- and output-oriented projections onto the frontier (when the firm is inefficient) are points in the region of the increasing returns. The firm is too large, when both projections are in the region of the diminishing returns to scale. A simple DEA model has to be solved to determine the scale efficiency of a firm as well as its appropriate scale characterization. Oriented models (both radial and nonradial) prioritize either the input conservation or the output expansion. As the models are based on the graph-hyperbolic distance function, directional distance function, geometric distance function, or Pareto–Koopmans efficiency, one looks for the output increase and input decrease simultaneously. The directional distance function, in particular, has emerged as the most popular analytical framework for measuring the efficiency when the objective is to expand the desirable or "good" output and reduce the undesirable of a "bad" output simultaneously. Finally, the chapter deals with the measurement of cost or profit efficiency in a market economy where the input and output prices are available.

16.1 Introduction

Data envelopment analysis (DEA) is a nonparametric method of evaluating efficiency in resource utilization. In this approach, the efficiency of a decision-making unit (DMU) is evaluated by comparing the outcome of the actual decision with what is deemed to be the best-achievable outcome. It is obvious that what constitutes the "best" outcome depends on the objective of the decision maker on one hand, and on the set of alternatives from which a particular decision is selected, on the other. The nature of decision making determines what are the choice variables and what are the parameters in a given context. For example, the manager of a primary healthcare center in a rural area might be assigned a given amount of resources (in the form of physicians and supplies) and the objective would be to immunize as many infants as possible. In this case, the efficiency would be measured by the ratio of the actual number of infants immunized and the maximum number possible given the resources. This is an output-oriented technical efficiency. A different example would be one where a landscaper has to irrigate a lawn of a given size and the efficiency lies in getting the task completed using the minimum necessary amount of water. This relates to an input-oriented technical efficiency. In the output-oriented case, realizing the full output potential is of a primary importance. In the input-oriented case, conserving the inputs has a priority over expanding the output. In many other cases, there are market prices of inputs reflecting the relative worth of individual inputs. In such cases, the objective may be to produce the target output at the minimum cost. This may actually involve increasing the quantity used of some input so long as the resulting increase in cost is more than the offset by the saving resulting from economizing on the use of a more-valuable input. It may be noted that cost minimization is a valid objective even for the public sector and other nonprofit agencies (such as schools and hospitals) because any cost saving ultimately releases the valuable resources for the production of other outputs. Finally, from the perspective of a business enterprise engaged in producing the outputs for a profit, the bottom line is the amount of profit earned. In this case, all the outputs and all the inputs are choice variables and the only constraint on the producer's behavior is that the input–output bundle selected must be such that it must be technologically possible to produce the planned output from the input bundle selected. There are economic theories of a producer's behavior corresponding to the alternative objectives. Correspondingly, there are appropriate DEA optimization problems that yield the relevant benchmarks for comparison with the actual outcome for evaluating the efficiency.

The present chapter is organized as follows. Section 16.2 introduces the concept of the production technology and defines the measures of output- and input-oriented technical efficiency. The corresponding DEA linear-programming (LP) models are also formulated. Section 16.3 considers the returns-to-scale properties of the technology in detail and explains how to measure scale efficiency. The most-productive scale size (MPSS) for a given input–output bundle is defined and the alternative ways to identify the nature of the returns to scale at a particular point on the efficient frontier of the technology set are described. Section 16.4 covers the nonradial measures of technical efficiency. Section 16.5 explains the concepts of the distance function, the directional distance function, and the geometric distance function. Section 16.6 deals with the question of invariance of different efficiency measures in light of data transformation. Section 16.6 provides the measures of technical efficiency in the presence of bad or undesirable outputs. Section 16.7 deals with the measurement of cost and profit efficiency. Section 16.8 is the conclusion.

16.2 Production Technology

At the core of productivity and efficiency, analysis is the concept of the production technology described by the production possibility set. Production is the process of converting inputs into outputs. A bundle of inputs ($x \in R_n^+$) is acquired by the producer from outside. It then goes through various parallel or sequential processes of transformation and ultimately exits the jurisdiction of the firm as a finished product in the form of an output bundle ($y \in R_m^+$). A pair of input–output bundles (x, y) is a feasible production plan if the output bundle y can be produced from the input bundle x. The production-possible set (T) includes all feasible production plans. Thus,

$$T = \{(x, y) : x \text{ can produce } y\}. \tag{16.1}$$

Often, the production-possibility set is defined by means of a *production correspondence* or a *transformation function*

$$F(x, y) = \alpha, \tag{16.2}$$

mapping from R_{m+n}^+ to the (0, 1) interval on the real line. The production-possibility set can then be expressed as

$$T = \{(x, y) : F(x, y) \leq 0\}. \tag{16.3}$$

It is assumed that the production correspondence is nonincreasing in the inputs and nondecreasing in the outputs. Thus, $F_i = \partial F/\partial x_i \leq 0$ for each input i and $F_j = \partial F/\partial y_j \geq 0$ for each output j. These are also known as free-disposability assumptions. They imply that if any input–output bundle is feasible, then, an increase in any input not accompanied by a decrease in another input or an increase in any output will not render the new input–output bundle infeasible. This rules out negative marginal productivity. Similarly, a decrease in any output quantity will not affect feasibility. Moreover, if $F(x^0, y^0) = 0$, then, (x^0, y^0) is a technically efficient input–output bundle.

In the single-output case, one uses the *production function*

$$y^* = f(x), \tag{16.4}$$

where y^* is the maximum quantity of the scalar output that can be produced from the input bundle, x. In this case,

$$T = \{(x, y) : y \leq f(x)\}. \tag{16.5}$$

In parametric analysis, one specifies an explicit form of the production function and uses statistical estimation techniques such as the maximum likelihood procedure to calibrate the parameters of the specified function using sample data of the inputs and outputs.* In DEA, one avoids any kind of functional specification and instead makes a number of quite general assumptions about the nature of the underlying production technology to construct the production-possibility set from sample data.

Let the data set $D = \{(x^j, y^j); j = 1, 2, \ldots, N\}$ be the set of observed input–output bundles of N firms from a particular industry. The following assumptions are made about the production technology:

1. Each observed input–output bundle is feasible.
2. The production-possibility set is convex.
3. Inputs are freely disposable. That is, if $(x^0, y^0) \in T$ and $x^1 \geq x^0$, then $(x^1, y^0) \in T$.
4. Outputs are freely disposable. That is, if $(x^0, y^0) \in T$ and $y^1 \leq y^0$, then $(x^0, y^1) \in T$.

There would, of course, be infinitely many sets satisfying these assumptions. In DEA, T is estimated by the set

$$S = \left\{ (x, y) : x \geq \sum_{j=1}^{N} \lambda_j x^j ; y \leq \sum_{j=1}^{N} \lambda_j y^j ; \sum_{j=1}^{N} \lambda_j = 1; \lambda_j \geq 0; (j = 1, 2, \ldots, N) \right\}. \tag{16.6}$$

It is the smallest convex set containing the observed data points and satisfying the free-disposability assumption.† It is also known as the free-disposal convex hull of the set D.

* The stochastic production function was introduced by Aigner et al. (1977) and Meeusen and van den Broeck (1977). For an excellent and comprehensive exposition of the stochastic frontier analysis (SFA), see Kumbhakar and Lovell (2000).

† In the DEA literature, it is common to refer to "minimum extrapolation" as if it is one of the assumptions about the technology. In reality, it is merely an "estimation" criterion, much like the maximum likelihood principle in SFA.

16.2.1 Technical Efficiency

The *output-oriented* technical efficiency* of a firm producing the output bundle y^0 from the input bundle x^0 is measured as

$$\tau_y(x^0, y^0) = \frac{1}{\varphi^*}, \tag{16.7}$$

where

$$\begin{aligned} \varphi^* &= \max \varphi \\ \text{s.t.} \quad & \sum_{j=1}^{N} \lambda_j y^j \geq \varphi y^0; \\ & \sum_{j=1}^{N} \lambda_j x^j \leq x^0; \\ & \sum_{j=1}^{N} \lambda_j = 1; \\ & \lambda_j \geq 0; \quad (j = 1, 2, \ldots, N); \quad \varphi \textit{ unrestricted.} \end{aligned} \tag{16.8}$$

Note that by convexity, $\sum_{j=1}^{N} \lambda_j = 1$ and $\lambda_j \geq 0$ ensures that the input–output bundle $\left(\sum_{j=1}^{N} \lambda_j x^j, \sum_{j=1}^{N} \lambda_j x^j\right)$ is feasible. Owing to the free disposability of inputs, $x^0 \geq \sum_{j=1}^{N} \lambda_j x^j \Rightarrow \left(x^0, \sum_{j=1}^{N} \lambda_j y^j\right)$ is also feasible. Finally, due to free disposability of outputs, $\varphi y^0 \leq \sum_{j=1}^{N} \lambda_j x^j \Rightarrow (x^0, \varphi y^0)$ is feasible. The optimal value of the objective function in Equation 16.8 shows the maximum rate by which *all outputs* of the firm can be expanded without any increase in any individual input. When different outputs can be expanded at different rates, φ^* is the lowest of these expansion factors. For example, in a two-output case, if one output can be expanded by a factor of 1.5 and the other can be expanded by a factor of 1.25, then, φ^* equals the lower of the two values. In this case, it is possible to expand the *output bundle* itself by at least 25% across the board and one output can be expanded even beyond that. The output-oriented technical efficiency is 0.80 implying that it is realizing only 80% of the potential output producible from its current input bundle.

An alternative measure of technical efficiency of the firm is its *input-oriented* technical efficiency

$$\begin{aligned} \tau_x(x^0, y^0) &= \min \theta \\ \text{s.t.} \quad & \sum_{j=1}^{N} \lambda_j y^j \geq y^0; \end{aligned}$$

* This is also known as Farrell efficiency named after Farrell (1957) who extended the earlier work by Debreu (1951) and Shephard (1953). However, the multiple-output LP formulation is due to Charnes, Cooper, and Rhodes (CCR) (1978). The model in Equation 16.8 is a generalization of the original CCR model by Banker, Charnes, and Cooper (BCC) (1984) that allows variable returns to scale.

$$\sum_{j=1}^{N} \lambda_j x^j \leq \theta x^0; \tag{16.9}$$

$$\sum_{j=1}^{N} \lambda_j = 1;$$

$$\lambda_j \geq 0; \quad (j = 1, 2, \ldots, N); \quad \theta \text{ unrestricted.}$$

The input-oriented technical efficiency of the firm shows the factor by which the entire input bundle can be scaled down without reducing any output.* In the multiple-input case, it may be possible to reduce some individual inputs even further. In general, the input- and output-oriented technical efficiency measures of a firm will be different.

When the production-possibility set is defined by a transformation function or production correspondence as in Equation 16.3, the *graph* of the technology is

$$G = \{(x, y) : F(x, y) = 0\}. \tag{16.10}$$

The nonparametric version of the graph would be

$$G = \{(x, y) : (x, y) \in T; \beta < 1 \Rightarrow (\beta x, y) \notin T; \alpha > 1 \Rightarrow (x, \alpha y) \notin T\}. \tag{16.11}$$

It is apparent that every $(x, y) \in G$ is technically efficient in both the input and the output orientation. The graph of the technology constitutes the frontier of the production-possibility set.

16.2.2 Constant Returns to Scale

The technology exhibits constant returns to scale (CRS) globally if

$$(x, y) \in T \Rightarrow (kx, ky) \in T \; \forall k \geq 0.$$

An implication of CRS is that any nonnegative radial expansion or contraction of a feasible input–output bundle is also a feasible input–output bundle. Under the CRS assumption, an empirical estimate of the production-possibility set is

$$S^C = \left\{ (x, y) : x \geq \sum_{j=1}^{N} \lambda_j x^j; y \leq \sum_{j=1}^{N} \lambda_j y^j; \lambda_j \geq 0; (j = 1, 2, \ldots, N) \right\}. \tag{16.12}$$

Note the absence of the restriction that the λs add up to unity. When only convexity is assumed, all the weighted averages of the observed input–output bundles are also feasible. It was necessary that the weights add up to 1. Thus, so long as $\sum_{j=1}^{N} \lambda_j = 1$ and each λ_j is nonnegative, $\left(\sum_{j=1}^{N} \lambda_j x^j, \sum_{j=1}^{N} \lambda_j y^j\right)$ is feasible. But with the added assumption of CRS, $\left(k \sum_{j=1}^{N} \lambda_j x^j, k \sum_{j=1}^{N} \lambda_j y^j\right)$ is also feasible for any $k \geq 0$. Now, consider the weights $\mu_j = k\lambda_j$, $k \geq 0$. CRS implies that $\left(\sum_{j=1}^{N} \mu_j x^j, \sum_{j=1}^{N} \mu_j y^j\right)$ is feasible. But $\sum_{j=1}^{N} \mu_j = k$ that can be any nonnegative number. This explains why the weights need not add up to 1 in Equation 16.12.

* As shown in a later section, it is the inverse of the Shephard (input) distance function.

The set S^C is sometimes described as the free-disposal conical hull of D.
The output-oriented CRS technical efficiency* is

$$\tau_y^C(x^0, y^0) = \frac{1}{\varphi_C^*}, \tag{16.13}$$

where

$$\begin{aligned} \varphi_C^* &= \max \varphi \\ \text{s.t.} \quad & \sum_{j=1}^{N} \lambda_j y^j \geq \varphi y^0; \\ & \sum_{j=1}^{N} \lambda_j x^j \leq x^0; \\ & \lambda_j \geq 0; \quad (j = 1, 2, \ldots, N); \quad \varphi \textit{ unrestricted.} \end{aligned} \tag{16.14}$$

Similarly, the input-oriented CRS technical efficiency is

$$\begin{aligned} \tau_x^C(x^0, y^0) &= \min \theta \\ \text{s.t.} \quad & \sum_{j=1}^{N} \lambda_j y^j \geq y^0; \\ & \sum_{j=1}^{N} \lambda_j x^j \leq \theta x^0; \\ & \lambda_j \geq 0; \quad (j = 1, 2, \ldots, N); \quad \theta \textit{ unrestricted.} \end{aligned} \tag{16.15}$$

It is easy to verify that under the CRS assumption, input- and output-oriented measures of technical efficiency are identical.

Figure 16.1a and b graphically explains the concepts of output- and input-oriented technical efficiency for variable returns to scale (VRS) in the one-input one-output case. In Figure 16.1a, the curve $y^* = f(x)$ represents the production function. Point A represents the input–output combination (x_A, y_A). The maximum output producible from input x_A is $y_A^* = f(x_A)$. The efficient input–output bundle (x_A, y_A^*) is shown by the point A^* on the frontier. The output-oriented technical efficiency of the firm A is measured as $\tau_y^A = y_A/y_A^* = Ax_A/A^*x_A$. Similarly, for firm B, the actual output produced from input x_B is y_B and the maximum producible is y_B^*. The output-oriented technical efficiency of firm B is $\tau_y^B = y_B/y_B^* = Bx_B/B^*x_B$. Further, the minimum quantity of the input needed for producing output y_A is $x_A^* = f^{-1}(y_A)$. Hence, the input-oriented technical efficiency of A is $\tau_x^A = x_A^*/x_A = Cy_A/Ay_A$. The corresponding input-oriented technical efficiency of B is $\tau_x^B = x_B^*/x_B = Dy_B/By_B$. It is clear that, in general, the input- and output-oriented technical efficiencies of the same firm will differ.

Figure 16.1b illustrates the case of CRS. Here, the production function is $f(x) = kx$. It is clear from the properties of similar triangles that Ox $\tau_x = \tau_y$ for every firm.

---

* This is the CCR model.

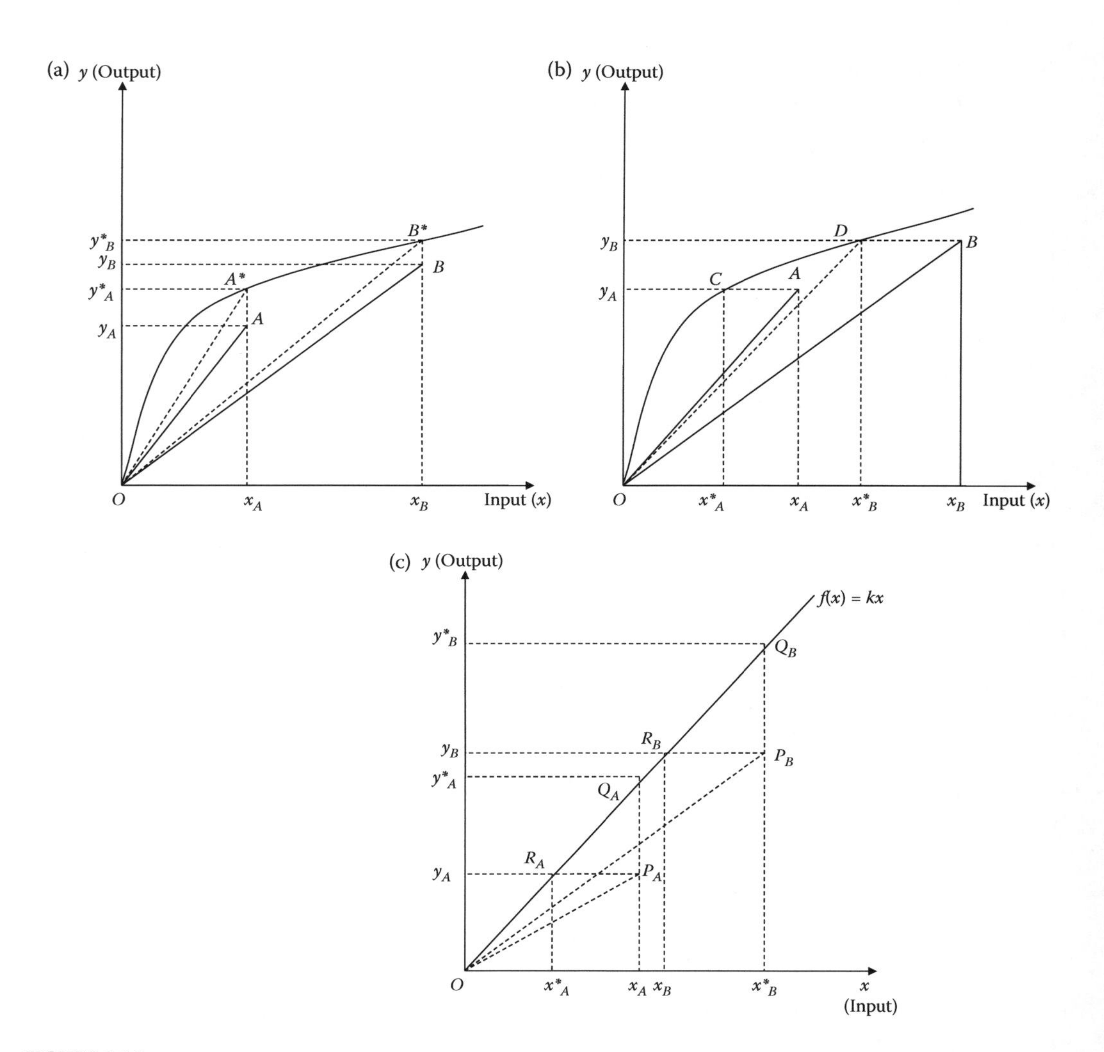

FIGURE 16.1
(a) Output-oriented technical efficiency. (b) Input-oriented technical efficiency. (c) Technical efficiency under CRS.

The dual of the maximization problem in Equation 16.14 is†

$$\begin{aligned} &\min u^{0'} x^0 \\ \text{s.t.} \quad & u^{0'} x^j - v^{0'} y^j \geq 0; \quad (j = 1, 2, \ldots, N); \\ & v^{0'} y^0 = 1; \\ & u^0, \quad v^0 \geq 0. \end{aligned} \tag{16.16}$$

Several points need special attention in this problem. First, x^j is the observed input vector and y^j is the corresponding output vector of firm j. All these are parameters. Also, (x^0, y^0) is one of the (x^j, y^j) bundles. The vectors u^0 and v^0 are the choice variables in this problem. These can

† This is the *multiplier form* of the model.

be interpreted as the *shadow prices* of the inputs and outputs. Second, the optimal values of these dual variables depend on the values of (x^0, y^0). That is the reason why they are superscripted. In Equation 16.16, these shadow prices of the inputs and outputs are chosen in such a way that evaluated at these prices:

a. The shadow value of the output bundle y^0 equals to unity.
b. The shadow value of any observed output bundle cannot exceed the shadow cost of the corresponding input bundle. This is true for the (x^0, y^0) bundle as well.

The problem in Equation 16.16 can be easily recast as the original linear-fractional functional-programming problem:

$$\begin{aligned} &\min \frac{u^{0\prime}x^0}{v^{0\prime}y^0} \\ \text{s.t.} \quad &\frac{u^{0\prime}x^j}{v^{0\prime}y^j} \geq 1; \quad (j = 1, 2, \ldots, N); \\ &u^0, \quad v^0 \geq 0. \end{aligned} \tag{16.17}$$

This can also be expressed as

$$\begin{aligned} &\max \frac{v^{0\prime}y^0}{u^{0\prime}x^0} \\ \text{s.t.} \quad &\frac{v^{0\prime}y^j}{u^{0\prime}x^j} \leq 1; \quad (j = 1, 2, \ldots, N); \\ &u^0, \quad v^0 \geq 0. \end{aligned} \tag{16.18}$$

This is the so-called *ratio form* of the DEA problem introduced by CCR in their pioneering 1978 paper.* The LP problems in Equations 16.14 and 16.15 above are generally called the CCR output- and input-oriented DEA models. In contrast, the previous models in Equations 16.8 or 16.9 are the corresponding BCC models named after Banker et al. (1984). The CCR models are known as the CRS problems. Because no specific assumption is made about the returns to scale in the BCC models, they are described as variable returns to scale (or VRS) problems.

16.2.3 Multipliers as Shadow Prices

It may be noted that the inverse of the output-oriented proportional expansion factor in Equation 16.14 is clearly a technical efficiency measure. In contrast, the ratio measure in Equation 16.18 is a total productivity measure. It is true that by the standard-duality results, they can be shown to be mathematically equivalent. But as noted by Førsund (2013), they are conceptually quite different. The Farrell efficiency measure directly relates to the frontier of the production-possibility set. But, there is no such obvious link in the case of the ratio measure. This is one reason as to why there is so much confusion about the interpretation of the aggregation weights in the CCR model in its multiplier form in the operations research (OR) literature and people have tried to impose arbitrary weight restrictions to avoid zero weights. However, as shown below, the weights come from the gradient of a supporting hyperplane at the efficient projection of an observed input–output bundle and,

* They use a normalization that was first considered in Charnes and Cooper (1968).

contrary to the popular belief, are far from arbitrary. As shown by Førsund (2013), the *ratios* of the multipliers associated with two inputs show the marginal rate of substitution between those inputs. Similarly, for a pair of outputs, the ratio of the multipliers shows the marginal rate of transformation between them.

16.2.4 Slacks

An inherent problem with inequality-constrained optimization is that more often than not, some of the constraints prove to be nonbinding at the optimal solution. This results in the presence of slacks in the input and/or the output constraints. As such, the presence of positive slacks at the optimal solution of an LP problem poses no particular problem. After all, when constraints are in the form of weak inequalities, it is only common to end up with slacks. In DEA, however, the presence of slacks in the output or input constraints has been a matter of concern right from the beginning. This is because of what they imply about the measured technical efficiency of the firm. Consider, for example, the following two-input one-output BCC input-oriented problem:

$$
\begin{aligned}
& \min \theta \\
\text{s.t.} \quad & \sum_{j=1}^{N} \lambda_j x_{1j} \leq \theta x_{10}; \\
& \sum_{j=1}^{N} \lambda_j x_{2j} \leq \theta x_{20}; \\
& \sum_{j=1}^{N} \lambda_j y_{2j} \geq y_0; \\
& \sum_{j=1}^{N} \lambda_j = 1; \quad \lambda_j \geq 0; \quad (j = 1, 2, \ldots, N).
\end{aligned}
\tag{16.19}
$$

Now, suppose that at the optimal solution, $\sum_{j=1}^{N} \lambda_j^* x_{1j} = x_{10}$ but $\sum_{j=1}^{N} \lambda_j^* x_{2j} = 0.5x_{20}$. Obviously, in this case, θ^* equals to unity and the firm under an evaluation is considered to be operating at 100% efficiency. This, of course, is quite difficult to accept given the fact that it can cut down its use of input 2 by half without reducing the output or increasing input 1. To correct this anomaly, CCR (1979) modified their original formulation of the problem so that a firm could be considered efficient only when θ^* (or φ^*, when it is an output-oriented model) was 1 *as well as all the input and output slacks were* 0. To enforce this added requirement, they included a very small penalty (ε) for the presence of any output or input slack in the objective function. A revised version of Equation 16.19 incorporating the slacks explicitly would be

$$
\begin{aligned}
& \min \theta - \varepsilon[s_1^- + s_2^- + s_1^+] \\
\text{s.t.} \quad & \sum_{j=1}^{N} \lambda_j x_{1j} + s_1^- = \theta x_{10}; \\
& \sum_{j=1}^{N} \lambda_j x_{2j} + s_2^- = \theta x_{20};
\end{aligned}
$$

$$\sum_{j=1}^{N} \lambda_j y_{2j} - s_1^+ = y_0; \tag{16.20}$$

$$\sum_{j=1}^{N} \lambda_j = 1;$$

$$s_1^-, s_2^-, s_1^+ \geq 0;$$

$$\lambda_j \geq 0; \quad (j = 1, 2, \ldots, N).$$

Here, (s_1^-, s_2^-) are the input slacks and s_1^+ is the output slack. CCR stipulated that ε should be a *non-Archemdian* (or *infinitesimal*) positive number to ensure that decreasing θ gets a preemptive priority over increasing the input or output slacks in the solution algorithm of the problem. But a practical question is: what numerical value should one use to actually solve the problem in Equation 16.20? No matter how small a value one can choose, it is always possible to pick one that is smaller. Thus, the minimization problem in Equation 16.20 cannot be actually solved. It can be seen, however, that irrespective of any numerical value of ε, the objective is first to minimize θ and then to maximize the sum of the slacks. In practice, the problem is solved in two steps. In step 1, one solves the problem in Equation 16.19 without any concern about the slacks. If there are multiple optimal solutions, then, there will be different vectors λ^* going with the same minimum value θ^*. The objective in the second stage is to select the optimal solution that maximizes the sum of the slacks through the following model:

$$\max \quad s_1^{-1} + s_2^- + s_1^+$$

$$\text{s.t.} \quad \sum_{j=1}^{N} \lambda_j x_{1j} + s_1^- = \theta^* x_{10};$$

$$\sum_{j=1}^{N} \lambda_j x_{2j} + s_2^- = \theta^* x_{20};$$

$$\sum_{j=1}^{N} \lambda_j y_{2j} - s_1^+ = y_0; \tag{16.21}$$

$$\sum_{j=1}^{N} \lambda_j = 1;$$

$$s_1^-, s_2^-, s_1^+ \geq 0;$$

$$\lambda_j \geq 0; \quad (j = 1, 2, \ldots, N).$$

It is important to note that in the second step of the problem θ^* is a parameter rather than a choice variable. The optimal value of θ obtained from Equation 16.19 along with those of the slacks obtained from Equation 16.21 constitutes the optimal solution for Equation 16.20 irrespective of an actual numerical value of ε.

The practical usefulness of this two-step procedure is not very clear. It certainly flags the presence of slacks at an optimal solution of Equation 16.19 even when the optimal value of θ is 1. But it does not provide a more-comprehensive measure of efficiency that incorporates penalties for the presence of slacks. The optimal values of the λs in Equation 16.19 define the technically efficient projection $\left(x_0^* = \sum_{j=1}^{N} \lambda_j^* x^j, y_0^* = \sum_{j=1}^{N} \lambda_j^* y^j\right)$ of the input–output bundle (x^0, y^0). When multiple optimal solutions exist for Equation 16.19, there are many such projections. The problem in step 2 helps to select among them. The CCR method provides no justification for the criterion of choice behind the problem in Equation 16.21.

An alternative to Equation 16.20 would be the so-called *additive model* that simply maximizes the sum of the input and output slacks and is formulated as follows:

$$\begin{aligned}
&\max \sum_{r=1}^{m} s_r^+ + \sum_{i=1}^{n} s_i^- \\
\text{s.t.}\quad &\sum_{j=1}^{N} \lambda_j y_{rj} - s_r^+ = y_{r0}; \quad (r = 1, 2, \ldots, m); \\
&\sum_{j=1}^{N} \lambda_j x_{ij} + s_i^- = x_{i0}; \quad (i = 1, 2, \ldots, n); \\
&\sum_{j=1}^{N} \lambda_j = 1; \\
&s_i^- \geq 0, \quad (i = 1, 2, \ldots, n); \\
&s_r^+ \geq 0, \quad (r = 1, 2, \ldots, m); \\
&\lambda_j \geq 0; \quad (j = 1, 2, \ldots, N).
\end{aligned} \tag{16.22}$$

Although sometimes used in the literature, this additive model is useless as a measure of inefficiency.* The objective function is the sum of input and output slacks that are measured in heterogeneous units and has no meaning. Its only usefulness lies in the fact that this will be 0 only when the θ^* in Equation 16.20 (or φ^* in an output-oriented model) equals to unity while all the input and output slacks are 0.

16.3 Scale Efficiency

While full technical efficiency requires a firm to produce the maximum output(s) from its observed input bundle, to be considered as scale efficient, the firm needs to operate at the scale where the average productivity reaches a maximum. Because the average productivity is a meaningful concept only when a single output is produced from a single input, the concept of scale efficiency is best described in the context of a one-input one-output technology. Consider a firm with input–output (x_0, y_0). Its average productivity is y_0/x_0. Clearly, if it is not technically efficient, then, it is possible to increase the output without changing the input or to lower the input without reducing the output. In either case, its productivity would increase. Now, suppose that the production function is $y^* = f(x)$

* However, there have been several revisions of the additive model (e.g., Cooper et al. 1999, 2011) incorporating weights for the input and output slacks that provide useful measures of (in)efficiency.

and the corresponding graph of the technology is

$$G = \{(x, y) : y = f(x); x \geq 0; y \geq 0\}. \tag{16.23}$$

For any $(x, y) \in G$, $AP(x) = y/x = f(x)/x$. Thus, if $(x_0, y_0) \in G$, $AP(x_0) = f(x_0)/x_0$. Because y_0 is the maximum output producible from input x_0, an increase in the average productivity is not possible so long as the input level does not change. There may exist other input levels, however, where the average productivity is higher. Let x^* be the input level where the average productivity attains a maximum. In that case, $dAP(x)/dx = xf'(x) - f(x)/x^2 = 0$ at the input level x^*. Frisch (1965) described the input level where the average productivity is maximum as the *technical optimal production scale* (TOPS).

Several points are to be noted:

1. At the technically optimal input level (x^*), locally CRS holds because $dAP(x)/dx|_{x=x^*} = 0$.
2. $AP(x) = f(x)/x \leq AP(x^*) = f(x^*)/x^*$ for all input levels (x).
3. At the input level (x^*), the marginal productivity and average productivity are equal. Thus, $f'(x^*) = f(x^*)/x^*$. This implies that $f(x^*) = f'(x^*) \cdot x^*$.

The scale efficiency of the firm operating at the input level x_0 is

$$SE(x_0) = \frac{AP(x_0)}{AP(x^*)} \leq 1. \tag{16.24}$$

More specifically,

$$SE(x_0) = \frac{f(x_0)/x_0}{f(x^*)/x^*} = \frac{f(x_0)}{x_0 f'(x^*)}. \tag{16.25}$$

Now, define $\delta \equiv f'(x^*)$ and consider a *pseudo* production function

$$y^{**} = r(x) = \delta x. \tag{16.26}$$

Then, the denominator in Equation 16.25 becomes $\delta x_0 = r(x_0)$. Therefore, an alternative measure of scale efficiency is

$$SE(x_0) = \frac{f(x_0)}{r(x_0)}. \tag{16.27}$$

Note that 2 and 3 above together imply that

$$f(x) \leq f'(x^*)x = \delta x = r(x),$$

and

$$f(x^*) \leq f'(x^*)x^* = \delta x^* = r(x^*).$$

In other words, $f(x) \leq r(x)$ for all x and $f(x) = r(x)$ at $x = x^*$. In particular, $f(x_0) \leq r(x_0)$. This ensures that $SE(x_0) \leq 1$.

The *pseudo* production function is a tangent to the graph of the technology and *would have been* the production frontier if the technology did exhibit CRS globally. Note that under VRS,

$y^* = f(x)$ is the true production function and the right measure of technical efficiency is $\tau_y = y_0/y_0^* = y_0/f(x_0)$. The other measure $\tau_y^C = y_0/y_0^{**} = y_0/r(x_0)$ is not the correct measure unless the technology exhibits CRS at all input levels. But this CRS efficiency measure, false as it is, does serve a useful purpose because the ratio of the CRS and VRS measures of technical efficiency is a valid measure of the scale efficiency. This can be shown as follows:

$$SE(x_0) = \frac{f(x_0)}{r(x_0)}$$

$$= \frac{y_0/r(x_0)}{y_0/f(x_0)} = \frac{\tau_y^C(x_0, y_0)}{\tau_y(x_0, y_0)}. \tag{16.28}$$

It should be noted that the expression in Equation 16.28 measures the *output-oriented* scale efficiency of the input level x_0. In a perfectly analogous manner, one can take the output level y_0 as given and measure the *input-oriented* scale efficiency

$$SE(y_0) = \frac{\tau_x^C(x_0, y_0)}{\tau_x(x_0, y_0)}. \tag{16.29}$$

In Figure 16.2, the curve $y^* = f(x)$ is the true VRS production function. At the input level x_0, the firm produces output $y_0 < f(x_0)$. This input–output combination is shown by the point A. If technical inefficiency is eliminated, then, it could move to the point B on the production function. Here, its average productivity would be $AP^*(x_0) = Bx_0/Ox_0$. But the maximum average productivity along the production function is attained at the point C where the input level is x^* and this maximum average productivity is $AP^*(x^*) = Cx^*/Ox^*$. The scale efficiency at input level x_0 is $SE(x_0) = AP^*(x_0)/AP^*(x^*)$. Now, consider the tangent line $y^{**} = r(x)$. The point D shows the maximum output, y_0^{**} that would have been producible from input x_0 if CRS holds. Now, the average productivity remains constant along the tangent line. Thus, comparing the average productivities at points B and C is equivalent to comparing the productivities at points B and D. But the average productivity at D would be $r(x_0)/x_0$ whereas at B it is $f(x_0)/x_0$. Thus, the scale efficiency at input level x_0 is $f(x_0)/r(x_0) = Bx_0/Dx_0$.

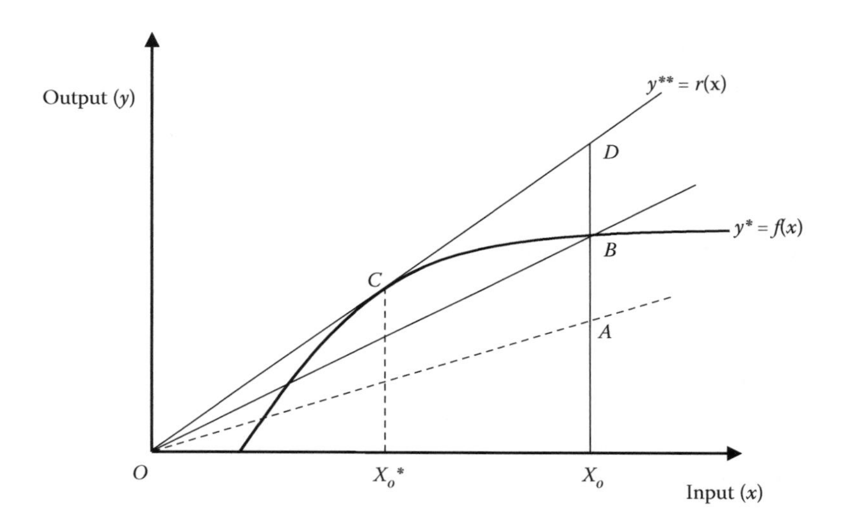

FIGURE 16.2
Scale efficiency.

16.3.1 Ray-Average Productivity and Returns to Scale

The single-input single-output case was useful for illustrative purposes but is of little relevance in real life because seldom if ever any output is produced from one input alone. We now consider a multiple-input single-output technology. The production function now shows the maximum *scalar* output producible from a *vector* of inputs. Consider an input–output combination (x^0, y_0) that lies in the graph, G. That is,

$$y_0 = f(x^0); \quad x^0 = (x_{10}, x_{20}, \ldots, x_{n0}).$$

Now, consider another bundle (x^1, y_1) also in the graph such that $x^1 = \beta x^0$. The two input bundles differ only in scale but not in input proportions. The vectors x^0 and x^1 lie on the same ray through the origin in the input space. If the bundle x^0 is considered to be one unit of a *composite input*, then, x^1 represents β units of the same input. If $\beta > 1$, then, the bundle x^1 is a radial expansion of the x^0 bundle. Now, suppose that $y_1 = \alpha y_0$. The *ray-average productivity* measured by the *output per unit of the composite input* at (x^0, y_0) is y_0 and at (x^1, y_1) it is $\alpha y_0/\beta$. If $\alpha > \beta > 1$, then, the *ray-average productivity* is increasing at (x^0, y_0) and we conclude that the locally increasing returns to scale (IRS) holds at this point on the graph. On the other hand, $1 < \alpha < \beta$ signifies the locally diminishing returns to scale (DRS). Finally, $\alpha = \beta$ implies CRS. Note that all these are the local characteristics of the technology and are evaluated as $\beta \to 1$ from above. The technology may exhibit increasing, constant, or DRS at different points on the graph. This is why it is described as VRS.

16.3.2 Most-Productive Scale Size

Banker (1984) generalized Frisch's concept of the technically optimal production scale to the multiple-output multiple-input case. A feasible input–output bundle (x^0, y^0) is an MPSS if for all nonnegative scalars (α, β) for which $(\beta x^0, \alpha y^0)$ is a feasible input–output combination, $\alpha/\beta \leq 1$. In other words, (x^0, y^0) is an MPSS only if there is no other feasible input–output bundle with the same mix of inputs and outputs but a higher *ray-average productivity*. It is obvious that no feasible input–output bundle can be an MPSS unless it is in the graph. Recall that if $(x^0, y^0) \in T$ but $\notin G$, then, there will exist either some $\beta < 1$ such that $(\beta x^0, y^0) \in T$ or some $\alpha > 1$ such that $(x^0, \alpha y^0) \in T$. In the former case, one gets $\alpha/\beta > 1$ for $\alpha = 1$. In the latter case, $\alpha/\beta > 1$ for $\beta = 1$.

The following lemma due to Ray (2010) shows that when the production-possibility set is convex, IRS holds at all scales smaller than the smallest MPSS. Similarly, DRS holds at all scales larger than the largest MPSS.

Lemma 16.1

For any convex productivity possibility set T, if there exist nonnegative scalars α and β such that $\alpha > \beta > 1$, and both $(\bar{x}, \bar{y})$ and $(\beta \bar{x}, \alpha \bar{y}) \in G$, then $\gamma > \delta$ for every γ and δ such that $1 < \delta < \beta$ and $(\delta \bar{x}, \gamma \bar{y}) \in G$.

Proof. Because both $(\bar{x}, \bar{y})$ and $(\beta \bar{x}, \alpha \bar{y})$ are feasible, by the convexity of T, for every $\lambda \in (0, 1)$, $((\lambda + (1 - \lambda)\beta)\bar{x}, (\lambda + (1 - \lambda)\alpha)\bar{y})$ is also feasible. Now, select λ such that $\lambda + (1 - \lambda)\beta = \delta$. Further, define $\mu = \lambda + (1 - \lambda)\alpha$. Using these notations, $(\delta \bar{x}, \mu \bar{y}) \in T$. But, because $(\delta \bar{x}, \gamma \bar{y}) \in G$, $\gamma \geq \mu$. However, because $\alpha > \beta$, $\mu > \delta$. Hence, $\gamma > \delta$. ■

An implication of this lemma is that, when the production-possibility set is convex, if the technology exhibits the locally DRS at a smaller input scale, then, it cannot exhibit the increasing returns at a bigger input scale. This is easily understood in the single-input single-output case. When both x and y are scalars, the average productivity at $(\bar{x}, \bar{y})$ is $\bar{y}/\bar{x}$ and at $(\beta\bar{x}, \alpha\bar{y})$ it is $(\alpha/\beta)(\bar{y}/\bar{x})$. Thus, when $\alpha > \beta$, the average productivity has increased. The above lemma implies that for every input level x in-between $\bar{x}$ and $\beta\bar{x}$, the average productivity is greater than $\bar{y}/\bar{x}$. Thus, the average productivity could not first decline and then increase as the input level increased from $\bar{x}$ to $\beta\bar{x}$.

Two results follow immediately. First, locally IRS holds at every input–output bundle $(x, y) \in G$ that is smaller than the smallest MPSS. Second, locally DRS holds at every input–output bundle $(x, y) \in G$ that is greater than the largest MPSS. To see this, let $x = bx^*$ and $y = ay^*$, where (x^*, y^*) is the smallest MPSS for the given input and output mix. Because (x, y) is not an MPSS, $a/b < 1$. Further, assume that $b < 1$. Define $\beta = 1/b$ (>1) and $\alpha = 1/a$. Then, $(x^*, y^*) = (\beta x, \alpha y)$ and $\alpha/\beta > 1$. Because the ray-average productivity is higher at a larger input scale, by virtue of the lemma, locally IRS holds at (x, y). Next, assume that $b > 1$. Again, because (x, y) is not an MPSS, $a/b < 1$. That is, the ray-average productivity has fallen as the input scale is increased from x^* to $x = bx^*$. Then, by virtue of the lemma, the ray-average product could not be any higher than a/b at a slightly greater input scale, $\bar{\bar{x}} = (1 + \varepsilon)x$. But, because (x, y) is not an MPSS, the ray-average product cannot remain constant as the input scale is slightly increased. Hence, the ray-average product must fall as the input scale is slightly increased from x. Thus, locally DRS holds at every $(x, y) \in G$, when x is larger than the largest MPSS.

16.3.3 Identifying the Nature of the Local Returns to Scale

16.3.3.1 Banker's Primal Approach

Banker (1984) developed the following important theorem that serves as a basis for identifying the nature of the local returns to scale at the input–output bundle (x^0, y^0) if it is on the VRS frontier and at its efficient projection if it is an interior point.*

Theorem 16.1

An input–output bundle (x^0, y^0) is an MPSS if and only if the optimal value of the objective function of a CCR–DEA model equals to unity for this input–output combination.

Proof. Consider the input-oriented CCR–DEA problem:

$$\begin{aligned} & \min \theta \\ \text{s.t.} \quad & \sum_{j=1}^{N} \lambda_j x^j \leq \theta x^0; \\ & \sum_{j=1}^{N} \lambda_j y^j \geq y^0; \\ & \lambda_j \geq 0, \quad (j = 1, 2, \ldots, N); \quad \theta \textit{ unrestricted}. \end{aligned} \tag{16.30}$$

A complete proof of this theorem requires us to show that (a) the optimal value θ^* must be unity if (x^0, y^0) *is* an MPSS and (b) θ^* cannot be unity if (x^0, y^0) *is not* an MPSS. ■

* See also Banker and Thrall (1992) and Banker et al. (2004).

First, assume that (x^0, y^0) is an MPSS but $\theta^* < 1$ in Equation 16.30, where the optimal solution is $(\theta^*, \lambda_j^*;\ j = 1, 2, \ldots, N)$. Then, the feasibility of the solution implies that

$$\begin{aligned} &\sum_{j=1}^{N} \lambda_j^* x^j \le \theta^* x^0; \\ &\sum_{j=1}^{N} \lambda_j^* y^j \ge y^0; \\ &\lambda_j^* \ge 0, \quad (j = 1, 2, \ldots, N). \end{aligned} \tag{16.31}$$

Now, define $k = \sum_{j=1}^{N} \lambda_j^*$, $\beta = \theta^*/k$, $\alpha = 1/k$, and $\mu_j = \lambda_j^*/k$. Clearly, $\sum_{j=1}^{N} \mu_j = 1$ and $\mu_j \ge 0$, where $(j = 1, 2, \ldots, N)$. Thus,

$$\begin{aligned} &\sum_{j=1}^{N} \mu_j x^j \le \beta x^0; \\ &\sum_{j=1}^{N} \mu_j y^j \ge \alpha y^0; \\ &\sum_{j=1}^{N} \mu_j = 1; \\ &\alpha, \beta > 0; \quad \mu_j \ge 0, \quad (j = 1, 2, \ldots, N). \end{aligned} \tag{16.32}$$

Therefore, $(\beta x^0, \alpha y^0)$ is a feasible input–output bundle under the VRS assumption. Further, $\alpha/\beta = 1/\theta^* > 1$ because θ^* has been assumed to be less than 1. This contradicts the assumption that (x^0, y^0) is an MPSS. Hence, part (a) is proven by contradiction.

Next, suppose that (x^0, y^0) *is not* an MPSS but θ^* is equal to 1. Because (x^0, y^0) *is not* an MPSS, there exist α, β, and $\mu_j (j = 1, 2, \ldots, N)$ such that

$$\begin{aligned} &\sum_{j=1}^{N} \mu_j x^j \le \beta x^0; \\ &\sum_{j=1}^{N} \mu_j y^j \ge \alpha y^0; \\ &\sum_{j=1}^{N} \mu_j = 1; \quad \frac{\alpha}{\beta} > 1; \\ &\alpha, \beta > 0; \quad \mu_j \ge 0, \quad (j = 1, 2, \ldots, N). \end{aligned} \tag{16.33}$$

Define $\lambda_j = \mu_j/\alpha$ $(j = 1, 2, \ldots, N)$. Then

$$\begin{aligned} &\sum_{j=1}^{N} \lambda_j x^j \leq \frac{\beta}{\alpha} x^0; \\ &\sum_{j=1}^{N} \lambda_j y^j \geq y^0; \\ &\lambda_j \geq 0, \quad (j = 1, 2, \ldots, N). \end{aligned} \tag{16.34}$$

This shows that $\theta = \beta/\alpha$ is a feasible value of the objective function in the minimization problem in Equation 16.30. But $\alpha/\beta > 1 \Rightarrow \beta/\alpha < 1$. This proves that the optimal value in Equation 16.30 cannot be $\theta^* = 1$ unless (x^0, y^0) is an MPSS. This completes the proof of the theorem.

This theorem only determines whether (x^0, y^0) is an MPSS or not. It does not say anything directly about the nature of the local returns to scale when it is not an MPSS. However, three important corollaries follow from the theorem:

1. If $k = \sum_{j=1}^{N} \lambda_j^* = 1$, then, (x^0, y^0) is an MPSS and CRS holds locally.
2. If $k = \sum_{j=1}^{N} \lambda_j^* < 1$, then, IRS holds locally at (x^0, y^0) or at its input-oriented efficient projection onto the VRS frontier if it is technically inefficient.
3. If $k = \sum_{j=1}^{N} \lambda_j^* > 1$, then, DRS holds locally at (x^0, y^0) or at its input-oriented efficient projection onto the VRS frontier if it is technically inefficient.

The intuition behind these corollaries is quite simple. When $k = 1$, the optimal solution from the CRS problem in Equation 16.30 is an optimal solution for the corresponding VRS problem. Because the CRS and VRS technical efficiency measures are identical, the scale efficiency equals to unity and (x^0, y^0) is an MPSS. Moreover, by virtue of part (a) of the theorem, θ^* equals to unity and (x^0, y^0) is on the frontier. If $k \neq 1$, then, the CRS input-oriented projection $(\theta^* x^0, y^0)$ is not a feasible solution for the corresponding VRS problem. But $(1/k)(\theta x^0, y^0)$ is both on the CRS and the VRS frontier. If $k < 1$, then, the input-oriented projection is to be scaled up to attain an MPSS and it lies in the IRS region. On the other hand, if $k > 1$, then, it is scaled down to the MPSS and the input-oriented projection falls in the DRS region on the VRS frontier.

A potential problem with this method of the returns-to-scale characterization is that there may be multiple optimal solutions to the DEA problem in Equation 16.30 with the sum of λs greater than 1 in some and less than 1 in the others. In that situation, conflicting conclusions would be drawn depending on which optimal solution was obtained. This requires a modification of corollaries (2) and (3) as follows:

2a. The locally IRS holds if $k = \sum_{j=1}^{N} \lambda_j^* < 1$ at all optimal solutions of the CRS–DEA problem in Equation 16.30.

3a. The locally diminishing returns hold if $k = \sum_{j=1}^{N} \lambda_j^* > 1$ at all optimal solutions of the CRS–DEA problem in Equation 16.30.

This can be implemented in two steps. In step 1, the DEA problem in Equation 16.30 is solved and the optimal value θ^* is determined. For (2a) above, in step 2, the following problem is solved:

$$\begin{aligned} &\max \sum_{j=1}^{N} \lambda_j \\ \text{s.t.}\quad &\sum_{j=1}^{N} \lambda_j x^j \leq \theta^* x^0; \\ &\sum_{j=1}^{N} \lambda_j x^j \geq y^0; \\ &\lambda_j \geq 0, \quad (j = 1, 2, \ldots, N). \end{aligned} \tag{16.35}$$

If the maximum value of the objective function is less than 1, then, it can be concluded that $k = \sum_{j=1}^{N} \lambda_j^* < 1$ at all optimal solutions of Equation 16.30. Similarly, to check for (3a), one minimizes the sum of λs in Equation 16.31 and if the minimum is greater than 1, then, one can conclude that DRS holds locally.

16.3.3.2 A Dual Approach

BCC (1984) offer an alternative method of identifying the local returns to scale from the following dual of the input-oriented VRS–DEA problem:

$$\begin{aligned} &\max v^{0\prime} y^0 - \delta_0 \\ \text{s.t.}\quad &v^{0\prime} y^j - \delta_0 \leq u^{0\prime} x^j, \quad (j = 1, 2, \ldots, N); \\ &u^{0\prime} x^0 = 1; \\ &u^0, v^0 \geq 0; \quad \delta_0 \textit{ unrestricted}. \end{aligned} \tag{16.36}$$

BCC have shown that

1. CRS holds at (x^0, y^0) if at the optimal solution of Equation 16.32 δ_0 is zero.
2. IRS holds at (x^0, y^0) if at the optimal solution of Equation 16.32 δ_0 is < 0.
3. DRS holds at (x^0, y^0) if at the optimal solution of Equation 16.32 δ_0 is > 0.

As in the case of Banker's approach, multiple optimal solutions pose a problem and conditions (2) and (3) have to be appropriately modified.

16.3.3.3 A Nesting Approach

Färe, Grosskopf, and Lovell (FGL) (1985) consider a technology that lies in-between CRS and the VRS technologies. They call it a nonincreasing returns to scale (NIRS) technology. Under the assumption of NIRS

$$(x^0, y^0) \in T \Rightarrow (kx^0, ky^0) \in T \quad \text{for any } k \in (0, 1).$$

The DEA estimate of an NIRS production-possibility set is

$$S^{NIRS} = \left\{ (x, y) : x \geq \sum_{1}^{N} \lambda_j x^j ; y \leq \sum_{1}^{N} \lambda_j y^j ; \sum_{1}^{N} \lambda_j \leq 1 ; \lambda_j \geq 0 ; j = 1, 2, \ldots, N \right\}. \quad (16.37)$$

It may be noted that the frontiers of the CRS and NIRS production-possibility sets coincide in the region of IRS. Similarly, the VRS and NIRS frontiers are identical in the DRS region. Therefore, when IRS holds at (x^0, y^0), in an input-oriented model, $\theta_*^C = \theta_*^{NIRS} < \theta_*^V$ where the superscripts C, N, and V refer to CRS, NIRS, and VRS. Similarly, $\theta_*^C < \theta_*^{NIRS} = \theta_*^V$ implies DRS. Of course, in the case of CRS, all three estimates of technical efficiency are equal to unity.

16.3.4 Identifying the Returns to Scale for an Inefficient Unit

The concept of the returns to scale is meaningful only when the relevant input–output bundle lies on the frontier of the production-possibility set. For an inefficient bundle, one must consider its efficient projection—either input- or output oriented. Unless similar returns to scale are found at both projections, one cannot conclusively determine the returns to scale at the observed input–output bundle.

The following DEA problem considered by Cooper et al. (1996) can be used not only to determine whether an input–output bundle (x^0, y^0) is an MPSS but also to identify the bundle (x^*, y^*) that is an MPSS for (x^0, y^0):

$$\begin{aligned} \text{Maximize} \quad & \rho = \frac{\alpha}{\beta} \\ \text{subject to} \quad & \sum_j \lambda_j x^j \leq \beta x^0 ; \\ & \sum_j \lambda_j y^j \geq \alpha y^0 ; \\ & \sum_j \lambda_j = 1 ; \\ & \alpha, \beta, \lambda_j (j = 1, 2, \ldots, N) \geq 0. \end{aligned} \quad (16.38)$$

Because (x^0, y^0) is assumed to be a feasible input–output bundle, $(\alpha = \beta = \rho = 1)$ is a feasible solution for this problem. Hence, the optimal value ρ^* is always greater than or equal to 1. When $\rho^* = \alpha^* / \beta^*$ exceeds unity, we know that (x^0, y^0) *is not an MPSS*. But we can also conclude that $(\beta^* x^0, \alpha^* y^0)$ *is an MPSS*.

As such, the objective function is nonlinear. However, it can be easily transformed into an LP problem. Define $t = 1/\beta$ and $\mu_j = t\lambda_j$ where $(j = 1, 2, \ldots, N)$. Note that the nonnegativity of β and λ_js ensures that t and μ_js are also nonnegative. Problem (5) can, therefore, be reformulated as the following LP problem:

$$\text{Maximize} \quad \rho$$

$$\text{subject to} \quad \sum_j \mu_j x^j \le x^0$$
$$\sum_j \mu_j y^j \ge \rho y^0; \tag{16.39}$$
$$\sum_j \mu_j = t;$$
$$t, \mu_j (j = 1, 2, \ldots, N) \ge 0.$$

From the optimal solution of this problem, we can derive $\beta^* = 1/t^*$ and $\alpha^* = \rho^*/t^*$. One can then infer the nature of the returns to scale from these values of α^* and β^*. It may be pointed out here that because the only restriction on t is the nonnegativity, Equation 16.27 is simply the output-oriented CCR–DEA problem and $1/\rho^*$ is the same as the output-oriented CRS technical efficiency $\tau_y^C(x^0, y^0)$.

When the bundle (x^0, y^0) is not an MPSS by itself, $\rho^* > 1$ so that $\alpha^* > \beta^*$. If the MPSS is unique, then, there are five different possibilities: (i) $1 < \beta^* < \alpha^*$; (ii) $\beta^* < \alpha^* < 1$; (iii) $\beta^* = 1 < \alpha^*$; (iv) $1 < \beta^* = \alpha^*$, and (v) $1 < \beta^* < \alpha^*$. When the MPSS is unique, for the case in (i) both the input- and output-oriented projections of the bundle (x^0, y^0) fall in the region of IRS. In this case, the unit is conclusively too small relative to its MPSS. Similarly, if $\beta^* < \alpha^* < 1$, then, both input- and output-oriented projections fall in the region of the DRS. The implication is that the unit is too large. When $\beta^* = 1 < \alpha^*$, the input scale corresponds to the MPSS but the output scale is too small. The opposite is true when $1 < \beta^* = \alpha^*$. Finally, in the intermediate case, where $\beta^* < 1 < \alpha^*$, the input scale is bigger than the MPSS and the output-oriented projection falls in the region of the DRS. At the same time, the input scale is smaller than the MPSS and the input-oriented projection falls in the region of IRS.

Zhu (2003) uses a single-input single-output example to partition the interior of the production-possibility set into six different regions for the returns-to-scale classification of inefficient production units.* In three out of these six regions, both input- and output-oriented efficient projections exhibit the same returns to scale: increasing, constant, or diminishing. In the remaining three, increasing returns at the input-oriented projection combine with constant or diminishing returns at the output-oriented projection, or constant returns at the input-oriented projection are associated with diminishing returns at the output-oriented projection. To correctly locate an inefficient unit in the appropriate region, one has to ascertain the returns to scale at both projections.

16.3.5 The Case of Multiple MPSS

Next, we consider the possibility of multiple MPSS. This is graphically depicted in Figure 16.3. Here, both C_1 and C_2 are MPSS and so are their convex combinations lying on the line segment connecting them. At C_1, (α_1^*, β_1^*) is the smallest MPSS. Similarly, (α_2^*, β_2^*) at C_2 is the largest MPSS. It is obvious that when Equation 16.39 has a unique optimal solution (in particular, t^* is unique), there cannot be multiple MPSS. For multiple optimal solutions, the largest $t^* = \sum_j \mu_j^*$ across all optimal solutions of Equation 16.39 corresponds to the smallest MPSS, β_1^*. Similarly, β_2^* corresponds to the smallest $t^* = \sum_j \mu_j^*$ at an optimal solution.

* See also the earlier paper by Seiford and Zhu (1999).

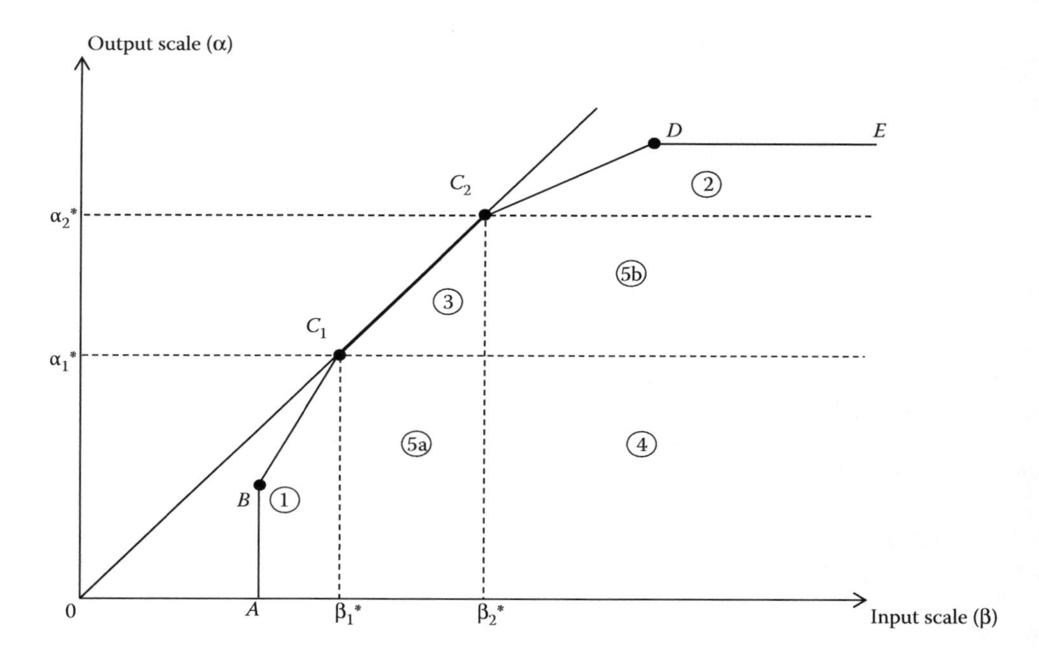

FIGURE 16.3
Multiple MPSS and the regions of the increasing, constant, decreasing, and the ambiguous returns to scale.

Note that across all optimal solutions, the value of the objective function is the same (ρ^*). Hence, $\beta_1^* = 1/t_1^*$, where

$$\begin{aligned} t_1^* &= \max \sum_j \mu_j \\ \text{s.t.} \quad & \sum_j \mu_j x^j \leq x^0; \\ & \sum_j \mu_j y^j \geq \rho^* y^0; \\ & \mu_j (j = 1, 2, \ldots, N) \geq 0. \end{aligned} \tag{16.40}$$

Similarly, $\beta_2^* = 1/t_2^*$, where

$$\begin{aligned} t_2^* &= \min \sum_j \mu_j \\ \text{s.t.} \quad & \sum_j \mu_j x^j \leq x^0; \\ & \sum_j \mu_j y^j \geq \rho^* y^0; \\ & \mu_j (j = 1, 2, \ldots, N) \geq 0. \end{aligned} \tag{16.41}$$

Once β_1^* and β_2^* have been determined from Equations 16.8 and 16.9, the corresponding values of α are readily obtained as $\alpha_1^* = \rho^*\beta_1^*$ and $\alpha_2^* = \rho^*\beta_2^*$.

As shown in Figure 16.3, the set of output–input scales (α, β) for which the input–output bundles $(\beta x^0, \alpha y^0)$ are feasible can be partitioned into six different regions defined below:

1. In region (1) toward the southwest of the smallest MPSS (C_1), $(\beta < \beta_1^*; \alpha < \alpha_1^*)$. When (x^0, y^0) falls in this region, $1 < \beta_1^* < \alpha_1^*$. Hence, IRS holds unambiguously.
2. In region (2) to the northeast of the largest MPSS (C_2), $(\beta_2^* < \beta; \alpha_2^* < \alpha)$. If (x^0, y^0) falls in this region, $\beta_1^* < \alpha_1^* < 1$. DRS holds unambiguously in this region.
3. In region (3), $\beta_1^* < \beta < \beta_2^*$ while $\alpha_1^* < \alpha < \alpha_2^*$. The points in this region lie between the smallest and the largest MPSS. It is interesting to note that even if the point $(\alpha = 1, \beta = 1)$ is not technically efficient and lies below the C_1C_2 line, both the input- and the output-oriented projection of the inefficient bundle will fall in the region of CRS. Thus, there is no scale inefficiency in this region even though there may be a technical inefficiency.
4. In region (4), $\beta_2^* < \beta; \alpha < \alpha_1^*$. When the actual input–output bundle lies here, $\beta_2^* < 1 < \alpha_1^*$. The input bundle x^0 is larger than the largest MPSS; hence, the output-oriented projection falls in the area of diminishing returns. At the same time, the actual output bundle is smaller than the smallest MPSS. Hence, IRS holds at the input-oriented projection. Thus, the returns to scale cannot be unambiguously defined at the actual input–output bundle.
5. In region (5a), $\beta_1^* < \beta < \beta_2^*$ but $\alpha < \alpha_1^*$. When the actual input–output bundle lies here, y^0 is smaller than the smallest MPSS and the input-oriented projection falls in the area of the increasing returns. At the same time, the actual input bundle lies between the smallest and the largest MPSS. Hence, CRS holds at the output-oriented projection. Here also, the returns-to-scale characterization depends on the orientation.
6. In region (5b), $\beta_2^* < \beta$ while $\alpha_1^* < \alpha < \alpha_2^*$. When the actual input–output bundle lies here, x^0 is larger than the largest MPSS. Hence, the output-oriented projection falls in the area of diminishing returns. At the same time, the actual output bundle lies between the smallest and the largest MPSS. Hence, CRS holds at the input-oriented projection. Here, the input bundle is too large. But the actual output bundle, if produced from the technically efficient input bundle, would correspond to an MPSS.

16.3.6 Output- or Input-Oriented?

Except in the case of globally CRS, output- and input-oriented technical efficiency measures would differ for the same firm. An important question is how to decide which measure is preferable. As a general rule, the answer depends on whether output augmentation is more important than input conservation in a specific context. In many situations, however, there is no clear-cut priority. A rule of thumb would then be to select the orientation that yields a *lower measure of efficiency* under the VRS assumption. The logic behind this criterion is the fact that the corresponding efficient projection would have a higher level of scale efficiency. This can be explained by a simple one-input one-output example. Consider a technically inefficient input–output combination (x_0, y_0). Now, suppose that the output-oriented efficient projection is $(x_0, \varphi^* y_0)$ while in the input-oriented projection it is $(\theta^* x_0, y_0)$. Thus, the corresponding technical efficiency measures are $\tau_y = 1/\varphi^*$ and $\tau_x = \theta^*$. Assume arbitrarily that $\tau_y < \tau_x$. This implies that $1/\varphi^* < \theta^*$ or $1/\theta^* < \varphi^*$. Therefore, $y_0/\theta^* x_0 < \varphi^* y_0/x_0$. This shows that that average productivity is higher at the output-oriented efficient projection than at the input-oriented projection of (x_0, y_0).

16.4 Nonradial Measures of Technical Efficiency

In an output-oriented analysis of technical efficiency, the objective is to produce the maximum output from a given quantity of inputs. For this, we first define the *(producible) output set* of any given input bundle. For the input bundle x^0, the output set

$$P(x^0) = \{y : (x^0, y) \in T\} \tag{16.42}$$

consists of all the output bundles that can be produced from x^0.

Because there are different output sets for different input bundles, the production-possibility set is equivalently characterized by a family of output sets. Each output set is a subset of the m-dimensional output space. The following properties of output sets follow from the relevant assumptions made about the production-possibility set:

(P1) If (x^j, y^j) is an actually observed input–output combination, then $y^j \in P(x^j)$
(P2) If $y^0 \in P(x^0)$ and if $x^1 \geq x^0$, then $y^0 \in P(x^1)$
(P3) If $y^0 \in P(x^0)$ and if $y^1 \leq y^0$, then $y^1 \in P(x^0)$
(P4) Each output set $P(x)$ is convex

The *output isoquant* of any input bundle x^0 can be defined as

$$\bar{P}(x^0) = \{y : y \in P(x^0) \text{ and } \lambda y \notin P(x^0) \text{ if } \lambda > 1\}. \tag{16.43}$$

Thus, if $y^0 \in \bar{P}(x^0)$, then, the output-oriented radial technical efficiency of the pair (x^0, y^0) equals to unity because it is not possible to increase *all* the outputs holding the input bundle unchanged. This does not, of course, rule out the possibility that some individual components of the y^0 output bundle can be increased.

The *efficient subset* of the output isoquant of x^0, on the other hand, is

$$P^*(x^0) = \{y : y \in P(x^0) \text{ and } y' \notin P(x^0) \text{ if } y' \geq y^0\}. \tag{16.44}$$

Thus, an output-oriented radial technically efficient projection of y^0 produced from x^0 onto $\bar{P}(x^0)$ may include slacks in individual outputs. But no such slacks may exist if the projection is onto $P^*(x^0)$. The radial measure of the output-oriented technical efficiency does not reflect any unutilized potential for increasing the individual outputs. Again, as shown below, a nonradial output-oriented measure does take account of all the potential increases in any component of the output bundle.

The problem of slacks in any optimal solution of a radial DEA model arises because we seek to expand all the outputs or contract all the inputs by the same proportion. In nonradial models, one allows the individual outputs to increase or the inputs to decrease at different rates. Färe and Lovell (1978) introduced the following input-oriented, *nonradial* measure of technical efficiency called the

Russell measure:

$$\rho_x(x^0, y^0) = \min \frac{1}{n} \sum_i \theta_i$$
$$\text{s.t.} \quad \sum_j \lambda_j y_{rj} \geq y_{r0}; \quad (r = 1, 2, \ldots, m);$$
$$\sum_j \lambda_j x_{ij} \leq \theta_i x_{i0}; \quad (i = 1, 2, \ldots, n); \tag{16.45}$$
$$\sum_j \lambda_j = 1; \quad \lambda_j \geq 0; \quad (j = 1, 2, \ldots, N).$$

When input slacks do exist at the optimal solution of a radial DEA model, the nonradial Russell measure in Equation 16.39 falls below the conventional measure obtained from an input-oriented BCC model (16.6). Because the radial projection is always a feasible solution for Equation 16.8, $\rho_x \leq \tau_x$. That is, the nonradial Russell measure of technical efficiency never exceeds the corresponding radial measure.

The analogous output-oriented nonradial VRS measure of technical efficiency is

$$RM_y(x^0, y^0) = \frac{1}{\rho_y},$$

where

$$\rho_y = \max \frac{1}{m} \sum_r \phi_r$$
$$\text{s.t.} \quad \sum_j \lambda_j y_{rj} \geq \phi_r y_{r0}; \quad (r = 1, 2, \ldots, m);$$
$$\sum_j \lambda_j x_{ij} \leq x_{i0}; \quad (I = 1, 2, \ldots, n); \tag{16.46}$$
$$\sum_j \lambda_j = 1; \quad \lambda_j \geq 0; \quad (j = 1, 2, \ldots, N).$$

While no input slacks can exist at the optimal solution of Equation 16.39, the presence of any output slack is not ruled out. Similarly, input slacks may remain at the optimal solution of Equation 16.40.

16.4.1 Graph-Efficiency Measures

All the technical efficiency measures considered above are either output- or input-oriented. Instead of focusing exclusively on increasing outputs or reducing inputs, one may wish to achieve both the objectives simultaneously. A problem with a graph-efficiency measure is that the benchmark input–output bundle selected on the frontier depends on the relative importance attached to input reduction vis-à-vis output expansion. In the extreme case of input orientation, output expansion is given a zero weight. Conversely, 100% weight assigned to an output expansion leads to the output-oriented projection. As discussed before, the relative importance of outputs and inputs is usually a matter of judgment by the analyst. Färe, Grosskopf, and Lovell (1988) introduced so-called *graph-hyperbolic* measure of efficiency is by selecting a benchmark on the frontier where the actual output bundle (y^0) is scaled up as δy^0 while the observed input bundle (x^0) is scaled down as $(1/\delta)x^0$. It is easy to see that in a single-output single-input case, both the actual input–output bundle (x^0, y^0) and its efficient projection $((1/\delta)x^0, \delta y^0)$ will lie on a rectangular hyperbola. This explains the name *graph hyperbolic* efficiency.

The VRS–DEA–LP problem for measuring the graph efficiency is

$$\begin{aligned} &\max \delta \\ \text{s.t.} \quad &\sum_{j=1}^{N} \lambda_j y^j \geq \delta y^0; \\ &\sum_{j=1}^{N} \lambda_j x^j \leq \tfrac{1}{\delta} x^0; \\ &\sum_{j=1}^{N} \lambda_j = 1; \\ &\lambda_j \geq 0, \quad (j = 1, 2, \ldots, N); \quad \delta \textit{ unrestricted}. \end{aligned} \tag{16.47}$$

A problem with the DEA problem (16.47) is that it is a non-LP problem. It is possible, however, to use a linear approximation for $1/\delta$ in the input constraint. Note that a first-order Taylor's series approximation to $f(\delta) = 1/\delta$ at the point of approximation δ_0 is

$$f(\delta) \simeq f(\delta_0) + (\delta - \delta_0) f'(\delta_0) = \frac{1}{\delta_0} - \frac{1}{\delta_0^2}(\delta - \delta_0).$$

Using $\delta_0 = 1$ as the point of approximation, $1/\delta \approx 2 - \delta$ and the problem in Equation 16.41 can be linearized as

$$\begin{aligned} &\max \delta \\ \text{s.t.} \quad &\sum_{j=1}^{N} \lambda_j y^j \geq \delta y^0; \\ &\sum_{j=1}^{N} \lambda_j x^j + \delta x^0 \leq 2x^0; \\ &\sum_{j=1}^{N} \lambda_j = 1; \\ &\lambda_j \geq 0, \quad (j = 1, 2, \ldots, N); \quad \delta \textit{ unrestricted}. \end{aligned} \tag{16.48}$$

It should be noted that if (x^0, y^0) is quite far from the frontier, then, $\delta_0 = 1$ may lead to a very poor approximation. This is likely to be true if either the input- or the output-oriented efficiency is low. In that case, the linear approximation may need to be applied iteratively with the optimal value of δ providing the point of approximation for each successive iteration.

In the case of CRS, the optimization problem in Equation 16.47 can be handled quite easily. Without the restriction $\sum_{j=1}^{N} \lambda_j = 1$, Equation 16.47 can be written as

$$\begin{aligned} & \max \delta^2 \\ \text{s.t.} \quad & \sum_{j=1}^{N} \delta\lambda_j y^j \geq \delta^2 y^0; \\ & \sum_{j=1}^{N} \delta\lambda_j x^j \leq x^0; \\ & \lambda_j \geq 0, \quad (j = 1, 2, \ldots, N); \quad \delta \textit{ unrestricted.} \end{aligned} \tag{16.49}$$

Defining $\varphi = \delta^2$ and $\mu_j = \delta\lambda_j$, the DEA problem in Equation 16.49 becomes the standard CCR output-oriented problem

$$\begin{aligned} & \max \varphi \\ \text{s.t.} \quad & \sum_{j=1}^{N} \mu_j y^j \geq \varphi y^0; \\ & \sum_{j=1}^{N} \mu_j x^j \leq x^0; \\ & \mu_j \geq 0, \quad (j = 1, 2, \ldots, N); \quad \varphi \textit{ unrestricted.} \end{aligned} \tag{16.50}$$

Note that φ should be constrained to be strictly positive. But given that $\varphi = 1$ is always a feasible solution, it would be a nonbinding constraint. The optimal value of δ can be computed as $\sqrt{\varphi}$.

16.4.2 Pareto–Koopmans Efficiency

An input–output combination (x^0, y^0) is not Pareto–Koopmans efficient if it violates either of the following efficiency postulates:

1. It is not possible to increase any output in the bundle y^0 without reducing any other output and/or without increasing any input in the bundle x^0; or
2. It is not possible to reduce at least any input in the bundle x^0 without increasing any other input and/or without reducing any output in the bundle y^0.

Clearly, unless $RM_x(x^0, y^0) = RM_y(x^0, y^0) = 1$, at least one of the two efficiency postulates is violated and (x^0, y^0) is not Pareto–Koopmans efficient. The input–output bundle (x^0, y^0) is Pareto–Koopmans efficient, when both of the following conditions hold:

1. $x^0 \in V^*(y^0)$
2. $y^0 \in P^*(x^0)$

Thus, the nonradial technical efficiency (whether input oriented or output oriented) by itself does not ensure the overall Pareto efficiency.

A nonradial Pareto–Koopmans measure of technical efficiency of the input–output pair (x^0, y^0) can be computed as

$$\gamma(x^0, y^0) = \min \frac{(1/n)\sum_i \theta_i}{(1/m)\sum_r \phi_r}$$

$$\text{s.t.} \quad \sum_{j=1}^{N} \lambda_j y_{rj} \geq \phi_r y_{r0}; \quad (r = 1, 2, \ldots, m);$$

$$\sum_{j=1}^{N} \lambda_j x_{ij} \leq \theta_i x_{i0}; \quad (i = 1, 2, \ldots, n); \tag{16.51}$$

$$\sum_{j=1}^{N} \lambda_j = 1; \quad \lambda_j \geq 0; \quad (j = 1, 2, \ldots, N).$$

Note that the efficient input–output projection (x^*, y^*) satisfies

$$x^* = \sum_{j=1}^{N} \lambda_j^* x^j \leq x^0 \quad \text{and} \quad y^* = \sum_{j=1}^{N} \lambda_j^* y^j \geq y^0.$$

Thus, (x^0, y^0) is Pareto–Koopmans efficient, if and only if $\phi_r^* = 1$ for each output r and $\theta_i^* = 1$ for each input i, implying that $\gamma(x^0, y^0) = 1$. We can visualize the Pareto–Koopmans global efficiency measure as the product of two factors. The first is the input-oriented component

$$\gamma_x = \frac{1}{n}\sum_i \theta_i \tag{16.52}$$

and the second is an output-oriented component

$$\gamma_y = \frac{1}{(1/m)\sum_r \phi_r}. \tag{16.53}$$

Thus,

$$\gamma(x^0, y^0) = \gamma_x \cdot \gamma_y. \tag{16.54}$$

16.4.3 A Slack-Based Measure of Efficiency

Tone (2001) introduced essentially the same measure of the overall efficiency and called it as a slack-based measure (SBM).* Tone's SBM is

$$\rho = \min \frac{1 - ((1/n)\sum_{i=1}^{n}(s_i^-/x_{i0}))}{1 + ((1/m)\sum_{r=1}^{m}(s_r^+/y_{r0}))}$$

$$\text{s.t.} \quad \sum_{j=1}^{N} \lambda_j x_{ij} + s_i^- = x_{i0}, \quad (i = 1, 2, \ldots, n);$$

* This is the same as the *extended Russell measure* of Pastor et al. (1999).

$$\sum_{j=1}^{N} \lambda_j y_{rj} + s_r^+ = y_{r0}, \quad (r = 1, 2, \ldots, m); \tag{16.55}$$

$$\sum_{j=1}^{N} \lambda_j = 1;$$

$$\lambda_j, s_i^-, s_r^+ \geq 0, \quad (j = 1, 2, \ldots, N;\ i = 1, 2, \ldots, n;\ r = 1, 2, \ldots, m).$$

In this formulation, s_i^- is the *total* slack (rather than the *radial* slack) in input i. Similarly, s_r^+ is the *total* slack in output r. Now, consider the benchmark bundle (x^*, y^*) where $x_i^* = \theta_i x_{i0} = x_i^0 - s_i^-$ for input i and $y_r^* = \varphi_r y_{r0} = y_r^0 + s_r^+$ for output r. It is obvious that $\theta_i = 1 - (s_i^-/x_{i0})$ and $\varphi_r = 1 + (s_r^+/y_{r0})$. That is, the objective function in Equation 16.51 is the same as that in Equation 16.55. Similarly, the constraints are the same in both problems except that the nonnegativity of the slacks implies that in the SBM, the θs cannot exceed unity and the φs cannot be lower than 1. No such constraints are implicit in Equation 16.51.

16.4.4 Linearization of the Pareto–Koopmans DEA Problem

The objective function in Equation 16.55 is nonlinear. Tone transformed this linear-fractional functional-programming problem into an LP problem by normalizing the denominator to unity. Alternatively, as shown in Ray (2004), one may replace the objective function by a linear approximation.

Define

$$\gamma(x^o, y^o) = f(\theta, \varphi). \tag{16.56}$$

Using $\theta_i^0 = 1$ for all i and $\phi_r^0 = 1$ for all r as the point of approximation,

$$f(\theta, \varphi) \approx 1 + \frac{1}{n} \sum_i \theta_i - \frac{1}{m} \sum_r \phi_r. \tag{16.57}$$

We may, therefore, replace the objective function in Equation 16.51 by Equation 16.57 and solve Equation 16.51 iteratively using the optimal solution from each iteration as the point of approximation for the next iteration until convergence. Once we obtain the optimal (θ^*, ϕ^*) from this problem, we evaluate

$$\gamma(x^0, y^0) = \frac{(1/n) \sum_i \theta_i^*}{(1/m) \sum_r \phi_r^*} \tag{16.58}$$

as a measure of the Pareto–Koopmans efficiency* of (x^0, y^0).

Apart from an overall measure, Equation 16.51 also provides information about the potential for reducing the individual inputs (θ_i^*) and increasing the individual outputs (φ_r^*). Also, a decomposition of Equation 16.51 into the input- and output-oriented components can be obtained from Equation 16.12.†

* For empirical applications of this Pareto–Koopmans measure, see Ray and Jeon (2009) and Ray and Ghose (2014).

† See Ray (2004) appendix to Chapter 2 for a proof of these properties.

16.5 Distance Functions

Shephard (1953) defined the *distance function* evaluated at any nonnegative input–output bundle (x, y) as

$$D(x, y) = \min \beta : \left(x, \frac{1}{\beta} y\right) \in T \tag{16.59}$$

It may be noted that the bundle (x, y) itself may not be feasible and may lie outside the production-possibility set. Shephard only assumed that every input–output bundle can be projected onto the frontier of the production-possibility set by appropriately scaling (up or down) the input or the output bundle. This was described as the *attainability* postulate.

If an output bundle y cannot be produced from the input bundle x, then, the attainability assumption ensures that any appropriate scaling down (without altering the mix of outputs) would result in a feasible bundle. That would imply a value of β greater than 1. On the other hand, if (x, y) is in the interior of the production-possibility set, then, the output bundle can be scaled up and still be producible from the input bundle x. In that case, β would be less than 1.

Two things emerge out of the above. First, an alternative characterization of the production-possibility set is

$$T = \{(x, y) : D(x, y) \leq 1\}. \tag{16.60}$$

Second, β in Equation 16.59 is simply the inverse of φ in Equation 16.8. Thus, the Shephard distance function is the same as the Farrell output-oriented technical efficiency, τ_y.

The distance function defined in Equation 16.59 is more accurately described as the *output distance function*

$$D^O(x, y) = \min \beta : \left(x, \frac{1}{\beta} y\right) \in T. \tag{16.61}$$

The *input distance function* can, analogously, be defined as

$$D^I(x, y) = \max \mu : \left(\frac{1}{\mu} x, y\right) \in T. \tag{16.62}$$

It is clear that $D^I(x, y)$ is the inverse of the input-oriented Farrell efficiency, τ_x. Moreover, under CRS, the *output* and *input* distance functions are inverses of each other. The following properties of the distance functions should be noted.

The *output* distance function, $D^o(x, y)$ is

1. Homogeneous of degree 1 in y
2. Increasing (nondecreasing) in y
3. Decreasing (nonincreasing) in x
4. Convex in y

Similarly, the *input* distance function, $D^I(x, y)$, is

1. Homogeneous of degree 1 in x
2. Increasing (nondecreasing) in x
3. Increasing (nondecreasing) in y
4. Convex in x

16.5.1 Directional Distance Function

Chambers, Chung, and Färe (CCF) (1996, 1998) introduced the Nerlove–Luenberger *directional distance function*:

$$\vec{D}(x, y; g^x, g^y) = \max \beta : (x + \beta g^x, y + \beta g^y) \in T. \tag{16.63}$$

Here, (g^x, g^y) is an arbitrary point that serves to define the direction along which the input–output bundle (x, y) is projected onto the frontier.* It is important to recognize that (g^x, g^y) need not be a feasible input–output bundle (or for that matter, even nonnegative!). Its only role is to define the direction along which the (x, y) bundle is to be projected onto the frontier.† Of course, if (x, y) is already on the frontier, then, β will be equal to 0. If one selects $(g^x = -x, g^y = y)$, then, the directional distance function becomes

$$\vec{D}(x, y; -x, y) = \max \beta : ((1 - \beta)x, (1 + \beta)y) \in T. \tag{16.64}$$

In this case, β is the maximum proportionate reduction in all the inputs that are simultaneously feasible with the same proportionate increase in all the outputs. For this reason, it is sometimes described as the *proportional* distance function. Further, if one selects $(g^x = 0, g^y = y)$, then, the directional distance function coincides with the (inverse of) the output-oriented Shephard distance function. On the other hand, $(g^x = -x, g^y = 0)$ leads to the input-oriented Shephard distance function.

The DEA–LP problem for measuring the directional distance function shown in Equation 16.61 above is

$$\begin{aligned}
&\vec{D}(x^0, y^0; -x^0, y^0) = \max\ \beta \\
\text{s.t.}\quad & \sum_{j=1}^{N} \lambda_j x^j + \beta x^0 \le x^0; \\
& \sum_{j=1}^{N} \lambda_j y^j - \beta y^0 \ge y^0; \\
& \sum_{j=1}^{N} \lambda_j = 1; \\
& \lambda_j \ge 0, \quad (j = 1, 2, \ldots, N); \quad \beta \textit{ unrestricted.}
\end{aligned} \tag{16.65}$$

As elsewhere, the restriction $\sum_{j=1}^{N} \lambda_j = 1$ is removed when CRS is assumed. It may be noted that even though β is unrestricted in principle, in practice, it can never exceed 1 because the target input bundle $(1 - \beta)x^0$ would otherwise become negative. Under the CRS assumption, the directional distance function, β, can be easily derived from the output expansion factor, φ, in the CCR–DEA problem. More specifically, $\varphi = (1 + \beta)/(1 - \beta)$. Alternatively, $\beta = (\varphi - 1)/(\varphi + 1)$. It should be noted that β is a measure of the technical *inefficiency*. Also, like the CCR/BCC–DEA measures, the CCF directional distance function is also a radial measure because all the inputs are scaled down by the factor $(1 - \beta)$ while all the outputs are scaled up by the factor $(1 + \beta)$. Individual output- and/or input slacks may exist at the optimal solution of the DEA–LP problem in Equation 16.65. Hence, a value of β equal to 0 does not guarantee Pareto–Koopmans efficiency.

* See also Färe and Grosskopf (2000).

† One may allow different values of β for the inputs and outputs. As shown by Aparicio et al. (2013), this leads to a modified directional distance function.

16.5.2 Geometric Distance Function

Portela and Thanassoulis (2005) introduced the geometric distance function that provides a nonradial Pareto–Koopmans measure of technical efficiency. Consider the input bundle $x^0 = (x_1^0, x_2^0, \ldots, x_n^0)$ and the output bundle $y^0 = (y_1^0, y_2^0, \ldots, y_m^0)$. The geometric distance function can be evaluated as

$$\begin{aligned}
&\gamma(x^0, y^0) = \min \frac{\prod_{i=1}^{n} (\theta_i)^{\alpha_i}}{\prod_{r=1}^{m} (\varphi_r)^{\beta_i}} \\
\text{s.t.} \quad &\sum_{j=1}^{N} \lambda_j x_i^j \leq \theta_i x_i^0, \quad (i = 1, 2, \ldots, n); \\
&\sum_{j=1}^{N} \lambda_j y_r^j \geq \varphi_r y_r^0, \quad (r = 1, 2, \ldots, m); \\
&\sum_{j=1}^{N} \lambda_j = 1; \\
&\theta_i \leq 1; \quad (i = 1, 2, \ldots, n); \\
&\varphi_r \geq 1, \quad (r = 1, 2, \ldots, m); \\
&\lambda_j \geq 0, \quad (j = 1, 2, \ldots, N).
\end{aligned} \tag{16.66}$$

The exponents α_i and β_r are, respectively, the predetermined weights assigned to the individual inputs and outputs. The weights are nonnegative and satisfy $\sum_i \alpha_i = \sum_r \beta_r = 1$. Portela and Thanassoulis (2005) set each α_i equal to $1/n$ and each β_r equal to $1/m$.

The only difference between the geometric distance function in Equation 16.63 above and the Pareto–Koopmans efficiency measure in Equation 16.51 is that one is a ratio of the arithmetic means of the θs and the φs while the other is a ratio of their respective geometric means. Also, Portela and Thanassoulis (2005) restrict the θs to be no greater than and the φs to be no less than unity. Removing these restrictions would allow a greater flexibility by allowing the inputs to increase if that would permit and an even greater increase in outputs or the outputs to decline if the inputs decline even more.

As in the case of Equation 16.51, the objective function in Equation 16.63 is also nonlinear. However, by taking its natural log, one gets $\ln \gamma = (1/n) \sum_{i=1}^{n} \ln \theta_i - (1/m) \sum_{r=1}^{m} \ln \varphi_r$ that can be linearized at $(\theta_i = 1; i = 1, 2, \ldots, n; \varphi_r = 1; r = 1, 2, \ldots, m)$ as $\ln \gamma \approx (1/n) \sum_{i=1}^{n} \theta_i - (1/m) \sum_{r=1}^{m} \phi_r$. One can set this up as an (approximate) linear objective function to iteratively solve the DEA optimization problem in Equation 16.63.

16.5.3 Data Transformation

In practical applications, the same input or output quantities can, often, be measured in alternative units. The cultivated area may be measured in acres or in hectares. Oil may be measured in gallons or in liters. The output may be measured in pounds or kilograms. A change in the unit of measurement is essentially a change in *scale*. An efficiency measure is *scale invariant* if a change in scale does not alter the measured efficiency of the same input–output bundle.*

---

* See Ali and Seiford (1990) and Lovell and Pastor (1995).

The CCR and BCC technical efficiency measures (both input and output oriented) are scale invariant. Consider the CCR output-oriented model first. Suppose that the scale of measurement of all (or some) of the individual inputs and the outputs is changed. Specifically, the new measure of input i is $\tilde{x}_i = a_i x_i$. Similarly, output r is measured as $\tilde{y}_r = b_r y_r$. For the transformed data, the CCR output-oriented DEA problem will be

$$\begin{aligned} &\max \varphi \\ \text{s.t.} \quad &\sum_{j=1}^{N} \lambda_j \tilde{x}_i^j \geq \tilde{x}_i^0, \quad (i = 1, 2, \ldots, n); \\ &\sum_{j=1}^{N} \lambda_j \tilde{y}_r^j \geq \varphi \tilde{y}_r^0, \quad (r = 1, 2, \ldots, m); \\ &\lambda_j \geq 0, \quad (j = 1, 2, \ldots, N) \quad \varphi \textit{ unrestricted.} \end{aligned} \tag{16.67}$$

But the input and output constraints are actually

$$\sum_{j=1}^{N} \lambda_j (a_i x_i^j) \geq a_i x_i^0 \quad (i = 1, 2, \ldots, n);$$

$$\sum_{j=1}^{N} \lambda_j (b_r y_r^j) \geq \varphi b_r y_r^0, \quad (r = 1, 2, \ldots, m).$$

Hence, the cancelation of common factors on both sides of the inequalities reduces the transformed DEA problem (16.64) to the original CCR output-oriented problem (16.14). For the BCC output-oriented problem, there is the additional restriction that the λs add up to unity. But data transformation does not affect that constraint in any way. Hence, the BCC output-oriented technical efficiency measure is also scale invariant. The proof of scale invariance of the input-oriented measures (both CCR and BCC) will be analogous.

A different kind of transformation known as *translation* of the origin is one where some constant is added to (or subtracted from) any input or output quantity of all firms in the sample. Data translation is common in applications where some input or output values are found to be negative in the data set. A constant is added to all the observations of that input or output to ensure the nonnegativity of the data.

We now show that the CCR–DEA efficiency measures (whether input or output oriented) are not translation invariant. For this, suppose that the transformed input–output data are $\tilde{x}_i^j = x_i^j + c_i (i = 1, 2, \ldots, n)$ and $\tilde{y}_r^j = y_r^j + d_r (r = 1, 2, \ldots, m)$. The CCR–DEA problem with the transformed data is actually

$$\begin{aligned} &\max \varphi \\ \text{s.t.} \quad &\sum_{j=1}^{N} \lambda_j (x_i^j + c_i) \geq x_i^0 + c_i \quad (i = 1, 2, \ldots, n); \end{aligned} \tag{16.68}$$

$$\sum_{j=1}^{N} \lambda_j (y_r^j + d_r) \geq \varphi(y_r^0 + d_r), \quad (r = 1, 2, \ldots, m);$$

$$\lambda_j \geq 0, \quad (j = 1, 2, \ldots, N) \quad \varphi \textit{ unrestricted}.$$

It is obvious that the additional term $\sum_{j-1}^{N} \lambda_j c_i$ on the left-hand side of the input constraint does not cancel with c_i on the right-hand side. Nor does $\sum_{j-1}^{N} \lambda_j d_r$ cancel out φd_r in the output constraints. Hence, the problem in Equation 16.68 will not have the same optimal solution as the original CCR problem in Equation 16.14. In a similar manner, it can be shown that the CCR input-oriented problem is also nontranslation invariant.

Next, consider the BCC problems where VRS is captured by the additional constraint that the λs add up to unity. Given that $\sum_{j=1}^{N} \lambda_j = 1, \sum_{j=1}^{N} \lambda_j c_i = c_i$. Thus, the input constraints in the output-oriented BCC problem with the transformed data are the same as the input constraints in the corresponding problem before data translation. However, the output constraints will differ unless each d_r equals to 0. That is, there is no output translation. We conclude, therefore, that the BCC output-oriented DEA model is invariant to *input translation*. Similarly, the BCC input-oriented DEA problem is invariant to *output translation*.

16.6 Weak Disposability and Bad Output

In many cases, production results in some *bad* or an undesirable output side by side with the *good* or desirable output. In manufacturing, the production of the desired output (such as industrial machinery or steel) leaves the firm with some industrial waste potentially damaging the environment. The generation of power at an electrical utility plant also results in the emission of smoke and polluting particulates in the atmosphere. Traditionally, productivity and efficiency analysts have focused solely on the quantity of the good output produced ignoring the bad output. A greater awareness of the environmental quality has prompted the researchers in recent times to rethink about their criterion of efficiency measurement. It is now recognized that one must include some penalty for the bad output produced to get a measure of the *net* output produced.

An important consideration in this context is: *if some outputs are undesirable, why would not the firm produce the desirable or good output alone without producing the bad output at the same time?* This relates to the concept of a *weak disposability*.

One of the critical assumptions made about the technology at the outset was that the outputs were freely disposable. Specifically, in the two-output case, it would imply that if the output bundle $y^0 = (y_1^0, y_2^0)$ can be produced from some input bundle x^0, then, any nonnegative output bundle $y^1 = (y_1^1, y_2^1) \neq y^0$ can also be produced from x^0 so long as $y_1^1 \leq y_1^0$ and $y_2^1 \leq y_2^0$. This would, of course, permit producing the bad output at zero level without reducing the good output! This, however, is not possible because bad outputs are weakly disposable.

As defined by Ball et al. (2001), the bad output (b) and the good (g) are weakly disposable if

$$(g^0, b^0) \in T \Rightarrow (kg^0, kb^0) \Rightarrow T | 0 \leq k \leq 1. \quad (16.69)$$

That is, the bad output can be reduced only if the good output is reduced proportionately. It is clear that some bad output will necessarily be produced if *any amount of the good output is produced.* Thus, $b = 0$ if and only if $g = 0$. Shephard and Färe (1974) characterized this as *null jointness.* Note that in this interpretation of the relationship between the good and the bad output, the two are produced as joint products. Ball et al. (2001) assume that while the bad output is weakly disposable (with the good output), the good output, however, is strongly disposable. With weak disposability of the bad output, an empirical estimate of the relevant technology set would be

$$T = \left\{ \begin{array}{l} (g, b; x) : g \leq k \sum_j \lambda_j g^j; \\ b = k \sum_j \lambda_j b^j; x \geq \sum_j \lambda_j x^j; \\ \sum_j \lambda_j = 1; k, \lambda_j \geq 0, (j = 1, 2, \ldots, N) \end{array} \right\} \tag{16.70}$$

Of course, under the CRS assumption, the restriction on the sum of the λs does not apply. It can be seen that some of the restrictions in Equation 16.66 are nonlinear. However, Färe et al. bypass this nonlinearity problem by setting k to unity. Assuming that the criterion of efficiency is a simultaneous increase in the good output and a decrease in the bad output, the relevant graph-hyperbolic output-oriented DEA problem is*

$$\begin{aligned} & \delta^* = \max \delta \\ \text{s.t.} \quad & \sum_{j=1}^{N} \lambda_j g_j \geq \delta g_0; \\ & \sum_{j=1}^{N} \lambda_j b_j = \frac{1}{\delta} b_0; \\ & \sum_{j=1}^{N} \lambda_j x_{ij} \leq x_{i0}, \quad (i = 1, 2, \ldots, n); \\ & \sum_{j=1}^{N} \lambda_j = 1; \\ & \lambda_j \geq 0, (j = 1, 2, \ldots, N); \quad \delta \textit{ unrestricted}. \end{aligned} \tag{16.71}$$

On the other hand, an output-oriented directional distance function would be

$$\begin{aligned} & \vec{D}(x_{10}, x_{20}; g_0, b_0) = \max \beta \\ \text{s.t.} \quad & \sum_{j=1}^{N} \lambda_j g_j \geq (1 + \beta) g_0; \\ & \sum_{j=1}^{N} \lambda_j b_j = (1 - \beta) b_0; \end{aligned} \tag{16.72}$$

* See Färe et al. (1989).

$$\sum_{j=1}^{N} \lambda_j x_{ij} \leq x_{i0}, \quad (i = 1, 2, \ldots, n);$$

$$\sum_{j=1}^{N} \lambda_j = 1;$$

$$\lambda_j \geq 0, (j = 1, 2, \ldots, N); \quad \beta \textit{ unrestricted.}$$

A problem with the specification such as Equations 16.71 or 16.72 is that the technological interdependence between the good and the bad output is not explicitly stated. For example, if the two outputs are joint in the sense that they must always increase or decrease together, then, how can the good output be treated as strongly disposable while the bad output is only weakly disposable? If a lower amount of power is being generated, then, where is the smoke coming from? One could argue that lower power generation is the result of an inefficient use of resources and the same amount of fossil fuel is being burnt so that the level of pollution is not reduced. In that case, the jointness is directly between the fossil fuel and pollution and only indirectly between the good and the bad output. This raises a question of the balance of materials.

Following Førsund (2009) and Murty et al. (2012), an alternative interpretation of the bad output would be that it is an unwanted by-product of some input or the inputs used for the production of the good output. In agricultural production, for example, fertilizers and chemical pesticides are used along with land, labor, and capital for crop production. But an unwanted consequence of using chemicals is ground-water contamination. Thus, the bad output can be reduced only if the polluting inputs are reduced as well. However, to the extent that there is room for input substitution, it may be possible to maintain the crop output level. In this case, weak disposability applies to the bad output and the polluting inputs rather than between the good and the bad output. The underlying production technology involves two separate subtechnologies. Suppose that there are two outputs g (good output) and b (bad output) produced from two inputs: x_1 and x_2. Both the inputs are used for the production of g but b is produced only from x_2. We can think of two production-possibility sets: $T_1 = \{(x_1, x_2; g) : g \text{ can be produced from } (x_1, x_2)\}$ and $T_2 = \{(x_2, b) : b \text{ can be produced from } x_2\}$.

The usual disposability assumptions are made about the good output and both the inputs in T_1. However, the bad output and the offending input (x_2) are assumed to be weakly disposable in T_2. An output-oriented directional distance function would be

$$\vec{D}(x_{10}, x_{20}; g_0, b_0) = \max \beta$$

$$\text{s.t.} \quad \sum_{j=1}^{N} \lambda_j g_j \geq (1 + \beta) g_0;$$

$$\sum_{j=1}^{N} \lambda_j b_j = (1 - \beta) b_0;$$

$$\sum_{j=1}^{N} \lambda_j x_{1j} \leq x_{10}; \tag{16.73}$$

$$\sum_{j=1}^{N}\lambda_j x_{2j} = (1-\beta)x_{20};$$

$$\sum_{j=1}^{N}\lambda_j = 1;$$

$$\lambda_j \geq 0, (j = 1, 2, \ldots, N); \quad \beta \textit{ unrestricted.}$$

Similarly, a graph-hyperbolic (output-oriented) measure of technical efficiency would be $\tau_{Graph} = 1/\delta^*$ where

$$
\begin{aligned}
\delta^* &= \max \delta \\
\text{s.t.} \quad & \sum_{j=1}^{N}\lambda_j g_j \geq \delta g_0; \\
& \sum_{j=1}^{N}\lambda_j b_j = \frac{1}{\delta} b_0; \\
& \sum_{j=1}^{N}\lambda_j x_{1j} \leq x_{10}; \qquad (16.74) \\
& \sum_{j=1}^{N}\lambda_j x_{2j} = \frac{1}{\delta} x_{20}; \\
& \sum_{j=1}^{N}\lambda_j = 1; \\
& \lambda_j \geq 0, \quad (j = 1, 2, \ldots, N); \quad \delta \textit{ unrestricted.}
\end{aligned}
$$

16.7 DEA with Market Prices

There is a widely held belief that DEA should be used only for the public sector and nonprofit organizations such as schools or municipal governments where market prices for the outputs are not always available. But for commercial firms that buy and sell their inputs and outputs at observable market prices, one should use parametrically specified econometric models rather than DEA. It should be noted, however, that the cost function relating the expenditure to input prices and output quantities, the revenue function relating the receipts to output prices and input quantities, or the profit function relating the net revenues to prices of the inputs and the outputs are all derived from the assumptions about the technology and the objectives of the firm. Econometrics and DEA are two alternative methods of calibrating the relationship between the prices, quantities, expenses, and revenues as they are relevant in a particular problem. The choice between the two alternative techniques should not depend on the availability or lack of information about the market prices of the inputs and the outputs.

16.7.1 DEA for Cost Minimization

Consider a producer using the input bundle x^0 to produce the output bundle y^0. Further, assume that the market price vector of the inputs is w^0 and the firm is a price taker in the input market. Then, its actual cost is $C_0 = w^{0\prime}x^0$. Clearly, because y^0 is being produced from x^0, (x^0, y^0) is a feasible input–output combination. That is $(x^0, y^0) \in T$, the question is whether $C_0 = w^{0\prime}x^0$ is the *minimum* cost of producing the output bundle y^0. The cost-minimization problem of the firm is to

$$\min w^{0\prime}x : (x, y^0) \in T. \tag{16.75}$$

Suppose x^* is the cost-minimizing input bundle and $C^* = w^{0\prime}x^*$ is the minimum cost. Given a reference technology, this minimum cost will depend on both the input price vector w^0 and the target output bundle y^0 and can be expressed as

$$C(w^0, y^0) = \min w^{0\prime}x : (x, y^0) \in T. \tag{16.76}$$

In production economics, the minimum cost function $C^* = C(w, y)$ is known as the dual-cost function.

Using the free-disposal convex hull, S (from Equation 16.6 above) as an estimate of the production-possibility set, the DEA problem for cost minimization can be set up as

$$\begin{aligned} C^* &= \min w^{0\prime}x \\ \text{s.t.} \quad &\sum_{j=1}^{N} \lambda_j x^j \le x; \\ &\sum_{j=1}^{N} \lambda_j y^j \ge y^0; \\ &\sum_{j=1}^{N} \lambda_j = 1; \\ &x \ge 0; \quad \lambda_j \ge 0 \quad (j = 1, 2, \ldots, N). \end{aligned} \tag{16.77}$$

If CRS is assumed, then, the constraint $\sum_{j=1}^{N} \lambda_j = 1$ is excluded. Like the λ_js, the optimal input vector x is also a (vector of) choice variable(s) in this LP problem. Note that at the optimal solution, there cannot be any slacks in any of the input constraints. To see that, define

$$\sum_{j=1}^{N} \lambda_j x^j = \bar{x}; \quad \sum_{j=1}^{N} \lambda_j y^j = \bar{y}.$$

By convexity, $(\bar{x}, \bar{y}) \in T$. Further, by the free disposability of outputs, $\bar{y} \ge y^0 \Rightarrow (\bar{x}, y^0) \in T$. Hence, the minimum cost cannot be any higher than $w^{0\prime}\bar{x}$. But if there is any input slack at the optimal solution (x^*), then, $w^{0\prime}x^0 > w^{0\prime}\bar{x}$. In that case, $w^{0\prime}x^*$ cannot be the optimal solution of the cost-minimization problem in Equation 16.73.

The (overall) cost efficiency of the firm can be measured as

$$\gamma = \frac{C(w^0, y^0)}{C_0} = \frac{w^{0\prime}x^*}{w^{0\prime}x^0}. \tag{16.78}$$

Farrell (1957) provides an interesting decomposition of the overall efficiency into two distinct components denoting the technical and allocative efficiency.

Consider the observed input–output bundle of the firm, (x^0, y^0) and its input-oriented technical efficiency: $\tau_x(x^0, y^0) = \beta = \min \theta : (\theta x^0, y^0) \in T$. Because $(x^0, y^0) \in T$, $\beta \leq 1$. Now, define $x_t^0 = \beta x^0$. It is the technically efficient projection of the observed input bundle x^0. The cost of this technically efficient input bundle is

$$C_t = w^{0'} x_t^0 = \beta w^{0'} x^0 = \beta C_0.$$

Obviously, $C_t/C_0 = \beta$ is the technical efficiency of the firm. Next, compare $C_t = w^{0'} x_t^0$ with the minimum cost $C^* = w^{0'} x^*$. It follows from the definition of a minimum, that $C^* \leq w^{0'} x$ over all the input bundles of x so long as $(x, y^0) \in T$. Because the bundle x_t^0 is one such bundle, $C^* = w^{0'} x^* \leq w^{0'} x_t^0 = C_t$. Farrell defined the ratio

$$\alpha \equiv \frac{w^{0'} x^*}{w^{0'} x_t^0} \text{ as an allocative efficiency.}$$

Hence, we have the decomposition

$$\frac{C^*}{C_0} = \left(\frac{C_t}{C_0}\right) \cdot \left(\frac{C^*}{C_t}\right) \tag{16.79}$$

or,

$$\gamma = (\beta) \cdot (\alpha). \tag{16.80}$$

As explained above, each of the three ratios, α, β, and γ, lies between 0 and 1. The measure of technical efficiency (β) shows the potential reduction in cost by a proportional reduction in all the inputs without changing the output. In contrast, allocative efficiency (α) measures the reduction in cost by changing the input mix and substituting a relatively less-expensive input for another that is (relatively) more expensive. The overall cost efficiency (γ) reflects the potential for cost reduction by scaling down the input bundle to the extent possible and then selecting a different input mix to take advantage of input substitution.

16.7.2 Profit Maximization

Finally, one can consider the problem of a profit-maximizing firm in a competitive market producing m outputs. The output price vector $p \gg 0$ is determined by market demand and supply and is not within the control of the firm. The firm merely selects the optimal input–output bundle that is feasible and maximizes the difference between the revenue and cost at the applicable market prices of the outputs and the inputs. Thus, conceptually, the maximum profit is

$$\pi^* = \max p'y - w'x : (x, y) \in T. \tag{16.81}$$

With reference to the empirically constructed set S, the relevant DEA problem becomes

$$\begin{aligned} & \max \pi = p'y - w'x \\ \text{s.t.} \quad & \sum_{j=1}^{N} \lambda_j y^j \geq y; \\ & \sum_{j=1}^{N} \lambda_j x^j \leq x; \\ & \sum_{j=1}^{N} \lambda_j = 1; \quad \lambda_j \geq 0 \quad (j = 1, 2, \ldots, N). \end{aligned} \tag{16.82}$$

Note that in this problem, the right-hand sides of both the input and output constraints are themselves the choice variables. Another important point is that for the profit-maximization problem (without any other constraint), *one must allow VRS*. If CRS is assumed, for every feasible (x^0, y^0) that yields the profit $\pi^0 = p'y^0 - w'x^0$, then, (tx^0, ty^0) is also feasible. Hence, the profit will be either unbounded or 0. In fact, locally diminishing returns must hold at the optimal point (x^*, y^*) on the frontier. Otherwise, if (kx^*, ky^*) is feasible for some $k > 1$, then, it is possible to increase the profit to

$$\pi = p'(ky^*) - w'(kx^*) = k(p'y^* - w'x^*)$$

and (x^*, y^*) cannot be the profit-maximizing bundle.

A common measure of profit efficiency is

$$\rho = \frac{\pi^0}{\pi^*} \tag{16.83}$$

where (x^0, y^0) is the actual input–output of the firm under evaluation and $\pi^0 = p'y^0 - w'x^0$ is its actual profit. It should be remembered, however, that if the actual profit is negative, then, the efficiency falls below zero. Additionally, if the maximum profit is also negative, then, the ratio exceeds unity. This may appear to be strange.

However, from the standpoint of pure economic theory, because *all* the inputs and outputs are being freely chosen, we are considering what is known as the long-run profit- maximization problem. If no input–output bundle yields a positive profit, then, the firm should have the option to shut down and earn zero profit. In the LP problem, a zero input–output bundle can be selected only if all λs are set to 0. But that would violate the summation constraint on the λs. So, if a firm is earning negative profit but is still in business, then, that is not a long-run solution in the strict sense and there must be some constraints that keep it from closing down.

16.8 Summing Up

Although it was introduced as an optimization problem in the OR/management science (MS) literature, DEA has grown into a nonparametric alternative to SFA for measurement of production efficiency. This chapter provides the neoclassical production economics behind the different formulations of DEA for the measurement of technical and, scale, cost, and profit efficiency. Given the limited scope of the overview, many important issues could not be addressed. The most important among them are (i) measurement and decomposition of the total factor-productivity growth over time, (ii) the role of exogenous (or contextual) factors in efficiency measurement, and (iii) bootstrapping for generating confidence intervals of DEA efficiency measures. The more-ambitious reader should refer to Ray (2004) and Cooper et al. (2000).

References

Aigner, D.J., C.A.K. Lovell, and P. Schmidt. 1977. Formulation and estimation of stochastic frontier production function models, *Journal of Econometrics*, 6:1, 21–37.

Ali, A.I. and L.M. Seiford. 1990. Translation invariance in data envelopment analysis, *Operations Research Letters*, 9, 403–405.

Aparicio, J., J.T. Pastor, and S.C. Ray. 2013. An overall measure of technical inefficiency at the firm and at the industry level: The "lost profit on outlay," *European Journal of Operational Research*, 226:1, 154–162.

Ball, V.E., R. Färe, S. Grosskopf, and R. Nehring 2001. Productivity of the U.S. agricultural sector: The case of undesirable outputs, in *New Developments in Productivity Analysis, Studies in Income and Wealth*, Vol. 63, C. Hulten, E. Dean, and M. Harper (Eds.), pp. 541–586. Chicago: University of Chicago Press.

Banker, R.D. 1984. Estimating the most productive scale size using data envelopment analysis, *European Journal of Operational Research*, 17:1, 35–44.

Banker, R.D., A. Charnes, and W.W. Cooper. 1984. Some models for estimating technical and scale inefficiencies in data envelopment analysis, *Management Science*, 30:9, 1078–1092.

Banker, R.D., W.W. Cooper, L.M. Seiford, R.M. Thrall, and J. Zhu. 2004. Returns to scale in different DEA models, *European Journal of Operational Research*, 154, 345–362.

Banker, R.D. and R.M. Thrall. 1992. Estimating most productive scale size using data envelopment analysis, *European Journal of Operational Research*, 62, 74–84.

Chambers, R.G., Y. Chung, and R. Färe. 1996. Benefit and distance functions, *Journal of Economic Theory*, 70, 407–419.

Chambers, R.G., Y. Chung, and R. Färe. 1998. Profit, directional distance functions, and Nerlovian efficiency, *Journal of Optimization Theory and Applications*, 98, 351–364.

Charnes, A. and W.W. Cooper. 1968. Programming with linear fractional functionals, *Naval Research Logistics Quarterly*, 15, 517–522.

Charnes, A., W.W. Cooper, and E. Rhodes. 1978. Measuring the efficiency of decision making units, *European Journal of Operational Research*, 2:6, 429–444.

Charnes, A., W.W. Cooper, and E. Rhodes. 1979. Short communication: Measuring the efficiency of decision making units, *European Journal of Operational Research*, 3:4, 339.

Cooper, W.W., S.K. Park, and J.T. Pastor. 1999. RAM: A range adjusted measure of inefficiency for use with additive models, and relations to other models and measures in DEA, *Journal of Productivity Analysis*, 11, 5–42.

Cooper, W.W., J.T. Pastor, F. Borras, J. Aparicio, and J.D. Pastor. 2011. BAM: A bounded adjusted measure of efficiency for use with bounded additive models, *Journal of Productivity Analysis*, 35, 85–94.

Cooper, W.W., L. Seiford, and K. Tone. 2000. *Data Envelopment Analysis: A Comprehensive Text with Uses, Example Applications, References and DEA-Solver Software*. Norwell, MA: Kluwer Academic Publishers.

Cooper, W.W., R.G. Thompson, and R.M. Thrall. 1996. Introduction: Extensions and new developments in DEA, *Annals of Operations Research*, 66, 3–45.

Debreu, G. 1951. The coefficient of resource utilization, *Econometrica*, 19:3, 273–292.

Färe, R. and S. Grosskopf. 2000. Theory and application of directional distance functions, *Journal of Productivity Analysis*, 13, 93–103.

Färe, R. and C.A.K. Lovell. 1978. Measuring the technical efficiency of production, *Journal of Economic Theory*, 19:1, 150–162.

Färe, R., S. Grosskopf, and C.A.K. Lovell. 1985. *The Measurement of Efficiency of Production*. Boston: Kluwer-Nijhoff.

Färe, R., S. Grosskopf, C.A.K. Lovell, and C. Pasurka. 1989. Multilateral productivity comparisons when some outputs are undesirable: A non-parametric approach, *Review of Economics and Statistics*, 71:1, 90–98.

Farrell, M.J. 1957. The measurement of technical efficiency, *Journal of the Royal Statistical Society Series A, General*, 120, Part 3, 253–281.

Førsund, F. 2009. Good modelling of bad outputs: Pollution and multiple-output production, *International Review of Environmental and Resource Economics*, 3:1, 1–38.

Førsund, F. 2013. Weight restrictions in DEA: Misplaced emphasis? *Journal of Productivity Analysis*, 40, 271–283.

Frisch, R. 1965. *Theory of Production.* Chicago: Rand McNally and Company.

Kumbhakar, S. and C.A.K. Lovell. 2000. *Stochastic Frontier Analysis*. New York: Cambridge University Press.

Lovell, C.A.K. and J.T. Pastor. 1995. Units invariant and translation invariant DEA models, *Operations Research Letters*, 18, 147–151.

Meeusen, W. and J. van den Broeck. 1977. Efficiency estimation from Cobb–Douglas production functions with composed errors, *International Economic Review*, 18:2, 435–444.

Murty, S., R. Russell, and S. Levkoff. 2012. On modeling pollution-generating technologies, *Journal of Environmental Economics and Management*, 64:2012, 117–135.

Pastor, J.T., J.L. Ruiz, and I. Sirvent. 1999. An enhanced DEA Russell-graph efficiency measure, *European Journal of Operational Research*, 115, 596–607.

Portela, M.C.A.S. and E. Thanassoulis. 2005. Profitability of a sample of Portuguese bank branches and its decomposition into technical and allocative components, *European Journal of Operational Research*, 162/3, 850–866.

Ray, S.C. 2004. *Data Envelopment Analysis: Theory and Techniques for Economics and Operations Research.* New York: Cambridge University Press.

Ray, S.C. 2010. A one-step procedure for returns to scale classification of decision making units in data envelopment analysis, *University of Connecticut Economics Working Paper*, 2010-07.

Ray, S.C. and A. Ghose. 2014. Production efficiency in Indian agriculture: An assessment of the post green revolution years, *Omega*, 44:2014, 58–69.

Ray, S.C. and Y. Jeon. 2009. Reputation and efficiency: A non-parametric assessment of America's top-rated MBA programs, *European Journal of Operational Research*, 189:2008, 245–268.

Seiford, L. and J. Zhu. 1999. An investigation of returns to scale in data envelopment analysis, *Omega, International Journal of Management Science*, 27, 1–11.

Shephard, R.W. 1953. *Cost and Production Functions*. Princeton: Princeton University Press.

Shephard, R.W. and R. Färe. 1974. The law of diminishing returns, *Zeitschrift für Nationalökonomie*, 34, 69–90.

Tone, K. 2001. A slacks-based measure of efficiency in data envelopment analysis, *European Journal of Operational Research*, 130, 498–509.

Zhu, J. 2003. *Quantitative Models for Performance Evaluation and Benchmarking: Data Envelopment Analysis with Spreadsheets and DEA Excel Solver*. Boston: Kluwer Academic Press.

Index

B

C

E

F

H

V

W

X

Y

Z

For Product Safety Concerns and Information please contact our EU representative GPSR@taylorandfrancis.com Taylor & Francis Verlag GmbH, Kaufingerstraße 24, 80331 München, Germany

Printed and bound by CPI Group (UK) Ltd, Croydon, CR0 4YY
08/05/2025
01864550-0001